Handbook of Cereals, Pulses, Roots, and Tubers

Handbook of Cereals, Pulses, Roots, and Tubers

Functionality, Health Benefits, and Applications

Edited by

SNEH PUNIA

ANIL KUMAR SIROHA

MANOJ KUMAR

CRC Press
Taylor & Francis Group
Boca Raton London New York

CRC Press is an imprint of the
Taylor & Francis Group, an **informa** business

CRC Press
Boca Raton and London
First edition published 2022

by CRC Press
6000 Broken Sound Parkway NW, Suite 300, Boca Raton, FL 33487-2742
and by CRC Press
2 Park Square, Milton Park, Abingdon, Oxon, OX14 4RN

Library of Congress Cataloging-in-Publication Data

Names: Punia, Sneh, editor. | Siroha, Anil Kumar, editor. | Kumar, Manoj (Chemical scientist), editor.
Title: Cereals, pulses, roots, and tubers : functionality, health benefits, and applications / edited by Sneh Punia, Anil Kumar Siroha, Manoj Kumar.
Description: First edition. | Boca Raton : CRC Press, 2022. | Includes bibliographical references and index.
Identifiers: LCCN 2021014714 (print) | LCCN 2021014715 (ebook) | ISBN 9780367692506 (hardback) | ISBN 9780367726027 (paperback) | ISBN 9781003155508 (ebook)
Subjects: LCSH: Cereal products. | Root crops. | Tubers. | Grain--Health aspects.
Classification: LCC TP434 .C48 2022 (print) | LCC TP434 (ebook) | DDC 664/.7--dc23
LC record available at https://lccn.loc.gov/2021014714
LC ebook record available at https://lccn.loc.gov/2021014715

ISBN: 978-0-367-69250-6 (hbk)
ISBN: 978-0-367-72602-7 (pbk)
ISBN: 978-1-003-15550-8 (ebk)

Typeset in Palatino
by KnowledgeWorks Global Ltd.

Contents

Section I   Cereals and Pseudocereals

Section III   Oilseed Crops

Section IV   Roots and Tubers

List of Abbreviations

α	alpha
β	beta
Γ	gamma
λ	lambda
%	Percentage
2D-DIGE	Two-dimensional differential gel electrophoresis
ΔHgel	Enthalpy of gelatinization
μg-	Microgram
4-MBG	4-methylthio-3-butenyl glucosinolate
a*	Redness
AA	Alpha amylase
AAPH	2,2'-Azobis(2-amidinopropane) dihydrochloride
AC	Amylose content
ACE	Angiotensin-converting enzyme
ACR	Acrylamide
AFLP	Amplified fragment length polymorphism
AG	Amyloglucosidase
AIA	α-amylase inhibitor activity
Ais	Amylase inhibitors
ALA	α-Linolenic acid
AM	Amylose
ANF	Antinutritional factor
ANN	Annealing
AP	Amylopectin
AR	Aldose reductase
AT	Autoclaving treatment
b*	Greenness
BA	Brabenderamylograph
BBI	Bowman-Birk inhibitor
B	Branching enzyme
BCAA	Branched chain amino acids
BD	Bulk density
BQPM	Boiled quality protein maize
BU	Brabender unit
BV	Breakdown viscosity
°C	Degree centigrade
Ca	Calcium
C	Colocassia
CFTRI	Central Food Technological Research Institute
CHD	Coronary heart disease
CHO	Carbohydrate
CIAE	Central Institute of Agricultural Engineering
CIP/IPC	International Potato Center
cP	Centipascal
C_p	Specific heat
CNVs	Copy number varieties

CPV	Cool paste viscosity
CQPM	Raw quality protein maize
CTI	Chymotrypsin inhibitor
Cu	Copper
cv.	Cultivar
CVD	Cardiovascular disease
Da	Dalton
DAD	Diode array detection
D	Dioscorea
DE	Diosgenin equivalent
DF	Dietary fiber
DHA	Docosahexaenoic acid
DHT	Dry heat treatment
DM	Dry matter
DNA	Deoxyribonucleic acid
DPPH	2,2-diphenyl-1-picryl-hydrazyl-hydrate
DS	Degree of substitution
DSC	Differential scanning calorimeter
DW	Dry weight
e.g.	Such as
EAAI	Essential amino acid index
EAAs	Essential amino acids
EFSA	European Food Safety Authority
ELISA	Enzyme-linked immunosorbent assay
EPA	Eicosapentaenoic Acid
ER	Estrogenic Receptor
et al.	*et alia* (and others)
FAO	Food and Agricultural Organization of the United Nations
FE	Ferric Equivalent
Fe	Iron
FP	Flaxseed Protein
FRAP	Ferric reducing antioxidant power
FTIR	Fourier transform infrared spectroscopy
FV	Final viscosity
Fw	Fresh weight
g	Gram
GA	Gallic acid
GAB	Guggenheim, Anderson, and Boer
GABA	Gamma-aminobutyric acid
GAE	Gallic acid equivalents
GC	Gel consistency
GFI	Good Food Institute
GHz	Gigahertz
GI	Glycemic index
GMOs	Genetically modified organisms
GPx	Glutathione peroxidase
GRAS	Generally recognized as safe
hr	Hour
HA	Hemagglutination activity
HAA	Hydrophobic amino acid
HCN	Hydrogen cyanide
HDL	High-density lipoprotein
HGP	Human Genome Project

HHP	High hydrolic pressure
HMT	Heat moisture treatment
HPLC	High performance liquid chromatography
HPLC	high-pressure liquid chromatography
HPP	High-pressure processing
HPV	Hot paste viscosity
HTST	High-temperature short time
i.e.	That is
ICAR	The Indian Council of Agricultural Research
ICAT	Isotope-coded affinity tag
IDF	Insoluble dietary fiber
IEC	ion-exchange chromatography
IFCT	The Indian Food Composition Tables
IgE	Immunoglobulin E
I	Interleukin
Ils	Ionic liquids
IP6	Myoinositol hexakisphosphate
ISO	International Organization for Standardization
iTRAQ	isobaric tag for relative and absolute quantitation
IU	Inhibitory unit
IYP	International Year of the Potato
J/g	Joule/gram
kDa	Kilo Dalton
Kg	Kilogram
kGy	Kilo Gray
K	Potassium
L	Length
L*	Lightness
LC	Liquid chromatography
LDL	Low-density lipoprotein
Lidar	light detection and ranging
MDA	Monodehydroascorbate
mg	Milligram
Mg	Magnesium
MHz	Megahertz
ml/g	Milliliter/gram
mm	Micrometers
MMT	Million per metric tons
Mn	Manganese
MPa	Mega pascal
MRM	Multiple reaction monitoring
MT	Microwave treatment
MTHFR	methylenetetrahydrofolate reductase
MS	mass spectrometry
MUFA	Monounsaturated fatty acids
NADES	Natural deep eutectic solvents
NaOH	Sodium hydroxide
NPBTs	New plant breeding technologies
NGF	Nerve growth factor
NHANES	National Health and Nutrition Examination Survey
NIN	National Institute of Nutrition
NMR	Nuclear magnetic resonance spectroscopy
NPY	Neuropeptide-Y

NS	Native starch
NSPs	Nonstarch polysaccharides
NTCDs	Non-transmissible Inveterate Maladies
ODM	Oligonucleotide directed mutagenesis
ORAC	Oxygen radical absorbance capacity
P	Para
PAH	Polycyclic aromatic hydrocarbons
PA	Phytic acid
PCR	polymerase chain reaction
PEF	Pulsed electric field
PHA	Phytohemagglutinin
Phy	Phytate
Pis	Protease inhibitors
PKD	Polycystic Kidney Disease
PSs	Porous starches
PT	Pasting temperature
PUFA	Polyunsaturated fatty acids
PV	Peak viscosity
QPM	Quality protein maize
RAPD	random amplified polymorphic detection
RDA	Recommended dietary allowances
Rd DM	RNA dependent DNA methylation
RDS	Rapidly digesting starch
RFLP	restriction fragment length polymorphism detection
RFOs	Raffinose family oligosaccharides
RNA	Ribonucleic acid
ROS	Reactive oxygen species
Rpm	Rotation per meter/minute
RQPM	Roasted quality protein maize
RS	Resistance starch
RVA	Rapid visco-analyzer
s	Second
SAUs	State Agricultural Universities
SCFA	Short chain fatty acid
SDF	Soluble dietary fiber
SDG	Secoisolariciresinol-di-glucoside
SDN	Site-Directed Nucleases
SDS-PAGE	Sodium dodecyl sulfate-polyacrylamide gel electrophoresis
SEC	size-exclusion chromatography
SEM	Scanning electron microscopy
SFA	Saturated fatty acids
SFC	Supercritical fluid chromatography
SILAC	stable isotope labeling with amino acids in cell culture
SME	Small and medium enterprise
SNPs	Single-nucleotide polymorphisms
SOD	Sulphuroxide dismutase
STRs	short tandem repeats
SSR	Simple sequence repeats
T	Thickness
T$_{end}$	Endset Temperature
T$_{onset}$	Onset temperature
TAI	Trypsin alpha amylase inhibitor
Tc	Conclusion temperature

TE	Trolox equivalent
TI	Trypsin inhibitor
TIA	Trypsin inhibitor activity
TIU	Trypsin inhibitor unit
To	Onset temperature
TPC	Total phenolic content
Tp	Peak temperature
TPTZ	2,4,6-Tri(2-pyridyl)-s-triazine
TV	Trough viscosity
U	Unit
UAVs	Unmanned Aerial Vehicles
UHPLC	ESI–MS/MS-ultrahigh performance liquid chromatography–electrospray ionization tandem mass spectrometry
UHP	Ultra high pressure
UNGAR	United Nation General Assembly Resolution
USDA	United States Department of Agriculture
UV	Ultraviolet
VLDL	Very low-density lipoprotein
W/V	Weight/volume
W	Width
XRD	X-ray diffraction
Zn	Zinc

Preface

Handbook of Cereals, Pulses, Roots, and Tubers: Functionality, Health Benefits, and Applications has been systematically divided into four major subsections: cereals and pseudocereals, pulses, oilseed crops, and roots and tubers.

Section I: Cereals and Pseudocereals

This section will highlight the nutritional aspects of the cereals, pulses, roots and tubers. Chapter 1 discusses the nutritional profile, processing, and contribution of wheat to healthy diet and health. The phenolic antioxidants and proteins are considered one of the main components that decides the overall health-promoting effects of any crop. These components are mostly affected by agronomic factors, climatic factors, and genomic organization. The effect of these external and internal factors on the antioxidant bioactives and amino acids will be discussed by Ali Yigit and Osman Erekul in Chapter 2. Biotechnology has recently contributed immensely in the improvement of the nutritional profile of the various crops and a general overview is presented in Chapter 3, contributed by Shourabh Joshi, Rakesh Kumar Prajapat, Pawan S. Mainkar, Vinod Kumar, Neeshu Joshi, Rahul Bharadwaj, Sunita Gupta, and N. K. Gupta.

The section will be followed by discussing the more specific and most important crops under the category of cereals and pseudocereals. Rice is most widely cultivated and consumed across the globe. The food properties of rice and its industrial processing will be discussed in Chapter 4, which is authored by Priya Dangi, Nisha Chaudhary, Ayushi Gupta, and Isha Garg. Oats are one of the emerging functional foods and are worth considering for their metabolism and health benefits in the human body (Chapter 5, authored by Sneh Punia, Anil Kumar Siroha, and Manoj Kumar). Moreover, millets are also gaining interest due to their health benefits and the advantage that millets can be cultivated with less water requirement with lesser disease incidence. Various minor millets include foxtail millet (Chapter 6), proso millets (Chapter 7), barnyard millet (Chapter 9), and finger millet (Chapter 10). The sources, characterization, and health benefits of β-Glucan is discussed in Chapter 8. Another component that we considered to be important are antinutritional factors (ANFs) present in the cereals. A comprehensive profile of the ANFs and their beneficial and harmful effects in the human body are reviewed in Chapter 11, authored by Vivek Saurabh, Sumit B. Urhe, Anurag Upadhyay, and Sampada Shankar. The potential health benefits and bioactive profile of maize and its products are presented in Chapter 12, by Florence A. Bello, Margaret O. Edet and Lawrence J. Iwok.

A smaller group of crops i.e., pseudocereals are currently gaining the interest of the scientific communities due to their exceptional nutritional profile. Herein, chia seeds (Chapter 14) and quinoa (Chapter 16) are discussed with respect to their potential application as functional foods by Nisha Chaudhary, Priya Dangi, Sunil Bishnoi, Rajesh Kumar, and Kusum Ruhlania. The health benefits of the amaranth and buckwheat grain is also highlighted in Chapter 13 and Chapter 15, respectively.

Section II: Pulses

Pulses are one of the important sources of protein for the vegetarian population. Hence, this section will comprehensively deliver the compositional, nutritional, and health benefits of the various pulses. The overall application of pulses is limited by the presence of ANFs. Hence, it is important to know various ANFs in the legumes and possible ways to overcome the problem of ANFs, which is mentioned in

Chapter 18 authored by Priya Dangi, Nisha Chaudhary, Deepali Gajwani, and Neha. The food properties and application of faba beans will be discussed in Chapter 20 by Nikita Wadhawan, Sagar M.Chavan, Seema Tanwar and N.K. Jain. Aderonke Olagunju, O.S. Omoba will discuss the processing, health benefits, and application of the lentils and pigeon peas in Chapter 19 and Chapter 21, respectively. Pinto beans (Chapter 17), kidney beans (Chapter 22), black gram (Chapter 23), rice beans (Chapter 24), adzuki beans (chapter 25), lupines (Chapter 26), and mung bean (Chapter 27) are also discussed in this reference book. An overall view of the pulses as nutraceutical and health-promoting effects are discussed in Chapter 28 authored by Nikita Wadhawan, Chavan Sagar Madhukar and Gaurav Wadhawan.

Section III: Oilseed Crops

This reference book will also contain a small section on very important oilseed crops. These crops (sesame, flaxseed, and soy beans) are not only an important source of oil, but also a very important candidate for the development of functional foods. An overview of sesame is discussed by Sneh Punia, Anil Kumar Siroha, and Manoj Kumar in Chapter 29. Various properties of flaxseed and its application as an important functional food ingredient are discussed in Chapter 30 by Alok Mishra and Amrita Poonia. The food processing aspects and utilization of the soy bean in foods is discussed in the last chapter of this section, Chapter 31, by Nikita Wadhawan, Sagar M. Chavan, N. K. Jain, and Seema Tanwar.

Section IV: Roots and Tubers

Roots and tubers are considered as one of the nature's buried treasures and have been important food crops for thousands of years. These crops are rich in nutrients and dietary fiber along with many health benefits for the human body. These crops have a crucial role in the food security, agriculture, and income generation for as many as 2.2 billion people in developing nations. Considering their importance, in this reference book will discuss the health benefits, processing, food properties, composition, food application, and ANFs in last five chapters. Chapter 32 will discuss the overall characteristics of roots and tubers, authored by Adeleke Omodunbi Ashogbon. The most important tuber crop—the potato—will be discussed in Chapter 33 by Kadiri Oseni, Olawoye Babatunde, and Oluwajuyitan Timilehin David with respect to its functional food properties, processing, and application. Other tuber and root crops discussed in this section are sweet potato and lotus seed flour and starch in Chapter 34 and Chapter 35, respectively. The last chapter (Chapter 36) of the discuss the various ANFs present in the roots and tuber crops and their possible minimization by different means.

Editors

Sneh Punia, Ph.D., is a researcher in the Department of Food, Nutrition and Packaging Sciences, at Clemson University, Clemson, USA. She previously worked as Assistant Professor (C) in the Department of Food Science and Technology, Chaudhary Devi Lal University, Sirsa. She was involved in the mandated research activities of the institution and has expertise on extraction and functional characterization of antioxidants and starch as well as their modifications and functional products.

She has presented her research in various national and international conferences and published more than 50 research papers/book chapters in national and international journals/books. She has published two edited and two authored books with CRC Press and Taylor & Francis Group. She also serves as the reviewer for various international journals. She is an active member of the Association of Food Scientists and Technologists (AFSTI), in Mysore, India.

Anil Kumar Siroha, Ph.D., is presently working as Assistant Professor (C) in the Department of Food Science and Technology, at Chaudhary Devi Lal University, in Sirsa. He has both master's and doctorate degrees in Food Science and Technology. His areas of interest include starch, starch modification, and the development of new products. He has published more than 20 research papers in national and international journals and several book chapters.

Dr. Siroha edited *Pearl Millet: Properties, Functionality and Applications* (CRC Press, 2020). He also serves as a reviewer for various national and international journals. He is an active member of the Association of Food Scientists and Technologists (AFSTI), in Mysore, India.

Manoj Kumar, Ph.D., is working as a scientist in the Chemical and Biochemical Processing Division at ICAR—Central Institute for Research on Cotton Technology, in Mumbai-400019, India. Dr. Kumar is involved in the mandated research activities of the institute and also acts as quality manager of the institute related to the National Accreditation Board for Testing and Calibration Laboratories (NABL). Dr. Kumar has expertise on extraction and functional characterization of proteins and other bioactive compounds from plant-based matrices. Currently, Dr. Kumar is working on a project funded by Science for Equity, Empowerment, and Development (SEED) Division of DST (Government of India). Dr. Kumar has been honored with several awards throughout his academic career including ICAR's national talent scholarship and junior research fellowship during undergraduate and postgraduate studies. He is an active member of the International Cotton Researchers Association, Indian Society of Cotton Improvement, Indian Fibre Society, and many other professional societies.

Contributors

Adeleke Omodunbi Ashogbon
Department of Chemical Sciences
Adekunle Ajasin University
Akungba-Akoko, Nigeria

Florence A. Bello
Department of Food Science and Technology
University of Uyo
Uyo, Nigeria

Rahul Bharadwaj
Department of Genetics and Plant Breeding,
    Agriculture Research Station
Mandore-Agriculture University-Jodhpur
Rajasthan, India

Sunil Bishnoi
Department of Food Technology
Guru Jambheshwar University of Science &
    Technology
Hisar, Harayana, India

Nisha Chaudhary
Department of Food Science and Technology
College of Agriculture, Nagaur
Agriculture University
Jodhpur, Rajasthan, India

Sagar M. Chavan
Department of Processing and Food
    Engineering
College of Technology and Engineering
Udaipur, Rajasthan, India

Priya Dangi
Department of Food Technology
Institute of Home Economics
University of Delhi
New Delhi, India

Margaret O. Edet
Department of Food Science and Technology
University of Uyo
Uyo, Nigeria

Osman Erekul
Department of Field Crops
Faculty of Agriculture
Aydın Adnan Menderes University
Aydın, Turkey

Deepali Gajwani
Department of Food Technology
Institute of Home Economics
University of Delhi
New Delhi, India

Isha Garg
Department of Food Technology
Institute of Home Economics
University of Delhi
New Delhi, India

Prixit Guleria
Department of Food Technology
Maharshi Dayanand University
Rohtak, Haryana, India

Antima Gupta
Department of Food Science and Technology
Punjab Agricultural University
Ludhiana, Punjab, India

Ayushi Gupta
Department of Food Technology
Institute of Home Economics
University of Delhi
New Delhi, India

N. K. Gupta
SKN Agriculture University
Jobner, Rajasthan, India

Sunita Gupta
SKN Agriculture University
Jobner, Rajasthan, India

Lawrence J. Iwok
Department of Food Science and Technology
University of Uyo
Uyo, Nigeria

N.K. Jain
Department of Processing and Food Engineering
College of Technology and Engineering
Udaipur, Rajasthan, India

Prachi Jain
Department of Food Technology
Bhaskaracharya College of Applied Sciences
Delhi University
Delhi, India

Alka Joshi
Division of Food Science and Postharvest
   Technology
ICAR—Indian Agricultural Research Institute
New Delhi, India

Neeshu Joshi
Department of Agronomy
Agriculture Research Sub-Station,
Sumerpur-Agriculture University
Jodhpur, Rajasthan, India

Riya Joshi
Department of Food Technology
Institute of Home Economics
University of Delhi
New Delhi, India

Shourabh Joshi
Department of Basic Sciences
College of Agriculture
Nagaur-Agriculture University
Jodhpur, Rajasthan, India

Oseni Kadiri
Department of Biochemistry
Edo State University Uzairue
Auchi, Nigeria

Priyanka Kajla
Guru Jambheshwar University of Science &
   Technology
Hisar, Haryana, India

Gurkirat Kaur
Electron Microscopy and Nanoscience Lab
Punjab Agricultural University
Ludhiana, Punjab, India

Gurwinder Kaur
Department of Food Science and Technology
I. K. Gujral Punjab Technical University
Kapurthala, Punjab, India

Dipanshi Kaushik
Guru Jambheshwar University of Science &
   Technology
Hisar, Haryana, India

Twinkle Kesharwani
Department of Food Technology
Bhaskaracharya College of Applied Sciences
Delhi University
New Delhi, India

Vikono Ksh
Division of Food Science and Postharvest
   Technology
ICAR—Indian Agricultural Research Institute
New Delhi, India

Manoj Kumar
Chemical and Biochemical Processing
   Division
ICAR—Central institute for Research on Cotton
   Technology
Mumbai, India

Rajesh Kumar
Department of Food Technology
Guru Jambheshwar University of Science &
   Technology
Hisar, Harayana, India

Vinod Kumar
Department of Biochemistry
College of Basic Sciences and Humanities
CCS Haryana Agricultural University
Hisar, Haryana, India

Parveen Kumari
Guru Jambheshwar University of Science &
   Technology
Hisar, Haryana, India

Roshanlal Yadav
Department of Food Technology
Bhaskaracharya College of Applied Sciences
University of Delhi
New Delhi, India

C. Lalmuanpuia
Department of Food Technology
Bhaskaracharya College of Applied Sciences
Delhi University
New Delhi, India

Chavan Sagar Madhukar
College of Technology and Engineering
Udaipur, Rajasthan, India

Pawan S. Mainkar
ICAR-Directorate of Onion and Garlic Research
Pune, Maharashtra, India

Alok Mishra
Department of Dairy Science and Food
  Technology
Institute of Agricultural Sciences
Banaras Hindu University
Varanasi, India

Manju Lata Mishra
Department of Food Science
Nutrition & Technology College of Home
  Science
CSK HPKV
Palampur, India

Shrestha Naudiyal
Guru Jambheshwar University of Science and
  Technology
Hisar, Haryana, India

Neha
Department of Food Technology
Institute of Home Economics
University of Delhi
New Delhi, India

Aderonke Ibidunni Olagunju
Department of Food Science and Technology
Federal University of Technology
Akure, Ondo State, Nigeria

Babatunde Olawoye
Department of Food Science and Technology
First Technical University
Ibadan, Oyo State, Nigeria

Timilehin David Oluwajuyitan
Department of Food Science and Technology
Federal University of Technology
Akure, Nigeria

Olufunmilayo Sade Omoba
Department of Food Science and Technology
Federal University of Technology
Akure, Ondo State, Nigeria

Amrita Poonia
Department of Dairy Science and Food
  Technology
Institute of Agricultural Sciences
Banaras Hindu University
Varanasi, India

Jayashree Potkule
Chemical and Biochemical Processing
  Division
ICAR-Central institute for Research on Cotton
  Technology
Mumbai, India

Saranya Prabha
Department of Food Technology
Institute of Home Economics
University of Delhi
New Delhi, India

Rakesh Kumar Prajapat
School of Agriculture
Suresh Gyan Vihar University
Jaipur, Rajasthan, India

Uma Prajapati
Division of Food Science and Postharvest
  Technology
ICAR-Indian Agricultural Research Institute
New Delhi, India

Sneh Punia
Department of Food Science and
  Technology
Chaudhary Devi Lal University
and
Sirsa, India
Department of Food, Nutrition and Packaging
  Sciences
Clemson University
Clemson, South Carolina, USA

Reetu
Department of Food Science and
  Technology
I. K. Gujral Punjab Technical University
Kapurthala, Punjab, India

Kusum Ruhlania
Guru Jambheshwar University of Science &
  Technology
Hisar, Haryana, India

J.K. Sahu
Centre for Rural Development and
    Technology
Indian Institute of Technology
New Delhi, India

Vivek Saurabh
Division of Food Science & Postharvest
    Technology
ICAR-Indian Agricultural Research Institute
New Delhi, India

Sampada Shankar
Division of Food Science & Postharvest
    Technology
ICAR-Indian Agricultural Research Institute
New Delhi, India

Neha Sharma
Department of Food Science and Technology
Punjab Agricultural University
Ludhiana, Punjab, India

Nitya Sharma
Centre for Rural Development and Technology
Indian Institute of Technology
New Delhi, India

Savita Sharma
Department of Food Science and Technology
Punjab Agricultural University
Ludhiana, Punjab, India

Sucheta Sharma
Department of Biochemistry
Punjab Agricultural University
Ludhiana, Punjab, India

Arashdeep Singh
Department of Food Science and Technology
Punjab Agricultural University
Ludhiana, Punjab, India

Barinderjit Singh
Department of Food Science and Technology
I. K. Gujral Punjab Technical University
Kapurthala, Punjab, India

Sukriti Singh
Maharishi Markenedeshvar College of Nursing
Maharishi Markenedeshvar (deemed to be)
    University
Mullana, India

and
Department Food Science and Technology
Guru Nanak Dev University
Amritsar, India

Prabhjot Singla
Department of Biochemistry
Punjab Agricultural University
Ludhiana, Punjab, India

Anil Kumar Siroha
Department of Food Science and Technology
Chaudhary Devi Lal University
Sirsa, Haryana, India

Seema Tanwar
Department of Processing and Food Engineering
College of Technology and Engineering
Udaipur, Rajasthan, India

Anurag Upadhyay
Department of Agronomy
Institute of Agricultural Sciences
Banaras Hindu University
Varanasi, India

Sumit B. Urhe
Division of Food Science & Postharvest
    Technology
ICAR-Indian Agricultural Research Institute
New Delhi, India

Gaurav Wadhawan
Pacific Medical College & Hospital
Udaipur, Rajasthan, India

Nikita Wadhawan
Department of Renewable Energy Engineering
College of Technology and Engineering
Udaipur, Rajasthan, India

Baljeet Singh Yadav
Department of Food Technology
Maharshi Dayanand University
Rohtak, Haryana, India

Ali Yigit
Department of Field Crops
Faculty of Agriculture
Aydın Adnan Menderes University
Aydın, Turkey

Section I

Cereals and Pseudocereals

1

Wheat: Contribution to Healthy Diet and Health

Nisha Chaudhary
Agriculture University, Jodhpur, Rajasthan, India

Priya Dangi
Institute of Home Economics, University of Delhi, New Delhi, India

Manju Lata Mishra
Himachal Pradesh Agriculture University, Palampur, Himachal Pradesh, India

Vinod Kumar
Chaudhary Charan Singh Haryana Agricultural University, Hisar, Haryana, India

CONTENTS

1.1 Introduction

The importance of wheat—on both economics and health—and its immense contribution to the diets of humans cannot be challenged. Wheat is vital source of protein and other micronutrients and contributes as a prime source of vegetable protein in the human diet when compared with other cereals (Shewry, 2009; Punia et al., 2017). In the current year (2019–20), global wheat production reached about 742.3 million tons (Statistica, 2020a), which makes it the second most dominant crop in terms of global production after corn. Wheat is incomparable in its range of cultivation—from 67°N in Scandinavia and Russia to 45°S in Argentina—considering the elevated regions in the tropics and subtropics (Feldman, 1995). This attribute has made wheat cultivation possible in distinct continents worldwide and its availability at global level. The satisfying taste, good price, availability, processing efficiency, and nutritive quality help make wheat a staple cereal all-round. Now, cereal consumption patterns are changing; the use of wheat is taking a downturn for consumers with increasing socioeconomic status. Wheat consumption in developing countries is more than in developed countries (Gayathri & Rashmi, 2016). Per capita consumption of wheat in Middle East countries exceeds 150 kg per year, whilst in South Asian countries, such as India, it is 66 kg to 70 kg per year. Areas such as Central America and Sub-Saharan Africa, which primarily rely up on other cereals and millets, withstand far less per capita wheat consumption. The average annual consumption per capita in Europe of cereal grains is 131 kg per year, with wheat making up the majority (108 kg per capita per year). In Asia, approximately half of the yearly cereal consumption is rice (European Food Information Council, 2015). Further, the demand of wheat is increasing in new markets ahead from its region of climatic adaptation and cultivation. The basis for this increase in demand for wheat is its peculiar visco-elastic properties that help form unique and variable food products. Subsequent increasing consumption of these products—along with industrialization and urbanization—created the convenience for production and consumption by replacing traditional eating patterns. Specifically, the exceptional characteristics and composition of gluten protein favor transformation of wheat in to products such as breads, cakes, cookies, pastries, pies, noodles, pasta, and a huge range of breakfast cereals and functional ingredients after processing.

Wheat is a refined reservoir of macronutrients (carbohydrates, protein, fat), micronutrients (sterols, minerals, vitamins, and phenolic compounds), cell wall polysaccharides, dietary fiber, and many biologically active minor elements. The starchy endosperm of wheat accounts for most of the scientific and technological concern for food processing up to and throughout the last century. The realization of the impact of bran and germ layers for maintaining human health spurred scientific interest to define the types, quantity, and promising physiological functions of various vitamins, minerals, and phytochemicals of whole grain and its food products. It is understood that processing food modifies its structure at different levels—this is important in controlling the metabolism of different nutrient compounds and the bioavailability of phytochemicals (Kong & Singh, 2008).

1.2 Dietary Wheat Species

The first wheat cultivation occurred nearly 10,000 years ago as part of the Neolithic Revolution, which was a transition phase from hunting and gathering of food to conventional agriculture. These earliest cultivated forms of wheat were diploid (genome AA, einkorn) and tetraploid (genome AABB, emmer), and their genetic fingerprints indicate their origin as the southeastern part of Turkey (Dubcovsky & Dvorak, 2007). About 9,000 years ago, cultivation expanded to the Near East during the first appearance of hexaploid bread wheat (Feldman, 2001). Farmers chose the earliest cultivated forms of wheat from wild populations. These were essentially landraces considering their superior yield and other typical features, and were an early and clearly nonscientific type of plant breeding. This domestic cultivation also contributed to the genetic traits selection, which separated these early crops from their wild relatives. Importantly, two related traits need to be mentioned here. First, is maturity: Shattering of the spike leads to seed loss at the time of harvesting, which is significant to insure seed dispersal in natural populations; the mutations at the Br (brittle rachis) locus is responsible for nonshattering traits (Nalam, Vales, Watson, Kianian, & Riera-Lizarazu, 2006). Secondly, a substantial trait is the change in hulled forms where the glumes stick tightly to the grain. A dominant mutant at the Q locus made it so that the effects of recessive mutations at the tenacious glume (Tg) locus were modified (Dubcovsky & Dvorak, 2007). Diploid, tetraploid, and hexaploid wheat contain a tough rachis (except for spelt forms of bread wheat). Various antiquated forms of einkorn, emmer, and spelt are hulled, yet modern types of tetraploid and hexaploid wheat are threshed freely. However, the domestication of the natural population developed einkorn and emmer wheat. Cultivation of bread wheat existed at these early times, emerging from the hybridization of cultivated emmer with the unrelated wild grass *Triticum tauschii*.

Feldman (2001) elegantly described and summarized the spread of wheat from its origin across the world. Wheat cultivation found its way into Europe through Anatolia to Greece by 8000 BCE, moving both northward via the Balkans to the Danube River and westward to Italy, France, and Spain during 7000 BCE. Eventually, wheat cultivation was spread to the UK and Scandinavia in 5000 BCE. Likewise, wheat reached central Asia via Iran and successively reached China and Africa by about 3000 BCE, initially via Egypt. Wheat cultivation made its way to Mexico by the Spaniards in 1529 CE and to Australia in 1788.

Triticum aestivum is the major wheat species cultivated throughout the world that is a hexaploid species and is usually known as "common" or "bread" wheat. Whereas, the noticeable part of total wheat production of world includes *Triticum turgidum* var. durum, which is a tetraploid species adapted to the arid conditions surrounding the Mediterranean Sea. Such wheat is utilized to make pasta, thus usually referred as "pasta" or "durum" wheat. Remaining wheat species are cultivated on limited areas by virtue of cultural consideration or for extending market in health foods. These wheat include einkorn (diploid *Triticum monococcum* var. monococcum), emmer (tetraploid *Triticum turgidum* var. dicoccum), and spelt that is being a cultivated form of hexaploid wheat (*Triticum aestivum* var. spelta). Spelt, emmer, and various forms of einkorn are distinct from bread and durum wheats due to their being hulled—as the glumes remain tightly closed over the grain and are not removed by threshing in such wild forms (Shewry, 2009).

1.3 Nutrient Contribution of Wheat in Diet

Wheat is very common, but its importance cannot be overlooked. Wheat is consumed by most of the population in the world. For this reason, its effects on health should be updated periodically. As we know, cereal grains (wheat, rice etc.) are the major source of carbohydrates that form the nutritional base that supports human life for most of the world's population; according to (FAO, 2001) wheat provides 25% of total plant-derived energy, more than 36% of protein, and about 22% of iron. The wheat carbohydrates, vitamins, minerals, phytochemicals and their relation to human nutrition are well documented in *The Wheat And Nutrition Series* by Jones, Peña, Korczak, & Braun (2017).

While it is growing, wheat is primarily affected by climatic condition, genotype, and nitrogen fertilization, which all directly affect the quality and quantity of protein content. Protein content and starch composition are the main influencers on the dietary importance of bread when using different types of whole wheat and flour (Al-Saleh & Brennan, 2012). Hard or soft endosperm wheat (*Triticum aestivum*) make up 95% of total production, while durum wheat (*Triticum durum*) accounts for only 5%. Durum wheat contains more protein i.e. 12–16%, while soft wheat contain 8–10% (USDA/NASS, 2001; Punia et al., 2019). In Table 1.1, the factors used for the conversion of nitrogen values to protein (per g N) given by Greenfield & Southgate (2003).

TABLE 1.1

Factors for the Conversion of Nitrogen Values to Protein (per g N) (Greenfield & Southgate (2003))

Foodstuff	Factor
Wheat whole	5.83
Wheat bran	6.31
Wheat embryo	5.80
Wheat endosperm	5.70

Soft wheat flours provide weak dough, which is good for making cakes, cookies, and crackers. Hard or durum wheat is milled to make semolina and pasta (Sayaslan, Seib, & Chung, 2006). Wheat is a highly adaptive crop in almost all regions of temperate environments and, combined with its genetic diversity, makes it an important and increasingly popular grain to feed people on large scale (Shewry, 2009). Although it's a chief source of energy-giving food, it provides some other essential nutrients like protein, vitamins, minerals, dietary fiber, and other components that are important for the growth and development of human health (Shewry & Hey, 2015). Most of the nutritional quality of the grains are affected by various processing methods; the utilization of highly refined products should not be used (Oghbaei & Prakash, 2016). Wheat grain is made up of three different sections: bran (outer layered fiber), germ (inner micronutrient rich), and the endosperm (loaded with carbohydrate). The bran is full of fibers, antioxidants, iron, zinc, copper, magnesium, B group vitamins, and phytonutrients. The germ holds B group vitamins, vitamin E, phytonutrients, antioxidants, and unsaturated fats. The endosperm is mostly covered with starchy carbohydrates, proteins, and few amount of vitamins and minerals (Gayathri & Rashmi, 2016). All the nutritional composition of the different types of wheat and the portion used for different purposes in the United States (USDA, National Nutrient Database for Standard Reference) and India (NIN, ICMR) are given in Tables 1.2 and 1.3, respectively.

1.4 Relationship of Wheat Composition to Diet and Health

1.4.1 Protein

Protein content present in the wheat flour determines the ability to process it into diverse foods since fully grown wheat grains constitute 8–20% of the total dry matter. Wheat proteins show high complexity and different interactions with each other, consequently making them hard to differentiate and classify according to their solubility and extractability in various solvents. Albumins and globulins of wheat endosperm correspond to 20–25% of total grain proteins (Žilić, Barać, Pešić, Dodig, & Ignjatović-Micić, 2011). Albumin wheat proteins are smallest in size followed by globulins. Gliadins and glutenins in wheat have complex high-molecular weight and act as storage proteins, which does not cause enzyme activity, but does produce spongy baked products due to the gas formed during the dough formation process. Physiologically, the enzymes or active proteins in wheat proteins belong to albumin and globulin groups. These proteins are concentrated in the seed coats, aleurone cells, and germ, with a lower concentration in the endosperm. The albumin and globulin portions cover about 25% of the total grain proteins (Belderok, Mesdag, & Donner, 2000).

1.4.1.1 Nutritional Quality of Protein

Protein quality depends on the degree of essential amino acid present in any foodstuff. If a food possesses all essential amino acids, then the food protein is considered of good and high quality. Essential amino acids cannot be synthesized on their own; a helper must be provided through diet. The essential amino acids are lysine, isoleucine, leucine, phenylalanine, tyrosine, threonine, tryptophan, valine, histidine, and methionine. The most limiting amino acid in wheat is lysine. The requirements for essential amino acids are higher in children for growth purposes, and needed in adult at lesser levels to maintain

TABLE 1.2

Nutritional Composition as per USDA Nutrient Database of Wheat for Different Purposes (USDA National Nutrient Database for Standard Reference, 2015)

Nutritional Composition	Wheat Flour, Whole-Grain	Wheat Flour, Whole-Grain, Soft Wheat	Wheat Bran, Crude	Wheat Germ, Crude	Wheat, Durum Triticum durum Desf.	Wheat, Hard White Triticum aestivum L.	Wheat, Soft White Triticum aestivum L.
PROXIMATE PRINCIPLES AND DIETARY FIBER *(All values are expressed per 100g edible portion)*							
Moisture/water (g)	10.74	12.42	9.89	11.12	10.94	9.57	10.42
Protein (N x 5.83/6.31*)(g)	13.21	9.61	15.55*	23.15	13.68	11.31	10.69
Ash (g)	1.58	1.53	5.79	4.21	1.78	1.52	1.54
Total Fat (g)	2.50	1.95	4.25	9.72	2.47	1.71	1.99
Total Dietary fiber (g)	10.7	13.1	42.8	13.2	-	12.2	12.7
Carbohydrate, by difference(g)	71.97	74.48	64.51	51.80	71.13	75.90	75.36
Energy (kcal)	340	332	216	360	339	342	340
WATER SOLUBLE VITAMINS (All values are expressed per 100g edible portion)							
Thiamine (B1) (mg)	0.502	0.297	0.523	1.882	0.419	0.387	0.410
Riboflavin (B2) (mg)	0.165	0.188	0.577	0.499	0.121	0.108	0.107
Niacin (B3) (mg)	4.957	5.347	13.578	6.813	6.738	4.381	4.766
Pantothenic Acid (B5) mg	0.603	1.011	2.181	2.257	0.935	0.954	0.850
Total (B6) (micro gram)	0.407	0.191	1.303	1.300	0.419	0.368	0.378
Total Folates (B9) microgram	44	28	79	281	43	38	41
FAT SOLUBLE VITAMINS							
Vitamin E (alpha Tocopherols) (mg)	0.71	0.53	1.49	-	-	1.01	1.01
Beta Tocopherols (mg)	0.23	-	-	-	-	-	-
Gamma Tocopherols (mg)	1.91	-	-	-	-	-	-
Alpha Tocotrienols (mg)	0.30	-	-	-	-	-	-
Gamma Tocotrienols (mg)	0.03	-	-	-	-	-	-
Phylloquinones (K1) micro gram	1.9	-	1.9	-	-	1.9	1.9
CAROTENOIDS (All values are expressed per 100g edible portion)							
Lutein+ Zeaxanthin	220	220	240	-	-	220	220
β-Carotene (microgram)	5	5	6	-	-	5	5
MINERALS AND TRACE ELEMENTS (All values are expressed per 100g edible portion)							
Calcium (Ca) (mg)	34	33	73	39	34	32	34
Iron (Fe) (mg)	3.60	3.71	10.57	6.26	3.52	4.56	5.37
Magnesium (Mg) (mg)	137	117	611	239	144	93	90
Phosphorus (P) (mg)	357	323	1013	842	508	355	402
Potassium (K) (mg)	363	394	1182	892	431	432	435
Sodium (Na) (mg)	2	3	2	12	2	2	2
Zinc (Zn) (mg)	2.60	2.96	7.27	12.29	4.16	3.33	3.46
Copper (Cu) (mg)	0.410	0.475	0.998	0.796	0.553	0.363	0.426
Manganese (Mn) (mg)	4.067	3.399	11.5	13.301	3.012	3.821	3.406
Selenium (Se) (microgram)	61.8	12.7	77.6	79.2	89.4	-	-
STARCH AND SUGARS (All values are expressed per 100g edible portion)							
Total Starch (g)	57.77	-	-	-	-	-	-
Fructose (g)	0.05	-	-	-	-	-	-
Glucose (g)	0.00	-	-	-	-	-	-
Sucrose (g)	0.36	-	-	-	-	-	-
Total Sugars (g)	0.41	1.02	0.41	-	-	0.41	-

(Continued)

TABLE 1.2 (Continued)

Nutritional Composition	Wheat Flour, Whole-Grain	Wheat Flour, Whole-Grain, Soft Wheat	Wheat Bran, Crude	Wheat Germ, Crude	Wheat, Durum Triticum durum Desf.	Wheat, Hard White Triticum aestivum L.	Wheat, Soft White Triticum aestivum L.
FATTY ACID PROFILE (All values are expressed per 100g edible portion)							
Total Saturated Fatty Acids (TSFA) (g)	0.430	0.430	0.630	1.665	0.454	0.277	0.368
Total Mono Unsaturated Fatty Acids (TMUFA) (g)	0.283	0.283	0.637	1.365	0.344	0.203	0.227
Total Poly Unsaturated Fatty Acids (TPUFA) (g)	1.167	1.167	2.212	6.010	0.978	0.750	0.837
AMINO ACID PROFILE(All values are expressed per 100g edible portion)							
Tryptophan (g)	0.174	-	0.282	0.317	0.176	-	-
Threonine (g)	0.367	-	0.500	0.968	0.366	-	-
Isoleucine (g)	0.443	-	0.486	0.847	0.533	-	-
Leucine (g)	0.898	-	0.928	1.571	0.934	-	-
Lysine (g)	0.359	-	0.600	1.468	0.303	-	-
Methionine (g)	0.228	-	0.234	0.468	0.221	-	-
Cystine(g)	0.275	-	0.371	0.456	0.286	-	-
Phenylalanine (g)	0.682	-	0.595	0.928	0.681	-	-
Tyrosine (g)	0.275	-	0.436	0.704	0.357	-	-
Valine (g)	0.564	-	0.726	1.198	0.594	-	-
Arginine (g)	0.648	-	1.087	-	0.483	-	-
Histidine(g)	0.357	-	0.430	0.643	0.322	-	-
Alanine (g)	0.489	-	0.765	-	0.427	-	-
Aspartic acid (g)	0.722	-	1.130	2.070	0.617	-	-
Glutamic acid (g)	4.328	-	2.874	3.995	4.743	-	-
Glycine (g)	0.569	-	0.898	1.424	0.495	-	-
Proline (g)	2.075	-	0.882	1.231	1.459	-	-
Serine (g)	0.620	-	0.684	1.102	0.667	-	-
POLYPHENOLS							
Caffeine (mg)	0.0	-	0.0	-	-	0.0	0.0

(- indicates lack of reliable data/NA.)

health (Shewry & Hey, 2016). Plant proteins are much more cost effective and efficient than meat protein to produce, but they are nutritionally inadequate due to their deficiency in essential amino acids (Bicar, Woodman-Clikeman, Sangtong, Peterson, Yang, Lee, et al., 2008).

1.4.1.2 Bioactive Proteins in Wheat

Plant proteins are not easily digested and absorbed and can cause allergic reactions by animals. The albumin and globulin proteins are involved in this abnormal immunological reaction, which causes type 1 hypersensitivity reactions mediated by allergen-specific immunoglobulin E (IgE). For instance, Baker's asthma occurs due to inhalation of flour particles in which water-soluble flour proteins bond to serum IgE (Šramkováa, Gregováb, & Šturdíka, 2009). Celiac disease (gluten-sensitive enteropathy) results in malabsorption and causes poor growth, diarrhea, and abdominal pain. The wheat proteins gliadins and glutenins are contributing factors, whereas thioredoxin (Trh) is involved in germination and seed development. Thioredoxin activates nitrogen and carbon during the reduction of the gliadin and glutenin storage proteins, and the disulphide proteins are inactivated that inhibit starch-degrading enzymes (Lindsay, 2002).

TABLE 1.3

Nutritional Composition of Wheat as Given by NIN, ICMR Wheat Varieties for Different Purposes (Longvah, Ananthan, Bhaskarachary, & Venkaiah, 2017)

Nutritional Composition	Wheat Flour, Refined (Triticum aestivum)	Wheat Flour, *Atta* (Triticum aestivum)	Wheat, Whole (Triticum aestivum)	Wheat, Bulgur (Triticum aestivum)	Wheat, Semolina (Triticum aestivum)	Wheat, Vermicelli (Triticum aestivum)	Wheat, Vermicelli, Roasted (Triticum aestivum)
PROXIMATE PRINCIPLES AND DIETARY FIBER (All values are expressed per 100g edible portion)							
Moisture	11.34±0.93	11.10±0.35	10.58±1.11	8.61±0.32	8.94±0.68	9.59±0.37	7.61±0.47
Protein	10.36±0.29	10.57±0.37	10.59±0.60	10.84±0.75	11.38±0.37	9.70±0.52	10.37±0.70
Ash	0.51±0.07	1.28±0.19	1.42±0.19	1.23±0.06	0.80±0.17	0.60±0.04	0.56±0.04
Total Fat	0.76±0.07	1.53±0.12	1.47±0.05	1.45±0.02	0.74±0.10	0.45±0.03	0.49±0.05
Total Dietary fiber	2.76±0.29	11.36±0.29	11.23±0.77	8.81±0.45	9.72±0.74	9.28±0.69	9.55±0.40
Insoluble dietary fiber	2.14±0.30	9.73±0.47	9.63±0.19	6.56±0.20	8.16±0.58	7.53±0.51	7.79±0.29
Soluble dietary fiber	0.62±0.14	1.63±0.64	1.60±0.75	2.25±0.38	1.55±0.18	1.75±0.24	1.76±0.18
Carbohydrate	74.27±0.92	64.17±0.32	64.72±1.74	69.06±0.74	68.43±0.99	70.39±0.61	71.42±0.71
Energy	1472±16	1340±7	1347±23	1430±6	1396±18	1392±8	1423±13
WATER SOLUBLE VITAMINS (All values are expressed per 100g edible portion)							
Thiamine (B1)	0.15±0.017	0.42±0.044	0.46±0.067	0.24±0.027	0.29±0.025	0.13±0.011	0.12±0.012
Riboflavin (B2)	0.06±0.008	0.15±0.010	0.15±0.041	0.12±0.004	0.04±0.004	0.01±0.003	0.01±0.002
Niacin (B3)	0.77±0.07	2.37±0.10	2.68±0.19	2.05±0.05	1.13±0.10	0.86±0.02	0.67±0.05
Pantothenic Acid (B5)	0.72±0.08	0.87±0.04	1.08±0.21	0.84±0.03	0.75±0.08	0.52±0.05	0.49±0.05
Total (B6)	0.08±0.008	0.25±0.032	0.26±0.036	0.24±0.011	0.11±0.010	0.03±0.004	0.03±0.001
Biotin (B7)	0.58±0.09	0.76±0.12	1.03±0.58	2.50±0.35	0.44±0.04	2.00±0.19	1.34±0.18
Total Folates (B9)	16.25±2.62	29.22±1.92	30.09±3.79	26.30±3.61	25.68±3.64	14.35±2.38	13.21±2.15
FAT SOLUBLE VITAMINS (All values are expressed per 100g edible portion)							
Ergocalciferol (D2)	6.73±0.96	13.43±1.77	17.49±3.51	6.27±0.31	8.19±0.81	4.06±0.35	3.21±0.21
Alpha Tocopherols	0.05±0.01	0.21±0.09	0.60±0.33	0.20±0.01	0.16±0.02	0.03±0.01	0.01±0.01
Beta Tocopherols	-	0.06±0.01	0.37±0.12	-	0.07±0.02	-	-
Alpha Tocotrienols	0.02±0.01	0.06±0.03	0.07±0.03	0.05±0.02	0.05±0.03	0.03±0.02	0.02±0.01
α-Tocopherol Equivalent	0.05±0.01	0.26±0.09	0.77±0.35	0.21±0.01	0.20±0.01	0.03±0.01	0.01±0.01
Phylloquinones (K1)	1.00±0.46	1.50±0.47	1.75±0.26	1.50±0.41	1.20±0.48	1.00±0.51	1.00±0.52
CAROTENOIDS (All values are expressed per 100g edible portion)							
Lutein	24.41±9.21	42.12±11.27	52.56±5.67	47.67±13.64	29.94±7.39	19.31±5.80	12.89±4.12
Zeaxanthin	1.30±0.72	1.31±0.69	1.47±0.68	1.06±0.80	1.13±0.66	0.89±0.57	1.08±0.42
β-Carotene	1.97±0.80	2.67±1.29	3.03±2.13	2.55±0.85	1.60±0.59	1.68±0.31	0.92±0.58
Total Carotenoids	270±69.0	284±31.9	287±40.5	191±15.1	276±29.9	39.54±5.33	21.33±5.71
MINERALS AND TRACE ELEMENTS (All values are expressed per 100g edible portion)							
Aluminium (Al)	0.94±0.33	1.54±0.53	0.55±0.23	0.43±0.16	0.64±0.19	1.15±0.38	1.20±0.16
Arsenic (As)	-	-	-	0.40±0.18	-	-	-
Cadmium (Cd)	0.001±0.000	0.001±0.001	0.002±0.001	0.001±0.001	0.002±0.001	0.001±0.001	0.001±0.000
Calcium (Ca)	20.40±2.46	30.94±3.65	39.36±5.65	27.09±1.62	29.38±2.11	19.42±1.74	22.63±3.46
Chromium (Cr)	0.005±0.002	0.006±0.005	0.006±0.003	0.007±0.002	0.006±0.003	0.006±0.005	0.007±0.002
Cobalt (Co)	0.001±0.001	0.006±0.003	0.003±0.002	0.001±0.000	0.003±0.002	0.002±0.001	0.003±0.002
Copper (Cu)	0.17±0.02	0.48±0.11	0.49±0.12	0.40±0.07	0.46±0.11	0.19±0.03	0.22±0.05
Iron (Fe)	1.77±0.38	4.10±0.67	3.97±0.78	3.86±0.34	2.98±0.34	2.02±0.41	2.09±0.42
Lead (Pb)	0.004±0.002	0.006±0.003	-	0.008±0.011	0.004±0.000	0.008±0.002	0.009±0.008

(Continued)

TABLE 1.3 (Continued)

Nutritional Composition	Wheat Flour, Refined (Triticum aestivum)	Wheat Flour, *Atta* (Triticum aestivum)	Wheat, Whole (Triticum aestivum)	Wheat, Bulgur (Triticum aestivum)	Wheat, Semolina (Triticum aestivum)	Wheat, Vermicelli (Triticum aestivum)	Wheat, Vermicelli, Roasted (Triticum aestivum)
Lithium (Li)	0.003±0.003	0.002±0.001	0.005±0.004	-	0.002±0.002	0.001±0.001	0.001±0.000
Magnesium (Mg)	30.69±2.77	125±11.5	125±14.8	116±14.0	37.89±3.71	34.18±5.28	39.03±7.18
Manganese (Mn)	0.63±0.09	2.98±0.36	3.19±0.59	1.95±0.13	1.98±0.18	0.67±0.11	0.75±0.15
Mercury (Hg)	-	-	-	2.01±0.12	2.67±3.69	-	-
Molebdeum (Mo)	0.013±0.003	0.022±0.009	0.073±0.030	0.029±0.013	0.018±0.011	0.031±0.016	0.019±0.004
Nickel (Ni)	0.005±0.003	0.021±0.015	0.014±0.005	0.018±0.016	0.008±0.006	0.008±0.002	0.010±0.005
Phosphorus (P)	110±9.8	315±41.4	315±41.8	245±27.9	119±8.5	99±12.94	107±21.6
Potassium (K)	148±7.0	311±38.3	366±59.6	330±33.4	284±26.5	163±26.7	177±30.3
Selenium (Se)	-	53.12±5.47	47.76±5.96	10.54±2.23	10.93±5.06	15.33±3.25	14.29±3.11
Sodium (Na)	1.54±0.48	2.04±0.31	2.50±0.20	2.09±0.34	2.31±0.61	2.71±0.79	3.43±0.14
Zinc (Zn)	0.88±0.07	2.85±0.32	2.85±0.65	1.97±0.25	2.13±0.32	0.83±0.12	0.88±0.22

STARCH AND SUGARS (All values are expressed per 100g edible portion)

Total Available CHO	71.82±1.07	58.62±2.68	59.30±1.86	61.74±3.52	59.85±2.99	56.99±1.91	56.69±2.10
Total Starch	70.03±1.01	56.82±2.69	57.53±1.86	60.54±3.62	58.20±2.95	55.31±1.93	54.55±2.05
Fructose	0.64±0.03	0.72±0.03	0.74±0.11	0.32±0.34	0.60±0.04	0.60±0.02	0.63±0.16
Glucose	0.75±0.02	0.78±0.05	0.73±0.08	0.55±0.13	0.55±0.03	0.58±0.02	1.01±0.02
Sucrose	0.40±0.05	0.30±0.02	0.30±0.09	0.33±0.08	0.50±0.04	0.50±0.02	0.50±0.05
Total Free Sugars	1.79±0.08	1.80±0.06	1.77±0.15	1.20±0.24	1.65±0.08	1.68±0.03	2.14±0.17

FATTY ACID PROFILE (All values are expressed per 100g edible portion)

Palmitic (C16:0)	91.24±1.50	191±5.6	176±7.4	179±4.2	81.63±4.28	59.43±4.42	46.97±4.72
Stearic (C18:0)	7.31±0.73	14.55±3.10	14.83±2.25	17.41±2.78	7.24±1.49	5.15±0.65	2.98±0.21
Oleic (C18:1n9)	50.64±2.98	149±7.5	141±9.4	152±5.8	67.34±3.25	36.06±2.43	28.84±1.22
Linoleic (C18:2n6)	325±6.8	697±19.4	616±22.1	657±9.0	306±3.0	209±17.4	162±12.8
α-Linolenic (C18:3n3)	17.45±1.21	44.93±1.64	38.51±3.88	42.69±3.34	19.21±0.84	11.26±0.74	8.55±0.50
Total Saturated Fatty Acids (TSFA)	98.55±1.87	206±8.2	191±8.0	196±2.6	88.87±5.16	64.59±4.94	49.95±4.74
Total Mono Unsaturated Fatty Acids (TMUFA)	50.64±2.98	149±7.5	141±9.4	152±5.8	67.34±3.25	36.06±2.43	28.84±1.22
Total Poly Unsaturated Fatty Acids (TPUFA)	343±7.8	742±19.2	654±23.7	700±11.3	325±2.4	220±17.8	170±13.1

AMINO ACID PROFILE (All values are expressed in g per 100g protein)

Histidine	1.95±0.23	2.56±0.25	2.65±0.31	2.23±0.25	2.38±0.27	1.76±0.11	1.63±0.23
Isoleucine	3.19±0.27	3.78±0.21	3.83±0.20	3.48±0.28	3.43±0.26	1.56±0.56	1.46±0.13
Luecine	6.22±0.46	6.13±0.48	6.81±0.33	6.61±0.66	6.71±0.59	5.23±1.23	5.63±0.24
Lysine	2.05±0.18	2.42±0.22	3.13±0.26	2.42±0.10	2.54±0.13	1.83±0.04	1.54±0.30
Methionine	1.64±0.20	1.77±0.08	1.75±0.21	1.62±0.36	1.57±0.23	1.17±0.14	1.15±0.08
Cystine	2.03±0.27	2.24±0.18	2.35±0.23	1.96±0.05	1.79±0.03	1.83±0.02	1.85±0.08
Phenylalanine	4.29±0.28	5.03±0.14	4.75±0.38	4.46±2.10	4.77±0.32	4.90±1.74	4.96±2.06
Threonine	2.34±0.08	2.58±0.14	3.01±0.17	2.46±0.34	2.71±0.15	2.26±0.16	2.25±0.28
Tryptophan	1.04±0.16	0.99±0.16	1.40±0.10	1.11±0.15	1.04±0.12	1.07±0.09	0.99±0.14
Valine	4.01±0.44	5.12±0.48	5.11±0.05	4.28±0.25	4.47±0.39	3.54±1.35	3.71±0.24

ORGANIC ACIDS (All values are expressed per 100g edible portion)

Total Oxalate	20.22±0.77	52.38±10.71	52.46±3.32	40.23±1.94	28.43±3.76	23.84±2.34	21.91±1.31
Soluble Oxalate	11.02±0.64	26.20±1.99	25.55±7.62	25.27±2.11	22.74±2.63	18.86±1.94	17.55±1.95

Nutritional Composition	Wheat Flour, Refined (Triticum aestivum)	Wheat Flour, *Atta* (Triticum aestivum)	Wheat, Whole (Triticum aestivum)	Wheat, Bulgur (Triticum aestivum)	Wheat, Semolina (Triticum aestivum)	Wheat, Vermicelli (Triticum aestivum)	Wheat, Vermicelli, Roasted (Triticum aestivum)
Insoluble Oxalate	9.20±0.57	24.18±1.35	26.96±7.58	14.96±0.69	5.90±0.59	5.57±0.78	5.66±0.77
Cis-Aconitic Acid	-	-	-	-	-	4.20±0.70	2.89±0.58
Fumaric Acid	1.07±0.07	1.50±0.13	1.61±0.39	1.57±0.27	1.04±0.02	1.20±0.20	1.05±0.03
Mallic Acid	0.02±0.01	0.02±0.01	0.03±0.03	0.02±0.01	0.02±0.01	0.02±0.01	0.02±0.01
Quinic Acid	5.57±1.78	6.49±1.98	6.16±2.71	6.42±2.91	5.81±1.71	4.86±2.13	4.13±1.90
Succinic Acid	1.05±0.06	1.15±0.12	1.48±0.58	1.54±0.42	1.09±0.09	0.97±0.77	0.83±0.38
Tartaric Acid	1.07±0.04	1.41±0.18	1.72±0.59	1.43±0.13	1.21±0.26	0.62±0.11	0.64±0.27
POLYPHENOLS (All values are expressed per 100g edible portion)							
Syringic acid	0.02±0.01	0.04±0.01	0.07±0.01	0.06±0.02	0.03±0.01	0.03±0.01	0.02±0.00
Sinapinic acid	0.06±0.02	0.13±0.02	0.12±0.02	0.11±0.01	0.04±0.01	0.03±0.01	0.02±0.01
Total polyphenols	5.17±0.24	13.98±2.54	14.33±1.76	9.53±1.40	6.50±1.30	5.55±0.34	7.37±0.57
3,4-Dihydroxy benzoic acid	0.23±0.09	0.68±0.19	0.71±0.14	0.68±0.10	0.25±0.05	0.21±0.03	0.20±0.04
Protocatechuic acid	0.09±0.01	0.62±0.11	0.74±0.35	0.71±0.12	0.12±0.01	0.04±0.03	0.02±0.01
Vanillic acid	0.13±0.01	0.26±0.08	0.40±0.31	0.40±0.09	0.18±0.05	0.15±0.05	0.07±0.03
Gallic acid	0.63±0.10	2.79±0.37	2.44±0.25	2.75±0.62	0.23±0.07	0.11±0.02	0.06±0.01
PCoumaric acid	0.06±0.02	0.38±0.09	0.67±0.15	0.63±0.07	0.05±0.02	0.04±0.01	0.02±0.01
Caffeic acid	0.07±0.02	0.78±0.07	0.68±0.12	0.66±0.13	0.14±0.02	0.02±0.01	0.03±0.01
Chlorogenic acid	0.16±0.02	0.71±0.13	0.76±0.11	0.73±0.13	0.28±0.07	0.24±0.03	0.03±0.02
Ferulic acid	0.04±0.01	0.14±0.01	0.15±0.02	0.15±0.01	0.05±0.02	0.03±0.01	0.03±0.02
Apigenin	-	-	-	-	0.03±0.01	-	-
Apigenin-6-C-gluoside	0.07±0.02	0.31±0.08	0.30±0.06	0.26±0.02	0.04±0.01	0.03±0.01	0.03±0.01
OLIGOSACCHARIDES, PHYTOSTEROLS, PHYTATES AND SAPONINS (All values are expressed per 100g edible portion)							
Oligosaccharides							
Raffinose	0.35±0.016	0.47±0.100	0.47±0.171	0.35±0.062	0.47±0.194	-	-
Stachyose	-	0.04±0.005	0.05±0.030	-	0.05±0.003	-	-
Phytosterols							
Campesterol	3.77±0.12	6.21±0.20	9.73±0.15	4.14±0.02	10.36±0.26	2.80±0.08	3.23±0.02
Stigmasterol	0.53±0.03	0.77±0.03	1.42±0.12	0.75±0.03	1.36±0.04	0.40±0.02	0.47±0.01
β-Sitosterol	25.63±0.90	36.60±2.69	55.25±2.06	25.41±0.18	54.28±2.53	22.37±0.28	25.58±0.15
Phytate	123±16.0	632±15.9	638±29.2	679±14.9	549±11.7	168±14.6	165±12.9
Total Saponin	-	-	-	-	-	-	-

1.4.2 Carbohydrates

1.4.2.1 Wheat as Carbohydrates: A Source of Energy

Wheat grain is approximately 80–85% carbohydrate (CHO) and provides a dense source of energy to humans. Polysaccharides are main source of dietary fiber, offer health benefits, and help prevent certain diseases (Stone & Morell, 2009). During the last decade, dietary carbohydrates have negatively impacted the population in terms of cardiovascular disease (CVD) risk at levels "worse than saturated fats," referencing high-protein diets for weight loss (Augustin, Kendall, Jenkins, Willett, Astrup, Barclay, et al., 2015). Carbohydrate's main function is to provide energy in the form of glucose to all cells and

TABLE 1.4

Glycemic Index of Wheat and Its Derived Products (Atkinson,
Foster-Powell, & C., 2008)

Wheat-Based Products	GI and Load Values
White wheat bread	75 ± 2
Whole wheat/whole-meal bread	74 ± 2
Specialty grain bread	53 ± 2
Unleavened wheat bread	70 ± 5
Wheat roti	62 ± 3
Chapatti	52 ± 4
Wheat flake biscuits	69 ± 2
Spaghetti, whole meal	48 ± 5
Spaghetti, white	49 ± 2

tissues. Excess carbohydrate is stored in the form of glycogen in the liver and has restrictive capacity to store it. Carbs are further used to make fat in the body. For utilization of fat, the body needs CHOs to metabolize. The main fuel of the brain and all body tissues is glucose—the brain alone uses 20% of the body's required energy and oxygen for the survival (Raichle & Gusnard, 2002). In cereal grain, all starch is contained in the endosperm, while the outer layer contains no starch. It is present in granule form, which is varied by its size, shape, and other properties. Starch (glycemic CHOs) is a polymer of glucose with $\alpha 1 \rightarrow 4$ and $1 \rightarrow 6$ linkages, which is hydrolysed by the enzyme amylases in humans. These glycemic CHOs can be broken down, absorbed as glucose in the small intestine, and provided in the bloodstream. CHOs in most grains are present in very small amounts (less than 1%) in the form of simple sugars and have less impact on the glycemic response (Jones, Peña, Korczak, & Braun, 2017). Glycemic index (GI) refers to the relative carbohydrate content of food compared with standards, such as glucose or white bread, based on their impact on blood glucose levels (50 g equivalent of carbohydrate) (Atkinson, Foster-Powell, & Brand-Miller, 2008). Low, medium, and high GI foods are ranged from <55, 55–69, and >70 respectively. Wheat flour blend with multiflour composition improves the GI of Indian bread (Nagaraju, Sobhana, Thappatla, Epparapalli, Kandlakunta, & Korrapati, 2020). Table 1.4 gives few examples of the glycemic index of wheat and its derivatives products.

In wheat grains, the undigested portion of starch is not absorbed in the small intestine—this is known as resistant starch, which may help in weight loss by delaying the digestion in the bowel. All cereals are rich source of nonstarch polysaccharides (NSP); Arabinoxylans are the major water-soluble NSP present in wheat (McKevith, 2004). Nonstarch polysaccharides and resistant oligosaccharides, lignin and its complex in plants, resistant starch and dextrins, and dietary fiber are mostly present in cereals, pulses, fruits, and vegetables. Consumption of such foods helps in the prevention of nutritional disorders like gut-related problems, cardiovascular diseases, type 2 diabetes, certain types of cancer, and obesity (Verma & Banerjee, 2010). A different method of lignan extraction gives different results. Syringaresinol lignan is abundantly present in triticale grain and triticale products, but its content varies with variety. Bran contains a maximum level of phytoestrogens followed by refined flour, which is less than 10 times in bran. Various methods like malting, fermentation, sprouting, and extrusion techniques can enhance the levels of these compounds (Makowskaa, Waskiewiczb, & Chudya, 2020).

1.4.2.2 Starch

Wheat starch contained varies from 60–75% of the total dry weight of the grain. It occurs in the form of granules in seed. Two types of starch granules are present in large (25–40 μm) lenticular and small (5–10 μm) spherical ones (Belderok, Mesdag, & Donner, 2000). Starch is primarily a polymer of glucose and chemically on breakdown gives amylose and amylopectin. The amylose molecular weight is around 1,500 glucose molecules. It is a linear α-(1,4)- linked glucose polymer with a degree of polymerization (DP) of 1,000 to 5,000 glucose units leading to a linear structure. The counterpart is branched (i.e. Amylopectin) and its unit chain gives 20–25 glucose molecules long. Its molecular weight is about 108. The ratio of amylose to amylopectin is relatively constant at 23. It is a much larger glucose polymer (DP 105–106) in which α-(1,4)-linked glucose polymers are connected by 5–6% α-(1,6)-linkages. Usually, wheat starch typically holds 20–30% amylose and 70–80% amylopectin (Konik-Rose, Thistleton, Chanvrier, Tan, Halley, Gidley, et al., 2007). The modification in the ratio amylose-amylopectin can be used for the production of starches with novel functional properties and enhanced wellbeing (Poole, Donovan, & Erenstein, 2020).

1.4.2.3 Dietary Fiber

Dietary fibers are nondigestible carbohydrates and lignin, which offers health benefits. The various sources of dietary fiber consist of whole grains, fruits, vegetables, and legumes like beans and peas. There are two types of soluble or insoluble fiber. Soluble fiber can be obtained from fruits, legumes, and oats. It lowers bad cholesterol (LDL) and can facilitate in the control of blood sugar whereas insoluble fibers are found in wheat, bran, vegetables, and fruits, which tends to control appetite, reduce risk of developing type 2 diabetes, and prevent constipation (Lane-Elliot, 2016). On U.S. food labels, inulin, fructo-oligosaccharides, and other oligosaccharides are classified as fiber. Oligosaccharides act as prebiotics, which is "a selectively fermented ingredient that allows specific changes, both in the composition and/or activity in the gastrointestinal microflora that confers benefits upon on host and their health" (Slavin, 2013). Utilization of fiber in the diet can help in the prevention of heart disease, blood lipids, cancer, glucose absorption, insulin secretion, and diverticular disease. A soluble dietary fiber is dissolved in water and comprises pectic substances and hydrocolloids, whereas insoluble dietary fiber do not dissolve in water and consist of cellulose, hemicellulose, and lignin like whole grains. Major components of wheat endosperm cell walls are arabinoxylans (AX) and (1→3),(1→4)-β-glucans. Arabinoxylan is the most favorable substrate for fermentative production of short-chain fatty acids (SCFAs) especially butyrate in the colon that improves bowel health and can lower the risk of cancer by different mechanisms (Šramková, Gregová, & Šturdík, 2009). Wheat flour, white and whole meal, contains 3.1 and 9 nonstarch polysaccharides g/100 g, respectively (Buttriss & Stokes, 2008). With nutritional benefits, whole-wheat flours contain significant amount of antinutritional factors, such as phytates, which reduces their bioavailability in the human gut; the inorganic phosphorus (Pi) is present in the mature seeds of cereal (40–80%) which helps form complexes with other minerals like Ca^{2+}, Fe^{3+}, Zn^{2+} and Mg^{2+} (Bilgiçli, İbanoğlu, & Herken, 2007).

1.4.3 Phytochemicals

The most important major group of phytochemicals in wheat grain are phenolics and terpenoids. They provide a wide range of health benefits but more studies by health agencies (like the FDA or EFSA) are needed to establish clear data. The plant sterols and stanol esters help in lowering cholesterol in blood and reduce the risk of cardiovascular diseases (Shewry & Hey, 2016). Phenolic compounds present in wheat raw flour are given in Table 1.5.

TABLE 1.5

Phenolic Constituents of Raw Wheat Flour (*Triticum spp.*) (Adefegha, Olasehinde, & Oboh, 2018)

Parameter	Raw Flour mg/g	Parameter	Raw Flour
Catechin	0.63 ± 0.02	Total phenol (GAE mg/g)	42.37 ± 1.75
Chlorogenic acid	1.95 ± 0.01	Total flavonoid (QE mg/g)	35.71 ± 0.87
Rutin	6.34 ± 0.05	Starch (g/100 g)	24.99 ± 0.67
Epicatechin	0.08 ± 0.04	Sugar (g/100 g)	3.68 ± 0.05
Quercetin	1.73 ± 0.01	Starch/sugar ratio	0.15
Kaempferol	0.59 ± 0.01	Amylose content (A) (g/100 g)	3.63 ± 0.60
		Amylopectin content (Am) (g/100 g)	21.36 ± 1.25

Babu, Ketanapalli, Beebi, and Kolluru (2018) reported on the total phenolic acid content in wheat bran, which is present about 4.5μg/g bran on wet basis (e.g., ferulic acid of hydroxycinnamic acids). These phenolic compounds greatly manipulated the flavor, texture, color and, most importantly, the nutritional properties of food.

1.4.4 Vitamins

Vitamins are vital organic substances that are not synthesized by the human body. Nutritionally, vitamins are an essential micronutrient for well-being and perform functions as precursors or coenzymes of niacin, thiamin, biotin, pantothenic acid, vitamin B6, vitamin B12, and folate. Vitamin A aids in vision and vitamin C in hydroxylation reactions, antioxidative defence systems, human genetic regulation, and genomic stability through the use of folic acid, vitamin B12, vitamin B6, niacin, and vitamins E and D (Paredes-López & Osuna–Castro, 2006). Vitamin A deficiency is becoming common in many countries. Wheat flour can act as vehicle for vitamin A through fortification and the consumption of vitamin-A fortified wheat flour may improve the deficiency at mass level. One fourth of the requirement of recommended dietary allowance for vulnerable groups can come from diversified flour products like baked goods, mixes, etc. The level of fortification affects the costs of product for the consumption (Klemm, West, Jr., Palmer, Johnson, Randall, Ranum, et al., 2010).

1.4.5 Methyl Donors in Wheat (Betaine, Choline)

In the homocysteine cycle, glycine betaine acts as a methyl donor, which is present as rich source in wheat. Choline (precursor of betaine) and trigonelline (the structural analogue of betaine and choline) are collectively known as methyl donors. Physiologically, betaine acts as an osmolyte, which protects cells, proteins, and enzymes from environmental stress (low water, high salinity, or extreme temperature). It helps to keep liver, heart, and kidney healthy (Craig, 2004). The different concentrations of choline and betaine in wheat are given in Table 1.6.

Betaine is solely obtained through diet in humans, although it can also be produced by reversible conversion of choline. The interrelationship of choline, betaine, and energy metabolism signify novel functions. Mthyl-deficient diets interrupt energy metabolism and protein synthesis in the liver, fatty liver, or muscle disorders. Folic acid supplementation additionally works as betaine; in folate deficiency, methionine load, or alcohol consumption, betaine is major determinant in plasma total homocysteine (tHcy). These two methyl donors help to elevate hypomethylation and tHcy. Choline is increased during pregnancy whereas betaine is decreased (Obeid, 2013). Aleurone-rich foods and whole-grain foods are related with improvement in the number of health biomarkers counting as decreased total plasma homocysteine and LDL cholesterol (Keaveney, Price, Hamill, Wallace, McNulty, Ward, et al., 2015).

TABLE 1.6

Choline and Betaine Concentrations in Wheat (Zeisel, Mei, Howe, & Holden, 2003)

Foodstuff	Total Choline (mg choline moiety/100 g food)	Betaine (mg/100 g food)
Wheat germ, toasted	152.08	1240.48
Wheat bran	74.39	1339.35
Wheat bread	26.53	201.41
White bread	12.17	93.20
Wheat cracker	31.80	198.71
Biscuit-plain	8.89	38.24

1.5 Progress in Improving Nutritional Quality

Wheat encompasses a wide range of nutritional components like iron (Fe), zinc (Zn), vitamins, and phenolic acids, which play a vital role in plant metabolism and human health. Low quantity in the endosperm and interventions with other components reduce the bioavailability of these nutritional components, which generates the requirement of nutritional quality enhancement by incorporating diverse biological tools. The nutritional quality improvement leads to a series of processes to ensure that the nutrients are bioavailable after consumption. The major process necessitates a genotypic and phenotypic characterization of pivotal biological processes or pathways, which are implicated in the assimilation, accumulation, biosynthesis, translocations and remobilization of chosen nutritional quality components like Fe, Zn, vitamins and phenolic acids in wheat grain (Borrill, Connorton, Balk, Miller, Sanders, & Uauy, 2014; Ma, Li, Zhang, Wang, Qin, Ding, et al., 2016; Meena, Abhishake, Punesh, Imran, Vinod, & Harcharan Singh, 2020). The process of biofortification is involved eventually, being the most sustainable approach to reach the nutritional requirements of the global community on an economical background. However, infusion of biofortification involves the genetic and phenotypic profile of respective crop, across distinct environments. The substantial progression for improving nutritional quality in wheat includes the deployment of assorted plans that consider conventional, technological, and transgenic approaches (Tiwari, Rawat, Neelam, Kumar, Randhawa, & Dhaliwal, 2010). Fundamental genetic and agronomic practices are utilized as conventional-based approaches in terms of agronomic biofortification, soil and foliar application, and genetic biofortification. These approaches associate germplasm screening to display the genetic variation for Fe and Zn levels of grain across various wheat genotypes grown in diverse environments (Cakmak, Pfeiffer, & McClafferty, 2010). In addition, a breakthrough has been gained to establish genetic variation of Fe and Zn across different wheat species, and, simultaneously, locus (QTL) Gpc-B1, which is an important quantitative trait from wild emmer wheat (*Triticum turgidum ssp. dicoccoides*), was detected and mapped on chromosome arm 6BS (Joppa, Du, Hart, & Hareland, 1997). Cloning of the gene of this locus was rendered that significantly improved Zn, Fe and protein accumulation by 12%, 18%, and 38%, respectively (Uauy, Distelfeld, Fahima, Blechl, & Dubcovsky, 2006). The Xuhw89 marker was found related Gpc-B1 locus including 0.1 cM genetic distance, which can be exploited to identify and select lines containing enriched levels of particular micronutrients in the wheat grain (Distelfeld, Cakmak, Peleg, Ozturk, Yazici, Budak, et al., 2007). Furthermore, numerous efforts have also been made in some wheat species to determine genetic variation in the levels of phenolic compounds. Advanced analytical technologies with high-throughput are employed in the character of technological-based approaches, such as ribonucleic acid sequencing (RNAseq), ribonucleic acid interference (RNAi), genomics, transcriptomics, and metabolomics, to determine and characterize candidate genes for improving nutritional quality (Chaudhary, Kumar, Sangwan, Pant, Saxena, Joshi, et al., 2020; Nathani, Kumar, Dhaliwal, Sircar, & Roy, 2020). This may incorporate genome editing-based approaches as well, like the CRISPR Cas9 approach, which was recently used in wheat (Zhang, Liang, Zong, Wang, Liu, Chen, et al., 2016). Transgenic-based approaches principally implicate the application of genetic modification for improving nutrient concentration in wheat grain. A limited headway has been achieved by integrating transgenic approaches in the pursuance of improving nutritional quality (Masuda, Aung, & Nishizawa, 2013). Technological applications have also played the key role in wheat improvement, which is exhibited by the success in enhancing the bioavailability of Fe and Zn and reducing the antinutrients like phytic acid and polyphenols. Noteworthy, such antinutrients inhibit Fe absorption by decreasing Fe bioavailability in the human body. Nevertheless, agronomic biofortification and application of nutritional enhancers were adopted as strategies to upgrade the bioavailability of micronutrients and phenolic acids (Laddomada, Caretto, & Mita, 2015; Velu, Ortiz-Monasterio, Cakmak, Hao, & Singh, 2014). Wheat endosperm lacks essential transporters for the translocation of Fe into the endosperm and there is a dearth of relevant research related to manipulating the transporter proteins to translocate additional Fe into wheat endosperm (Balk, Connorton, Wan, Lovegrove, Moore, Uauy, et al., 2019). Similarly, literature is deficient in demonstrating the translocation and transporters of phenolic acids and vitamins into the endosperm. Continuous attempts were initiated to address such related challenges by applying discrete conventional, technological and transgenic approaches. However, a sufficient breakthrough has been obtained to fathom primitive mechanism of assimilation, translocation and bio-synthesis of micronutrients into wheat grain (Borrill, Connorton, Balk, Miller, Sanders, & Uauy, 2014).

1.6 The Impact of Processing on Wheat Grain Components With Regard to Health Benefits

Processing is an indispensable step to transform cereal grains into food that is a safe and appealing final product for human consumption. Processing can aid to minimize lurking hazardous compounds such as pesticides, mycotoxins and heavy metals, while also favoring the production of food products with tempting and unique properties. The basic character of wheat grain is to be processed into various end products as flour, semolina, and varying bakery products; it gains most of the importance in the human diet. Such processing combines various systematic steps, each influencing both or any one of the composition and physical-chemical characteristics of its varied components, which consecutively deduce the techno-functional quality and the nutritional or health promoting properties of the end product. Starch and gluten proteins are the key components to describe peculiar textural properties of wheat foods present in the starchy endosperm, hence, this is most correlated with refined flour or semolina. Nevertheless, the health properties of wheat-based products are mainly linked with their dietary fibers and bioactive compounds in the grain peripheral layers—largely in the aleurone layer that is considered in the bran fraction during milling (Tosi, Hidalgo, & Lullien-Pellerin, 2020). Various milling combinations impact profoundly the relative abundance of the various grain components in the initial processing outcomes (wheat flours/semolina) and, consequently, in the secondary processing outcomes (bread, pasta, biscuits, breakfast cereals or snacks etc.) (Hemery, Rouau, Lullien-Pellerin, Barron, & Abecassis, 2007). Further, the bioavailability of various grain components also are affected by procedures followed under processing steps in terms of dough making, microbial fermentation, extrusion, and baking. Recently, consumption of whole-grain products have been linked with better health benefits (Cooper, Kable, Marco, De Leon, Rust, Baker, et al., 2017; Kristensen, Toubro, Jensen, Ross, Riboldi, Petronio, et al., 2012; Nelson, Mathai, Ashton, Donkor, Vasiljevic, Mamilla, et al., 2016). Numerous other studies poorly described the whole-grain products with regard to their composition (Brouns, van Rooy, Shewry, Rustgi, & Jonkers, 2019; Thielecke & Nugent, 2018), therefore, this contradiction still remains under discussion.

1.6.1 Alkylresorcinols

Alkylresorcinols are phenolic lipids, chemically defined as 1,3-dihydroxy-5-n-alkylbenzenes containing an odd number of carbon atoms constituted between 17 and 25, gives a mixture of alkylresorcinol homologues in specific proportions in accordance with the cereal, which is being enriched in 19 and 21 carbon homologues for wheat (Ross, Shepherd, Schüpphaus, Sinclair, Alfaro, Kamal-Eldin, et al., 2003). Certain *in vitro* studies related to antioxidant properties (Kozubek & Tyman, 1999), chronic diseases inhibition properties (Zhu, Soroka, & Sang, 2012), or glycerol-3 phosphate deshydrogenase activity inhibition properties of alkylresorcinols were identified. Whereas, *in vivo* efficiency estimation remains difficult to understand due to association of alkylresorcinol consumption in cereal products with other potentially active molecules. The accumulation of total alkylresorcinol content in wheat grains varies between 54 and 1,489 µg/g (d.m.) while a mean content around 500–700 µg/g (d.m.) in accordance with the species, cultivars and growing environment with contents (Andersson, Kamal-Eldin, & Aman, 2010; Ross, et al., 2003). Alkylresorcinols are generally present at the frontier among the outer cuticle of the testa and the inner cuticle of the pericarp in wheat grain (Landberg, Marklund, Kamal-eldin, & Åman, 2014). Consequently, their determination during fractionation sourced out of the aleurone layer and the outer pericarp that is just less than 4% of the total grain mass (Barron, Samson, Lullien-Pellerin, & Rouau, 2011). During the progression of milling, external tissues inclusive of the aleurone layer are separated from the starchy endosperm, which contains the maximum alkylresorcinol concentrated (three to five times higher than in grains) in the bran and shorts fractions. Hence, only limited amounts obtained in refined flours or semolina and subsequent final products (Ross, et al., 2003). Therefore, the way to increase the alkylresorcinol content in flour can be accomplished by adding ground shorts or bran, or fortification of recovered fractions after pearling of grain that can be added between 5% and 10% of the total grain mass (Bordiga, Locatelli, Travaglia, & Arlorio, 2016).

1.6.2 Tocols

Tocols comprise two types of amphipathic and lipo-soluble molecules, namely tocopherols and tocotrienols, which exhibit a polar chromanol ring and a hydrophobic 16-carbon side chain interrelated to a phytyl or an isoprenoid chain, respectively (Tiwari & Cummins, 2009). This side chain is fully composed of tocopherols whilst it constitutes three double bonds in tocotrienols. As per the discrimination in the number and position of methyl groups in the chromanol ring, four different forms of these molecules have been identified as α-, β-, γ-, and δ-. Tocols show antioxidant properties by means of scavenging lipid peroxyl radicals and quenching the singlet or reactive oxygen and nitrogen species (Kamal-Eldin & Appelqvist, 1996). Whilst, β-tocotrienol corresponds the vitamin E activity and out of this 5% belongs to α-Tocopherol. Total quantity of tocol in wheat grains varies between 30–88 μg/g (d.m), constituting β -tocotrienol followed by α-tocopherol being the major components, depending on genotypes and environment (Lampi, Nurmi, Ollilainen, & Piironen, 2008). Germ enriched wheat milling fractions generally contain α-tocopherol whereas bran and flours fractions are enriched in β-tocotrienol (Piironen, Syvaoja, Varo, Salminen, & Koivistoinen, 1986). The order of total tocol content presence in decreasing manner is germ> bran> flour. Tocols are light and temperature sensitive, which induces losses along processing (Andersson, Dinberg, Aman, & Landberg, 2014; Tiwari & Cummins, 2009). The amount of vitamin E in milled wheat products relies mainly on the extraction rate of the flour (reduced to 50% from whole grain to white flour), milling method, and storage (Nielsen & Hansen, 2008). Milling via stone-mill brought about the total loss of 24% during storage of vitamin E, which included significant amount of germ, whereas this loss became 50% in case of roller-milled wheat flour, which was devoid of germ and bran as it worked as an antioxidant. Oxidation of vitamin E was designated as the main reason behind the losses during the subsequent processing steps similar to the preparation of gruels and porridges involving processes like extrusion cooking and drum-drying that sabotage a large fraction of vitamin E in white flour (Håkansson & Jägerstad, 1990). It was reported that the ratio of tocotrienols to tocopherols increased after extrusion cooking, suggesting the tocotrienols as the primary residual isomers of vitamin E (Zielinski, Kozlowska, & Lewczuk, 2001). In addition, a higher ratio of tocotrienols to tocopherols in the diet is evident in metabolic regulation (Tosi, Hidalgo, & Lullien-Pellerin, 2020).

1.6.3 Short-Chain Carbohydrates

Fractans (fructooligosaccharides or FOS) are the predominant low molecular mass carbohydrate fraction of wheat flour, consisting of three to five fructose units with different structures, sometimes including a single glucose unit (Roberfroid, 2005). Limited quantity of galactooligosaccharides (GOS), such as raffinose and stachyose, are also present (Huynh, Lachlan, Mather, Wallwork, Graham, Welch, et al., 2008). Fructan content in wheat grains was observed to range between 7–29 mg/g as per their genotypes, therefore adequate selection can control this character (Huynh et al. 2008). Higher concentration of fractan (34–40 mg/g) is actually present in bran and shorts fractions obtained from milling than white flour and germ portion (14–25 mg/g) confined to wheat cultivars and climatic conditions, (Haska et al. 2008; Knudsen 1997). In the case of exceeding extraction rates, the endosperm may comprise approximately half of the total fructan present. The processing step also alters the amount of fractans, as in bread making that depicted noticeable variation on fructans content (Gélinas, McKinnon, & Gagnon, 2016a). The process of dough mixing reduced fructans by about 20%, whether including or not including the baker's yeast. Dough fermentation accompanied the higher reduction, by degrading up to 80% of wheat grain fructans during a 180-minute period, while the process of baking did not impact the result. Moreover, bread making did not lead to a change in the chain length of fructans. During the course of investigation on pasta making with regard to variation in fractans content, Gélinas, McKinnon, and Gagnon (2016a) disclosed that the drying temperature of 40°C or 80°C could not change the composition of fractans significantly; on the contrary, boiling (cooking) of pasta leached about 40–50% of fructans into the water irrespective of the cooking time. FOS and GOS come under the category of FODMAPs (fermentable oligo-, di- and monosaccharides and polyols). This term is allocated to the human indigestible molecules that lack corresponding hydrolytic enzymes to be absorbed in the intestinal lumen, though act as a prebiotic substrate for bacteria in the gut (Gibson & Shepherd, 2005).

1.6.4 Sterols

Sterols and their saturated forms, stanols, are composed of four cyclic compounds that contain a cyclopentane perhydrophenanthrene nucleus, have a presence of hydroxyl group at position 3 of the A-ring, and a placement of a side chain at carbon 17. Phytosterols were found at lower levels in wheat bran and germ in total- and low-density serum lipoprotein (Andersson, Skinner, Ellegard, Welch, Bingham, Bingham, et al., 2004). The variation of accumulation of total phytosterol content in wheat grains in 26 genotypes, three growing seasons, and four locations was observed between 700 to 928 μg/g (Nurmi, Lampi, Nyström, Turunen, & Piironen, 2010).

Nyström, Paasonen, Lampi, and Piironen (2007) analyzed the commercial milling fractions for total sterol concentration; germ was the most enriched fraction followed by fine and coarse brans. A similar enrichment of the bran fractions were also observed with steryl ferulate forms. The effect of processing was seen in the sterols' bioavailability by their release from the food matrix. The particle-size reduction by centrifugal milling from an average of 97 μm to 47 μm of wheat bran, minimally increased the amount of sterols (about 5%) (Nyström, Lampi, Rita, Aura, Oksman-Caldentey, & Piironen, 2007). The sterol content was reduced by the application of thermal treatments, such as roasting or microwave heating. Achievement of sterols of cereal products decreased dramatically by the sole addition of water (without enzymes). The soaking of wheat bran resulted in deprived sterol content, possibly by the viscous structure formation of arabinoxylan hydrates, which block the hydrophobic sterols inside.

1.6.5 Phytic Acid and Minerals (Iron, Magnesium, and Zinc)

Phytic acid is a myo-inositol 1,2,3,4,5,6-hexakis dihydrogen phosphate that makes up the prime phosphate storage in cereal grains. The concentration of phytic acid in wheat grains ranges between 12–18 mg/g (Barron, Samson, Lullien-Pellerin, & Rouau, 2011), reside mainly in phytin globoids within protein storage vacuoles of aleurone cells, and includes a major portion of magnesium (Mg) and Fe (O'Dell, De Boland, & Koirtyohann, 1972). The amount of phytic acid in the aleurone layer is usually found from 95–190 mg/g (Barron, Samson, Lullien-Pellerin, & Rouau, 2011), therefore deposition of phytic acid can be used as a monitoring marker in aleurone estimation in milling (Hemery, Lullien-Pellerin, Rouau, Abecassis, Samson, Åman, et al., 2009).

Phytic acid has long been contemplated as an antinutritional factor because of the formation of insoluble complexes with dietary cations, specifically Mg, Fe, Zn and calcium (Ca); as a result mineral absorption in humans is impaired (Zandomeneghi, Festa, Carbonaro, Galleschi, Lenzi, & Calucci, 2000). Wheat grain contains ample amounts of these minerals Mg, Fe, Zn, and some trace elements, which are essential for good preventive nutrition. Aleurone contains the most cereal grain minerals, therefore the mineral accumulated bran fraction is achieved after milling. Cubadda, Aureli, Raggi, and Carcea (2009) described varied extent of mineral loss during milling of durum wheat grains. After milling, semolina retained the highest concentrations nearby 77–85% of that in grain (dry weight basis), accompanied by Ca (54–60%), copper (Cu, 49–53%) potassium (K) and P (42–47%), Fe (36–38%), and Mg and Zn (32–36%). In wheat grain, accumulation of Fe mainly takes place in aleurone cells in the form of either soluble or insoluble phytate salts as per the nature of the bonding. Insoluble forms are not available to iron 11 transporters in the human gut, but it is evident that soluble salts as monoferric phytate (MFP) as source of Fe, may be a bioavailable (Sandberg, Brune, Carlsson, Hallberg, Skoglund, & Rossander-Hulthén, 1999).

The studies carried out on white flour and whole grain in relation to convergence of Fe showed that Fe complexed with NA/DMA in white flour was four to five times higher than the whole grain (Eagling, Wawer, Shewry, Zhao, & Fairweather-Tait, 2014; Eagling et al. 2014b). Further, the endosperm contains significant lower content of Fe than the bran, but the lower phytic acid content of the endosperm leads to the better source of Fe in bioavailable form. Aleurone layer is also embodied of Mg and Zn largely as a phytate salts, while Mg having been colocalized within globoid crystal with P and K. Speciation and localization studies on Zn showed its accumulation in endosperm as well, in association with small cysteine-rich proteins (Persson, de Bang, Pedas, Kutman, Cakmak, Andersen, et al., 2016; Persson et al. 2016). The chelating activity of phytic acid in hindering bioaccessibility of Fe, Mg, and Zn, severely cut down the bioavailability of such minerals in wheat products (Das, Raychaudhuri, & Chakraborty, 2012).

Interestingly, breeding of wheat remains a viable option in reducing phytic acid, whilst processing may also function as the most effective strategy to solubilize Fe, Mg, and Zn from phytate salts in wheat flour (Magallanes-López, Hernandez-Espinosa, Velu, Posadas-Romano, Ordoñez-Villegas, Crossa, et al., 2017). The process of sourdough carries drastic degradation of phytate by the activation of wheat endogenous phytase through microbial acidification of the dough. Leenhardt, Levrat-Verny, Chanliaud, and Rémésy (2005) displayed that the 4-hour sourdough fermentation reduced the phytic acid to 70%, followed by five times increment in soluble Mg, while Rodriguez-Ramiro, Brearley, Bruggraber, Perfecto, Shewry, and Fairweather-Tait (2017) reported almost complete destruction of phytic acid after 36-hour sourdough fermentation. Germination of wheat substantially increase the phytase activity, which results in the drastic increase in the availability of Zn, but not much for Fe (Liu, Qiu, & Beta, 2010). The process of extrusion cooking (a high-temperature for a short-time), used for the production of various breakfast cereals and salty and sweet cereal snacks, was shown to enhance the mineral availability efficiently. Extrusion is supposed to hydrolyse the phytic acid and mineral complex to release phosphate molecules with 13–35% reduction in phytate content (Andersson, Hedlund, Jonsson, & Svensson, 1981). The presence of tannins and fibers also evidenced to impair the absorption of minerals from wheat grain that is subjected to form insoluble complexes with divalent ions in the gastrointestinal tract. Shear forces and elevated temperature in extrusion cooking lead to the accountable destruction of polyphenols and may carry out the modification of fiber components and their chelating properties, which ultimately contribute to enhance the bioavailability of minerals (Singh, Gamlath, & Wakeling, 2007; Wang, Klopfenstein, & Ponte, 1993).

1.6.6 Vitamin B Complex

Vitamin B is a water-soluble vitamin in wheat grains, occuring in various forms as thiamine (B1), riboflavin (B2), niacin (B3), pantothenic acid (B5), pyridoxine (B6), and folate (B9). They act as coenzymes or their precursors in human body, and also as components implicated in genetic regulation and genomic stability. Wheat grain is a vital source of the vitamin-B complex, though a high genetic difference in their content between and within wheat species have been proclaimed. Bran and germ parts of wheat are usually rich in B vitamins. The aleurone layer is specifically full of vitamin B3 containing about 171–741 mg/kg (dry matter) and B9 possessing 4.0–6.0 mg/kg (fresh weight basis) (Ndolo, Fulcher, & Beta, 2015). White flour contains significantly lower vitamin B content which is 1.46–2.19 of B1, 0.43–0.58 of B2, and 0.28–0.52 mg/kg (dry matter.) of B6 when compared with whole meal encompassing 2.24–4.16 of B1, 0.75–0.96, of B2, and 1.31–2.58 mg/kg (dry matter) of B6 (Batifoulier, Verny, Chanliaud, Rémésy, & Demigné, 2006). These data indicate the occurrence of considerable vitamin retardation during milling; therefore, it is recommended to consume the high end amount of whole-meal products for the fulfilment of the recommended dietary allowance for the B vitamins. Usually, the classical breadmaking process exhibited a significant loss of B vitamins. Nurit et al. (2016) explained that the bread making caused a noticeable destruction of vitamins B1, B5, and B6, although substantial increase in vitamin B2. Contrasting variations were demonstrated by vitamin B3, where nicotinic acid was reduced and nicotinamide was increased. Batifoulier, Verny, Chanliaud, Rémésy, and Demigné (2005) proposed resolution to this obstacle by illustrating a long yeast fermentation process, which led to an escalation of B1 and B2 concentrations by virtue of yeast metabolism. Further, studies depicted the influence of processing on vitamin B9, which is vital for normal human metabolic function. Germination is one the processes that appeared to have the highest impact on folate content of wheat flours. Hefni and Witthöft (2012) determined the folate content of wheat grain after germination was 0.58 mg/kg and of 0.14 mg/kg (dry matter), respectively, yet a three-to-six-fold increase in total folate was observed in both studies.

The bread-making process did not seem to affect the flour native folates (B9) content as Gujska and Majewska (2005) stated that the flour-native folate content was increased from flour to proofed dough by the action of fermenting yeast. On the contrary, the folate content was reduced upon baking and attained the values near to the native flour.

1.6.7 Dietary Fiber

Dietary fiber is the consumable homologous carbohydrate from plant parts that is resistant to digestion and absorption in the small intestine of humans, along with complete or partial fermentation in the large intestine, and a bolstered physiological effect or benefit to health (AACC, 2001). Dietary fiber of mature wheat grain is represented by the cell-wall polysaccharides, which are extensively present in the grain. Lignin is a phenolic polymer present only in the pericarp or seed coat, while bran tissues are enriched with fractans and endosperm cells contains some portion. Resistant starch are present only in minute quantity in the endosperm cells (Stone & Morell, 2009). The nonstarch polysaccharides are the part of cell wall and are the prime component of dietary fiber, accounting for approximately 11% (dry weight) of the mature wheat grain (Andersson, Dinberg, Aman, & Landberg, 2014). The content and composition of cell-wall nonstarch polysaccharides vary among the grain tissues as in endosperm. The outer layer of starchy endosperm contains at least 50% of cell wall material in the form of 60% arabinoxylan, 25% cellulose, and ~10% lignin (Shewry & Hey, 2015; Stone & Morell, 2009). The cell walls of mature grain comprised of about 35–40% of the aleurone layer and accounts mainly 65% arabinoxylan and 30% β-glucan (Bacic & Stone, 1981). The starchy endosperm cells are structured of thinner walls (2–3% dry weight), consisting of 70% arabinoxylan and 20% β-glucan (Stone & Morell, 2009). Arabinoxylan and β-glucan exist in varied proportion in different parts of whole grain, where β-glucan has been reported in frequent abundance close to the germ. However, in endosperm, arabinose in arabinoxylan was observed increasing from outward to inward, being a complex polymer of aromatic alcohols found closely associated with cell-wall polysaccharides (Davin, Jourdes, Patten, Kim, Vassão, & Lewis, 2008; Saulnier, Robert, Grintchenko, Jamme, Bouchet, & Guillon, 2009). Lignin is insoluble and cements and anchor the cellulose microfibrils and other matrix polysaccharides, consequently, making very rigid cell walls. Such characteristics make it highly resistant to bacterial degradation in the large intestine. Resistant starches (3% in cereal grain) made up of starch entrapped in the food matrix escape the digestion in small intestine and become a convenient substrate by colonic microorganisms for fermentation, which generates short-chain fatty acids and has a positive effect on human health (Topping, Bajka, Bird, Clarke, Cobiac, Conlon, et al., 2008).

The whole grain contains the uneven composition of dietary fiber within distinct grain tissues. Milling can bring about a striking effect on the quantity of dietary fiber and flour composition. Wholemeal wheat flour possesses nearly 13% (dry weight basis) of total dietary fiber, arabinoxylan represents the half of it including ~0.57% water-soluble portion. White flour contains 3.5% of total dietary fiber, out of that about 75% is arabinoxylan along with the same quantity of water-soluble portion (Shewry & Hey, 2015). Therefore, the extraction rate of milling affects the dietary fiber composition of flour, and therefore, the efficacy of bran and endosperm tissue break-up. Milling parameters also disclose the particle size production, which plays a key role in determining fiber properties. Zhu, Huang, Peng, Qian, and Zhou (2010) investigated the effects of ultrafine grinding on the physicochemical properties of wheat bran dietary fiber. It was revealed that the decrease in particle size significantly contracted the hydration properties of wheat bran dietary fiber in terms of water-holding capacity, water retention, and swelling capacity and a redistribution of fiber components took place from insoluble to soluble fractions. Ultra-fine grinding also enhanced the antioxidant capacity of wheat bran, apparently owing to the greater exposure of the phenolic acids linked to fibers (Rosa et al. 2013). Extrusion cooking technology carried out the notable alteration in physicochemical properties of wheat dietary fiber by combining high temperature, pressure, and shear force. Successively producing highly expanded and low-density products in the form of ready-to-eat breakfast cereals and snacks. Extrusion, specifically at highest extrusion speed has been utilized to intensify the solubility of dietary fiber in wheat bran successfully (Rashid, Rakha, Anjum, Ahmed, & Sohail, 2015). Cleemput, Booij, Hessing, Gruppen, and Delcour (1997) also observed mixing and baking led to substantial rise (7–12%) in water extractable nonstarch polysaccharides, whilst bottom levels (0–5 percent) of water-unextractable nonstarch polysaccharides solubilized during fermentation. Further, fermentation withdrew clear changes in molecular weight distribution of arabinoxylan (AX) without modifying the arabinose to xylose (A/X) ratio. Comino, Collins, Lahnstein, and Gidley (2016) also stated an enhancement of about 18.5% in total solubilized nonstarch polysaccharides including 12.5% and 6% increase in water extractable arabinoxylan and β-glucan, respectively occurred during the bread-making process. The breadmaking process, and peculiar fermenting inoculum and fermentation parameter, fosters a significant impact on the other dietary fiber, particularly fructans and resistant starch. Yeast leavening culminates the large reduction of fructans present in flour in dough and bread (Gélinas, McKinnon, & Gagnon, 2016b). On the contrary, sourdough process tend to produce higher quantity of resistant starch (Scazzina, Del Rio, Pellegrini, & Brighenti, 2009), possibly due to the presence of organic acids evolved during fermentation that could expedite the debranching of the amylopectin moieties while baking. The making and cooking of pasta and noodles also alters the amount and composition of dietary fiber. The process of pasta making draws out reduced glycemic responses in final products, inducing the metabolic advantages of low glycemic index food (Monge, Cortassa, Fiocchi, Mussino, & Carta, 1990). Processing methods of cereal products, depending on wet-heat and extrusion cooking are being evaluated for their capacity to cause the formation of amylose-lipid complexes, which results in a novel form of resistant starch (Panyoo & Emmambux, 2017). Alkaline and/or boiling (100°C) conditions could not considerably affected the dietary fiber comprising of arabinoxylan and β-glucan in the production of yellow alkaline noodles (Comino, Collins, Lahnstein, & Gidley, 2016).

1.7 Application of Omics Technology in Nutritional Quality Improvement of Wheat

The Omics technologies as important tool in food and nutritional science have gained recent attention (Chaudhary, et al., 2020; Vyas, Singh, Singh, Kumar, & Dhaliwal, 2018). The inclusion of Omics techniques in high throughput screening studies not only provide greater opportunities of fast analysis but also offer the advantage of comprehensive study of large sample size with multiple parameters. The integration of Omics technologies including proteomics, genomics, transcriptomics, and metabolomics with other such approaches is of great importance in studies of molecular interactions of different kinds. The nutritional bioavailability of phytochemicals is a serious concern and it is affected by complex interactions of particular nutrient with genomic, molecular, and environmental factors (Bohn, McDougall, Alegría, Alminger, Arrigoni, Aura, et al., 2015; Desmarchelier & Borel, 2017). With a tremendous shift in global interest toward food quality over food quantity, poor bioavailability of various phytochemicals including micronutrients and vitamins is inviting various interventions to address this issue. Application of one or other Omics techniques is central to implementation of these interventions. This is giving rise to concepts of personalized nutrition, nutrigenomics, metagenomics, and several others new terms. The complex nature of foods further necessitates the use of this technology in comprehensive analysis of nutritional bioavailability and associated factors. Considering the dependence of the majority of the world's population on wheat as part of their diet, improving the nutritional bioavailability of phytochemicals in grain is reported in different studies. However, variation in responses, parameters studied, sample size, analytical procedures, and other associated factors between different studies is a major cause of concern. Therefore, comprehensive studies of maximum parameters using integrated Omics technologies might be significant in understanding and improving the nutritional bioavailability in this very important crop.

1.8 Negative Impact of Wheat Consumption on Health

The various disorders associated with wheat consumption in human health are explored in the sections that follow.

1.8.1 Wheat Allergy

Allergies are hypersensitive responses to foreign components; generally, those are proteins that are linked with the evolution of a specific category of antibody called IgE. Symptoms after eating wheat products cover atopic dermatitis, urticaria (hives or nettle rash), respiratory problems, and gastrointestinal symptoms. A number of assorted proteins have been involved for such conditions including α-amylase inhibitors and gluten proteins constituted of gliadin and glutenin (Matsuo, Yokooji, & Taogoshi, 2015). Zuidmeer, Goldhahn, Rona, Gislason, Madsen, Summers, et al. (2008) described a meticulous review of the pervasiveness of allergies by plant foods and took into account the data from 15 studies with regard to wheat allergy. Consequently, a group of subjects was formed ranging between approximately 500 and 10,000, who were diagnosed by skin prick test, serum IgE analysis, and perception of wheat allergy analyzed by questionnaires. In the four studies, where children of 3–14 years of age were tested with a diet based on wheat exhibited a mean predominance of 0.25% (7 in 2,807 subjects) alongside a range from 0–0.5%. Two groups investigated in the UK showed 0.2% prevalences in children 9–12 years of age and 0.3% in children of 6 years (Venter, Pereira, Grundy, Clayton, Arshad, & Dean, 2006a; Venter, Pereira, Grundy, Clayton, Roberts, Higgins, et al., 2006b). Wheat sensitivity was found prevalent (1.2 prrcent) in UK adolescents of 15 years as revealed by the responses of skin prick test, in comparison to UK children (0.2–0.6%) of 4–11 years (Pereira, Venter, Grundy, Clayton, Arshad, & Dean, 2005; Venter, Pereira, Grundy, Clayton, Arshad, & Dean, 2006a; Venter, et al., 2006b). Adults of many other countries showed high level of sensitization to 3.6% on the basis of IgE estimation to wheat in the serum. It was also reported that infants have wheat allergies that commonly later disappear. This outcome is evident considering the data on perception of wheat allergy, which is likely to be greater than sensitization to dietary challenge in children than adults. Out of the 36 studies under meta-analysis, only six applied the double-blind food challenges that demonstrated the prevalence of wheat allergy within the limits when compared with allergies reported by other plant foods such as fruit, nuts, vegetables, soy, and sesame (Zuidmeer, et al., 2008).

1.8.2 Wheat-Dependent Exercise-Induced Anaphylaxis (WDEIA)

Wheat-dependent exercise-induced anaphylaxis (WDEIA) is a precisely characterized type of wheat allergy. This kind of allergic response is triggered by the ingestion of wheat- and crustacean-based food most commonly, accompanied by physical exercise (Beaudouin et al. 2006). Yokooji, Okamura, Chinuki, Morita, Harada, Hiragun, et al. (2015) studied WDEIA in detail and recognized its two forms. Conventional (CO) WDEIA is defined as the prevailing form, contemplated to be sensitized by the gastrointestinal tract, considering the ω-5 gliadin as a major allergen (Morita, Matsuo, Mihara, Morimoto, Savage, & Tatham, 2003; Palosuo, Varjonen, Kekki, Klemola, Kalkkinen, Alenius, et al., 2001). Another type of WDEIA seems to be sensitized through the skin and/or mucosa via hydrolyzed wheat protein (HWP) as an ingredient of a soap. γ-gliadin has appeared as the dominant sensitizing agent in HWPWDEIA and exposure to soap or consumption of wheat may account for the reactions (Yokooji, Kurihara, Murakami, Chinuki, Takahashi, Morita, et al., 2013). The outbreak of food-dependent exercise-induced anaphylaxis, including WDEIA, has been reported up to 0.017% among Japanese children (Aihara, Takahashi, Kotoyori, Mitsuda, Ito, Aihara, et al., 2001). During the investigation of 935 Japanese adults for wheat allergy, which was including WDEIA by using questionnaires, skin prick tests, and estimation of ω-5 gliadin specific IgE, only two allergic subjects (0.21%) were identified (Morita, Chinuki, Takahashi, Nabika, Yamasaki, & Shiwaku, 2012).

1.8.3 Celiac Disease

Celiac disease is an autoimmune condition, which affects the small intestine and gives rise to malabsorption, fatigue, weight loss, abdominal pain, vomiting and diarrhea. As a result, patients with celiac disease (CD) encounter nutrient deficiencies that can cause iron anemia and folate deficiency. However, this disease may also be asymptomatic or outbreak with only mild symptoms in some individuals. The role of wheat gluten proteins (gliadins and glutenins) is well established in triggering CD (Gilissen, van der Meer, & Smulders, 2014). Presently, 31 short peptide sequences have been defined as celiac toxic named as "epitops" are found in wheat gluten proteins and several barley and rye proteins (Sollid, Qiao, Anderson, Gianfrani, & Koning, 2012). However, originally CD has been taken as a pediatric condition, but is now identified in different age groups and extensive screening has shown a significant level of undiagnosed CD in adults. Likewise, a study carried out in Cambridge (UK) including 7,550 participants revealed that 1.2% of adults between 45–76 years of age were serologically positive (West, Logan, Hill, Lloyd, Lewis, Hubbard, et al., 2003). In an another study, 16,847 adults aged 50 years or more in Minnesota depicted 0.8% undiagnosed CD (Godfrey, Brantner, Brinjikji, Christensen, Brogan, Van Dyke, et al., 2010). Eventually, the occurrence of CD in Europe and countries with high European ancestry as the United States and Australia is widely predicted at nearly 1% of the population, whist substantial variation takes place from as low as 0.2% to more than 5% between different countries (Godfrey, et al., 2010; Rubio-Tapia, Ludvigsson, Brantner, Murray, & Everhart, 2012; Walker, Murray, Ronkainen, Aro, Storskrubb, D'Amato, et al., 2010). An escalating predominance of CD has been attributed to changes in infant feeding in Sweden (Myléus, Ivarsson, Webb, Danielsson, Hernell, Högberg, et al., 2009) while Lohi, Mustalahti, Kaukinen, Laurila, Collin, Rissanen, et al. (2007) stated a two-fold enhancement (from 1.05% to 1.99%) in CD in adults between 1978–1980 and 2000–2001 in Finland. Green and Cellier (2007) noted that CD is about two times as prevalent in women than men, in consideration with higher prevalence of other autoimmune diseases. It was also reported that the prevalence in women decreased after about age 65 (Green, Stavropoulos, Panagi, Goldstein, McMahon, Absan, et al., 2001). Commonly, it has been observed that even when the initial serological screening of CD is confirmed by small bowel biopsy, patients may not feel differences in bodyweight or other symptoms. However, the association of CD with multiplied risk of a range of other disorders cannot be ignored, which suggests the treatment of this condition even in the absence of symptoms (Godfrey, et al., 2010; P. H. Green & Cellier, 2007; Solaymani-Dodaran, West, & Logan, 2007).

1.8.4 Nonceliac Gluten Sensitivity (NCGS)

A growing number of patients have reported symptoms linked with wheat consumption, other than the classical allergic or autoimmune responses; this new condition was therefore termed as nonceliac gluten sensitivity (NCGS). The extent of symptoms of NCGS varies widely, such as gastrointestinal symptoms, pains in muscles and joints, tiredness, headache, dermatitis, depression, anxiety and anemia. It has also not been postulated definitely whether NCGS produces a single syndrome or a combination of several conditions (Sapone, Bai, Ciacci, Dolinsek, Green, Hadjivassiliou, et al., 2012). In addition, the role of gluten has not been defined clearly for NCGS, which indicates the action of other grain components for the outbreak of related symptoms. So, it may be defined as nonceliac wheat sensitivity (NCWS) more appropriately. The pathogenesis of NCGS/NCWS is not comprehended, yet it is likely to present a mixture of factors considering the stimulation of the innate immune system. This shortcoming of perception causes a difficulty in diagnosis, mostly due to the lack of validated diagnostic criteria and the high nocebo and placebo effects of gluten (Diez-Sampedro, Olenick, Maltseva, & Flowers, 2019). Recently, the recommendation of a gluten-free diet followed by a double-blind, placebo-controlled gluten challenge, featuring a variation of 30% or more in one to three main symptoms, contained a positive result in both phases (Catassi, Elli, Bonaz, Bouma, Carroccio, Castillejo, et al., 2015). The accurate prevalence of NCGS/NCWS will not be obtained until these criteria are precisely employed. Therefore, the prevalence stated in prior published investigations is likely to be larger than the true values. Further, a survey of was conducted in the Sheffield area of the UK including 1,002 adults, 13% identified with self-reported gluten sensitivity (GS) while successive study with 200 GS patients showed the prevalence of CD by 7% and NCGS by 93% (Aziz, Lewis, Hadjivassiliou, Winfield, Rugg, Kelsall, et al., 2014). The ratio of females to males of NCGS patients in the latter studies was approximately 4:1 (Aziz, et al., 2014; Volta, Bardella, Calabrò, Troncone, & Corazza, 2014).

1.8.5 Schizophrenia and Autism Spectrum Disorder

The relationship of ingestion of wheat and milk with schizophrenia and autism spectrum disorder have been investigated extensively. Patients with both types of conditions improved by interventions with gluten-free and casein-free diets (Whiteley, Haracopos, Knivsberg, Reichelt, Parlar, Jacobsen, et al., 2010;

Whiteley, Shattock, Knivsberg, Seim, Reichelt, Todd, et al., 2013). The speculation has been made that the release of neuroactive peptides after digestion of wheat gluten is responsible for such neurological effects and induced the theory of gluteomorphins (Dohan, Harper, Clark, Rodrigue, & Zigas, 1984). Suggestively, these peptides are considered as opioid peptides, released in the gastrointestinal tract after gluten digestion, which are subsequently absorbed into the bloodstream and transmit neurological effects and "addictive" properties (Shewry & Hey, 2016).

1.8.6 FODMAPs and Gastrointestinal Disorders

As is common with diverse plant foods, wheat also possesses little amount of fermentable carbohydrates known as FODMAPs (fermentable, oligo-, di-, monosaccharides and polyols). Fractans are found most abundantly about 2% (dry weight), sucrose about 0.5-1.5% (dry weight) and raffinose about 0.2–0.7% dry weight as FODMAPs (Shewry & Hey, 2016). Irritable bowel disease (IBS) and inflammatory bowel disease may be improved by the presence of a low-FODMAP diet, as fermentation in the colon is reduced (Muir & Gibson, 2013). Again, self-reported NCGS patients showed a considerable improvement of symptoms under a tightly controlled intervention trial using a low-FODMAP diet (Biesiekierski, Peters, Newnham, Rosella, Muir, & Gibson, 2013). Therefore, the low quantities of FODMAPs in gluten-free products may responsible for improvements in the condition of IBS and NCGS patients on gluten-free diets (Muir & Gibson, 2013).

1.9 Potential Harm of a Gluten-Free Diet

A study in 2015 reviewed the records of patients of celiac disease found that 11% of the subjects had avoided gluten at some point without a celiac disease diagnosis. The reasons revealed behind the adoption of gluten-free diet included lactose intolerance and irritable bowel syndrome (Tanpowpong, Broder-Fingert, Katz, & Camargo, 2015). Similarly, a study that contained 579 children and adolescents as subjects reported that 7.4% were avoiding gluten without a diagnosis of celiac disease. The study disclosed that the strongest predictors of participants consuming the gluten-free diet featured irritability, family history of celiac disease, diarrhea, bowel movement changes, and autism (Tanpowpong, Broder-Fingert, Katz, & Camargo, 2012). It is evident that there are considerable drawbacks of following the gluten-free diet. For instance, gluten-free processed grain products, as in breads, cereals, and crackers, generally contain lower amounts of dietary fiber, vitamins, and minerals (Missbach, Schwingshackl, Billmann, Mystek, Hickelsberger, Bauer, et al., 2015). Such diets are suspected to raise the risks for nutritional deficiencies, particularly in B vitamins, Fe, and other trace minerals, as wheat (particularly wheat bran and germ) is rich source of dietary fiber, vitamin B complex, minerals, and phytochemicals, including phenolic acids and betaine, which produce enormous health benefits (Shewry & Hey, 2016). Persons, who consume a gluten-free diet also may fail to adhere to recommendations related to daily servings of cereal grain products. Furthermore, an investigation found that gluten-free bread and bakery products were 267% more expensive than gluten-containing breads on an average, while gluten-free cereals were 205% more expensive than gluten-containing cereals grains (Missbach, et al., 2015).

The impact of wheat among on the general public's health is escalating. Specifically, in the United States, one-third of adults wished to cut down or exclude gluten consumption. The sale of gluten-free products in 2015 amounted to nearly $2.79 billion U.S. In accordance with the industry experts, in 2014, the United States was the leading market in terms of gluten-free food retail sales. Italy and the United Kingdom rounded off the dominant three consumer markets. Therefore, a valuable market for gluten-free foods was turned up, and forecasted to excel to nearby $7.59 billion in the United States by 2020 (Statistica, 2020b). As several studies found, gluten-free foods may be depleted in essential nutrients and micronutrients needed for daily basic nutrition levels in comparison to conventional diets (Pellegrini & Agostoni, 2015; Wu, Neal, Trevena, Crino, Stuart-Smith, Faulkner-Hogg, et al., 2015) and food scientists have recognized the challenge of enhancement in the nutritional quality and health benefits of gluten-free diets (Capriles, dos Santos, & Arêas, 2016).

1.10 Conclusion

Wheat consumption is increasing globally, even in the countries or regions where climates are not as suitable for wheat production. Enormous options in terms of wheat-based foods contribute to various essential and healthful ingredients to the human diet, as protein, B vitamins, dietary fiber, minerals, and phytochemicals (Punia & Sandhu, 2016). These components may also range widely in quantity and composition by virtue of variation in genotype, environment, and the extent and type of processing. Dietary fiber is specifically important as a constituent of food, being associated in reducing the risk of CVD, type 2 diabetes, and numerous forms of cancer. There is limited research on integrating selected Omics technologies for improving the bioavailability and stability of selected nutritional constituents, thus improving the nutritional characteristics of wheat. The respective research outputs should provide data that can be easily utilized to improve nutritional quality.

The last few years have seen increasing cases of various diseases related to wheat, including wheat allergy, celiac disease, wheat-dependent exercise-induced anaphylaxis, nonceliac gluten sensitivity, schizophrenia, autism spectrum disorder, FODMAPs and gastrointestinal disorder. However, adverse reactions to wheat were well understood only a decade before and some sort of related diagnoses are still a challenge full of complications. To monitor the prevalence of the above stated gluten disorders, researchers will need to observe a large population on a gluten-free diet to monitor for signs and reasons of the increasing rates of disease.

REFERENCES

AACC. (2001). The Definition of Dietary Fiber. *Cereal Foods World*, *46*(3), 112–126.

Adefegha, S. A., Olasehinde, T. A., & Oboh, G. (2018). Pasting alters glycemic index, antioxidant activities, and starch hydrolyzing enzyme inhibitory properties of whole wheat flour. *Food Science & Nutrition*, *6*, 1591–1600.

Aihara, Y., Takahashi, Y., Kotoyori, T., Mitsuda, T., Ito, R., Aihara, M., Ikezawa, Z., & Yokota, S. (2001). Frequency of food-dependent, exercise-induced anaphylaxis in Japanese junior-high-school students. *Journal of Allergy and Clinical Immunology*, *108*(6), 1035–1039.

Al-Saleh, A., & Brennan, C. S. (2012). Bread wheat quality: some physical, chemical and rheological characteristics of Syrian and English bread wheat samples. *Foods*, *1*, 3–17.

Andersson, A. A., Kamal-Eldin, A., & Aman, P. (2010). Effects of environment and variety on alkylresorcinols in wheat in the HEALTHGRAIN diversity screen. *Journal of Agricultural Food Chemistry*, *58*(17), 9299–9305.

Andersson, A. A. M., Dinberg, L., Aman, P., & Landberg, R. (2014). Recent findings on certain bioactive components in whole grain wheat and rye. *Journal of Cereal Science*, *59*, 294–311.

Andersson, S. W., Skinner, J., Ellegard, L., Welch, A. A., Bingham, S. A., Bingham, S., Mulligan, A., Andersson, H., & Khaw, K. T. (2004). Intake of dietary plant sterols is inversely related to serum cholesterol concentration in men and women in the EPIC Norfolk population: a cross-sectional study. *European Journal of Clinical Nutrition*, *58*(10), 1378–1385.

Andersson, Y., Hedlund, B., Jonsson, L., & Svensson, S. (1981). Extrusion cooking of a high-fiber cereal product with crispbread character. *Cereal Chemistry*, *58*, 370–374.

Atkinson, F. S., Foster-Powell, K., & Brand-Miller, J. C. (2008). International tables of glycemic index and glycemic load values. *Diabetes Care*, *31*, 2281–2283.

Augustin, L. S., Kendall, C. W., Jenkins, D. J., Willett, W. C., Astrup, A., Barclay, A. W., Björck, I., Brand-Miller, J. C., Brighenti, F., Buyken, A. E., Ceriello, A., La Vecchia, C., Livesey, G., Liu, S., Riccardi, G., Rizkalla, S. W., Sievenpiper, J. L., Trichopoulou, A., Wolever, T. M., Baer-Sinnott, S., & Poli, A. (2015). Glycemic index, glycemic load and glycemic response: An International Scientific Consensus Summit from the International Carbohydrate Quality Consortium (ICQC). *Nutrition, Metabolism & Cardiovascular Diseases*, *25*(9), 795–815.

Aziz, I., Lewis, N. R., Hadjivassiliou, M., Winfield, S. N., Rugg, N., Kelsall, A., Newrick, L., & Sanders, D. S. (2014). A UK study assessing the population prevalence of self-reported gluten sensitivity and referral characteristics to secondary care. *European Journal of Gastroenterology and Hepatology*, *26*(1), 33–39.

Babu, C. R., Ketanapalli, H., Beebi, S. K., & Kolluru, V. C. (2018). Wheat bran-composition and nutritional quality: A review. *Advances in Biotechnology & Microbiology, 9,* 21–27.

Bacic, A., & Stone, B. (1981). Chemistry and organization of aleurone cell wall components from wheat and barley. *Functional Plant Biology, 8*(5), 475–495.

Balk, J., Connorton, J. M., Wan, Y., Lovegrove, A., Moore, K. L., Uauy, C., Sharp, P. A., & Shewry, P. R. (2019). Improving wheat as a source of iron and zinc for global nutrition. *Nutrition bulletin, 44*(1), 53–59.

Barron, C., Samson, M. F., Lullien-Pellerin, V., & Rouau, X. (2011). Wheat grain tissue proportions in milling fractions using biochemical marker measurements: Application to different wheat cultivars. *Journal of Cereal Science, 53*(3), 306–311.

Batifoulier, F., Verny, M. A., Chanliaud, E., Rémésy, C., & Demigné, C. (2005). Effect of different breadmaking methods on thiamine, riboflavin and pyridoxine contents of wheat bread. *Journal of Cereal Science, 42,* 101–108.

Batifoulier, F., Verny, M. A., Chanliaud, E., Rémésy, C., & Demigné, C. (2006). Variability of B vitamin concentrations in wheat grain, milling fractions and bread products. *European Journal of Agronomy, 25*(2), 163–169.

Belderok, B., Mesdag, H., & Donner, D. A. (2000). *Bread-Making quality of wheat—A century of breeding in Europe.* New York: Springer.

Bicar, E. H., Woodman-Clikeman, W., Sangtong, V., Peterson, J. M., Yang, S. S., Lee, M., & Scott, M. P. (2008). Transgenic maize endosperm containing a milk protein has improved amino acid balance. *Transgenic Research, 17,* 59–71.

Biesiekierski, J. R., Peters, S. L., Newnham, E. D., Rosella, O., Muir, J. G., & Gibson, P. R. (2013). No effects of gluten in patients with self-reported non-celiac gluten sensitivity after dietary reduction of fermentable, poorly absorbed, short-chain carbohydrates. *Gastroenterology, 145*(2), 320–328, e321-323.

Bilgiçli, N., İbanoğlu, S., & Herken, E. N. (2007). Effect of dietary fibre addition on the selected nutritional properties of cookies. *Journal of Food Engineering, 78,* 86–89.

Bohn, T., McDougall, G. J., Alegría, A., Alminger, M., Arrigoni, E., Aura, A. M., & Martínez-Cuesta, M. C. (2015). Mind the gap—deficits in our knowledge of aspects impacting the bioavailability of phytochemicals and their metabolites—a position paper focusing on carotenoids and polyphenols. *Molecular Nutrition & Food Research, 59*(7), 1307–1323.

Bordiga, M., Locatelli, M., Travaglia, F., & Arlorio, M. (2016). Alkylresorcinol content in whole grains and pearled fractions of wheat and barley. *Journal of Cereal Science, 70,* 38–46.

Borrill, P., Connorton, J. M., Balk, J., Miller, A. J., Sanders, D., & Uauy, C. (2014). Biofortification of wheat grain with iron and zinc: integrating novel genomic resources and knowledge from model crops. *Front Plant Sci, 5,* 53.

Beaudouin, E., Renaudin, J. M., Morisset, M., Codreanu, F., Kanny, G., Moneret-Vautrin, D. A. (2006). Food-dependent exercise-induced anaphylaxis--update and current data. *Europian Annals of Allergy & Clinical Immunology, 38*(2), 45–51.

Brouns, F., van Rooy, G., Shewry, P., Rustgi, S., & Jonkers, D. (2019). Adverse reactions to wheat or wheat components. *Comprehensive Reviews in Food Science and Food Safety, 18*(5), 1437–1452.

Buttriss, J. L., & Stokes, C. S. (2008). Dietary fibre and health: an overview. *British Nutrition Foundation Nutrition Bulletin, 33,* 186–200.

Cakmak, I., Pfeiffer, W. H., & McClafferty, B. (2010). REVIEW: Biofortification of durum wheat with zinc and iron. *Cereal Chemistry, 87*(1), 10–20.

Capriles, V. D., dos Santos, F. G., & Arêas, J. A. G. (2016). Gluten-free breadmaking: Improving nutritional and bioactive compounds. *Journal of Cereal Science, 67,* 83–91.

Catassi, C., Elli, L., Bonaz, B., Bouma, G., Carroccio, A., Castillejo, G., Cellier, C., Cristofori, F., de Magistris, L., Dolinsek, J., Dieterich, W., Francavilla, R., Hadjivassiliou, M., Holtmeier, W., Körner, U., Leffler, D. A., Lundin, K. E. A., Mazzarella, G., Mulder, C. J., Pellegrini, N., Rostami, K., Sanders, D., Skodje, G. I., Schuppan, D., Ullrich, R., Volta, U., Williams, M., Zevallos, V. F., Zopf, Y., & Fasano, A. (2015). Diagnosis of non-celiac gluten sensitivity (NCGS): The Salerno experts' criteria. *Nutrients, 7*(6), 4966–4977.

Chaudhary, N., Kumar, V., Sangwan, P., Pant, N. C., Saxena, A., Joshi, S., & Yadav, A. N. (2020). Personalized nutrition and -omics. In *Reference Module in Food Science*: Elsevier.

Cleemput, G., Booij, C., Hessing, M., Gruppen, H., & Delcour, J. A. (1997). Solubilisation and changes in molecular weight distribution of arabinoxylans and protein in wheat flours during bread-making, and the effects of endogenous arabinoxylan hydrolysing enzymes. *Journal of Cereal Science, 26,* 55–66.

Comino, P., Collins, H., Lahnstein, J., & Gidley, M. J. (2016). Effects of diverse food processing conditions on the structure and solubility of wheat, barley and rye endosperm dietary fibre. *Journal of Food Engineering, 169,* 228–237.

Cooper, D. N., Kable, M. E., Marco, M. L., De Leon, A., Rust, B., Baker, J. E., Horn, W., Burnett, D., & Keim, N. L. (2017). The effects of moderate whole grain consumption on fasting glucose and lipids, gastrointestinal symptoms, and microbiota. *Nutrients, 9*(2).

Craig, S. A. (2004). Betaine in human nutrition. *American Journal of Clinical Nutrition, 80,* 539–549.

Cubadda, F., Aureli, F., Raggi, A., & Carcea, M. (2009). Effect of milling, pasta making and cooking on minerals in durum wheat. *Journal of Cereal Science, 49,* 92–97.

Das, A., Raychaudhuri, U., & Chakraborty, R. (2012). Cereal based functional food of Indian subcontinent: a review. *Journal of Food Science and Technology, 49*(6), 665–672.

Davin, L. B., Jourdes, M., Patten, A. M., Kim, K.-W., Vassão, D. G., & Lewis, N. G. (2008). Dissection of lignin macromolecular configuration and assembly: Comparison to related biochemical processes in allyl/propenyl phenol and lignan biosynthesis. *Natural Product Reports, 25*(6), 1015–1090.

Desmarchelier, C., & Borel, P. (2017). Overview of carotenoid bioavailability determinants: From dietary factors to host genetic variations. *Trends in Food Science & Technology, 69,* 270–280.

Diez-Sampedro, A., Olenick, M., Maltseva, T., & Flowers, M. (2019). A gluten-free diet, not an appropriate choice without a medical diagnosis. *Journal of Nutrition and Metabolism, 2019,* 2438934.

Distelfeld, A., Cakmak, I., Peleg, Z., Ozturk, L., Yazici, A. M., Budak, H., Saranga, Y., & Fahima, T. (2007). Multiple QTL-effects of wheat Gpc-B1 locus on grain protein and micronutrient concentrations. *Physiologia Plantarum, 129*(3), 635–643.

Dohan, F. C., Harper, E. H., Clark, M. H., Rodrigue, R. B., & Zigas, V. (1984). Is schizophrenia rare if grain is rare? *Biological psychiatry, 19*(3), 385–399.

Dubcovsky, J., & Dvorak, J. (2007). Genome plasticity a key factor in the success of polyploidy wheat under domestication. *Science, 316,* 1862–1866.

Eagling, T., Wawer, A. A., Shewry, P. R., Zhao, F. J., & Fairweather-Tait, S. J. (2014). Iron bioavailability in two commercial cultivars of wheat: comparison between wholegrain and white flour and the effects of nicotianamine and 2'-deoxymugineic acid on iron uptake into Caco-2 cells. *Journal of Agricultural and Food Chemistry, 62*(42), 10320–10325.

European Food Information Council. (2015). Whole Grain Fact Sheet. In). Available online: https://www.eufic.org/en/whats-in-food/article/whole-grains-updated-2015

Greenfield, H. & Southgate, D. A. T. (2003). *Food Composition data-Production, management and use* (second edition). In Review of mthods of analysis, (pp. 103). Food & Agriculture Organization of the United Nations, Rome.

Feldman, M. (1995). Wheats. In J. Smartt & N. Simmonds (Eds.), *Evolution of Crop Plants).* Harlow, UK: Longman Scientific and Technical.

Feldman, M. (2001). *Origin of cultivated wheat.* In A. P. Bonjean & W. J. Angus (Eds.), (pp. 3–56). Paris, France: Lavoisier Publishing.

Gayathri, D., & Rashmi, B. S. (2016). Critical Analysis of Wheat as Food. *Maternal and Pediatric Nutrition, 2*(2), 1–3.

Gélinas, P., McKinnon, C., & Gagnon, F. (2016a). Fructans, water-soluble fibre and fermentable sugars in bread and pasta made with ancient and modern wheat. *International Journal of Food Science and Technology, 51*(3), 555–564.

Gélinas, P., McKinnon, C., & Gagnon, F. (2016b). Fructans, water-soluble fibre and fermentable sugars in bread and pasta made with ancient and modern wheat. *International Journal of Food Science & Technology, 51*(3), 555–564.

Gibson, P. R., & Shepherd, S. J. (2005). Personal view: Food for thought–Western lifestyle and susceptibility to Crohn's disease. The FODMAP hypothesis. *Aliment Pharmacological Therories, 21*(12), 1399–1409.

Gilissen, L. J. W. J., van der Meer, I. M., & Smulders, M. J. M. (2014). Reducing the incidence of allergy and intolerance to cereals. *Journal of Cereal Science, 59,* 337–353.

Godfrey, J. D., Brantner, T. L., Brinjikji, W., Christensen, K. N., Brogan, D. L., Van Dyke, C. T., Lahr, B. D., Larson, J. J., Rubio-Tapia, A., Melton, L. J., 3rd, Zinsmeister, A. R., Kyle, R. A., & Murray, J. A. (2010). Morbidity and mortality among older individuals with undiagnosed celiac disease. *Gastroenterology, 139*(3), 763–769.

Green, P. H., & Cellier, C. (2007). Celiac disease. *New England Journal of Medicine, 357*(17), 1731–1743.

Green, P. H. R., Stavropoulos, S. N., Panagi, S. G., Goldstein, S. L., McMahon, D. J., Absan, H., & Neugut, A. I. (2001). Characteristics of adult celiac disease in the USA: results of a national survey. *American Journal of Gastroenterology, 96*(1), 126–131.

Gujska, E., & Majewska, K. (2005). Effect of baking process on added folic acid and endogenous folates stability in wheat and rye breads. *Plant Foods Human Nutrition, 60*(2), 37–42.

Håkansson, B., & Jägerstad, M. (1990). The effect of thermal inactivation of lipoxygenase on the stability of vitamin E in wheat. *Journal of Cereal Science, 12*, 177–185.

Haska, L., Nyman, M. & Andersson, R. (2008). Distribution and characterisation of fructan in wheat milling fractions. *Journal of Cereal Science, 48*, 768–774.

Hefni, M., & Witthöft, C. M. (2012). Enhancement of the folate content in Egyptian pita bread. *Food Nutrition and Research, 56*.

Hemery, Y., Lullien-Pellerin, V., Rouau, X., Abecassis, J., Samson, M.-F., Åman, P., von Reding, W., Spoerndli, C., & Barron, C. (2009). Biochemical markers: Efficient tools for the assessment of wheat grain tissue proportions in milling fractions. *Journal of Cereal Science, 49*(1), 55–64.

Hemery, Y., Rouau, X., Lullien-Pellerin, V., Barron, C., & Abecassis, J. (2007). Dry processes to develop wheat fractions and products with enhanced nutritional quality. *Journal of Cereal Science, 46*, 327–347.

Huynh, B.-L., Lachlan, P., Mather, D. E., Wallwork, H., Graham, R. D., Welch, R. M., & Stangoulis, J. C. R. (2008). Genotypic variation in wheat grain fructan content revealed by a simplified HPLC method. *Journal of Cereal Science, 48*, 369–378.

Jones, J. M., Peña, R. J., Korczak, R., & Braun, H. J. (2017). The wheat and nutrition series: A compilation of studies on wheat and health. *Cereal Foods World*, 1-174 (Papers published by the International Maize and What Improvement Center (CIMMYT) as a part of special series)

Joppa, L. R., Du, C., Hart, G. E., & Hareland, G. A. (1997). Mapping gene(s) for grain protein in tetraploid wheat (Triticum turgidum L.) Using a population of recombinant inbred chromosome lines. *Crop Science, 37*(5), cropsci1997.0011183X003700050030x.

Kamal-Eldin, A., & Appelqvist, L. A. (1996). The chemistry and antioxidant properties of tocopherols and tocotrienols. *Lipids, 31*(7), 671–701.

Keaveney, E. M., Price, R. K., Hamill, L. L., Wallace, J. M., McNulty, H., Ward, M., Strain, J. J., Ueland, P. M., Molloy, A. M., Piironen, V., von Reding, W., Shewry, P. R., Ward, J. L., & Welch, R. W. (2015). Postprandial plasma betaine and other methyl donor-related responses after consumption of minimally processed wheat bran or wheat aleurone, or wheat aleurone incorporated into bread. *British Journal of Nutrition, 113*(3), 445–453.

Klemm, R. D. W., West, K. P., Jr. Palmer, A. C., Johnson, Q., Randall, P., Ranum, P., & Northrop-Clewes, C. (2010). Vitamin A fortification of wheat flour: Considerations and current recommendations. *Food and Nutrition Bulletin, 31*, 547–561.

Kong, F., & Singh, R. P. (2008). Disintegration of solid foods in human stomach. *Journal of Food Science, 73*(5), R67–80.

Konik-Rose, C., Thistleton, J., Chanvrier, H., Tan, I., Halley, P., Gidley, M., Kosar-Hashemi, B., Wang, H., Larroque, O., Ikea, J., McMaugh, S., Regina, A., Rahman, S., Morell, M., & Li, Z. (2007). Effects of starch synthase IIa gene dosage on grain, protein and starch in endosperm of wheat. *Theoretical and Applied Genetics, 115*(8), 1053–1065.

Kozubek, A., & Tyman, J. H. P. (1999). Resorcinolic lipids, the natural non-isoprenoid phenolic amphiphiles and their biological activity. *Chemical Reviews, 99*(1), 1–26.

Kristensen, M., Toubro, S., Jensen, M. G., Ross, A. B., Riboldi, G., Petronio, M., Bügel, S., Tetens, I., & Astrup, A. (2012). Whole grain compared with refined wheat decreases the percentage of body fat following a 12-week, energy-restricted dietary intervention in postmenopausal women. *Journal of Nutrition, 142*(4), 710–716.

Laddomada, B., Caretto, S., & Mita, G. (2015). Wheat bran phenolic acids: Bioavailability and stability in whole wheat-based foods. *Molecules (Basel, Switzerland), 20*(9), 15666–15685.

Lampi, A. M., Nurmi, T., Ollilainen, V., & Piironen, V. (2008). Tocopherols and tocotrienols in wheat geno-types in the HEALTHGRAIN Diversity Screen. *Journal of Agricultural and Food Chemistry, 56*(21), 9716–9721.

Landberg, R., Marklund, M., Kamal-eldin, A., & Åman, P. (2014). An update on alkylresorcinols -Occurrence, bioavailability, bioactivity and utility as biomarkers. *Journal of Functional Foods, 7*, 77–89.

Lane-Elliot, A. (2016). The role of fiber. *University of Michigan health system.*

Leenhardt, F., Levrat-Verny, M. A., Chanliaud, E., & Rémésy, C. (2005). Moderate decrease of pH by sour-dough fermentation is sufficient to reduce phytate content of whole wheat flour through endogenous phytase activity. *Journal of Agricultural and Food Chemistry, 53*(1), 98–102.

Lindsay, D. G. (2002). The challenges facing scientists in the development of foods in Europe using biotech-nology. *Phytochemicals Revision, 1*, 101–111.

Liu, Q., Qiu, Y., & Beta, T. (2010). Comparison of antioxidant activities of different colored wheat grains and analysis of phenolic compounds. *Journal of Agricultural and Food Chemistry, 58*(16), 9235–9241.

Lohi, S., Mustalahti, K., Kaukinen, K., Laurila, K., Collin, P., Rissanen, H., Lohi, O., Bravi, E., Gasparin, M., Reunanen, A., & Mäki, M. (2007). Increasing prevalence of coeliac disease over time. *Aliment Pharmacology Therapeutices, 26*(9), 1217–1225.

Longvah, T., Ananthan, R., Bhaskarachary, K., & Venkaiah, K. (2017). Indian Food Composition Tables. Indian Council of Medical Research, National Institute of Nutrition, Department of Health Research, Ministry of Health and Family Welfare, Government of India.

Ma, D., Li, Y., Zhang, J., Wang, C., Qin, H., Ding, H., Xie, Y., & Guo, T. (2016). Accumulation of phenolic compounds and expression profiles of phenolic acid biosynthesis-related genes in developing grains of white, purple, and red wheat. *Frontiers in Plant Science, 7*, 528.

Magallanes-López, A. M., Hernandez-Espinosa, N., Velu, G., Posadas-Romano, G., Ordoñez-Villegas, V. M. G., Crossa, J., Ammar, K., & Guzmán, C. (2017). Variability in iron, zinc and phytic acid content in a worldwide collection of commercial durum wheat cultivars and the effect of reduced irrigation on these traits. *Food Chemistry, 237*, 499–505.

Makowskaa, A., Waskiewiczb, A., & Chudya, S. (2020). Lignans in triticale grain and triticale products. *Journal of Cereal Science, 93*.

Masuda, H., Aung, M. S., & Nishizawa, N. K. (2013). Iron biofortification of rice using different transgenic approaches. *Rice (N Y), 6*(1), 40.

Matsuo, H., Yokooji, T., & Taogoshi, T. (2015). Common food allergens and their IgE-binding epitopes. *Allergology International, 64*(4), 332–343.

McKevith, B. (2004). British Nutrition Foundation. *Nutrition bulletin, 29*, 111–142.

Meena, V., Abhishake, S., Punesh, S., Imran, S., Vinod, K., & Harcharan Singh, D. (2020). Phytase medi-ated beneficial impact on nutritional quality of biofortified wheat genotypes. *Current Nutrition & Food Science, 16*, 1–11.

Missbach, B., Schwingshackl, L., Billmann, A., Mystek, A., Hickelsberger, M., Bauer, G., & König, J. (2015). Gluten-free food database: the nutritional quality and cost of packaged gluten-free foods. *PeerJ, 3*, e1337.

Monge, L., Cortassa, G., Fiocchi, F., Mussino, G., & Carta, Q. (1990). Glyco-insulinaemic response, diges-tion and intestinal absorption of the starch contained in two types of spaghetti. *Diabetes, Nutrition and Metabolism, 3*, 239–246.

Morita, E., Chinuki, Y., Takahashi, H., Nabika, T., Yamasaki, M., & Shiwaku, K. (2012). Prevalence of wheat allergy in Japanese adults. *Allergology International, 61*(1), 101–105.

Morita, E., Matsuo, H., Mihara, S., Morimoto, K., Savage, A. W., & Tatham, A. S. (2003). Fast omega-glia-din is a major allergen in wheat-dependent exercise-induced anaphylaxis. *Journal of Dermatological Science, 33*(2), 99–104.

Muir, J. G., & Gibson, P. R. (2013). The low FODMAP diet for treatment of irritable bowel syndrome and other gastrointestinal disorders. *Gastroenterology & Hepatology, 9*(7), 450–452.

Myléus, A., Ivarsson, A., Webb, C., Danielsson, L., Hernell, O., Högberg, L., Karlsson, E., Lagerqvist, C., Norström, F., Rosén, A., Sandström, O., Stenhammar, L., Stenlund, H., Wall, S., & Carlsson, A. (2009). Celiac disease revealed in 3% of Swedish 12-year-olds born during an epidemic. *Journal of Pediatric Gastroenterology and Nutrition, 49*(2), 170–176.

Nagaraju, R., Sobhana, P. P., Thappatla, D., Epparapalli, S., Kandlakunta, B., & Korrapati, D. (2020). Glycemic index and sensory evaluation of whole grain based multigrain Indian breads (rotis). *Preventive Nutrition and Food Science, 25*, 194–202.

Nalam, V. J., Vales, M. I., Watson, C. J., Kianian, S. F., & Riera-Lizarazu, O. (2006). Map-based analysis of genes affecting the brittle rachis character in tetraploid wheat (Triticum turgidum L.). *Theoretical Applications in Genetics, 112*(2), 373–381.

Nathani, S., Kumar, V., Dhaliwal, H. S., Sircar, D., & Roy, P. (2020). Biological application of a fluorescent zinc sensing probe for the analysis of zinc bioavailability using Caco-2 cells as an in-vitro cellular model. *Journal of Fluorescence, 30*(6), 1553–1565.

Ndolo, V. U., Fulcher, R. G., & Beta, T. (2015). Application of LC-MS-MS to identify niacin in aleurone layers of yellow corn, barley and wheat kernels. *Journal of Cereal Science, 65*, 88–95.

Nelson, K., Mathai, M. L., Ashton, J. F., Donkor, O. N., Vasiljevic, T., Mamilla, R., & Stojanovska, L. (2016). Effects of malted and non-malted whole-grain wheat on metabolic and inflammatory biomarkers in overweight/obese adults: a randomised crossover pilot study. *Food Chemistry, 194*, 495–502.

Nielsen, M. M., & Hansen, Å. (2008). Stability of vitamin E in wheat flour and whole wheat flour during storage. *Cereal Chemistry, 85*(6), 716–720.

Nurit, E., Lyan, B., Pujos-Guillot, E., Branlard, G., & Piquet, A. (2016). Change in B and E vitamin and lutein, b-sitosterol contents in industrial milling fractions and during toasted bread production. *Journal of Cereal Science, 69*, 290–296.

Nurmi, T., Lampi, A.-M., Nyström, L., Turunen, M., & Piironen, V. (2010). Effects of genotype and environment on steryl ferulates in wheat and rye in the HEALTHGRAIN diversity screen. *Journal of Agricultural and Food Chemistry, 58*(17), 9332–9340.

Nyström, L., Lampi, A. M., Rita, H., Aura, A. M., Oksman-Caldentey, K. M., & Piironen, V. (2007). Effects of processing on availability of total plant sterols, steryl ferulates and steryl glycosides from wheat and rye bran. *Journal of Agricultural and Food Chemistry, 55*(22), 9059–9065.

Nyström, L., Paasonen, A., Lampi, A. M., & Piironen, V. (2007). Total plant sterols, steryl ferulates and steryl glycosides in milling fractions of wheat and rye. *Journal of Cereal Science, 45*, 106–115.

O'Dell, B. L., De Boland, A. R., & Koirtyohann, S. R. (1972). Distribution of phytate and nutritionally important elements among the morphological components of cereal grains. *Journal of Agricultural and Food Chemistry, 20*(3), 718–723.

Obeid, R. (2013). The metabolic burden of methyl donor deficiency with focus on the betaine homocysteine methyl transferase pathway. *Nutrients, 5*, 3481–3495.

Oghbaei, M., & Prakash, J. (2016). Effect of primary processing of cereals and legumes on its nutritional quality: A comprehensive review. *Cogent Food & Agriculture, 2*(1).

Palosuo, K., Varjonen, E., Kekki, O.-M., Klemola, T., Kalkkinen, N., Alenius, H., & Reunala, T. (2001). Wheat ω-5 gliadin is a major allergen in children with immediate allergy to ingested wheat. *Journal of Allergy and Clinical Immunology, 108*(4), 634–638.

Panyoo, A. E., & Emmambux, M. N. (2017). Amylose–lipid complex production and potential health benefits: A mini-review. *Starch-Stärke, 69*(7-8), 1600203.

Paredes-López, O., & Osuna–Castro, J. A. (2006). Functional foods and biotechnology. In K. Shetty, G. Paliyath, A. L. Pometto & R. E. Levin (Eds.), New York: Marcel Dekker Inclusive.

Pellegrini, N., & Agostoni, C. (2015). Nutritional aspects of gluten-free products. *Journal of the Science of Food and Agriculture, 95*(12), 2380–2385.

Pereira, B., Venter, C., Grundy, J., Clayton, C. B., Arshad, S. H., & Dean, T. (2005). Prevalence of sensitization to food allergens, reported adverse reaction to foods, food avoidance, and food hypersensitivity among teenagers. *Journal of Allergy and Clinical Immunology, 116*(4), 884–892.

Persson, D. P., de Bang, T. C., Pedas, P. R., Kutman, U. B., Cakmak, I., Andersen, B., Finnie, C., Schjoerring, J. K., & Husted, S. (2016). Molecular speciation and tissue compartmentation of zinc in durum wheat grains with contrasting nutritional status. *New Phytology, 211*(4), 1255–1265.

Piironen, V., Syvaoja, E. L., Varo, P., Salminen, K., & Koivistoinen, P. (1986). Tocopherols and tocotrienols in cereal products from Finland. *Cereal Chemistry, 63*(2), 78–81.

Poole, N., Donovan, J., & Erenstein, O. (2020). Agri-nutrition research: Revisiting the contribution of maize and wheat to human nutrition and health. *Food Policy*, 101976.

Punia, S., & Sandhu, K. S. (2016). Physicochemical and antioxidant properties of different milling fractions of Indian wheat cultivars. *International Journal of Pharma And Bio Sciences, 7*(1), 61–66.

Punia, S., Sandhu, K. S., & Siroha, A. K. (2019). Difference in protein content of wheat (Triticum aestivum L.): Effect on functional, pasting, color and antioxidant properties. *Journal of the Saudi Society of Agricultural Sciences, 18*(4), 378–384.

Punia, S., Sandhu, K. S., & Siroha, Sharma, S. (2017). Comparative studies of color, pasting and antioxidant properties of wheat cultivars as affected by toasting and roasting. *Nutrafoods*, 16, 95–102.

Raichle, M. E., & Gusnard, D. A. (2002). Appraising the brain's energy budget. *Proceedings of the National Academy of Sciences of the United States of America*, 99, 10237–10239.

Rashid, S., Rakha, A., Anjum, F. M., Ahmed, W., & Sohail, M. (2015). Effects of extrusion cooking on the dietary fibre content and Water Solubility Index of wheat bran extrudates. *International Journal of Food Science & Technology*, 50(7), 1533–1537.

Roberfroid, M. B. (2005). Introducing inulin-type fructans. *British Journal of Nutrition*, 93 Suppl 1, S13–25.

Rodriguez-Ramiro, I., Brearley, C. A., Bruggraber, S. F., Perfecto, A., Shewry, P., & Fairweather-Tait, S. (2017). Assessment of iron bioavailability from different bread making processes using an in vitro intestinal cell model. *Food Chemistry*, 228, 91–98.

Rosa, N. N., Barron, C., Gaiani, C., Dufour, C., & Micard, V. (2013). Ultra-fine grinding increases the antioxidant capacity of wheat bran. *Journal of Cereal Science*, 57, 84–90.

Ross, A. B., Shepherd, M. J., Schüpphaus, M., Sinclair, V., Alfaro, B., Kamal-Eldin, A., & Åman, P. (2003). Alkylresorcinols in cereals and cereal products. *Journal of Agricultural and Food Chemistry*, 51(14), 4111–4118.

Rubio-Tapia, A., Ludvigsson, J. F., Brantner, T. L., Murray, J. A., & Everhart, J. E. (2012). The prevalence of celiac disease in the United States. *American Journal of Gastroenterology*, 107(10), 1538–1544; quiz 1537, 1545.

Sandberg, A. S., Brune, M., Carlsson, N. G., Hallberg, L., Skoglund, E., & Rossander-Hulthén, L. (1999). Inositol phosphates with different numbers of phosphate groups influence iron absorption in humans. *American Journal of Clinical Nutrition*, 70, 240–246.

Sapone, A., Bai, J. C., Ciacci, C., Dolinsek, J., Green, P. H., Hadjivassiliou, M., Kaukinen, K., Rostami, K., Sanders, D. S., Schumann, M., Ullrich, R., Villalta, D., Volta, U., Catassi, C., & Fasano, A. (2012). Spectrum of gluten-related disorders: consensus on new nomenclature and classification. *BMC Medicine*, 10, 13.

Saulnier, L., Robert, P., Grintchenko, M., Jamme, F., Bouchet, B., & Guillon, F. (2009). Wheat endosperm cell walls: Spatial heterogeneity of polysaccharide structure and composition using micro-scale enzymatic fingerprinting and FT-IR microspectroscopy. *Journal of Cereal Science*, 50, 312–317.

Sayaslan, A., Seib, P. A., & Chung, O. K. (2006). Wet-milling properties of waxy wheat flours by two laboratory methods. *Journal of Food Engineering*, 72, 167–178.

Scazzina, F., Del Rio, D., Pellegrini, N., & Brighenti, F. (2009). Sourdough bread: Starch digestibility and postprandial glycemic response. *Journal of Cereal Science*, 49, 419–421.

Shewry, P. R. (2009). Wheat. *Journal of Experimental Botany*, 60(6), 1537–15553.

Shewry, P. R., & Hey, S. J. (2015). The contribution of wheat to human diet and health. *Food and Energy Security*, 4(3), 178–202.

Shewry, P. R., & Hey, S. J. (2016). Do we need to worry about eating wheat? *Nutrition Bulletin*, 41(1), 6–13.

Singh, S., Gamlath, S., & Wakeling, L. (2007). Nutritional aspects of food extrusion: a review. *International Journal of Food Science & Technology*, 42(8), 916–929.

Slavin, J. (2013). Fiber and prebiotics: mechanisms and health benefits. *Nutrients*, 5, 1417–1435.

Solaymani-Dodaran, M., West, J., & Logan, R. F. A. (2007). Long-term mortality in people with celiac disease diagnosed in childhood compared with adulthood: A population-based cohort study. *Official Journal of the American College of Gastroenterology*, 102(4).

Sollid, L. M., Qiao, S.-W., Anderson, R. P., Gianfrani, C., & Koning, F. (2012). Nomenclature and listing of celiac disease relevant gluten T-cell epitopes restricted by HLA-DQ molecules. *Immunogenetics*, 64(6), 455–460.

Šramkováa Z., Gregováb, E., & Šturdíka, E. (2009). Chemical composition and nutritional quality of wheat grain. *Acta Chimica Slovaca*, 2, 115–138.

Statistica (2020a). Global wheat production from 2011/2012 to 2019/2020 (in million metric tons). Available online: https://www.statista.com/statistics/267268/production-of-wheat-worldwide-since-1990/, Release date: Feb, 2020.

Statistica (2020b). U.S. Gluten-free Foods MarketStatistics & Facts. In: Statista Research Department.

Stone, B., & Morell, M. K. (2009). Carbohydrates. In K. Khan & P. R. Shewry (Eds.), *Wheat: Chemistry and Technology*, vol. Fourth (pp. 299–362). MN: American Association of Cereal Chemists, St Paul.

Tanpowpong, P., Broder-Fingert, S., Katz, A. J., & Camargo, C. A. (2012). Predictors of gluten avoidance and implementation of a gluten-free diet in children and adolescents without confirmed celiac disease. *The Journal of Pediatrics*, *161*(3), 471–475.

Tanpowpong, P., Broder-Fingert, S., Katz, A. J., & Camargo, C. A., Jr. (2015). Predictors of dietary gluten avoidance in adults without a prior diagnosis of celiac disease. *Nutrition*, *31*(1), 236–238.

Thielecke, F., & Nugent, A. P. (2018). Contaminants in grain—a major risk for whole grain safety? *Nutrients*, *10*(9), 1213.

Tiwari, U., & Cummins, E. (2009). Nutritional importance and effect of processing on tocols in cereals. *Trends in Food Science and Technology*, *20*, 511–520.

Tiwari, V. K., Rawat, N., Neelam, K., Kumar, S., Randhawa, G. S., & Dhaliwal, H. S. (2010). Substitutions of 2S and 7U chromosomes of Aegilops kotschyi in wheat enhance grain iron and zinc concentration. *Theoretical and Applied Genetics*, *121*(2), 259–269.

Topping, D. L., Bajka, B. H., Bird, A. R., Clarke, J. M., Cobiac, L., Conlon, M. A., Morell, M. K., & Toden, S. (2008). Resistant starches as a vehicle for delivering health benefits to the human large bowel *Microbial Ecology in Health and Disease*, *20*, 103–108.

Tosi, P., Hidalgo, A., & Lullien-Pellerin, V. (2020). The impact of processing on wheat grain components associated with health benefits. In G. Igresias, T. M. Ikeda & C. Guzmán (Eds.), *Wheat Quality for Improving Processing and Human Health*, (pp. 387–420): Springer.

Uauy, C., Distelfeld, A., Fahima, T., Blechl, A., & Dubcovsky, J. (2006). A NAC Gene regulating senescence improves grain protein, zinc, and iron content in wheat. *Science*, *314*(5803), 1298–1301.

USDA National Nutrient Database for Standard Reference. (2015). Composition of Foods Raw, Processed, Prepared.Vol. 28. Washington, DC: U.S. Department of Agriculture.

USDA/NASS. (2001). *Agricultural statistics*. Washington, DC: U.S. Department of Agriculture.

Velu, G., Ortiz-Monasterio, I., Cakmak, I., Hao, Y., & Singh, R. P. (2014). Biofortification strategies to increase grain zinc and iron concentrations in wheat. *Journal of Cereal Science*, *59*(3), 365–372.

Venter, C., Pereira, B., Grundy, J., Clayton, C. B., Arshad, S. H., & Dean, T. (2006a). Prevalence of sensitization reported and objectively assessed food hypersensitivity amongst six-year-old children: a population-based study. *Pediatric Allergy Immunology*, *17*(5), 356–363.

Venter, C., Pereira, B., Grundy, J., Clayton, C. B., Roberts, G., Higgins, B., & Dean, T. (2006b). Incidence of parentally reported and clinically diagnosed food hypersensitivity in the first year of life. *Journal of Allergy and Clinical Immunology*, *117*(5), 1118–1124.

Verma, A. K., & Banerjee, R. (2010). Dietary fibre as functional ingredient in meat products: a novel approach for healthy living – a review. *Journal of Food Science and Technology*, *47*, 247–257.

Volta, U., Bardella, M. T., Calabrò, A., Troncone, R., & Corazza, G. R. (2014). An Italian prospective multi-center survey on patients suspected of having non-celiac gluten sensitivity. *BMC Medicine*, *12*, 85.

Vyas, P., Singh, D., Singh, N., Kumar, V., & Dhaliwal, H. S. (2018). Nutrigenomics: Advances, opportunities and challenges in understanding the nutrient-gene interactions. *Current Nutrition & Food Science*, *14*(2), 104–115.

Walker, M. M., Murray, J. A., Ronkainen, J., Aro, P., Storskrubb, T., D'Amato, M., Lahr, B., Talley, N. J., & Agreus, L. (2010). Detection of celiac disease and lymphocytic enteropathy by parallel serology and histopathology in a population-based study. *Gastroenterology*, *139*(1), 112–119.

Wang, W.-M., Klopfenstein, C. F., & Ponte, J. G. J. (1993). Effects of twin-screw extrusion on the physical properties of dietary fiber and other components of whole wheat and wheat bran and on the baking quality of the wheat bran. *Cereal Chemistry*, *70*, 707–711.

West, J., Logan, R. F., Hill, P. G., Lloyd, A., Lewis, S., Hubbard, R., Reader, R., Holmes, G. K., & Khaw, K. T. (2003). Seroprevalence, correlates, and characteristics of undetected coeliac disease in England. *Gut*, *52*(7), 960–965.

Whiteley, P., Haracopos, D., Knivsberg, A. M., Reichelt, K. L., Parlar, S., Jacobsen, J., Seim, A., Pedersen, L., Schondel, M., & Shattock, P. (2010). The ScanBrit randomised, controlled, single-blind study of a gluten- and casein-free dietary intervention for children with autism spectrum disorders. *Nutritional Neuroscience*, *13*(2), 87–100.

Whiteley, P., Shattock, P., Knivsberg, A.-M., Seim, A., Reichelt, K. L., Todd, L., Carr, K., & Hooper, M. (2013). Gluten- and casein-free dietary intervention for autism spectrum conditions. *Frontiers in Human Neuroscience*, *6*, 344–344.

Wu, J. H., Neal, B., Trevena, H., Crino, M., Stuart-Smith, W., Faulkner-Hogg, K., Yu Louie, J. C., & Dunford, E. (2015). Are gluten-free foods healthier than non-gluten-free foods? An evaluation of supermarket products in Australia. *British Journal of Nutrition, 114*(3), 448–454.

Yokooji, T., Kurihara, S., Murakami, T., Chinuki, Y., Takahashi, H., Morita, E., Harada, S., Ishii, K., Hiragun, M., Hide, M., & Matsuo, H. (2013). Characterization of causative allergens for wheat-dependent exercise-induced anaphylaxis sensitized with hydrolyzed wheat proteins in facial soap. *Allergology International, 62*(4), 435–445.

Yokooji, T., Okamura, Y., Chinuki, Y., Morita, E., Harada, S., Hiragun, M., Hide, M., & Matsuo, H. (2015). Prevalences of specific IgE to wheat gliadin components in patients with wheat-dependent exercise-induced anaphylaxis. *Allergology International, 64*(2), 206–208.

Zandomeneghi, M., Festa, C., Carbonaro, L., Galleschi, L., Lenzi, A., & Calucci, L. (2000). Front-surface absorbance spectra of wheat flour: Determination of carotenoids. *Journal of Agricultural and Food Chemistry, 48*(6), 2216–2221.

Zeisel, S. H., Mei, H. M., Howe, J. C., & Holden, J. M. (2003). Concentrations of choline-containing compounds and betaine in common foods. *Journal of Nutrition and Metabolism, 133*, 1302–1307.

Zhang, Y., Liang, Z., Zong, Y., Wang, Y., Liu, J., Chen, K., Qiu, J. L., & Gao, C. (2016). Efficient and transgene-free genome editing in wheat through transient expression of CRISPR/Cas9 DNA or RNA. *Nature Communications, 7*, 12617.

Zhu, K.-X., Huang, S., Peng, W., Qian, H., & Zhou, H. (2010). Effect of ultrafine grinding on hydration and antioxidant properties of wheat bran dietary fiber. *Food Research International, 43*, 943–948.

Zhu, Y., Soroka, D. N., & Sang, S. (2012). Synthesis and inhibitory activities against colon cancer cell growth and proteasome of alkylresorcinols. *Journal of Agricultural and Food Chemistry, 60*(35), 8624–8631.

Zielinski, H., Kozlowska, H., & Lewczuk, B. (2001). Bioactive compounds in the cereal grains before and after hydrothermal processing. *Innovative Food Science and Emerging Technologies., 2*(3), 159–169.

Žilić, S., Barać, M., Pešić, M., Dodig, D., & Ignjatović-Micić, D. (2011). Characterization of proteins from grain of different bread and durum wheat genotypes. *International Journal of Molecular Sciences, 12*, 5878–5894.

Zuidmeer, L., Goldhahn, K., Rona, R. J., Gislason, D., Madsen, C., Summers, C., Sodergren, E., Dahlstrom, J., Lindner, T., Sigurdardottir, S. T., McBride, D., & Keil, T. (2008). The prevalence of plant food allergies: a systematic review. *Journal of Allergy and Clinical Immunology, 121*(5), 1210–1218 e1214.

2

Agronomic, Environmental and Genotype Effects on Phenols, Antioxidants and Amino Acid Content of Wheat

Ali Yigit and Osman Erekul
Aydın Adnan Menderes University, Turkey

CONTENTS

2.1 Introduction

Wheat is one of the world's cereal crops with more than 700 million tons being harvested from 214 million hectare area (FAO, 2018). The hexaploid wheat (*Triticum aestivum*) originated from domesticated tetrapoid (*Triticum turgidum* spp *dicoccum*) and the diploid donor of the D genome (*Aegilops tauschii*) 7,000–9,500 years ago. *Aegilops tauschii* has been considered the main contributor to the bread-making properties of bread wheat. Due to the D genome hexaploid, wheat is unique because of its gluten that allows the elastic properties of the dough in bread and bakery products (Chantret et al., 2005; Shewry 2009).

Bread wheat supplies 90–95% of the wheat production consumed as flour (whole-grain or refined) used for bread types and bakery products.

Wheat is one of the most consumed cereal species with a major contribution to the human diet. It is a good source of carbohydrates and protein and also contains phytochemicals, which have a significant health benefit (Punia et al., 2017a; Punia and Sandhu, 2016). Wheat grain is usually ground into white flour—without bran—to make bread, cake, pasta, noodles, or other bakery products. Recently, whole wheat consumption has increased because of consumer awareness of its health benefits (Wrigley, 2009).

Phenolic compounds and antioxidants are the main chemical components that directly react with the oxygen reactive species (ROS) or single oxygen molecules that protect cells from DNA damage. Whole-grain products are well known to be rich in phenolic compounds—primarily concentrated in the outer layers of grain—such as aleurone, testa, and pericarp (Martini et al., 2015). The aleurone layer is a part of the endosperm that separated along with the bran in the milling process. After harvesting grain, antioxidant and phenolic compounds are mainly affected by postharvest processing techniques (Lin, Guo and Mennel, 2008). Flour milling causes a reduction of phytochemicals in the milled products.

In general, white flour is preferred by consumers in bakery products, which means the bran that is rich in phenolic compounds is removed. Phenolic compounds in grain dramatically decrease with increased refinement from whole grain to flour because of the distribution of phenolics found in the outer layers of grain. Besides postharvest management, genotype, agronomic management (organic and conventional production, sowing date, fertilization, or pest management), and environmental factors affect antioxidant activity (Mazzoncini et al., 2015). Environmental factors (i.e. drought, solar radiation, precipitation, soil properties) are believed to major determinants of antioxidant activity of wheat grain that are difficult to control and may vary greatly from year to year.

Wheat grain is an important protein source in the human diet; protein quality as well as protein content is highly desirable. Besides protein quality, amino acids should be thought of as nutrients in the daily intake for a healthy diet. Wheat proteins contain low levels of some amino acids, especially lysine and threonine, but are rich in glutamin and prolin (Abdel-Aal and Hucl, 2002). These limited amino acids determine protein quality, which changes the amino acid composition of the grain.

Amino acid content depends on genotypic and environmental factors, such as nitrogen application time and concentration in the soil, availability of soil moisture, and temperature in grain filling periods (Qabaha, 2010). However, wheat protein content highly depends on genotype and environmental factors, and it is known that variation in protein content significantly modifies the amino acid profile of wheat. In recent years, scientists have shared the same opinion that climate changes threaten world crop production and food supply. In the view of climate change, drought, salinity, and heat stress are the major abiotic components for wheat productivity and cause adverse effects for wheat yield and quality (El Sabagh et al., 2019). On the other hand, water availability and rainfall are major factors limiting wheat production, which is mainly dependent on rainfall during wheat growing periods around the world. Changes in optimal conditions throughout wheat production can affect wheat health and nutritional properties from sowing to harvest. With climate change already affecting crop productivity across the globe and with the increasing consumer awareness of healthy food, a renewed focus is needed on both food safety and nutritional properties of grains. With the need to focus greater attention on health and nutrient values of wheat with agricultural perspective, this chapter aims to provide useful information about antioxidant and amino acid changes due to agricultural factors.

2.2 Wheat Grain Structure

Cereals are an important source of nutrients in most human diets in both developed and developing countries, providing a major proportion of dietary energy. Cereals are edible seeds that belong to *Graminea* family. Worldwide, wheat and rice are the most important and most produced crops, accounting for more than half of the world's cereal production. Cereal grains show wide variation in size and shape and changes among cereal grains. One major difference is the hull surrounding the grain when the grain is harvested or processed (Blakeney et al., 2009).

Among the cereals, wheat is an important source of minerals, proteins, vitamins, dietary fibers, and contains essential nutrients. Whole-wheat grain has three main parts as percentage of grain (Figure 2.1):

- Endosperm (80–85%)
- Bran (13–17%)
- Germ (embryo, 2–3%)

The endosperm is the largest part of the grain and is mainly made up of starch. The aleurone layer surrounds the starchy endosperm and consists of cubic-shaped thick-walled cells. It is rich in both lipids and proteins. While the testa is only one to two layers of cells thick, the pericarp is multilayered. After maturity, the pericarp consists of dry empty cells. The hairs on the end of the grain are known as the "brush" (Blakeney et al., 2009). The endosperm contains food reserves to use during seed germination, and it is rich in starch. Apart from carbohydrates, the endosperm comprises fats (1.5%) and protein (13%) approximately. Albumin, globulin, glutenin, and gliadins are proteins that form gluten during

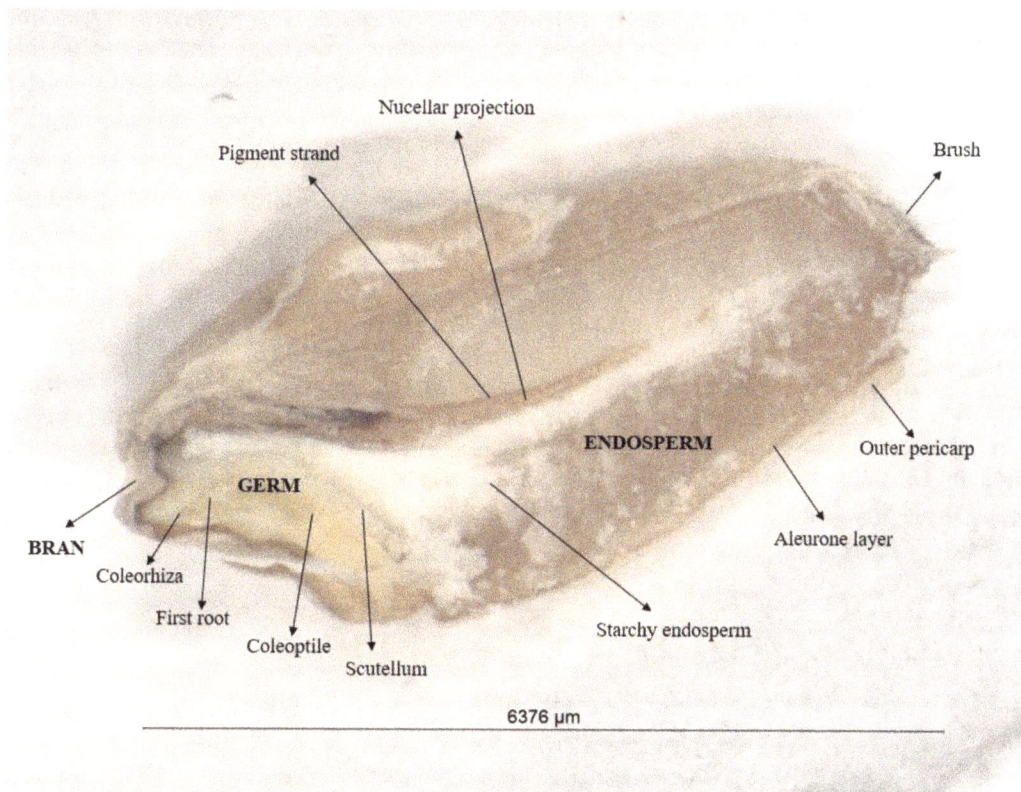

FIGURE 2.1 Microscopic anatomy of wheat grain (original, Yigit, A), morphological features based on and modified from Bechtel et al. (2009).

the dough-making process. The endosperm contains low levels of minerals and fibers—0.5% and 1.5%, respectively.

The outer layers of grain bran contain water-insoluble fiber and half of the bran comprises fiber components (53%). Proteins and carbohydrates each represent 16% (d.m.) of bran. The mineral content (ash) is rather high (7.2%) in bran compared to other parts of the grain (Šramkováa, Gregováb and Sturdik, 2009). Bran contains vitamins and minerals, but it is separated from the endosperm during the milling process. This process leads to reduced health and nutritional benefits in grain.

The germ is the embryo or sprouting part of the seed. It consists of about 10–15% lipids, 26–35% protein, 17% sugar, 1.5–4.5% fiber and about 4% minerals. It also has significant amounts of bioactive compounds, such as phytosterols, tocopherols, carotenoids, thiamin, and riboflavin. However, wheat germ has unsaturated germ oil that is removed for flour quality in terms of increase in acidity (Brandolini and Hidalgo, 2012).

2.2.1 Wheat Grain Development

Maturation of wheat grain, described as the grain-filling period, is the most sensitive development stage in wheat cultivation because yield and quality can change by many factors (i.e. heat stress, drought, deficit irrigation, diseases, pests, soil fertility).

During maturation periods, changes occur in different phases: step I—development of the endosperm nucleus; step II—cellularization of the endosperm cells; step III—termination of meristematic activity; step IV—achieve maximum weight; and step V—reach approximately 14% moisture at harvest maturity (Bechtel et al., 2009).

Endosperm start to develop with a diploid cell, followed by repeated division of triploid nuclei and the gradual formation of cell walls. Next, cell expansion begins with starch and protein accumulation and water content increases. Accumulation of starch and protein in each grain is determined early in grain fill, and the final size of the cells is influenced by water availability and duration of the grain-filling period. In this growth stage, heat stress (daily maximum temperature >30°C) accelerates the rate of grain-filling whereas filling duration is shortened.

Under heat stress and deficit water conditions in the plant roots, the supply of photoassimilates may be limited and decrease synthesis of starch, protein, and nutritional properties of grain. In the next step, water intake stops before and the kernel begins to dehydrate. Late in maturity, the formation of a waxy layer breaks the input of sugars and amino acids into the grain. Protein and starch deposition stops and the grain reaches maximum dry weight. Finally, grain water content reaches 10–15% at which time it is ready to harvest (Dupont and Altenbach, 2003; Farooq et al., 2011).

Yield, quality, and the nutritional value of wheat grain depends on the conditions during the grain-filling period. In this period, reserved assimilates move from the flag leaf, stem, and ear parts of plant and are sensitive to biotic and abiotic stress conditions. Identifying and understanding how the nutritional value of grain is affected by the factors and processes that regulate bioactive compounds and amino-acid accumulation will aid in the development and selection of higher nutritional value varieties during these stages.

2.3 Antioxidant Activity and Phenolic Content of Wheat Grain

2.3.1 Reactive Oxygen Species (ROS) and Bioactive Compounds

Phenolic compounds are common dietary phytochemicals and have several different functions in plants, such as providing resistance and defense against microbial infections. They are attributed to have protective effect against chronic diseases suggested in epidemiological evidence. Antioxidants have beneficial health effects and are believed to directly react with and quench free radicals (ROS), reducing peroxides and stimulating the antioxidant defense enzyme system (Li et al., 2006; Sonntag et al., 2020). Free radicals naturally occur inside the body to perform metabolic activities. But the level of these free radicals leading to an imbalance with the existing antioxidants increases due to certain stress factors (Narwal et al., 2014).

Air pollutants, soil salinity, drought (deficit water availability), high temperature, and biotic stress (pests and disease) conditions can all target DNA (mutation, breaking of strands, protein DNA cross-links), while lipids can also cause modified amino acids and inactivation of enzymes in the human body (Das and Roychoudhury, 2014).

2.3.2 Phenolic Acid Composition of Wheat Grain

Plant phenolics are mainly found in fruits, vegetables, wines, and teas. However, many phenolic compounds are also in fruits and vegetables, and are also reported in cereals. Phenolic acids have an important role in oxidative stress in the body by maintaining a balance between oxidants and antioxidants. Phenolic acids are composed of benzoic and cinnamic acids, which are found in all cereals. Hydroxybenzoic acids and hydroxycinnamic acids are two classes of phenolic acids found in cereal grains. Hydroxybenzoic acids are: protocatechuic, *p*-Hydroxybenzoic, salicylic, and syringic acids. Hydroxycinnamic acids are: ferulic, ceffeic, *p*-Coumaric and cinnamic acids. These are found in wheat grain (Dykes and Rooney, 2007; Punia et al. 2017).

Phenolic compounds are higher in bran compared to the inner layer (endosperm) of grain that contains 17% of the total phenol content. The bran coatings are made up of the aleurone, the hyaline layer, and the inner and outer pericarp. Bran is a key factor in identifying whole-grain benefits and contains various amount of phenolic acids. Phenolic acid composition in whole grains change with a wide range for ferulic acid from 4.5 to 1,270 g/100 g, sinapic acid from 1.3 to 6.3 g/100 g, *p*-coumaric acid from 0.2 to 37.2 g/100 g, and total phenol content from 350 to 1,505 g/100 g values (Călinoiu and Vodnar, 2008).

Phenolic acid composition is mainly influenced by cereal variety and milling process. Wheat bran is clearly highest in dietary fiber and phenolic acid content compared to whole grain.

Bran obtained from bread wheat genotypes has significantly higher amounts of phenolic acids compared to durum wheat genotypes. Ferulic acid had the highest value in the bran of bread and durum wheats with its concentration ranging between from 5502.05–9160.92 mg/kg, accounting for about 95–99% of total grain ferulic acid. White flour contained very low amounts of phenolic acid composition compared to bran (Žilić et al., 2012).

Phenolic acid composition changes among fractions of different grains and varieties. In some cases, even the distribution of phenolic acids within grain fractions varied widely; on average, sinapic acid, in wheat varieties was highly concentrated in the germ (403 µg/g), followed by the aleurone layer (245 µg/g), and pericarp layer (200 µg/g). The aleurone layer is the most important fraction because of its high ferulic acid concentration. Ferulic acid contributed between 45–50% to the total amount of phenolics in wheat. Phenolic acid composition in whole grain differed from the highest to lowest ferulic (689 µg/g), vanillic (94 µg/g), sinapic (66 µg/g), *p*-coumaric (54 µg/g), and syringic acid (31 µg/g), respectively (Ndolo and Beta, 2014).

Whole grains commonly include phenolic acids, including vanillic acid, caffeic acid, ferulic acid, *p*-coumaric acid, and syringic acid. The concentration of phenolic acids varies among the cereals. Predominant components in wheat are ferulic, vanillic, syringic, and sinapic acids. The profile of phenolic acids were similar between wheat genotypes with ferulic acid predominantly (72–85%). The concentration of total ferulic acid levels ranged from 455 µg/g to 621 µg/g, which has the highest proportion among the phenolic acids; values of total vanillic acid ranged from 8.4 µg/g to 12.7 µg/g), syringic acid values ranged from 8.9 µg/g to 17.8 µg/g), and *p*-coumaric acid values ranged from 10.4 µg/g to 14.10 µg/g (Li, Shewry and Ward, 2008).

Vaher et al. (2010) detected phenolic acids in spring and winter wheat varieties. They also reported that ferulic acid has the highest value followed by sinapic, syringic, vanillic, and p-coumaric acids. Ferulic acid values ranged between 14.4 µg/g (flour) to 268.9 µg/g (bran) with a wide range. The value of ferulic and sinapic acids of winter wheat variety was two times higher than the spring variety. Ferulic acid constitutes 40% of total phenolic acids in wheat bran.

Mazzoncini et al. (2015) also suggest that phenolic acid content and total phenolic contents were found higher in bran fraction than white flour. They have obtained seven phenolic acids from both bran and flour milling products: protocatechuic acid, gallic acid, vanillic acid, *p*-hydroxybenzoic acid, syringic acid, *p*-coumaric acid, caffeic acid, and also ferulic acid detected only in bran fraction. In bran fraction they detected the highest ferulic acid values ranged from 103.47–149.96 µmol/100 g compared to white flour values (1.72–4.95 µmol/100 g). With the exception of *p*-hydroxybenzoic acid, the bran fraction had a higher content of vanillic acid, syringic acid, gallic acid, *p*-coumaric acid, and ferulic acid than white flour.

Another study also reported ferulic, *p*-hydroxybenzoic, *p*-coumaric, syringic, and vanillic acids detected in all bran fraction of wheat varieties. Similar to other studies, ferulic acid was the predominant phenolic acid in all samples and constituted approximately 53.47–66.68% (d.m.) of total phenolic acids. The level of *p*-coumaric, *p*-hydroxybenzoic, vanillic, syringic, and ferulic acid ranged 5.81–8.60 µg/g, 8.89–18.98 µg/g, 14.45–33.11 µg/g, 36.45–55.70 µg/g and 130.06–146.38 µg/g hard red winter wheat bran, respectively. The study also suggested the effects of growing conditions primarily on ferulic acid content and phenolic acid composition (Zhou, Yin and Yu, 2005).

Mpofu, Sapirstein and Beta (2006) obtained highly significant and strong correlations between total phenolic content and antioxidant activity ($r = 0.73$), between total phenolic and ferulic acid content ($r = 0.84$) and also between antioxidant activity and ferulic acid content ($r = 0.72$). In their study, ferulic acid was found as the main phenolic acid representing approximately 63% of total content of phenolic acids. The mean concentration of ferulic acid ranged from 371–441 µg/g. Thus, it is shown that ferulic acid is the main phenolic acid that affects total phenolic acid composition and antioxidant activity properties of wheat grain.

Overall, ferulic acid is the primary component and has a great impact on phenolic acid composition compared to other phenolic compounds. Wheat grain has significant levels of antioxidants, and phenolic compounds seem to have a great impact and potential on health benefits. To understand changes of phenolic composition and antioxidant properties of wheat grain, it is necessary to determine which factors change antioxidant activity of grain. In the next section, we will discuss the factors that affect antioxidant activity of wheat.

2.3.3 Factors Effecting Wheat Antioxidant Activity

Wheat antioxidant researchers encourage consumption of wheat-based foods and food products rich in natural antioxidants, especially those produced from whole-grain products. Whole grain and its products have attracted the attention of consumers due totheir health-promoting properties in recent years. However, wheat antioxidants need to be approached in multidisciplinary studies to advance knowledge on wheat antioxidants. Therefore, it can be understood how agricultural factors may alter antioxidant activity of wheat grain (Yu, 2008).

Grain structure and antioxidant properties of wheat may vary widely by many factors during the wheat-growing period. Many factors can lead to high variability in grain structure and chemical composition of wheat, such as stress conditions in the germination period, nutrient deficiencies, fertilization, climate and soil conditions, limited water availability, and high temperatures during maturity periods. Many crop scientists focus on which factors are responsible for antioxidant activity of wheat grain.

Wheat antioxidants are available in different proportions and amounts in milling fractions of grain. Zhou, Laux, and Yu (2004) analyzed wheat grain and fractions for their antioxidant properties. They compared aleurone, bran, and whole grain for antioxidant properties and total phenolic acid composition. In the study, they found that aleurone had the highest radical scavenging activity, total phenolic content, and phenolic acid concentration followed by bran and other grain parts. Wheat grain milling fractions differed from one another in their total phenolic content. The highest value of total phenolic content of 4.04 mg of gallic acid equivalent (GAE) of wheat was obtained in the aleurone, while it had the lowest total phenolic content value of 1.8 mg of GAE/g of wheat. The results demonstrate that antioxidants, including phenolic acids, are concentrated in the aleurone fraction of wheat bran.

Antioxidant activity and phytochemical profiles of 11 diverse wheat genotypes are determined by Adom, Sorrells and Liu (2003). In the study, total phenolic content values ranged from 709.8–860.0 μmol GAE/100 g, total antioxidant activity values ranged from 37.6–46.4 μmol of vitamin C/g and total flavonoid content values ranged from 105.8–141.8 μmol of catechin equivalent/100 g of wheat. Wheat varieties showed significant differences for antioxidant properties in the study. They also reported a high correlation between total antioxidant activity and total phenolic content ($r^2 = 0.983$).

Analogously, 46 bread wheat varieties commonly grown in Turkey changed in a wide range in terms of total phenolic content and antioxidant activities. In the study, total phenolic content of varieties ranged from 102.4–211.8 μg GAE/g (whole wheat flour) and values of varieties changed in a wide range and intensely between 140–200 μg GAE/g. In all evaluated varieties, Tosunbey (26.3%), Momtchill (25.9%), and Selimiye (23.7%) showed significant DPPH radical scavenging activity indicating these varieties as having good phenolic content compared to other varieties. Wheat varieties showed statistically significant differences and the genotypic factor had an important impact on antioxidant properties of wheat grain. Furthermore, significant and positive correlation was found between total phenolic content and antioxidant activity ($p < 0.01$); wheat varieties had higher total phenolic content and antioxidant activity can be used to improve health effects of commonly grown varieties in wheat-breeding programs (Yiğit, 2015).

Boukid et al. (2019) reported the possibility of breeding genetically different and phenolic-rich wheat grain in their study. Significant differences were observed in terms of free and bound phenolic content of durum varieties. Bound phenolic acids had greater impact than free phenolics on total phenolic content of grain. Total phenolic content of durum varieties ranged between 177.48–272.62 mg GAE/100 g and modern durum wheat varieties had the highest phenolic content; the lowest values were obtained from landraces. They observed total phenolic content of wheat is a complex trait influenced by both genotype and environmental factors. Ferulic acid content had the highest value among phenolic acids so it has greater impact on total phenolic content and antioxidant properties of wheat grain. Total ferulic acid recorded a high genetic diversity among the durum wheat varieties that indicated phenolic acid composition mainly change by genotype. Antioxidant activity of wheat varieties were significantly affected by variability in phenolic acid composition. Modern wheat genotypes had a high total phenolic content and antioxidant activity compared to other old and landrace varieties.

Eight Canadian wheat varieties are examined for phenolic content and antioxidant activity in wheat milling fractions. Total phenolic content of wheat fractions differed significantly with the values of 1.300–5.300 mg/kg based on wheat genotypes and fractions. Total phenolic content decreased into inner parts of grain and confirmed that phenolic compounds are mostly located in the outer layers of grain. Significant differences observed in antioxidant activity of wheat samples were observed in the 5%, and 10% fractions had higher values compared to other parts. Antioxidant activity of wheat fractions ranged between 10.3% and 26.0%; inhibition values varied approximately two-fold between wheat samples. Total phenolic content and antioxidant activity of wheat-grain fractions were highly correlated ($r^2 = 0.94$), whcih emphasizes evidence that phenolic compounds have antioxidant activity in wheat grain. Genotype and milling fractions are the main factors that affect antioxidant activity and phenolic composition, but they are mainly influenced by growing environment (Beta et al., 2005).

Wild, feral, and domesticated diploid 15 *Triticum* accessions were evaluated for their phenolic acids and antioxidant activities. In the present study, total phenolic content showed a significant difference ($p \leq 0.05$) between *Triticum* accessions. The highest total phenolic content obtained from *Triticum urartu* (ID1277 genotype) with 831 mg/kg (d.m.) value, and the lowest value (510 mg/kg d.m.) obtained from *Triticum durum* (Dylan genotype). Phenolic acids ranked from most to least abundant in the bound form were ferulic (557.9 mg GAE/kg dm), *p*-coumaric (43.2 mg GAE/kg dm), syringic (4.9 mg GAE/kg dm), vanillic (3.9 mg GAE/kg dm), and *p*-hydroxybenzoic (1.5 mg GAE/kg dm) acids. *Triticum urartu* had the highest concentration of total phenolic acids. The total phenolic acids had the lowest values reported from einkorn, emmer, durum and bread wheat samples. However, the genotype was the factor that highlighted significant differences in statistical results among wheat species. *Triticum thaoudar* and *Triticum urartu* species generally had the highest antioxidant values (2.2 and 2.0 mmol Trolox equivalent [TE]/kg d.m., respectively) and *Triticum durum* and *Triticum monococcum* had the lowest antioxidant activity value (0.7 mmol TE/kg d.m.). The results of the study clearly demonstrated that wild wheats have higher phenolic content, phenolic acids, and antioxidant activity than other wheat species (Yılmaz, Brandolini and Hidalgo, 2015).

Free and bound phenolic acids, dietary fiber, and antioxidant activity of four Indian wheat varieties were evaluated by Revanappa and Salimath (2009). They observed significant differences in free, bound and total phenolic content among the wheat varieties. Free and bound phenolic acid content values ranged between 946.6 µg/g flour and 711.2 µg/g flour. The obtained highest total phenolic acid observed in MACS-2679 variety (also had the highest bound phenolic acid) with 940.6 µg/g flour value while the lowest value 711.2 µg/g flour obtained from GW-322 (also had the lowest free and bound phenolic acids) variety. The researches also reported correlation between phenolic acid content and antioxidant activity, which provided evidence that wheat grain antioxidant activity derives from phenolic compounds. The radical scavenging (%) of free and bound phenolic extracts is affected by phenolic acid composition and genotype. The antioxidant activity of bound phenolic acid extracts were found to be higher compared to free phenolic acids and ferulic acid, which was found to be the most important phenolic acid for antioxidant properties of wheat grain.

Wheat genotype and growing conditions (environment factors) can strongly influence grain antioxidants and their composition. Statistically significant differences were obtained between wheat varieties for total phenolic content, antioxidant activity, and phenolic acid concentration. Also, differences among genotypes at each location were also highly significant for all the parameters. Mean total phenolic content ranged from 1,709–2,009 µg/g and total antioxidant activity values ranged between 13.21% and 15.06% (DPPH, discoloration). The mean concentration of ferulic acid ranged from 371–441 µg/g between genotypes and locations, and ferulic acid was found to be the predominant phenolic acid representing approximately 63% of total phenolic content. In the study, significant and strong correlations were obtained between total phenolic content and antioxidant acitivity ($r^2 = 0.73$), between total phenolic content and ferulic acid concentration ($r^2 = 0.84$), and between antioxidant acitivity and ferulic acid content ($r^2 = 0.72$), so it can be explained that phenolic content and ferulic acid composition determine antioxidant properties of wheat grain (Mpofu, Sapirstein and Beta, 2006).

Genotype differences for total phenolic content, antioxidant activity, and phenolic acids would be possible to select for their quantitive traits in breeding programs. However, environmental factors should be taken into consideration and genotype and environment interactions should be evaluated with further studies (Mpofu, Sapirstein and Beta, 2006).

Zhou and Yu (2004) studied on antioxidant properties of wheat grown in different locations and they determined effects of growing conditions, including high-temperature stress in maturity periods and solar radiation on antioxidant activity of wheat bran extracts. In the study, antioxidant activity of wheat bran extracts differed significantly and approximately radical scavenging activity (% DPPH remaining) values changed from 25–45% under different locations. The Walsh location (Colorado, USA) contained the greatest level of total phenolic content (3.05 mg/g of bran) followed by the results 2.29 mg/g, 2.46 mg/g, 2.49 mg/g, and 2.97 mg/g values from other locations. The DPPH radical scavenging activity from the four non-irrigated locations was significantly correlated to both total solar radiation and daily average solar radiation. However, there is no significant correlation between total phenolic content of bran extracts and any tested growing conditions, including total or daily average solar radiation or daily high temperature. In addition to temperature changes, significant positive correlation found between radical scavenging activity and solar radiation, so it can be claimed that total and daily average solar radiations may alter antioxidant activity of wheat grain.

The studies clearly show that wheat genotype, environment, and their interactions are known to strongly influence the activity of grain antioxidants. Environmental effects are thought to have greater impact than genotypic effects because it is controlled by many factors. Temperature changes and precipitation amount from anthesis to maturity are mainly responsible for the grain composition development and variation of biochemical compounds.

The HEALTHGRAIN project aimed to improve consumer health by increasing the intake of bioactive compounds in grains and developing health-promoting and high quality cereal foods. For this purpose, 26 bread wheat cultivars grown in six different ecological conditions in France, United Kingdom, Poland, and Hungary to determine free, conjugated, and bound phenolic acid concentration. Free phenolic acid consisted only 1–2% of the total phenolic acids but the amounts of free phenolics showed the greatest variation in both growing seasons and locations. Bound phenolic acids have the greatest contribution and comprise 75–80% of total phenolic acids; mean values of bound phenolic acids ranged from 536–745 µg/g dm. Bound phenolic acids had less variation due to the environmental changes than other phenolic acids. The mean amounts of total phenolic acids ranged from 707–900 µg/g dm and were significantly affected by genotype and growing location. Environmental effects were larger than genotypic differences for phenolic acid composition. In the project, it is expressed that bound phenolic acids were less influenced by environment while more genotypic differences were noticeable. It is also suggested that genotypes with higher and more stable contents of phenolic acids can be selected to improve health effects of wheat grain in wheat-breeding programs (Fernandez-Orozco et al., 2010).

Genotype (G), environment (E), and the GxE interaction on antioxidant properties were investigated in 10 wheat varieties grown in different locations by Lv et al. (2013). The authors pointed out genotype, growing environment, or their interaction have influence on nutrient value of wheat grain. The environment had a significant effect on radical scavenging properties of hard wheat bran. Genotype and environmental conditions individually caused changes on antioxidant activity of wheat flour while the effect of the GxE interaction was not detected. They found significant and strong positive correlations between radical scavenging activity and overall average temperature, average high air temperature, and average low temperature. The correlation results suggested that higher air temperature during maturity periods was related to increased antioxidant activity of wheat. As a result, the growing environment has a stronger effect while genotype had less effect on antioxidant activity. The authros also claimed that it is possible to select the optimal environment and genotype to improve healthy compounds of wheat in breeding programs.

Nitrogen (N) is an essential nutritional elements of wheat, and N fertilizers are commonly used in conventional agriculture systems for plant fertilization, increase in yield, and quality properties. Applying an optimal nitrogen dose is fundamental to obtain better plant growth, yield, and quality results. Cereals respond more efficiently to nitrogen fertilization than other crops. Excessive nitrogen can lead to increase disease susceptibility and loss of nitrogen from soil that causes environmental damages. It also causes major changes in grain chemical composition and end-use quality (Kong et al., 2017).

Influence of nitrogen fertilization and maturity periods on total phenolic concentration of wheat was investigated by Stumpf et al. (2015). During grain development, phenolic compounds reached the highest level (0.14 mg/GAE/g dm) at the late milk stage, while phenolic content decreased about 20% in both medium milk and dough maturity stages. N fertilization caused inverse effects in different fractions of phenolic compounds. In the study, the most remarkable point made by the authors was that most phenolic compounds are mainly located in the outer layers of grain, so it is expected that increasing proportion bran fraction in grain may cause higher phenolic compounds. However the smaller grains with higher surfaces had the lowest thousand-grain mass and total phenolic content results. As a result of their study, it is emphasized that excessive nitrogen application can also reduce the bioavailable phenolic compounds and less health-promoting effects of wheat grain.

Organic products are preferred by consumers and an eco-friendly and environmental sustainability crop system. Organic products generally have a lower nutritional value, especially in terms of proteins, but have higher concentration of secondary metabolites. This may be due to exposure of the antioxidant defense system to pest attacks and stress conditions. Mazzoncini et al. (2015) focused on how organic and conventional cropping systems influence antioxidant properties of wheat. In the study, total phenolic acid and phenolic acid composition were higher in bran than white flour and bran fraction radical scavenging activity had a higher antioxidant activity than white flour. The organic cultivation system didn't reduce total phenolic acid, but caused an increase in antioxidant activity in the bran. In bran and white flour, ferulic acid was higher in the conventional system compared to organic cultivation, but the bran fraction had higher phenolic acid, and ferulic acid content decreased in white flour. Organic crop systems were not effective on most phenolic acids in comparison with conventional system. Antioxidant activity wasn't significantly affected by crop systems in both radicals. The researches also reported that both wheat and other crops showed no significant differences between the two cultivation systems in terms of phenolic-acid concentration. In general, bran was confirmed as rich in phenolics and antioxidant activity. Nitrogen deficiency may cause different composition of phenolic acids and higher antioxidant activity found in the bran from the organic system. This can be explained as stress conditions that may have an effect on composition of secondary metabolites in plants.

A study was conducted to determine genotype, weather conditions, and crop system effects on antioxidant activity and phenolic acid content of wheat with colored grain. In the study, six wheat genotypes with different grain colors (blue aleurone, purple pericarp, and yellow endosperm) were grown in organic and conventional crop systems. Genotype and its interactions with crop year and crop system significantly affected grain yield, total phenolic content, phenolic acids, total carotenoid, and total anthocyanin content, but total antioxidant activity was mostly affected by genotype. The research showed that the highest antioxidant activity values were 226.7 mg/kg dm and 212.7 mg/kg (d.m.) obtained from blue aleurone and purple pericarp, respectively. Total antioxidant activity value was found to be higher in the blue aleurone colored genotype compared to the standard and yellow endosperm. Crop season 2016 marked by slightly higher-than-average temperature and lower precipitation during grain filling caused higher antioxidant compounds and wheat genotypes had higher antioxidant activity due to being exposed to higher weather-stress conditions. The organic cultivation system had the highest antioxidant activity and phenolic acids, but its wheat yield decreased compared to the conventional system. It is suggested wheat genotypes grown in organic crop systems were much more damaged by the pests, so the authors suggested that higher biotic stress may cause higher concentrations of antioxidant compounds in organic cultivated wheat genotypes (Zrcková et al., 2018).

2.4 Protein Quality of Wheat

Plants are the major harvesters of solar energy and producers of human food supply as well as the primary source of carbon, vitamins, minerals, protein, essential amino acids, and fatty acids. Plants provide approximately 65% of edible protein on a global basis. The world's nourishment depends on 20 different plant crops, which are divided into cereals, vegetables, legumes, fruits, and nuts. In all the world, cereal grains and food legumes (also oil-seed legumes) are the most important groups to provide protein and amino acid supply in daily nutrition in humans (Young and Pellett, 1994).

The human body uses protein to build and repair tissue. Proteins provide energy, enzymes, hormones, skin, blood and bone formation. The amount of daily protein intake is important for a heathy life; the effects of having access to quality protein are listed below (FAO, 2013):

- **Short-term effects:** Growth and tissue repair, immune function and host defense system, muscle and skeletal formation, mental performance and antioxidant system
- **Long-term effects:** Linear growth and aging, age-related functional losses, nutrition-related chronic diseases (cancer, hypertension, etc.)

Among the cereals, wheat flour is used widely in food and the baking industry. Its functionality is significantly different from other cereals because its unique properties of protein allows it to produce visco-elastic dough. Wheat proteins are divided into gluten proteins and a heterogeneous group of nongluten proteins. They are classified to protein fractions based on solubility in various solvents listed below:

- Albumins (soluable in water)
- Globulins (insoluable in pure water, but soluable in NaCl solutions)
- Gliadins (soluable in ethyl alcohol)
- Glutenin (soluable in dilute acid or sodium hydroxide solutions)

Gluten proteins are a generally accepted protein fraction that determine the bread-making quality potential of wheat flour. Gluten proteins consist of glutenin and gliadin protein fractions—both of which have high levels of proline and glutamine. Wheat gluten proteins provide a visco-elastic form when kneading flour with water, allowing the dough to hold gas during fermentation to result in the typical structure of bread (Veraverbeke and Delcour, 2002; Šramkováa, Gregováb and Sturdik, 2009).

Wheat protein content is a major component of wheat grain quality—its high protein value is the major emphasis for nutritional enhancement and improved product quality. Wheat protein content varies from about 7–22%—the greater part of this variation depends on nongenetic factors and the environmental impacts that make breeding for high-protein content difficult (Shewry, 2007).

Grain protein content is inversely related to grain size and grain yield (Erekul and Köhn, 2006). This inverse effect causes difficulties in improving high-protein content wheat varieties. Breeders and farmers mainly make a selection between high-yield and high-quality wheat varieties. Heat stress causes an increase in protein content while grain yield and functionality of protein significantly decrease, affecting the bread-making quality of wheat. Wheat grain protein content substantially increases when heat stress occurs in the early grain-filling period. Synthesis of glutenin decreases while synthesis of glutenin fraction increase in heat stress conditions, so increasing the gliadin/glutenin ratio mainly relates to lower bread-making performance (Farooq et al., 2011).

Protein quality is associated with the proportion of essential amino acids, which cannot be synthesized by animals and must be provided in daily nourishment. If only one of these amino acids is limited in the diet, it can result in poor growth of livestock and humans. The use of protein degenerates and metabolism and productivity decline because of the lack or absence of these amino acids. The most critical essential amino acids are lysine and methionine in human and animal nutrition. The contents of amino acids are largely determined by the endosperm, which consists of about 80% of grain weight. The aleurone and embryo tissues of grains contain higher essential amino-acid content compared to other parts of the grain (Shewry, 2007). Lysine is the most limiting amino acid in wheat. The nutritional quality is determined by the higher protein and limiting amino acid content, especially lysine. Protein content can be modified and increased in wheat breeding programs, nevertheless, a negative correlation exists between lysine expressed as a percent of protein and protein content in wheat genotypes. Therefore lysine can be described as a primary measure of protein quality in wheat grain (Anjum et al., 2005; Noberbekova, Suleimenov and Zhapayev, 2018).

Wheat grain quality is affected by genetic and environmental factors. Cultivar selection, climate conditions, and agricultural practices are the main causes of changes in protein composition. High-temperature climate conditions increase leaf and plant senescence and reduce the duration of all development stages

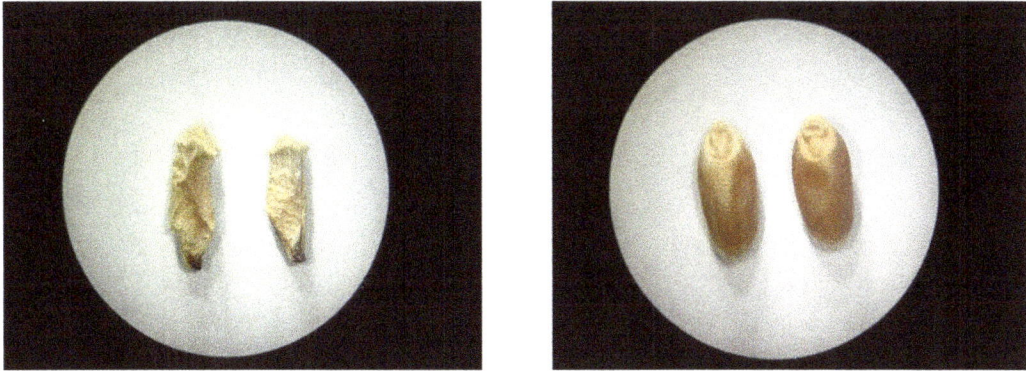

FIGURE 2.2 Grain size differences in heat-stress conditions (left) and optimal growing conditions (right) (original, Yigit, A).

in wheat. Heat-stress effects start with a shortage of grain filling and reduction of grain weight and directly cause a decrease of yield and quality components (Figure 2.2). Heat stress, occurring for even a few days in maturity periods can have a major impact on protein composition. Heat stress also decreases essential amino acid content in grain associated with increasing protein content, which is related to nutritional quality of protein (Castro et al., 2007).

Protein and amino acid composition is influenced by many factor from sowing to harvest. In all development stages, wheat growth encounters many stress factors, so it is important to apply optimal management practices and fertilizers, protection from pests and diseases, supply suitable irrigation, and suitable climate and soil conditions. With the need to focus greater attention on quality of protein, a brief discussion and summarized results on how amino acid composition responds to different environment, genotype, and agricultural factors follows in the next section.

2.5 Factors Affecting Amino Acid Composition of Wheat

Ancient wheat species are an important source of genes related to health benefit effects, nutritional value, and resistance to diseases for commercial wheat varieties in wheat breeding programs. A study was conducted to evaluate the nutritional protein value in emmer wheat [*Triticum dicoccum* (Schrank) Schuebl]. Six varieties of emmer from different genetic resources were grown in different locations. Environmental factors only affected methionine content, while valine, leucine, tyrosine, and phenylalanine amino acid values obviously changed between wheat species. Lysine was found to be the most limiting essential amino acid in emmer wheat, which achieved quite low values of a 0.38 and 0.43 chemical score. Leucine was found the third limiting amino acid, and a proportion of leucine wasn't influenced by location or wheat species. It is suggested that crude protein is not only a crucial factor of grain quality, but it is also necessary to know the structure of amino acid composition in wheat. The proportion of amino acids in emmer wheat grains is higher than the proportion of amino acids in grains of modern varieties. Emmer wheat contains the same amino acids as commercial wheat varieties, however, it is characterized by higher protein in grain and higher content of amino acids (Konvalina et al., 2008; 2011).

Genotype is the most important factor affecting the amino acid composition of wheat. It affects grain size, weight, and protein content of grain controlled by genotypic features. Fifty bread wheat varieties were collected from different locations in Turkey and the study aimed to determine the genotype effects on amino acid composition of wheat grain. In the study, it was found that amino acid composition showed statistically significant different results in all varieties. Glutamic acid and prolin are the most abundant amino acids with 2.20 g/100 g and 1.30 g/100 g mean values, respectively. While methionine and lysine had the lowest values respectively 0.267 g/100 g and 0.392 g/100 g and were described as one the most limiting amino acids in wheat grain. Wheat protein content and 17 amino acids showed statistically significant positive correlation coefficient values (the lowest in lysine $r = 0.206^{**}$ [$p < 0.01$],

Recovery in flour (%)

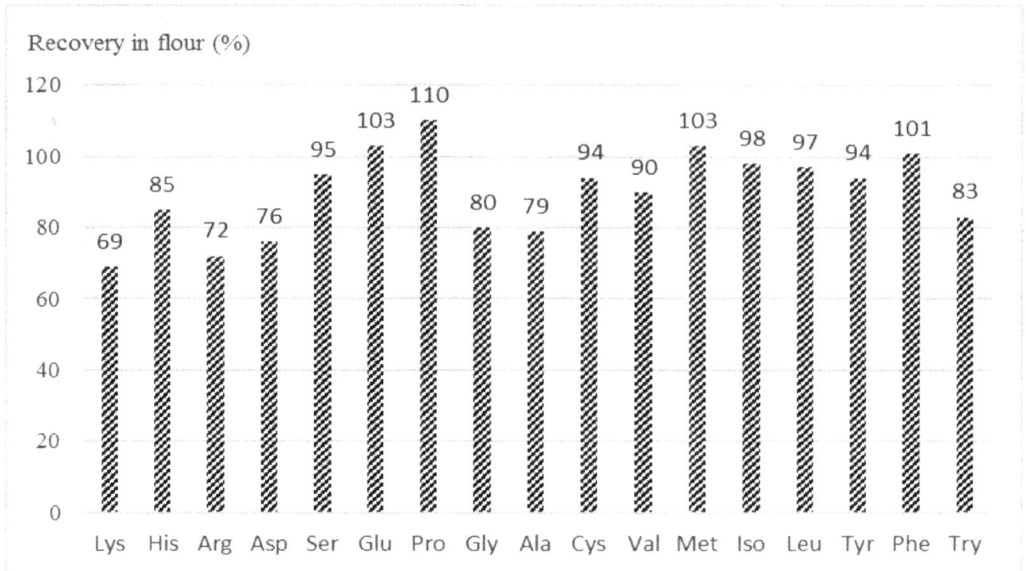

FIGURE 2.3   Recovery of amino acids after milling from wheat grain to flour (modified from Cornell, 2012).

the highest in leucine $r = 0.943^{**}$ [$p < 0.01$]) and the wheat grain protein content increase caused better nutritional quality with increasing amino acid values in grain (Yiğit, 2015).

Wheat grain parts contain different amounts of amino acids and amino acid composition of grain is affected by the milling process. Glutamine and proline levels reduced by half, arginine tripled, and asparagine, alanine, histidine, lysine, and glysine were double those in wheat flour. Whole wheat and endosperm have a similar amino acid composition. After milling process of hard red wheat grain to flour glutamic acid, proline, methionine, phenylalanine, serine, isoleucine, and leucine amino acids recovered with high level (>95%) compared to other amino acids (Fig 2.3).

Whole wheat grain and endosperm are very rich in glutamine and proline amino acids while lysine, histidine, cysteine, methionine, tyrosine, and tryptophan content is very low and limited. Wheat flour protein content is similar to whole wheat, but removal of the germ causes a decrease in milling process (Cornell, 2012).

Nitrogen fertilization directly affects grain yield and protein quality in wheat. Amino acids are the major forms in which N is remobilized from wheat leaves to the grain in maturity periods. A study was conducted to determine the effects of nitrogen fertilization on amino acid composition of wheat grain. The variation in amino acid composition was investigated by applying different nitrogen doses in common wheat varieties in Mediterranean climate conditions. Aspartic acid, serine, and lysine amino acids resulted in significant increases in response to the nitrogen application in comparison with others (Table 2.1).

Nitrogen fertilization caused an increase in all evaluated amino acids and this situation is mainly attributed to the decreased grain expansion and lower kernel weight in the N-deficient treatment that led to a reduced proportion of amino acids in grain. It is also emphasized that glutamic acid, arginine, alanine, valine, and leucine amino acids were found in higher values compared to others. Lysine, methionine, and threonine are limiting amino acids with the lowest values in amino acid composition of wheat in the region (Öncan-Sümer, 2008).

Irrigation is a key application improving grain yield and quality, especially in arid and hot climate conditions. Conversely, nitrogen application is often known as the most limiting factor in terms of wheat yield and quality. Increased nitrogen application as well as fertilizer timing and type is crucial for optimal plant growth. Zhang et al. (2017) aimed to determine the effects of nitrogen fertilization doses and irrigation treatments on protein and the amino acid composition of wheat. Nitrogen application caused a significant increase in grain protein content while no significant protein content response to irrigation was observed.

TABLE 2.1

Amino Acid Composition Change Response to Excessive and Deficient Nitrogen Application (modified from Öncan-Sümer [2008])

Amino Acids (mol/100 mol)	N Deficiency (0 kg/ha)	Excessive N (240 kg/ha)	Increase (%)
Aspartic acid	4.22	6.23	47.6
Glutamic acid	13.70	18.63	35.9
Serine	3.12	4.74	51.9
Histidine	1.63	2.06	26.3
Arginine	6.80	8.75	28.6
Glycine	1.41	1.71	21.2
Threonine	2.36	3.36	42.3
Alanine	16.18	20.84	28.8
Tryptophan	2.63	3.43	30.4
Methionine	1.24	1.57	26.6
Valine	6.62	7.94	19.9
Phenylalanine	2.69	3.67	36.4
Isoleucine	4.02	4.66	15.9
Leucine	8.55	10.26	20.0
Lysine	1.05	1.63	55.2

Total amino acid (essential and nonessential) values were not significantly affected by irrigation. N application caused an increase in the amino acid composition of wheat. Leucine and phenylalanine accounted for a large proportion of the essential amino acid content. The highest total amino acid, essential and nonessential values, was obtained with the combination of 300 kg/ha N and irrigation of 750 m³/ha. A significant increase in essential amino acid values was observed following the higher nitrogen doses. Essential and nonessential amino acid content increased linearly with an increase in protein content.

Environmental conditions and water availability are crucial factors for protein quality during the grain-filling period. High temperature during maturity combined with water stress has favorable effects on protein content, and even short periods of very high temperature (>35°C) have negative effects on grain quality (Erekul et al., 2009).

The amino acid content of wheat is influenced by environmental conditions in comparison to genotype and genotype-environment interactions. Amino acid composition changed significantly in all durum wheat genotypes with the exception of arginine and cysteine. Methionine and lysine essential amino acids had the highest percentage (13.9% and 18.5%) of variation for all genotypes, respectively. Irrigation significantly prolonged the grain-filling duration, ranging from 28 days under rain-fed conditions. to 37.7 days under irrigated conditions. Amino acid composition varied significantly between environments; the highest values were found in irrigated conditions while the lowest values were in dry rain-fed conditions for all amino acids. Protein content significantly decreased in environmental conditions caused by a longer grain-filling duration, while the concentration of all amino acids significantly increased. Glutamine, phenylalanine, and proline content increased with the shortened grain-filling duration and faster leaf senescence caused by high temperature and water availability. On the contrary, the increase in the grain-filling duration period diminished the other amino acids and, thus, the baking quality conditions of the product. Conditions that create a shorter maturity period, such as high temperature and drought, affect the balance of protein fractions and amino acid composition in wheat grain (Moral et al., 2007).

2.6 Concluding Remarks

This chapter has pointed out that how the health and nourishment properties of wheat respond directly to agricultural factors. Wheat is cultivated in larger areas across the world in a diverse range of climates and environments, so its yield and quality properties are controlled by many factors. Wheat grain structure

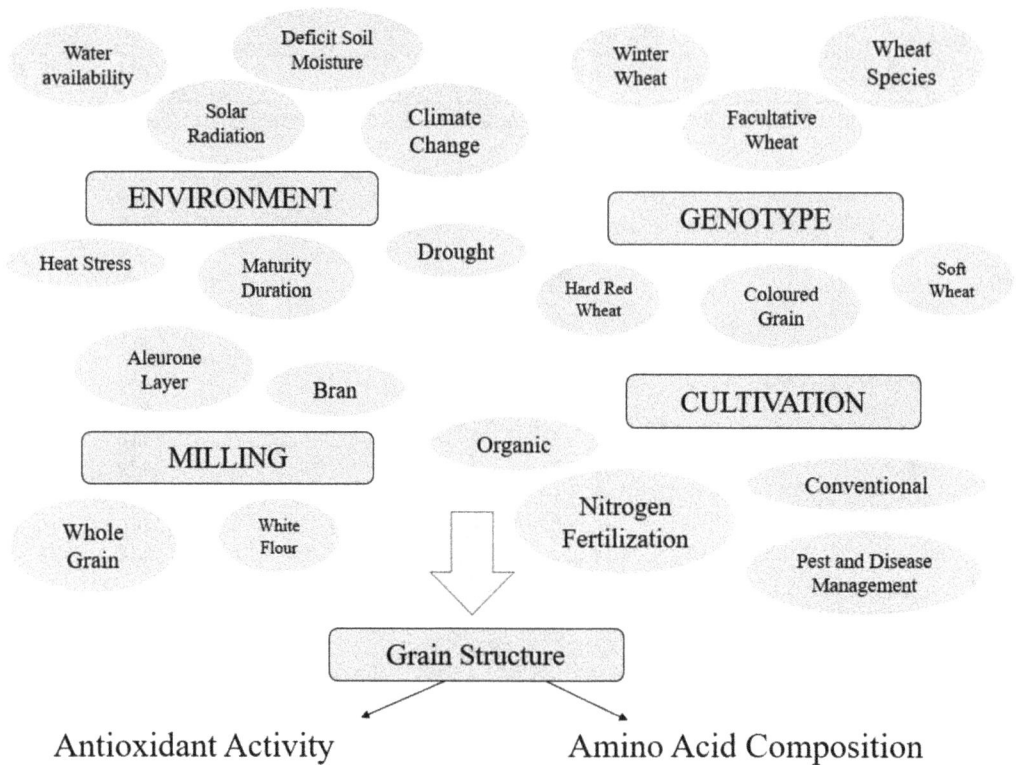

FIGURE 2.4   Schematic design of factors.

has unique parts that provide healthy food and products for human consumption, but this structure is affected by many factors from sowing to harvest. The outer layers of wheat grain have health promoting biochemical compounds, but their separation in the milling process leads to reduced benefits of grain (Figure 2.4).

The health benefits and antioxidant activity of wheat is based on phenolic acid composition, so ferulic acid comes to the forefront compared to other phenolic acids. Conversely, limited essential amino acids (lysine and methionine) certainly play a major role in protein quality and nourishment benefits. Taking this into account, factors were evaluated that contribute to a better understanding of the biochemical changes and mechanisms by which genetic, environmental, and industrial processes can modify grain composition, revealing the health potential of wheat grain. Further studies are essential between breeders, agronomists, physiologists, and cereal chemists with interdisciplinary studies on the effects of agricultural factors on health and nourishment issues.

REFERENCES

Abdel-Aal, E. S. M., P. Hucl 2002. Amino acid composition and in vitro protein digestibility of selected ancient wheats and their end products. *Journal of Food Composition and Analysis*, 15: 737–747.

Adom, K. K., M. E. Sorrells, R. H. Liu 2003. Phytochemical profiles and antioxidant activity of wheat varieties. *Journal of Agricultural and Food Chemistry*, 51, 7825–7834.

Anjum, F. M., I. Ahmad, M. S. Butt, M. A. Sheikh, I. Pasha 2005. Amino acid composition of spring wheats and losses of lysine during chapati baking. *Journal of Food Composition and Analysis*, 18: 523–532.

Bechtel, D. B., J. Abecassis, P. R. Shewry, A. D. Evers 2009. Development, Structure, and Mechanical Properties of the Wheat Grain. In: K. Khan and P. R. Shewry (Eds.), *Wheat Chemistry and Technology*. St. Paul, M. N. AACC International Inc., 51–80.

Beta, T., S. Nam, J. E. Dexter, H. D. Sapirstein 2005. Phenolic content and antioxidant activity of pearled wheat and roller-milled fractions. *Cereal Chemistry*, 82 (4): 390–393.

Blakeney, A. B., R. L. Cracknell, G. B. Crosbie, S. P. Jefferies, D. M. Miskelly, L. O'Brien, J. F. Panozzo, D. A. I. Suter, V. Solah, T. Watts, T. Westcott, R. M. Williams 2009. Understanding Australian wheat quality: Fundamentals of wheat quality. *Wheat Quality Objectives Group*.

Boukid, F., M. Dall'Asta, L. Bresciani, P. Mena, D. Del Rio, L. Calani, R. Sayar, Y. Weon Seo, I. Yacoubi, M. Mejri 2019. Phenolic profile and antioxidant capacity of landraces, old and modern Tunisian durum wheat. *European Food Research and Technology*, 245 (1): 73–82.

Brandolini, A., A. Hidalgo 2012. Wheat germ: not only a by-product. *International Journal of Food Sciences and Nutrition*, 63 (S1): 71–74.

Călinoiu, L. F., D. C. Vodnar 2008. Whole grains and phenolic acids: A review on bioactivity, functionality, health benefits and bioavailability. *Nutrients*, 10 (11): 1615.

Castro, M., C. J. Peterson, M. Dalla Rizza, P. Diaz Dellavalle, D. Vázquez, V. Ibáñez, A. Ross 2007. Influence of heat stress on wheat grain characteristics and protein molecular weight distribution. In: Buck, H. T., J. E., Nisi and N. Salomón (Eds.), *Wheat Production in Stressed Environments*. Springer: 365–371.

Chantret, N., J. Salse, F. Sabot, S. Rahman, A. Bellec, B. Laubin, I. Dubois, C. Dossat, P. Sourdille, P. Joudrier, M. F. Gautier, L. Cattolico, M. Beckert, S. Aubourg, J. Weissenbach, M. Caboche, M. Bernard, P. Leroy, B. Chalhoub 2005. Molecular basis of evolutionary events that shaped the hardness locus in diploid and polyploid wheat species (*Triticum* and *Aegilops*). *The Plant Cell*, 17: 1033–1045.

Cornell, H. J 2012. The chemistry and biochemistry of wheat. In: Cauvain, S.P. (Ed.), *Bread Making, Improving Quality*, 2nd ed. Cambridge, UK: Woodhead Publishing.

Das, K., A. Roychoudhury 2014. Reactive oxygen species (ROS) and response of antioxidants as ROS-scavengers during environmental stress in plants. *Frontiers in Environmental Science*, 2 (53): 1–13.

Dupont, F. M., S. B. Altenbach 2003. Molecular and biochemical impacts of environmental factors on wheat grain development and protein synthesis. *Journal of Cereal Science*, 38: 133–146.

Dykes, L., L.W. Rooney 2007. Phenolic compounds in cereal grains and their health benefits. *Cereal Foods World*, 52: 105–111.

El Sabagh, A., A. Hossain, C. Barutçular, M. S. Islam, S. I. Awan, A. Galal, A. Iqbal, O. Sytar, M. Yıldırım, R. S. Meena, S. Fahad, U. Najeeb, O. Konuşkan, R. A. Habib, A. Llanes, S. Hussain, M. Farooq, M. Hasanuzzaman, K. H. Abdel-aal, Y. Hafez, F. Çığ, H. Saneoka 2019. Wheat (*Triticum aestivum* L.) production under drought and heat stress-adverse effects, mechanisms and mitigation: a review. *Applied Ecology and Environmental Research*, 17 (4): 8307–8332.

Erekul, O, W. Köhn 2006. Effect of weather and soil conditions on yield components and bread-making quality of winter wheat (*Triticum aestivum* L.) and winter triticale (*Triticosecale* Wittm.) Varieties in North-East Germany. *Journal of Agronomy and Crop Science*, 192: 452–464.

Erekul, O., T. Kautz, F. Ellmer, I. Turgut 2009. Yield and bread making quality of different wheat (*Triticum aestivum* L.) genotypes grown in Western Turkey. *Archives of Agronomy and Soil Science*, 55(2): 169–182.

FAO, 2013. Dietary protein quality in human nutrition: Findings and recommendations of the 2011 FAO expert consultation on protein quality evaluation in human nutrition. *FAO and Food Nutrition Paper*, Chapter 4.

FAO, 2018. Food and Agriculture Organization Corporate Statistical Database (FAOSTAT), Crop Statistics. http://www.fao.org/faostat/en/#data/QC (accessed September 10, 2020).

Farooq, M., H. Bramley, J. A. Palta, H. M. Siddique 2011. Heat stress in wheat during reproductive and grain-filling phases. *Critical Reviews in Plant Sciences*, 30: 1–17.

Fernandez-Orosco, R., L. Li, C. Herflett, P. R. Shewry, J. L. Ward 2010. Effects of environment and genotype on phenolic acids in wheat in the HEALTHGRAIN diversity screen. *Journal of Agricultural and Food Chemistry*, 58: 9341–9352.

Kong, L., Y. Xie, L. Hu, J. Si, Z. Wang 2017. Excessive nitrogen application dampens antioxidant capacity and grain filling in wheat as revealed by metabolic and physiological analyses. *Scientific Reports*, 7, 43363, doi: 10.1038/srep43363.

Konvalina, P., Jr. J. Moudrý, Z. Stehno, J. Moudrý 2008. Amino acid composition of emmer landraces grain. *Lucrări Ştiinţifice seria Agronomie*, 51: 241–249.

Konvalina, P., I. Capouchová, Z. Stehno, Jr. J. Moudrý, J. Moudrý 2011. Composition of essential amino acids in emmer wheat landraces and old and modern varieties of bread wheat. *Journal of Food, Agriculture & Environment*, 9(3/4): 193–197.

Li, W., T. Beta, S. Sun, H. Corke 2006. Protein characteristics of Chinese black-grained wheat. *Food Chemistry*, 98: 463–472.

Li, L., P. R. Shewry, J. L. Ward 2008. Phenolic acids in wheat varieties in the HEALTHGRAIN divesity screen. *Journal of Agricultural and Food Chemistry*, 56: 9732–9739.

Lin, C., G. Guo, D. L. Mennel 2008. Effects of postharvest treatments, food formulation, and processing conditions on wheat antioxidant properties. In: L. Yu (Ed.), *Wheat Antioxidants*. Hoboken, N.J.: John Wiley & Sons Inc., 73–87.

Lv, J., Y. Lu, Y. Niu, M. Whent, M. F. Ramadan, J. Costa 2013. Effect of genotype, enviroment, and their interaction on phytochemical compositions and antioxidant properties of soft winter wheat flour. *Food Chemistry*, 138: 454–462.

Martini, D., F. Taddei, R. Ciccoritti, M. Pasquini, I. Nicoletti, D. Corradini, M. G. D'Egidio 2015. Variation of total antioxidant activity and of phenolic acid, total phenolics and yellow coloured pigments in durum wheat (*Triticum turgidum* L. var. *durum*) as a function of genotype, crop year and growing area. *Journal of Cereal Science*, 65: 175–185.

Mazzoncini, M., D. Antichi, N. Silvestri, G. Ciantelli, C. Sgherri 2015. Organically vs conventionally grown winter wheat: Effects on grain yield, technologically quality, and on phenolic composition and antioxidant properties of bran and refined flour. *Food Chemistry*, 175: 445–451.

Moral, L. F., Y. Rharrabti, V. Martos, C. Royo 2007. Environmentally induced changes in amino acid composition in the grain of durum wheat grown under different water and temperature regimes in a Mediterranean environment. *Journal of Agricultural and Food Chemistry*, 55: 8144–8151.

Mpofu, A., H. D. Sapirstein, T. Beta 2006. Genotype and environmental variation in phenolic content, phenolic acid composition and antioxidant activity of hard spring wheat. *Journal of Agricultural and Food Chemistry*, 54: 1265–1270.

Narwal, S., V. Thakur, S. Sheoran, S. Dahiya, S. Jaswal, R. K. Gupta 2014. Antioxidant activity and phenolic content of the Indian wheat varieties. *Journal of Plant Biochemistry and Biotechnology*, 23(1): 11–17.

Ndolo, V. U., T. Beta 2014. Comparative studies on composition and distribution of phenolic acids in cereal grain botanical fractions. *Cereal Chemistry*, 91(5): 522–530.

Noberbekova, N. K., Y. T. Suleimenov, R. K. Zhapayev 2018. Influence of fertilizing with nitrogen fertilizer on the content of amino acids in sweet sorghum grain. *Agriculture and Food Sciences Research*, 5(2): 64–67.

Öncan Sümer, F 2008. Ekmeklik Buğday (*Triticum aestivum* L.) Çeşitlerinde Bitki Sıklığı ve Azot Dozlarının Verim, Verim Unsurları, Agronomik ve Kalite Özellikleri Üzerine Etkileri ve Özellikler Arası İlişkiler. AydınAdnan Menderes University Graduate School of Natural and Applied Sciences, *Ph.D. thesis*.

Punia, S., K. S. Sandhu, A. K. Siroha 2017. Difference in protein content of wheat (*Triticum aestivum* L.): Effect on functional, pasting, color and antioxidant properties. *Journal of Saudi Society of Agricultural Sciences*, doi: https://doi.org/10.1016/j.jssas.2017.12.005.

Punia, S., K. S. Sandhu, S. Sharma (2017a). Comparative studies of color, pasting and antioxidant properties of wheat cultivars as affected by toasting and roasting. *Nutrafoods*, 16, 95–1021.

Punia, S., & K. S. Sandhu (2016). Physicochemical and antioxidant properties of different milling fractions of Indian wheat cultivars. *International Journal of Pharma and Bio Sciences*, 7(1), 61–66.

Qabaha, K 2010. Development and optimization of a microwave-assited protein hydrolysis method to permit amino acid profiling of cultivated and wild wheats and to relate the amino acid to grain mineral concentrations. Sabancı University, Graduate School of Engineering and Natural Sciences, Ph.D. thesis.

Revanappa, S. B., P. V. Salimath 2009. Phenolic acid profiles and antioxidant activities of different wheat (*Triticum aestivum* L.) varieties. *Journal of Food Biochemistry*, 35: 759–775.

Shewry, P. R 2007. Improving the protein content and composition of cereal grain. *Journal of Cereal Science*, 46: 239–250.

Shewry, P. R 2009. Wheat. *Journal of Experimental Botany*, 60(6): 1537–1553.

Sonntag, F., D. Bunzel, S. Kulling, I. Porath, F. Pach, E. Pawelzik, I. Smit, M. Naumann 2020. Effect of potassium fertilization on the concentration of antioxidants in two cocktail tomato cultivars. *Journal of Applied Botany and Food Quality*, 93: 34–43. https://doi.org/10.5073/JABFQ.2020.093.005.

Šramkováa, Z., E. Gregováb, E. Šturdík 2009. Chemical composition and nutritional quality of wheat grain. *Acta Chimica Slovaca*, 2(1): 115–138.

Stumpf, B., F. Yan, B. Honermeier 2015. Nitrogen fertilization and maturity influence the phenolic concentration of wheat grain (*Triticum aestivum*). *Journal of Plant Nutrition and Soil Science*, 178: 118–125.

Vaher, M., K. Matso, T. Levandi, K. Helmja, M. Kaljurand 2010. Phenolic compounds and the antioxidant activity of the bran, flour and whole grain of different wheat varieties. *Procedia Chemistry*, 2: 76–82.

Veraverbeke, W. S., J. A. Delcour 2002. Wheat protein composition and properties of wheat glutenin in relation to bread making functionality. *Critical Reviews in Food Science and Nutrition*, 42 (3): 179–208.

Wrigley, C. W 2009. Wheat: A unique grain for the World. In: K. Khan and P. R. Shewry (Eds.), *Wheat Chemistry and Technology*. St. Paul, M.N.: AACC International Inc., 1–17.

Yılmaz, V. A., A. Brandolini, A. Hidalgo 2015. Phenolic acids and antioxidant activity of wild, feral and domesticated diploid wheats. *Journal of Cereal Science*, 64: 168–175.

Yiğit, A 2015. Türkiye'de Yaygın Olarak Yetiştirilen Ekmeklik Buğday (*Triticum aestivum* L.) Çeşitlerinin, Protein, Aminoasit Dağılımı ve Antioksidan Aktivitelerinin Belirlenmesi. Aydın Adnan Menderes University Graduate School of Natural and Applied Sciences, *Masters thesis.*

Young, V. R., P. L. Pellett 1994. Plant proteins in relation to human protein and amino acid nutrition. *The American Journal of Clinical Nutrition*, 59: 1203S–12S.

Yu, L 2008. Overview and Prospective. In L. Yu (Ed.), *Wheat Antioxidants*. Hoboken, New Jersey.: John Wiley & Sons Inc., 1–5.

Zhang, P., G. Ma, C. Wang, H. Lu, S. Li, Y. Xie, D. Ma, Y. Zhu, T. Guo 2017. Effect of irrigation and nitrogen application on grain amino acid composition and protein quality in winter wheat. *Plos One*, 12(6): https://doi.org/10.1371/journal.pone.0178494.

Zhou, K., L. Yu 2004. Antioxidant properties of bran extracts from Trego wheat grown at different locations. *Journal of Agricultural and Food Chemistry*, 52: 1112–1117.

Zhou, K., J. J. Laux, L. Yu 2004. Comparison of Swiss red wheat grain and fractions for their antioxidant properties. *Journal of Agricultural and Food Chemistry*, 52: 1118–1123.

Zhou, K., J. Yin, L. Yu 2005. Phenolic acid, tocopherol and carotenoid compositions, and antioxidant functions of hard red winter wheat bran. *Journal of Agricultural and Food Chemistry*, 53: 3916–3922.

Žilić, S., A. Serpen, G. Akıllıoğlu, M. Janković, V. Gökmen 2012. Distributions in phenolic compounds, yellow pigments and oxidative enzymes in wheat grains and their relation to antioxidant capacity of bran and debranned flour. *Journal of Cereal Science*, 56: 652–658.

Zrcková, M., I. Capouchová, M. Eliášová, L. Paznocht, K. Pazderů, P. Dvořák, P. Konvalina, M. Orsák, Z. Štěrba 2018. The effect of genotype, weather conditions and cropping system on antioxidant activity and content of selected antioxidant compounds in wheat with coloured grain. *Plant Soil Environement*, 11: 530–538.

3

Biotechnological Advancements to Explore Crop Based Studies on Nutritional Aspects

Shourabh Joshi
Nagaur-Agriculture University, Jodhpur, India

Rakesh Kumar Prajapat
Suresh Gyan Vihar University, Jaipur, India

Pawan S. Mainkar
ICAR-Directorate of Onion and Garlic Research, Pune, Maharashtra, India

Vinod Kumar
CCS Haryana Agricultural University, Hisar, Haryana, India

Neeshu Joshi
Sumerpur-Agriculture University, Jodhpur, India

Rahul Bharadwaj
Mandore-Agriculture University, Jodhpur, India

Sunita Gupta
PME, SKN Agriculture University, Jobner, Rajasthan, India

N.K. Gupta
SKN Agriculture University, Jobner, Rajasthan, India

CONTENTS

3.1 Introduction

The world's population is 6.2 billion and is expected to reach about 9.6 billion by 2050, resulting in an expected 70% increase in food demand. Worldwide, more than 800 million people are hungry, and more than 2 billion are facing malnutrition (FAO 2019a). Insufficient and low caloric food causes serious health issues, and malnutrition worsens the scenario by affecting physical and mental development and causing unacceptably high numbers of premature death (Development Initiatives 2018). Reducing these problems and achieving "zero hunger and improved nutrition" was among the 17 Sustainable Development Goal 2 which was established by the United Nations in 2015, requires major revolutions in global food systems. The level and composition of available nutrients vary from crop to crop and among crops, too. Improving the nutritional quality of staple crops seems to be an effective and straight-forward solution to the problem. While the Green Revolution significantly increased the yield potential of wheat and rice, avoiding severe food crises (Pingali 2012), its major focus on these two crops led to an increase in selective pressure from pests and diseases (Tilman *et al.* 2002) and lacked a focus to improve the nutritive value of crops. Lower productivity, marginal environments, exposure to extreme environments, and limited genetic resources are the key hurdles in crop improvement programs. To meet these challenges, scientists have been successful in identifying suitable landraces within the wild relatives that can tolerate biotic and abiotic stress (Gilliham *et al.* 2017). The integration of modern biotechnological tools with plant-breeding methods in a sustainable manner has the potential to fulfil the goal of attaining food security for present and future generations. The 21st century breeding teams—with their advancement in digital technologies and ability to share data sets in seconds—have enormous potential to explore molecular traits and tinker with them at the single nucleotide level. Development of emerging genomics platforms facilitates genomics-assisted breeding with higher precision and efficiency for crop improvement (Varshney *et al.* 2005). During the last decade, rapid development of thousands of SSR and millions of SNP markers could make it possible to generate saturated genetic maps and identify novel marker-trait association in crops that were earlier considered to be orphan crops. High throughput phenotyping and genotyping platforms further speed up the crop improvement process.

3.2 Molecular Breeding: A Potential Tool in Development of Crop-Based Research

As molecular biology grew, scientific interest deepened toward the identification of novel heritable variation, characterization, and annotation of the genes with available sequence information in the public domain. Molecular breeding allows direct selection of targeted genes/quantitative trait loci (QTLs) by using DNA-based molecular markers (Moose and Mumm 2008). Molecular breeding offers important advantages in comparison to phenotypic selection: they are rapid, can be performed at the seedling stage, offer identification of codominant alleles, and they are cheaper. Even though marker-assisted selection revolutionized the process of identification of functional variants with specific genes/QTLs, its implication is limited in the public domain (Collard and MacKill 2008). Molecular breeding can help to identify landraces with higher nutritive value that perform better under biotic and abiotic stress situations (Kumar *et al.* 2019). Modern breeding aimed to explore the unrevealed genetic diversity among the wild ancestors, which were eventually lost during domestication, to utilize them for betterment of cultivated species (Tanksley and McCouch 1997, Zamir 2001). But the major concern is "linkage drag," which is the association of negative effects of the genetic region (gene/QTLs) of interest present in wild relatives.

This negatively associated region will lower the performance of the gene/QTLs of interest so they cannot be utilized directly in a crop improvement breeding program. Hence, exploration of genetic diversity and identification of suitable marker systems are the prime objectives. DNA markers make it feasible to map QTLs in segregating populations, and showed that transfer of these genomic regions from wild germplasm to cultivated species has the capability to improve yield as well as other traits of interest e.g. for rice (Septiningshi *et al.* 2003) and pepper (Rao *et al.* 2003).

3.3 Genetic Engineering: A Tool to Incorporate Demand Driven Genes in Crops

Genetic engineering provides tools to transfer novel genes in such genetic background where they could not be transferred in crops by natural crossing/breeding methods. Despite the fact that the outcome of the technology comes after extensive experimental testing, evaluation of its composition, considering safety concerns, performance of important agronomic traits, and investigation of the crop's effect on the environment prior to its release in the marketplace, rigorous regulatory barriers and other issues have stalled the full potential of this time-tested technology. However, some of the important successful examples are discussed here. In 1996, Bt corn was introduced to fight the European corn borer (*Ostrinia nubilalis*) and was described as "the most important technological advancement in insect pest management since the development of synthetic insecticides" (Obrycki *et al.* 2001). Herbicide tolerant soybeans (Roundup Ready) were effective against glyphosate-based herbicide (Fernandez-Cornejo *et al.* 2000). Crop varieties were developed against infection caused by viruses or insect pests (Food and Drug Administration 2001). Plants were also genetically engineered to withstand various abiotic stressors, including cold, heat, drought, and ill-drained soil conditions, such as salinity and aluminum contamination (Thomashow 2001). Metabolic pathways (starch biosynthesis and nitrogen assimilation) were also modified to increase yield (Oscarson 2000). A successful attempt was also made to prevent spoilage of tomato fruits by increasing their shelf-life with the incorporation of a polygalacturonase-producing gene, which is a pectin-degrading enzyme found in ripened tomatoes, which slows down the ripening process (Redenbaugh *et al.* 1993). Foods and feeds with higher micronutrients, protein, starch content, and oil composition also developed. A new strain of potatoes containing 30–60% more starch has been developed by inserting a bacterial gene for an enzyme in the starch biosynthetic pathway. These potatoes with higher starch absorb less fat during frying (Muller-Rober *et al.* 1992). A peanut oil with higher oleic acid content was also developed (McKinney 2000). One of the most popular stories among genetically engineered crops is the genetically modified rice known as Golden Rice and its ability to synthesize higher amount beta-carotene, by incorporation of genes coding three enzymes of the phytoene synthase pathway (two genes from a daffodil and one from the bacteria *Erwinia uredovora*) (Ye *et al.* 2000). It could decrease malnutrition and blindness associated with vitamin A deficiency particularly in the marginal regions where rice is staple food.

3.4 New Plant Breeding Technologies (NPBTS)

New plant breeding technologies (NPBTs) refers not only to genetic modification using techniques of modern biotechnology tools, but it also includes other techniques as site-directed nucleases (SDN includes ZFN-1/2/3 and CRISPER systems), oligonucleotide directed mutagenesis (ODM), Cis genesis, RNA-dependent DNA methylation (Rd DM) reverse breeding, and agro infiltration. NPBTs are less prone to create multiple unknown, unintentional mutations across the genome. Unlike GMOs, many of the new breeding techniques will result in no foreign DNA in the product. NPBTs, which cover genetically modified organisms (GMOs) and gene-edited crops, could be a milestone in crop improvement programs (Zaidi *et al.* 2019). It could contribute to higher crop yields, more nutritious foods, higher storage and processing of quality foods, lower use of chemical fertilizers and pesticides, better crop resilience to climate stress, and reduced postharvest losses (Bailey-Serres *et al.* 2019; Eshed and Lippman 2019).

3.5 Omics-Based Technologies & Their Impact on Nutritional Studies

Omics—modern technologies of biology like genomics, transcriptomics, proteomics, and metabolomics—primarily aim to characterize and evaluate pools of biological molecules (genes, mRNA, protein, and metabolites) that interpret the structure, role, and elements of life forms. Novel Omics advances and bioinformatics strategies investigate the complex relationship between diet and the digestive system. The application of advanced technologies in nutritional research (nutrigenomics, nutritranscriptomics, nutriproteomics, and nutrimetabolomics) provides an opportunity to investigate dietary relationships (Figure 3.1). Genomics is the field of biology that entails structure, function, evolution, sequencing, and genome editing and mapping of the genes present in the genome; the study of the interaction between nutrition, health, and genome of an organism is called nutrigenomics (Davis and Hord 2005). Transcriptomics is the study of a total set of RNA transcript produced by the genome in a specific cell, while nutritranscriptomics studies the diet-gene interaction and alterations in the gene expression. Proteomics is the study of the whole complement of proteins (proteome), and nutriproteomics is the study of the relationship between dietary supplements and the individual's proteome. Metabolomics is the study of all the metabolites synthesized by a living being (Fiehn 2002), while the study of the interaction between the metabolite and dietary changes that lead to integrative action for metabolite analysis in nutritional studies is called nutrimetabolomics. These emerging technologies help to study the molecular functioning of an individual's body, which might be helpful in giving personalized nutrition (Chaudhary *et al.* 2021).

Understanding the relationship between nutrition and human science is significant and can encourage the development of nutritional research that might improve our wellness status and future well-being by decreasing hazard variables for nontransmittable infections, such as cardiovascular infections, cancer, corpulence, and metabolic disorders. The physiological impacts of diet and bioactive compounds are one of the most important objectives of current nutritional research as it might lead to the disclosure of novel biomarkers. Advances in high throughput next-generation sequencing, mass spectrometry, nuclear magnetic resonance (NMR), X-ray crystallography, and microarray, allow massive genome and gene expression analysis, metabolomic profiling, proteomic study, and in-depth analysis of pathological scenarios. Therefore, these platforms are important for the identification and analysis of new nutritional biomarkers and bioactive compounds that distinguish the atomic marks of dietary supplements and nonnutrients driving to a particular phenotype, and to give nutritional recommendations for personalized well-being support and malady anticipation.

3.5.1 Genomics

Genomics explores the full hereditary complements of a life form, centering on the structure, function, mapping, evolution, and genome editing of the individual. Genomics also works to sequence and examine

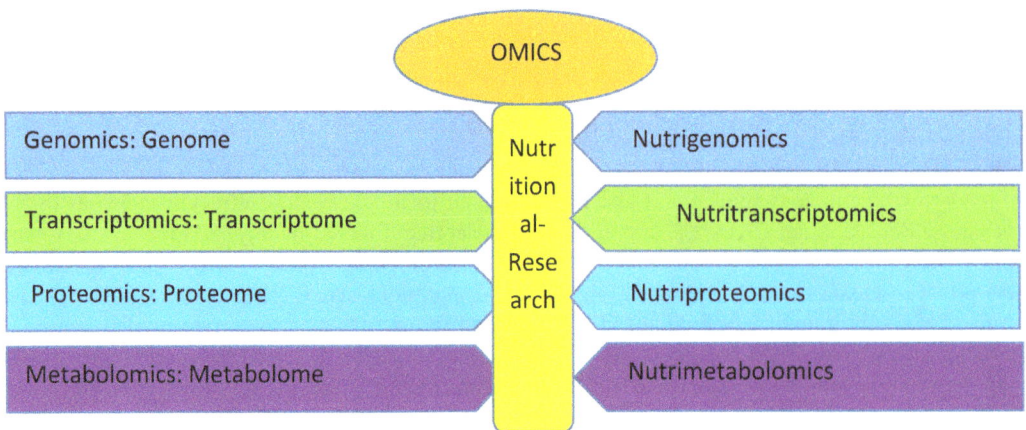

FIGURE 3.1 Modern analytical Omics technologies.

genomes through the use of progressed DNA-sequencing methods and bioinformatics to collect and analyze the work and structure of whole genomes. Functional genomics utilizes the tremendous amount of
information delivered from sequencing to analyze the functions and interaction of genes. The genome-
wide investigation is the key approach that functional genomics considers high throughput strategies,
instead of the gene-by-gene conventional approach. It resolves the variations in the genome between
individuals and correlates these variations with all of the phenotypic parameters. Structural genomics-
based approaches involve the modelling-based approach and high throughput experimental methods to
determine the three-dimensional structures of every single protein encoded by a given species' genome.
Epigenomics refers to the study of complete sets of epigenetic modifications on the hereditary materials in a cell or tissue at a given time. These are the reversible modifications on genetic materials (DNA
methylations) or histone proteins that change the gene expression without altering the hereditary materials (Russell 2010, Sanhueza and Valenzuela 2012). Metagenomics (environmental genomics) refers to the
study of genetic materials recovered directly from environmental samples. It has provided a powerful lens
to look over the microbial world that can revolutionize the perception of the entire world. Nutrigenomics is
the study of the impact of bioactive nourishment on gene expression of a person, and includes investigation
into nutritional components that act on the genome. It is a major undertaking to progress our understanding of sustenance and genomics. Looking into the zone of sustenance requires atomic instruments to view
just a few of the reactions from a food eaten by individuals in less-connected population groups, helping to
interpret how the bioactive compound of a particular diet may increase or suppress the expression of genes
(Dauncey 2012, Cozzolino and Cominetti 2013, Liu and Qian 2011). From the Human Genome Project
(HGP) data, it's now known that people share a 99.9% identity between their genomes and that mere 0.1%
of the gene sequence made a particular contrast between their tallness, eye color, hair, weight, nutritional
necessities, and the chance of creating nontransmissible inveterate maladies (NTCDs) (Ong and Rogero
2009, Cozzolino and Cominetti 2013). Single-nucleotide polymorphisms (SNPs) account for 90% of all
human hereditary variety and can alter the works of the "housekeeping genes" included within the cell,
creating a chance of creating disease (Norheim *et al.* 2012, Dauncey 2012). The methylenetetrahydrofolate
reductase (MTHFR) gene common C677T polymorphism is the classic illustration of the diet-SNP interaction. This polymorphism declines the activity of MTHFR enzyme and results in the reduced use of folic
acid to change homocysteine into methionine, and consequently, into the S-adenosyl-methionine (SAM).
SAM is a methyl donor necessary for the upkeep of methylation of cytosine in hereditary material and
control of gene expression in metabolic responses. The action of the MTHFR enzyme-catalyzed response
can be changed depending on the sum of two basic supplements, such as folic acid (folate) as a substrate for
MTHFR, and cofactor of MTHFR, riboflavin. The physical astuteness and stability of the genome, called
"genome health," is decided by an endless supply of specific supplements. The "genome well-being nutrigenomics" concept considers how the lack or overabundance of sustenance can cause genome changes
at the chromosomal level. The most objective is to characterize the ideal dietary admissions and tissue
culture medium concentration to preserve harm to the genome at its least conceivable level *in vivo* and *in
vitro*. This becomes imperative because increased harm to the genome is among the elemental causes of
infertility, neurodegenerative infections, developmental weakness, and cancer. The use of particular supplements in people with particular gene variations might result in improved resistance toward these major
health issues. It's important we being to see diet in terms of its effect on genome-protective supplements
(Fenech *et al.* 2010). Characterizing the perfect concentration of micronutrients required to preserve cells
in a genomically unfaltering state remains one of the foremost challenges for nutrigenomics work. This
challenge shows the necessity for contrasting genetic establishments. Nutrigenomics applications are far-
reaching in the area of nutritional science and has the potential to be an important tool for nutritionists,
dietitians, specialists or any platform that involves nutrition therapy for the treatment of infection. It may
conceivably shed light on potential diet-related maladies as well as plan for dietary techniques and the
antagonistic or useful impacts of a diet or supplements (Palou, 2007). The field of nutrigenomics might not
have evolved without high-throughput Omics (transcriptomic, proteomic, and metabolomic) innovations.
The fast headway of these Omics advances, such as nucleic acids, can be analyzed with tall throughput
DNA sequencing, highly refined mass spectrometry of protein and metabolites, and the strides made in
DNA microarray chips. Moreover, there's the advance of computer innovation for handling vastly large
amounts of organic information.

3.5.2 Transcriptomics

Transcriptomics is the study of the total set of RNA transcripts, counting coding and noncoding RNA, delivered by the genome in a particular cell utilizing high-throughput strategies. Major challenges in atomic science lie in understanding how the same genome can grant rise to diverse cell sorts and how gene expression is controlled. Transcriptomics advances give a wide account of which cellular forms are dynamic and which are torpid. The most widely used modern methods within the field of transcriptomics for transcriptional profiling, such as hybridization-based microarray strategy, can identify the expression of thousands of qualities at a time. These challenges transitioned transcriptome investigation to a sequencing-based procedure (RNA-seq) to rise as an elective to the microarray. The RNA sequencing strategy (RNA-Seq) is based on the high-throughput sequencing procedures that uncover nearness and amount of all the transcripts in a biological sample at a given time. Among all the Omics, transcriptomics is the most broadly utilized in nutritional investigation as it explores gene expression changes in reaction to dietary intercessions and dietary constituents in creature tissue at a characterized nutritional state (Wickramasinghe *et al.* 2012). Presently, distinctive transcriptomic platforms can be utilized to secure profitable transcriptome information. Microarray innovation could be a hybridization-based high-throughput transcriptomics examination strategy broadly utilized for profiling transcriptomes in a certain tissue (Yue *et al.* 2015, Wang *et al.* 2017). The essential guideline of microarray innovation includes hybridization of cDNA tests with spotted particular short or long oligonucleotide probes, or cDNA (Mill operator *et al.* 2009, Chon *et al.* 2011). This innovation can be used for the investigation of gene expression profiling and genotyping for point mutations, SNPs, and short tandem repeats (STRs). The microarray procedure has a few specialized restrictions in spite of the fact that it can create high-resolution information in a brief period of time; its reliance on the available information of the genomic groupings and the high abundance and low expression of certain transcripts may make high background noise due to nonspecific hybridization. Moreover, it does not permit for the discovery of transcripts from repeated sequences that present in dynamic range (Bradford *et al.* 2010, Wickramasinghe *et al.* 2014). A novel transcriptomic approach, RNA-Seq, has illustrated exceptional explanatory potential relative to microarray innovation for gene expression profiling. It has been utilized within the circumstances of nutrient-gene intelligence, permitting for synchronous distinguishing proof of all gene expressions in reaction to particular nutrients, particular diets, and conjointly physiological conditions, such as vitality confinement, vitamin and mineral insufficiencies, and maladies (Wickramasinghe *et al.* 2012, Kulahoglu and Brautigam 2014). This approach helps to create gigantic amounts of information and features an incredible advantage to speed up our use of this information, helping nutritionists to handle the molecular mechanism of nutrition for progressing nutritional investigation.

3.5.3 Proteomics

Proteomics covers the entire protein complement of a cell tissue expressed by genome under a particular set of conditions. Protein functions as effector molecules and their expression level is dependent on the corresponding RNA transcript synthesis and also on host translational control and regulation. In this way, proteome could be exceedingly dynamic and continually changing in reaction to the cellular state or the outside natural conditions experienced by a living being. Consequently, proteomics would be considered as the most appropriate information to characterize an organic framework (Cox and Mann 2007) to depict the basic differing qualities of proteins and the relationship of these qualities with the natural forms required to utilize the diverse proteomics-based advances. Proteomics innovations included the progress in mass spectrometry (LC-MS-MS, MALDI-TOF/TOF and SELDI-TOF-MS). Protein fractionation procedures and bioinformatics examination are utilized in different investigations, such as disclosure and discovery of diagnostic markers, identification of candidates for immunization generation, change of expression designs in reaction to distinctive signals, understanding pathogenicity mechanism, and illustration of useful protein pathways in numerous diseases. A few ordinary strategies exist for the characterization of proteins like ion-exchange chromatography (IEC), size-exclusion chromatography (SEC), high-pressure liquid chromatography (HPLC), and affinity chromatography, but for particular protein examination, two high-throughput procedures can be utilized: enzyme-linked immunosorbent assay (ELISA) and western blotting. Be that as it may, all these strategies are perhaps confined to a selective number of proteins and additionally are incapable in characterizing protein expression levels (Kurien and Scofield 2006). For the complex protein samples, sodium dodecyl sulfate-polyacrylamide gel electrophoresis (SDS-PAGE) and 2D-DIGE (two-dimensional differential gel electrophoresis), can be utilized (Marouga *et al.* 2005). Innovations have been created counting progresses in mass spectrometry, multiple reactions monitoring, and multiplexed immunoassays as an alternative to ordinary gel-based strategies of protein profiling. The mass spectrometry (MS) has been created to dissect the complex protein blends with higher affectability (Yates 2011). Edman corruption was created to decide the amino acid sequence of a particular protein (Smith 2001). For quantitative proteomics investigation, stable isotope labeling with amino acids in cell culture (SILAC), isotope-coded affinity tag (ICAT) labeling, and isobaric tag for relative and absolute quantitation (iTRAQ) strategies have been created (Shiio and Aebersold 2006, Kroksveen *et al.* 2015). Two high-throughput methods, such as nuclear magnetic resonance (NMR) spectroscopy and X-ray crystallography that decides proteins three-dimensional (3D) structure, that can help find proteins' biological functions (Wiese *et al.* 2007, Smyth and Martin 2000). Dietary proteomics (nutriproteomics) could be a budding research area, utilizing the elements of all the proteomic strategies to characterize molecular and cellular changes in protein expression and function, while additionally representing the interaction of food supplements with the proteins. As dietary supplements are displayed within the complex mixtures, the bioavailability and functions of each supplement can be modified by the nearness of other supplements with proteins through post translational modifications (Ganesh and Hettiarachchy 2012). Utilizing the proteomic methods in nutritional research to discover the cause and impact in living frameworks and to find and create biomarkers that can offer assistance in disease regulation, can additionally evaluate the security and usefulness of unused nutrients (Keusch 2006). Nutriproteomics coordinates with nutrigenomics and nutrimetabolomics and can help in evaluating people's well-being, diet-derived impacts on human digestion, and make a difference in creating personalized dietary proposals by anticipating diet-related infections such as cancer, diabetes, or neurodegenerative illnesses (Ovesna *et al.* 2008). Several applications of proteomics in sustenance research exist: Biomarker revelation and advancement; a quantifiable indicator of a particular organic state (especially one pertinent to the hazard of withdrawal) or nearness or stage of disease (Keshishian *et al.* 2009); nutraceutical disclosure and improvement; bioactive nourishment, which can cause the avoidance or treatment of diseases; and preclinical sustenance and dietary investigation. Previously, dietary research focused on dietary deficiencies in different populations; however, today, dietary research centers on the advancement of well-being by focusing on diet and dietary components that may anticipate the onset of infections, and how to best develop a prevention plan (Kussmann *et al.* 2006).

3.5.4 Metabolomics

Metabolomics is the analytical profiling technique for systematic identification and quantification of all endogenous and exogenous metabolites (metabolome) present in biological systems in a high-throughput manner. It faces a significant analytical challenge, unlike genomics and proteomics techniques, since it tries to measure and quantify small molecules (<1kDa) that have different physical properties (Kuehnbaum and Britz-McKibbin 2013). The metabolome is downstream of transcriptome and proteome and is detailed to be complementary to genomics, transcriptomics, and proteomics. Metabolomics focuses on the phenotype, which is sensitive to epigenetic and hereditary adjustments (McKillop and Flatt 2011, Sun *et al.* 2013, Tsoukalas *et al.* 2017, Nalbantoglu *et al.* 2019). A few natural components like pharmaceuticals, lifestyle variables, basic illness, and the upstream impact of the genome and proteome all influence the composition of metabolites. The plenitude of metabolites in organic frameworks is frequently directly related to pathogenic components and is portrayed as proximal causes of infection (Gerszten and Wang 2008). Gene expression and protein concentration might show the potential for physiological changes, however, metabolite concentrations and their dynamic changes in cells and tissues are the real end-points of the physiological regulatory process. Metabolomics has several applications in well-being and disease, including precision medication, personalized medication, epidemiologic populace studies, single-cell metabolic phenotyping, metabolome-wide association studies (MWAS), exactness metabolomics, and with other Omics, such as integrator Omics, biotechnology, and bioengineering. Nutrimetabolomics is the application of metabolomics tools within the field of nourishment investigation to research the metabolic reactions of people and dietary mediations and to distinguish metabolic diseases impacted by the nutrient metabolome (Zhang *et al.* 2012, Lemieux *et al.* 2014). The identification and measurement of metabolites used as food admissions biomarkers might offer assistance to survey nutritional interventions and approve dietary studies. The progressed innovation and the development of more studies within the field of nutrimetabolomics have made a difference in the understanding of the health-diet relationship. The biomarkers of utilization ought to be characterized by their specificity, so they are not affected by the addition of comparable nourishment. Particular biomarkers for specific nourishment permitted a distant understanding of their effect on metabolic pathways control. Foods that are similar but have distinctive chemical compositions may have diverse impacts on the metabolome (Barnes *et al.* 2010). Nutrimetabolomics can speed up our ability to distinguish metabolic infections that are impacted by supplements and create a focus on diet-based medicines. Currently, the broadly utilized apparatus for metabolomics tests is proton nuclear magnetic resonance (NMR) innovation.

3.6 Recent Advancements in Biotechnological-Led Nutritional Research

3.6.1 High-Throughput Phenotyping Technologies

The term phenotype was originally proposed by Johannsen. He stated that all "types" of organisms, distinguishable by direct inspection or only by finer methods of measuring or description, may be characterized as *"phenotypes"*(Johannsen 1911). Simply, a phenotype is the "observable characters of an organism." The phenotype is the outcome of the interaction between genotype and environment (Flood *et al.* 2016). Phenomics is a rapidly emerging concept of genetics, which aims to characterize phenotypes in a precise way and establish a connection with the corresponding genes or alleles (NIFA-NSF Phenomics Report 2011).

Advances in genome sequencing technologies have garnered major attention toward phenotyping in recent years. In plant breeding, the conventional method of recording data is used on a large scale for various attributes, such as short duration, short height, bold seed, biotic and abiotic stress tolerance, high seed yield, high oil/protein content, seed color, plant type, etc. These data are recorded on thousands of breeding lines or germplasm. Collecting of phenotypic data in such a way may hamper an established crop breeding program as sometimes the data collection results in ambiguous information. However, with the inclusion of advanced technologies, plant phenotyping can maturing gradually but still lack precise and unambiguous data (Mahlein 2016, Minervini et al. 2015). Hence, in the present era of genome sequencing, further advances in phenotyping methods may be of great significance for a plant breeder for establishing a sound breeding program through upgrading the selection effectiveness and accelerating genetic gains.

3.6.1.1 Satellite Imaging

Satellite images are essentially eyes in the sky. Images collected through satellites may help farmers in crop forecasting, identify crop stress, and prepare different types of maps related soil and water properties. Although the use of satellite imaging in plant breeding programs is comparatively low and breeding lines and germplasm are evaluated in very small-sized plots, this technology may still have a deciding role in the evaluation of multilocation yield trials. Thus, in view of the reducing costs for satellite imaging, we can look forward to its increased use in plant breeding (Chawade et al. 2019).

3.6.1.2 Unmanned Aerial Vehicles (UAVS)

Unmanned aerial vehicles (UAV), more commonly known as drones, are a class of aircraft that can fly without a human pilot on board. In general, UAVs are classified into four groups: parachutes, blimps, rotocopters, and fixed wing systems (Sankaran et al. 2015). These vehicles principally fly at an altitude between 10 meters and 250 meters and can come up with remarkably higher resolution photography compared to satellites.

3.6.1.3 Proximal Phenotyping

Proximal phenotyping is carried out with proximal sensors—a number of technologies that employ a sensor close to, or in direct contact with the object. For instance, a red, green, and blue (rgb) image, infrared imaging, thermal imaging, fluorescence imaging, light detection and ranging (Lidar), near-infrared imaging, hyperspectral imaging, spectroscopy and so many other high –throughput phenotyping sensors and imaging technologies are used for phenotyping. These sensors are assembled with the camera, navigation unit, and other necessary equipment required for the research objective. This assembled unit is framed on manually or machine-driven instruments for recording observations. A proximal sensing cart with three infrared thermometers, two digital cameras, a GPS receiver for positioning, and two data loggers was custom-made on two bicycles and a metal frame (White et al. 2013). This was used in breeding nurseries of wheat, barley, camelina, and cotton (Chawade et al. 2019). Likewise, a modified model of Primary sclerosing cholangitis, including infrared thermometer, ultrasonic transducer, multispectral reflectance sensor, GPS receiver, altitude heading and reference system, RGB cameras, data logger, and weather station was also evolved (Thompson et al. 2018). Proximal phenotyping is unquestionably of great importance in plant breeding experiments where observations like leaf area index, counting plants, plant height, early vigor, plant maturity, plant chlorophyll content, canopy temperature, nitrogen content, etc. can be collected with high efficiency at low cost.

3.6.2 High-Throughput Genotyping Technologies

The genotype is the total hereditary makeup of a life form. The innovation of recognizing the little hereditary contrasts within the genotype of a living being that can lead to major changes within the phenotype called is genotyping. Genotyping can recognize DNA arrangement contrasts at three diverse levels: Single nucleotide polymorphisms (contrasts within the DNA sequence at the single nucleotide level); copy number varieties (CNVs, an increase or decrease in the number of duplicates of a gene or a locus-specific DNA repeat component); and structural variations (caused by chromosomal rearrangement that frequently cover megabases of DNA). Genotyping decides the variations within the organism's DNA sequence utilizing organic measures and compares it with a reference sequence to uncover the allelic variations acquired from their parents. SNP genotyping is performed utilizing the conventional strategies, *in situ* hybridization, restriction fragment length polymorphism detection (RFLP), simple sequence repeats (SSR), random amplified polymorphic detection (RAPD), amplified fragment length polymorphism (AFLP), polymerase chain reaction (PCR), and DNA sequencing. However several technological limitations influence the genotyping process—nearly all genotyping is fractional and little division of an individual's genotype is decided. Genotyping is the most vital innovation in the investigation of genes and their variations related with the infection. Numerous high-throughput advances for genotyping have developed in recent years for the synchronous genotyping for a couple to a few thousands markers in hundreds to thousands of individuals. Microsatellite genotyping uses simple sequence repeats of a short nucleotide motif redundancy (1–5 nucleotides) and can occur in culminate reiteration. The SNPs are biallelic, dinucleotide repeats (CA) are the foremost commonly utilized markers for human studies, and tri- allelic/tetra-allelic SNP repeats are frequently utilized markers. These repeats have appeared to be exceptionally abundant and exceedingly polymorphic within the eukaryotic genomes (Litt and Luty, 1989). Microsatellite genotyping has been the most broadly utilized strategy of choice for numerous recent linkage analysis studies, particularly genome scans outlined to recognize genes underlying common diseases. The microsatellites utilized have critical, diverse frequencies of different alleles, with high heterozygosity, so the information contains a generally high sum of data when examining the family-based transmission of alleles. Among all the variations, SNPs are the most copious (>90% of all contracts between the individuals) form of hereditary variation within the genomes; subsequently, snips are the most excellent hereditary variation asset for population studies and genome mapping (Frohlich *et al.* 2004). They are thought to be useful discerning differences in biological forms among individuals. Genomic determination utilizing the SNP markers may be an effective modern tool for hereditary selection (Seidel 2009). Genetic association studies utilizing SNP markers are anticipated to show proof of hereditary factors capable of complex phenotypes like chronic infections and responses to different nutritional components. There are a few focal points of SNP variations as markers, such as most of the SNP markers are found within the coding region of DNA, that subsequently can influence protein function specifically. They are more reasonable for high-throughput hereditary analysis than microsatellites. They are more suitable for long-term determination markers as they are steadily inherited compared to other DNA markers. SNPs are broader and generate potential markers close to the locus of more interest than other sorts of polymorphism. In this manner, as compared to multiallelic marker data, the SNP marker is much lower; five SNP markers give rise to as much data as one microsatellite marker (Beuzen *et al.* 2000, Saxena *et al.* 2017). The intriguing part of SNPs has been shown in the improvement of these SNP genotyping high-throughput strategies: 1) allele discrimination methods, which can distinguish single-base pair differences used to separate genotypes, mutations, and polymorphisms by comparing the fluorescence signal obtained by utilizing allele-specific, dye-labeled probes; 2) primer extension includes the hybridization of a probe to the bases upstream of the SNP nucleotide that, after a mini-sequencing response where DNA polymerase expands the hybridized primer, adds a complementary base to the SNP nucleotide (Syvanen 2001); 3) dynamic allele-specific hybridization (DASH), which measures a quantifiable change in Tm in DNA that comes from the instability of mismatched base pairs and can identify changes (Howell *et al.* 1999); 4) pyrosequencing—a next-generation real-time DNA sequencing innovation that quantitatively measures the incorporated nucleotide. It has <250 bases in a read that limits their capacity for entire genome sequencing. However, it can produce real-time results and has the potential to be enormously scaled up, which makes them a reasonable choice for sequencing small regions to perform SNP genotyping. It is suited to distinguish numerous SNPs in an exceedingly polymorphic major histocompatibility complex region of the genome (Harbron and Rapley, 2004); and 5) high-throughput chemistry measures, such as Flap endonuclease (FEN), which catalyzes structure-specific cleavage that is profoundly delicate to mismatches and has a high degree of specificity utilized to interrogate SNPs (Olivier, 2005), and oligonucleotide ligation assay, which examines an SNP by hybridizing two probes specifically over the SNP polymorphic location where DNA ligase interceded ligation can happen if the allele-specific probes coordinate the target DNA.

3.6.3 Chip-Based Breeding & Genotyping Platforms

Feeding a huge population (9 billion by 2050) completely depends on the major crops, however, the recent productive gain developments in major crops are insufficient to bolster a worldwide population (Ray *et al.* 2012). Production is mainly challenged by climate change (altered rainfall, floods/drought, extreme weather), but changes in severity and distribution of pathogens and pests is anticipated to surge in the future (Abberton *et al.* 2016). However, advanced genomics and biotechnological methods have had large effects on pharmaceutical and public well-being and have also given the potential for another Green Revolution—a moment that would come from the cumulative effect of new scientific methodologies and products with routine breeding methods (Borlaug, 2000). Advancement and application of molecular markers in crop hereditary qualities have picked up momentous consideration, beginning with restriction fragment length polymorphisms (RFLPs) (Tanksley *et al.* 1989) and, recently, single nucleotide polymorphism (SNP) markers based on next-generation sequencing innovations (Varshney *et al.* 2009). The use of crop-specific simple sequences repeat (SSR) markers, as they are copious in number and profoundly polymorphic, were effectively utilized and encouraged the advancement of high-density maps for diverse crops such as apples (Khan *et al.* 2012), potatoes (Sharma *et al.* 2013), pears (Wu *et al.* 2014), common wheat, rice, grain, maize and other crops was a major landmark. Future practical breeding platforms ought to receive computerized genotyping advances like SNP arrays and GBS advances, target functional polymorphisms supporting economic characteristics, and have desirable prediction precision for quantitative traits with widespread applications in hereditary foundations in crops. Crop breeding chips and genotyping platforms will give phenomenal openings to hasten the advancement of cultivars with desired yield potential, quality, and improved adjustment to moderate the impacts of climate change. The prospects of crop genomics are changing expeditiously, and the current imperative question is raised: Can all the genomics data including SNPs, InDels, CNVs, and epigenetic variations be contained in a single chip? (A DNA chip is a little piece of silicon glass to which a huge number of manufactured oligonucleotides [single-stranded DNA] have been chemically bonded). The

Eureka digital research platform outlined a low-density genotyping assay for the synchronous discovery of SNPs, InDels, CNVs, and methylation, but the downstream examination pipeline for allele calling can be a challenge in polyploid crops. So also, the Affymetrix Human SNP cluster 6.0 can distinguish 1.8 million polymorphisms, of which 960,000 are CNV (Nishida *et al.* 2008). Numerous chips are accessible for genome-wide genotyping in any crop with a certain sum of inclination in each—based on the hereditary background of the germplasm and goals, researchers have to choose the finest one. The concept of breeding chips is more widespread as these chips are based on functional genes approved over hereditary backgrounds. Subsequently, the choice of technology includes more extensive worthiness to handle the benefits from the post-genomics time. The future of crop genotyping is questionable—either sequencing will supplant all genotyping platforms or genotyping platforms will proceed. Whole-genome resequencing information for hereditary studies will be utilized in major crops such as wheat, maize, rice, cotton, soybean, and a few vegetables within the coming years since whole genome sequencing (WGS) from a much bigger gene pool is presently developing. However, genotyping platforms still need to advance in minor crops and crops with complex and huge genomes.

3.6.4 Successful Examples of Biofortified Crops

Malnutrition is more prevalent in developing countries where diets are of poor nutritional quality (deficient in protein, minerals, vitamins, and essential amino acids) and this is a major cause of poor health and higher disease susceptibility among the population. The deficiencies of micronutrients i.e. Fe, Zn and vitamin A could lead to irreparable damage to the body in the form of blindness, stunted growth, mental retardation, learning disabilities, low working capacity, and even premature death. Malnutrition has serious consequences on socioeconomic structure and the Gross Domestic Product of the affected countries. The biotechnological advances and various crop improvement strategies have contributed to a great extent in the nutritional improvement of important food crops worldwide. In one such major initiative, HarvestPlus along with its partners, is developing and promoting new biofortified varieties of staple food crops with higher amounts of vitamin A, iron or zinc.

India, the largest democracy and developing country, and where a majority of the population is dependent on agriculture, has taken the issue of malnutrition as a challenge and begun several initiatives. The National Agricultural Research System including ICAR and State Agricultural Universities (SAUs) in India have made a significant progress in the development of many biofortified varieties of cereals, pulses, oilseeds, vegetables, and fruits in last decade (Table 3.1). To highlight the important biofortified varieties and their characteristics, the Indian Council of Agricultural Research published a booklet entitled *Biofortified Varieties: Sustainable Way to Alleviate Malnutrition* in 2017 (Yadava *et al.* 2017).

The Prime Minister of India, Shri Narendra Modi, dedicated 17 biofortified varieties of eight crops to the nation on the occasion of 75th Anniversary of the Food and Agriculture Organization (FAO) of the United Nations, on October 16, 2020, which is celebrated as World Food Day. The biofortified varieties are 1.5 to 3 times more nutritious than the traditional varieties. The biofortified varieties were of huge importance for ensuring good income for the farmers along with opening new avenues of entrepreneurship for them. A recent study examined the biofortified wheat cultivars for their quality parameters and zinc bioavailability (Saxena *et al.* 2020, Nathani *et al.* 2010).

3.6.5 Future Perspectives

At the onset of the Green Revolution in India, a major focus was set on increased food grain production and was successfully achieved with the help of conventional breeding strategies. Next, quality became the key concern rather than mere quantity of the food produced. In order to meet current and future challenges, molecular breeding took the place of conventional breeding—with its own advantages mentioned in earlier in this chapter—and biotechnology became a key scientific tool in development of biofortified crops.

In the last decade, next-generation sequencing platforms generate huge amounts of genetic information at the allelic level, helping to identify candidate genes and discover novel SNPs, which facilitates genomic mapping of important qualitative genes/QTLs, the further identification of functional markers,

TABLE 3.1

Successful Examples of Biofortified Varieties of Different Crops Developed By Public Sector Research Institutes in India in Last Decade

S. No.	Crop and Variety	Biofortified/Improved for	Developing Institute	Notified Year
Rice				
1	CR Dhan 315	Zinc	ICAR-NRRI, Cuttack	2020
2	DRR Dhan 45	Zinc (contains 22.6 ppm in polished grains in comparison to 12.0–16.0 ppm in popular varieties)	ICAR-IIRR, Hyderabad	2016
Wheat				
3	HD 3298	Protein (12.12%), iron (43.1 ppm)	ICAR-IARI, New Delhi	2020
4	HI 1633	Protein, iron, zinc	ICAR-IARI, RS Indore	2020
5	DBW 303	Protein, iron	IARI-IIWBR, Karnal	2020
6	DDW 48	Zinc (>40 ppm), protein (12.1%)	IARI-IIWBR, Karnal	2020
7	WB 02	High iron (40 ppm) and zinc (42 ppm) in comparison to 28.0–32.0 ppm iron and 32.0 ppm zinc in popular varieties	ICAR-Indian Institute of Wheat and Barley Research, Karnal	2017
8	HPBW 01	Iron (40 ppm), zinc (40.6 ppm) in comparison to 28.0–32.0 ppm iron and 32.0 ppm zinc in popular varieties	Punjab Agricultural University, Ludhiana	2017
Finger Millet				
9	CFMV 1 (Indravathi)	Calcium, iron, zinc	ICAR-IIMR, Hyderabad	2020
10	CFMV 2 (Gira)	Calcium, iron, zinc	ICAR-IIMR, Hyderabad	2020
Small Millet				
11	CCLMV1 (JAICAR Sama 1)	Protein (14.4%), iron (58 ppm)	ICAR-IIMR, Hyderabad	2020
Mustard				
12	Pusa Mustard 32	Low erucic acid		
13	Pusa Mustard 30	Low erucic acid (<2%) as compared to 40% erucic acid in popular varieties	ICAR-Indian Agricultural Research Institute, New Delhi	2013
14	Pusa Double Zero Mustard 31	Low erucic acid (<2.0%) in oil and also contains <30.0 ppm glucosinolates in seed meal as compared to >120.0 in popular varieties	ICAR-Indian Agricultural Research Institute, New Delhi	2016
Groundnut				
15	Girnar 4	Increased oleic acid (~80% oleic acid content as compared to 40–50% in popular varieties)	ICAR-Directorate of Groundnut Research, Junagadh in collaboration with the ICRISAT	2019
16	Girnar 5	Increased oleic acid (~80% oleic acid content as compared to 40–50% in popular varieties)	ICAR-Directorate of Groundnut Research, Junagadh in collaboration with the ICRISAT	2019

(Continued)

TABLE 3.1 (Continued)

S. No.	Crop and Variety	Biofortified/Improved for	Developing Institute	Notified Year
Pearl Millet				
17	HHB 311	Iron (83 ppm)	CCS Haryana Agricultural University, Hisar, INDIA	2020
18	HHB 299	Iron (73 ppm) and zinc (41 ppm) as compared to 45.0–50.0 ppm iron and 30.0–35.0 ppm zinc in popular varieties/hybrids	CCS Haryana Agricultural University, Hisar, INDIA in collaboration with ICRISAT, Patancheru	2017
19	AHB 1200	Iron (73.0 ppm)	VNMKV, Parbhani in collaboration with ICRISAT, Patancheru	
Lentils				
20	Pusa Ageti Masoor	Contains 65.0 ppm iron as compared to 55.0 ppm iron in popular varieties	ICAR-Indian Agricultural Research Institute, New Delhi	2017
Sweet Potato				
21	Sree Neelima	Enriched with anthocyanin	ICAR-CTCRI, Sreekariyam, Thiruvananthapuram	2015
22	DA 340	Enriched with anthocyanin	ICAR-CTCRI, Sreekariyam, Thiruvananthapuram	2020
Potato				
23	Bhu Sona	High β-carotene (14.0 mg/100 g) content as compared to 2.0–3.0 mg/100 g β-carotene in popular varieties	ICAR-Central Tuber Crops Research Institute, Thiruvananthapuram, Kerala	2017
Pomegranate				
24	Solapur Lal	Higher iron (5.6–6.1 mg/100g), zinc (0.64–0.69 mg/100g) and vitamin C (19.4–19.8 mg/100 g) in fresh arils in comparison to 2.7–3.2 mg/100g, 0.50–0.54 mg/100g and 14.2–14.6 mg/100g, respectively in popular variety Ganesh	ICAR-National Research Centre on Pomegranate, Pune	2017
Maize Hybrids				
25	Pusa Vivek QPM9 Improved	High provitamin-A (8.15 ppm), lysine (2.67%), and tryptophan (0.74%) as compared to 1.0–2.0 ppm provitamin-A, 1.5–2.0% lysine and 0.3–0.4% tryptophan content in popular hybrids	ICAR-Indian Agricultural Research Institute, New Delhi	2017
26	Pusa HM4 Improved	Contains higher amounts of tryptophan (0.91%) and lysine (3.62%) amino acids	ICAR-Indian Agricultural Research Institute, New Delhi	2017
27	Pusa HM8 Improved	Contains higher amounts of tryptophan (1.06%) and lysine (4.18%)	ICAR-Indian Agricultural Research Institute, New Delhi	2017

and development of new varieties with better nutritive values. High-throughput genotyping platforms are not only cost effective, but also generate huge data sets. Similarly, high-throughput phenotyping techniques precisely generate uniform information for multiple parameters. As genome sequencing becomes a reality, crops earlier considered as "orphan crops" are now "genomic-resource rich." A multi-omics approach, along with modern bioinformatics tools, will lead to better understanding of complex interactions, which will be important to understand individual nutrition requirement-led research with diversified nutritive sources.

REFERENCES

Abberton, M., Batley, J., Bentley, A., Bryant, J., Cai, H., Cockram, J., Costa de Oliveira, A., Cseke, L. J., Dempewolf, H., De Pace, C. and Edwards, D. 2016. Global agricultural intensification during climate change: a role for genomics. *Plant Biotechnology Journal* 14, no. 4: 1095–1098.

Bailey-Serres, J., J. E. Parker, J. E., E. A. Ainsworth, E. A., G.E.D. Oldroyd, G. E. D., and J. I. Schroeder, J. I.. 2019. Genetic strategies for improving crop yields. *Nature* 575: 109–188.

Barnes, V. M., Teles, R., Trivedi, H. M., Devizio, W., Xu, T., Lee, D.P. and Guo, L. 2010. Assessment of the effects of dentifrice on periodontal disease biomarkers in gingival crevicular fluid. *Journal of Periodontology* 81, no. 9: 1273–1279.

Beuzen, N. D., Stear, M. J. and Chang, K. C. 2000. Molecular markers and their use in animal breeding. *The Veterinary Journal* 160, no. 1: 42–52.

Borlaug, N.E. 2000. Ending world hunger. The promise of biotechnology and the threat of antiscience zealotry. *Plant Physiology* 124, no. 2: 487–490.

Bradford, J. R., Hey, Y., Yates, T., Li, Y., Pepper, S. D., Miller, C. J. 2010. A comparison of massively parallel nucleotide sequencing with oligonucleotide microarrays for global transcription profiling. *BMC Genomics* 11: 282.

Chaudhary, N., Kumar, .V., Sangwan, P., Pant, N. C., Saxena, A., Joshi, S. and Yadav, A. N. 2021. Personalized nutrition and -Omics. *Comprehensive Foodomics*, 495–507. https://doi.org/10.1016/B978-0-08-100596-5.22880-1.

Chawade, A., Ham, J.V., Blomquist, H., Bagge, O., Alexandersson, E., and Ortiz, R. 2019. High-throughput field-phenotyping tools for plant breeding and precision agriculture. *Agronomy* 9, 258.

Chon, D. H., Rome, M. N., Kim, H. S. and Park, C. 2011. Investigating the mechanism of sludge reduction in activated sludge with an anaerobic side-stream reactor. *Water Science Technology* 63(1): 93–99. https://doi.org/10.2166/wst.2011.015

Collard, B. C. Y and Mackill, D. J. 2008. Marker-assisted selection: an approach for precision plant breeding in the twenty-first century. *Phil. Trans. R. Soc. B* 363: 557–572.

Collard, B. C. Y., Beredo, J. C., Lenaerts, B., Mendoza, R., Santelices, R., Lopena, V., Verdeprado, H., Raghavan, C., Gregorio, G. B., Vial, L., Demont, M., Biswas, P. S., Iftekharuddaula, K.M., Rahman, M. A., Cobb, J. N., Islam, M. R. 2017. Revisiting rice breeding methods—evaluating the use of rapid generation advance (RGA) for routine rice breeding. *Plant Production Science* 20: 337–352. https://doi.org/10.1080/1343943X.2017.1391705

Corthesy-Theulaz, I., Den Dunnen, J.T., Ferre, P., Geurts, J.M.W., Muller, M., Van Belzen, N., *et al.* 2005. Nutrigenomics: the impact of biomics technology on nutrition research. *Annals of Nutrition and Metabolism* 49, no. 6: 355–65.

Cox, J. and Mann, M. 2007. Is proteomics the new genomics? *Cell* 130, no. 3: 395–398.

Cozzolino, S. M. F. and Cominetti, C. 2013. Biochemical and Physiological Bases of Nutrition in Different Stages of Life in Health and Disease,. 1ˢᵗ edition. São Paulo, Brazil: Monole.

Dauncey, M. J. 2012. Recent advances in nutrition, genes and brain health. *Proceedings of the Nutrition Society*. 71, no. 4: 581–591.

Davis, C. D. and Hord, N. G. 2005. Nutritional "*omics*" technologies for elucidating the role(s) of bioactive food components in colon cancer prevention1. *The Journal of Nutrition* 135: 2694–97.

Development Initiatives. 2018. Global Nutrition Report 2018. Bristol: Development Initiatives.

Eshed, Y., and Z. B. Lippman, Z. B.. 2019. Revolutions in agriculture chart a course for targeted breeding of old and new crops. *Science* 366: eaax0025.

FAO. 2019. The State of Food Security and Nutrition in the World. Rome: Food and Agriculture Organization of the United Nations.

Fenech, M.F. 2010. Dietary reference values of individual micronutrients and nutriomes for genome damage prevention: current status and a road map to the future. *The American Journal of Clinical Nutrition* 91, no. 5: 1438S–1454S.

Fernandez-Cornejo, J., McBride, W. D., Klotz-Ingram, C., Jans, S. & Brooks, N. 2000. Genetically engineered crops for pest management in U.S. agriculture: Farm level effects. Agricultural Economics Report. No. 786: Resource Economics Division, Economic Research Service, U.S. Department of Agriculture, Washington, D.C.

Fiehn, O. 2002. Metabolomics-the link between genotypes and phenotypes. *Plant Mol. Biol.* 48(1–2): 155–171.

Flood, P. J., Kruijer, W., Schnabel, S. K., van der Shoor, R., Jalink, H., Snel, J. F. H., Harbinson, J. and Aarts, M. G. M. 2016. Phenomics for photosynthesis, growth and reflectance in *Arabidopsis thaliana* reveals circadian and long-term fluctuations in heritability. *Plant Methods* 12: 14.

Food and Drug Administration. 2001. Performance Improvement. Compendium of HHS Agency FY 2000: Evaluations completed and in progress. U.S. Department of Health & Human services.

Fröhlich, C. & Mettenleiter, M. 2004. Terrestrial laser scanning-new perspectives in 3D surveying. *International Archives of Photogrammetry, Remote Sensing and Spatial Information Sciences* 36.

Fu, W. J., Stromberg, A. J., Viele, K., Carroll, R. J. and Wu, G. 2010. Statistics and bioinformatics in nutritional sciences: analysis of complex data in the era of systems biology. *The Journal of Nutritional Biochemistry* 21, no. 7: 561–572.

Ganesh, V. and Hettiarachchy, N. S. 2012. Nutriproteomics: a promising tool to link diet and diseases in nutritional research. *BiochimicaetBiophysicaActa (BBA)-Proteins and Proteomics* 1824, no. 10: 1107–1117.

Gehlenborg, N., O'Donoghue, S. I., Baliga, N. S., Goesmann, A., Hibbs, M. A., Kitano, H., Kohlbacher, O., Neuweger, H., Schneider, R., Tenenbaum, D., Gavin, A. C. 2010. Visualization of omics data for systems biology. *Nature Methods* 7: S56–S68.

Gerszten, R. E. and Wang, T. J. 2008. The search for new cardiovascular biomarkers. *Nature* 451, no. 7181: 949–952.

Gilliham, M., Able, J. A. and Roy, S.J. 2017. Translating knowledge about abiotic stress tolerance to breeding programmes. *Plant Journal* 90: 898–917.

Harbron, S. and Rapley, R. 2004. Molecular Analysis and Genome Discovery. London: John Wiley & Sons Ltd. ISBN 978-0-471-49919-0.

Howell, W. M., Jobs, M., Gyllensten, U. and Brookes, A. J. 1999. Dynamic allele-specific hybridization. A new method for scoring single nucleotide polymorphisms. *Nature Biotechnology* 17, no. 1: 87–8.

Johannsen, W. 1911. The genotype conception of heredity. *The American Naturalist* 45: 129–159.

Keshishian, H., Addona, T., Burgess, M., Mani, D. R., Shi, X., Kuhn, E., and Carr, S. A. 2009. Quantification of cardiovascular biomarkers in patient plasma by targeted mass spectrometry and stable isotope dilution. *Molecular & Cellular Proteomics* 8, no. 10: 2339–2349.

Keusch, G.T. 2006. What do -omics mean for the science and policy of the nutritional sciences? *The American Journal of Clinical Nutrition* 83, no. 2: 520S–522S.

Khan, M. A., Han, Y., Zhao, Y. F., Troggio, M. and Korban, S.S. 2012. A multi-population consensus genetic map reveals inconsistent marker order among maps likely attributed to structural variations in the apple genome. *PloS one* 7, no. 11: e47864.

Kroksveen, A. C., Jaffe, J. D., Aasebo, E., Barsnes, H., Bjorlykke, Y., Franciotta, D. 2015. Quantitative proteomics suggests decrease in the secretogranin-1 cerebrospinal fluid levels during the disease course of multiple sclerosis. *Proteomics* 15, no. 19: 3361–3369.

Kuehnbaum, N. L. and Britz-McKibbin, P. 2013. New advances in separation science for metabolomics: resolving chemical diversity in a post-genomic era. *Chemical Reviews* 113, no. 4: 2437–68.

Kulahoglu, C. and Brautigam, A. 2014. Quantitative transcriptome analysis using RNA-seq. In: Staiger Dorothee S. (E, edditor.), Plant Circadian Networks. New York: Humana. P. 71–91.

Kumar V., Joshi, S., Pant, N. C., Sangwan, P., Yadav, A. N., Saxena, A. and Singh, D. 2019. Molecular approaches for combating multiple abiotic stresses in crops of arid and semi-arid region. In: Singh S., Upadhyay S., Pandey A., Kumar S. (Eds.), Molecular Approaches in Plant Biology and Environmental Challenges. Energy, Environment, and Sustainability. Singapore: Springer, Singapore.

Kurien, B. and Scofield, R. 2006. Western blotting. *Methods (San Diego, CA)*. 38, no. 4: 283–293.

Kussmann, M., Raymond, F. and Affolter, M. 2006. OMICS-driven biomarker discovery in nutrition and health. *Journal of Biotechnology* 124: 758–787.

LeMieux, M. J., Aljawadi, A. and Moustaid-Moussa, N. 2014. Nutrimetabolomics. *Advances in Nutrition* 5: 792–794.

Litt, M. and Luty, J.A. 1989. A hypervariable microsatellite revealed by in vitro amplification of a dinucleotide repeat within the cardiac muscle actin gene. *American Journal of Human Genetics* 44, no. 3: 397.

Liu, B. and Qian, S. B. 2011. Translational regulation in nutrigenomics. *American Society for Nutrition* 2: 511–519.

Mahlein, A. K. 2016. Plant disease detection by imaging sensors—parallels and specific demands for precision agriculture and plant phenotyping. *Plant Diseases* 100: 241–251.

Marouga, R., David, S., Hawkins, E. 2005. The development of the DIGE system: 2D fluorescence difference gel analysis technology. *Analytical and Bioanalytical Chemistry* 382, no. 3: 669–678.

McKillop, A. M. and Flatt, P. R. 2011. Emerging applications of metabolomic and genomic profiling in diabetic clinical medicine. *Diabetes Care* 34: 2624–2630

McKinney, S. 2000. Biotech for product developers. Accessed 12/19/2001. Available from: http://www.food-productdesign.com/archive/2000/ 0500pp.html. Accessed Dec. 19, 2001.

Minervini, M.; Scharr, H.; Tsaftaris, S. A. 2015. Image analysis: The new bottleneck in plant phenotyping [applications corner]. *IEEE Signal Processing Magazine* 32, 126–131.

Moose, S. P., Mumm, R. H. 2008. Molecular plant breeding as the foundation for 21st century crop improvement. *Plant Physiology* 147: 969–977.

Muller-Rober, B., Sonnewald, U. & Willmitzer, L. 1992. Inhibition of the ADP-glucose phyrophosphorylase in transgenic potatoes leads to sugarstoring tubers and influences tuber formation and expression of tuber storage protein genes. *EMBO Journal.* 11: 1229–1238.

Nalbantoglu, S., Abu-Asab, M., Suy, S., Collins, S. and Amri, H. 2019. Metabolomics-based bio signatures of prostate cancer in patients following radiotherapy. *Omics: A Journal of Integrative Biology* 23, no. 4: 214–223.

Nathani, S., Kumar, V., Dhaliwal, H. S., Sircar, D., Roy, P. 2010. Biological application of a fluorescent zinc sensing probe for the analysis of zinc bioavailability using Caco-2 cells as an in-vitro cellular model. *Journal of Fluorescence* 30, no. 6: 1553–65.

NIFA-NSF Phenomics Workshop Report. 2011. Phenomics:: Genotype to Phenotype. p. 12.

Nishida, N., Koike, A., Tajima, A., Ogasawara, Y., Ishibashi, Y., Uehara, Y. and Tokunaga, K. 2008. Evaluating the performance of Affymetrix SNP Array 6.0 platform with 400 Japanese individuals. *BMC Genomics* 9, no. 1: 431.

Norheim F, Gjelstad IM, Hjorth M, Vinknes KJ, Langleite TM, Holen T, Jensen J, Dalen KT, Karlsen AS, Kielland A, Rustan AC, Drevon CA. Molecular nutrition research: the modern way of performing nutritional science. *Nutrients* 4(12):1898–944. 10.3390/nu4121898.

Obrycki, J. J., Losey, J. E., Taylor, O. R. & Jesse, L. C. H. 2001. Transgenic insecticidal corn: beyond insecticidal toxicity to ecological complexity. *BioScience* 51: 353–361.

Olivier, M. 2005. The invader assay for SNP genotyping. *Mutation Research* 573, no. 1–2: 103–10.

Ong, T. P. and Rogero, M. M. 2009. Nutrigenomics: importance of nutrient-gene interaction for health promotion. *Journal of the ABESO*, no. 40.

Oscarson, P. 2000. The strategy of the wheat plant in acclimating growth and grain production to nitrogen availability. *Journal of Experimental Botany* 51: 1921–1929.

Ovesná, J., Slabý, O., Toussaint, O., Kodíček, M., Maršík, P., Pouchová, V. and Vaněk, T. 2008. High throughput 'omics' approaches to assess the effects of phytochemicals in human health studies. *British Journal of Nutrition* 99, no. E-S1: ES127–ES134.

Ovesna, J., Slaby, O., Toussaint, O., Kodíček, M., Marsik, P., Pouchova, V. and Vanek, T. 2008. High throughput 'omics' approaches to assess the effects of phytochemicals in human health studies. *The British journal of nutrition.* 99 E Suppl 1. ES127–34. 10.1017/S0007114508965818.

Palou, A. 2007. From nutrigenomics to personalised nutrition. *Genes & Nutrition* 2, no. 1: 5–7.

Pingali, P. L. 2012. Green revolution: impacts, limits, and the path ahead. *PProceedings of the National Academy of Sciences of the United States of Americaroc Natl Acad Sci USA* 109: 12302–12308. https://doi. org/10.1073/pnas.091295310.

Rao, G. U., Ben Chaim, A., Borovsky, Y. and Paran, I. 2003. Mapping of yield-related QTLs in pepper in an interspecific cross of *C. annuum* and *C. frutescens. Theoretical and Applied Genetics* 106: 1457–1466.

Ray, D. K., Ramankutty, N., Mueller, N. D., West, P. C. and Foley, J. A. 2012. Recent patterns of crop yield growth and stagnation. *Nature Communications* 3, no. 1: 1–7.

Redenbaugh, K., Berner, T., Emlay, D., Frankos, B., Hiatt, W., Houck, C., Kramer, M., Malyj, L., Marineau, B., Rachman, N., Rudenko, L., Sanders, R., Sheehy, R. and Wixtrom, R. 1993. Regulatory issues for commercialization of tomatoes with an antisense polygalacturonase gene. *In Vitro Cellular and Developmental Biology* 29P: 17–26.

Russell, P. J. 2010. I Genetics: A Molecular Approach, (3rd ed.). San Francisco: Pearson Benjamin Cummings. ISBN 978-0-321-56976-9.

Sales, N. M. R., Pelegrini, P. B. and Goersch, M. C. 2014. Nutrigenomics: definitions and advances of this new science. *Journal of Nutrition And Metabolism.* Article ID 202759. https://doi.org/ 10.1155/ 2014/202759

Sanhueza, C. and Valenzuela, B. 2012. Nutrigenomics: revealing molecular aspects of a personalized nutrition. *Revista Chilena de Nutrición* 39, no. 1: 71–85.

Sankaran, S., Khot, L.R., Espinoza, C. Z., Jarolmasjed, S., Sathuvalli, V. R., Vandemark, G. J., Miklas, P. N., Carter, A. H., Pumphrey, M. O., Knowles, N. R. 2015. Low-altitude, high-resolution aerial imaging systems forrow and field crop phenotyping: A review. *European Journal of Agronomy* 70, 112–123.

Saxena, A., Verma, M., Singh, B., Sangwan, P., Yadav, A. N., Dhaliwal, H.S., Kumar, V. 2020. Characteristics of an acidic phytase from aspergillus aculeatus APF1 for dephytinization of biofortified wheat genotypes. *Applied Biochemistry and Biotechnology* 191, no. 2: 679–694.

Saxena, R. K., Kale, S.M., Kumar, V., Parupali, S., Joshi, S., Singh, V., Garg, V., Das, R. R., Sharma, M., Yamini, K. N., Ghanta, A., Rathore, A., Sameerkumar, C. V., Saxena, K. B., Varshney, R. K. 2017. Genotyping-by-sequencing of three mapping populations for identification of candidate genomic regions for resistance to sterility mosaic disease in pigeonpea. *Scientific Reports* 7, 1813.

Seidel, G.E. 2009. Brief introduction to whole-genome selection in cattle using single nucleotide polymorphisms. *Reproduction, Fertility and Development* 22, no. 1: 138–144.

Septiningsih, E. M., Prasetiyono, J., Lubis, E., Tai, T. H. and Tjubaryat, T. 2003. Identification of quantitative trait loci for yield and yield components in an advanced backcross population derived from the *Oryza sativa* variety IR 64 and the wild relative *O. rufipogon*. *Theoretical and Applied Genetics* 107: 1419–1432.

Sharma, S. K., Bolser, D., de Boer, J., Sønderkær, M., Amoros, W., Carboni, M. F. and Eguiluz, M. 2013. Construction of reference chromosome-scale pseudomolecules for potato: integrating the potato genome with genetic and physical maps. *G3: Genes, Genomes, Genetics* 3, no. 11: 2031–2047.

Shiio, Y., Aebersold, R. 2006. Quantitative proteome analysis using isotope-coded affinity tags and mass spectrometry. *Nature Protocols* 1, no. 1: 139–145.

Smith, J.B. 2001. Peptide sequencing by Edman degradation. In: Encyclopedia of Life Sciences. Hoboken, NJ: Wiley-Blackwell., Hoboken, NJ, USA

Smyth, M.S., Martin, J.H. 2000. X ray crystallography. *Molecular Pathology* 53, no. 1: 8–14.

Sun, J., Beger, D. R. and Schnackenberg, K. L. 2013. Metabolomics as a tool for personalizing medicine: *Personalized Medicine.* 10: 149–161.

Syvanen, A.C. 2001. Accessing genetic variation: genotyping single nucleotide polymorphisms. *Nature Reviews Genetics* 2, no. 12: 930–42.

Tanksley, S. D. and McCouch, S.R. 1997. Seed banks and molecular maps: Unlocking genetic potential from the wild. *Science* 277: 1063–1066.

Tanksley, S. D., Young, N. D., Paterson, A. H. and Bonierbale, M.W. 1989. RFLP mapping in plant breeding: new tools for an old science. *Biotechnology* 7, no. 3: 257–264.

Thomashow, M. F. 2001. So what's new in the field of plant cold acclimation? Lots! *Plant Physiology* 125: 89–93.

Thompson, A. L., Thorp, K. R., Conley, M., Andrade-Sanchez, P., Heun, J. T., Dyer, J. M. and White, J. W. 2018. Deploying a proximal sensing cart to identify drought-adaptive traits in upland cotton for high-throughput phenotyping. *Frontiers in Plant Science* 9: 507.

Tilman, D., Cassman, K. G., Matson, P. A., Naylor, R. and Polasky, S. 2002. Agricultural sustainability and intensive production practices. *Nature* 418: 671–677.

Tsoukalas, D., Alegakis, A., Fragkiadaki, P., Papakonstantinou, E., Nikitovic, D., Karataraki, A. and Tsatsakis, A. M. 2017. Application of metabolomics: Focus on the quantification of organic acids in healthy adults. *International Journal of Molecular Medicine* 40, no. 1: 112–120.

Varshney, R. K., Graner, A. and Sorrells, M. E. 2005. Genomics- assisted breed-ing for crop improvement. *Trends in Plant Science* 10: 621–630.

Varshney, R. K., Nayak, S. N., May, G. D. and Jackson, S. A. 2009. Next-generation sequencing technologies and their implications for crop genetics and breeding. *Trends in Biotechnology* 27, no. 9: 522–530.

Wang, X., Chen, M., Zhong, M., Hu, Z., Qiu, L., Rajagopalan, S., Fossett, N.G., Chen, L.C. and Ying, Z. 2017. Exposure to Concentrated Ambient PM2.5 Shortens Lifespan and Induces Inflammation-Associated Signaling and Oxidative Stress in Drosophila. *Toxicol. Sci.* 156(1): 199–207.

White, J. W., Conley, M. M. and Flexible, A. 2013. Low-cost cart for proximal sensing. *Crop Science* 53, 1646–1649.

Wickramasinghe, S., Cánovas, A., Rincón, G. and Medrano, J.F. 2014. RNA-sequencing: a tool to explore new frontiers in animal genetics. *Livestock Science* 166: 206–216.

Wickramasinghe, S., Rincon, G., Islas-Trejo, A. and Medrano, J.F. 2012. Transcriptional profiling of bovine milk using RNA sequencing. *BMC Genomics* 13:45.

Wiese, S., Reidegeld, K. A., Meyer, H. E. and Warscheid, B. 2007. Protein labeling by iTRAQ: A new tool for quantitative mass spectrometry in proteome research. *Proteomics* **7**, no. 3: 340–350.

Wu, J., Li, L. T., Li, M., Khan, M. A., Li, X. G., Chen, H, and Zhang, S. L. 2014. High-density genetic linkage map construction and identification of fruit-related QTLs in pear using SNP and SSR markers. *Journal of Experimental Botany* 65, no. 20: 5771–5781.

Xu, Y. and Crouch, J. H. 2008. Marker-assisted selection in plant breeding: From publications to practice. *Crop Science* 48, no. 2: 391–407.

Yadava D. K., P. R. Choudhury, Hossain, F. and Kumar, D. 2017. *Biofortified Varieties: Sustainable Way to Alleviate Malnutrition.* New Delhi: Indian Council of Agricultural Research., New Delhi.

Yates, Iii J.R. 2011. A century of mass spectrometry: from atoms to proteomes.; *Nature Methods* 8, no. 8: 633–637.

Ye, X., Al-Babili, S., Kloti, A., Zhang, J., Lucca, P., Beyer, P. and Potrykus, I. 2000. Engineering the pro-vitamin A (beta-carotene) biosynthetic pathway into (carotenoid-free) rice endosperm. *Science* 287: 303–305.

Yue, L., Shonkoff, E. T. and Dunton, G. F. 2015. The acute relationships between affect, physical feeling states, and physical activity in daily life: a review of current evidence. *Frontiers in Psychology* 6: 1975. 10.3389/fpsyg.2015.01975

Zaidi, S. S., H. Vanderschuren, M. Qaim, M. M. Mahfouz, A. Kohli, S. Mansoor, and M. Tester. 2019. New plant breeding technologies for food security. *Science* 363: 1390–1391

Zamir, D. 2001. Improving plant breeding with exotic genetic libraries. *Nature Reviews Genetics* 2: 983–989.

Zhang, A., Sun, H., Wang, P., Han, Y. and Wang, X. 2012. Modern analytical techniques in metabolomics analysis. *Analyst* 137, no. 2: 293–300.

Zhang, X., Yap, Y. and Wei, D. 2008. Novel omics technologies in nutrition research. *Biotechnology Advances* 26: 169–76.

4

Rice Processing and Properties

Priya Dangi,
Institute of Home Economics, University of Delhi, New Delhi, India

Nisha Chaudhary
College of Agriculture, Nagaur, Agriculture University, Jodhpur, India

Ayushi Gupta and Isha Garg
Institute of Home Economics, University of Delhi, New Delhi, India

CONTENTS

4.1 Introduction

Rice *(Oryza sativa L.)* is considered as a semiaquatic annual grass chiefly grown as a *kharif* season crop. The majority of the rice cultivated in the world belongs to two main species: *Oryza sativa* (Asian rice) and *Oryza glaberrima* (African rice). The former species is produced and utilized on a larger scale in comparison to the *Oryza glaberrima* species, which is restricted to the region of West Africa. The widely cultivated Asian rice includes three subspecies: *indica, javanica, and japonica,* which vary in kernel size and dimensions.

According to FAO (2020), the global rice production reached 501.3 million tons in the marketing year 2019–20 with India, Pakistan, Thailand, Vietnam, and the United States as leading exporters. India contributed roughly 112 million tons of rice to the global production in this marketing year (Singh, 2019). Rice production is further expected to rise to 550 million tons worldwide by the year 2035 (You, You, Yue, Mun, & Lee, 2017). Around two-third of the world's population eat rice as their staple food. The consumption of rice in some Asian countries (Bangladesh, Burma, Cambodia, China, India, Indonesia, Korea, Laos, the Philippines, Sri Lanka, Thailand, and Vietnam) is relatively high, contributing about 75% of their daily calorie intake (FAO, 2001). Rice is predominantly consumed in the form of fully milled white rice, which comprises regularly milled as well as parboiled rice and only a small fraction is consumed as brown rice. It is also used in the preparation of processed foods, such as noodles, puffed rice, fermented sweet rice, snack foods, and beverages (such as beer, wine, sake, and vinegar).

The process of converting paddy into well-milled, edible, silky-white rice, involves a series of steps such as parboiling, drying, and milling that must be carried out with utmost care in order to produce high-quality rice. The market value of rice depends largely on its physical (moisture content, grain dimensions, weight, density, and color), mechanical (grain strength and elasticity), thermal (specific heat, thermal conductivity, thermal diffusivity, coefficient of expansion), and biochemical properties. Understanding the physical characteristics of a rice kernel is of paramount importance as it influences the design and dimensions of the operating equipment and helps in optimizing the storage and processing conditions. In cases, where the machineries are improperly designed and operations are executed inappropriately, cracked and broken rice kernels are produced, which fetches a low marketing price. Substantially, the percentage of whole grain rice obtained is one critical parameter, since long whole grains command higher market prices than the broken ones. The cooking and eating quality of rice is essential for establishing the economic value of rice. The preference of rice varies greatly among people; some people prefer long and flaky rice, others prefer short and sticky rice. The texture, aroma, flavor and a number of other parameters play a crucial role in visualizing consumer acceptability and deciding the cost of the rice (Ghadge & Prasad, 2012).

4.2 Rice Properties

4.2.1 Physical Properties of Rice

The physical properties are the properties of the grain either individually or in mass that are physical in nature but are not specifically included among other categories such as cooking, chemical, and physico-chemical properties. Unlike other cereals, rice is consumed as whole grain, therefore, knowledge of the physical properties, such as grain dimensions, hardness, grain friction, size, bulk density, true density, angle of internal friction and static coefficient of friction, shape, uniformity and general appearance is essential in handling, storage, and processing of rice. Moisture content, variety of rice and degree of milling often leads to difference in the physical properties (Bhattacharya, 2011).

4.2.1.1 Grain Dimensions, Weight and Uniformity

Since rice is produced and marketed according to its size and shape, the physical dimensions, weight and uniformity are of prime importance. The shape of rice is observed to be cylindrical with three dimensions: length (L), width (W) and thickness (T). USDA (2020) recognized seven classes of milled rice based on length/width ratios and termed them as long-grain, medium-grain, short-grain, mixed, second head, screenings, and brewers. Rice was classified on the basis of grain length as extra-long (>7.5 mm), long (6.61–7.5 mm), medium (5.51–6.60 mm), or short (<5.50 mm) and on the basis of grain shape depending upon the length-width ratio as slender (>3), medium (2–3), or bold (<1) (B.O Juliano, 1993). Kernel dimensions are primary quality factors in many areas of processing, drying, handling equipment, breeding, marketing, and grading.

The shape of rice can be expressed in the terms of sphericity (Φ). Reddy and Chakraverty (2004) defined sphericity as the ratio of the surface area of a sphere (having the same volume as that of rice) to the surface area of rice. Ghadge and Prasad (2012) computed sphericity (S_p) as

$$S_p = \frac{(LWT)1/3}{L} \times 100$$

Where
L is length, W is width, and T is thickness.
The aspect ratio (R_a) is used for the classification of the shape of rice and is calculated as:

$$R_a = \frac{W}{L} \times 100$$

Where
L is length, and W is width.

4.2.1.2 Color

Color is used as one of the important criteria of quality in all varieties of rice. The assessment, however, is performed on well-milled head rice. Color is influenced by such things as smut, which imparts a grey or red color to rice seed, giving the milled rice a rosy color (Luh, 1991). Rice varieties are classed as either light (straw) or dark (gold) hulled. Although hull color is not of major importance in the production of regular white milled rice, it is influential in the production of parboiled rice. Varieties with light colored hulls are generally preferred by parboilers as they tend to produce a lighter colored product than dark hulled varieties processed under similar parboiling conditions (Bett-Garber, Champagne, Thomson, & Lea, 2012).

4.2.1.3 Bulk density, true density and porosity

Knowledge of grain's bulk density, true density and porosity is useful in sizing grain hoppers and making correct facilities for storage. These properties of the grain can affect the rate of heat and mass transfer of moisture during the aeration and drying processes (Jouki, Emam-Djomeh, & Khazaei, 2012). Varieties differ in their fraction of high density grain due to differences, both in the degree of filling of the hull by caryopsis and in actual caryopsis density.

Reddy and Chakraverty (2004) defined bulk density as the ratio of mass of the paddy to its total (bulk) volume. It is determined by filling a circular container of known volume with paddy and calculated by the given formula:

$$\rho_b = \frac{M}{V}$$

Where
  ρ_b [kgm^{-3}] is bulk density,
  M [kg] is mass of the paddy sample, and
  V [m^3] is volume of the container.

The true density (tρ) is the ratio of mass of the paddy to its true volume. It is determined using the Toluene displacement method.

The porosity (ε) of the paddy is the ratio of the volume of internal pores in between the paddy to its bulk volume. It is determined using following relationship:

$$\varepsilon = \frac{(1-\rho_b)}{\rho_b} \times 100$$

Where
  ε [%] is porosity,
  ρ_b [kgm^{-3}] is bulk density, and
  ρ_t [kgm^{-3}] is true density.

A grain bed with low porosity will have greater resistance to water vapor escape during the drying process, which may lead to the need for higher power to drive the aeration fans.

A study of rice properties done at CFTRI summarized that density remains constant at 1.452 g/ml in all rice varieties. Bulk density varies appreciably in rice (0.777–0.847) and angle of repose is relatively constant in different varieties of rice (average 37.5 degrees). With increasing moisture content, density decreases linearly. However, bulk density decreases twice as fast and the porosity increases, owing to a concurrent progressive increase in the frictional property, which decreases the degree of grain packing. Density of rice increases slightly with milling, but bulk density, porosity, and angle of repose is markedly affected by the degree of milling.

4.2.1.4 Coefficient of Friction and Angle of Repose

Ghadge and Prasad (2012) defined the angle of repose (φ) as the angle in degrees with the horizontal at which the material will start forming a heap vertically, which can be determined using relationship:

$$\varphi = \frac{tan^{-1}(2H)}{D}$$

Where
  H is height of material, and
  D is distance between material and horizontal.

The friction coefficient is defined as the ratio between the friction force (force due to the resistance of movement) and the normal force on the surface of the material used in the wall. For biological products, two types of friction coefficients are considered, the static coefficient determined by the force capable to initiate the movement, and the dynamic coefficient determined by the force needed to maintain the movement of the grains in contact with the wall surface, which depends on the type and nature of the material in contact. The static coefficient of friction is used to determine the angle at which chutes must be positioned to achieve consistent flow of materials through the chute. Such information is useful in sizing motor requirements for grain transportation and handling (Jouki, Emam-Djomeh, & Khazaei, 2012).

4.2.2 Mechanical Properties

Mechanical properties of grains such as rice have a significant impact on the energy demand of grinding mills. Some rice varieties observe severe breakage losses during processing, making it quite important to determine their engineering properties in order to optimize the design of machinery used for milling these varieties (Kruszelnicka, Marczuk, Kasner, Bałdowska-Witos, Piotrowska, Flizikowski, et al., 2020). Considering the process of grinding, e.g., by means of grinding machines or roller mills, permanent deformation (fragmentation) occurs after exceeding the load value corresponding to the compressive strength limit. Strength is closely related to the power necessary to cause the strain and the grinded material cross-section field (hence being dependent on its geometric features). Thus, material fragmentation occurs upon application of appropriate forces, which in the system of grinding machines and roller mills, is performed by the rotary motion of rollers (Kunze & Calderwood, 2004). In such a case, the force is a direct effect of torque, which in turn is related to the power of the devices affecting the energy demand for grain processing. In general, the higher force applied corresponds to higher power, that is, the machine energy is needed.

Grain strength depends on the type of material, especially its internal structure (porosity), moisture, components of the grain, and biological properties. The internal structure of the grain endosperm and tegmen have an impact on the strength properties and the energy needed for grinding. The endosperm, which is characterized by higher glassiness, is usually harder, thus, for permanent deformation; it is necessary to use higher forces, which result in increased energy demands as compared to materials whose endosperm is less glassy. The glassiness of the endosperm also has an influence on the material fragmentation efficiency and the size of particles after division; the higher the glassiness, the easier it is to separate the endosperm from bran, breaking the grain into smaller parts (Kruszelnicka, et al., 2020).

4.2.3 Thermal Properties

Knowing the thermal properties of rice is important for rice processing or storage. The information about these properties plays an essential role in order to yield high-quality rice.

4.2.3.1 Specific Heat

Specific heat refers to the heat required by a substance of unit mass to raise its temperature by 1°C. The specific heat, C_p, is given by the formula:

$$C_p = \frac{Q}{m\Delta T}$$

Where
Q is amount of heat,
m is mass of substance, and
ΔT is the change in temperature.

The specific heat of rice in the temperature range of 10-28°C, as reported by Kunze and Calderwood (2004) is 1.84 J/g°C. Generally, an increase in the moisture content of cereal grains results in increase in its specific heat (Chakraverty, Mujumdar, & Ramaswamy, 2003).

4.2.3.2 Thermal Conductivity

Thermal conductivity refers to the rate at which heat gets transferred through a unit cross-sectional area of material by conduction when the temperature gradient is perpendicular to area.

The steady-state method is based on Fourier's law of heat conduction:

$$q = -kA\frac{\Delta T}{\Delta x}$$

Where
    q is the rate of heat transfer,
    k is thermal conductivity,
    A is area of cross-sectional surface,
    ΔT is temperature difference, and
    Δx is the distance between the two ends.

Thermal conductivity increases with increase in temperature, moisture content, and bulk density of rice. Yang, Siebenmorgen, Thielen, and Cnossen (2003) reported that an increase in the moisture content of a sample at a given temperature increases its thermal conductivity. As an example, the average thermal conductivity of Bengal rice at 61°C and 9.2% moisture content was reported to be 0.112 W/(mK), which rose to 0.126 W/(mK) when the moisture content was increased to 17% at the same temperature. Furthermore, the average thermal conductivity at 12.1% moisture content and 24°C (at glassy state) was reported to be 0.102 W/(mK), which rose to 0.111 W/(mK) when the temperature was increased to 61°C at the same moisture content (rubbery state).

4.2.3.3 Thermal Diffusivity

Thermal diffusivity refers to how quickly the heat can diffuse through a substance under transient conduction of heat transfer. It can be computed by the formula:

$$\alpha = \frac{k}{\left(\rho\, C_p\right)}$$

Where
    α is thermal diffusivity,
    k is thermal conductivity,
    ρ is density, and
    C_p is specific heat.

4.2.3.4 Coefficient of Thermal Expansion

The rice grains expand and shrink during heating and cooling, respectively, although the changes are minute. The coefficient of thermal expansion was reported to be 4.62×10^{-4} for rubbery state, which was much more than 0.87×10^{-4} for glassy state. This is of importance for rubbery-glassy state transition, which has an effect on rice fissuring (Perdon, Siebenmorgen, & Mauromoustakos, 2000).

4.2.4 Biochemical Properties

The composition of grain makes it a palatable food of high energy value, which leads nutritionists to have a major interest in the composition of the kernel. Moisture content is one of the critical parameters that affects shelf life of rice. If maintained and stored properly, dry rice can be maintained for years while only a few days are required for wet rice to spoil. Rough rice having moisture content of 13% is commonly accepted as a safe level for storage for less than 6 months, while 12% or less moisture is recommended for long-term storage.

The majority of fat is centralized in the bran layer of rice, which is removed during the milling process (Punia et al., 2021a, b). Lipids are also known to influence viscoelastic properties by forming inclusion complexes with the helical structure of amylose. It was observed that defatting rice starch reduced both gelatinization temperature and gel viscosity of starch (Juliano & Tuaño, 2018).

The protein content of milled rice is low in comparison with other cereals, although the whole rice grain content ranged from 7.0% to 10.8% of which 70–80% is the glutelin. Protein, as the other major constituent of rice, has not been thought to strongly influence cooking and eating qualities. When differences in gross protein content were examined in relation to texture of cooked rice, only a weak relationship was found; the higher protein rice was somewhat less tender than low-protein rice because commonly eaten rice generally contains about 7% protein and does not fluctuate widely from this level.

Rough rice has higher fiber and ash content, but lower protein and available carbohydrates than brown rice. The important B-vitamins in rice are thiamine, riboflavin, and niacin. Modern milling removes most of these vitamins because they are found largely in the bran and germ. The mineral composition of the rice grain depends considerably on the availability of soil nutrients during crop growth and on the diverse sampling, preparation, and analytical methods used by various investigators. Minerals are generally present in higher levels in brown than in milled rice. A considerable portion of the rice caryopsis ash is accounted by phosphorus. Potassium, magnesium and silicon are also present in large amounts in brown and milled rice. By contrast, silica is the major element in hull ash (Bett-Garber, Champagne, Thomson, & Lea, 2012).

4.3 Rice Processing

The harvested paddy contains undesirable, inedible portions that need to be separated to obtain edible white rice. Husk or hull, which comprises approximately one-fifth weight of the paddy, is a woody, siliceous, inedible covering that, when removed, results in brown rice. This brown rice is further enclosed by a fibrous and fatty bran layer that poses difficulty in cooking, causing a demand to remove this layer by the process of abrasion. The paddy undergoes a series of steps in the milling process that affects the quality of the edible rice produced in terms of its cooking, nutritional, and eating quality. Milling is termed as a process that removes foreign material, husk, bran, germ, and broken kernels to yield high-quality white rice grains. The milling yield of rice is affected by the variety of rice, degree of milling (quantity of bran that is removed in the rice), and grain breakage (Bhattacharya, 2011).

Bhattacharya and Ali (2015) described the unit operations and various types of equipment involved in rice milling in Table 4.1.

4.3.1 Drying

When rice is harvested, it contains a high level of moisture that is not suitable for storage purposes. Drying is one essential process that must be performed as soon as rice is harvested to prevent its spoilage during storage. The ideal moisture content for storage is 12% for long-term and 14% for short-term. Rice grains are hygroscopic in nature and exchange water with air depending upon the vapor pressure of both the grain and air (Bhattacharya & Ali, 2015). During drying, a moisture gradient develops in the rice grain where the center has comparatively higher moisture content than the surface. The drying rate at the surface of the grain is initially rapid, but later slows because of the rate of moisture travelling from center to surface. If the surrounding air is heated, the vapor pressure between grain and air increases, which in turn increases the rate of internal moisture content. One should be very careful while drying rice grains to achieve a minimum of broken kernels and a high-quality white rice after milling (Kunze & Calderwood, 2004).

Drying of rice kernels can be done either by natural or an in-storage drying process. In natural drying, the just-harvested wet paddy is left on the field for several days to expose it to natural air and sun, drying it to a level of 15–18%. This method is rather inexpensive, but cracking of grain and breakage during milling are the major drawbacks. Grains should be dried in thin layers and tossed regularly to prevent uneven drying. The grain should not be over-dried, i.e., it should not be dried to less than 13–14%. In-storage drying makes use of forced natural air or slightly heated air to dry the grains in a storage bin, which consists of a ducting system from where forced air is passed by a fan. If the weather is humid and the relative humidity is higher than 70%, the air should be slightly heated to reduce its relative humidity for effective drying. This heat is called supplemental heat. The heat provided should raise the

TABLE 4.1

Equipment and Their Principles Used in Milling of Rice

Milling Stage	Equipment	Principle of Equipment
Cleaning	Scalper	The paddy is placed in a rotating drum made of perforated metal sieve at an elevated height. The paddy falls through the perforated openings and the larger particles, like straw, remain on the sieve and are discharged at the end.
	Paddy cleaner	The paddy is fed into a column where it is aspirated as it goes through a series of decks covered with perforated steel sheets. Objects larger or wider than the paddy are removed on the top deck, whereas the objects of the same size or shorter than the paddy are removed on the lower deck.
	Drum-type cleaner	The paddy is first passed through a scalper to remove larger particles, after which the paddy is put on vibrating sieves. A fan pulls an air stream through the paddy to remove chaff and dust. Larger particles are removed on the top sieve due to the presence of the large perforations.
	Destoner	The paddy is separated from stones and other impurities on the basis of density differences between them. Air is pulled from under the perforated metal sheet deck through rice. Dense objects remain on deck while the vibrations of the deck cause lighter rice to fall under force of gravity and discharge.
	Magnetic separator	The paddy is made to move under permanent magnets such that the metal particles get attached to it. A magnetic drum separator can also be used where ceramic magnets produce strong metallic fields, which attach the iron particles to the drum's surface.
	Thickness grader	The grains are put in a cylinder consisting of rectangular slots. These grains are continuously allowed to tumble. Thin grains fall in this slot and get discharged, whereas the normal grains remaining on the top are discharged separately after being conveyed to the end of the cylinder.
De-husking	Disc sheller	The disc sheller consists of two iron discs placed horizontally, whose inner surface is coated with a layer of abrasive emery. The paddy gets evenly distributed over the disc surface during rotation and gets aligned vertically due to centrifugal force and friction of disc. The length-wise caught grains get de-husked by pressure and shearing action.
	Centrifugal sheller	High speed impeller discs are placed vertically with a radial blade in a metallic casing along with a hard rubber ring fixed on the inner side. The laterally fed paddy is rotated by blades moving outward in radial direction at high speed by centrifugal force. Frictional force causes the tip of grains to collide with the hard rubber ring at an angle, resulting in de-husking due to the impact force.
	Rubber roll sheller	The sheller consists of two rubber-covered rollers rotating in opposite directions, where one roll moves at about 25% higher speed. The paddy grains that fall between the two rolls get de-husked as a result of shearing and frictional forces due to differences in the peripheral speed of rolls.
Husk separation	Oscillating sieves	Two self-cleaning sieves are equipped in a plansifter, wherein one is finely perforated for bran and dust removal, while the other is largely perforated for broken grain collection. The sieve overflow contains husk, brown rice, and unshelled paddy.
	Husk aspirator	Air is forced through a mixture of husk, brown rice, and unshelled paddy, whereby air lifts the husks and discharges it.
Paddy separation	Compartment type separator	The oscillating compartment assembly consists of one or several stacked decks in a zig-zag channel. The surface of the table is made to oscillate in a perpendicular direction to the grain feed, which throws grain sideways, causing the unshelled/unhusked paddy to move up to the inclined slope and brown rice to move down the slope. In this manner, streams of pure paddy and pure brown rice are separated.
	Tray-type separator	The indented trays are made to oscillate at an incline. Brown rice, though having a higher density, ends up getting caught in indents and eventually is moved to the higher end by the upward oscillation motion. Due to the repelling effect of the smooth surface of brown rice, the paddy floats on top of brown rice and then slides down toward the lower end of tray. This is where it is sent back to the sheller.

(Continued)

TABLE 4.1 (Continued)

Milling Stage	Equipment	Principle of Equipment
De-branning/ whitening/ polishing	Abrasive polishers	The abrasive polisher removes bran by the swirling motion of grains between cone and screen, which is enhanced by rubber brakes. The grains are made to come in contact with a rotating emery surface that cuts and abrades bran from the surface of the rice grain.
	Friction polishers	The rice is fed into the equipment under slight pressure. A strong air stream is blown through a hollow shaft and slit of the cylinder that separates bran due to friction.
	Water jet polishers	Almost completely milled rice is fed into the water jet polisher. A strong air flow and water mist enters the chamber where rice moves in circular motion, rubbing against each other at high speed and under high pressure. The humidification softens the grain surface and the frictional force and pressure from the air removes dust, bran, and aleurone layer.
Grading	Oscillating sieve	Single or double perforated steel screens kept horizontally at an angle of 4° to 12° is made to vibrate to and fro by vertical eccentric drive. A single screen separates the rice into head rice and large broken rice at the upper, and small broken rice at the lower section. Whereas milled rice is separated into head rice, large and small broken rice through individual screens when double screens are used.
	Plansifter	Steel cables suspend the perforated screens (single or double) and give a swinging motion by eccentric drive. The different sized openings separate the milled rice into head, large broken rice, small broken rice, or more fractions on the basis of number of screens and openings.
	Indented cylinder	Rice is fed at the raised end into the revolving cylinder consisting of indentations. While moving downward, the broken grains fall into indents and get trapped, while the head rice slides down. The broken grains fall from the indents when they are inverted while rotating at a higher point and are collected in the collecting tray.
	Indented disc separator	Cast iron discs with indentations on both sides are arranged radially. These rotate in a bed of rice and the broken grains get trapped in the indentations depending on their size; these fall out when inverted down and get collected in a collecting tray.
Color sorting	Optical sorter	The rice grains pass through a photo-detector system, i.e., the sensor, which detects the light deflected by rice and compares with the standard color. The rice with color different from standard color is blown out by a strong stream of pressurized air.

temperature maximum by 5–8°C, reducing the relative humidity to about 60%. Care must be taken while dealing with supplemental heat because if the air becomes too dry, i.e., if the relative humidity becomes less than 40%, then the grain would become over-dried resulting in cracks in the grain. The grain must not be dried below 12% moisture (Bhattacharya & Ali, 2015).

4.3.2 Cleaning

Once the paddy is harvested, it undergoes a cleaning process to remove any impurities present like sand, stones, straw, weed seeds, and foreign materials like metal or glass pieces. The importance of this step lies in obtaining cleaned rice, which can improve the efficiency of the milling process.

4.3.3 De-husking

After the cleaning process, the paddy is de-husked ensuring the least damage possible to the bran and brown rice grain. This process marks the splitting of the paddy into brown rice and husk, in addition, a number of other materials like unshelled paddy, broken rice, bran, and germ are also obtained depending on the type of sheller used. Along with husk, these by-products are separated from the brown rice on the basis of their size, density, and frictional properties by using oscillating sieves and husk aspirators.

4.3.4 Paddy Separation

This step ensures the separation of the unshelled paddy from the brown rice and paddy mixture, which is then again sent back to the sheller for de-husking. Compartment-type and tray-type separators are the two types of paddy separators that work on the principle of density and coefficient of friction of constituents.

4.3.5 De-branning/Whitening/Polishing

The outer (and also sometimes inner) bran layers are removed from the surface of brown rice to yield debranned white rice, which is essential for easy cooking and better digestion. However, if the bran is excessively removed, it leads to the reduction of nutritional quality of rice. The germ and fine broken grains are also removed in this step.

4.3.6 Grading

Different fractions of rice grains differ in their length and are thus graded for determining their market value. The whole unbroken grains and the broken grains with at least three-fourth length compared to the whole grains are graded as head rice. Broken rice that is one-eighth or less in length of whole rice is graded as fine broken and the remaining broken grains are classified as either large broken or medium broken.

4.3.7 Color Sorting

The discolored grains are sorted from the milled rice grains in this step. An optical sorter consisting of a photo-detector is used to complete this process.

4.4 Parboiling

The term par-boiling refers to partially boiled (or partially cooked) rice. This method of treatment of rice originated in India for storage and conservation purposes and has been widely practiced since ancient times. About a fifth of all rice is parboiled before milling, and 90% of all parboiled rice is produced in South Asia (Bhullar & Bhullar, 2013).

The parboiling process refers to the hydrothermal treatment given to the paddy where it is allowed to be pre-cooked within the husk without affecting its size and shape. During this process, a crystalline form of starch present in the paddy is changed into an amorphous state as a result of the gelatinization process. This treatment is beneficial for coarse and medium rice with a soft structure as it is prone to breakage during the milling process. The major objectives of parboiling are to increase the total and head yield of the paddy, prevent nutrient loss during milling, salvage wet or damaged paddy, and prepare the rice according to the requirements of consumers (Bhattacharya, 2011).

4.4.1 Parboiling Process

The process of parboiling has three basic steps: soaking, steaming, and drying. In general, the paddy is soaked in water for a short period of time, then heated once or twice by steam, and lastly dried before milling.

4.4.1.1 Soaking

The fundamental aim of this step is to hydrate the paddy enough to promote the gelatinization process upon heating. This can be achieved simply by soaking the paddy in water for a defined period. The duration of soaking is dependent upon the temperature of the water. Low temperature conditions (<60–65°C) require very little attention to the soaking time as the equilibrium moisture level will never exceed 30–32% (wet basis) thereby avoiding the risk of over-soaking of paddy. However, under these conditions, it will take longer to achieve hydration, which can lead to fermentation and the development of off-flavors. At high temperatures, the rate of hydration increases exponentially, which promotes the soaking process and minimizes the risk of fermentation—both of which help bring about the gelatinization of starch. As hydration continues, the moisture content of the kernel exceeds 30–32% (wet basis) and the husk is no longer able to hold the expanded structure and bursts, resulting in leaching and deformation of the grain.

Practically, the process of soaking can be accomplished in three ways:

1. Soaking can be done at a temperature greater than 75°C, where the soaking time is carefully monitored. Once the moisture content of grain reaches 30–32%, the water is drained off and the paddy undergoes a tempering stage to equalize its moisture content.
2. Soaking the paddy at ~70°C with strict regulation over time.
3. Soaking begins at ~75°C and the batch is gradually allowed to cool naturally during soaking.

4.4.1.2 Steaming or Heating

Steaming is done to gelatinize the hydrated starch. Adequately and uniformly hydrated starch can undergo the gelatinization in as little as 2 minutes of steaming at atmospheric pressure. In cases where over-imbibition may occur, chances of splitting of the husk during steaming are very common. Besides steaming, other methods such as mild heating with hot water or sand and ohmic and microwave heating can be successfully used for carrying out the gelatinization process.

4.4.1.3 Drying

During the soaking stage, the water uptake by paddy increases the moisture level to approximately 35–38%. For safe storage and efficient milling process, this moisture content needs to be lowered significantly to 12–14% moisture. Drying the paddy in an effective manner is of paramount importance as if it is not done properly, it may lead to cracks after the milling process and head rice yield will decrease considerably. Drying can be accomplished either in the sun or with hot air by following a two-stage process. In the first stage, the paddy is dried to a moisture content of 16% instantaneously, followed by its tempering for about 4 or 8 hours in the second stage, which will relax the steep moisture gradient developed in the paddy (Bhattacharya, 2011).

4.4.2 Methods of Parboiling

Depending upon the scalability of the process, traditional and modern methods are employed to obtain parboiled rice. The traditional method makes use of pottery or a boiler for direct or indirect heating and practices either a single or double steaming process. Agricultural residues serve as the main energy sources for carrying out local parboiling processes. Nontraditional/modern methods are highly energy and capital intensive, and applicable only to large scale operations (Roy, Shimizu, Shiina, & Kimura, 2006).

4.4.2.1 Non-commercial Methods

4.4.2.1.1 Soak-Drain-Cook Process

The most common and widely used process includes soaking, draining, cooking, and drying. The paddy is soaked in water at a suitable temperature, varying from ambient (2–3 days) to about 70°C (3–4 hours) to confer saturation (approximately 30% moisture, wet basis). The water is drained off and the paddy is then steamed, or otherwise heated by infrared or microwave (rarely) or some other form of heating, to cook (or to gelatinize) the starch. The steaming can either be carried out at atmospheric pressure, i.e., open steaming, or under elevated pressure (0.5–2.0 kg/cm^2 gauge pressure).

4.4.2.1.2 Low-Moisture Parboiling

Low-moisture parboiling is characterized by partially soaking the paddy followed by high-pressure steaming. In this process, the paddy is not soaked to saturation, but only partially soaked or even simply wetted and later exposed to high-pressure steaming (1–3 kg/cm² gauge) that brings about the desired gelatinization.

4.4.2.1.3 Dry-Heat Parboiling

The initial step of soaking follows the same pathway as in the soak-drain-cook process to attain the saturation level. Afterwards, instead of being steamed, the paddy is subjected to conduction heating. As a result, the soaked rice is gelatinized as well as dried simultaneously.

4.4.2.2 Commercial Method: CFTRI (Hot-Soaking) Process

The commercial process was developed by Central Food Technological Research Institute (CFTRI), Mysore. The process exposes the wet paddy to reduced pressure conditions for a short time initially, followed by steeping the paddy in 75–85°C water for 2–3 hours. The water is then drained off and the paddy is heated under reduced pressure in a steam-jacketed vessel with live injected steam. The water discharge valve is kept open in order to remove condensed water during steaming. The parboiled paddy is dried either under sun or by mechanical driers and taken for further processing (Bhattacharya, 2011).

4.4.3 Effects of Parboiling

As a consequence of the parboiling process, numerous physical, chemical, and sensorial changes appear in the raw grain and its constituents as presented in Table 4.2. The typical changes taking place during different stages of parboiling are described further in detail below.

4.4.3.1 Changes During Soaking

4.4.3.1.1 Enzymic Activity

The soaking stage marks the initiation of the germination process in the grain and its extent relies greatly upon temperature and presence of air and light. Soaking the paddy at room temperature for extended periods creates an anaerobic condition, and as a result, seeds die out. However, some of the enzymes remain active to convert sucrose into reducing sugar and assist in *de novo* production of sugars and amino acids that are partly responsible for the discoloration of parboiled rice. This can be prevented by soaking the paddy at a high temperature (>70°C), which would kill the seed quickly but simultaneously inactivate the enzymes.

4.4.3.1.2 Migration of Water-Soluble Molecules from Outer Regions to Endosperm

During soaking, the water-soluble components present in the husk and bran layers solubilize in the water and enter the endosperm region. As a result, milled parboiled rice is enriched with B-vitamins, sugars, and certain minerals as compared to milled raw rice.

TABLE 4.2

Comparison of Quality Characteristics of Raw Milled Rice and Parboiled Milled Rice

Property of Rice Grains	Raw Rice	Parboiled Rice
Appearance	Opaque and white	Translucent with faint amber (yellow-brown) color
Dimensions	Comparatively longer rice kernels	Slightly shorter but broader than raw rice grains
Milling efficiency	Yield of head rice is low with a larger proportion broken	Increased yield with less number of broken pieces
Nutritional profile	Vitamin-B content is comparatively low	High vitamin-B content
Cooking behavior and eating quality	Soft, rough, and sticky	Firmer, fluffier, and less sticky
Recovery of bran oil	Comparatively low	High

4.4.3.2 Changes During Steaming

The gelatinization process that takes place during steaming is pivotal for improved head rice recovery and enhanced retention of thiamine after the milling process. Interestingly, these changes are brought about by disruption of starch, protein, and fat in the rice.

4.4.3.2.1 Changes in Starch

Soaking the paddy at high temperature exclusively, or at an ambient temperature followed by steaming, induces swelling of the starch molecules as a result of water uptake (gelatinization). This gelatinization results in loss of birefringence and the destruction of A-type starch granules. A proportion of gelatinized starch undergoes re-association with lipid molecules, which explains the quintessential properties of parboiled rice: Its firm texture in comparison to raw milled rice due to the diminished hydration at high-temperature conditions.

4.4.3.2.2 Changes in Fat

Rice bran oil is one of the valuable by-products obtained during milling of rice. In raw rice, the oil is in the form of distinct globules centered below the grain surface in the aleurone layer. During the parboiling process, these fat globules are disrupted and oil is released that penetrates easily into the soft bran tissue, forming a band on the surface. The oil does not seep into the tough endosperm. This phenomenon clearly explains the higher oil content of parboiled bran in comparison to raw bran.

4.4.3.2.3 Changes in Protein

Rao and Juliano (1970) reported the rupturing of protein bodies in rice grains as a result of the steaming process. This is due to the reduced solubility of protein and its extent is determined by the severity of the process.

4.4.4 Effect of Parboiling on Milling Quality

Amongst the several factors, the cracking of the kernel is one of the main factors for breakage during raw rice milling. Cracks may develop as a result of delayed harvesting, threshing, or a rapid drying process. Rice breakage is related to milling conditions, particularly by the relative humidity, temperature, and extent of milling. Parboiling aids in starch gelatinization, which fills up the void spaces in the rice, reducing the breakage of kernels during milling. The most advantageous aspects of parboiling are the increase in the head yield of rice during polishing, the polish percentage, and breakage over time; parboiled rice takes longer than raw rice to attain the same degree of polishing. Parboiled rice requires three to four times as much abrasive load as raw rice for the same level of polishing. Regarding the color of rice, the polishing need for parboiled rice is less as compared to raw rice. For example, if consumers favor 80% bran removal to achieve the parboiled rice, this rice would require polishing of 3% whereas raw rice would need to be polished to 4% for the same quantity of bran removal (Bhattacharya, 2011).

4.4.5 Advantages and Disadvantages of Parboiling

The process offers numerous advantages and disadvantages, which are described below:

- **Advantages:**
  - The de-husking of rice becomes easier and the grain becomes tougher, reducing the losses in milling.
  - The nutritive value of parboiled rice increases as the water-soluble vitamins and mineral salts, present in the hull and bran coat, are solubilized in water and transported to the endosperm. The riboflavin and thiamine content is four times higher in parboiled rice than in milled rice.
  - Parboiled rice has comparatively lower moisture content (10–11%) than milled rice, which corresponds to its better storage.
  - The starch grains embedded in a proteinaceous matrix are gelatinized and expanded until they fill up the surrounding void spaces, thereby minimizing the occurrence of cracks and fissures on milling.

- **Disadvantages:**
  - Most of the naturally occurring enzymes in the grain are inactivated during the steaming stage.
  - Natural antioxidants present in grain are also destroyed.
  - The process may affect the quality of milled rice and bran by promoting oxidative rancidity.
  - The heating step causes discoloration of grains.
  - Due to defective steeping, parboiling sometimes causes an unpleasant smell in rice.
  - Sometimes during the treatment, over-imbibition and deformation of the grain may occur, which can be restricted by keeping the water and heat separated.

4.5 Quality Analysis of Rice

4.5.1 Cooking Quality

The major proportion of milled rice is consumed as table rice, and only a slight part is converted into various other rice products. Table rice is specifically cooked in water until it turns soft by utilizing different modes, such as using a pressure cooker, double boilers, or microwave. Overall, the cooking quality of rice is mainly affected by the rice variety and its milling quality, ageing, and the cooking method employed, which in turn influences the sensory and economic considerations of the grain.

4.5.1.1 Changes Taking Place During Cooking Process

- Rice grains are hydrated readily by absorbing roughly twice its weight of water.
- Hydrated starch granules in the presence of heat undergo the gelatinization process.
- Rice grains become soft, easily digestible, and fit for consumption.
- Expansion of the grain takes place in all the directions, chiefly longitudinally.
- Inflation of the grain volume (both true and bulk volume).
- Semi-translucent or translucent uncooked grains change to an opaque color.
- Dissolved solids and suspended particles leach from the grain into the excess of water during cooking.
- Considerable sections of grains burst either along the ventral or dorsal edge of the grain.

4.5.1.2 Factors Affecting Cooking Quality

Numerous internal factors of rice (moisture content, amylose and amylopectin ratio, and type of starch granules), processing parameters (degree of milling, precooking, water/rice ratio during cooking, cooking methods, gel consistency, and gelatinization temperature, cooking losses and cooking time) and post-processing conditions (drying and storage conditions) are the factors of great importance that influence the cooking and textural characteristics of rice. Soft-textured cooked rice is obtained from those having high water-binding capacity, which is greatly influenced by degree of milling and is indirectly associated with the profit gained by the farmers and rice milling industry. Consequently, it is obligatory to select a suitable degree of milling that depreciates the levels of losses and boosts cooking/eating qualities (Mohapatra & Bal, 2006).

4.5.1.2.1 Amylose to Amylopectin Ratio

The texture of rice is determined by its amylose content. A fluffy, non-sticky, flaky, and dry texture is obtained by cooking rice varieties whose amylose content is more than 25%, which makes it capable of absorbing more water. Amylose content of rice is related to the water absorption and volume expansion of rice. To cook high-amylose rice, more water is needed to obtain the optimum texture as compared to low-amylose varieties. Waxy starches being devoid in amylose and rich in amylopectin absorb less water and cook to a sticky and pasty texture (Frei, Siddhuraju, & Becker, 2003). Furthermore, Perez, Juliano,

Bourne, and Morales (1993) reported that amylose content is correlated positively to hardness whilst negatively to the stickiness value of cooked milled rice.

4.5.1.2.2 Ageing

Rice undergoes a change in its cooking behavior as it ages; there is a remarkable difference between recently harvested rice and rice that has been stored for some time. Just after harvesting, rice (commonly known as new rice in South Asia) cooks to a somewhat lumpy, sticky, and moist mass. However, as it ages, it cooks comparatively drier, fluffier, and free-flowing (Bhattacharya & Ali, 2015). The former is preferred by the people who eat rice with chopsticks since the rice grains cling together. The texture of rice can be characterized as sticky/adhesive or firm/hard/tender. Soaking is an important process, which if not done, can lead to difficulty in cooking the rice. It is a common observation that when rice is cooked in excess water, starch solids leach out of the rice grains, however, the case is different when they are cooked in limited water, where these solids get redeposited. These solids, called gruel, make the cooked rice sticky. The parboiled rice is hard, so usually less starch is leached into the cooking water, thus preventing the problem of sticky or pasty rice. Short rice grains have a stickier texture upon cooking than medium or long rice grains (Elbashir, 2005).

4.5.1.2.3 Gelatinization Temperature

The gelatinization temperature is directly related to the cooking time; rice takes longer to cook when its gelatinization temperature is high (Bhattacharya & Sowbhagya, 1971).

4.5.1.2.4 Water Uptake

Water uptake is related to the surface area of rice grains per unit weight; small and slender grains cook comparatively quicker than large and round grains. At ambient temperature, the water uptake is inversely related to amylose content and gelatinization temperature of the rice. However, chalkiness of the rice grain promotes water uptake, which reduces the palatability of cooked rice (Bhattacharya, 2011).

4.5.1.2.5 Cooking Losses

Depletion of nutrients takes place when rice is washed in excess water prior to cooking due to discarding the gruel, which is also resultant of cooking rice in excess water. Parboiled rice has less vitamin loss than raw milled rice. Elbashir (2005) observed a significant loss in starch, amylose, and amylopectin content in cooked rice.

4.5.1.2.6 Grain Elongation

Linear elongation in rice grains without significant increase in girth is considered a desirable characteristic of high-quality cooking rice. The gelatinization temperature is positively related to grain elongation. Nonetheless, more elongation in rice would likely indicate low amylose content (Perez, Juliano, Bourne, & Morales, 1993).

4.5.2 Eating Quality

Eating quality is defined as the sensory perception of the aroma, whiteness, gloss, flavor, tenderness (or hardness), and cohesiveness (or stickiness) of the cooked rice. It is a function of milling quality; adequately milled rice has more consumer acceptance for its eating quality than brown and under-milled rice. Eating quality is evaluated by three major physicochemical characteristics of the starch—amylose content, gel consistency, and gelatinization temperature—and can also be evaluated on the basis of grain quality considering its size, appearance, and shape (Ahmed, Tanweer, Kabir, & Latif, 2020).

4.5.2.1 Grain Quality

The chalkiness and translucency of rice is influenced by the blurriness or opacity of its endosperm. Cruz, Kumar, Kaushik, and Khush (1989) mentioned that customer acceptability decreases with the increase of chalkiness in the grain.

4.5.2.2 Amylose Content (AC)

Amylose content (AC) is considered to be a major predictor of rice eating quality as it has been associated with mechanical textural attributes such as hardness and stickiness (Custodio, Cuevas, Ynion, Laborte, Velasco, & Demont, 2019). Waxy rice has near zero amylose, and is used for special foods such as desserts and snacks. High amylose cultivars (>25%) are common in *Indica* rice, and are dry and fluffy on cooking, often becoming hard after cooling. Low amylose cultivars (15–20%) are soft and sticky, and include nearly all-temperate *japonica* cultivars. Intermediate amylose (20–25%) rice is soft but not sticky, and is widely preferred by most consumers (Phing Lau, Latif, Rafii, Ismail, & Puteh, 2016).

4.5.2.3 Gel Consistency (GC)

Gel consistency is a measure of firmness of cooked rice as it tempers the tendency of the cooked rice to harden after cooling. Within the same amylose group, varieties with a softer gel consistency are preferred where the cooked rice has a higher degree of tenderness. Harder gel consistency is associated with harder cooked rice and this feature is particularly evident in high amylose rice. Hard cooked rice also tends to be less sticky. Rice varieties can be grouped into three GC classes: high (hard and very flaky texture), medium (flaky but softer rice), and low (soft and non-flaky rice). Rice with the same amylose content can be classified as hard gel consistency (26–40 mm), medium gel consistency (41–60 mm), or soft gel consistency (61–100 mm) (de Oliveira, Pegoraro, & Viana, 2020). GC is reported to be affected by milling (lipid content), protein content, ageing of milled rice (fat oxidation), and rice flour particle size (efficiency of dispersion) (Perez, Juliano, Bourne, & Morales, 1993).

4.5.2.4 Aroma

Aroma is a value-added character to rice, since it is a preferred trait by consumers. Rarely, however, do consumers describe the aroma of rice beyond the subjectivity of "with fragrance," "no fragrance," or "bad or unpleasant smell." Good aroma tends to be associated with pleasant aromatics found in Jasmine and Basmati rice types. Most often, the volatile compound 2-acetyl-1-pyrroline (2-AP) is found in relatively high concentrations in these aromatic rice varieties, lending a popcorn-like roasted smell. The aroma of 2-AP is also associated with a milky and sweet nutty smell and exhibits a low threshold value in water (Custodio, Cuevas, Ynion, Laborte, Velasco, & Demont, 2019). However, how these volatiles contribute to the aroma of rice is unclear; hence defining aromatic rice is still incomplete.

4.6 Conclusion

With world rice production expecting to increase, there is a need to pay attention to its various properties and processing techniques to ensure high quality. Physical properties of rice are important for efficient handling, storage, and processing of rice grains. The physical properties, namely, bulk density, true density, and porosity, are used to determine the size of hoppers and the design of storage facilities. Low porosity grains require higher power to drive aeration fans for drying. The static coefficient of friction is used in sizing motor requirements for rice grain handling and transportation. Mechanical properties of rice have a significant impact on the energy demand used in designing machinery and, optimizing them to prevent grain breakage. High glassiness corresponds to easier separation of endosperm. The coefficient of thermal expansion is an important property with respect to rice fissuring. Milling of paddy is a very essential process, which renders the rice edible, giving it a characteristic texture, flavor, and color—all important for consumer acceptability. Head rice recovery and number of broken grains determines the efficiency of milling process. A hydrothermal treatment called parboiling is given to the paddy to prevent grain breakage and increase the head rice yield, along with making the de-husking process easier, enhancing the nutritional content, and making the rice less sticky during cooking. Consumers judge the quality of rice based on its cooking and eating ability—the better the final rice quality, the higher its market value. The cooking quality of rice is improved by high amylose content, ageing, and low gelatinization temperature. The eating quality of rice is determined by sensory perceptions, such as aroma, tenderness, cohesiveness, flavor etc., which are affected majorly by grain chalkiness, amylose content, and gelatinization temperature. Since rice is consumed as a whole grain and in variety of processed products, understanding its key attributes and milling process is of prime importance to obtain a product of excellent cooking and eating quality with wide consumer acceptability.

REFERENCES

Ahmed, F., Tanweer, F., Kabir, M., & Latif, A. (2020). Rice quality: Biochemical composition, eating quality, and cooking quality. In A. Costa de Oliveira, C. Pegoraro & V. V. Ebeling (Eds.), *The Future of Rice Demand: Quality Beyond Productivity*, (pp. 3–24): Springer, Cham.

Bett-Garber, K. L., Champagne, E. T., Thomson, J. L., & Lea, J. (2012). Relating raw rice colour and composition to cooked rice colour. *Journal of the Science of Food and Agriculture*, 92(2), 283–291.

Bhattacharya, K. R. (2011). Rice Quality: A Guide to Rice Properties and Analysis. In K. R. Bhattacharya (Ed.): Woodhead Publishing Series.

Bhattacharya, K. R., & Ali, S. Z. (2015). *An Introduction to Rice-Grain Technology, 1st. ed*. New York: WPI Publishing.

Bhattacharya, K. R., & Sowbhagya, C. M. (1971). Water uptake by rice during cooking. *Cereal Science Today*, 16(12), 420–424.

Bhullar, G. S., & Bhullar, N. K. (2013). Agricultural Sustainability: Progress and Prospects in Crop Research. Elsevier Inc.

Chakraverty, A., Mujumdar, A. S., & Ramaswamy, H. S. (2003). *Handbook of Postharvest Technology: Cereals, Fruits, Vegetables, Tea, and Spices*: CRC Press.

Cruz, N. d. l., Kumar, I., Kaushik, R. P., & Khush, G. S. (1989). Effect of temperature during grain development on stability of cooking quality components in Rice. *Japanese Journal of Breeding*, 39(3), 299–306.

Custodio, M. C., Cuevas, R. P., Ynion, J., Laborte, A. G., Velasco, M. L., & Demont, M. (2019). Rice quality: How is it defined by consumers, industry, food scientists, and geneticists? *Trends in Food Science & Technology*, 92, 122–137.

de Oliveira, A. C., Pegoraro, C., & Viana, V. E. (2020). *The Future of Rice Demand: Quality Beyond Productivity*: Springer.

Dhull, S. B., Punia, S., Kumar, M., Singh, S., & Singh, P. (2020). Effect of different modifications (physical and chemical) on morphological, pasting, and rheological properties of black rice (Oryza sativa L. Indica) starch: A comparative study. *Starch-Stärke*, 2000098.

Elbashir, L. T. M. (2005). *Physiochemical properties and cooking quality of long and short rice (Oryza sativa) grains*. University of Khartoum, Khartoum (Sudan).

FAO. (2001). *Food Balance Sheet*. Rome: Food and Agricultural Organization of the United Nations.

FAO. (2020). Cereal markets to remain well supplied in 2020/21. In: *FAO Cereal Supply and Demand Brief*. Food and Agriculture Organization of the United Nations.

Frei, M., Siddhuraju, P., & Becker, K. (2003). Studies on the in vitro starch digestibility and the glycemic index of six different indigenous rice cultivars from the Philippines. *Food Chemistry*, 83(3), 395–402.

Ghadge, P., & Prasad, K. (2012). Some physical properties of rice kernels: Variety PR-106. *Journal Food Process Technology*, 3(8), 1000175.

Jouki, M., Emam-Djomeh, Z., & Khazaei, N. (2012). Physical properties of whole rye seed *(Secale cereal)*. *International Journal of Food Engineering*, 8(4).

Juliano, B. O. (1993). *Rice in Human Nutrition*: Food and Agriculture Organization of the United Nations.

Juliano, B. O., & Tuaño, A. (2018). Gross structure and composition of the rice grain. In J. Bao (Ed.), *Rice*, (pp. 31–53): Woodhead Publishing and AACC International Press.

Juliano, B. O., & Tuaño, A. (2019). Gross structure and composition of the rice grain. In *Rice*, (pp. 31–53).

Kruszelnicka, W., Marczuk, A., Kasner, R., Bałdowska-Witos, P., Piotrowska, K., Flizikowski, J., & Tomporowski, A. (2020). Mechanical and processing properties of rice grains. *Sustainability*, 12(2), 552.

Kunze, O. R., & Calderwood, D. L. (2004). *Rice: Chemistry and Technology, 3rd ed*. American Association of Cereal Chemists.

Luh, B. S. (1991). *Rice: Utilization* (Vol. 2): Springer US.

Mohapatra, D., & Bal, S. (2006). Cooking quality and instrumental textural attributes of cooked rice for different milling fractions. *Journal of Food Engineering, 73*(3), 253–259.

Perdon, A., Siebenmorgen, T., & Mauromoustakos, A. (2000). Glassy state transition and rice drying: Development of a brown rice state diagram 1. *Cereal Chemistry, 77*, 708–713.

Perez, C. M., Juliano, B. O., Bourne, M. C., & Morales, A. A. (1993). Hardness of cooked milled rice by instrumental and sensory methods *Journal of Texture Studies, 24*(1), 81–94.

Punia, S., Kumar, M., Sandhu, K. S., & Whiteside, W. S. (2021a). Rice-bran oil: An emerging source of functional oil. *Journal of Food Processing and Preservation*, e15318.

Punia, S., Kumar, M., Siroha, A. K., & Purewal, S. S. (2021b). Rice bran oil: Emerging trends in extraction, health benefit, and its industrial application. *Rice Science, 28*(3), 2.

Phing Lau, W. C., Latif, M. A., Rafii, M. Y., Ismail, M. R., & Puteh, A. (2016). Advances to improve the eating and cooking qualities of rice by marker-assisted breeding. *Critical Reviews in Biotechnology, 36*(1), 87–98.

Rao, S. N. R., & Juliano, B. O. (1970). Effect of parboiling on some physicochemical properties of rice. *J Agric Food Chem, 18*(2), 289–294.

Reddy, B. S., & Chakraverty, A. (2004). Physical properties of raw and parboiled paddy. *Biosystems Engineering, 88*(4), 461–466.

Roy, P., Shimizu, N., Shiina, T., & Kimura, T. (2006). Energy consumption and cost analysis of local parboiling processes. *Journal of Food Engineering*, 646–655.

Singh, S. K. (2019). India: Grain and Feed Annual. In: M. Wallace (Ed.), Global Agricultural Information Week. USDA Foreign Agriculture Service.

USDA. (2020). Rice Inspection Handbook. Washington, DC: United States Department of Agriculture.

Yang, W., Siebenmorgen, T. J., Thielen, T. P. H., & Cnossen, A. G. (2003). Effect of glass transition on thermal conductivity of rough rice. *Biosystems Engineering, 84*(2), 193–200.

You, K. Y., You, L. L., Yue, C. S., Mun, H. K., & Lee, C. Y. (2017). Physical and chemical characterization of rice using microwave and laboratory methods. In: Amanullah and Fahad, S. (Ed.), *Rice- Technology and Production*. IntechOpen, pp. 81–99).

5

Oat *(Avena Sativa): Functional Components*

Sneh Punia
Chaudhary Devi Lal University, Sirsa, India
Clemson University, Clemson, SC, USA

Anil Kumar Siroha
Chaudhary Devi Lal University, Sirsa

Manoj Kumar
ICAR-Central Institute for Research on Cotton Technology, Mumbai, India

CONTENTS

5.1 Introduction

Oats belongs to the *Poaceae* family and are commonly known as *Avena sativa*. According to the Food and Agriculture Organization of the United Nations (FAO) (2018), globally, oats occupy an area of more than 10 million hectares (ha) with total production of more than 25 million tons. Regarding annual production and area harvested for oats production, this crop is generally estimated as a minor crop and

TABLE 5.1

Composition of Oat Grain

Component	Availability	References
Starch	~60%	Doehlert et al., 2013; Berski et al. (2011)
Protein	Total: 11–15%	Robert et al. (1985)
	~9 to 15%	Zhu, 2017
	Globulins: 80% of total protein	Lasztity (1996)
	Prolamins: 15% of total protein	Klose et al. (2009)
	Glutelin: 5–66%of total protein	
	Albumin: 1–12% of total protein	
Lipids	~3–11%	Zhu, 2017
	5–9%	Flander et al. (2007)
β-glucan	~2–8%	Zhu, 2017
	2.3 and 3.2%	Andersson and B¨ orjesdotter (2011)
	2.3–85%	Flander et al. (2007)
	3.9 to 6.8%%	Wood (1994)

mostly used as animal feed, and to a smaller extent, for food. In 2009, livestock feeding (horses, cattle, sheep, and poultry) was the primary use (70%) of oats. Because of the unique bioactive compounds in oats, the use of this beneficial crop as animal feed has begun to gradually decline and interest has been developed in exploring oats as healthy and functional food for human consumption (Punia et al., 2020; Sandhu et al., 2017; Ahmad et al., 2010). As reported by Jing and Hu (2012), about 70 species of oats have been explored around the world, among them *Avena sativa* and *Avena nuda,* commonly known as hulled and naked oats, respectively, are the most cultivated oat species.

Oat grains comprise plentiful nutrients i.e. proteins (9–15%) with a higher lysine content than wheat and corn, fat (3–11%), beta-glucans (β-glucans, 2–8%) (Zhu, 2017), and starch (60%) (Rasane et al., 2015; Decker et al., 2014) (Table 5.1). Like all monocot cereal grains, oats contain relatively high amount of soluble fibers, protein, and fat and have less starch when compared to other cereal grain; they also contain one-third more protein, nearly four times more fat, and have less starch (Meydani, 2009). In comparison to wheat, oat grains have 2–5 times more lipids (5–9%) and are rich in hydrolytic enzymes (Ovando-Martínez et al., 2013). Oats also contain trace elements such as tocols, folic acid, manganese, iron, copper, carotenoids, alkaloids, niacin, amino acids (having sulphur), phytates, resorcinols, and lignins (Peterson, 2001).

Oats contain fat soluble vitamin E (20–30 mg kg^{-1}) which is found to be an important antioxidant with free-radical scavenging properties. Among vitamins, both α-tocopherol and α-tocotrienol account for 86–91%, with a lesser amount of niacin (0.032%), thiamine (0.002%), and riboflavin (0.001%). Being a rich source of phenolic compounds, it contains hydroxybenzoic acid, ferulic, caffeic, p-coumaric, and vanillic acids as major phenolics. Zhu (2017) observed small amounts of flavonoids like vitexin, glycosylvitexin, apigenin, tricin, glycosylvitexin, and isovitexin in oats. In recent years, the influence of functional components of oats has been appraised, as these components have antioxidant potential and therapeutic benefits including anti-inflammatory, antiproliferative, and antigenotoxic effects.

5.2 Functional Compounds

The health benefits associated with oat consumption have been attributed to their distinctive polyphenolic compounds in either free or in bound form (Adom & Liu, 2002). Another study conducted by Rasane et al. (2015) also reported that anticancerous and cholesterol lowering effects of oats associated with β-glucan (dietary fibres), protein, lipid and starch, and phytochemicals present in the whole oat grain.

The most important functional compounds reported in oats are phenolic acids, flavonoids, β-glucans, avenanthramides, fat soluble vitamin E, and carotenoids (Ryan et al., 2007).

5.2.1 Phenolic Acids

Polyphenols are functional molecules and secondary metabolites, including phenolic acid and flavonoids in plants, that benefit humans through actions on the immune system. Structurally, they contain one or more benzene rings with hydroxyl, carbonyl, and carboxylic acid groups (Abuajah et al., 2015). They are derivatives of cinnamic and benzoic acids and are categorized as hydroxycinnamic acid and hydroxybenzoic acid. Oats contain a number of polyphenolic compounds i.e. ferulic, caffeic, cinnamic, p-coumaric, and vanillic (Cai et al., 2012). The functional components present in oats are presented in Table 5.2. Researchers have reported many phenolics found in oat groats, hull, and flour. Vogel (1961) reported o-coumaric, p-hydroxybenzoic, p-coumaric, sinapic, ferulic, vanillic, and syringic acid are the phenolic acids present in hulls. In groats, phenolic acids present were ferulic, o-coumaric, p-hydroxybenzoic p-coumaric, caffeic, and catechol (Durkee & Thivierge, 1977; Tian & White, 1994), whereas in flour, trans-p-coumaric, vanillic, p-hydroxybenzoic, protocatechuic, syringic, caffeic, and ferulic acid were present (Sosulski et al. (1982).

TABLE 5.2

Functional Components of Oats

Components	Amount
Total Phenolic Content	87 mg/kg (Sosulski et al., 1982)
	1.18 mg/g (Irakali et al. 2012)
Total Phenolic Acids	137.4 µg/g (Liu et al. 2004)
Hydroxybenzoic Acid	
Gallic Acid	1.71 µg/g (Irakali et al. 2012)
O-Hydroxybenzoic Acid	1.75 µg/g (Irakali et al. 2012)
Salicylic Acid	4.56 µg/g (Irakali et al. 2012)
Vanillic	7.64 µg/g (Irakali et al. 2012)
Syringic Acid	11.31 µg/g (Irakali et al. 2012)
Protocatechuic Acid	1.15 µg/g (Irakali et al. 2012)
Hydroxycinnamic Acid	
Ferulic Acid	290–300 mg/kg (Durkee and Thivrege, 1977)
	75 mg/kg (Sosulski et al.1982)
	185 µmol (Adom and Liu, 2002)
	250 mg/kg (Mattilla & Hellström, 2005)
	12.13 µg/g (Irakali et al., 2012)
P-Coumaric Acid	4.19 µg/g (Irakali et al., 2012)
Cinnamic Acid	2.63 µg/g (Irakali et al., 2012)
Sinapic Acid	6.44 µg/g (Irakali et al., 2012)
	55 mg/kg (Cui and Liu, 2013)
Caffeic Acid	2.67 µg/g (Irakali et al., 2012)
Carotenoids	
Lutein	0.23 mg/kg (Panfili et al., 2004).
Zeacanthin	0.12 mg/kg (Panfili et al., 2004).
Carotene	0.01 mg/kg (Panfili et al., 2004).
Total Tocopherol	50–75 mg/kg (Peterson 2001; Hammon 1983)
	19–30 mg/kg (Peterson and Qureshi, 1993)
	13–25 mg/kg (Bryngelsson et al., 2002b)
Avenanthramides	Avenanthramide A- 25 to 47 mg/kg
	Avenanthramide B- 21 to 43 mg/kg
	Avenanthramide C- 28 to 62 mg/kg(Dimberg et al., 1996)
	40–132 mg/kg (Dimberg et al., 1993)

In oats, ferulic acid is reported in abundance followed by syringic acid, *chlorogenic* acid, *caffeic* acid, p-coumaric acid, p-hydroxybenzoic acid, and protocatechuic acid (Cui & Liu., 2013).

Sosulski et al. (1982) studied nine different types of phenolic content in oat flours and reported that ferulic acid constituted the greatest percentage (75 mg/kg) of the acids. They reported that bound form of phenolic acid comprised 66.3% of total phenolic content, whereas free and soluble forms were 10% and 23%, respectively. Adom and Liu (2002) evaluated bound, soluble, and free forms of total ferulic acid with 97.8%, 1.8%, and 0.4%, respectively, in whole oat flour. Durkee and Thivierge (1977), and Weidner et al. (1999), reported ferulic acid levels within range of 12.13 μg/g and 290–300 mg/kg. Many researchers have reported that phenolic compounds are mostly concentrated in bran fraction. Mattila et al. (2005) reported that bran fraction contained phenolic acid concentrations of 250 mg/kg, which is higher than as reported by Sosulski et al. (1982) in de-branned flour (71.4 mg/kg). Both cultivar and location affects individual phenolic concentrations (Emmons & Peterson, 2001). Dimberg et al. (1996) observed that the differences in the phenolic compounds may be due to the storage length and with contents increasing with increase in storage time.

5.2.1.1 Metabolism

Phenolic compounds present as glycoside and aglycones containing a number of sugar molecules. Glucose is reported to be the most commonly used sugar, followed by galactose>rhamnose>xylose> arabinose. Mannose, fructose, and galacturonic acids are not used very often. Aglycones, which are bound to sugars (β-glycosides), are absorbed in the duodenum after removal of sugar molecules by β-glucosidase enzymes and lactase (also known as lactase phlorizin hydrolase [LPH]) enzymes (Di Carlo et al., 1999). For absorption, β-glucosidase and LPH enzymes release the aglycone into the lumen by a diffusion mechanism (Figure 5.1). Phenolic compounds, which are bound to rhamnose, reach the gut and are hydrolyzed by rhamnosidases before they are absorbed (Manach et al., 2004). Although hydroxycinnamic acids' (HCA) potential health benefits depend on their bioavailability, these acids are common in the human diet. These acids esterified to organic acids, sugars, and fats. Human tissue does not have esterases to break ester links, so HCA's metabolizes in the large intestine microbiome, although most of the absorption has been completed in the small intestine (almost one-third) (Clifford, 2000; Erk et al., 2014). Some HCAs, such as ellagitannins, are resistant to the action of LPH (present in lining of the villi) and cannot be absorbed in the small intestine. Accordingly, the HCAs reached the large intestine and cleaved by microbiota into aglycones, which are again metabolized by microbiota into many hydroxyphenylacetic acids (Landete, 2011). When an aglycone has been absorbed in the intestinal tract, it undergoes some degree of phase II metabolism (as methylation, sulfation, and glucuronidation) at the enterocyte level. Then, these products by the portal vein enter the blood stream, and then the liver. In the

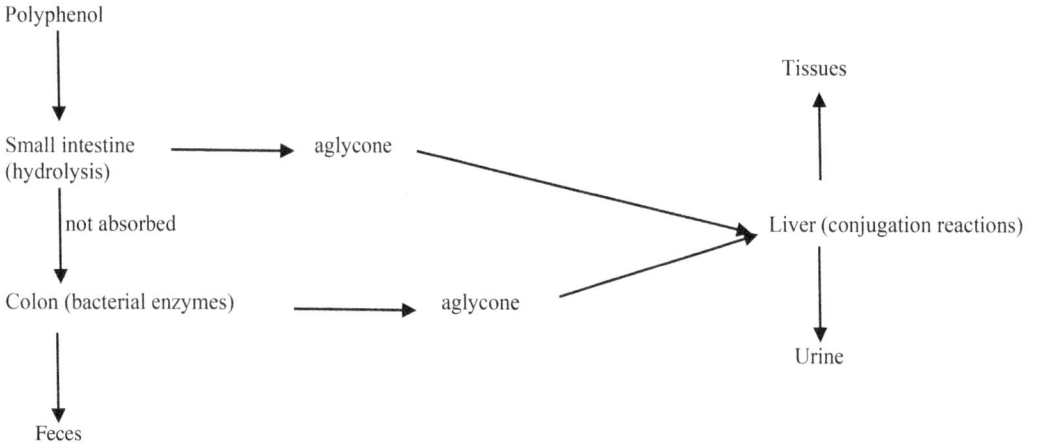

FIGURE 5.1 Metabolism of polyphenols (phenolic acid and flavonoids) in human gut.

liver, they may go through phase II metabolism or undergo conjugation where they would be transported back to the bloodstream until they're secreted into urine (Stalmach et al., 2010). A fraction of conjugated products are excreted back into the intestinal tract where they are converted into deconjugated compounds by gut microbial enzymes before being reabsorbed again (Cardona et al., 2013). The metabolites which are unabsorbed are excreted through feces.

Bioavailability of phenolic acids from food sources are reported by many researchers. Chen et al. (2004) reported that p-coumaric acid is the most bioavailable phenolic acid among the oat phenolic compounds. On the basis of their pharmacokinetic profile, they reported the maximum plasma concentrations (1.55 μmol/L) of p-coumaric acid. They further reported that this acid is 10% more bioavailable than others phenolics present in oats.

5.2.2 Flavonoids

Oats are not considered to be appreciable source of flavonoids. Flavonoids constitute a major portion of phenolics compounds in oats and consist of an aromatic ring combined to another aromatic ring (Clifford, 2000). Flavonoids in oats include flavone apigenein and its 6-O- and 8-O glucoside as well as flavone quercetin and its 3-O rutinoside (Collins & Webster, 1986). Peterson (2001) identified major flavonoids as apigenin, luteolin, tricin, kaempferol, and quercetin in oats. Eggum and Gullord (1983) reported tannin levels between 0.85% and 1.48% in their study among 12 diverse oat cultivars.

5.2.2.1 Metabolism

The general description of the metabolic pathway of flavonoids is shown in Figure 5.1. Most of the flavonoids, except catechins, are present as β-glycosides having association with sugars (Hollman, 2004). They are ingested as glycosides with aglycone and transferred in the intestinal lumen, where flavonoid glycosides are hydrolyzed into aglycones by LPH or β-glucosidase (Day et al., 2000). Mostly ingested glycosidic flavonoids are not absorbed by the small intestine, but reach the large intestine where they are exposed to hydrolysis and fermentation by microbiomes. Within the intestinal epithelium, glycosidic flavonoids are subjected to phase I metabolism (by cytochrome P450 monooxygenases) and the produced metabolites are transported to the liver via the hepatic portal vein. Within the liver, they again pass through phase (I and II) metabolism through which more polar compounds are formed. The egestion of flavonoids from the body is via the kidney, via bile in urine, or it is transferred back into the lumen of the intestines. Via the biliary route, flavonoids are excreted into the duodenum, where microbial enzymes act upon them so they are reabsorbed and undergo enterohepatic recycling (Lampe & Chang, 2007). A fraction reaching the colon are exposed to colonic microbiota (metabolic reactor) where unabsorbed flavonoids are catabolized into phenolic and aromatic acids, which become bioavailable (Williamson & Clifford, 2010).

5.2.2.2 Health Benefits

Collectively, health benefits of functional components of oats are summarized in Table 5.3. It is evident that a diet enriched with antioxidants is associated with a lower risk of degenerative and cardiovascular diseases. Polyphenolic compounds play a pivotal roles in mediating their properties and are characterized by the presence of aromatic rings bearing one or more hydroxyl moieties (Leiro et al., 2004). Recently FDA (U.S.) has claimed an association between the consumption of oatmeal, bran, and flour and a reduced risk of coronary heart disease (Mazza, 1998). Oats as a healthful food with physiological health benefits (i.e. hypocholestrolemic and hypoglycaemic effects) has potential of reducing cancer and hypertension. Oat polyphenols possess antiproliferative, anti-inflammatory, and anti-itching properties along with protection against cardiovascular diseases, skin irritation, and colon cancer (Meydani, 2009). Polyphenols present in healthy foods readily metabolized into phenolic acids and aldehydes by intestinal microflora, and these metabolites are reported to have anti-inflammatory properties (Rios et al., 2003).

TABLE 5.3

Health Benefits of Functional Compounds of Oats

Compounds	Health benefits
Phenolic compounds	
	Have hypocholestrolemic effect and hypoglycemic effect, has potential of reducing cancer and hypertension
	Possess antiproliferative, anti-inflammatory, and anti-itching properties
	Anti-inflammatory effect
β-glucan	
	Stimulates wound healing
	Reducing plasma cholesterol
	Controlling blood glucose concentration
	Stimulate humoral and cellular immunity
	Reduces psychophysical stress
	Enhances immune response
	Attenuate chronic fatigue syndrome
	Metabolically regulate diabetes
	Inhibits the development of cancer
tocols	
	Antiaging ability, anticarcinogenic potential and also equally good in preventing chronic disease
	Behaves as cholesterol synthesis inhibitors
AVNs	
	Antipathogenic property
	Anti-inflammatory
	Protects pancreatic β-cells from injury
Phytosterols	
	Cholesterol-lowering potential
Carotenoids	
	Carotenoids exert anticarcinogenic properties, reducing cardiovascular diseases, enhancing immune response and cell defence against free radicals
	Maintaining eye health

5.2.3 Beta-glucan (β-glucan)

Oat β-glucan comprises the polysaccharide $(1\rightarrow3)$, $(1\rightarrow4)$-β-D-glucan, is commonly known as β-glucan (Daou & Zhang, 2012), and is principally found in the aleurone cell walls and endosperm in oat grains (Bhatty, 1993). The importance of β-glucan reducing plasma cholesterol has been associated with water-soluble mixed linkage (1,3)(1,4)-b-D-glucans (MLG) as soluble fibers (Colleoni-Sirghie et al., 2003), which is the principal component and contributes approximately 85% of endospermic cell walls (Miller et al., 1995). Soluble fibers have gel-forming properties, which increase viscosity of intestinal chyme by disturbing micelle formation, which may inhibit cholesterol absorption, slow cholesterol transfer across the unstirred layer, and increase bile acid excretion by inhibiting bile acid reabsorption. The solubility of β-glucan in oats is due to $(1\rightarrow3)$ linkage, varies with detection methods, and depends on cultivars. Zhu (2017) and Andersson and Börjesdotter (2011) reported β-glucan of oats between 2–8% and 2.3–3.2%, in their respective studies. Wood (1994) reported the β-glucan within in the range of 3.9–6.8% in 11 oat cultivars.

5.2.3.1 Metabolism

The general description of the metabolic pathway of β-glucan is shown in Figure 5.2. Oat fibers possess gastrointestinal (GT) effects. Soluble fiber contributes due to its high solubility and as a

β-glucan

↓

Stomach (depolymerization only)

↓

Small intestine (viscous)

↓

Colon (fermentation)

↓ butyric acid, acetic acid
stimulates growth of colonic bacteria

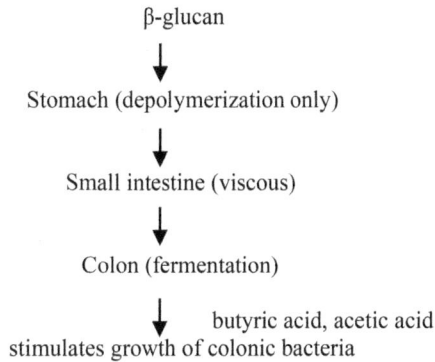

FIGURE 5.2   Metabolism of β-glucan.

substrate for fermentation in the colon, whereas insoluble fiber has a bulking effect. After ingestion, dietary fibers absorb water and increase in size in the gastric tract, ultimately affecting
satiety. The solubility of the fibers is independent of size or hydrothermal treatment given to
the oat grains (Daou & Zhang, 2012). Since humans do not have specific enzymes to hydrolyze
β-glucan, it remains intact in the small intestine. However, viscosity is increased by increased
mucin throughout the small intestine (Mälkki & Virtanen, 2001). Lazaridou and Biliaderis (2007)
reported depolymerization of β-glucan in the gastrointestinal tract. A study conducted by Ulmius
et al. (2012) concluded that the increasing volume of β-glucan is independent of bile acids or
enzymes (pepsin) as they did not observe any binding of fibers with bile or enzymes; instead, viscosity is dependent on pH. From the small intestine, viscous dietary fibers reach the large bowel
where they are fermented and metamorphose into acetic, propionic, and butyric acids. Casterline et
al. (1997) reported that dietary fiber has high fermentability and yields higher amounts of butyric
acid when compared with other cereals. The health attributes of butyric acid have also been noted
due to its anticarcinogenic properties.

5.2.3.2 Health Benefits

Being a rich source of β-glucan, oats have increased attention in recent years as they are considered
a bioactive component with the ability to reduce blood cholesterol (AbuMweis et al., 2010), lessen the
elevated blood glucose concentration (Behall et al., 2006), enhance immune response, and have anticancer effects (Chan et al., 2009). Anderson (1986) reported oats as a rich source of insoluble fiber, reducing
intestinal transit time. β-glucan has two important traits: it possesses high viscosity at low concentration
due to increased gastric and intestinal contents and has favorable effects on the colon, by enhancing
production of microbial mass (Mälkki & Virtanen, 2001).

5.2.4 Tocopherols

Naturally, tocols (vitamin E) exist in eight forms, i.e. α-,β-,Y- and δ-tocopherol, and α-,β-,Y- and
δ- tocotrienol, with α-tocopherol and α-tocotrienol in abundance (Barnes, 1983a, b). As reported by
many researchers, oats have only α and β tocol forms, but Lásztity et al. (1980) reported eight forms.
With α-tocotrienol as the predominant homologue, researchers found total tocols (tocopherol + tocotrienol) within the range of 20–30 mg kg^{-1} in oat cultivars (Peterson & Qureshi, 1993). Peterson (2001)
reported that α-tocotrienol and α-tocopherol constitute about 91% of total tocols with values in the range
of 15–50 mg/kg. Bryngelsson et al. (2002a, b) studied tocol levels for eight oat cultivars and reported
the values ranging from 13.8–25.3 mg/kg. The germ and endosperm have the majority constituting 96%
of the tocols, whereas the hull had 4%. In oat groat and hulls, the α-tocotrienol of 13% and 70%, and
α-tocopherol accounting for 63% and 19% of the tocols (Bryngelsson et al., 2002a).

Tocols (as dl-α-tocopheryl acetate)

↓   hydrolysis

Stomach

↓   digestion

Bile acid ————————→   ←———————— Pancreatic enzymes

Small intestine (Micelles)

↓

Enterocytes

↓

Chylomicrons

↓

Lymph

↓

Plasma

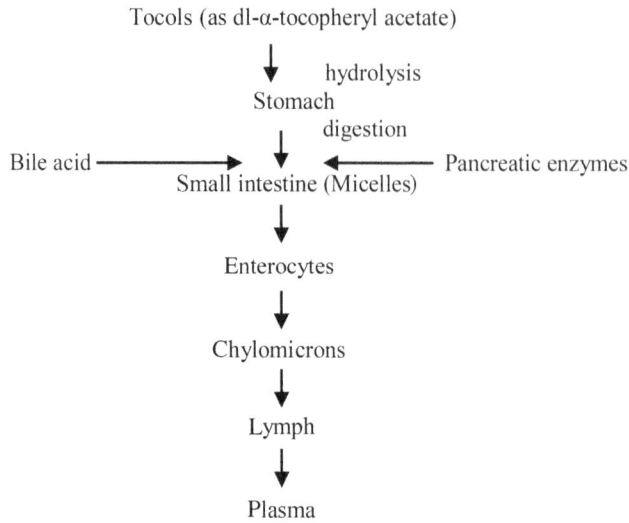

FIGURE 5.3 Metabolism of tocols.

5.2.4.1 Metabolism

The mechanism of tocol absorption and other fat-soluble vitamins is quite similar and shown in Figure 5.3. Tocols are generally taken up as dl-α-tocopheryl acetate in the human diet (Jakobsen et al., 1995). Before absorption in the small intestine, tocols must be hydrolyzed either into free alcohols or with emulsified fat (Gallo-Torres, 1970). Borel et al. (2001) reported that absorption to tocol acetates is associated with lipid-rich foods, whereas before the absorption process, bile acids and pancreatic enzymes are necessary. They further concluded that tocols are firstly hydrolyzed with gastric lipase in the stomach and are thereafter digested by pancreatic and other digestive enzymes secreted in intestine (Burton et al., 1988). Absorption of tocols is accelerated by medium-sized triglycerides, inhibited by polyunsaturated fatty acids (PUFA), and mainly occurs in the upper and middle intestine (Traber et al., 1993). Further micelles (clusters of monoglycerides, bile acids, fatty acids, etc.) are transferred into enterocytes (absorptive epithelial cells of small intestine) from where they reach the lymph system within chylomicrons (transports lipids to tissues), and then in plasma where they are processed by lipoprotein lipase (Bjørneboe et al., 1986). It is reported that some α and β tocopherol are transferred directly to tissues through chylomicron and mediated by lipoprotein lipase LPL (Traber et al., 1993). As a result of LPL activity, tocols and lipids are circulated into adipose and muscular tissue and, to some extent, brain tissue. It is believed that biological activities (inhibition of lipid peroxidation in biological membranes) of tocol are from its antioxidant action (Burton & Traber, 1990). α-tocopherol is reported to have the highest biological activity (Barnes & Taylor, 1981) and antioxidant activity (Serbinova et al., 1991) of others tocols.

5.2.4.2 Health Benefits

Vitamin E has antiaging ability, anticarcinogenic potential, and is also equally good in preventing chronic disease, stroke, and cardiovascular disease (Packer & Fuchs, 1993). Tocotrienols behave as cholesterol synthesis inhibitors (Pearce et al., 1992; Wang et al., 1993) due to antioxidant and free-radical scavenging activity.

5.2.5 Avenanthramides

Oats have low-molecular-weight phenolic molecules called avenanthramides (AVNs) of approximately 40 different types. Structurally, avenanthramides consist of an anthranilic and hydroxycinnamic acid derivatives (Collins & Mullin, 1988; Collins, 1989; Bratt et al., 2003; Wise, 2011) that are not present in other cereal grains. AVNs were the first identified and characterized alkaloids present in groats and hulls of oats. Oats contain at least 25 different AVNs and N-acyl anthranilate alkaloids. The three major AVNs reported in oats are AVA-A, AVA-B and AVA-C, which contain hydroxyanthranilic and hydroxy-cinnamic acids, such as p-coumaric, caffeic, and ferulic acids (Collins, 1989; Peterson et al., 2002; Gani et al., 2012). Oats contain 21–132 mg/kg of AVN-B and 25–47 mg/kg and 28–62 mg/kg of AVN-A and AVN-C C (Dimberg et al., 1996). Peterson (2001) reported that oat grain contains 54 mg/kg, 36 mg/kg, and 52 mg/kg of AVN- A, B and C, respectively. Li et al. (2017) reported total concentration of AVN in the range between 22.1–471.2 mg kg^{-1}, which is higher than what was reported by Chu et al. (2013) and lower than reported by Tong et al. (2014). Some researchers reported that among 40 AVA, only 15 AVA have been found in oat grains and 2p, 2c, and 2f (2 indicating 5-hydroxyanthranilic acid and p, c, and f indicating the hydroxycinnamic acids p-coumaric, caffeic or ferulic acids) are most abundant (Cui & Liu, 2013; Boz, 2015).

5.2.5.1 Metabolism

When AVN are ingested in high doses, they become bioavailable during passage through the gastrointestinal tract. In plasma, most of the AVNs are metabolized into hydroxycinnamic acids. Hydroxycinnamic acids are again metabolized to smaller hydroxybenzoic acids by reduction, methylation, and sulfation (Schär et al., 2018).

5.2.5.2 Health Benefits

Avenanthramide compounds are fabricated by the action of pathogens and have an antipathogenic property (Collins & Mullin, 1988; Collins, 1989). Emmons and Peterson (2001) reported that a diet supplemented with AVN increased superoxide dismutase (an enzyme with antioxidant defense systems against O_2) activity in the liver, kidneys and skeletal muscle and enhanced glutathione peroxidase (an enzyme with antioxidant defense systems against H_2O_2) activity in heart and skeletal muscles. The antioxidant potential of AVNs is approximately 30 times greater than that of oats' other phenolic compounds, such as caffeic and vanillin acids (Pundir et al., 2013). AVNs are reported to have the ability to suppress ICAM-1, which ultimately reduces inflammatory cytokine production by inhibiting monocytes adhesion to the aorta's endothelial cells (Cui & Liu., 2013). Lv et al. (2009) reported that dihydroavenanthramide (DHAv), an AVN analog, protects pancreatic β-cells from injury.

5.2.6 Phytosterols

Plant sterols are found in all foods of plant origin and function as structural components of the cell membrane. Phytosterols, having the same basic structure as cholesterol, are abundant in cereals. β-sitosterol (65%), campesterol (30%), and stigmasterol (3%) are commonly occurring phytosterols in the human diet (Weihrauch & Gardner, 1978). Määttä et al. (1999) reported β-sitosterol, as the most abundant phytosterol in oats, followed by campesterol, D5-avenasterol, and stigmasterol.

5.2.6.1 Metabolism

Dietary intake is the only way to get plasma phytosterols. After ingestion, only 5% sterols are absorbed by the human or mammal body (Salen et al., 1970). The human body does not have the potential to synthesize phytosterols, therefore, they are taken exclusively from food into intestinal absorption. They are firstly incorporated into micelles and then taken up by enterocytes and incorporated into chylomicrons at much lower concentrations. Inside the enterocytes, phytosterols interfere with the esterification and incorporation of cholesterol into chylomicrons (Smet et al., 2012). From chylomicrons, enterocytes enter blood circulation and are then taken up by the liver. Inside the liver, they are metabolized into cholesterol and other metabolites by the action of several enzymes; a key enzyme called cholesterol 7∝-hydroxylase is found in bile acids and is rapidly secreted into bile by hepatic ATP-binding cassette (ABC) proteins, G5/G8 transporters (Ogbe et al., 2015). Phytosterols as a liver constituent are then secreted into the blood from the liver and then taken up by cells that may store them either in cytoplasm or in the cell membrane. However, phytosterols do not affect cell the membrane's physical or biological properties (Trautwein et al., 2003).

5.2.6.2 Health Benefits

According to a study conducted by Ikeda et al. (1988), plant sterols have cholesterol-lowering potential by inhibiting absorption cholesterol from the intestinal tract. From bile salt micelles, sterols have ability to displace cholesterol and restrict solubility of cholesterol, which has an inhibitory effect on cholesterol absorption in plasma (Heinemann et al., 1991; Wester, (2000).

5.2.7 Carotenoids

Carotenoids are pigmented compounds (responsible for the orange, red, and yellow color of various fruits and vegetables) of lipophilic nature (Britton & Hornero-Mendez, 1996). Among all carotenoids xanthophyll is found in the highest quality in cereal grains. Xanthophylls includes lutein, zeaxanthin, and β-cryptoxanthin found in appreciable amount, but α- and β-carotene in small amounts (Mellado-Ortega & Hornero-Méndez, 2015). Amongst the cereal crops, oats have a relatively low carotenoid content. Panfili et al. (2004) reported that lutein is the major carotenoid of oats with lesser amounts of zeaxanthin and α- and β-carotene; values of lutein were 0.23 mg/kg, zeaxanthin was 0.12 mg/kg, and α- and β-carotenes were 0.01 mg/kg.

5.2.7.1 Metabolism

Generally, carotenoids are soluble in fat and insoluble in water and aqueous media (content of the human gut). Both lutein and zeaxanthin (having hydroxyl groups) are polar molecules whereas α-, β-carotene, and lycopene are nonpolar (Abdel-Aal et al., 2013). It is strongly suggested that the bioavailability of carotenoids is increased by consumption of fats and oils. First, carotenoids are solubilized in micelles by bile salts, taken up by epithelial cells of the jejunum, and further incorporated into the chylomicron (transports lipids to tissues) (Figure 5.4). Thereafter, the chylomicron is secreted into the lymph system and circulated within the body through vascular systems (Nagao, 2014). Bile plays a crucial role in absorption of carotenoids and their subsequent cleavage to retinol (El-Gorab & Underwood, 1973; El-Gorab, et al., 1975).

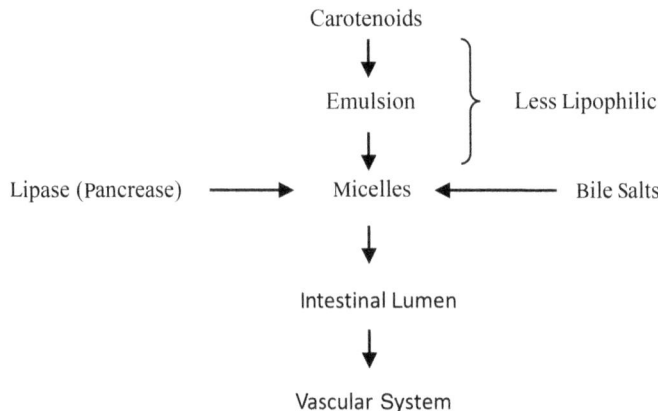

FIGURE 5.4 Metabolism of carotenoids.

5.2.7.2 Health Benefits

Ingested pigmented carotenoids exert antioxidant and anticarcinogenic properties, reducing cardiovascular diseases and enhancing immune response and cell defense against free radicals (Britton & Hornero-Mendez, 1996). Cui & Liu (2013) reported that although α- and β-carotene and β-cryptoxanthin possess provitamin A activity, β-carotene is the leading precursor of vitamin A. It is evident from research that after absorption, both zeaxanthin and lutein (yellow pigment) are moved into the retina and transported to macula where lutein is metamorphosed into mesozeaxanthin. Collectively, for maintaining eye health, zeaxanthin and mesozeaxanthin play a very important role (Mozaffarieh et al., 2003).

5.3 Conclusion

Being nutritional sources of dietary fiber, protein, and carbohydrates, oats play an important role in prevention of cardiovascular diseases, diabetes, inflammation, and cancer. These health attributes provided by oats could be due to the presence of a functional compound, such as β-glucan, phenolic compounds, etc. These functional compounds provide various chances for incorporating oats in healthy food products. To investigate the mechanisms of action of these functional compounds in disease prevention, it is necessary to know the factors that determine releasing behavior, their absorption, and their bioavailability in the organism. Further investigations are needed to determine the bioavailability of functional compounds and the mechanisms that contribute to the different absorption and elimination profiles of various components.

REFERENCES

Abdel-Aal, E. S. M., Akhtar, H., Zaheer, K., & Ali, R. 2013. Dietary sources of lutein and zeaxanthin carotenoids and their role in eye health. *Nutrients*, 5(4):1169–1185.

Abuajah, C. I., Ogbonna, A. C., & Osuji, C. M. 2015. Functional components and medicinal properties of food: a review. *Journal of Food Science and Technology*, 52(5):2522–2529.

AbuMweis, S. S., Jew, S., & Ames, N. P. 2010. β-glucan from barley and its lipid-lowering capacity: a meta-analysis of randomized, controlled trials. *European journal of clinical nutrition*, 64(12): 1472–1480.

Adom, K. K., & Liu, R. H. 2002. Antioxidant activity of grains. *Journal of Agricultural and Food Chemistry*, 50(21):6182–6187.

Ahmad, A., Anjum, F. M., Zahoor, T., Nawaz, H., & Ahmed, Z. (2010). Extraction and characterization of β-d-glucan from oat for industrial utilization. *International Journal of Biological Macromolecules*, 46(3): 304–309.

Anderson, J. W. (1986). Fiber and health: an overview. *Nutrition Today*, 21(6):22–26.

Andersson, A. A., & Börjesdotter, D. (2011). Effects of environment and variety on content and molecular weight of β-glucan in oats. *Journal of Cereal Science*, 54(1):122–128.

Barnes, P.J. (1983a). Non-saponifiable lipids in cereals. In: *Lipids in Cereal Technology*. Barnes, P.J. ed. Academic Press: New York. pp 33–55.

Barnes, P.J. (1983b). Cereal tocopherols. In: *Progress in Cereal Chemistry and Technology, Proc. 7th World Cereal and Bread Congr.* Holas, J.; Kratochvil, J. eds. Elsevier: Amsterdam. pp 1095–1100.

Barnes, P.J. & Taylor, P.W. (1981). Gamma-tocopherol in Occurrence of avenanthramides and hydroxycinnamoyl- barley germ. *Phytochemistry*, 20,1753–1754.

Behall, K. M., Scholfield, D. J., Hallfrisch, J. G., & Liljeberg-Elmståhl, H. G. 2006. Consumption of both resistant starch and β-glucan improves postprandial plasma glucose and insulin in women. *Diabetes care*, 29(5):976–981.

Berski, W., Ptaszek, A., Ptaszek, P., Ziobro, R., Kowalski, G., Grzesik, M., & Achremowicz, B. J. C. P. (2011). Pasting and rheological properties of oat starch and its derivatives. *Carbohydrate Polymers*, 83(2), 665–671.

Bhatty, R.S. 1993. Non-malting uses of barley, In: *Barley: Chemistry and Technology*. MacGregor, A.W.; Bhatty, R.S. (Eds).

Bjørneboe, A., Bjørneboe, G. E. A., Bodd, E., Hagen, B. F., Kveseth, N., & Drevon, C. A. 1986. Transport and distribution of α-tocopherol in lymph, serum and liver cells in rats. *Biochimica et Biophysica Acta (BBA)-Molecular Cell Research*, *889*(3):310–315.

Borel, P., Pasquier, B., Armand, M., Tyssandier, V., Grolier, P., Alexandre-Gouabau, M. C., & Lairon, D. 2001. Processing of vitamin A and E in the human gastrointestinal tract. *American Journal of Physiology-Gastrointestinal and Liver Physiology*, 280(1):G95–G103.

Boz, H. 2015. Phenolic amides (avenanthramides) in oats-a review. *Czech Journal of Food Sciences*, 33(5): 399–404.

Bratt, K., Sunnerheim, K., Bryngelsson, S., Fagerlund, A., Engman, L., Andersson, R. E., & Dimberg, L. H. (2003). Avenanthramides in oats (Avena sativa L.) and structure– antioxidant activity relationships. *Journal of agricultural and food chemistry*, 51(3), 594–600.

Britton, G. E. O. R. G. E., & Hornero-Mendez, D. A. M. A. S. O. (1996). Carotenoids and colour in fruit and vegetables. In *Proceedings-Phytochemical Society of Europe* (Vol. 41, Pp. 11–28). Oxford University Press Inc.

Bryngelsson, S., Mannerstedt-Fogelfors, B., Kamal-Eldin, A., Andersson, R., & Dimberg, L. H. (2002a). Lipids and antioxidants in groats and hulls of Swedish oats (Avena sativa L). *Journal of the Science of Food and Agriculture*, 82(6), 606–614.

Bryngelsson, S., Dimberg, L. H., & Kamal-Eldin, A. (2002b). Effects of commercial processing on levels of antioxidants in oats (Avena sativa L.). *Journal of Agricultural and Food Chemistry*, 50(7), 1890–1896.

Burton, G. W., Ingold, K. U., Foster, D. O., Cheng, S. C., Webb, A., & Hughes, L. (1988). Comparison of free α-tocopherol and α-tocopheryl acetate as sources of vitamin E in rats and humans. *Lipids*, 23(9), 834–840.

Burton, G. W., & Traber, M. G. (1990). Vitamin E: antioxidant activity, biokinetics, and bioavailability. *Annual review of nutrition*, 10(1), 357–382.

Cai, S., Wang, O., Wu, W., Zhu, S., Zhou, F., Ji, B., & Cheng, Q. (2012). Comparative study of the effects of solid-state fermentation with three filamentous fungi on the total phenolics content (TPC), flavonoids, and antioxidant activities of subfractions from oats (Avena sativa L.). *Journal of Agricultural and Food Chemistry*, 60(1), 507–513.

Cardona, F., Andrés-Lacueva, C., Tulipani, S., Tinahones, F. J., & Queipo-Ortuño, M. I. (2013). Benefits of polyphenols on gut microbiota and implications in human health. *The Journal of nutritional biochemistry*, 24(8), 1415–1422.

Casterline, J. L., Oles, C. J., & Ku, Y. (1997). In vitro fermentation of various food fiber fractions. *Journal of Agricultural and Food Chemistry*, 45(7), 2463–2467.

Chan, G. C. F., Chan, W. K., & Sze, D. M. Y. (2009). The effects of β-glucan on human immune and cancer cells. *Journal of hematology & oncology*, 2(1), 1–11.

Chen, C. Y., Milbury, P. E., Kwak, H. K., Collins, F. W., Samuel, P., & Blumberg, J. B. (2004). Avenanthramides and phenolic acids from oats are bioavailable and act synergistically with vitamin C to enhance hamster and human LDL resistance to oxidation. *The Journal of nutrition*, 134(6), 1459–1466.

Chu, Y. F., Wise, M. L., Gulvady, A. A., Chang, T., Kendra, D. F., Van Klinken, B. J. W., & O'Shea, M. (2013). In vitro antioxidant capacity and anti-inflammatory activity of seven common oats. *Food chemistry*, 139(1-4), 426–431.

Clifford, M. N. (2000). Chlorogenic acids and other cinnamates–nature, occurrence, dietary burden, absorption and metabolism. *Journal of the Science of Food and Agriculture*, 80(7), 1033–1043.

Colleoni-Sirghie, M., Fulton, D. B., & White, P. J. (2003). Structural features of water soluble (1, 3)(1, 4)-β-D-glucans from high-β-glucan and traditional oat lines. *Carbohydrate Polymers*, 54(2), 237–249.

Collins, F. W. (1989). Oat phenolics: avenanthramides, novel substituted N-cinnamoylanthranilate alkaloids from oat groats and hulls. *Journal of Agricultural and Food Chemistry*, 37(1), 60–66.

Collins, F. W., & Mullin, W. J. (1988). High-performance liquid chromatographic determination of avenanthramides, N-aroylanthranilic acid alkaloids from oats. *Journal of Chromatography A*, 445, 363–370.

Collins, F. W., & Webster, W. H. (1986). Oat phenolics: structure, occurrence, and function. In *Oats: Chemistry and Technology. American Association of Cereal Chemists, Inc.* pp 227–295.

Cui, S., & Liu, R. H. (2013). Health Benefits of Oat Phytochemicals. *Oats Nutrition and Technology*, 171–194.

Daou, C., & Zhang, H. (2012). Oat beta-glucan: its role in health promotion and prevention of diseases. *Comprehensive reviews in food science and food safety*, 11(4), 355–365.

Day, A. J., Cañada, F. J., Díaz, J. C., Kroon, P. A., Mclauchlan, R., Faulds, C. B., ... & Williamson, G. (2000). Dietary flavonoid and isoflavone glycosides are hydrolysed by the lactase site of lactase phlorizin hydrolase. *FEBS letters*, *468*(2–3), 166–170.

Decker, E. A., Rose, D. J., & Stewart, D. (2014). Processing of oats and the impact of processing operations on nutrition and health benefits. *British Journal of Nutrition*, *112*(S2), S58–S64.

Di Carlo, G., Mascolo, N., Izzo, A. A., & Capasso, F. (1999). Flavonoids: old and new aspects of a class of natural therapeutic drugs. *Life sciences*, *65*(4), 337–353.

Dimberg, L. H., Molteberg, E. L., Solheim, R., & Frølich, W. (1996). Variation in oat groats due to variety, storage and heat treatment. I: Phenolic compounds. *Journal of Cereal Science*, *24*(3), 263–272.

Dimberg, L. H., Theander, O. L. O. F., & Lingnert, H. A. N. S. (1993). Avenanthramides: a group of phenolic antioxidants in oats. *Cereal Chemistry*, *70*(6), 637–641.

Doehlert, D. C., Simsek, S., Thavarajah, D., Thavarajah, P., & Ohm, J. B. (2013). Detailed composition analyses of diverse oat genotype kernels grown in different environments in North Dakota. *Cereal Chemistry*, *90*(6), 572–578.

Durkee, A. B., & Thivierge, P. A. (1977). Ferulic acid and other phenolics in oat seeds (Avena sativa L. var Hinoat). *Journal of Food Science*, *42*(2), 551–552.

Eggum, B. O., & Gullord, M. (1983). The nutritional quality of some oat varieties cultivated in Norway. *Plant Foods for Human Nutrition*, *32*(1), 67–73.

El-Gorab, M., & Underwood, B. A. (1973). Solubilization of β-carotene and retinol into aqueous solutions of mixed micelles. *Biochimica et Biophysica Acta (BBA)-Lipids and Lipid Metabolism*, *306*(1), 58–66.

El-Gorab, M. I., Underwood, B. A., & Loerch, J. D. (1975). The roles of bile salts in the uptake of beta-carotene and retinol by rat everted gut sacs. *Biochimica et Biophysica Acta*, *401*(2), 265–277.

Emmons, C. L., & Peterson, D. M. (2001). Antioxidant activity and phenolic content of oat as affected by cultivar and location. *Crop Science*, *41*(6), 1676–1681.

Erk, T., Hauser, J., Williamson, G., Renouf, M., Steiling, H., Dionisi, F., & Richling, E. (2014). Structure–and dose–absorption relationships of coffee polyphenols. *Biofactors*, *40*(1), 103–112.

FAO—Food and Agriculture Organization of the United Nations (2018). Available from FAOSTAT Statistics database-agriculture, Rome, Italy.

Flander, L., Salmenkallio-Marttila, M., Suortti, T., & Autio, K. (2007). Optimization of ingredients and baking process for improved wholemeal oat bread quality. *LWT-Food Science and Technology*, *40*(5), 860–870.

Gallo-Torres, H. E. (1970). Obligatory role of bile for the intestinal absorption of vitamin E. *Lipids*, *5*(4), 379–384.

Gani, A., Wani, S. M., Masoodi, F. A., & Hameed, G. (2012). Whole-grain cereal bioactive compounds and their health benefits: a review. *Journal of Food Science and Technology*, *3*(3), 146–56.

Heinemann, T., Kullak-Ublick, G. A., Pietruck, B., & Von Bergmann, K. (1991). Mechanisms of action of plant sterols on inhibition of cholesterol absorption. *European Journal of Clinical Pharmacology*, *40*(1), S59–S63.

Hollman, P. C. (2004). Absorption, bioavailability, and metabolism of flavonoids. *Pharmaceutical biology*, *42*(sup1), 74–83.

Ikeda, I., Tanaka, K., Sugano, M., Vahouny, G. V., & Gallo, L. L. (1988). Inhibition of cholesterol absorption in rats by plant sterols. *Journal of Lipid Research*, *29*(12), 1573–1582.

Irakli, M. N., Samanidou, V. F., Biliaderis, C. G., & Papadoyannis, I. N. (2012). Development and validation of an HPLC-method for determination of free and bound phenolic acids in cereals after solid-phase extraction. *Food Chemistry*, *134*(3), 1624–1632.

Jakobsen, K., Engberg, R.M., Andersen, J. O., Jensen, S. K., Henckel, P., Bertelsen, G., Skibsted, L. H., & Jensen, C. (1995). Supplementation of broiler diets all-rac-α- or a mixture of natural RRR-α-, Υ-,δ -tocopheryl acetate. 1. Effect on Vitamin E status of broilers in vivo and at slaughter. *Poultry Science*, 74, 1984–1994.

Jing, P., & Hu, X. (2012). Nutraceutical properties and health benefits of oats. *Cereals and pulses: Nutraceutical properties and health benefits*, 21–36.

Klose, C., Schehl, B. D., & Arendt, E. K. (2009). Fundamental study on protein changes taking place during malting of oats. *Journal of Cereal Science*, 49(1), 83–91

Lampe, J. W. & Chang, J. L. (2007). Interindividual differences in phytochemical metabolism and disposition. In *Seminars in cancer biology, Academic Press*. 17 (5), 347–353.

Landete, J. M. (2011). Ellagitannins, ellagic acid and their derived metabolites: a review about source, metabolism, functions and health. *Food research international, 44*(5), 1150–1160.

Lásztity, R., Berndorfer-Kraszner, E., Huszár, M., Inglett, G., & Munck, L. (1980). On the presence and distribution of some bioactive agents in oat varieties. *Cereals for food and beverages. Recent progress in cereal chemistry and technology,* 429–445.

Lásztity, R. (1996). Oat proteins. The chemistry of cereal proteins, 275–292.

Lazaridou, A., & Biliaderis, C. G. (2007). Molecular aspects of cereal β-glucan functionality: Physical properties, technological applications and physiological effects. *Journal of cereal science, 46*(2), 101–118.

Leiro, J., Alvarez, E., Arranz, J. A., Laguna, R., Uriarte, E., & Orallo, F. (2004). Effects of cis-resveratrol on inflammatory murine macrophages: antioxidant activity and down-regulation of inflammatory genes. *Journal of leukocyte biology, 75*(6), 1156–1165.

Li, X. P., Li, M. Y., Ling, A. J., Hu, X. Z., Ma, Z., Liu, L., & Li, Y. X. (2017). Effects of genotype and environment on avenanthramides and antioxidant activity of oats grown in northwestern China. *Journal of Cereal Science, 73,* 130–137.

Liu, L., Zubik, L., Collins, F. W., Marko, M., & Meydani, M. (2004). The antiatherogenic potential of oat phenolic compounds. *Atherosclerosis, 175*(1), 39–49.

Lv, N., Song, M. Y., Lee, Y. R., Choi, H. N., Kwon, K. B., Park, J. W., & Park, B. H. (2009). Dihydroavenanthramide D protects pancreatic β-cells from cytokine and streptozotocin toxicity. *Biochemical and biophysical research communications, 387*(1), 97–102.

Määttä, K., Lampi, A. M., Petterson, J., Fogelfors, B. M., Piironen, V., & Kamal-Eldin, A. (1999). Phytosterol content in seven oat cultivars grown at three locations in Sweden. *Journal of the Science of Food and Agriculture, 79*(7), 1021–1027.

Mälkki, Y., & Virtanen, E. (2001). Gastrointestinal effects of oat bran and oat gum: a review. *LWT-Food Science and Technology, 34*(6), 337–347.

Manach, C., Scalbert, A., Morand, C., Rémésy, C., & Jiménez, L. (2004). Polyphenols: food sources and bioavailability. *The American journal of clinical nutrition, 79*(5), 727–747.

Mattila, P., Pihlava, J. M., & Hellström, J. (2005). Contents of phenolic acids, alkyl-and alkenylresorcinols, and avenanthramides in commercial grain products. *Journal of Agricultural and Food Chemistry, 53*(21), 8290–8295.

Mazza, G. (1998). (Ed.), *Functional foods: biochemical and processing aspects* (Vol. 1). CRC Press.

Mellado-Ortega, E., & Hornero-Méndez, D. (2015). Carotenoids in cereals: an ancient resource with present and future applications. *Phytochemistry reviews, 14*(6), 873–890.

Meydani, M. (2009). Potential health benefits of avenanthramides of oats. *Nutrition Reviews, 67*(12), 731–735.

Miller, S. S., Fulcher, R. G., Sen, A., & Arnason, J. T. (1995). Oat endosperm cell walls: I. Isolation, composition, and comparison with other tissues. *Cereal chemistry, 72*(5), 421–427.

Mozaffarieh, M., Sacu, S., & Wedrich, A. (2003). The role of the carotenoids, lutein and zeaxanthin, in protecting against age-related macular degeneration: a review based on controversial evidence. *Nutrition journal, 2*(1), 1–8.

Nagao, A. (2014). Bioavailability of dietary carotenoids: Intestinal absorption and metabolism. *Japan Agricultural Research Quarterly: JARQ, 48*(4), 385–391.

Ogbe, R. J., Ochalefu, D. O., Mafulul, S. G., & Olaniru, O. B. (2015). A review on dietary phytosterols: Their occurrence, metabolism and health benefits. *Asian Journal of Plant Science & Research, 5*(4), 10–21.

Ovando-Martínez, M., Whitney, K., Reuhs, B. L., Doehlert, D. C., & Simsek, S. (2013). Effect of hydrothermal treatment on physicochemical and digestibility properties of oat starch. *Food Research International, 52*(1), 17–25.

Packer, L., & Fuchs, J. (1993). *Vitamin E in health and disease.* Marcel Dekker: New York.

Panfili, G., Fratianni, A., & Irano, M. (2004). Improved normal-phase high-performance liquid chromatography procedure for the determination of carotenoids in cereals. *Journal of Agricultural and Food Chemistry, 52*(21), 6373–6377.

Pearce, B. C., Parker, R. A., Deason, M. E., Qureshi, A. A., & Wright, J. K. (1992). Hypocholesterolemic activity of synthetic and natural tocotrienols. *Journal of medicinal chemistry, 35*(20), 3595–3606.

Peterson, D. M. (2001). Oat antioxidants. *Journal of cereal science, 33*(2), 115–129.

Peterson, D. M., Hahn, M. J., & Emmons, C. L. (2002). Oat avenanthramides exhibit antioxidant activities in vitro. *Food chemistry, 79*(4), 473–478.

Peterson, D. M., & Qureshi, A. A. (1993). Genotypes and environment effects on tocols of barley and oats. *Cereal Chemistry*, 70(2), 157–162.

Pundir, P., Wang, X., & Kulka, M. (2013). Asthma in the 21st Century—Unexpected Applications of Ancient Treatments. *Using Old Solutions to New Problems: Natural Drug Discovery in the 21st Century*, 277.

Punia, S., Sandhu, K. S., Dhull, S. B., Siroha, A. K., Purewal, S. S., Kaur, M., & Kidwai, M. K. (2020). Oat starch: Physico-chemical, morphological, rheological characteristics and its applications-A review. *International journal of biological macromolecules*, 154, 493–498.

Rasane, P., Jha, A., Sabikhi, L., Kumar, A., & Unnikrishnan, V. S. (2015). Nutritional advantages of oats and opportunities for its processing as value added foods-a review. *Journal of food science and technology*, 52(2), 662–675.

Rios, L. Y., Gonthier, M. P., Rémésy, C., Mila, I., Lapierre, C., Lazarus, S. A., ... & Scalbert, A. (2003). Chocolate intake increases urinary excretion of polyphenol-derived phenolic acids in healthy human subjects. *The American Journal of Clinical Nutrition*, 77(4):912–918.

Ryan, D., Kendall, M., & Robards, K. (2007). Bioactivity of oats as it relates to cardiovascular disease. *Nutrition research reviews*, 20(2), 147–162.

Salen, G., Ahrens, E. H., & Grundy, S. M. (1970). Metabolism of β-sitosterol in man. *The Journal of clinical investigation*, 49(5), 952–967.

Sandhu, K. S., Godara, P., Kaur, M., & Punia, S. (2017). Effect of toasting on physical, functional and antioxidant properties of flour from oat (Avena sativa L.) cultivars. *Journal of the Saudi Society of Agricultural Sciences*, 16(2), 197–203.

Schär, M. Y., Corona, G., Soycan, G., Dine, C., Kristek, A., Alsharif, S. N., & Spencer, J. P. (2018). Excretion of avenanthramides, phenolic acids and their major metabolites following intake of oat bran. *Molecular nutrition & food research*, 62(2), 1700499.

Serbinova, E., Kagan, V., Han, D., & Packer, L. (1991). Free radical recycling and intramembrane mobility in the antioxidant properties of alpha-tocopherol and alpha-tocotrienol. *Free Radical Biology and Medicine*, 10(5), 263–275.

Smet, E. D., Mensink, R. P., & Plat, J. (2012). Effects of plant sterols and stanols on intestinal cholesterol metabolism: suggested mechanisms from past to present. *Molecular nutrition & food research*, 56(7), 1058–1072.

Sosulski, F., Krygier, K., & Hogge, L. (1982). Free, esterified, and insoluble-bound phenolic acids. 3. Composition of phenolic acids in cereal and potato flours. *Journal of Agricultural and Food Chemistry*, 30(2), 337–340.

Stalmach, A., Steiling, H., Williamson, G., & Crozier, A. (2010). Bioavailability of chlorogenic acids following acute ingestion of coffee by humans with an ileostomy. *Archives of Biochemistry and Biophysics*, 501(1), 98–105.

Tian, L. L., & White, P. J. (1994). Antioxidant activity of oat extract in soybean and cottonseed oils. *Journal of the American Oil Chemists' Society*, 71(10), 1079–1086.

Trautwein, E. A., Duchateau, G. S., Lin, Y., Mel'nikov, S. M., Molhuizen, H. O., & Ntanios, F. Y. (2003). Proposed mechanisms of cholesterol-lowering action of plant sterols. *European Journal of Lipid Science and Technology*, 105(3–4), 171–185.

Tong, L. T., Liu, L. Y., Zhong, K., Yan, W. A. N. G., Guo, L. N., & Zhou, S. M. (2014). Effects of cultivar on phenolic content and antioxidant activity of naked oat in China. *Journal of Integrative Agriculture*, 13(8), 1809–1816.

Traber, M. G., Cohn, W., & Muller, D. P. R. (1993). Absorption, transport and delivery to tissues. *Vitamin E in health and disease*, 35–51.

Ulmius, M., Adapa, S., Önning, G., & Nilsson, L. (2012). Gastrointestinal conditions influence the solution behaviour of cereal β-glucans in vitro. *Food chemistry*, 130(3), 536–540.

Vogel, J. J. (1961). *Studies on the diet in relation to dental caries in the cotton rat...* University of Wisconsin–Madison.

Wang, L., Newman, R. K., Newman, C. W., Jackson, L. L., & Hofer, P. J. (1993). Tocotrienol and fatty acid composition of barley oil and their effects on lipid metabolism. *Plant Foods for Human Nutrition*, 43(1), 9–17.

Weidner, S., Amarowicz, R., Karamać, M., & Dąbrowski, G. (1999). Phenolic acids in caryopses of two cultivars of wheat, rye and triticale that display different resistance to pre-harvest sprouting. *European Food Research and Technology*, 210(2), 109–113.

Weihrauch, J. L., & Gardner, J. M. (1978). Sterol content of foods of plant origin. *Journal of the American Dietetic Association*, *73*(1), 39–47.

Wester, I. (2000). Cholesterol-lowering effect of plant sterols. *European Journal of Lipid Science and Technology*, *102*(1), 37–44.

Williamson, G., & Clifford, M. N. (2010). Colonic metabolites of berry polyphenols: the missing link to biological activity?. *British Journal of Nutrition*, *104*(S3), S48–S66.

Wise, M. L. (2011). Effect of chemical systemic acquired resistance elicitors on avenanthramide biosynthesis in oat (Avena sativa). *Journal of agricultural and food chemistry*, *59*(13), 7028–7038.

Wood, P. J. (1994). Evaluation of oat bran as a soluble fibre source. Characterization of oat β-glucan and its effects on glycemic response. *Carbohydrate Polymers*, *25*:331–336,

Zhu, F. (2017). Structures, properties, modifications, and uses of oat starch. *Food chemistry*, *229*, 329–340.

6

Postharvest Processing of Foxtail Millet and its Potential as an Alternative Protein Source

Nitya Sharma and J. K. Sahu
Centre for Rural Development and Technology, Indian Institute of Technology, New Delhi, India

CONTENTS

6.1 Introduction

Foxtail millet, also known as Italian or German millet, is one of the oldest cultivated crops with a history going back approximately 8,000 years in areas like India, Africa, and China. Due to its ease of cultivation (especially in drier and cooler regions) and higher yield as compared to other millet varieties, foxtail millet has been identified as a major millet in terms of its worldwide production. It grows about 2–5 feet tall in sandy and loamy soils and produces about 1 ton of forage on 2.5 inches of moisture. Foxtail millet has a short harvesting period of 75–90 days, which includes generation time from planting to flowering (5–8 weeks) and planting to seed maturity (8–15 weeks), producing hundreds of seeds per inflorescence. Further, the morphology of foxtail millet grains comprises 13.5% (w/w) of husk and 1.5–2% (w/w) of bran and germ, where in the digestible and nondigestible portion constitutes about 79% and 21%, respectively (Sharma & Niranjan, 2018).

Owing to the presence of a pertinent amount of nutritional components and various health benefits (Figure 6.1), foxtail millet has attracted a lot of attention, with research focusing on exploiting its unique agronomic features and understanding its role in environmental security and potential economic benefit the crop can offer farmers. The postharvest research focus for this crop has mainly been on suitable processing techniques, the effect of pretreatment and processing operations on nutrient and antinutrient content, flour functionality, and the development of new food products, as well as their packaging and storability. To understand the effect of these mechanism on human health, characterization and comprehensive analysis of the foxtail millet constituents must be explored.

Extensive research has been carried out on the molecular studies of foxtail millet starch, as it majorly contributes to the postprocessing consumer acceptance for foxtail millet products. The least explored foxtail millet protein is not only the second major component of the foxtail millet flour, but also contains a rich content of EAAs, including methionine. This protein has shown to have therapeutic functions as an essential dietary component (Sachdev, Goomer and Singh, 2020). Thus, it can be concluded that foxtail

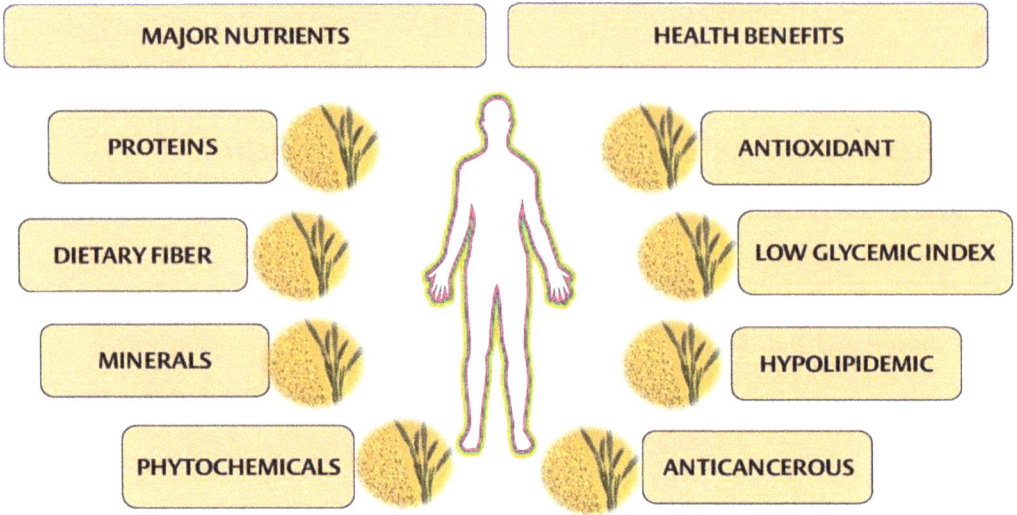

FIGURE 6.1   Major nutrients and health benefits of Foxtail millet.

millet protein isolates/concentrates can be used as a sustainable and low-cost plant protein alternative, which in combination with other plant proteins, can even mimic a typical composition of an animal-protein source.

Foxtail millet is a tractable model crop due to its comparatively short growing cycle and excellent abiotic stress tolerance characteristics. Thus, with the occurrence of adversities in the environment and modern agriculture, crops like foxtail millet need to be introduced and studied extensively for their potential and response to various processing conditions. This chapter discusses foxtail millet grains under three main themes: its nutritional and medicinal profile, the effect of various postharvest processes on its constituents and properties, and its potential, utilization, and importance as an alternative protein source.

6.2 Nutritional Value and Pharmacological Uses of Foxtail Millets

Foxtail millet grains are one of the most digestible and nonallergenic grains available with a significant potential for improving human health. Table 6.1 shows the complete nutritional profile of foxtail millet grains. The proximate composition of foxtail millet grains comprises 10–12% protein, 4.7–6.6% fat, 60.6–64.5% carbohydrates, and 2.29–2.7% lysine. Here, carbohydrate content accounts for 75% of the total weight of nutritional components, consisting of reducing sugar, starch, cellulose, and other minor components, with starch being the main form of carbohydrate. Starch from underutilized sources (millets) are suitable for numerous food and nonfood industrial applications (Punia et al., 2021). Further, foxtail millet is a rich source of nutrients with a wide range of amino acids present in it. It constitutes eight EAAs, including isoleucine, leucine, phenylalanine, threonine, and tryptophan, along with a varying content of lysine, methionine, and valine. Foxtail millet grains are also characterized by substantial amounts of minerals and fatty acids, including, unsaturated fatty acids like linoleic, oleic, and linolenic acids and saturated fatty acids like palmitic, and stearic acids. The foxtail millet bran accounts for 67% of linoleic acid, and its yellow pigment (content in the range of 5.4–19.6 mg/kg) is mainly due to the presence of zeaxanthin, cryptoxanthin, and xanthophyl. Although, these pigments are highly sensitive to light and acids, they are also highly thermostable and resistant to oxidoreduction. Verma et al. (2020) has also reported the presence of antioxidants like carotenoids (173 µg/100 g) and tocopherol (˜1.3 mg/100 g) in foxtail millet flour, with a total antioxidant capacity of 5.0 mM TE/g. According to the review, vitamins also form an exceptional component of foxtail millet grains, with substantial amounts of vitamin A (1.9 mg/kg), which is approximately twice as found in maize; vitamin B1 (5.7 mg/kg), which is around 1.7 times as in

TABLE 6.1

Nutrient Profile of Foxtail Millet

Carbohydrates (g 100 g^{-1})	60.9	Vitamin profile (mg 100 g^{-1}):	
Protein (g 100 g^{-1})	12.3	Thiamine	0.59
Fat (g 100 g^{-1})	4.3	Niacin	3.2
Crude fibre (g 100 g^{-1})	8.0	Riboflavin	0.11
Mineral matter (g 100 g^{-1})	3.3	Vit A (carotene)	32.0
Calcium (mg 100 g^{-1})	31	Folic acid	15.0
Phosphorus (mg 100 g^{-1})	290	Vit B5	0.82
Iron (mg 100 g^{-1})	2.8	Vit E	31.0
Amylose (%)	17.5	Essential amino acids (mg g^{-1} of N):	
Amylopectin (%)	82.5	Arginine	220
Micronutrient profile (mg 100 g^{-1}):		Histidine	130
Magnesium	81	Lysine	140
Sodium	4.6	Tryptophan	60
Potassium	250	Phenyl alanine	420
Copper	1.40	Methionine	180
Manganese	0.60	Cystine	100
Zinc	2.4	Threonine	190
Sulphur	171	Leucine	1040
Chloride	37	Isoleucine	480
Molybdenum	0.070	Valine	430

rice and maize; and vitamin E (43.5 mg/kg) and folic acid (2.37 mg/kg), which are also higher than rice, wheat, and maize (Verma et al., 2020).

Although, the antinutrients in foxtail millet grains, such as phytic acid and tannin, can be easily tailored or reduced to negligible levels by using suitable processing methods, they have shown to possess various medicinal uses on consumption. For instance, it helps to improve the appetite, acts as a good astringent, is a diuretic, and helps to improve the action of the gall bladder. Apart from this, foxtail millet grains have a low glycemic index, which helps in controlling blood sugar levels and cholesterol. Its high levels of resistant starch and antioxidants also add the power to reduce inflammation and potentially promotes anticancer, antiaging, and other benefits. Since it is naturally gluten free, thus it also improves the overall digestive health.

6.3 Postharvest Processing of Foxtail Millet

Post-harvest processing of foxtail millets has been proven to be an effective method in reducing phytochemicals with antinutritional effects and improving the bioavailability of nutrients. Apart from conventional processing techniques like de-hulling, milling, steaming, cooking, germination, and malting, various advanced processing techniques like extrusion and high-pressure processing have been used to optimize the properties of foxtail millet grains. The effect of such postharvest processing methods on the various constituents of foxtail millet grains has been extensively discussed by Sharma and Niranjan (2018). Thus, this chapter focusses on some significant recent advances (Table 6.2). This section will discuss the postharvest processing methods used for foxtail millet grains.

6.3.1 Conventional Postharvest Processing

Foxtail millet are small seeded grains with a characteristic thick pericarp layer that constitutes the inedible portion of grains. The foxtail millet husk is a distinct entity that not only poses a major challenge in primary processing of foxtail millets, but also discourages its consumption. Therefore, de-hulling and

TABLE 6.2

Postharvest Processing Effects on the Constituents of Foxtail Millet (Recent Advances)

Process	Process Variables	Parameters	Effect	Reference
De-hulling and milling	*De-hulling*: Break rolls of a laboratory scale mill were adjusted to separate the husk from endosperm with little damage to husk. *Milling*: Separated endosperm was milled on reduction rolls to produce flour fractions: C1, C2, C3 = endospermic fractions BDR = bran duster flour SRF = C1+C2+C3+BDR Control = De-husked grains pulverized to 250µ size	Nutrient composition, carbohydrate digestibility, protein digestibility, total starch, pasting profile, swelling power and solubility index, total carotenoids, antioxidant activity	Nutritional quality of endospermic fractions was comparable to control flour. The BDR fraction contained the highest protein, dietary fiber, minerals such as iron, phytic acid, total phenolic compound (TPC), and carotenoid contents. Control had the highest antioxidant activity. Carbohydrate and protein digestibility was highest for C3 and C2 fractions, respectively. The swelling power and solubility index was highest for C1 and control sample at 96°C, respectively. The BDR fraction showed the highest gelatinization temperature. C1 fraction showed the highest peak viscosity.	Kumar et al., 2016
Milling	Pulverized in mixer and passed through 0.88 mm sieve.	Proximate composition, rapid visco-analyzer (RVA) pasting profile, thermal properties.	Protein, fat, crude fiber, carbohydrate, and ash content were 11.04%, 3.86%, 6.61%, 63.21% and 3.3%, respectively. Peak viscosity was ~2270 cP. Gelatinization temperatures T_o, T_p & T_c were 73.52°C, 78.01°C, and 91.40°C, respectively.	Kharat et al., 2019
Dehulling, steaming and cooking	*De-hulling*: Passed through de-huller of 2 kW three times removing 90% of hull. *Steaming*: 100°C for 30 minutes with 1:5 (w/w) seed and water ratio. *Cooking*: 100°C for 30 minutes with 1:20 (w/w) seed and water ratio.	Total phenolic and flavonoid content, phenolic acid, phytic acid content, antioxidant activity	De-hulling decreased total phenolic content (TPC) and increased total flavonoid content (TFC). Steaming and cooking further decreased TPC and TFC. Cooking increased TPC, phytic acid, and cinnamic acid content with higher radical scavenging capacity.	Zhang et al., 2017
Cooking	*Roasting*: 150°C for 35 minutes in an Isotemp forced-air oven. *Boiling*: 1:10 roasted millet and water ratio was boiled for 30 minutes. *Freeze drying*: Frozen at −40°C for 24 hours, then placed in vacuum freeze-dryer at −40°C for 48 hours.	Volatile aroma composition, sensory properties	57 aroma compounds were identified. Pyrazines and several unsaturated aldehydes were observed during roasting and boiling, respectively. Freeze-drying after boiling decreased complexity of volatile compounds. Correlation of odor-active compounds with sensory analysis indicated that pyrazine and dienal compounds were responsible for "popcorn-like" and "boiled rice" odors, respectively.	Bi et al., 2019

(Continued)

TABLE 6.2 (Continued)

Process	Process Variables	Parameters	Effect	Reference
Germination and heat treatment	*Washing and Soaking*: 12 hours at room temperature. *Germination*: 30°C for 3 days at 90% relative humidity (RH) in climate incubator. *Drying*: 40°C for 24 hours. *Grinding*: In lab grinder and passed through 40-mesh sieve. *Boiling*: 1:5 flour and water ratio was boiled for 30 minutes and drained before drying. *Microwave treatment*: Ground flour was evenly spread to a 0.5 cm thickness in a plastic disc and heated using a household microwave oven at 100 W for 30 minutes.	Amino acid profile, *in vitro* gastrointestinal digestion of protein Isolate and digest fractionation, peptide and protein content, chemical antioxidant capacity, chemical anti-inflammatory activities, cell based antioxidant activity	Protein digest fractions containing low-molecular weight peptides (<3 kDa) and most hydrophobic subfractions (F4) abundant in random coil structures were responsible for antioxidant and anti-inflammatory bioactivity. Seven novel peptides were identified from F4 subfractions derived from boiled germinated millet. All seven peptides reduced reactive oxygen species production, increased glutathione content, and superoxide dismutase activity in Caco-2 cells.	Hu et al., 2020
High-pressure soaking and germination	25 g grains were suspended in 150 ml of deionized water and sealed in polyethylene pouches (3 cm × 20 cm). High pressure soaking was done at three levels of pressure (200 MPa, 400 MPa, and 600 MPa), four levels of temperature (20°C, 40°C, 60°C, and 80°C) & four levels of time (30, 60, 90, and 120 minutes).	Absorption, gelatinization and biochemical properties	Increasing pressure and temperature helped in attaining higher moisture content by increasing water diffusion rates until starch was partially gelatinized. Germination and high-pressure processing improved nutritional quality, including protein content, total phenolic content, and antioxidant activity, and decreased antinutritional (phytic acid and tannin content) content. The optimum conditions for high-pressure soaking were 400 MPa at 60°C for 60 minutes.	Sharma et al., 2018
Germination and high-pressure processing	*Germination*: Grains were soaked in deionized water for 15 hours and then placed in a controlled-environment growth cabinet for 72 hours at 25 ± 2°C & 50% RH. *High-pressure soaking*: Temperature-controlled pressure vessel system was used with 400 MPa pressures for 60 minutes at 60°C.	Functional and moisture sorption properties	Germinated grains when subjected to high-pressure soaking had higher values of porosity, occluded air content, water and oil absorption capacity, dispersibility, swelling capacity, wettability, flowability, gelatinization enthalpy, emulsification activity, and stability; and lower values of bulk density, insolubility index, foaming capacity, and stability. Bound water % decreased for nongerminated flour increased for germinated flour, with increase in temperature, confirming decrease in availability of active sites.	Sharma et al., 2018a

(Continued)

TABLE 6.2 (Continued)

Process	Process Variables	Parameters	Effect	Reference
Drying	*Hot-air drying (HAD)*: 95°C for 2 h. *Microwave drying (MD)*: 350 W, 490 W, and 700 W at 2450 Hz. *Freeze drying (FD)*: Frozen at −80°C and placed in a vacuum freeze dryer at −40°C to −45°C for 72 hours.	Rehydration rate, cellular structure, paste viscosity, degree of starch crystallization, % relative crystallinity, sensory properties	FD maintained inherent morphology, volume and porous nature of cooked millet with high rehydration capacity. MD and HAD led to formation of more V-type crystalline structure of amylose-lipid complexes, affecting the swelling capacity of starch.	Wang, Chen & Guo, 2017
Extrusion	Corotating twin-screw extruder with L:D ratio of 8:1 and screw diameter of 35 mm, pitch of 60 mm and cylindrical die of 3 mm diameter was used. Feed rate was 4 kg h^{-1} and cutter speed was set at 150 rpm. Feed moisture (X_1) levels were 15%, 17.5%, and 20% (w.b.), temperature (X_2) levels were 110°C, 120°C, and 130°C, and screw speed (X_3) levels were 300 rpm, 350 rpm, and 400 rpm.	Specific mechanical energy (SME), physical properties, cellular structure	Minimum SME was 22.14 at 130°C, 17.5% feed moisture, and 300 screw rpm. Maximum was 34.14 kJ kg^{-1} at 110°C, 17.5% feed moisture, and 400 screw rpm. Extrudates had an expansion ratio of 4.41 at 130°C, 17.5% feed moisture, and 400 screw rpm. Bulk density range was 71.2–172.6 kg/m^3, water absorption index, and water solubility index ranges were 3.51–4.18 (g/g) and 9.37–18.54%, respectively. Cellular structures of extrudates showed higher relative porosity in both core and outer layer with numerous open cells separated by a thin wall.	Kharat et al., 2019

pulverization/grinding/milling is the foremost postharvest process to make foxtail millet grains consumable. Much effort is being put forth by researchers to evaluate and further develop efficient primary processing machineries for small millets like foxtail millets. For instance, Durairaj et al. (2019) developed a double-chamber centrifugal de-huller, by assessing an existing set of primary processing machineries, including a grader, abrasive-roller de-huller, and de-stoner. They identifed subsequent areas of improvement to enhance the efficiency and utility of the processing machines, along with the quality of output obtained in small millets like foxtail millets. The developed centrifugal de-huller showed the following improvements: (1) 10% increased efficiency, (2) 4–5% fewer broken grains, (3) improved quality in terms of vitamins, minerals and soluble fiber due to better retention of the major bran and endosperm, (4) a versatile scale of operation ranging between 50–500 kg/h that meets requirements at the village through small- and medium-enterprise levels, and (5) found suitable for de-hulling all types of small millets. Similarly, a performance evaluation was done for millet processing machinery established at the Millet Processing Centre in the tribal area of Tamil Nadu (India) (Ambrose et al., 2017). The de-stoner, de-huller and polisher for foxtail millet were found to be 90%, 86%, and 86% efficient, respectively, with a processing charge of Rs. 6 per kg, leaving approximately 86% savings in cost to the farmers of the area.

Foxtail millet, as compared to other millet varieties like finger and pearl millet, is the richest source of protein, crude fiber, and minerals with values of 11.04%, 6.61%, and 3.3%, respectively (Kharat et al., 2019). These nutrients are unevenly distributed within the millet morphology, which creates a space for the preparation of different nutrient-rich fractions from millet. Kumar et al. (2016) explored this possibility of producing different biocomponent-rich foxtail millet fractions using a roller mill and bypassing the de-husking step. The nutrient evaluation of foxtail millet flour fractions indicated that the bran duster flour (BDR) had the highest protein (25.36 g/100 g), total fiber (13.12 g/100 g), and iron content (17.40 g/100 g), and the lowest carbohydrate and protein digestibility of 75.73% and 80.78%, respectively. In terms of functional properties, fractions with maximum foxtail millet endospermic fractions showed, (1) the highest swelling power and solubility index, which further increased by 3.6 times after cooking, and (2) the lowest gelatinization temperatures, due to the absence of bran fractions. As per Kharat et al. (2019), on comparing the pasting properties of different millet starches, the foxtail millet starch stood comparable to hard wheat flour but with better paste stability. Furthermore, the nutraceutical potential including phytic acid and total carotenoid content was highest for bran-rich fractions, with a value of 0.92 g/100 g and 3.89 µg/g, respectively. The total phenolic content was also highest for the bran-rich fractions (1.25 mg GA E/g) (Kumar et al., 2016), with the highest amount of cinnamic acid (Zhang et al., 2017). While the antioxidant activity measured in terms of ferric reducing antioxidant power (FRAP) and 2,2-diphenyl-1-picryl-hydrazyl-hydrate (DPPH) activity was highest for the whole foxtail millet grain flour, i.e., 80.0 mM Fe^{2+}equiv/g and 1.25 mg GA E/g, respectively (Kumar et al., 2016), it decreased with processing in the following order: dehulling>cooking>steaming (Zhang et al., 2017).

Cooking of foxtail millet grain flour, that included roasting, boiling, and freeze-drying also had a significant impact on its aroma profile. Bi et al. (2019) revealed that roasting significantly increased the contents of pyrazines, boiling caused an increase in the contents of unsaturated aldehydes, alcohol, and benzene derivatives, and freeze-drying after boiling reduced the contents of volatile compounds, thus decreasing the complexity of flavors. Further, on correlating the odor-active profile data with descriptive sensory analysis, it was established that the compound 2-ethyl-3,5-dimethyl pyrazine contributed "popcorn-like," "boiled beans," and "smoky" aroma descriptions, while (E,E)-2,4-decadienal was responsible for a "boiled rice" aroma. Apart from imparting peculiar odors, *in vitro* gastrointestinal digestion of cooked (germinated and boiled) foxtail millet grains synthesized bioactive peptides with a potential to modulate oxidative stress and inflammation (Hu et al., 2020).

Thus, foxtail millet can be conveniently consumed as a rice alternative after de-husking and cooking or can be converted to different nutrient-rich flour fractions without de-husking and separation to potentially be used to prepare designer foods.

6.3.2 Advanced Postharvest Processing

The exceptional physicochemical and nutritional properties of foxtail millet grains make them a suitable candidate for large-scale utilization and consumption. This may involve processing of foxtail millet grains using conventional as well as advanced methods like hot-air drying, freeze-drying, microwave treatment, high-pressure processing, and extrusion (Table 6.2). These advanced postharvest processing methods when applied either alone or in combination with convention processing methods, have the potential to release bioactive compounds, inactivate or optimize antinutritional factors, modulate amino acid composition, and increase nutrient bioaccessibility. For instance, Hu et al. (2020) combined the method of germination and microwave treatment to process foxtail millet grains and found that *in vitro* gastrointestinal digestion of processed grains synthesized seven peptides that significantly reduced the reactive oxygen species production and increased glutathione content and superoxide dismutase activity in Caco-2 cells, along with two other peptides that reduced nitric oxide and tumor necrosis factor-α to 42.29% and 44.07%, respectively, and interleukin-6 to 56.59% and 43.45%, respectively. Similarly, Sharma et al. (2018, 2018a) combined the method of germination with high-pressure processing, and studied its effect on water absorption, gelatinization, biochemical, functional, and moisture sorption properties.

The study conducted by Sharma et al. (2018) revealed that all operational parameters of high-pressure treatment (i.e. pressure, temperature, and soaking time) had a significant effect on the quality of germinated as well as nongerminated foxtail millet grains, with a predominant effect of pressure and temperature. The foxtail millet grains treated at an optimum condition of 400 MPa pressure, 60°C temperature and 60 minutes of soaking time not only attained higher moisture content and partially gelatinized starch due to the increase in water diffusion rates, but also achieved significant improvements in the nutritional quality of the flour, protein content, total phenolic content, and antioxidant activity, along with a significant decrease in the antinutritional (phytic acid and tannin content) content. Furthermore, Sharma et al. (2018a) confirmed that combining germination and high-pressure processing significantly improved the functional properties of foxtail millet flour, including porosity, occluded air content, water and oil absorption capacity, dispersibility, swelling capacity, wettability, flowability, gelatinization enthalpy, emulsification activity, and stability. It also affected the sorption properties of the flour, where the moisture sorption isotherms showed a sigmoidal type II shape fitting satisfactorily to the GAB model, and the percentage of bound water increased with temperature, thus, confirming an increase in the availability of active sites with total sorption ability. The study also concluded that a correlation between moisture sorption, thermodynamic properties, spreading pressures, and glass transition temperatures can obtain conditions for the maximum stability of foxtail millet flour during storage.

Foxtail millet is a potential candidate for healthy instant foods like dried ready-to-eat cereal products. For this category of food products, the method and conditions of drying play an important role to deliver the required quality. The traditional convective hot-air drying has usually led to quality deterioration, long rehydration times, and inferior eating quality. Thus, Wang et al. (2017) analyzed the effect of freeze-drying and microwave-drying techniques as an alternative on the eating quality of gelatinized millet (α-millet) products. On comparison, the freeze-drying method maintained the inherent morphology, volume, and porous nature of cooked foxtail millet so that it possessed high rehydration capacity, along with a minimum formation of more V-type crystalline structure (probably amylose-lipid complexes) that adversely affected the swelling capacity of starch therein. Likewise, extrusion cooking can be effectively used to produce similar dried ready-to-eat millet flour-based products, especially whole grain foxtail millet flour, which produced the best extrudate characteristics as compared to other millet varieties (Kharat et al., 2019).

Therefore, processing can be effectively used to produce various foxtail millet-based products with tailored properties as desired. However, there is a need to research this area further to enable the selection of the correct processing technique and explore the potential of foxtail millet grains as an ingredient for various heath benefitting foods.

6.4 Potential of Foxtail Millet as an Alternative Protein Source

Meat is a high biological value source of protein and has been consumed by humans since the prehistoric era. Exploring the potential of foxtail millet as an alternative protein source can help to meet the higher protein requirements of an expanding world population, which is currently about 7.8 billion and is estimated by the United Nations to reach more than 9 billion by 2050 (www.gfi.org). As overall demand for meat increases, there is an urgent need to increase supply of protein from sustainable alternative sources. Moreover, it has been found that protein production from animals has a greater impact on the environment as compared to plant-based protein productions, like greenhouse gas emissions, excess use of land, water, and nitrogen, and loss of terrestrial and aquatic biodiversity. Other factors that add to the list of reasons to discourage animal-based protein production, include global poverty, animal welfare, and human health. In the present scenario, total protein consumption from animal protein accounts for a majority of 55–73%, with only a smaller contribution of about 24–39% of the protein intake coming from plant origin mainly derived from cereals (Kumar et al., 2017). Food producers, processors, technologists, and policy makers with a sustainability objective are hoping to see consumers make a shift toward plant-based meat-protein alternatives. Furthermore, plant-based protein sources are easily available and cheaper than the animal-based protein sources and can play an important role in food and nutrition security in underdeveloped and developing communities.

Albumin, globulin, cross-linked prolamin, b-prolamin, and glutelin form the major protein fractions of millets, the content of each varying with the variety of millets. Further, millet proteins have shown the presence of high quantity EAAs, including sulfur-containing amino acids such as cysteine and methionine. Bioactive peptides from millets are usually smaller than 10 kDa and may exist naturally or are derived from their native/precursor protein during hydrolysis by digestive or proteolytic enzymes. These bioactive peptides have a positive influence on body functions that exert various distinct biochemical effects on human health, such as antimicrobial, antioxidant, antihypertensive, ACE-inhibitory, antiproliferative/anticancer, and antidiabetic effects. Protein from foxtail millet grains has already been identified as an essential dietary component, however, the therapeutic functions extend the possibility of using this component for physiological benefits on humans. Sachdev et al. (2020) has anthologized diverse literature that suggests that foxtail millet protein fractions could be a receptacle of fabricated functional foods (e.g. extenders) and drug delivery systems with novel properties; for example, bioactive compounds encapsulated in nanosized protein particles. Foxtail millet can also be used as a sustainable and low-cost source of protein isolates/concentrates, which can mirror a typical composition of an animal protein source when combined with other plant- or animal-based protein sources. Studies reveal that processing foxtail millet causes either a nonsignificant change or a significant increase in the protein content, thus suggesting its potential as a valuable protein source (Sharma and Niranjan, 2018).

Various challenges need to be addressed to make foxtail millet protein fractions a commercially viable ingredient. These challenges may include the *in vivo* studies of the food products with alternative protein, along with the sensory and consumer studies. The consumers may be ready to shift from animal proteins to alternative plant proteins, but this shift needs to be strategized to make it permanent. Thus, improving the cooking quality and optimization of foxtail millet protein as an ingredient is necessary. Further, for achieving the desired quality of foxtail millet protein concentrate, various considerations must be taken into account, especially postprocessing technical aspects, such as protein content, amino acid composition, digestibility, bioavailability, allergenicity, toxicity due to chemicals/enzymes/microbes involved, and functionality at time of manufacturing (Sá et al., 2020). Employing a cost-effective alpha amylose (AA) fortification strategy may be a key point to improve the quality of foxtail millet protein that can be further utilized to address the problem of malnutrition. Thus, foxtail millet protein fractions can be effectively used to gradually replace animal protein foods from the human diet.

6.5 Conclusion

Foxtail millet is a crop recognized for its high nutrient value, climate compatibility, and potential to adapt to low-input agricultural practices. It is a versatile and important food source that has recently gained a lot of attention by the researchers. Owing to the yielding potential of foxtail millet grains in difficult environments, it is a cereal suitable to address food and nutrition security. Whole foxtail millet grains processed using advanced techniques like drying, roasting, germination, and fermentation can produce products that can be preferred over other grains due to its positive effect on nutritional value, health benefits, and palatability. The present literature gives a clear understanding of the effect of processing on various properties of foxtail millet grains, but further research and development, particularly on its potential as a protein alternative and establishing its pharmacological properties, is necessary to exploit its full potential.

REFERENCES

Ambrose, D. C., Annamalai, S. J. K., Naik, R., Dubey, A. K., & Chakraborthy, S. (2017). Performance studies on millet processing machinery for tribal livelihood promotion. *J Appl Nat Sci*, 9(3), 1796–1800.

Bi, S., Wang, A., Wang, Y., Xu, X., Luo, D., Shen, Q., & Wu, J. (2019). Effect of cooking on aroma profiles of Chinese foxtail millet (*Setaria italica*) and correlation with sensory quality. *Food Chemistry, 289*, 680–692.

Durairaj, M., Gurumurthy, G., Nachimuthu, V., Muniappan, K., & Balasubramanian, S. (2019). Dehulled small millets: The promising nutricereals for improving the nutrition of children. *Maternal & Child Nutrition, 15*, e12791.

Hu, S., Yuan, J., Gao, J., Wu, Y., Meng, X., Tong, P., & Chen, H. (2020). Antioxidant and Anti-Inflammatory Potential of Peptides Derived from In Vitro Gastrointestinal Digestion of Germinated and Heat-Treated Foxtail Millet (*Setaria italica*) Proteins. *Journal of Agricultural and Food Chemistry, 68*(35), 9415–9426.

Kharat, S., Medina-Meza, I. G., Kowalski, R. J., Hosamani, A., Ramachandra, C. T., Hiregoudar, S., & Ganjyal, G. M. (2019). Extrusion processing characteristics of whole grain flours of select major millets (foxtail, finger, and pearl). *Food and Bioproducts Processing, 114*, 60–71.

Kumar, H. V., Gattupalli, N., Babu, S. C., & Bhatia, A. (2020). Climate-SMART Small Millets (CSSM): A way to ensure sustainable nutritional security. In *Global Climate Change: Resilient and Smart Agriculture* (pp. 137–154). Springer, Singapore.

Kumar, K. V. P., Dharmaraj, U., Sakhare, S. D., & Inamdar, A. A. (2016). Flour functionality and nutritional characteristics of different roller milled streams of foxtail millet (*Setaria italica*). *LWT, 73*, 274–279.

Kumar, P., Chatli, M. K., Mehta, N., Singh, P., Malav, O. P., & Verma, A. K. (2017). Meat analogues: Health promising sustainable meat substitutes. *Critical Reviews in Food Science and Nutrition, 57*(5), 923–932.

Punia, S., Kumar, M., Siroha, A. K., Kennedy, J. F., Dhull, S. B., & Whiteside, W. S. (2021). Pearl millet grain as an emerging source of starch: A review on its structure, physicochemical properties, functionalization, and industrial applications. *Carbohydrate Polymers*, 117776.

Sá, A. G. A., Moreno, Y. M. F., & Carciofi, B. A. M. (2020). Plant proteins as high-quality nutritional source for human diet. *Trends in Food Science & Technology, 97*, 170–184.

Sachdev, N., Goomer, S., & Singh, L. R. (2020). Foxtail Millet: A potential crop to meet future demand scenario for alternative sustainable protein. *Journal of the Science of Food and Agriculture, 101*(3), 831–842. DOI: 10.1002/jsfa.10716.

Sharma, N., & Niranjan, K. (2018). Foxtail millet: Properties, processing, health benefits, and uses. *Food Reviews International, 34*(4), 329–363.

Sharma, N., Goyal, S. K., Alam, T., Fatma, S., & Niranjan, K. (2018a). Effect of germination on the functional and moisture sorption properties of high–pressure-processed foxtail millet grain flour. *Food and Bioprocess Technology, 11*(1), 209–222.

Sharma, N., Goyal, S. K., Alam, T., Fatma, S., Chaoruangrit, A., & Niranjan, K. (2018). Effect of high pressure soaking on water absorption, gelatinization, and biochemical properties of germinated and non-germinated foxtail millet grains. *Journal of Cereal Science, 83*, 162–170.

Verma, K. C., Joshi, N., Rana, A. S., & Bhatt, D. (2020). Quality parameters and medicinal uses of foxtail millet (*Setaria italica* L.): A review. *Journal of Pharmacognosy and Phytochemistry, 9*(4), 1036–1038.

Wang, R., Chen, C., & Guo, S. (2017). Effects of drying methods on starch crystallinity of gelatinized foxtail millet (α-millet) and its eating quality. *Journal of Food Engineering, 207*, 81–89.

Zhang, L., Li, J., Han, F., Ding, Z., & Fan, L. (2017). Effects of different processing methods on the antioxidant activity of 6 cultivars of foxtail millet. *Journal of Food Quality, 2017*.

7

Proso Millet Flour and Starch: Properties and Their Applications

Anil Kumar Siroha
Chaudhary Devi Lal University, Sirsa, India

Sneh Punia
Chaudhary Devi Lal University, Sirsa, India
Clemson University, Clemson, SC, USA

CONTENTS

7.1 Introduction

Proso millet (*Panicum miliaceum* L.) belongs to family *Poaceae* and is one of the oldest cultivated and first domesticated crops. It originated in about 1,000 years ago and was extensively cultivated in Asia, Australia, North America, Europe, and Africa (Yang et al., 2018). The worldwide production of millet is 31,019,370 tons and India is the top producer of millet (11,640,000 tons) followed by Niger (3,856,344 tons)

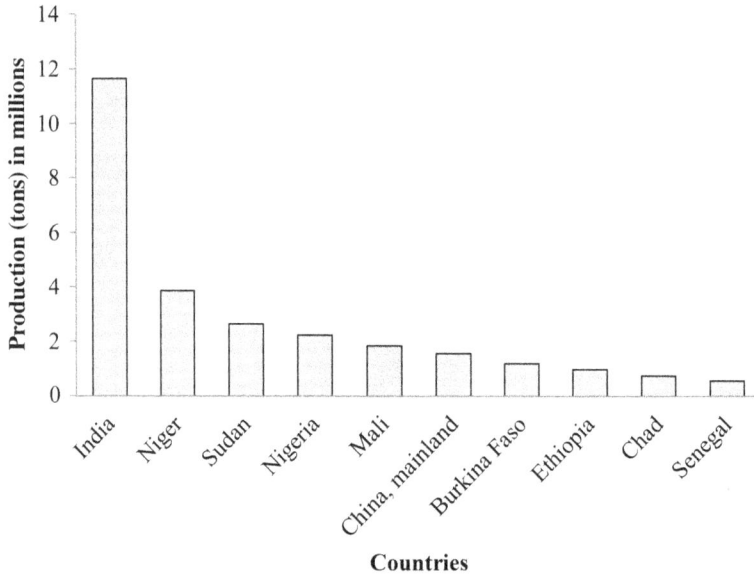

and Nigeria (2,240,744 tons) (Figure 7.1) (FAO, 2018). Among different millet varieties (finger, foxtail, little, and pearl millet), proso millet is the only millet variety grown on a commercial scale in the United States. The majority of this crop is used as bird food, but recently there has been increased interest in proso millet as a food for humans due to the rapidly growing gluten-free food market (McDonald et al., 2003).

Starch is the major nutrient component in proso millet grains (62–68%) followed by protein (10.6–13.6%), fat (3.3–3.8%), fiber (19.4%), and minerals (2.8–3.3%) (Yanez et al., 1991; Devisetti et al., 2014; Vali Pasha et al., 2018) (Table 7.1). The essential amino acid index (EAAI) of proso millet is reported to be higher (51%) than that of wheat (Kalinova & Moudry, 2006). Proso millet-based products have a lower glycemic index (GI) than corn, which means proso millet appears to be a good ingredient for producing low-GI products (McSweeney et al., 2017).

The use of grain flours in food formulations is dependent on flour functionality. The water absorption capacity plays an important role in the development of food products because it influences to a large extent their interaction with water. The protein/nitrogen solubility provides useful information on

TABLE 7.1

Nutritional Constituent of Proso Millets (g/100 g)

Protein	Carbohydrates/Starch	Fat	Crude Fiber	Mineral matter	Reference
8.3	65.9	1.4	9.0	2.6	FAO, 1995
11.0	56.1	3.5	9.0	3.6	Devi et al., 2014
13.6	70	3.8	19.4	3.3	Devisetti et al., 2014
8.3	65.0	1.4	15.0	2.6	Saha et al., 2016
3.7	54.0	0.12	9.7		Sharma et al., 2017
11.6	51.8	3.4		3.0	Vali Pasha et al., 2018
13.6	71.9	3.32		1.17	Gulati et al., 2018
11.68	61.80	4.18	16.61	2.72	Mustač et al., 2020
8.28	78.51	3.24		1.25	Kumar et al., 2020

effective utilization (Devisetti et al., 2014). The successful performance of flour as a food ingredient depends on functional characteristics, such as foaming, emulsification, gelation, water and oil absorption capacities, and the viscosity contributed to the end product (Adebowale & Lawal, 2004).

Starch is the major reserve carbohydrate found in the plants and it is major source of energy for human food (Punia et al., 2021). Starch is made up of two main components: Amylose and amylopectin. The physicochemical and functional properties of starch depend on the amylose and amylopectin starch content, the amylose/amylopectin ratio, and the inner structure of starch granules (Li et al., 2014). Proso millet starch has an A-type crystalline structure, which is mainly found in cereal starches. Starch granules have a bimodal distribution with two basic shapes and sizes, small spherical and large polygonal, and starch granule sizes range from 1.8–13.5 µm (Yanez et al., 1991). Peak viscosity (PV), hot paste viscosity (HPV) and cool paste viscosity (CPV) ranged from 2,215–3,787 cP, 1,444–2,177 cP, and 2,577–3,373 cP, respectively for different varieties of proso millet starch (Li et al., 2020). Differential scanning calorimeter (DSC) parameters showed that the onset temperature (T_o) of proso millet starches ranged from 73.1–76.4°C, peak temperature (T_p) ranged from 78.0–81.5°C, and their gelatinization enthalpy varied from 0.81 Jg^{-1} to −4.48 Jg^{-1}, respectively (Kim et al., 2012). Under different temperatures and pH values, stability of native starch varies unfavorably. To improve properties and functionality (viscosity, solubility, texture, and heat stability), native starches are modified.

Starch is frequently modified with physical, chemical, and enzymatic methods to enhance different functional properties due to the instability of starches during heating, shearing, acidity and the subsequent storage, starch is generally modified to increase the application of starches in food and nonfood industries (Lin et al., 2019). Physical modification has gained widespread acceptance because they do not need chemical reagents to modify starches. The required functional characteristics, such as heat tolerance, texture, adhesion, and solubility, may be acquired by using different modification procedures, which have many applications (Mohammed & Bin, 2020).

Processing is commonly done to enhance the quality of the grains by converting them into edible form (Jaybhaye et al., 2014). Processing of millets starts at the time of postharvesting with separation of the grains from the panicles and removal of impurities. To improve the nutritional, and sensorial characteristics, there are some processing technologies that are used in manufacturing of food products such as physical (milling, decortication, cooking, roasting, blanching, extrusion, and popping), chemical (acid treatment), and biological processes (fermentation, germination) (Saleh et al., 2013; Dias-Martins et al., 2018). Processing is commonly done to enhance the quality of the grains by converting them into edible form (Jaybhaye et al., 2014).

Cereal-based foods have been consumed as a staple food worldwide in various forms, *i.e.*, bakery products, noodles and pasta, snack foods, breakfast cereals, and others (Tebben et al., 2018; Xu et al., 2019). Millet grains are gluten-free, nonacid-forming (Ramashia et al., 2019), easy to digest with low GI (Chandrasekara et al., 2012). Its low-GI food property is reported to be a good choice for people with celiac disease (disease caused by gluten-containing cereal protein ingestion) and diabetes, as consumption of the grain assists in the regulation of blood glucose levels (Jideani & Jideani, 2011). Millets are now considered superior to wheat and rice as the whole grains and their products provide protein necessary for development, antioxidants, minerals, and vitamins (Taylor & Emmanbux, 2008; Purewal et al., 2019). This chapter will summarize the flour, starch, starch modification, and applications of proso millet in the food sector.

7.2 Nutritional Composition

Nutritional quality of food is a key element in maintaining human overall physical well-being (Saleh et al., 2013). Therefore, solving the problem of deep-rooted food insecurity and malnutrition, dietary quality should be taken into consideration (Singh & Raghuvanshi, 2012). Proso millet is rich in protein, mineral substances, and vitamins, and its nutritive parameters are comparable or better than common cereals. Moreover, the quantities of nutrients in millet are similar to the recommended ratio of protein, saccharide, and lipids (Kalinova & Moudry, 2006).

7.2.1 Carbohydrate Content

Millet carbohydrate in milled grains consists of free sugars (2–3%), nonstarchy polysaccharides (15–20%), and starch (60–75%) and among the free sugars, glucose, fructose, and sucrose are prominent (Chauhan et al., 2018). Polysaccharides constitute the main part (97.1%) of total saccharides (64.5%) in proso millet (Beaker, 1994). Sucrose (0.66% d.m.) and raffinose (0.08% d.m.) are present from soluble saccharides (FAO, 1995). The monosaccharides glucose, fructose, and galactose are found in traces (Becker and Lorenz, 1978). The disaccharide content was primarily sucrose at a proportion of 0.5–0.9% of the total dry weight, which is comparable to other cereal grains (Becker et al., 1977; Becker & Lorenz, 1978). The major edible carbohydrate of millet is starch. For nutritional purposes, starch is generally classified into rapidly digestible starch (RDS), slowly digestible starch (SDS), and resistant starch (RS), depending on the rate and extent of digestion (Englyst et al., 1992). Concerning the relationship between GI and the digestibility of starch, literature studies showed that GI is positively correlated with the amount of RDS, while SDS is the key component for low-GI cereal foods (Englyst et al., 1999). Therefore, increasing the amount of SDS and RS, or decreasing the amount of RDS in diets is essential to the development of low-GI food products for improved health (Luo & Zhang, 2018).

7.2.2 Protein

The protein content in proso millet is around 11% (dry basis) (Kalinova and Moudry, 2006) with higher amount of essential amino acids (leucine, isoleucine, methionine), as compared to wheat. Hence, the protein quality of proso (EAAI) is higher (51%) compared to wheat. Lorenz and Dilsaver (1980) conducted a study on proximate composition and nutritive value of the proso-millet flours and observed that millet flour had higher ash, crude fat, and protein content as compared to wheat flour. Proso millet protein contains limiting amino acids (lycine and threonine), which improved the biological value of the protein (Nishizawa et al., 1989), which elevated the plasma level of high-density lipoprotein (HDL) in rats and mice (Nishizawa et al., 1990; Nishizawa & Fudamoto, 1995). Millets contain a high amount of lecithin, which provides excellent support for nervous system health by helping to restore nerve cell function, regenerate myelin fiber, and intensify brain-cell metabolism (Habiyaremye et al., 2017). It is observed that different varieties of proso millet are particularly rich in glutamate (Glu), leucine (Leu), and alanine (Ala) (averaging 32.55 mg/g, 16.19 mg/g, and 12.25 mg/g, respectively) (Shen et al., 2018).

7.2.3 Lipid

Lipids are a minor constituent of millet grains and help in various reactions in the biological system. Lipid content of proso millet varied from 0.12–4.18 g/100 g (Table 7.1) with free (62.2%), bound (27.8%), and structural lipids (10.0%) in total lipids (Sridhar & Lakhshminarayana, 1994. In the free lipids, hydrocarbons, sterols esters, triglycerides, diglycerides, and fatty acids are present whereas bound lipids, monogalactosyl diglycerides, digalactosyl diglycerides, phosphatidyl, ethanol amine, phosphadyl serine, and phosphatidyl choline are described (Lorenz & Hwang, 1986). Linolenic acid and oleic acid are the two dominant fatty acids in different proso millet varieties, and their average contents were 61.74% and 22.16%, respectively (Shen et al., 2018). These results were consistent with Annor et al. (2015), who reported linolenic acid (65.5%) and oleic acid (20.0%) in millets.

7.2.4 Vitamins and Minerals

Millets are rich sources of vitamin E and B-complex vitamins (except vitamin B 12). Vitamin B—such as niacin, folacin, riboflavin, and thiamine—and phosphorus are present in millets and play a key role in energy synthesis in the body (Sarita & Singh, 2016). Millets are recognized nutritionally for being a good source of minerals including magnesium, manganese, and phosphorus (Shashi et al., 2007). Vali Pasha et al. (2018) reported mineral content of proso millet and observed calcium (0.18 g/kg), phosphorus (4.94 g/kg), sodium (0.58 g/kg), potassium (4.99 g/kg), and magnesium (2.53 g/kg), respectively. The main portion of mineral compounds is contained in the pericarp, aleurone layer, and germ. The mineral content of proso millet grain ranged from 1.5–4.2% (Jasovskij, 1987; Ravindran, 1991; Kalinva, 2002), which is higher than wheat (1.5–2%). Proso millet is poor in calcium content, but its seed contains higher amounts of phosphorus content and it is almost free from sodium (Kalinova, 2007).

7.2.5 Dietary Fiber

Dietary fiber refers to the nondigestible carbohydrate, which is impervious to enzymatic assimilation and absorption in the small intestine of humans. Dietary fibers are classified into two categories according to their water solubility: insoluble dietary fiber (IDF) and soluble dietary fiber (SDF) (Lattimer & Haub, 2010; Dhingra et al., 2012). Millets contain both insoluble and soluble dietary fiber and have comparable or even higher total dietary fiber than other cereals.

7.3 Flour

Proso millet is rich in protein, mineral substances, and vitamins, and its nutritive parameters are comparable or better than common cereals. Moreover, the quantities of nutrients in millet are very similar to the recommended ratio of protein, saccharide, and lipids (Kalinova and Moudry, 2006). Proso millet contains protein (9.4–14.80 g/100 g), fat (1.2–4.5 g/100 g), ash (0.6–3.3 g/100 g) and carbohydrates (70.0–74.0 g/100 g) with various other essential minerals and phytochemicals (Bagdi et al., 2011; Devisetti et al., 2014). The nutritional value of millet flours vary due to differences in geographical region, variety and growth conditions (Kaur et al., 2007). The high ash and high carbohydrate content of the flour indicates that these millets could serve as a significant source of minerals and energy for consumers (Kaushal et al., 2012).

Water absorption capacity (WAC) and oil absorption capacity (OAC) are important physicochemical properties (Elleuch et al., 2011). WAC and OAC of proso millet flour varied from 74.0–196.0 g/100 g and 63–95.5 ml/100 ml (Devisetti et al., 2014; Thilagavathi et al., 2015; Bora & Das, 2018; Kumar et al., 2020). WAC represents the ability of a product to associate with water under conditions where water is limited (Singh, 2001). The high WAC of flour may prove useful in products where viscosity is required, such as in soups and gravies (Kaushal et al., 2012). The oil-absorbing mechanism involves capillarity interaction, which allows the absorbed oil to be retained. Adebowale and Lawal (2004) reported that flours with good OAC are potentially useful in flavor retention, improvement of palatability, and extension of shelf life, particularly in bakery or meat products where fat absorption is desired.

Emulsion activity (EA), which represents the ability of flour to emulsify oil. Devisetti et al. (2014) observed EA of proso millet flour 4.9 ml/100 ml. The EA reflects the ability and capacity of a protein to aid in the formation of an emulsion and is related to the protein's ability to absorb oil and water in an emulsion (Du et al., 2014). Foaming capacity of proso millet flour varied from 0.20–8.7 ml/100 ml (Devisetti et al., 2014; Kumar et al., 2020). Foam formation and stability generally depend on the interfacial film formed by proteins, which keeps air bubbles in suspension and slows the rate of coalescence. Foaming properties are dependent on the proteins as well as on other components, such as carbohydrates (Ma et al., 2011).

The bulk density (BD) of proso millet flour varied from 698–830 g/l (Devisetti et al., 2014; Bora and Das, 2018; Kumar et al., 2020). The BD is used to determine the capacity of storage and transport, while the true density is useful to design proper separation equipment (Brooker et al., 1992; Kachru et al., 1994). The BD of flour plays an important role in weaning food formulation, that is, reducing the BD of the flour is probably helpful to the formulation of weaning foods (Milán-Carrillo et al., 2000).

7.4 Starch

Starch is the major storage polysaccharide of higher plants and it is deposited in partially crystalline granules varying in morphology and molecular structure between and within plant species (Blazek & Copeland, 2008). Starch is composed of two main components: amylose and amylopectin. Amylose is smaller with few branches, whereas amylopectin is larger with many branches. The branches in amylopectin are arranged in a clustered fashion. The tightly branched unit in the amylopectin is termed a building block, which has an internal chain length of 1–3 glucosyl residues (Pérez & Bertoft, 2010). Major factors influencing the physicochemical properties of starch are amylose content and particle size (Zhang et al., 2016). Bhat and Riar (2016) reported that shape and particle size indicated a considerable effect on the functional properties of starch due to variations in distributional pattern of starch granule.

7.4.1 Physicochemical Properties

Swelling power (SP) and solubility provide information about the degree of the interaction between starch chains within both the amorphous and crystalline zones (Singh et al., 2003). During heating of starch molecules in excess water, the crystalline structure of starch molecules are disrupted due to the breakdown of the hydrogen bond, and the exposed hydroxyl group of starch bonds with the water molecules, increasing swelling and the solubility power of the starch (Wani et al., 2012). SP and solubility of proso millet starch is observed 24.99% –27% and 8.5%–34.88%, respectively (Wu et al., 2014; Manjot & Adedeji, 2016). SP can be influenced strongly by the amylopectin molecular structure (Tester et al., 1993); the higher SP of proso millet starch can be used as thickener and binder agent.

7.4.2 Thermal Properties

When heated with sufficient water over a range of temperatures, granular starch undergoes an order-disorder phase transition termed gelatinization. Water uptake by the amorphous region, radial swelling of the granules, breakdown of the crystalline region with the disruption of double helices, and the leaching of starch molecules characterize gelatinization (Hoover, 2001). The cooking quality of starches are measured by its gelatinization, which has an important role in food processing as starches with a low gelatinization temperature showed better cooking quality (Waters et al., 2006). To obtain consistent results of starch gelatinization temperature and enthalpy of gelatinization (ΔHgel, the starch sample must contain at least two times (w/w) of water. Without a sufficient amount of water, the starch gelatinization peak broadens and shifts to a higher temperature (Donovan, 1979). Proso millet starch had an onset (T_o), peak (T_p), and conclusion (T_c) temperature of 66.81–72.7°C; 72.79–76.55°C, and 78.30–82.44°C, respectively (Table 7.2) (Yanez et al., 1991; Chao et al., 2014; Wen et al., 2014; Wu et al., 2014). Noda et al. (1998) postulated that the T_o, T_p, and T_c values were influenced by the molecular architecture of the crystalline region, which corresponds to the distribution of amylopectin shorter chains. The starch melting temperature range (T_c-T_o) gives an indication of the homogeneity and quality of the amylopectin crystals (Annor et al., 2014). A narrow melting range indicates amylopectin crystals with more homogeneous quality and uniform stability and vice versa (Ratnayake et al., 2001). ΔHgel is an indicator of the

TABLE 7.2

Thermal Properties of Proso Millet Starch

Starch:Water Ratio (w/w)	Heating Rate (°C/min)	Gelatinization Parameters				References
		T_o (°C)	Tp (°C)	Tc (°C)	ΔHgel (J/g)	
1:3	10	67.8–69.0	72.5–73.9	80.3–81.8	3.51–3.57	Yanez et al., 1991
1:3	10	73.1–76.4	78.0–81.5	79.3–86.0	−0.01 to −4.48	Kim et al., 2012
1:3	5	68.4	72.0		4.0	Annor et al., 2014
1:3	10	67.9–73.5	74.6–77.1	80.4–81.3	10.37–13.06	Chao et al., 2014
1:2	10	68.65	71.37	80.4	16.10	Sun et al., 2014
1:2	10	68.56	74.53	82.43	5.16	Wu et al., 2014
1:3	10	66.81–70.01	72.79–76.55	78.30–82.44	10.40–14.46	Wen et al., 2014
1:2	10	72.93	78.61	94.55	3.83	Singh & Adedeji, 2017
1:4	10	64.16	68.45	79.09	10.58	Li et al., 2018

loss of molecular double-helical order within the granule (Cooke & Gidley, 1992; Tester & Morrison, 1990). The lower ΔHgel suggests a lower percentage of organized arrangements or a lower stability of the crystals (Chiotelli & Meste, 2002).

7.4.3 Pasting Properties

Pasting properties provide important information about the cooking behavior of starches during heating and cooling cycles (Siroha et al., 2020). For determination of pasting properties, starch is heated in an excess of water under a constant shearing rate and the temperature profile is programmed according to the method chosen. Pasting properties are affected by water content, shear rate, temperature profile, starch type, and concentration of starch used for determination. A typical curve of pasting profile is shown in Figure 7.2. Chao et al. (2014) reported that waxy proso millet starches had higher average PV (3304 cP), trough viscosity (TV, 2,422 cP), and breakdown viscosity (BV, 882 cP), but lower setback viscosity (SV, 267 cP) and pasting temperature (PT, 63.65°C) than those of nonwaxy proso millet (PV 2,134 cP; TV, 1,646 cP; BV, 488 cP; SV, 1,102 cP; PT, 63.80°C) and maize (PV 2,059 cP; TV, 1,619 cP; BV, 440 cP; SV, 1,161 cP; PT, 63.80°C). PV, TV, BV, final viscosity (FV), and PT is observed 2,807–3,827 cP, 1,061– 2,122 cP, 1,705–1,746 cP, 2,694–2,876 cP and 57.40–76.03°C, respectively for proso millet starch

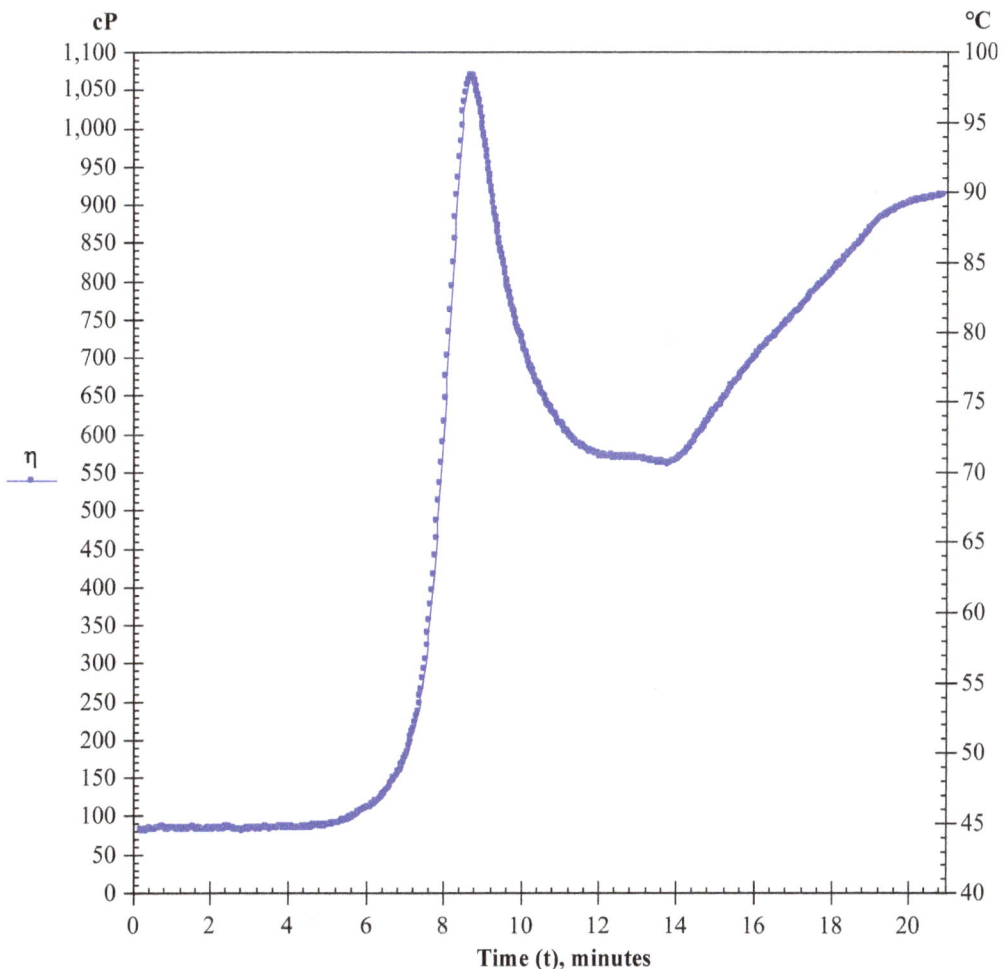

FIGURE 7.2   Pasting graph of millet starch.

TABLE 7.3

Pasting Properties of Proso Millet Starch

Starch (%)	Peak Viscosity (PV)	Breakdown Viscosity (BV)	SV	Unit	Pasting Temperature (PT, °C)	References
12	988–1,445	672–1,049	275–324	RVU	77.4–77.6	Kim et al., 2012
8	2059–3,515	440–967	197–1161	cP	63.60–63.80	Chao et al., 2014
12	2,822	1854	501	cP	76.0	Sun et al., 2014
8	219–457	79–240	115–201	BU		Wen et al., 2014
14	4.60	2.60	1.69	Pa.s	79.23	Singh &Adedeji, 2017
12	2,807	1,746	1,634	cP		Li et al., 2018
14	3,827	1,705	754	cP	76.03	Zheng et al., 2020

(Table 7.3) (Li et al., 2018; Zheng et al., 2020). Amylose content is believed to have a marked influence on the breakdown viscosity (a measure of the susceptibility of cooked starch granules to disintegration) and the setback viscosity, which is a measure of the recrystallization of gelatinized starch during cooling) (Lee et al., 1995). High amylose content has also been suggested as the major factor contributing to the nonexistence of a peak, a high stability during heating, or a high setback during cooling (Lii & Chang, 1981; Jin et al., 1994).

It is difficult to directly compare pasting results of proso millet starch with other commercial starch like corn, rice, or potato due to the different instruments used and experimental conditions employed for measuring pasting properties.

7.4.4 X-ray Diffraction Pattern

Starch granules possess a semi-crystalline structure corresponding to different polymorphic forms. X-ray diffraction (XRD) is used to study the presence and characteristics of that crystalline structure (Punia, 2020a). Arrangement of the amylopectin helices are supposed to be responsible for starch crystallinity, whereas amylose is associated to amorphous regions (Zobel, 1988; Singh et al., 2006). The crystalline parts of starch always show sharp peaks, whereas amorphous parts of starch are dispersive (Gernat et al., 1990). Starch granules had a semi-crystalline structure related to different polymorphic forms and based on this, starch can be differentiated into three types, namely A, B and C (Buléon et al., 1998). Proso millet starches exhibited double peaks at 17° and 18° and a single peak at 23°, a characteristic A-type crystallinity pattern shown by cereal starches (Kim et al., 2012; Sun et al., 2014; Wen et al., 2014). A-type corresponds to an orthorhombic structure while B-type corresponds to a hexagonal one. Water plays an essential role in the formation of the crystalline structures present in starch (Rodriguez-Garcia et al., 2021). In the A-type crystal, the double helices are closely packed into a monoclinic unit cell (with dimensions a = 20.83 Å, b = 11.45 Å, c = 10.58 Å, space group B2) containing eight water molecules (Popov et al., 2009). Degree of crystallinity is observed 23.69–45.2% (Kim et al., 2012; Wu et al., 2014; Wen et al., 2014). The differences in relative crystallinity among starches can be attributed to the following: (1) crystal size, (2) amount of crystalline regions (influenced by amylopectin content and amylopectin chain length), (3) orientation of the double helices within the crystalline domains, and (4) extent of interaction between double helices (Hoover & Ratnayake, 2002).

7.4.5 Morphological Properties

The size and shape of starch granules are important in relation to their technological properties, including the viscosity of their pastes (Wojeicchowski et al., 2018). Scanning electron microscopy (SEM) of proso millet starch granules show bimodal distribution, small spherical, large polygonal, rarely elliptical shapes with indentations on the surface, and smooth edges (Yanez et al., 1991; Kim et al., 2012; Chao et al., 2014; Annor et al., 2014). These pores are normal structures related to the genetic make-up of the source plant and are not artefacts of isolation, preparation, or observation techniques (Hoover et al., 1996). The physico-chemical, functional, and nutritional properties of starches are affected by the shape and size of the starch granules; larger granules are responsible for high paste viscosity while small granules had higher digestibility (Bello-Pérez et al., 2010). Starch granules range in size from 1.8–24 μm for proso millet starch (Yanez et al., 1991; Annor et al., 2014; Wu et al., 2014). The particle size of starch powder has significance in handling, packing, and product formulation and in various foods or other miscellaneous applications (Igathinathane et al., 2010). Starch granule size influences water absorption, solubility, and swelling (Hedayati et al., 2016). Small granules have a high surface area that can lead to high water-absorption capacity (Lindeboom et al., 2004). Agnes et al. (2017) reported small starch granules exhibited higher solubility and increased water absorption capacity. Smaller starch granules with a diameter similar to lipid micelles (approximately 2 μm) can be applied as fat mimetics (McClements et al., 2017).

7.5 Starch Modification

Regardless of starch key functional roles, native starch sometimes affects several quality characteristics of food products and renders instability of paste under shear, acid, or freezing conditions and poor paste clarity (Shaikh et al., 2015). Therefore, the applications of starch in various industries can be increased by adopting different physical, chemical, and enzymatic modification methods (Rafiq et al., 2016).

7.5.1 Physical Modification

To produce more natural food components, there is an interest to improve the properties of native starches without using chemicals for modification (Ortega-Ojeda & Eliasson, 2001). Chemicals are not required in physical modification of starch; therefore, physical modification is preferred. Different physical modification i.e. heat moisture treatment (HMT), autoclaving treatment (AT), and microwave treatment (MT) of proso millet starch is reported by Zheng et al. (2020). It is observed that after physical modifications, amylose content and resistant starch content increased while reverse is observed for PV, BV, SV, and PT as compared to native starch. X-ray diffractometry showed that the relative crystallinity of the HMT sample decreased by 20.88%, and the crystalline peaks disappeared from the AT and MT sample patterns. Singh and Adedeji (2017) also evaluated the properties of HMT-modified proso millet starch. Swelling power, solubility, PV, BV, FV, and SV decreased while gelatinization temperature increased after HMT. Modifications influenced the thermal and pasting properties, with low breakdown implying that starches were more stable during continued shearing and heating after modifications. Low SV after modification could enable the starches to be used in canned and frozen foods.

Ultra-high pressure (UHP) treatment (150 MPa, 300 MPa, 450 MPa, and 600 MPa for 15 minutes) for proso millet starch is reported by Li et al. (2018). The XRD patterns showed that the UHP at 600 MPa converted the A-type crystals to B-type. The DSC data revealed an increase in gelatinization temperatures and a decrease in gelatinization enthalpy after treatment at 150–450 MPa. Pasting property analysis showed that TV, FV, PT, and peak time were decreased; however, PV and BV were decreased with the increase in pressure up to 600 MPa. UHP treatment resulted in a higher swelling capacity and solubility at a relatively low temperature.

Sun et al. (2014) reported the effect of dry heat treatment (DHT) (130°C for 2 or 4 hours) on physico-chemical and structural properties of proso millet starch. PV and BV viscosity decrease while TV, SV, and FV increase after heat treatment. Increase in gelatinization temperature is observed as compared to native starch. The relative crystallinity decreased with longer heating time, and these changes may suggest a decrease in the crystalline areas due to DHT. The gel structure became more compact and the particles are plumper after DHT.

7.5.2 Chemical Modification

Chemical modification is a classical way to effectively improve the functionalities of starch. Native starches contain free hydroxyl groups in the 2, 3 and 6 carbons of the glucose molecule, making them highly reactive. This allows them to be modified by different chemical treatments and thus regulate their properties (Bao et al., 2003). Chemical modifications incorporate new functional groups without affecting the morphology or size distribution of the granules, therefore, making them fit for various industrial applications (Punia, 2020b).

In the industry, acid-modified starch is prepared with dilute hydrochloric acid (HCl) or sulfuric acid (H_2SO_4) at 25–55°C for various time periods (Lin et al., 2005). Acid modification of starch is a chemical modification process involving hydrolysis of starch using hydrochloric acid, which breaks the glycosidic linkages of α-glucan chains, changing the structure and characteristics of native starch (Hoover, 2000). The typical procedure for manufacture of acid-thinned starch involves treating a concentrated starch slurry at a temperature lower than the gelatinization temperature of the starch with mineral acid for a period of time (Wurzburg, 1986). Singh and Adedeji (2017) reported acid hydrolysis of proso millet starch. Starch was modified using HCl (0.14 mol/L) for 8 hours and 50°C; physicochemical as well as pasting and thermal properties were evaluated. Swelling power, onset temperature, and peak temperature are decreased while reverse is observed for solubility power. AM reduced the amylose content and water binding capacity of starch, and also improved the clarity of the paste. After modification, drastic changes are observed in the pasting properties of modified starches. Acid hydrolysis is widely used for production of starch gum candies, paper, cationic, and amphoteric starches (Wurzburg, 1986).

7.6 Processing

Technology used for converting the grain into an edible form and thereby enhancing its quality is known as processing. Processing of cereals and millets plays significant role during its utilization as food. Millet grains, before consumption, are usually processed (Adebiyi et al., 2018) to improve their edible, nutritional, and sensorial characteristics by using traditional processing techniques (Nazni & Devi, 2016). Negative changes in millet are not avoidable because industrial method of processing are not well developed compared to other cereal (Singh & Sarita, 2016). Physical (milling, decortication, cooking, roasting, blanching, extrusion and popping), chemical, and biological processes (fermentation, germination) are used for production of food (Dias-Martins et al., 2018).

7.6.1 Milling

Milling is the most common traditional processing method that converts dried and moistened cereal grains into flour by using wooden or stone mortar and pestle. Milling or grinding is normally practiced in developing countries mostly by women (Young, 1999). The milling process of grains consists of sorting, cleaning, hulling, branning, and kilning for further processing (Rasane et al., 2015). Millet grains contain greater proportions of husk and bran, requiring de-husking and de-branning prior to consumption. De-husking of millet in centrifugal sheller followed by de-branning in a huller yields the grain of satisfactory quality (Hadimani & Malleshi, 1993). Good quality of proso millet flour was obtained by de-hulling in a barley pearler with the grain then sent into Quadrumat Jr. mill (Lorenz et al., 1980). Decortication, milling, and sieving of millet grain are mostly carried out manually; therefore, there is a need for convenient and motorized milling technology so millet grain production can become a consistent source for industrial food uses at commercial scale (Saleh et al., 2013).

7.6.2 Decortication/De-hulling

The decortication process, also known as de-hulling, is the removal of the bran (pericarp and germ) of the millet grain, in order to promote improvements in the quality attributes, such as palatability, grain coloration, reduction of phytic acid, and fat content (Hama et al., 2011). Decortication had no effect on the protein and fat content of proso millet; however, it significantly decreased the content of crude fiber, dietary fiber, minerals, total phenol content and antioxidant capacity (Bagdi et al. (2011). Chandrasekara et al. (2012) reported the effect of de-hulling on antioxidant properties of proso millet. Total phenolic content, 2,2-diphenyl-1-picryl-hydrazyl-hydrate (DPPH) scavenging activity, hydroxyl radicals scavenging activity, superoxide radical scavenging capacity, hydrogen peroxide scavenging, and oxygen radical absorption capacity decreased after de-hulling. As phenolic compounds are found to be concentrated in the outermost layers, the bran fractions obtained as milling by-products may be used as a natural source of antioxidants and as a value-added product in the preparation of functional food ingredients and/or for enrichment of certain products (Kundgol et al., 2013).

7.6.3 Fermentation

Fermentation is a metabolic process that converts complex material into a simpler form with the help of microorganisms. It is the most effective and oldest method of processing and preserving foods. Nutritional enrichment by increasing protein content, mineral extractability, palatability, and decrease in cooking time have also been reported (Badau et al., 2005 Osman, 2011; Ranasalva & Visvanathan, 2014; Jay et al., 2005; Sasikumar, 2014). Wang et al. (2019) reported the fermentation conditions to improve the sensory quality of broomcorn millet sour porridge. The strains combination of *L. brevis* L1, *A. aceti* A1 and *S. cerevisiae* E4 (1:1:1, v/v/v) are used to ferment broomcorn millet. Optimum fermentation conditions were observed: Fermentation time of 30 hours at a temperature of 30°C.

7.6.4 Germination/Malting

Germination or malting of cereal grains may result in some biochemical modifications and produce malt with improved nutritional quality that can be used in various traditional recipes. It has been found that germination of proso millet grains increased the free amino acids and total sugars and decreased the dry weight and starch content. Increases in lysine, tryptophan, and non-protein nitrogen were also noticed (Parameswaran & Sadasivam, 1994). Germinated grains are suitable for preparing baby foods in many developed countries. Mixes consisting of 70% of either popped or malted proso millet flour, 15% roasted soy meal, and 15% roasted peanut meal were successfully used to prepare five infant food products (sweet gruel, salty gruel, halwa, burfi, and biscuits) in India (Srivastava et al., 2001).

7.6.5 Roasting

Roasting/toasting is a rapid processing method that uses a dry-heat treatment for a short time. Roasting is a simple and a commonly used household technology, which has been reported to improve the edibility and digestibility of grains, reduce their antinutrient properties, and prevents the loss of nutritious components (Huffman & Martin, 1994). Roasting enhanced the antioxidant properties of proso millet significantly by increasing its content of total phenolic content. Therefore, roasted proso millet is a potentially good source of nutraceuticals for food formulations (Fei et al., 2018). Obadina et al. (2016) reported effect of roasting on nutritional properties of pearl millet. Roasting reduced the protein, crude fiber, moisture content, lysine, methionine, potassium, phosphorus, calcium, magnesium, and phenolic contents of the pearl millet flour, their composition reduced as the roasting temperature increased. Roasting increased the ash, carbohydrate, energy value, and iron contents of the pearl millet flours but had no effect on the fat, threonine, and glycine contents.

7.7 Applications

Consumers, food manufacturers and health professionals are uniquely influenced by the growing popularity of the gluten-free diet. Consumer expectations have urged the food industry to continuously adjust and improve the formulations and processing techniques used in gluten-free product manufacturing (El Khoury et al., 2018). Mustač et al. (2020) prepared bread by using proso millet bran to increase the nutritional profile of bread. Millet bran is an edible ingredient with high content of dietary fiber, micronutrients, and bioactive compounds (Bagdi et al., 2011). McSweeney et al. (2017) reported different products (muffin, couscous, extruded snack, and porridge) by using refined proso millet and corn flour. These products can be consumed by people with celiac or gluten sensitivities. Gluten-free pasta using proso millet and different hydrocolloids (guar gum, xanthan, and sodium alginate) is prepared by Romero et al., 2017. Wang et al. (2019) prepare sour porridge using broomcorn millet (proso millet). Three strains, namely, *Lactobacillus brevis* L1, *Acetobacter aceti* A1, and *Saccharomyces cerevisiae* E4, were used as inocula for the fermentation of broomcorn millet sour porridge. Proso millet protein could be promisingly used in food and pharmaceutical applications for the encapsulation of lipophilic components (Wang et al., 2018).

7.8 Conclusion

With food industries challenged to produce health-promoting food products, their focus has turned to less exploited ingredients like millets. This article highlights the flour, starch, nutrition composition, and applications of proso millet. Proso millet is suitable to address food and nutrition security as it contains all essential nutrients and requires fewer costs to grow and maintain soil fertility. Millet contains about 60–75% starch, which can be easily isolated and used in various food applications. Starch modification is used to improve the properties of native starch, so that its application in foods can be explored. Processing technologies can be used to improve nutrient content and enhance the diet quality of millets. Generally, processing (decortication/de-hulling, fermentation, germination, and roasting) alters the grain quality and improves the nutritional availability and storage stability of the flour and its products. The majority of proso millet is used as bird food but has seen a recent increased interest in its use in human food due to the rapidly growing gluten-free food market.

REFERENCES

Adebiyi, J. A., Obadina, A. O., Adebo, O. A., & Kayitesi, E. (2018). Fermented and malted millet products in Africa: Expedition from traditional/ethnic foods to industrial value-added products. *Critical Reviews in Food Science and Nutrition*, *58*(3), 463–474.

Adebowale, K. O. and Lawal, O. S. 2004. Comparative study of the functional properties of bambarra groundnut (*Voandzeia subterranean*), jack bean (*Canavalia ensiformis*) and mucuna bean (*Mucuna pruriens*) flours. *Food Research International* 37: 355–365.

Agnes, A. C., Felix, E. C. and Ugochukwu, N. T. 2017. Morphology, rheology and functional properties of starch from cassava, sweet potato and cocoyam. *Asian Journal of Biology*: 1–13.

Annor, G., Marcone, M., Corredig, M., Bertoft, E. and Seetharaman, K. 2015. Effects of the amount and type of fatty acids present in millets on their *in vitro* starch digestibility and expected glycemic index (eGI). *Journal of Cereal Science* 64: 76–81.

Annor, G. A., Marcone, M., Bertoft, E. and Seetharaman, K. 2014. Physical and molecular characterization of millet starches. *Cereal Chemistry* 91: 286–292.

Badau, M. H., Nkama, I. and Jideani, I. A. 2005. Phytic acid content and hydrochloric acid extractability of minerals in pearl millet as affected by germination time and cultivar. *Food Chemistry* 92: 425–435.

Bagdi, A., Balázs, G., Schmidt, J., Szatmári, M., Schoenlechner, R., Berghofer, E. and Tömösközia, S. J. A. A. 2011. Protein characterization and nutrient composition of Hungarian proso millet varieties and the effect of decortication. *Acta Alimentaria* 40: 128–141.

Bao, J., Xing, J., Phillips, D. l. and Corke, H. 2003. Physical properties of octenyl succinic anhydride modified rice, wheat, and potato starches. *Journal of Agricultural & Food Chemistry* 51: 2283–2287.

Beaker, H.G. (1994). Buchweizen, Dinkel, Gerste, Hafer, Hirse, und Reis-die Sachal-und Spelzgetreide und ihre Bedeutung Fur die Emahrung. *AID –Ver brauchrdienst* 39, 123–130.

Becker, R. and Lorenz, K. 1978. Saccharides in proso and foxtail millets. *Journal of Food Science* 43: 1412–1414.

Becker, R., Lorenz, K. and Saunders, R. M. 1977. Saccharides of maturing triticale, wheat, and rye. *Journal of Agricultural & Food Chemistry* 25: 1115–1118.

Bello-Pérez, L. A., Sánchez-Rivera, M. M., Núñez-Santiago, C., Rodríguez-Ambriz, S. L. and Román-Gutierrez, A. D. 2010. Effect of the pearled in the isolation and the morphological, physicochemical and rheological characteristics of barley starch. *Carbohydrate Polymers* 81: 63–69.

Bhat, F. M. and Riar, C. S. 2016. Effect of amylose, particle size and morphology on the functionality of starches of traditional rice cultivars. *International Journal of Biological Macromolecules* 92: 637–644.

Blazek, J. and Copeland, L. 2008. Pasting and swelling properties of wheat flour and starch in relation to amylose content. *Carbohydrate Polymers* 71: 380–387.

Bora, P. and Das, P. 2018. Some physical and functional properties of Proso Millet (*Panicum miliaceum* L.) grown in Assam. *International Journal of Pure & Applied Bioscience* 6: 1188–1194.

Brooker, D. B., Bakker-Arkema, F. and Hall, C. W. 1992. Drying and Storage of Grains and Oilseeds. Springer, NY: Van Nostrand Reinhold.

Buléon, A., Colonna, P., Planchot, V. and Ball, S. 1998. Starch granules: Structure and biosynthesis. *International Journal of Biological Macromolecules* 23: 85–112.

Chandrasekara, A., Naczk, M. and Shahidi, F. 2012. Effect of processing on the antioxidant activity of millet grains. *Food Chemistry* 133: 1–9.

Chao, G., Gao, J., Liu, R., Wang, L., Li, C., Wang, et al. 2014. Starch physicochemical properties of waxy proso millet (*Panicum Miliaceum* L.). *Starch/Stärke* 66: 1005–1012.

Chauhan, M., Sonawane, S. K. and Arya, S. S. 2018. Nutritional and nutraceutical properties of millets: A review. *Clinical Journal of Nutrition & Dietetics* 1: 1–10.

Chiotelli, E. and Meste, M. L. 2002. Effect of small and large wheat starch granules on thermo mechanical behaviour of starch. *Cereal Chemistry* 79: 286–293.

Cooke, D. and Gidley, M. J. 1992. Loss of crystalline and molecular order during starch gelatinisation: Origin of the enthalpic transition. *Carbohydrate Research* 227: 103–112.

Devi, P. B., Vijayabharathi, R., Sathyabama, S., Malleshi, N. G. and Priyadarisini, V. B. 2014. Health benefits of finger millet (*Eleusine coracana* L.) polyphenols and dietary fiber: A review. *Journal of Food Science & Technology* 51(6), 1021–1040.

Devisetti, R., Yadahally, S. N. and Bhattacharya, S. 2014. Nutrients and antinutrients in foxtail and proso millet milled fractions: Evaluation of their flour functionality. *LWT-Food Science & Technology* 59: 889–895.

Dhingra, D., Michael, M., Rajput, H. and Patil, R.T. 2012. Dietary fiber in foods: A review. *Journal of Food Science & Technology* 49: 255–266.

Dias-Martins, A. M., Pessanha, K. L. F., Pacheco, S., Rodrigues, J. A. S. and Carvalho, C. W. P. 2018. Potential use of pearl millet (*Pennisetum glaucum* [L.] R. Br.) in Brazil: Food security, processing, health benefits and nutritional products. *Food Research International* 109: 175–186.

Donovan, J. W. 1979. Phase-transitions of the starch-water system. *Biopolymers* 18: 263–275.

Du, S. K., Jiang, H., Yu, X. and Jane, J. L. 2014. Physicochemical and functional properties of whole legume flour. *LWT-Food Science & Technology* 55: 308–313.

El Khoury, D., Balfour-Ducharme, S. and Joye, I. J. 2018. A review on the gluten-free diet: Technological and nutritional challenges. *Nutrients* 10: 1410.

Elleuch, M., Bedigian, D., Roiseux, O., Besbes, S., Blecker, C. and Attia, H. 2011. Dietary fibre and fibre-rich by-products of food processing: Characterisation, technological functionality and commercial applications: A review. *Food Chemistry* 124:411–421.

Englyst, H. N., Kingman, S. M. and Cummings, J. H. 1992. Classification and measurement of nutritionally important starch fractions. *European Journal of Clinical Nutrition* 46: S33–S50.

Englyst, K. N., Englyst, H. N., Hudson, G. J., Cole, T. J. and Cummings, J. H. 1999. Rapidly available glucose in foods: An in vitro measurement that reflects the glycemic response. *The American Journal of Clinical Nutrition* 69: 448–454.

FAO (Food and Agricultural Organization of the United Nations) 2018. The statistical division. http://faostat. fao.org/beta/en/#data/QC (Accessed Oct. 9, 2020).

FAO, 1995. Sorghum and Millets in Human Nutrition. Rome: Food and Agriculture Organization of The United Nations, p. 184.

Fei, H., Lu, Z., Wenlong, D. and Aike, L. 2018. Effect of roasting on phenolics content and antioxidant activity of proso millet. *International Journal of Food Engineering* 4: 110–116.

Geervani, P. and Eggum, B.O. 1989. Nutrient composition and protein quality of minor millets. *Plant Foods for Human Nutrition* 39: 201–208.

Gernat, C., Radosta, S., Damaschun, G. and Schierbaum, F. 1990. Supramolecular structure of legume starches revealed by X-ray scattering. *Starch/Stärke* 42: 175–178.

Gulati, P., Zhou, Y., Elowsky, C. and Rose, D. J. 2018. Microstructural changes to proso millet protein bodies upon cooking and digestion. *Journal of Cereal Science* 80: 80–86.

Habiyaremye, C., Matanguihan, J. B., D'Alpoim Guedes, J., Ganjyal, G. M., Whiteman, M. R., Kidwell, K. K. and Murphy, K. M. 2017. Proso millet (*Panicum miliaceum* L.) and its potential for cultivation in the Pacific Northwest, US: A review. *Frontiers in Plant Science* 7: 1961.

Hadimani, N. A. and Malleshi, N. G. 1993. Studies on milling, physico-chemical properties, nutrient composition and dietary fibre content of millets. *Journal of Food Science & Technology* 30: 17–20.

Hama, F., Icard-Vernière, C., Guyot, J. P., Picq, C., Diawara, B. and Mouquet-Rivier, C. 2011. Changes in micro- and macronutrient composition of pearl millet and white sorghum during in field versus laboratory decortication. *Journal of Cereal Science* 54: 425–433.

Hedayati, S., Shahidi, F., Koocheki, A., Farahnaky, A. and Majzoobi, M. 2016. Functional properties of granular cold-water swelling maize starch: Effect of sucrose and glucose. *International Journal of Food Science & Technology* 51: 2416–2423.

Hegde, P. S. and Chandra, T. S. 2005. ESR spectroscopic study reveals higher free radical quenching potential in kodo millet (*Paspalum scrobiculatum*) compared to other millets. *Food Chemistry* 92: 177–182.

Hoover, R. 2000. Acid-treated starches. *Food Reviews International* 16: 369–392.

Hoover, R. 2001. Composition, molecular structure, and physicochemical properties of tuber and root starches: A review. *Carbohydrate Polymers* 45: 253–267.

Hoover, R. and Ratnayake, W. S. 2002. Starch characteristics of black bean, chick pea, lentil, navy bean and pinto bean cultivars grown in Canada. *Food Chemistry* 78: 489–498.

Hoover, R., Swamidas, G., Kok, L. S. and Vasanthan, T. 1996. Composition and physicochemical properties of starch from pearl millet grains. *Food Chemistry* 56: 355–367.

Huffman, S. L. and Martin, L. H. 1994. First feedings: Optimal feeding of infant and toddlers. *Nutrition Review* 14: 127–159.

Igathinathane, C., Tumuluru, J. S., Sokhansanj, S., Bi, X., Lim, C. J., Melin, S. and Mohammad, E. 2010. Simple and inexpensive method of wood pellets macro-porosity measurement. *Bioresource Technology* 101: 6528–6537.

Jasovskij, I.V. 1987. Selection and seed production of proso millet. *Agropromizdat, Moskva*, 255 (Russian).

Jay, J. M., Loessner, M. J., and Golden, D. A. 2005. Modern Food Microbiology, 7th ed. Springer, India.

Jaybhaye, R. V., Pardeshi, I. L., Vengaiah, P. C. and Srivastav, P. P. 2014. Processing and technology for millet based food products: A review. *Journal of Ready to Eat Food* 1: 32–48.

Jideani, A. I. and Jideani, V. A. 2011. Traditional and possible technology uses of *Digitaria exillis* (Acha) and *Digitaria ibrua*: A review. *Plant Foods for Human Nutrition* 54: 362–373.

Jin, M., Wu, J. and Wu, X. 1994. A study on the properties of starches used for starch-noodle making. In: Xie, G., Ma, Z. (Eds.) *Proceedings of 1994 International Symposium and Exhibition on New Approaches in the production of Food Stuffs and Intermediate Products from Cereal Grains and Oil Seeds*. CCOA: Beijing, p. 488–496.

Kachru, R. P., Gupta, R. K. and Alam, A. 1994. Physico-Chemical Constituents and Engineering Properties of Food Crops, 1st ed.. Jodhpur, India: Scientific Publishers.

Kalinová, J. (2002). Comparison of productivity and quality in common buck-wheat and proso millet (in Czech). PhD Thesis, University of South Bohemia, Ceske Budejovice, pp. 175.

Kalinova, J. 2007. Nutritionally important components of proso millet (*Panicum miliaceum* L.). *Food* 1: 91–100.

Kalinova, J. and Moudry, J. 2006. Content and quality of protein in proso millet (*Panicum miliaceum* L.) varieties. *Plant Foods for Human Nutrition* 61: 43–47.

Kaur, M., Sandhu, K. S. and Singh, N. 2007. Comparative study of the functional, thermal and pasting properties of flours from different field pea (*Pisum sativum* L.) and pigeon pea (*Cajanus cajan* L.) cultivars. *Food Chemistry* 104: 259–267.

Kaushal, P., Kumar, V. and Sharma, H. K. 2012. Comparative study of physicochemical, functional, antinutritional and pasting properties of taro (*Colocasia esculenta*), rice (*Oryza sativa*) flour, pigeonpea (*Cajanus cajan*) flour and their blends. *LWT-Food Science & Technology* 48: 59–68.

Khatkar, B. S., Rajneesh, B. and Yadav, B. S. 2013. Physicochemical, functional, thermal and pasting properties of starches isolated from pearl millet cultivars. *International Food Research Journal* 20: 1555–1561.

Kim, S. K., Choi, H. J., Kang, D. K. and Kim, H. Y. 2012. Starch properties of native proso millet (*Panicum miliaceum* L.). *Agronomy Research* 10: 311–318.

Kumar, S. R., Sadiq, M. B. and Anal, A. K. 2020. Comparative study of physicochemical and functional properties of pan and microwave cooked underutilized millets (proso and little). *LWT-Food Science & Technology* 128: 109465.

Kundgol, N. G., Kasturiba, B., Math, K. K., Kamatar, M. Y. and Usha, M. 2013. Impact of decortication on chemical composition, antioxidant content and antioxidant activity of little millet landraces. *International Journal of Engineering Science* 2: 1705–1720

Lattimer, J. M., and Haub, M. D. 2010. Effects of dietary fiber and its components on metabolic health. *Nutrients* 2: 1266–1289.

Lee, M. H., Hettiarachchy, N. S., McNew, R. W. and Gnanasambandam, R. 1995. Physicochemical properties of calcium-fortified rice. *Cereal Chemistry* 72: 352–355.

Li, K., Zhang, T., Narayanamoorthy, S., Jin, C., Sui, Z., Li, et al. 2020. Diversity analysis of starch physicochemical properties in 95 proso millet (*Panicum miliaceum* L.) accessions. *Food Chemistry* 324: 126863.

Li, W., Gao, J., Saleh, A. S., Tian, X., Wang, P., Jiang, H. and Zhang, G. 2018. The modifications in physicochemical and functional properties of proso millet starch after ultra-high pressure (UHP) process. *Starch/Stärke* 70: 1700235.

Li, W., Xiao, X., Zhang, W., Zheng, J., Luo, Q., Ouyang, S. and Zhang, G. 2014. Compositional, morphological, structural and physicochemical properties of starches from seven naked barley cultivars grown in China. *Food Research International* 58:7–14.

Lii, C. Y. and Chang, S. M. 1981. Characterization of red bean (*Phaseolus radiatus* var. Aurea) starch and its noodle quality. *Journal of Food Science* 46: 78–81.

Lin, C. L., Lin, J. H., Lin, J. J. and Chang, Y. H. 2019. Progressive alterations in crystalline structure of starches during heat-moisture treatment with varying iterations and holding times. *International Journal of Biological Macromolecules* 135: 472–480.

Lin, J. H., Lii, C. Y. and Chang, Y. H. 2005. Change of granular and molecular structures of waxy maize and potato starches after treated in alcohols with or without hydrochloric acid. *Carbohydrate Polymers* 59: 507–515.

Lindeboom, N., Chang, P. R. and Tyler, R. T. 2004. Analytical, biochemical and physicochemical aspects of starch granule size, with emphasis on small granule starches: A review. *Starch/Stärke* 56: 89–99.

Lorenz, K. and Hwang, Y. S. 1986. Lipids in proso millet (*Panicum miliaceum*) flours and brans. *Cereal Chemistry* 63: 387–390.

Lorenz, K., Dilsaver, W. and Bates, L. 1980. Proso millets. Milling characteristics, proximate compositions, nutritive value of flours. *Cereal Chemistry* 57: 16–20.

Luo, K. and Zhang, G. 2018. Nutritional property of starch in a whole-grain-like structural form. *Journal of Cereal Science* 79: 113–117.

Ma, Z., Boye, J. I., Simpson, B. K., Prasher, S. O., Monpetit, D. and Malcolmson, L. 2011. Thermal processing effects on the functional properties and microstructure of lentil, chickpea, and pea flours. *Food Research International* 44: 2534–2544.

Manjot, S. and Adedeji, A. A. 2016. Physicochemical and functional properties of proso millet starch. In: *2016 ASABE Annual International Meeting, Orlando, Florida, USA, July 17–20, 2016*. American Society of Agricultural and Biological Engineers.

McClements, D. J., Chung, C. and Wu, B. C. 2017. Structural design approaches for creating fat droplet and starch granule mimetics. *Food & Function* 8: 498–510.

McDonald, S.K., Hofsteen, L. and Downey, L. 2003. Crop Profile for Proso Millet in Colorado. USDA Crop Profiles. Regional IPM Centers. Available at: http://www. ipmcenters.org/CropProfiles/.

McSweeney, M. B., Seetharaman, K., Dan Ramdath, D. and Duizer, L. M. 2017. Chemical and physical characteristics of proso millet (*Panicum miliaceum*)-based products. *Cereal Chemistry* 94: 357–362.

Milán-Carrillo, J., Reyes-Moreno, C., Armienta-Rodelo, E., Carábez-Trejo, A. and Mora-Escobedo, R. (2000). Physicochemical and nutritional characteristics of extruded flours from fresh and hardened chickpeas (*Cicer arietinum* L). *LWT-Food Science & Technology* 33: 117–123.

Mohammed, O. and Bin, X. 2020. Review on the physicochemical properties, modifications, and applications of starches and its common modified forms used in noodle products. *Food Hydrocolloids* 112: 106286.

Mustač, N. Č., Novotni, D., Habuš, M., Drakula, S., Nanjara, L., Voučko, B., et al. 2020. Storage stability, micronisation, and application of nutrient-dense fraction of proso millet bran in gluten-free bread. *Journal of Cereal Science* 91: 102864.

Nazni, P. and Devi, S. R. 2016. Effect of processing on the characteristics changes in barnyard and foxtail millet. *Journal of Food Processing & Technology* 7: 1–9.

Nishizawa, N. and Fudamoto, Y. 1995. The elevation of plasma concentration of high-density lipoprotein cholesterol in mice fed with protein from proso millet. *Bioscience, Biotechnology & Biochemistry* 59: 333–335.

Nishizawa, N., Oikawa, M. and Hareyama, S. I. 1990. Effect of dietary protein from proso millet on the plasma cholesterol metabolism in rats. *Agricultural & Biological Chemistry* 54: 229–230.

Nishizawa, N., Oikawa, M., Nakamura, M. and Hareyama, S. 1989. Effect of lysine and threonine supplement on biological value of proso millet protein. *Nutrition Reports International (USA)* 40: 239–245.

Noda, T., Takahata, Y., Sato, T., Suda, I., Morishita, T., Ishiguro, K. and Yamakawa, O. 1998. Relationships between chain length distribution of amylopectin and gelatinization properties within the same botanical origin for sweet potato and buckwheat. *Carbohydrate Polymers* 37: 153–158.

Obadina, A., Ishola, I. O., Adekoya, I. O., Soares, A. G., de Carvalho, C. W. P. and Barboza, H. T. 2016. Nutritional and physico-chemical properties of flour from native and roasted whole grain pearl millet (*Pennisetum glaucum* [L.] R. Br.). *Journal of Cereal Science* 70: 247–252.

Ortega-Ojeda, F. E. and Eliasson, A. C. 2001. Gelatinisation and retrogradation behaviour of some starch mixtures. *Starch/Stärke* 53: 520–529.

Osman, M. A. 2011. Effect of traditional fermentation process on the nutrient and antinutrient contents of pearl millet during preparation of Lohoh. *Journal of the Saudi Society of Agricultural Sciences* 10: 1–6.

Parameswaran, K. P. and Sadasivam, S. 1994. Changes in the carbohydrates and nitrogenous components during germination of proso millet, *Panicum miliaceum*. *Plant Foods for Human Nutrition* 45: 97–102.

Pérez, S. and Bertoft, E. 2010. The molecular structures of starch components and their contribution to the architecture of starch granules: A comprehensive review. *Starch/Stärke* 62: 389–420.

Popov, D., Buléon, A., Burghammer, M., Chanzy, H., Montesanti, N., Putaux, J. L., et al. 2009. Crystal structure of A-amylose: A revisit from synchrotron microdiffraction analysis of single crystals. *Macromolecules* 42: 1167–1174.

Punia, S. 2020a. Barley starch: Structure, properties and *in vitro* digestibility-A review. *International Journal of Biological Macromolecules* 155: 868–875.

Punia, S. 2020b. Barley starch modifications: Physical, chemical and enzymatic—A review. *International Journal of Biological Macromolecules* 144: 578–585.

Punia, S., Kumar, M., Siroha, A. K., Kennedy, J. F., Dhull, S. B., & Whiteside, W. S. (2021). Pearl millet grain as an emerging source of starch: A review on its structure, physicochemical properties, functionalization, and industrial applications. *Carbohydrate Polymers*, 117776.

Purewal, S.S., Sandhu, K.S., Salar, R.K. and Kaur, P. 2019. Fermented pearl millet: a product with enhanced bioactive compounds and DNA damage protection activity. *Journal of Food Measurement & Characterization* 13: 1479–1488.

Rafiq, S. I., Singh, S. and Saxena, D. C. 2016. Effect of heat-moisture and acid treatment on physicochemical, pasting, thermal and morphological properties of Horse Chestnut (*Aesculus indica*) starch. *Food Hydrocolloids* 57: 103–113.

Ramashia, S. E., Anyasi, T. A., Gwata, E. T., Meddows-Taylor, S. and Jideani, A. I. O. 2019. Processing, nutritional composition and health benefits of finger millet in sub-saharan Africa. *Food Science & Technology* 39: 253–266.

Ranasalva, N. and Visvanathan, R. 2014. Development of cookies and bread from cooked and fermented pearl millet flour. *African Journal of Food Science* 8: 330–336.

Rasane, P., Jha, A., Sabikhi, L., Kumar, A. and Unnikrishnan, V. S. 2015. Nutritional advantages of oats and opportunities for processing as value-added foods - Review. *Journal of Food Science & Technology* 52: 662–675.

Ratnayake, W., Hoover, R., Shahidi, F., Perera, C. and Jane, J. 2001. Composition, molecular structure, and physicochemical properties of starches from four field pea (*Pisum sativum* L.) cultivars. *Food Chemistry* 74: 189–202.

Ravindran, G. 1991. Studies on millets: Proximate composition, mineral composition, and phytate and oxalate contents. *Food Chemistry* 39: 99–107.

Rodriguez-Garcia, M. E., Hernandez-Landaverde, M. A., Delgado, J. M., Ramirez-Gutierrez, C. F., Ramirez-Cardona, M., Millan-Malo, B. M. and Londoño-Restrepo, S. M. 2021. Crystalline structures of the main components of starch. *Current Opinion in Food Science* 37: 107–111.

Romero, H. M., Santra, D., Rose, D. and Zhang, Y. 2017. Dough rheological properties and texture of gluten-free pasta based on proso millet flour. *Journal of Cereal Science* 74: 238–243.

Saha, D., Gowda, M. C., Arya, L., Verma, M. and Bansal, K. C. 2016. Genetic and genomic resources of small millets. *Critical Reviews in Plant Sciences* 35: 56–79.

Saleh, A. S., Zhang, Q., Chen, J. and Shen, Q. 2013. Millet grains: Nutritional quality, processing, and potential health benefits. *Comprehensive Reviews in Food Science & Food Safety* 12: 281–295.

Sarita, E. S. and Singh, E. 2016. Potential of millets: nutrients composition and health benefits. *Journal of Scientific & Innovative Research* 5: 46–50.

Sasikumar, R. 2014. Fermentation technologies in food production. In: Lai, W. F. (Ed.), Progress in Biotechnology for Food Applications. OMICS Group, California, USA.

Shaikh, M., Ali, T. M. and Hasnain, A. 2015. Post succinylation effects on morphological, functional and textural characteristics of acid-thinned pearl millet starches. *Journal of Cereal Science* 63: 57–63.

Sharma, S., Sharma, N., Handa, S. and Pathania, S. 2017. Evaluation of health potential of nutritionally enriched Kodo millet (*Eleusine coracana*) grown in Himachal Pradesh, India. *Food Chemistry* 214: 162–168.

Shashi, B. K., Sharan, S., Hittalamani, S., Shankar, A. G. and Nagarathna, T. K. 2007. Micronutrient composition, anti micronutrient factors and bioaccessibility of iron in different finger millet (*Eleucine coracana*) genotype Karnataka. *Journal of Agriculture Science* 20: 583–585.

Shen, R., Ma, Y., Jiang, L., Dong, J., Zhu, Y. and Ren, G. 2018. Chemical composition, antioxidant, and antiproliferative activities of nine Chinese proso millet varieties. *Food & Agricultural Immunology* 29: 625–637.

Singh, U. (2001). Functional properties of grain legume flours. *Journal of Food Science and Technology*, 38(3), 191–199.

Singh, E. and Sarita, A. 2016. Nutraceutical and food processing properties of millets: A review. *Austin Journal of Nutrition & Food Sciences* 4: 1077.

Singh, J., McCarthy, O. J. and Singh, H. 2006. Physico-chemical and morphological characteristics of New Zealand Taewa (Maori potato) starches. *Carbohydrate Polymers* 64: 569–581.

Singh, M. and Adedeji, A. A. 2017. Characterization of hydrothermal and acid modified proso millet starch. *LWT-Food Science & Technology* 79: 21–26.

Singh, N., Singh, J., Kaur, L., Sodhi, N. S. and Gill, B. S. 2003. Morphological, thermal and rheological properties of starches from different botanical sources. *Food Chemistry* 81: 219–231.

Singh, P. and Raghuvanshi, R. S. 2012. Finger millet for food and nutritional security. *African Journal of Food Science* 6: 77–84.

Siroha, A. K., Punía, S., Sandhu, K. S. and Karwasra, B. L. 2020. Physicochemical, pasting, and rheological properties of pearl millet starches from different cultivars and their relations. *Acta Alimentaria* 49: 49–59.

Sridhar, R. and Lakshminarayana, G. 1994. Contents of total lipids and lipid classes and composition of fatty acids in small millets: Foxtail (*Setaria italica*), proso (*Panicum miliaceum*), and finger (*Eleusine coracana*). *Cereal Chemistry* 71: 355–359.

Srivastava, S., Thathola, A. and Batra, A. 2001. Development and nutritional evaluation of proso millet-based convenience mix for infants and children. *Journal of Food Science & Technology (Mysore)* 38: 480–483.

Sun, Q., Gong, M., Li, Y. and Xiong, L. 2014. Effect of dry heat treatment on the physicochemical properties and structure of proso millet flour and starch. *Carbohydrate Polymers* 110: 128–134.

Taylor, J. R. N. and Emmambux, M. N. 2008. Millets. In: A. Arend, and F. D. Bello (Eds.), *Handbook of gluten free cereal*.

Tebben, L., Shen, Y. and Li, Y. 2018. Improvers and functional ingredients in whole wheat bread: A review of their effects on dough properties and bread quality. *Trends in Food Science & Technology* 81: 10–24.

Tester, R. F. and Morrison, W. R. 1990. Swelling and gelatinization of cereal starches. *Cereal Chemistry* 67: 558–563.

Tester, R. F., Morrison, W. R. and Schulman, A. H. 1993. Swelling and gelatinization of cereal starches. V. Risø mutants of Bomi and Carlsberg II barley cultivars. *Journal of Cereal Science* 17: 1–9.

Thilagavathi, T., Banumathi, P., Kanchana, S. and Ilamaran, M. 2015. Effect of heat moisture treatment on functional and phytochemical properties of native and modified millet flours. *Plant Archives* 15: 15–21.

Vali Pasha, K., Ratnavathi, C. V., Ajani, J., Raju, D., Manoj Kumar, S. and Beedu, S. R. 2018. Proximate, mineral composition and antioxidant activity of traditional small millets cultivated and consumed in Rayalaseema region of south India. *Journal of the Science of Food & Agriculture* 98: 652–660.

Wang, L., Gulati, P., Santra, D., Rose, D. and Zhang, Y. 2018. Nanoparticles prepared by proso millet protein as novel curcumin delivery system. *Food Chemistry* 240: 1039–1046.

Wang, Q., Liu, C., Jing, Y. P., Fan, S. H. and Cai, J. 2019. Evaluation of fermentation conditions to improve the sensory quality of broomcorn millet sour porridge. *LWT-Food Science & Technology* 104: 165–172.

Wani, A. A., Singh, P., Shah, M. A., Schweiggert-Weisz, U., Gul, K. and Wani, I. A. 2012. Rice starch diversity: Effects on structural, morphological, thermal, and physicochemical properties—A review. *Comprehensive Reviews in Food Science & Food Safety* 11(5): 417–436.

Waters, D. L., Henry, R. J., Reinke, R. F. and Fitzgerald, M. A. 2006. Gelatinization temperature of rice explained by polymorphisms in starch synthase. *Plant Biotechnology Journal* 4: 115–122.

Wen, Y., Liu, J., Meng, X., Zhang, D. and Zhao, G. 2014. Characterization of proso millet starches from different geographical origins of China. *Food Science & Biotechnology* 23: 1371–1377.

Wojeicchowski, J. P., Siqueira, G. L. D. A. D., Lacerda, L. G., Schnitzler, E. and Demiate, I. M. 2018. Physicochemical, structural and thermal properties of oxidized, acetylated and dual-modified common bean (*Phaseolus vulgaris* L.) starch. *Food Science & Technology* 38: 318–327.

Wu, Y., Lin, Q., Cui, T. and Xiao, H. 2014. Structural and physical properties of starches isolated from six varieties of millet grown in China. *International Journal of Food Properties* 17: 2344–2360.

Wurzburg, O. B. 1986. Modified Starches-Properties and Uses. CRC Press Inc.

Xu, J., Wang, W. and Li, Y. 2019. Dough properties, bread quality, and associated interactions with added phenolic compounds: A review. *Journal of Functional Foods* 52: 629–639.

Yanez, G. A., Walker, C. E. and Nelson, L. A. 1991. Some chemical and physical properties of proso millet (*Panicum milliaceum*) starch. *Journal of Cereal Science* 13: 299–305.

Yang, Q., Zhang, P., Qu, Y., Gao, X., Liang, J., Yang, P. and Feng, B. 2018. Comparison of physicochemical properties and cooking edibility of waxy and non-waxy proso millet (*Panicum miliaceum* L.). *Food Chemistry* 257: 271–278.

Young, R. 1999. Finger miller processing in East Africa. *Vegetation History & Archaeobotany* 8: 31–34.

Zhang, Y., Zhu, K., He, S., Tan, L. and Kong, X. 2016. Characterizations of high purity starches isolated from five different jackfruit cultivars. *Food Hydrocolloids* 52: 785–794.

Zheng, M. Z., Xiao, Y., Yang, S., Liu, H. M., Liu, M. H., Yaqoob, S., et al. 2020. Effects of heat–moisture, autoclaving, and microwave treatments on physicochemical properties of proso millet starch. *Food Science & Nutrition* 8: 735–743.

Zobel, H. F. 1988. Starch crystal transformations and their industrial importance. *Starch/Stärke* 40: 1–7.

8

Beta-glucan (β-glucan)

Priya Dangi
Institute of Home Economics, University of Delhi, New Delhi, India

Nisha Chaudhary
College of Agriculture, Nagaur, Agriculture University, Jodhpur, India

Riya Joshi and Saranya Prabha
Institute of Home Economics, University of Delhi, New Delhi, India

CONTENTS

8.1 Introduction

β-glucan is a polymer of D-glucose and is a major component of the cell walls of yeast, fungi, some bacteria, and cereals such as barley, oats, wheat, rye, sorghum, and rice. Structurally, β-glucan in cereals contains D-glucopyranosyl units joined together by β(1–3) and β(1–4) glycosidic linkages (Lazaridou,

Biliaderis, & Izydorczyk, 2007). β-glucan displays vast diversity in molecular features, such as molecular weight, size, and ratio of β(1–4) to β(1–3) linkages and trisaccharides to tetrasaccharides, which determine solubility, viscosity, gelation, and many other properties. Oat β-glucan exhibits the typical ratio of 2:1 for trisaccharides to tetrasaccharides, while barley and wheat correspond to 3:1 and 4:1, respectively (Cui, Wood, Blackwell, & Nikiforuk, 2000).

An interesting fact about β-glucan is that it possesses wide clinical applications, including immune-modulating agents, reducing hypertension, and antitumor and anticancer activities. The United States Food and Drug Administration (FDA) has accepted the efficiency of oat and barley β-glucan in lowering the risk of coronary heart diseases (CHDs). Consuming at least 3 g of oat β-glucan per day helps lower low density lipoproteins (LDLs) and reduce the risk of CHDs (Othman, Moghadasian, & Jones, 2011). Furthermore, β-glucan from oats is capable of stimulating the growth of some probiotic bacteria, such as *Bifidobacterium* and *Lactobacillus rhamosus* strains, thus, it is considered as a potent prebiotic (Kontula et al., 1998). It also manifests antimicrobial effects against *E. coli* and *B. subtilis* (Rahar, Swami, Nagpal, Nagpal, & Shah, 2011). For these reasons, processed food products such as breakfast cereals, baked products, and dairy and meat products are fortified with β-glucan fractions (Lazaridou et al., 2007).

The physical attributes of β-glucan, such as solubility, viscosity and hydration, are the prime determinants governing its functionality and bioactivity in food products. For instance, the solubility behavior of β-glucan is dependent on its degree of polymerization and presence of functional groups, such as hydroxyl or carboxylic acids. Soluble β-glucans are found to be more effective immunity regulators than insoluble β-glucans (Xiao, Trincado, & Murtaugh, 2004). Interestingly, the insoluble fraction of β-glucan tends to hydrate by absorbing water and swelling (Hromádková et al., 2003). Modification at the molecular level results in structural reformation affecting the bioavailability of β-glucan (Zhang & Kim, 2012). Since β-glucan in nature is an insoluble bioactive compound, different modification processes were developed to maximize its utilization (Ahmad, Mustafa, & Che Man, 2015). Modification can be achieved by physical (use of heat, ultrasonic waves, microwaves, or irradiation), chemical (by the addition of carboxymethyl, sulphur, or phosphorous groups) and biological (employing β-glucanase enzyme) methods. It is the physiochemistry of β-glucan that gives rise to functional properties like thickening, stabilizing, emulsification, and gelation, which are applied in food industry for the production of various kinds of products like soups, sauces, and beverages (Dawkins and Nnanna 1995).

8.2 Sources and Diversity Among β-glucan

Classically, β-glucan can be classified into various forms depending on their chemical structure, origin, and nature of linkage.

- Linear (1–3) β-glucans are polymers of glucose linked through 1–3 glycosidic bonds. They are commonly found in capsules of some rhizobial species.
- Curdlan is a linear unbranched (1–3) β-glucan that does not solubilize in water, alcohol, or nearly any organic solvent. It commonly originates from bacterial species like *Agrobacterium* and *Rhizobium* species (Kenyon, Esch, & Buller, 2005).
- Paramylon originates from the species of euglenoids and haptophytes and is a nondispersible, crystalline form arising from linear (1–3) β-glucan.
- Linear (1–3) β-glucan found in fungi and lichen is classified into *pachyman* and *lichen glucans*.
- The *Mycolaminarin* family of water soluble, side-chain branched (1–3) (1–6) β-glucan serve as carbohydrate reserves in species of Chromistan oomycete like *Pythium* (Perret et al., 1992)
- *Leucosin* is a family of water soluble, side-chain branched (1–3) (1–6) β-glucan resembling *Chyrsolaminarin* generally found in diatoms (Wustman, Gretz, & Hoagland, 1997).

- Cyclic (1–3) (1–6) β-glucan produced by legume symbionts are water soluble. Some examples are *Bradyrhizobium japonicum, Rhizobium loti, Azospirillum braslense*, and *Azorhizobium calinnodans* (Miller, Gore, Johnson, Benesi, & Reinhold, 1990).
- Sulfated linear (1–3)(1–4) β-glucan are found in the cell walls of *Kappachycus alvarezii* (Lechat, Amat, Mazoyer, Buléon, & Lahaye, 2001).

Amongst cereals, the most common and widely available sources of β-glucan are oats and barley (Punia et al., 2020; Punia, 2019; Sandhu et al., 2017). In oats, it is centralized mainly in three regions—the cell walls of starchy endosperm, the cell wall of the sub-aleurone layer, and the inner cells of the aleurone layer. Total β-glucan content of barley grain ranges from 2.5–11.3% by weight of the kernel on an average, but it may go as high as 13–17%. The β-glucan content of oat is highly variable, ranging from 2.2–7.8%. The level of β-glucan in rye, wheat, sorghum, and rice was found to be much lower: 1.2–2.9%, 0.41–.4%, 0.1–1.0%, and 0.04% respectively. In general, both genotypic and environmental factors govern the β-glucan content of cereal grains; however, in the case of oats and barley, genotypic factors play a dominant role. The difference in β-glucan content within the same genotype has also been seen, and this difference is attributed to environmental conditions, particularly growth site or moisture content. Availability of water during the maturation period affects the final β-glucan content significantly as moist conditions during the preharvest stage are responsible for lower β-glucan in comparison to dry conditions. Moreover, a higher temperature during growth of oats and barley grain is seen to favor a higher amount of β-glucan (Lazaridou et al., 2007).

8.3 Extraction and Purification of β-glucan

Because of numerous health benefits offered by oat β-glucan, it is essential to utilize its full potential through extraction and purification. Research aims to find most economic methods of extracting high purity β-glucan from cereals and their by-products for incorporation in food. High β-glucan content indicates the presence of high dietary fiber in cereal grains but, it should be noted that the soluble component of β-glucan is responsible for a larger part of the beneficial activity on health. Generally, a higher amount of soluble β-glucan is found in oats compared to barley (Lazaridou et al., 2007). Dry and wet separation processing are the two major techniques used to extract β-glucan.

8.3.1 Dry Separation

A dry separation process is generally applied because of its simplicity and low price. Sibakov et al. (2011) developed a fractionation process (Figure 8.1) that involves extraction of oat lipids by supercritical carbon dioxide (SC^-CO_2) followed by sequential milling and bran separation processes. Since the lipids present in oat endosperm remain associated with starch and protein, the removal of lipids in the preliminary step facilitates the milling process and makes the dry separation process more effective than if high-lipid containing oats were used as a raw material. This process yield a β-glucan concentrate with up to 34% of β-glucan and a molecular weight ranging between 1,000 and 2,000 kDa (Heneen et al., 2009).

8.3.2 Wet Separation

The wet separation process is not usually preferred due to its complexity, however, it may result in extraction of 20–70% of β-glucan in the final product (Benito-Román, Alonso, & Lucas, 2011). The beginning of the process is marked with the hydrolyzation of oats and barley either as whole or cracked, meal or flour with water at acidic, neutral, or alkaline conditions (Figure 8.2). Once the β-glucan is solubilized, the slurry is centrifuged and β-glucan is precipitated with the assistance of ethanol. This β-glucan extract should be purified to remove other substances, such as starch and protein, that emerged

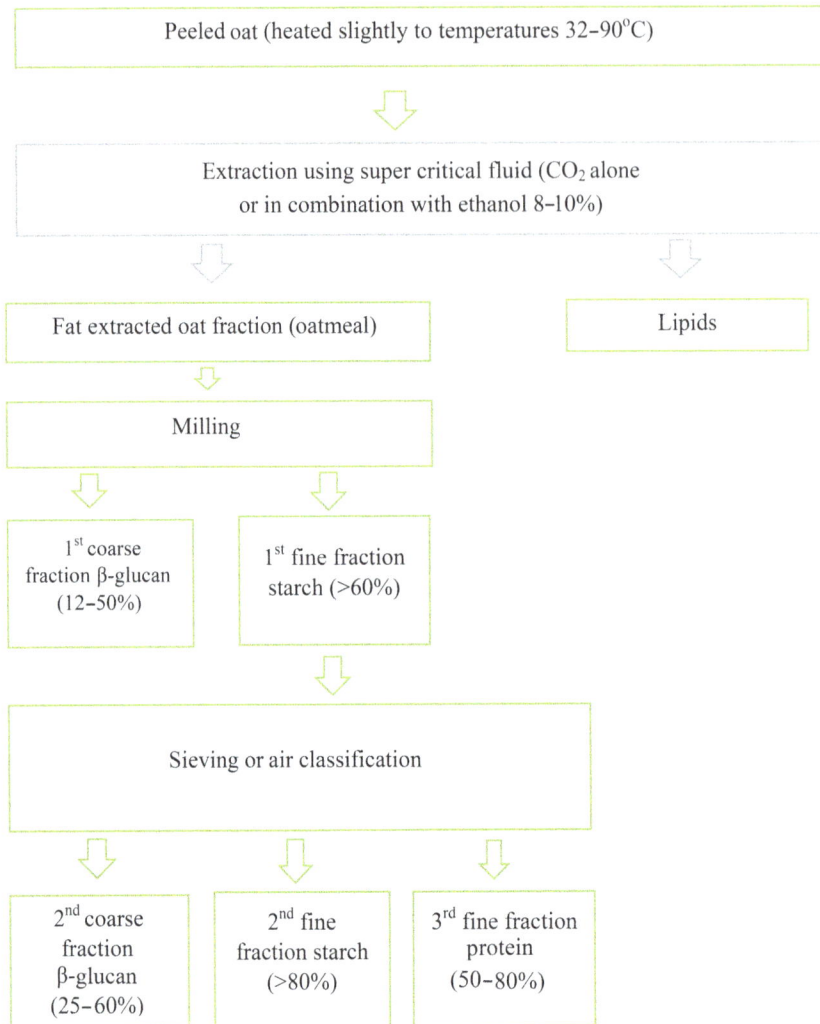

FIGURE 8.1   Flowchart representing dry extraction of β-glucan.

with the β-glucan during crude extraction. The use of ethanol ensures the inhibition of few indigenous enzymes and thus removes any dispersed sugars, proteins, and other nonpolar molecules. Thermally stable α-amylase enzyme could be useful in degrading the starch molecules that come alongwith the extraction of nonstarch polysaccharides (NSP) (Hu, Zhao, Zhao, & Zheng, 2014). Experimental conditions maintained during extraction (such as type of processing, pH, solvent used, temperature, duration of extraction) and pretreatments (heating and drying), as well as the presence of enzymes (endogenous or from contaminating microorganisms), the method of milling, and particle size all influence the content of β-glucan extracted and its native structure (Bricker & Tollison, 2011; Lazaridou et al., 2007)

The native structure of β-glucan is highly susceptible to alterations which affects its functionality. When β-glucan is solubilized in the aqueous system, the endogenous β-glucanases of the grain simultaneously become active and begin to degrade the β-glucan, fragmenting it by the virtue of the shear generated from mixing and centrifugation (Bhatty 1995). Consequently, the molecular weight and viscosity of β-glucan is reduced, decreasing its overall potential in lowering cholesterol, glucose, and prebiotic activities (Wood 2007). Recent studies have concluded that the bioactivity of β-glucan can be preserved to a greater extent by inactivating the enzymes during the purification process, possibly by heating, ethanol precipitation, or use of hydrochloric acid (HCl) or trichloroacetic acid (TCA). Amongst these

```
┌─────────────────────────────────────────────────────┐
│                  Milled whole oat flour               │
└─────────────────────────────────────────────────────┘
                          │
                          ▼
┌─────────────────────────────────────────────────────┐
│           Reflux with 80% ethanol for about 6 hours   │
└─────────────────────────────────────────────────────┘
                          │
                          ▼
┌─────────────────────────────────────────────────────┐
│   Heat the mix stirring continuously in the hot plate │
│   after adding sodium hydroxide at 45°C for 90 minutes│
└─────────────────────────────────────────────────────┘
                          │
                          ▼
┌─────────────────────────────────────────────────────┐
│  Centrifuge the mix, collect the supernatant, and     │
│              divide it into three portions            │
└─────────────────────────────────────────────────────┘
```

Acid treatment	Alkaline treatment	Enzyme treatment
Citric acid is added till pH 3.5 is achieved	pH is adjusted to 10 with sodium bicarbonate	pH is adjusted to 7 with citric acid
	Centrifugation	α-amylase is added at 40°C for 3 hours
Supernatant is mixed with ethanol (80%) in 1:2 ratio and hold for 15 minutes		Centrifugation
Centrifugation		Supernatant treated with protease enzyme and incubated at 37°C for 3 hours

```
┌─────────────────────────────────────────────────────┐
│     Pellets were separated and dried in vacuum oven   │
└─────────────────────────────────────────────────────┘
```

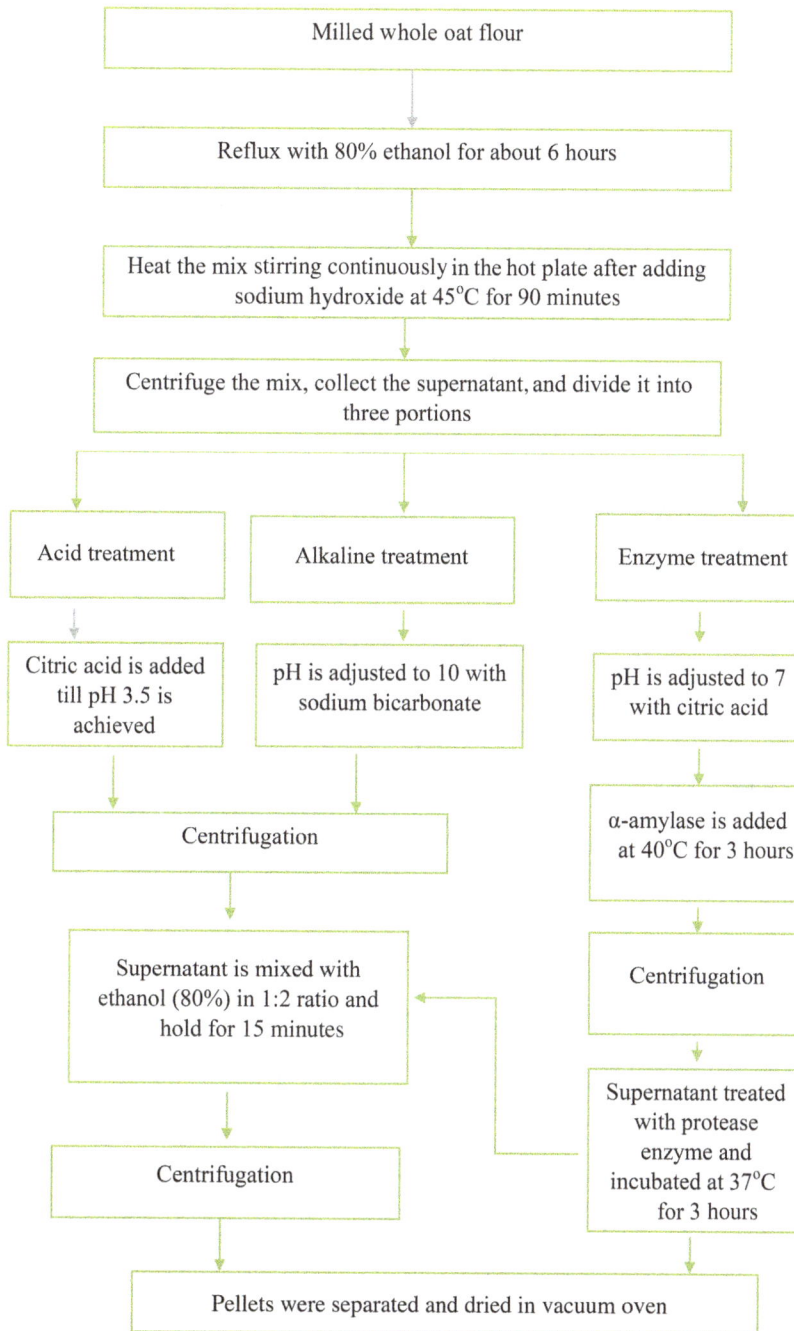

FIGURE 8.2 Flowchart representing wet extraction of β-glucan.

methods, ethanol reflux is the one most effective and widely used method (Bhatty 1995; Temelli 1997; Rimstein, Stenberg, Andersson, Andersson, & Aman, 2003). The large scale applicability of the wet separation procedure is further limited by its high cost; large amounts of water and solvent are required, which has impeded the widespread application of β-glucan in the food industry.

Removal/precipitation of adhered starch is necessary for maximum yield of purified β-glucan and could be carried out at the time of nonstarch polysaccharide extraction using various methods, including acid, alkali and water extraction. Use of acid could degrade β-glucan (Li, Cui, & Wang, 2011; Wood 2011). Although sodium hydroxide (NaOH) may further alleviate the percentage of β-glucan purified to 80–100%, it would yield decreased molecular size (Lazaridou et al., 2007). Convincingly, hot water is the most preferred method. Temelli (1997) indicated that the recovery rate of β-glucan was directly correlated with temperature: the higher the temperature, the greater the extraction. This is the main reason for using thermal stable α-amylase to hydrolyze starch during water extraction. For further purification, precipitation with ethanol and ammonium sulphate is used followed by dialysis or ultrafiltration. Dialysis could separate amino acids, peptides, and glucose from β-glucan, increasing the purity and viscosity of the final product. Dialysis could provide highly purified and viscous product than other methods, however, it is time consuming and inappropriate for large-scale production. In order to scale up the process, other techniques like ultrafiltration, deproteinization by active carbon followed by alcohol precipitation, isoelectric precipitation followed by solvent extraction, and an alcohol extraction method coupled with ammonium sulfate could be employed.

8.4 Characterization of β-glucan

Different species and varieties of oats exhibit different structures and properties that are affected by growing and storage conditions. *In vitro* studies demonstrate that the primary determinant of blood-glucose and cholesterol-lowering properties of oat β-glucan is its ability to increase the viscosity of digesta in the gastrointestinal (GI) tract. Therefore, the chemical structure, molecular weight, the rate and extent of dissolution and solution rheology of oat β-glucan are key factors in determining the physiological function of oat-containing foods (Wang and Ellis 2014).

8.4.1 Rheological Behavior

β-glucan exhibits Newtonian behavior and displays high viscosity even at low concentrations. However, at very low concentrations, the viscosity of the solution does not change with a change in shear rate (Wang and Ellis 2014). At a concentration above 2 g/L, it exhibits pseudoplastic or shear thinning behavior where the viscosity decreases with an increase in the shear rate (Dongowski, Knappe-Drzikova et al. 2005). The rheological properties of oat β-glucan are complex and are influenced by the sources, pretreatment given, solvent concentration, temperature and pH (Hu et al., 2014).

Wang, Xu, Liu, Wang, and Xie (2008) stated that the presence of weakly acidic and alkaline conditions can decrease the viscosity of oat β-glucan when compared to neutral conditions. Johansson, Karesoja, Ekholm, Virkki, and Tenhu (2008) compared the viscosities of equal-sized oat and barley β-glucans and the viscosity of oat β-glucan proved higher, suggesting that there were small structural differences between oats and barley β-glucan, such as the ratio of tri- and tetrasyl units. When compared at equivalent β-glucan concentrations, viscosity of oat β-glucan came out to be 100 times higher than barley β-glucan.

8.4.2 Solubility

Molecular weight and conformation of β-glucan are significant as they offer insight into other physical properties like solubility, viscosity, viscoelasticity, and gelatinization. The term "solubility" in the case of polysaccharides refers to the property where polysaccharides (in solid form) disperse in a liquid medium. In addition to varietal differences, the solubility of β-glucan from oat bran varies largely with processing and storage conditions along with the structural characteristics of a polysaccharide (Mälkki et al., 1992). Beer, Wood, Weisz, and Fillion (1997) reported that only 12–33% of total β-glucan present in bran and rolled oats was solubilized, whereas Henrion, Francey, Lê, and Lamothe (2019) reported a slightly higher solubility value of 39% in a commercial oat bran. Incorporation of oat bran into an extruded cereal showed an increase in solubility of β-glucan from 39–67%. An increase in extrusion temperature and mechanical disruption lead to complete solubilization of β-glucan. Microscopic observations revealed that the process of extrusion disrupted the cell-wall structure, which aided in the release of β-glucan and its dispersion into the food matrix. Loss of solubility of β-glucan during frozen storage is supposed to be caused by the reorganization of β-glucan chains due to intermolecular interactions (mostly by hydrogen bond formation), which leads to an increased ordered structure (Lazaridou et al., 2007). Presence of substitution groups like COOH increases solubility. Solubility can be expressed as the percentage of a dissolved fraction relative to the total amount of polysaccharide in the original mixture. Sometimes, solubility and extractability can be used interchangeably. The presence of endogenous enzymes like β-glucanase cause depolymerization and brings about an increase in the solubility of the polymer. It was found experimentally that with a decrease in molecular weight from 22 million Da to 4 million Da, there was an increase in solubility. However, further reduction in molecular weight to 1.2 million Da showed decreased solubility.

8.5 Modifications of β-glucan

Native β-glucan being high molecular weight compound exhibits high viscosity which interferes with its food application. This prompts the need to modify the structure of β-glucan by adopting physical, chemical, enzymatic, mechanical and other processing methods as presented in Table 8.1.

8.5.1 Physical Methods

The physical treatments applied to β-glucan focuses on reorganizing its molecular arrangement and rheology. Various techniques such as use of low temperature (like refrigeration, freezing), high temperature

TABLE 8.1

Overview of Modification Methods Employed to Modify β-glucan

Treatment	Modified β-glucan	Applications
Drying	Reduced water holding capacity	Used in preparation of high fiber pasta
Kilning and flaking	Increased molecular weight and viscosity (Wang 2014)	Used for preparation of cereal products like oat flakes, flour, groats (Ames et al., 2015)
Extrusion	Reduced molecular weight and increased apparent viscosity and solubility (Tosh et al., 2010)	Used in preparation of maize bran/oat flour extruded breakfast cereals (Holguín-Acuña et al., 2008)
Cooking	Increased amount of soluble β-glucan (Johansson, Tuomainen, Anttila, Rita, & Virkki, 2007)	Used in preparation of porridge (Beer et al., 1997)
Freezing	Lower molecular weight, reduced viscosity and water-binding capacity	Used in beverages (Biörklund, van Rees et al. 2005)
Homogenization	Increased viscosity and bioactivity (Bangari 2011)	Used in preparation of milk fortified with β-glucan (Bangari, 2011)
Gamma radiation	Decreased viscosity and increased antioxidant activity (Shah, Masoodi, Gani, & Ashwar, 2015)	Used in preparation of feed for chickens (broilers) (Campbell, Sosulski, Classen, & Ballance, 1987)
Microwaves	Increased molecular weight	Used in preparation of gluten-free β-glucan fortified bread with improved loaf volume (Pérez-Quirce, Ronda, Lazaridou, & Biliaderis, 2017)
Substitution with ascorbic acid	Reduced molecular weight and viscosity	In beverages (Kivelä et al., 2009)
Acetylation	Increased entrapment of bile acids, reduced viscosity, hardness and adhesiveness	Used as acetylated oat starch (Berski et al., 2011) or for making strong gels (de Souza et al., 2015)
Enzymatic	Decrease in molecular weight promotes solubility and increased bioactivity (Roubroeks et al., 2001)	Used in producing feed for broiler chicks (addition of β-glucanase showed increase in weight gain and greater digestibility in chicks) (Wood et al., 1989)

treatments (such as flaking, kilning, cooking) or drying are commonly used methods for modification purpose (Kenji, Takae, & Takafumi, 2013)

- Kilning is a hydrothermal treatment used at industrial scale to produce shelf stable cereal products as it aims at deactivating β-glucanase and other enzymes that might cause degradation or rancidity in the product (Ames, Storsley, & Tosh, 2015)
- Drying is one of the most ancient techniques for extending shelf life and to improve the quality of food. Studies showed that drying has varying effect on β-glucan at various temperatures as in no significant effects on water holding capacity were observed at temperature 25 and 50°C but at higher temperatures of 75 and 100°C, the sample showed loss in water holding capacity (Kumar and Kalita 2017).
- Flaking process applies thermal and mechanical shear to the cereals, thereby influencing the molecular arrangement and interactions among β-glucan and other compounds. To obtain good quality flaked cereals, steaming is considered as a necessary step prior to flaking owing to its capacity to inactivate enzymes like lipases and glucanases (Yokoyama, Knuckles, Wood, & Inglett, 2002).
- Low temperature processing reduces molecular weight, solubility and/or extractability of cereal β-glucan as freezing is not able to deactivate enzyme β-glucanase. This enzyme causes reduction in molecular weight as they break the linkage (Ames et al., 2015).

8.5.2 Chemical Methods

It is the availability of the hydroxyl group in the structure of β-glucan that renders it a site for coupling with other functional groups. This connection allows alterations in the molecular framework, and thus, it can suitably be applied to different food products (Ahmad et al., 2015). A range of chemical reagents, such as organic acid (ascorbic acid), mineral acid (phosphoric acid), and transition metals (iron) possess the ability to break apart the glycosidic bonds between β-glucan molecules. This act depreciates its molecular weight and viscosity can successfully be employed. Acid treatment can be given to β-glucan by either using chloroacetic acid in 2-propanol or alkali water with pH more than 12 or by incorporating sulfo-hydrophilic groups. Kivelä, Nyström, Salovaara, and Sontag-Strohm (2009) observed that the addition of ascorbic acid and its oxidation products resulted in destabilization of β-glucan. Conversely, an improvement in the solubility of β-glucan in cold water has been observed after etherifying primary and secondary alcohol groups present on carbohydrates (Zeković, Kwiatkowski, Vrvić, Jakovljević, & Moran, 2005; Zhu, Du, & Xu, 2016). Furthermore, sulphated β-glucan has a high yield with substantial water solubility and bioactive characteristics, such as antitumor, antithrombotic, and immunity booster (Wang et al., 2017)

Besides this, another interesting interplay may arise from the conjugation of the β-glucan molecule with various functional biomolecules, such as proteins, amino acids, organic acids, and polyphenols. The maillard reaction is one of the most striking reactions that modify the viscoelastic nature of β-glucan and is desirable for certain products (Sun et al., 2018).

8.5.3 Enzymatic Methods

The use of enzymes (β-1,3-glucanase, cellulase, lichenase) brings about the debranching, depolymerization, and de-esterification in the structure of β-glucan. Depolymerization is significant in minimizing the viscous character of β-glucan solution (McCleary 1986) and changes its structure from a semiflexible chain into extended random coils (Roubroeks, Andersson, Mastromauro, Christensen, & Åman, 2001)

8.5.4 Miscellaneous

β-glucan can be modified using gamma irradiations through electron beams, X-rays, gamma rays, or by applying mechanical stress (Methacanon, Weerawatsophon, Tanjak, Rachtawee, & Prathumpai, 2011).

Shah, Ahmad, et al. (2015) reported that gamma irradiated β-glucan manifested modified antioxidant and antiproliferative activity, enhancing the functional value of food products. Microwave energy can also be suitably applied to cause modifications in β-glucan chains and enhance their physiochemical characteristics. Tao and Xu (2008) produced water-soluble modified β-glucan from water insoluble β-glucan originated from *Pleurotus tuber regium* using microwaves. High-pressure homogenization is another technique that can reduce molecular weight, viscosity and shear thinning behavior of the sample.

8.6 Health Implications of β-glucan

More health impacts of β-glucan have been discovered ever since its first health benefit was reported. β-glucan is now used to treat many diseases and disorders as discussed in the sections that follow.

8.6.1 Effect on the Gastrointestinal Tract

Drzikova, Dongowski, Gebhardt, and Habel (2005) have shown that β-glucan has the ability to prevent colorectal cancer by increasing the bulkiness in stool and helping to prevent constipation. This is, perhaps, because as oats are digested, β-glucan units take up water and swell, resulting in increased viscosity and giving the feeling of satiety. Since the human small intestine lacks enzymes for the degradation of β-glucan, it remains as whole and maintains viscosity (Mälkki and Virtanen 2001). Moreover, it was observed that a change in pH caused reaggregation of β-glucan units, thereby increasing the size ranging between $200–700 \times 10^6$ g/mol (Ulmius, Adapa, Önning, & Nilsson, 2012). Simultaneously, oat β-glucan acts as prebiotic as it stimulates the growth of bacterial strains in the colon (Gibson and Roberfroid 1995).

8.6.2 Effect on Cholesterol Level and Cardiovascular Diseases (CVD)

Scientific evidence has shown that oat β-glucan has significant cholesterol lowering effects that can also be linked to reduced cardiovascular diseases (CVDs). Amongst the various theories that propose to explain the reduction of cholesterol level by β-glucan, the most widely accepted one is that β-glucan's increased viscosity helps in capturing micelles containing bile acids and prevents them from being reabsorbed by lumen. This results in their elimination from the body with the feces (Ellegard and Andersson 2007). Furthermore, β-glucan reduces LDLs and total serum cholesterol by entrapping dietary cholesterol and preventing its reabsorption by the small intestine, which is otherwise a prime initiate of cardiovascular problems (Butt, Tahir-Nadeem, Khan, Shabir, & Butt, 2008). Conclusively, hydrolysed β-glucan exhibits cholesterol lowering effects to a greater extent than the native one (Bae, Kim, Lee, & Lee, 2010)

8.6.3 Antidiabetes Activity

Oat β-glucan has shown beneficial effects in patients with type 2 diabetes. Granfeldt, Hagander, and Björck (1995)) reported that consumption of flaked oats (muesli) and oat flakes (oat porridge) by diabetic patients resulted in lower glucose and insulin responses. Tappy and colleagues also reported a significant decrease in plasma glucose levels with increase in concentration of cooked oat bran in breakfast cereal (Tappy, Gügolz, & Würsch, 1996). The drop in insulin and glucose responses is specifically attributed to the viscosity of β-glucan (Wood et al., 1994).

8.6.4 Anticancerous Effects

β-glucan showed anticarcinogenic properties and found its usage in radiotherapy where it acts as a guard and prevents damage to normal cells (Bashir and Choi 2017). β-glucan exhibits antitumorous and anticancerous activity by the virtue of activating macrophages that attack tumor cells and destroy them. It also modulates lymphocytes, neutrophils, and natural killer (NK) cell activity alongwith components of the innate immune system that help prevent tumorous growth (Hong et al., 2004). The potency of β-glucan in preventing tumors depends upon type of tumor, genetics of the host animal, the dose, route, and timing of administration, as well as tumor load (Yang, Jang, Radhakrishnan, Kim, & Song, 2008). Recent research showed that as orally ingested β-glucan boosts exogenously administered antibodies, it can be used in combination with specific monoclonal antibodies that activate or bind to tumor-enabling neutrophils to encounter the tumor cells (Hong et al., 2004).

8.6.5 Antimicrobial Activity

Cereal derived β-glucan has shown effective resistance against viral, bacterial, fungal, and parasitic pathogens, which allows integration into the diet to stimulate the immune system (Liang et al., 1998). Oat β-glucan ingested orally, either alone or in combination with sucrose, resulted in positive effects on susceptibility to the herpes simplex pneumonia type 1 respiratory infection (Davis et al., 2004). Moreover, it was observed that an intraperitoneal injection of β-glucan improved immunity response against bacterial pathogens (Rodríguez, Chamorro, Novoa, & Figueras, 2009). Yun, Estrada, Van Kessel, Park, and Laarveld (2003) reported that when given β-glucan, immunosuppressed mice infected with oocysts of *Eimeria vermiformis* showed greater immunity response and reduced mortality. The efficacy of oat β-glucan as an antimicrobial agent, by the cause of cellular and antigen-specific humoral immunity, has been seen against *Staphylococcus aureus, Escherichia coli* and *Bacillus subtilis*.

8.6.6 Effects on Skin

The antioxidant activity of β-glucan is used by dermatologists to treat various skin problems, including skin aging (Singh and Agarwal 2009). Oat β-glucan is used for reducing fine lines and wrinkles (Pillai, Redmond, & Röding, 2005). Furthermore, by its moisturizing and anti-irritant property, it has found usage in many cosmetics and personal care products (Zulli, Suter, Biltz, & Nissen, 1998).

8.7 Applications of β-glucan in Food

Oat β-glucan is a valuable functional ingredient that has many health-promoting properties. Its promising physiochemical characteristics and functionality favor its incorporation in various food systems, such as sauces, soups, beverages, and many other products to increase its regular intake. Moreover, increasing consumer demand for natural, healthy, and safe food is another driving force for exploiting β-glucan use in processed food products (Dawkins and Nnanna 1995). β-glucan as a functional ingredient not only has an application in the food industry, but also in the cosmetic and pharmaceutical industries. The various functional properties of β-glucan are described in the sections that follow.

8.7.1 Thickening Agent

β-glucan possesses the capability to be used as a thickening agent, making it a good substitute for pectin, gum arabic, xanthan gum, alginates, and carboxymethyl-cellulose in a traditional beverage (Ahmad, Anjum, Zahoor, Nawaz, & Dilshad, 2012). Trogh et al. (2004) observed that the addition of β-glucan into flour would result in improved bread loaf volume. Johansson (2006) revealed that the viscosity of barley β-glucan was higher than that of oat β-glucan, specifically due to the greater proportion of insoluble β-glucan in barley.

8.7.2 Fat Mimetics

Fat mimetics are classically protein- or carbohydrate-derived compounds that imitate the functionality of true fats in terms of their physical, textural, and organoleptic characteristics (Shaltout and Youssef 2007). Oatrim is one such product isolated from hydrolyzed oat flour or bran containing β-glucan; it can be used as a replacement for shortening in the preparation of oatmeal raisin cookies. The resulted product is low in fat and calories while simultaneously enriched with soluble fiber (Inglett, Warner, & Newman, 1994). Z-trim gel is another similar product obtained from amalgam of wheat bran, corn, oat hulls, and some other insoluble fiber sources like soybean, rice, peas, etc. This gel, in combination with oatrim, allows production of food with different textures such as low calorie snacks, cheese, some baked products, and hamburger and other meat products. The exploitation of β-glucan is not restricted to cereal-based products, rather it has found its application in the dairy and beverage industry also. In recent studies, incorporation of cereal β-glucan in fruit juice beverages and soups has received greater customer satisfaction (Ahmad et al., 2012) In the dairy industry, β-glucan is used in preparation of low-fat ice cream and yogurt (Tudorica, Jones, Kuri, & Brennan, 2004), where its association with whey protein increases the nutritional status as well (Temelli, Bansema, & Stobbe, 2004).

8.7.3 Emulsifying Agent

Reformulation is one of the most effective tools to improve the fat content of different food products and develop healthy and functional foods. To produce healthy and functional meat products, oat bran proves to be a potentially useful ingredient. Reformulation with oat bran gives an added advantage to the product as it contains soluble dietary fiber and β-glucan that together exhibits emulsifying properties (Arendt and Zannini 2013). Lazaridou and Biliaderis (2007) reported that a wide range of hydrogels with varying molecular weights and characteristics have been prepared from cereal β-glucan under various isothermal conditions (5–45°C), molecular sizes, and polymer concentrations. These gels not only offer emulsifying properties, they also provide thermal stability and appropriate texture and color to the product. Oat bran seems to be a potential replacement in the development of structured emulsions and gels with solid-like properties. Kontogiorgos, Biliaderis, Kiosseoglou, and Doxastakis (2004) assessed the creaming and rheological aspects of a concentrated egg yolk emulsion stabilized by β-glucan obtained from barley and oats individually. It was concluded that high and low molecular weight β-glucans showed stabilization of oil in water emulsion by different ways. The former showed stability due to increased viscosity while the later stabilized because of network formation in continuous phase.

8.7.4 Prebiotics

The dairy sector offers huge potential to exploit β-glucan to develop new nutraceutical products. Bekers et al. (2001) developed a dairy-based functional food containing oat β-glucan with a symbiotic association of probiotics and prebiotics, whereby dietary fiber is released during the fermentation process. Furthermore, development of oat β-glucan incorporated yogurt also showed promising results, where inclusion of β-glucan improved the body and texture of unsweetened yogurt. However, in case of sweetened yogurt, the presence of sweeteners increased the apparent viscosity of such products (Fernández-Garía, McGregor, & Traylor, 1998).

8.8 Safety Concerns

For a short period of time, β-glucan might be safe for most adults when ingested orally. β-glucan in injectable form containing microparticles is generally not safe, as it may cause headache, nausea, dizziness, vomiting, diarrhea, chills, fever, rash, pain at the injection site, back and joint pain, high or low blood pressure, flushing, decreased number of white blood cells, and increased urine. β-glucan consumption is known to cause thickening of skin on hands and feet of people suffering from AIDS. There is not enough information about the effects of β-glucan when applied to the skin. Potential side effects of oral consumption of β-glucan are not very well known (Rahar et al., 2011).

REFERENCES

Ahmad, A., et al. (2012). "Beta glucan: A valuable functional ingredient in foods." *Critical Reviews in Food Science and Nutrition* **52**: 201–212. doi: 10.1080/10408398.2010.499806

Ahmad, N. H., et al. (2015). "Microbial polysaccharides and their modification approaches: A review." *International Journal of Food Properties* **18**(2): 332–347. doi: 10.1080/10942912.2012.693561

Ames, N., et al. (2015). "Effects of processing on physicochemical properties and efficacy of β-glucan from oat and barley." *Cereal Foods World* **60**(1): 4–8.

Arendt, E. and E. Zannini (2013). Cereal Grains for the Food and Beverage Industries. Woodhead Publisher.

Bae, I. Y., et al. (2010). "Effect of enzymatic hydrolysis on cholesterol-lowering activity of oat β-glucan." *New Biotechnology* **27**(1): 85–88.

Bangari, S. (2011). *Effects of oat beta glucan on the stability and textural properties of beta glucan fortified milk beverage. (Master of Science)*, University of Wisconsin, Stout.

Bashir, K. M. I. and J. S. Choi (2017). "Clinical and physiological perspectives of β-Glucans: The past, present, and future." *International Journal of Molecular Sciences* **18**(9): 1906–1952.

Beer, M. U., et al. (1997). "Effect of cooking and storage on the amount and molecular weight of (1→3) (1→4)-β-d-glucan extracted from oat products by an in-vitro digestion system." *Cereal Chemistry* **74**(6): 705–709. doi: https://doi.org/10.1094/CCHEM.1997.74.6.705

Bekers, M., et al. (2001). "Oats and fat-free milk based functional food product." *Food Biotechnology* **15**(1): 1–12. doi: 10.1081/FBT-100103890

Benito-Román, et al. (2011). "Optimization of the β-glucan extraction conditions from different waxy barley cultivars." *Journal of Cereal Science* **53**(3): 271–276. doi: 10.1016/j.jcs.2011.01.003

Berski, W., et al. (2011). "Pasting and rheological properties of oat starch and its derivatives." *Carbohydrate Polymers* **83**(2): 665–671.

Bhatty, R. S. (1995). "Laboratory and pilot plant extraction and purification of β-glucans from hull-less barley and oat brans." *Journal of Cereal Science* **22**(2): 163–170.

Biörklund, M., et al. (2005). "Changes in serum lipids and postprandial glucose and insulin concentrations after consumption of beverages with β-glucans from oats or barley: A randomised dose-controlled trial." *European Journal of Clinical Nutrition* **59**(11): 1272–1281.

Bricker, J., & Tollison, S. (2011). "Comparison of motivational interviewing with acceptance and commitment therapy: a conceptual and clinical review." *Behavioural and Cognitive Psychotherapy* **39**(5): 541–559. doi: 10.1017/s1352465810000901

Butt, M. S., et al. (2008). "Oat: unique among the cereals." *European Journal of Nutrition* **47**(2): 68–79. doi: 10.1007/s00394-008-0698-7

Campbell, G. L., et al. (1987). "Nutritive value of irradiated and β-glucanase-treated wild oat groats *(Avena fatua L.)* for broiler chickens." *Animal Feed Science and Technology* **16**(4): 243–252.

Chen, C., et al. (2014). "Chain conformation and anti-tumor activity of derivatives of polysaccharide from Rhizoma Panacis Japonici." *Carbohydrate Polymers* **105**(1): 308–316.

Cui, S., et al. (2000). "Physicochemical properties and structural characterization by two-dimensional NMR spectroscopy of wheat β-D-glucan—comparison with other cereal β-D-glucans." *Carbohydrate Polymers* **41**(3): 249–258. doi: 10.1016/S0144-8617(99)00143-5

Davis, J. M., et al. (2004). "Effects of moderate exercise and oat β-glucan on innate immune function and susceptibility to respiratory infection." *American Journal of Physiology-Regulatory, Integrative and Comparative Physiology* **286**(2): R366–R372.

Dawkins, N. L. and I. A. Nnanna (1995). "Studies on oat gum [(1→3, 1→4)-β-D-glucan]: Composition, molecular weight estimation and rheological properties." *Food Hydrocolloids* **9**(1): 1–7.

de Souza, N. L., et al. (2015). "Functional, thermal and rheological properties of oat β-glucan modified by acetylation." *Food Chemistry* **178**: 243–250. doi: https://doi.org/10.1016/j.foodchem.2015.01.079

Dongowski, G., et al. (2005). "Rheological behaviour of β-glucan preparations from oat products." *Food Chemistry* **93**: 279–291.

Drzikova, B., et al. (2005). "The composition of dietary fibre-rich extrudates from oat affects bile acid binding and fermentation in vitro." *Food Chemistry* **90**(1–2): 181–192.

Ellegard, L. and H. Andersson (2007). "Oat bran rapidly increases bile acid excretion and bile acid synthesis: an ileostomy study." *European Journal of Clinical Nutrition* **61**(8): 938–945.

Fernández-Garía, et al. (1998). "The addition of oat fiber and natural alternative sweeteners in the manufacture of plain yogurt." *Journal of Dairy Science* **81**(3): 655–663. doi: https://doi.org/10.3168/jds. S0022-0302(98)75620-6

Gibson, G. R. and M. B. Roberfroid (1995). "Dietary modulation of the human colonic microbiota: Introducing the concept of prebiotics." *Journal of Nutrition* **125**(6): 1401–1412.

Granfeldt, Y., et al. (1995). "Metabolic responses to starch in oat and wheat products. On the importance of food structure, incomplete gelatinization or presence of viscous dietary fibre." *European Journal of Clinical Nutrition* **49**(3): 189–199.

Heneen, W. K., et al. (2009). "The distribution of oil in the oat grain." *Plant Signaling & Behavior* **4**(1): 55–56. doi: 10.4161/psb.4.1.7313

Henrion, M., et al. (2019). "Cereal B-glucans: The impact of processing and how it affects physiological responses." *Nutrients* **11**(8): 1729. doi: 10.3390/nu11081729

Holguín-Acuña, A. L., et al. (2008). "Maize bran/oat flour extruded breakfast cereal: A novel source of complex polysaccharides and an antioxidant." *Food Chemistry* **111**(3): 654–657. doi: https://doi.org/10.1016/j.foodchem.2008.04.034

Hong, F., et al. (2004). "Mechanism by which orally administered β-1, 3-glucans enhance the tumoricidal activity of antitumor monoclonal antibodies in murine tumor models." *The Journal of Immunology* **173**(2): 797–806.

Hromádková, Z., et al. (2003). "Influence of the drying method on the physical properties and immunomodulatory activity of the particulate (1→3)-β-D-glucan from *Saccharomyces cerevisiae.*" *Carbohydrate Polymers* **51**(1): 9–15.

Hu, X., et al. (2014). "Structure and characteristics of β-glucan in cereal: A review." *Journal of Food Processing and Preservation* **39**(6): 3145–3153. doi: 10.1111/jfpp.12384

Inglett, G. E., et al. (1994). "Sensory and nutritional evaluations of oatrim." *Cereal Foods World* **39**(10): 755–759.

Johansson, L. (2006). Structural analyses of (1->3),(1->4)-B-D-glucan of oats and barley. *Department of Applied Chemistry and Microbiology.* Helsinki, University of Helsinki.

Johansson, L., et al. (2008). "Comparison of the solution properties of (1→3),(1→4)-β-d-glucans extracted from oats and barley." *LWT - Food Science and Technology* **41**(1): 180–184. doi: 10.1016/j.lwt.2007.01.012

Johansson, L., et al. (2007). "Effect of processing on the extractability of oat β-glucan." *Food Chemistry* **105**(4): 1439–1445. doi: https://doi.org/10.1016/j.foodchem.2007.05.021

Kenji, I., et al. (2013). "The use of Lentinan for treating gastric cancer." *Anti-Cancer Agents in Medicinal Chemistry* **13**(5): 681–688. doi: http://dx.doi.org/10.2174/1871520611313050002

Kenyon, W., et al. (2005). "The curdlan-type exopolysaccharide produced by Cellulomonas flavigena KU forms part of an extracellular glycocalyx involved in cellulose degradation." *Antonie van Leeuwenhoek* **87**(2): 143–148. doi: 10.1007/s10482-004-2346-4

Kivelä, R., et al. (2009). "Role of oxidative cleavage and acid hydrolysis of oat beta-glucan in modelled beverage conditions." *Journal of Cereal Science* **50**(2): 190–197.

Kontogiorgos, V., et al. (2004). "Stability and rheology of egg-yolk-stabilized concentrated emulsions containing cereal b-glucans of varying molecular size." *Food Hydrocolloids* **18**(6): 987–998. doi: 10.1016/j.foodhyd.2004.04.003

Kontula, P., et al. (1998). "The colonization of a simulator of the human intestinal microbial ecosystem by a probiotic strain fed on a fermented oat bran product: effects on the gastrointestinal microbiota." *Applied Microbiology and Biotechnology* **50**(2): 246–252. doi: 10.1007/s002530051284

Kumar, D. and P. Kalita (2017). "Reducing postharvest losses during storage of grain crops to strengthen food security in developing countries." *Foods* **6**(1): 8.

Lazaridou, A. and C. G. Biliaderis (2007). "Molecular aspects of cereal β-glucan functionality: Physical properties, technological applications and physiological effects." *Journal of Cereal Science* **46**(2): 101–118.

Lazaridou, A., et al. (2007). Cereal B-glucans: Structures, physical properties and physiological functions. *Functional Food Carbohydrates* C. G. Biliaderis and M. S. Izydorczyk, CRC Press: 1–72.

Lechat, H., et al. (2001). "Structure and distribution of glucomannan and sulfated glucan in the cell walls of the red alga *Kappaphycus alvarezii* (Gigartinales, Rhodophyta)." *Journal of Phycology* **36**: 891–902. doi: 10.1046/j.1529-8817.2000.00056.x

Li, W., et al. (2011). "Studies of aggregation behaviours of cereal [beta]-glucans in dilute aqueous solutions by light scattering: Part I. Structure effects." *Food Hydrocolloids* **25**(2): 189–195.

Liang, J., et al. (1998). "Enhanced clearance of a multiple antibiotic resistant *Staphylococcus aureus* in rats treated with PGG-glucan is associated with increased leukocyte counts and increased neutrophil oxidative burst activity." *International Journal of Immunopharmacology* **20**(11): 595–614.

Mälkki, Y. and E. Virtanen (2001). "Gastrointestinal effects of oat bran and oat gum: A review." *LWT—Food Science and Technology* **34**(6): 337–347.

Mälkki, Y., et al. (1992). "Oat bran concentrates. Physical properties of beta-glucan and hypocholesterolemic effects in rats." *Cereal Chemistry* **69**(6): 647–653.

McCleary, B. V. (1986). "Enzymatic modification of plant polysaccharides." *International Journal of Biological Macromolecules* **8**(6): 349–354.

Methacanon, P., et al. (2011). "Interleukin-8 stimulating activity of low molecular weight β-glucan depolymerized by γ-irradiation." *Carbohydrate Polymers* **86**(2): 574–580. doi: https://doi.org/10.1016/j.carbpol.2011.04.075

Miller, K. J., et al. (1990). "Cell-associated oligosaccharides of *Bradyrhizobium* spp." *Journal of Bacteriology* **172**(1): 136–142. doi: 10.1128/jb.172.1.136-142.1990

Othman, R., et al. (2011). "Cholesterol-lowering effects of oat b-glucan." *Nutrition Reviews* **69**(6): 299–309. doi: 10.1111/j.1753-4887.2011.00401.x

Pérez-Quirce, S., et al. (2017). "Effect of microwave radiation pretreatment of rice flour on gluten-free bread-making and molecular size of β-glucans in the fortified breads." *Food and Bioprocess Technology* **10**(8): 1412–1421.

Perret, J., et al. (1992). "Effect of growth conditions on the structure of β-d-glucans from Phytophthora parasitica dastur, a phytopathogenic fungus." *Carbohydrate Polymers* **17**(3): 231–236. doi: https://doi.org/10.1016/0144-8617(92)90009-F

Pillai, R., et al. (2005). "Anti-wrinkle therapy: Significant new findings in the non-invasive cosmetic treatment of skin wrinkles with B-glucan." *International Journal of Cosmetic Science* **27**: 292–292. doi: 10.1111/j.1463-1318.2005.00268_3.x

Punia, S., Sandhu, K. S., Dhull, S. B., Siroha, A. K., Purewal, S. S., Kaur, M., & Kidwai, M. K. (2020). "Oat starch: Physico-chemical, morphological, rheological characteristics and its applications-A review." *International Journal of Biological Macromolecules, 154*, 493–498.

Punia, S. (2019). "Barley starch: Structure, properties and in vitro digestibility—A review." *International Journal of Biological Macromolecules*, 155, 868–875.

Rahar, S., et al. (2011). "Preparation, characterization, and biological properties of β-glucans." *Journal of Advanced Pharmaceutical Technology & Research* **2**(2): 94–103. doi: 10.4103/2231-4040.82953

Rimstein, L., et al. (2003). "Determination of β-glucan molecular weight using SEC with calcofluor detection in cereal extracts." *Cereal Chemistry* **80**(4): 485–490.

Rodríguez, I., et al. (2009). "β-Glucan administration enhances disease resistance and some innate immune responses in zebrafish (Danio rerio)." *Fish & Shellfish Immunology* **27**(2): 369–373.

Roubroeks, J. P., et al. (2001). "Molecular weight, structure and shape of oat (1→3),(1→4)-β-d-glucan fractions obtained by enzymatic degradation with (1→4)-β-d-glucan 4-glucanohydrolase from Trichoderma reesei." *Carbohydrate Polymers* **46**(3): 275–285. doi: https://doi.org/10.1016/S0144-8617(00)00329-5

Sandhu, K. S., Godara, P., Kaur, M., & Punia, S. (2017). "Effect of toasting on physical, functional and antioxidant properties of flour from oat (Avena sativa L.) cultivars." *Journal of the Saudi Society of Agricultural Sciences, 16*(2), 197–203.

Shah, A., et al. (2015). "Effect of γ-irradiation on structure and nutraceutical potential of β-d-glucan from barley (Hordeum vulgare)." *International Journal of Biological Macromolecules* **72**: 1168–1175. doi: https://doi.org/10.1016/j.ijbiomac.2014.08.056

Shah, A., et al. (2015). "Effect of γ-irradiation on antioxidant and antiproliferative properties of oat β-glucan." *Radiation Physics and Chemistry* **117**: 120–127. doi: https://doi.org/10.1016/j.radphyschem.2015.06.022

Shaltout, O. E. and M. M. Youssef (2007). "Fat replacers and their applications in food products: A review." *Alexandria Journal of Food Science and Technology* **4**(1): 29–44.

Sibakov, J., et al. (2011). "Lipid removal enhances separation of oat grain cell wall material from starch and protein." *Journal of Cereal Science* **54**(1): 104–109. doi: https://doi.org/10.1016/j.jcs.2011.04.003

Singh, R. and R. Agarwal (2009). "Cosmeceuticals and silibinin. "*Clinics in Dermatology* **27**(5): 479–484.

Sun, T., et al. (2018). "Maillard reaction of oat β-glucan and the rheological property of its amino acid/peptide conjugates." *Food Hydrocolloids, 76*, 30–34. doi: https://doi.org/10.1016/j.foodhyd.2017.07.025

Tao, Y. and W. Xu (2008). "Microwave-assisted solubilization and solution properties of hyperbranched polysaccharide." *Carbohydrate Research* **343**(18): 3071–3078.

Tappy, L., et al. (1996). "Effects of breakfast cereals containing various amounts of β-glucan fibers on plasma glucose and insulin responses in NIDDM subjects." *Diabetes Care* **19**(8): 831–834.

Temelli, F. (1997). "Extraction and functional properties of barley β-glucan as affected by temperature and pH." *Journal of Food Science* **62**(6): 1194–1201.

Temelli, F., et al. (2004). "Development of an orange-flavored barley β-glucan beverage with added whey protein isolate." *Journal of Food Science* **69**(7): 237–242.

Tosh, S. M., et al. (2010). "Processing affects the physicochemical properties of β-glucan in oat bran cereal." *Journal of Agricultural and Food Chemistry* **58**(13): 7723–7730.

Trogh, I., et al. (2004). "The combined use of hull-less barley flour and xylanase as a strategy for wheat/hull-less barley flour breads with increased arabinoxylan and (1→3,1→4)-β-D-glucan levels." *Journal of Cereal Science* **40**(3): 257–267. doi: https://doi.org/10.1016/j.jcs.2004.08.008

Tudorica, C. M., et al. (2004). "The effects of refined barley β-glucan on the physico-structural properties of low-fat dairy products: curd yield, microstructure, texture and rheology." *Journal of the Science of Food and Agriculture* **84**(10): 1159–1169.

Ulmius, M., et al. (2012). "Gastrointestinal conditions influence the solution behaviour of cereal β-glucans in vitro." *Food Chemistry* **130**(3): 536–540. doi: https://doi.org/10.1016/j.foodchem.2011.07.066

Wang, Q. and P. Ellis (2014). "Oat β-glucan: Physico-chemical characteristics in relation to its blood-glucose and cholesterol-lowering properties." *British Journal of Nutrition* **112** (2): S4–S13.

Wang, X. (2014). *The effects of processing on the nutritional characteristics of oat fibre.* Master's thesis. Department of Human Nutritional Science: University of Manitoba.

Wang, H. B., et al. (2008). "Rheological properties of β-glucans from oats." *Trans. China Soc. Agric. Eng* **24**: 31–35.

Wang, Q., et al. (2017). "β-Glucans: Relationships between modification, conformation and functional activities." *Molecules* **22**(2). doi: 10.3390/molecules22020257

Wijesinghe, W. A. J. P. and Y.-J. Jeon (2012). "Biological activities and potential industrial applications of fucose rich sulfated polysaccharides and fucoidans isolated from brown seaweeds: A review." *Carbohydrate Polymers* **88**(1): 13–20.

Wood, P. J. (2007). "Cereal β-glucans in diet and health." *Journal of Cereal Science* **46**(3): 230–238.

Wood, P. J. (2011). "Oat β-glucan: Properties and function, oats: Chemistry and Technology." *Cereal Chemistry.*

Wood, P. J., et al. (1989). "Physiological effects of beta-D-glucan rich fractions from oats." *Cereal Foods World*

Wood, P. J., et al. (1994). "Effect of dose and modification of viscous properties of oat gum on plasma glucose and insulin following an oral glucose load." *British Journal of Nutrition* **72**(5): 731–743.

Wustman, B. A., et al. (1997). "Extracellular matrix assembly in Diatoms (Bacillariophyceae) (I. A Model of adhesives based on chemical characterization and localization of polysaccharides from the marine diatom *Achnanthes longipes* and other diatoms)." *Plant Physiology* **113**(4): 1059. doi: 10.1104/pp.113.4.1059

Xiao, Z., Trincado, C. A., & Murtaugh, M. P. (2004). "β-Glucan enhancement of T cell IFNγ response in swine." *Veterinary Immunology and Immunopathology* **102**(3): 315–320.

Yang, J.-L., et al. (2008). "β-glucan suppresses LPS-stimulated NO production through the down-regulation of iNOS expression and NFκB transactivation in RAW 264.7 macrophages." *Food Science and Biotechnology* **17**(1): 106–113.

Yokoyama, W. H., et al. (2002). Food processing reduces size of soluble cereal β-glucan polymers without loss of cholesterol-reducing properties. *Bioactive Compounds in Foods* American Chemical Society: Vol. 816, 105–116.

Yun, C.-H., et al. (2003). "β-Glucan, extracted from oat, enhances disease resistance against bacterial and parasitic infections." *FEMS Immunology & Medical Microbiology* **35**(1): 67–75.

Zeković, D. B., et al. (2005). "Natural and modified (1→3)-β-D-glucans in health promotion and disease alleviation." *Critical Reviews in Biotechnology* **25**(4): 205–230. doi: 10.1080/07388550500376166

Zhang, M., & Kim, J. A. (2012). "Effect of molecular size and modification pattern on the internalization of water soluble β-(1→ 3)-(1→ 4)-glucan by primary murine macrophages." *The International Journal of Biochemistry & Cell Biology* **44**(6): 914–927.

Zhu, Fengmei, et al. (2016). "A critical review on production and industrial applications of beta-glucans." *Food Hydrocolloids* **52**: 275–288. doi: https://doi.org/10.1016/j.foodhyd.2015.07.003

Zulli, F., et al. (1998). "Improving skin function with CM-glucan, a biological response modifier from yeast." *International Journal of Cosmetic Science* **20**: 79–86. doi: 10.1046/j.1467-2494.1998.171740.x

9

Barnyard Millet—Composition, Properties, Health Benefits, and Food Applications

Parveen Kumari, Priyanka Kajla and Dipanshi Kaushik
Guru Jambheshwar University of Science & Technology, Hisar, Harayan, India

CONTENTS

9.1 Introduction

Barnyard millet is the oldest cultivated millet. There are two primary types of barnyard millet: *Echinochloa esculenta*, a Japanese millet, and *Echinochloa frumentacea*, an Indian barnyard millet (Figure 9.1). The Indian farm millet is also called Billion Dollar Grass. *Echinochloa frumentacea* belongs to the *Poacea* family and it is a self-pollinating crop. *Echinochloa frumentacea* is the hardest millet and is generally known by a few names like sanwa and jhangora (Hindi), shyama (Sanskrit), oodalu (Kannada), kuthiravaali (Tamil), udalu and kodisama (Telugu), shamul (Marathi), sama (Gujarati), shamula (Bengali), and swank (Punjabi). It is grown in different regions, including India, China, Africa, East Indies, Japan, the United States, and Malaysia. In India, it is mainly cultivated in Orissa, Bihar, Madhya Pradesh, Tamil Nadu, Punjab, Maharashtra, Gujarat, and on the slopes of Uttarakhand.

In India, barnyard millet is grown in Himalayan territory in the north to the Deccan region of the south. This underutilized kharif millet grows well under rainfed conditions in slopes up to an elevation of 2,000 meters above sea level. It is usually found in the hilly areas and undulating fields of sloping, minimal, or ancestral zones, where scarcely any choices exist for crop improvement. These two types of barnyard millet are typically scattered and grow well in water-logged conditions, growing viably into paddy. Barnyard (*Echinochloa frumentacea*), an important minor millet that has emerged as an extremely useful crop for food production. The nutritional profile of farm millet per 100 grams is: 10.1% protein, 8.7% water, 3.9% fat, 6.7% unrefined fiber, 68.8% starch, 398 kcal energy, total dietary fiber content 12.5% out of which 4.2% is soluble while 8.4% is insoluble (Ugare et al., 2014).

FIGURE 9.1 Barnyard millet—different colors and shape of the panicle. (Source: AICRP, 2017)

9.2 Origin and Taxonomy

Barnyard millet is in the class *Echinochloa*, of the *Poaceae* family (Clayton and Renvoize, 2006). The variety *Echinochloa* comprises about 250 enduring species that are broadly cultivated in different agroclimatic conditions around the globe (Bajwa et al., 2015). The scarcity of information about the various species of *Echinochloa* leads to difficulty in identifying its morphological characteristics as well as its aggregate versatility due to its insignificant interspecific-intraspecific variation (Chauhan and Johnson, 2011). Thirty-five species have been identified to date by morphological, cytological, and subatomic marker examination for their taxa and phylogenetic relationship (Yamaguchi et al., 2005). Different species of *Echinochloa* include *E. crus-galli* (allohexaploid), *E. colona* (allohexaploid), and *E. oryzicola* (allotetraploid), while the rest are identified as weeds. *E. crus-galli* is the fastest growing weed in rice fields and is found in more than in 60 nations due to its capability to grow even in extreme conditions; it can grow in hypoxic conditions and can penetrate a depth of 100 mm, it has extensive natural resistance, lavish seed production, and imitates the characteristics of rice (Barrett, 1983) (Table 9.1).

Most of its species are not commercially cultivatable but are grown as minor millets by a few ranchers around the globe. *E. frumentacea,* commonly referred to as Indian farm millet, is derived from a

TABLE 9.1

Popular Indian Varieties of Barnyard Millet

State	Varieties
Uttar Pradesh	DHBM-93-3, VL 172 and VL 207, Anurag, Kanchan, VL 29
Tamil Nadu	VL 181, VL 29, CO 1, CO 2, DHBM-93-3
Gujarat	DHBM-93-3, VL-172, Gujarat Banti- 1
Karnataka	RAU 11, DHBM-93-3, VL 172, VL 181, DHB-93-2
Bihar	VL Madira 181
Uttarakhand	VL 172, PRJ 1, VL 207, VL 29, DHBM-93-3, PRS 1

Source: AICRP, 2017.

wild species *E. colona* (L.) and shows an equal cultivation trend in India and Africa (Doggett, 1989; Upadhyaya et al., 2014). Both wild and developed *Echinochloa* species do not contain the same morphological and developmental characteristics.

9.3 Nutritional Profile

Barnyard millet has the best nutritional profile among all millets. Farm millet contains adequate amount of iron, quality protein, fat, dietary fiber, calcium, magnesium, and zinc. The starch content of farm millet ranges between 51.5–62.0 g/100 g, which is significantly less than that of other millets. The total dietary fiber content of farm millet ranges 8.1–16.3% (Table 9.2).

This proportion of starch to total dietary fiber assures a slow rise of postprandial blood glucose levels, decreases LDL cholesterol, and removes other undesirable fatty substances from the body. Several clinical studies also reported similar findings, affirming that incorporation of this millet in the diet is helpful in curing diabetes and cardiovascular disorders. Research studies have demonstrated that barnyard millet contains higher qualitative and quantitative protein as compared to other millets and oats.

TABLE 9.2

Nutritional Profile of Barnyard Millet

Nutritional Component	Amount
Moisture	8.74%
Protein	10.1%
Fat	3.9%
Palmitic acid	10.80 mg/100 g
Oleic acid	53.80 mg/100 g
Linoleic acid	34.90 mg/100 g
Crude fiber	6.7%
Carbohydrate	68.8%
Total dietary fiber	12.5%
Insoluble fiber	8.4%
Soluble fiber	4.2%
Total minerals	2.1%
Phosphorus	281 mg/100 g
Iron	5 mg/100 g
Magnesium	83 mg/100 g
Calcium	19 mg/100 g
Thiamine	0.33 mg/100 g
Riboflavin	0.10 mg/100 g
Niacin	4.20 mg/100 g
Phenolic compounds	
Total phenols	0.8 mg/g
Flavonoids	0.6 mg/g
Total carotenoids	36.7–50.8 mg/100 g
Bioactive compounds	
GABA (gamma-Aminobutyric acid)	11.5–12.3%
β-glucan	5.0–6.0%
Antinutritional factors (Phytic acid)	3.30–3.70 mg/g

Source: Saleh et al., 2013; Chandel et al., 2014; Ugare et al., 2014; Kaur and Sharma, 2020; IFCT 2017.

Barnyard millet is a reasonably good source of iron, ranging 15.6–18.6 mg/100 g, which is significantly higher than oats and millet (Saleh et al., 2013; Vanniarajan et al., 2018). Phytic acid content is lower, possibly due to de-hulling of the millet, which further leads to the decerease in the bioavailability of minerals among all major millets. These nutritional characteristics makes millet a super food for individuals suffering from various health issues, as well as a good candidate for combating malnutrition. Barnyard millet contains double the amount of polyphenols and carotenoids than other millets, which are well known to have superb health benefits (Panwar et al., 2016). Likewise, tannins, steroids, sugars, flavonoids, alkaloids, glycosides, and other phenolic compounds found in farm millet are known to possess different health benefits like anticancer-causing, antimicrobial, and anti-inflammatory qualities as well as the treatment of radiation related infections (Sharma et al., 2016; Sayani and Chatterjee, 2017). These nutritional components make farm millet an appropriate food for the nourishment of all sections of the society.

9.4 Utilization

Farm millet is traditionally used as substitute for rice in Indian Himalayan areas. Rice millet is de-hulled and processed in different products like barnyard millet porridge, and is consumed in Uttarakhand, while in south India, this millet is utilized in traditional dishes, such as dosa, idli, and chakli. In addition to these products, millet is used in the preparation of bread rolls, desserts, noodles, rusks, and popped items (Ugare, 2008), however, to scale-up this process will require substantial industry contribution to market the items universally. Barnyard millet is appropriate in the diets of diabetics, babies, and pregnant women due to its high iron content. Despite its potential, the lack of processed products is the major reason this crop is so underutilized. Therefore, to satisfy the health conscious consumers as well as maximize the utilization of this millet, more value-added and functional products need to be developed. Expanded use of minor millets not only provides the consumer with quality and assorted food, but is also significant for restoring farm millet production. Farm millet is a significant wellspring of grain in the Himalayan district. The leaves of farm millet are wide, and the plant grows quickly, subsequently delivering voluminous feed. Barnyard grain is an exceptional choice for making animal feed.

9.5 Therapeutic Advantages

Barnyard millet is known to lower blood glucose and cholesterol levels, therefore it is beneficial for diabetic and cardiovascular patients, while it can also be a boon for those that are gluten intolerant.

9.5.1 Diabetes

Despite the presence of starch, barnyard millet is suitable for people with diabetes because of its low sugar content, making it easy to incorporate into their diet. Millet phenolics, alpha-glucosidase, and pancreatic amylases lower the postprandial increase in blood glucose level by restricting the enzymatic hydrolysis of complex sugars (Shobana et al., 2009). The presence of inhibitors like aldose reductase lowers the risk of cataract and other eye-related problems common to diabetes. De-hulled and processed millet is known to reduce the risk of diabetes (Ugare et al., 2014).

9.5.2 Cardiovascular Diseases

Obesity and smoking increase the risk of coronary illness and stroke. Currently, a major portion of the world's population faces a high risk of cardiovascular disease. Rodents fed with native and modified starches of barnyard millet showed insignificant alterations in serum cholesterol and blood glucose level in comparison to rice and other millets (Kumari and Thayumanavan, 1997).

9.5.3 Gluten Intolerance

Medically, gluten intolerance is called celiac disease. The particular explanation of celiac isn't clear, yet it has a genetic component. Gluten affects the internal lining of the digestive tract, which causes an allergic reaction in the body. Therefore, those allergic or sensitive to gluten are advised to supplement their diet with gluten-free grains like millets, pseudocereal, etc. (Rao et al., 2018).

9.6 Processing and Food Applications

Processing of millets is generally carried out by traditional methods before its consumption and preparation into various foods to improve nutritive and organoleptic properties. Processing reduces antinutrient contents, which improves the bioavailability of essential micronutrients. The millet and its processed products are good resources to combat malnutrition (Nazni and Devi, 2016).

A major step of barnyard millet grain processing includes de-hulling, which improves its nutritional quality further (Lohani et al. 2012). De-hulling is traditionally carried out by pounding and is very labor-intensive and time consuming. The small size and firm covering of the palea and lemma make manual grain de-hulling even more difficult and laborious. A small, lightweight threshing machine of low cost was designed by Singh et al. (2003) to perform both de-husking and threshing and was found most suitable for small farmers in hilly areas. The threshing capacity of the machine was 40–60 kg/h and de-husking capacity was 5–6 kg/h for barnyard millet grain. Machines with a processing (de-husking) capacity of 100 kg/h are also available in the market for large-scale industry. Another affordable machine was developed by an Indian researcher in collaboration with Canadian researchers and had a de-husking efficiency of 98% with damaged grains of less than 2%. A similar machine operated with a single-phase electric motor of 1-horsepower was also manufactured by the ICAR-Central Institute of Agricultural Engineering in Bhopal, India. Its grinding and de-husking capacity was 100 kg/h at a moisture content of 10–12%. It significantly reduced the time and labor load for processing farm millet.

Millet-based biscuits were developed by incorporating refined wheat flour and barnyard/foxtail millet flour in a ratio of 55:45 and were subjected to sensory analysis and acceptability tests by diabetic subjects and trained panels. Their nutritional profile and glycemic index were compared with control samples prepared using refined wheat flour. All three categories of biscuits were found acceptable by panelists and can be stored in polythene bags (thermally sealed) at room temperature up to 60 days. The glycemic index of barnyard millet biscuits was 68 and the lowest of 50.8 was found in those made from foxtail millet. Total dietary fiber, crude fiber, and ash content were reported higher in millet-incorporated biscuits in comparison to refined wheat-flour biscuits (Anju and Sarita, 2010).

Barnyard millet contains adequate amounts of amylose rich starch, which is utilized for making biodegradable films. The addition of borage seed oil in farm millet starch improves the properties (rigidity, dampness, stretchability, and water penetrability) of the biodegradable film. These biofilms are safe against different pathogenic microorganisms (Cao et al., 2017). The concentrate *E. colona* along with an amalgamation of silver nanoparticles (AgNPs) are utilized for making ecoblends used in the field of medication and farming. Many researchers revealed that the polyphenolic extract of *E. esculenta* had antimutagenic activity against 3-(5-nitro-2-furyl) acrylic corrosive strains of *Salmonella typhimurium*. The seed of *E. crus-galli* has a novel antifungal peptide, EcAMP1, which is very effective against different phytopathogens, such as *alternaria, botrytis, trichoderma* and *fusarium*. This antifungal peptide has α-helical clip structure that binds to the conidia of the pathogens and hinders the stretching of the hyphae without hampering the cytoplasm (Nolde et al., 2011). This characteristic can be used for designing innovative proteins and novel antimicrobials by drug and agribusiness industries. Barnyard millet produces superior and voluminous fodder to oat and rice as it contains higher calcium and protein content (Yabuno 1987). It also contains a fair amount of digestible fiber and total digestible nutrients (National Research Council 1996).

In addition to these applications, farm millet was found to have a higher yield of straw, approximately 6.3 tons/ha (Bandyopadhyay, 2009). Millet contains a good measure of protein, soluble fiber, debris, and fat in the range 7.6%, 23%, 12% and 2.0%, respectively. In addition to its good quantity, the quality of the feed is also superior in terms of good nitrogen fixation properties and good absorbability, making it a quality feed for domestic applications (Sood et al., 2015).

A study by Sharma et al., (2016) optimized the conditions of germination to enhance the bioavailability of various nutritive bioactive components in barnyard millet. A significant increment was noticed in phenolic compounds, total flavonoids, antioxidant activity, dietary fiber content, GABA content, mineral content, and in γ-amino butyric acid, while phytic acid was reported in germinated barnyard millet.

Polishing of de-husked barnyard millet can be done using a rice polisher. Maximum nutritional retention along with optimum polishing was attained at moisture level of 8–10% and the best recommended milling time for farm millet was 3 minutes. A decrease was noticed in protein, fiber, fat, and ash content with an increase in milling time and moisture content as it influenced bran removal. The degree of polishing negatively affected the milling yield and head recovery. The moisture level of millet plays a major role not only in storage, but also in influencing quality parameters in milling, de-husking, polishing, and in development of new machineries (Lohani et al., 2012). Processing of barnyard millet is generally carried out in a similar manner to rice (parboiling, de-hulling, cooking) (Surekha et al., 2013). However, various processing techniques influence nutritional, functional, pasting properties, and antinutrient levels in farm millet (Nazni and Devi, 2016).

Barnyard millet can be utilized in preparation of antinutrient value-added products and can serve as a nutritional ingredient in preparation of baby food, dietary foods, and other nutritive products (Surekha et al., 2013; Vijayakumar et al., 2009). Barnyard millet flour has a greater compatibility with other flours and can be utilized successfully in preparation of valued-added and novel food products without altering native taste and flavor (Surekha et al., 2013).

9.7 Conclusions

Barnyard millet is a well-known minor millet that grows under extreme environmental conditions. Despite its dietary and agronomical advantages, this millet is has gone virtually unrecognized as a crop and has had little consideration from specialists across the globe. Additionally, it has great cooking and sensory qualities. Being a rich source of important bioactive compounds, including antioxidants, millet may serve as a natural treasure of nutraceutical and other health promoting compounds. Millet also contains a significant amount of some non-nutritive compounds, such as phytate, tannin, and other important phytochemicals. Barnyard millet can be utilized in various aspects of the food and pharmaceutical industries; its heat-treated de-hulled grains have a positive effect on lipid, serum, and blood-glucose levels. In both developed and developing countries, millets are gaining in popularity as a food source as well as in the production of biofilms and biofuels. Along with the traditional applications in preparing porridge, chapatti, and as animal feed, millet can have better utilization in the preparation of therapeutic food due to its low glycemic index, nutraceutical and functional food qualities. Owing to its superb nutritional profile, different processed and value-added food products need to be developed to awaken consumers about its nutritional and therapeutic advantages.

REFERENCES

All India Coordinated Research Project on Small Millets. 2017. Proceedings of the 28th Annual Group Meeting. 14-15 April, 2017 at UAS, Bengaluru.

Anju, T. and Sarita, S. 2010. Suitability of foxtail millet (*Setaria italica*) and barnyard millet (*Echinochloa frumentacea*) for development of low glycemic index biscuits. *Malaysian Journal of Nutrition.* 361–368.

Aoki, D. and Yamaguchi, H. 2008. Genetic relationship between *Echinochloa crus-galli* and *Echinochloa oryzicola* accessions inferred from internal transcribed spacer and chloroplast DNA sequences. *Weed Biology and Management.* 8:233–242.

Bajwa, A., Jabran, K., Shahid, M., Ali, H. H., Chauhan, B., and Ehsanullah. 2015. Eco-biology and management of *Echinochloa crus-galli. Crop Protection.* 75:151–162.

Bandyopadhyay, B. B. 1999. Genotypic differences in relation to climatic adaptation of two cultivated barnyard millet species. *Indian Journal of Genetics.* 105–108.

Bandyopadhyay, B. B. 2009. Evaluation of barnyard millet cultivars for fodder yield under single and double cut treatments at higher elevation of hills. *Agricultural Science Digest*. 29: 66–68.

Barrett, S. H. 1983. Crop mimicry in weeds. *Economic Botany*. 37, 255–282. doi: 10.1007/bf02858881

Cao, T. L., Yang, S. Y., and Song, K. B. 2017. Characterization of barnyard millet starch films containing borage seed oil. *Coatings*. 183.

Chandel, G., Meena, R. K. and Dubey, M. 2014. Nutritional properties of minor millets: neglected cereals with potentials to combat malnutrition. *Current Science*. 107:1109–1111.

Chauhan, B., and Johnson, D. E. 2011. Ecological studies on *Echinochloa crusgalli* and the implications for weed management in direct-seeded rice. *Crop Protection*. 30:1385–1391.

Clayton, W. D., and Renvoize, S. A. 2006. Genera Graminum: Grasses of the world in Kew Bulletin Additional Series XIII, Royal Botanical Gardens Kew. Chicago, IL: University of Chicago Press.

Doggett, H. 1989. "Small millets-a selective overview," in *Small Millets in Global Agriculture*, eds A. Seetharam, K. W. Riley, and G. Harinarayana, (Oxford: Oxford), 3–18.

Gupta, A., Mahajan, V., Gupta, H., Singh, H., & and Bisht, G. 2009. Barnyard millet (*Echinochloa frumentacea*) germplasm, easy de-hulling type. *Indian Journal of Plant Genetic Resources*. 281–317.

IFCT—Indian Food Composition Tables 2017. National Institute of Nutrition. Telangana, India.

Kaur, H., and Sharma, S. 2020. An overview of Barnyard millet (Echinochloa frumentacea). *Journal of Pharmacognosy and Phytochemistry*. 819–822.

Kumari, K. S. and Thayumanavan, B. 1997. Comparative study of resistant starch from minor millets on intestinal responses, blood glucose, serum cholesterol and triglycerides in rats. *Journal of Food Science and Technology*. 75:296–302.

Lohani, U. C., Pandey, J. P. and Shahi N. C. 2012. Effect of degree of polishing on milling characteristics and proximate compositions of barnyard millet (*Echinochloa frumentacea*). *Food Bioprocess Technology*. 5:1113–1119.

National Research Council, 1996. Lost Crops of Africa Vol. I: Grains. Board on Science and Technology for International Development. Washington, D.C: National Academy Press.

Nazni, P. and Devi, S. R. 2016. Effect of Processing on the characteristics changes in barnyard and foxtail millet. *Journal of Food Processing and Technology*. 7:3. doi: http://dx.doi.org/10.4172/2157-7110.1000566

Nolde, S., Vassilevski, A. A., Rogozhin, E. A., Barinov, N. A., Balashova, T. A., Samsonova, O. V. 2011. Disulfide-stabilized helical hairpin structure and activity of a novel antifungal peptide EcAMP1 from seeds of barnyard grass (*Echinochloa crus-galli*). *Journal of Biological Chemistry*. 286: 25145–25153.

Panwar, P., Dubey, A., and Verma, A. K. 2016. Evaluation of nutraceutical and antinutritional properties in barnyard and finger millet varieties grown in Himalayan region. *Journal of Food Science and Technology*. 53(6): 2779–2787.

Prabhakar, C. G. and Prabhu, B. 2017. Improved Production Technology for Barnyard Millet. AICRP, Bengaluru, India.

Rao, B. D., Ananthan, R., Hariprasanna, K., Bhatt, V., Rajeshwari, K., Sharma, S. 2018. Nutritional and Health Benefits of Nutri Cereals. ICAR—Indian Institute of Millets Research (IIMR), Hyderabad, India.

Saleh, A., Zhang, Q., Chen, J. and Shen, Q. 2013. Millet grains: Nutritional quality, processing, and potential health benefits. *Comprehensive Reviews in Food Science and Food Safety*. 12: 281–295.

Sayani, R., and Chatterjee, A. 2017. Nutritional and biological importance of the weed *Echinochloa colona*: A review. *International Journal of Food Science and Biotechnology*. 2: 31–37

Sharma, A., Sood, S., Agrawal, P. K., Kant, L., Bhatt, J. C., and Pattanayak, A. 2016. Detection and assessment of nutraceuticals in methanolic extract of finger (*Eleusine coracana*) and barnyard millet (*Echinochloa frumentacea*). *Asian Journal of Chemistry*. 28:1633–1637

Sharma, S., Saxena, D. C. and Riar C. S. 2016. Analysing the effect of germination on phenolics, dietary fibres, minerals and γ-amino butyric acid contents of barnyard millet (*Echinochloa frumentaceae*). *Food Bioscience*. 13:60–68.

Shobana, S., Sreerama, Y. N. and Malleshi, N. G. 2009. Composition and enzyme inhibitory properties of finger millet (*Eleusine coracana* L.) seed coat phenolics: Mode of inhibition of α-glucosidase and pancreatic amylase. *Food Chemistry* 115:1268–1273.

Singh, K. P., Kundu, S. and Gupta, H. S. 2003. Development of higher capacity thresher for ragi/kodo. In: Recent Trend in Millet Processing and Utilization. CCSHAU, Hissar, India, pp. 109–116.

Sood, S., khulbe, R., & Gupta, A. K. 2015. Barnyard millet—a potential food and feed crop of future. *Plant Breeding*. 135–147.

Surekha, N., Ravikumar, S. N., Mythri, S., and Rohini, D. 2013. Barnyard millet (*Echinochloa frumenta-cea* Link) cookies: Development, value addition, consumer acceptability, nutritional and shelf life evaluation. *IOSR Journal of Environmental Sciences Toxicology and Food Technology.* 7:01–10. doi: 10.9790/2402-0730110

Ugare, R., 2008. Health benefits, storage quality and value addition of barnyard millet (*Echinochloa fru-mentacaea* Link). *M. Sc. Thesis,* Department of Food Science and Nutrition, College of Rural Home Science, University of Agricultural Sciences, Dharwad.

Ugare, R., Chimmad, B., Naik, R., Bharati, P. and Itagi, S. 2014. Glycemic index and significance of barn-yard millet (Echinochloa frumentacae) in type II diabetics. *Journal of Food Science and Technology.* 51:392–395.

Upadhyaya, H., Dwivedi, S. L., Singh, S. K., Sube, S., Vetriventhan, M. and Sharma, S. 2014. Forming core collections in barnyard, kodo, and little millets using morphoagronomic descriptors. *Crop Science.* 54: 2673–2682

Vanniarajan, C., Anand, G., Kanchana, S., Arun Giridhari, V. and Renganathan, V. G. 2018. A short duration high yielding culture—barnyard millet ACM 10145. *Agricultural Science Digest—A Research Journal.* 8: 123–126.

Veena, B., Chimmad, B. V., Naik, R. K. and Shantakumar, G. 2004. Development of barnyard millet based traditional foods. *Karnataka Journal of Agricultural Sciences.* 17:522–527.

Vijayakumar, T. P., Mohankumar, J. B., and Jaganmohan, R. 2009. Quality evaluation of chapati from millet flour blend incorporated composite flour. *Indian Journal of Nutrition and Dietetics.* 46: 144–155.

Yabuno, T. 1987. Japanese barnyard millet (*Echinochloa utilis,* Poaceae) in Japan. *Economic Botany.* 41: 484–493.

Yamaguchi, H., Utano, A. Y. A., Yasuda, K., Yano, A., and Soejima, A. 2005. A molecular phylogeny of wild and cultivated *Echinochloa* in East Asia inferred from non-coding region sequences of trnT-L-F. *Weed Biology Management.* 5: 210–218. doi: 10.1111/j.1445-6664.2005.00185.x

10

Finger Millet (Eleusine coracana L.)— Properties and Health Benefits

Parveen Kumari, Priyanka Kajla and Shrestha Naudiyal
Guru Jambheshwar University of Science and Technology, Hisar, Haryana, India

CONTENTS

10.1 Introduction

Millet belonging to the family *Poaceae* are generally small-sized, annual crops that easily adapt to tropical and arid climates with potential to grow even in areas of low fertility (Ramashia et al., 2019). Pearl millet, finger millet, brown top millet, barnyard millet, foxtail millet, sorghum, proso millet, little millet, and kodo millet are some of the generally grown millets in India (Gopalan et al., 2009; Punia et al., 2021). World millet production accounts for 30.73 million tons, out of which 11.42 million tons are produced in India, accounting for 37% of total world production (FAO, 2018). The use of millet in developing countries is more common in Africa's Nigeria, Namibia, and Uganda. Outside Africa, consumption of millet is also observed in some of the Asian countries, specially India, China, and Myanmar. Millet consumption is inconsequential in developed countries. The use of millet grain as animal feed is also insignificant, less than 2 million tons are used as feed in animals worldwide. Commercially, finger millet is also used in preparation of alcoholic beverages, for example, beer in Zimbabwe. Food scientists have studied the effect of incorporating different millet flour into composite flour, but the commercial utilization is still limited (FAO, 2018).

Finger millet (*Eleusine coracana* L.) is one of the ancient crops of India; in Indian literature this millet is known as "nrtta-kondaka," which means "dancing grain" (Achaya, 2009). It is generally known as "ragi" or "mandua" and is widely grown in different states of India (Gopalan et al., 2009) and Africa (Shukla and Srivastava, 2011). Finger millet occupies the largest cultivation area in India and is sixth in production among the major cereals while it ranks 4th among millets in the world (Upadhyaya et al., 2007). India is the largest producer of all the millets out of which finger millet accounts for about 85% of total millet production. This millet is grown on an area of 1.19 million hectares having 1.98 million tons production and 1,661 kg/ha productivity (Divya, 2011).

Finger millet acts as a staple food for people from low income groups as it is a calorie-dense and protein-rich crop (O'Kennedy et al., 2006). Owing to its high nutritional profile, its consumption pattern is increasing daily. Among all the millet, finger millet is good source of calcium and phosphorus, which help in development of bones, blood clotting, normal heart function, and maintaining energy metabolism. In addition to these, finger millet is also a good source of other minerals like iron, zinc, potassium, and sodium (David et al., 2014; Udeh et al., 2017). Finger millet contains adequate amounts of dietary fiber and polyphenols. The low glycemic index of finger millet enables gradual food digestion by keeping the blood sugar level at a normal rate. Other benefits of its high dietary fiber include the reduced risk of diabetes mellitus, cholesterol lowering properties, improved fecal bulk, transit of gastrointestinal content, and a lower risk of colon cancer (Lebovitz, 2001; Lopez et al., 1999; Tharanathan and Mahadevamma, 2003). Phytochemicals are found in substantial amounts on the outer layer of the grains and also contribute to its diverse health benefits (Rao and Muralikrishna, 2007; Viswanath et al. 2009; Hegde et al., 2005; Chethan et al., 2008). Food allergies and intolerance are a growing concern and call for products that serve certain customer requirements. Some people are allergic to food containing gluten and some have celiac disease, which makes it hard for them to digest gluten protein present in wheat, this we've seen a considerable increase in the gluten-free food market. Finger millet is an entirely gluten-free millet. It is becoming famous among gluten-free products and is gaining popularity in a gluten-free diet. The rising trend among customers for a gluten-free lifestyle to enhance health and gain more control over their diet has led to the high adoption of gluten-free foods. Finger millet acts as a perfect alternative to people with many health-related problems. Therefore, these factors are positively correlating with an increased trend in consumption of finger millets.

Finger millet (also known as ragi) is a small-seeded cereal grain having a spherical, oval, or globular shape ranging in size from 1–1.8 mm. It also features a wide variety of colors, which range from light brown to brick red with the latter more commonly found. The pericarp, which has little nutritional significance, is not attached to the seed coat, making for effortless removal by rubbing and soaking grains in water or by process of threshing (Gull et al., 2016). Ragi's seed coat or testa contains five layers, which make it unique when compared to other millets, and is also very tightly adhered to the aleurone layer and endosperm. Thus, it is usually milled with an intact seed coat, which may be a possible reason for its high dietary fiber content (Shobana et al., 2013). The starchy endosperm is further divided into corneous regions (organized starch granules) and floury endosperm (loosely packed starch granules) (McDonough et al., 1986).

10.2 Nutritional Composition

Finger millet is a storehouse of numerous health profiting nutrients and phytochemicals (Table 10.1). These nutrients are discussed next.

10.2.1 Carbohydrates

Finger millet contains 72% carbohydrates, which further comprises 11.5% dietary fiber (Gopalan et al., 2009), 1.04% free sugars, and 65.5% starchy and nonstarchy polysaccharides (Malleshi, et al., 1986). Starch is composed of amylose and amylopectin and the amylose content is lower at around 16% (Wankhede et al., 1979).

10.2.2 Proteins

Protein content ranges from 6–8% and varies by variety. The major fraction of finger millet proteins are prolamins (Virupaksha et al., 1975) and the essential amino acid composition is as follows: Arginine (300 mg/g N), histidine (130 mg/g N), lysine (220 mg/g N), tryptophan (100 mg/g N), phenylalanine (310 mg/g N), leucine (690 mg/g N), tyrosine (220 mg/g N), methionine (210 mg/g N), cystine (140 mg/g N), valine (480 mg/g N), threonine (240 mg/g N), and isoleucine (400 mg/g N) (Gopalan et al., 2009).

10.2.3 Lipids

Total lipid content in finger millet is about 1.5%. The fatty acids commonly found are palmitic acids (25%), linoleic (25%), and oleic (49%) (Mahadevappa and Raina, 1978).

TABLE 10.1

Nutritional Composition of Finger Millet

S. No	Nutrients	Amount (g/100g)
1.	Fat (g)	1.5
2.	Protein (g)	8.7
3.	Starch (g)	72.0
4.	Total Dietary fibre (g)	19.6
5.	Insoluble	16.1
	Soluble	3.5
6.	Total minerals (g)	2.2

Source: (Shobana, 2009; Gopalan et al., 2009)

10.2.4 Micronutrients

Finger millet contains adequate amounts of potassium (408 mg/100 g), iron (3.9 mg/100 g), calcium (344 mg/100 g), phosphorus (283 mg/100 g), and other microminerals and vitamins (Gopalan et al., 2009). The bioavailability of microminerals can be improved by using methods such as germination and fermentation (Sripriya et al., 1997).

10.2.5 Phenolic Compounds

Finger millet is a versatile source of polyphenolics, which are abundantly present in its outer seed coat. Polyphenolics in finger millet exist in both free and bound form, out of which major bound phenolics includes ferulic acid (64–96%), p-coumaric acid (50–99%), and proanthocyanidins (Rao and Muralikrishna, 2002; Chethan and Malleshi, 2007; Dykes and Rooney, 2007).

10.3 Bioactive Compounds in Finger Millet

In recent years, several studies have been carried out on bioactive compounds present in finger millet. The studies revealed that dietary fiber, microminerals, and polyphenols are the major bioactive compounds present in finger millet, which exhibits both nutritional and pharmaceutical properties. These properties are utilized in the management of various diseases.

10.3.1 Dietary Fiber

Dietary fiber (DF) is a major bioactive compound known to contribute to a healthy and quality life. It significantly influences the rate and level of blood glucose, which is altered after consumption of carbohydrates. Nonstarch polysaccharides (NSPs), modified starches, non-α-glucan oligosaccharide, polyols, and resistance starches are some categories of primary DF (Fardet, 2010; Lafiandra et al., 2014). Water-soluble and insoluble polysaccharides, cellulose, pectin, hemicellulose A and B, and lignin constitute a major portion of polysaccharides present in this millet, and glucose, xylose, arabinose, and galactose are the main sugars along with mannose and rhamnose present in lesser amounts. Arabinoxylans present in this millet are found to have high antioxidant activities credited to its higher level of polyphenols and bound ferulic acids (Nirmala et al., 2000; Amadou et al., 2013) (Table 10.2).

10.3.2 Minerals

Finger millet consists of various minerals and contains adequate amounts of calcium, iron, and other important microminerals, such as zinc, potassium, and sodium. These contents explain finger millet's association with various health benefits (Table 10.3).

TABLE 10.2

Therapeutic Effect of Dietary Fiber

Dietary Fiber Component	Amount (%)	Health Benefits	Reference
Total dietary fiber	22.0	Blood glucose lowering property, reduce risk of diabetes mellitus, cholesterol lowering property, improved fecal bulk, colon cancer management, transit of gastrointestinal contents, enhanced fecal transit time, and fermentation in gut.	(Shobana and Malleshi, 2007); (Rao and Muralikrishna, 2007); (Baron, 1998; Lebovitz, 2001); (Gopalan, 1981); (Lopez et al., 1999); (Tharanathan and Mahadevamma, 2003)
Insoluble	19.7		
Soluble	2.5		
Water soluble (NSP)	0.13		
Hemicellulose A	1.4		
Hemicellulose B	1.9		

TABLE 10.3

Mineral Content of Finger Millet and Its Health Benefits

Minerals	Amount (%)	Health Benefits	Reference
Calcium	317.1–398.0	Essential for tooth and bone formation, helps in blood clotting, and normal heart function	(David et al., 2014); (Udeh et al., 2017)
Iron	2.12–3.9	Helps in the formation of hemoglobin, can be used to improve anemic condition	
Zinc	1.44–2.13	Helps in the activation of various enzymes, also aids in wound healing	
Potassium	294.0–1070.0	Important for regulation of electrolyte and water balance, also responsible for proper functioning of muscles	
Sodium	0.6–0.9	Helps in transmission of nerve impulses, also maintains osmotic balance of cells	

10.3.3 Phenolic Compounds

Phenolic compounds are the most widely versatile groups of phytochemicals that are found in plant foods, constituting a significant part of the human diet. Several studies have reported that finger millet exhibits a vast variety of phenolic compounds, including phenolic acids and flavonoids.

10.3.3.1 Phenolic Acid

Major polyphenolic acids include hydroxybenzoic and hydroxycinnamic acids. Hydroxybenzoic acids accounts for about 70–71% of the phenolic acids and are the free form acids that include gallic, gentisic, p-hydroxybenzoic, protocatechuic, vanillic, syringic, ferulic, caffeic, and p-coumaric acids. Hydroxycinnamic acids make up the remaining 30% of the phenolic acids and are bound form acids that includes sinapic, caffeic, p-coumaric, chlorogenic, transcinnamic, and transferulic acids (Udeh et al., 2017).

10.3.3.2 Flavonoids

Catechin, quercetin, epigallocatechin, gallocatechin, and epicatechin are the main of flavonoids reported in this finger millets (Udeh et al., 2017) (Table 10.4).

10.4 Antinutritional Factors

Antinutrients are substances commonly produced in plants during their natural metabolism process that are responsible for decreased nutrient digestion, absorption, and utilization and may simulate other negative health effects in humans and animals after ingestion (Akande et al., 2010). Some of the antinutritional factors existing in ragi include tannins, protease inhibitors, NSPs, glucans, oxalates, and phytates. All of these can directly or indirectly influence the digestibility of nutrients.

10.4.1 Tannins

Tannins are a group of non-nitrogenous phenolic constituents and are chemically classified into two vast categories: hydrolysable and condensed. Condensed tannins on hydrolysis yield flavans, while the hydrolysable tannins on hydrolysis yield gallic acid. Finger millet contains huge percentages of tannins ranging from 0.04–3.74% of catechin equivalents (Rao and Deosthale, 1988). Ramachandra et al. (1997) reported the tannin content of ragi ranges from 0.04–3.47% with lower percentages of tannins in white varieties of finger millet (0.05%) when compared to brown and dark-brown varieties (0.61%). Gunashree et al. (2014) reported (18.75 mg CAE/100 g) in raw ragi and found a significant reduction of 73.6% tannins after soaking and pressure-cooking the sample. Experiments in animal diets found tannins were

TABLE 10.4

Therapeutic Effects of Phenolic Compounds

Phenolic Acid (Hydroxybenzoic Acids)	Amount (mg/g)	Health Benefits	Reference
Gentisic acid	4.5	Inhibition of collagen glycation, crosslinking; Inhibition of malt and pancreatic amylases; Inhibition of aldose reductase, wound healing property, and intestinal α-glucosidase	(Hithamani and Srinivasan, 2014); (Chethan et al., 2008); (Dykes and Rooney, 2007); (Shahidi and Chandrasekara, 2013); (Rao and Muralikrishna, 2007); (Viswanath et al., 2009); (Monnier, 1990); (Hegde et al., 2005); (Rohn et al., 2002); (Chethan et al., 2008)
Vanillic acid	20.0		
Gallic acid	3.91–30.0		
Syringic acid	10.0–60.0		
Salicylic acid	5.12–413.0		
Protocatechuic acid	119.8–405.0		
p-hydroxybenzoic acid	6.3–370.0		
(Hydroxycinnamic acid)			
Chlorogenic acid	-		
Caffeic acid	5.9–10.4		
Sinapic acid	11.0–24.8		
p-coumaric acid	1.81–41.1		
Transcinnamic acid	35–100.0		
Transferulic acid	41–405.0		
Flavonoids			
Quercetin	3		
Catechin, epigallocatechin, luteolin, myricetin, apigenin, gallocatechin, kempherol, narigenin, diadzein, epicatechin, procyanidin	-		

responsible for decreased feed intake, growth rate, reduced net metabolizable energy, and protein digestibility. Tannins impair nutrient digestibility and cause nitrogen retention, thus causing growth depression in poultry (Kumar et al., 2016).

10.4.2 Protease Inhibitors

Protein inhibitors affect protein digestion by making digestive enzymes—trypsin and chymotrypsin—unavailable. It is reported that the presence of a single functional protein inhibitor in finger millet is responsible for inactivation of both amylase inhibitor and trypsin inhibitor. Further, it is concluded that finger millet had more antitryptic activity than antichymotryptic activity. However, in some reports, the chymotrypsin inhibitory unit values attained from different finger millets were comparable (Kumar et al., 2016). Shivaraj et al. (1992) isolated two inhibitors from ragi and named them chymotrypsin inhibitor (CTI) and trypsin alpha amylase inhibitor (TAI). Gunashree et al. (2014) reported 102.6 mg/g of trypsin inhibitors in raw ragi and a 61.5% reduction in trypsin inhibitors when ragi were roasted and then pressure-cooked.

10.4.3 Nonstarch Polysaccharides (NSPs)

Nonstarch polysaccharides (NSPs) are defined as polymeric carbohydrates and they differ in composition and structure from amylose and amylopectin; they contain glycosidic bonds in addition to (1–4) and (1–6) bonds present in starch. NSPs also have high molecular weight ranging from 8–9 billion. Millets have NSPs, such as beta glucans, phytates, oxalates, etc., and all of them may directly or indirectly affect the digestibility of nutrients in millets. It is reported that NSPs in broiler feed may be responsible for growth depression and a decreased in feed conversion. It was further concluded that ragi contained 6.2–7.2% pentosan content, while another report found millet contains more hexoses than pentosans.

Kumar et al. (2016) and Gunashree et al. (2014) reported that raw ragi had elevated phytic acid of 685 mg/100 g while substantial reduction of 61.5% was observed in a roasted sample.

10.5 Health Benefits

Numerous studies have been conducted that reveal ragi as a storehouse of health benefits. Some of these health-beneficial properties are discussed as in the sections that follow.

10.5.1 Antioxidant Activity

Phenolic acids and their derivatives, flavonoids and tannins, can behave as reducing agents and prevent oxidation via free-radical termination, metal chelation, and singlet oxygen quenching mechanism, preventing excessive oxidation that can result in aging and cancer. Phenolic compounds act as antioxidants, as they have the ability to donate hydrogen atoms from hyroxyl groups on benzene rings to electron-deficient free radicals, and, in turn, form a stabilized and less reactive phenoxyl radical (Devi et al., 2014). Several studies support that finger millet possesses high antioxidant activity. Viswanath et al. (2009) reported that diadzene, gallic, coumaric, syringic, and vanillic acids were identified as major phenolic acids from the extracted phenolics of the seed coat and whole flour of finger millet. Antioxidant activity in the seed coat extract was higher (86%) compared to (27%) in the whole flour extract, which was determined by the β-carotene–linoleic acid assay. Veenashri and Muralikrishna (2011) reported that the bran xylo-oligosaccharide (XO) mixture from finger millet displayed a higher antioxidant activity of about 70% at 60 mg concentration, which is higher than the antioxidant activity exhibited by rice, wheat, or maize bran XO mixtures (70% at 1,000 mg concentration) in 1,1-Di phenyl-2 picryl-hydrazyl (DPPH) and FRAP (ferric reducing antioxidant power) assays. Chandrasekara and Shahidi (2011) inferred that insoluble bound phenolic fractions showed lower oxygen radical absorbance capacity than free phenolic fractions. The greater antioxidant capacity of the free phenolic fraction of finger millet is due to the high total phenolic content and presence of flavonoids, such as catechin, gallocatechin, epicatechin, and procyanidin dimmer.

10.5.2 Blood Glucose Lowering Property

Diets rich in millet are usually recommended for diabetics as they can help controls postprandial blood glucose surge and reduce chronic vascular complications (Baron, 1998; Lebovitz, 2001). Carbohydrates present in finger millet are more slowly digested and assimilated than those present in other cereals (Kavitha and Prema, 1995). Regular consumption of finger millet is recognized to reduce gastrointestinal tract disorders (Tovey, 1994) and the risk of diabetes mellitus (Gopalan, 1981) due to its high polyphenol and dietary fiber content (Chethan et al., 2008). Partial inhibition of amylase and α-glucosidase during enzymatic hydrolysis of complex carbohydrates and delay of the absorption of glucose ultimately lead to control of postprandial blood glucose levels—all of this occurring due to the presence of phenolics (Shobana et al., 2009). Beneficial effects of dietary fiber include slower gastric emptying and formation of un-absorbable complexes with accessible carbohydrates in the gut lumen, both of which might result in the reduction of quantity absorbed and delayed absorption of carbohydrates (Kawai et al., 1987; Rasmussen et al., 1991). Ragi flour and ragi starch gave the lowest glycemic response in a study of glucose levels in the blood of six normal males, one diabetic male, and one diabetic female after consumption of cooked rice, parboiled rice, wheat, ragi flour, rice starch, and ragi starch. Supplementation of ragi in the diet for a month showed a higher reduction of fasting and postprandial glucose levels than supplementation with other millets (Geetha and Parvathi, 1990).

10.5.3 Inhibition of Collagen Glycation and Crosslinking

High oxidative stress and hyperglycemia contribute to hastened accumulation of glycation end products and the crosslinking of collagen in diabetes mellitus (Monnier, 1990). Nonenzymatic glycosylation and crosslinking of collagen is caused by free radicals whereas antioxidative conditions and free radical scavengers inhibit these reactions (Fu et al., 1992). The effects of the methanolic extracts of finger millet on glycation and crosslinking of collagen were studied by Hegde et al. (2002) and it was found that the collagen incubated with glucose (50 mM) and 3 mg methanolic extract inhibited glycation. This may be due to the presence of natural antioxidants mainly of polyphenolic nature and other phytochemicals in the seed coat of the grain, giving finger millet the potential for a therapeutic role in dietary supplements for the prevention of glycation-provoked complications, as in diabetes or aging.

10.5.4 Cholesterol Lowering Property

Cholesterol lowering properties of a ragi diet on male albino rats for 8 weeks reported lower serum cholesterol level (65 mg/dl) as compared to casein diet fed animals (95 mg/dl) (Pore and Magar, 1976). Another similar study concluded that the lowering of cholesterol in Alloxan-induced diabetic rats with a finger millet-incorporated diet (55% in the basal diet) for 28 days resulted in a 13% reduction in serum cholesterol levels when compared to the control animals (Hegde et al., 2005). The study by Shobana et al. (2010) also reported lower serum cholesterol (43%) and triacyl glycerol (62%) and lower atherogenic index (1.82) as compared to the diabetic controls (5.3) and concluded this may be due to the presence of soluble dietary fiber trapping fatty substances in the gastrointestinal tract, reducing cholesterol levels in the blood and lessening the risk of heart disease (Lopez et al., 1999).

10.5.5 Wound Healing Property

One researcher studied the antioxidant effects of finger millet on the dermal wound healing process in diabetes-induced rats and inferred that feeding diabetic animals with finger millet for 4 weeks regulated the glucose levels and enhanced the antioxidant status, which accelerated the dermal wound healing, increased the expression of nerve growth factor, boosted epithelialization, increased synthesis of collagen, and activated fibroblasts and mast cells—all of which could be due to the structure, antioxidative mechanism, and the synergistic effects of various phenolic compounds Rajasekaran et al. (2004). It is indicated that the phenolic antioxidants present in finger millet partially protected the insulin-producing cells from alloxan-mediated cell damage, and hence encouraged the healing process. A finger millet flour aqueous paste was applied to wounded rats for 16 days and noted a significant increase in protein and collagen and a decrease in lipid peroxides. Wound rate contraction was 90% as compared to 75% for control untreated rats and complete wound closure occurred after 13 days when compared to 16 days for untreated rats (Hegde et., 2005).

10.5.6 Improvement in Hemoglobin Status

Germinated finger millet-based food was fed to children in rural Tanzania for 6 months. The food comprised finger millet flour, kidney beans, ground peanuts, and dried mangoes (at a ratio of 75:10:10:5) and it was concluded that supplementing children's diets with the germinated finger millet-based food showed a general improvement in hemoglobin status, which may be due to the presence of goosignificantd amounts of iron in finer millet (Tatala et al., 2007).

10.5.7 Inhibition of Malt Amylase, Pancreatic Amylase, and Intestinal α-glucosidase

Polyphenols are widely known to restrict the activity of digestive enzymes such as amylase, pepsin, trypsin, glucosidase, and lipases (Rohn et al., 2002). Polyphenols may act as inhibitors of amylase and glucosidase, leading to a reduction in postprandial hyperglycemia. Polyphenols in millets may affect the amylase activity in several ways, such as by competing with the substrate to bind to the active site of the enzyme or by disrupting the catalytic process. The mode of inhibition also depends on the substrate specificity of the enzyme, hence, finger millet phenolics can be used as amylase and α-glucosidase inhibitors for modulation of the carbohydrate breakdown and regulation of the glycemic index of foods. This can reduce diabetes mellitus also may help regulate the glucose uptake from the intestinal lumen by restricting carbohydrate digestion and absorption, leading to glucose homeostasis (Bailey, 2001).

10.5.8 Inhibition of Aldose Reductase

In a study that investigated the mode of inhibition of aldose reductase (AR) by finger millet polyphenols in patients with diabetes-induced cataracts (caused by the accumulation of sorbitol), the cataracts were mediated by the action of AR enzyme (Chethan et al., 2008). The characteristic factor for AR-mediated sugar-induced cataracts is nonenzymatic glycation (the binding of glucose to a protein molecule), which is induced during diabetes. Crude phenolic extracts from finger millet consist of phenolics, such as gallic, p-coumaric, vanillic, protocatechuic, syringic, ferulic, p- benzoic, trans-cinnamic acids, and quercetin. The mode of inhibition could be due to the prevention of enzymatic conversion of glyceraldehyde to glycerol and glucose to sorbitol thus restoring the depletion of NADPH (nicotinamide adenine dinucleotide phosphate) levels. The structure and function from analytical studies shows that phenolics with a hydroxy group at the 4th position are important for inhibition of AR and the presence of a neighboring O-methyl group in phenolics also denatures the AR activity. Quercetin is the most powerful AR inhibitory component among the other polyphenolic constituents as it displays noncompetitive inhibition on AR enzymes.

10.5.9 Benefits of High Dietary Fiber Content

According to Bingham (1987) and Marlett (1990), dietary fiber in finger millet is higher than many other cereals and is divided into two categories according to water solubility; each category is responsible for its different therapeutic effects. Water-soluble fiber (SDF) consists of nonstarchy polysaccharides, namely β-glucan and arabinoxylan, while water-insoluble fiber (IDF) consists of lignin, cellulose, and hemicelluloses. Some other health benefits related with high fiber foods are as improved fecal bulk, reduction of blood lipids, prevention of colon cancer, barrier to digestion, mobility of gastrointestinal contents, enhanced fecal transit time and fermentability characteristics (Tharanathan and Mahadevamma, 2003). Water soluble fiber has the ability to absorb water and form a gel like structure, and is entirely fermented in the large intestine, thus bringing many desirable metabolic effects. These fractions also trap fatty substances in the gastrointestinal tract, further reduce cholesterol level in blood, and lower the threat of heart diseases. Lopez et al. (1999) concluded that resistant starch (RS) is a functional fiber fraction present in ragi. RS flees the enzymatic digestion process, providing fermentable sugars to colonic bacteria and further producing desirable metabolites. These metabolites in the colon, such as short-chain fatty acids, stabilize colonic cell proliferation and help prevent colon cancer (Gee et al., 1992; Englyst et al., 1992). The insoluble fiber has a crucial effect on gastrointestinal transit time, it binds with water and bulks up stool, and speeds up intestinal transit. It binds with some carcinogens and further lessens the contact time for fecal mutagens to interact with the intestinal epithelium and also amends the digestive microflora activity, leading to a transformation or reduction in the production of mutagens (Thebaudin et al., 1997).

10.6 Processing and Utilization

Finger millet can be utilized by subjecting it to various processing method that are discussed in this section.

10.6.1 Soaking

In this method, millet is soaked in water at a temperature of 30–60°C and left undisturbed for 12–24 hours. Afterward, the millet is rinsed, cleaned, and dried at a temperature of 60°C in an oven. Next, the millets are processed into flour. This processing method helps in reducing antinutritional factors and thereby increasing microminerals (Banusha and Vasantharuba, 2013; Saleh et al., 2013).

10.6.2 Milling

Finger millet cannot be polished and cooked like rice because it has a delicate endosperm with a whole and intact seed coat; the grain needs to be milled for preparation of flour. Whole meal finger millet flour is darker, less desirable, and not attractive, hence a refined flour is prepared. This flour is free from the seed coat and has a higher appeal, but it also has high glycemic response due to its high starch and low dietary fiber content. Mineral and vitamin content are also lower, and, therefore, it can be observed that whole flour is better than the refined flour in terms of nutritional content (Shobana, 2009). Milling of millet can be done using traditional mortar and pestle method as well as by use of modern machinery like stone mills, burr mills, hammer mills, and ball mills. Removal of phytic acid and tannins during milling leads to the increase in bioavailability of divalent minerals, such as calcium, iron, and magnesium (Singh and Raghuvanshi, 2012).

10.6.3 Germination

Germination is a simple, easy, and inexpensive processing method. Millets are soaked for 2–24 hours and spread over a wet cloth for another 24–48 hours; the millet can also be incubated for 48 hours at 30°C (Shimray et al., 2012). This processing method helps in the breakdown of complex compounds into simple ones, thus enhancing its nutritional and functional properties (Pushparaj and Urooj, 2011).

10.6.4 Malting

Usually barley is preferred for the production of malt, but finger millet has been studied and shown good malting characteristics, too. The small size of finger millet is advantageous and can lead to even germination as well as even kilning (Sastri, 1939). The processing of finger millet includes steeping it for 12 hours and germination for 2–3 days; next, the grain is dried, its roots are removed and, is further followed by kilning, milling, moist conditioning, and sieving to produce the final product of refined finger millet malt flour (Malleshi, 2007). Malt flour is an amylase-rich food because it contains significant amounts of alpha-amylases. It is used in the production of infant foods, cakes, and confections.

10.6.5 Decortication

This process is done in a different way for finger millet when compared with other cereals as finger millet's seed coat is tightly intact with endosperm, preventing the use of the normal method of decortication. For the easy separation of the outer seed cover and to harden the soft endosperm layer during decortication, finger millet is subjected to hydrothermal processing (Malleshi, 2006). Decorticated millet has the ability to be cooked like rice. The by-product of decortication, which is the outer seed layer rich in phenolic compounds, is the insoluble dietary fiber and minerals that can be further utilized in different food formulations (Shobana, 2009).

10.6.6 Popping

A high temperature short time (HTST) process is widely used to formulate processed products. The HTST treatment of millet uses sand as heating media. This millet can be popped using an air popper instrument, which produces a good quality product with enjoyable flavor and aroma. During popping, millet starch is gelatinized and the endosperm bursts, resulting in the production of popped products. This processing method improves the insoluble DF of millet products and reduces antinutrient factors (Choudhury et al., 2011; Sarkar et al., 2015).

10.6.7 Roasting

Roasting is similar to popping, but differs in terms of volume expansion of the products. Roasted millet has an appealing color and aroma. This processing method decreases the antinutritional factors more than other processing methods. In addition, it reduces the moisture content, which in turn improves the shelf-stability of millet (Singh and Raghuvanshi, 2012; Thapliyal and Singh, 2015).

10.6.8 Fermentation

Fermentation is the most common processing method for finger millet. This process has been used traditionally and has been continuously applied in the production of alcoholic and nonalcoholic beverages and foods at household and small-scale cottage industry levels (Osungbaro, 2009). The process of fermentation of millet can be carried out at room temperature for 24–72 hours, depending on the type of product to be produced (Blandino et al., 2003). During fermentation, microorganisms are breaking down complex nutritional components into different bioproducts, such as antibiotics, organic acid, amino acid, peptides, and the nutritional and functional properties of the fermented millet (Ranasalva and Visvanathan, 2014; Rasane et al., 2015).

Finger millet flour is widely utilized in the preparation of unleavened flat breads, or roti, in Indian households. Whole-grain popped finger millet is widely consumed in Karnataka as a snack popularly known as "hurihittu" that consists of a vegetable or milk protein source, oil seeds, and jaggery (a type of cane sugar) (Malleshi, 2007). A mildly alcoholic fermented beverage known as "kodo ko jaanr" is consumed in the Sikkim, Darjeeling, Bhutan, and Nepal regions (Thapa and Tamang, 2004). Newer products include noodles (Shukla and Srivastava, 2011), vermicelli (Sudha et al., 1998), biscuits (Krishnan et al., 2011), pasta products (Krishnan and Prabhasankar, 2010), papads (Vidyavati, et al., 2004), muffins (Rajiv et al., 2011), finger millet soup mixes (Guha and Malleshi, 2006) and many more.

10.7 Conclusions

Finger millet—with its nutritional superiority over millets and cereals—can serve as a major contributor to food security, especially for low-income populations. It is a storehouse of multiple nutrients, such as carbohydrates, proteins, lipids, micronutrients (calcium, phosphorus, potassium, iron), and phytochemicals (phenolic compounds and flavonoids), all of which account for numerous health benefits. Millet's significant source of DF and phenolic compounds offers different therapeutic advantages, such as anti-ulcer, antimicrobial, antioxidant, anti-inflammatory, anticancer and, atherosclerogenic effects. Finger millet utilization can be enhanced in rural and urban areas by using proper processing techniques and formulating various value-added products due to its gluten-free nature and well-balanced protein profile. Finger millet has immense potential for commercial exploitation beyond the traditional applications of nutraceutical and functional foods. It can be further concluded that finger millet is versatile and many traditional as well as newer products can be formulated using different processing techniques that will not only decrease antinutritional factors but also yield different products that may have great market potential.

REFERENCES

Achaya, K. T. 2009. The Illustrated Food of India A–Z. New Delhi, India: Oxford University Press.

Akande, K. E., Doma, U. D, Agu, H. O. and Adamu, H. M. (2010). Major antinutrients found in plant protein sources: Their effect on nutrition. *Pakistan Journal of Nutrition.* 9(8): 827–832

Amadou, I., Gounga, M. E., Le, G. W. 2013. Millets: Nutritional composition, some health benefits and processing – A review. *Food Science and Nutrition.* 25: 501–508.

Gunashree, B.S., Kumar, S. R., Roobini, R. and Venkateswaran. G. 2014. Nutrients and antinutrients of ragi and wheat as influenced by traditional processes. *International Journal of Current Microbiology and Applied Sciences.* 3(7): 720–736.

Bailey, C. J. (2001). New approaches to the pharmacotherapy of diabetes. In: Pickup JC, William G, (Eds.), Text Book of Diabetes, 3rd ed. UK: Blackwell Science Ltd.

Banusha, S. and Vasantharuba, S. 2013. Effect of malting on nutritional contents of finger millet and mung bean. *American-Eurasian Journal of Agriculture and Environmental Science.* 13(12): 1642–1646.

Baron, A. D. 1998. Postprandial hyperglycemia and alpha-glucosidase inhibitors. *Diabetes Research and Clinical Practice.* 40: S51–S55.

Bingham, S. 1987. Definitions and intakes of dietary fiber. *American Journal of Clinical Nutrition.* 45:1226–1231

Blandino, A., Al-Aseeri, M. E., Pandiella, S. S., Cantero, D. and Webb, C. 2003. Cereal-based fermented foods and beverages. *Food Research International.* 36(6): 527–543.

Chandrasekara, A. and Shahidi, F. 2011. Determination of antioxidant activity in free and hydrolyzed fractions of millet grains and characterization of their phenolic profiles by HPLC-DAD-ESI-MSn. *Journal of Functional Foods.* 3: 144–158.

Chethan, S., Dharmesh, S.M. and Malleshi, N. G. 2008. Inhibition of aldose reductase from cataracted eye lenses by finger millet (*Eleusine coracana*) polyphenols. *Bioorganic and Medicinal Chemistry.* 16:10085–10090.

Chethan, S., Sreerama, Y.N. and Malleshi, N. G. 2008. Mode of inhibition of finger millet malt amylases by the millet phenolics. *Food Chemistry.* 111:187–191.

Chethan, S. and Malleshi, N. G. 2007. Finger millet polyphenols: Characterization and their nutraceutical potential. *American Journal of Food Technology.* 2(7):582–592.

Choudhury, M., Das, P. and Baroova, B. 2011. Nutritional evaluation of popped and malted indigenous millet of Assam. *Food Science and Technology (Campinas).* 48(6): 706–711.

David, B. M., Michael, A., Doyinsola, O., Patrick, I. and Abayomi, O. 2014. Proximate composition, mineral and phytochemical constituents of *Eleusine coracana* (finger millet). *International Journal of Advanced Chemistry.* 2 (2): 171–174.

Devi, P. B., Vijayabharathi, R., Sathyabama, S., Malleshi, N.G. and Priyadarisini, V. B. 2014. Health benefits of finger millet (*Eleusine coracana* L.) polyphenols and dietary fiber: A review. *Journal of Food Science and Technology.* 51(6):1021–40. 10.1007/s13197-011-0584-9

Divya, G. M. 2011. Growth and instability analysis of finger millet crop in Karnataka. Unpublished Master's thesis, University of Agricultural Sciences, Bengaluru, India.

Dykes, L. and Rooney, L. W. 2007. Phenolic compounds in cereal grains and their health benefits. *Cereal Foods World.* 52 (3): 105–111.

Englyst, H. N., Kingman, S. M. and Cummings, J. H. 1992. Classification and measurement of nutritionally important starch fractions. *European Journal of Clinical Nutrition.* 46: S33–S50.

FAO, 2018. Ethiopia: Report on feed inventory and feed balance. FAO, Food and Agriculture Organization of the United Nations, Rome, Italy.

Fardet, A. 2010. New hypotheses for the health-protective mechanisms of whole-grain cereals: What is beyond fibre? *Nutrition Research Review.* 23: 65–134.

Fu, M. X., Knecht, K. J. W., Thorpe, S. R. and Baynes, J. W. 1992. Role of oxygen in cross linking and chemical modification of collagen by glucose. *Diabetes.* 41:42–48.

Gee, J. M., Johnson, I. T. and Lind, L. 1992. Physiological properties of resistant starch. *European Journal of Clinical Nutrition.* 46: S125–S131.

Geetha, C. and Parvathi, E. P. 1990. Hypoglycaemic effect of millet incorporated breakfast on selected non-insulin dependent diabetic patients. *Indian Journal of Nutrition Dietetics.* 27:316–320.

Gopalan, C. 1981. Carbohydrates in diabetic diet. India: Bulletin of Nutrition Foundation, p. 3.

Gopalan, C., Rama, S. B. V. and Balasubramanian, S. C. 2009. Nutritive value of Indian foods. National Institute of Nutrition, Indian Council of Medical Research, Hyderabad, India.

Guha, M. and Malleshi, N. G. (2006). A process for preparation of pre-cooked cereals and vegetables based foods suitable for use as instant soup mix or of similar type foods. *Indian Patent* 257/NF/06.

Gull, A., Gulzar Ahmad, N., Prasad, K. and Kumar, P. 2016. Technological, processing and nutritional approach of finger millet (*Eleusine coracana*)—A Mini Review. *Journal of Food Processing and Technology.* 7: 593

Hegde, P. S., Chandrakasan, G. and Chandra, T.S. 2002. Inhibition of collagen glycation and cross linking in vitro by methanolic extracts of Finger millet (*Eleusine coracana*) and Kodo millet (*Paspalum scrobiculatum*). *Journal of Nutritional Biochemistry.* 13:517–521.

Hegde, P. S., Anitha, B. and Chandra, T. S. 2005. In vivo effect of whole grain flour of finger millet (*Eleusine coracana*) and kodo millet (*Paspalum scrobiculatum*) on rat dermal wound healing. *Indian Journal of Experimental Biology.* 43(3): 254–258.

Hegde, P. S., Rajasekaran, N. S. and Chandra, T. S. 2005. Effects of the antioxidant properties of millet species on oxidative stress and glycemic status in alloxan-induced rats. *Nutrition Research.* 25: 1109–1120.

Hithamani, G. and Srinivasan, K. 2014. Effect of domestic processing on the polyphenol content and bioaccessibility in finger millet (*Eleusine coracana*) and pearl millet (*Pennisetum glaucum*). *Food Chemistry.* 164: 55–62.

Hulse, J. H., Laing, E. M. and Pearson, O. E. 1980. Sorghum and the Millets: Their Composition and Nutritive Value. London: Academic Press.

Kavitha, M. S. and Prema, L. 1995. Post prandial blood glucose response to meals containing different CHO in diabetics. *Indian Journal of Nutrition Dietetics.* 32:123–126.

Kawai, K., Murayama, Y., Okuda, Y. and Yamashita, K. 1987. Post prandial glucose, insulin and glucagon responses to meals with different nutrient compositions in NIDDM. *Endocrinologia Jponica.* 34(5):745–753.

Krishnan, M. and Prabhasankar, P. 2010. Studies on pasting, microstructure, sensory, and nutritional profile of pasta influenced by sprouted finger millet (*Eleucina coracana*) and green banana (*Musa paradisiaca*) flours. *Journal of Texture Studies.* 41: 825–841.

Krishnan, R., Dharmaraj, U., Sai Manohar, R. and Malleshi, N. G. 2011. Quality characteristics of biscuits prepared from finger millet seed coat based composite flour. *Food Chemistry.* 129(2): 499–506.

Kumar, S. I., Babu, C. G., Reddy, V. C. and Swathi, B. 2016. Anti-nutritional factors in finger millet. *Journal of Nutrition and Food Sciences.* 6: 491.

Kurien, P. P. and Desikachar, H. S. R. 1966. Preparation of refined white flour from ragi using a laboratory mill. *Journal of Food Science and Technology.* 3: 56–58.

Lafiandra, D., Riccardi, G. and Shewry, P. R. 2014. Review—Improving cereal grain carbohydrates for diet and health. *Journal of Cereal Science.* 59: 312–326.

Lebovitz, H. E. 2001. Effect of the postprandial state on non traditional risk factors. *American Journal of Cardiology.* 88:204–205

Lopez, H. W., Levrat, M. A., Guy, C., Messanger, A., Demigne, C. and Remesy, C. 1999. Effects of soluble corn bran arabinoxylans on cecal digestion, lipid metabolism, and mineral balance (Ca, Mg) in rats. *Journal of Nutritional Biochemistry.* 10:500–509.

Mahadevappa, V. G. and Raina, P. L. 1978. Lipid profile and fatty acid composition of finger millet (*Eleusine coracana*). *Journal of Food Science and Technology.* 15: 100–102.

Malleshi, N. G., Desikachar, H. S. R. and Tharanathan, R. N. 1986. Free sugars and non-starchy polysaccharides of finger millet (*Eleusine coracana*), pearl millet (*Pennisetum typhoideum*), foxtail millet (*Setaria italica*) and their malts. *Food Chemistry.* 20: 253–261.

Malleshi, N. G. 2006. Decorticated finger millet (*Eleusine coracana*) and process for preparation of decorticated finger millet. United States Patent 2006/7029720 B2.

Malleshi, N. G. 2007. Nutritional and technological features of ragi (finger millet) and processing for value addition. In K. T. Krishne Gowda and A. Seetharam (Eds.), Food Uses of Small Millets and Avenues for Further Processing and Value Addition. Indian Council of Agricultural Research. UAS, GKVK, Bangalore, India.

Marlett, J. A. 1990. Analysis of dietary fiber in human foods. In: Kritchevsky, D., Bonfield, C., Anderson, J. W., (Eds.), Dietary Fibre: Chemistry, Physiology and Health Effects. New York: Plenum. pp. 31–48.

McDonough, C. M., Rooney, L. W. and Earp, C. F. 1986. Structural characteristics of *Eleusine coracana* (Finger millet) using scanning electron and fluorescence microscopy. *Food Microstructure.* 5: 247–256.

Monnier, V. M. 1990. Nonenzymatic glycosylation, the Maillard reaction and the aging process. *The Journal of Gerontology.* 45:105–111.

Nirmala, M., Subba Rao, M. V. S. S. T. and Muralikrishna, G. 2000. Carbohydrates and their degrading enzymes from native and malted finger millet (Ragi, *Eleusine coracana*, Indaf-15). *Food Chemistry.* 69: 175–180.

O'Kennedy, M. M., Grootboom, A. and Shewry, P. R. 2006. Harnessing sorghum and millet biotechnology for food and health. *Journal of Cereal Sciences.* 44 (3): 224–235

Osungbaro, T. O. 2009. Physical and nutritive properties of fermented cereal foods. *African Journal of Food Science.* 3(2): 23–27.

Pore, M. S. and Magar, N. G. 1976. Effect of ragi feeding on serum cholesterol level. *Indian Journal of Medical Research.* 64(6): 909–914.

Punia, S., Kumar, M., Siroha, A. K., Kennedy, J. F., Dhull, S. B. andWhiteside, W. S. (2021). Pearl millet grain as an emerging source of starch: A review on its structure, physicochemical properties, functionalization, and industrial applications. *Carbohydrate Polymers*, 117776.

Pushparaj, F. S. and Urooj, A. 2011. Influence of processing of dietary fibre, tannin and in vitro protein digestibility of pearl millet. *Food and Nutrition Sciences.* 2(08): 895–900.

Rajiv, J., Chandrashekhar, S., Dasappa, I. and Rao, G. V. 2011. Effect of replacement of wheat flour with finger millet flour (*Eleusine corcana*) on the batter microscopy, rheology and quality characteristics of muffins. *Journal of Texture Studies.* 42 (6). DOI: 10.1111/j.1745-4603.2011.00309.x

Rajasekaran, N. S., Nithya, M., Rose, C. and Chandra, T. S. 2004. The effect of finger millet feeding on the early responses during the process of wound healing in diabetic rats. *Biochimica et Biophysica Acta.* 1689:190–201.

Ramachandra, G., Virupaksha, T. K. and Shadaksharaswamy, M. 1997. Relationship between tannin levels and in vitro—protein digestibility of finger millet (*Eleusine coracana* Gaertn). *Journal of Agricultural Food Chemistry.* 25:1101–1104.

Ramananthan, M. K. and Gopalan, C. 1957. Effect of different cereals on blood sugar levels. *Indian Journal of Medical Sciences.* 45:255–262.

Ramashia, S. E., Anyasi, T. A., Gwata, E. T., Meddows-Taylor, S. and Jideani A. I. O. 2019. Processing, nutritional composition and health benefits of finger millet in sub-saharan Africa. *Food Science and Technology.* 39:253–266. doi: https://doi.org/10.1590/fst.25017

Ranasalva, N. and Visvanathan, R. 2014. Development of bread from fermented pearl millet flour. *Food Processing & Technology.* 5(5): 2–5.

Rao, P. V. and Deosthale, Y. G. 1988. In vitro availability of iron and zinc in white and colouredra (Eleusine coracana): Role of tannin and phytate. *Plant Foods for Human Nutrition.* 38: 35–41.

Rao, R.S.P. and Muralikrishna, G. 2007. Structural characteristics of water-soluble feruloyl arabinoxylans from rice (*Oryza sativa*) and ragi (finger millet, *Eleusine coracana*). Variations upon malting. *Food Chemistry.* 104: 1160–1170.

Rao, S., M. V. S. S. T. and Muralikrishna, G. (2002). Evaluation of the antioxidant properties of free and bound phenolic acids from native and malted finger millet (Ragi, *Eleusine coracana* Indaf-15). *Journal of Agricultural and Food Chemistry.* 50: 889–892.

Rasane, P., Jha, A., Sabikhi, L., Kumar, A. and Unnikrishnan, V. S. 2015. Nutritional advantages of oats and opportunities for processing as value-added foods—Review. *Journal of Food Science and Technology.* 52(2): 662–675.

Rasmussen, O., Winther, C. and Hermansen, K. 1991. Glycemic responses to different types of bread in IDDM patients: Studies at constant insulinaemia. *European Journal of Clinical Nutrition.* 45(2):97–103.

Rohn, S., Rawel, H. M. and Kroll, J. 2002. Inhibitory effects of plant phenols on the activity of selected enzymes. *Journal of Agricultural Food Chemistry.* 50:3566–3571.

Saleh, S. M., Zhang, Q., Chen, J. and Shen, Q. 2013. Millet grains, nutritional quality, processing and potential health benefits. *Comprehensive Reviews in Food Science and Technology.* 12(3): 281–295.

Sarkar, P., Lohith Kumar, D. H., Dhumal, C., Panigrahi, S. S. and Choudhary, R. 2015. Traditional and ayurvedic foods of Indian origin. *Journal of Ethnic Foods.* 2(3): 97–109.

Sastri, B. N. (1939). Ragi, *Eleucine coracana* Gaertn—A new raw material for the mating industry. *Current Science.* 1:34–35.

Shahidi, F. and Chandrasekara, A. (2013). Millet grain phenolics and their role in disease risk reduction and health promotion: a review. *Journal of Functional Foods.* 5: 570–581.

Shimray, C. A., Gupta, S. and Venkateswara Rao, G. 2012. Effect of native and germinated finger millet flour on rheological and sensory characteristics of biscuits. *International Journal of Food Science & Technology.* 47(11): 2413–2420.

Shivaraj, B., Narayana Rao, H. and Patabiraman, N. 1992. Natural plant enzyme inhibitors. Isolation of a trypsin/α-amylase inhibitor and a chymotrypsin/trypsin-inhibitor from ragi (Eleusine coracana) grain by affinity chromatography and study of their properties. *Journal of Science of Food and Agriculture.* 33: 1080–1091

Shobana, S., Krishnaswamy, K., Sudha, V., Malleshi, N. G., Anjana, R. M., Palaniappan, L. and Mohan, V. 2013. Finger millet (ragi, *Eleusine coracana* L.): A review of its nutritional properties, processing, and plausible health benefits. *Advances in Food and Nutrition Research.* 69:1043–4526

Shobana, S., Sreerama, Y. N. and Malleshi, N. G. 2009. Composition and enzyme inhibitory properties of finger millet (*Eleusine coracana* L.) seed coat phenolics: Mode of inhibition of α-glucosidase and pancreatic amylase. *Food Chemistry.* 115:1268–1273.

Shobana, S. and Malleshi, N. G. 2007. Preparation and functional properties of decorticated finger millet (*Eleusine coracana*). *Journal of Food Engineering.* 79: 529–538.

Shobana, S. 2009. Investigations on the carbohydrate digestibility of finger millet with special reference to the influence of its seed coat constituents. Ph.D. Thesis. University of Mysore, Mysore.

Shobana, S., Harsha, M. R., Platel, K., Srinivasan, K. and Malleshi, N. G. 2010. Amelioration of hyperglycemia and its associated complications by finger millet (*Eleusine coracana* L.) seed coat matter in streptozotocin induced diabetic rats. *British Journal of Nutrition.* 104: 1787–1795.

Shobana, S., Sreerama, Y. N. and Malleshi, N. G. 2009. Composition and enzyme inhibitory properties of finger millet (*Eleusine coracana* L.) seed coat phenolics: Mode of inhibition of a-glucosidase and a-amylase. *Food Chemistry.* 115: 1268–1273.

Shukla, K. and Srivastava, S. (2011). Evaluation of finger millet incorporated noodles for nutritive value and glycemic index. Journal of Food Science and Technology. 51(3):527–534.

Shukla, K. and Srivastava, S. 2011. Quality characteristics of finger millet based baby food preparation as affected by its varieties and processing techniques. *Journal of Functional and Environmental Botany.* 1: 77–84.

Singh, P. and Raghuvanshi, S. 2012. Finger millet for food and nutritional security. *African Journal of Food Science.* 6(4): 77–84.

Sripriya, G., Antony, U. and Chandra, T. S. 1997. Changes in carbohydrate, free aminoacids, organic acids, phytate and HCL extractability of minerals during germination and fermentation of finger millet (*Eleusine coracana*). *Food Chemistry.* 58(4): 345–350.

Subba Rao, M.V.S.S.T. and Muralikrishna, G. 2001. Non-starch polysaccharides and bound phenolic acids from native and malted finger millet (Ragi, *Eleusine coracana*, Indaf-15). *Food Chemistry.* 72: 187–192.

Sudha, M. L., Vetrimani, R. and Rahim, A. (1998). Quality of vermicelli from finger millet (Eleusine coracana) and its blend with different milled wheat fractions. *Food Research International*, 31(2), 99–104.

Tatala, S., Ndossi, G., Ash, D. and Mamiro, P. 2007. Effect of germination of finger millet on nutritional value of foods and effect of food supplement on nutrition and anaemia status in Tanzanian children. *Tanzania Health Research Bulletin.* 9(2): 77–86.

Thapa, S. and Tamang, J. P. 2004. Product characterization of kodo ko jaanr: Fermented finger millet beverage of the Himalayas. *Food Microbiology.* 21(5): 617–622.

Thapliyal, V. and Singh, K. 2015. Finger millet: potential millet for food security and power house of nutrients. *International or Research in Agriculture and Forestry.* 2(2): 22–33.

Tharanathan, R. N. and Mahadevamma, S. 2003. Grain legumes—a boon to human nutrition. *Trends in Food Science and Technology.* 14:507–518.

Thebaudin, J. Y., Lefebvre, A. C., Harrington, M. and Bourgeois, C. M. 1997. Dietary fibre: Nutritional and technological interest. *Trends in Food Science and Technology.* 8:41–48.

Tovey, F. 1994. Diet and duodenal ulcer. *Journal of Gastroenterology and Hepatology.* 9:177–185.

Udeh, H. O., Duodu, K. G. and Jideani, A. I. O. 2017. Finger millet bioactive compounds, bioaccessibility, and potential health effects—a review. *Czech Journal of Food Sciences.* 35: 7–17.

Upadhyaya, H. D., Gowda, C. L. L. and Reddy, V. G. 2007. Morphological diversity in finger millet germplasm introduced from Southern and Eastern Africa. *Journal of SAT Agricultural Research.* 3(1): 1–3

Veenashri, B. R. and Muralikrishna, G. 2011. In vitro anti-oxidant activity of xylo-oligosaccharides derived from cereal and millet brans—A comparative study. *Food Chemistry.* 126: 1475–1481.

Vidyavati, H. G., Mushtari, J., Vijayakumari, Gokavi, S. S. and Begum, S. 2004. Utilization of finger millet in the preparation of papad. *Journal of Food Science and Technology.* 41: 379–382.

Virupaksha, T. K., Ramachandra, G. and Nagaraju, D. 1975. Seed proteins of finger millet and their amino acid composition. *Journal of the Science of Food and Agriculture.* 26: 1237–1246.

Viswanath, V., Urooj, A. and Malleshi, N. G. 2009. Evaluation of antioxidant and antimicrobial properties of finger millet (*Eleusine coracana*). *Food Chemistry*, 114: 340–346.

Wankhede, D. B., Shehnaj, A. and Raghavendra Rao, M. R. 1979. Preparation and physicochemical properties of starches and their fractions from finger millet (*Eleusine coracana*) and foxtail millet (*Setaria italica*). *Starch.* 31: 153–159.

11

Antinutritional Factors in Cereals

Vivek Saurabh and Sumit B. Urhe
ICAR-Indian Agricultural Research Institute, New Delhi, India

Anurag Upadhyay
Banaras Hindu University, Varanasi, India

Sampada Shankar
ICAR-Indian Agricultural Research Institute, New Delhi, India

CONTENTS

11.1 Introduction

Cereal is a Latin word derived from *Ceres*, referring to the goddess of harvest and agriculture. The crops covered under cereals are mostly grass belonging to the family *Gramineae* or *Poaceae* grown for edible starchy grains. As cereals are the chief source for edible food grain, it owns higher production and acreage all over the world and continuously fulfils the demand of the world's population, making it a staple food crop for humans. Moreover, cereals are used as animal feed for the production of animal-based food and other materials. The major crops that belong to this group are rice, wheat, maize, sorghum, barley, oat, and millet, which is grown by several countries in a very large amount. Among these crops, rice and wheat alone share almost 50% of world cereal crop production. The major cereal grain producing countries are China, the United States, India, Russia, Brazil, Indonesia, Argentina, France, Ukraine, Canada, Bangladesh, Vietnam, Pakistan, Australia, and Germany.

Cereals are classified as carbohydrate-rich foods as they contain the highest share of carbohydrate (approximately 75%) of whole grain. Excluding carbohydrate, cereals contain approximately 6–15% protein, i.e. gliadins and glutenins in wheat, glutelin or oryzenin in rice, zein or prolamin in maize, hordeins and glutelins in barley, albumins and globulins in oats. Cereals contain lipids in lower amounts that are mostly present in the germ of cereal grains varying from 1–3% in barley, rice, and wheat, to 5–9% in corn and 5–10% in oats on a dry matter (DM) basis. Cereals are also an important source of vitamins like thiamin, riboflavin, niacin, and vitamin E. Whole-grain cereals also contain considerable amounts of iron, magnesium, and zinc, as well as lower levels of many trace elements, e.g. selenium. As well as containing a range of phytochemicals, which may provide some of the health benefits seen among populations consuming diets based on plant foods, cereals also contain several antinutrients that create hurdles to the easy absorption and bioavailability of phytonutrients known as antinutrients or antinutritional factors (ANFs). The variation in the degree of ANFs depends on the type of food and the method of preparation. The major ANFs present in cereals includes phytic acid (PA), protease inhibitors (PI), amylase inhibitors (AI), saponins, tannins, and lectins. Antinutrients are highly biologically active secondary metabolites, produced in plants naturally, but affecting absorption, digestion, and utilization of phytonutrient elements in the body of different animals. The reduced utilization of nutrients present in cereals causes "hidden hunger" in humans in developing and underdeveloped countries where cereals are the main source of food and nutrition, which has become a big concern (Handa et al., 2017; Nadeem et al., 2010). Apart from harmful effects, ANFs also show several positive effects in control and treatment of disease like diabetes, cancer, etc. ANFs can be eliminated through application of biological and physicochemical approaches such as soaking, germination, milling, high pressure processing (HPP) or high hydrostatic pressure (HHP), extrusion, microwave (MW) heating, etc. These approaches are very useful due to their easiness and effectiveness in the elimination of these ANFs.

ANFs are perceived as a threat to nutrient bioavailability, however, they do contain several therapeutic properties. Most food sources containing ANFs are usually not consumed in their raw form. They undergo the simple method of domestication (like soaking and cooking) before consumption. So, the ANF content in the unprocessed foods may not in any way reflect the actual contents being consumed. As such, more emphasis should be on the assessment of PA in ready-to-use foods, rather than its contents in raw foods (Abdulwaliyu et al., 2019).

11.2 Cereals

11.2.1 Rice

Rice (*Oryza sativa* L., 2n = 24) is a self-pollinated C_3 crop and an important staple food crop. About 90% of total world production is contributed by Asian regions that supply the food for more than 60% of the world population, especially for the Asian continent. Most of the world's rice is grown in the tropics, which include countries of South and Southeast Asia, West Africa, and Central and South America. Important rice growing countries are India, China, Indonesia, Thailand, and Bangladesh. It is considered a high-energy or high-caloric food. Rice contains more carbohydrates (78.2 g/100 g), less protein (6.8 g/100 g), fat (0.5 g/100 g), crude fiber (0.2 g/100 g), mineral matter (0.6 g/100 g), calcium (10.0 mg/100 g), and phosphorus (160.0 mg/100 g). It is used for preparing puffed rice, rice flakes, rice wafers, canned rice, and also for starch and brewing industry. The rice bran obtained as a byproduct of milling is used for the extraction of edible rice bran oil. However, rice also contains several ANFs in the different parts of grains. The ANFs present in unstabilized rice bran is about 4% phytate, 22.58 activity/mg hemagglutinin, and 8.44 activity/mg trypsin inhibitor (Khan et al., 2009). Rice contains approximately 9.5–10.8 mg/g phytate (Dost & Tokul, 2006; Lestienne et al., 2005) and rice flour contains about 93.70 mg/100 g PA, 172 mg/100 g polyphenol, and 9.8 mg/100 g oxalate (Kaushal et al., 2012).

11.2.2 Wheat

Wheat (*Triticum aestivum* L., 2n = 42) is a self-pollinated C_3 crop. It is a chief crop used for the preparation of bread and *chapatis*. Wheat is usually processed to prepare *Dalia* or flour. Thereafter, flour is also used to produce a wide range of products such as *roti, chapatti, paratha, puri, dosa, noodles, upma, halwa, kheer, payasam, laddoo*, bread, buns, cake, biscuits, flakes, etc. Moreover, it is also used for the preparation of breakfast cereals, animal feed, dextrose, and alcohol. Wheat is the most widely cultivated among all the cereals. It is commonly grown in North and Latin America, Europe, West Asia, South Asia, North Africa, South Africa, East Africa, and Australia. Wheat grain contains about 60–68% starch, 8–15% protein, 1.5–2.0% fat, 2.0–2.5% cellulose, 1.5–2.0% minerals, and relatively higher amounts of protein, niacin, and thiamin. ANFs are also present in wheat. ANFs are mostly found in the outer layer and bran. A higher amount of phenolic compounds and more than 80% of PA are situated in the bran. The PA, saponins, and total phenolic content present in wheat bran is 3.68 mg/g, 1.16 mg Diosgenin equivalent (DE)/g, and 99.11–188.61 mg gallic acid equivalent (GAE)/100 g respectively (López-Perea et al., 2019). Dost & Tokul (2006) estimated a higher level (20–24.3 mg/g) of PA by using UV-Vis and high-performance liquid chromatography (HPLC) methods in grain. Moreover, tannin (2020 µg/g DW) and trypsin inhibitor (46.30 unit/g) was found in mature wheat grain (Singh et al., 2012).

11.2.3 Maize

Maize (*Zea mays* L., 2n = 20) is a cross-pollinated crop with C_4 mechanism of photosynthesis. Maize is known as the "Queen of cereals" due to its very high yield potential. It is considered as a miracle crop. Maize is consumed after being roasted, boiled, or heated at different physiological maturity stages. Moreover, it is used for starch preparation at the commercial level. The use of flakes, bread, buns, rusks, and biscuits etc. is becoming popular. Baby corn is used as a vegetable for salads, manchurian, etc. Maize is used as an important ingredient in animal feed. It is a basic ingredient of swine and poultry rations. Maize stover is used as fodder for animals. It is commonly grown in the United States, China, Brazil, Mexico, India, Romania, Philippines, and Indonesia. Maize grain contains about 71.5% starch, 1.97% sugar, 10.3% protein, 4.8% fat, and 1.44% ash. The protein content varies between 6–15% in different strains. PA, polyphenols, trypsin inhibitors (TIs), α-amylase inhibitors (α-AIs), oxalate, and tannin are generally present in maize. In the two genotypes (Var–113 and TL–98B–6225–9×TL617) of maize, 1,047 mg/100 g and 87.16 mg/100 g PA; 460.50 mg/100 g and 363.70 mg/100 g polyphenols were reported, respectively (Sokrab et al., 2014). The amount of ANF content in crop varies with the genotype of the crop. In yellow maize, 9.87 mg/g DM phytates and in raw quality protein maize, 0.291% phytate has reported. The presence of oxalate (0.128%) and tannin (0.002%) in raw quality protein maize; TI, (3.65 TIU/mg), and α-AI (917 AIU/mg) in yellow maize has also been reported (Chukwuma, 2016; Ejigui et al., 2005).

11.2.4 Sorghum

Sorghum or jowar (*Sorghum bicolor* L. Moench, 2n = 20) is a self-pollinated (cross-pollinated under specific conditions) C_4 plant. It is fourth in importance among the world's leading cereals. Millions of people in Africa and Asia depend on sorghum as a staple food. It has the capacity to withstand drought, grows relatively quickly, and produces not only good yields of grain but also very large quantities of fodder. It is generally consumed as chapati or boiled and cooked like rice and popped grains. Moreover, it is also used as poultry and animal feed for cattle and pigs. The leading sorghum grower countries are the United States, India, China, Nigeria, Sudan, and Argentina. Sorghum grain contains about 70% carbohydrates, 10–12% protein, and 3% fat. The phosphorus concentration observed in bran is higher than observed in the whole grain. Phenolic compounds, phytate, and total phosphorus are mainly deposited in the grain bran and aleurone layers (López-Perea et al., 2019). Sorghum contains about 925 mg/100 g DM phytate (Lestienne et al., 2005).

11.2.5 Barley

Barley (*Hordeum vulgare*, 2n = 14) is a self-pollinated C_3 plant. It is the fifth most important cereal in the terms of production in the world. Of total barley produced in the world, about 2–3% is used for human consumption, 65% for animal feed, and 30% for alcohol and vinegar production. Moreover, it is also used for the preparation of energy drink and cookies. It is mostly growing in Russia, France, Germany, Ukraine, and Canada. Barley grains contains about 11.5% protein, 74% carbohydrate, 1.3% fat, 3.9% crude fiber, and 1.5% ash. On the nutritional point of view, some undesired compounds, such as PA, are present in the grain of barley. Dost & Tokul (2006) estimated 21.5–24.6 mg/g of PA in the grain. It also contains about 0.5% α-galactoside content (Andersen et al., 2005). Barley contains a very high amount of β-glucans. The concentration of β-glucans generally range between 3.9–4.9% (Zhang et al., 2001), but may reach up to 10% (Izydorczyk et al., 2000). The season and environmental conditions also affect the accumulation of β-glucans in barley grains. In Lithuania, lower accumulation (1.64%) of β-glucans was recorded in a spring crop compared to a winter season crop (2.89%) (Alijošius et al., 2016). Additionally, in China, in the cultivars Sumei 21 and QB25 reported the range of 2.98–8.62% β-glucans out of 164 cultivars (Zhang et al., 2002). PA was found mainly in the aleurone layer and it contains about 80% of total phytate, the rest is in the germ. The amount of phytate in 100 genotypes or cultivars of barley ranged from 3.85 mg g^{-1} to 9.85 mg g^{-1} in China (Dai et al., 2007). The presence of the total phenolic compounds, tannin, saponin, and PA present in Egyptian barley germplasm has been seen up to 7.50 mg g^{-1} total phenolic compound, 1.98 mg/g tannin, 9.21 mg/g saponin, and 1.87 mg/g vicine present in the different genotypes (Mohamed et al., 2010). Apart from grain, PA (2.32 mg/g) and saponins (1.96 mg DE/g) are also present in barley husk.

11.2.6 Pearl Millet

Pearl millet (*Pennisetum glaucum* L., 2n = 14) is a highly cross-pollinated C_4 plant. It is a staple food that supplies a major proportion of calories and protein to large segments of the populations living in the semiarid tropical regions of Africa and Asia. The crop is supremely adapted to heat and aridity, and production is likely to increase as the world gets hotter and drier. It is mostly grown in Asia followed by Europe and America. In the United States, Australia, South Africa and South America, pearl millet is grown as a forage crop and used to feed the cattle. Pearl millet contains about 67% carbohydrate, 9–21% protein, 2–3% fat, and some amount of minerals like iron, zinc, magnesium, copper, manganese, potassium, and phosphorous. Mature kernels are rich in vitamin A but deficient in vitamins B and C. The ANFs present in the grain are distributed unequally, the amount of phytate present in the germ, endosperm, and bran was reported at 752 mg/100 g, 86 mg/100 g, and 278 mg/100 g, respectively (Lestienne et al., 2005).

11.2.7 Finger Millet

Finger millet (*Eleusine coracana*, 2n = 36) is a self-pollinated C_3 plant also known as ragi or raji in different parts of India. This is the sixth most important cereal crop in the world. It is consumed as flour or used to prepare malt and fermented beverages. Moreover, it is also used to prepare several cousins like *idli, dosa, laddu, bhakri, ambil, papad, ragi mudde,* etc. It is grown mainly in the eastern and central part of Africa, India, parts of China, Sri Lanka, Taiwan, and parts of Indonesia and Guam. Finger Millet contains about 65–75% carbohydrate, 5–8% protein, 1.3% fat, and 2.5–3.5% minerals. However, the millet also contains phytates (0.51%), polyphenols, tannins (0.53%), trypsin inhibitors (0.47%), and dietary fiber, which were once considered antinutrients due to their metal chelating and enzyme inhibition activities. Finger millet contains about 51.67 mg/100 g phytate, 0.47 mg/100 g trypsin and 53.33 mg/100 g tannin (Abioye, 2018).

11.2.8 Oats

Oats (*Avena sativa* L., 2n = 42) are self-pollinated plants, equipped with the C_3 mechanism of photosynthesis. It is consumed due to its high nutritive value and richness of dietary fiber and also used as feed to ruminant animals. Moreover, oatmeal, cookies, bread, and other food materials are prepared from oats. Most of the oat cultivation areas are in Russia, the United States, Canada, Poland, China, France, and Australia. Oats contains 67% carbohydrate, 6.5% fat, 13% protein with a good amount of vitamins, minerals, and fiber. Apart from the health promoting components, some ANFs such as PA (4.5 mg/g), saponin (1514.7 µg/g), and tannins (1.7 mg/g) are found in oats (Kaur et al., 2019). In another study, Dost & Tokul (2006) reported higher PA (11.5–12.5 mg/g) in oat, which is higher than rice and about half of the PA content present in wheat.

11.3 Strategies to Reduce Antinutritional Factors

11.3.1 Soaking

Soaking is a simple technological treatment that is often used for the preparation of food at home. Soaking is a traditional practice that helps wash cereals, remove dirt and chemical substances, disinfect surfaces, and overcome some nutritional hurdles or ANFs. In this method, grains are dipped in a liquid solution for some duration. Soaking exaggerates the elimination of water soluble antinutrients from foods through leaching. Moreover, it also permits the water to scatter in the protein fraction and starch granules, which provides protein denaturation, starch gelatinization, and softens the texture of the food. The softened texture of the food reduces the cooking time and helps to enhance the bioavailability of minerals from the food (Hotz & Gibson, 2007). Generally, cereal grains like rice, wheat, corn, barley, sorghum, millets, and oats are soaked followed by sprouting or germination by humans for consumption as food. It is also applied as a pretreatment to other processing methods, such as fermentation and germination, for removal of ANFs from a food source (Kumar et al., 2010). Soaking also provides significant benefits in terms of protein digestibility, removal of seed coats and any further residue present on the surface of the grains (Adeleke et al., 2017). Soaking generally reduces the concentration of antinutrient phytochemicals like phytate, tannins, etc. Due to abovementioned advantages, it was suggested that wheat should be soaked for 12–24 hours before being consumed as food (Hendek Ertop and Bektaş, 2018). At the time of soaking, solvent leached out phytate or endogenous/exogenous phytase enzymes hydrolyse phytate, resulting into enhanced *in vitro* solubility of minerals such as zinc and iron (from 2–23%) (Lestienne et al., 2005; Vashishth et al., 2017). The efficiency of leaching depends upon the grain and type of solvent used that diffuses easily in the pores of vegetal materials and allows the grain to leach out bioactive compounds (López-Perea et al., 2019). The phytate to iron (Phy/Fe) molar ratio is used as a parameter to estimate the availability of minerals. The inverse proportion can be seen in the molar ratio and bioavailability—the lower the molar ratio, the higher the bioavailability. Soaking significantly reduces the molar ratio of phytate to zinc (Phy/Zn) in millet, maize, sorghum, and rice from 18.5, 40.6, 62.8, and 36.3 to 16.4, 41.4, 58.5, and 45.8, respectively, but no significant effect of soaking observed in Fe availability (Lestienne et al., 2005). Apart from water, lactic acid shows a significant reduction in phytate content. Soaking in 2.5% lactic acid for 48 hours reduced about 24.4% and 29.9% of phytate-phosphorus in corn and wheat, respectively (Vötterl et al., 2019).

11.3.2 Germination

Germination or the biological activation of grain is a bioprocess by which a plant grows from a seed. During germination, several biochemical changes occur in the grain. Development of the autotrophic seedling by dry seed after water absorption is known as germination. It includes four stages, i.e. imbibition (water absorption), initiation in enzymatic activity, transfer of reserve food from endosperm to embryo, and radicle emergence followed by seedling growth. During germination, several hydrolytic enzymes produced from the scutellum and aleurone layer of seed/grain (Szewińska et al., 2016). Germinated seeds are considered a good source of ascorbic acid, riboflavin, choline, thiamine, tocopherols, and pantothenic acid. They are also used as a biological and environmentally friendly processing technique to overcome ANFs, enhancing the sensory quality and bioavailability of nutrients. Germination is one of the easiest methods to enhance the bioavailability of nutrients in the human body. The important factors responsible for the germination of any seeds are temperature and humidity. In addition to these, nutrients are required for its growth and development (Sangronis & Machado, 2007). For a good germination, the grain should be washed and soaked in water for 12–48 hours at ambient temperature. After gaining 30–35% water absorption, the water should be drained and the grain then formed a heap to let them germinate. Germination can be used successfully to make low-phytate ingredients with high endogenous phosphatase activities the phytase enzyme, which reduces phytate content and results in decreased PA concentration in the cereal. Consequently, it leads to higher absorption of phosphorus by monogastric animals as well as a reduction in phosphate pollution (Centeno et al., 2003). For barley, soaking followed by germination for 3–5 days enhanced the phytase up to 3,151 U/kg and acid phosphatase 3,151 U/g (212% and 634%, respectively) with a concomitant decrease in phytate phosphorus content (up to 58%) (Centeno et al., 2003). Germination has been attributed to the removal of polyphenols by leaching into soaking water as well as enzymatic hydrolysis. The reduction in polyphenol content (up to 75%) was found in wheat as compared to soaking, MW, and other treatments (Singh et al., 2017). Finger millet seeds were presoaked then germinated at room temperature. Total phenolic and tannin content decreased significantly from 33.33–61.66% (Abioye, 2018). Germination of two corn genotype (Var–113 [high phytate] and TL–98B–6225–9·TL617 [low phytate]) for 6 days, significantly reduced 84% of PA while it increased 138.27% polyphenol contents. After 6 days of germination, higher extractable calcium and iron was reported in Var–113; while, higher phosphorus, and magnesium were noticed and TL–98B–6225–9·TL617 (Sokrab, Mohamed Ahmed, et al., 2012).

11.3.3 Fermentation

Fermentation is a complex metabolic activity associated with many chemical and metabolic changes by which the complex molecules are converted into free sugars, alcohol, vitamins, CO_2, and enhance the bioavailability of minerals, which is results from the breakdown of some of the antinutritional compounds. It is one of the oldest and simplest techniques and is widely used in the food processing industry to develop fermented products, preserve products, remove undesirable components, such as PA, tannins, and polyphenols, and this also helps to enhance the absorption of nutrition in humans (Simwaka et al., 2017). Fermentation leads to degeneration of grain compounds, particularly starch and soluble sugars by endogenous and exogenous enzymes (Hassan et al., 2006). The bacterial strains such as *Leuconostoc, Streptococcus, Pediococcus, Micrococcus, Bacillus,* and *Lactobacillus* spp. are commonly used for the fermentation of grains (Singh et al., 2015). This process enriches the nutritional value of grains by increasing the amount of essential amino acids such as lysine, methionine, and tryptophan (Mohapatra et al., 2019). The bacteria like *Lactobacillus plantarum* used in fermentation excrete lysine, thereby enhancing the quality of proteins in corn (Newman & Sands, 1984). The grains of two corn genotypes (Var–113 and TL–98B–6225–9·TL617) were germinated for 6 days followed by 14 days of fermentation. Fermentation significantly increased calcium and phosphorus extractability and reduced PA and polyphenol content with the increase in days of fermentation (Sokrab, Ahmed, et al., 2012). Thereafter, another experiment was conducted on the same genotype of corn grain flour. The corn flour fermented for 14 days which increased the level of extractable calcium (94.73%), phosphorus (76.55%), and iron (84.93%) (Sokrab et al., 2014). The reduction in phytate and PA levels was also observed in finger millet and pearl millet (Eltayeb et al., 2008; Sripriya et al., 1997). The lactic acid fermentation of corn flour for 96 hours at 30°C successfully reduced phytate, Tis, and α-AI by 61.5%, 41.7%, and 16.6%, respectively (Ejigui et al., 2005). Likewise, different species of lactic acid bacteria were used for the fermentation of corn paste (Kutukutu). Consequently, bacterial species *L. brevis* G25 *L. plantarum* A6 reduced the tannin content by 98.8% and *L. buchneri* M11 reduced the phytate content by 95.5% as well as improved the bioavailability of magnesium and iron by 50.5% and 70.6%, respectively (Roger et al., 2015). A reduction of about 24.3% in phytate levels was noted in yeast fermented corn cultivars (Cui et al., 2012). Two pearl millet cultivars (Gazira and Gadarif) fermented after processing for 12–24 hours significantly reduced PA, polyphenol, and tannin content (Eltayeb et al., 2008). Phytate reduction was also recorded in both fermented and malted sorghum up to 20–21% and 19–33% after 2 and 3 days, respectively (Onyango et al., 2013).

11.3.4 Milling

Milling is a very important postharvest practice of grains. It is an act or process in which the grains are de-hulled, split, polished, or ground into flour of varying particle size. There is two milling process: (1) The whole grain is converted into flour without subtracting any parts, and (2) the whole grain undergoes differential milling to separate the grain into different parts. For example, wheat could be milled as whole wheat flour or undergo roller milling to yield multiple products as refined wheat flour, bran, germ, semolina, etc. The cereals after harvesting from parent plants possess a protective coating layer and indigestible substances on it that is rich in antinutrients. The primary aims of milling are the removal of the protective layer and to facilitate easy processing. Milling leads to a nutritional difference in the whole and processed grains. It removes all the nutrients present in the outer layer of grains, including dietary fiber, ANFs such as PA, tannin, polyphenol, and some enzyme inhibitors like TI, as well as minerals and some vitamins (Oghbaei & Prakash, 2016).

11.3.5 Roasting

Roasting is a dry heat treatment for a shorter duration used to induce the development of color, taste, and flavor; it also changes the chemical composition, modifying nutritional value and shelf life. It is carried out at above 100°C with average moisture content between 15–22% of the grain. Roasting helps in moisture reduction as a pretreatment of grains to remove husks, making water absorption easier and reducing cooking time. Moreover, heating destroys the naturally occurring heat liable ANFs so, its use in the reduction of antinutrients in plant based foods enhances the nutritional value (Oboh et al., 2010).

Roasted quality protein maize (RQPM) had a higher amount of microelements (iron, zinc, copper, manganese, and selenium) than boiled quality protein maize (BQPM) and raw quality protein maize (CQPM). However, boiling significantly reduced phytate content by 9.62% while roasting had a 5.84% phytate reduction compared with the raw. In BQPM, oxalate reduced significantly by 7.03% while in RQPM, only 3.13% oxalate reduction was achieved. Both cooking methods (boiling and roasting) had a similar reduction (50%) efficacy for tannin. The results demonstrated that boiling was more effective in the reduction of ANFs than roasting, but the nutrient loss was higher in the boiling treatment (Chukwuma, 2016). The study of Tiwari et al. (2014) revealed that roasting not only reduced antinutritional factors like PA and total polyphenols, but also caused the least loss (non-significant) in iron and zinc contents as well as antioxidants. A significant decrease was observed in PA content. The maximum reduction in PA was from 728–410 mg/100 g (43.68%) for a treatment time of 60 seconds was found to be comparable to fermentation (45.32%). Reduction in polyphenols from 352.64–338.26 mg/100 g and from 340.13–337.79 mg/100 g was reported after 30 seconds and 60 seconds, respectively. The iron content of heat treated pearl millet flour for 30 and 60 seconds decreased from 6.86–6.80 mg/100 g; 6.86–6.79 mg/100 g, respectively. The zinc content reduced from 3.059–3.005 mg/100 g; and 3.059–2.985 mg/100 g during roasting for 30 and 60 seconds, respectively (Tiwari et al., 2014). Roasting reduced the 3.13% oxalate and 50% of tannin content present in quality protein maize (QPM) (Chukwuma, 2016).

11.3.6 Boiling and Cooking

Boiling and cooking are traditionally used hydrothermal practices applied also at the household level. These hydrothermal processes successfully reduce the ANFs from various cereals and pulses. In QPM, boiling significantly reduced phytate content by 9.62%, oxalate by 7.03%, and tannin by 50% (Chukwuma, 2016). Cooking of maize greatly reduced PA from 473–353.0 mg/100 g, polyphenol from 533–343 mg/100 g, and tannin from 673–562 mg/100 g. When the cooking of maize was preceded by fermentation, a further reduction in phytate was observed in polyphenols of 473–211 mg/100g (50%) and tannin also followed a trend similar to phytate (Awada et al., 2005)

11.3.7 Extrusion

Extrusion is a method of development of weaning food. This applies force on pumpable materials through a small opening or die. The pressure and heat are generated by rotating screw assembly and heated coils present in the barrel. As a fully automated technology, extrusion has control over applied pressure, product temperature, screw speeds, and die shape. Products processed through high-temperature short time (HTST) (>100°C) treatment, by which raw material gets cooked, kneaded, mixed, and forced to take the desired shape from the die opening. The heat liable ANFs (phytate, tannins, saponins, and oxalates) present in the cereal grains or flour get destroyed by the application of the high temperature during extrusion processing. Short-time processing conserves nutrients and ensures better nutrient availability. Commercially, the extrusion process was used to develop new food products from cereals, like pasta, baby foods, animal foods, breakfast cereals, and starch modification, etc. Besides the application of extrusion processing to formulate nutritive foods, it has several other benefits like starch gelatinization, enhancement of soluble dietary fibers, reduction in lipid peroxidation, and lowering ANFs (Nikmaram et al., 2017). Extrusion is being considered the best method to abolish α-AIs, trypsin, and chymotrypsin and hemagglutinin activity without modifying the protein content in food products. Extrusion processing is a suitable technique for improving the quality of cereal bran for use in food preparation. It was applied on bran of wheat, rice, barley, and oats. The extrusion process reduced the PA by 54.51%, polyphenol by 73.38%, oxalates by 36.84%, and trypsin inhibitor by 72.39%. The heat treatment caused the highest reduction in polyphenols followed by TIs, PA, and oxalates. The highest reduction in antinutrients was observed at 140°C and 20% moisture content. Treatment also significantly increases bulk density compared to raw bran, increases redness, and decreases yellowness of bran after extrusion treatment (Kaur et al., 2015). Extrusion cooking also reduced PA content by nearly 50% in a maize-common bean blend (Maseta et al., 2016). *Uji* (thin lactic fermented porridge) was developed by blending maize and finger millet in a single screw extruder having barrel temperatures of 150–180°C with a screw speed of 200 rpm. Tannin content decreased from 1677 mg/100 g in the raw blend to 697 mg/100 g after extrusion of the unfermented blend and further to 551 mg/100 g after fermentation and extrusion, but phytate content was unaffected after treatment (Onyango et al., 2005).

11.3.8 Microwave

Microwaves (MW) are the electromagnetic spectrum whose frequencies range from 300 MHz to 300 GHz and contains magnetic, electric, perpendicular, and oscillating fields. MWs are generally used as energy carriers using a frequency range from 0.915–2.450 GHz. MW instruments work on the phenomena of dielectric heating. This frequency produces a thermal energy collision and dipole movements in polar molecules of the food sample (Kumar et al., 2020). MWs increases the temperature above 100°C, which effectively destroys the ANFs. MW is getting popularity attributable to its high efficiency, fast processing, and economic feasibility. MWs have an HTST treatment, so it breaks down antinutritional factors without disturbing the other nutritional qualities and there may also be a vitamin retention aspect to MW cooking. However, due to the shorter time and a smaller amount of water, MW cooking affects the nutritional content more than that of conventional methods (Abbas & Ahmad, 2018). Application of MW heating at 2,450 MHz for 2.5 minutes on edible cereal bran resulted in 53.85%, 57.21%, 65.00%, and 100% reduction in PA, polyphenols, oxalates, and saponins, respectively (Kaur et al., 2012).

11.3.9 Irradiation

The application of ionizing radiation to food acts as a preservation treatment to reduce microbial load, inhibit germination, disinfestation, and decontamination (Tresina et al., 2017). Gamma (γ) and electron beam irradiation are the most widely used forms of radiation for food. Irradiation treatment alone or in combination with other processing operations has been shown to eliminate ANFs present in cereals and legumes (Bhat & Sridhar, 2008; Tresina & Mohan, 2011) by the formation of free radicles. Free radicles are responsible for the structural breakdown of ANFs. In phytate, free radicles cause the breakdown of phytate into lower inositol phosphates by forming cleavage in the phytate ring. The electron beam has several distinct advantages compared to γ irradiation, such as short treatment time, no residuals in products, no need for porous packaging, precision control, and elimination of the need to use rinse water (Shawrang et al., 2011). Irradiation doses 10 kGy or above are mostly applied to grains like cereals and pulses to reduce ANFs and for disinfection. The effect of ionizing radiation doses on the food grains initiates chemical changes that include the formation of radiolytic products, oxidation, the reduction of carbonyl groups to hydroxyl derivatives, destruction of double bonds, decreases in aromatic and heterocyclic compounds, free radical formation, and enzyme degradation (Machaiah & Pednekar, 2002; Siddhuraju et al., 2002). All these changes due to irradiation depend on the application dose and time of exposure. Maize and sorghum treated with 2 kGy doses of γ irradiation significantly reduced the concentration of PA and tannins (Hassan et al., 2009). The combined treatment of radiation and cooking significantly reduced the phytate content of whole pearl millet flour from 761.61–293.40 mg/100 g (Mohamed et al., 2010). Jan et al. (2020) reported irradiation treatment on brown rice based weaning food treated by irradiation at 0 kGy, 2 kGy, 4 kGy, 8 kGy, and 10 kGy doses. Irradiation also signifies a rehydration ratio, solubility index, decreased browning index, antioxidant activity, and iron and calcium content improvement on the cereal based foods. However, microbial load, organoleptic properties, and ANF content decreased with increasing irradiation doses (Siddhuraju et al., 2002).

Phytate and tannins are the important ANFs of sorghum. Treatment of 10–30 kGy of γ irradiation successfully reduced 28–86% tannin content and 39–90% of phytate content as compared to nontreated grains (Shawrang et al., 2011).

11.4 Adverse and Beneficial Effects of ANFs on Health

11.4.1 Phytic Acid and Phytate

Phosphorus is a macronutrient for the plant, which is essential, and intended for the development of the seed/grain and successful seedling growth. In the seed, a large share of phosphorus stored in the form of *Myo*-inositol 1,2,3,4,5,6–hexakisphosphate (InsP6), known as PA. It contains about 50–80% proportion of total phosphorus and more than 1% of total dry weight of seed (Dai et al., 2007; Hendek Ertop & Bektaş, 2018; Nikmaram et al., 2017). PA is mainly present in plants as a phosphate ester of inositol or inositol polyphosphate of its salt form known as phytate, which accumulates in the seeds or grains during the ripening period. These are generally present in the aleurone layer and pericarp of wheat and endosperm of maize. The concentration and distribution of PA in a crop vary from the crop species, varieties, environmental condition, soil and agricultural practices, maturity stage, harvesting, age of grains, and processing techniques. At physiological pH, the phosphate is partially ionized. It is reported that applying higher amounts of phosphoric fertilizer enhance phosphate (PO_4^{3-}) absorption by the roots. Further, it esterifies through hydroxyl group to the carbon chain (C–O–P) as a single phosphate ester or attaches to another phosphate by an energy rich pyrophosphate bond, consequently enhancing the accumulation of the phytate concentration in a crop (Abdulwaliyu et al., 2019).

PAs are chemically active in a very wide range of pH and six reactive (negatively charged) phosphate groups make it a very effective chelator. Due to its ability to form chelates, it reacts and binds several divalent and trivalent mineral ions, such as Zn^{2+}, $Fe^{2+/3+}$, Ca^{2+}, Mg^{2+} present in grains, consequently, it decreases their solubility, functionality, digestibility, intestinal absorption of minerals, affects enzymatic degradation and peptic digestion in nonruminant animals and their bioavailability. These minerals are required daily for effective functionality of the basic chemistry of life and due to reduced bioavailability of these minerals and enzymatic inhibition, PA considered as an ANF (Dai et al., 2007; Gemede & Ratta, 2014; Panwar et al., 2016). The mineral deficiencies caused by PA can be predicted by the molar ratio between PA and minerals, which should be below critical levels. An increase in the molar ratio above critical level affects the absorption and bioavailability of minerals (Sokrab et al., 2014). The inhibitory effect of phytate on zinc absorption can also be predicted *in vitro* by the molar ratio of Phy/Zn. Davies and Olpin (1979) showed that molar ratios above 10–15 progressively inhibited zinc absorption and were associated with suboptimal zinc status. In the same way, Saha et al. (1994) showed that absorption of ratio labelled iron in rats decreased significantly when the molar ratios of Phy/Fe were above 14 in wheat-flour-based diets containing 0.19–1.85% of phytate (Lestienne et al., 2005). Calcium (Ca) itself does not interact with Zn strongly, but there are multiple interactions between Zn, Ca, and phytate. When Ca content is higher, it stimulates fiber to bind Zn through coprecipitation and formation of insoluble combinations of Ca, Zn, and PA in the gastrointestinal (GI) tract.

The presence of phytate concentration above 0.25% is considered lethal for health (Maseta et al., 2016). The Recommended Daily Intake (RDI) value of phytate varied by country; for example, 180 mg RDI/ day has been specified for Sweden, whereas the UK and the United States accept 631–746 mg RDI/ day (Nissar et al., 2017). Phytate can be hydrolyzed by phytase enzyme into a bioavailable form, such as lower *myo*-inositol phosphates and phosphate, but phytase enzyme does not produce naturally in nonruminant animals and humans so they remain unavailable or poorly utilized in the body (Centeno et al., 2003). Furthermore, processing methods such as fermentation, germination, appertization, and extrusion were found very effective in the reduction of PA content in grains and processed products, enhancing nutrient bioavailability. Despite the antinutrient effect of PA, it also has some health benefits like anticarcinogenic and antioxidant properties that can help prevent several coronary diseases and various types of cancers (Udomkun et al., 2019). In the human body, several life-threatening diseases like diabetes, cancer, cardiovascular diseases, and neurodegenerative diseases originate due to oxidative stress. Antioxidants have potential to reduce the incidence of oxidative stress in body. PA has a very high chelating potential as well as exhibits antioxidant properties. This natural antioxidant (PA) can downregulate the series of chain reactions, inhibits the formation of iron catalyzed hydroxyl radical (OH^-), linoleic acid auto-oxidation and Fe(II)/ascorbate-induced lipid peroxidation in human colonic epithelial cells, oxidative stress through the formation of mineral chelates, and boost the immune system (Abdulwaliyu et al., 2019). So, the consumption of PA (below critical level) is considered beneficial for health.

11.4.2 Polyphenols

Phenolic compounds belong to a specific group of secondary metabolites present in plants and widely distributed in the plant kingdom. Phenolic compounds contain hydroxyl groups and aromatic rings. These are broadly categorized as phenolic acids, flavonoids, coumarins, and tannins. The predominant phenolic acids present derive from benzoic and cinnamic acids in plants (Van Hung, 2016). Generally, the presence of a phenolic compound is concentrated on the outer layer of seed or grain and found least in the endosperm portion. Therefore, the extracts from the bran always contained higher total phenolic compounds as compared to the endosperm. For instance, finger millet contains nearly 90% of polyphenols, which are concentrated in the seed coat tissue, hence the outer portion has the highest polyphenol content (Oghbaei & Prakash, 2012). The presence of tannins is very limited to cereals such as finger millet, sorghum, and barley. In cereals, the most common phenolic content is phenolic acid, which is mostly (about 90%) present in bound form. These are generally categorized into two types i) hydroxybenzoic acid and ii) hydroxycinnamic acid derivatives. The presence of p-hydroxybenzoic, chlorogenic, vanillic, caffeic, syringic, p-coumaric, ferulic, and sinapic acids has been seen in raw as well as processed cereals. The transferulic acid was found to be dominant phenolic acid in raw brown rice, wheat, oat, corn, and sorghum; p-hydroxybenzoic is the major acid in barley and oats; and diazene is the dominant acid in the millet (Van Hung, 2016). Existence of polyphenolic pigments in pericarp, aleurone and endosperm areas in pearl millet may cause the development of unpleasant gray color and taste to the finished product (Rani et al., 2018).

A class of polyphenols that interferes with numerous enzymatic systems in humans and most notably those that control thyroid hormone synthesis, which enhances the incidence of goitrogen in humans are known as goitrogenic polyphenols (C-glycosylflavones), such as glucosyl vitexin, glucosyl orientin, and vitexin. C-glycosylflavones are stable to hydrolysis and are biologically active both in plant as well as dietary components. This can lead to the development of a goiter, i.e. a swelling of the neck or larynx resulting from enlargement of the thyroid gland (Boncompagni et al., 2018).

Tannins are a widely available, higher molecular weight (500–3,000 Da), water-soluble polyphenol that is present in two forms i.e. hydrolysable and condensed tannins in plants. Based on the chemical structure, it is classified into hydrolysable tannins composed of polyhydroxy alcohol esterified with gallic or ellagic acid and the condensed tannins, which are flavonoid-based polymers. It is highly stable against heat and bitter in taste. During the process of digestion in ruminant animals, tannins hydrolyze into a toxic substance. Being a polyphenolic compound and containing hydroxyl and other groups, it binds or forms precipitate or strong reversible or irreversible complexes with proteins that impair protease access to labile peptide bonds and various other organic compounds, such as alkaloids and amino acids, which further reduces their digestibility in both humans and animals. It also results in the production of off-colored products. These tannins usually occur in food products, which inhibit the enzymatic activity of amylase, lipase, trypsin, and chymotrypsin, thus decreasing the bioavailability of protein and carbohydrates and interfering with the absorption of vitamins such as vitamin B_{12} and minerals such as iron and Zn, finally affecting the digestive system (Abbas & Ahmad, 2018). Although, PA is the primary chelator for iron and zinc in whole pearl millet flour, these also bind with polyphenols and form iron and zinc complexes (Krishnan & Meera, 2018). The major inhibitory effect on iron absorption was due to iron-binding galloyl groups rather than phenolic catechol groups that affected iron absorption to a minor extent. Most food polyphenols could actively inhibit dietary non-heme iron absorption. The correlation of iron bioaccessibility with the particular phenolic group instead of total polyphenol content appears to be more appropriate (Krishnan & Meera, 2018). The chemical structure of polyphenols determines their rate and extent of GI tract absorption and the nature of metabolites present in plasma. The antinutritional effect of tannins depends on chemical structure and dosage—the acceptable total daily intake of tannic acid for a man is 560 mg/100 g (Udomkun et al., 2019).

Improvement in protein digestibility after processing could be attributed to the reduction or elimination of polyphenolic compounds or promote structural changes of protein such as globulin, thereby increasing chain flexibility and accessibility to proteases (Martín-Cabrejas et al., 2009). The profiles and quantities of polyphenols and tannins in foods are affected by processing due to their highly reactive nature, which may affect their antioxidant activity and the nutritional value of foods. Selection of suitable processing methods for polyphenol reduction is also a very important step. Milling and refining of flour is a suitable method for wheat flour because polyphenols in wheat grain are principally contained in the outer layers (aleurone cells, seed coat) and separated during processing (Krishnan & Meera, 2018). The presence of polyphenol content in the food can be reduced by several methods such as soaking, blanching, germination, acid treatment, fermentation, and exogenous and endogenous enzyme formation during processing (Nadeem et al., 2010). Tiwari et al. (2014) observed that increasing the duration of fermentation activate polyphenol oxidase enzymes subsequently decreased polyphenols in pearl millet. Similarly, fermentation significantly reduces polyphenol content in corn (Sokrab et al., 2014) and sorghum (Abdelhaleem et al., 2008) as well as heat treatment and fermentation in pearl millet flour (Tiwari et al., 2014). Nonetheless, a contradiction was reported in a study about germination and change in concentration of polyphenol. Sokrab et al. (2012) found that the germination of pearl millet significantly increased the polyphenol concentration. The increment in polyphenols could be a result of solubilization of condensed tannin when the seeds were soaked in water and the polyphenols migrated to the outer layer as a result of germination, as indicated by the browning of the germinated seeds. No polyphenols were observed on mineral extractability. Processing at a high temperature (140°C) and maintained moisture level (20%) significantly reduced polyphenol levels from 64.84% to 80.74% in wheat, rice, barley, and oat cereal bran (Kaur et al., 2015).

Although free phenolics and tannins are not desirable for human consumption due to several detrimental effects on nutritional availability and health, they do have several health benefitting effects such as antioxidant, anticarcinogenic, anti-inflammatory, antiarteriosclerotic, antiallergenic, antiatherogenic, antimicrobial, antithrombotic activities, and cardioprotective and vasodilatory effects. They also play an important role in combating oxidative stress in the human body by maintaining a balance. The health benefits from phenolic compound are mainly attributed to their ability to scavenge free radicals or reactive oxygen species (i.e., superoxide anion, hydroxyl radicals, and peroxy radicals), form chelates with pro-oxidant metals, thus preventing degenerative diseases such as heart disease and cancer, low–density lipoprotein oxidation, and DNA strand scission or enhancing immune function (Hung, 2016).

11.4.3 Enzyme Inhibitors

Enzymes are the biological catalysts that accelerate chemical reactions and improve digestion in the gut. Amylase and protease such as trypsin and chymotrypsin are the important enzymes that catalyze the hydrolysis of starch and protein, which speeds up the digestion in the digestive system. Amylase and protease have several others characteristics and uses. For example, protease is used by the food industry to control viscosity, elasticity, cohesion, emulsification, foam stability and whip ability, flavor development, texture modification, solubility, digestibility and extractability, and it catalyzes many physiological and pathophysiological processes such as food digestion, tissue remodeling, and host defense (Akande et al., 2010; Bora, 2014). TIs possess the characteristics of an insecticide (Abbas & Ahmad, 2018). Food products that originated from plants and contain various water-soluble protein that exhibits a growth-retarding property attributed to inhibition of enzymatic activity in the GI tract of animals are known as enzymatic inhibitors. These are considered as ANFs, such as α-AIs and PI. α-AIs impair growth, causing pancreatic hyperplasia and a metabolic disturbance of sulphur and amino acid utilization that creates a problem in digestion due to accumulation of nondigested polysaccharides in the intestine. PIs are classified in to two types i) Kunitz and ii) Bowman-Birk. Kunitz especially acts on trypsin, while Bowman-Birk inhibits both trypsin and chymotrypsin. TIs are polypeptides that form well-characterized stable complexes with trypsin on a one-to-one molar ratio, obstructing the enzymatic action. An imbalance may generate between endogenous PI and proteases due to some physiopathological conditions such as uncontrolled proteolysis, which can lead to tissue damage such as inflammation, hypertension, gastric ulcer, tumor growth, and metastasis, while TIs could cause gastric distress and lead to pancreatic hypertrophy or hyperplasia due to nonhydrolysis of proteins. Despite that, it is also reported that pancreatic hyperactivity increases production of trypsin and chymotrypsin with consequent loss of cystine and methionine (Gilani et al., 2012; Akande et al., 2010). Wheat contains amylase-trypsin inhibitors (ATIs), which are grouped into i) wheat monomeric AIs, ii) wheat dimeric AI and iii) wheat tetrameric AIs. Bread wheat has nearly twice as much trypsin inhibitor activity (TIA) as other wheat. ATIs of wheat are found responsible for disease in a broader spectrum such as Baker's asthma, celiac disease, nonceliac wheat sensitivity, wheat allergy, and irritable bowel syndrome (Abbas & Ahmad, 2018). Compared with legumes, ANF levels in cereal are quite low. Although digestive enzyme (protease and amylase) inhibitors have been identified in most cereals, they do not pose serious nutritional problems, however, barley contains higher levels of TIA compared to other cereals (Bora, 2014). For example, total protein inhibitor in barley normally is about 0.45 g/kg, whereas in defatted soy flour it is more than 32 g/kg (Abbas & Ahmad, 2018).

Enzyme inhibitors are heat labile, so they can be decomposed by heat treatment. Cooking generally inactivates heat-sensitive factors as a result of denaturation of these heat labile proteins. Extrusion cooking is a high temperature short time treatment. Application of extrusion cooking of 140°C and 20% moisture content reported very effective in reduction of TIs to 71.17%, 73.11%, 73.07%, and 72.21% in wheat, rice, barley, and oat bran respectively (Kaur et al., 2015). Treatment of moist heat is very effective in the reduction of PI in grains because of uniform distribution of heat in the PI medium (Akande et al., 2010), but there is no significant effect of germination (Sandarani & Kulathunga, 2019).

The abovementioned details represent the various harmful effects of enzyme inhibitors on human health. But in some research, health benefitting effects are also have been discussed. With the various antinutritional properties of ANFs, some health benefitting properties also have been reported. TIs have been implicated in reducing protein digestibility and in pancreatic hypertrophy (Akande et al., 2010). In the pharmaceutical industry, PI is used as antiviral drugs for the treatment of HIV/AIDS and suppression of carcinogenesis in *in vivo* and *in vitro* conditions, however, the mechanism is still unclear (Sandarani & Kulathunga, 2019). Amylase inhibitors are used for the treatment of diabetes as it interferes and slows down the digestion process, altering blood sugar levels and insulin in the human body (Abbas & Ahmad, 2018). Moreover, ATIs have been identified as one of the main triggers of nonceliac wheat sensitivity.

11.4.4 Lectins

Lectins are plant-derived proteinous compounds, abundant in nature, and present in plants—especially in seeds. It takes part in the biological recognition phenomena involving cells and proteins, therefore, they improve resistance, act as defense mechanisms, and protect plants against other plants, bacteria, fungi, or insects. This protein offers the ability of binding specific carbohydrate moieties on the glycocalyx or cell surface of erythrocytes and agglutinate erythrocytes without any alteration in their covalent structure. The ability to bind and agglutinate red blood cells is well known and used for blood typing, hence the lectins are commonly called hemagglutinins, but lectins and hemagglutinins are specifically classified on the basis of known or unknown sugar specificity agglutination with red blood cells. The proteins bind with known sugar specificity called lectins and the unknown once called as hemagglutinins. Hemagglutinins and lectins contain at least one noncatalytic domain that exhibits reversible binding to specific monosaccharides or oligosaccharides. This causes cauterization of the membrane glycoreceptor, transmembrane signalling, and cell aggregation similar to physiological agonists of platelets (Buul & Brouns, 2014; Signorello et al., 2020). These are stable in a wide range of physiological pH and they are highly resistant to proteolysis. Because of their ability to bind to various cell types and cause damage to several organs, lectins are widely recognized as antinutrients within food. Most lectins are resistant to heat and the effects of digestive enzymes and are able to bind to several tissues and organs *in vitro* and *in vivo*. Sometimes, consumptions of lectin in small quantity (0.5%) may show toxic and fatal effects. Postconsumption, lectins bind to intestinal cells and may cause some damage i.e. lesions on an internal layer of the GI tract, impart nutrient absorption in the GI tract, growth inhibition, diarrhea, nausea, bloating, and vomiting (Shi et al., 2018). Although, numerous studies show the harmful effects of lectins from pulses, there is no evidence of the fatal effects of lectins from cereals (Bora, 2014; Nikmaram et al., 2017). Lectin is present in many cereal grains such as wheat, barley, oats, corn, and rice, but the highest lectin reported was wheat germ agglutinin (WGA) having concentration up to 0.5 g/kg (Peumans & Van Damme, 1996). In the body, WGA binds to N-glycolylneuraminic acid of the glycocalyx that leads to disturbance in immunity by inducing a proinflammatory immune response (Punder & Pruimboom, 2013). WGA binds to glycocalyx and modifies its structure, which affects the aerobic mechanism, it causes oxidative stress in platelets, decreases energy availability and metabolic activity of platelets, causing the release of lactic dehydrogenase to increased and lead to the death of platelets (Signorello et al., 2020). WGA also caused hyperplastic and hypertrophic growth in the small intestines, hypertrophic growth of the pancreas, and thymus atrophy was reported by Punder & Pruimboom (2013). The effect of wheat lectins is broadly reviewed by Van Buul & Brouns (2014). They found that the grain processed through cooking, baking, and extrusion cooking do not cause negative impacts on human health. On the contrary, it is suggested to consume cereals and whole grain products that contain lectin and WGA. Whole grain contains fiber (β-glucan and arabinoxylan) and phytochemicals (phenolics, sterols, tocols, and vitamins) in the aleurone layer of the bran. Regular consumption of whole grain reduces the risks for type 2 diabetes, heart disease, syndrome X related events, obesity, some types of cancer, and cardiovascular disease (Van Buul & Brouns, 2014).

11.4.5 Saponins

Saponins are higher molecular weighted secondary metabolites widely present in the plant kingdom. Similar to tannins, saponins also take part in the defense mechanism of plants and protection against attack by pathogenic microbes such as fungus. Moreover, it is also used in pharmaceuticals. Saponins are composed of aglycone (fat-soluble nucleus) that is classified on the basis of backbone structure into triterpenoid, neutral, or alkaloid steroid having one or more sugar chains (Hassan et al., 2010). Backbone structures are mostly linked to two polar glycosidic side chains, but sometimes three side chains occur. The most common saccharide side chains consist of one to three glucose, rhamnose, arabinose, xylose, or glucuronic acid monomers (Wiesner et al., 2017). The special foaming characteristic was responsible for the name saponin, as it is derived from the Latin word *sapo*, meaning soap. Due to the presence of both hydrophilic and hydrophobic regions, saponins spontaneously form micelles in solutions, thus it is called a natural detergent.

The presence of saponins in higher concentration in food makes it a limiting factor for consumption due to its bitter taste and irritation in the throat. It is therefore compulsory to remove saponins before consumption. Saponins possess various biological activities such as hemolytic activity, anti-inflammatory activity, antimicrobial activity, molluscicide activity, cytotoxicity, and antitumor activity.

Biological activities are affected by factors such as the nucleus type of saponin, number of sugar side chains, and type of functional groups (Hassan et al., 2010). In the presence of bile, the hydrophobic portions intertwine with the bile and large micelles are formed that are too large to be absorbed across the gut wall (Franks, 2004). Saponins can also interact directly with proteins and lipids and form insoluble complexes that prevent absorption in the GI tract, e.g. cholesterol. Moreover, saponins also form complexes with zinc and iron and thus affect their bioavailability. Overall, saponins aid growth of the body, reduce the bioavailability of nutrients and minerals, decrease enzyme activity, and affect protein digestibility by inhibiting various digestive enzymes such as trypsin and chymotrypsin.

Production of saponins in plants mostly belong to dicotyledonous groups, but a few monocots are present that produce the same secondary metabolites i.e. oats, pearl millet, quinoa, and sorghum. Oats (*Avena* spp.) produces both steroidal (avenacosides) and triterpenoid saponins (avenacins) in their leaves and roots, respectively (Osbourn et al., 2003). The triterpenoid saponin content in oats and oat products ranges from 1.77–18.20 μg/g (Hu & Sang, 2020). Saponin content in quinoa (whole) seeds ranges from 0.03–2.05% (Mir et al., 2018). Heat processing of saponin-containing grains into food or feed does not degrade saponin content significantly and the absorption of intact saponins into the GI tract is quite low (Wiesner et al., 2017). De-hulling and washing were found quite effective in saponin reduction at a primary level (Mir et al., 2018).

Saponins have some health-promising properties, like reduction of serum cholesterol by binding, anti-carcinogenic, antimicrobial, immune modulating, and anti-inflammatory effects (Isaiah, 2020). Saponins present in sorghum exhibits antimicrobial activity and inhibitory effect on *S. aureus* (Gram-positive organism) but not against Gram-negative organisms and fungi (*Escherichia coli, Staphylococcus aureus,* and *Candida albicans*) (Soetan et al., 2006). It is found beneficial for animals. It is used in feed to reduce ruminal methane production by repression of methanogenic microorganisms and to induce therapeutic anti-inflammatory and immune-stimulating effects in livestock animals (Wiesner et al., 2017).

11.5 Conclusion

Cereals are the chief source of staple food in the world to fulfil daily energy and nutritional requirements. The presence of a group of particular secondary metabolites interferes with the absorption and uptake in the GI tract, and at higher concentrations, causes inflammation and several degenerative diseases. Application of higher-dose of fertilizer increases ANF accumulation in plants, which contributes to micronutrient deficiencies and subsequently "hidden hunger," a nutritional problem peculiar to underdeveloped and developing countries. So, it becomes a necessity to reduce the concentration of ANFs present in food grains. The accumulation of ANFs depends on genetic constituents, environmental condition, and cultural practices such as nutrient management. Several physical and biological methods are able to effectively reduce ANF concentration from food material. On the contrary, regular consumption of whole-grain products containing ANFs shows several health benefitting effects, such as reduction of risks for type 2 diabetes, heart disease, syndrome X related events, weight gain, and some types of cancers. Keeping these benefits in mind, several dietary guidelines emphasize the importance of grain foods, particularly whole grains in the diet. The consumption of food containing ANFs up to a certain level is beneficial, otherwise it becomes harmful at higher concentrations. So, it is necessary to maintain an optimum concentration of ANFs by using suitable processing method(s), crop species, and cultural practices. Moreover, genetic modification of staple food crops containing a higher level of ANFs and lower levels of micronutrients can be an effective and permanent solution.

REFERENCES

Abbas, Y., & Ahmad, A. (2018). Impact of processing on nutritional and antinutritional factors of legumes: A review. *Annals of Food Science and Technology, 19*(2), 199–215.

Abdelhaleem, W. H., El Tinay, A. H., Mustafa, A. I., & Babiker, E. E. (2008). Effect of fermentation, malt-pretreatment and cooking on antinutritional factors and protein digestibility of sorghum cultivars. *Pakistan Journal of Nutrition, 7*(2), 335–341. https://doi.org/10.3923/pjn.2008.335.341

Abioye, V. F., Ogunlakin, G. O., & Taiwo, G. (2018). Effect of germination on anti-oxidant activity, total phenols, flavonoids and anti-nutritional content of finger millet flour. *Journal of Food Processing and Technology, 9*(2), 1–5. https://doi.org/10.4172/2157-7110.1000719

Abdulwaliyu, I., Arekemase, S. O., Adudu, J. A., Batari, M. L., Egbule, M. N., & Okoduwa, S. I. R. (2019). Investigation of the medicinal significance of phytic acid as an indispensable anti-nutrient in diseases. *Clinical Nutrition Experimental, 28*, 42–61. https://doi.org/10.1016/j.yclnex.2019.10.002

Adeleke, O., Adiamo, O. Q., Fawale, O. S., & Olamiti, G. (2017). Effect of soaking and boiling on anti-nutritional factors, oligosaccharide contents and protein digestibility of newly developed Bambara groundnut cultivars. *Turkish Journal of Agriculture—Food Science and Technology, 5*(9), 1006. https://doi.org/10.24925/turjaf.v5i9.1006-1014.949

Akande, K. E., Doma, U. D., Agu, H. O., & Adamu, H. M. (2010). Major antinutrients found in plant protein sources: Their effect on nutrition. *Pakistan Journal of Nutrition, 9*(8), 827–832. https://doi.org/10.3923/pjn.2010.827.832

Alijošius, S., Švirmickas, G. J., Kliševiciute, V., Gruźauskas, R., Šašyte, V., Racevičiute-Stupeliene, A., Daukšiene, A., & Dailidavičiene, J. (2016). The chemical composition of different barley varieties grown in Lithuania. *Veterinarija Ir Zootechnika, 73*(95), 9–13.

Andersen, K. E., Bjergegaard, C., Møller, P., Sørensen, J. C., & Sørensen, H. (2005). Compositional variations for α-galactosides in different species of Leguminosae, Brassicaceae, and barley: A chemotaxonomic study based on chemometrics and high-performance capillary electrophoresis. *Journal of Agricultural and Food Chemistry, 53*(14), 5809–5817. https://doi.org/10.1021/jf040471v

Awada, S. H., Hady, A., Hassan, A. B., Ali, I., & Babiker, E. E. (2005). Antinutritional factors content and availability of protein, starch and mineral of maize (Zeamays linnaus) and lentil (Lens culinaris) as influenced by domestic processing. *Journal of Food Technology, 3*(4), 523–528.

Bhat, R., & Sridhar, K. R. (2008). Nutritional quality evaluation of electron beam-irradiated lotus (Nelumbo nucifera) seeds. *Food Chemistry, 107*(1), 174–184. https://doi.org/10.1016/j.foodchem.2007.08.002

Boncompagni, E., Orozco-Arroyo, G., Cominelli, E., Gangashetty, P. I., Grando, S., Tenutse Kwaku Zu, T., Daminati, M. G., Nielsen, E., & Sparvoli, F. (2018). Antinutritional factors in pearl millet grains: Phytate and goitrogens content variability and molecular characterization of genes involved in their pathways. *PLoS ONE, 13*(6), 1–30. https://doi.org/10.1371/journal.pone.0198394

Bora, P. (2014). Anti-Nutritional factors in foods and their effects. *Journal of Academia and Industrial Research, 3*(6), 285–290.

Centeno, C., Viveros, A., Brenes, A., Lozano, A., & De La Cuadra, C. (2003). Effect of several germination conditions on total P, phytate P, phytase, acid phosphatase activities and inositol phosphate esters in spring and winter wheat. *The Journal of Agricultural Science, 141*(3–4), 313–321. https://doi.org/10.1017/S0021859603003666

Chukwuma, O. E. (2016). Effect of the traditional cooking methods (boiling and roasting) on the nutritional profile of quality protein maize. *Journal of Food and Nutrition Sciences, 4*(2), 34. https://doi.org/10.11648/j.jfns.20160402.12

Cui, L., Li, D., & Liu, C. (2012). Effect of fermentation on the nutritive value of maize. *International Journal of Food Science & Technology, 47*(4), 755–760. https://doi.org/10.1111/j.1365-2621.2011.02904.x

Dai, F., Wang, J., Zhang, S., Xu, Z., & Zhang, G. (2007). *Food Chemistry Genotypic and environmental variation in phytic acid content and its relation to protein content and malt quality in barley. 105*, 606–611. https://doi.org/10.1016/j.foodchem.2007.04.019

Davies, N. T., & Olpin, S. E. (1979). Studies on the phytate: Zinc molar contents in diets as a determinant of Zn availability to young rats. *British Journal of Nutrition, 41*(3), 591–603. https://doi.org/10.1079/bjn19790074

Dost, K., & Tokul, O. (2006). Determination of phytic acid in wheat and wheat products by reverse phase high performance liquid chromatography. *Analytica Chimica Acta, 558*(1–2), 22–27. https://doi.org/10.1016/j.aca.2005.11.035

Ejigui, J., Savoie, L., Marin, J., & Desrosiers, T. (2005). Beneficial changes and drawbacks of a traditional fermentation process on chemical composition and antinutritional factors of yellow maize (Zea mays). *Journal of Biological Sciences, 5*(5), 590–596. https://doi.org/10.3923/jbs.2005.590.596

Eltayeb, M. M., Hassan, A. B., Mohamed, G. A., & Babiker, E. E. (2008). Effect of processing followed by fermentation on HCl extractability of Ca, P, Fe and Zn of pearl millet (*Pennisetum glaucum* L.) cultivars. *International Journal of Agricultural Research, 3*(5), 349–356. https://doi.org/10.3923/ijar.2008.349.356

Franks, P. A. (2004). Nutraceuticals from grains. In Encyclopedia of Grain Science. Academic Press, pp. 312–318. https://doi.org/10.1016/B0-12-765490-9/00105-1

Gemede, H. F., & Ratta, N. (2014). Antinutritional factors in plant foods : Potential health benefits and adverse effects. *International Journal of Nutrition and Food Sciences*. *3*(4), 284–289. https://doi.org/10.11648/j.ijnfs.20140304.18

Gilani, G. S., Xiao, C. W., & Cockell, K. A. (2012). Impact of antinutritional factors in food proteins on the digestibility of protein and the bioavailability of amino acids and on protein quality. *British Journal of Nutrition, 108*(S2), S315–S332. https://doi.org/10.1017/S0007114512002371

Handa, V., Kumar, V., Panghal, A., Suri, S., & Kaur, J. (2017). Effect of soaking and germination on physico-chemical and functional attributes of horsegram flour. *Journal of Food Science and Technology*, *54*(13), 4229–4239. https://doi.org/10.1007/s13197-017-2892-1

Hassan, A. B., Ahmed, I. A. M., Osman, N. M., Eltayeb, M. M., Osman, G. A., & Babiker, E.E. (2006). Effect of processing treatments followed by fermentation on protein content and digestibility of pearl millet (Pennisetum typhoideum) cultivars. *Pakistan Journal of Nutrition*, *5*(1), 86–89. https://doi.org/10.3923/pjn.2006.86.89

Hassan, A. B., Osman, G. A. M., Rushdi, M. A. H., Eltayeb, M. M., & Diab, E. E. (2009). Effect of gamma irradiation on the nutritional quality of maize cultivars (Zea mays) and sorghum (Sorghum bicolor) grains. *Pakistan Journal of Nutrition*, *8*(2), 167–171. https://doi.org/10.3923/pjn.2009.167.171

Hassan, S. M., Haq, A. U., Byrd, J. A., Berhow, M. A., Cartwright, A. L., & Bailey, C.A. (2010). Haemolytic and antimicrobial activities of saponin-rich extracts from guar meal. *Food Chemistry*, *119*(2), 600–605. https://doi.org/10.1016/j.foodchem.2009.06.066

Hendek Ertop, M., & Bektaş, M. (2018). Enhancement of bioavailable micronutrients and reduction of antinutrients in foods with some processes. *Food and Health*, *4*(3), 159–165. https://doi.org/10.3153/FH18016

Hotz, C., & Gibson, R. S. (2007). Traditional food-processing and preparation practices to enhance the bio-availability of micronutrients in plant-based diets. *Journal of Nutrition*, *137*(4), 1097–1100. https://doi.org/10.1093/jn/137.4.1097

Hu, C., & Sang, S. (2020). Triterpenoid saponins in oat bran and their levels in commercial oat products. *Journal of Agricultural and Food Chemistry*, *68*(23), 6381–6389. https://doi.org/10.1021/acs.jafc.0c02520

Isaiah, O. O. (2020). The synergistic interaction of phenolic compounds in pearl millets with respect to antioxidant and antimicrobial properties. *American Journal of Food Science and Health*, *6*(3), 80–88. http://www.aiscience.org/journal/paperInfo/ajfsh?paperId=4954

Izydorczyk, M. S., Storsley, J., Labossiere, D., MacGregor, A. W., & Rossnagel, B. G. (2000). Variation in total and soluble β-glucan content in hulless barley: Effects of thermal, physical, and enzymic treatments. *Journal of Agricultural and Food Chemistry*, *48*(4), 982–989. https://doi.org/10.1021/jf991102f

Jan, A., Sood, M., Younis, K., & Islam, R. U. (2020). Brown rice based weaning food treated with gamma irradiation evaluated during storage. *Radiation Physics and Chemistry*, *177*, 109158. https://doi.org/10.1016/j.radphyschem.2020.109158

Kaur, Satinder, Sharma, S., Dar, B. N., & Singh, B. (2012). Optimization of process for reduction of antinutritional factors in edible cereal brans. *Food Science and Technology International*, *18*(5), 445–454. https://doi.org/10.1177/1082013211428236

Kaur, Satinder, Sharma, S., Singh, B., & Dar, B. N. (2015). Effect of extrusion variables (temperature, moisture) on the antinutrient components of cereal brans. *Journal of Food Science and Technology*, *52*(3), 1670–1676. https://doi.org/10.1007/s13197-013-1118-4

Kaur, Sukhdeep, Bhardwaj, R. D., Kapoor, R., & Grewal, S. K. (2019). Biochemical characterization of oat (*Avena sativa* L.) genotypes with high nutritional potential. *LWT*, *110*, 32–39. https://doi.org/10.1016/j.lwt.2019.04.063

Kaushal, P., Kumar, V., & Sharma, H. K. (2012). Comparative study of physicochemical, functional, antinutritional and pasting properties of taro (*Colocasia esculenta*), rice (*Oryza sativa*) flour, pigeonpea (*Cajanus cajan*) flour and their blends. *LWT—Food Science and Technology*, *48*(1), 59–68. https://doi.org/10.1016/j.lwt.2012.02.028

Khan, S. H., Butt, M. S., Anjum, F. M., & Jamil, A. (2009). Antinutritional appraisal and protein extraction from differently stabilized rice bran. *Pakistan Journal of Nutrition*, *8*(8), 1281–1286. https://doi.org/10.3923/pjn.2009.1281.1286

Krishnan, R., & Meera, M. S. (2018). Pearl millet minerals: Effect of processing on bioaccessibility. *Journal of Food Science and Technology*, *55*(9), 3362–3372. https://doi.org/10.1007/s13197-018-3305-9

Kumar, M., Tomar, M., Saurabh, V., Mahajan, T., Punia, S., del Mar Contreras, M., Rudra, S. G., Kaur, C., & Kennedy, J. F. (2020). Emerging trends in pectin extraction and its anti-microbial functionalization using natural bioactives for application in food packaging. *Trends in Food Science & Technology, 105*, 223–237. https://doi.org/10.1016/j.tifs.2020.09.009

Kumar, V., Sinha, A. K., Makkar, H. P. S., & Becker, K. (2010). Dietary roles of phytate and phytase in human nutrition: A review. *Food Chemistry, 120*(4), 945–959. https://doi.org/10.1016/j.foodchem.2009.11.052

Lestienne, I., Icard-Vernière, C., Mouquet, C., Picq, C., & Trèche, S. (2005). Effects of soaking whole cereal and legume seeds on iron, zinc and phytate contents. *Food Chemistry, 89*(3), 421–425. https://doi.org/10.1016/j.foodchem.2004.03.040

López-Perea, P., Guzmán-Ortiz, F. A., Román-Gutiérrez, A. D., Castro-Rosas, J., Gómez-Aldapa, C. A., Rodríguez-Marín, M. L., Falfán-Cortés, R. N., González-Olivares, L. G., & Torruco-Uco, J. G. (2019). Bioactive compounds and antioxidant activity of wheat bran and barley husk in the extracts with different polarity. *International Journal of Food Properties, 22*(1), 646–658. https://doi.org/10.1080/10942912.2019.1600543

Machaiah, J. P., & Pednekar, M. D. (2002). Carbohydrate composition of low dose radiation-processed legumes and reduction in flatulence factors. *Food Chemistry, 79*(3), 293–301. https://doi.org/10.1016/S0308-8146(02)00142-5

Martín-Cabrejas, M. A., Aguilera, Y., Pedrosa, M. M., Cuadrado, C., Hernández, T., Díaz, S., & Esteban, R. M. (2009). The impact of dehydration process on antinutrients and protein digestibility of some legume flours. *Food Chemistry, 114*(3), 1063–1068. https://doi.org/10.1016/j.foodchem.2008.10.070

Maseta, E., C. E. Mosha, T., Laswai, H., & N. Nyaruhucha, C. (2016). Nutritional quality, mycotoxins and antinuntritional factors in quality protein maize-based supplementary foods for children in Tanzania. *International Journal of Sciences, 2*(07), 37–44. https://doi.org/10.18483/ijsci.1082

Mir, N. A., Riar, C. S., & Singh, S. (2018). Nutritional constituents of pseudo cereals and their potential use in food systems: A review. *Trends in Food Science and Technology, 75*, 170–180. https://doi.org/10.1016/j.tifs.2018.03.016

Mohamed, A. A., Matter, M. A., & Rady, M. R. (2010). Assessment of some barley germplasms based on RAPD analysis and anti-nutritional factors. *Journal of Crop Science and Biotechnology, 13*(2), 61–68. https://doi.org/10.1007/s12892-010-0010-2

Mohamed, E. A., Abdelraheem Ali, N., Ahmed, S. H., Mohamed Ahmed, I. A., & Babiker, E. E. (2010). Effect of radiation process on antinutrients and HCl extractability of calcium, phosphorus and iron during processing and storage. *Radiation Physics and Chemistry, 79*(7), 791–796. https://doi.org/10.1016/j.radphyschem.2010.01.018

Mohapatra, D., Patel, A. S., Kar, A., Deshpande, S. S., & Tripathi, M. K. (2019). Effect of different processing conditions on proximate composition, anti-oxidants, anti-nutrients and amino acid profile of grain sorghum. *Food Chemistry, 271*, 129–135. https://doi.org/10.1016/j.foodchem.2018.07.196

Nadeem, M., Anjum, F. M., Amir, R. M., Khan, M. R., Hussain, S., & Javed, M. S. (2010). An overview of anti-nutritional factors in cereal grains with special reference to wheat—A review. *Pakistan Journal of Food Science, 20*(1–4), 54–61.

Newman, R. K., & Sands, D. C. (1984). Nutritive value of corn fermented with lysine excreting lactobacilli. *Nutrition Reports International, 30*(6), 1287–1293.

Nikmaram, N., Leong, S. Y., Koubaa, M., Zhu, Z., Barba, F. J., Greiner, R., Oey, I., & Roohinejad, S. (2017). Effect of extrusion on the anti-nutritional factors of food products: An overview. *Food Control, 79*, 62–73. https://doi.org/10.1016/j.foodcont.2017.03.027

Nissar, N., Wani, S. M., Hameed, O. Bin, Wani, T. A., & Ahmad, M. (2017). Influence of paddy (Oryza sativa) sprouting on antioxidant activity, nutritional and anti-nutritional properties. *Journal of Food Measurement and Characterization, 11*(4), 1844–1850. https://doi.org/10.1007/s11694-017-9566-6

Oboh, G., Ademiluyi, A. O., & Akindahunsi, A. A. (2010). The effect of roasting on the nutritional and antioxidant properties of yellow and white maize varieties. *International Journal of Food Science & Technology, 45*(6), 1236–1242. https://doi.org/10.1111/j.1365-2621.2010.02263.x

Oghbaei, M., & Prakash, J. (2012). Bioaccessible nutrients and bioactive components from fortified products prepared using finger millet (Eleusine coracana). *Journal of the Science of Food and Agriculture, 92*(11), 2281–2290. https://doi.org/10.1002/jsfa.5622

Oghbaei, M., & Prakash, J. (2016). Effect of primary processing of cereals and legumes on its nutritional quality: A comprehensive review. *Cogent Food and Agriculture, 36*(1), 1–14. https://doi.org/10.1080/23311932.2015.1136015

Onyango, C. A., Ochanda, S. O., Mwasaru, M. A., Ochieng, J. K., Mathooko, F. M., & Kinyuru, J. N. (2013). Effects of malting and fermentation on anti-nutrient reduction and protein digestibility of red sorghum, white sorghum and pearl millet. *Journal of Food Research*, 2(1), 41. https://doi.org/10.5539/jfr.v2n1p41

Onyango, C., Noetzold, H., Ziems, A., Hofmann, T., Bley, T., & Henle, T. (2005). Digestibility and antinutrient properties of acidified and extruded maize-finger millet blend in the production of uji. *LWT—Food Science and Technology*, 38(7), 697–707. https://doi.org/10.1016/j.lwt.2004.09.010

Osbourn, A. E., Qi, X., Townsend, B., & Qin, B. (2003). Dissecting plant secondary metabolism - Constitutive chemical defences in cereals. *New Phytologist*, 159(1), 101–108. John Wiley & Sons, Ltd. https://doi.org/10.1046/j.1469-8137.2003.00759.x

Panwar, P., Dubey, A., & Verma, A. K. (2016). Evaluation of nutraceutical and antinutritional properties in barnyard and finger millet varieties grown in Himalayan region. *Journal of Food Science and Technology*, 53(6), 2779–2787. https://doi.org/10.1007/s13197-016-2250-8

Peumans, W. J., & Van Damme, E. J. M. (1996). Prevalence, biological activity and genetic manipulation of lectins in foods. *Trends in Food Science & Technology*, 7(4), 132–138. https://doi.org/10.1016/0924-2244(96)10015-7

Punder, K. de, & Pruimboom, L. (2013). The dietary intake of wheat and other cereal grains and their role in inflammation. *Nutrients*, 5(3), 771–787. https://doi.org/10.3390/nu5030771.

Rani, S., Singh, R., Sehrawat, R., Kaur, B. P., & Upadhyay, A. (2018). Pearl millet processing: a review. *Nutrition and Food Science*, 48(1), 30–44. https://doi.org/10.1108/NFS-04-2017-0070

Roger, T., Léopold, T. N., & Funtong, M. C. M. (2015). Nutritional properties and antinutritional factors of corn paste (Kutukutu) fermented by different strains of lactic acid bacteria. *International Journal of Food Science*, 2015. https://doi.org/10.1155/2015/502910

Saha, P. R., Weaver, C. M., & Mason, A. C. (1994). Mineral bioavailability in rats from intrinsically labeled whole wheat flour of various phytate levels. *Journal of Agricultural and Food Chemistry*, 42(11), 2531–2535. https://doi.org/10.1021/jf00047a029

Sandarani, M. D. J. C., & Kulathunga, K. A. A. V. (2019). A brief review: Lectins, protease inhibitors and saponins in cereals and legumes. *Asian Food Science Journal*. https://doi.org/10.9734/afsj/2019/v10i430044

Sangronis, E., & Machado, C. J. (2007). Influence of germination on the nutritional quality of *Phaseolus vulgaris* and *Cajanus cajan*. *LWT - Food Science and Technology*, 40(1), 116–120. https://doi.org/10.1016/j.lwt.2005.08.003

Shawrang, P., Sadeghi, A. A., Behgar, M., Zareshahi, H., & Shahhoseini, G. (2011). Study of chemical compositions, anti-nutritional contents and digestibility of electron beam irradiated sorghum grains. *Food Chemistry*, 125(2), 376–379. https://doi.org/10.1016/j.foodchem.2010.09.010

Shi, L., Arntfield, S. D., & Nickerson, M. (2018). Changes in levels of phytic acid, lectins and oxalates during soaking and cooking of Canadian pulses. *Food Research International*, 107(March), 660–668. https://doi.org/10.1016/j.foodres.2018.02.056

Siddhuraju, P., Makkar, H. P., & Becker, K. (2002). The effect of ionising radiation on antinutritional factors and the nutritional value of plant materials with reference to human and animal food. *Food Chemistry*, 78(2), 187–205. https://doi.org/10.1016/S0308-8146(01)00398-3

Signorello, M. G., Ravera, S., & Leoncini, G. (2020). Lectin-induced oxidative stress in human platelets. *Redox Biology*, 32, 101456. https://doi.org/10.1016/j.redox.2020.101456

Simwaka, J. E., Chamba, M. V. M., Huiming, Z., Masamba, K. G., & Luo, Y. (2017). Effect of fermentation on physicochemical and antinutritional factors of complementary foods from millet, sorghum, pumpkin and amaranth seed flours. *International Food Research Journal*, 24(5), 1869–1879.

Singh, A. k., Rehal, J., Kaur, A., & Jyot, G. (2015). Enhancement of attributes of cereals by germination and fermentation: A review. *Critical Reviews in Food Science and Nutrition*, 55(11), 1575–1589. https://doi.org/10.1080/10408398.2012.706661

Singh, B., Singh, J. P., Shevkani, K., Singh, N., & Kaur, A. (2017). Bioactive constituents in pulses and their health benefits. *Journal of Food Science and Technology*, 54(4), 858–870). Springer. https://doi.org/10.1007/s13197-016-2391-9

Singh, S., Gupta, A. K., & Kaur, N. (2012). Influence of drought and sowing time on protein composition, antinutrients, and mineral contents of wheat. *The Scientific World Journal*, 2012. https://doi.org/10.1100/2012/485751

Soetan, K. O., Oyekunle, M. A., Aiyelaagbe, O. O., & Fafunso, M.A. (2006). Evaluation of the antimicrobial activity of saponins extract of Sorghum Bicolor L. Moench. *African Journal of Biotechnology*. https://doi.org/10.5897/AJB06.252

Sokrab, A. M., Ahmed, I. A. M., & Babiker, E. E. (2012). Effect of malting and fermentation on antinutrients, and total and extractable minerals of high and low phytate corn genotypes. *International Journal of Food Science and Technology*, 47(5), 1037–1043. https://doi.org/10.1111/j.1365-2621.2012.02938.x

Sokrab, A. M., Mohamed Ahmed, I. A., & Babiker, E. E. (2012). Effect of germination on antinutritional factors, total, and extractable minerals of high and low phytate corn (Zea mays L.) genotypes. *Journal of the Saudi Society of Agricultural Sciences*, 11(2), 123–128. https://doi.org/10.1016/j.jssas.2012.02.002

Sokrab, A. M., Mohamed Ahmed, I. A., Babiker, E. E., Ahmed, I. A. M., Babiker, E. E., Mohamed Ahmed, I. A., Babiker, E. E., & Ahmed, I. A. M. (2014). Effect of fermentation on antinutrients, and total and extractable minerals of high and low phytate corn genotypes. *Journal of Food Science and Technology*, 51(10), 2608–2615. https://doi.org/10.1007/s13197-012-0787-8

Sripriya, G., Antony, U., & Chandra, T. S. (1997). Changes in carbohydrate, free amino acids, organic acids, phytate and HCl extractability of minerals during germination and fermentation of finger millet (Eleusine coracana). *Food Chemistry*, 58(4), 345–350. https://doi.org/10.1016/S0308-8146(96)00206-3

Szewińska, J., Simińska, J., & Bielawski, W. (2016). The roles of cysteine proteases and phytocystatins in development and germination of cereal seeds. *Journal of Plant Physiology*, 207, 0–21. https://doi.org/10.1016/j.jplph.2016.09.008

Tiwari, A., Jha, S. K., Pal, R. K., Sethi, S., & Krishan, L. (2014). Effect of pre-milling treatments on storage stability of pearl millet flour. *Journal of Food Processing and Preservation*, 38(3), 1215–1223. https://doi.org/10.1111/jfpp.12082

Tresina, P. S., & Mohan, V. R. (2011). Effect of gamma irradiation on physicochemical properties, proximate composition, vitamins and antinutritional factors of the tribal pulse *Vigna unguiculata* subsp. unguiculata. *International Journal of Food Science and Technology*, 46(8), 1739–1746. https://doi.org/10.1111/j.1365-2621.2011.02678.x

Tresina, P. S., Paulpriya, K., Mohan, V. R., & Jeeva, S. (2017). Effect of gamma irradiation on the nutritional and antinutritional qualities of Vigna aconitifolia (Jacq.) Marechal: An underutilized food legume. *Biocatalysis and Agricultural Biotechnology*, 10, 30–37. https://doi.org/10.1016/j.bcab.2017.02.002

Udomkun, P., Tirawattanawanich, C., Ilukor, J., Sridonpai, P., Njukwe, E., Nimbona, P., & Vanlauwe, B. (2019). Promoting the use of locally produced crops in making cereal-legume-based composite flours: An assessment of nutrient, antinutrient, mineral molar ratios, and aflatoxin content. *Food Chemistry*, 286(February), 651–658. https://doi.org/10.1016/j.foodchem.2019.02.055

Van Buul, V. J., & Brouns, F. J. P. H. (2014). Health effects of wheat lectins: A review. *Journal of Cereal Science*, 59(2), 112–117. https://doi.org/10.1016/j.jcs.2014.01.010.

Van Hung, P. (2016). Phenolic compounds of cereals and their antioxidant capacity. *Critical Reviews in Food Science and Nutrition*, 56(1), 25–35. https://doi.org/10.1080/10408398.2012.708909

Vashishth, A., Ram, S., & Beniwal, V. (2017). Cereal phytases and their importance in improvement of micronutrients bioavailability. *3 Biotech*, 7(1), 1–7. https://doi.org/10.1007/s13205-017-0698-5

Vötterl, J. C., Zebeli, Q., Hennig-Pauka, I., & Metzler-Zebeli, B.U. (2019). Soaking in lactic acid lowers the phytate-phosphorus content and increases the resistant starch in wheat and corn grains. *Animal Feed Science and Technology*. https://doi.org/10.1016/j.anifeedsci.2019.04.013

Wiesner, M., Hanschen, F. S., Maul, R., Neugart, S., Schreiner, M., & Baldermann, S. (2017). Nutritional quality of plants for food and fodder. In Encyclopedia of Applied Plant Sciences. Elsevier, pp. 285–291. https://doi.org/10.1016/B978-0-12-394807-6.00128-3

Zhang, G., Chen, J., Wang, J., & Ding, S. (2001). Cultivar and environmental effects on (1→ 3, 1→ 4) β-D-glucan and protein content in malting barley. *Journal of Cereal Science*, 34(3), 295–301. https://doi.org/10.1006/jcrs.2001.0414

Zhang, Guoping, Junmei, W., & Jinxin, C. (2002). Analysis of β-glucan content in barley cultivars from different locations of China. *Food Chemistry*, 79(2), 251–254. https://doi.org/10.1016/S0308-8146(02)00127-9

12

Nutritional Value, Bioactive Compounds, and Potential Health Benefits of Maize and Maize-Based Food Products

Florence A. Bello, Margaret O. Edet and Lawrence J. Iwok
University of Uyo, Uyo, Akwa Ibom State, Nigeria

CONTENTS

12.1 Introduction

Maize (*Zea mays* L.), also known as corn is an annual cereal crop belonging to the maideas tribe and the family of *Gramineae*. It has considerable social, cultural, and historical relevance since pre-Hispanic times (Dominguez-Hernandez et al. 2018). Maize is an important grain for the world's economy and food security as it was originated in Mexico more than 6,000 years ago (González-Ortega et al. 2017). It was introduced into Europe in the 16th century, then spread to Africa and Far East Asia and it is now considered as one of the most important cereals in the world due to its highest genetic yield potential (Orhun 2013; Chaves-López 2020).

Maize is the third most important crop after wheat and rice with more cultivation in many countries than any other crop in the world (Orhun 2013). Its production is estimated to be 1.2 billion metric tons

TABLE 12.1

World Maize Production (Thousand Metric Tons)

Countries	2016/17	2017/18	2018/19	2019/20	2020/21
United States	384,778	371,096	364,262	345,894	378,466
China	263,613	259,071	257,330	260,770	260,000
Brazil	98,500	82,000	101,000	102,000	110,000
European Union	61,884	62,007	64,362	66,665	66,300
Argentina	41,000	32,000	51,000	50,000	50,000
Ukraine	27,969	24,115	35,805	35,887	38,500
India	25,900	28,753	27,715	28,636	28,000
Mexico	27,575	27,569	27,600	25,000	28,000
South Africa	17,551	13,104	11,824	16,250	14,000
Russia	15,305	13,201	11,415	14,275	15,000
Canada	13,889	14,095	13,885	13,404	14,000
Indonesia	10,900	11,900	12,000	12,000	12,000
Nigeria	11,548	10,420	11,000	11,000	11,500
Philippines	8,087	8,084	7,608	8,030	8,200
Ethiopia	7,847	8,007	8,350	8,500	8,600
Serbia	7,600	4,000	7,000	7,700	7,200
Others	103,678	109,199	111,461	106,826	112,614
Subtotal	742,846	707,525	759,355	766,943	783,914
World Total	1,127,624	1,078,621	1,123,617	1,112,837	1,162,380

Source: USDA (2020).

in 2020/2021 (Table 12.1). The United States is the largest producer of maize in the world, contributing 378.5 million metric tons representing approximately 33% of the total world production. It is also known as the queen grain of the Americas and it is the driver of the U.S. economy (Shah et al. 2016). China and Brazil ranked second and third with production of 260 million metric tons and 110 million metric tons, respectively. South Africa, Nigeria, and Ethiopia are the largest producers of maize in Africa. The total consumption of maize in the world is estimated to rise to 1.16 billion metric tons in the year 2020/21 with 851 million metric tons accounted for food, seed, industrial, feed, and waste as presented in Table 12.2.

More than 80% of the maize harvested in developing countries is used as food (FAOSTAT, 2016). It is consumed as maize on the cob (green) after boiling or roasting (Bello and Oluwalana 2017). Since fresh maize sugar gradually converts into starch after harvest (Bello and Badejo 2017), maize kernels are allowed to mature and dry before harvest for the preparation of various kinds of foods. In contrast to wheat and rice in industrialized countries, a larger proportion of maize is utilized as livestock feed and production of biofuel, with less than 10% directly used for food and food ingredients.

The major parts of maize kernel include the endosperm, germ, and bran. Beyond its role as a major source of food, the health benefits of maize not only come from basic nutrients such as carbohydrates, vitamins, and minerals, but also from their unique bioactive compounds that are known to enhance health and prevent various degenerative diseases (Žilić et al. 2012; Serna-Saldivar 2016). Diverse maize varieties are cultivated all over the world as a result of differences in genotypes. Maize can be classified based on different kernel characteristics such as flint (*Zea mays* var. *indurate*), dent (*Zea mays* var. *indentata*), waxy (*Zea mays* var. *ceratina*), pop (*Zea mays* var. *everta*), and sweet maize (*Zea mays* var. *saccharata* and *Zea mays* var. *rugosa*), which differ significantly in physicochemical properties and end-use quality. Also in terms of pericarp color, which includes white, yellow, red, blue, purple, red, and multicolor (Figure 12.1), pigmented maize is excellently rich in phenolic compounds, carotenoids, and anthocyanins, which have antioxidant and bioactive properties (Saikaew et al. 2018).

Maize-based foods are characterized with different attributes across each country based on diverse processing techniques and forms of consumption (Mensah et al. 2013). The starchy grain is used for the development of varieties of indigenous and industrial foods (maize meal, flour, flakes, popcorn, tortillas, bread,

TABLE 12.2

World Maize Total Consumption (Thousand Metric Tons)

Countries	2016/17	2017/18	2018/19	2019/20	2020/21
United States	313,785	313,981	310,446	301,385	313,704
China	255,000	263,000	274,000	277,000	279,000
European Union	74,000	76,500	88,000	81,900	88,500
Brazil	60,500	63,500	67,000	69,000	70,000
Mexico	40,400	42,500	44,100	44,200	45,250
India	24,900	26,700	28,500	28,000	28,200
Egypt	15,100	15,900	16,200	17,100	17,300
Japan	15,200	15,600	16,000	16,000	16,000
Vietnam	13,000	13,600	14,300	14,950	15,500
Canada	12,949	13,985	15,087	14,200	14,800
Argentina	11,200	12,400	13,800	13,500	15,000
Indonesia	12,300	12,400	12,900	12,600	13,000
South Africa	12,663	12,230	12,500	12,600	12,300
Nigeria	11,400	11,100	11,300	11,400	11,800
Korea, South	9,435	10,000	10,947	11,400	11,850
Iran	9,300	9,800	10,300	9,900	11,400
Others	171,626	178,194	180,980	186,234	194,447
Subtotal	774,006	775,677	833,814	822,842	851,031
World Total	1,087,791	1,089,658	1,144,260	1,124,227	1,164,735

Source: USDA (2020).

cookies, complementary etc.) with unique distinctive flavors not duplicated by any other cereal. The processing methods include grinding (wet or dry), roasting, boiling, cooking, fermentation, nixtamalization (alkali cooking), etc. The consumption of whole maize grain products has been linked to the reduced risk of chronic diseases including cardiovascular disease, type 2 diabetes, obesity, some cancers. and with the improvement of digestive tract health (Fung et al. 2002; Tighe et al. 2010; Mourouti et al. 2016).

Nutritional quality of maize also depends upon the processing method used for food preparations. The preparation techniques of most indigenous foods have led to loss of a large portion of proteins, lipids, minerals and vitamins that are present at the pericarp and germ of maize. In the past decades, much research effort has been undertaken to improve the nutritional quality of maize for human consumption.

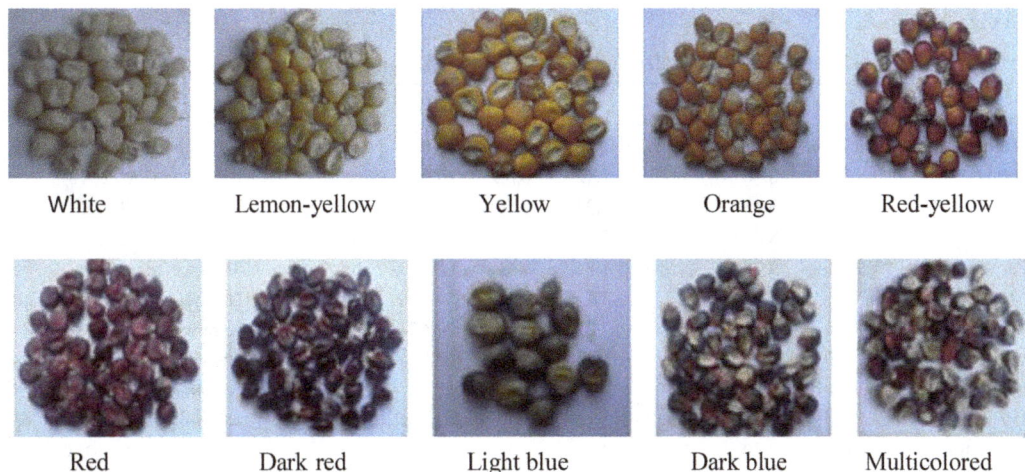

White	Lemon-yellow	Yellow	Orange	Red-yellow

Red	Dark red	Light blue	Dark blue	Multicolored

FIGURE 12.1 Maize kernel colors (see Zilic et al., 2012).

The prevalence of degenerative diseases has fostered increasing research interest in the breeding of new maize varieties with a remarkable genetic variability, and both GMO and regular types are planted throughout the globe (Serna-Saldivar 2016; Suri and Tanumihardjo 2016). The nutritional improvement of maize-based food ingredients with higher nutrients crops is encouraging. Thus, this chapter aims to discuss the major nutritional composition and bioactive compounds in maize and their health-promoting effects, in order to better understand the nutritional and health potential of maize and consequently improve its consumption.

12.2 Compositional Characteristic of Maize Kernel

12.2.1 Macronutrients of Maize Kernel

The composition of maize kernel according to the kernel color is presented in Table 12.3. Carbohydrates, especially starch, are the major chemical substance in maize and mostly located in the endosperm fraction of the kernel. Nigerian white maize was reported to contain 87.96% carbohydrate content, which was the highest value among the maize varieties (Bello and Udo 2018). This value was actually higher than 67.71% discovered in Ethiopian white maize of BH 660 variety (Demeke 2018). Plant varieties, climate, and cultivation affect the proportion of digestible and indigestible starch amounts in maize composition (Beckles and Thitisaksakul 2014). Low moisture contents are important for cereals and they provide good resistance to spoilage by microorganisms and insects during storage.

Protein is the second-most-abundant constituent of maize kernel found in the endosperm and located within subcellular bodies (protein bodies). Maize varieties differ in protein content ranging between 6.36% (Bello and Udo 2018) and 13.13% (Mutlu et al. 2017). It consists almost entirely of a prolamine-rich protein fraction known as zein, which accounts for about 60% of the total grain proteins. Several million people, particularly in developing countries, derive their protein and calorie requirements from maize. Protein from maize kernel is deficient in essential amino acids, lysine, and tryptophan, leading to harmful consequences such as growth retardation, protein energy malnutrition, anemia, pellagra, free radical damage, etc. (Graham et al. 1990). Protein content varies depending on growing conditions and genetic and environmental factors. The nitrogen content of soil also affects the protein content and therefore the usage of fertilizers and their amount play a very important role in the differences among plant material in respect to protein content (Agama-Acevedo et al. 2011).

The fat contents of maize varieties varied from 3.72–7.43%. The purple maize varieties of Argentina and Cote d'Ivoire contained higher fat contents (Akaffou et al. 2018; Mansilla et al. 2020). Maize oil is an important source of essential fatty acids. These essential fatty acids help maintain the body cells,

TABLE 12.3

Proximate Composition (% d.m.) of Maize According to Kernel Colors

	Moisture	Protein	Fat	Ash	Fibre	CHO	Countries	References
White	8.76	6.63	3.72	1.18	0.51	87.96	Nigeria	Bello and Udo (2018)
Yellow	9.28	6.36	4.67	1.12	0.52	87.33	Nigeria	Bello and Udo (2018)
Blue	9.39	13.13	4.30	1.34	2.68	63.94	Turkey	Mutlu et al. (2017)
Yellow	10.6	8.44	4.27	2.40	ND	ND	India	Trehan et al. (2018)
Purple	9.17	9.88	4.53	1.84	ND	ND	India	Trehan et al. (2018)
Blue	ND	11.21	5.13	ND	ND	58.76	Mexico	Uriarte-Aceves et al. (2015)
Yellow	11.55	13.11	5.44	2.33	1.45	77.67	Bangladesh	Kabir et al. (2019)
White	11.45	12.32	5.33	1.37	1.82	67.71	Ethiopia	Demeke (2018)
Yellow	ND	8.85	5.30	2.000	1.45	77.85	Côte d'Ivoire	Akaffou et al. (2018)
Deep purple	ND	10.09	6.21	1.89	ND	74.66	Côte d'Ivoire	Akaffou et al. (2018)
Red-yellow	ND	9.10	5.5	ND	8.30	77.00	Italy	Bacchetti et al. (2013)
Purple	ND	10.52	7.43	1.96	ND	ND	Argentina	Mansilla et al. (2020)

ND, Not determined; *CHO*, carbohydrate; *dm*, dry matter.

protect the nervous system, and also help fight against bad cholesterol (Bressani et al. 1990). Maize germ contains about 45–50% of oil that is used in cooking, salads, and is obtained from the wet milling process (Orthoefer et al. 2003). The oil contains 14% saturated fatty acids, 30% monounsaturated fatty acids, and 56% polyunsaturated fatty acids. The lipids present in the endosperm contain more saturated fatty acids than do germ lipids. The germ is composed of the embryo, the living organ of the grain, and the scutellum, which nourishes the embryo. The germ is characterized by high fat and protein content (approximately 33% and 18%, respectively) and low starch content (approximately 8%). Germ oil is relatively stable due to its high levels of natural antioxidants (tocols) and is considered healthy because of its high concentration of oleic and linoleic acids. However, the fat content can change dependiong on variety, soil conditions, and climatic effects (Uriarte-Aceves et al. 2015).

Dietary fiber is an important component of maize kernel and is defined as the portion of food derived from plant cell, which is resistant to hydrolysis or digestion by the elementary enzyme system in human beings. Maize kernels contain 0.51–8.30% crude fiber content with Italian maize variety (Roccacontrada Giallo) having the highest crude fiber content (Bacchetti et al. 2013). Some of the bacteria in the large intestine degrades some components of the fiber-releasing products that can be absorbed into the body and also used as a source of energy. Crude fiber is the residue remaining after the treatment with hot sulphuric acid, alkali, and alcohol. The major component of crude fiber is a polysaccharide called cellulose and a part of dietary fiber found especially in the pericarp layer (86%) in the kernel coat (Mutlu et al. 2017). It has been demonstrated that fiber, particularly its insoluble fraction, helps normalize intestinal obstipation, accelerating and increasing the fecal bulk (Akaffou et al. 2018).

Maize kernel contains 1.12–2.40% ash content. In all the varieties reviewed, yellow maize had the highest ash content (Akaffou et al. 2018; Trehan et al. 2018; Kabir et al. 2019), except Nigerian maize (Bello and Udo, 2018). Ash content in a food substance indicates inorganic remains after the organic matter has been burnt away and provides an estimate of the total mineral content of a food (Kidist, 2018). The variability in ash content of maize varieties may be due to genetics and environmental factors (Ndukwe et al. 2015).

12.2.2 Micronutrients of Maize Kernel

Most micronutrients are concentrated in the outer layers of the grain. Minerals are essential for body functions. They are classified as macrominerals or microminerals depending on their dietary requirement. Minerals play an important part in biological processes. The recommended dietary allowance and adequate intake are generally used to quantify suggested daily intake of minerals (Akaffou et al. 2018). The concentration levels of macroelements: potassium (K), magnesium (Mg), phosphorus (P), and calcium (Ca) in maize kernels are in appreciable quantities with the exception of sodium (Na) as shown in Table 12.4. Cote d'Ivoire's deep purple maize variety contains higher potassium content (317.18 mg/100 g) as reported by Akaffou et al. (2018). Iron (Fe) and zinc (Zn) contents (0.32 mg/100 g and 0.10 mg/100 g, respectively) are insignificant in all the varieties of maize with very low concentration reported for Nigerian white maize (Oboh et al. 2010). The factors responsible for the differences in mineral composition of maize kernels could be varieties, colors, soil, stage of maturity, environmental factors, fertilizers application etc. (Kabir et al. 2019).

TABLE 12.4

Mineral Composition (mg/100 d.m.) of Maize According to Kernel Colors

	K	Mg	P	Ca	Na	Fe	Zn	Cu	Mn	Countries	References
Yellow	307.29	98.17	127.25	12.41	2.81	2.21	2.2	0.61	0.61	Côte d'Ivoire	Akaffou et al. (2018)
Deep purple	317.18	92.32	125.52	10.92	3.17	2.58	2.26	0.44	0.52	Côte d'Ivoire	Akaffou et al. (2018)
Yellow	254.63	483.47	ND	8.77	1.85	1.04	0.05	ND	ND	Nigeria	Oboh et al. (2010)
White	312.5	425.4	ND	1.18	2.16	0.32	0.10	ND	ND	Nigeria	Oboh et al. (2010)
Yellow	ND	ND	ND	43.48	ND	2.15	1.57	ND	ND	Canada	Ejigui et al. (2005)
White	ND	ND	ND	31.25	ND	2.39	2.52	ND	ND	Ethiopia	Demeke (2018)
Yellow	ND	ND	ND	48.10	ND	3.18	2.92	ND	ND	Ethiopia	Demeke (2018)

ND, Not determined; *dm,* dry matter.

Vitamins play a crucial role in proper metabolic functions and therefore, their daily intake is essential since they cannot be synthesized at adequate amounts (or at all) in the human body. These essential micronutrients are divided into two subcategories according to their solubility: the water-soluble vitamin C and the group of B vitamins (thiamin-B_1, riboflavin-B_2, niacin-B_3, pantothenic acid-B_5, pyridoxine-B_6, biotin, folic acid, and cobalamin-B_{12}), and the fat-soluble vitamins A, D, E, and K (Gan et al. 2017). Maize kernel is widely used to prepare foods often containing low levels of vitamins that have low bioavailability. It is reported to contain low concentrations (0.17 mg/100 g and 1.61 mg/100 g) of vitamins A and C, respectively (Noah, 2017). Thiamin, riboflavin, and niacin of white maize kernel according to Okafor et al. (2018) were 0.32 mg/100 g, 0.14 mg/100 g and 1.94 mg/100 g, respectively. In developing countries, where there is high child and adult mortality, Zn, Fe, and vitamin A are among the top 10 risk factors contributing to the burden of disease (WHO, 2020).

Micronutrient bioavailability is therefore one of the main factors defining the amounts of micronutrients absorbed and the micronutrient status of individuals. Once provitamin A carotenoids are consumed, they are absorbed and converted into retinol or vitamin A. Many factors such as age, gender, genetic factors, health status, the food matrix, and the amounts and types of carotenoids in the meal affect the bioavailability or efficiency with which provitamin A carotenoids are absorbed in the intestine (Hess et al. 2005). There is considerable evidence that provitamin A, Fe, and Zn mutually enhance their bioavailabilities (García-Casal et al. 1998; Hess et al. 2005).

12.3 Bioactive Compounds in Maize and Their Potential Health Benefits

Maize grains contain significant amounts of non-nutrient phytochemicals. Phytochemicals are the most important bioactive chemical compounds naturally present in plants and are associated with promoting health and reducing risk of major chronic diseases beyond its role as a major source of food (Bacchetti et al. 2013; Sheng et al. 2018; Mansilla et al. 2020). Maize is an essential source of various major phytochemicals such as carotenoids, phenolic compounds, and phytosterols (Kopsell et al. 2009). Bioactive compositions vary among different types of maize around the world based on kernel colors (Moreira et al. 2015). Anthocyanins are concentrated in red, blue, purple, and black maize, carotenoids are concentrated in yellow and red maize, and phytosterols are concentrated in the kernel part of color (Luo and Wang 2012). Like many other grain products, bioactive compounds in maize are also distributed mainly in the kernel and bran (Adom and Liu 2002; Pang et al. 2018).

12.3.1 Phenolics

Phenolics are naturally occurring compounds that possess one or more aromatic rings and hydroxyl groups. They are the most widely distributed category of phytochemicals in the plant kingdom (Saxena et al. 2013). Phenolic compounds include a wide range of groups, such as phenolic acids, flavonoids, and tannins (Amador-Rodríguez et al. 2019). It has been found that phenolic compounds in plants occur in free and bound forms. A large portion of phenolic present in grains were proved to exist in bound form, while this phenomenon is quite opposite in fruits and vegetables (Acosta-Estrada et al. 2014; Bacchetti et al. 2013; Urias-Peraldí et al. 2013). The content of phenolic compounds in grains varies extensively and is dependent on genotype, the part of the grain sampled, grain handling, and grain processing (Ragaee et al. 2011).

Different maize varieties (dent corn, baby corn, popcorn, and sweet corn) contain a number of phenolic acids (*p*-hydroxybenzoic, vanillic, syringic, caffeic, *o*-coumaric, *p*-coumaric, and ferulic acids) (Prasanthi et al. 2017). The concentrations of phenolic acids in different maize kernel according to Žilić et al. (2012) in increasing order of concentration were ferulic acid > *p*-coumaric acid > *o*-coumaric acid (Table 12.5a). The authors reported highest ferulic acid (4,521.26 µg/g) in Serbian light blue popcorn while *p*-coumaric acid (560.77 µg/g) and *o*-coumaric acid (39.74 µg/g) were the highest in France orange flint maize. Ferulic acid (4-hydroxy-3-methoxycinnamic acid) and *p*-coumaric acid, which are hydroxycinnamic acid derivatives, are reported to be the predominant phenolic acids detected in maize (Adom and Liu 2002). Due to its diverse health benefits (antioxidant, anti-inflammatory, and anticancer activities) ferulic acid is regarded as one of the most important phenolic acids and receiving increasingly

TABLE 12.5a

Concentration of Phenolic Acids (µg/g d.m.) in Maize According to Kernel Colors

	Ferulic Acid	*p*-coumaric	*o*-coumaric	Kernel Type	Genotype Name	Country of Origin
White	2777.97	264.55	22.13	Semi dent	ZPL1	Mexico
Lemon-yellow	3274.65	265.26	24.30	Semi dent-semi flint	ZPL-2	USA
Yellow	2859.81	461.09	16.32	Dent	ZPL-3	USA
Orange	2899.11	560.77	39.74	Flint	ZPL-4	France
Dark red	2835.57	236.58	0.83	Popcorn	ZPP-1 selfed	Serbia
Light blue	4521.26	333.29	27.09	Popcorn	ZPP-2	Serbia
Multicolored	2586.21	186.25	33.68	Dent	ZPP-3	Netherland

Source: Žilić et al. (2012)

more attention (Kroon and Williamson 1999). The antioxidant and cardiovascular protective properties of *p*-coumaric acid makes it to be of great interest (Jyoti and Prince, 2013). The phenolic contents and their compositions in maize are dependent on varieties, pigments and also varied by cultivation, fertilizer application, ripeness, and postharvest conditions.

Anthocyanin and its derivatives are water soluble glycosides of polyhydroxy and polymethoxy derivates of 2-phenylbenzopyrylium or flavylium salts responsible for the purple, blue, and red colors in vegetal tissues, belonging to the class of flavonoids (Escribano-Bailón et al. 2004; Castañeda-Ovando et al. 2009; Lemos et al. 2012). It is located in the pericarp of the maize endosperm, having a great impact on the kernel color and could be separated into fractions for development of functional colorants or food ingredients. Phenolic compound of different maize kernel types and colors is presented in Table 12.5b. Žilić et al. (2012) reported higher total anthocyanin contents (6,960.70 mg CGE/100 g and 5,971.50 mg/100 g) in dark red maize and dark blue maize, respectively. Other researchers reported low anthocyanin contents in purple maize grown in Argentina (Mansilla et al. 2020), Côte d'Ivoire (Akaffou et al. 2018), and Thailand (Saikaew et al. 2018), while no detection was reported for Mexican white maize. A wide range of beneficial biological activities of anthocyanin pigments in maize and anthocyanin-rich foods has been published as potential agents in anti-inflammatory, anticarcinogenic, and antioxidant activities including vital roles it plays as cardiovascular disease prevention, obesity control, and diabetes alleviation properties, antimutagenic activity, anti-inflammatory, hypoglycemic and antihypertensive properties (Tsuda et al. 2003; Konczack and Zhang 2004; Lopez-Martinez et al. 2009; He and Giusti 2010; Cavalcanti et al. 2011; Norberto et al. 2013; González-Muñoz et al. 2013). Anthocyanins are also involved in ultraviolet (UV) scavenging, fertility, antimicrobial, and antimutagenic effects as part of its presumed health-promoting properies (Cevallos-Casals and Cisneros-Zevallos 2003; Kuhnen et al., 2011).

TABLE 12.5b

Concentration of Phenolic in Maize According to Kernel Colors on Dry Matter Basis

	TPC (mg GAE/100g)	TAC (mg GAE/100g)	TFC (mg CGE/100g)	Tannin (%)	Country	References
Purple	1023.00	172.00	ND	ND	Argentina	Mansilla et al. (2020)
White	319.00	ND*	ND*	ND	Argentina	Mansilla et al. (2020)
Yellow	116.51	17.04	35.41	60.45	Côte d'Ivoire	Akaffou et al. (2018)
Deep purple	253.80	178.54	190.95	52.89	Côte d'Ivoire	Akaffou et al. (2018)
White	ND	ND*	2480.64	ND	Mexico	Žilić et al. (2012)
Dark red	ND	6960.70	2705.40	ND	Serbia	Žilić et al. (2012)
Dark blue	ND	5971.50	3074.20	ND	USA	Žilić et al. (2012)
Yellow	300.25	ND	89.40	ND	Nigeria	Oboh et al. (2010)
Blue	13478.20	9154.30	ND	ND	Turkey	Mutlu et al. (2017)
Purple	326.85	115.05	ND	ND	Thailand	Saikaew et al. (2018)

ND, Not determined; *ND**, Not detected; *TPC*, Total phenol content; *TAC*, Total anthocyanin content; *TFC*, Total flavonoid content.

TABLE 12.6

Concentration of Carotenoids (mg/kg d.m.) in Maize According to Kernel Colors

	Zea-xanthin	Lutein	β-crypto-xanthin	α-carotene	β-carotene	Total carotene	Country	References
Orange	14.03	3.83	6.89	0.88	3.13	ND	Zambia	Mugode et al. (2014)
Red	1.94	0.26	0.041	0.053	0.39	ND	Italy	Bacchetti et al. (2013)
Yellow	2.15	0.32	0.035	0.038	0.27	ND	Italy	Bacchetti et al. (2013)
White	ND	ND*	ND	ND	0.21	ND	Mexico	Žilić et al. (2012)
Light blue	ND	0.01	ND	ND	ND*	ND	USA	Žilić et al. (2012)
Yellow	3.90	2.44	0.04	ND	ND	0.06	South Brazil	Kuhnen et al. (2011)
White	0.04	0.03	ND	ND	ND*	ND	South Brazil	Kuhnen et al. (2011)
Orange	3.00	1.35	0.04	ND	ND	0.05	South Brazil	Kuhnen et al. (2011)

ND, Not determined; *ND**, Not detected.

12.3.2 Carotenoids

Carotenoids are fat-soluble compounds present in plants, algae, and photosynthetic bacteria belonging to a family of red, orange, and yellow pigments, which are chemically constituted of 40 units of isoprene (Patil et al. 2009; Shah et al. 2016). Carotenoids generally categorized into two groups: carotenes, which lack oxygen functions (α-carotene and β-carotene,) and xanthophylls, which contain more than one oxygen atom (zeaxanthin, lutein, β-cryptoxanthin). α-carotene, β-carotene and β-cryptoxanthin are important precursors of vitamin A in humans while lutein and zeaxanthin are nonprovitamin A (Safawo et al. 2010). The predominant carotenoids found in corn are zeaxanthin and lutein as it was reported for purple, yellow, and orange maize kernels by Kuhnen et al. (2011) as presented in Table 12.6. The highest zeaxanthin (14.03 mg/kg) and all-trans β-carotene (3.13 mg/kg) were reported for Zambian biofortified orange maize (Mugode et al. 2014). Xanthophylls are generally more polar than carotenes due to the presence of hydroxyl group. They have provitamin A activity to metabolize to vitamin A in the human body. Studies by Singh et al. (2011) reported the presence of higher level of xanthophylls in some maize samples to be responsible for the higher yellow pigment content. Consumption of carotenoids are associated with a lower risk of various degenerative diseases, as antioxidants work as regulators of the human immune system in preventing cardiovascular diseases and age-related diseases (Messias et al. 2013). The role of the hydroxyl groups by di-acetylation extracted from maize protein residues was investigated and suggested that hydroxyl groups were responsible for the antitumor activity of lutein and zeaxanthin (Sun and Yao 2007).

12.3.3 Phytosterols

Phytosterols (plant sterols) are the essential components of plant cell walls and membranes (Piironen 2000). More than 250 different phytosterols have been found so far and are classified into three based on their methyl group numbers at carbon-4 position: 4-desmethylsterols (simple sterols), 4-monomethylsterols, and 4, 4-dimethylsterols. According to Verleyen et al. (2002), maize oil is very rich in phytosterols. Sitosterol, campesterol, and stigmasterol are mostly consumed phytosterols from maize oil. Their distribution varies in different fractions of maize kernel such as the endosperm, pericarp, and germ (Harrabi et al. 2008). The health benefits of dietary phytosterols is the inhibition of cholesterol absorption through intestine and subsequent compensatory stimulation of the synthesis of cholesterol, resulting in the enhanced elimination of cholesterol in stools (Luo and Wang 2012).

12.3.4 Tocopherols

Vitamin E is common name for eight naturally occurring compounds, lipid-soluble antioxidants with two distinct groups of four tocopherols (α, β, τ, γ), and four tocotrienols (α-T_3, β-T_3, τ-T_3, γ-T_3) (DellaPenna and Pogson 2006; Xu et al. 2019). Tocopherols are essential micronutrients, but unable to be synthesized by the human body (Anđelković et al. 2018). Among cereals, maize grain has the highest content of tocopherols and the two predominant isomers are γ-tocopherol and α-tocopherol located in the embryo (Rocheford et al. 2002). Although γ-tocopherol is naturally most abundant in maize, α-tocopherol has higher biological activity and is considered more desirable for consumption. Anđelković et al. (2018) reported higher value (27.20 µg/g) in Serbian dark orange maize while 6.08, 4.95 µg/g and 4.87 µg/g were discovered in yellow, red and dark red, respectively. The recommended daily allowance for vitamin E for both adult men and women is reported to be 15 mg/d (Otten et al. 2006).

12.4 Processing Technologies of Maize and Their Effects on Nutrients and Bioactive Compounds

Processing of grains into a wide variety of food products before consumption is of primary importance of increasing food palatability, enhancing food digestion, reducing foodborne diseases, and increasing shelf life (Xu et al. 2019). However, processing and mode of preparation sometimes negatively affects the contributions of bioactive compounds in maize for human nutrition and health as well as its sensory characteristics such as taste, color, and texture (Ayatse et al. 1983; Das et al. 2017) as summarized in Table 12.7. Maize is processed in different ways and forms throughout the world according to people's cultures through thermal and nonthermal methods. These include boiling, roasting (whole maize on the cob), popping, cooking, lime cooking (nixtamalization), milling, soaking, germination, fermentation, etc. Therefore, choosing appropriate processing methods would enhance the bioavailability and bioaccessibility of nutritive and health-promoting value of the maize grains (Singh et al. 2016).

12.4.1 Boiling and Roasting

Freshly harvested maize is consumed directly as maize on the cob after boiling in water. Roasting is a technique known to produce healthy crunchy maize as reported by Oboh et al. (2010) to enhance fat, carbohydrate, sodium, potassium, calcium, and magnesium. However, the same researchers also reported the negative impact of roasting on nutritional composition (especially protein, ash, total dietary fiber, and iron) and bioactive compounds (phenol and flavonoid content). Roasting was also found to reduce β-carotene and pro-vitamin A contents in different yellow maize hybrids (Alamu et al. 2014), and total anthocyanin and total polyphenol contents of Peruvian purple maize (Mrad et al. 2014).

12.4.2 Drying

Drying of maize locally after harvesting is done in a different way depending on local uses and customs to a safer level of moisture content. Maize on the cob is stored with or without husks or as shelled grain that is often packaged in bags (Pixley et al. 2013). Depending on the culture, the husks are removed and maize on cobs are hung or placed under the sun to dry. Other farmers shell the ears and dry the grain under the sun on a mat, in a wide basket, or on a cement surface. The dried maize kernels are processed into flour, grits, bread, or transformed into crunchy products such as corn flakes, popcorn, and snacks (Mrad et al. 2014). Drying and storage temperature, oxygen availability, and light conditions are the determining factors that justify the stability of carotenoid compounds in maize. An increase in levels of provitamin A carotenoid levels has shown to reduce the losses encountered during maize drying (Burt et al. 2010).

12.4.3 Steeping and Sprouting

Many research findings have shown the positive impacts of steeping (soaking) on nutritional compositions. Okafor et al. (2018) reported increase in crude protein (10.32–12.43%), fat (7.60–9.21%), ash (0.66–0.94%), and crude fiber content (0.22–0.29%) after steeping yellow maize in water for 48 hours. The authors also recorded an increase in mineral composition (K, P, Na) with significant higher contents in Fe (3.68–20.89 mg/100 g), vitamins B_1, B_2, and B_3 as well as amino acids especially lysine content

TABLE 12.7

Effect of Processing Methods on the Nutrients and Bioactive Compounds of Maize

Methods of Processing	Retention (Positive Effect)	Losses (Negative Effect)	References
Steaming	Significant increase in protein and ash contents. Increase in minerals (Fe, Zn, Cu, Mn, K, P, and Ca). Enhancement in total carotenoids and β-cryptoxanthin,	Reduction in fat, total dietary fiber, carbohydrate, and energy. Losses in lutein, zeaxanthin, α-cryptoxanthin, and phenolic acid (4-OH benzoic, vanillic, syringic, *o*-coumaric, *p*-coumaric, caffeic, ferulic).	Prasanthi et al. (2017)
Popping (without oil)	Significant increase in carbohydrate, energy (K, Ca, Mg, Na and Cu). Improvement in zeaxanthin concentration.	Significant reduction in protein and fat contents. Reduction in minerals (Fe, Zn, P, Mn). Higher losses in total carotenoids, lutein, α- and β- cryptoxanthin, phenolics.	Prasanthi et al. (2017)
Popping (with oil)	Significant increase in fat and carbohydrate energy, Mg, Zn, and Na.	Significant reduction in protein, dietary fiber. Losses in minerals (Fe, Cu, Na, K, P, Ca, and Mn). Higher losses in total carotenoids, lutein, α- and β- cryptoxanthin, and phenolics.	Prasanthi et al. (2017)
Boiling (in water) (maize grits)	Not reported.	Significant reduction in proximate composition and energy value. Losses in carotenoids and xanthophylls. Reduction in total phenolic contents and phenolic acids.	Prasanthi et al. (2017)
Roasting	Significant increase in fat, carbohydrate, Na, K, Ca and Mg.	Significant reduction in protein, fiber, ash, and Fe content. Losses in phenol and flavonoid content.	Oboh et al. (2010)
Roasting	Not reported.	Reduction in β-carotene and provitamin A content.	Alamu et al. (2014)
Roasting	Not reported.	Reduction in total anthocyanin and total polyphenol content.	Mrad et al. (2014)
Milling	Significant increase in crude fiber and Fe content.	Losses in fat and Zn content.	Greffeuille et al. (2010)
Milling	Enhancement in zeaxanthin, β- cryptoxanthin and β-carotene content.	Not reported.	Pillay et al. (2011)
Cooking (dough)	Significant increase in crude fiber and Fe content.	Losses in fat and Zn content.	Greffeuille et al. (2010)
Precooked (grains)	Not reported.	Reduction in fat, fiber, Fe, Zn content.	Greffeuille et al. (2010)
Steeping	Significant increase in protein, fat, ash, fiber, minerals (K, P, Na, Fe, and Ca). Enhancement of vitamins (B_1, B_2, and B_3) and amino acids (lysine, tryptophan, isoleucine, methionine, phenylalanine, leucine, and valine).	Significant reduction in carbohydrate.	Okafor et al. (2017)
Souring	Significant increase in ash, fiber, protein, Ca, Mg, and retention of fat. Increase in vitamins and amino acids.	Significant reduction in carbohydrate, K, P, Na, Fe and Na.	Prasanthi et al. (2017)

(Continued)

TABLE 12.7 (Continued)

Methods of Processing	Retention (Positive Effect)	Losses (Negative Effect)	References
Fermentation	Significant increase in Fe content.	Losses in fat, fiber and Zn.	Greffeuille et al. (2010)
Germination	Significant increase in protein, ash, energy, Ca, Mg, K, and P.	Significant reduction in fat, fiber, carbohydrate, and Fe content.	Bello and Udo (2018)
Traditional nixtamalization	Not reported.	Significant reduction in concentration level of total anthocyanin content, total soluble phenol content, and antioxidant activity.	Amador-Rodríguez et al. (2019)

(11.85–39.21 mg/g) were improved. No losses were reported except reduction in carbohydrate content from 80.16–74.21%.

Sprouting (germination) is the process in which a plant emerges from a seed or spore and begins growth. The germination phase begins after the kernels have absorbed enough water to start enzyme production and starch hydrolysis. The sprouting process unlocks many nutrients, which are in bound forms, increases the bioavailability of nutrients, energy density, and acceptability (Mtebe et al. 1991). It is widely functioned particularly in breaking down of certain antinutrients, such as phytate and protease inhibitors (Steiner et al. 2007). Sprouting is also known to increase the phenolics and flavonoids in a natural way (Laila and Murtaza 2014; Paucar-Menacho et al. 2016). Bello and Udo (2018) reported an improvement in protein, ash, energy, Ca, Mg, K, and P of local white maize sprouted for 72 hours while losses were recorded for crude fat, fiber, carbohydrate, and Fe.

12.4.4 Fermentation

Fermentation is one of the oldest known food processing methods and its history stretches back to the Neolithic period, as indicated by archaeological findings of clay tools for cheese making (Tsafrakidou et al. 2020). It is recognized as a natural way to preserve foods, improving the digestibility, destroying undesirable components, and inhibiting undesirable microorganisms (Marshall and Mejia 2012). Fermentation processes have two systems, submerged fermentation, which consists of a liquid medium, and solid-state fermentation, which consists of growth of microbes on solid particles (Sandhu et al. 2017). Submerged fermentation (steeping maize in water until soft) has been the major processing method of maize prior to wet-milling. Fermentation is reported to have a positive influence on phenolic content of cereals, but the degree of influence depended on microorganism species (Kariluoto et al. 2006; Acosta-Estrada et al. 2014; Sandhu et al. 2017). It enhances nutritive value through the synthesis of certain amino acids, improves availability of B group vitamins and minerals, and improves sensory attributes by contributing to the sweet and sour taste and specific aroma of maize products (Ejigui et al. 2005; Ferri et al. 2016; Peyer et al. 2016). The improvement of bioactive compounds during fermentation of maize has been attributed to the action of native enzymes and the presence of water and oxygen (Achi and Ukwuru 2015). These factors actually promote a series of simultaneous reactions that affect the structure and solubility of bioactive compounds (Konopka et al. 2014). Prasanthi et al. (2017) reported the enhancement of crude protein, ash, fiber, Ca, Mg, vitamins B_1, B_2, B_3 and the improvement of amino acids. The researchers observed losses in carbohydrate and Fe content as well as in some macroelements (K, P, and Na).

12.4.5 Milling

The milling process of maize kernels is categorized into dry and wet milling. Dry milling of maize is done to produce grits of different particle sizes from grains of harder endosperm, which are used in the development of food products such as breakfast cereals, extruded products, snacks, and tortillas (Singh et al. 2014). The objective of dry milling of maize into grit is to get maximum recovery of grit with minimum contamination from the hull, germ, and tip cap. Two different processes are used for industrial dry milling of maize grains. The nondegerming process grinds grains without the separation of germ. The product (meal) of this process has comparatively shorter shelf-life, as the germ is retained. This oil in presence of oxygen and lipolytic enzymes is prone to oxidative and hydrolytic rancidity. Hence, it is necessary to remove the germ from maize to produce maize products with much lower fat content and greater shelf-life. The second process includes tempering and degerming steps, which remove most of the germ and hull thereby leaving the endosperm as free of oil and fiber as possible, and also recovering maximum yield of endosperm and germ as large clean particles. The grains are cleaned to remove dirt, broken kernels and extraneous plant materials and then passed through a Beall degerminator or fluted roller mills to separate the germ and hull. The cleaned grains are tempered by conditioning with water (cold or hot) or steam for varying lengths of time depending upon the target products. The essence of conditioning is to make the hull germ tough and endosperm friable (mellow). The grit is dried to around 15% moisture content, cooled, and graded to obtain fractions of different particle sizes, ranging from large hominy grits to fine flour. These fractions differ in composition and end-use suitability (Shevkani et al. 2014). Traditionally in India, maize is milled in stone mills to get whole meal (or atta) that is used in the production of roti (unleavened flat bread) (Singh et al. 2019). Most of the phenolic compounds present in corn bran is in bound form. Mamatha et al. (2012) expressed that carotenoid (lutein and zeaxanthin) content in white corn is not affected by drying and milling.

Wet-milling of maize is done to produce starch and other valuable byproducts such as gluten, germ, and bran (Singh et al. 2019). Steeping of maize in a solution of sulfurous and lactic acid is done to soften the kernels and loosen the bran, facilitate the separation of starch from protein, and prevent fermentation. Steeped water is drained and grains are coarsely ground to free the germ from the endosperm and hull. The germ portion is separated, dried, and expelled for oil extraction while the fiber and starch suspension is then fine-milled and separated by screening, centrifugation, and washing. Centrifugation with the aid of battery of hydrocyclones successively purifies cornstarch (reduces protein and lipid content) and affects physicochemical and functional properties of the starch (Singh et al. 2014). The dried starch can be modified or used to develop valuable products like fructose syrups, dextrose, sorbitol, and dextrins through chemical processes, enzymatic processes, or a combination. According to Mamatha et al. (2012), milling enhanced the retention of carotenoids in maize products. Pillay et al. (2011) reported that milling maize grain into mealie meal (South African foods) resulted in a higher retention of zeaxanthin (115.9–136.4%), β-cryptoxanthin (118.9–137.2%) and β-carotene (105.6–134.3%). According to these authors, the observed higher retention could be as a result of increased the concentrations of available carotenoids due to the greater breakdown of the maize kernel matrix during the production of mealie meal. Milling also led to an increase in crude fiber and Fe content of maize flour and its cooked dough in the production of owo (Benin foods) while losses were reported for lipid and Zn content (Greffeuille et al. 2011). Optimization of milling conditions can therefore help in minimizing the loss of the nutrients and maximizing removal of antinutrients and toxins such as mycotoxins.

12.4.6 Lime Cooking and Steeping

Nixtamalization is the traditional form of processing maize in Mexico and Central America where maize grain is converted to dough (masa), and ultimately to corn flour, snacks, and tortillas. The maize kernels are lime-cooked with 1% calcium hydroxide solution in a proportion of 1:2 (maize: solution) at 92–95°C for 45 minutes and then steeped overnight and the resulting nixtamal is washed and stone-ground into masa using a commercial mill (Colín-Chávez et al. 2020). Conventional processing equipment is employed with raw maize and dry finished product. The obtained product may be dried for commercial sales of further processed consumer food product. (Gwirtz and Garcia-Casal 2014). The traditional nix-tamalization process has been reported to have negative effect as it reduced the total polyphenol content, antioxidant activity, and anthocyanin content in corn and corn products (De la Parra et al. 2007; Mora-Rochín et al. 2016). Significant losses of total anthocyanin, which decreased from 759.55 mg C3G/kg to 252.53 mg C3G/kg (66%), and also total soluble phenolic content from 154 mg GAE/100g to 46.33 mg GAE/100g were reported in nixtamalized Creole blue maize flours (Amador-Rodríguez et al. 2019). The authors attributed the significant losses to the combined effect of alkaline and thermal processing during nixtamalization, as well as to the physical losses of the pericarp and the leaching of phenolics into the cooking liquor as confirmed the findings of Del Pozo-Insfran et al. (2006). High-energy milling has been reported to be a possible soluble of retaining the bioactive compounds in nixtamalized maize products (Amador-Rodríguez et al. 2019).

12.5 Possible Approaches to Improve Nutritional and Bioactive Compounds in Maize and Maize-based Food Products

Poor diet is the leading cause of mortality and morbidity worldwide, exceeding the burdens attributable to many other major global health challenges (Afshin et al. 2019). Protein-energy malnutrition (PEM), micronutrient (MN) deficiencies, and food insecurity continue to be health challenges worldwide. Globally, more than 1 billion people, especially women and children, are estimated to lack sufficient dietary energy availability, and at least twice as many suffer MN deficiencies (Rautiainen et al. 2016; Waller et al. 2020). The inadequate supply of lysine and tryptophan results in weakness, slow growth, loss of appetite, depression, and anxiety (Yadava et al. 2018). The human body cannot synthesize essential amino acids like lysine and tryptophan, and vitamins such as vitamin-A, thus they are supplemented through food (Prasanna et al. 2020). It is estimated that 45% of death in children (<5 years) is associated with malnutrition (Black et al. 2013).

Those affected by MN (hidden hunger) predominately reside in low-resource regions, food insecure, and low-income households (Horton 2008). Vitamin-A deficiency causes visual disablement, diminishing body growth, reproductive disorders and weak immune system in humans (Lonzano-Alejo et al. 2007). MNs commonly lacking among populations worldwide include iron, vitamin A, and iodine (Dunn et al. 2014; WHO 2020). Iron deficiency can lead to anemia, resulting in an increased risk of developmental issues, infection, and death (Horton 2008). The developed nations have had tremendous success in alleviating nutrient deficiencies through dietary diversification, improved public health care, food fortification, and supplementation. In developing countries, these strategies are often too expensive and difficult to sustain, especially in rural areas. Poor people from rural areas typically consume what they plant and are dependent upon a small number of staple crops for the vast majority of their nutrition (Dichi and Miglioranza 2013).

Consumption of maize grains poor in essential nutrients results in health problems especially in children and pregnant women (Bouis 2018). The human body cannot synthesize essential amino acids like lysine and tryptophan, and vitamins such as vitamin A, thus they are supplemented through food (Prasanna et al. 2020). The content of micronutrients such as vitamin A and essential microminerals (Fe and Zn) are low in maize. Thus, consumption of maize as a primary diet staple cannot help people live a healthy life. Developing micronutrient-enriched corn (fortification) is necessary for increasing micronutrient content and bioavailability (Xu et al. 2019). Fortification has been implemented around the world based on local and regional practices in the areas of maize-based food production (Ranum, 2014).

12.5.1 Biofortification (Endogenous Fortification)

Agronomists have been working diligently to improve the intrinsic nutrient content and yield of staple crops in developing countries through biofortification (De Valençaet al. 2017; Nikhil and Salakinkop 2018; Ngigi et al. 2019). Biofortification is an effective strategy aimed to increasing content and/or bioavailability of essential nutrients in crops during plant growth (Bouis et al. 2011). Biofortification is an umbrella term pertaining to four major endogenous fortification strategies including the use of fertilizers (agronomic), conventional (crop breeding), transgenic (genetic engineering), or mutagenesis (Zhu et al. 2007; Bouis et al. 2011; Miller and Welch 2013; Meena et al. 2016; Haskell et al. 2017). The maize kernel has become a vector for biofortification of macronutrients and micronutrients in traditional maize varieties for researchers, for example, Harvest Plus is a global research institute working on biofortification via maize breeding to generate micronutrient-enriched corn to help reduce the micronutrient malnutrition (Ortiz-Monasterio et al. 2007).

Nutritional goals for biofortification include increased vitamin and mineral content, elevated amino acid and antioxidant values, and favorable fatty acid composition in edible parts of crops (Hirschi 2009). Among cereals, maize possesses the greatest natural genetic variation for carotenoids, suitable for provitamin A biofortification (Menkir et al., 2008). The yellow maize contains carotenoid isoforms, α-carotene, β-carotene, β-cryptoxanthin, zeaxanthin, and lutein (Cazzonelli and Pagson, 2010). However, maize is not only a good source for provitamin A carotenoids but also a source of vitamin E, tocochromanols and phenolic compounds, which have antioxidant properties (Muzhingi et al., 2017).

Biofortification of maize hybrids rich in provitamin A and essential amino acids have been success-fully achieved via breeding and released for cultivation (Prasanna et al. 2020). Biofortified maize cul-tivars (first generation), were enhanced for single nutrient, but new scientific achievements are directed to second generation of biofortified maize genotypes, with increased content of combination of more micro- and macronutrients (Gupta et al. 2015). Simultaneous breeding for multinutritional traits in maize, could increase utilization of maize grain in nutrient rich meals instead of using synthetic carot-enoids and tocopherols.

O'Hare et al. (2015) reported significant increase of zeaxanthin concentration in sweet corn kernels at least 10-fold to 25 μg/g fresh weight (FW) through conventional breeding and selection. Dark orange grain was recommended by Anđelković et al. (2018) in their studies as a good source for multinutrient biofortification for provitamin A and vitamin E (β-carotene and α-tocopherol, respectively). The bio-fortified sweet maize hybrids at different stages of kernel development was reported to contain higher lysine, tryptophan, β-carotene, β-cryptoxanthin, and provitamin A contents over traditional yellow sweet corn (Mehta et al. 2020). Besides sweetness, accumulation of carotenoids and essential amino acids makes it more nutritious.

12.5.2 Fortification of Maize-based Foods with Legumes and Other Protein-rich Ingredients (Exogenous Fortification)

In developing and underdeveloped countries, fortification of staple foods are the most effective ways to upgrade the nutritional status of the population. The relatively poor protein quality of maize demands its fortification with protein rich food in order to achieve a nutritional balanced diet and bioavailabil-ity. This entails the partial substitution of cereals with other nutrient dense foods such as legumes to diversify as well as upgrade the nutritive value of indigenous agricultural food products (Temba et al. 2016). Diversifying diets with legumes is a cheaper, surer, and more sustainable way to supply a range of nutrients to the body and combat malnutrition and associated health problems than other approaches that target only a single or a few nutritional factors (Ojiewo et al. 2015).

Food-to-food fortification with varieties of ingredients from different sources, especially legumes, have been explored to enhance nutritive value in local maize-based foods (Table 12.8). Ikya et al. (2013) reported that fortification of traditional maize food (*agidi*) with full fat soy flour improved the protein content and 30% full fat soy flour supplementation was generally accepted by the panelists. *Agidi* is a thick gel produced from fermented maize paste and is a popular food widely consumed in Nigeria and some West African countries. The production of gruel from germinated maize flour fortified with soy bean and *Moringa oleifera* was carried out by Klang et al. (2019) to obtain the desired flow and nutritional value. The authors reported significant increase in protein, Fe, and Ca content from 6.82% to 12.00%, 1.79 mg/100 g to 6.70 mg/100 g, and 3.00 mg/100 g to 4.50 mg/100 g, respectively. This approach could therefore help to fight against child malnutrition.

The prevalence of Fe and Zn deficiency in the diet of children under age 5 in South Africa was attributed to high intake of maize porridge (Motadi et al. 2015). In view of this, Netshiheni et al. (2019) successfully carried out an improved traditional maize porridge blending with *Moringa oleifera* leaves and termite (*Macrotermes falciger*) powder. The authors reported higher values of protein, Ca, Fe, and Zn content. The fortified porridge has potential to contribute significantly to the dietary intake of many households in the Limpopo province of South Africa. Nutritional and sensory evaluation of different maize-meal porridge (MMP) developed for three districts in the Eastern province of Zambia were reported by Alamu et al. (2016). The three fortificants (soybean flour, groundnut paste, and powdered milk) added separately to MMP as discussed by the authors significantly increase the protein contents of MMP. The consumer acceptance of powdered milk fortified MMP was most preferred by the panel-ists. *Abari*, a maize-based food product that looks like *moinmoin*, fortified with white kidney bean flour had increased protein content and was acceptable at a 30% substitution level with kidney bean flour (Olanipekun et al. 2015). Protein content, amino acid, antioxidant property, and *in vitro* protein digest-ibility of Mexican tortillas were enhanced through fortification with soy and defatted chia (León-López et al. 2019). *Kokoro* is a widely accepted local snack prepared from maize flour in Southwestern Nigeria and its relatively low cost of production made it is cheaper compared to other snacks. *Kokoro* fortified

TABLE 12.8

Studies on Fortification of Local Maize-Based Foods with Increase in Protein/Amino Acids, Microelements, Bioactive, and Sensory Evaluation

Fortification Ingredients	Maize Food Type	Protein/Amino Acids, Microelements and Sensory	Country	References
Full fat soy flour	*Agidi*	Increase in protein (8.92–20.77%). Sensory (Yes with 30% full fat soy flour substitution).	Nigeria	Ikya et al. (2013)
Defatted chia flour	Tortillas	Significant protein increase (8.88–14.33%). Enhancement in lysine (2.44–3.42 g/100 g protein), in vitro protein digestibility (75.90–83.22%), antioxidant activity (5,985–12,499).	Mexico	León-López et al. (2019)
Soybean flour	Emergency food	Significant protein increase (13.38–15.26%)	Indonesia	Aini et al. (2018)
Tempeh flour	Emergency food	Significant protein increase (12.69–16.25%)	Indonesia	Aini et al. (2018)
Soybean and yellow ginger flour	*Kokoro*	Increase in protein (8.32–15.91%), lysine (3.16– 3.24 g/100 g).	Nigeria	Fasasi and Alokun (2013)
Moringa oleifera leaves and termite powder	Porridge	Significant increase in protein (10.00–21.20%), Fe (7.70–27.90 mg/100 g), Zn (3.40–7.60 mg/100 g). Sensory (Yes, with no general acceptability).	South Africa	Netshiheni et al. (2019)
Soybean flour and *Moringa oleifere* leaf powder	Gruel	Increase in protein (6.82–12.00%), Fe (1.79–6.70 mg/100 g).	Cameroon	Klang et al. (2019)
Soybean flour	Porridge	Increase in protein (2.50–12.66%). Sensory (Yes, not preferred).	Zambia	Alamu et al. (2016)
Groundnut paste	Porridge	Increase in protein (2.50–12.11%). Sensory (Yes, not preferred).	Zambia	Alamu et al. (2016)
Powdered milk	Porridge	Increase in protein (2.50–12.05%). Sensory (Yes, most preferred).	Zambia	Alamu et al. (2016)
White kidney bean	*Abari*	Increase in protein (5.90–7.81%). Sensory (Yes with 30% kidney bean flour substitution).	Nigeria	Olanipekun et al. (2015)
Soybean and tigernut milk	*Ekoki*	Increase in protein (10.52–16.74%), β-carotene (32.8–149.20 mg/100 g). Sensory (Yes, white maize *ekoki* was most preferred without fortification).	Nigeria	Enidiok et al. (2017)

with soybean and ginger flours was reported to contain higher protein, mineral, and vitamin A contents but lower lysine content as a result of yellow ginger flour incorporation (Fasasi and Alokun 2013).

Studies have shown maize among cereals as a base for complementary food formulation, which is enhanced with ingredients rich in nutrients and bioactive compounds that would potentially address the problem of protein-energy malnutrition with further reduction on the risk of cardiovascular and degenerative diseases in developing countries. The few of out of many reported findings on maize-based complementary foods is presented in Table 12.9. The nutritional demands of infants necessitate the introduction of complementary foods that are readily consumed and digested by the young child and that provide additional nutrition to meet all the growing child's needs from six months up to 18-24 months of age (UNICEF, 2018).

Maize flour fortified with soybean and pigeon pea as reported by Aderonke et al. (2014) has shown to enhance the protein and mineral (Ca, Zn, and Fe) contents of formulated complementary food. Incorporation of defatted fluted pumpkin flour has been reported to significantly increased the protein content from 8.19%–17.60% (Ikujenlola et al. 2013). Amankwah et al. (2009) incorporated rice, soybean,

TABLE 12.9

Studies on Fortification of Maize-Based Complementary Foods with Increase in Protein/Amino Acids, Microelements, Bioactive, and Sensory Evaluation

Fortification Ingredients	Protein/Amino Acids, Microelements and Sensory	Country	References
Soybean and pigeon pea flours	Increase in protein (12.61–20.75%), Fe (4.81–10.22 mg/100 g), Zn (2.75–6.77%). No sensory	Nigeria	Aderonke et al. (2014)
Defatted fluted pumpkin flour	Increase in protein (8.19–17.60%). No sensory	Nigeria	Ikujenlola et al. (2013)
Defatted sesame and mushroom flours	Increase in protein (10.64– 25.32), Fe (1.25–2.54 mg/100 g), lysine (3.70–6.17 g/100 cP). Sensory: Yes, 100% quality protein maize was preferred without fortification.	Nigeria	Ikujenlola et al. (2013)
Sorghum and mungbean flours	Increase in protein (13.99–17.19%). Sensory: Yes, 30% mungbean flour substitution).	Nigeria	Onwurafor et al. (2017)
Mungbean malt flour	Protein increase (7.89–22.68%). Enhancement in lysine (0.82–5.44 g/g protein). No sensory.	Nigeria	Onwurafor et al. (2020)
Rice, soybean, and fishmeal flours	Increase in protein (11.19–19.13%). Fe (16.00–71.00 mg/100 g). Sensory: Yes, with 74% of panelists suggested removal of all the fortification ingredients.	Ghana	Amankwah et al. (2009)
Garden pea flour	Increase in protein (6.24–12.04%). Sensory: Yes, with 10% garden pea flour inclusion.	Nigeria	Ukeyima et al. (2019)
Cowpea and sweet potato (co-fermentation)	Increase in protein (6.70–10.70 g/100 g), lysine (0.82–1.03 g/100 g protein). Reduction in Fe (3.80–3.20 mg/100 g) and Zn (0.85–0.19 mg/100 g). Increase in β- carotene (986.30–1782.50 µg/100 g) and β-carotenoid (164.30–297.00 µg/100 g). No sensory.	Nigeria	Oyarekua (2013)
Soybean, peanut, and *Moringa oleifera* powder	Increase in protein (10.40–17.59%). No sensory.	Nigeria	Shiriki et al. (2015)
Pea and barley flours	Increase in protein (13.00–18.00%). Sensory: Yes, with 35% pea flour and 10% barley flour inclusion.	Ethiopia	Fikiru et al. (2017)
Pigeon pea (co-fermentation)	Increase in protein (14.95–22.79%), reduction in Fe (15.90–14.24 mg/100 g). Enhancement in lysine (19.98–93.95 mg/g). No sensory.	Nigeria	Okafor et al. (2018)
Pea and anchote flours	Increase in protein (14.92–20.99%), Fe (11.48–12.61 mg/100 g), Zn (2.73–3.00 mg/100 g). No sensory.	Ethiopia	Gemede (2020)
Moringa oleifera seed flour	Increase in protein (15.27–17.00%), Fe (45.69–67.36 mg/100 g), Zn (14.57–23.68 mg/100 g). No sensory.	Nigeria	Adeoti and Osundahunsi (2017)
Plantain and soybean flours	Increase in protein (10.64–12.95%), Fe (ND), Zn (0.81–3.54 mg/100 g), Vit. A (0.17–1.18 mg/100 g), Vit. C (1.61–7.43 mg/100 g). Sensory: Yes, with 15% plantain flour and 15% soybean flour inclusion.	Nigeria	Noah (2017)
Soybean and sweet potato flours	Increase in protein (8.51–12.20%), Fe (3.65–4.61 mg/100 g), Zn (3.70–4.30 mg/100 g). Sensory: Yes, 5% soybean flour and 5% sweet potato flour inclusion.	Nigeria	Okoye and Egbujie (2018)

and fish meal into fermented maize flour and an increase in protein and Fe contents were reported. The authors claimed that 74% of panelists suggested removal of the fortificants (rice, soybean, and fish meal) from the product. Garden pea flour has shown to have a great potential in enhancing the protein content (6.24–12.04%) of maize-based complementary diet with 10% inclusion (Ukeyima et al. 2019).

Studies have established that cofermentation of maize with cowpea and sweet potato (Oyarekua 2013), and maize and pigeon pea (Okafor et al. 2018) have significantly improved the quality of protein, lysine, Fe and β-carotene contents. Fikiru et al. (2017) also showed its application in an effort to increase the macronutrient in an infant feeding in Ethiopia with the fortification of maize with pea and barley grain malt. The researchers reported enhancement in protein content and sensory evaluation showed acceptability of 35% pea and 10% barley grain malt. The maize fortified with sorghum and roasted mungbean flours yielded diets with higher protein content as reported by Onwurafor et al. (2017). Higher protein and lysine content was also recorded in maize flour fortified with mungbean malt (Onwurafor et al. 2020). The research has shown that mungbeans are rich in protein and essential amino acids (Mubarak 2005; Skylas et al. 2018).

The potential of maize flour in the formulation of bakery products have been on the increase due to its gluten-free content (Hera et al. 2013). Celiac disease is an autoimmune disease related to an intolerance of gluten, which affects adults and children and is known to be treated by avoiding gluten-containing food (Crowe 2008; Belorio et al. 2019). Maize-based bakery foods have been nutritionally and bioactively enhanced through food-to-food fortification as summarized in Table 12.10. Cookies are ready-to-eat, convenient and inexpensive baked products produced from wheat and other ingredients. (Adeyeye et al. 2017). Use of maize in cookie production has been demonstrated by Awolu et al. (2017).

TABLE 12.10

Studies on Fortification of Maize-Based Cookies/Biscuits and Bread with Increase in Protein/Amino Acids, Microelements, Bioactive, and Sensory Evaluation

Fortification Ingredients	Maize Food Type	Protein/Amino Acids, Microelements and Sensory	Country	References
Fenugreek flour	Biscuits	Increase in protein (11.56–15.00%). No sensory	Egypt	Hussein et al. (2011)
Soy protein isolate	Cookies	Increase in protein (6.21–29.11%). Sensory: Yes, up to 20% supplementation.	Nigeria	Adeyeye et al. (2017)
Bambara groundnut flour	Cookies	Increase in protein (13.47–19.26%), Fe (3.20–5.10 mg/100 g). Sensory: Yes, 25% supplementation.	Nigeria	Akpapunam and Darbe (1994)
Soya flour	Cookies	Increase in protein (9.78–36.47%). Sensory: Yes, 10% supplementation.	India	Mishra et al. (2012)
Soybean and tigernut flours	Cookies	Increase in protein (10.90–16.40%), lysine (1244.25–1332.25 mg/100 g). Sensory: Yes, 20% soy flour supplementation.	Nigeria	Awolu et al. (2017)
Acorn flour	Biscuits	Increase in protein (35.80–36.50 g/kg). Sensory: Yes, 40% supplementation.	Poland	Korus et al. (2017)
Hemp flour	Biscuits	Increase in protein (50.00–79.20 g/kg). Sensory: Yes, supplementation with hemp flour was not preferred.	Poland	Korus et al. (2017)
Soybean flour	Bread	Increase in protein (4.10–7.06%). No sensory.	Ethiopia	Mesfin and Shimelis (2013)
Cassava and pigeon pea flours	Bread	Increase in protein (3.65–5.91%). Sensory: Yes, supplementation with 30% cassava flour and 30% pigeon pea flour.	Nigeria	Olunlade et al. (2013)
Cassava and soybean flours	Bread	Protein increase (3.65–6.66%). Sensory: Yes, supplementation with 30% cassava flour and 10% soybean flour.	Nigeria	Olunlade et al. (2013)

The researchers fortified maize flour with soybean and tigernut flours, producing cookies richer in protein and lysine content with acceptable level of 20% soy flour inclusion. Soy proteins are unique among plant proteins by virtue of their relatively high biological value and presence of essential lysine, which is a limiting amino acid in maize kernel (Lang et al. 1999). Soy flour, soy isolate, and soy concentrate are three protein-rich products of soybean (Mishra et al. 2012). Maize-soy protein isolate flour cookie was reported to contain high protein content (Adeyeye et al. 2017).

There has been an interest in developing gluten-free products using alternative gluten-free raw materials with higher nutritional values similar to that of wheat products (Capriles et al. 2016; Gómez and Martínez 2016). Fortification of maize flour with acorn and hemp in separate biscuit productions were reported by Korus et al. (2017). The two ingredients enhanced the protein contents but the maize-hemp biscuit had a higher value. Upon the sensory attribute, the hemp fortified cookie was rejected due to greenish color and bitter taste. The maize-acorn biscuit was accepted with higher flavonoids and total phenolic contents. The productions of two bread samples from maize, cassava, and soybean flours, and also maize, cassava and pigeon pea flours were reported to contain high protein content and the bread was accepted at 30% cassava flour and 10% for each soybean flour and pigeon pea flour (Olunlade et al. 2013). Mesfin and Shimelis (2013) also reported increase in protein content of maize-soybean flour bread.

12.6 Conclusion and Future Prospects

Food utilization of maize kernel irrespective of varieties and colors has been on the increase across the continents as the world population becomes larger day by day. Due to good agronomic practices, maize has the highest genetic yield potential among the cereals. Different technologies of maize processing from traditional to industrial levels across diverse cultures impact positively or negatively the products' nutrients and phytochemicals, which are vital to human health. Fortification has shown to be a better approach in nutritional value, bioactive compounds, and sensory attributes enhancement, which, therefore, has the potential to alleviate the prevalent challenges of protein-energy and micronutrient malnutrition in developing and underdeveloped countries of the world as well as consumer acceptability of maize-based food products.

Food-to-food fortification is a processing technology with a great potential to increase the protein, microminerals, carotenoids, and phenolic contents of maize-based food products. With increasing cultivation and consumption of maize all over the world as well as consumer desire for nutritionally rich and healthy foods, food-to-food fortification has a great potential to be a better solution to macro- and micromalnutrition in a near future. The technology is low cost and affordable, which would encourage farmers in cultivating local crops that are rich in protein, micronutrients, and bioactive compounds. Besides the protein-rich source foods, fruit and vegetable byproducts, plant ingredients and extracts, and whole grain-like seeds should be incorporated for the development of functional maize-based foods with enhanced dietary fiber content as well as bioactive compounds with health promoting effects.

REFERENCES

Achi, O. K., and M. Ukwuru, M. 2015. Cereal-based fermented foods of Africa as functional foods. *International Journal of Applied Microbiology* 2:71–83.

Acosta-Estrada, B. A., Gutiérrez-Uribe, J. A., and S. O. Serna-Saldívar. 2014. Bound phenolics in foods, a review. *Food Chemistry* 152:46–55.

Adeoti, O. A., and O. F. Osundahunsi. 2017. Nutritional characteristics of maize-based complementary food enriched with fermented and germinated *Moringa Oleifera* seed flour. *International Journal of Food Science, Nutrition and Dietetics* 6(2), 350–357. http://dx.doi.org/10.19070/2326-3350-1700062

Aderonke, A. M., Fashakin, J. B., and S. I. Ibironke. 2014. Determination of mineral contents, proximate composition and functional properties of complementary diets prepared from maize, soybean and pigeon pea. *American Journal of Nutrition and Food Science* 1(3):53–56. https://doi.org/10.12966/ajnfs.07.01.2014

Adeyeye, S. A. O., Adebayo-Oyetoro, A. O., and S. A. Omoniyi. 2017. Quality and sensory properties of maize flour cookies enriched with soy protein isolate. *Cogent Food & Agriculture* 3:1278827. http://dx.doi.org /10.1080/23311932.2017.1278827

Adom, K. K., and R. H. Liu. 2002. Antioxidant activity of grains. *Journal of Agricultural and Food Chemistry*, 50, 6182–6187. http://dx.doi.org/10.1021/jf0205099

Afshin, A., Sur, P. J., and K. A. Fay. 2019. Health effects of dietary risks in 195 countries, 1990–2017: a systematic analysis for the global burden of disease study 2017. *The Lancet* 1:958–972. https://doi. org/10.1016/S0140-6736(19)30041-8

Agama-Acevedo, E., Salinas-Moreno, Y., Pacheco-Vargas, G., and L. A. Bello-Pérez. 2011. Características físicas y químicas de dos razas de maíz azul: morfología del almidón. *Revista Mexicana de Ciencias Agrícolas* 2(3):317–329.

Aini, N., Prihananto, V., Wijonarko, G., Sustriawan, B., Dinayati, M., and F. Aprianti. 2018. Formulation and characterization of emergency food based on instant corn flour supplemented by instant tempeh (or soybean) flour. *International Food Research Journal* 25(1): 287–292.

Akaffou, F. A., Koffi, D. M., Cisse, M. and S. L. Niamké. 2018. Physicochemical and functional properties of flours from three purple maize varieties named "Violet de Katiola" in Côte d'Ivoire. *Asian Food Science Journal* 4(4):1–10. https://doi.org/10.9734/AFSJ/2018/44034

Akpapunam, M. A., and J. W. Darbe 1994. Chemical composition and functional properties of blends of maize and Bambara groundnut flours for cookies production. *Plant Foods for Human Nutrition* 46:147–155.

Alamu, E. O., Maziya-Dixon, B., Popoola, I., Gondwe, T., and D. Chikoye. 2016. Nutritional evaluation and consumer preference of legume fortified maize-meal porridge. *Journal of Food and Nutrition Research* 4(10):664–670. https://doi.org/10.12691/jfnr-4-10-6

Alamu, O. E., Menkir, A., Maziya-Dixon, B., and O. Olaofe. 2014. Effects of husk and harvest time on carotenoid content and acceptability of roasted fresh cobs of orange maize hybrids. *Food Science & Nutrition* 2(6):811–820. https://doi.org/10.1002/fsn3.179

Amador-Rodríguez, K. Y., Martínez-Bustos, F. and H. Silos-Espino. 2019. Effect of high-energy milling on bioactive compounds and antioxidant capacity in nixtamalized creole corn flours. *Plant Foods for Human Nutrition* https://doi.org/10.1007/s11130-019-00727-9

Amankwah, E. A., Barimah, J., Nuamah, A. K. M., Oldham, J. H., and C. O. Nnaji. 2009. Formulation of weaning food from fermented maize, rice, soybean and fishmeal. *Pakistan Journal of Nutrition* 8(11): 1747–1752.

Anđelković, V., Masarović, J. Srebrić, M., and S. M. Drinić. 2018. Pigmented maize—A potential source of β-carotene and α-tocopherol. *Journal of Engineering and Processing Management* 10(2):1–7. https:// doi.org/10.7251/JEPM181002001A

Awolu, O. O., Omoba, O. S., Olawoye, O., and M. Dairo. 2017. Optimization of production and quality evaluation of maize-based snack supplemented with soybean and tiger-nut (*Cyperus esculenta*) flour. *Food Science and Nutrition* 5(1):3–13. https://doi.org/10.1002/fsn3.359

Ayatse, J. O., Eka, O. U., and E. T. Ifon. 1983. Chemical evaluation of the effect of roasting on the nutritive value of maize. *Food Chemistry* 12:135–147.

Bacchetti, T., Masciangelo, S., Micheletti, A., and G. J. Ferretti. 2013. Carotenoids, phenolic compounds and antioxidant capacity of five local Italian corn (*Zea mays* L.) Kernels. *Journal of Nutrition and Food Sciences* 3:6. 10.4172/2155-9600.1000237

Beckles, D. M., and M. Thitisaksakul. 2014. How environmental stress affects starch composition and functionality in cereal endosperm. *Starch-Starke* 66(1–2):58–71.

Bello, F. A. and A. A. Badejo. 2017. Combined effects of packaging film and temperatures on the nutritional composition of stored fresh maize (*Zea mays*) on the cob. *American Journal of Food Science and Technology* 5(1):23–30. https://doi.org/10.12691/ajfst-5-1-5

Bello, F. A. and I. B. Oluwalana. 2017. Impact of modified atmosphere packaging on nutritive values and sensory qualities of fresh maize (*Zea mays* L.) under tropical ambient storage condition. *International Journal of Nutrition and Food Sciences* 6(1):19–24. https://doi.org/10.11648/j.ijnfs.20170601.14

Bello, F. A., and J. A. Udo. 2018. Effect of processing methods on the nutritional composition and functional properties of flours from white and yellow local maize varieties. *Journal of Advances in Food Science and Technology* 5(1):1–7.

Belorio, M., Sahagún, M., and M. Gómez. 2019. Influence of flour particle size distribution on the quality of maize gluten-free cookies. *Foods* 8:83. https://doi.org/10.3390/foods8020083

Black, R. E., Victora, C. G., and Walker, S. P., et al. 2013. Maternal and child undernutrition and overweight in low-income and middle-income countries. *Lancet* 382, 427–451.

Bouis, H. E., Hotz, C., and McClafferty, B., et al. 2011. Biofortification: A new tool to reduce micronutrient malnutrition. *Food Nutrition Bulletin* 32(1 Suppl):S31–S40.

Bouis, H., 2018. Reducing mineral and vitamin deficiencies through biofortification: Progress under HarvestPlus. In H.K. Biesalski and R. Birner (Eds.), Hidden Hunger: Strategies to Improve Nutrition Quality, 118:112–122. Basel Switzerland: Karger Publishers.

Bressani, R., Benavides, V., Acevedo, E., and M. A. Ortiz. 1990. Changes in selected nutriment content and in protein quality of common and quality maize during tortilla preparation. *Food Revue. Cereal Chemistry* 6(67):515–518.

Burt, A. J., Grainger, C. M., Young, J. C., Shelp, B. J., and E. A. Lee. 2010. Impact of postharvest handling on carotenoid concentration and composition in high-carotenoid maize (*Zea mays* L.) kernels. *Journal of Agricultural & Food Chemistry* 58(14):8286–92.

Capriles, V. D., dos Santos, F. G., and J. A. G. Arêas. 2016. Gluten-free breadmaking: improving nutritional and bioactive compounds. *Journal of Cereal Science* 67:83–91.

Castañeda-Ovando, A., Pacheo-Hernández, de L., Páez- Hernández, M. E., Rodríguez, J. A. and Gálan-Vidal, C. A. 2009. Chemical studies of anthocyanins: A review. *Food Chemistry* 113(4): 859–871.

Cavalcanti, R. N., Santos, D. T., and M. A. A. Meireles. 2011. Non-thermal stabilization mechanisms of antho-cyanins in model and food systems-an overview. *Food Research International* 44:499–509. https://doi.org/10.1016/j.foodres.2010.12.007

Cazzonelli, C. I., and B. J. Pagson. 2010. Source to sink regulation of carotenoid biosynthesis in plants. *Trends in Plant Science* 15:266–274. https://doi.org/10.1016/j.tplants.2010.02.003

Cevallos-Casals, B. A., and L. Cisneros-Zevallos. 2003. Stoichiometric and kinetic studies of phenolic anti-oxidants from Andean purple corn and red-fleshed sweetpotato. *Journal of Agricultural and Food Chemistry* 51:3313–3319.

Chaves-López, C., Rossi, C., Maggio, F., Paparella, A., and A. Serio. 2020. Changes occurring in spontane-ous maize fermentation: an overview. *Fermentation* 6:36. https://doi.org/10.3390/fermentation6010036

Colín-Chávez, C., Virgen-Ortiz, J. J., Serrano-Rubio, L. L., Martínez-Téllez, M. A., and M. Astier. 2020. Comparison of nutritional properties and bioactive compounds between industrial and artisan fresh tortillas from maize landraces. *Current Research in Food Science* 3:189–194. https://doi.org/10.1016/j.crfs.2020.05.004

Crowe, S. E. 2008. Celiac disease. In Nutritional and Gastrointestinal Disease. USA: Humana Press Inc. Totowa, NJ, pp 123–147.

Das, A. K., Bhattacharya, S., and V. Singh. 2017. Bioactives-retained non-glutinous noodles from nixtamal-ized dent and flint maize. *Food Chemistry* 217:125–132. https://doi.org/10.1016/j.foodchem.2016.08.061

De la Parra, C., Serna Saldivar, S. O., and R. H. Liu. 2007. Effect of processing on the phytochemical pro-files and antioxidant activity of corn for production of masa, tortillas, and tortilla chips. *Journal of Agriculture & Food Chemistry* 55:4177–4183. https://doi.org/10.1021/jf063487p

De Valença, A. W., Bake, A., Brouwer, I. D., and K. E. Giller. 2017. Agronomic biofortification of crops to fight hidden hunger in sub-Saharan Africa. *Global Food Security* 12:8–14. https://doi.org/10.1016/j.gfs.2016.12.001

Del Pozo-Insfran, D., Brenes, C. H., Serna Saldivar, S. O. and S. T. Talcott, 2006. Polyphenolic and antioxidant content of white and blue corn (*Zea mays* L.) products. *Food Research International* 39(6):696–703. https://doi.org/10.1016/j.foodres.2006.01.014

DellaPenna, D., and B. J. Pogson. 2006. Vitamin synthesis in plants: Tocopherols and carotenoids. *Annual Review in Plant Biology* 57:711–738. https://doi.org/10.1146/annurev.arplant.56.032604.144301.

Demeke, K. H. 2018. Nutritional quality evaluation of seven maize varieties grown in Ethiopia. *Biochemistry and Molecular Biology* 3(2):45–48. https://doi.org/10.11648/j.bmb.20180302.11

Dichi, I., and L. H. S. Miglioranza. 2013. Fortification of corn flour-derived products. In V.R. Preedy et al. (Eds.), Handbook of Food Fortification and Health: From Concepts to Public Health Applications Volume 1. Totowa : Humana Press Inc. pp. 149–158. https://doi.org/10.1007/978-1-4614-7076-2_12

Dominguez-Hernandez, M. E., Zepeda-Bautista, R., Valderrama-Bravo, M. D. C., Dominguez-Hernandez, E., and C. Hernandez-Aguilar. 2018. Sustainability assessment of traditional maize (*Zea mays* L.) agroeco-system in Sierra Norte of Puebla, Mexico. *The Agroecology of Sustainable Food Systems* 42(4):383–406. https://doi.org/10.1080/21683565.2017.1382426

Dunn, M. L., Jain, V., and B. P. Klein. 2014. Stability of key micronutrients added to fortified maize flours and corn meal. *Annals of the New York Academy of Sciences* 1312:15–25. https://doi.org/10.1111/nyas.12310

Ejigui, J., Savoie, L., Marin J., and T. Desrosiers. 2005. Beneficial changes and drawbacks of a traditional fermentation process on chemical composition and antinutritional factors of yellow maize (*Zea mays*). *Journal of Biological Sciences* 5(5):590–596.

Enidiok, S. E., Bello, F. A., and I. I. Udo I. I. 2017. Nutritional evaluation and sensory attributes of *ekoki* produced from maize and enhanced with soy and tigernut milks. *Journal of Advances in Food Science & Technology* 4(4):131–136.

Escribano-Bailón, M.T., Santos-Buelga, C., and J. C. R. Rivas-Gonzalo. 2004. Anthocyanins in cereals. *Journal of Chromatograph* 1054:129–141.

FAOSTAT. 2016. Food Balance Sheet, Available from http://faostat.fao.org/site/345/default.aspx accessed on Nov. 30, 2016.

Fasasi, O. S., and O. A. Alokun. 2013. Physicochemical properties, vitamins, antioxidant activities and amino acid composition of ginger spiced maize snack "kokoro" enriched with soy flour (a Nigeria based snack). *Agricultural Sciences* 4(5B):73–77. https://doi.org/10.4236/as.2013.45B014

Ferri, M. D., Serrazanetti, I., Tassoni, A., Baldissarri, M., and A. Gianotti. 2016. Improving the functional and sensorial profile of cereal-based fermented foods by selecting Lactobacillus plantarum strains via a metabolomics approach. *Food Research International* 89:1095–105.

Fikiru, O., Bultosa, G., Forsido, S. F., and M. Temesgen. 2017. Nutritional quality and sensory acceptability of complementary food blended from maize (*Zea mays*), roasted pea (*Pisum sativum*), and malted barley (*Hordium vulgare*) *Food Science and Nutrition* 5(2):173–181. https://doi.org/10.1002/fsn3.376

Fung, T. T., Hu, F. B., Pereira, M. A. et al. 2002. Whole-grain intake and the risk of type 2 diabetes: A prospective study in men. *American Journal of Clinical Nutrition* 76:535–540.

Gan, R. Y., Li, H. B., Gunaratne, A., Sui, Z. Q., and H. Corke. 2017. Effects of fermented edible seeds and their products on human health: bioactive components and bioactivities. *Comprehensive Review Food Science & Food Safety* 16: 489–531.

García-Casal, M. N., Layrisse, M., Solano, L., and E. Tropper. 1998. Vitamin A and β-Carotene can improve nonheme iron absorption from rice, wheat and corn by humans. *Journal of Nutrition* 128(3):646–50. https://doi.org/10.1093/jn/128.3.646

Gemede, H. F. 2020. Nutritional and antinutritional evaluation of complementary foods formulated from maize, pea, and anchote flours. *Food Science and Nutrition* 8:2156–2164. https://doi.org/10.1002/fsn3.1516

Gómez, M., and M. M. Martínez. 2016. Changing flour functionality through physical treatments for the production of gluten-free baking goods. *Journal of Cereal Science* 67:68–74.

Gonzalez-Muñoz, A., Quesille-Villalobos, A. M., Fuentealba, C., Shetty, K., and L. G. Ranilla. 2013. Potential of Chilean native corn (*Zea mays* L.) accessions as natural sources of phenolic antioxidants and in vitro bioactivity for hyperglycemia and hypertension management. *Journal of Agricultural and Food Chemistry* 61:10995–11007. https://doi.org/10.1021/jf403237p

González-Ortega, E., Piñeyro-Nelson, A., Gómez-Hernández, E., et al. 2017. Pervasive presence of transgenes and glyphosate in maize-derived food in Mexico, *The Agroecology of Sustainable Food Systems* 41(9-10):1146–1161. https://doi.org/10.1080/21683565.2017.1372841

Graham, G. G., Lembake, J., and E. Morales. 1990. Quality-protein maize as the sole source of dietary protein and fat in rapidly growing young children. *Pediatrics* 85:85–91.

Greffeuille, V., Kayodé, P., Icard-Vernière, C., Gnimadi, M., Rochette, I., and C. Mouquet-Rivier. 2010. Changes in iron, zinc and chelating agents during traditional African processing of maize: Effect of iron contamination on bioaccessibility. *Food Chemistry* 126:1800–1807. https://doi.org/10.1016/j.foodchem.2010.12.087

Gupta, H. S., Hossain, F., and V. Muthusamy. 2015. Biofortification of maize: An Indian perspective. *Indian Journal of Genetics and Plant Breeding* 75(1):1–22. https://doi.org/10.5958/0975-6906.2015.00001.2

Gwirtz, J. A., and Garcia-Casal, M. N. 2014. Processing maize flour and corn meal food products. *Annals of the New York Academy of Sciences* 1312:66–75. https://doi.org/10.1111/nyas.12299

Harrabi, S., St-Amand, A., Sakouhi, F., et al. 2008. Phytostanols and phytosterols distributions in corn kernel. *Food Chemistry* 111:115–120. http://dx.doi.org/10.1016/j.foodchem.2008.03.044

Haskell, M., Tanumihardjo, S., Palmer, A., Melse-Boonstra, A., Talsma, E., and B. Burri. 2017. Effect of regular consumption of provitamin A biofortified staple crops on vitamin A status in populations in low-income countries. *African Journal of Food, Agriculture, Nutrition and Development* 17(02):11865–11878. https://doi.org/10.18697/ajfand.78.HarvestPlus02

He, J., and M. M. Giusti. 2010. Anthocyanins: natural colorants with health-promoting properties. *Annual Review of Food Science and Technology* 1:163–187.

Hera, E. D. L., Talegón, M., Caballero, P., and M. Gómez. 2013. Influence of maize flour particle size on gluten-free bread making. *Journal of Science of Food and Agriculture* 93:924–932.

Hess, S. Y., Thurnham, D. I., and R. F. Hurrell. 2005. *I*nfluence of provitamin A carotenoids on iron, zinc and vitamin A status. HarvestPlus Technical Monograph, vol. 6. International Food Policy Research Institute (IFPRI) and International Center for Tropical Agriculture (CIAT), Washington, DC, and California.

Hirschi, K. D. 2009. Nutrient biofortification of food crops. *Annual Review of Nutrition* 29:401–421. https://doi.org/10.1146/annurev-nutr-080508-141143

Horton, R. 2008. Maternal and child undernutrition: An urgent opportunity. *Lancet* 371:179.

Hussein, A. M. S., Abd El-Azeem, A. S., Hegazy, A. M., Afifi, A. A., & Gamal, H. R. (2011). Physiochemical, Sensory and Nutritional Properties of corn-fenugreek Flour composite biscuits. *Austutrilian Journal of Basic and Applied Science*, 5(4), 84–95.

Ikujenlola, A. V., Oguntuase, S. O., and S. V. Omosuli. 2013. Physico-chemical properties of complementary food from malted quality protein maize (*Zea mays* L.) and defatted fluted pumpkin flour (*Telfairia occidentalis* Hook, F). *Food and Public Health* 3(6): 323–328. https://doi.org/10.5923/j.fph.20130306.09

Ikya, J. K., Gernah, D. I., and I. A. Sengev. 2013. Proximate composition, nutritive and sensory properties of fermented maize, and full fat soy flour blends for "agidi" production. *African Journal of Food Science* 7(12): 446–450. https://doi.org/10.5897/AJFS09.224

Jyoti, R.A., and Prince, P. S. M. 2013. Preventive effects of p-coumaric acid on lysosomal dysfunction and myocardial infarct size in experimentally induced myocardial infarction. *European Journal of Pharmacology* 699:33–39.

Kabir, S. H., Das, A. K., Rahman, M. S., et al. 2019. Effect of genotype on proximate composition and biological yield of maize (*Zea mays* L.). *Archives of Agriculture and Environmental Science* 4(2):185–189. https://dx.doi.org/10.26832/24566632.2019.040209

Kariluoto, S., Aittamaa, M., Korhola, M., et al. 2006. Effects of yeasts and bacteria on the levels of folates in rye sourdoughs. *International Journal of Food Microbiology* 106:137–143.

Kidist, H. D. 2018. Nutritional quality evaluation of seven maize varieties grown in Ethiopia. *Biochemistry and Molecular Biology* 3(2):45–48. https://doi.org/10.11648/j.bmb.20180302.11

Klang, J. M., Tene, S. T., Doungmo Fombasso, A. D., Tsopzong, A. B. T., and H. M. Womeni. 2019. Application of germinated corn flour on the reduction of flow velocities of the gruels made from corn, soybean, *Moringa oleifera* leaf powder and cassava. *Journal of Food Processing and Technology*. 10(7):1000800.

Konczack, I., and W. Zhang. 2004. Anthocyanins-more than nature's colours. *Journal of Biomedical and Biotechnology* 5:239–240.

Konopka, I., Tanska, M., Faron, A., and S. Czaplicki. 2014. Release of free ferulic acid and changes in antioxidant properties during the wheat and rye bread making process. *Food Science and Biotechnology* 23:831.

Kopsell, D. A., Armel, G. R., Mueller, T. C., et al. 2009. Increase in nutritionally important sweet corn kernel carotenoids following mesotrione and atrazine applications. *Journal of Agricultural and Food Chemistry* 57: 6362-6368. http://dx.doi.org/10.1021/jf9013313

Korus, A., Gumul, D., Krystyjan, M., Juszczak, L., and J. Korus. 2017. Evaluation of the quality, nutritional value and antioxidant activity of gluten-free biscuits made from corn-acorn flour or corn-hemp flour composites. *European Food Research Technology* https://doi.org/10.1007/s00217-017-2853-y

Kroon, P. A., and G. Williamson 1999. Hydroxycinnamates in plants and food: Current and future perspectives. *Journal of the Science of Food and Agriculture* 79:355–361.

Kuhnen, S., Lemos, P. M., Campestrini, L. H., et al. 2011. Carotenoid and anthocyanin contents of grains of Brazilian maize landraces. *Journal of Science and Food Agriculture* 91:1548–1553.

Laila, O., and I. Murtaza. 2014. Seed sprouting: A way to health promoting treasure. *International Journal of Current Research and Review* 6(23):70–74.

Lang, V., Bellise, F., Alamowitch, C., et al. 1999. Varying the protein source in mixed meal modifies glucose, insulin and glucagon kinetics in healthy men, has weak effects on subjective satiety and fails to affect food intake. *European Journal of Clinical Nutrition* 53:959–965.

Lemos, M. A., Aliyu, M. M., and G. Hungerford. 2012. Observation of the location and form of anthocyanin in purple potato using time-resolved fluorescence. *Innovative Food Science and Emerging Technology* 16:61–68.

León-López, L., Reyes-Moreno, C., Ley-Osuna, A.H. et al. 2019. Improvement of nutritional and nutraceutical value of nixtamalized maize tortillas by addition of extruded chia flour. *Biotecnia* XXI (3):56–66.

Lonzano-Alejo, N., Carrillo, V. G., Pixley, K., and N. P. Rojas. 2007. Physical properties and carotenoid content of maize kernels and its nixtamalized snacks. *Innovation in Food Science and Emerging Technology* 8:385–389.

Lopez-Martinez, L., Oliart-Ros, R., Valerio-Alfaro, G., et al. 2009. Antioxidant activity, phenolic compounds and anthocyanins content of eighteen strains of Mexican maize, *LWT-Food Science and Technology* 42:1187–1192.

Luo, Y., and Q. Wang. 2012. Bioactive compounds in corn. In L. L. Yu, R. Tsao, and F. Shahidi. (Eds.), *Cereals and Pulses: Nutraceutical Properties and Health Benefits*, 1st Ed. New Jersey: John Wiley & Sons, Inc., pp. 1–19

Mamatha, B. S., Arunkumar, R., and V. Baskaran. 2012. Effect of processing on major carotenoid levels in corn (Zea mays) and selected vegetables: Bioavailability of lutein and zeaxanthin from processed corn in mice. *Food Bioprocess Technology* 5:1355–1363.

Mansilla, P. S., Nazar, M. C., and G. T. Pérez. 2020. Flour functional properties of purple maize (*Zea mays* L.) from Argentina. Influence of environmental growing conditions. *International Journal of Biological Macromolecules* 146:311–319. https://doi.org/10.1016/j.ijbiomac.2019.12.246

Marshall, E., and D. Mejia. 2012. Traditional fermented food and beverages for improved livelihoods. In Food and Agriculture Organization of the United Nations. Rome, Italy, pp. 1–79.

Meena, N. L., Gupta, O. P., and S. K. Sharma. 2016. Nutritional enhancers/promoters in biofortification. In U. Singh et al.(Eds.), Biofortification of Food Crops. India: Springer India, pp. 349–357.

Mehta, B. K., and Muthusamy, V., et al. 2020. Composition analysis of lysine, tryptophan and provitamin-A during different stages of kernel development in biofortified sweet corn. *Journal of Food Composition and Analysis* 94: 103625. https://doi.org/10.1016/j.jfca.2020.103625

Menkir, A., Liu, W., White, W., Maziya-Dixon, B., and T. Rocheford. 2008. Carotenoid diversity in tropical-adapted yellow maize inbred lines. *Food Chemistry* 109(3):521–529.

Mensah, J. O., and Aidoo, R., A. N. Teye. 2013. Analysis of street food consumption across various income groups in the Kumasi metropolis of Ghana. *International Review of Management and Business Research* 2(4):951.

Mesfin, W., and A. Shimelis. 2013. Effect of soybean/cassava flour blend on the proximate composition of Ethiopian traditional bread prepared from quality protein maize. *African Journal of Food, Agriculture, Nutrition and Development* 13:7985–8003.

Messias, R. S., Galli, V., Silva, S. D. A., Schirmer, M. A., and C. V. Rombaldi. 2013. Micronutrient and functional compounds biofortification of maize grains. *Critical Review of Food Science and Nutrition* 10.1080/10408398.2011.649314

Miller, B. D. D., and R. M. Welch. 2013. Food system strategies for preventing micronutrient malnutrition. *Food Policy* 42:115–128.

Mishra, V., Puranik, V., Akhtar, N., and G. K. Rai. 2012. Development and compositional analysis of protein rich soyabean-maize flour blended cookies. *Journal of Food Processing and Technology* 3:9 https://doi.org/10.4172/2157-7110.1000182

Mora-Rochín, S., Gaxiola-Cuevas, N., Gutiérrez-Uribe, J. A. et al. 2016. Effect of traditional nixtamalization on anthocyanin content and profile in Mexican blue maize (*Zea mays* L.) landraces. *LWT—Food Science & Technology* 68:563–569. https://doi.org/10.1016/j.lwt.2016.01.009

Motadi, A. S., Mbhenyane, X. G., Mbhatsani, H. V., Mabapa, N. S., and R. L. Mamabolo. 2015. Prevalence of zinc deficiency among preschool children aged 3-5 years in Vhembe district, Limpopo province, South Africa. *Journal of Nutrition and Food Science* 5(6).

Mourouti, N., Kontogianni, M.D., Papavagelis, C., et al. 2016. Whole grain consumption and breast cancer: A case-control study in women, *Journal of American College of Nutrition* 35:143–149.

Mrad, R., Debs, E., Saliba, R., Maroun, R. G., and N. Louka. 2014. Multiple optimization of chemical and textural properties of roasted expanded purple maize using response surface methodology. *Journal of Cereal Science* 60:397–e405. http://dx.doi.org/10.1016/j.jcs.2014.05.005

Mtebe, K., Ndabikunze, B., Bangu, N. T. A., and E. Mwemezi. 1991. Effect of cereal germination on the energy density and nutrient content of Togwa. In: A. Westby, and P. J. A. Reilly (Eds.), *Traditional African Foods Quality and Nutrition*, Stockholm: IFS, pp 67–72.

Mubarak, A. E. 2005. Nutritional composition and anti-nutritional factors of mung bean seeds (*Phaseolus aureus*) as affected by some home traditional processes. *Food Chemistry* 89:489–495.

Mugode, L., Ha, B., Kaunda, A., et al. 2014. Carotenoid retention of biofortified provitamin A maize (Zea mays L.) after Zambian traditional methods of milling, cooking and storage. *Journal of Agricultural and Food Chemistry* 62:6317–6325. http://dx.doi.org/10.1021/jf501233f

Mutlu, C., Arslan-Tontul, S., Candal, C., Kilic, O., and E. Mustafa. 2017. Physicochemical, thermal, and sensory properties of blue corn (*Zea mays* L.). *Journal of Food Science* https://doi.org/10.1111/1750-3841.14014

Muzhingi, T., Palacios, N., Miranda, A., Cabrera, M. L., Yeum, K. J. and G. Tang. 2017. Genetic variation of carotenoids, vitamin E and phenolic compounds in biofortified maize. *Journal of the Science of Food and Agriculture* 97(3):793–801. https://doi.org/10.1002/jsfa.7798

Mwemezi, E. 1991. Effect of cereal germination on the energy density and nutrient content of Togwa. In A. Westby and P. J. A. Reilly (Eds.), Traditional African Foods Quality and Nutrition. Stockholm.:IFS, pp. 67–72.

Ndukwe O. K., Edeoga H. O., and G. Omosun. 2015. Varietal differences in some nutritional composition of ten maize (*Zea mays* L.) varieties grown in Nigeria. *International Journal of Academic Research and Reflection* 5:1–11.

Netshiheni, K. R. 2019. Nutritional and sensory properties of instant maize porridge fortified with *Moringa oleifera* leaves and termite (*Macrotermes falciger*) powders. *Nutrition and Food Science* 49 (4):654–667. https://doi.org/10.1108/NFS-07-2018-0200

Ngigi, P. B., Lachat, C., Masinde, P.W., *et al.* 2019. Agronomic biofortification of maize and beans in Kenya through selenium fertilization. *Environmental Geochemistry Health* 41:2577–2591. https://doi.org/10.1007/s10653-019-00309-3

Nikhil, K., and S. Salakinkop. 2018. Agronomic biofortification of maize with zinc and iron micronutrients. *Modern Concepts & Developments in Agronomy* 1(4):000522. https://doi.org/10.31031/MCDA.2018.01.000522

Noah, A. A. 2017. Nutrient composition and sensory evaluation of complementary food made from maize, plantain soybean blends. *International Journal of Current Microbiology and Applied Sciences* 6(12): 5421–5428. https://doi.org/10.20546/ijcmas.2017.612.507

Norberto, S., Silva, S., Meireles, M., et al. 2013. Blueberry anthocyanins in health promotion: A metabolic overview. *Journal of Functional Foods* 5:1518–1528.

O'Hare, T. J., Fanning, K. J., and I. F. Martin. 2015. Zeaxanthin biofortification of sweet-corn and factors affecting zeaxanthin accumulation and colour change. *Archives of Biochemistry and Biophysics* 572:184–187. http://dx.doi.org/10.1016/j.abb.2015.01.015

Oboh, G., Ademiluyi, A. O., and A. A. Akindahunsi. 2010. The effect of roasting on the nutritional and antioxidant properties of yellow and white maize varieties. *International Journal of Food Science and Technology* 45:1236–1242. https://doi.org/10.1111/j.1365-2621.2010.02263.x

Ojiewo, C., Keatinge, D. H., Hughes, J., et al. 2015. The role of vegetables and legumes in assuring food, nutrition, and income security for vulnerable groups in sub-Saharan Africa. *World Medical and Health Policy* 7:187–210.

Okafor, U. I., Omemu, A. M., Obadina, A. O., Bankole, M. O., and S. A. O. Adeyeye. 2018. Nutritional composition and antinutritional properties of maize ogi cofermented with pigeon pea. *Food Science and Nutrition* 6:424–439. https://doi.org/10.1002/fsn3.571

Okoye, J. I., and A. E. Egbujie. 2018. Nutritional and sensory properties of maize-based complementary foods fortified with soybean and sweet potato flours. *Discourse Journal of Agriculture and Food Sciences*, 6(3):17–24.

Olanipekun, O. T., Olapade, O. A., Suleiman, P., and S. O. Ojo. 2015. Nutrient and sensory analysis of abari made from composite mixture of kidney bean flour and maize flour. *Sky Journal of Food Science* 4(2): 019–023.

Olunlade, B. A., Ayanleke, O. D., and A. A. Adeola. 2013. Influence of Soybean or Pigeon pea on the proximate composition and sensory attributes of bread from maize-cassava flour blends. *Direct Research Journal of Agriculture and Food Science* 1(6):73–79.

Onwurafor, E. U., Umego, E. C., Uzodinma, E. O., and E. D. Samuel. 2017. Chemical, functional, pasting and sensory properties of sorghum-maize-mungbean malt complementary food. *Pakistan Journal of Nutrition* 16(11):826–834. 10.3923/pjn.2017.826.834

Onwurafor, E., Ani, J., Uchegbu, N., and G. Ziegler. 2020. Quality evaluation of fermented maize-based complementary foods as affected by amylase-rich mungbean malt. *Journal of Food Stability* 3(1):26–37. https://doi.org/10.36400/J.Food.Stab.3.1.2020-0025

Orhun, G.E. 2013. Maize for life. *International Journal of Food Science and Nutrition Engineering* 3(2):13–16. https://doi.org/10.5923/j.food.20130302.01

Orthoefer, F., Eastman, J., and G. List. 2003. Corn oil: Composition, processing and utilization. In P. J. White, L. A. Johnson (Eds.), Corn: Chemistry and technology, 2nd ed. St. Paul, MN: American Association of Cereal Chemists, pp. 671–693.

Ortiz-Monasterio, J. I., Palacios-Rojas, N., Meng, E., Pixley, K., Trethowan, R., and R. J. Peña. 2007. Enhancing the mineral and vitamin content of wheat and maize through plant breeding. *Journal of Cereal Science* 46:293–307.

Otten, J. J., Hellwig, J. P., Meyers, L. D., and I. O. D. R. I. Medicine. 2006. Dietary reference intakes: The essential guide to nutrient requirements. *American Journal of Clinical Nutrition* 85:924.

Oyarekua, M. A. 2013. Effect of co-fermentation on nutritive quality and pasting properties of maize/cowpea/sweet potato as complementary food. *African Journal of Food, Agriculture, Nutrition and Development* 13(1):7171–7719.

Patil, B. S., Jayaprakasha, G. K., Murthy, K. N. C., and A. Vikram. 2009. Bioactive compounds: Historical perspectives, opportunities, and challenges. *Journal of Agricultural & Food Chemistry* 57:8142–8160.

Paucar-Menacho, L. M., Martınez-Villaluenga, C., Duenas, M., Frias, J., and E. Penas. 2016. Optimization of germination time and temperature to maximize the content of bioactive compounds and the antioxidant activity of purple corn (*Zea mays* L.) by response surface methodology. *LWT—Food Science & Technology* 76:236–44.

Peyer, L. C., Zannini, E., and E. K. Arendt, 2016. Lactic acid bacteria as sensory biomodulators for fermented cereal-based beverages. *Trends in Food Science & Technology* 54:17–25.

Piironen, V., Lindsay, D. G., Miettinen, T. A., Toivo, J., and A. Lampi. 2000. Plant sterols: Biosynthesis, biological function and their importance to human nutrition. *Journal of the Science of Food and Agriculture* 80, 839–966.

Pillay, K., Siwela, M., Derera, J., and F. J. Veldman. 2011. Provitamin A carotenoids in biofortified maize and their retention during processing and preparation of South African maize foods. *Journal of Food Science Technology vol. 51, pp. 634–644.* https://doi.org/10.1007/s13197-011-0559-x

Pixley, K., Rojas, N. P. Babu, R., Mutale, R., Surles, R., and E. Simpungwe. 2013. Biofortification of maize with Provitamin A carotenoids. In S. A. Tanumihardjo (Ed.), Carotenoids and Human Health, Nutrition and Health. New York: Springer Science+Business Media, pp. 271–292. http://doi.org/10.1007/978-1-62703-203-2_17

Prasanna, B. M., Palacios-Rojas, N., Hossain, F., et al. 2020 Molecular breeding for nutritionally enriched maize: status and prospects. *Frontiers Genetics* https://doi.org/10.3389/fgene.2019.01392

Prasanthi, P. S., Naveena, N., Rao, M. V., and K. Bhaskarachary 2017. Compositional variability of nutrients and phytochemicals in corn after processing. *Journal of Food Science and Technology* 54:1080–1090. https://doi.org/10.1007/s13197-017-2547-2

Ragaee, S., Guzar, I., Dhull, N., and K. Seetharaman. 2011. Effects of fiber addition on antioxidant capacity and nutritional quality of wheat bread. *LWT—Food Science and Technology* 44:2147–2153. https://doi.org/10.1016/j.lwt.2011.06.016

Ranum, P., Peña-Rosas, J. P., and M. N. Garcia-Casal. 2014. Global maize production, utilization, and consumption. *Annals of the New York Academy of Sciences* 1312:105–12.

Rautiainen, S., Manson, J. E., Lichtenstein, A. H., and H. D. Sesso. 2016. Dietary supplements and disease prevention—a global overview. *Nature Reviews Endocrinology* 12:407–420. https://doi.org/10.1038/nrendo.2016.54

Rocheford, T. R., Wong, J. C., Egesel C. O., and R. J. Lambert. 2002. Enhancement of vitamin E levels in corn. *Journal of the American College and Nutrition* 21:191S–198S. https://doi.org/10.1080/07315724.2002.10719265

Safawo, T., Senthi, N., Raveendran, M., et al. 2010. Exploitation of natural variability in maize for β-Carotene content using HPLC and gene specific markers. *Journal of Plant Breeding* 1:548–555.

Saikaew, K., Lertrat, K., Ketthaisong, D., Meenune, M., and R. Tangwongchai. 2018. Influence of variety and maturity on bioactive compounds and antioxidant activity of purple waxy corn (*Zea mays* L. var. *cerazh*)

Sandhu, K. S., Punia, S., and M. Kaur. 2017. Fermentation of cereals: A tool to enhance bioactive compounds. In S. K. Gahlawat et al. (Eds.), In the book: Plant Biotechnology: Recent Advancements and Developments.Chapter 8 Springer, Singapore. pp. 157–170. https://doi.org/10.1007/978-981-10-4732-9_8

Saxena, M., Saxena, J. Nema, R., Singh, D., and A. Gupta. 2013. Phytochemistry of medicinal plants. *Journal of Pharmacognosy and Phytochemistry* 1(6):168–182.

Serna-Saldivar, S. O. 2016. Cereal Grains: Properties, Processing, and Nutritional Attributes. Boca Raton, FL: CRC Press (Taylor & Francis Group).

Serna-Saldivar, S. O. 2016. Non-wheat Foods. In Encyclopedia of Food Grains, 2nd ed. Waltham, MA: Elsevier Academic Press. pp. 97–109. http://dx.doi.org/10.1016/B978-0-12-394437-5.00126-1

Shah, T. R., Prasad, K., and P. Kumar. 2016. Maize-a potential source of human nutrition and health: A review. *Cogent Food and Agriculture* 2:1166995. https://doi.org/10.1080/23311932.2016.1166995

Sheng, S., Li, T., and R. Liu. 2018. Corn phytochemicals and their health benefits. *Food Science and Human Wellness* 7:185–195. https://doi.org/10.1016/j.fshw.2018.09.003

Shevkani, K., Kaur, A., Singh, G., Singh, B., and N. Singh. 2014. Composition, rheological and extrusion behaviour of fractions produced by three successive reduction dry milling of corn. *Food Bioprocess Technology* 7:1414–1423.

Shiriki, D., Igyor, M. A., and D. I. Gernah. 2015. Nutritional evaluation of complementary food formulations from maize, soybean and peanut fortified with *Moringa oleifera* leaf powder. *Food and Nutrition Sciences* 6(5): 494–500. http://dx.doi.org/10.4236/fns.2015.65051

Singh, A., Sharma, V., Banerjee, R., Sharma, S., and A. Kuila. 2016. Perspectives of cell-wall degrading enzymes in cereal polishing. *Food Biosciences* 15:81–6.

Singh, N., Shevkani, K., Kaur, A., Thakur, S., Parmar, N., and A. S. Virdi. 2014. Characteristics of starch obtained at different stages of purification during commercial wet milling of maize. *Starch* 66:668–677.

Singh, N., Singh, S., and K. Shevkani. 2011. Maize: Composition, bioactive constituents, and unleavened bread. In V. R. Preedy, R. R., Watson and V. B. Patel (Eds.), Flour and Breads and Their Fortification in Health and Disease Prevention. London: London Academic Press, Elsevier, pp. 89–99.

Singh, N., Singh, S., and K. Shevkani. 2019. Maize: Composition, bioactive constituents, and unleavened bread. In V. R. Preedy and R. R., Watson (Eds.), Flour and Breads and Their Fortification in Health and Disease Prevention, 2nd ed. London: London Academic Press, Elsevier, pp. 111–120 https://doi.org/10.1016/B978-0-12-814639-2.00009-5

Skylas, D. J., Molloy, M. P., Willows, R. D., Salman, H., Blanchard, C. L., and K. J. Quail. 2018. Effect of processing on mungbean (*Vigna radiata*) flour nutritional properties and protein composition. *Journal of Agricultural Science* 10(11). https://doi.org/10.5539/jas.v10n11p16

Steiner, T., Mosenthin, R., Zimmermann, B., Greiner, R., and S. Roth. 2007. Distribution of total phosphorus, phytate phosphorus and phytase activity in legume seeds, cereals and cereal byproducts as influenced by harvest year and cultivar. *Animal Feed Science and Technology* 133:320–334.

Sun, Z., and Yao, H. Y. 2007. The influence of di-acetylation of the hydroxyl groups on the anti-tumor proliferation activity of lutein and zeaxanthin. *Asian Pacific Journal of Clinical Nutrition* 16(Suppl 1): 447–452.

Suri, D. J., and Tanumihardjo, S. A. 2016. Effects of different processing methods on the micronutrient and phytochemical contents of maize: From A to Z. *Comprehensive Reviews in Food Science and Food Safety* 15:912–926. https://doi.org/10.1111/1541-4337.12216

Temba, M. C., Njobeh, P. B., Adebo, O. A., et al. 2016. The role of compositing cereals with legumes to alleviate protein energy malnutrition in Africa. *International Journal of Food Science and Technology* 51:543–554. https://doi.org/10.1111/ijfs.13035

Tighe, P., Duthie, G. Vaughan, N. et al. 2010. Effect of increased consumption of whole-grain foods on blood pressure and other cardiovascular risk markers in healthy middle-aged persons: A randomized controlled trial. *American Journal of Clinical Nutrition* 92:733–740.

Trehan, S., Singh, N., and K. Amritpa. 2018. Characteristics of white, yellow, purple corn accessions: Phenolic profile, textural, rheological properties and muffin making potential. *Journal of Food Science and Technology vol. 55, pp. 2334–2343.* https://doi.org/10.1007/s13197-018-3171-5

Tsafrakidou, P., Michaelidou, A., and C. G. Biliaderis. 2020. Fermented cereal-based products: Nutritional aspects, possible impact on gut microbiota and health implications. *Foods* 9:734. http://doi.org/10.3390/foods9060

Tsuda, T., Horio, F., Uchida, K., H., Aoki, H., and T. Osawa. 2003. Dietary cyanidin 3-O-beta-D-glucoside-rich purple corn color prevents obesity and ameliorates hyperglycemia in mice. *Journal of Nutrition* vol. 133(7) pp. 2125–2130.

Ukeyima, M. T., Acham, I. O., and B. A. Amaechi. 2019. Quality evaluation of maize based complimentary food supplemented with garden peas. *Asian Food Science Journal* 12(1):1–10. https://doi.org/10.9734/AFSJ/2019/v12i130074

UNICEF (2020). Companion to breast feed after 6, Retrieved from: http://www.unicef.org/programme/breast-feeding/food.htm.Nutritional

UNICEF. 2020. Improving young children's diets during the complementary feeding period. Retrieved from https://www.unscn.org/en/news-events/recent-news?idnews=2030

Uriarte-Aceves, P. M., Cuevas-Rodríguez, E. O., Gutiérrez-Dorado, R., et al. 2015. Physical, compositional, and wet-milling characteristics of Mexican blue maize (*Zea mays* L.) Landrace. *Cereal Chemistry* 92(5):491–496.

Urias-Peraldí, M., Gutiérrez-Uribe, J., R. Preciado-Ortiz. 2013. Nutraceutical profiles of improved blue maize (*Zea mays*) hybrids for subtropical regions. *Field Crops Research* 141:69–76.

Verleyen, T., Forcades, M., Verhe, R., et al. 2002. Analysis of free and esterified sterols in vegetable oils. *Journal of the American Oil Chemists' Society* 79:117–122. https://dx.doi.org/10.1007/s11746-002-0444-3

Waller, A. W., Dominguez-Uscanga, A., Barrera, E. L., Andrade, J. E., and J. M. Andrade. 2020. Stakeholder's perceptions of Mexico's federal corn flour fortification program: A qualitative study. *Nutrients* 12:433. https://doi.org/10.3390/nu12020433

WHO. 2020. Micronutrient survey manual. Geneva: Centers for Disease Control and Prevention, Nutrition International, UNICEF,.

Xi, P., and R. H. Liu. 2016. Whole food approach for type 2 diabetes prevention. *Molecular Nutrition and Food Research* 60:1819–1836.

Xu, J., Li, Y., and W. Wang. 2019. Corn. In. J. Wang et al. (Ed.), Bioactive Factors and Processing Technology for Cereal Foods. Springer Singapore, pp. 33–53. https://doi.org/10.1007/978-981-13-6167-8_3

Yadava, D. K., Hossain, F., and T. Mohapatra. 2018. Nutritional security through crop biofortification in India: Status & future prospects. *Indian Journal of Medical Research* 148:621–631.

Zhu, C., Naqvi, S., Gomez-Galera, S., Pelacho, A. M., Capell, T., and P. Christou. 2007. Transgenic strategies for the nutritional enhancement of plants. *Trends in Plant Sciences* 12:548–55.

Žilić, S., Serpen, A., Akıllıoğlu, G., Gökmen, V., and J. Vančetović. 2012. Phenolic compounds, carotenoids, anthocyanins, and antioxidant capacity of colored maize (*Zea Mays* L.) kernels. *Journal of Agricultural and Food Chemistry* 60(5):1224–1231. https://doi.org/10.1021/jf204367z

13

Grain Amaranth: Processing, Health Benefits and Applications

Babatunde Olawoye
First Technical University, Ibadan, Oyo State, Nigeria

Oseni Kadiri
Edo State University Uzairue, Auchi, Edo State, Nigeria

Timilehin David Oluwajuyitan
Federal University of Technology Akure, Nigeria

CONTENTS

13.1 Introduction

Recently, there was a threat to human survival owing to the increase in the gap between food supply and the world population, which led to a dramatic increase in the number of people suffering from chronic undernutrition from 515 million in 2014 to a projected value of 800 million in 2030 (FAO, 2020). The surge in undernutrition cases across the world could be attributed to the fact that cereal (which served as a major carbon source to human) had other industrial applications apart from its production and cultivation crises. This issue had led researchers to focus on the search for an available underutilized crop with high nutritional characteristics that could serve as an alternative source of energy (Olawoye & Gbadamosi, 2020b; Olawoye & Gbadamosi, 2017). The underutilized crops with high nutritional and nutraceutical properties that fall into this category are pseudo-cereal especially grain amaranth; an Andean crop usually cultivated for its leaf that is utilized as a vegetable while the grain is usually discarded.

Grain amaranth is best recognized by its leaves that are twice or three times as long as they are wide and often have pointed leaf tips (Olawoye & Kadiri, 2016). Although the crop was one of the staple foods in the precolonial era, the cultivation and production of grain amaranth were given less consideration and thus currently it can be classified as a forgotten, neglected, and alternative crop of great nutritional value (Berghofer & Schoenlechner, 2002). Amaranth species are mostly cultivated for their leaves such as Nigeria *Amaranthus viridis*, while other species cultivated mainly for their seeds include *Amaranthus caudatus* of Peru origin, *Amaranthus cruentus*, *Amaranthus hypochrondriacus* of Mexican origin, as well varieties from other Andean countries. Amaranth is a pseudo-cereal multipurpose crop with good potential for exploitation as grain, vegetable, and fodder. The nutritional value of pseudo-cereals and grain amaranth in particular is mainly connected to their exceptionally high protein content as compared to true cereals.

Furthermore, grain amaranth has an exceptional quality protein with a high amino acid balance of cysteine, lysine, and methionine when compared with cereal grains in which lysine is deficient (Mota et al., 2016). Apart from its high protein content, amaranth contains sufficient levels of important micronutrients such as minerals and vitamins and substantial amounts of other bioactive components such as saponins, phytosterols, squalene, and polyphenols (Belton & Taylor, 2002; Schoenlechner et al., 2010; Taylor et al., 2016) that convey additional health benefits in the management of degenerating diseases. Owing to this, and coupled with the fact that it's gluten-free, grain amaranth has caught the attention of the world as a potential raw material in the food and pharmaceutical industry.

13.2 History and Production of Grain Amaranth

The genus *Amaranthus* belongs to the family *Amaranthaceae* with 50-70 species (Costea & Demason, 2001). While 15 of these species are native to Africa, a majority of these species are native to the Americas (Das, 2016). There is a dearth of knowledge as regards the interspecific relationship and taxonomic position of the genus *Amaranthus* due basically to its high phenotypic variation, which is mostly based on the precept of its inflorescence characteristics (Sauer, 1955). In classifying *Amaranthus*, Sauer (1955) grouped this plant into two subgenera. More recent taxonomy study classifies *Amaranthus* into *Albersia*, *Amaranthus* and *Acnida* (Costea & Demason, 2001). The genome-wide molecular markers have been partly used in reproducing these result. Conversely, the subgenus, *Achida*, was split into two phylogenetics through a polyploidization event within its subgenus (Stetter & Schmid, 2017). The *Amaranthus* subgenus consists of the hybridus complex which includes the three amaranth grain species and their two potential wild ancestors, *A. quitensis and A.hybridus.*

The first documented cultivation of grain amaranth was in the Americas with archaeological records of seeds collected from Northern Argentina dating back to the initial mid-Holocene (8000–7000 BP; Arreguez et al., 2013). Seeds from the Tehuacan cave in Central America have been recorded as the oldest grain found to date with seeds of *A. cruentus* and *A. hypochondriacus* dating back 6,000 and 1,500 years, respectively (Brenner et al., 2000).

Amaranth grains were an important part of the Aztec civilization, as seen in archaeological evidence, and were grown mainly in Mexico in the 1400s. There was however a decline in amaranth production after its introduction to other continents such as Asia, Africa, and Europe. Distribution pattern of genotypes of *A. hypochondriacus* suggest that the wild (dark seeded) and domesticated (light) were most likely introduced to Asia in the early 1700s from Latin America by sea and land routes (Sauer, 1993). There have been several hypotheses suggesting the domestication of the three species of grain amaranth over the last 50 years.

Single domestication and multiple domestication models were among models suggested by Sauer (1967) resulting in all the three grain amaranths. Reports based on the meiotic behavior of allozymes analyses, DNA polymorphism and hybrid chromosomes provided supporting evidence in favor of the hypothesis that suggest that *A. hybridus* is the ancestor common to the three cultivated species of grain amaranth (Pandey, 1999). Findings based on genotyping-by-sequencing and simple sequence repeats agrees that the ancestor of all three grain amaranth was the *A. hybridus* though this is not fully agreed upon if this is from a single or independent domestication event (Kietlinski et al., 2014; Stetter & Schmid, 2017; Stetter et al., 2017).

Population genetic analysis and phylogenetic analysis have been used to show the distinct geography grouping of *A. caudatus*, *A. hybridus*, and *A. quitensis* from Central America and *A. hybridus* from South America (Stetter & Schmid, 2017). However, it is uncertain the role and status of *A. quitensis*, the closest and wild of the previously discussed species. It is most likely that *A. quitensis* contributed to the domestication of the South American *A. caudatus* through gene flow or is an intermediate of *A. hybridus* domestication (Stetter et al., 2017).

13.3 Chemical and Nutritional Components of Grain Amaranth

The chemical and nutritional component present in grain amaranth varies according to varieties, species, geographical origin and climate conditions. Table 13.1 shows the chemical composition of grain amaranth compared to other cereal. When compared with cereal grain, grain amaranth has a higher nutritional value, especially protein, with a significant higher lysine content. Next are the chemical and nutritional components present in grain amaranth.

13.3.1 Protein

Generally, the protein content (16–18%) in pseudo-cereal is especially higher than common cereal grains (<14%) such as maize, wheat, millet, or sorghum, however, lower than legumes and oilseeds. About 65% of grain amaranth protein is found in the seed coat and the germ, while the rest is located at the endosperm. Among pseudo-cereal crops, grain amaranth had the highest protein content followed by quinoa, while buckwheat had the lowest protein content. In quality, amaranth protein had been reported to be very good by assay (Berghofer & Schoenlechner, 2002). Amaranth protein when compared with casein had a slightly low protein-efficiency ratio, which however increases and competes with casein after cooking. In terms of amino acid composition, the quality of protein in amaranth is on par with casein (an animal source of protein), which makes it unusual for plant source protein (Martínez-Villaluenga et al., 2020). Contrary to other common cereal and pseudo-cereal grains, grain amaranth protein is mainly made up of albumins and globulins with little or no prolamin proteins (a storage protein that served as the major component in the protein fraction of gluten) (Menegassi et al., 2011). Prolamin protein is the main storage protein in cereal grain and it's reported to be the toxic protein causing celiac disease; an immune-mediated disease caused by the inability of consumers of gluten products to ingest gluten (Ruales & Nair, 1993). In bakery products, gluten is very important owing to its ability to form a film that traps in CO_2 produced during the fermentation of sugar (Olawoye et al., 2020a). Moreover, it's aggregated into an extensible three-dimensional protein-starch matrix. The absence of this gluten in grain amaranth makes it a suitable raw material in the production of a gluten-free diet. Apart, from its suitability in the management of celiac disease, the *in vivo* and *in vitro* allergic study of amaranth protein, as well as its extract, revealed that antigen-specific Immunoglobin-E production was inhibited by increasing Th1 cytokine responses (Niro et al., 2019).

The dominance of grain amaranth over other pseudo-cereals and cereal grains is its excellent balance of amino acids, especially essential amino acids (Table 13.2), which is required being sourced from the plant as the human body cannot manufacture or synthesis it (Mota et al., 2016). Among the essential amino acids, grain amaranth protein contains an appreciable higher percentage of lysine compared to

TABLE 13.1

Nutritional Composition of Amaranth Grain as Compared to Cereals (g/100 g dry matter)

Grain Type	Fat	Protein	Ash	Crude Fiber	Carbohydrate
Amaranth	10.75	14.7	2.61	7.27	65.27
Wheat	2.9	16	1.8	2.6	74.1
Rice	2.2	9.1	7.2	10.2	71.2
Corn	2.2	11.1	1.7	2.1	80.2

Source: Repo-Carrasco-Valencia, *et al.* (2010).

TABLE 13.2

Contents of Essential Amino Acids in Amaranth Compared to
Wheat and Rice Grains

Amino Acids (g/100 g)	Amaranth	Rice	Wheat
Arginine[b]	8.2	6.3	4.8
Cystine[b]	2.3	2.5	2.2
Histidine[a]	2.4	2.2	2
Isoleucine[a]	3.2	3.5	4.3
Leucine[a]	5.4	7.5	6.7
*Lysine[a]	6	3.2	2.8
Methionine[a]	3.8	3.6	1.3
Phenylalanine[a]	3.7	4.8	4.9
Valine[a]	3.8	5.1	4.6
Threonine[a]	3.3	3.2	2.9
Tryptophane[a]	1.1	1.1	1.2
Tyrosine[b]	2.7	2.6	3.7

[a] Indispensable amino acids.
[b] Conditional indispensable amino acids.
*Limiting amino acids in cereals.
Source: Brinegar *et al.* 2006.

cereal grains, especially wheat grain in which its lysine content is three times higher. Clouse et al. (2016) revealed that the higher lysine content in amaranth could be attributed to its high globulins and albumins, which account to 44–77% of the total protein. Grain amaranth had been reported by Bucaro Segura & Bressani (2002) to be limited in threonine followed by leucine. The protein composition in grain amaranth is close to the protein recommended to adults by the Food and Agriculture Organization of the United Nations (FAO) and this made amaranth to be rated higher than other seeds such as barley, maize, wheat and soy beans (Silva-Sanchez et al., 2008). When grain amaranth is made into flour and mixed with corn flour in the production of cookies, bread, or tortillas, the products made from the blended flour almost reach 100 scores because they complement each other in relation to amino acids (Wu et al., 2017).

13.3.2 Carbohydrate

Starch, which carbohydrate is mostly referred to, is the main macromolecule in plants and occurs in granular form with diverse sizes and shapes. It forms the bulk of physiological energy in humans and is located at the perisperm of the grain amaranth either as a single entity or in compound structure (Olawoye & Gbadamosi, 2020a). A report by Kong et al. (2009) revealed that grain amaranth starch is embedded with a protein matrix thereby forming a complex with the protein and hindering the hydrolysis of the starch by enzymes, which in turn, decreases the digestibility of the starch. Carbohydrate content of grain amaranth varied between 60–62%, a value which is less than common cereal grains. A report on the diameter of grain amaranth starch by Mir et al. (2018), revealed that the diameter of the amaranth starch granule is smaller when compared to that of wheat or maize. The small granule size exhibited by grain amaranth makes it exhibit a low gelatinization temperature. The gelatinization temperature of grain amaranth ranges between 54.2–61.9°C, a value which is quite low compared to the gelatinization temperature of quinoa (57–64°C) (Inouchi et al., 1999). The gelatinization properties of grain amaranth are not only dependent on the granule sizes but also varies with respect to the ultrastructure of the starch as well as the degree of crystallinity of the starch (Roa et al., 2015). Aside from the starch content in grain amaranth, there is also another carbohydrate found in amaranth in small amounts containing 2.1% monosaccharides, 2.4% disaccharides, 2.3–3.7% and 2.7–3.8% pentosans. A comparison between amaranth starch and other common cereal starch revealed that amaranth starch had higher water and oil absorption capacity as well as swelling capacity (Coelho et al., 2018). This property made starch isolated from grain amaranth suitable for food application where high binding capacity is required.

13.3.3 Dietary Fiber

Dietary fiber is the food component that escapes digestion at the gastrointestinal tract but is converted to short-chain fatty acid at the large intestine and hence, is considered as a health-promoting food component (Olawoye et al., 2020b). Grain amaranth contains dietary fiber that ranged between 8–15% and it is considered to be high. A research carried out by Zhu (2020) revealed that the total dietary fiber found in grain amaranth was 13.4% of which insoluble fiber was 11% and soluble fiber was 2.4% of the total dietary fiber. The value of the insoluble fiber is comparable with common cereal grains such as rye and wheat grain, but contains a less soluble fiber when compared with these common cereals (Lamothe et al., 2015). A study carried out on the dietary fiber content of grain amaranth revealed that the insoluble fiber is made up of homogalacturonans dispersed with RG-1 stretches containing galactan and arabian side chains and cellulose. Among pseudo-cereal grains, amaranth has a solution fraction of dietary fiber that differs from and is higher than buckwheat and quinoa (Mustafa et al., 2011).

13.3.4 Lipids

With respect to fat, grain amaranth contains the highest lipid content when compared to other pseudo-cereals as well as most cereal grain. The fat content, however, varies among different varieties and is characterized by higher concentration (75%) of unsaturated fatty acids, especially linoleic acid, oleic acid, and palmitic acid. Although present in low amounts, grain amaranth also consists of 3.9% stearic acid and 0.7% of linolenic acid (Kim et al., 2006). The higher concentration of polyunsaturated fatty acid in grain amaranth poses a positive effect in the management of cardiovascular diseases. Aside from the unsaturated fatty acids, other lipids such as sterols, phospholipids, tocopherol, triglycerides and glycolipids (Dus-Zuchowska et al., 2019). It also contains a higher concentration of unsaturated triterpene, also known as squalene, a lipid usually found in the liver of deep-sea animals such as sharks. This unsaturated open-chain triterpene had found application in the cosmetic and pharmaceutical industries. With respect to the quality of the oil as well as a fat concentration up to 8.6% in some amaranth varieties, grain amaranth had been dubbed pseudo-oilseed and was considered for cultivation solely for its oil content. Also, squalene had been reported to act as protective in the treatment of cancer and also reduce the side effects induced by chemotherapy (Pina-Rodriguez & Akoh, 2009).

13.3.5 Vitamin and Minerals

Although vitamin and minerals are present in grain amaranth and other pseudo-cereals in small amount, they, however, played an important role in the physiological functioning of the body. Souci et al. (1994) reported a higher vitamin (thiamine) content in grain amaranth than wheat. Also, grain amaranth is particularly rich in vitamin E and folic acid and is also an excellent source of riboflavin as well as vitamin C (Morales et al., 2020). Folic acid reported in amaranth is in the region of 82ug/100 g, a value which is twice the value obtained for wheat (43 ug.100 g). Vitamin E is present in amaranth in the form of tocopherol as well as tocotrienols (Morales et al., 2020). This vitamin has an antioxidant effect and had been reported to increase oil stability in amaranth. Tocopherol occurs in amaranth in two forms, alpha-tocopherol and beta-tocopherol. Alpha-tocopherol, which had the highest antioxidant effect, is found in the seed to be the most abundant of the total tocopherol (100–129 mg/kg) extracted using supercritical fluid extraction according to Bruni et al. (2002).

Amaranth has been reported to be an excellent source of minerals such as potassium, magnesium, and calcium as well as other important minerals. The values found are in grain amaranth are reported to be twice as high than that found in common or other cereal grain. Among the pseudo-cereal groups, amaranth was reported to have the highest mineral composition followed by quinoa and buckwheat. The high calcium found in amaranth makes it suitable for the formulation of food products and in the treatment and management of celiac disease (Gliszczynska-Swiglo et al., 2018). This is due to the fact people with celiac disease complication also had the prevalence of osteoporosis and osteopenia diseases.

13.3.6 Phytochemicals

To guard against the invasion of microbes, pest, and insects, plants synthesize a secondary metabolite also known as phytochemicals that serve as defense mechanisms. The phytochemicals are bioactive in nature and act as antioxidants, which helps in the management of degenerating diseases (Morales et al., 2020). The phytochemicals present in grain amaranth include, but are not limited to, phenolic acids, flavonoids, sterols, etc. Phenolic acids are the secondary metabolic derivatives of cinnamic and benzoic acids, which exist in pseudo-cereal and grain amaranth in particular. In a report of Olawoye & Kadiri (2016), it was revealed that grain amaranth flour is made up of polyphenol which has antioxidant effects. Three main polyphenols had been isolated from different varieties of amaranth and they include nicotiflorin, rutin, and isoquercitrin. Among the polyphenols, isoquercitrin was found to be the least with the value ranging between 0.3–0.5 ug/g followed by nicotiflorin whose value ranged between 4.8–7.2 ug/g, and finally, rutin is found to be the highest of the polyphenols and is found in the concentration of 4.0–10.1 ug/g (Morales et al., 2020). The degradation of this polyphenol in the intestine yielded aglyconic molecules in the presence of a beta-glucosidase enzyme. Nicotoflorin on the other hand had been reported to act as a protective in reducing brain or memory disfunction. Different phenolic acids had been reported to be present in grain amaranth with the main phenolic acids being ferulic acid, p-hydroxybenzoic acid, and caffeic acid (Mir et al., 2018). Flavonoids identified in amaranth include several groups of flavonols, flavones, flavonones, flavanones, and anthocyanins. The presence of these phytochemicals in grain amaranth made the pseudo-cereal an interesting raw material in the food and pharmaceutical industries.

13.4 Processing and Utilization of Grain Amaranth

Grain amaranth processing varies based on the end-use product and country. The grains are commonly consumed as traditional food products in many countries such as *Alegria* (Mexico), *chapatti* and *laddoos* (India), *satto* (Nepal) and *bollos* (Peru). Recently, a wide range of modern food products have been developed from grain amaranth by processing (sorting, cleaning, and milling) it into flour suitable for both infants and adults, these include extruded products (infant formula, noodles, pasta, flakes, and breakfast cereals), baked products (bread, snacks, cakes and biscuits), drinks (beverage, yoghurt), and distilled spirits/vinegar.

It was also observed in Tagwira et al. (2006) that the consumption of grain amaranth among nursing mothers increased the rate of breast milk production. The grain is also reported to contain oil rich in squalene, an organic compound mainly found in marine sharks that is reported to exhibit wound-healing, anticholesterol, anti-inflammatory, and anticancer effects thereby enhancing its application in food, cosmetic, and pharmaceutical industries (He & Corke, 2003; Karamać et al., 2019; Aderibigbe et al., 2020). Owing to its unique properties (gluten-free), it provides an alternative food protein source to people suffering from celiac disease and this is attributed to the fact that grain amaranth protein contains albumins and globulins in which prolamins, the toxin responsible for celiac diseases, is very low (Aderibigbe et al., 2020; Martínez-Villaluenga et al., 2020).

Grain amaranth can also be processed into flour and used in enrichment/fortification as cereal substitute in countries with comparative disadvantage in the production of wheat, barley, and other cereal crops. It also worthy of note that grain amaranth is a rich source of sulphur containing amino acids (methionine and cysteine) thereby making it a good combination with cereals that are limited in methionine and cysteine for development of novel food products. For instance, Diaz et al. (2013) developed a gluten-free extruded maize-based snack with 20% inclusion of amaranth grain flour. Likewise, Martinez et al. (2014) reported on improved technology and sensory quality of bread wheat pasta developed with the addition of amaranth grain flour.

In Africa, Nigeria in particular, the leaves of amaranth vegetables are another part of the crop that is commonly consumed raw, as cooked green leafy vegetables, or used in soup making called *Tete* by the Yorubas in Nigeria (Olawoye & Gbadamosi, 2020). Value addition in the use of amaranth as a vegetable includes drying, milling into powder form, and packaging in transparent polythene bags for use in the preparation of sauces during the scarcity of vegetables in the dry season to prolong its shelf life (Aderibigbe et al., 2020).

Literature studies have shown that secondary metabolites (phenolic compounds) such as flavonols, hydroxycinnamic acids, and benzoic acids from amaranth leaves extracts play a defending role against several diseases, such as cardiovascular disease, cancer, cataracts, atherosclerosis, retinopathy, arthritis, emphysema, and neurodegenerative in human beings (Kraujalis & Venskutonis, 2013; Longato et al., 2017; Molina et al., 2018; Sarker & Oba, 2019).

Meanwhile, due to climate change and shortage in rainfall as well as an increase in the cost of animal proteins, grain amaranth protein is used as alternative protein sources in animal feed formulation (Manyelo et al., 2020). A positive increase in animal production has been reported on poultry, pigs, sheep, and cattle feed on amaranth feed meals (Gresta et al., 2020). Recently, Manyelo et al. (2020) reported that grain amaranth can also be used as a potential source of protein and methionine in monogastric nutrition with no negative implication on animals productivity. Kambashi et al. (2014) reported an increase in protein and fat contents of rabbit diets developed with increasing level (0%, 16%, and 32%) of amaranth. Likewise, Longato et al. (2017) reported that up to 10% inclusion of amaranth in broiler diets significantly improved broiler performance and growth indices.

13.5 Effect of Processing Techniques/Methods on Grain Amaranth

Pseudo-cereals are usually not eaten raw but are processed before being eaten. Cooking is a typical processing method for pseudo-cereals before consumption. However, there is occasional loss of soluble nutrients such as minerals, phenolic compounds, and amino acids. In a study by Repo-Carrasco-Valencia et al. (2010), iron (Fe) content decreases significantly in amaranth grains during boiling. The wet procedure adopted for this process was said to be responsible for this significant loss of mineral content. Zinc content was however not affected.

The cooking effect on amaranth seeds of *A.cruentus* and *A. caudatus* was investigated by Gamel et al. (2006). Phenolic content was observed to reduce from 5.16–3.53 g/kg for raw amaranth and cooked amaranth, while from 5.24–3.96 g/kg for raw and processed amaranth, respectively for amaranth subjected to steaming processes. In a related study by Queiroz et al. (2009), the phenolic content in *Amaranthus cruentus* grains was observed to reduce from 31.7 mg to 24.10 mg of Gallic acid equivalent/g. The antioxidant activity was however observed to increase from 55.42–79.52% in raw to cooked seed.

Puffing or popping is a common way of improving the flavor of process amaranth grains. This process gives amaranth grain a pleasant flavor. Iron and calcium content decreases significantly in amaranth grain subjected to popping. The popping process did not significantly affect some essential minerals such as manganese (Mn), potassium (K), phosphorous (P), magnesium (Mg), copper (Cu), calcium (Ca), and iron (Fe) (Gamel et al., 2004). In a study by Pedersen et al. (2010), a significant difference in Ca, Fe, and K was reported though no significant difference was observed for Mg and Zn content after popping process. A higher amount of Ca and Cu in the outer layer of the amaranth grain was said to be responsible for this observation.

During heat treatment, about 90% of tryptophan was significantly depleted in a study by Amere et al. (2015). Tyrosine was another aromatic acid that was significantly affected. The nonenzymatic browning reaction was given as the most likely reason for the decrease recorded in amino acids levels after heat treatment (Tovar et al., 1989).

The popping process resulted in a reduction in the total phenolic content of *Amaranthus* grain (*A. cruentus*) from 31.7-to 22.71 mg GAE/g Queiroz et al. (2009). A similar trend was reported in a related study by Muyonga et al. (2014). Popping was reported to reduce the phenolic content from 3.42 – 2.99 mg

GAE/g though there was no significance in the result of raw and popped grains. Yanez et al. (1986) also reported a significant decrease in the total phenolic content of grain exposed to popping process. The popping process was however reported to increase total flavonoid content of *A. hypochondriacus* and *A. cruentus*, respectively.

Lysine, aspartic acid, glutamic acid, and glycine content were amino acids not affected after amaranth was processed using an extrusion process at different temperatures and moisture contents (Chávez-Jauregui *et al.*, 2000). The extrusion process however affected the phenolic content of amaranth seeds (Repo-Carrasco-Valencia et al., 2009). Phenolic acid decarboxylation was attributed to this observation. There is quite a paucity of data as regards the effect of extrusion on amaranth seed's total phenolic content.

Certain amino acid contents such as serine, tyrosine, lysine, leucine, and aspartic increases when *A. caudatus* and *A. cruentus* seed grains were germinated (Gamel et al. 2005). Phenylalanine, arginine, threonine, and tyrosine, however, decrease in both species of amaranth investigated. Valine was recorded as the highest amino acid increased while tyrosine was the least affected in the same studies by these authors. Iron, manganese, copper, sodium, and magnesium levels were not significantly affected (Gamel et al. 2005). Decomposition of tannins or phytate, which binds minerals by enzyme activity such as phytase, is responsibly attributed to these findings. Increases in soluble phenolic compounds have been reported in germinating seeds and sprouts. Increased water content during the germination process, which is dependent on temperature and time, have been implicated for this process (Mbithi et al., 2001; Guo et al., 2012).

Free and bound phenolic content of amaranth grains increased when germination conditions were optimized in a study by Perales-Sánchez et al. (2014). Antioxidant activities were also reported to increase. Phenolic compounds were however reduced for seeds dried at high temperature (60°C or 90°C) compared to those dried at low temperature (30°C).

Bioactive compound composition can be modified by the fermentation process depending on conditions such as pH and temperature (Nufer et al., 2009). In a related study by Amere et al. (2015), virtually all amino acids increased in pigmented *Amaranthus* grains when subjected to fermentation. Proline, glutamic acid, and tyrosine remained unchanged while arginine decreased significantly. Sripriya et al. (1997) and Pranoto et al. (2013) attributed the hydrolysis of protein during fermentation to be responsible for the significant phenylalanine and lysine increase when compared to other amino acids.

Antioxidant and total phenolic content of *Amaranthus* grain increased significantly upon fermentation (Alvarez-Jubete et al. 2010). Increased activity of endogenous hydrolytic enzymes was linked to this effect. Toasting is a processing method that uses dry heat for a shorter duration in improving the texture, crispiness, color, flavor, shelf life, and volume of amaranth grains. There was a reduction in the iron content of kiwicha (*Amaranthus caudatus*) after toasting, though values for zinc and calcium were unaffected (Repo-Carrasco-Valencia et al., 2009). Toasting of raw amaranth improves its protein quality and digestibility. Bressani et al. (1992) however reported a decrease in the lysine content of *Amaranthus caudatus* from 18.9 to 3.0 after the roasting process.

In studies by Queiroz et al. (2009), there was no significant increase in the total phenolic content of *A. cruentus* and *A. hypochondriacus*, nonetheless, flavonoid content was increased. Muyonga et al. (2014) reported 62.3% and 47.1% increase in the flavonoid content of *A. cruentus* and *A. hypochondriacus*, respectively. Deactivation of endogenous oxidative enzymes thereby inhibiting enzymatic oxidation, which causes antioxidant loss in raw plant materials, was associated with the flavonoid content increase as reported by Queiroz et al. (2009) and Muyonga et al. (2014).

In related studies, the antioxidant activities of *A. cruentus* and *A. hypochondriacus* were increased by 62.3% and 42.8% after the toasting process (Kähkönen et al., 1999; Lopez-Martinez et al., 2009). An observed increase in flavonoid was associated with increased antioxidant activity.

13.6 Amaranth Starch and Application

Starch is the most abundant macromolecule found in cereals, pseudo-cereals as well as some legume grains. It is made of glucose residue joined linearly together by an alpha 1,4 glycosidic bond to form amylose and branched at the alpha 1,6 glycosidic bond to form amylopectin (Bet et al., 2018). Usually, in starch crops, amylopectin formed 75% of the starch composition while the lesser part of the starch is made up of amylose. The ratio of amylose to amylopectin goes in a long way to determine the structural, functional, digestive properties as well as industrial application of the starch (Jan et al., 2015). The branching of the glucose unit in the formation of amylopectin does not occur randomly, but occurs or is arranged in clusters leading to double helix formations. The helices organize together to form a crystal that is separated by amorphous regions as found in amylose starch. Grain amaranth starch exists in very minute granules that ranged between 0.8–2.5 um in diameter and is regarded as the smallest ever granule size recorded when compared to commercial starch granules such as rice starch (3–8 um) and potato starch granule (15–100 um) (Choi et al., 2004). With respect to shape, a grain amaranth starch granule is either spherical or polygonal. The functionality of the starch is influenced by the size of the starch granules. According to (Li & Zhu, 2017), amaranth starch has a molar mass averaging at 11.8×106 g/mole, which is higher when compared to wheat starch (5.5×106 g/mol), however, it had a lower value (17.4×106 g/mol) than waxy maize starch. Amaranth starch when examined under X-ray diffractometry revealed it exhibited an A-type X-ray diffraction pattern with reflection angles at 15.3°, 17.1°, 18.0°, 20.0°, and 23.5° 2θ as well as 35% relative starch crystallinity (Tang et al., 2002).

As said earlier, the gelatinization properties of amaranth starch are dependent on a variety of factors such as degree of starch crystallinity and size and structure of the starch granule. A report on the gelatinization behavior of the starch revealed that grain amaranth starch gelatinized at a low temperature with an onset temperature range of 46.1–57.3°C, peak temperature ranging between 54.3–61.8°C, and conclusion temperature ranging 66.2–68.55°C (Inouchi et al., 1999). These gelatinization temperatures are similar to that of potato and wheat starches but are lower than corn starch. The low gelatinization temperature of grain amaranth had been related to the granule size of the starch; starch with small granule size tends to gelatinize at a low temperature and vice versa. The gelatinization enthalpy of amaranth starch is 7.3–10.5 j/g, a value low compared to rice starch, potato starch, wheat, and corn starch, which were 14.2–16.3 j/g, 18.8 j/g, 12.1 j/g, and 17.2–20.5 j/g, respectively (Olawoye, 2016). Although the gelatinization temperature of amaranth is similar to that of wheat, they differ in pasting behavior. Amaranth starch had higher water absorption capacity, swelling power, excellent freeze-thawing, and retrogradation stability than corn and wheat starches (Alonso-Miravalles & O'Mahony, 2018). These properties have been attributed to its low amylose content. Amaranth starch had found an application as a thickener in soups, cookies, gluten-free diets, pastas, muffins, breakfast cereals, and also in the formulation of healthy foods (Fasuan & Akanbi, 2018). Other potential commercial and industrial application of amaranth starch include biodegradable films, cosmetics, and laundry starch as well as in paper coatings.

13.7 Future Application and Utilization of Grain Amaranth

Considering the high cost of animal protein in many developing countries, in the near future, it will become a necessity for food industries to extract protein from amaranth grains, a cheap and high source of protein (Martínez-Cruz et al., 2014), and apply industrial processes. Amaranth grain proteins can be further purified into bioactive peptides (Montoya-Rodriguez et al., 2015). Plant-derived bioactive peptides are currently replacing the use of synthetic drugs in the management of health challenges due to side effects (Martínez-Maqueda et al., 2012). Hence, the development of food and nutraceuticals from grains amaranth in the nearest future is imperative to enact health-promoting benefits such as antioxidants, antidiabetics, anticancer, and antihypertensive properties with no side effects on human consumption.

Another future use of grain amaranth is in the field of genetic engineering. FAO reported that about 70% of the human diet must be derived from cereal, legumes, fruits, and vegetable (Mandal & Mandal, 2000). However, some cereals are deficient in lysine and tryptophan, while legumes are deficient in methionine and cysteine. Amarantine, a protein gene extracted from grains amaranth (pseudo-cereal), has been reported to contain a substantial amount of lysine and tryptophan (Martínez-Cruz et al., 2014). Hence, amaranthine could be isolated, purified, characterized, and introduced into legumes to improve their nutritional value, which can help reduce the cases of protein malnutrition. Grains amaranth starch could also be used in the production of biodegradable film as well as in nanotechnology.

13.8 Conclusion

Grain amaranth under the class of pseudo-cereal is an underutilized crop that is rich nutritionally and also contains phytochemicals with health-promoting properties. The uniqueness of the seed is in its high-quality protein and lipids, which are higher than other common cereals, coupled with the fact that its protein lacks gluten, a protein known to induce celiac disease. The lack of gluten protein makes it an ideal raw material in the formulation of gluten-free diets. Apart from its high-quality protein, its oil is a rich source of squalene, which also conveys health benefits. These unique properties of the grain should facilitate its utilization as food or a functional food ingredient in human diets.

REFERENCES

Aderibigbe, O. R., Ezekiel, O. O., Owolade, S. O., Korese, J. K., Sturm, B., & Hensel, O. (2020). Exploring the potentials of underutilized grain amaranth (Amaranthus spp.) along the value chain for food and nutrition security: A review. Critical Reviews in Food Science and Nutrition. https://doi.org/10.1080/10408398.2020.1825323

Alonso-Miravalles, L., & O'Mahony, J. A. (2018). Composition, protein profile and rheological properties of pseudocereal-based protein-rich ingredients. *Foods*, *7*(5). https://doi.org/10.3390/foods7050073

Alvarez-Jubete, L., Wijngaard, H., Arendt, E. K., & Gallagher, E. (2010). Polyphenol composition and in vitro antioxidant activity of amaranth, quinoa buckwheat and wheat as affected by sprouting and baking. *Food Chemistry*, *119*(2), 770–778.

Amere, E., Mouquet-Rivier, C., Servent, A., Morel, G., Adish, A., & Haki, G. D. (2015). Protein quality of amaranth grains cultivated in Ethiopia as affected by popping and fermentation. *Food and Nutrition Sciences*, *6*(01), 38.

Arreguez, G. A., Martínez, J. G., & Ponessa, G. (2013). *Amaranthus hybridus* L. ssp. hybridus in an archaeological site from the initial mid-Holocene in the Southern Argentinian Puna. *Quaternary International*, *307*, 81–85.

Belton, P. S., & Taylor, J. R. (2002). Pseudocereals and Less Common Cereals: Grain Properties and Utilization Potential. Springer.

Berghofer, E., & Schoenlechner, R. (2002). Grain amaranth. In Pseudocereals and Less Common Cereals: Grain Properties and Utilization Potential. Springer, pp. 219–260.

Bet, C. D., de Oliveira, C. S., Colman, T. A. D., Marinho, M. T., Lacerda, L. G., Ramos, A. P., & Schnitzler, E. (2018). Organic amaranth starch: A study of its technological properties after heat-moisture treatment. *Food Chemistry*, *264*, 435–442. https://doi.org/10.1016/j.foodchem.2018.05.021

Brenner, D. M., Baltensperger, D. D., Kulakow, P. A., Lehmann, J. W., Myers, R. L., Slabbert, M. M., & Sleugh, B. B. (2000). Genetic resources and breeding of Amaranthus. *Plant Breeding Reviews*, *19*, 227–285.

Bressani, R., Sánchez-Marroquín, A., & Morales, E. (1992). Chemical composition of grain amaranth cultivars and effects of processing on their nutritional quality. *Food Reviews International*, *8*(1), 23–49.

Brinegar, C., Sine, B., and Nwokocha, L. (2006). High-cysteine 2S seed storage proteins from quinoa (*Chenopodium quinoa*). *Journal Agriculture and Food Chemistry*. *44*: 1621–1623

Bruni, R., Guerrini, A., Scalia, S., Romagnoli, C., & Sacchetti, G. (2002). Rapid techniques for the extraction of vitamin E isomers from *Amaranthus caudatus* seeds: Ultrasonic and supercritical fluid extraction. *Phytochemical Analysis*, *13*(5), 257–261. https://doi.org/10.1002/pca.651

Bucaro Segura, M. E., & Bressani, R. (2002). [Protein fraction distribution in milling and screened physical fractions of grain amaranth]. *Archivos Latinoamericanos de Nutrición*, *52*(2), 167–171. https://www.ncbi.nlm.nih.gov/pubmed/12184151 (*Distribucion de la proteina en fracciones fisicas de la molienda y tamizado del grano de amaranto.*)

Chávez-Jáuregui, R. N., Silva, M. E. M. P., & Arêas, J. A. G. (2000). Extrusion cooking process for amaranth (*Amaranthus caudatus* L.). *Journal of Food Science*, *65*(6), 1009–1015.

Choi, H., Kim, W., & Shin, M. (2004). Properties of Korean amaranth starch compared to waxy millet and waxy sorghum starches. *Starch—Stärke*, *56*(10), 469–477. https://doi.org/https://doi.org/10.1002/star.200300273

Clouse, J. W., Adhikary, D., Page, J. T., Ramaraj, T., Deyholos, M. K., Udall, J. A., Fairbanks, D. J., Jellen, E. N., & Maughan, P.J. (2016). The amaranth genome: Genome, transcriptome, and physical map assembly. *Plant Genome*, *9*(1). https://doi.org/10.3835/plantgenome2015.07.0062

Coelho, L. M., Silva, P. M., Martins, J. T., Pinheiro, A. C., & Vicente, A. A. (2018). Emerging opportunities in exploring the nutritional/functional value of amaranth. *Food and Function, 9*(11), 5499–5512. https://doi.org/10.1039/c8fo01422a

Costea, M., & DeMason, D. A. (2001). Stem morphology and anatomy in *Amaranthus* L. (Amaranthaceae), taxonomic significance. *Journal of the Torrey Botanical Society,* 254–281.

Das, S. (2016). Amaranthus: A Promising Crop of Future. Springer Nature.

Diaz, J. M. R., Kirjoranta, S., Tenitz, S., Penttilä, P. A., Serimaa, R., Lampi, A. M. and Jouppila, K. (2013). Use of amaranth, quinoa and kañiwa in extruded corn-based snacks. *Journal of Cereal Science, 58*(1), 59–67.

Dus-Zuchowska, M., Walkowiak, J., Morawska, A., Krzyzanowska-Jankowska, P., Miskiewicz-Chotnicka, A., Przyslawski, J., & Lisowska, A. (2019). Amaranth oil increases total and LDL cholesterol levels without influencing early markers of atherosclerosis in an overweight and obese population: A randomized double-blind cross-over study in comparison with rapeseed oil supplementation. *Nutrients, 11*(12). https://doi.org/10.3390/nu11123069

FAO, I., UNICEF, WFP & WHO. (2020). The State of Food Security and Nutrition in the World 2020. Transforming food systems for affordable healthy diets. In. Rome, Italy: FAO.

Fasuan, T. O., & Akanbi, C. T. (2018). Application of osmotic pressure in modification of Amaranthus viridis starch. *LWT, 96,* 182–192. https://doi.org/10.1016/j.lwt.2018.05.036

Gamel, T. H., Linssen, J. P., Alink, G. M., Mosallem, A. S., & Shekib, L. A. (2004). Nutritional study of raw and popped seed proteins of *Amaranthus caudatus* L and *Amaranthus cruentus* L. *Journal of the Science of Food and Agriculture, 84*(10), 1153–1158.

Gamel, T. H., Linssen, J. P., Mesallem, A. S., Damir, A. A. and Shekib, L. A. (2005). Effect of seed treatments on the chemical composition and properties of two amaranth species: starch and protein. *Journal of the Science of Food and Agriculture, 85*(2), 319–327.

Gamel, T H., Linssen, J. P., Mesallam, A. S., Damir, A. A. and Shekib, L. A. (2006). Seed treatments affect functional and antinutritional properties of amaranth flours. *Journal of the Science of Food and Agriculture, 86*(7), 1095–1102.

Gliszczynska-Swiglo, A., Klimczak, I., & Rybicka, I. (2018). Chemometric analysis of minerals in gluten-free products. *Journal of the Science of Food and Agriculture, 98*(8), 3041–3048. https://doi.org/10.1002/jsfa.8803

Gresta, F., Meineri, G., Oteri, M., Santonoceto, C., Lo Presti, V., Costale, A., & Chiofalo, B. (2020). Productive and qualitative traits of *Amaranthus cruentus* L.: An unconventional healthy ingredient in animal feed. *Animals, 10*(8), 1428

Guo, X., Li, T., Tang, K., & Liu, R. H. (2012). Effect of germination on phytochemical profiles and antioxidant activity of mung bean sprouts (*Vigna radiata*). *Journal of Agricultural and Food Chemistry, 60*(44), 11050–11055.

He, H. P., & Corke, H. (2003). Oil and squalene in amaranthus grain and leaf. *Journal of Agricultural and Food Chemistry, 51*(27), 7913–7920.

Inouchi, N., Nishi, K., Tanaka, S., Asai, M., Kawase, Y., Hata, Y., Konishi, Y., Yue, S., & Fuwa, H. (1999). Characterization of amaranth and quinoa starches. *Journal of Applied Glycoscience, 46*(3), 233–240.

Jan, R., Saxena, D., & Singh, S. (2015). Physico-chemical and textural property of starch isolated from Chenopodium (*Chenopodium album*) grains. *Cogent Food and Agriculture, 1*(1), 1095052.

Kähkönen, M. P., Hopia, A. I., Vuorela, H. J., Rauha, J. P., Pihlaja, K., Kujala, T. S., & Heinonen, M. (1999). Antioxidant activity of plant extracts containing phenolic compounds. *Journal of Agricultural and Food Chemistry, 47*(10), 3954–3962.

Kambashi, B., Picron, P., Boudry, C., Thewis, A., Kiatoko, H., & Bindelle, J. (2014). Nutritive value of tropical forage plants fed to pigs in the Western provinces of the Democratic Republic of the Congo. *Animal Feed Science and Technology, 191,* 47–56.

Karamać, M., Gai, F., Longato, E., Meineri, G., Janiak, M. A., Amarowicz, R., & Peiretti, P. G. (2019). Antioxidant activity and phenolic composition of amaranth (*Amaranthus caudatus*) during plant growth. *Antioxidants, 8*(6), 173.

Kietlinski, K. D., Jimenez, F., Jellen, E. N., Maughan, P. J., Smith, S. M., & Pratt, D. B. (2014). Relationships between the Weedy Amaranthus hybridus (Amaranthaceae) and the Grain Amaranths. *Crop Science, 54*(1), 220–228. doi:https://doi.org/10.2135/cropsci2013.03.0173

Kim, H. K., Kim, M. J., & Shin, D. H. (2006). Improvement of lipid profile by amaranth (*Amaranthus esculantus*) supplementation in streptozotocin-induced diabetic rats. *Annals of Nutrition and Metabolism, 50*(3), 277–281. https://doi.org/10.1159/000091686

Kong, X., Corke, H., & Bertoft, E. (2009). Fine structure characterization of amylopectins from grain amaranth starch. *Carbohydrate Research*, *344*(13), 1701–1708. https://doi.org/10.1016/j.carres. 2009.05.032

Kraujalis, P., & Venskutonis, P. R. (2013). Supercritical carbon dioxide extraction of squalene and tocopherols from amaranth and assessment of extracts antioxidant activity. *The Journal of Supercritical Fluids*, *80*, 78–85.

Lamothe, L. M., Srichuwong, S., Reuhs, B. L., & Hamaker, B. R. (2015). Quinoa (*Chenopodium quinoa* W.) and amaranth (*Amaranthus caudatus* L.) provide dietary fibres high in pectic substances and xyloglucans. *Food Chemistry*, *167*, 490–496. https://doi.org/10.1016/j.foodchem.2014.07.022

Li, G., & Zhu, F. (2017). Molecular structure of quinoa starch. *Carbohydrate Polymers*, *158*, 124–132. https:// doi.org/10.1016/j.carbpol.2016.12.001

Longato, E., Meineri, G., & Peiretti, P. G. (2017). The effect of *Amaranthus caudatus* supplementation to diets containing linseed oil on oxidative status, blood serum metabolites, growth performance and meat quality characteristics in broilers. *Animal Science Papers and Reports*, *35*(1), 71–86.

Lopez-Martinez, L. X., Oliart-Ros, R. M., Valerio-Alfaro, G., Lee, C. H., Parkin, K. L., & Garcia, H. S. (2009). Antioxidant activity, phenolic compounds and anthocyanins content of eighteen strains of Mexican maize. *LWT—Food Science and Technology*, *42*(6), 1187–1192.

Mandal, S., & Mandal, R. (2000). Seed storage proteins and approaches for improvement of their nutritional quality by genetic engineering. *Current Science*, *79*, 576–589.

Manyelo, T. G., Sebola, N. A., van Rensburg, E. J., & Mabelebele, M. (2020). The probable use of Genus amaranthus as feed material for monogastric animals. *Animals*, 10(9), 1504.

Martinez, C. S., Ribotta, P. D., Anon, M. C., & Leon, A. E. (2014) Effect of amaranth flour (*Amaranthus mantegazzianus*) on the technological and sensory quality of bread wheat pasta. *Food Science and Technology International*, *20*: 127–135

Martínez-Cruz, O., Cabrera-Chávez, F., & Paredes-López, O. (2014). Biochemical characteristics, and nutraceutical and technological uses of amaranth globulins. In S. D. Milford (Ed.), *Globulins: Biochemistry, Production and Role in Immunity*. New York, NY: Nova Science Publishers, pp. 41–70.

Martínez-Maqueda, D., Miralles, B., Recio, I., & Hernández-Ledesma, B. (2012). Antihypertensive peptides from food proteins: a review. *Food & Function*, *3*(4), 350–361. https://doi.org/10.1039/c2fo10192k

Martínez-Villaluenga, C., Peñas, E., & Hernández-Ledesma, B. (2020). Pseudocereal grains: Nutritional value, health benefits and current applications for the development of gluten-free foods. *Food and Chemical Toxicology*, *137*, 111178. https://doi.org/10.1016/j.fct.2020.111178

Mbithi, S., Van Camp, J., Rodriguez, R., & Huyghebaert, A. (2001). Effects of sprouting on nutrient and antinutrient composition of kidney beans (*Phaseolus vulgaris* var. Rose coco). *European Food Research and Technology*, *212*(2), 188–191.

Menegassi, B., Pilosof, A. M., & Arêas, J. A. (2011). Comparison of properties of native and extruded amaranth (*Amaranthus cruentus* L.–BRS Alegria) flour. *LWT–Food Science and Technology*, *44*(9), 1915–1921.

Molina, E., González-Redondo, P., Moreno-Rojas, R., Montero-Quintero, K., & Sánchez-Urdaneta, A. (2018). Effect of the inclusion of *Amaranthus dubius* in diets on carcass characteristics and meat quality of fattening rabbits. *Journal of Applied Animal Research*, *46*(1), 218–223.

Montoya-Rodríguez, A., Gómez-Favela, M. A., Reyes-Moreno, C., Milán-Carrillo, J., & González de Mejía, E. (2015). Identification of bioactive peptide sequences from amaranth (*Amaranthus hypochondriacus*) seed proteins and their potential role in the prevention of chronic diseases. *Comprehensive Reviews in Food Science and Food Safety*, *14*, 139–158.

Morales, D., Miguel, M., & Garces-Rimon, M. (2020). Pseudocereals: A novel source of biologically active peptides. *Critical Reviews in Food Science and Nutrition*, 1–8. https://doi.org/10.1080/10408398.2020 .1761774

Mota, C., Santos, M., Mauro, R., Samman, N., Matos, A. S., Torres, D., & Castanheira, I. (2016). Protein content and amino acids profile of pseudocereals. *Food Chemistry*, *193*, 55–61. https://doi.org/10.1016/j. foodchem.2014.11.043

Mustafa, A. F., Seguin, P., & Gelinas, B. (2011). Chemical composition, dietary fibre, tannins and minerals of grain amaranth genotypes. *International Journal of Food Sciences and Nutrition*, *62*(7), 750–754. https://doi.org/10.3109/09637486.2011.575770

Muyonga, J. H., Andabati, B., & Ssepuuya, G. (2014). Effect of heat processing on selected grain amaranth physicochemical properties. *Food Science & nutrition*, *2*(1), 9–16.

Niro, S., D'Agostino, A., Fratianni, A., Cinquanta, L., & Panfili, G. (2019). Gluten-free alternative grains: Nutritional evaluation and bioactive compounds. *Foods*, *8*(6). https://doi.org/10.3390/foods8060208

Nufer, K. R., Ismail, B., & Hayes, K. D. (2009). The effects of processing and extraction conditions on content, profile, and stability of isoflavones in a soymilk system. *Journal of Agricultural and Food Chemistry*, *57*(4), 1213–1218.

Olawoye, B. (2016). Effect of processing methods on the chemical composition, functional and anti-oxidant properties of *Amaranthus virides* seed. Master's thesis, Department of Food Science and Technology, Obafemi Awolowo University, Ife, Nigeria.

Olawoye, B. T., & Gbadamosi, S. O. (2017). Effect of different treatments on in vitro protein digestibility, antinutrients, antioxidant properties and mineral composition of *Amaranthus viridis* seed. *Cogent Food & Agriculture*, *3*(1), 1296402.

Olawoye, B., & Gbadamosi, S. O. (2020a). Digestion kinetics of native and modified cardaba banana starch: A biphasic approach. *International Journal of Biological Macromolecules*, *154*, 31–38. https://doi.org/10.1016/j.ijbiomac.2020.03.089

Olawoye, B., & Gbadamosi, S. O. (2020b). Influence of processing on the physiochemical, functional and pasting properties of Nigerian *Amaranthus viridis* seed flour: A multivariate analysis approach. *SN Applied Sciences*, *2*(4), 607. https://doi.org/10.1007/s42452-020-2418-8

Olawoye, B., & Kadiri, O. (2016). Optimization and response surface modelling of antioxidant activities of Amaranthus virides seed flour extract. *Annals Journal of Food Science and Technology*, *17*, 114–123.

Olawoye, B., Gbadamosi, S. O., Otemuyiwa, I. O., & Akanbi, C. T. (2020a). Gluten-free cookies with low glycemic index and glycemic load: Optimization of the process variables via response surface methodology and artificial neural network. *Heliyon*, *6*(10), e05117. https://doi.org/10.1016/j.heliyon.2020.e05117

Olawoye, B., Gbadamosi, S. O., Otemuyiwa, I. O., & Akanbi, C. T. (2020b). Improving the resistant starch in succinate anhydride-modified cardaba banana starch: A chemometrics approach. *Journal of Food Processing and Preservation*, *44*(9), e14686. https://doi.org/10.1111/jfpp.14686

Pandey, R. M. (1999). Evolution and improvement of cultivated amaranths with reference to genome relationship among *A. cruentus*, *A. powellii* and *A. retroflexus*. *Genetic Resources and Crop Evolution*, *46*(3), 219–224.

Pedersen, H. A., Steffensen, S. K., Christophersen, C., Mortensen, A. G., Jørgensen, L. N., Niveyro, S., et al. (2010). Synthesis and quantitation of six phenolic amides in *Amaranthus* spp. *Journal of Agricultural and Food Chemistry*, *58*(10), 6306–6311.

Perales-Sánchez, J. X., Reyes-Moreno, C., Gómez-Favela, M. A., Milán-Carrillo, J., Cuevas-Rodríguez, E. O., Valdez-Ortiz, A., & Gutiérrez-Dorado, R. (2014). Increasing the antioxidant activity, total phenolic and flavonoid contents by optimizing the germination conditions of amaranth seeds. *Plant Foods for Human Nutrition*, *69*(3), 196–202.

Pina-Rodriguez, A. M., & Akoh, C. C. (2009). Synthesis and characterization of a structured lipid from amaranth oil as a partial fat substitute in milk-based infant formula. *Journal of Agriculture and Food Chemistry*, *57*(15), 6748–6756. https://doi.org/10.1021/jf901048x

Pranoto, Y., Anggrahini, S., & Efendi, Z. (2013). Effect of natural and Lactobacillus plantarum fermentation on in-vitro protein and starch digestibilities of sorghum flour. *Food Bioscience*, *2*, 46–52.

Queiroz, Y. S. D., Soares, R. A. M., Capriles, V. D., Torres, E. A. F. D. S., & Áreas, J. A. G. (2009). Efeito do processamento na atividade antioxidante do grão de amaranto (*Amaranthus cruentus* L. BRS-Alegria). *Archivos Latinoamericanos de Nutrición*.

Repo-Carrasco-Valencia, R., de La Cruz, A. A., Alvarez, J. C. I., & Kallio, H. (2009). Chemical and functional characterization of kaniwa (*Chenopodium pallidicaule*) grain, extrudate and bran. *Plant Foods for Human Nutrition*, *64*(2), 94–101.

Repo-Carrasco-Valencia, R., Hellström, J. K., Pihlava, J. M., & Mattila, P. H. (2010). Flavonoids and other phenolic compounds in Andean indigenous grains: Quinoa (*Chenopodium quinoa*), kañiwa (*Chenopodium pallidicaule*) and kiwicha (*Amaranthus caudatus*). *Food Chemistry*, *120*(1), 128–133.

Roa, D. F., Baeza, R. I., & Tolaba, M. P. (2015). Effect of ball milling energy on rheological and thermal properties of amaranth flour. *Journal of Food Science and Technology*, *52*(12), 8389–8394. https://doi.org/10.1007/s13197-015-1976-z

Ruales, J., & Nair, B. M. (1993). Content of fat, vitamins and minerals in quinoa (*Chenopodium quinoa*, Willd) seeds. *Food Chemistry*, *48*(2), 131–136. https://doi.org/10.1016/0308-8146(93)90047-J

Sarker, U., & Oba, S. (2019). Nutraceuticals, antioxidant pigments, and phytochemicals in the leaves of *Amaranthus spinosus* and *Amaranthus viridis* weedy species. *Scientific Reports, 9*(1), 1–10.

Sauer, J. (1955). Revision of the dioecious amaranths. *Madrono, 13*(1), 5–46.

Sauer, J. D. (1967). The grain amaranths and their relatives: A revised taxonomic and geographic survey. *Annals of the Missouri Botanical Garden, 54*(2), 103–137.

Sauer, J. D. (1993). Amaranthaceae-amaranth family. In Historical Geography of Crop Plants: A Select Roster. Boca Raton, FL: CRC Press, pp. 9–14.

Schoenlechner, R., Drausinger, J., Ottenschlaeger, V., Jurackova, K., & Berghofer, E. (2010). Functional properties of gluten-free pasta produced from amaranth, quinoa and buckwheat. *Plant Foods for Human Nutrition, 65*(4), 339–349. https://doi.org/10.1007/s11130-010-0194-0

Silva-Sanchez, C., de la Rosa, A. P., Leon-Galvan, M. F., de Lumen, B. O., de Leon-Rodriguez, A., & de Mejia, E. G. (2008). Bioactive peptides in amaranth (*Amaranthus hypochondriacus*) seed. *Journal of Agriculture and Food Chemistry, 56*(4), 1233–1240. https://doi.org/10.1021/jf072911z

Souci SW, Fachmann W, Kraut H (1994) *Food Composition and Nutrition Tables*. Wissenschaft Verlag, Stuttgart

Sripriya, G., Antony, U., & Chandra, T. S. (1997). Changes in carbohydrate, free amino acids, organic acids, phytate and HCl extractability of minerals during germination and fermentation of finger millet (*Eleusine coracana*). *Food Chemistry, 58*(4), 345–350.

Stetter, M. G., & Schmid, K. J. (2017). Analysis of phylogenetic relationships and genome size evolution of the Amaranthus genus using GBS indicates the ancestors of an ancient crop. *Molecular Phylogenetics and Evolution, 109*, 80–92.

Stetter, M. G., Müller, T., & Schmid, K. J. (2017). Genomic and phenotypic evidence for an incomplete domestication of South American grain amaranth (*Amaranthus caudatus*). *Molecular Ecology, 26*(3), 871–886.

Tagwira, M., Tagwira, F., Dugger, R., & Okum, B. (2006). Using grain amaranth to fight malnutrition. RUFORUM working document, *1*, 201–206.

Tang, H., Watanabe, K., & Mitsunaga, T. (2002). Characterization of storage starches from quinoa, barley and adzuki seeds. *Carbohydrate Polymers, 49*(1), 13–22. https://doi.org/10.1016/S0144-8617(01)00292-2

Taylor, J. R. N., Taylor, J., Campanella, O. H., & Hamaker, B. R. (2016). Functionality of the storage proteins in gluten-free cereals and pseudocereals in dough systems. *Journal of Cereal Science, 67*, 22–34. https://doi.org/10.1016/j.jcs.2015.09.003

Tovar, L. R., Brito, E., Takahashi, T., Miyazawa, T., Soriano, J., & Fujimoto, K. (1989). Dry heat popping of amaranth seed might damage some of its essential amino acids. *Plant Foods for Human Nutrition, 39*(4), 299–309.

Wu, T., Taylor, C., Nebl, T., Ng, K., & Bennett, L. E. (2017). Effects of chemical composition and baking on in vitro digestibility of proteins in breads made from selected gluten-containing and gluten-free flours. *Food Chemistry, 233*, 514–524. https://doi.org/10.1016/j.foodchem.2017.04.158

Yanez, G. A., Messinger, J.K., Walker, C. E., & Rupnow, J. H. (1986). *Amaranthus hypochondriacus*: Starch isolation and partial characterization. *Cereal Chemistry, 63*(3), 273–276.

Zhu, F. (2020). Dietary fiber polysaccharides of amaranth, buckwheat and quinoa grains: A review of chemical structure, biological functions and food uses. *Carbohydrate Polymers, 248*, 116819. https://doi.org/10.1016/j.carbpol.2020.116819

14

Chia Seeds—A Renewable Source as a Functional Food

Nisha Chaudhary
Agriculture University, Jodhpur, Rajasthan, India

Priya Dangi
Institute of Home Economics, University of Delhi, New Delhi, India

Rajesh Kumar and Sunil Bishnoi
Guru Jambheshwar University of Science & Technology, Hisar, Haryana, India

CONTENTS

14.1 Introduction

The word chia (*Salvia hispanica L.*) is derived from a Spanish word "chian" or "chien" that means "oily." Mexico and Guatemala are regarded as the origin of chia; it has been an important component of human food for about 5,500 years. Chia being a herbaceous plant, was traditionally utilized by the Aztec and Maya people to prepare folk medicines, food, and canvases (Armstrong, 2003). Chia is an oilseed, regarded as the powerhouse of omega-3 fatty acids, containing superior quality protein, higher quantity of dietary fiber, vitamins, minerals, and is a vital source of bioactive and polyphenolic compounds (Punia & Dhull, 2019; Cahill, 2003). Chia possess a considerable amount of polyunsaturated omega-3 fatty acids, which can help prevent inflammation, increase cognitive performance, and minimize the level of cholesterol in the human body. Chia holds the richness of polyphenols derived from caffeic acid that acts as an antioxidant to protect the body from free radicals, aging, and chronological diseases. Further, high concentration levels of carbohydrate-based dietary fiber are present in chia, associated in the reduction of inflammation, cholesterol, and irregular bowel function. Considering the excellent phytochemical content and eminent nutritional and functional properties, chia is gaining increasing importance worldwide. As after the recognition of its nutraceutical benefits, chia is now called as "the seed of 21st century," "new gold," or "super nutrient" (Dick, Costa, Gomaa, et al., 2015). Recent studies show the constituents of chia seeds are ascribed a beneficial effect on the enhancement of the blood lipid profile by means of their hypotensive, hypoglycaemic, antimicrobial, and immunostimulatory belongings (Kulczyński, Kobus-Cisowska, Taczanowski, Kmiecik, & Gramza-M., 2019).

In recent years, consumers are concentrating more on the health benefits in addition to nutrition from food in an effort to preserve a healthy lifestyle. Whilst all kinds of foods are functional in one way or another, functional foods have turned into one of the rapidly expanding areas of the food industry because of the increasing awareness of people working to prevent lethal diseases like cancer, diabetes mellitus, and cardiovascular disease. Functional foods are explained as "the food or food components that manifest efficiency in protecting from diseases and attaining a healthier life by administering additional benefits on human physiology and metabolic functions apart from basic nutritional requirements of the body" (Berner & O'Donnell, 1998; Hasler & Brown, 2009). Likewise, chia is admired as a functional food by virtue of its usage in the food industry and its extensive health-promoting properties. Chia also covers huge physicochemical and functional characteristics that conclude its suitability for the food industry. In food processing operations of various foods, chia can serve as good thickening agent, gel-forming agent, chelating agent, foam enhancer, emulsifier, suspending agent, and rehydration factor. Accordingly, this can be conceptualized that novel food products enriched with various vital nutrients such as omega-3 fatty acids, protein, soluble/insoluble fiber and phenolic compounds could be incorporated into frozen foods, bakery products, beverages, desserts, sweets, pasta products, and various meat products and eventually fabricated at industrial level and utilized commercially (Coelho & Myriam de las Mercedes, 2015; Inglett, Chen, & Liu, 2014; Pintado, Herrero, Jiménez-Colmenero, & Ruiz-Capillas, 2016).

14.2 History

Chia is also known as Mexican chia or salba chia, which is the species of flowering plant that belongs to the mint family *Lamiaceae*. These seeds are grown for being used as edible seeds. This plant is native to Mexico and Guatemala, where it was an extensively vital crop for pre-Columbian Aztecs and many Mesoamerican Indian cultures (Petruzzello, 2020). Since 1500 BC, the seeds have been the part of human nutrition (Cahill, 2003). Traditionally pre-Columbian societies have used the chia seeds in the form of food, folk medicines, primary cosmetics, and a part of religious rituals in addition to its culinary application. The diets during pre-Columbian regimes were very rich and nutritious as today. The Aztecs incorporated chia in food in the year 3500 BC for the first time. Between the years 1500–900 BC, it was already acknowledged as a cash crop in the center of Mexico. Chia, was known as "running food" during that time, as it was recognized to boost energy and endurance (Craig, 2006).

Chia seeds constituted a significant portion of the diet collectively with beans, corn (maize), squash, and amaranth for the indigenous peoples. Aztecs used to roast the seeds and grind them into a flour for different food preparations, while warriors and messengers depended majorly on whole seeds for securing the nourishment during long journeys. Chia seed was given cultural and religious significance and Spanish conquistadors banned its cultivation and interchanged it with other foreign grains, like wheat and barley. Just before the late 20th century, the plant was predominantly overlooked as a food crop, though it did achieved some popularity in the United States in the 1980s as an element of the terracotta novelties called "chia pets." Later, agricultural engineer Wayne Coates accomplished a huge promotion of this plant in the early 1990s, hence chia was recognized for its efficacy as an alternative crop and a health-promoting food (Petruzzello, 2020).

The chia seed has been widely used in many countries as food or medicine for thousands of years because of its colossal nutritional and therapeutic aptitude (Fernandez, Ayerza, Coates, Vidueiros, & Pallaro, 2006). Chia seeds are acclaimed for their health benefits, therefore are now grown commercially in various countries, such as Argentina, Australia, Bolivia, Peru, and the United States (Petruzzello, 2020). These days, chia seeds are utilized in diverse industrial fields as functional food, animal feed, and cosmetics. In the food and beverage industry, chia seed is recognized in the form of super packaged food and as a backing product in cereals, muesli bars, peanut oil, and beverages (Dinçoğlu & Yeşildemir, 2019).

Presently, regardless of the fact that a considerable proportion of chia is consumed now in South America, expectedly the demand for chia products is escalating in the UK, Brazil, Chile, Spain, and India. According to the Novel Food Authorization (The European Union Novel Food), the production of various new products including chia seeds has started, though the chia seed was banned for incorporating in food in the European Union until 2009. After authorization, the application of chia seeds was approved for baked goods (up to 10%), breakfast cereals (up to 10%), nuts, fruit, and different seed blends (up to 10%) and chia seed prepacks (on the basis of up to 15 g consumption per day) (Dinçoğlu & Yeşildemir, 2019).

14.3 Structure and Composition

Chia is an annual herbaceous plant and it can attain the height of nearly 1 meter (3 feet). The lime-green leaves of the chia plant are oppositely arranged, containing serrated margins. The spikes of small blue, purple, or white flowers on plant, shows a high rate of self-pollination. Chia is very suitable to grow in desert zones as the plant requires lesser irrigation and grows properly in sandy loam soils, yet it shows sensitivity toward frost and day length. The plant repels insects, pests, and disease, so is thus considered as a good candidate for organic production (Petruzzello, 2020). The small, flat, and oval chia seeds are about 2.0–2.5 mm long, 1.2–1.5 mm in width and 0.8–1.0 mm thick. These seeds feature a shiny, patchy, or speckled seed coat that varies from dark brown to grey-white in color, but commonly found in black. As per the sensitivity of the chia plant to daylight, it produces black and white seeds with speckles on the surface. The black and white seeds differ slightly in their physicochemical properties. The white seeds are comparatively larger, thicker, and broader than the black seeds. The average moisture content also varies for black and white seeds, containing 7.2% and 6.6%, respectively. White chia seeds accommodate the higher oil yield than black seeds, which is nearly 33.8% and 32.7%, respectively. Further, the composition of protein and fatty acid of both the seeds varies noticeably (Capitani, Spotorno, Nolasco, & Tomás, 2012; Ixtaina, Nolasco, & Tomas, 2008). However, both kind of chia seeds are a rich source of fatty acids, dietary fiber, protein, minerals, and bioactive compounds. These can be used easily as whole, milled, or ground. Oil extracted from chia seeds finds application in various food preparations in the form of dressings, sauces, gravies, etc. A mucilaginous gel is produced when chia seeds are soaked in water.

14.4 Nutritional Composition of Chia

14.4.1 Storage Proteins

Proteins, peptides, and amino acids, being different matrices, are necessary cell components that enabe normal function of the organism. The protein content of chia seed is in the range of 16–26%, depending on the geographical area where the crop was harvested. The content of proteins in chia seeds is 16–26%, most of them being prolamins (538 g/kg of crude protein), followed by glutelins (230 g/kg of crude protein), globulins (70 g/kg of crude protein), and albumins (39 g/kg of crude protein) (Ayerza & Coates, 2011). Patients suffering from celiac disease can consume chia seeds because they do not contain gluten proteins (Uribe, Perez, Kauil, Rubio, & Alcocer., 2011). Chia seeds contain more proteins than rice, maize, barley, or oats seeds (Grancieri, Martino, & Gonzalez de Mejia, 2019).

According to the data of the US Department of Agriculture (USDA database 2018), chia seeds contain 18 amino acids, including 7 exogenous amino acids, which are considered to be indispensable. The study by Olivos-Lugo, Valdivia-López, & Tecante (2010) revealed that glutamic acid, which is responsible for proper functioning of the brain, is the predominant amino acid in chia seeds. The amino acid content of the protein obtained from chia is more complete than the protein from other grains that are limited in terms of two or more amino acids (Ayerza & Coates, 2011). Olivos-Lugo, Valdivia-López, & Tecante (2010) reported the amino acid profile, chemical score, and *in vitro* digestibility tests. They showed that the protein contained high amounts of glutamic acid (123 g/kg of raw protein), arginine (80.6 g/kg of raw protein), and aspartic acid (61.3 g/kg of raw protein). The amount of proteins in chia seeds depends mainly upon environmental and agronomical factors (Ullah, Nadeem, Khalique, Imran, Mehmood, Javid, et al., 2016). The US Department of Agriculture has confirmed that chia seeds contain some exogenous amino acids (arginine, leucine, phenylalanine, valine, and lysine) and some endogenous amino acids (glutamic and aspartic acid, alanine, serine, and glycine) (Table 14.1). For example, the content of amino acid serine is 1.05 g/100 g, glutamic acid 3.50 g/100 g, glycine 0.95 g/100 g, alanine 1.05 g/100 g, lysine 0.97 g/100 g, and histidine 0.53 g/100 g (Ullah, et al., 2016). The absence of the protein gluten makes chia seeds highly valued to patients suffering from celiac disease. Moreover, food rich in proteins is highly recommended to people who are trying to lose weight.

Globulins are widely distributed in nature; they are particularly common as main storage proteins in dicotyledon seeds such as pea, soybean, and bean (Kumar & Khatkar, 2017). Chia seeds fit in this distribution. Sandoval-Oliveros & Paredes-López (2013) determined the sedimentation coefficient of chia globulin and

TABLE 14.1

Amino Acid Composition of Chia (*Salvia hispanica L.*)

Amino Acids	Value (g/100 g of protein)
Alanine	4.1
Arginine	9
Aspartic acid	8.9
Cysteine	1.4
Glutamic Acid	15.2
Glycine	5.1
Proline	4.2
Serine	5.8
Tyrosine	2.5
Essential Amino acids	
Hystidine	2.5
Isoleucine	3.3
Leucine	6.5
Lysine	4.5
Methionine	1.2
Phenylalanine	4.8
Threonine	3.6
Tryptophan	0.6
Valine	4.9

Source: Berner and O'Donnell (1998), Sandoval-Oliveros and Paredes-López (2013), Orona-Tamayo, Valverde, Nieto-Rendon, and Paredes-Lopez (2015).

found that 11S globulin is the principal component, followed by 7S-like proteins; however, other minor components such as 6S and 19S with unusual sedimentation coefficients can be formed as intermediates derived from 11S globulins. This result is supported by two-dimensional gel electrophoresis for globulin fraction in a broad pH range (3–10). The protein map shows a denaturing globulin profile that confirms a high heterogeneity with molecular masses ranging from 11 kDa to 70 kDa and focused over the range of pI 4–9. In this proteomic map, a chain of six abundant proteins was present at approximately 29–35 kDa with a range of pI between 4.5 and 7.0. This globulin shows acid pH tendencies due to the presence of the acid 11S subunit. The 11S globulin consists of a single band of 383 kDa, comprising a hexameric conformation of about 50–60 kDa; however, under reducing conditions, these monomers are resolved into acidic (approximately 30 kDa) and basic (approximately 20 kDa) subunits. Protein identification confirmed the presence of peptides belonging to the 11S protein with less proportion of 7S protein peptides (Sandoval-Oliveros & Paredes-López, 2013) confirming the abundant presence of 11S subunit containing disulfide bonds in its structure.

It is important to study the amino acid composition of chia proteins due to its high quality for human health. The protein composition of chia seed flour is an excellent source of sulfur amino acids (cysteine and methionine), arginine, aspartic, and glutamic amino acids (Olivos-Lugo, Valdivia-López, & Tecante, 2010). Globulin, the highest protein fraction, also has significant amounts of aromatic and sulfur amino acids, glutamic and aspartic acids, as well as threonine and histidine. Hence, the flour and the isolated globulins from chia seeds contain the essential amino acids for human nutrition, which are histidine, isoleucine, leucine, lysine, methionine, phenylalanine, threonine, tryptophan, and valine. The amino acids in chia seeds are important for human metabolic activities. Glutamic acid has the potential to stimulate the central nervous system, is involved in its immunologic performance, and enhances athletic endurance (McCormack, Hoffman, Pruna, Jajtner, Townsend, Stout, et al., 2015; Segura-Campos, Ciau-Solís, Rosado-Rubio, Chel-Guerrero, & Betancur-Ancona, 2014). Aspartic acid stimulates the hormonal regulation for the proper functionality of the nervous system (Brosnan & Brosnan, 2013), while other amino acids like arginine apparently protect against cardiovascular diseases (Böger, 2007). Sulfur amino acids may be involved in the functionality of the tertiary and also quaternary structures of the proteins. These results suggest that the quality of chia proteins and their amino acids is similar or higher than other important cereals and oilseeds. The presence of these biomolecules in chia seeds represents an important nutraceutical contribution to the daily diet (Sandoval-Oliveros & Paredes-López, 2013).

Functional properties of proteins from chia have been explored. Hernandez-Jardon (2007) reported that the proteins have the ability to form elastic films when glycerol or sorbitol is incorporated as plasticizer, with a molecular weight between 4 kDa and 50 kDa. Mechanical properties reflect the advantages of the films: they showed good properties of tensile strength, elastic modulus, and strength in comparison with synthetic films. The films obtained from chia protein presented very low permeability to oxygen (16.27 cm^3 μm/m^2 d kPa) compared with other films made with synthetic (LDPE 57.85 cm^3 μm/m^2 d kPa) and natural macromolecules (concentrate of whey protein 31.3 cm^3 μm/m^2 d kPa) (Alvarado-Suárez, 2008). This property is of paramount importance, since oxygen is a key factor that might cause oxidation, thus, initiating several objectionable changes in food matrices such as odor, color, flavor, and nutrient deterioration; therefore, obtaining films with proper barrier properties against oxygen can help to improve food quality and to extend the shelf life of foods.

Furthermore, differential scanning calorimetry (DSC) has also been used to study the thermal properties of the protein fractions of chia. Olivos-Lugo, Valdivia-López, & Tecante (2010) reported on a DSC study of the main protein fractions and a protein isolate of chia seed from the state of Jalisco, Mexico. Globulins exhibited a denaturation temperature (Td) of 125°C, which makes them the fraction with the greatest thermal stability. This value might indicate the presence of a high number of hydrophobic interactions leading to an increase in enthalpy change of denaturation (ΔHd) because this is a process with significant energy absorption. Glutelins and globulins behaved in a similar way, which confirms that their structure is mainly stabilized by hydrophobic interactions as reported by Del Angel, Martínez, & López (2003). Albumins denatured over a temperature range of approximately 40°C, starting at 61°C and ending at 101°C, while for prolamins, globulins, and glutelins the interval was approximately 20°C. When denaturation occurs over a narrow temperature range, the transition is considered highly cooperative (Bove, Ma, & Harwalkar, 1997). On the other side, Sandoval-Oliveros & Paredes-López (2013) and Paredes-Lopez (1991) report denaturation temperatures of 103°C, 105°C, 85.6°C, and 91°C for crude albumins, globulins, prolamins, and glutelins, respectively, and concluded that albumins and globulins have relatively good thermal stability. Selected globulin peptides characterized by mass spectrometry showed to be homologous to sesame proteins.

14.4.2 Fibers

Fiber is one of the most important components of healthy diet. Intake of adequate amount of dietary fiber is associated with the prevention of cardiovascular diseases like stroke, myocardial infarction, vascular diseases, obesity, hypertension, hyperglycemia, and hyperlipidemia. Dietary fibers cannot be digested and absorbed by the small intestine but get fermented in the large intestine. On the basis of their physiochemical properties and functions, dietary fibers are classified in two forms—insoluble fibers, which exhibit bulking action, and soluble fibers which become fermented partially/completely in the colon (Suri, Passi, & Manchanda [2015] and Anderson, Baird, Davis, Ferreri, Knudtson, & Koraym [2009]). The total dietary fiber content of chia seeds ranges from 36–40 g per 100 g, which is much higher than that present in several grains, vegetables, and fruits such as corns, carrot, spinach, banana, pear, apple, kiwi (Ovando, Rubio, Guerrero, & Ancona, 2009). The content of fiber in chia seeds depends on the region of cultivation and climate. Chia seeds contain about twice as much fiber as bran and 4–5 times more than almonds, soy, quinoa, or amaranth (Pizarro, Almeida, Coelho, Samman, Hubinger, & Chang, 2014). Insoluble and soluble dietary fiber varies from 23–46% and 2.5–7.1%, respectively (Coates & Ayerza, 2009).The American Dietetic Association established the preferable ratio between insoluble and soluble dietary fiber fractions at 3:1 (Ayerza & Coates, 2011).

Chia contains about 5% mucilage, which can also act as soluble fiber. Chia mucilage has approximately 48% total sugar content, 4% protein, 8% ash, and 1% fat. The fiber is a polysaccharide with a high molecular weight, and its basic structure is a tetrasaccharide with 4-O-methyl-a-D-glucoronopyranosyl residues that branch b-D-xylopyranosyl on the main chain structure. The monosaccharide composition is 16% D-xylose1D-mannose, 2% D-arabinose, 6% D-glucose, 3% galacturonic acid, and 12% glucuronic acid. The insoluble dietary fiber of chia is capable of retaining water several times of its weight during hydration, providing bulk and prolonging gastrointestinal transit time. Increased gastrointestinal time is directly related to gradual increase in postprandial blood glucose levels and decrease in insulin resistance over a period of time (Muñoz, Cobos, Díaz, & Aguilera, 2012). Considering the recommendations for dietary fiber intake for adults (25–30 g/day), consumption of 15 g chia seed per day receives approximately 20% of daily fiber intake. Therefore, chia seed could contribute to health particularly in its fiber type and content.

14.4.3 Vitamins and Minerals

14.4.3.1 Vitamins

Vitamins are necessary for normal function of an organism. An adequate supply of these elements enables optimal control of the amount of hormones, growth regulators, and differentiation of cells and tissues. It also protects the organism from oxidative stress. Vitamins cannot be synthesized by the body and thus necessary to consume in diet from natural foods (Rendón-Villalobos, Ortíz-Sánchez, Solorza-Feria, & Trujillo-Hernández, 2012). Chia seed is a good source of B-complex vitamins, such as B6 with 0.1 mg/100 g, thiamine with 0.7 mg/100 g, riboflavin with 0.2 mg/100 g, and niacin with 7.2 mg/100 g of seed. Vitamin C is also present with 5.4 mg/100 g of seed (Craig, 2003). Silva 2014) reported the average content of vitamin E in chia of 8.1 mg/100 g, with the appreciable quantities of γ-tocopherol.

14.4.3.2 Minerals

Minerals, like vitamins, are not synthesized by the body but are necessary for maintaining the body in optimal health. It is therefore necessary to use external sources such as food, nutritional supplements, and absorption through the skin, to ensure an adequate supply of minerals (Rendón-Villalobos, Ortíz-Sánchez, Solorza-Feria, & Trujillo-Hernández, 2012). Chia seed's mineral composition varies between:

103–260 mg/100 g for sodium; 726–826 mg/100 g for potassium; 456–590 mg/100 g for calcium; 9–12 mg/100 g for iron; 77–449 mg/100 g for magnesium; 604–919 mg/100 g for phosphorus; Zn, 5–6.5 mg/100 g for zinc; 1.7–1.9 mg/100 g for copper and 1.4–3.8 mg/100 g for manganese in different studies (Jin, Nieman, Sha, Xie, Qiu, & Jia, 2012). It's observed that chia seeds have more calcium, phosphorus, magnesium, zinc, iron, and copper than many seeds. However, fiber components of chia seed can affect bioavailability of these minerals (Capitani, Spotorno, Nolasco, & Tomás, 2012). It is claimed that together with its alpha-linolenic acid (ALA) and fiber composition, chia seed's high mineral content provides beneficial effects on health (Chicco, D'Alessandro, Hein, Oliva, & Lombardo, 2008). On the other hand, mineral composition of seed oil extract can vary depending on extraction method. Metals, especially copper and iron are not desired in oils because of their pro-oxidant nature. This is why chia seed oils obtained by pressing or solvent extraction may/can be preferred (Ixtaina, Martínez, Spotorno, Mateo, Maestri, Diehl, et al., 2011).

14.4.4 Fatty Acid Composition

Lipids are bioactive substances, which the human body needs to accumulate energy, form structural elements of cell membranes, and regulate physiological functions. If there are no enzymatic systems capable of forming double bonds at positions n-3 and n-6, the organism cannot synthesize fatty acids, such as omega-3 (ω-3) alpha-linolenic acid and ω-6 alpha linoleic acid. Therefore, it is necessary to provide the organism with a supply of lipids in food. Chia seeds contain 25–40% of fat, most of which is in the form of polyunsaturated fatty acids, such as ω-3 alpha-linolenic acid and ω-6 alpha-linoleic acid (Marineli, Lenquiste, Moraes, & Marostica, 2015).The yield of total lipids is directly related to the method used, as demonstrated in the work by Ixtaina, Nolasco, & Tomas (2008) where the percentages found for the seeds studied ranged from 26.70% to 33.60% and from 20.30% to 24.80%, in the extraction with n-hexane solvent and pressure, respectively. The use of a nonpolar solvent, such as n-hexane, can lead to greater extraction of total lipids (Tanamati, Oliveira, Visentainer, Matsushita, & de Souza, 2005). The most important characteristic of chia is its high content of polyunsaturated fatty acids (PUFAs). The seed has around 25–40% fat that comprises 55–60% linolenic acid (ω-3) and 18–20% linoleic acid (ω-6) (Table 14.2), which are essential fatty acids for good health (Ayerza & Coates, 2011). Chia contains the highest percentage of any plant source of ALA. This fatty acid is the precursor for the long-chain PUFAs considered as essential fatty acids because the human body cannot produce them. Chia seeds contain high concentrations of PUFAs that provide potent lipid antioxidants (Ayerza & Coates, 2011; Marineli, Lenquiste, Moraes, & Marostica, 2015). Chia seed is more suitable due to its n-6/n-3 rate rather than another ALA source. Nevertheless it should be remembered that n-6/n-3 rate of chia oil is unbalanced. This doesn't mean to consume it as the only oil source in diet, rather, "since the extent to which ALA is converted into physiologically essential long chain n-3 FA, in particular eicosapentaenoic

TABLE 14.2

Fatty Acid Composition of Chia Seeds (*Salvia hispanica L.*)

Saturated Fatty Acids(g/100 g)		Monounsaturated Fatty Acids(g/100 g)		Polyunsaturated Fatty Acids(g/100 g)	
Myristic acid	0.030 g	Myristoleic acid	0.030 g	Linoleic acid	5.835 g
Pentadecylic acid	0.030 g	Palmitoleic acid	0.029 g	α-Linolenic acid	17.830 g
Palmitic acid	2.170 g	Vaccenic acid	2.203 g	Total Polyunsaturated	23.665 g
Margaric acid	0.063 g	Paullinic acid	0.046 g	fatty acids (g/100 g)	
Stearic acid	0.912 g	Total Monounsaturated	2.309 g		
Arachidic acid	0.093 g	fatty acids (g/100 g)			
Behenic acid	0.032 g				
Total Saturated fatty acids (g/100 g)	3.330 g				

Source: USDA national Nutrient Database for Standard Reference Release, 2011.

acid (EPA, 20:5 n–3) and docosahexaenoic acid (DHA, 22:6 n–3), is very limited" (Schuchardt & Hahn, 2013), chia and other seeds can be a good substitute for people who don't eat animal omega-3 sources. Furthermore, a higher content of ALA and LA (lenoleic acid) results in low oxidative stability and shorter shelf life of the oil. However, in the dark, the combined addition of ascorbyl palmitate and tocopherol significantly reduced lipid oxidation and improved oil shelf life (at least up to 300 storage days) (Bodoira, Penci, Ribotta, & Martínez, 2017). The addition of the different antioxidants increased the induction time of chia seed oil. The best effects were recorded in chia seed oil with the addition of ascorbyl palmitate, rosemary, and its blend with green tea extract (Ixtaina, et al., 2011). Besides adding antioxidants, there is new alternative to improve the chia oil stability, like oil in water emulsions, as a delivery system (Julio, Ixtaina, Fernández, Sánchez, Wagner, Nolasco, et al., 2015).

14.4.5 Total Polyphenolic Contents and Their Antioxidant Activity

Among the diversity of compounds that can be present in all the varieties of chia seed, antioxidants are without a doubt some of the most important. Antioxidants present in chia are of phenolic nature and can be in free form or bonded to sugars by glycosidic linkages, which increases their solubility in water. The identity of the main antioxidants as well as their antioxidant activity is well documented in the literature. The most important phenolic compounds include chlorogenic and caffeic acids, and the flavanols myricetin, quercetin, and kaempferol (Ayerza & Coates, 2011). The analyzed extracts of chia seeds did not contain the anthocyanin group. In both chia seeds, i.e., Jalisco and Sinaloa, the flavonol group is present in the highest amount: Sinaloa seeds contain 0.590 mg/mL crude extract and 0.650 mg/mL hydrolyzed extract, while Jalisco seeds contain 0.379 mg/mL crude extract and 0.427 mg/mL hydrolyzed extract. It is more likely that the main phenolic compounds of crude extracts are a mixture of glycosides of quercetin and kaempferol, in which the aglycon form of quercetin and kaempferolin predominates in the hydrolyzed extracts. Low amounts of chlorogenic and caffeic acids are present only in the nonhydrolyzed crude extracts, probably because these acids are sensitive to hydrolysis conditions. The antioxidant activity of chia extracts was evaluated screening the antioxidant activity of 2,20-azino-bis-3-ethylbenzothiazoline-6-sulphonic acid radical cation (ABTS+ radical), β-carotene linoleic acid model system (β-CLAMS), and *in vitro* liposome peroxidation system assays.

The results show that crude extracts have an antioxidant activity similar to the commercial antioxidant Trolox used as a reference (Reyes-Caudillo, Tecante, & Valdivia-Lopez, 2008). In the same work, the authors reported for the first time that different concentration of polyphenols from the hydrolyzed and crude chia seed extracts showed antioxidant effect in a model water-in-oil food emulsion. In a more recent study (Martínez-Cruz & Paredes-López, 2014), chia seed (*S. hispanica L.*) was analyzed for total phenolic compounds, antioxidant activity, and quantification of phenolic acids and isoflavones by ultra high performance liquid chromatography. The phenolic compounds identified and quantified in the chia extracts were rosmarinic, protocatechuic, caffeic, and gallic acids, and daidzin. Additionally, ferulic acid, glycitin, genistin, glycitein, and genistein were also detected. Among the reported compounds, rosmarinic acid was the most abundant antioxidant (0.9267 mg/g). In summary, this study shows chia seed to have a high antioxidant capacity and represents a novel isoflavone source that could be incorporated into human diet. Chia seeds have numerous antioxidant compounds, such as vitamins, polyphenols, and peptides. These compounds can inhibit the activation of the NF-κB transcription factor *in vitro*, thus reducing the inflammatory and carcinogenic processes (Aggarwal & Shishodia, 2006; Ellulu, 2017) and protecting against the attack of reactive oxygen species (ROS) or nitrogen (Kampa, Nistikaki, Tsaousis, Maliaraki, Notas, & Castanas, 2002).

These antioxidant actions can protect the organism from pathologies, like neurological diseases, inflammation, immunodeficiency, ischemic heart disease, strokes, Alzheimer's and Parkinson's diseases, and cancer (Marcinek & Krejpcio, 2017). It has been demonstrated that rats fed a high-fat diet including chia seeds for 6 weeks or 12 weeks experienced a decrease of thiol levels and plasma catalase and glutathione peroxidase activities, while liver levels of the glutathione reductase became enhanced (Marineli, Lenquiste, Moraes, & Marostica, 2015). Rats that received a long-term sucrose-rich diet and were fed chia seeds, returned to the same activities of antioxidant enzymes catalase, superoxide dismutase, and glutathione reductase as control values (de Souza Ferreira, de Sousa Fomes, Santo da Silva, & Rosa, 2015).

In addition, increases in superoxide dismutase and IL-10 plasma concentrations were observed when Wistar rats consumed chia seed flour plus a high-fat diet for 35 days in comparison with a control group (calcium carbonate) (da Silva, Toledo, Grancieri, de Castro Moreira, Medina, Silva, et al., 2019). In healthy humans who had received chia seeds (12 weeks), a better plasma antioxidant activity was observed compared to hypertensive (Toscano, da Silva, Toscano, de Almeida, da Cruz Santos, Vázquez-Ovando, et al., 2010) or overweight patients (Nieman, Cayea, Austin, Henson, McAnulty, & Jin, 2009). It has been demonstrated that germinated chia showed increased protein quality, as measured by a protein efficiency ratio (PER). The amount of the gamma-aminobutyric acid (GABA), total phenolic content, and antioxidant activity increased even more in the flour of germinated seeds (Gómez-Favela, Gutiérrez-Dorado, Cuevas-Rodríguez, Canizalez-Román, del Rosario León-Sicairos, Milán-Carrillo, et al., 2017), as well as in normal chia flour (Grancieri, Martino, & Gonzalez de Mejia, 2019; Martínez-Cruz & Paredes-López, 2014). The albumin and globulin fractions showed a high antiradical activity against 2,2-diphenyl-1-picryl-hydrazyl-hydrate (DPPH) and prolamin as well as globulin ability to chelate ferrous ions (KnezHrnčič, Ivanovski, Cör, & Knez, 2020).

14.5 Therapeutic Aspects of Chia

The therapeutic potential of chia in terms of its use as nutraceutical and pharmaceutical is barely known to the world. Scientific studies have proven the effectiveness of chia in disease such as hypertension, dyslipidemia, diabetes, anxiety, weak immunity, depression, etc. These therapeutic benefits are attributed by the presence of bioactive and nutritional substances. Use of chia (*Salvia hispanica*) as nutraceutical and dietary supplement are reported by several researchers (De Falco, Amato, & Lanzotti, 2017; Marcinek & Krejpcio, 2017). Various health benefits studied by these authors includes prevention of myocardial infraction and strokes, inhibition of blood clotting, cholesterol reduction, epilepsy and stress prevention, blood glucose stabilization, blood pressure lowering, antineoplastic, laxative and analgesic properties, and higher content of omega-3, iron, fiber, calcium, and magnesium. Various studies concluded that more research is needed to prove safety of chia.

14.5.1 Antiproliferative Effects

Foods containing omega-3 and omega-6 fatty acids are proven to have antiproliferative effects and, thus, retard the growth of tumor cells (Bishnoi, 2017). Rosas-Ramírez, Fragoso-Serrano, Escandón-Rivera, Vargas-Ramírez, Reyes-Grajeda, and Soriano-García (2017) studied the effect of administration of chia seed to Wistar rats and found significant decrease in tumor weight by increasing eicosapentaenoic acid (EPA), which have direct apoptotic effect by encouraging rate of programmed cell death. The study depicted the beneficial effects of chia on proliferation of HeLa, HepG2, and MCF7 cancerous cells.

14.5.2 Cardioprotective Effects

Cardio protective effects of chia was studied by Ayerza & Coates (2007) and exhibited its effectiveness due to presence of omega-3 fatty acids, which prevent hypertension by blocking calcium and sodium channel abnormalities and improving parasympathetic tone, heart rate variability, and ventricular arrhythmia protection. Presence of ALA and EPA in the diet is also important for numerous physiological functions (Pawlosky, Hibbeln, Lin, & Salem, 2003). Ayerza & Coates (2007) depicted the cardioprotective effects of feeding chia seeds to male Wistar rats by increasing levels of HDL cholesterol and lowering the omega-6 to omega-3 ratio. The rats suffering from dyslipidemia were fed chia seed in their diet, also resulted in decreased visceral adiposity and decreased LDL cholesterol levels. The feeding of 50 g/day chia seed to 12 healthy subjects for 30 days was found to be significant for decreasing blood pressure without any side effect (Vertommen, Van de Sompel, Loenders, Van der Velpen, & De Leeuw, 2005). In the study by Rahman, Nadeem, Ahmad, et al. (2015), the addition of olein fraction from chia oil in different factions in milk fat were found to possess antioxidant potential by improved the DPPH free-radical scavenging activity.

Antioxidant activity along with angiotensin-converting enzyme inhibitory activity is shown by low molecular-weight peptides, which are produced by hydrolysis of chia (Segura-Campos, Salazar-Vega, Chel-Guerrero, & Betancur-Ancona, 2013). A comparative study to check inhibitory effect on angiotensin-converting enzyme by *Salvia hispanica* L., *Phaseoluslunatus* L. and *Phaseolus vulgaris* L. was conducted by Salazar-Vega, Segura-Campos, Chel-Guerrero, et al. (2012) and resulted that protein hydrolysates from chia seeds had greater antioxidant potential that ultimately proved their antihypertensive activity. High antiradical activity against 2,2′-azinobis (3-ethylbenzothiazoline6-sulfonic acid), 2,2-diphenyl-1-picrylhydrazyl, and angiotensin-converting enzyme activity by chia was also reported by (Orona-Tamayo, Valverde, Nieto-Rendon, & Paredes-Lopez, 2015).

14.5.3 Diabetes

Vuksan, Choleva, Jovanovski, Jenkins, Au-Yeung, Dias, et al. (2017) found the effectiveness of chia for type 2 diabetes by virtue of increased weight loss in the subjects. Chia seeds were found effective for satiety and thus for obese patients as well (Jenkins, Brissette, Jovanovski, Au-Yeung, Ho, Zurbau, et al., 2016). Ingestion of 37 g chia seeds per day for 12 weeks was found to be significant in reduction of systolic blood pressure and C-reactive protein concentration in blood plasma in 20 type 2 diabetes patients. It was found that there was double increase of ALA and EPA in the plasma of subjects with chia seeds. A significant reduction of triacylglycerols, C-reactive protein concentrations, and insulin resistance in group with chia-based diet was reported by (Guevara-Cruz, Tovar, Aguilar-Salinas, Medina-Vera, Gil-Zenteno, Hernández-Viveros, et al., 2012). The problem of strokes and heart attacks are more prominent in type 2 patients and their prevention by use of chia seeds are attributed to their anticoagulant and anti-inflammatory properties (Vuksan, Whitham, Sievenpiper, Jenkins, Rogovik, Bazinet, et al., 2007). However Nieman, Cayea, Austin, Henson, McAnulty, & Jin (2009) conducted study on 76 adults for 12 weeks and found no significant reduction in body weight, blood pressure, and blood sugar levels. Presence of active ingredients in chia seeds provides numerous health benefits but medical efficiency and their safety is still needs further scientific research.

14.5.4 Effect on the Immune System

Studies suggested that chia possess significant immune stimulant activity. The thymus weight and serum IgE concentration as an indicator of immunity stimulant activity was studied on 23-days-old weaning male Wistar rats (Fernandez, Vidueiros, Ayerza, Coates, & Pallaro, 2008). The results of the study depicted that administration of chia in any form to diet of subjects resulted in significant increase in IgE concentration. Usually flaxseed and marine products are used for their immune stimulant activity, but they result in fishy flavor, diarrhea, and gastrointestinal-tract problems (Yadav, Khatak, Singhania, & Bishnoi, 2020). These problems were not reported with chia, thus, giving it the upper hand as an ingredient for boosting immunity of individual (Ayerza & Coates, 2007).

14.5.5 Antiobesity Activities

Researchers had investigated the antiobesity effects of chia on animals, including the effects of chia seed and its oil on obese rats. Improvement in insulin and glucose tolerance, cardiac and hepatic fibrosis, inflammation and stearoyl-CoA 9-desaturase (SCD) activity inhibition was reported with the use of chia. It is proven that SCD inhibition eventually protects from murine obesity, cellular lipid accumulation, and insulin resistance (Poudyal, Panchal, Waanders, Ward, & Brown, 2012).

In a similar study by Marineli, Moura, Moraes, Lenquiste, Lollo, Morato, et al. (2015), chia was found to be associated to aid glucose and insulin tolerance in obese rats. In the study, subjects in one group were fed a high-fructose diet with inclusion of soybean oil, lard, and fructose and in other groups, chia seeds and oil were fed. The benefits of feeding chia seed was found in terms of reduced plasma thiobarbituric acid reactive substances, reduced plasma 8-isoprostane, and increased plasma glutathione levels in comparison to the high-fructose group. Inclusion of chia in the diet of Wistar rats was found to improve 18:0 to 16:0 fatty acid hepatic ratios and decreases SCD-1 index in the Fortino, Oliva, Rodriguez, Lombardo,

& Chicco (2017), but the study by Guevara-Cruz, et al. (2012) found no significant difference in cholesterol (both LDL and HDL) in patients with metabolic syndrome.

14.5.6 Angiotensin I-converting Enzyme Inhibitory Peptides of Chia

There is always a search for natural bioactive peptides that may prevent humans from side effects of angiotensin I-converting enzymes (ACE-I). The hydrolysate and ultra-filtered fractions of chia were found to have 58.46% and 69.31% ACE-I inhibition by Segura-Campos, Ciau-Solís, Rosado-Rubio, Chel-Guerrero, & Betancur-Ancona (2014). The gel filtration chromatography of chia amino acids showed the ACE-I inhibition ability of chia by obstructing the angiotensin II generation. Hydrolysed chia proteins have been used in white bread and carrot cream and have shown significant ACE-I inhibition potential (Suri, Passi, & Manchanda, 2015).

14.5.7 Other Therapeutic Aspects of Chia

Chia has been used in skin curative formulation (4% chia oil) and used by patients with pruritus affected by end stage renal disease and healthy volunteers having xerotic pruritus. Use for 8 weeks significantly improved the skin hydration, lichen simplex chronicus, and prurigo nodularis in all the patients (Jeong, Park, Park, & Hwan Kim, 2010).

The leaves of chia can be used as insect repellent because it contains an essential oil that contains β-caryophyllene, globulol, γ-muroleno, β-pinene, α-humoleno, germacren-B, and widdrol having strong insect repellent characteristics (Ullah, Nadeem, & Imran, 2017). From recent years, chia seeds are being used in a number of food products including biscuits, pasta, yogurt, cake, and other bakery and confectionary products (Borneo, Aguirre, & León, 2010). Chia oil is also being used in butter oil for enhancement of omega-3 fatty acids (Nadeem, Ajmal, Rahman, & Ayaz, 2015) and in mucilage as functional coating for improving functional properties (Munoz, Aguilera, Rodriguez-Turienzo, Cobos, & Diaz, 2012).

The allergens and other safety issues associated with use of chia are a matter of research. The extracts from commercial products of chia were analyzed by SDS-PAGE and immunoblot and identification of proteins by MS/MS to determine potential allergens such as lectin, elongation factor and 11S globulin. There is potential for an allergic reaction from chia seed and its oil through an IgE reaction (Jimenez, Pastor Vargas, De las Heras, Sanz Maroto, Vivanco, & Sastre, 2015).

14.6 Efficient Uses of Chia

Chia can be incorporated after processing in the form of flour, roasted, or after soaking and also after extraction of oil, the oil cake and oil can be utilized further. Various forms of chia have been used as a nutrient, blended with other foods and beverages. There are enormous studies suggesting that consumption of chia seed and its products can serve as an important alternative to improve human health and subsequently act as functional food in their diet.

14.6.1 Applications in the Food Industry

The excellent water holding capacity of chia seed—27 times its own weight—makes it feasible to use in varied food preparations (Muñoz, Cobos, Díaz, & Aguilera, 2012). Numerous studies demonstrate the emulsification ability of chia seeds. The oil retention and water holding capacities of chia seed have been found to be more than usual commercial thickeners; the added feature of carrying functional properties also gets incorporated into the food when used as whole chia seed or flour (Falco, Amato, & Lanzotti, 2017). Chia mucilage is readily utilized as a foam stabilizer, emulsifier, or binder in the food industry (Muñoz, Cobos, Díaz, & Aguilera, 2012). Chia gum can be used as a flavoring in the food industry, increasing the sensation of food flavor in the mouth due to the presence of appropriate fat absorption capacity. This gum remains stable even at elevated temperature up to 244°C, which indicates the promising property of chia seed gum to incorporate in high-value food formulations (Timilsena, Adhikari,

Kasapis, & Adhikari, 2016). Chia gum generally contains 26.2% fat and two fractions can be achieved under fat extraction in the form of gum with fat (FCG) and gum partially defatted (PDCG). The PDCG can be added in yogurt, sauces, and pastries by virtue of its high viscosity, while FCG can be used to emulsify and stabilize sauces, mayonnaise, and various meat products (Segura-Campos, Ciau-Solís, Rosado-Rubio, Chel-Guerrero, & Betancur-Ancona, 2014).

Good textural properties are important features of bakery products, which requires the ingredient characteristics like water holding, absorption, and emulsion capacity. When the other fiber sources as soybean, wheat, wheat hulls, and maze are compared with the fiber-rich fractions of chia, chia provides higher water holding, emulsifying activity (53.26 mL/100 mL), emulsion stability and absorption. Therefore, chia can be easily used in bread, sweets, desserts, and cookies (Capitani, Spotorno, Nolasco, & Tomás, 2012; Vázquez-Ovando, Rosado-Rubio, Chel-Guerrero, & Betancur-Ancona, 2009). Coelho & Myriam de las Mercedes (2015) produced two types of broth during a study (1) by adding 7.8% chia flour with 0.9% fat and (2) 11% chia flour with 1% fat, both resulted in the reduced loaf volume of bread, presumably due to the gluten-free composition of chia and weak interactions of wheat flour proteins (gliadin and glutenin) with chia fiber. The addition of chia flour boosted the nutritional profile of the bread by increased ash content, fiber, and lipid profile and reduced carbohydrate content. Sensory attributes of chia in different bakery products also have been studied in which most of the panelists were observed as a "relatively lover" or "very lover" for the breads with chia seeds and 83% of panelist remarked to purchase this product. Chia seeds are rich in pulp content, thus the breads fortified with chia seeds enhance the feeling of satiety and act as a prebiotic. The use of whole chia seeds with oat wheat flour and chia seeds with oat bran by replacing wheat flour maintained the textural quality of cookies. Further, bakery application with chia and oat combination can subsidize the nutritional value of resulted bakery products as well (Inglett, Chen, & Liu, 2014). Another study demonstrated the role of chia mucilage as a fat replacer in foods. During cake making, fat was replaced with chia mucilage up to 25%, which ultimately perpetuated the nutritional composition with sensory and textural properties of the cake (Felisberto, Wahanik, Gomes-Ruffi, Clerici, Chang, & Steelet al., 2015). The nutritional composition of chia seeds in terms of low carbohydrate, high protein, and insoluble fiber contents accounts for its suitability in bakery products for diabetic and obese patients. Patients with gluten sensitivity, celiac disease, or other wheat related sensitivities can easily incorporate chia in their diets being gluten-free (Dinçoğlu & Yeşildemir, 2019).

Chia seed oil also finds its application in sports drinks, corn flakes, chocolate, and meat products (Falco, Amato, & Lanzotti, 2017). Chia mucilage found its usage in ice cream production as an emulsifier and stabilizer and rendered productive influence on the ice cream quality, but caused darkness in ice-cream color (Campos, Ruivo, Scapim, Madrona, & Bergamasco, 2016). In the commercial production of marmalades and sauces, chia seeds can be incorporated due to fine gelling properties of chia seeds. In an investigation by Özbek, Şahin-Yeşilçubuk, and Demirel (2019), strawberry marmalade was prepared by fortifying 5% chia seeds, which resulted in the increased phenolic components (15.45%) and diet fiber content (168%) and decreased energy value (48%). Consequently, this strawberry marmalade with added chia seed was sugar-free and can be designated as a new functional food due to its high antioxidant properties, dietary fiber content, and its n-3 values.

14.6.2 Use in Edible Film Production

Hydrocolloids of chia mucilage can be processed into an edible film very conveniently. A novel edible film was prepared that was based on chia mucilage hydrocolloid (1% w/v) with different concentrations of glycerol as a plasticizer. This edible film can protect the food product being transparent in appearance and exhibited excellent absorption of ultraviolet light, fair thermal stability, and high water solubility (Dick, Costa, Gomaa, Subirade, Rios, & Flôres, 2015).

14.7 Negative Effects and Safe Consumption Levels

In 2009, European Parliament and Council of Europe gave the approval to chia seeds as a Novel Food. Up to now, any sort of investigation related to chia seeds consumption in any form did not reveal any adverse toxic effects, allergic, or antinutritional effects. The regulations (2011) under the food safety and standards act (2006) do not administer any information regarding the safe consumption level or prohibition of chia seeds. Hence, chia seeds may be contemplated as safe for human consumption and have an assuring future as a functional food (Capitani, Spotorno, Nolasco, & Tomás, 2012; Petruzzello, 2020). As per the US Dietary Guidelines, chia seeds can be taken as crucial edibles, therefore consumption should be in limits, not surpassing 48 g/day. However, subsequent clinical trials on animals and humans are needed in large numbers to discover the safety aspects of chia seeds.

14.8 Future Perspectives

The recent increased interest in functional foods has steered people toward foods with significant health benefits like chia. Therefore, scientists are also taking it as a subject of many studies in relation to functionality, processing technologies, usage, and health well-being. Chia is now becoming part of the novel foodstuffs with health-promoting benefits. Likewise, the rich amount of fiber content of chia seeds allows its recommendation for diabetic people and patients of hypercholesterolaemia. The supplementation of chia in the daily diet is important by virtue of its high content of omega-3 fatty acids. An enormous usage of chia seeds is proposed for a diverse group of dishes because of its range of physicochemical and technofunctional qualities. Chia produces attractive looking gels when soaked with water and the seeds can be used in the form of flour or whole seeds. Presently, chia seeds find their usefulness worldwide in cereal products, like breakfast cereals, nutri-cereal bars, wafers, chips, rice crisps, cookies, breads, cakes, noodles, pasta, and in many Indian traditional sweets. The utilization of chia seeds to produce different kinds of dairy products, fruit and vegetables products, or meat stuffs contains great perspectives since many culinary recipes can be prepared with chia seeds. However, the industrial application of chia seeds is hindered by the mentioned legal status of the seed in European countries as a novel food ingredient, which means that a food business operator has to obtain the official premarket approval for every chia product not mentioned in the European Union list of authorized novel foods and novel food ingredients. This regulation is established as an effect of Regulation (EU) 2015/2283 of the European Parliament and regulation on novel foods of the Council (November 25, 2015), amending Regulation (EU) No 1169/2011 of the European Parliament and of the Council and repealing Regulation (EC) No 258/97 of the European Parliament and of the Council and Commission Regulation (EC) No 1852/2001 (OJ 327, 11.12.2015, p.1).

Chia seeds can be the key component of health-improving food by virtue of abundant nutritional and nutraceutical value including the biological and technological potential. However, still wide-ranging human clinical trial-based studies need to be performed to define its safety, mechanisms of action, and true potential.

14.9 Conclusions

Chia seeds have been part of humankind since pre-Colombian time by the Aztecs as a part of foodstuff, medicine, and religious ceremonies. Chia seeds are abundantly rich in dietary fiber (insoluble and soluble), essential amino acids, omega-3 fatty acids, and bioactive compounds. Chia have been considered to play many vital physiochemical and health promoting properties, which is evident in its suitability in the food industry. Chia produces important rheological and technofunctional properties being good thickening agent, gel former, chelating agent, foam enhancer, emulsifier, clarifying agent, stabilizer, and rehydrating agent. Such properties can be utilized at the commercial level for new product development enriched with vital nutrients like omega-3, protein, soluble-insoluble fiber, and phytochemicals. Chia seeds can be fortified in the production of frozen food products, baked products, beverages, confectionary products, baby foods, pasta, sausages, etc. Chia seeds may contribute in the prevention, treatment, and management of numerous health-related problems, and likewise in immunity enhancement and cardiovascular diseases. Chia also helps balance blood glucose levels through its ability to slow down carbohydrate digestion due to the presence of dietary fiber. Researches (*in vivo* and *in vitro*) reveal its safety for human consumption in certain limited amounts that deliver enormous health benefits. However, there is still a huge scope of research left to be done on chia seeds with regard to the food industrial applications and nutraceutical properties. The long lists of chia functionalities designate it as a functional food that can boost the health of masses at multitudinal levels.

REFERENCES

Aggarwal, B. B., & Shishodia, S. (2006). Molecular targets of dietary agents for prevention and therapy of cancer. *Biochemical Pharmacology, 71*, 1397–1421.

Alvarado-Suaˊrez, L. A. (2008). *Estudiosobrelaspropiedadesmecaˊnicas y funcionales de biopeliˊculasformadas con proteiˊna de semilla de limoˊnmexicano (Citrus aurantifolia)*. Meˊxico: Facultad de Quiˊmica, UNAM.

Anderson, J. W., Baird, P., Davis, R. H., Ferreri, S., Knudtson, M., & Koraym, A. (2009). Health benefits of dietary fiber. *Nutrition Reviews, 67*(4), 188–205.

Armstrong, D. (2003). Application for approval of whole chia (*Salvia hispanica L*) seed and ground whole chia as novel food ingredients. Northern Ireland.

Ayerza, R., & Coates, W. (2011). Protein content, oil content and fatty acid profiles as potential criteria to determine the origin of commercially grown chia (*Salvia hispanica L.*). *Journal of Industrial Crops and Products 34*, 1366–1371.

Ayerza, R. J., & Coates, W. (2007). Seed yield, oil content and fatty acid composition of three botanical sources of ω-3 fatty acid planted in the Yungas ecosystem of tropical Argentina. *Tropical Science, 47*(4), 183–187.

Berner, L. A., & O'Donnell, J. A. (1998). Functional foods and health claims legislation: Applications to dairy foods. *International Dairy Journal, 8*(5/6), 355–562.

Bishnoi, S. (2017). Herbs as Functional Foods. Functional Foods: Sources and Health Benefits. Scientific Publishers, Jodhpur, 141–172.

Bodoira, R. M., Penci, M. C., Ribotta, P. D., & Martínez, M. L. (2017). Chia (*Salvia hispanica L.*) oil stability: Study of the effect of natural antioxidants. *LWT—Food Science and Technology, 75*, 107–113.

Böger, R. H. (2007). The pharmacodynamics of L-arginine. *Journal of Nutrition, 137*, 1650–1655.

Borneo, R., Aguirre, A., & León, A. E. (2010). Chia (*Salvia hispanica L*) gel can be used as egg or oil replacer in cake formulations. *Journal of the American Dietetic Association, 110*(6), 946–949.

Bove, J. I., Ma, C. Y., & Harwalkar, V. R. (1997). Coagulation of proteins. *Food Proteins and Their Applications, 80*(25).

Brosnan, J. T., & Brosnan, M. E. (2013). Glutamate: A truly functional amino acid. *Amino Acids, 45*, 413–418.

Cahill, J. P. (2003). Ethnobotany of chia, *Salvia hispanica L.* (Lamiaceae). *Economic Botany, 57*(4), 604–618.

Campos, B. E., Ruivo, T., Scapim, M. S., Madrona, G. S., & Bergamasco, R. D. C. (2016). Optimization of the mucilage extraction process from chia seeds and application in ice cream as a stabilizer and emulsifier. *LWT—Food Science and Technology, 65*, 874–883.

Capitani, M. I., Spotorno, V., Nolasco, S., & Tomás, M. (2012). Physicochemical and functional characterization of by-products from chia (*Salvia hispanica L.*) seeds of Argentina. *LWT—Food Science and Technology, 45*(1), 94–102.

Chicco, A. G., D'Alessandro, M. E., Hein, G. J., Oliva, M. E., & Lombardo, Y. B. (2008). Dietary chia seed (*Salvia hispanica L.*) rich in α-linolenic acid improves adiposity and normalises hyper triacylglycerolaemia and insulin resistance in dyslipaemic rats. *British Journal of Nutrition, 101*(41-50).

Coates, W., & Ayerza, R. (2009). Chia (Salvia hispanica L.) seeds as an omega-3 fatty acid source for feeding pigs: Effects on fatty acid composition and fat stability of the meat and internal fat, growth performance and meat sensory characteristics. *Journal of Animal Science, 87*, 3798–3804.

Coelho, M. S., & Myriam de las Mercedes, S.-M. (2015). Effects of substituting chia (*Salvia hispanica L.*) flour or seeds for wheat flour on the quality of the bread. *LWT—Food Science Technology, 60*(2), 729–736.

Craig, R. (2006). Application for approval of whole chia (Salvia hispanica L) seed and ground whole chia as novel food ingredients. Northern Ireland: Craig & Sons Ltd.

da Silva, B. P., Toledo, R. C. L., Grancieri, M., de Castro Moreira, M. E., Medina, N. R., Silva, R. R., & Martino, H. S. D. (2019). Effects of chia (*Salvia hispanica L.*) on calcium bioavailability and inflammation in Wistar rats. *Food Research International, 116*, 592–599.

De Falco, B., Amato, M., & Lanzotti, V. (2017). Chia seeds products: An overview. *Phytochemistry Reviews, 16*(4), 745–760.

de Souza Ferreira, C., de Sousa Fomes, L. D. F., Santo da Silva, G. E., & Rosa, G. (2015). Effect of chia seed (*Salvia hispanica L.*) consumption on cardiovascular risk factors in humans: A systematic review. *Nutricionhospitalaria, 32*, 1909–1918.

Del Angel, S. S., Martínez, E. M., & López, M. A. V. (2003). Study of denaturation of corn proteins during storage using differential scanning calorimetry. *Food Chemistry, 83*(4), 531–540.

Dick, M., Costa, T., Gomaa, A. I., Subirade, M., Rios, A., & Flôres, S. H. (2015). Edible film production from chia seed mucilage: Effect of glycerol concentration on its physicochemical and mechanical properties. *Carbohydrate Polymers, 130*, 198–205.

Dinçoğlu, A. H., & Yeşildemir, Ö. (2019). A renewable source as a functional food: Chia seed. *Nutrition and Food Science, 15*(4), 327–337.

Ellulu, M. S. (2017). Obesity, cardiovascular disease, and role of vitamin C on inflammation: A review of facts and underlying mechanisms. *Inflammopharmacology 25*(3), 313–328.

Falco, B. d., Amato, M., & Lanzotti, V. (2017). Chia seeds products: An overview. *Phytochemistry Reviews, 16*, 745–760.

Felisberto, M. H. F., Wahanik, A. L., Gomes-Ruffi, C. R., Clerici, M. T. P. S., Chang, Y., & Steel, C. (2015). Use of chia (Salvia hispanica L.) mucilage gel to reduce fat in pound cakes. *LWT—Food Science and Technology, 63*, 1049–1055.

Fernandez, I., Vidueiros, S. M., Ayerza, R., Coates, W., & Pallaro, A. (2008). Impact of chia (*Salvia hispanica L.*) on the immune system: Preliminary study. *Proceedings of the Nutrition Society, 67*.

Fernandez, I. R., Ayerza, W., Coates, S. M., Vidueiros, N., & Pallaro, A. N. (2006). Nutritional characteristics of chia. *Actualización en Nutrición, 7*, 23–25.

Fortino, M. A., Oliva, M. E., Rodriguez, S., Lombardo, Y. B., & Chicco, A. (2017). Could post weaning dietary chia seed mitigate the development of dyslipidemia, liver steatosis and altered glucose homeostasis in offspring exposed to a sucrose-rich diet from utero to adulthood? *Prostaglandins Leukotrienes and Essential Fatty Acids, 116*, 19–26.

Gómez-Favela, M. A., Gutiérrez-Dorado, R., Cuevas-Rodríguez, E. O., Canizalez-Román, V. A., del Rosario León-Sicairos, C., Milán-Carrillo, J., & Reyes-Moreno, C. (2017). Improvement of chia seeds with anti-oxidant activity, GABA, essential amino acids, and dietary fiber by controlled germination bioprocess. *Plant Foods for Human Nutrition, 72*(4), 345–352.

Grancieri, M., Martino, H. S. D., & Gonzalez de Mejia, E. (2019). Chia seed (*Salvia hispanica L.*) as a source of proteins and bioactive peptides with health benefits: A review. *Comprehensive Reviews in Food Science and Food Safety, 18*(2), 480–499.

Guevara-Cruz, M., Tovar, A. R., Aguilar-Salinas, C. A., Medina-Vera, I., Gil-Zenteno, L., Hernández-Viveros, I., & Torres, N. (2012). A dietary pattern including nopal, chia seed, soy protein, and oat reduces serum triglycerides and glucose intolerance in patients with metabolic syndrome. *Journal of Nutrition, 142*(1), 64–69.

Hasler, C. M., & Brown, A. C. (2009). Position of the American Dietetic Association: Functional foods. *Journal of American Dietetic Association, 109*(4), 735–746.

Hernandez-Jardon, G. (2007). *Proteı́nas de chia (Salvia hispanica L.): Estudio para valorarsuspropie-dadescomoformadoras de pelı́culas.* Me´xico: Facultad de Quı́mica, UNAM.

Inglett, G. E., Chen, D., & Liu, S. (2014). Physical properties of sugar cookies containing chia-oat composites. *Journal of Science Food Agriculture, 94*(15), 3226–3233.

Ixtaina, V. Y., Martínez, M. L., Spotorno, V., Mateo, C. M., Maestri, D. M., Diehl, B. W. K., Nolasco, S. M., & Tomás, M. C. (2011). Characterization of chia seed oils obtained by pressing and solvent extraction. *Journal of Food Composition and Analysis, 24*, 166–174.

Ixtaina, V. Y., Nolasco, S. M., & Tomas, M. C. (2008). Physical properties of chia (*Salvia hispanica L.*) seeds. *Industrial Crops and Products, 289*(3), 286–293.

Jenkins, A. L., Brissette, C., Jovanovski, E., Au-Yeung, F., Ho, H. V. T., Zurbau, A., Sievenpiper, J., & Vuksan, V. (2016). Effect of salba-chia (salvia hispanica l), an ancient seed, in the treatment of overweight and obese patients with type 2 diabetes: A double-blind, parallel, randomized controlled trial. *The FASEB Journal, 30*(1), 126.

Jeong, S. K., Park, H. J., Park, B. D., & Hwan Kim, H. (2010). Effectiveness of topical chia seed oil on pruritus of end-stage renal disease (ESRD) patients and healthy volunteers. *Annals of Dermatology, 22*(2).

Jimenez, G. S., Pastor Vargas, C., De las Heras, M., Sanz Maroto, A., Vivanco, F., & Sastre, J. (2015). Allergen characterization of chia seeds (*Salvia hispanica*), a new allergenic food. *Journal of Investigational Allergology and Clinical Immunology*, *25*, 55–56.

Jin, F., Nieman, D. C., Sha, W., Xie, G., Qiu, Y., & Jia, W. (2012). Supplementation of milled chia seeds increases plasma ALA and EPA in postmenopausal women. *Plant Foods for Human Nutrition*, *67*(2), 105–110.

Julio, L. M., Ixtaina, V. Y., Fernández, M. A., Sánchez, R. M. T., Wagner, J. R., Nolasco, S. M., & Tomás, M. C. (2015). Chia seed oil-in-water emulsions as potential delivery systems of ω-3 fatty acids. *Journal of Food Engineering*, *162*(48-55).

Kampa, M., Nistikaki, A., Tsaousis, V., Maliaraki, N., Notas, G., & Castanas, E. (2002). A new automated method for the determination of the Total Antioxidant Capacity (TAC) of human plasma, based on the crocin bleaching assay. *BMC Clinical Pathology*, *2*(1), 1–16.

KnezHrnčič, M., Ivanovski, M., Cör, D., & Knez, Ž. (2020). Chia seeds (*Salvia Hispanica L.*): An overview— Phytochemical profile, isolation methods, and application. *Molecules (Basel, Switzerland)*, *25*(1), 11.

Kulczyński, B., Kobus-Cisowska, J., Taczanowski, M., Kmiecik, D., & Gramza-Michalowska, A. (2019). The chemical composition and nutritional value of chia seeds-current state of knowledge. *Nutrients*, *11*(6), 1242.

Kumar, R., & Khatkar, B. S. (2017). Thermal, pasting and morphological properties of starch granules of wheat (*Triticum aestivum L.*) varieties. *Journal of Food Science and Technology*, *54*(8), 2403–2410.

Marcinek, K., & Krejpcio, Z. (2017). Chia seeds (*Salvia hispanica L.*): Health promoting properties and thera-peutic applications-a review. *Roczniki Państwowego Zakładu Higieny*, *68*(2), 123–128.

Marineli, R., Lenquiste, S. A., Moraes, E. A., & Marostica, M. R., Jr. (2015). Antioxidant potential of dietary chia seed and oil (*Salvia hispanica L.*) in diet-induced obese rats. *Food Research International*, *76*, 666–674.

Marineli, R., Moura, C., Moraes, E., Lenquiste, S., Lollo, P. C., Morato, P., Amaya- Farfan, J., & Marostica, M. R. (2015). Chia (Salvia hispanica L.) enhances HSP, PGC-1α expressions and improves glucose toler-ance in diet-induced obese rats. *Nutrition*, *51*, 740–748.

Martínez-Cruz, O., & Paredes-López, O. (2014). Phytochemical profile and nutraceutical potential of chia seeds (*Salvia hispanica L.*) by ultrahigh performance liquid chromatography. *Journal of Chromatography*, *1346*, 43–48.

McCormack, W. P., Hoffman, J. R., Pruna, G. J., Jajtner, A. R., Townsend, J. R., Stout, J. R., & Fukuda, D. H. (2015). Effects of L-alanyl-L-glutamine ingestion on one-hour run performance. *Journal of the American College of Nutrition*, *34*(6), 488–496.

Muñoz, L., Cobos, A., Díaz, O., & Aguilera, J. (2012). Chia seeds: Microstructure, mucilage extraction and hydration. *Journal of Food Engineering*, *108*, 216–224.

Munoz, L. A., Aguilera, J. M., Rodriguez-Turienzo, L., Cobos, A., & Diaz, O. (2012). Characterization and microstructure of films made from mucilage of salvia hispanica and whey protein concentrate. *Journal of Food Engineering*, *111*, 511–518.

Nadeem, M., Ajmal, M., Rahman, F., & Ayaz, M. (2015). Analytical characterization of butter oil enriched with omega-3 and 6 fatty acids through chia (*Salvia hispanica L.*) seed oil. *Pakistan Journal of Analytical & Environmental Chemistry*, *16*(2), 68–71.

Nieman, D. C., Cayea, E. J., Austin, M. D., Henson, D. A., McAnulty, S. R., & Jin, F. (2009). Chia seed does not promote weight loss or alter disease risk factors in overweight adults. *Nutrition Research*, *29*(6), 414–418.

Olivos-Lugo, B. L., Valdivia-López, M. Á., & Tecante, A. (2010). Thermal and physicochemical properties and nutritional value of the protein fraction of Mexican chia seed (*Salvia hispanica L.*). *Food Science and Technology International*, *16*(1), 89–96.

Orona-Tamayo, D., Valverde, M. E., Nieto-Rendon, B., & Paredes-Lopez, O. (2015). Inhibitory activity of chia (*Salvia hispanica L.*) protein fractions against angiotensin I-converting enzyme and antioxidant capac-ity. *Food Science & Technology*, *64*, 236–242.

Ovando, J. V., Rubio, G. R., Guerrero, L. C., & Ancona, D. B. (2009). Physicochemical properties of fibrous fraction from chia (*Salvia hispanica L.*). *LWT—Journal of Food Science and Technology*, *42*, 168–173.

Özbek, T., Şahin-Yeşilçubuk, N., & Demirel, B. (2019). Quality and nutritional value of functional strawberry marmalade enriched with chia seed (*alvia hispanica L.*). *Journal of Food Quality*, *2019*, 2391931.

Paredes-Lopez, O. (1991). Safflower proteins for food use. In B. J. F. Hudson (Ed.), *Development in Food Proteins*, vol. 7. London, UK: Elsevier, pp. 1–33.

Pawlosky, R., Hibbeln, J., Lin, Y., & Salem, N. (2003). N-3 fatty acid metabolism in women. *British Journal of Nutrition, 90*, 993–994.

Petruzzello, M. (2020). Chia. In, Encyclopedia Britannica. Available online at: https://www.britannica.com/plant/chia. Accessed on October 27, 2020.

Pintado, T., Herrero, A. M., Jiménez-Colmenero, F., & Ruiz-Capillas, C. (2016). Strategies for incorporation of chia (*Salvia hispanica L.*) in frankfurters as a health-promoting ingredient. *Meat Science, 114*, 75–84.

Pizarro, P. L., Almeida, E. L., Coelho, A. S., Samman, N. C., Hubinger, M. D., & Chang, Y. K. (2014). Functional bread with n-3 alpha linolenic acid from whole chia (*Salvia hispanica L.*) flour. *Journal of Food Science and Technolgy, 52(7), 4475–82*.

Poudyal, H., Panchal, S. K., Waanders, J., Ward, L., & Brown, L. (2012). Lipid redistribution by α-linolenic acid-rich chia seed inhibits and induces cardiac and hepatic protection in diet-induced obese rats. *Journal of Nutrition and Biochemistry, 24*, 153–162.

Punia, S., & Dhull, S. B. (2019). Chia seed (*Salvia hispanica L.*) mucilage (a heteropolysaccharide): Functional, thermal, rheological behaviour and its utilization. *International Journal of Biological Macromolecules, 140*, 1084–1090.

Rahman, U., Nadeem, M., Ahmad, S., Azeem, M. W., & Tayyab, M. (2015). Fractionation of chia oil to enhance omega 3 & 6 fatty acids: Oxidative stability of fractions. *Food Science and Biotechnology, 25(1), 41–47*.

Rendón-Villalobos, R., Ortíz-Sánchez, A., Solorza-Feria, J., & Trujillo-Hernández, C. A. (2012). Formulation, physicochemical, nutritional and sensorial evaluation of corn tortillas supplemented with chia seed (*Salvia hispanica L.*). *Czech Journal of Food Sciences, 30*(2), 118–125.

Reyes-Caudillo, E., Tecante, A., & Valdivia-Lopez, M. A. (2008). Dietary fibre content and antioxidant activity of phenolic compounds present in Mexican chia seeds. *Food Chemistry, 107*(2), 656–663.

Rosas-Ramírez, D. G., Fragoso-Serrano, M., Escandón-Rivera, S., Vargas-Ramírez, A. L., Reyes-Grajeda, J. P., & Soriano-García, M. (2017). Resistance-modifying activity in vinblastine-resistant human breast cancer cells by oligosaccharides obtained from mucilage of chia seeds (*Salvia hispanica*). *Phytotherapy Research, 31*(6), 906–914.

Salazar-Vega, I. M., Segura-Campos, M. R., Chel-Guerrero, L. A., & Betancur-Ancona, D. A. (2012). Antihypertensive and antioxidant effects of functional foods containing chia (*Salvia hispanica*) protein hydrolysates. In: B. Valdez (Ed.), *Scientific, Health and Social Aspects of the Food Industry*. Rijeka, Croatia, 381–398.

Sandoval-Oliveros, M. R., & Paredes-López, O. (2013). Isolation and characterization of proteins from chia seeds (*Salvia hispanica* L.). *Journal of Agricultural and Food Chemistry, 61*(1), 193–201.

Schuchardt, J. P., & Hahn, A. (2013). Bioavailability of long-chain omega-3 fatty acids. *Prostaglandins, Leukotrienes and Essential Fatty Acids, 89*(1), 1–8.

Segura-Campos, M. R., Ciau-Solís, N., Rosado-Rubio, G., Chel-Guerrero, L., & Betancur-Ancona, D. (2014). Chemical and functional properties of chia seed (*Salvia hispanica* L.) gum. *International Journal of Food Science, 2014*, 241053.

Segura-Campos, M. R., Salazar-Vega, I. M., Chel-Guerrero, L. A., & Betancur-Ancona, D. A. (2013). Biological potential of chia (*Salvia hispanica L.*) protein hydrolysates and their incorporation into functional foods. *LWT—Food Science and Technology, 50*(2), 723–731.

Silva, A. S. (2014). Chia flour supplementation reduces blood pressure in hypertensive subjects. *Plant Foods for Human Nutrition, 64*(4), 392–398.

Suri, S., Passi, S. J., & Manchanda, S. C. (2015). Effect of prebiotic—guar gum supplementation among dyslipidemic patients with or without hyperglycemia. *International Journal of Advanced Technology in Engineering and Science, 3*(11), 30–40.

Tanamati, A., Oliveira, C. C., Visentainer, J. V., Matsushita, M., & de Souza, N. E. (2005). Comparative study of total lipids in beef using chlorinated solvent and low-toxicity solvent methods. *Journal of the American Oil Chemists' Society, 82*(6), 393–397.

Timilsena, Y. P., Adhikari, R., Kasapis, S., & Adhikari, B. (2016). Molecular and functional characteristics of purified gum from Australian chia seeds. *Carbohydrate Polymers, 136*, 128–136.

Toscano, L. T., da Silva, C. S. O., Toscano, L. T., de Almeida, A. E. M., da Cruz Santos, A., Vázquez-Ovando, J. A., Rosado-Rubio, J. G., Chel-Guerrero, L. A., & Betancur-Ancona, D. A. (2010). Dry processing of chía (*Salvia hispanica L.*) flour: Chemical characterization of fiber and protein. *CyTA—Journal of Food, 8*(2), 117–127.

Ullah, R., Nadeem, M., & Imran, M. (2017). Omega-3 fatty acids and oxidative stability of ice cream supplemented with olein fraction of chia (*Salvia hispanica L.*) oil. *Lipids in Health and Disease, 16*(1), 34.

Ullah, R., Nadeem, M., Khalique, A., Imran, M., Mehmood, S., Javid, A., & Hussain, J. (2016). Nutritional and therapeutic perspectives of Chia (*Salvia hispanica L.*): A review. *Journal of Food Science and Technology, 53*(4), 1750–1758.

USDA National Nutrient Database for Standard Reference, Release 28. 2011. Basic Report 12006, Seeds- Chia Seeds, dried. Release date: January 2016. Available online at: http: //www.ars.usda.gov/ba/bhnrc/ndl

Uribe, J. A. R., Perez, J. I. N., Kauil, H. C., Rubio, G. R., & Alcocer, C. G. (2011). Extraction of oil from chia seeds with supercritical CO_2. *The Journal of Supercritical Fluids, 56*(2), 174–178.

Vázquez-Ovando, A., Rosado-Rubio, G., Chel-Guerrero, L., & Betancur-Ancona, D. (2009). Physicochemical properties of a fibrous fraction from chia (*Salvia hispanica L.*). *LWT—Food Science and Technology, 42*(1), 168–173.

Vertommen, J., Van de Sompel, A. M., Loenders, M., Van der Velpen, C., & De Leeuw, I. (2005). Efficacy and safety of 1 month supplementation of SALBA (*Salvia Hispanica Alba*) grain to diet of normal adults on body parameters, blood pressure, serum lipids, minerals status and haematological parameters. Results of a pilot study. *The 23th International Symposium on Diabetes and Nutrition of the European Association for the Study of Diabetes.*

Vuksan, V., Choleva, L., Jovanovski, E., Jenkins, A. L., Au-Yeung, F., Dias, A. G., & Duvnjak, L. (2017). Comparison of flax (*Linum usitatissimum*) and Salba-chia (*Salvia hispanica L.*) seeds on postprandial glycemia and satiety in healthy individuals: A randomized, controlled, crossover study. *European Journal of Clinical Nutrition, 71*(2), 234–238.

Vuksan, V., Whitham, D., Sievenpiper, J. L., Jenkins, A. L., Rogovik, A. L., Bazinet, R. P., Vidgen, E., & Hanna, A. (2007). Supplementation of conventional therapy with the novel grain Salba (*Salvia hispanica L.*) improves major and emerging cardiovascular risk factors in type 2 diabetes: Results of a randomized controlled trial. *Diabetes Care, 30*(11), 2804–2810.

Yadav, M., Khatak, A., Singhania, N., & Bishnoi, S. (2020). Comparative analysis of various processing on total phenolic content and antioxidant activity of flaxseed. *International Journal of Chemical Studies, 8*(4), 3738–3744.

15

Buckwheat

Nutritional Composition, Health Benefits, and Applications

Jayashree Potkule
ICAR—Central Institute for Research on Cotton Technology, Mumbai, India

Sneh Punia
Chaudhary Devi Lal University, Sirsa, India
Clemson University, Clemson, SC, USA

Manoj Kumar
ICAR—Central Institute for Research on Cotton Technology, Mumbai, India

CONTENTS

15.1 Introduction

Buckwheat (*Fagopyrum spp.*) is a historical pseudocereal belongings to the *Polygonaceae* family (Cheng et al., 2020). Its global production was reported more than 2.9 million tons in 2018 (FAOSTAT, 2020). Buckwheat classified into *Fagopyrum esculentum* (common buckwheat), *Fagopyrum tataricum* (Tartary buckwheat) and *Fagopyrum cymosum* (cymosum buckwheat). Common buckwheat and Tartary buckwheat are the most widely grown and consumed species from 26 known species of Fagopyrum.

Buckwheat contains high nutrient value and health benefits. Buckwheat contains bioactive compounds such as flavonoids, polysaccharides, fatty acids, amino acids, proteins, fagopyrins, iminosugars, starch, dietary fiber, minerals, and vitamins (Huda et al., 2020). Buckwheat also contains microelements and macroelements such as sodium, copper, potassium, zinc, iron, magnesium, manganese, and calcium (Huda et al., 2020). The nutritional composition and phytochemical profile imparts various health benefits to buckwheat including cardioprotective, hepatoprotective, neuroprotective, antioxidant, antimicrobial, anti-inflammatory, antifatigue, antidiabetic, cognition-improvement, cholesterol, and blood pressure lowering (Zhou et al., 2019; Huda et al., 2020). Sucrose and calcium chloride treatment improves the nutritive value, antioxidant activity, and health benefits of buckwheat by enhancing accumulation of flavonoids, polyphenols, γ-aminobutyric acid and vitamins (Sim et al., 2020). Phosphorus fertilization improves the growth, quality, and development of Tartary buckwheat (Zhang et al., 2019).

Buckwheat is a versatile, nutritious, and low-cost food commodity, which increases its applications in foods. Cooking methods greatly affect the chemical composition, structure, rheological, and functional properties of buckwheat flour. Extrusion cooking technology improves the morphology, structure, crystallinity, rheological, physicochemical, and functional properties of modified buckwheat (Cheng et al., 2020). Buckwheat is a good source of bioactive compounds and nutrition, providing various pharmaceutical and health benefits. Hence, due to its health-promoting factors, buckwheat has various food applications. This chapter discusses the nutritional profile, biological activities/health benefits, and industrial application of buckwheat.

15.2 Nutritional Profile of Buckwheat

Due to its high nutritional value, many researchers worked on buckwheat including proximate analysis, its composition, rheology, bioactive compounds, and physiochemical and functional properties. Bonafaccia, Marocchini, and Kreft (2003) studied chemical composition of bran, flour and grains of common buckwheat and found 10.6–21.6% protein, 40.7–78.4% starch, 1.82–4.08% ash, 2.34–7.20% fat, 6.77–27.38% total dietary fiber (0.78–0.91% soluble and 5.89–26.6% insoluble dietary fiber). The nutritional composition of Tartary buckwheat constitutes 10.3–25.3% protein, 37.6–79.4% starch, 1.8–4.97% ash, 2.45–7.35% fat, 6.29–25.97% total dietary fiber (0.52–1.18% soluble and 5.77–25.43% insoluble dietary fiber). They also analyzed vitamins, fatty acids, color, and amino acid composition of both common and Tartary buckwheat (Sytar et al., 2016). Bhinder et al. (2019) reported the proximate composition (moisture, ash, fat protein, and starch content), vitamins, minerals, amino acid profile, and functional properties of Tartary buckwheat varieties. The nutritional components of the buckwheat are discussed in detail in next section.

15.2.1 Carbohydrates/polysaccharides

Buckwheat polysaccharides include starch and nonstarch polysaccharides. Buckwheat contains 70–91% starch depending on their milling process, variety, and geographical location (Škrabanja & Kreft, 2016). Acid, alkali, and hot water are all used in the extraction of polysaccharide from buckwheat. For industrial and laboratory purposes, hot water extraction is the most preferable and commonly used method. The nonconventional techniques including ultrasound and microwave-assisted extraction methods have been proven to enhance the yields of polysaccharides. Factors such as raw material, temperature, time, and solvent used in the extraction method affected the polysaccharide yield (Ji et al., 2019). Jindal and Saxena, (2015) extracted starch from common buckwheat at different alkali concentrations and obtain a maximum 45.3% yield at 0.2% alkali concentration with good water and oil binding capacity, solubility, and swelling power. Sequential extraction of buckwheat waste with water, ammonium oxalate, and sodium hydroxide contains 3.2–6.3% total polysaccharides in the husk and 7.6–12.2% in straw with monosaccharide composition such as of arabinose, rhamnose, xylose, glucose, mannose, and galactose as well as other polysaccharides like inositol and uronic acids (Zemnukhova et al., 2004). Gao et al. (2020) reported resistant and native starch of common buckwheat was suitable raw material in starch-based food products and development of other innovative food products. Starch from buckwheat has many applications in food and food products as well as in pharma fields due to their biological activities. Biological activities of buckwheat depend on the chemical composition, molecular weight and chain confirmations (Ji et al., 2019).

15.2.2 Protein

Buckwheat has higher biological value due to its well-balanced amino acid profile of protein and is rich in essential amino acids like lysine and arginine (Ji et al., 2019; Cheng et al., 2020). Tomotake et al. (2002) extracted buckwheat protein by alkali extraction and obtained 65.8% protein yield by isoelectric point precipitation. They also studied the physiochemical and functional properties of buckwheat protein, and the results suggested that the protein solubility of buckwheat increases at pH 3, reduced at pH 4, and again increased above pH 5. Buckwheat protein also shows good water and oil holding capacity as well as emulsifying stability and viscosity. Better functional properties of buckwheat protein enhance their application in food products. Tang, (2007) extracted buckwheat protein using ultrasound-assisted extraction with spray and freeze-drying methods and studied their *in vitro* digestibility and functional properties. Results suggested that functional and nutritional values and applications of protein depend on their processing conditions.

Buckwheat protein mainly contains four different fractions including albumin, glutelin, prolamins, and globulins. Guo & Yao (2006) extracted fractions of buckwheat protein and obtained 43.8% albumin that was rich in threonine, histidine, valine, lysine, phenylalanine, isoleucine, and leucine. Prolamin (10.5%) is rich in threonine, isoleucine, histidine, and valine. Glutelin (14.6%) was found rich in histidine, valine, threonine, leucine, isoleucine, and globulin (7.82%) with lysine and methionine. Buckwheat has antioxidant, antitumor, antimicrobial, antidiabetic, hypotensive, and hypocholesterol activity as well as protease inhibitory activity due to their bioactive peptides. Trypsin inhibitor of buckwheat was effective against Gram-positive and Gram-negative bacteria and fungi and reduces the growth of various types of cancer cells (Zhou et al., 2015a).

15.2.3 Dietary Fiber

Nonstarch polysaccharide components are known as dietary fiber and are mainly categorized as soluble or insoluble fiber. Buckwheat is a good source of dietary fibers. Fiber fractions of buckwheat were composed of 39% cellulose, 1.8% pectin, 20% lignin, and 39% hemicellulose (Zhang, Wang, Tan, et al., 2020). Buckwheat hull contains higher amount of dietary fiber than grains (Mackèla, Andriekus, and Venskutonis, 2017). Organic solvents were used for defatting buckwheat flour and was precipitated or hydrolyzed using solvents/enzymes to remove protein. In the next step, amylase was used to remove starch, and finally, nonstarch polysaccharide/dietary fiber was recovered using 96% ethanol. Microwave/ultrasound treatment enhances the yields of dietary fiber. Buckwheat grains contained 2.9% insoluble and 2.4% soluble dietary fiber. Similar dietary fiber composition was found in both common and Tartary buckwheat (Wefers and Bunzel, 2015; Zhu, 2020). Dietary fiber mainly consist xyloglucans, pectins, and arabinogalactans. 6.7–9.1% total dietary fiber found in Canadian buckwheat vary from 6.7–9.1% (Izydorczyk et al., 2014). High fiber content in diet lowers the risk of cardiovascular diseases, obesity, diabetes, and type of cancers in humans (Wefers and Bunzel, 2015; Zhu, 2020).

15.2.4 Fatty Acids

Fat/lipids are recovered by defatting buckwheat flour using organic solvents such as n-hexane, methanol, and acetone. Golijan et al. (2019) studied lipid fractions of buckwheat grains and found 5.36% and 3.43% total lipid content in conventional and organic buckwheat grains than maize and spelt collected in the same year. They also reported fatty acid content of buckwheat, maize, and spelt, such as triacylglycerol, linolenic acid, stearic acid, palmitoleic acid, myristic acid, palmitic acid, oleic acid, behenic acid, arachidic acid, eicosenoic acid, and unsaturated and saturated fatty acids. Similarly, Sinkovic, Kokalj, and Vidrih (2020) determined the fatty acid composition of bran, hull, light flour, and whole grains of both common and Tartary buckwheat using gas chromatography. Common buckwheat was composed of total fatty acids 48–62 g/kg in bran, 20–22 g/kg in whole grains, 7–9 g/kg in light flour, and 2–5 g/kg in the hull. Tartary buckwheat was composed of 40–52 g/kg total fatty acid in bran, 20–24 g/kg in whole grains, 6–8 g/kg in light flour, and 3–6 g/kg in the hull. Common buckwheat contained 37–47 g/kg linoleic acid and 20–36 g/kg oleic acid and Tartary buckwheat contained 35–44 g/kg linoleic acid and 26–40 g/kg oleic acid. Lauric acid, myristic acid, palmitic acid, stearic acid, palmitoleic acid, and saturated and unsaturated fatty acids were found in both common and Tartary buckwheat. Both common and Tartary buckwheat has similar fatty acid composition (Sinkovic, Kokalj, and Vidrih, 2020). Treatment with superheated steam leads to the suppression of unsaturated fatty acids, changes in glycerophospholipid and glycerolipids metabolism, inactivation of lipase, lipid hydrolytic rancidity retardation, and maintenance of nutrition of lipids during storage of common buckwheat (Wang et al., 2020).

15.2.5 Phenolic Compounds

Phenolic compounds consist of phenolic acids and flavonoids, which are an important bioactive compound in buckwheat. Flavonoids and their subgroups were detected in seed, root, fruit, flower, leaves, stem, seedling, sprouted seed, seed husk, and seed coat (Huda et al., 2020). Flavonoids are classified into subgroups such as flavanones, flavones anthocyanins, proanthocyanidins, fagopyrins, flavonolignans, flavanols, and isoflavones. Factors like cultivated species, organ, growth stage, area, and season affect the flavonoid content of buckwheat (Huda et al., 2020). Acid, alkaline, and enzyme treatments are required to extract phenolic acids and flavonoids from the cell wall of buckwheat seeds in free and bound form (Lee et al., 2016). Dzah et al. (2020) extracted polyphenols by hot water, subcritical water, ultrasound, and ultrasound-subcritical water extraction methods, purified by chromatography techniques, and determined polyphenols using High performance liquid chromatography (HPLC) and liquid chromatography-mass spectroscopy (LC-MS). Results suggest that the extraction method affected the yield and composition of polyphenols. Similarly, Kalinová, Vrchotová, and Tříska (2019) extracted phenolic compounds, analyzed using HPLC and finally detected by UV visible spectrophotometer at 350 nm and 220 nm by using standards as reference. Rutin, catechin, epicatechin, epicatechin gallate, hyperoside, isoquercitrin, isovitexin, and procyanidin B2 content was obtain from the seed coat. Rutin, isovitexin, vitexin, isoquercitrin, quercetin, quercitrin, orientin, hyperoside, isoorientin, epicatechin, catechin, epicatechin gallate, vanillin, vanillic acid, procyanidin B2, and protocatechuic acid content were obtained from hulls. Rutin comprises 90% of the total phenolic compound and detected in all buckwheat varieties. Hence, rutin is the main flavanol in buckwheat grain extracted using acetone, methanol, ethanol, and other solvents (Lee et al., 2016). Some phenols also found in other parts like endosperm, cotyledon, and groats of buckwheat achenes. Roasting reduces the antioxidant activity and total phenol and flavonoid content in both common buckwheat and Tartary buckwheat (Ma et al., 2020). The phenolic compounds of buckwheat have a protective effect on the immune system, inhibit peroxidation of low density lipoprotein (LDL), and delay aging (Kalinová, Vrchotová and Tříska, 2019).

15.2.6 Vitamins

Vitamins also play an important role in nutritional value of buckwheat. Buckwheat contains vitamin B (B_1 [thiamine], B_2 [riboflavin], B_3 [niacin], B_5 [panthothenic acid], and B_6 [pyridoxine]), vitamin C (ascorbic acid), vitamin E (tocopherols), and vitamin K. Common buckwheat contains a higher vitamin E content than Tartary buckwheat (Huda et al., 2020). Germination of buckwheat seed also increases the level of vitamins like vitamin B1 and C (Zhou et al., 2015b). Tartary buckwheat contains high vitamin B and α-tocopherol content than common buckwheat (Ahmed et al., 2013). Gamma (γ)-tocopherol (117.8 µg/g), δ-tocopherol (7.3 µg/g), and α-tocopherol (2.1 µg/g) of vitamin E exist in buckwheat seeds (Li, 2019). Vitamins like thiamine, vitamin B6, folate, and niacin work on the transformation of carbohydrates into energy development of the brain and its functions, formation of red blood cells, digestive system, nerves, and skin (Huda et al., 2020).

15.2.7 Minerals

Minerals play important role in physiological functions. Many researchers conclude that buckwheat is a good source of essential minerals (Ahmed et al., 2013). Trace elements such as potassium (K), copper (Cu), calcium (Ca), phosphorus (P), magnesium (Mg), barium (Ba), selenium (S), iodine (I), platinum (Pt), boron (B), iron (Fe), cobalt (Co), and zinc (Zn) present in different parts of buckwheat. Excessive amount of trace elements are found in the seed coat and seed of buckwheat (Huda et al., 2020). Buckwheat flour contains Zn, manganese (Mn), Mg, and Cu in higher amounts and Ca in lower amounts than other flours (Ahmed et al., 2013). Fertilization affected on the physiochemical characteristics of buckwheat and enhances their food and nonfood applications (Zhang et al., 2019). Minerals like P are used in bone and teeth formation, Zn is essential in immune system, Cu helps in Fe absorption with collagen synthesis, Mn helps in development of connective tissues and bones. Mg regulates nervous system and myocardium functions, myocardial infarction, arteriosclerosis resistance, and hypertension prevention (Li, 2019).

Fagopyrum species also contain some other phytochemicals like steroids, terpenoids alkaloids, tannins, anthraquinones, and coumarins (Huda et al., 2020). Some examples of bioactive compound of buckwheat are shown in Tables 15.1 and 15.2.

15.3 Biological Activities of Buckwheat and its Health Benefits

15.3.1 Antioxidant Activity

Antioxidant activity of buckwheat depends on the phenolic compound content of rutin and quercetin. Different flavonoids of common buckwheat and Tartary buckwheat contributed and differently affected the antioxidant activity of buckwheat (Lee et al., 2016). Bączek et al. (2020) found 98.0 μmol Trolox/g, 12.88 μmol Trolox/g, and 5.86 μmol Trolox/g of antioxidant activity using Trolox equivalent antioxidant capacity (ABTS), ferric reducing antioxidant power (FRAP), and ABTS (PCL) using ACW (commercial kit use as a standard in antioxidant activity) as standard methods after digestion of buckwheat bread. A 3% sucrose treatment and a 7.5 mM CaCl2 treatment enhance the nutritional value, antioxidant activity, and health benefits in buckwheat sprouts (Sim et al., 2020). Buckwheat hull is a byproduct of processing and it contains a higher antioxidant activity than grains (Kalinová, Vrchotová, and Tříska, 2019). Buckwheat extract has high antioxidant activity, which lowers the risk of cardiovascular disease, aging, and types of cancer (Inglett et al., 2011).

15.3.2 Cholesterol Lowering/hypolipidemic Activity

Buckwheat and other proteins are use in the diet for many animal studies. Buckwheat protein decreases plasma cholesterol, significantly decreases hepatic cholesterol as compared to soybean and casein in rats, and improves cholesterol metabolism (Kayashita, Shimaoka, and Nakajyoh, 1995). Tomotake et al. (2007) obtained 65.8% and 45.8% protein from common and Tartary buckwheat, respectively, using alkali extraction and isoelectric precipitation. Consumption of Common buckwheat and Tartary buckwheat protein reduces 32% and 25% serum cholesterol after 13 days and shows a 62% and 43% lithogenic index reduction after 27 days by enhancing excretion of sterols and bile acids in mice and rat fed with cholesterol. Similarly, Zhou et al. (2018) reported buckwheat protein was beneficial for cholesterol metabolism and regulates composition of microbiota in gut. They observed larger decreases in plasma total cholesterol (TC) and triglyceride (TG) content and increases in excretion of free fatty acid chains and bile acid in mice than mice fed with casein. Buckwheat protein stimulated the growth of *Bifidobacterium, Enterococcus,* and *Lactobacillus* and inhibited growth of *Escherichia coli* also. Chen et al. (2020) reported levels of TG, TC, and LDL cholesterol increased and high density lipoprotein cholesterol (HDL-C) decreased in *Lactobacillus plantarum* and *Saccharomyces cerevisiae* fermented buckwheat sprouts than nonfermented. They concluded fermentation of buckwheat sprouts grown in maifanite mineral water affected on the hypolipidemic property.

TABLE 15.1

Extraction of Bioactive Compounds from Buckwheat

Biomolecules	Source	Extraction Method	Optimum Conditions		Result	References
Carbohydrates	Common buckwheat	Water extraction	Water: raw material	30:1	16.50% yield obtained at optimum condition	Chai et al., 2007
			Temperature	70°C		
			Time	3 hours		
	Common buckwheat	Alkali extraction	Water: raw material	50:1	25.16% yield obtained by alkali extraction	Chai et al., 2008
			Temperature	80°C		
			Time	1 hours		
			Alkali	0.3 M NaOH		
	Tartary buckwheat	Ultrasound assisted extraction	Water: raw material	40:1	7.23% yield obtained at optimum condition	Zhang et al., 2012
			Temperature	50°C		
			Time	30 minutes		
			Ultrasonic power	400W		
Protein	Tartary buckwheat	Enzymatic (alcalase) hydrolysis	Temperature	46.16°C	Experimental protein hydrolysis value 57.71% found significant to the predicted value 57.69%	Tao et al., 2019
			Time	20.8 minutes		
			Enzyme	40.61U/g		
Polyphenol	Tartary buckwheat	Ultrasound assisted extraction	Water: raw material	3000: 50	19.3 mg/g yield with 6.6 mgGAE/g total free phenol content	Dzah et al., 2020
			Temperature	50°C		
			Time	20 minutes		
			Ultrasonic power	90W		
			Frequency	20kHz		
		Hot water extraction	Water: raw material	3000: 50	4.2 mg/g yield with lower free phenol content than other methods	
			Temperature	80°C		
			Time	60 minutes		
		Subcritical water extraction	Water: raw material	3000: 50	53.3 mg/g yield with 7.9 mgGAE/g total free phenol content	
			Temperature	220°C		
			Time	60 minutes		
		Ultrasound assisted-subcritical water extraction	Water: raw material	3000: 50	31.8 mg/g yield with 6.8 mgGAE/g total free phenol content	
			Temperature	50°C and 220°C		
			Time	20 and 60 minutes		
Flavonoid	Common buckwheat	Ultrasound assisted extraction with deep eutectic solvent	Water: raw material		Above 97% yield using solid phase extraction	Rois Mansur et al., 2019
			Temperature	56°C		
			Time	40 minutes		
			Ultrasonic power	520 and 700W		
			Frequency	40kHz		

TABLE 15.2

Extraction of Polyphenol from Buckwheat

Source	Phenolic Compound	Subgroups	Concentration	Result	References
Common buckwheat flour	Total phenol		313 mg/100 g	Presence of phenolic compound in buckwheat improves the antioxidant activity	Quettier-Deleu, et al., 2000
	Flavonoids		9,8 mg/100 g		
		Rutin	2.27 mg/100 g		
		Quercetin	0.15 mg/100 g		
		Hyperoside	0.19 mg/100 g		
	Flavonol		220 mg/100 g		
	Proanthocyanidin		159.3 mg/100 g		
Common buckwheat hull	Total phenol		333 mg/100 g		
	Flavonoids		45.6 mg/100 g		
		Rutin	5.20 mg/100 g		
		Quercetin	0.60 mg/100 g		
		Hyperoside	1.61 mg/100 g		
	Flavonol		169 mg/100 g		
	Proanthocyanidin		137 mg/100 g		
Commercial Tartary buckwheat tea	Flavonoid	Rutin	Whole plant: 0.23–0.96 mg/g Whole bran: 2.09–3.83 mg/g Whole embryo:13.6–15 mg/g		Peng et al., 2015
		Quercetin	Whole plant: 26.2–30.2 mg/g Whole bran: 4.78–17.4 mg/g Whole embryo: 5.22–13.6 mg/g		
Tartary buckwheat	Flavonoid	flavan3-ols	Catechin-7-O-glucoside, epigallocatechin-3-gallate, epigallocatechin, and epicatechin	Flavonoid extracted with different extraction techniques and determined by HPLC and LC-MS	Dzah et al., 2020
		Flavonols	Rutin, kaempferol, kaempferol-3-O-glucoside, kaempferol-3-rutinoside, quercetin-3-O-glucuronide hyperin, and quercetin		
		Flavones	Isovitexin, vitexin, isoorientin, and orientin		
		Anthocyanins	Cyanidin 3-O-galactoside, cyanidin-3-O-rutinoside, and cyanidin-3-O-glucoside		
Buckwheat fermented with agaricus strains	Total phenol content		Agaricus blazei: 3.92–18 mg/g Agaricus bisporus: 2.17–6 mg/g Agaricus bisporus: 3.68–5.41 mg/g in 0 to 21 days of fermentation	Total phenolic content increases with fermentation time	Kang et al., 2017

15.3.3 Anticancer Activity

Buckwheat polysaccharides shows anticancer activity against cancer cells such as liver cancer, breast cancer, prostate cancer, and leukemic THP-1 cells in humans by inducing differentiation of cells, DNA damage, cell cycle arrest, disruption of mitochondrial membrane, production of nitric oxide, and immunotherapy (Zhu, 2020). Trypsin inhibitor from recombinant buckwheat induces hepatic cell lines by apoptosis and was studied using electrophoresis of DNA, 3-(4,5-dimethylthiazol-2-yl)-2,5-diphenyl tetrazolium bromide (MTT), flow cytometry, cytochrome C measurement, morphological analysis of nuclei, and caspase activation *in vivo* and *in vitro*. Results show fragmentation of DNA, reduction of cell viability due to apoptosis, dysfunction of mitochondria and caspase 3, 8, and 9 activation. Anticancer activity was also dependent on the time and concentration of buckwheat (Bai et al., 2015). Buckwheat (TBWSP31) antitumor proteins arrested the cell cycle in the G_0/G_1 phase and inhibited cell growth in the G_0/G_1 to S phase of the cell cycle. They also involve in the participation of protein factors (bcl-2 and fas are death effector molecules which regulates apoptosis or cell death) and bcl-2 and fas induced apoptosis in breast cancer Bcap37 cells. Morphological changes caused by apoptosis such as detachment of culture, shrinkage of cells, change in shape, blebbling of plasma membrane, absence of microvilli, and apoptotic bodies formation were observed by using scanning electron microscopy and inverted microscopy (Guo et al., 2010). Peptides isolated from buckwheat have antiproliferation activity against leukemia cells, breast cancer, lung cancer, gastric cancer, colon cancer, liver embryonic, HeLa, HepG2, and MCF cancer cell lines (Giménez-Bastida, & Zieliński, 2015).

15.3.4 Antidiabetic Activity

Deficiency of insulin or defective activity of insulin increase glucose levels in plasma, which causes diabetes mellitus (Giménez-Bastida, & Zieliński, 2015). Regulation of glucose, insulin, and the lipid profile improves the outcome of diabetic individuals. Qiu et al. (2016) used Tartary buckwheat protein to minimize risk factors in type 2 diabetes mellitus such as insulin resistance, lipid profile, and fasting glucose by replacing daily staple food with buckwheat protein in 165 patients. After 4 weeks, diet information and blood samples were collected, revealing that buckwheat protein lowered the (2.46–2.39 Ln mU/L) fasting insulin content, (5.08–4.79 mmol/L) TC, and (3–2.8 mmol/L) LDL-C than the systematic diet plan. The overall study shows buckwheat protein improves fasting glucose, lipid profile, and insulin resistance in type 2 diabetes mellitus patients. Stringer et al. (2013) studied the effects of consumption of buckwheat flour on the postprandial response of insulin, glucose, and gastrointestional hormones. Results showed consumption of buckwheat flour alters the postprandial plasma glucagon-like peptide-1, pancreatic polypeptide, glucose-dependent insulinotropic peptide, and no change was observed in insulin and glucose concentrations in healthy and type 2 diabetes mellitus individuals.

15.3.5 Hypoglycaemic Activity

Buckwheat has hypoglycemic activity, Tartary buckwheat (nonstarch) polysaccharide was isolated and purified by using a sephadex column. Tartary buckwheat polysaccharide-II shows higher inhibition of alpha-D-glucosidase than Tartary buckwheat polysaccharide-I and crude polysaccharide, buts it was significant with acarbose used as positive control and its dose-dependant activity (Kaur et al., 2015).

15.3.6 Anti-inflammatory Activity

Nuclear factor kappa B plays important role in inflammation by regulating interleukin and necrosis factors. It was an important mediator in cyclooxygenase-2 and nitric oxide synthase. Phenolic extract of buckwheat reduces lipopolysaccharides (LPS) and induces nuclear kappa factor B (Giménez-Bastida, & Zieliński, 2015). The lipopolysaccharide inducement of protective activity of buckwheat shows anti-inflammatory activity against human colon cancer in mice. Cytokines like interleukin 6 (IL-6) and tumour necrosis factor alpha (TNF-α) up regulated in liver and spleen of mice after LPS administration (Ishii et al., 2008). Nam et al. (2017) studied anti-inflammatory activity of high flavonoid-containing common (rutin, *C*-glycosyl flavones and quercetin-3-*O*-robinobioside) and Tartary buckwheat sprouts (rutin) with high flavonoid content for evaluation of LPS, which stimulates macrophages. Results suggest that Tartary buckwheat LPS induces peritoneal macrophages and inhibits mediators like cytokines (IL-6 and TNF-α) and nitric oxide in LPS stimulated RAW 264.7.

15.3.7 Other Biological Activities

Buckwheat also contains other biological activities such as cardioprotective, hepatoprotective, neuroprotective, antimicrobial, anti-inflammatory, immunomodulation, anti-fatigue, wound healing, and cognition-improvement (Ji et al., 2019; Zhou et al., 2019; Huda et al., 2020). Giménez-Bastida, & Zieliński (2015) discuss the bioactive compounds, biological activities, and health benefits of buckwheat.

Immunomodulatory activity was studied for the enhancement of defense mechanisms and reduction of toxicity against various infectious agents. Immunosuppression was of studied *in vivo* by the administration of 50 mg/kg and 100 mg/kg dosage of buckwheat polysaccharides in mice and results suggest buckwheat polysaccharides increase leukocyte counts in blood, thymus, and spleen indices in immunosuppressed mice. This result concluded that buckwheat can act as potential immunomodulator (Oh et al., 2018).

Different extracts of common buckwheat stems, seeds, and other parts have neuroprotective activity through butylcolinesterase, acetylcholisterase, antioxidant, and tyrosinase inhibitory activities (Giménez-Bastida, & Zieliński, 2015). Some examples of biological activities of buckwheat are shown in Table 15.3.

15.4 Food Applications

Buckwheat has various food applications like biscuits, cakes, breads, crepes, cookies, casseroles, porridge, noodles, pancakes, soups, muffins, and other products. Mostly, buckwheat is used in flour form with other cereals or by replacing wheat in home-made food like noodles, pancakes, bread, pasta, and biscuits (Cheng et al., 2020). Good functional properties of buckwheat improve its texture and bioactive activities like antimicrobial and antioxidant activities, which increase the shelf life of buckwheat breads (Zhu, 2020). Bioaccessibility of food provides important information about nutrient efficiency of food products. Zieliński, Szawara-Nowak, and Wronkowska (2020) prepared buckwheat biscuits after lactic acid and *Rhizopus oligosporus* fermentation of the buckwheat flour. They observed bioaccessibility of antiadvanced glycation end products (anti AGEs activity), antioxidant activity, and total phenolic content of the biscuits after *in vitro* digestibility was improved.

Wronkowska et al. (2010) used cornstarch with 10%, 20% 30%, and 40% buckwheat flour in 2% sunflower oil, 8% yeast, 1% salt, 5% sugar and water to make gluten-free bread. Buckwheat bread shows high antioxidant activity by DPPH (4.1 μmol Trolox/g), ABTS (2.5 μmol Trolox/g), and reducing activity (1.5 μmol Trolox/g) in relation with total phenols (1.22 FAE/g), high sensory quality (7.1 units), and trace elements (Mg, P, and K). Buckwheat breads are beneficial to individuals suffer with celiac disease. Sakač et al. (2011) used rice and buckwheat flour (light or whole grain) in combination of 90:10, 80:20, and 70:30 in gluten-free bread making and investigated their phenolic content and antioxidant activity. Whole-grain buckwheat flour increased antioxidant activity and phenolic content more than light buckwheat flour. Bączek et al. (2020) found 5.6 mg GAE/g of total phenolic content after digestion of buckwheat bread. Buckwheat honey reduces cholesterol in blood and improves the heart health. Małgorzata et al. (2020) found antimicrobial activity of 20 samples of buckwheat honey against of *S. aureus, S. enterica, E. coli,* and *K. pneumoniae* using H_2O_2, minimum inhibitory and bacterial concentrations. The antioxidant activity of buckwheat honey was found to be similar to the commercial honey using FRAP and DPPH assay (Małgorzata et al., 2020). Deng et al. (2018) reported the antioxidant level of buckwheat honey was greater than the manuka honey. Buckwheat honey shows antibacterial activity against *Pseudomonas aeruginosa* and *Staphylococcus aureus*. The addition of buckwheat flour in different concentration improves the quality, antioxidant activity, and free and bound phenolic content in crust and flavonoids like flavones, flavan-3-ols, and flavonols in the crumb of buckwheat loaves (Verardo et al., 2018).

15.5 Conclusion

Buckwheat is a good source of nutrition due to the presence of bioactive compounds such as protein, carbohydrates, polyphenols, vitamins, and minerals. Biological activities of buckwheat include cardioprotective, hepatoprotective, neuroprotective, antioxidant, anti-inflammatory, antimicrobial, antifatigue,

TABLE 15.3

Biological Activities of Buckwheat

Biological Activity	Source	Method	Results	References
Antioxidant	Tartary buckwheat grout	ABTS, FRAP, and DPPH	14–42 mmol TE/g	Lee et al., 2016
	Tartary buckwheat hull		25–55 mmol TE/g	
	Common buckwheat grout		78–146 mmol TE/g	
	Common buckwheat hull		30–80 mmol TE/g	
	Commercial Tartary buckwheat tea (plant, bran, and embryo)	DPPH	214.14–294.44 µmol trolox equivalent/g dry weight	Peng et al., 2015
		ABTS	120.46–383.8 µmol trolox equivalent/g dry weight	
	Buckwheat sprouts fermented with *Lactobacillus plantarum* and *Saccharomyces cerevisiae*	DPPH and OH	Antioxidant activity of fermented buckwheat sprouts were better than unfermented sprouts	Chen et al., (2020)
	Tartary buckwheat	Cell antioxidant activity by using ABTS, DPPH, TEAC, and FRAP	Subcritical water extraction: 300 µg/ml Ultrasound assisted extraction: 250 µg/ml Hot water extraction: 200 µg/ml Ultrasound-subcritical water extraction: 150 µg/ml	Dzah et al., 2020
Anticancer	Buckwheat polysaccharides	Differential induction, cytokine, phagocytosis and superoxide anion assay	Inhibits differentiation, growth, and proliferation of leukemic cells (THP-1 cells) in humans	Wu & Lee, 2011
	Tartary buckwheat	MTT assay	High cytotoxicity obtained in subcritical water and ultrasound-subcritical water extraction followed by ultrasound assisted extraction and lower cytotoxicity obtained in hot water extraction of buckwheat in HepG2 cancer cells	Dzah et al., 2020
Anti-inflammatory	Buckwheat bread alone and in combination with TNF-α	Cell migration, membrane potential of mitochondria, cell cycle, and intestinal inflammation	Consumption of buckwheat prevents inflammatory diseases.	Giménez-Bastida et al., (2018)
Antidiabetic	Tartary buckwheat	Oral cute toxicity, fasting glucose level in blood, oral glucose tolerance, immunohistochemical analysis, plasma Glucagon, C-peptide, triglyceries and cholesterol analysis	D-chiro-inositol lowers the glucose level in plasma, enhances glucose tolerance and decreases insulin level, plasma C-peptide, glucagon, triglyceride, cholesterol, and urea were lower in diabetic mice than control.	Yao et al., 2008

antidiabetic, cognition-improvement, cholesterol and blood pressure lowering, and wound healing properties that increase their importance and health benefits in food products. This study provides beneficial information on buckwheat as a novel, valuable, and healthy food product.

REFERENCES

Ahmed, A., Khalid, N., Ahmad, A., Abbasi, N. A., Latif, M. S. Z., and Randhawa, M. A. 2013. Phytochemicals and biofunctional properties of buckwheat: A review. *The Journal of Agricultural Science*, 152(03): 349–369.

Bączek, N., Jarmułowicz, A., Wronkowska, M., and Monika Haros, C. 2020. Assessment of the glycaemic index, content of bioactive compounds, and their *in vitro* bioaccessibility in oat-buckwheat breads. *Food Chemistry*, 127199.

Bai, C.-Z., Feng, M.-L., Hao, X.-L., Zhao, Z.-J., Li, Y.-Y., and Wang, Z.-H. 2015. Anti-tumoral effects of a trypsin inhibitor derived from buckwheat *in vitro* and *in vivo*. *Molecular Medicine Reports*, 12(2):1777–1782.

Bhinder, S., Kaur, A., Singh, B., Yadav, M. P., and Singh, N. 2019. Proximate composition, amino acid profile, pasting and process characteristics of flour from different Tartary buckwheat varieties. *Food Research International*, 108946.

Bonafaccia, G., Marocchini, M., and Kreft, I. 2003. Composition and technological properties of the flour and bran from common and Tartary buckwheat. *Food Chemistry*, 80(1): 9–15.

Chai, R .J., Ma, J. H., and Xu D. Q. 2007. Study on extraction of water soluble polysaccharide from buckwheat. *Science and Technology of Food Industry*, 28:163–164.

Chai, R. J., Ma, J. H., and Xu D. Q. 2008. Isolation of buckwheat polysaccharide by alkali method. *Forest By-Product and Speciality in China*, 5:32-33.

Chen, T., Piao, M., Ehsanur Rahman, S. M., Zhang, L., and Deng, Y. 2020. Influence of fermentation on antioxidant and hypolipidemic properties of maifanite mineral water-cultured common buckwheat sprouts. *Food Chemistry*, 321:126741.

Cheng, W., Gao, L., Wu, D., Gao, C., Meng, L., Feng, X., and Tang, X. 2020. Effect of improved extrusion cooking technology on structure, physiochemical and nutritional characteristics of physically modified buckwheat flour: Its potential use as food ingredients. *LWT*, 109872.

Deng, J., Liu, R., Lu, Q., Hao, P., Xu, A., Zhang, J., and Tan, J. 2018. Biochemical properties, antibacterial and cellular antioxidant activities of buckwheat honey in comparison to manuka honey. *Food Chemistry*, 252:243–249.

Dzah, S. C., Duan, Y., Zhang, H., Antwi Authur, D., and Ma, H. 2020. Ultrasound-, subcritical water- and ultrasound assisted subcritical water-derived Tartary buckwheat polyphenols show superior antioxidant activity and cytotoxicity in human liver carcinoma cells. *Food Research International*, 137,109598.

FAOSTAT database: FAOSTAT (2020). Production/yield quantities of buckwheat in World + (Total) 2018. http://www.fao.org/faostat/en/#data/QC/visualize, Accessed date: 10 August 2020, 2020.

Gao, L., Wang, H., Wan, C., Leng, J., Wang, P., Yang, P., and Gao, J. 2020. Structural, pasting and thermal properties of common buckwheat (*Fagopyrum esculentum* Moench) starches affected by molecular structure. *International Journal of Biological Macromolecules*, 156:120–126.

Giménez-Bastida, J. A., and Zieliński, H. 2015. Buckwheat as a Functional Food and Its effects on health. *Journal of Agricultural and Food Chemistry*, 63(36):7896–7913.

Giménez-Bastida, J. A., Laparra-Llopis, J. M., Baczek, N., and Zielinski, H. 2018. Buckwheat and buckwheat enriched products exert an anti-inflammatory effect on the myofibroblasts of colon CCD-18Co. *Food and Function*, 9(6):3387–3397.

Golijan, J., Milinčić, D. D., Petronijević, R., Pešić, M. B., Barać, M. B., Sečanski, M., and Kostić, A. Ž. 2019. The fatty acid and triacylglycerol profiles of conventionally and organically produced grains of maize, spelt and buckwheat. *Journal of Cereal Science*, 102845.

Guo, X., & Yao, H., 2006. Fractionation and characterization of Tartary buckwheat flour proteins. *Food Chemistry*, 98(1):90–94.

Guo, X., Zhu, K., Zhang, H., and Yao, H. 2010. Anti-tumor activity of a novel protein obtained from Tartary buckwheat. *International Journal of Molecular Sciences*, 11(12):5201–5211.

Huda, M. N., Lu, S., Jahan, T., Ding, M., Jha, R., Zhang, K., and Zhou, M. 2020. Treasure from garden: Bioactive compounds of buckwheat. *Food Chemistry*, 127653.

Inglett, G. E., Chen, D., Berhow, M., and Lee, S. 2011. Antioxidant activity of commercial buckwheat flours and their free and bound phenolic compositions. *Food Chemistry*, 125(3):923–929.

Ishii, S., Katsumura, T., Shiozuka, C., Ooyauchi, K., Kawasaki, K., Takigawa, S., and Ohba, K. 2008. Anti-inflammatory effect of buckwheat sprouts in lipopolysaccharide-activated human colon cancer cells and mice. *Bioscience, Biotechnology, and Biochemistry*, 72(12):3148–3157.

Izydorczyk, M. S., McMillan, T., Bazin, S., Kletke, J., Dushnicky, L., and Dexter, J. 2014. Canadian buckwheat: A unique, useful and under-utilized crop. *Canadian Journal of Plant Science*, 94(3):509–524.

Ji, X., Han, L., Liu, F., Yin, S., Peng, Q., and Wang, M. 2019. A mini-review of isolation, chemical properties and bioactivities of polysaccharides from buckwheat *(Fagopyrum Mill)*. *International Journal of Biological Macromolecules*, 127:204–209.

Jindal, N., and Saxena, D. C. 2015. Process standardization for isolation of starch from buckwheat *(Fagopyrum esculentum)* flour. *Journal of Agricultural Engineering and Food*, 25–28.

Kalinová, J. P., Vrchotová, N., and Tříska, J. 2019. Phenolics levels in different parts of common buckwheat *(Fagopyrum esculentum)* achenes. *Journal of Cereal Science*, 85:243–248.

Kang, M., Zhai, F.-H., Li, X.-X., Cao, J.-L., and Han, J.-R. 2017. Total phenolic contents and antioxidant properties of buckwheat fermented by three strains of *Agaricus. Journal of Cereal Science*, 73:138–142.

Kaur, M., Sandhu, K. S., Arora, A., and Sharma, A. 2015. Gluten free biscuits prepared from buckwheat flour by incorporation of various gums: Physicochemical and sensory properties. *LWT—Food Science and Technology*, 62(1):628–632.

Kayashita, J., Shimaoka, I., and Nakajyoh, M. 1995. Hypocholesterolemic effect of buckwheat protein extract in rats fed cholesterol enriched diets. *Nutrition Research*, 15(5):691–698.

Lee, L.-S., Choi, E.-J., Kim, C.-H., Sung, J.-M., Kim, Y.-B., Seo, D.-H., and Park, J.-D. 2016. Contribution of flavonoids to the antioxidant properties of common and Tartary buckwheat. *Journal of Cereal Science*, 68:181–186.

Li, H. 2019. Buckwheat. In J. Wang, B. Sun, & R. Tsao (Eds.), *Bioactive Factors and Processing 779 Technology for Cereal Foods*. Springer Nature Singapore Pte Ltd, 137–150.

Ma, Q., Zhao, Y., Wang, H.-L., Li, J., Yang, Q.-H., Gao, L.-C., and Feng, B.-L. 2020. Comparative study on the effects of buckwheat by roasting: Antioxidant properties, nutrients, pasting, and thermal properties. *Journal of Cereal Science*, 103041.

Mackèla, I., Andriekus, T., and Venskutonis, P. R. 2017. Biorefining of buckwheat *(Fagopyrum esculentum)* hulls by using supercritical fluid, Soxhlet, pressurized liquid and enzyme-assisted extraction methods. *Journal of Food Engineering*, 213:38–46.

Małgorzata, D., Dorota, G.-L., Sylwia, S., Monika, T., Sabina, B., and Ireneusz, K. 2020. Physicochemical quality parameters, antibacterial properties and cellular antioxidant activity of Polish buckwheat honey. *Food Bioscience*, 100538

Nam, T. G., Lim, T.-G., Lee, B. H., Lim, S., Kang, H., Eom, S. H., and Kim, D.-O. 2017. Comparison of anti-inflammatory effects of flavonoid-rich common and Tartary buckwheat sprout extracts in lipo-polysaccharide-stimulated RAW 264.7 and peritoneal macrophages. *Oxidative Medicine and Cellular Longevity*, 9658030.

Oh, M.-J., Choi, H.-D., Ha, S. K., Choi, I., and Park, H.-Y. 2018. Immunomodulatory effects of polysaccharide fraction isolated from *Fagopyrum esculentum* on innate immune system. *Biochemical and Biophysical Research Communications*, 496(4):1210–1216.

Peng, L. X., Zou, L., Wang, J. B., Zhao, J. L., Xiang D. B., and Zhao, G. 2015. Flavonoids, antioxidant activity and aroma compounds analysis from different kinds of Tartary buckwheat tea. *Indian Journal of Pharmaceutical Science*, 77(6):661–667.

Qiu, J., Liu, Y., and Yue, Y. 2016. Dietary Tartary buckwheat intake attenuates insulin resistance and improves lipid profiles in patients with type 2 diabetes: A randomized controlled trial. *Nutrition Research*, 36(12).

Quettier-Deleu, C., Gressier, B., Vasseur, J., Dine, T., Brunet, C., Luyckx, M., and Trotin, F. 2000. Phenolic compounds and antioxidant activities of buckwheat *(Fagopyrum esculentum Moench)* hulls and flour. *Journal of Ethnopharmacology*, 72(1-2):35–42.

Rois Mansur, A., Song, N.-E., Won Jang, H., Lim, T.-G., Yoo, M., and Gyu Nam, T. 2019. Optimizing the ultrasound-assisted deep eutectic solvent extraction of flavonoids in common buckwheat sprouts. *Food Chemistry*, 293, 438–445.

Sakač, M., Torbica, A., Sedej, I., and Hadnađev, M. 2011. Influence of breadmaking on antioxidant capacity of gluten free breads based on rice and buckwheat flours. *Food Research International*, 44(9):2806–2813.

Sim, U., Sung, J., Lee, H., Heo, H., Sang Jeong, H., and Lee, J. 2020. Effect of calcium chloride and sucrose on the composition of bioactive compounds and antioxidant activities in buckwheat sprouts. *Food Chemistry*, 312, 126075.

Sinkovic, L., Kokalj, D., and Vidrih, R. 2020. Vladimir Meglic Milling fractions fatty acid composition of common (*Fagopyrum esculentum* Moench) and Tartary (*Fagopyrum tataricum* (L.) Gaertn) buckwheat. *Journal of Stored Products Research*, 85:101551.

Škrabanja, V., and Kreft, I. 2016. Nutritional value of buckwheat proteins and starch. *Molecular Breeding and Nutritional Aspects of Buckwheat*, 169–176.

Stringer, D. M., Taylor, C. G., Appah, P., Blewett, H., and Zahradka, P. 2013. Consumption of buckwheat modulates the post-prandial response of selected gastrointestinal satiety hormones in individuals with type 2 diabetes mellitus. *Metabolism*, 62(7):1021–1031.

Sytar O., Brestic M., Zivcak M., and Tran L.P. 2016. The contribution of buckwheat genetic resources to health and dietary diversity. *Current Genomics*, 17(3):193–206.

Tang, C. H. 2007. Functional properties and *invitro* digestibility of buckwheat protein products: Influence of processing. *Journal of Food Engineering*, 82:568–576.

Tao, T., Pan, D., Zheng, Y. Y., and Ma, T. M. 2019. Optimization of hydrolyzed crude extract from Tartary buckwheat protein and analysis of its hypoglycemic activity *in vitro*. *IOP Conference Series: Earth and Environmental Science*, 295, 032065.

Tomotake, H., Shimaoka, I., Kayashita, J., Nakajoh, M., and Kato, N. 2002. Physicochemical and functional properties of buckwheat protein product. *Journal of Agricultural and Food Chemistry*, 50(7):2125–2129.

Tomotake, H., Yamamoto, N., Kitabayashi, H., Kawakami, A., Kayashita, J., Ohinata, H., and Kato, N. 2007. Preparation of Tartary buckwheat protein product and its improving effect on cholesterol metabolism in rats and mice fed cholesterol-enriched diet. *Journal of Food Science*, 72(7):S528–S533.

Torbica, A., Hadnađev, M., and Dapčević Hadnađev, T. 2012. Rice and buckwheat flour characterisation and its relation to cookie quality. *Food Research International*, 48(1):277–283.

Verardo, V., Glicerina, V., Cocci, E., Garrido Frenich, A., Romani, S., and Fiorenza Caboni, M. 2018. Determination of free and bound phenolic compounds and their antioxidant activity in buckwheat bread loaf, crust and crumb. *LWT—Food Science and Technology*, 87:217–224.

Wang, L., Wang, L., Qiu, J., and Li, Z. 2020. Effects of superheated steam processing on common buckwheat grains: Lipase inactivation and its association with lipidomics profile during storage. *Journal of Cereal Science*, 103057.

Wefers, D., and Bunzel, M. 2015. Characterization of dietary fiber polysaccharides from dehulled common buckwheat (*Fagopyrum esculentum*) seeds. *Cereal Chemistry Journal*, 92(6):598–603.

Wronkowska, M., Zielińska, D., Szawara-Nowak, D., Troszyńska, A., and Soral-Śmietana, M. 2010. Antioxidative and reducing capacity, macroelements content and sensorial properties of buckwheat-enhanced gluten-free bread. *International Journal of Food Science & Technology*, 45(10):1993–2000.

Wu, S.-C., and Lee, B.-H. 2011. Buckwheat polysaccharide exerts antiproliferative effects in THP-1 human leukemia cells by inducing differentiation. *Journal of Medicinal Food*, 14(1–2):26–33.

Yao, Y., Shan, F., Bian, J., Chen, F., Wang, M., And Ren, G. 2008. D-chiro-Inositol-enriched Tartary buckwheat bran extract lowers the blood glucose level in KK-Ay mice. *Journal of Agricultural and Food Chemistry*, 56:10027–10031.

Zemnukhova, L. A., Tomshich, S. V., Shkorina, E. D., and Klykov, A. G. 2004. Polysaccharides from buckwheat production wastes. *Russian Journal of Applied Chemistry*, 77(7):1178–1181.

Zhang, D., Wang, L., Tan, B., and Zhang, W. 2020. Dietary fiber extracted from different types of whole grains and beans: A comparative study. *International Journal of Food Science & Technology*, 55(5):2188–2196.

Zhang, W., Yang, Q., Xia, M., Bai, W., Wang, P., Gao, X., and Gao, J. 2019. Effects of phosphate fertiliser on the physicochemical properties of Tartary buckwheat (*Fagopyrum tataricum (L.) Gaertn.*) starch. *Food Chemistry*, 125543.

Zhang, Z.-L., Zhou, M.-L., Tang, Y., Li, F.-L., Tang, Y.-X., Shao, J.-R., and Wu, Y.-M. 2012. Bioactive compounds in functional buckwheat food. *Food Research International*, 49(1), 389–395.

Zhou, X., Hao, T., Zhou, Y., Tang, W., Xiao, Y., Meng, X., (2015b). Relationships between antioxidant compounds and antioxidant activities of Tartary buckwheat during germination. *Journal of Food Science and Technology*, 52(4):2458–2463.

Zhou, X., Wen, L., Li, Z., Zhou, Y., Chen, Y., and Lu, Y. 2015a. Advance on the benefits of bioactive peptides from buckwheat. *Phytochemistry Reviews*, 14(3):381–388.

Zhou, X.-L., Yan, B.-B., Xiao, Y., Zhou, Y.-M., and Liu, T.-Y. 2018. Tartary buckwheat protein prevented dyslipidemia in high-fat diet-fed mice associated with gut microbiota changes. *Food and Chemical Toxicology*, 119:296–301.

Zhou, Y., Ma, Y., Li, L., and Yang, X. 2019. Purification, characterization, and functional properties of a novel glycoprotein from Tartary buckwheat *(Fagopyrum tartaricum)* seed. *Food Chemistry*, 125671.

Zhu, F. (2020). Dietary fiber polysaccharides of amaranth, buckwheat and quinoa grains: A review of chemical structure, biological functions and food uses. *Carbohydrate Polymers*, 116819.

Zieliński, H., Szawara-Nowak, D., and Wronkowska, M. 2020. Bioaccessibility of anti-AGEs activity, antioxidant capacity and phenolics from water biscuits prepared from fermented buckwheat flours. *LWT*, 109051.

16

Functional Potential of Quinoa

Nisha Chaudhary
Agriculture University, Jodhpur, Rajasthan, India

Priya Dangi
University of Delhi, New Delhi, India

Rajesh Kumar, Sunil Bishnoi and Kusum Ruhlania
Guru Jambheshwar University of Science & Technology, Hisar, Haryana, India

CONTENTS

16.1 Introduction

Quinoa (*Chenopodium quinoa* Willd.) is known as superfood. It is an annual herbaceous crop from the family *Chenopodiaceae*, the same as of spinach and beets. The genus *Chenopodium* grows worldwide, including nearly 250 identified species. This species is native of the Andean region of South America and was domesticated by the regional people of mainly Peru and Bolivia about 3,000 years ago. Quinoa has been known as "mother seed" by Incas and was considered a sacred plant. It possesses excellent quantity of proteins and quality as well in terms of an extraordinary balance of essential amino acids resembling the biological value of protein in milk (Farro, 2008; Jancurová, Minarovičová, & Dandár, 2009; Spehar, 2007). Overall, this plant exhibits marvellous nutritional composition and bizarre growing conditions as it requires sandy-loam to loamy-sand soils with a pH range between 6.0 and 8.5. It can bear high temperatures up to 35°C during cultivation (Farro, 2008). In spite of having cereal characteristics, like its botanical aspects of the occurrence of a panicle-type inflorescence, it does not belong to the *Gramineae* family—this is the overall reasoning behind its designation as a "pseudocereal" (Farro, 2008; A. V. Vega-Gálvez et al., 2010). In comparison to cereals, quinoa contains more protein and has a balanced ratio of essential amino acids, lipid content, dietary fiber, vitamins B1, B2, B6, C, E, and minerals, particularly calcium, phosphorus, iron, and zinc. Quinoa, is especially rich in lysine and sulphur-containing amino acid, which makes it a more complete protein than most vegetables. Moreover, the essential amino acid composition of quinoa is recognized as close to the ideal protein equilibrium recommended by the Food and Agriculture Organization of the United Nations (FAO) and is similar to the biological value of protein in milk (Filho et al., 2017). Quinoa also contains antinutritional factors like saponins, oxalates, phytic acid, trypsin inhibitors, and tannins in the seed coat of the grain in higher concentrations, with saponins as the most prevalent among all. Saponins induce a bitter taste in quinoa-based foods. Saponins are soluble in water and can be readily removed by washing with water and by applying abrasion for removing the seed coat (Borges, Bonomo, Paula, Oliveira, & Cesário, 2010; Comai, Bailoni, Zancato, Costa, & Allegri, 2007).

   In addition to being a potential nutritional food crop, it is gluten-free, which enables a greater scope to provide a variety of nutritious food products for patients with wheat sensitivity, gluten intolerance, and celiac disease. The absence of gliadins, one of the gluten forming proteins of wheat, and the presence of analogous protein fractions to gliadin (as present in various millets like barley and rye), proves the adequacy of quinoa for being the base material for preparing gluten-free food products, a mandatory food option for celiac patients. Celiac disease causes chronic inflammation of the small intestinal mucosa, ultimately leading to villous atrophy, diarrhea, and weight loss. The related studies have noted the increasing prevalence of celiac disease worldwide, varying from 1:100 to 1:300 persons out of a healthy population (Filho et al., 2017; Niro, D'Agostino, Fratianni, Cinquanta, & Panfili, 2019). The peculiar characteristics of quinoa are the reasons behind its increasing popularity among people of developed countries, especially from those who covet substitute foods with high nutritional value. In order to fulfil such international demands, cultivation and exports have been increased from Andean countries primarily (Farro, 2008). Despite its having such magnificent attributes, the majority of the world's population is not fully aware of the benefits of quinoa, which accounts for the limited marketing worldwide. Therefore, intense studies are required to have a better understanding regarding this efficient "pseudocereal," for identifying its major components, illustrating its nutritional and functional advantages for human health, and for removing the effects of antinutritional factors (Alvarez-Jubete, Arendt, & Gallagher, 2010).

16.2 History and Origin

Quinoa is a food plant belonging to the family *Chenopodiaceae* and genus *Chenopodium*. It has been cultivated for thousands of years back as a food in the Andean regions of Chile, Ecuador, Peru, and Bolivia. Its cultivation is the oldest ever recorded. Ancient populations recognized the extraordinary nutritional importance of quinoa, which they named "golden grain" and designated as a sacred food (Farro, 2008).

The Inca regarded Quinoa as "the mother of all grains" and considered it as an offering from their gods. Other than food, quinoa was utilized for medicinal purposes. It was a dominant staple for the Inca population and is still a prime source of food for their native descendants as the Quechua and Aymara people. Legends stated that the Inca emperor ceremoniously planted quinoa seeds initially every year. Primarily, quinoa seeds were mixed in soup preparations, used as a basic cereal food, and also fermented to prepare beer or chichi (traditional Andes' drink) (Vega-Gálvez et al., 2010). After the outstanding traditional importance as a food in the highlands of the Andes in South America, presently, quinoa has achieved immense popularity as a health food/superfood in North America, Europe, Australia, Japan, and India. Now, quinoa is cultivated or tested in 95 countries around the world (Singh, Singh, & Singh, 2016).

Quinoa is not considered as a true grain because it is a dicot plant (typical cereal grains are monocots) and is instead a fruit. Ultimately, it has been known as a pseudocereal or even a pseudo-oilseed due to its unique composition and exceptional balance of protein and fat (Cusack, 1984).

16.3 Nutritional Properties of Quinoa

16.3.1 Carbohydrates

Carbohydrates consist of one of the largest groups of organic compounds established in nature, and collectively with proteins, these forms the main components of living organisms, being the most abundant and economical source of energy for humans. Carbohydrates can be classified according to their degree of polymerization in three groups: mono-, oligo-, and polysaccharides (Mahan & Escott-Stump, 2010). Starch is the major carbohydrate polymer of quinoa and is present between 52–69% (d.m.). Total dietary fiber is near the value found in cereals (7–9.7% d.m.), wherein the embryo contains higher levels than those in perisperm. The soluble fiber content is reported to range from 1.3–6.1% (d.m.). Quinoa still presents around 3% simple sugars, mostly maltose, followed by D-galactose and D-ribose, plus low levels of fructose and glucose (Abugoch James, 2009). Dietary fiber's indigestibility in the small intestine has positive effects. Thus, more content of fibers in quinoa can enhance digestibility by assisting the absorption of the other nutritional components present in quinoa in large intestine (Ogungbenle, 2003). Quinoa carbohydrates can be considered as nutraceuticals because they have beneficial hypoglycemic effects and induce reduction of free fatty acids. Previous studies revealed that in some varieties of quinoa, insoluble and total dietary fiber are decreased due to the extrusion process. On the contrary, soluble dietary fiber content improved throughout extrusion. During extrusion, high temperature and shear stress may disrupt chemical bonds and create smaller and more soluble particles. There is, therefore, a transformation of some of the components of the insoluble fiber into soluble fiber during extrusion (Repo-Carrasco, Espinoza, & Jacobsen, 2003).

Starch is the main biopolymer component of plants, naturally occurring as granules of many types, shapes, and sizes. Quinoa starch granules may be found as single entities or collective forms of spherical of oval complex structures and have a polygonal shape with diameter 0.6–2.2 μm, comparatively smaller than other cereal grains. The extremely small size can be beneficially exploited and used in blends with synthetic polymers in the preparation of biodegradable packaging. Quinoa starch has excellent freeze-thaw stability and can be used as an ideal thickener for soups, condiments, and sauces. It can be used in capacities that require a low gelation temperature, stability during storage at low temperatures, and in some other functions where resistance to retrogradation is required. It is also used to make creamy and smooth textures similar to fats (Abugoch James, 2009; A. V. Vega-Gálvez et al., 2010).

Quinoa starch has an average molar mass of 11.3×106 g/mol, i.e. more than wheat starch and lower than rice and waxy maize starch (Park, Ibáñez, & Shoemaker, 2007). With a minimum degree of polymerization of 4,600 glucose units, maximum of 161,000 and 70,000 average weights, it is a highly branched structure. The length of the chain depends on the botanical source ranging from 500–6,000 glucose units. The amylose content of quinoa starch varies between 3–22%, which is more than some barley and rice varieties but lower than wheat or maize (Abugoch James, 2009). As described by Tang, Watanabe, and Mitsunaga (2002), amylose has an average degree of polymerization of 920, much lower than that of barley (1,660). According to Tari, Annapure, Singhal, and Kulkarni (2003) quinoa starch contains 77.5% amylopectin, which is higher than some wheat and rice varieties. Due to higher content of amylopectin, it has a length distribution similar to waxy starch amylopectin, 317 branching and average polymerization degrees of 6,700 glucose units per fraction. Quinoa amylopectin has small number of longer chain 13–20 units in comparison with starches from other sources (Abugoch James, 2009; Tang et al., 2002).

X-ray diffraction (XRD) has been used to explain the structure and crystallinity of the granules of starches. On the basis of biological origin, the content and proportion of amylose and amylopectin, the branching pattern of amylopectin, and the starch granules show three types of diffractions linked with dissimilar crystalline polymorphic forms: Type A (cereal grains), Type B (tubers), and Type C (crystals A and B simultaneously present on the same granule) (Kumar & Khatkar, 2017; Lopez-Rubio, Flanagan, Gilbert, & Gidley, 2008). Quinoa starches have a diffraction pattern of X-rays with reflections in 15.3 degrees, 17 degrees, 18 degrees, 20 degrees, and 23.4 degrees, typical of cereal starches (type A). Relative degree of crystallinity is between 35–43%, which is higher than barely starches and lower than amaranth and buckwheat starches. The relative crystallinity is similar to that of waxy maize and barley starches (Abugoch James, 2009; Tang et al., 2002). Differential scanning calorimetry (DSC) thermograms of quinoa starches showed two types of thermal transitions, starch gelatinization and other one for the amylose-lipid complex. The gelatinization properties of starches depends on the botanical source, varieties, and other factors like the size of granules and proportion of granule types and type of crystalline organization of the starch granule. The peak, initial, and final temperatures of quinoa starches were 62.5°C, 54.5°C and 71.3°C, respectively (Tang et al., 2002). According to Abugoch James (2009), enthalpy change was 10.3 J/g, and 0.186 J/g. Quinoa starch gelatinizes at low temperatures, reported between 62–67°C. The gelatinization temperature of quinoa starch is lesser than amaranth and waxy barley starch and greater than wheat and rice upon some extent. Compared to wheat and barley starches, quinoa starch has high maximum viscosity, water absorption capacity (WAC), and greater swelling power. Quinoa starch shows higher paste stability because of a lower pasting temperature and a lower rate of starch breakdown i.e., increased shear resistance during heating. Starch viscosity in a Rapid Visco Analyzer (RVA) revealed similar pattern as cereal and root starch paste. It shows excellent stability even under freezing and retrogradation processes (Abugoch James, 2009; Tang et al., 2002).

16.3.2 Proteins

Quinoa seed got attention as a novel food item because of its nutritional value and quality of its protein content. Quinoa protein has an amino acid composition near to the ideal protein recommended by the FAO and is analogous to milk. Previous literature on quinoa protein proved a quality similar to casein and other milk proteins for quinoa protein (Koziol, 1992; Miranda et al., 2011). The absence of gluten forming proteins (gliadins present in wheat) and other fractions similar to gliadins, makes it more appropriate to use quinoa protein in the production of gluten-free food products, a proper nutritional diet for patients with celiac disease (Almeida & Sá, 2009; Borges et al., 2010). The regulation of metabolic processes, enzyme formation, energy requirements, and building and maintenance of body tissue are the main functions of dietary protein. Proteins in the form of lipoprotein are involved in the carrying of phospholipids, cholesterol, and vitamins that are fat soluble. According to Koziol (1992), on dry matter basis, quinoa seed contains 13.8–16.5% protein content, with an average of 15%. In comparison to other cereal grains, quinoa protein content is higher and near wheat protein content on a dry matter basis (Comai, Bertazzo, et al., 2007).

Quinoa Protein vacuoles are mostly concentrated and enclosed by the embryo surrounding the perisperm, which also contains other nutritional components such as lipids (Ando et al., 2002; Taylor & Parker, 2002). The storage proteins of quinoa grains contain albumin (2S) 35% and globular chenopedin (globulins 11S) 37%, these are the major portion of protein. Prolamins are present in low concentration and it varies according to different varieties (Abugoch James, 2009). Globulins are soluble in dilute saline solutions and insoluble in water achieved in a crystalline state (Thanapornpoonpong, Vearasilp, Pawelzik, & Gorinstein, 2008). The second structure of globulin consists of 20% of α-helices, 35% of β-sheets, and 45% of a periodic structure (Drzewiecki, Delgado-Licon, Haruenkit, et al., 2003). Additionally, it contains a hexamer structure made up of six pairs of small basic (B) and larger acidic (A) polypeptides with a molecular mass of 22–23 kDa and 32–39 kDa, correspondingly linked by a single disulfide bond. These subunits are connected by noncovalent interactions. Quinoa globulins are rich in glutamic acid, asparagine/aspartic acid, arginine, serine, leucine, and glycine, excluding methionine and cysteine (sulphur containing amino acids). In the subunits of globulins, A is considerably higher in glycine, methionine, and histidine and lower in alanine, cysteine, leucine, and tyrosine compared to the B subunit (Brinegar & Goundan, 1993). The functional properties of globulins and structure are dependent on pH and most stable at a range of 6–9, in which they have hexamer structure. Globulins lose their tertiary structure at an acidic pH (Marcone, Yada, Aroonkamonsri, & Kakuda, 1997). In addition, these can undergo subunit dissociation at a higher pH because of electrostatic repulsion. The structure of quinoa proteins, especially globulins, can be influenced by extraction conditions, such as temperature, pH, ionic strength, and enzymatic modifications (Abugoch James, 2009; Ruiz, 2016a).

The 2S-type of albumin is the second most abundant protein estimated to account for 35% of the total protein of quinoa (Brinegar, Sine, & Nwokocha, 1996). The albumin is readily soluble in water and can be extracted from a suspension in water (Osborne, 1924). In addition, the secondary structure of quinoa, albumin, is made up of 4% of α-helices, 50% of β-sheets, and 46% of aperiodic structure (Drzewiecki, Delgado-Licon, Haruenkit, et al., 2003). Albumin protein has a heterodimer structure consisting of about 30–40 and 60–90 residues providing polypeptides with a molecular mass of about 8–9 kDa associated by two disulphide bonds (Brinegar et al., 1996). The 2S albumin has been found to show significant levels of polymorphism. Such polymorphic variants lead to various properties (Sun, Leung, & Tomic, 1987). Quinoa albumin contains high levels of cysteine (15.6 mol%), arginine, histidine, and lysine whereas, it has comparatively low content of methionine (0.6 mol%) (Brinegar et al., 1996). While 2S albumin types are recognized as allergens in various nuts such as hazelnut, they show some kinds of antifungal activity by displaying synergistic activity and promote the permeabilization of the plasma membrane of phytopathogens (Garino et al., 2010). Furthermore, it shows a high rate of nonprotein tryptophan, that is more easily absorbed and can contribute to increasing the availability of this amino acid in the brain, thereby influencing the synthesis of the serotonin neurotransmitter. Quinoa protein contains lower amounts of prolamins (0.5–7% of total protein), making quinoa a better gluten-free alternative (Thanapornpoonpong et al., 2008).The nutritional quality of the protein is determined by the proportion of the essential amino acids, namely those that cannot be synthesized by animals and must, therefore, be supplied in the diet. Quinoa contains a very good amino acid profile, including appropriate concentrations of essential amino acids desirable for development and maintenance of metabolic activities with required bioavailability (Miranda et al., 2010). Above all, the lysine content of its protein is virtually twice that of wheat, 25% more than rice, and is also higher than that suggested by the FAO/ World Health Organization (WHO) prototype concerning essential and nonessential amino acids needed for usual metabolic activities (Aluko & Monu, 2003; Comai, Bertazzo, et al., 2007; WHO, 2007). As per the given recommendations of FAO/WHO, quinoa would fulfil more than 150% of the daily essential amino acid needs of school children and also more than 200% of the daily essential amino acid requirements of adults, due to no deficiency of any essential amino acid (Abugoch James, 2009; Miranda et al., 2010). The protein digestibility and bioavailability of amino acids in quinoa differs according to the variety and handling that the grains are given, and even more significantly with cooking (Abugoch James, 2009; Comai, Bertazzo, et al., 2007; Koziol, 1992).

Quinoa protein solubility is also connected to the hydrophilic-hydrophobic balance of the protein and its interface with the solvent. Surface hydrophobicity is a useful attribute for the analysis of the denaturation and accumulation of proteins often calculated by the first incline of fluorescence intensity against protein concentration plots. Past research on this demonstrated that pH and heating time could significantly influence quinoa globulins surface hydrophobicity, which enhanced with heating at 100°C for 5 minutes at all pH conditions (6.5, 8.5, and 10.5) due to protein unfolding in the tertiary structure (Mäkinen, Zannini, Koehler, & Arendt, 2016).

16.3.3 Lipids

Due to the quality and quantity of quinoa lipid fraction, quinoa has been considered as an alternative oilseed crop. Quinoa seed contains about 6.9% lipids, which is more compared to other cereal grains. According to previous studies, iodine value of quinoa oil is 54%, which proves that it presents a higher amount of unsaturated fatty acids. Quinoa oil is highly composed of unsaturated fatty acids that provide health benefits as compared to saturated fatty acids. Its acid value and peroxide values are 0.5% and 2.44%, respectively, which depicts that these lipids are stable and not oxidized early when exposed to heat and oxygen. The molecular weight or chain length of fatty acids can be determined by saponification value, which is the number of milligrams of potassium hydroxide (KOH) required to saponify 1g of lipid or fat under specific conditions (Ogungbenle, 2003). According to Lindeboom, Chang, Falk, and Tyler (2005), quinoa lipids have a 192% saponification value, which is lower than butter (240%) and soy bean oil (194%). Palmitic acid is the major saturated fatty acid, comprising around 10% of the total fatty acid content. It contains about 87% of unsaturated fatty acids including oleic acid, linoleic acid, and linolenic acid as shown in Table 16.1. Linoleic acid is the main group that comprises 52% of the total lipid fatty acids. Seed germ or embryo fatty-acid content is higher that the perisperm and seed coat. If we calculate total fatty acid content in quinoa (6.5%), we find it is lower than wheat germ oil and higher than corn oil (Ando et al., 2002). Commercialized oils crops are much higher in total fatty-acid content, varying from 20% in olive oil to 60% in walnut oil. In quinoa oil, saturated fatty acid is 11% of the total lipids i.e. palmitic (10%) and stearic acid 0.8%. According to Hu, Manson, and Willett (2001), stearic acid was reported to reduce the levels of high density lipoprotein, the good cholesterols that defend against cardiovascular disease (CVD). The monounsaturated fatty acids level in quinoa oil is roughly 25%, less than that of olive oil and canola. According to metabolic studies, rising intake of linoleic acid may develop lipid profiles, lower cholesterol (Chait et al., 1993), and improve insulin sensitivity (Lovejoy, 1999).

Quinoa oil contains 52% linoleic acid, which is lower than corn oil (59%) and wheat germ oil (55%). This value is more than walnut oil (51%) and much greater than canola, flax seed oil, and olive oil. Alpha (α)-linoleic acid (ALA), i.e. n-3 fatty acid, is a functional food component that is gaining attention from food consumers and processors. This is important to infants and teenagers for the development of the nervous system and brain. Alpha-linoleic acid is precursor for the docosahexaenoic acid (DHA) and eicosapentaenoic acid (EPA) by the unsaturation chain making longer pathways (Cunnane & Thompson, 1995). Both these acids enhance brain and nerve functioning and improve immunity. Transformation of linoleic acid to arachidonic acid is inhibited by α-linoleic acid because inflammation is controlled

TABLE 16.1

Fatty Acid Composition of Quinoa Seed According to Part of Grains

Fatty Acid	Whole Grain	Perisperm	Embryo
Myristic (C14:0)	0.2	0.1	0.2
Palmitic (C16:0)	10.2	10.8	9.5
Stearic (C18:0)	0.8	0.7	0.9
Oileic (C18:1)	24.9	29.5	19.7
Linoleic (C18:2)	52.5	49.0	56.4
Linolenic (C18:3)	10.1	8.7	11.7
Others	1.4	1.2	1.6

Source: Ando et al., 2002; USDA, 2011.

in the human body. Unnecessary inflammation can cause obesity, diabetes, and cardiovascular disease (Wymann & Solinas, 2013). Quinoa oil contains about 10% ALA that is less than that of flax oil (58%) and comparatively higher than wheat germ oil, coconut oil, and canola oil. The ratio of ALA to linoleic acid, 1:10 to 1:5, is considered to prevent or control heart disease (Hu et al., 2001). The ratio of ALA and linoleic acid in quinoa oil is close to 1:5. The composition of quinoa oil is well balanced and advantageous to an individual's well-being.

16.3.4 Minerals

Minerals are inorganic compounds beneficial for important functions of the body and are not produced by living beings. Imbalances may occur due to a low bioavailability of minerals. Among the well-known minerals are calcium, phosphorus, iron, potassium, sulphur, sodium, magnesium, zinc, copper, selenium, and chromium. These micronutrients are necessary to consume through diet to supply the daily needs and to enhance the bioavailability of the minerals (Vega-Gálvez et al., 2010). Ash content of quinoa is 3.4%, which is more than wheat, rice, and other cereals. Therefore quinoa seed contain higher amount of minerals. It contains calcium, iron, and magnesium (0.26%), more than wheat and corn. Several authors have reported large quantities of iron and calcium in quinoa. Brend, Galili, Badani, et al. (2012) stated large difference in the concentration of minerals in quinoa seeds. The variation in the values was due to genotypes, regions and environment, and applied fertilizers. According to Ogungbenle (2003), potassium is the most abundant mineral in quinoa grain, followed by magnesium and phosphorus, while iron confirmed the lowest value. Ando et al. (2002) analyzed the mineral content in quinoa grain, and reported higher levels of calcium, phosphorus, iron, potassium, magnesium, and zinc among the minerals analyzed. As reported by Alvarez-Jubete et al. (2010), calcium, magnesium, and iron are the major minerals lacking in gluten-free foods. Although quinoa grains contain more iron than other cereal grain, the bioavailability is affected to some extent by antinutritional factors like saponins and phytic acids. The process of cleaning and washing with water for saponin removal influences the mineral content. A considerable reduction is reported by Ruales and Nair (1993): 46% in calcium, 28% in iron, and 27% in manganese. An unnecessary quantity of dietary phytate has a negative effect on the mineral balance since it binds with inorganic multivalent minerals, such as Fe^{3+}, Zn^{2+}, Ca^{2+} e Mg^{2+}, making them less bioavailable for absorption. Distribution of minerals in grains was reported by Konishi, Hirano, Tsuboi, and Wada (2004) using energy dispersive X-raya combined with scanning electron microscopy (SEM). According to their study, potassium, magnesium, and phosphorus were located in the embryonic tissues. Since phytin cells are present in protein bodies of embryonic cells, it is recommended that phosphorous is allocated to phytic acid and that potassium and magnesium will help structure the phytates. Pericarp contains calcium and potassium where the cell wall is thickly developed.

16.3.5 Vitamins

Cereal-based foods are one of the major sources of vitamin B complex. Bread and rice contains 45% of thiamin, 30% of riboflavin, 28% of niacin, 14% of pyridoxine, and 19% of folate (Khan & Shewry, 2009). Vitamin B complex provides important function for the human body, while folate is known to affect the health of the neural system. According to Yang et al. (2010), the rate of neural tube defects were reduced in the United States by the promotion of folate-rich cereal food products. In quinoa, riboflavin (B2) is approximately 0.35mg/100 g and folate (B9) is 0.781mg/100 g, significantly higher than wheat, rice, and barley and even the recommended daily intake (RDI). Quinoa pyridoxine (B6) content is 0.487mg/100 g, which is 25% of the RDI as shown in Table 16.2. Folate, or folic acid, helps in embryo development and production of red blood cells. Quinoa is an important source of folate. Tocopherols or vitamin E is the chain-breaking antioxidant that protects the human body from free radical oxidation (Kline, Lawson, Yu, & Sanders, 2007). Vitamin E content in quinoa oil makes it is more stable during processing and storage, but it still contains high amounts of unsaturated fatty acids (Ng, Anderson, Coker, & Ondrus, 2007).

TABLE 16.2

Vitamin Content of Quinoa

Vitamins	Value in Quinoa Seed (mg/100 g)	Recommended Daily Intake (RDI) (mg/100 g)
Vitamin B1 (Thiamin)	0.31	1.5
Vitamin B2 (Riboflavin)	0.34	1.7
Vitamin B3 (Niacin)	1.24	20
Vitamin B6 (Pyridoxine)	0.487	2
Vitamin B9 (Folic acid)	0.781	0.4
Vitamin C (Ascorbic acid)	4.0	60
Vitamin E (Tocopherols)	5.37	30

Source: USDA, 2011.

16.3.6 Phytochemicals

16.3.6.1 Phenolic Compounds

Phenolic compounds are located in the outer layer of the grains and are major components of phytochemicals present in cereals. These compounds generally contain at least one phenol ring. According to the number of phenol rings, phenolic compounds are divided into simple phenols and polyphenols. As described by Khan and Shewry (2009), phenolic acids with one aromatic ring and one or more hydroxyl groups are simple phenols, while polyphenols includes phenolic acid dehydromers, lignins, tannins, and flavonoids that have three or more phenol rings. Guo and Beta (2013) revealed that antioxidant activity of phenolic compounds were higher, determined by methods of 2, 2-diphenyl-1-picylhydrazyl (DPPH) and ferric reducing antioxidant power (FRAP). Gallic acid equivalent (GAE) can be used to total phenolic content of the samples. Quinoa contains 71.7 mg GAE/100 g of total phenolic acid, which is more than the content of wheat, amaranth, and other cereals (Alvarez-Jubete et al., 2010). According to varieties and the color of seeds, quinoa seed showed variations in phenolic compounds and flavonoid content. Red quinoa contains 90% more total flavonids than simple grains (Brend et al., 2012).

16.3.6.2 Phenolic Acid

As described earlier, phenolic acids are phenolic compounds with a single phenol ring. These are bound to cell wall structural components and sugars by the ether, acetyl, or ester linkages (Yuan, Wang, & Yao, 2005). A few phenolic acids are free acids. Gallic acid and rosmarinic acid both have the highest potential of antioxidant activity (Soobrattee, Neergheen, Luximon-Ramma, Aruoma, & Bahorun, 2005). These acids have different functions in plants such as enzyme activity, microbial tenancy, nutrient uptake, and protection against pathogenic microbes (Kroon & Williamson, 1999). Some phenolic acids in cereals do not provide high antioxidant activity like ferulic acid, but these acids cross links for the making of cell walls and dietary fiber (Yadav, Moreau, & Hicks, 2007). Mattila, Pihlava, and Hellström (2005) analyzed the phenolic acids in quinoa through a high-performance liquid chromatography method. Table 16.3 shows the type and content of phenolic acids in quinoa in comparison with other cereals.

16.3.6.3 Flavonoids

Flavonoids are also a group of phenolic compounds containing 2-phenyl1-1,4-benzopyrone as a structural difference and can be classified into flavones, isoflavons, anthocyanidins, and proanthcyanins. Flavonoids like epicatechin and quercetin provide antioxidant activity and reduce the risk of coronary heart disease and diabetes (Giménez, Moreno, López-Caballero, et al., 2013). Quinoa has quercetin glycosides 43.4 μmol/100 g on dry weight basis and 41% μmol/100 g on dry weight basis of kaempferol glycosides (Alvarez-Jubete et al., 2010). A group of flavonoids called isoflavones act as phytoestrogens in humans and animals and stimulate osteoprogerin secretion and provide antioxidant properties (Li, Kong, Ahmad,

TABLE 16.3

Comparison of Phenolic Acid Content in Quinoa with Other Cereal Grains

Phenolic Acid	Quinoa (μg/g)	Rice (μg/g)	Wheat (μg/g)	Corn(μg/g)
Ferulic acid	430	240	870	380
Caffic acid	40	Not detected	35	25
Vanilic acid	43.5	7.8	14.9	4.6
Hydroxybenzoic acid	76.8	15	7.5	5.7
Gallic acid	320	Not detected	Not detected	Not detected
Cinamic acid	10	Not detected	Not detected	Not detected

Source: Mattila et al., 2005; Pa'sko et al., 2009.

Bao, & Sarkar, 2013). According to Vega-Gálvez et al. (2010), daidzein and genistein are the main isoflavones found in quinoa grains. Previous literature has reported that quinoa contains 102.4 mg/100 g (d.m.) of anthocyanin content, which is higher than soy bean, sorghum, rice, and amaranth (Pa'sko et al., 2009). These flavonoids provide color to the plants ranging from red to violet. A small amount of anthocyanin intake may lower blood pressure (BP) and reduce fat levels and inflammation in heart patients.

16.3.6.4 Carotenoids

Carotenoids are the precursor of the vitamin A. These are antioxidants as well as a colorant component of plants (Dini, Tenore, & Dini, 2010). Carotenoid varies among quinoa, with sweet quinoa having 0.4mg/10 g of carotenoids and bitter quinoa containing 0.08 mg/10 g (Hidalgo & Brandolini, 2008). These values are higher than similar types of cereals and grains. Carotenoids can satiate singlet or free oxygen molecules, which are most reactive molecules in light-induced oxygen. Carotenoids can be beneficial for eye health as reported by Alves-Rodrigues and Shao (2004).

16.4 Antinutritional Factors

Antinutritional factors are the chemical compound known as "secondary Metabolites." These compounds are associated with foods of plant origin. These antinutritional factors limit wider food utilization, causing negative effects to human health, reduce the bioavailability of minerals, and interfere with digestibility. Different types of antinutritional factors are identified in vegetables. In quinoa seeds, the major antinutritional factors are saponins, phytic acid, tannins, trypsin inhibitor, nitrates, and oxalates.

16.4.1 Saponins

Saponins are the primary antinutritional factor in quinoa and are located in the outer layer of quinoa seeds. They are a group of triterpenoid glycosides, act as a defense factor against insects, and herbivorous. They are bitter in taste and foam in water. Saponins are detergent-like molecules that are soluble in water or methanol. They are derivatives of the four main triterpenes or sterols in quinoa called sapogenins: hederagenin, oleanolic acid, phytolaccagenic acid, and 30-*O*-methyl-espergulagenate (Ridout, Price, DuPont, Parker, & Fentwick, 1991).The sugars present are: glucose, arabinose, and galactose (Farro, 2008). On the basis of its saponin content, quinoa is classified into two varieties: sweet (<0.11%) and bitter (>0.11%).

Saponins have toxic properties that alter cell membrane permeability, causing damage to intestinal mucosal cells. The toxicity depends on the saponin structure and sensitivity of the exposed organism. The main negative effect of saponin consumption is the reduction in mineral and vitamin absorption (Jancurová et al., 2009). The toxic effect of saponins limit the utility of quinoa. Now, they are considered health-promoting compounds as a result of their hypocholesterolemic, antiallergic, fungicide, and antioxidant properties. Saponins have many industrial applications such as use in soaps, shampoos, detergents, beer, toothpaste, and fire extinguishers (Vega-Gálvez et al., 2010). They are also used in pharmaceutical industry because of their ability to modify the small intestine permeability, which helps in absorption of certain drugs (Francis, Kerem, Makkar, & Becker, 2002). Due to its bitter characteristic, this substance is removed from the grain by different methods for both consumer acceptance and to improve its sensory quality. Saponins are removed through wet methods (washing in cold water) and dry methods (toasting, extrusion, rubbing of grains, or mechanical abrasion) or combination of both methods. On a household level, the wet method is mostly used because of the high water solubility of saponins. This method has both economic and ecological drawbacks because of the higher consumption of water for washing the grain and for drying (Farro, 2008; Jancurová et al., 2009). On a commercial level, abrasive de-hulling (dry method) is used for the removal of saponins, but the main drawback of this method is the loss of vitamins and mineral content with the layer of bran. For this reason, the wet method combined with polishing is used to reduce saponin content of quinoa grain (Borges et al., 2010).

16.4.2 Phytic Acid

The principle storage form of phosphorus is phytic acid, which has a saturated cyclic structure. Quinoa seed contains 0.7–1.2% phytate values, which is close to wheat, barley grains, rye, lentil, and faba beans (Filho et al., 2017). In quinoa, phytic acid is not only present in the outer layers of grain, but is uniformly distributed in the endosperm as well.

Phytates are capable of forming insoluble complexes with bivalent minerals such as Ca^{2+}, Mg^{2+}, Fe^{2+}, and Zn^{2+} therefore compromising the bioavailability of these minerals in the gastrointestinal tract. Compared to other cereal grains, quinoa is rich in iron so the phytates could alter the bioavailability of this mineral. Some treatments, like soaking, germination, and lactic acid fermentation of quinoa seeds, resulted in a reduced phytate content and the improved solubility of iron (Gupta, Gangoliya, & Singh, 2015). The most effective treatment for reducing phytic acid are soaking, germination, and fermentation. There is an approximately 30% reduction in the content of phytic acid in quinoa seeds by the brushing and rinsing method. The content of phytic acid in nonprocessed quinoa grains was 10.4 mg/g, while in processed grains it was 7.8 mg/g compared with other cereal grains such as wheat (8.7 mg/g), rye (7.7 mg/g), lentils (8.4 mg/g), and faba beans (8.0 mg/g) (Filho et al., 2017).

16.4.3 Tannins

Tannins are bitter, an astringent natural substance that is part of polyphenol group. The molecular weight of tannins ranges from 500 to more than 3,000. They are heat stable and cause biological effects by precipitating and binding proteins and other compounds such as alkaloids and amino acids (Filho et al., 2017). In humans and animals they decrease the digestibility of proteins by either decreasing the bioavailability of protein or increasing fecal nitrogen. Other negative effects tannins cause in foods are changes in color and decreased palatability due to enzymatic browning and astringency. They also inhibit the activities of chymotrypsin, amylase, and lipase, decrease the protein quality of foods, and interfere with iron absorption (Santos, 2006).

In quinoa, the tannins are present in small amounts (0.53%), which is lower than rice (1.3%). Tannins can be decreased by cleaning and rinsing with water. The concentration of tannins were higher in peel (0.92%) as compared to bran and flour. After peeling, the amount of tannins in bran present were 46–50% of the total tannins in quinoa seeds (Jancurová et al., 2009). Research revealed that the process of washing and cooking of quinoa seeds may reduce the tannins and improve its digestibility (Borges et al., 2010).

16.4.4 Trypsin Inhibitor

Trypsin inhibitors are protease inhibitors that form stable complexes with proteolytic enzymes. They reduce the activity of enzyme trypsin in the intestinal tract, thus preventing protein digestion. They are not resistant to heat, so can be easily removed by both in home and industrial methods of food preparation (Lopes, Dessimoni, Da Silva, Vieira, & Pinto, 2009). In quinoa seeds, the amount of protease inhibitors is less than 50 ppm. The amount of trypsin inhibitor in quinoa seed is very small compared to other cereal grains, and therefore do not cause much concern (Vega-Gálvez et al., 2010). Research revealed that the concentration of trypsin inhibitor in eight varieties of quinoa ranged from 1.36–5.04 UTI mg^{-1}, which is less than in beans (12.9–42.8 UTI mg^{-1}), soy bean (24.5–41.5 UTI mg^{-1}), and lentils (17.8 UTI mg^{-1}) (Jancurová et al., 2009).The embryo layer of quinoa seeds contains the highest concentration of phytates and trypsin inhibitors (Ando et al., 2002). In comparison to the whole grain, quinoa seeds embryo contained 60% of the total phytate and 89% of the trypsin inhibitory activity.

16.4.5 Nitrates

Nitrate is a polyatomic ion (NO_3^-) present in all plants and is an important source of nitrogen for plant growth. Excessive use of nitrate fertilizer can be harmful for plants because of its accumulation in edible parts of plants. High concentrations are present in leaves, mainly in mesophyll, and the maximum accumulation is found in the petioles and stems. Nitrites can react with secondary amines to form carcinogenic N-nitrosamines. Food containing nitrate causes an increase of gastric and bladder cancer risk in humans. Nitrates and nitrite alone have limited carcinogenic effect in humans, but when combined with amines, form carcinogenic N-nitrous compounds (Santos, 2006). In infants they can also cause methaemoglobinaemia.

According to the WHO, the acceptable daily intake (ADI) of nitrate for humans is 3.7 mg/kg of body weight. Therefore, the consumption of nitrate should not be more than 222 mg for a 60 kg adult, which is similar to the intake of 351 g of quinoa wholemeal 'BRS Piabiru'(Lopes et al., 2009). Thus, quinoa has no adverse effects to human health.

16.4.6 Oxalates

Oxalate is a dianion ($C_2O_4^{-2}$) toxic substance. It is mostly found in vegetables such as beets, tomatoes, and spinach and also in nuts and cocoa. Oxalic acid can bind with minerals to form insoluble calcium and iron oxalates. A higher intake of oxalates causes the increase in urinary oxalate, which is a risk factor for predisposed to kidney stones. Oxalic acid causes a number of antinutrient effects in humans, reduces the absorption of minerals in the gastrointestinal tract, causes irritation in the gut, decreases blood clotting, and can injury the urinary tract (Konishi et al., 2004). The concentration of oxalates changes with soil quality, climate, and level of maturity. In seeds, the concentration of oxalate is higher in the bran, so the process of de-hulling reduces the concentration of oxalate in horse gram (Jancurová et al., 2009).

16.5 Functional Properties of Quinoa

Functional properties are generally described as the behavior of food compounds during preparation, cooking, formulation, etc., that affect finished food products in terms of taste and appearance. Functional properties include foaming capacity and stability, swelling capacity, water absorption capacity, bulk density, oil absorption capacity, emulsion activity and stability, dextrinization, gelatinization, jelling, plasticity, retention of moisture, sensory attributes, cohesiveness, and adhesiveness. Functional properties are basically physicochemical properties of food that exhibit due to the complex formation between food composition, structure, and matrixes. These properties are also influenced by other factors such as compounds of foods like carbohydrates, fats, protein, moisture, and starch. For the formulation and manufacture of food products, functional properties play an important role related to the polymer interaction of molecules like protein or starch. Other factors such as pH, charge density, hydrophilic or hydrophobic ratio, ionic forces, temperature, size of interaction particles, and changes of environment directly affect the functional properties of food.

Some functional properties of quinoa have been described such as oil and water absorption, foaming stability, emulsion stability and capacity, viscosity, gelation, water holding capacity (WHC), water imbibing capacity (WIC) and others.

16.5.1 Solubility

Solubility works on the principle of protein hydrophilic-hydrophobic balance and interaction with solvents. Quinoa flour protein solubility is heat-labile and pH dependent (Dakhili, Abdolalizadeh, Hosseini, et al., 2019). It was found that lowest solubility was observed at 6 pH and highest at 10 pH, where the solubility range is 15–52% (Aluko & Monu, 2003). The solubility of quinoa flour was found to be 5.4–15.6% while cooking. The solubility values of quinoa flour in the acid pH region imply that the protein may be useful in the formulation of beverages, dehydrated soups and sauces, and low-acid foods.

Food industries are focusing on quinoa protein isolates (QPI) due to their suitable functional and physicochemical properties. Solubility of QPI is lower (~5%) at acidic pH (4–6) and higher solubility (~100%) in alkaline range of pH nearby 10.0 (Elsohaimy, Refaay, & Zaytoun, 2015). At higher pH of 11.0, albumins are more soluble (~90%) than globulins (~50%) (Tömösközi et al., 2008). Higher solubility at alkaline pH are attributed to the ionization of the carboxyl groups and deprotonation of the amine groups, which results in the improvement of the protein-solvent interactions.

16.5.2 Water Holding Capacity

The functional property of WHC is important due to its association with the mouth feel and textural properties of foods. WHC is related to hydration that allows the quinoa flour to retain water under a centrifugal gravity force or is considered as physically entrapped and bound in capillaries by hydrodynamic interaction. WHC is expressed as increases in weight and quinoa flour has the capability to retain 147% of its weight of water (Pellegrinia et al., 2018). WHC decreases with a rising temperature due to heating, which causes the denaturation of bounding (Ogungbenle, Oshodi, & Oladimeji, 2009). Presence of salt decreases the WHC from 147–79.5%, the inhabitance of a 0.5–10% salt concentration interrupts water holding force with matrix effects due to the masking of charges caused by electrostatic interaction and hydration between protein molecules. High salt concentrations indicate a low WHC, which is important for dried quinoa flour and enhances its storage stability (Pellegrinia et al., 2018). A sigmoid sorption isotherm was observed while measuring the WHC of starch granules of quinoa (Tang et al., 2002). Lindeboom et al. (2005) obtained 49.5–93% water binding capacity for quinoa starch while quinoa proteins exhibit 3.94 ± 0.06 ml/g (Ashraf, Saeed, Sayeed, & Ali, 2012).

16.5.3 Gelation

Gelation contributes toand is involved in the texture, sensorial perception, water holding ability, and flavor attributes of food. Covalent and noncovalent (hydrophobic, hydrogen, and electrostatic) interactions are very crucial in gel formation. The gelation ability of quinoa protein depends greatly on pH values, i.e. lower pH (<9) provide semisolid gelled foods, whereas higher pH (>10) provide a weaker gel that could be used in foods.

This property is also influenced by the presence of salt. It was found that lowest value of a quinoa gelation concentration was 16% (w/v) in distilled water; the addition of salt reduced the concentration to 10–14%. Different salts have different impacts on gelation. Generally, quinoa flour has low gel forming properties, however, the addition of 0.5% KCl salt can improve its gel forming properties (Ogungbenle et al., 2009).

16.5.4 Foaming Stability and Capacity

Foaming stability and capacity are related to surface tension. Quinoa flour has low values for this property. In a study by Ogungbenle et al. (2009), foaming properties were affected by the addition of salts such as KCl, NaCl, and CH3COONa salts with a positive impact of up to 10% of the concentration being safe toward human health. Foaming capacity and stability were measured by El Sohaimy, Mohamed, Shehata, et al., (2018) as 14.33% and 9.63%, respectively, for quinoa flour. For QPI, protein hydrophobicity, moisture retention ability, and interfacial layer properties are important factors for foaming properties (Nadathur, Wanasundara, & Scanlin, 2016) and quinoa proteins were found to have a higher foaming capacity and stability than soy bean protein, but lower than wheat and egg albumin (Abugoch, Romero, Tapia, Silva, & Rivera, 2008). Optimum foaming properties for QPI were observed at a higher pH (Aluko & Monu, 2003) and exhibit a foaming capacity of 58.37 ± 2.14% for 0.1% protein concentration to 78.62 ± 2.54% for 3% protein concentration with an average of 69.28% and foam stability of 54.54 ± 15.31% after 60 minutes (Elsohaimy et al., 2015). In a study by Chauhan, Eskin, and Mills (1999), it was reported that removal of saponin leads to a decrease in the foaming properties, but it can be improved by increasing the protein concentration and enzymatic hydrolysis. The improved foaming properties of QPI make them suitable to be used in different food products and industrial applications.

16.5.5 Emulsifying Capacity and Stability

Emulsifying capacity and stability are vital functional properties that effect the food behavior of products. Emulsion stability is the ability of an emulsion to resist changes and remain stable at particular conditions for a definite time period, while emulsifying capacity is the maximum extent of oil emulsified through dispersion under standard conditions. El Sohaimy et al. (2018) reported that quinoa flour has 100.40% of emulsifying capacity and 45% of emulsify stability. A report compared this value with yam bean flour and it was observed that quinoa flour has a higher value than yam bean, soy flour, and pigeon pea flour by 10–20%, 18%, and 49%, respectively. This study found that quinoa flour is an excellent source for binder formulation and stabilization of colloidal food, better than soy flour, pigeon flour, and yam flour. Emulsification also depends on the salt concentration and it decreases with salt addition (Ogungbenle et al., 2009). This property is very important for food formulation like sauces, desserts, beverages, and sausages. Foam expansion and stability for quinoa was studied by Aluko and Monu (2003). They found smaller foam expansion (<20%) for protein concentrate but higher values (>160%) for protein hydrolysate. Foam stability was better with protein concentrate. The main factors on which the emulsifying properties of QPI depend includes their ability to increase hydrophobicity, prevent coalescence of droplets, and lower surface tension. QPI possess higher emulsifying capacity and stability when compared with soy bean, wheat, and pearl millet (Elsohaimy et al., 2015). The functional properties of QPI can be affected by desaponification, enzymatic hydrolysis, and the presence of polysaccharides that make cross linkages with proteins.

16.5.6 Thermal Properties

Differential scanning calorimetry (DSC) analysis is used to determine the thermal properties such as denaturation temperature and enthalpy of denaturation. Denaturation temperature relates to thermal stability while enthalpy of denaturation reflects the extent of the ordered structure (Shevkani, Kaur, Kumar, & Singh, 2015). The pH conditions and presence of salts influence the thermal stability while the protein aggregation and disruption of bonds between globulin's acidic and basic subunits are the main factors that determine the enthalpy of denaturation. Denaturation enthalpy values for quinoa proteins are found to be higher than those for whey proteins, but are comparable to those for soy and sunflower globulins (González-Pérez & Vereijken, 2007; Kinsella, Damodaran, & German, 1985). In the studies, it was observed that with an increase in temperature, the unfolding of the secondary structure in the quinoa protein lead to better solubility, emulsifying activities and gel formation with regular structure (Mäkinen, Hager, & Arendt, 2014).

To determine the thermal properties of quinoa starch, a 10% (w/w) quinoa starch suspension was examined for viscosity changes over the temperature range of 55–95°C and found 1960 Brabender Units (BU) at 70°C, which is higher than wheat (910 BU, 45°C), and amaranth (580 BU, 74°C) and was comparable to waxy maize (1870 BU, 80°C) (Praznik, Mundigler, Kogler, Pelzl, & Huber, 1999). The properties of quinoa starch make it a novel ingredient that can be used in different food formulations (Watanabe, Peng, Tang, & Mitsunaga, 2007).

16.6 Processing Techniques of Quinoa

After harvesting, some processing steps are generally followed to make quinoa palatable or consumable. For fulfilling this purpose, grains must undergo some industrial processing for de-hulling or removal of the outer layer of grain to facilitate the removal of antinutrients present in the outer layer of quinoa seeds and to improve its sensory quality. Such processes cause a loss of beneficial nutrients of the grain, thus, to overcome the nutrient loss, some traditional and modern processing methods are recommended to enhance the bioavailability of nutrients, minimize nutrient loss, and to develop variety in food products of quinoa (Naga Sai Srujana, Anila Kumari, Jessie Suneetha, & Prathyusha, 2019), which are discussed next. After some initial processing, quinoa can be used to make bakery products, pastas, breakfast cereals, and other snacks.

16.6.1 Germination

The process of germination for quinoa results in an increase of oleic acid, but a decrease in linoleic acid nonpolar lipids, glycolipids, and polar lipids. A germination time of 72 hours lead to the ratio of omega-3 and omega-6 to 1:0.25 in glycolipids (Park & Naofumi, 2004). Antioxidant activity increases two-fold by germinating quinoa seeds for 3 days. As a result, phenolic acids increased 8.57 times and flavonoids multiplied 4.4 times (Carciochi, Dimitrov, & Galván D'Alessandro, 2013). Successive processes of oven-drying after germination exhibited improvement in phenolic content and antioxidant activity enhancement in quinoa seeds (Felipe, Guamis, Quevedo, & Trujillo, 2003). Soaking and germination of quinoa seeds enhance the iron solubility by two to four times.

16.6.2 Fermentation

Fermentation was achieved by the presence of natural microorganisms in the seeds or via inoculation with *Saccharomyces cerevisiae* strains, which are generally used for baking and brewing. This gave rise to diminishing ascorbic acid and tocopherol content, while phenolic compounds and antioxidant activity were increased (Carciochi, Galván-D'Alessandro, & Vandendriessche, 2016).

16.6.3 Malting

The application of malting under moderate thermal treatment is regarded as an effective tool to augment antioxidants in quinoa seeds (Carciochi et al., 2016).

16.6.4 Dehydration

Quinoa dehydration from 70–80°C illustrated vitamin E content enhancement, while dehydration treatment at 40°C, 50°C, and 80°C led to improved antioxidant capacity by following the temperature/drying time equivalent processes. The obvious reduction in 10% proteins, 12% fat, and 27% fiber and ash content each were obtained after dehydration of quinoa (Miranda et al., 2010).

16.6.5 Heat Processing

The application of heat to quinoa seeds resulted in reduced antinutritional factors (Silva, Pompeu, Olavo, Gonçalves, & Spehar, 2015). Heat processing is a crucial step needed for complete digestion of proteins present in foods. Such processes substantially alter the structure of protein, and subsequently, their resistance to digestion.

16.6.6 Extrusion Processing

Extrusion processing leads to antinutrient inactivation, aflatoxin destruction, and increases the digestibility of fiber (Saalia & Phillips, 2011). Insoluble dietary fiber is transformed into soluble dietary fiber by the application of extrusion treatment. Moreover, resistant starch and enzyme-resistant glucans were formed as a result of transglycosidation (Repo-Carrasco, Espinoza, & Jacobsen, 2003). Excelling extrusion temperatures during quinoa processing caused unsaturated fatty acid oxidation subjected to decreased amount of unsaturated fatty acids and omitted antinutrients like lectins and antitrypsin inhibitors, which raise protein digestibility (Chandran, 2015).

16.6.7 Roasting

Roasting quinoa seed flour at 177°C for 15 minutes, 30 minutes, and 45 minutes caused the escalated peak and final viscosity in cake batter. Also, it resulted in crippled starch-protein interactions and reduced the swelling of the starch granules, which led to granule rupturing and declined geometric mean diameters along with higher roasting temperatures (Rothschild et al., 2015).

16.6.8 Baking

Baking of red and yellow quinoa seeds multiplied the total phenolic content and antioxidant activity, apparently by virtue of the Maillard reaction during thermal processing (Brend et al., 2012).

16.6.9 High Pressure-High Temperature Processing

High pressure-high temperature processing of quinoa products resulted in a two-to-four-time increase in iron solubility after soaking and germination, a three-to-five-time increase after fermentation, and a five-to-eight-time increase after fermentation of the germinated flour. High pressure-high temperature processing multiplied the magnesium content by five to eight times after subsequent cooking and baking, while the amount of copper was decreased by 28% (Ruiz, 2016b).

16.7 Allergenicity of Quinoa

Quinoa is not connected with the any sort of allergenicity in general, being gluten-free due to its negligible to zero amount of prolamin content. Quinoa differs from the plant family that includes gluten containing wheat, barley, oats, and rye, making it a good staple food for people with allergies and sensitivities related to wheat or gluten proteins (Asao & Watanabe, 2010). Quinoa is a highly nutritious option for patients with celiac disease, who generally do not find healthful food alternatives. However, Zevallos, Ellis, Suligoj, et al. (2012) reported the celiactoxic epitopes in certain quinoa cultivars that can produce immune responses in certain patients with celiac disease.

In addition, when a food is introduced to human body for the first time, it can trigger an adverse reaction. Consumption of quinoa can mislead the immune system and might offer a confused immune response to quinoa proteins as a harmful substance, leading to the onset of an allergic reaction. It is common, however, that the improper preparation of quinoa can cause some digestion problems. Therefore, saponin removal is the most crucial step to be considered before consumption of quinoa (Singh et al., 2016).

16.8 Usage as Food

Quinoa has been the most functional and versatile grain of the Andes, used for human consumption in terms of whole grain, semolina, raw flour, toasted flour, flakes, puffed grain, instant powdered foods, etc. Quinoa can be transformed into many different traditional or innovative recipes by incorporating various processing/cooking methods. These days, quinoa is sold in many supermarkets and health food counters as an individual "superfood."

In this array of quinoa preparations, the grains can be mixed with legumes like broad beans and kidney beans to enhance the dietary quality, particularly in the form of school breakfasts for children. Such ready-to-eat breakfast food products are generally processed or semiprocessed, which also includes cereals in preparations. These are commonly granular, puffed, flaked, and shredded cereals consumed after mixing with hot liquid. Reconstituted baby foods are also prepared in a similar way. Practically all flour-based industry products can be prepared from whole quinoa and flour. Preliminary trials in the Andean region and at other places have demonstrated the practicability of mixing quinoa flour to bread at levels of 10%, 15%, 20%, and even as much as up to 40% in the preparation of noodles, up to 60%

in making of sponge cake, and up to 70% for preparing biscuits. One of the prime benefits for the food industry to utilize quinoa is its gluten-free properties that meet the growing international demand for such products (Food and Agriculture Organization of the United Nations, 2013).

Quinoa is marketed by many brands in raw but preprocessed form, which are already rinsed before packaging; otherwise, quinoa needs to be rinsed under running water to remove bitterness due to antinutrient called saponin. Saponin is abundantly found in the outer layer of grain and important to remove before consumption. Quinoa can be used in salads by mixing cooked chilled quinoa with pinto beans, scallions, and coriander. Quinoa can be an excellent substitute for rice in some vegetable recipe requiring stuffing. Since 16% of our daily iron requirement can be fulfilled by quinoa compared to brown rice as well as other vitamins and minerals in good quantity. Quinoa can be used to prepare patties to make veggie burgers. Quinoa can replace bread crumbs in a meatloaf recipe, and used for binding ingredients together. Quinoa provides a nutty flavor and crispy texture to bread (Singh et al., 2016).

Presently, high-quality foods containing a good quantity of protein are needed and the quinoa embryo is concentrated with up to 45% protein content. After separation from the seed in concentrated form, it can be added to children's food to help with undernourishment or it can be fortified in foods for adults with special nutritional needs like pregnant women. Thus, there is huge number of quinoa-based processed and semiprocessed foods available in the market, however they are generally unaffordable or costly to most of the population (Food and Agriculture Organization of the United Nations, 2013).

16.9 Conclusions

Quinoa is a potential crop that has been considered a complete food due to its high quality protein and exquisite amino acid composition. The abundance of the essential amino acid lysine, makes it outright grain in all cereals and vegetables. Further, quinoa does not contain gluten proteins, so it can be regarded as prime complete food for people with celiac disease or an allergy related to wheat. The fatty acid composition of the seed is of high quality and possess high nutritional value. It is an excellent source of vital minerals like iron, magnesium, copper, phosphorus, potassium, and zinc, also provides good amounts of fiber and vitamins (E and B). Quinoa contains phytochemicals like polyphenols, phytosterols, and flavonoids owing its nutraceutical benefits. Quinoa seeds contain some antinutritional factors, but these are removed during manufacturing process and home preparation methods. However, some antinutritional factors as well as their breakdown products may possess beneficial health effects if present in small amounts. Antinutritional factors can be easily extracted by following some simple processing techniques at the household as well as industrial level. The addition of some easy steps during processing of quinoa leads to increased digestion and absorption of protein and depreciation in antinutritional contents such as saponins, phytates, and tannins. In addition, the bioavailability of minerals (magnesium, iron, and zinc) present in quinoa increases, which are essential for certain physiological and neurological functions. The highly rated nutritional value and promising functional properties of quinoa make it most popular among novel functional food manufacturers and consumers. Quinoa it can be easily processed and fortified into breakfast foods, snack foods, beverages, baked food products, soups, desserts, and many other forms, and can be consumed as staple food.

REFERENCES

Abugoch James, L. E. (2009). Quinoa (*Chenopodium quinoa* Willd.): Composition, chemistry, nutritional and functional properties. *Advances in Food and Nutrition Research, 58*, 1–31.

Abugoch, L. E., Romero, N., Tapia, C. A., Silva, J., & Rivera, M. (2008). Study of some physicochemical and functional properties of quinoa (*Chenopodium quinoa* Willd.) protein isolates. *Journal of Agricultural and Food Chemistry, 56*(12), 4745–4750.

Almeida, S. G., & Sá, W. A. C. (2009). Amaranto (*Amaranthus* ssp) quinoa (Chenopodium Quinoa) alimentos alternativos para doentes celíacos. *Ensaios e Ciência: C. Biológicas, Agrárias e da Saúde, XIII*(1), 77–92.

Aluko, R. E., & Monu, E. (2003). Functional and bioactive properties of quinoa seed protein hydrolysates. *Journal of Food Science 68*(4), 1254–1258.

Alvarez-Jubete, L., Arendt, E. K., & Gallagher, E. (2010). Nutritive value of pseudocereals and their increasing use as functional gluten-free ingredients. *Trends in Food Science & Technology, 21*(2), 106–113.

Alves-Rodrigues, A., & Shao, A. (2004). The science behind lutein. *Toxicology Letter, 150*(1), 57–83.

Ando, H., Chen, Y., Tang, H., Shimizu, M., Watanabe, K., & Mitsunaga, T. (2002). Food components in fractions of quinoa seed. *Food Science and Technology Research, 8*(1), 80–84.

Asao, M., & Watanabe, K. (2010). Functional and bioactive properties of quinoa and amaranth. *Food Science & Technical Research, 16*, 163–168.

Ashraf, S., Saeed, S. M. G., Sayeed, S. A., & Ali, R. (2012). Impact of microwave treatment on the functionality of cereals and legumes. *International Journal of Agriculture and Biology, 14*(3).

Borges, J. T. S., Bonomo, R. C., Paula, C. D., Oliveira, L. C., & Cesário, M. C. (2010). Physicochemical and nutritional characteristics and uses of Quinoa (*Chenopodium quinoa* Willd.). *Temas Agrários, 15*(1), 9–23.

Brend, Y., Galili, L., Badani, H., Hovav, R., & Galili, S. (2012). Total phenolic content and antioxidant activity of red and yellow quinoa (*Chenopodium quinoa* Willd.) seeds as affected by baking and cooking conditions. *Food and Nutrition Sciences, 3*, 1150–1155.

Brinegar, C., & Goundan, S. (1993). Isolation and characterization of chenopodin, the 11S seed storage protein of quinoa (*Chenopodium quinoa*). *Journal of Agricultural and Food Chemistry, 41*(2), 182–185.

Brinegar, C., Sine, B., & Nwokocha, L. (1996). High-cysteine 2S seed storage proteins from quinoa (*Chenopodium quinoa*). *Journal of Agricultural and Food Chemistry, 44*(7), 1621–1623.

Carciochi, R. A., Dimitrov, K., & Galván D´Alessandro, L. (2013). Effect of malting conditions on phenolic content, Maillard reaction products formation, and antioxidant activity of quinoa seeds. *Journal of Food Science and Technology*, 1–8.

Carciochi, R. A., Galván-D'Alessandro, L., & Vandendriessche, P. (2016). Effect of germination and fermentation process on the antioxidant compounds of quinoa seeds. *Plant Foods for Human Nutrition, 71*(4), 361–367.

Chait, A., Brunzell, J. D., Denke, M. A., Eisenberg, D., Ernst, N. D., Franklin, F. A., & Ginsberg, H. (1993). Rationale of the diet-heart statement of the American Heart Association. *Report of Nutrition Committee, 88*(6), 3008–3029.

Chandran, S. (2015). Effect of extrusion on physico-chemical properties of quinoa-cassava extrudates fortified with cranberry concentrate. Master's Thesis, The State University of New Jersey.

Chauhan, G., Eskin, N., & Mills, P. (1999). Effect of saponin extraction on the nutritional quality of quinoa (*Chenopodium quinoa* Willd.) proteins. *Journal of Food Science and Technology, 2*, 123–126.

Comai, S., Bertazzo, A., Bailoni, L., Zancato, M., Costa, C. V. L., & Allegri, G. (2007). The content of proteic and nonproteic (free and protein-bound) tryptophan in quinoa and cereal flours. *Food Chemistry, 100*(4), 1350–1355.

Cunnane, S. C., & Thompson, L. U. (1995). The effect of dietary alpha-linolenic acid on blood lipids and lipoproteins in humans. In J. G. Chamberlain & G. J. Nelson (Eds.), *Flaxseed in Human Nutrition*. Champaign, IL: AOAC Press, 1–458.

Cusack, D. F. (1984). Quinua: Grain of the Incas. *Ecologist, 14*, 21–31.

Dakhili, S., Abdolalizadeh, L., Hosseini, S. M., Shojaee-Aliabadi, S., & Mirmoghtadaie, L. (2019). Quinoa protein: Composition, structure and functional properties. *Food Chemistry, 299*, 125–161.

Dini, I., Tenore, G. C., & Dini, A. (2010). Antioxidant compound contents and antioxidant activity before and after cooking in sweet and bitter Chenopodium quinoa seeds. *LWT—Food Science and Technology 43*(3), 447–451.

Drzewiecki, J., Delgado-Licon, E., Haruenkit, R., Pawelzik, E., Martin-Belloso, O., Park, Y.-S., et al. (2003). Identification and differences of total proteins and their soluble fractions in some pseudocereals based on electrophoretic patterns. *Journal of Agricultural and Food Chemistry, 51*(26), 7798–7804.

Elsohaimy, S., Refaay, T., & Zaytoun, M. (2015). Physicochemical and functional properties of quinoa protein isolate. *Annals of Agricultural Sciences, 60*(2), 297–305.

Farro, P. C. A. (2008). *Desenvolvimento de filmes biodegradáveis a partir de derivados do grão de quinoa (Chenopodium quinoa Willdenow) da variedade "Real."* Campinas: Faculdade de Engenharia de Alimentos, Universidade Estadual de Campinas.

Felipe, C. X., Guamis, L. B., Quevedo, T. J. M., & Trujillo, M. A. J. (2003). Liquid product of vegetable origin as milk substitute. *Patent Application*, EP1338206 A1

Filho, A. M., Pirozi, M. R., Borges, J. T., Pinheiro Sant'Ana, H. M., Chaves, J. B., & Coimbra, J. S. (2017). Quinoa: Nutritional, functional, and antinutritional aspects. *Critical Reviews in Food Science & Nutrition, 57*(8), 1618–1630. doi: 10.1080/10408398.2014.1001811

Food and Agriculture Organization of the United Nations. (2013). Quinoa—A future sown thousand years ago. *2013 International Year of Quinoa Secretariat*. Retrieved January 6, 2021, from http://www.fao. org/quinoa-2013/what-is-quinoa/use/en/?no_mobile=1

Francis, G., Kerem, Z., Makkar, H., & Becker, K. (2002). The biological action of saponins in animal systems: A review. *British Journal of Nutrition, 88*(6), 587–605.

Garino, C., Zuidmeer, L., Marsh, J., Lovegrove, A., Morati, M., Versteeg, S., & van Ree, R. (2010). Isolation, cloning, and characterization of the 2S albumin: A new allergen from hazelnut. *Molecular Nutrition & Food Research 54*(9), 1257–1265.

Giménez, B., Moreno, S., López-Caballero, M. E., Montero, P., & Gómez-Guillén, M. C. (2013). Antioxidant properties of green tea extract incorporated to fish gelatin films after simulated gastrointestinal enzymatic digestion. *LWT—Food Science and Technology, 53*(2), 445–451.

González-Pérez, S., & Vereijken, J. M. (2007). Sunflower proteins: Overview of their physicochemical, structural and functional properties. *Journal of the Science of Food and Agriculture 87*(12), 2173–2191.

Guo, W. W, & Beta, T. (2013). Phenolic acid composition and antioxidant potential of insoluble and soluble dietary fibre extracts derived from select whole-grain cereals. *Food Research International, 51*(2), 518–525.

Gupta, R. K, Gangoliya, S. S., & Singh, N. K. (2015). Reduction of phytic acid and enhancement of bioavailable micronutrients in food grains. *Journal of Food Science & Technology, 52*(2), 676–684.

Hidalgo, A., & Brandolini, A. (2008). Protein, ash, lutein and tocols distribution in einkorn (*Triticum monococcum* L. subsp. *monococcum*) seed fractions. *Food Chemistry, 107*(1), 444–448.

Hu, F. B., Manson, J. E., & Willett, W. C. (2001). Types of dietary fat and risk of coronary heart disease: A critical review. *Journal of the American College of Nutrition, 20*(1), 5–19.

Jancurová, M., Minarovičová, L., & Dandár, A. (2009). Quinoa: A review. *Czech Journal of Food Sciences, 27*(2), 71–79.

Khan, K., & Shewry, P. R. (2009). *Wheat: Chemistry and technology*. Saint Paul, MN: AACC International.

Kinsella, J. E., Damodaran, S., & German, B. (1985). Physicochemical and functional properties of oilseed proteins with emphasis on soy proteins. *New Protein Foods (USA)*.

Kline, K., Lawson, K. A., Yu, W., & Sanders, B. G. (2007). Vitamin E and cancer. *Vitamins and Hormones 76*, 435–461.

Konishi, Y., Hirano, S., Tsuboi, H., & Wada, M. (2004). Distribution of minerals in quinoa (*Chenopodium quinoa* Willd.) seeds. *Bioscience, Biotechnology and Biochemistry, 68*(1), 231–234.

Koziol, M. J. (1992). Chemical composition and nutritional evaluation of quinoa (*Chenopodium quinoa* Willd.). *Journal of Food Composition and analysis, 5*(1), 35–68.

Kroon, P. A., & Williamson, G. (1999). Hydroxycinnamates in plants and food: Current and future perspectives. *Journal of the Science of Food and Agriculture, 79*(3), 355.

Kumar, R., & Khatkar, B. S. (2017). Thermal, pasting and morphological properties of starch granules of wheat (*Triticum aestivum* L.) varieties. *Journal of Food Science and Technology 54*, 2403–2410.

Li, Y., Kong, D., Ahmad, A., Bao, B., & Sarkar, F. H. (2013). Antioxidant function of isoflavone and 3,3-diindolylmethane: Are they important for cancer prevention and therapy?. *Antioxidants & Redox Signaling, 19*(2), 139–150.

Lindeboom, N., Chang, P. R., Falk, K. C., & Tyler, R. T. (2005). Characteristics of starch from eight quinoa lines. *Cereal Chemistry, 82*(2), 216–222.

Lopes, C. O., Dessimoni, G. V., Da Silva, M. C., Vieira, G., & Pinto, N. A. V. D. (2009). Aproveitamento, composição nutricional e antinutricional da farinha de quinoa (*Chenoipodium quinoa*). *Alimentos e Nutrição, 20*(4), 669–675.

Lopez-Rubio, A., Flanagan, B. M., Gilbert, E. P., & Gidley, M. J. (2008). A novel approach for calculating starch crystallinity and its correlation with double helix content: A combined XRD and NMR study. *Biopolymers, 89*(9), 761–768.

Lovejoy, J. C. (1999). Dietary fatty acids and insulin resistance. *Current Atherosclerosis Reports, 1*(3), 215–220.

Mahan, L.K., & Escott-Stump, S. (Eds.). (2010). *Krause: Alimentos, nutrição e dietoterapia*. (Vol. *12*). Rio de Janeiro: Elsevier.

Mäkinen, O. E., Hager, A. S., & Arendt, E. K. (2014). Localisation and development of proteolytic activities in quinoa (*Chenopodium quinoa*) seeds during germination and early seedling growth. *Journal of Cereal Science 60*(3), 484–489.

Mäkinen, O. E., Zannini, E., Koehler, P., & Arendt, E. K. (2016). Heat-denaturation and aggregation of quinoa (*Chenopodium quinoa*) globulins as affected by the pH value. *Food Chemistry, 196*, 17–24.

Marcone, M. F., Yada, R. Y., Aroonkamonsri, W., & Kakuda, Y. (1997). Physico-chemical properties of puri-fied isoforms of the 12S seed globulin from mustard seed (*Brassica alba*). *Bioscience, Biotechnology, and Biochemistry 61*(1), 65–74.

Mattila, P., Pihlava, J. M., & Hellström, J. (2005). Contents of phenolic acids, alkyl- and alkenylresorcin-ols, and avenanthramides in commercial grain products. *Journal of Agricultural and Food Chemistry 53*(21), 8290–8295.

Miranda, M., Vega-Gálvez, A., López, J., Parada, G., Sanders, M., Aranda, M., Di Scala, K. (2010). Impact of air-drying temperature on nutritional properties, total phenolic content and antioxidant capacity of qui-noa seeds (*Chenopodium quinoa* Willd.). *Industrial Crops and Products, 32*(3), 258–263. doi: 10.1016/j. indcrop.2010.04.019

Miranda, M., Vega-Gálvez, A., Uribe, E., López, J., Marínez, E., Rodríguez, M. J., Scala, K. D. (2011). Physico-chemical analysis, antioxidant capacity and vitamins of six ecotypes of chilean quinoa (Chenopodium quinoa Willd.). *Procedia Food Science, Special Issue of the 11th International Congress on Engineering and Food*, 1 (2011):1439–1466.

Nadathur, S., Wanasundara, J. P., & Scanlin, L. (Eds.). (2016). *Sustainable Protein Sources*. Academic Press.

Naga Sai Srujana, M., Anila Kumari, B., Jessie Suneetha, W., & Prathyusha, P. (2019). Processing technolo-gies and health benefits of quinoa. *The Pharma Innovation Journal, 8*(5), 155–160.

Ng, S. C., Anderson, A., Coker, J., & Ondrus, M. (2007). Characterization of lipid oxidation products in qui-noa (*Chenopodium quinoa*). *Food Chemistry, 101*(1), 185–192.

Niro, S., D'Agostino, A., Fratianni, A., Cinquanta, L., & Panfili, G. (2019). Gluten-Free Alternative Grains: Nutritional Evaluation and Bioactive Compounds. *Foods, 8*(6), 208. doi: 10.3390/foods8060208

Ogungbenle, H. N., Oshodi, A. A., & Oladimeji, M. O. (2009). The proximate and effect of salt applications on some functional properties of quinoa (*Chenopodium quinoa*) flour. *Pakistani Journal of Nutrition, 8*(1), 49–52.

Ogungbenle, N. H. (2003). Nutritional evaluation and functional properties of quinoa (*Chenopodium quinoa*) flour. *International Journal of Food Sciences and Nutrition, 54*(2), 153–158.

Osborne, T. B. (1924). *The Vegetable Proteins*. London: Kluwer Academic Press.

Pa´sko, P., Barto´n, H., Zagrodzki, P., Gorinstein, S., Fołta, M., & Zachwieja, Z. (2009). Anthocyanins, total polyphenols and antioxidant activity in amaranth and quinoa seeds and sprouts during their growth. *Food Chemistry 115*(3), 994–998.

Park, I. M., Ibáñez, A. M., & Shoemaker, C. F. (2007). Rice starch molecular size and its relationship with amylose content. *Starch, 59*(2), 69–77.

Park, S. H., & Naofumi, M. (2004). Changes of bound lipids and composition of fatty acids in germination of quinoa seeds. *Food Science & Technology, 10*(3), 303–306.

Pellegrinia, M., Lucas-Gonzalesb, R., Riccia, A., Fontechac, J., Fernández-Lópezb, J., Pérez-Álvarezb, J. A., & Viuda-Martosb, M. (2018). Chemical, fatty acid, polyphenolic profile, techno-functional and antioxi-dant properties of flours obtained from quinoa (*Chenopodium quinoa* Willd.) seeds. *Industrial Crops & Products, 111*, 38–46.

Praznik, W., Mundigler, N., Kogler, A., Pelzl, B., & Huber, A. (1999). Molecular background of technological properties of selected starches. *Starch/Starke 51*, 197–211.

Repo-Carrasco, R., Espinoza, C., & Jacobsen, S. E. (2003). Nutritional value and use of the Andean crops quinoa (*Chenopodium quinoa*) and kañiwa (*Chenopodium pallidicaule*). *Food Reviews International 19*(1-2), 179–189.

Ridout, C., Price, L. R., DuPont, M. S., Parker, M.L., & Fentwick, G. R. (1991). Quinoa saponins-anlaysis and preliminary investigations into effects of reduction by processing. *Journal of Science Food & Agriculture, 54*, 165–176.

Rothschild, J., Rosentrater, K. A., Onwulata, C., Singh, M., Menutti, L., Jambazian, P., & Omary, M. B. (2015). Influence of quinoa roasting on sensory and physicochemical properties of allergen-free, gluten-free cakes. *International Journal of Food Science & Technology, 50*(8), 1873–1881. doi: https://doi.org/10.1111/ijfs.12837

Ruales, J., & Nair, B. M. (1993). Content of fat, vitamins and minerals in quinoa (*Chenopodium quinoa* Willd.) seeds. *Food Chemistry, 48*(2), 131–136.

Ruiz, G. A. (2016a). Exploring Novel Food Proteins and Processing Technologies. Wageningen University.

Ruiz, G. A. (2016b). Exploring novel food proteins and processing technologies: a case study on quinoa pro-tein and high pressure-high temperature processing. Ph.D. Thesis, Wageningen University & Research, Netherlands.

Saalia, F., & Phillips, R. D. (2011). Degradation of aflatoxins by extrusion cooking: Effects on nutritional quality of extrudates. *LWT—Food Science and Technology, 44*(6), 1496–1501.

Santos, M. A. T. (2006). Efeito do cozimento sobre alguns fatores antinutricionais em folhas debrocou, couve-flor e couve. *Ciência e Agrotecnologia, 30*(2), 294–301.

Shevkani, K., Kaur, A., Kumar, S., & Singh, N. (2015). Cowpea protein isolates: Functional properties and application in gluten-free rice muffins. *LWT—Food Science and Technology, 63*(2), 927–933.

Silva, J. A., Pompeu, G., Olavo, C., Gonçalves, D. B., & Spehar, C. R. (2015). The importance of heat against antinutritional factors from Chenopodium quinoa seeds. *Food Science and Technology, 35*(1), 74–82.

Singh, S., Singh, R., & Singh, K. V. (2016). Quinoa (*Chenopodium quinoa* Willd.), functional superfood for today's world: A Review. *World Scientific News, 58*, 84–96.

El Sohaimy, S. A., Mohamed, S. E., Shehata, M. G., Mehany, T., & Zaitoun, M. A. (2018). Compositional analysis and functional characteristics of quinoa flour. *Annual Research & Review in Biology 22*(1), 1–11.

Soobrattee, M. A., Neergheen, V. S., Luximon-Ramma, A., Aruoma, O. I., & Bahorun, T. (2005). Phenolics as potential antioxidant therapeutic agents: Mechanism and actions. *Mutation Research, 579*(1–2), 200–213. doi: 10.1016/j.mrfmmm.2005.03.023

Spehar, C. R. (2007). Quinoa: Alternativa para a diversificação agrícola e alimentar. Planaltina, Brasil, Embrapa Cerrados.

Sun, S. S., Leung, F. W., & Tomic, J. C. (1987). Brazil nut (Bertholletia excelsa HBK) proteins: Fractionation, composition, and identification of a sulfur-rich protein. *Journal of Agricultural and Food Chemistry 35*(2), 232–235.

Tang, H., Watanabe, K., & Mitsunaga, T. (2002). Characterization of storage starches from quinoa, barley and adzuki seeds. *Carbohydrate Polymer, 49*(1), 13–22.

Tari, T. A., Annapure, U. S., Singhal, R. S., & Kulkarni, P. R. (2003). Starch-based spherical aggregates: Screening of small granule sized starches for entrapment of a model flavouring compound, vanillin. *Carbohydrate Polymers 53*(1), 45–51.

Taylor, J. R., & Parker, M. L. (2002). Quinoa. In: Belton P. and Taylor J. (Eds.), *Pseudocereals and Less Common cereals—Grain Properties and Utilization Potential*: Springer, 93–119.

Thanapornpoonpong, S. N., Vearasilp, S., Pawelzik, E., & Gorinstein, S. (2008). Influence of various nitrogen applications on protein and amino acid profiles of amaranth and quinoa. *Journal of Agricultural and Food Chemistry, 56*(23), 11464–11470.

Tömösközi, S., Gyenge, L., Pelcéder, Á., Varga, J., Abonyi, T., & Lásztity, R. (2008). Functional properties of protein preparations from amaranth seeds in model system. *European Food Research and Technology, 226*(6), 1343–1348.

USDA. (2011). USDA National Nutrient Database for Standard Reference, Release 18, Nutrient Data Laboratory: U.S. Department of Agriculture, Agricultural Research Service.

Vega-Gálvez, A., Miranda, M., Vergara, J., Uribe, E., Puente, I., & Martínez, E. A. (2010). Nutritional facts and potential of quinoa (*Chenopodium quinoa* Willd.) an ancient Andean grain: A review. *Journal of Science Food & Agriculture, 90*, 2541–2547.

Vega-Gálvez, A. V., Miranda, M., Vergara, J., Uribe, E., Puente, L., & Martínez, E. A. (2010). Nutrition facts and functional potential of quinoa (*Chenopodium quinoa* Willd.), an ancient Andean grain: a review. *Journal of the Science of Food and Agriculture 90*, 2541–2547.

Watanabe, K., Peng, L., Tang, H., & Mitsunaga, T. (2007). Molecular structural characteristics of quinoa starch. *Food Science and Technology Research, 13*(1), 73–76.

WHO. (2007). World Health Organization Report of a joint WHO meeting. Geneva.

Wymann, M. P., & Solinas, G. (2013). Inhibition of phosphoinositide 3-kinase attenuates inflammation, obesity, and cardiovascular risk factors. *Annals of the New York Academy of Sciences, 1280*(1), 44–47.

Yadav, M. P., Moreau, R. A, & Hicks, K. B. (2007). Phenolic acids, lipids, and proteins associated with purified corn fiber arabinoxylans. *Journal of Agricultural and Food Chemistry, 55*(3), 943–947.

Yang, Q., Cogswell, M. E., Hamner, H. C., Carriquiry, A., Bailey, L. B., Pfeiffer, C. M., & Berry, R. J. (2010). Folic acid source, usual intake, and folate and Vitamin B-12 status in US adults: National health and nutrition examination survey (Nhanes) 2003–2006. *The American Journal of Clinical Nutrition, 91*(1), 64–72.

Yuan, X. P., Wang, J., & Yao, H. (2005). Antioxidant activity of feruloylated oligosaccharides from wheat bran. *Food Chemistry, 90*(4), 759–764.

Zevallos, V. F., Ellis, H. J., Suligoj, T., Herencia, L. I., & Ciclitira, P. J. (2012). Variable activation of immune response by quinoa (*Chenopodium quinoa* Willd.) prolamins in celiac disease. *The American Journal of Clinical Nutrition, 96*(2), 337–344.

Section II

Pulses

17

Pinto Beans As An Important Agricultural Crop and Its Health Benefits

Twinkle Kesharwani and C. Lalmuanpuia
Bhaskaracharya College of Applied Sciences, Delhi University, New Delhi, India

CONTENTS

17.1 Introduction

Legumes or pulse seeds, also referred as poor man's meat, are important sources of many nutrients, especially proteins (Tharanathan and Mahadevamma, 2003). The pinto bean (*Phaseolus vulgaris* L.) is an ecologically important pulse seed used throughout the world for its nutritional composition. It is a medium-sized, oval-shaped, brown-skinned bean (Ganesan and Xu, 2017). The name "pinto bean" is derived from the Spanish word meaning "painted" because of its mottled, brown, pink, and beige appearance (Stone and Stone, 1988; Geil and Anderson, 1994). It belongs to the group of common beans (*Phaseolus vulgaris*) that also include black beans, borlotti beans, cranberry (roman) beans, flageolet beans, haricot beans, kidney beans, mexican beans, navy beans, pea beans, pink beans, string beans, small white beans, and yellow beans (Hosfield et al., 2004; Shimelis and Rakshit, 2005; He et al., 2015; OECD, 2019).

Generally, the health-promoting effects of pinto beans are linked with its macronutrients but enough evidence has been shown by many researchers that its non-nutritional components also play an important role to improve human health (Singh et al., 2017). Bioactive compounds constitute this non-nutritional content and contain bioactive phytochemicals, dietary fiber, resistant starches, lectins, phenolic compounds, phytates, etc. Cotyledons are a good source of macronutrients (mainly proteins), while the bioactive compounds are mainly concentrated in hulls or seed coat. As compared to macronutrients, the bioactive compounds are present in small quantities but play a key role in metabolism, defense mechanism, energy reserve, and protection against multiple diseases (Kabagambe et al. 2005; Rochfort and Panozzo 2007; Singh et al., 2017). Although pinto beans also contain some of the antinutritional and toxic constituents, but their nutritional value can be improved as some simple processing steps like soaking and cooking can remove these antinutrients compounds effectively. The incorporation of pinto beans in various food applications will help in better utilization of its potential properties as a functional food.

17.2 Cultivation and Production of Pinto Beans

The pinto bean is an herbaceous vine belonging to order *Fabales*, family *Fabaceae*, genus *Phaseolus* L. and species *Phaseolus vulgaris* L. (OECD, 2015). Based on the archeological, botanical, chronological, and linguistic evidence, it originated and was domesticated in America (Gepts and Dpbouk, 1991; Papa and Gepts, 2003; Papa et al., 2005; Hayat et al., 2013). Because of its edibility and nutritional characteristics, it is now grown and consumed worldwide (Arkcoll and Clement, 1989; Shimelis and Rakshit, 2005).

P. vulgaris has lateral roots with irregular root nodules, hairy stems, trifoliolate leaves, hairy leaflets with small stipules, and white, pink, or violet bisexual flowers. The hairs present in stems helps in disease and insect resistance. The seed pods are narrow and enclose a maximum of 12 seeds per pod, but more commonly four to six seeds per pods are present (OECD, 2015). Depending on the cultivar, the seeds vary in color and size from 150–900 g per thousand seeds (Purseglove, 1968; Pillemer and Tingey, 1978; Mmbaga and Steadman, 1992; Brink and Belay, 2006; Wortmann, 2006).

In developing countries, dry beans are often produced by small farms in subtropical and tropical regions (Broughton et al., 2003; OECD, 2019), while in developed countries depending on the variety, row planting mechanized cultivation is commonly practiced (Liebenberg, 2009). Typically dry beans are planted on level land, but in some areas where the water table is high, sowing on ridges or hills may also be practiced (Wortmann, 2006). About 150,000–400,000 thousand seeds are sown for every 10,000 square meters of area either in rows or broadcast, but when intercropped, the sowing rate is reduced (Wortmann, 2006; OECD, 2015). The optimum soil temperature for seed germination is 22–30°C while the minimum requirement is 12°C for germination (OECD, 2015). Flowering begins after 4–6 weeks of sowing depending on the variety (Wortmann, 2006). Harvesting time varies according to the use of the crop. For dry beans, harvesting is done when the seeds have matured (approximate moisture content 50%) and the pods become yellow (Purseglove, 1968; Wortmann, 2006). In the case of small farms, harvesting and threshing may be done by hand while the process may be mechanized at the commercial level (Liebenberg, 2009). Loss of moisture prior to threshing can lead to postharvest losses to a major extent (FAO, 1999) while preharvesting loss of moisture can cause undesirable splitting of beans (Liebenberg, 2009; OECD, 2015). After harvesting, based on the requirements, seeds may be sorted by the specific variety or seeds of various varieties are mixed together (Wortmann, 2006; FAO, 1999).

More than 50% of common beans produced are consumed in the home, while the rest is value added and processed as bean paste, baked beans, canned beans, cereal products, frozen beans, packaged dry beans, puffed snacks, texturized vegetable protein, meat analogues, soups, and bean flours (Câmara et al., 2013; OECD, 2019).

17.3 Composition of Pinto Beans

17.3.1 Carbohydrate

The carbohydrate content of pinto beans varies considerably according to its variety. Table 17.1 represents the composition of pinto beans (g/100 g) dry weight basis. The average carbohydrate content of pinto beans ranges from 70.55% (OECD, 2019), which mainly consists of complex starch (50%) and nonstarch polysaccharides with small quantities of mono- and disaccharides (Geil and Anderson, 1994; Ganesan and Xu, 2017). Additionally, dietary fiber also constitutes about 17–23% of total carbohydrate in pinto beans (Shiga et al., 2009). Most of the fibers are insoluble in nature and consist of cellulose, hemicellulose, pentosan, pectin, and lignin (Ganesan and Xu, 2017). Oligosaccharides like raffinose (0.6%) and verbascose (0.15%) are also present in small amounts (Sgarbieri, 1984).

17.3.2 Protein

Protein content of pinto beans ranges from 23.27–26.72% of dry weight of beans, which are rich in almost all essential amino acids except methionine, tryptophan, and cystine (OECD, 2019). Table 17.2 shows the amino acid composition (per 100 g dry weight basis) of pinto beans. Complementation of dry beans with cereals and other good quality proteins can considerably improve the protein quality of pinto beans (Geil and Anderson, 1994). The principle protein fractions of pinto beans consist of globulin and albumin, which are responsible for about 50% and 10% of total protein content, respectively (Ganesan and Xu, 2017), with small quantities of prolamine and glutelin (Adebowale, et al. 2007). Like other

TABLE 17.1

Composition of Pinto Beans (g/100 g) Dry Weight Basis

Moisture (g/100 g)	Carbohydrate, (g/100 g)	Protein (g/100 g)	Total Lipid (fat) (g/100 g)	Ash (g/100 g)	Total Dietary Fiber, (g/100 g)
11.33	70.55	24.16	1.39	3.90	17.5

Data from: OECD. (2019). Safety Assessment of Foods and Feeds Derived from Transgenic Crops, Vol. 3.

TABLE 17.2

Amino Acid Content of Pinto Beans (g/100 g) Dry Weight Basis

Amino Acid	Content per 100 g	Amino Acid	Content per 100 g	Amino Acid	Content per 100 g
Alanine	0.98	Histidine	0.63	Proline	1.21
Arginine	1.24	Isoleucine	0.98	Serine	1.32
Aspartic acid	2.56	Leucine	1.76	Threonine	0.91
Cystine	0.21	Lysine	1.53	Tryptophan	0.27
Glutamic acid	3.41	Methionine	0.29	Tyrosine	0.48
Gylcine	0.90	Phenylalanine	1.23	Valine	1.13

Data from: OECD. (2019). Safety Assessment of Foods and Feeds Derived from Transgenic Crops, Vol. 3.

TABLE 17.3

Fatty Acid Composition of Pinto Beans (g/100 g) Dry Weight Basis

Total Saturated	Total Monounsaturated	Total Polyunsaturated
0.208	0.203	0.361

Data from: OECD. (2019). Safety Assessment of Foods and Feeds Derived from Transgenic Crops, Vol. 3.

pulses, protein fractions of pinto beans can be used to prepare gluten-free products, edible biodegradable films, and formulate different foods (Shevkani and Singh 2015; Santiago-Ramos et al., 2018).

17.3.3 Lipids

Pinto beans contains only a small quantity of lipids, which mainly contains unsaturated fatty acids (Ganesan and Xu, 2017). The composition of saturated and unsaturated fatty acids present in pinto beans is shown in Table 17.3. Because of their plant origin, the lipids of pinto beans are cholesterol-free, which is recommended for reducing the risk of many chronic diseases (Geil and Anderson, 1994).

17.3.4 Vitamins

Pinto beans are a relatively good source of B-group vitamins like thiamin, riboflavin, niacin, pyridoxine, pantothenic acid, and folate (Geil and Anderson, 1994), but are a poor source of Vitamin C and fat-soluble vitamins due to their low lipid content (Sgarbieri, 1984). However commercial and household processing lead to a considerable loss of these water-soluble vitamins. The principle vitamins present in pinto beans are shown in Table 17.4.

17.3.5 Minerals

Pinto beans are good source of certain minerals like iron, calcium, magnesium, sodium, and potassium (Câmara et al., 2013). Table 17.5 represents the mineral composition of pinto beans in a 100 g edible portion. However, the bioavailability of these minerals are lower as compared to animal sources due to the presence of several components like phytic acid, fiber, and phenolic compounds that reacts with minerals present in beans, decreasing its bioavailability (Sgarbieri, 1984; Geil and Anderson, 1994). A study on pinto beans by Saari et al., (2006) also showed that pinto beans are an excellent source of copper with high concentration and bioavailability. Another study by Petry et al., (2015) has elaborated that pinto beans can also act as a carrier for biofortification of iron to combat iron deficiency.

TABLE 17.4

Vitamin Composition (mg/kg, dry weight basis) of Pinto Beans

Thiamine	Riboflavin	Niacin	Pantothenic Acid	Pyridoxine	Folate
8.6–9.9	1.4–2.3	17.8	3.1–4.4	4.8	4.6

Data from: OECD. (2019). Safety Assessment of Foods and Feeds Derived from Transgenic Crops, Vol. 3.

TABLE 17.5

Mineral Composition of Pinto Beans (mg/100 g edible portion)

Calcium	Iron	Potassium	Magnesium	Sodium
71	2.57	96	43	286

Data from: Ganesan, K., & Xu, B. (2017). Polyphenol-rich dry common beans (Phaseolus vulgaris L.) and their health benefits. International Journal of Molecular Sciences.

Apart from the nutrients mentioned above, pinto beans also contain several phytochemicals and antinutrients in low amounts including phenolic compounds, lectins, tannins, phytates, trypsin inhibitors, and oligosaccharides, some of which show protective properties against multiple degenerative diseases (Câmara et al., 2013; Ganesan and Xu, 2017).

17.3.6 Phenolic Compounds

Principle phenolic components present in pinto beans are phenolic acids and flavonols (OECD, 2019). Phenolic acids in pinto beans primarily contain p-coumaric acid (4.9 mg/100 g), sinapic acid (7.8 mg/100 g), gallic acid (8.7 mg/100 g), and ferulic acid (18.0 mg/100 g) (Luthria and Pastor-Corrales, 2006; Câmara et al., 2013). Kaempferol is the principle flavonol present in pinto beans (OECD, 2019), which is found in two forms: Kaempferol 3-O-glucoside (14.8 mg/100 g) and Kaempferol 3-O-acetylglucoside (3.0 mg/100 g) (Xu and Chang, 2009; Yang et al., 2018). In edible beans, these compounds are mainly found in free form and can be extracted by hydrophilic solvent water mixtures (Los et al., 2018). Some reports also suggest that in common beans, phenolic acids are present in bound or conjugated forms while flavonoids are present in free form (Chen et al., 2015). Processing steps or pretreatment can affect the amount of phenolic content in pinto beans in both a positive and negative manner. A considerable reduction in total phenolic content can be observed when pinto beans are subjected to thermal processing while soaking in hot or cold water prior to cooking leads to an increase in the total phenolic content of pinto beans (Akillioglu and Karakaya, 2010; Yang et al., 2018).

17.3.7 Oligosaccharides

Several raffinose family oligosaccharides (RFOs), like raffinose and verbascose, are present in pinto beans and show prebiotic activity (Los et al., 2018). These oligosaccharides are not digested because of a lack of enzyme α-galactosidase in the human digestive system and act as a substrate for microbial fermentation (Geil and Anderson, 1994). Consumption of these compounds is thought to be beneficial because they stimulate the activity of several beneficial microbiota like lactobacilli and bifidobacterial in the colon (Chen et al., 2016; Los et al., 2018). Sometimes, these compounds also lead to flatulence or gas production (Cristofaro et al., 1974). However simple household preparations like soaking and cooking can remove these flatulence-causing oligosaccharides effectively. (Vidal-Valverde et al., 1993).

17.4 Antinutritional Factors

17.4.1 Lectins

Lectins, also known as hemagglutinins, are proteins present in pinto beans and other varieties of common beans (Gupta, 1987). Lectins are sugar-binding molecules that show high carbohydrate selectivity (Los et al., 2018). Lectins have the ability to agglutinate red blood cells (Yin et al., 2015). However, the lectin content of common beans can be effectively reduced by processing steps like steam cooking and extrusion at 82–85°C (Kelkar et al., 2012).

17.4.2 Tannins

Tannins are polyphenols that can interfere with the bioavailability of certain nutrients (Reed, 1995). Tannins can form complexes by binding with the protein or mineral content of beans, leading to reduced digestion and absorption of these nutrients (Junk-Knievel et al., 2008; Los et al., 2018). Cooking cannot remove the tannins, but more than 50% of tannins can be removed by washing the beans (Ziena et al., 1991).

17.4.3 Phytic Acid/Phytates

Phytic acid and phytates are generally considered antioxidants because of their ability to chelate metal cations (Nikmaram et al., 2017). But in the case of beans, they interfere with the absorption of minerals like iron, magnisium, and calcium (making them unavailable), which can also be considered an antinutrient compound (Los et al., 2018). The phytic acid content of pinto beans varies considerably in different part of the bean. In whole pinto beans, the amount of phytic acid varies from 6.1–23.8 mg/g, while de-hulled beans contain about 8.71 mg/g of phytic acid(OECD, 2019).

17.4.4 Trypsin Inhibitors

Trypsin inhibitors are protease inhibitors that cause metabolic disturbance (Adeyemo and Onilud, 2013). Trypsin inhibitors are responsible for slow digestion of proteins, which leads to the excretion as well as limited bioavailability of amino acids (Nikmaram et al., 2017). The amount and adverse effects of trypsin inhibitors can be reduced considerably by the application of thermal processing (Li et al., 2014).

17.5 Effect of Processing and Storage Conditions on Stability of Pinto Beans

Processing and storage conditions can influence the constituents as well as final quality of pinto beans. Pinto beans can be safely stored under 14% or less moisture content (wet basis) for 20 weeks below 20°C without affecting its nutritional and sensory quality. Storage at 30°C or above can have a considerable effect on seed germination, darkening of the seed coat, as well as on microbial stability. Beans with moisture content more than 14% (wet basis) must be dried to the safe moisture limit within 3 weeks for its prolonged storage, otherwise microbial load and quality loss will increase with increase in moisture and storage temperature (Rani et al., 2013). Longer storage periods can also lower the total phenolic content while increasing the total proanthocyanidin content of pinto beans (Martín-Cabrejas et al., 1997).

Thermal processing leads to the loss of heat-sensitive nutrients and also removes some of the antinutritional factors. Thermal treatment can alter the protein structure and cause inactivation of antinutritional factors in pinto and other beans, resulting in increased biological value as well as improved digestibility of the bean proteins (Hayat et al., 2013). Lectins are relatively heat stable, but cooking at around 100°C can inactivate them significantly (Coffey et al., 1992; Kelkar et al., 2012). The total lectin activity (also referred as hemagglutination activity [HA]) is used as an indicator to check the effectiveness of thermal processing (He et al., 2014). The total phenolic content of pinto beans can reduce significantly with the application of heat (Xu and Chang 2009). Akillioglu and Karakaya (2010) have suggested that soaking in water prior to cooking may help retain the phenolic compounds of pinto beans during thermal processing.

Nonthermal treatment such as gamma radiation can be used for extending the shelf life and reducing pathogenic microorganisms. Some studies have demonstrated that gamma treatment can reduce the allergenicity due to lectins present in the pinto beans without reducing its nutritional value (Kasera et al., 2012; Vaz et al., 2013; He et al., 2015). The high hydrolic pressure (HHP) treatment on pinto beans does not change the HA of phytohemagglutinin (PHA) at treatment of 150 MPa, however, at about 50 MPa, its unfolding process is initiated to molten globule state. High-pressure treatment at about 450 MPa can induce aggregation and rearrangements of protein, with a significant decrease in lectin activity (Liu et al., 2013).

Germination can reduce the total hydroxybenzoic content as compared to raw beans, while the total hydroxycinnamic and total flavonoid content will increase after germination (Xue et al., 2016; Duenas et al., 2016). In another study by Audu and Aremu (2011), it was observed that sprouting and fermentation can enhance the total essential amino acid content in pinto beans while it was reduced during roasting. They have concluded that the proportion of sulphur-containing amino acids were also improved significantly in cooked, fermented, roasted, and sprouted seeds while boiling had a negative effect.

17.6 Functional Properties of Pinto Beans

The bioactive compounds present in pinto bean are responsible for its functional properties. Some of the important functional properties of pinto beans are discussed next.

17.6.1 Antioxidant Activity

Pinto beans show excellent antioxidant properties due to the phenolic compound present within it that exerts a reducing effect against free radicals and other oxidative agents (Oomah et al., 2005). Apart from this, the polyphenol present in beans also chelates metal ions that otherwise act as pro-oxidants and enhance the rate of oxidation (Ganesan and Xu, 2017). But because of this, sometimes these phenolic compounds also interfere with the absorption and bioavailability of minerals present in beans. These phenolic compounds show maximum antioxidant properties during the digestion and absorption process because the acidic environment of the intestines helps the release and solubility of polyphenols (Akillioglu and Karakaya, 2010).

17.6.2 Antimutagenic Properties

Apart from their antioxidant activity, polyphenols present in pinto beans such as catechins, lectin-free fractions, and phenolic acids also act as antimutagens because of their effectivity against mutagenic compounds like nitrosamines, polycyclic aromatic hydrocarbons (PAH) and mycotoxins (De Mejia et al., 1999; Cardador-Martínez et al., 2002; Ganesan and Xu, 2017). The effectivity of polyphenols as antimutagens will depend on several factors such as number, position of phenolic-OH groups and degree of alkylation or acetylation (Hayat et al., 2013). Alkylation or acetylation can block these phenolic hydroxyl groups and decrease the antimutagenic properties of polyphenols (Edenharder and Tang 1997; Elizabeth et al., 2007).

17.6.3 Anticarcinogenic Properties

The anticarcinogenic properties of pinto beans are due to the ability of its substances, particularly phenolic and other bioactive compounds, to inhibit and interact with potent carcinogens (Ganesan and Xu, 2017). Although pinto beans show some anticarcinogenic properties, exposure in this field is lacking. Future studies should be focused on this area for better understanding and utilization of pinto beans as anticarcinogenic agents.

17.6.4 Antidiabetic Properties

Dietary fiber, resistant starch, phytosterols, and total phenolics content of pinto beans are responsible for its antidiabetic activity (Ganesan & Xu, 2017). These phenolics include anthocyanins, catechin, delphinidin glucoside, epicatechin, malvidin glucoside, myricetin 3-O-arabinoside, O-coumaric acid, petunidin glucoside, syringic acid, and vanillic acid, which are mainly concentrated into the seed coat of pinto beans and can be reduced by de-hulling (Mojica et al., 2015; Mojica et al., 2017). In addition to these phenolic compounds, enzyme α-amylase in common beans also shows inhibitory effects against absorption and metabolism of carbohydrates (He et al., 2015).

17.6.5 Anti-inflammatory Properties

The phenolic acids, short chain fatty acid precursors, and bioactive peptide fractions present in pinto beans are reported to have anti-inflammatory properties (Oseguera-Toledo et al., 2011). These anti-inflammatory substances can inhibit proinflammatory activity by modulating the immune response and inflammatory mediators without any cytotoxicity (Reverri et al., 2015; Gabriele et al., 2016).

17.6.6 Antiobesity and Cardio-protective Properties

Flavonoids and phytosterols present in beans can help minimize the risk of obesity and cardiovascular disease (CVD). Soluble fiber present in pinto beans can reduce cholesterol levels. The hypocholesteromic action of soluble fiber is due to its ability to change the pattern of absorption and reabsorption of cholesterol and bile acid (Geil and Anderson, 1994). In the colon, these soluble fibers are subjected to fermentation, which converts them into short chain fatty acids and leads to the reduction or inhibition of hepatic cholesterol (Anderson and Gustafson, 1988; Geil & Anderson, 1994). Resistant starch and dietary fibers associated with pinto beans are also reported to have a positive effect against obesity as it can increase satiety, change the utilization of fat and glucose, and control appetite, thus managing the metabolic process (Hayat et al., 2013).

17.7 Health Benefits of Pinto Beans

Pinto beans contain a considerable quantity of bioactive compounds that plays a key role in improving human nutrition and the management and prevention of many degenerative diseases. Some of the health benefits of pinto beans are discussed next and Table 17.6 summarizes the bioactive constituents of pinto beans with their functional properties and health benefits.

17.7.1 Oxidative Stress

Free radicals and other oxidative agents, which are produced by certain biochemical and physiological processes, are detrimental to human health as it can cause oxidative stress (Câmara et al., 2013). Oxidative stress is the condition when there is an imbalance between the production of free radicals in the cell and the ability of the body to detoxify these free radicals that can cause damage to our body system as well as cardiovascular, neurological, respiratory and degenerative diseases (Pizzino et al., 2017). Because of their high phenolic content, pinto beans exert excellent antioxidant properties that can be utilized against these reactive oxygen species (Câmara et al., 2013). Xu and Chang (2012) also reported that along with soy beans and black beans, pinto beans exerted the maximum antioxidant activity among 13 legumes. De-hulling can significantly reduce the phenolic content of pinto beans (Cao et al., 2010), so for the better utilization of its antioxidant properties, pinto beans should be consumed whole.

17.7.2 Diabetes Mellitus

Clinical studies show that regular consumption of pinto beans not only prevents the occurrence of diabetes but it is also advantageous in control and regulation of diabetes mellitus (Venn and Mann, 2004). According to Ambigaipalan et al., 2011, pinto beans are effective in reducing postprandial glycemic responses because of their slow digestibility as compared to other beans. The low digestibility of bean carbohydrates is because of the production of short chain fatty acids as well as their high content of amylase and resistant starch (Campos-Vega et al., 2010; Hayat et al., 2013). Pinto beans are also rich source of dietary fiber and phenolics, which show protective properties against diabetes and other CVDs (Geil and Anderson, 1994). Because of their phenolic content, common beans show better inhibition of many enzymes including α-glucosidase, α-amylase, and dipeptidyl peptidase, which are responsible for antihyperglycemic activities (Ganesan and Xu, 2017). Regular intake of more than three servings of beans per week can decrease the risk of diabetes by at least 25% (Campos-Vega et al., 2010).

17.7.3 Obesity

Obesity is a serious health condition that increases the threat of diabetes, cardiovascular, and other degenerative diseases. Pinto beans are high in fiber and are low glycemic index carbohydrates that are able to influence body weight, blood insulin, and blood glucose positively (Livesey et al., 2008; Câmara et al., 2013). Moreover the α-amylase inhibitors can interfere with starch digestion and cause energy restriction that leads to utilization of body fat (Obiro et al., 2008; Hayat et al., 2013). Extracts of pinto beans should be utilized as a therapeutic agent or dietary adjunct to control extra body weight and risk of obesity without the loss of body proteins (Pusztai et al., 1998; Spadafranca et al., 2013).

TABLE 17.6

Functional Properties and Health Benefits Associated with Pinto Beans

Bioactive Constituent	Functional Property	Health Benefits	References
Polyphenols	Antidiabetic activity	Slow digestibility of carbohydrates	Spadafranca et al., 2013
	Antioxidant activities	Inhibits oxidative agents	Oomah et al., 2010
	Anticarcinogenic properties	Cancer cell proliferation	Hayat et al., 2013 Xu & Chang, 2012
	Anti-inflammatory activity	Modulation of inflammatory cells or proinflammatory mediators	Oomah et al., 2010
	Antimutagenic properties	Inhibits metabolism of the toxicants or mutagens by interacting with mutagens.	De Mejia et al., 1999
	Hypoglycemic and antiobesity effect	Interferes with starch digestion, cause energy restriction and better utilization of fats	Maccioni et al., 2010
Phenolic acids	Antidiabetic property	Helpful in prevention and management of diabetes	Mojica et al., 2015
	Antimutagenic activity	Inhibition of potent mutagens by scavenging activities of phenolics	Cardador-Martínez et al., 2002; Frassinetti et al., 2015
	Antioxidant activities	Inhibits free radicals, chelates metal ions	Frassinetti et al., 2015 Ganesan and Xu, 2017
	Anti-inflammatory activity	Inhibits proinflammatory mediators	Oseguera-Toledo et al., 2011
Tannins	Anticarcinogenic activity	Chemoprotective effect against breast cancer	Ganesan and Xu, 2017
(+)-catechin,	Antimutagenic activity	Prevents colon and breast cancers	Cardador-Martínez et al., 2006
	Antioxidant activities	Effective against oxidative stress	Cardador-Martínez et al., 2006
	Antidiabetic activity	Protects against type 2 diabetes and related disorders	Mojica et al., 2015
Dietary fiber & resistant starch	Antidiabetic and hypoglycemic activity	Interferes with starch digestion and reducing postprandial glycemic responses	Ambigaipalan et al., 2011; Ganesan and Xu, 2017
	Antiobesity properties	Prevention and regulation of obesity and lifestyle related disorders	Hayat et al., 2013
	Cardio-protective activity	Improves lipid profiles associated with cardiovascular disease (CVD), reducing risk for CVD	Berrios et al., 2010; Wang and Toews, 2011 Geil and Anderson, 1994
	Anticarcinogenic activity	Inhibits tumor formation	Hayat et al., 2013
	Improvement in metabolic activity	Slows digestibility and subjected to microbial fermentation	Los et al., 2018; Marinangeli et al., 2020
		Higher production of SCFA and altered pH improve gastrointestinal health	Campos-Vega et al., 2010
Bioactive peptide fractions	Anti-inflammatory activity	Modulation of immune response and inflammatory mediators without causing any cytotoxicity	Gabriele et al., 2016
Short chain fatty acids	Hypocholesteromic activity	Reduction HDL and LDL cholesterol levels	Finley et al., 2007
	Anti-inflammatory activity	Acts as immunomodulatory agent to protect from acute and severe inflammation	Oseguera-Toledo et al., 2011
α-amylase inhibitors	Antidiabetic properties	Inhibitory action against absorption and metabolism of carbohydrates	He et al., 2015
Phytates and protease inhibitors	Anticarcinogenic properties	Inhibitory actions against colon cancer	Harland and Morris, 1995; Hayat et al., 2013
Lectin	Anticarcinogenic activity	Binds with cell membranes or cell receptors of cancer causing cells	De Mejía and Prisecaru, 2005
Lectin-free fractions	Antioxidant activities	Free radical scavenging and ROS generation inhibition	Frassinetti et al., 2015
	Antimutagenic activity	Inhibits mutagenic compounds	Frassinetti et al., 2015

17.7.4 Inflammation

Inflammation is a condition when body receptors can recognize any type of infection or tissue damage (Nathan, 2002). When inflammation is not repaired or treated properly, it can result in the occurrence of number of disorders including chronic inflammation (Dinarello, 2010). According to Middleton et al. (2000), the drugs commonly used for the treatment of acute inflammation (principally steroidal anti-inflammatory drugs) are less effective against chronic inflammation and may results in multiple side effects. However, bioactive constituents of foods such as flavonoids and vitamins, can counteract these adverse effects by the modulation of inflammatory cells, proinflammatory enzymes, or proinflammatory mediators (García-Lafuente et al., 2009; Câmara et al., 2013). Pinto beans show excellent anti-inflammatory effects because of their antioxidative and anti-inflammatory properties. According to Oomah et al. (2010), the hull extract of pinto beans can inhibit proinflammatory mediators effectively. So pinto bean extracts can be utilized for development of immunomodulatory or anti-inflammatory agents (He et al., 2015).

17.7.5 Metabolic Syndrome and Cardiovascular Health

Metabolic syndrome represents a set of metabolic conditions that can further develop the risk of CVD (Câmara et al., 2013). These conditions include increased triglycerides and total cholesterol, high levels of low density and very low density lipoproteins (LDL), low levels of high density lipoproteins (HDL), as well as elevated blood pressure and glucose levels (Finley et al., 2007). Many studies reported that the incorporation of whole pinto beans or their hulls in a diet high in saturated fats can improve lipid profiles significantly by lowering the level of LDL and increasing the level of HDL (McCrory et al., 2010; Nguyen et al., 2019). Winham et al., 2007 concluded that regular consumption of a half-cup of pinto beans for 8 weeks can reduce the serum total cholesterol and low density cholesterol concentrations by more than 8%. Bioactive compounds including slow digesting and nondigesting carbohydrates (eg; dietary fibers and resistance starches) present in common beans are subjected to microbial fermentation to improve metabolic activities and reduce the risk of development of metabolic and CVD (Los et al., 2018; Marinangeli et al., 2020).

17.7.6 Cancer

Cancer is a genetic disease caused by a number of genetic and environmental factors. Some studies have shown that regular consumption of pinto beans can lower the risk of development of certain types of cancers (Singh et al., 2017). According to Hayat et al., 2013, the anticancerous effects of common beans are due to the macro- and microconstituents including lectins, dietary fiber, phenolic compounds, phytic acids, resistance starch, etc. Lectins bind with cell membranes or cell receptors of cancer-causing cells (De Mejía and Prisecaru, 2005). Because of their selective targeting to cancerous cells, lectins can be utilized as a carrier for drug delivery against cancer-causing cells (Khopade et al., 1998). Tannins present in pinto beans shows protective effects against breast cancer (Ganesan and Xu, 2017). Finley et al. (2007) have shown in his study that consumption of beans can lower the risk of development of colon cancer due to the increased colonial short chain fatty acid formation, which may reduce HDL and LDL cholesterol levels.

17.8 Conclusion and Future Prospectives

Currently, many researches are focused on the replacement of energy-rich foods with pulses for management and prevention of many genetic and lifestyle-related diseases. Because of its low-fat, fiber-rich, and high-protein content, pinto beans are recommended as one of the major food constituents in a regular diet. Polyphenols and other bioactive compounds from pinto beans are of particular interest because of their biological properties including antioxidants, antibacterial, anti- inflammatory, antidiabetic, anti-obesity, antimutagenic, and anticarcinogenic properties. Although further studies are required to specify any direct correlation between constituents of pinto beans and prevention or control of various genetic and lifestyle-related diseases, its incorporation in different food formulations will definitely help to utilize its properties as a functional food.

Moreover, future research and clinical trials have wide potential in the study of pinto beans, including the role and direct impact of its bioactive constituents on regulation and management of various health related diseases, the effect of different processing steps on quality and properties of its constituents, and the formulation of nutraceuticals and functional foods by the incorporation of pinto beans as a major ingredient.

REFERENCES

Adebowale, Y. A., Adeyemi, I. A., Oshodi, A. A., and Niranjan, K. (2007). Isolation, fractionation and characterization of proteins from Mucuna bean. *Food Chem, 104*, 287–299.

Adeyemo, S. M. and Onilude, A. A. (2013). Enzymatic reduction of antinutritional factors in fermenting soybeans by *Lactobacillus plantarum* isolates from fermenting cereals. *Niger Food J, 31,* 84–90.

Akillioglu, H. G., and Karakaya, S. (2010). Changes in total phenols, total flavonoids, and antioxidant activities of common beans and pinto beans after soaking, cooking, and in vitro digestion process. *Food Sci Biotechnol, 19*(3), 633–639. https://doi.org/10.1007/s10068-010-0089-8

Ambigaipalan, P., Hoover, R., Donner, E., Liu, Q., Jaiswal, S., Chibbar, R., Nantanga, K. K. M., and Seetharaman, K. (2011). Structure of faba bean, black bean and pinto bean starches at different levels of granule organization and their physicochemical properties. *Food Res Int, 44*, 2962–2974.

Anderson, J. W., and Gustafson, N. J. (1988). Hypocholesterolemic effects of oat and bean products. *Am J Clin Nutr, 48,* 749–753.

Arkcoll, D. B., and Clement, C. R. (1989). Potential new crops from Amazon. In Wickens, G. F. S., Haq, N. and Day, P. (Eds.), *New Crops for Food Industry*. London: Chapman and Hall,. 150–165.

Audu, S., and Aremu, M. (2011). Nutritional composition of raw and processed pinto bean (*Phaseolus vulgaris* L.) grown in Nigeria. *J Food Ag Enviro, 9*, 72–80. https://www.researchgate.net/publication/287562235

Berrios JDJ, Morales P, Ca´mara M, and Sa´nchez-Mata MC (2010) Carbohydrate composition of raw and extruded pulse flours. *Food Res Int, 43*, 531–536.

Brink, M., and Belay, G. (2006). Plant resources of tropical Africa 1: cereals and pulses (pp. 54–57). *Wageningen*, The Netherlands: PROTA Foundation.

Broughton, W. J. et al. (2003), Beans (*Phaseolus* Spp.)—model food legumes, *Plant Soil, 252*, 55–128.

Câmara, C. R. S., Urrea, C.A., and Schlegel, V. (2013). Pinto beans (*Phaseolus vulgaris* l.) as a functional food: Implications on human health. *Agriculture (Switzerland), 3*(1), 90–111. https://doi.org/10.3390/agriculture3010090

Campos-Vega., R., Loarca-Pina, G., and Oomah, B. D. (2010). Minor components of pulses and their potential impact on human health. *Food Res Int, 43*, 461–582.

Cao, J. J., Gregoire, B. R., Sheng, X., and Liuzzi, J. P. (2010). Pinto bean hull extract supplementation favorably affects markers of bone metabolism and bone structure in mice. *Food Res Int, 43*, 560–566.

Cardador-Martínez, A., Albores, A., Bah, M., Calderón-Salinas, V., Castaño-Tostado, E., Guevara-González, R., Shimada-Miyasaka, A., and Loarca-Piña, G. (2006). Relationship among antimutagenic, antioxidant and enzymatic activities of methanolic extract from common beans (*Phaseolus vulgaris* L). *Plant Foods Hum Nutr, 61*, 161–168.

Cardador-Martínez, A., Castano-Tostado, E., and Loarca-Pina, G. (2002). Antimutagenic activity of natural phenolic compounds present in the common bean (*Phaseolus vulgaris*) against aflatoxin B1. *Food Addit Cont, 19*(1), 62–69.

Chen, P. X., Bozzo, G.G., Freixas-Coutin, J. A., Marcone, M. F., Pauls, P. K., Tang, Y., Zhang, B., Liu, R., and Tsao, R. (2015). Free and conjugated phenolic compounds and their antioxidant activities in regular and nondarkening cranberry bean (*Phaseolus vulgaris* L.) seed coats. *J Funct Foods, 18*:1047–1056.

Chen, Y., McGee, R., Vandemark, G., Brick, M., and Thompson, H. (2016). Dietary fiber analysis of four pulses using AOAC 2011.25: Implications for human health. *Nutrients, 8*:829.

Coffey, D., Uebersax, M., Hosfield, G., and Bennink, M. (1992). Stability of red kidney bean lectin. *J Food Biochem, 16*(1), 43–57.

Cristofaro, E., Mottu, F., and Wuhrmann, J. J. (1974) Involvement of the Raffinose family of oligosaccharides in flatulence. In: Sipple, H. L. and McNutt, K.W. (Eds.), Sugars in Nutrition, London: Academic Press, 313–336.

De Mejía, E. G., andPrisecaru, V. I. (2005). Lectins as bioactive plant proteins: A potential in cancer treatment. *Crit Rev Food Sci*, *45*(6), 425–445.

De Mejia, E. G., Castano-Tostado, E., and Loarca-Pina, G. (1999). Antimutagenic effects of natural phenolic compounds in beans. *Mut Res, 441,* 1–9.

Dinarello, C. (2010). Anti-inflammatory agents: Present and future. *Cell*, *140*, 935–50.

Duenas, M., Sarmento, T., Aguilera, Y., Benítez, V., Molla, E., Esteban, R. M., and Martín-Cabrejas, M. A. (2016). Impact of cooking and germination on phenolic composition and dietary fibre fractions in dark beans (*Phaseolus vulgaris* L.) and lentils (*Lens culinaris* L.). *LWT—Food Sci Technol*, *66*, 72–78. https://doi.org/10.1016/j.lwt.2015.10.025

Edenharder, R., and Tang, X. (1997). Inhibition of mutagenicity of 2-nitrofluorene, 3-nitrofluorene and 1-nitropyrene by flavonoids, coumarins, quinones and other phenolic compounds. *Food Chem Toxicol, 35,* 357–372.

Elizabeth, R.N., Annete, H., Francisco, G. R., Javier, I. F., Graciela, Z., and Alberto, G. J. (2007). Antioxidant and antimutagenic activity of phenolic compounds in three different colour groups of common bean cultivars (*Phaseolus vulgaris*). *Food Chem*, *103*, 521–527.

FAO (1999). Phaseolus Bean: Post-Harvest Operations. Food and Agriculture Organization of the United Nations.

Finley, J. W., Burrell, J. B., and Reeves, P. G. (2007). Pinto bean consumption changes SCFA profiles in fecal fermentations, bacterial populations of the lower bowel, and lipid profiles in blood of humans. *J Nutr*, *137*(11), 2391–2398. https://doi.org/10.1093/JN/137.11.2391

Frassinetti, S., Gabriele, M., Caltavuturo, L., Longo, V. and Pucci, L. (2015). Antimutagenic and antioxidant activity of a selected lectin-free common bean (*Phaseolus vulgaris* L.) in two cell-based models. *Plant Foods Hum Nutr*, *70*, 35–41.

Gabriele, M., Pucci, L., La Marca, M., Lucchesi, D., Della Croce, C. M., Longo, V., and Lubrano, V. (2016). A fermented bean flour extract down-regulates LOX-1, CHOP, and ICAM-1 in HMEC-1 stimulated by ox-LDL. *Cell Mol Biol Lett*, *21*, 10–21.

Ganesan, K., and Xu, B. (2017). Polyphenol-rich dry common beans (*Phaseolus vulgaris* L.) and their health benefits. *Int J Mol Sci*, *18*(11), 2331. https://doi.org/10.3390/ijms18112331

García-Lafuente, A., Guillamón, E., Villares, A., Rostagno, M. A., and Martínez, J. A. (2009). Flavonoids as anti-inflammatory agents: Implications in cancer and cardiovascular disease. *Inflamm Res*, *58*, 537–552.

Geil, P. B., and Anderson, J. W. (1994). Nutrition and health implications of dry beans: A review, *J Am Coll Nutr*, *13*:6, 549–558. doi: 10.1080/07315724.1994.10718446

Gepts, P., and Dpbouk, D. (1991). Origin, domestication, and evolution of the common bean (Phaseolus vulgaris L.). In Van Schoonhoven, A. & Voyset, O. (Eds.), Common Beans: Research for Crop Improvement. Wallingford, England: CAB International, 7–53.

Gupta, Y. P. (1987). Anti-nutritional and toxic factors in food legumes: A review, *Plant Foods Hum Nutr*, *37*(3), 201–228. http://link.springer.com/article/10.1007%2FBF01091786#.

Harland, B. F., and Morris, E .R. (1995). Phytate: A good or a bad food component? *Nutr Res, 15,* 733–754.

Hayat, I., Ahmad, A., Masud, T., Ahmed, A., and Bashir, S. (2013). Nutritional and health perspectives of beans (*Phaseolus vulgaris* L.): An overview, *Crit Rev Food Sci Nutr,* doi: 10.1080/10408398.2011.596639

He, S., Shi, J., Ma, Y., Xue, S. J., Zhang, H., and Zhao, S. (2014). Kinetics for the thermal stability of lectin from black turtle bean. *J Food Eng, 142,* 132–137.

He, S., Simpson, B. K., Sun, H., Ngadi M.O., Ma, Y., and Huang, T. (2015). *Phaseolus vulgaris* lectins: A systematic review of characteristics and health implications, *Crit Rev Food Sci Nutr,* doi: 10.1080/10408398.2015.1096234

Hosfield, G. L., Varner, G.V., Uebersax, M. A., and Kelly, J. D. (2004). Registration of "Merlot" small red bean. *Crop Sci*, *44*, 351–352.

Junk-Knievel, D. C., Vandenberg, A., and. Bett, K. E. (2008). Slow darkening in pinto bean (*Phaseolus vulgaris* L.) seed coats is controlled by a single major gene, *Crop Sci*, *48*(1), 189–193.

Kabagambe, E. K., Baylin, A., Ruiz-Narvarez, E., Siles, X., and Campos, H. (2005). Decreased consumption of dried mature beans is positively associated with urbanization and nonfatal acute myocardial infarction. *J Nutr, 135,* 1770–1775.

Kasera, R., Singh, A. B., Kumar, R., Lavasa, S., Prasad, K. N., and Arora, N. (2012). Effect of thermal processing and γ-irradiation on allergenicity of legume proteins. *Food Chem Toxicol, 50*(10), 3456–3461.

Kelkar, S., Siddiq, M., Harte, J. B., Dolan, K. D., Nyombaire, G., and Suniaga, H. (2012). Use of low-temperature extrusion for reducing phytohemagglutinin activity (PHA) and oligosaccharides in beans (*Phaseolus vulgaris* L.) cv. navy and pinto. *Food Chem, 133,* 1636–1639.

Khopade, A., Nandakumar, K., and Jainb, N. (1998). Lectin-functionalized multiple emulsions for improved cancer therapy. *J. Drug Target, 6*(4), 285–292.

Li, H., Qiu, J., Liu, C., Ren, C., and Li, Z. (2014). Milling characteristics and distribution of phytic acid, minerals, and some nutrients in oat (*Avena sativa* L.), *J Cereal Sci, 60,* 549–554.

Liebenberg, A. J. (ed.) (2009), *Dry bean production. Department: Agriculture, Forestry and Fisheries, Republic of South Africa.* Available at: http://www.nda.agric.za/docs/drybean/drybean.pdf

Liu, C., Zhao, M., Sun, W., and Ren, J. (2013). Effects of high hydrostatic pressure treatments on haemagglutination activity and structural conformations of phytohemagglutinin from red kidney bean (*Phaseolus vulgaris*). *Food Chem, 136*(3), 1358–1363.

Livesey, G., Taylor, R., Hulshof, T., and Howlett, J. (2008). Glycemic response and health—A systematic review and meta-analysis: Relations between dietary glycemic properties and health outcomes. *Am J Clin Nutr, 87,* 258S–268S.

Los, F. G. B., Zielinski, A. A. F., Wojeicchowski, J. P., Nogueira, A., and Demiate, I. M. (2018). Beans (*Phaseolus vulgaris* L.): Whole seeds with complex chemical composition. *Curr Op Food Sci, 19,* 63–71. https://doi.org/10.1016/j.cofs.2018.01.010

Luthria, D. L., and Pastor-Corrales, M. A. (2006). Phenolic acids content of fifteen dry edible bean (*Phaseolus vulgaris* L.) varieties. *J Food Compos Anal, 19,* 205–211.

Maccioni, P., Colombo, G., Riva, A., Morazzoni, P., Bombardelli, E., Gessa, G. L., and Carai, M. A. (2010) Reducing effect of a *Phaseolus vulgaris* dry extract on operant self-administration of a chocolate-flavored beverage in rats. *Br J Nutr, 104,* 624–628.

Marinangeli, C. P. F., Harding, S. V., Zafron, M., and Rideout, T. C. (2020). A systematic review of the effect of dietary pulses on microbial populations inhabiting the human gut. *Beneficial Microbes, 11*(5), 457–468. https://doi.org/10.3920/BM2020.0028

Martín-Cabrejas, M. A., Esteban, R. M., Perez, P., Maina, G., and Waldron, K. W. (1997). Changes in physicochemical properties of dry beans (*Phaseolus vulgaris* L.) during long-term storage. *J Ag Food Chem, 45*(8), 3223–3227. https://doi.org/10.1021/jf970069z

McCrory, M. A., Hamaker, B. R., Lovejoy, J. C., and Eichelsdoerfer, P. E. (2010). Pulse consumption, satiety, and weight management. *Adv Nutr, 1,* 17–30.

Middleton, E., Kandaswami, C., and Theoharides, T. C. (2000). The effects of plant flavonoids on mammalian cells: Implications for inflammation, heart disease, and cancer. *Pharmacol Rev, 52,* 673–751.

Mmbaga, M. T., and J. R. Steadman (1992), Nonspecific resistance to rust in pubescent and glabrous common bean genotypes, *Phytopathology, 82,* 1283–87.

Mojica, L., Berhow, M., and Gonzalez de Mejia, E. (2017). Black bean anthocyanin-rich extracts as food colorants: Physicochemical stability and antidiabetes potential. *Food Chem, 229,* 628–639.

Mojica, L., Meyer, A., Berhow, M. A., and González de Mejía, E. (2015). Bean cultivars (*Phaseolus vulgaris* L.) have similar high antioxidant capacity, in vitro inhibition of α-amylase and α-glucosidase while diverse phenolic composition and concentration. *Food Res Int, 69,* 38–48.

Nathan, C. (2002). Points of control in inflammation. *Nature, 420,* 846–852.

Nguyen, A. T., Althwab, S., Qiu, H., Zbasnik, R., Urrea, C., Carr, T. P., and Schlegel, V. (2019). Pinto beans (*Phaseolus vulgaris* L.) lower non-HDL Cholesterol in hamsters fed a diet rich in saturated fat and act on genes involved in cholesterol homeostasis. *J Nutr, 149*(6), 996–1003. https://doi.org/10.1093/jn/nxz044

Nikmaram, N., Leong, S. Y., Koubaa, M., Zhu, Z., Barba, F. J., Greiner, R., Oey, I., and Roohinejad, S. (2017). Effect of extrusion on the anti-nutritional factors of food products: An overview. *Food Control, 79,* 62–73.

Obiro, W. C., Tao, Z., and Bo, J. (2008). The nutraceutical role of the *Phaseolus vulgaris* α-amylase inhibitor. *Br J Nutr, 100,* 1–12.

OECD (2015). Consensus Document on the Biology of Common Bean (Phaseolus vulgaris L.), *Series on Harmonisation of Regulatory Oversight in Biotechnology, No.59.* Paris: OECD. Available at: http://www.oecd.org/officialdocuments/publicdisplaydocumentpdf/?cote=env/jm/mono(2015)47&doclanguage=en

OECD. (2019). Safety Assessment of Foods and Feeds Derived from Transgenic Crops, Vol. 3: *Common Bean, Rice, Cowpea and Apple Compositional Considerations.* Novel Food and Feed Safety series, OECD Publishing, Paris: OECD. https://doi.org/10.1787/f04f3c98-en

Oomah, B. D., Cardador-Martinez, A., and Loarca-Piña, G. (2005). Phenolics and antioxidative activities in common beans (*Phaseolus vulgaris* L). *J Sci Food Agric, 85*, 935–942.

Oomah, B. D., Corbé, A., and Balasubramanian, P. (2010). Antioxidant and anti-inflammatory activities of bean (*Phaseolus vulgaris* L.) hulls. *J Agric Food Chem, 58*, 8225–8230.

Oseguera-Toledo, M. E., De Mejia, E. G., Dia, V. P. & Amaya-Llano, S. L. (2011). Common bean (*Phaseolus vulgaris* L.) hydrolysates inhibit inflammation in LPS-induced macrophages through suppression of NF-κB pathways. *Food Chem, 127*, 1175–1185.

Papa, R., and Gepts, P. (2003). Asymmetry of gene flow and differential geographical structure of molecular diversity in wild and domesticated common bean (*Phaseolus vulgaris* L.) from Mesoamerica. *Theor Appl Genet, 106*, 239–250.

Papa, R., Acosta, J., Delgado-Salinas, A., and Gepts, P. (2005). A genome-wide analysis of differentiation between wild and domesticated *Phaseolus vulgaris* from Mesoamerica. *Theor Appl Genet, 111*, 1147–1158.

Petry, N., Boy, E., Wirth, J. P., and Hurrell, R. F. (2015). Review: The potential of the common bean (*Phaseolus vulgaris*) as a vehicle for iron biofortification. *Nutrients, 7*(2), 1144–1173. https://doi.org/10.3390/nu7021144

Pillemer, E. A., and W.M. Tingey (1978). Hooked trichomes and resistance of *Phaseolus vulgaris* to *Empoasca Fabae* (Harris), *Entomologia Experimentalis et Applicata, 24* (1) 83–94. doi: 10.1111/j.1570-7458.1978.tb02758.x

Pizzino, G., Irrera, N., Cucinotta, M., Pallio, G., Mannino, F., Arcoraci, V., Squadrito, F., Altavilla, D., and Bitto, A. (2017). Oxidative stress: Harms and benefits for human health. *Oxidative Med Cell Longev,* 2017. https://doi.org/10.1155/2017/8416763

Purseglove, J. W. (1968). *Tropical Crops: Dicotyledons. London: Longmans,* 719.

Pusztai, A., Grant, G., Buchan, W., Bardocz, S., De Carvalho, A., and Ewen, S. (1998). Lipid accumulation in obese Zucker rats is reduced by inclusion of raw kidney bean (*Phaseolus vulgaris*) in the diet. *Br J Nutr, 79*(2), 213–221.

Rani, P. R., Chelladurai, V., Jayas, D. S., White, N. D. G., and Kavitha-Abirami, C. V. (2013). Storage studies on pinto beans under different moisture contents and temperature regimes. *J Stored Prod Res, 52*, 78–85. https://doi.org/10.1016/j.jspr.2012.11.003

Reed, J. D. (1995). Nutritional toxicology of tannins and related polyphenols in forage legumes, *J Animal Sci, 73*, 1516–1528. https://www.ncbi.nlm.nih.gov/pubmed/7665384.

Reverri, E. J., Randolph, J. M., Steinberg, F. M., Kappagoda, C. T., Edirisinghe, I., and Burton-Freeman, B. M. (2015) Black beans, fiber, and antioxidant capacity pilot study: Examination of whole foods vs. functional components on postprandial metabolic, oxidative stress, and inflammation in adults with metabolic syndrome. *Nutrients, 7*, 6139–6154.

Rochfort, S., and Panozzo, J. (2007) Phytochemicals for health, the role of pulses. *J Agric Food Chem, 55*, 7981–7994.

Saari, J. T., Reeves, P. G., Johnson, W. T., and Johnson, L. A. K. (2006). Pinto beans are a source of highly bioavailable copper in rats. *J Nutr, 136*(12), 2999–3004. https://doi.org/10.1093/jn/136.12.2999

Santiago-Ramos, D., Figueroa-Ca´rdenas, J., De, D., Ve´les-Medina, JJ., and Salazar, R. (2018). Physicochemical properties of nixtamalized black bean (Phaseolus vulgaris L.) flours. Food Chem, 240, 456–462. http://dx. doi.org/10.1016/j.foodchem.2017.07.156.

Sgarbieri, V. C. (1984). Composition and nutritive value of beans (*Phaseolus vulgaris* L.). *World Rev Nutr Diet, Nutritional Value of Cereal Products, Beans and Starches series, 60*, 132–198.

Shevkani, K., and Singh, N. (2015). Relationship between protein characteristics and film-forming properties of kidney bean, field pea and amaranth protein isolates. *Int J Food Sci Technol, 50*, 1033–1043.

Shiga, T. M., B. R. Cordenunsi, and F.M. Lajolo (2009). Effect of cooking on non-starch polysaccharides of hard-to-cook beans. *Carb Polymers, 76*, 100–109. http://www.sciencedirect.com/science/article/pii/S0144861708004669.

Shimelis, E. A., and Rakshit S. K. (2005). Proximate composition and physico-chemical properties of improved dry bean (*Phaseolus vulgaris* L.) varieties grown in Ethiopia. *LWT, 38*, 331–338.

Singh, B., Singh, J. P., Shevkani, K., Singh, N., and Kaur, A. (2017). Bioactive constituents in pulses and their health benefits. *J Food Sci Technol, 54*(4), 858–870. https://doi.org/10.1007/s13197-016-2391-9

Spadafranca, A., Rinelli, S., Riva, A., Morazzoni, P., Magni, P., Bertoli, S., and Battezzati, A. (2013). *Phaseolus vulgaris* extract affects glycometabolic and appetite control in healthy human subjects. *Br J Nutr, 109*(10), 1789–1795.

Stone, S., and Stone, M. (1988). *The Brilliant Bean.* New York: Bantam Books.

Tharanathan, R. N., andMahadevamma, S. (2003). Grain legumes—a boon to human nutrition. *Trends Food Sci Technol, 14,* 507–518.

Vaz, A F., Souza, M. P., Vieira, L. D., Aguiar, J. S., Silva, T. G., Medeiros, P. L., Melo, A. M., Silva-Lucca, R. A., Santana, L. A., and Oliva, M. L. (2013). High doses of gamma radiation suppress allergic effect induced by food lectin. *Radiat Phys Chem, 85,* 218–226.

Venn, B. J., andMann, J. I. (2004). Cereal grains, legumes and diabetes. *Eur J Clin Nutr, 58,* 1443–1461.

Vidal-Valverde, C., Frias, J., and Valverde, S. (1993). Changes in the carbohydrate composition of legumes after cooking and soaking. *J Am Diet Assoc, 93,* 547–550.

Wang N., and Toews R. (2011) Certain physicochemical and functional properties of fibre fractions from pulses. *Food Res Int, 44,* 2515–2523.

Winham, D. M., Hutchins, A. M., and Johnston, C. S. (2007). Pinto bean consumption reduces biomarkers for heart disease risk. *J Am Coll Nutr, 26,* 243–249.

Wortmann, C.S. (2006), *Phaseolus Vulgaris L.* (common Bean); Record from PROTA4U. Brink, M. & Belay, G. (Editors). *PROTA (Plant Resources of Tropical Africa / Cereals and pulses/Céréales et légumes secs),* Wageningen, Netherlands

Xu, B., and Chang, S. K. C. (2009), Total phenolic, phenolic acid, anthocyanin, flavan-3-ol, and flavonol profiles and antioxidant properties of pinto and black beans (*Phaseolus vulgaris* L.) as affected by thermal processing, *J Ag Food Chem, 57*(11), 4754–4764. doi:10.1021/jf900695s

Xu, B., and Chang, S. K. C. (2012). Comparative study on antiproliferation properties and cellular antioxidant activities of commonly consumed food legumes against nine human cancer cell lines. *Food Chem, 134,* 1287–1296.

Xue, Z., Wang, C., Zhai, L., Yu, W., Chang, H., Kou, X., and Zhou, F. (2016). Bioactive compounds and antioxidant activity of mung bean (*Vigna radiata* L.), soybean (*Glycine max* L.) and black bean during the germination process. *Czech J Food Sci, 34*(1), 68–78. https://doi.org/10.17221/434/2015-CJFS

Yang, Q. Q., Gan, R. Y., Ge, Y. Y., Zhang, D., and Corke, H. (2018). Polyphenols in common beans (*Phaseolus vulgaris* L.): Chemistry, Analysis, and Factors Affecting Composition.: Common bean polyphenols.... *Comp Rev Food Sci Food Safety, 17*(6), 1518–1539. https://doi.org/10.1111/1541-4337.12391

Yin, C., Wong, J. H., and Ng, T. B. (2015). Isolation of a hemagglutinin with potent antiproliferative activity and a large antifungal defensin from *Phaseolus vulgaris* cv Hokkaido large pinto beans. *J Agric Food Chem, 63,* 5439–5448.

Ziena, H. M., Youssef, M., and El-Mahdy, A. R. (1991). Amino acid composition and some anti-nutritional factors of cooked faba beans (Medamnins): Effects of cooking temperature and time, *J Food Sci, 56*(5), 1347–1349. Available at: http://onlinelibrary.wiley.com/doi/10.1111/j.1365-2621.1991.tb04769.x/pdf.

18

Antinutritional Factors in Legumes

Priya Dangi
University of Delhi, New Delhi, India

Nisha Chaudhary
Agriculture University, Jodhpur, India

Deepali Gajwani and Neha
University of Delhi, New Delhi, India

CONTENTS

18.1 Introduction

Legumes are quite popular and are highly consumed, next only to cereals. They are a pool of good quality protein, potassium, fiber, and B-vitamins. The outstanding nutritional profile in terms of protein and fiber and being low in fat, makes legumes a valuable and cheap source of nutrition that offer numerous physiological benefits to the human body (Punia et al., 2019, 2020). They are known to lower the glycemic index of diabetic patients and aid in the prevention of certain metabolic diseases such as cardiovascular diseases and cancer (Hangen and Bennink 2002). Despite their preventive roles, legumes in their raw state contain certain toxic substances, or antinutritional factors, that limit their utilization as they exhibit adverse effects on digestibility, enzyme activity, nutrition, and health of the individual (Liener 2006). Table 18.1 presents different types of antinutritional factors along with their sources and health effects. For this reason, botanists, agriculturists, and food scientists are establishing breeding experiments to minimize the amount of these factors. Despite gaining achievements in reducing considerable

TABLE 18.1

Antinutritional Factors, Their Sources and Adverse Health Effects

Antinutrients	Sources	Adverse Health Effects	References
Trypsin inhibitors	Red gram, Bengal gram, Soy beans	Retards growth, reduces protein digestibility, enlarges pancreas.	Samtiya et al. (2020)
Cyanogenic glycosides	Lima beans	Suppresses growth, acute toxic, and can cause death.	Lawley et al. (2008)
Goitrogens	Soy beans, lentils	Interferes with thyroid hormone and affects brain and neurological development.	Mondal et al. (2016)
Lectins and hemagglutinins	Soy beans, field beans, white beans, double beans, etc.	Leads to poor growth, impairment in nutrient absorption, and may lead to hypertrophy and hyperplasia of pancreas.	Samtiya et al. (2020)
Phenolics and tannins	Red kidney beans, black gram, soy beans	Interferes with protein availability and digestibility, affects absorption of iron and vitamin-B. May cause loss of appetite, body weight, cardiac problems etc.	
Phytic acid	Legumes	Interferes with mineral bioavailability and inhibits the activity of digestive enzymes; May be associated with zinc deficiency in crop plants.	
Saponins	Soy beans, jack beans, sword beans, etc.	Affects the integrity of intestinal epithelial cells and interferes with the absorption of vitamin A, vitamin E, and lipids; May lead to nausea and vomiting.	
Alkaloids	Lupins	Affects the central nervous system, digestive system, immune system, etc.	Veerabahu et al. (2015)
Toxic amino acids	Found in legumes and mainly in seeds such as jack beans	Responsible for neurotoxic disturbances, hallucinations, diarrhea, paralysis, liver cirrhosis, etc.	
Antivitamins	Soy beans, kidney beans, alfalfa, and field peas	Causes destruction of carotene and interferes with the absorption of calcium, phosphorus, niacin, and pyridoxine.	Hamid et al. (2017)
Miscellaneous	Lentils, red kidney beans, and white beans	Oxalates are known to decrease calcium absorption; Oligosaccharides may cause flatulence.	Veerabahu et al. (2015)

amounts of the antinutrients in legumes, this method has not gained importance due to the fact that these secondary metabolites are developed in plants as a part of attaining protection against predators, such as herbivores, insects, pathogens, and also as a means of surviving harsh growing conditions (Owusu-Apenten 2003). Subsequently, the best way to eradicate this problem is to develop effective strategies that have the potential to reduce or remove these factors ahead of their utilization for human and animal consumption.

Such antinutritional factors in legumes can be eliminated or inactivated by traditional treatments and different processing methods including cooking, soaking, germination, fermentation, sterilization, dehulling etc. The fundamental objective of all these techniques is to eliminate the antinutritional factors from legumes, enhance protein digestibility, thereby upgrading the nutritional quality of legumes (Elkowicz and Sosulski 2006).

18.2 Classification and Health Effects of Antinutritional Factors

Antinutritional factors are defined as the poisonous compounds that restrict the availability of nutrients in the human body. Primarily, these antinutrients on the basis of heat stability can be categorized into two groups: heat-labile and heat-stable factors. Heat labile factors are highly sensitive to temperature conditions and are likely to undergo losses at high temperatures. Examples may include protease inhibitors, lectins, goitrogens, amylase inhibitors, and antivitamin factors. Heat stable antinutritional factors, on the other hand, are the stable compounds that resist degradation at higher temperature conditions. Alkaloids, cyanogenic glycosides, tannins, nonprotein amino acids, saponins, flavones, isoflavones, flatus producing oligosaccharides, and pyrimidine glycosides are heat-stable antinutritional factors (Veerabahu et al. 2015).

Depending upon the chemical structure, these factors can be classified into different types as described next.

18.2.1 Protease Inhibitors

Trypsin inhibitors belong to broad class of protease inhibitors possessing the ability to impede the proteolytic activity of certain enzymes such as trypsin and chymotrypsin. Protease inhibitors are found throughout the plant kingdom among legumes such as red gram, Bengal gram, double beans, cow peas, soy beans, and peas. Jain et al. (2009) revealed that the inhibitors obtained from soy beans, field beans, kidney beans, and Bengal gram suppress bovine pancreatic proteinases specifically, while inhibitors from cow peas and red gram are highly active against human chymotrypsin.

These proteases belong to two families: Kunitz and Bowman-Birk. Both of these types of inhibitors are present in soy beans, where the former has a molecular weight of about 21.5 KDa and the latter is of about 8 KDa. The Kunitz inhibitors possess two disulphide linkages and are particularly active against trypsin. However, the Bowman-Birk inhibitors have higher proportions of disulphide linkages (particularly seven) and shows their inhibiting activities toward trypsin and chymotrypsin simultaneously. In lima beans, cow peas, and lentils, the protease inhibitors have been characterized as the members of Bowman-Birk family (Guillamón et al. 2008).

Protease inhibitors constrain enzymatic activity by acting on serine proteases, such as trypsin and chymotrypsin. These serine protease inhibitors, such as trypsin inhibitors, lead to the formation of an irreversible stable trypsin enzyme-trypsin inhibitor complex that causes a decrease of trypsin in the intestine and interferes with the digestibility of dietary proteins. This reduces protein utilization in the body and consequently slows down animal growth. Additionally, this complex stimulates the intestine to release cholecystokinin, which profoundly increases the size of the pancreas and leads to pancreatic hypertrophy and hyperplasia (Savage and Morrison 2003).

18.2.2 Cyanogenic Glycosides

Cyanogenic glycosides are a group of nitrile-containing secondary compounds of plant origin. They are common in certain families such as *Fabaceae*, *Rosaceae*, and *Leguminosae*. These cyanogenic glycosides yield hydrogen cyanide when the food is chewed or digested. This act of chewing or digestion causes hydrolysis of the food substance by the enzyme present in the food, causing cyanide to be released (Bolarinwa et al. 2016). There are about 25 known cyanogenic glycosides and about 2,500 known species of plants that produce these toxic compounds. In fruits such as apples, apricots, peaches, plums, and cherries, cyanogenic glycosides are generally present in the edible parts (Lawley et al. 2008). Cyanogenic glycosides are also present in cassava, lima beans, and chickpeas. Most legumes except lima beans contain cyanide content in the range 10–20 mg/100 g, which is considered safe. Lima beans have been reported to yield 210–312 mg/100 g hydrogen cyanide, which had been known to cause serious outbreaks of human poisoning and intoxication (Srilakshmi 2003). The symptoms of cyanide poisoning in humans include vomiting, stomach ache, diarrhea, convulsion, rapid breathing, drop in blood pressure, raised pulse rate, dizziness, confusion, twitching, and in extreme cases death can also occur (Bolarinwa et al. 2016).

The toxicity of cyanogenic glycosides is dependent upon the release of hydrogen cyanide after plant consumption. When the cell structure of the plant gets damaged, as in cases of predation by animals or during the preparation or processing, the β-glycosidase enzyme is released and comes in contact with its substrate cyanogenic glycoside. This leads to the disintegration of glycoside to sugar and cyanohydrin, which then decomposes to yield hydrogen cyanide. Hydrogen cyanide is usually formed in order to protect the plants from predation (Lawley et al. 2008). The cyanide ions have adverse effects on the body such as slowed growth by interfering with certain essential amino acids and cultivation of nutrients in body as well as the potential to cause toxicity, neuropathy, and death (Osuntokun 2007).

18.2.3 Goitrogenic Factors

Goitrogens are the substances that lead to an increase in size of thyroid gland since they interfere with the normal production of thyroid hormone. Thyroid hormone plays an important role in many physiological functions, as it is very critical for brain and neurological development of the fetus and child (Mondal, et al. 2016). Iodine is the main constituent of the thyroid hormone (T_3 and T_4) and its deficiency is one of the causes of hypothyroidism in children and adults. Pearl millet, soy beans, ground nuts, lentils, and cassava are attributed to cause thyroid dysfunction as thiocyanates, isothiocyanates, and their derivatives are present in them. They also contain certain complex polyphenols that can also adversely affect iodine utilization by the thyroid gland (Davies 2007; Babiker et al. 2020). The thiocyanate ion, an inorganic substrate and a pseudohalide, exhibits similarity to iodide in its chemical behavior. Subsequently, it blocks the sodium iodide symporter, which conclusively blocks the thyroid gland to take up iodide, resulting in reduced thyroid hormone production (Tonacchera et al. 2005).

18.2.4 Lectins and Hemagglutinins

Lectins and hemagglutinins are the proteins or glycoproteins that are widely distributed in nature and are present in many plants that are consumed by humans and animals (Lawley et al. 2008). They are characterized by their highly specific sugar-binding activity, which allows them to easily attach to red blood cells and cause agglutination (Hamid et al. 2013). It was observed that a protein fraction of the castor bean named "ricin" was capable of agglutinating erythrocytes (red blood cells) and hence termed as phytohemagglutinins (Veerabahu, Tresina et al. 2016). Later, this hemagglutinating activity was detected in more than 800 different plants species out of which more than 600 belongs to the family *Leguminosae*. Some of the edible legumes such as navy beans, lentils, garden peas, soy beans, field beans, white beans, double beans, horse gram, and red kidney beans contain phytohemagglutinins. Although, the amounts and specificity vary widely among different sources, the concentration of lectin is found to be highest in red kidney beans (Lawley et al. 2008).

Lectins are proteins that are resistant to stomach acid and digestive enzymes, hence they do not break easily. Lectins may bind to the intestinal epithelial cells and cause damage to the gut by causing lesions and disruption of the microvilli lining of the digestive tract, which exposes the bacterial population to the blood stream and leads to the severe impairment in the nutrient absorption across the intestinal wall. This alteration in gut function can cause colitis, Crohn's disease, celiac sprue, and irritable bowel syndrome (Lawley et al. 2008). Besides agglutinating red blood cells, lectins exhibit certain unusual biological and chemical properties such as interaction with specific blood groups, mitogenesis, promotion of cell adhesion, inhibition of fungal growth, and insulin-like effect on fat cells.

18.2.5 Phenolics and Tannins

Tannins are water-soluble phenolic compounds present in high concentrations in the seed coat. They are found in abundance in agricultural crops including cereals, pulses, and forages. Pulses like red kidney beans, black gram, and soy beans contain high amount of polyphenolic compounds. They produce an astringent reaction in the mouth and are used for tanning animal hides into leather. They have been classified into two groups depending upon their chemical structure: condensed tannins and hydrolysable tannins. Condensed tannins are widely distributed in plants, they pass through the digestive tract unchanged, and are generally nontoxic, however, large quantities can lead to gastroenteritis problems (Duffus and Duffus 1991). Hydrolysable tannins are present in small amount in plants. Being simple derivatives of gallic acid, they can be classified depending upon the products obtained after hydrolysis such as gallotannins and ellagitannins (Pizzi 2019).

Tannins interfere with the digestibility of proteins and carbohydrates and slows growth as it lowers the activity of digestive enzymes such as α-amylase, trypsin, chymotrypsin, and lipase. Tannins also affect protein availability as they form a less digestive complex with proteins with the help of hydrogen bonding and hydrophobic interactions. The precipitation of the protein-tannin complex is affected by the pH, ionic strength, and molecular size of tannins. The inclusion of tannin phenolics into the precipitate increases with the increase in molecular size of the tannins until the molecular weight surpasses 5,000 Daltons. After this, the tannins become insoluble and ultimately lose all their protein precipitating capacity. Tannins may also cause disruption of the mucosa of the digestive tract, interfering with the absorption of vitamins and minerals. Furthermore, they may bind with iron irreversibly and interfere with its absorption (Lampart-Szczapa et al. 2003; Veerabahu et al. 2015).

18.2.6 Phytic Acid

Phytic acid is usually present in cereals or legumes in the form of salt known as phytate at a concentration of 0.4–6.4% by weight. Chemically, they are myoinositol 1, 2, 3, 4, 5, 6-hexakis dihydrogen phosphate (InsP6), which serves as the primary phosphate reserve and accounts for 60–90% of total phosphorous in most of the seeds. In monocotyledons such as wheat and rice, it is associated with aleurone or bran layer, allowing easy removal upon milling, whereas in dicotyledons, such as legumes, nuts, and oilseeds, it is found closely associated with protein, making its removal through milling difficult (Owusu-Apenten 2003). Phytic acid is involved in various important plants functions such as DNA repair, chromatin remodeling, endocytosis, nuclear messenger, RNA export, and potential hormone signaling for plant and seed growth (Veerabahu et al. 2015).

Being a negatively charged structure, it attaches to positively charged metal ions such as zinc, iron, magnesium, and calcium to form complexes, and owing to its chelating property, it reduces minerals absorption and their bioavailability in our body (Grases et al. 2017). Furthermore, it disturbs the activity of the enzymes involved in the protein degradation mechanism in the small intestine and stomach, mostly affecting population groups like infants, pregnant and lactating women consuming large portions of cereal-based foods (Marie and Sandberg 2006).

18.2.7 Saponins

Saponins are largely a group of secondary plant metabolites that are non-volatile and exhibit surface active characteristics (Samtiya et al. 2020). They encompasses a steroidal or triterpene aglycone (sapgenin) unit associated with one, two, or three saccharide chains via ester and/or ether linkages, resulting in a molecule of varying size and complexity. Sugars like galactose, arabinose, xylose, and glucose are most commonly linked with an aglycone. They are present in a number of legumes like soy beans, sword beans (*Canavalia gladiata*), and jack beans (*Canavalia ensiformis*). Minor proportions of saponins are also found in root crops (potato, yams, and alliums) and in oats, tea, and medicinal plants (ginseng). The range of saponins found in grain legumes lies between 0.5–5% of dry weight, with soy beans having the highest dietary source i.e. 5.6% (Owusu-Apenten 2003). High concentration of saponins can develop a bitter taste and astringency in products. Soya saponins are the most commonly found saponins in legumes and exhibit three different forms i.e. A, B, and E depending on the chemical structure of the aglycone (Veerabahu et al. 2015).

The ability of saponins to produce foam or lather in aqueous solutions and to hemolyse red blood cells is due to their amphiphilic nature, which is due to the lyophilic and lyophobic behavior of the carbohydrates and aglycone moiety, respectively (Owusu-Apenten 2003). Human digestive enzymes cannot completely hydrolyse saponins, which can cause severe implications in gastrointestinal digestion (Amin et al. 2011), leading to problems like nausea and vomiting. They form large mixed micelles in the human body by interacting with bile juice and cholesterol, thus, saponins show low cholesterol levels in some animal species, but the hypocholesterolemic effects of saponins in humans is under research. Collaterally, they are known to form insoluble complexes with 3-b-hydroxysteroids (Veerabahu et al. 2015). Saponins isolated from *Bulbostermma paniculatum* and *Pentapamax leschenaultia* affect human spermatozoa, by inhibiting acrosin activity, which leads to an extreme damage to the spermal plasma membrane (Zheng et al. 1994).

18.2.8 Alkaloids

Alkaloids are heat-stable antinutrient factors produced by plants as a defense mechanism against herbivores. Alkaloids are found in considerable amounts in lupins, while some are devoid of alkaloids such as faba beans, peas, and oilseeds. Lupanine (700 mg/g total alkaloids), sparteine (300 mg/g total alkaloids), albine (150 mg/g total alkaloids), and 13-a-hydrolupanine (80 mg/g total alkaloids) are the major alkaloids of *Lupinus albus*. The alkaloids are known to affect the central nervous system, digestive system, reproductive system, and the immune system in an individual (Veerabahu et al. 2015). Besides this, the other most commonly found alkaloids include nicotine (tobacco), cocaine (leaves of coca plant), quinine (cinchona bark), morphine (dried latex of the opium poppy), and solanine (unripe potatoes and potato sprouts) (Hamid et al. 2017).

18.2.9 Toxic Amino Acids

18.2.9.1 Mimosine

It is an unusual amino acid found in *Leucaena leucocephala*, comprising 3–5% (dry weight) of the protein. It retards the growth of cattle when it comprises more than half of their diet. In humans, hair loss is observed when leaves, pods, and seeds of *L. leucocephala* are consumed in the diet.

18.2.9.2 Djenkolic Acid

Djenkolic acid is a sulphur-containing amino acid present in *Pithecolobium lobatum* in a free state comprising 1–4% (dry weight) of protein. On consumption of seeds, blood along with, needle like clusters of djenkolic acid may appear in urine and ultimately lead to kidney failure.

18.2.9.3 Canavanine and Canaline

Canavanine and canaline are the primary metabolites of the *Canavalia* species, where the former is a toxic arginine anti-metabolite and latter is a toxic nonprotein amino acid. Canavinine affects the nitric oxide pathway, affecting peristalsis movement because of its structural similarity with arginine. Similarly, canaline affects the ornithine cycle as it is a structural analogue of citrulline. Additionally, canaline also reacts with aldehydes (vitamin B6) and with keto acids to form oxime.

18.2.9.4 3, 4-Dihydroxyphenylalanine (L-Dopa)

A non-protein amino acid, 3, 4-dihydroxyphenylalanine (l-dopa) was first isolated from *Vicia faba*, and later found in other genus like *Mucuna*. L-dopa in *Mucuna* is extracted and used for the treatment of Parkinson's disease in humans, though it is associated with certain side effects such as toxic confusional state, hallucination, and gastrointestinal disturbance i.e. nausea, vomiting, and anemia. L-dopa is capable of causing favism in an individual with glucose-6-phosphate dehydrogenase deficiency in erythrocytes. Moreover, it was reported that the L-dopa in its oxidized form conjugates with sulphydryl groups of proteins and eventually forms a protein-bound 5-S-cysteinyldopa cross-link. These cross-linkages bring about the polymerization of proteins, and as a result, lower the protein digestibility (Veerabahu et al. 2015).

18.2.10 Antivitamins

Antivitamins are heat-labile antinutritional factors most commonly found in leguminous plants. As an example, antivitamin A factor is present in soy beans and possess the capability to destroy carotene. Another factor called antivitamin D is present in soy beans and tends to decrease calcium and phosphorous absorption. Antivitamin E is present in seeds like kidney beans, soy beans, alfalfa, and field peas. Antithiamine factor called thiaminase is present in cottonseed, linseed, mung beans, and mustard seeds (Hamid et al. 2017).

18.2.11 Miscellaneous

18.2.11.1 Oxalate

The highest amount of oxalate salt is present in Anasazi beans and the lowest is in black-eyed beans. Other legumes that contain oxalate salts are lentils, red kidney beans, and white beans. Oxalate binds to calcium, forming calcium-oxalate crystals in monogastric animals. As a result, the complex formed decreases calcium absorption and availability in body, and increases chances of disease such as rickets and osteomalacia. It also cause renal stones around renal tubes through precipitation of calcium crystals.

18.2.11.2 Pyrimidine Glycosides

Pyrimidine glycosides are composed of one molecule of glucose linked to one pyrimidine nucleotide, as in the vicine and covicine present in *Vicia faba*. They reduce glutathione and glucose-6-phosphate activity, causing hemolytic anemia due to the biochemical abnormalities of blood cells.

18.2.11.3 Oligosaccharides

Oligosaccharides are the flatulence-causing factors present in legumes seeds. These have 1, 2, or 3-galactose units joined to α-1,6 galactosidic linkages. Monogastric animals lack α-1,6 galactosidase activity, so they cannot hydrolyse oligosaccharides and then are acted upon by microorganisms present in the large intestine, which produces flatus gases like hydrogen, carbon dioxide, and small amounts of methane. Also they cause abnormal growling, diarrhea, and discomfort (Veerabahu et al. 2015).

18.3 Identification and Quantification of Antinutritional Factors in Legumes

The foremost step in the quantification of antinutrients present in legumes is to extract them from the sample. Each group of antinutrients can be separated specifically by employing different reagents such as trichloroacetic acid, sulphuric acid, or hydrochloric acid for phytates while methanol, ethanol, or aqueous acetone alone or in combination favor the separation of polyphenols. Numerous novel techniques such as gas liquid chromatography, ion exchange chromatography, high-performance liquid chromatography either alone or in conjugation with mass spectroscopic techniques can also be employed for identification and quantification purposes. These analytical techniques are described in detail next.

18.3.1 Thin Layer Chromatography (TLC)

Thin layer chromatography offers an elementary and inexpensive procedure to identify the different components present in a sample mixture at a given time. It was suitably used for the qualitative analysis of polyphenolic constituents present in legumes by employing competent absorbents and reagents. As an example, hydrophilic flavonoids were partitioned exclusively using polyamide and microcrystalline cellulose as stationary phases. Contrarily, apolar flavonoids (flavons and isoflavonoids) are widely separated using stationary silicone gel. Saponins were observed by Sharma et al. (2012) as white spots against the pink background on TLC plates wherein n-butanol: water: acetic acid (84:14:7) was used as a solvent system. TLC can be considered as an ideal tool for the initial fractionation of antinutrients in legumes, and can further be used for monitoring fractions obtained from column chromatography.

18.3.2 Gas Chromatography (GC)

Gas chromatography is the preferred method for quantifying low molecular-weight alcohols, aromatic acids, and simple aglycones. The technique is based on evaporating the sample mixture; the molecules are differentially separated, depending on their volatility, and noted by a detector. The two fundamental ways that allow the dissociation of the mixture are chemical and electron ionization. Advancements in technology have led to the coupling of GC column to mass spectrometry (MS) ion sources directly along with different analyzers, such as low resolution quadrupols, ion traps, or time of flight (Tof) analyzers. The process is highly advantageous as it is capable of identifying the isomers of the target product as well. However, the limited applicability of GC in quantifying saponins is due to its ability to isolate and quantify the aglycone portion specifically, which makes the hydrolysis and derivatization of saponins a prerequisite step. This step is essentially required for enabling the evaporation and separation of compounds and allows certain modifications around the glycosidic portion of the molecule. Additionally, the majority of phenolic acid derivatives and flavonoids require substitution with trimethylsilyl (TMS) and methyl groups (CH3) on their polar heads (hydroxyl and/or carboxyl) to facilitate the volatization process (Stobiecki and Makkar 2004).

18.3.3 High-performance Liquid Chromatography (HPLC)

High-performance liquid chromatography is helpful in characterizing polyphenolic compounds by using chemically modified silicone consisting of hydrocarbon chains as stationary phase. Depending on the composition of the sample, the elution of fractions can be carried out in an isocratic or gradient manner. The isocratic elution is generally employed in various legumes and vegetables, for example, polyphenolic constituents of legumes and isoflavonoids from soy bean are comprised of the same group of molecules and can be effectively separated in this manner (Stobiecki and Makkar 2004). Concurrently, Carmona et al. (1991) well determined the concentration of total polyphenols as the accumulation of tannins and non-tannins in common white and black bean varieties. Being highly accurate and precise, this method offers numerous advantages over other methods. The phytates present in soy beans were efficiently separated by Kwanyuen and Burton (2005) by following a simple HPLC protocol involving a nominal sampling procedure.

18.3.4 Affinity Chromatography

Lectins display an interesting behavior whereby they occur in multiple forms that differ among each other only in their physical and chemical attributes to a slight extent, and exhibit quite similar biological functions. This peculiar homogeneity of lectins makes their purification a cumbersome process that can't be achieved like other antinutrients. Because these compounds exhibit a unique affinity for specific sugar residues, this property can be successfully used for purifying large amounts of lectins in a single stage of affinity chromatography. Applying this process, concanavalin A, a type of proteinaceous lectin, can be purified by the virtue of its ability to bind specific sugar moiety. Moreover, the glycoproteins can also be separated from non-glycosylated proteins (Stobiecki and Makkar 2004).

18.3.5 UV-VIS Spectrophotometry

UV-VIS spectrophotometry has remained one of the most quintessential techniques for the determination of various antinutrients in beans. Wang et al. (1998) analyzed the content of trypsin inhibitors present in field peas and grass peas that are responsible for decreasing protein digestibility and causing enlargement of the pancreas. Furthermore, this key technique was also utilized for measuring the saponin content by using oleanolic acid as a reference—the change in color produced as a result of the interaction between saponins and vanillin or anisaldehyde was quantified. The only drawback of using this technique is that sometimes the colored reactions, being non-specific, can lead to misinterpretation of the content of actual antinutrients (Stobiecki and Makkar 2004).

18.4 Elimination of Antinutritional Factors from Legumes

Regardless of being synthesized as secondary metabolites within the plant systems and offering defensive roles against attack by insects and animals, antinutritional factors—when consumed by humans—pose a severe impact on the absorption and utilization of carbohydrates, proteins, vitamins, and minerals by the body. This arouses a need for eliminating these elements from legumes prior to consumption. It is the structure and chemical properties of antinutrients, specifically their sensitivity towards heat, that determines which physical process will be the most compelling in their reduction or elimination, which will curtail their detrimental effects on human health. Conventionally, an amalgam of two approaches is preferred over a single method (Owusu-Apenten 2003).

18.4.1 Soaking

Soaking is a fairly reliable method to remove hydrosolvable antinutrients, wherein the legumes are soaked either in distilled or saline water for a defined period. During the course of soaking, water-soluble antinutrients along with enzyme inhibitors become readily dispersed in the surrounding water, and certain endogenous enzymes such as phytase is activated (Verde et al. 1992). Concurrently, water is diffused into the protein and starch molecules, expediting the process of gelatinization and softening the texture of beans, which results in a significant reduction in cooking time. Udensi et al. (2009) reported a reduction of 27.9% and 36% phytic acid content in *Mucana flaggelipes* during soaking for 6 hours and 24 hours, respectively, at room temperature. According to Frias et al. (2000), soaking of cow pea flour for 16 hours reduced 26% and 28% of stachyose and raffinose concentrations, respectively. An additional advantage of soaking is that it supplies moist conditions to the nuts, grains, and other edible seeds, thereby aiding in the germination and fermentation processes. Besides the several advantages to offer, soaking has its limitations due to the occurrence of certain metabolic reactions in the product in the presence of water.

18.4.2 Fermentation

Fermentation is one method to reduce the toxic substances in pulses. Fermentation has almost removed all the trypsin inhibitor activity from cow peas, thus, fermentation is one of the best methods for the effective elimination of inhibitors as compared to other treatments such as soaking, cooking, or germination (Ibrahim et al. 2002). Fermentation provides optimum pH conditions for the enzymatic degradation of phytate, which is present in the form of complexes with iron, zinc, calcium, magnesium, and protein in cereals. This reduction in phytate may increase the amount of soluble iron, zinc, and calcium by several times (Gupta et al. 2015). Fermentation also leads to the reduction in total phenolics and tannins and that might have been caused by polyphenol oxidase or fermentation microflora (Reddy and Pierson 1994).

Oligosaccharides are also shown to be reduced by the natural fermentation process (62.68%). Lentil flour has also shown complete removal of lectins after 72 hours of fermentation at 42°C having a flour concentration of 72 gl⁻¹ (Mohan et al. 2016). Fermentation is also one of the best ways to remove hydrogen cyanide (Lawley et al. 2008).

18.4.3 Germination

Germination is another widely accepted method for reducing the content of antinutritional elements and improving protein digestibility and availability in plant-based foods. During the germination process, enzymes become highly active, which catalyzes the reactions and produces absorbable polypeptides and essential amino acids in significant amounts. The process also triggers phytase enzyme, which decomposes phytate salts and thus reduces the phytic acid content present in the legumes (Laxmi 2015). The level of reduction is dependent upon the type of bean employed and the germinating conditions (Mohan et al. 2016). Furthermore, germination of lentils for 6 days also eliminates oligosaccharides, particularly raffinose (Verde et al. 1992).

18.4.4 Milling

Antinutritional factors such as phytic acid, tannins, polyphenols, and enzyme inhibitors are localized in the outer layer of bran. The process of differential milling is applied to various pulses in order to remove the husk and occasionally the bran layers. This results in an edible portion free of impurities that can be converted into a powder form with varying particle sizes. During the process of milling, with the removal of these antinutrients, certain minerals are also lost thereby reducing the nutritional value of milled legumes (Samtiya et al. 2020). Egounlety and Aworh (2003) suggested that the process of dehulling raw cow peas, ground beans, and soy beans has significantly reduced their tannin levels from 223 mg/100 g, 152 mg/100 g, and 68 mg/100 g, respectively, to non-detectable levels.

18.4.5 Heat Processing

An appreciable amount of complex sugars is present in beans that are non-digestible by digestive enzymes; as a result, consumption of such beans may lead to certain gastric problems in humans. To avoid this problem, the best way to consume these beans is to soak and then cook by boiling or in a pressure cooker. Processing by heat is the most-practiced method and is highly suitable for removing protein-based antinutrients (trypsin or chymotrypsin inhibitors) because of their ability to undergo denaturation during heating. The extent of the removal of antinutritional elements from legumes are influenced by factors such as temperature of processing, duration of heating, and moisture content.

18.4.5.1 Cooking

Consumption of legume seeds in excess amounts can cause toxicity as an effect of the presence of undesirable antinutritional factors in raw seeds. Cooking legumes for up to 90 minutes can almost eliminate antinutrients like trypsin and hemagglutinins, while polyphenolic compounds decrease up to only a certain limit (Srilakshmi 2003). Sometimes, cooking preceded by soaking of seeds has a greater effect on decreasing the percentage of phytic acid and, thus, improves the mineral availability (Demir and Elgün 2014). In comparison to simple cooking, pressure cooking reduces antinutritional factors to a larger extent as in the case of mung beans and black gram (Samtiya et al. 2020). Moreover, L-DOPA being soluble in water, can be easily removed by soaking and then boiling the legumes in clean water (Veerabahu et al. 2015).

18.4.5.2 Autoclaving

Autoclaving is a microbiological operation based on heating under high-pressure conditions to achieve sterilization. The same principle can be utilized for removing antinutrients and simultaneously reduce cooking time. When jack beans were autoclaved for 30 minutes at 125°C and 15 pounds of pressure, thermo-labile inhibiting substances such as cyanogenic glycosides, saponins, terpenoids, and alkaloids were not detected (Akande and Fabiyi 2010). This method is highly useful in inactivating antivitamin D, antivitamin E, and antivitamin K (Hamid et al. 2017).

18.5 Conclusions

Scientific evidence has shown that varied types of antinutrients are available in legumes and more precisely in the seed coat layer of beans. These compounds are produced as secondary metabolites and play defensive roles against attack by predators. Regardless of their protective action, these compounds, when ingested as a part of a regular diet, diminish the digestibility and availability of proteins, restrict the uptake of minerals and vitamins in the body, and can cause gastric problems, paralysis, renal failure, hemophilia, and many other severe diseases. Trypsin inhibitors, goitrogens, tannins, oxalate, antivitamins, lectins and haemagglutinin, toxic amino acids, cyanogenic glycosides, saponins, etc. are the various antinutritional compounds present in pulses such as Bengal gram, cow peas, soy beans, field beans, kidney beans, double beans, peas, etc. Many analytical techniques such as HPLC, TLC, GC, and spectrophotometric techniques can be utilized for their identification and quantification in the respective sources. Antinutrients exhibit ill-effects on human health and need to be removed either partially or wholly by applying various processing treatments such as milling, soaking, germination, fermentation, or heat processing prior to their consumption. Usually, a combination of two or more methods is preferred over a single method as single method may not be sufficient to remove or reduce antinutritional factors in pulses. These methods do not just eliminate the antinutrients—they carry out various metabolic reactions in the legumes, making them highly digestible and improving their sensory characteristics as well.

REFERENCES

Akande, K. E. and E. F. Fabiyi (2010). "Effect of processing methods on some antinutritional factors in legume seeds for poultry feeding." *International Journal of Poultry Science* **9**(10): 996–1001.

Amin, H. A. S., et al. (2011). "Comparative studies of acidic and enzymatic hydrolysis for production of soyasapogenols from soybean saponin." *Biocatalysis and Biotransformation* **29**(6): 311–319.

Babiker, A., et al. (2020). "The role of micronutrients in thyroid dysfunction." *Sudanese Journal of Paediatrics* **20**(1): 13–19.

Bolarinwa, I. F., et al., Eds. (2016). A review of cyanogenic glycosides in edible plants. In: *Toxicology—New Aspects to This Scientific Conundrum*. Intech. doi: 10.5772/64886

Carmona, A., et al. (1991). "Comparison of extraction methods and assay procedures for the determination of the apparent tannin content of common beans." *Journal of the Science of Food and Agriculture* **56**(3): 291–301.

Davies, N. T. (2007). "Anti-nutrient factors affecting mineral utilization." *Proceedings of the Nutrition Society* **38**(1): 121–128.

Demir, M. K. and A. Elgün (2014). "Comparison of autoclave, microwave, IR and UV-C stabilization of whole wheat flour branny fractions upon the nutritional properties of whole wheat bread." *Journal of Food Science and Technology* **51**(1): 59–66.

Duffus, C. M. and J. H. Duffus (1991). "Introduction and overview." In: J. P. F. D'Mello, C. M. Duffus and J. H. Duffus (Eds.), Toxic Substances in Crop Plants. Woodhead Publishing: 1–20.

Egounlety, M. and O. C. Aworh (2003). "Effect of soaking, dehulling, cooking and fermentation with *Rhizopus oligosporus* on the oligosaccharides, trypsin inhibitor, phytic acid and tannins of soybean (*Glycine max* Merr.), cowpea (*Vigna unguiculata* L. Walp) and groundbean (*Macrotyloma geocarpa Harms*)." *Journal of Food Engineering* **56**(2): 249–254.

Elkowicz, K. and F. Sosulski (2006). "Antinutritive factors in eleven legumes and their air-classified protein and starch fractions." *Journal of Food Science* **47**: 1301–1304.

Frias, J., et al. (2000). "Influence of processing on available carbohydrate content and antinutritional factors of chickpeas." *European Food Research and Technology* **210**(5): 340–345.

Grases, F., et al. (2017). Dietary Phytate and Interactions with Mineral Nutrients. Springer.

Guillamón, E., et al. (2008). "The trypsin inhibitors present in seed of different grain legume species and cultivar." *Food Chemistry* **107**(1): 68–74.

Gupta, R. K., et al. (2015). "Reduction of phytic acid and enhancement of bioavailable micronutrients in food grains." *Journal of Food Science and Technology* **52**(2): 676–684.

Hamid, H., et al. (2017). "Anti-nutritional factors, their adverse effects and need for adequate processing to reduce them in food." *AgricInternational* **4**(1): 56–60.

Hamid, R., et al. (2013). "Lectins: Proteins with diverse applications." *Journal of Applied Pharmaceutical Science* **3**: 93–103.

Hangen, L. and M. R. Bennink (2002). "Consumption of black beans and navy beans *(Phaseolus vulgaris)* reduced azoxymethane-induced colon cancer in rats." *Nutrition and Cancer* **44**(1): 60–65.

Ibrahim, S. S., et al. (2002). "Effect of soaking, germination, cooking and fermentation on antinutritional factors in cowpeas." *Nahrung* **46**(2): 92–95.

Jain, A. K., et al. (2009). "Antinutritional factors and their detoxification in pulses—A review." *Agricultural Reviews* **30**(1): 64–70.

Kwanyuen, P. and J. W. Burton (2005). "A simple and rapid procedure for phytate determination in soybeans and soy products." *Journal of the American Oil Chemists' Society* **82**(2): 81–85.

Lampart-Szczapa, E., et al. (2003). "Chemical composition and antibacterial activities of lupin seeds extracts." *Food/Nahrung* **47**(5): 286–290.

Lawley, R., et al. (2008). The Food Safety Hazard Guidebook. Royal Society of Chemistry.

Laxmi, N. (2015). "The impact of malting on nutritional composition of Foxtail millet, wheat and chickpea." *Journal of Nutrition & Food Sciences* **5**(5): 1–3.

Liener, I. E. (2006). "Legume toxins in relation to protein digestibility—A review." *Journal of Food Science* **41**: 1076–1081.

Marie, A. and A.-S. Sandberg (2006). "Phytate reduction in oats during malting." *Journal of Food Science* **57**: 994–997.

Mohan, V. R., et al. (2016). "Antinutritional factors in legume seeds: Characteristics and determination." In: *Encyclopedia of Food and Health.* Academic Press, 211–220.

Mondal, C., et al. (2016). "Studies on goitrogenic/antithyroidal potentiality of thiocyanate, catechin and after concomitant exposure of thiocyanate-catechin." *International Journal of Pharmaceutical and Clinical Research* **8**(1): 108–116.

Osuntokun, B. O. (2007). "Cassava diet and cyanide metabolism in Wistar rats." *British Journal of Nutrition* **24**(3): 797–800.

Owusu-Apenten, R. (2003). "Antinutritional factors in food legumes and effects of processing." In: The Role of Food, Agriculture, Forestry and Fisheries in Human Nutrition. Oxford, UK, Encyclopedia of Life Support Systems (EOLSS): 82–116.

Pizzi, A. (2019). "Tannins: Prospectives and actual industrial applications." *Biomolecules* **9**(8): 344–373.

Punia, S., Dhull, S. B., Sandhu, K. S., & Kaur, M. (2019). "Faba bean (*Vicia faba*) starch: Structure, properties, and in vitro digestibility—A review." *Legume Science*, **1**(1), e18.

Punia, S., Dhull, S. B., Sandhu, K. S., Kaur, M., & Purewal, S. S. (2020). "Kidney bean (*Phaseolus vulgaris*) starch: A review." *Legume Science*, **2**(3), e52.

Reddy, N. R. and M. D. Pierson (1994). "Reduction in antinutritional and toxic components in plant foods by fermentation." *Food Research International* **27**(3): 281–290.

Samtiya, M., et al. (2020). "Plant food anti-nutritional factors and their reduction strategies: An overview." *Food Production, Processing and Nutrition* **2**(1): 6.

Savage, G. P. and S. C. Morrison (2003). "Trypsin inhibitor." In: B. Caballero (Ed.), Encyclopedia of Food Sciences and Nutrition. Oxford, Academic Press: 5878–5884.

Sharma, O. P., et al. (2012). "An improved method for thin layer chromatographic analysis of saponins." *Food Chemistry* **132**(1): 671–674.

Srilakshmi, B. (2003). Food Science. New Age Publishers.

Stobiecki, M. and H. P. S. Makkar (2004). "Recent advances in analytical methods for identification and quantification of phenolic compounds." In: Proceedings of the 4th International Workshop on Antinutritional Factors in Legume Seeds and Oilseeds, Recent Advances of Research in Antinutritional Factors in Legume Seeds and Oilseeds. Toledo, Spain: Wageningen Academic Publishers, 11–28.

Tonacchera, M., et al. (2005). "Relative potencies and additivity of perchlorate, thiocyanate, nitrate, and iodide on the inhibition of radioactive iodide uptake by the human sodium iodide symporter." *Thyroid* **14**: 1012–1019.

Udensi, E. A., et al. (2009). "Effects of processing methods on the levels of some antinutritional factors in *Mucuna flagellipes*." *Nigerian Food Journal* **26**: 47–50.

Veerabahu, R. M., et al. (2015). "Antinutritional factors in legume seeds: Characteristics and determination." In: B. Caballero, P. Finglas and F. Toldra (Eds.), The Encyclopedia of Food and Health. Oxford: Academic Press (Elsevier).

Veerabahu, R. M., et al. (2016). *Antinutritional factors in legume seeds: Characteristics and determination. The Encyclopedia of Food and Health*. B. Caballero, P. Finglas and F. Toldra, Oxford: Academic Press. (Elsevier).

Verde, C. V., et al. (1992). "Effect of processing on the soluble carbohydrate content of Lentils." *Journal of Food Protection* **55**(4): 301–303.

Wang, X., et al. (1998). "Trypsin inhibitor activity in Field pea *(Pisum sativum L.)* and Grass pea *(Lathyrus sativus L.)*." *Journal of Agricultural and Food Chemistry* **46**(7): 2620–2623.

Zheng, X., et al. (1994). "Inhibition of acrosin by protein C inhibitor and localization of protein C inhibitor to spermatozoa." *American Journal of Physiology* **267**(2 Pt 1): C466–472.

19

Processing, Nutritional Composition, and Health Benefits of Lentils

Aderonke Ibidunni Olagunju and Olufunmilayo Sade Omoba
Federal University of Technology, Akure, Ondo State, Nigeria

CONTENTS

19.1 Introduction

Pulses, according to Food and Agriculture Organization of the United Nations (FAO), refer to plants harvested for their dry seed. Pulses belong to the *Fabacea* or *Leguminosae* family and they are the world's third largest group of plants. The seeds of pulses can be regarded as multipurpose and used for human consumption as well as animal fodder in addition to serving as a point of pride for the farmer due to the crop's ability to fix atmospheric nitrogen and increase the overall fertility of soil (Kouris-Blazos and Belski, 2016; Punia et al., 2019, 2020). Pulses are small in size and highly nutritious, pulse crops include lentils, lupines, beans (kidney beans, navy beans, faba beans, cow peas) and peas (chickpeas, pigeon peas, black-eye peas). They are uniquely rich sources of protein (17–30% by dry weight and doubling the amount in cereals) and essential amino acids. In addition to high protein, they also possess significant amounts of carbohydrates, micronutrients, and dietary fiber and low fat content, which has been proven to be effective at maintaining reduced low-density lipoprotein (LDL) cholesterol levels as well lowering blood pressure (FAO, 2016).

Lentils (*Lens culinaris*) is an important pulse of significance which has been cultivated and consumed since ancient time with usage limited in developed countries. It is grown in more than 70 countries and consumed worldwide in whole, dehulled, and split forms (Nosworthy et al., 2017). Lentils have been tagged one of the most nutritious and health promoting foods known to man (Faris and Attlee, 2017); it remains one of the world's healthiest foods and the most desired pulse in many regions because it has an excellent nutritional quality. Lentils are nutrient dense foods, listed among the soft seed-coated pulses that require shorter cooking time, therefore, have smaller nutrient loss compared to significant losses observed in pulses with a hard seed coat (Satya et al., 2010). Lentils are second to soy beans in terms of heat processing due to their fast cooking time (23–26 minutes) in comparison to >70 minutes for other pulses (Jood et al., 1998; Khazaei et al., 2019), and Canada presently stands as the world's leader in lentil production and export (Ramdath et al., 2020). India follows closely as the second largest producer and the largest consumer of lentils worldwide as they also import lentil to complement the large production that is majorly dedicated to domestic consumption (FAO, 2019).

There are several varieties of lentils, the seed coat exhibits different colors namely red, green, and brown or tan; however, the cotyledon exhibited is either red or yellow, which could be a result of the presence of anthocyanins in the seed (Jati et al., 2013). The two varieties showed significant differences in their nutritional composition and protein digestibility (Gharibzahedi et al., 2012, Jarpa-Parra, 2018). Lentils are subjected to different processing methods, including conventional cooking, de-hulling, and fractionation into components, namely, flour, starch, and protein fractions that have potential as ingredients for use in food formulations for the development of novel food products.

19.2 Nutritional Composition of Lentils

Lentils are a good source of carbohydrates (59–70%) as reported by several authors (Shams et al., 2008; Gharibzahedi et al., 2012; Ramdath et al., 2020). Certain carbohydrate components (especially oligosaccharides) have a functional significance by serving as selective promoters for the growth of beneficial gut microbes (prebiotics) that aid improvement of gut health, microbial balance restoration, and prevention of intestinal diseases (Fooks et al., 1999). Starch occupies most of the carbohydrate mass in pulses and is similar for lentils; moreover, Hoover et al. (2010) reported a high starch yield (47%) from lentil seeds when compared to the starch yield from several other pulse grains.

Lentils are also uniquely rich in high-quality protein and the seeds represent an excellent source of plant-based protein that serves as a viable alternative to animal and soy protein for food processing formulations (Khazaei et al., 2019). Similarly, the high protein content makes it a great food source for people in developing countries to meet their nutritional requirement (Abraham, 2015); this is due to its affordability compared to the high cost of animal protein products. Lentil contains approximately twice the amount of protein present in most whole-grain cereal such as wheat, oat, barley, and rice (Samaranayaka, 2017). Genetic diversity accounts for a great range of variation in protein content (18–36%) reported for lentil seeds (Khazaei et al., 2019). Nonetheless, the proteins in lentils are predominantly globulin (47%) and an appropriate quantity of albumin (Lombardi-Boccia et al., 2013). The seed proteins confer functionality in different food formulations, which can however, be characterized by subjecting the seed flour to different processing techniques. The composition of amino acids strongly influences the protein quality of lentil seeds (Kahraman, 2016), interestingly, lentils have a good proportion of amino acids with methionine and cysteine being the limiting amino acids. Data provided by the United States Dry Pea and Lentil Council webinar (Table 19.1) shows the amino acid composition of lentils compared closely with that of soy beans and eggs (which are important protein food sources).

Lentils are relatively low in fat, thus, they contribute low energy content. Several reports have confirmed the low-fat content in lentils ranging from 1.0–1.4% (Ryan et al., 2007; USDA, 2010). The fat is unevenly distributed over the fatty fractions as saturated fatty acids (SFA), monounsaturated fatty acids (MUFA), and polyunsaturated fatty acids (PUFA), contents averaging 16.7%, 23.7%, and 58.8% respectively.

Lentils are significant sources of dietary fiber, mostly insoluble dietary fiber, which accounts for approximately 97% of the total dietary fiber content. The dietary fiber represents a complex carbohydrate comprising cellulose, hemicellulose, and lignin.

19.3 Phytochemicals and Antioxidant Properties in Lentils

Pulses are generally known to contain a plethora of bioactive phytochemicals and lentils are no exception. Several studies reported lentils to contain higher total phenolic content and total flavonoids content than several other pulses such as green peas, chickpeas, cow peas, yellow peas, and mung beans

TABLE 19.1

Amino Acid Composition of Lentil Compared to Peas, Chickpeas, Soy beans, Eggs, Whole Wheat, and Brown Rice

Amino Acid	Lentil	Pea	Chickpea	Soy Bean	Egg	Whole Wheat	Brown Rice
Alanine	4.2	4.4	4.3	5.2	-	3.7	5.8
Arginine	7.7	8.9	9.4	8.6	6.2	4.9	7.6
Aspartic acid	11.1	11.8	11.8	14.0	11.0	5.5	9.4
Cystine	1.3	1.5	1.3	1.8	2.3	2.1	1.2
Glutamic acid	15.5	17.1	17.5	21.6	12.6	32.8	20.4
Glycine	4.1	4.4	4.2	5.2	4.2	4.3	4.9
Histidine	2.8	2.4	2.8	3.0	2.4	2.7	2.5
Isoleucine	4.3	4.1	4.3	5.4	6.6	3.4	4.2
Leucine	7.3	7.2	7.1	9.1	8.8	6.8	8.3
Lysine	7.0	7.2	6.7	7.4	5.3	2.7	3.8
Methionine	0.9	1.0	1.3	1.5	3.2	1.7	2.3
Phenylalanine	4.9	4.6	5.4	5.8	5.8	5.2	5.2
Proline	4.2	4.1	4.1	6.5	4.2	15.7	4.7
Serine	4.6	4.4	5.0	6.5	6.9	4.7	5.2
Threonine	3.6	3.6	3.7	4.8	5.0	2.8	3.7
Tryptophan	0.9	1.1	1.0	1.6	1.7	1.3	1.3
Tyrosine	2.7	2.9	2.5	4.2	4.2	2.1	3.8
Valine	5.0	4.7	4.2	5.6	7.2	4.3	5.9

Adapted from Samaranayaka (2017).

(Xu and Chang, 2007; Han and Baik, 2008; Xu and Chang, 2010; Zou et al., 2011). An array of polyphenolic substances has been identified in lentil extract, making lentils a significant dietary source of extractable phenolics. Existing reports have detected phenolic compounds (both phenolic acids and flavonoids) belonging to hydroxybenzoic acids and hydroxycinnamic acids, namely ρ-hydroxybenzoic, trans-ρ-coumaric, trans-ferulic and sinapic acids, in the extractable fraction of both red and green lentils (Amarowicz et al., 2009; Amarowicz et al., 2010; Alshikh et al., 2015). Zou et al. (2011) also reported the presence of several flavonols, flavan-3-ols, and condensed tannins (proanthocyanidins) in lentils, with the predominant being kaempferol glycoside, quercetin diglycoside, catechin, epicatechin, prodelphinidin dimer, and digallate procyanidin. Other flavonoid derivatives (flavan-3-ol monomers, dimer, and trimers) were also detected in a study by Alshikh et al. (2015) and a review report by Ganesan and Xu (2017) provided a detailed list of polyphenol compounds in lentils. These polyphenols confer potential health-promoting benefits exerted by the antioxidant, cardioprotective, anti-inflammatory, nephroprotective, antidiabetic, antiobesity, hypolipidemic, and chemopreventive activities of lentil seed and products. The different groups of phenolic compounds in lentils compared to other pulses are responsible for the greater antioxidant potentials and health-promoting effects in lentils than in other pulse crops.

Studies have shown that polyphenols make the most important contribution to antioxidant activity in relation to the human diet (Manach et al., 2004). Polyphenols act as reducing agents, hydrogen donating antioxidants, iron chelators, and singlet oxygen quenchers, which are the main properties of antioxidants, thus, the polyphenols naturally act as antioxidants in preventing the formation of reactive oxygen species. The high polyphenolic content of lentil suggests it is a good candidate for antioxidant activity. A study by Alshikh et al. (2015) reported that lentil flour showed potent reducing power and significant free-radical scavenging activity. Reports from several other authors corroborated the antioxidant potential of lentils and an earlier study confirmed the greater antioxidant activity in lentils than in different peas (Duenas et al., 2006). Another study by Xu and Chang (2008) showed that lentils had higher 2, 2-diphenyl-1-picylhydrazyl (DPPH) radical scavenging capacity and oxygen radical absorbing capacity than peas (green peas, yellow peas, and chickpeas). Recent studies established the contributions of total phenolic, flavonoid, and tannin content as well as kaempferol derivative, hydroxybenzoic compounds, protocatechuic, vanillic acid, aldehyde ρ-hydroxybenzoic, trans-ferulic acid and trans-ρ-coumaric acid in lentils to its antioxidant potential (Gharachorloo et al., 2012; Świeca and Baraniak, 2013; Talukdar and Talukdar, 2014; Elaloui et al., 2017).

19.4 Health Benefits of Lentils

Owing to the presence of phenolic compounds and antioxidant properties in lentils, there exists an array of health benefits associated with these components available to consumers of lentil seed, products and extracts. *In vitro* and *in vivo* studies have shown that lentils possess nutraceutical potentials and confer several health benefits relating to the maintenance of human well-being (Shams et al., 2008; Al-Tibi et al., 2010; Świeca et al., 2013; Peñas et al., 2015; Bolsinger et al., 2017)

Several studies have also shown that lentils possess components that have the potential to address a number of disease conditions (Ganesan and Xu, 2017). The phytosterols present in lentils contribute anti-inflammatory and hypolipidemic effects, whereas, the resistant starch contributes to glycemic load reduction. The proteins, on the other hand, contribute to blood pressure lowering potential while the phytochemicals (flavonoids, phytic acid, phytosterols, lectins, and saponins) have anticancer effects (Faris et al., 2013).

Diabetes refers to a group of metabolic diseases associated with defective insulin secretion or poor ability of insulin-sensitive tissues to be appropriately responsive to insulin. It is characterized by abnormal elevation of blood glucose levels usually as a result of the consumption of carbohydrate-rich meal. Interestingly, research output is a pointer to the fact that inclusion of lentil in the regime of a diabetic subjects may be useful in the management of the disease. Lentil is a rich source of β-glucan (Faris et al., 2013); this important fiber component is functional in inducing certain physiological effects that confer a positive health impact, capable of lowering cholesterol levels and, thus, reducing the risk for cardiovascular disease (Earnshaw et al., 2017; Roudi et al., 2017). Lentils are also rich in oligosaccharides, especially resistant starch, which has physiological effects similar to those of dietary fiber. Resistant starch possesses the ability to modulate postprandial blood-glucose levels by escaping enzymatic hydrolysis in the small intestine and passing into the colon where it ferments into butyrate-rich short chain fatty acids (SCFA) by naturally present bacteria. The SCFA is used as energy (Topping and Clifton, 2001). A study by Shams et al. (2008) attested to the fact that cooked lentils have the potential for improving glycemic control in type 2 diabetic patients, thus, making it a possible dietary hypoglycemic agent. Al-Tibi et al. (2010) also reported a significant decrease in serum blood glucose of diabetic rats upon administration of cooked lentils. The consumption of lentils has been associated with low postprandial blood glucose response attributable to the significant α-glucosidase inhibitory activity of lentils, as well the presence of higher amounts of dietary fiber, protein, and phenolic compounds in whole lentils (Tosh et al., 2013; Ramdath et al., 2020).

Consumption of lentils have also been reported to display associated benefits in antiobesity activity. Studies showed that consumption of lentils resulted in hypolipidemic and antihypercholesterolemic activities (Mahmoud, 2011, Bazzano et al., 2011, Vohra et al., 2016).

Regular consumption of lentil has shown the potential to reduce the risk of cardiovascular and coronary disease with reported cardioprotective activity (Ganesan and Xu, 2017).

Lectins are known bioactive compounds found in lentils; lectins act as therapeutic agents and they exhibit anticancer properties by binding to cancer cell membranes, eventually inhibiting the growth of tumors (De Mejia Prisecaru, 2005). Other polyphenol compounds in lentils have also been reported to be greatly responsible for the reduction of the incidence of various types of cancer, in addition, lentil bioactive compounds showed chemoprotective potential (Rodríguez-Juan et al., 2000; Adebamowo et al., 2005; Perabo et al., 2008; Caccialupi et al., 2010; Busambwa et al., 2014; Chan et al., 2015).

19.5 Lentil Processing

Processing is a vital aspect of food production and traditionally involves the separation of the inedible portion from the edible portion. However, lentils can also be subjected to unconventional technological processes to separate the protein, carbohydrate, and other fractions (Thavarajah et al., 2016).

Processing can be carried out using dry or wet methods, however, the aim of processing pulses generally is to improve the nutrient content as well as extract functional ingredients or fractions for subsequent use for value addition or product development. Therefore, processing can be divided into three levels namely primary, secondary, and tertiary processing. Primary processing involves cleaning, grading, and packaging processes. The basic principle of primary processing uses a series of mechanical separation techniques with the aim of obtaining clean, wholesome lentil seeds (Vandenberg, 2009). Secondary processing involves unit operations such as decortication, splitting, sorting, and polishing of the whole or split seeds. It may also involve subjecting seeds to thermal processing to obtain a semifinished product packaged in cans or jars. Soaking and cooking are techniques to promote edibility of the seeds as well shorten antinutritional factors (ANFs) that may limit availability of nutrients. De-hulling is also carried out to reduce soaking and cooking time and reduce antinutrients while adding value to the seed. Finally, tertiary processing involves milling of whole or decorticated seeds; it also entails fractionation of the seeds into protein and starch components for use as functional ingredients in food products.

Microwave and other traditional cooking methods have effect on the nutritional composition and antinutritional factors of lentils. Hefnawy (2011) recommended microwave cooking for lentils as the technique improved nutritional quality, reduced cooking time whereas other cooking treatment significantly decreased mineral contents and content of some amino acids (viz lysine, tryptophan, total aromatic and sulphur-containing amino acids).

In another study, Morales et al. (2015) evaluated the effect of extrusion processing on a new snack product from lentil flour, they reported an increase in soluble fiber fraction, antioxidant activity and majority of the polyphenolic fractions but a marked decrease in the total tocopherol.

Lentils have also been sprouted and incorporated (at 5 and 10%) into wheat flour to produce bread, the results from the study showed that the incorporation of germinated lentil modified the quality of the dough and the flour as well increased the content of phenolic acids and flavonoids in the bread (Anguilar et al., 2019).

19.6 Bioactive Functional Groups in Lentils

Lentils (*Lens culinaris*) are examples of pulses. In addition to their rich nutritional compositions and potential health benefits, they contain some bioactive functional groups usually regarded as ANFs, which reduce their nutrient utilization. ANFs are toxic phytochemical substances or undesirable chemicals found in abundance in these crops, they are secondary metabolites, and are biologically active. These group of natural compounds are capable of causing harmful effects by decreasing the utilization of nutrients, micronutrients, and/or food consumption (Shanthakumari et al., 2008; Gemede and Ratta, 2014). In lentils, these ANFs includes polyphenols (phytate and tannins), trypsin inhibitors, and oligosaccharides (Pal et al., 2017). They have been reported to hamper many biochemical reactions by themselves or through the activities of their metabolic products (Singh, et al., 2017; Bresciani and Marti, 2019). Among such biochemical reactions of great nutritional importance is the inhibition of digestive enzymes and interference with protein digestibility, resulting in low protein digestibility and consequently poor bioavailability of proteins.

19.6.1 Phytate in Lentils

Phytate, also known as inositol hexaphosphate (IP_6), occurs naturally in plants and is the main storage form of phosphorus (P), inositol, and inorganic phosphate ions in plants. Phosphorus bound to inositol is mainly referred to as phytate (comprising inositols bound with four to six phosphate groups) and it makes up about 50–80% of the total phosphorus in plants (Delia et al., 2011). It comprises 1–5% of the weight of pulses (such as lentils), cereals, nuts, and oil seeds. Factors that influences the phytate content of these plants include the cultivars and soil types, climate and irrigation conditions, as well as processing and location (Delia et al., 2011). The major location of phytate in cereals is in the aleurone and the germ, while they exist in the endosperm or cotyledon (protein bodies) in legumes (Schlemmer et al., 2009). This makes separation difficult through simple processing techniques, such as milling (Sinha and Khare, 2017).

As reported by Fredlund et al. (2006) and Humer et al. (2014), phytate is a highly negatively charged ion (due to the presence of phosphate groups), and therefore its presence in lentils (when consumed) negatively influences the bioavailability of mineral cations (Zn^{2+}, $Fe^{2+/3+}$, Ca^{2+}, Mg^{2+}, Mn^{2+} and Cu^{2+}), by forming insoluble complexes with the cations. Also, they alter protein structure by forming indigestible complexes with proteins (at varied pH) and inhibit digestive enzymes, causing reduction in the solubility of proteins, activity of enzymes, and proteolytic digestibility. Studies had revealed that different processing techniques result in significant reductions in lentil phytate. Thavarajah et al. (2009) and Dellavalle et al. (2013) reported that decortication (removal of seed coat) of lentils prior to cooking reduced the phytate content by more than half (50%) and lentils exhibited increased iron (Fe) bioavailability. Similarly, Ghavidel and Prakash (2007) reported that germination and de-hulling reduced the phytate content of lentils and was corroborated by the findings of Pal et al. (2017). The reduced phytate content observed in germinated lentils might be attributed to the increased activity of endogenous phytase and some other phosphatases that hydrolyze phytate, releasing low inositol phosphates (phosphate, inositol, and micronutrients) required for the support of the developing seedling. Germination, therefore, is regarded as the most effective way of dephosphorylation. The sprouting process (an act of soaking in water and allowing seeds to germinate and sprout) significantly ($p<0.05$) reduced phytate in lentil seeds with the percentage decrease ranging from 45.85–73.76% as sprouting days increased (Fouad and Rehab, 2015). Hefnawy (2011) also reported a significant reduction in phytate by cooking, which corroborates the report of Wang et al. (2009).

19.6.2 Tannin in Lentils

Tannins are a class of astringent, polyphenolic biological molecules that fastens to and precipitate proteins and many other macromolecules. Tannins are responsible for discoloration and astringent taste in seeds. They are also able to chelate with metal ions (de Jesus et al. 2012) and bind to proteins through hydrogen bounding and hydrophobic interaction, thus resulting in reduced nutritional properties (Singh et al., 2017). According to Wang et al. (2009), tannins were significantly reduced by cooking, and this was corroborated by the findings of Hefnawy (2011). Fouad and Rehab (2015) also reported that the sprouting process caused a significant reduction in tannin content of lentils, the reduction observed corresponds to increased sprouting days. Several authors ascribed the reduction of tannins after soaking to their water-soluble nature, resulting in leaching into the water during soaking (Shimelis and Rakshit, 2007; Kalpanadevi and Mohan, 2013; Fouad and Rehab 2015). In addition to leaching, tannin could be further reduced during germination by the formation of a hydrophobic association of tannins with seed proteins and enzymes, as well the binding of polyphenols with carbohydrate or proteins (Khandelwal et al., 2010; Rusydi and Azrina, 2012). Pal et al. (2017) reported that germination and de-hulling significantly ($p<0.05$) reduced tannin content of lentils and agrees with the previous findings of Ghavidel and Prakash (2007) in cow peas, chickpeas, green gram, and lentils.

   The effectiveness of de-hulling in reducing the tannin content of lentils is corroborated in the earlier findings of Wang (2008) that showed dehulling improved the palatability and taste (astringent) of lentils, which was attributed to the presence of tannins in the seed coat (hull). Similarly, the reduction of tannins after heat treatment (boiling, cooking, microwave, and autoclaving) might be attributed to the heat labile property of tannins and the further degradation as a result of heat (Khattab and Arntfield, 2009). Barros et al. (2012) further reported that the decrease in tannins during cooking might be due to the interaction of tannins with starch, especially amylose, to form an insoluble complex, consequently increasing resistant starch content and decreasing starch digestibility. Fermentation as a processing technique also reduces tannin content, this reduction has been attributed to the activity of polyphenol oxidase or fermentation flora (Gunenc et al., 2017; Polat et al., 2020).

19.6.3 Enzyme (Protease) Inhibitors in Lentils

Enzyme inhibitors are molecules that bind to enzymes at their active site, thereby inhibiting their activity (Lopina, 2017). Lentils are a rich source of dietary fiber and enzyme inhibitors (Aslani et al., 2015) and the enzyme inhibitors (protease inhibitors) have been characterized to belong to the Bowman-Birk family (Clemente and Arqués 2014). The protease inhibitors consist of two distinct binding loops responsible for the inhibition of two enzyme molecules that may be the same (trypsin- like) or different enzymes (trypsin/chymotrypsin-like). The Bowman-Birk-like trypsin/chymotrypsin inhibitors in lentils inhibit the activity of these digestive enzymes (*in vitro* and *in vivo*), they intensely inhibit the activity of pancreatic enzymes (trypsin and chymotrypsin), and consequently decrease the digestion and absorption of proteins by forming indigestible complexes (Gemede and Ratta, 2014). Present in soy beans are the original Bowman-Birk inhibitors (BBI) capable of inhibiting only trypsin and chymotrypsin in combination. The trypsin inhibitors (TI) though present in most legumes, vary significantly. According to Shi et al. (2017), TI could range as low as 3.16–4.92 TIU/mg in lentils to as high as 15.18–20.83 TIU/mg in common beans. Several authors have reported on the use of diverse food processing methods to inactivate or reduce TIs in lentils (Świeca and Baraniak, 2014; Shi et al., 2017; Avilés-Gaxiola, et al., 2018). Shi et al. (2017) reported that de-hulling significantly increased the TIs in lentils. This was attributed to the presence of TIs in the cotyledon; removal of the hull, therefore, increases the concentration of TIs. Also, roasting (80°C/1 minute) has been reported to effectively reduce TI activity in lentils by about 95.6% (Avilés-Gaxiola, et al., 2018). This agrees with the report of Qayyum et al. (2012) that TI reduction is a function of heat treatment, which might be attributed to the heat sensitivity of TIs. Soaking followed by cooking resulted in 100% reduction in TI activities in lentils, which might be attributed to the leaching of TIs into the water used for soaking (Shi et al., 2017; Avilés-Gaxiola, et al., 2018). According Sharif et al. (2014), soaking alone proved to be more effective than cooking only. Another report found that lentils, when soaked for 12 hours, blanched at 70°C for 9 minutes, and then cooked at 116°C for 42 minutes (a canning process) completely removed TI activity (Hefnawy, 2011; Avilés-Gaxiola, et al., 2018). Thermal treatment (cooking, boiling, autoclaving, microwaving, and roasting), which is easily done at home and at industrial levels, is an effective method to inactivate TIs in lentils.

19.6.4 Oligosaccharides in Lentils

Oligosaccharides of lentils are composed of raffinose, ciceritol, stachyose (the main oligosaccharide in lentils), and verbascose (a minor oligosaccharide in lentils). The utilization of lentils is limited because of the presence of oligosaccharides (flatus-producing carbohydrates). They are referred to as flatus-producing carbohydrates due to the presence of α-galactosides bonds. Alpha (α) galactosidase is absent in humans, thus making the breakdown of α-galactosides bonds impossible, leaving the oligosaccharides (carbohydrates) undigested in the intestine. The undigested carbohydrates undergoes anaerobic fermentation with soluble dietary fiber producing SCFA. These fermentation substrates encourage the growth of lactobacilli and bifido-bacteria and the reduction of enterobacteria in the intestinal microflora. Fermentation also releases H_2, CO_2, and CH_4 (Han and Baik, 2006; Njoumi et al. 2019), causing bloating and flatulence. Several authors reported that considerable quantities of flatus-producing carbohydrates can be removed by common processing methods (Han and Baik, 2006; Hefnawy, 2011; Njoumi et al. 2019). According to Hefnawy (2011), cooking significantly reduced raffinose and starchyose, whereas verbascose was removed totally. The decrease observed could be accredited to their dispersion into cooking water, which agrees with the findings of Ruperéz (1998), who observed a 35.9% loss of starchyose after pressure cooking of soaked lentils. Han and Baik (2006) also reported a reduction in total oligosaccharide content of lentils by 22.9–50.1% after soaking in water for 12 hours; 55.9–58.7% was observed after soaking with ultrasound for 1.5 hours, whereas 34.5–55.9% reduction was observed after soaking with high hydrostatic pressure for 1 hour. The authors concluded that the use of ultrasound during soaking assisted in the reduction of oligosaccharides in lentils, particularly when soaking time is reduced. Abdel-Gawad (1993) also observed a 28% decrease in oligosaccharide content of lentils after soaking in tap water for 12 hours and a 37% decrease after soaking in sodium bicarbonate solution for 12 hours. This was further corroborated with the report of Njoumi et al. (2019), where a 10% loss in total oligosaccharides was observed in lentils after 16 hours of soaking. The loss might be attributed to the leaching out of water-soluble raffinose, stachyose, and verbascose into the soaking water or their enzymatic degradation into lower molecular-weight sugars (Vidal-Valverde et al., 2002; Han and Baik, 2006; Coffigniez et al., 2018).

19.7 Health Importance of Antinutritional Factors in Lentils

ANFs are useful to humans in spite of the various reported hazards. Studies have revealed that dietary phytate found in lentils can cause a reduction in the rate of starch digestion and glycemic response (Lee et al., 2006; Kumar et al., 2010). Phytate inhibits the α-amylase enzyme and consequently culminates in incomplete digestion of starch, thus reducing starch digestibility (Lee et al., 2015). Phytate effectively reduced blood glucose *in vivo* (Lee et al., 2006). Kumar et al. (2010) attributed the effect of phytate on blood glucose to its influence on calcium channel activity as it deters serine threonine protein phosphatase activity and this opens intracellular calcium channels, enforcing the release of insulin. Therefore, it would be suitable in the control of carbohydrate metabolism disorders. Evidence-based studies had also shown that phytate protects against different forms of cancer (Khatiwada et al., 2012; Kapral et al., 2017) through numerous ways of involving antioxidative properties, distruption of cellular signal transduction, inhibition of cell cycle, and improvement of natural killer cells activity (Kumar et al., 2010). Phytate displays a specific target against cancer cells without affecting the normal cells, this is an advantage over existing cancer drugs.

Tannin in lentils, due to their high antioxidant activity, exhibit possible beneficial cardiovascular and anticancer activities (Chan et al., 2015; Ganesan and Xu, 2017). BBI have also been studied as cancer risk reducing factors (Souza et al., 2014; Iqbal et al., 2017). Oligosaccharides in lentils act as probiotics and have other numerous health benefits including anti-inflammatory, antitumour, and antiulcerative properties (Uddin et al., 2015; Singh et al., 2017; tan et al., 2018).

19.8 Conclusion

Lentils are an important pulse seed consumed either as a whole grain or in decorticated form and its processing results in a variety of products with useful applications in food formulation. Lentils are uniquely rich in high-quality protein with a good proportion of amino acids. Lentils possess bioactive components that have the potential to address a number of disease conditions including diabetes, obesity, and other cardiovascular diseases.

REFERENCES

Abdel-Gawad, A. S. (1993). Effect of domestic processing on oligosaccharide content of some dry legume seeds. Food Chemistry, 46, 25–31.

Abraham, R. (2015). Lentil (*Lens culinaris* Medikus) current status and future prospect of production in Ethiopia. Advances in Plants Agriculture Research, 2, 00040. doi: 10.15406/apar.2015.02.00040

Adebamowo, C. A., Cho, E., Sampson, L., Katan, M. B., Spiegelman, D., Willett, W. C., Holmes, M. D. (2005). Dietary flavonols and flavonol-rich foods intake and the risk of breast cancer. International Journal of Cancer, 114, 628–633.

Aguilar, C. H., Pacheco, A. D., Tenango, M. P., Bravo, M. D., Soto-Hernandez, M., Orea, A. C. (2019). Lentil sprouts: A nutraceutical alternative for the elaboration of bread. Journal of Food Science & Technology, Mysore, 57, doi: 10.1007/s13197-019-04215-5.

Alshikh, N., de Carmago, A. N., Shahidi, F. (2015). Phenolics of selected lentil cultivars: Antioxidant activities and inhibition of low-density lipoprotein and DNA damage. Journal of Functional Foods, 18, 1022–1038.

Al-Tibi, A. M. H., Takruri, H. R., Ahmad, M. N. (2010). Effect of dehulling and cooking of lentils (*Lens Culinaris*, L.) on serum glucose and lipoprotein levels in streptozotocin-induced diabetic rats. Malaysian Journal of Nutrition, 16, 409–418.

Amarowicz, R., Estrella, I., Hernández, T., Robredo, S., Troszyńska, A., Kosińska, A., Pegg, R. B. (2010). Free radical-scavenging capacity, antioxidant activity, and phenolic composition of green lentil (*Lens culinaris*). Food Chemistry, 121, 705–711.

Aslani, Z., Alipour, B., Bahadoran, Z., Bagherzadeh, F., Mirmiran, P. (2015) Effect of lentil sprouts on glycemic control in overweight and obese patients with type 2 diabetes. International Journal of Nutrition & Food Science, 4, 10–14.

Avilés-Gaxiola, S., Chuck-Hernández, C., Saldívar, S. O. (2018). Inactivation methods of trypsin inhibitor in legumes: A review. Journal of Food Science, 83, 17–29.

Barros, F., Awika, J. M., Rooney, L. W. (2012). Interaction of tannins and other sorghum phenolic compounds with starch and effects on in vitro starch digestibility. Journal of Agriculture & Food Chemistry, 60, 11609–11617.

Bazzano, L. A., Thompson, A. M., Tees, M. T., Nguyen, C. H., Winham, D. M. (2011). Non-soy legume consumption lowers cholesterol levels: A meta-analysis of randomized controlled trials. Nutrition, Metabolism & Cardiovascular Diseases, 21, 94–103.

Bolsinger, J., Landstrom, M., Pronczuk, A., Auerbach, A., Hayes, K. C. (2017). Low glycemic load diets protect against metabolic syndrome and type 2 diabetes mellitus in the male Nile rat. The Journal of Nutritional Biochemistry, 42, 134–148.

Bresciani, A., Marti, A. (2019). Using pulses in baked products: Lights, shadows, and potential solutions, Foods, 8. doi: 10.3390/foods8100451.

Busambwa, K., Miller-Cebert, R., Aboagye, L., Dalrymple, L., Boateng, J., Shackelford, L., Verghese, M. (2014). Inhibitory effect of lentils, green split and yellow peas (sprouted and non-sprouted) on azoxy-methane-induced aberrant crypt foci in Fisher 344 male rats. Internal Journal of Cancer Research, 10, 27–36.

Caccialupi, P., Ceci, L. R., Siciliano, R. A., Pignone, D., Clemente, A., Sonnante, G. (2010). Bowman-Birk inhibitors in lentil: Heterologous expression, functional characterization and antiproliferative properties in human colon cancer cells. Food Chemistry, 120, 1058–1066.

Chan, Y. S., Yu, H., Xia, L., Ng, T. B. (2015). Lectin from green speckled lentil seeds (*Lens culinaris*) triggered apoptosis in nasopharyngeal carcinoma cell lines. Chinese Medicine, 10, 25, doi: 10.1186/s13020-015-0057-6.

Clemente, A., Arqués, M. C. (2014). Bowman-Birk inhibitors from legumes as colorectal chemopreventive agents. World Journal of Gastroenterology, 20, 10305–10315.

Coffigniez, F., Briffaz, A., Mestres, C., Alter, P., Durand, N., Bohuon, P. (2018). Multi-response modeling of reaction diffusion to explain alpha-galactoside behavior during the soaking-cooking process in cowpea. Food Chemistry, 242, 279–287.

de Jesus, N. Z., Falcão, H. S., Gomes, I. F., Leite, T. J., Lima, G. R., Barbosa-Filho, J. M., Tavares, J. F., da Silva, M. S., de Athayde-Filho, P. F., Batista, L. M. (2012). Tannins, peptic ulcers and related mechanisms. International Journal of Molecular Sciences, 13, 3203–3228.

De Mejia, E. G., Prisecaru, V. I. (2005). Lectins as bioactive plant proteins: A potential in cancer treatment. Critical Reviews in Food Science & Nutrition, 45, 425–445.

Delia, E., Tafaj, M., & Männer, K. (2011). Total phosphorus, phytate and phytase activity of some cereals grown in Albania and used in non-ruminant feed rations. Biotechnology & Biotechnological Equipment, 25, 2587–2590.

Dellavalle, D. M., Vanderberg, A. Glahn, R. P. (2013). Seed coat removal improves iron bioavailability in cooked lentils: Studies using an in vitro digestion/Caco-2 cell culture model. Journal of Agriculture & Food Chemistry, 61, 34, 8084–8089.

Duenas, M., Hernandez, T., Estrella, I. (2006). Assessment of in vitro antioxidant capacity of the seed coat and the cotyledon of legumes in relation to their phenolic contents. Food Chemistry, 98, 95–103.

Earnshaw, S. R., McDade, C. L., Chu, Y., Fleige, L.E., Sievenpiper, J. L. (2017). Cost-effectiveness of maintaining daily intake of oat β-glucan for coronary heart disease primary prevention. Clinical Therapeutics, 39, 804–818.

Elaloui, M., Ghazghazi, H., Ennajah, A., Manaa, S., Guezmir, W., Karray, N. B., Laamouri, A. (2017). Phenolic profile, antioxidant capacity of five Ziziphus spina-christi (L.) Willd provenances and their allelopathic effects on Trigonella foenum-graecum L. and Lens culinaris L. seeds. Natural Product Research, 31, 1209–1213.

Faris, M. A. E, Attlee, A. (2017). Chapter 3-Lentils (Lens culinaris, L.): A novel functional food. In: Exploring the nutrition and health benefits of functional foods, 1st Ed., USA IGI Global, 42–72.

Faris, M. A. I. E.; Takruri, H.R.; Issa, A. Y. (2013). Role of lentils (*Lens culinaris* L.) in human health and nutrition: A review. Mediterranean Journal of Nutrition & Metabolism, 6, 3–16.

FAO (2016). Food and Agriculture Organization of the United Nations. Pulses: Nutritious Seed for a Sustainable Future, Rome, FAO/OCCP, 1–189.

FAO. (2019). The global economy of pulses. Rawal, V. & Navarro, D. K. (Eds.). Rome: FAO, 1–174.

Fooks, L.J., Fuller R., Gibson G.R. (1999). Prebiotics, probiotics and human gut microbiology. International Dairy Journal, 9, 53–61.

Fouad, A. A., Rehab, F. M. A. (2015). Effect of germination time on proximate analysis, bioactive compounds and antioxidant activity of lentil (*Lens culinaris* Medik.) sprouts. Acta Scientiarum Polonorum Technologia Alimentaria, 14, 233–246.

Fredlund, K., Isaksson, M., Rossander-Hulthén, L., Almgren, A., Sandberg, A. S. (2006). Absorption of zinc and retention of calcium: Dose-dependent inhibition by phytate. Journal of Trace elements in Medicine & Biology, 20, 49–57.

Ganesan, K., Xu, B. (2017). Polyphenol-rich lentils and their health promoting effects. International Journal of Molecular Sciences, 18, 2390. doi: 10.3390/ijms18112390

Gemede, H. F., Ratta, N. (2014). Antinutritional factors in plant foods: Potential health benefits and adverse effects. Global Advanced Research in Food Science & Technology, 3,103–117.

Gharachorloo, M., Tarzi, B. G., Baharinia, M., Hemaci, A. H. (2012). Antioxidant activity and phenolic content of germinated lentil (*Lens culinaris*). Journal of Medicinal Plants Research, 6, 4562–4566.

Gharibzahedi, S.M., Mousari, S. M., Jafari, F. M., Faraji, K. (2012). Proximate composition, mineral content, and fatty acids profile of two varieties of lentil seeds cultivated in Iran. Chemistry of Natural Compound, 47, 976–978.

Ghavidel, R. A., Prakash, J. (2007). The impact of germination and dehulling on nutrients, antinutrients, in vitro iron and calcium bioavailability and in vitro starch and protein digestibility of some legume seeds. LWT—Food Science and Technology, 40, 1292–1299.

Gunenc, A., Yeung, M. H., Lavergne, C., Bertinato, J., Hosseinian, F. (2017). Enhancements of antioxidant activity and mineral solubility of germinated wrinkled lentils during fermentation in kefir. Journal of Functional Foods, 32, 72–79.

Han, H., Baik, B-K. (2008). Antioxidant activity and phenolic content of lentils (*Lens culinaris*), chickpeas (*Cicer arietinum* L.), peas (*Pisum sativum* L.) and soybeans (*Glycine max*), and their quantitative changes during processing. International Journal of Food Science and Technology, 43, 1971–1978.

Han, I. H., Baik, B-K (2006). Oligosaccharide content and composition of legumes and their reduction by soaking, cooking, ultrasound and high hydrostatic pressure. Cereal Chemistry, 83, 428–433.

Hefnawy, T. H. (2011). Effect of processing methods on nutritional composition and anti-nutritional factors in lentils (*Lens culinaris*). Annals of Agricultural Science, 56, 57–61.

Hoover, R., Hughes, T., Chung, H., Liu, Q. (2010). Composition, molecular structure, properties, and modification of pulse starches: A review. Food Research International, 43, 399–413.

Humer, E., Schwarz, C., Schedle, K. (2014). Phytate in pig and poultry nutrition. Journal of Animal Physiology & Animal Nutrition, 99, 605–625.

Iqbal, J., Abbasi, B.A., Mahmood, T., Kanwal, S., Ali, B., Shah, S. A. (2017). Plant-derived anticancer agents: A green anticancer approach. Asian Pacific Journal of Tropical Biomedicine, 7, 1129–1150.

Jarpa-Parra, M. (2018). Lentil protein: A review of functional properties and food application. An overview of lentil protein functionality. International Journal of Food Science & Technology, 53, 892–903.

Jati, I. R. A. P., Vadivel, V., Biesalski, H. K. (2013). Antioxidant activity of anthocyanins in common legume grains. In: Bioactive Food as Dietary Interventions for Liver and Gastrointestinal Diseases. Watson, R.R., Preedy, V.R. (Ed.), Academic Press, 485–497. doi.10.1016/B978-0-12-39715-8.00007-5.

Jood, S., Bishnoi, S., Sharma, A. (1998). Chemical analysis and physicochemical properties of chickpea and lentil cultivars. Nahrung/Food, 42, 71–74.

Kahraman, A. (2016). Nutritional components and amino acids in lentil varieties. Selcuk Journal of Agriculture & Food Sciences, 30, 34–38.

Kalpanadevi, V., Mohan, V. R. (2013). Effect of processing on antinutrients and in vitro protein digestibility of the underutilized legume, Vigna unguiculata (L.) Walp subsp. unguiculata. LWT—Food Science & Technology, 51, 455–461.

Kapral, M., Wawszczyk, J., Jesse, K., Paul-Samojedny, M., Kusmierz, D., Weglarz, L. (2017). Inositol hexaphosphate inhibits proliferation and induces apoptosis of colon cancer cells by suppressing the AKT/mTOR signaling pathway. Molecules, 22, 1657. https://doi.org/10.3390/molecules22101657

Khandelwal, S., Udipi, S. A., Ghugre, P. (2010). Polyphenols and tannins in Indian pulses: Effect of soaking, germination and pressure cooking. Food Research International, 43, 526–530.

Khatiwada, J., Davis, S., Williams, L. L. (2012). Synergistic effects of green tea catechin and phytic acid increases the cytotoxic effects of human colonic adenocarcinoma cell lines. International Journal of Cancer Research, 8, 49–62.

Khattab, R. Y., Arntfield, S. D. (2009). Nutritional quality of legume seeds as affected by some physical treatments 2. Antinutritional factors. LWT—Food Science &Technology, 42, 1113–1118.

Khazaei, H., Subedi, M., Nikerson, M., Martínez-Villaluenga, C., Frias, J., Vandeberg, A. (2019). Seed protein of lentils: Current status, progress, and food applications. Foods, 8, 391. doi: 10.3390/foods8090391

Kouris-Blazos, A., Belski, R. (2016). Health benefits of legumes and pulses with a focus on Australian sweet lupins. Asia Pacific Journal of Clinical Nutrition, 25, 1–17.

Kumar, V., Sinha, A. K., Makkar, H., Becker, K. (2010). Dietary roles of phytate and phytase in human nutrition: A review. Food Chemistry, 120, 945–959.

Lee, S-K., Park, H-J., Chun, H-K., Cho, S-Y., Cho, S-M. (2006). Dietary pytic acid lowers the blood glucose level in diabetic KK mice. Nutrition Research, 26, 474–479.

Lee, H., Loh, S., Bong, C., Sarbini, S., Yiu, P. H. (2015). Impact of phytic acid on nutrient bioaccessibility and antioxidant properties of dehusked rice. Journal of Food Science & Technology, 52, 7806–7816.

Lombardi-Boccia, G., Ruggeri, S., Aguzzi, A., Cappelloni, M. (2013). Globulins enhance in vitro iron but not zinc dialysability: A study on six legume species. Journal of Trace Elements in Medicine & Biology, 17, 1–5.

Lopina, O. D. (2017). Enzyme inhibitors and activators. In: Enzyme Inhibitors and Activators, Senturk, M. (Ed.), Rijeka: InTech, 243–257. doi: 10.5772/67248

Mahmoud, N. E. (2011). The semi-modified diets as antioxidants, hypolipidemic and hypocholesterolemic agents. Rome: FAO. Available from: http://agris.fao.org/aos/records/EG2012000695 (accessed on September 14, 2020)

Manach, C., Scalbert, A., Morand, C., Rémésy, C., Jiménez, L. (2004). Polyphenols: Food sources and bioavailability. The American Journal of Clinical Nutrition, 79, 727–747.

Morales, P., Cebadera-Miranda, L., Cámara, R. M., Reis, F. S., Barros, L., Berrios, J. D., Ferreira, I. C., Cámara, M. (2015). Lentil flour formulations to develop new snack-type products by extrusion processing: phytochemicals and antioxidant capacity. Journal of Functional Foods, 19, 537–544.

Njoumi, S., Amiot, M.J, Rochette, I., Bellagha, S., Mouquet-Rivier, C. (2019). Soaking and cooking modify the alpha-galacto-oligosaccharide and dietary fibre content in five Mediterranean legumes. International Journal of Food Sciences and Nutrition, 70, 551–561.

Nosworthy, M. G., Neufeld, J., Frohlich, P., Young, G., Malcolmson, L., House, J. D. (2017). Determination of the protein quality of cooked Canadian pulses. Food Science & Nutrition, 5, 896–903.

Pal, R. S., Bhartiya, A., Yadav, P., Kant, L., Mishra, K. K., Aditya, J. P., Pattanayak, A. (2017). Effect of dehulling, germination and cooking on nutrients, anti-nutrients, fatty acid composition and antioxidant properties in lentil (*Lens culinaris*). Journal of Food Science & Technology, 54, 909–920.

Peñas, E., Limón, R. I., Martínez-Villaluenga, C., Restani, P., Pihlanto, A., Frias, J. (2015). Impact of elicitation on antioxidant and potential antihypertensive properties of lentil sprouts. Plant Foods for Human Nutrition, 70, 401–407.

Perabo, F. G., Von Löw, E. C., Ellinger, J., von Rücker, A., Müller, S. C. (2008). Soy isoflavone genistein in prevention and treatment of prostate cancer. Prostate Cancer & Prostatic Diseases, 11, 6–12.

Polat, H., Capar, T. D., Inanir, C., Ekici, L., Yalcin, H. (2020). Formulation of functional crackers enriched with germinated lentil extract: A Response Surface Methodology Box-Behnken Design. LWT—Food Science & Technology, 123. doi: 10.1016/j.lwt.2020.109065

Punia, S., Dhull, S. B., Sandhu, K. S., Kaur, M. (2019). Faba bean (Vicia faba) starch: Structure, properties, and in vitro digestibility—A review. Legume Science, *1*(1), e18.

Punia, S., Dhull, S. B., Sandhu, K. S., Kaur, M., Purewal, S. S. (2020). Kidney bean (Phaseolus vulgaris) starch: A review. Legume Science, *2*(3), e52.

Qayyum, M. M. N., Butt, M. S., Anjum, F. M., Nawaz, H. (2012) Composition analysis of some selected legumes for protein isolates recovery, The Journal of Animal & Plant Sciences, 22, 1156–1162.

Ramdath, D. D., Lu, Z-H., Maharaj, P. L., Winberg, J., Brummer, Y., Hawke, A. (2020). Proximate analysis and nutritional evaluation of twenty Canadian lentils by principal component and cluster analyses. Foods, 9, 175. doi: 10.3390/foods9020175

Rodríguez-Juan, C., Pérez-Blas, M., Suárez-García, E., López-Suárez, J. C., Múzquiz, M., Cuadrado, C., Martín-Villa, J.M. (2000). Lens culinaris, Phaseolus vulgaris and Vicia faba lectins specifically trigger IL-8 production by the human colon carcinoma cell line Caco-2. Cytokine, 12, 1284–1287.

Roudi, R., Mohammadi, S. R., Roudbary, M., Mohsenzadegan, M. (2017). Lung cancer and β-glucans: Review of potential therapeutic applications. Investigational New Drugs, 35, 509–517.

Rupéréz, P. (1998). Oligosaccharides in raw and processed legumes. Zeitschrift für Lebensmittel-Untersuchung und -Forschung A, 206, 130–133.

Rusydi, M. R. M., Azrina, A. (2012). Effect of germination on total phenolic, tannin and phytic acid contents in soy bean and peanut. International Food Research Journal, 19, 673–677.

Ryan, E., Galvin, K., O'Connor, T. P., Maguire, A. R., O'Brien, N. M. (2007). Phytosterol, squalene, tocopherol content and fatty acid profile of selected seeds, grains, and legumes. Plant Foods for Human Nutrition, 62, 85–91.

Samaranayaka, A. (2017). Lentil: Revival of poor man's meat. In: Sustainable Protein Sources. Sudarshan, P.N., Janitha, P. D., Wanasundara, L. S. (Eds.), Academic Press, 185–196. doi: 10.1016/B978-0-12-802778-3.00011-1

Satya, S., Kaushik, G., Naik, S. N. (2010). Processing of food legumes: a boon to human nutrition. Mediterranean Journal of Nutrition & Metabolism, 3, 183–195.

Schlemmer, U., Frølich, W., Prieto, R.M., Grases, F. (2009). Phytate in foods and significance for humans: Food sources, intake, processing, bioavailability, protective role and analysis. Molecular Nutrition & Food Research, 53, S330–375.

Shams, H., Tahbaz, F., Entezari, M., Abadi, A. (2008). Effects of cooked lentils on glycemic control and blood lipids of patients with type 2 diabetes. ARYA Atherosclerosis, 4, 1–5.

Shanthakumari, S., Mohan, V., Britto, J. (2008). Nutritional evaluation and elimination of toxic principles in wild yam (Dioscorea spp.). Tropical and Subtropical Agroecosystems, 8, 319–325.

Sharif, H. R., Zhong, F., Anjum, F. M., Khan, M. I., Sharif, M. K., Khan, M. A., Haider, J., Shah, F. H. (2014). Effect of soaking and microwave pre-treatments on nutritional profile and cooking quality of different lentil cultivars. Pakistan Journal of Science, 24, 186–194.

Shi, L., Mu, K., Arntfield, S. D., Nickerson, M.T. (2017). Changes in levels of enzyme inhibitors during soaking and cooking for pulses available in Canada. Journal of Food Science & Technology, 54, 1014–1022.

Shimelis, E. A., Rakshit, S. K. (2007). Effect of processing on antinutrients and in vitro protein digestibility of kidney bean (*Phaseolus vulgaris* L.) varieties grown in East Africa. Food Chemistry, 103, 161–172.

Singh, B., Singh, J. P., Shevkani, K., Singh, N., Kaur, A. (2017). Bioactive constituents in pulses and their health benefits. Journal of Food Science & Technology, 54, 858–870.

Sinha, K., Khare, V. (2017). Review on: Antinutritional factors in vegetable crops. The Pharma Innovation Journal, 6, 353–358.

Souza, L. C., Camargo, R., Demasi, M., Santana, J. M., de Sá, C. M., de Freitas, S. M. (2014). Effects of an anticarcinogenic Bowman-Birk protease inhibitor on purified 20S proteasome and MCF -7 breast cancer cells. PLOS ONE, 9: e86600.

Świeca, M., Baraniak, B. (2014). Influence of elicitation with H_2O_2 on phenolics content, antioxidant potential and nutritional quality of *Lens culinaris* sprouts. Journal of the Science of Food & Agriculture, 94, 489–496.

Świeca, M., Baraniak, B., Gawlik-Dziki, U. (2013). In vitro digestibility and starch content, predicted glycemic index and potential in vitro anti-diabetic effect of lentil sprouts obtained by different germination techniques. Food Chemistry, 138, 1414–1420.

Talukdar, D., Talukdar, T. (2014). Coordinated response of sulfate transport, cysteine biosynthesis, and glutathione-mediated antioxidant defense in lentil (*Lens culinaris* Medik.) genotypes exposed to arsenic. Protoplasma, 251, 839–855.

Tan, H., Chen, W., Liu, Q., Yang, G., Li, K. (2018). Pectin oligosaccharides ameliorate colon cancer by regulating oxidative stress- and inflammation-activated signaling pathways, Frontiers in Immunology, 9, 1–13.

Thavarajah, D., Thavarajah, P., Johnson, C. R., Kumar, S. (2016). Lentil (*Lens culinaris* Medikus): A whole food rich in prebiotic carbohydrates to combat global obesity. In: Grain Legumes, Goyal A. (Ed.), IntechOpen, 35–53. doi.10.5772/62567.

Thavarajah, P., Thavarajah, D., Vanderberg, A. (2009). Low Phytic acid Lentils (*Lens culinaris* L.): A potential solution for increased micronutrient bioavailability. Journal of Agriculture & Food Chemistry, 57, 9044–9049.

Topping, D. L., Clifton, P.M. (2001). Short-chain fatty acids and human colonic function: Roles of resistant starch and nonstarch polysaccharides. Physiological Reviews, 81, 1031–1064.

Tosh, S., Farnworth, E., Brummer, Y., Duncan, A., Wright, A., Boye, J., Marcotte, M., Benali, M. (2013). Nutritional profile and carbohydrate characterization of spray-dried lentil, pea and chickpea ingredients. Foods, 2, 338. doi: 10.3390/foods2030338

Uddin, M., Sarker, M., Islam, Z., Ferdosh, S., Akanda, M., Haque, J., Yunus, K. B. (2015). Phytosterols and their extraction from various plant matrices using supercritical carbon dioxide: A review. Journal of the Science of Food & Agriculture, 95, 1385–1394.

USDA (2010). United States Department of Agriculture National Nutrient Database for Standard Reference, Release 23. Retrieved from: http://www.ars.usda.gov/research/publications/publications.htm?seq_no_115=243584 (accessed August 2020).

Vandenberg A. 2009. Postharvest processing and value addition. In: Erskine W, Muehlbauer F J, Sarker A, Sharma B, (Eds.), The Lentil: Botany, Production and Uses. CAB International, UK, 391.

Vidal-Valverde, C., Frias, J., Sierra, I., Blazquez I. F., Lambein, F., Kuo, Y. H. (2002) New functional legume foods by germination: Effect on the nutritive value of beans, lentils and peas. European Food Research & Technology, 215, 472–477.

Vidal-Valverde, C., Sierra, I., Frias, J., Prodanov, M., Sotomayor, C., Hedley, C., Urbano, G. (2002). Nutritional evaluation of lentil flours obtained after short-time soaking process. European Food Research & Technology, 215, 138–144.

Vohra, K., Gupta, V. K., Dureja, H., Garg, V. (2016). Antihyperlipidemic activity of *Lens culinaris* Medikus seeds in Triton WR-1339 induced hyperlipidemic rats. Journal of Pharmacognosy & Natural Products, 2, 117. doi: 10.4172/2472-0992.1000117

Wang, N. (2008). Effect of variety and crude protein content on dehulling quality and on the resulting chemical composition of red lentil (*Lens culinaris*). Journal of the Science of Food & Agriculture, 88, 885–890.

Wang, N., Hatcher, D.W., Toews, R., Gawalko, E.J. (2009). Influence of cooking and dehulling on nutritional composition of several varieties of lentils (*Lens culinaris*). Food Science & Technology, 42, 842–848.

Xu, B., Chang, S. K. (2008). Effect of soaking, boiling, and steaming on total phenolic content and antioxidant activities of cool season food legumes. Food Chemistry, 110, 1–13.

Xu, B., Chang, S. K. (2010). Phenolic substance characterization and chemical and cell-based antioxidant activities of 11 lentils grown in the Northern United States. Journal of Agriculture & Food Chemistry, 58, 1509–1517.

Xu, B.J., Chang, S. K. C. (2007). Comparative analysis of phenolic composition, antioxidant capacity, and color of cool season legumes and other selected food legumes. Journal of Food Science, 72, S167–S177.

Zou, Y., Chang, S. K. C., Gu, Y., Qian, S. Y. (2011). Antioxidant activity and phenolic compositions of lentil (*Lens culinaris* var. Morton) extract and its fractions. Journal of Agriculture & Food Chemistry, 59, 2268–2276.

20

Faba Bean Properties, Functionality, and its Applications

Nikita Wadhawan, Sagar M. Chavan, Seema Tanwar and N. K. Jain
College of Technology and Engineering, Udaipur, Rajasthan, India

CONTENTS

20.1 History

Faba beans belong to the *Fabaceae* family and they are originated by the *Vicia* L. genus, which is one of the oldest agricultural commodities grown. Faba beans date back to prehistoric times in the Middle East (Multari et. al, 2015). Caracuta et al. (2016) have identified seeds of a potential ancestor of faba bean adjacent to Mount Carmel, Israel. The remains were carbon dated to 14,000 years BP (before present). Moreover, Caracuta et al. (2015) have determined that faba beans were already domesticated about 10,200 years BP in the Lower Galilee area of Israel. Faba beans are considered one of the earliest domesticated crops in light of numerous archaeological findings in Eurasia and Africa, which date back to the early Neolithic era (Duc et al., 2015). Faba beans are cultivated in more than 58 countries worldwide. However, a few findings suggest the origin of faba beans to be Central Asia. The Chinese used them for food almost 5,000 years ago and they were cultivated by Egyptians 3,000 year ago (Askar, 1986).

20.2 Introduction

Faba beans, also known as horse beans, are a legume crop grown primarily for human consumption. Some varieties of seeds are useful as animal feed. Faba beans are grown mainly for its protein content (20–41%) and starch, as well as for fodder for animals. It is an annual legume whose botanical name is *Vicia Forba* L. Faba beans are known by many names, each indicating a particular subgroup, including fava beans, broad beans, horse beans, windoor beans, tick beans, Bakela, Boby Kurmouvje, Faveira, Fulmasri, Feve, Yeshil Bakla, etc. After soy beans and peas, faba beans are the most demanded legume in the world. The main faba-beans producing countries are China, some in Europe like United Kingdom, France, Austria, Czech Republic, Estonia or Denmark, and, Ethiopia, Egypt, and Australia.

World production of dry faba bean seeds from 1999–2003 amounted to 3.9 million tons per year from 2.60 million ha. Egypt is the leading country in consumption of faba beans and about 75% of the daily per capita protein intake of its population comes from cereals and beans. While China (114,000 ton per year) and Morocco (112,000 ton per year) are the largest producers, the production of the beans in tropical Africa and Asia is negligible. The world export shows a growing trend from 1998 to 2012, amounting to 475,000 tons. The main areas exporting faba beans include Australia (201,000 tons), the UK (114,000 tons), China (63,000 tons), France (53,000 tons), Italy (169,000 tons), and Spain (52,000 tons). The export from African and Asia are however very less (FAOStat, 2009). *V. Faba* is an annual herb with coarse and upright stems that are unbranched and measure 0.3–2 meters tall with one hollow stem at the base. The leaves and alternate pinnate consist of two to six leaflets each up to 8 cm long, unlike most other members of the genus.

Faba bean production worldwide was 4 million tons in the year 2014, offering a 1,807 kg/ha yield that same year. In East and West Asia, faba bean production was 37.3 and 1.5% of the world's production in 2014. After that, most of the production was found in African and European countries.

20.3 Morphology

Faba beans plants can measure up to 2 meters tall. They have hollow, coarse, and unbranched stems. Its leaves are alternate, up to 8 cm long, pinnately compound, with two to six leaflets, and are without tendrils or with very rudimentary ones. Faba beans can explore in soil up to 90 cm with strong roots. Its flower size is up to 3–4 cm long and is white with black or dark purple spots. The fruit is 10 cm long and 1–2 cm in diameter with a cylindrical shape. Their pods are green at first, then turn into dark brown or black at maturity. Germination of faba bean seeds takes 10–14 days in optimum growing conditions (Etemadi et al., 2015), but in dry conditions or when soil temperature is cold, it will take much longer. Faba bean plants grow one node per week on average. Its plant can grow 90–130 cm tall due the relatively strong stems that grow upright depending on the genotype. When the plant is around 30 cm tall, at the 8–10 node growth stage, the faba beans produce their first flower in the month of June. Flowers and pods appear about 20 cm above the ground. Faba beans pods will produce 25% flowers, which each have three to six seeds (Etemadi et al., 2015). Irrigation, soil fertility, time of planting, and type of management practices can significantly reduce the number of aborted flowers, thus improving seed/pod yield. Faba beans have been identified for their efficient nitrogen (N) fixing capacities, which is the highest among the cool season legumes (Mekkei, 2014). Reports indicated that faba beans can fix 50–330 kg N·hm-2 (Galloway et al., 2004; Etemadi et al., 2018d) depending on cultivation management and environmental conditions (Hu and Schmidhalter, 2005).

Faba beans may play a role in enhancing land productivity of small and marginal farmers who own pieces of land less than 2 ha and are facing challenges from increased competition for water, rising farm costs, and climate change. Under global warming and climate change scenarios, faba beans are one of the best performing crops due to their unique ability to excel under almost all type of climatic conditions coupled with their wide adaptability to a range of soil environments (Singh et al., 2012). As a nitrogen-fixing plant, faba beans are capable of fixing atmospheric nitrogen, which results in increased residual soil nitrogen for use by subsequent crops and in green manure, having the potential to fix free nitrogen up to 150–300 kg/ha. Faba beans are still a minor crop in India and its potential still remains unexploited. During the reproductive phases of flower and pod development, the optimal temperature for plant growth is 15–20°C. Faba bean tolerance of frost is better compared to other grain legumes. Faba bean flowers will abort if temperatures exceed 27°C and are also particularly sensitive to hot, dry conditions during podding. There are two subspecies commonly grown:

- Vicia faba varieties major (broad beans) produces large seeds (650–850 g/1000 seeds). It is cultivated mainly for human consumption, through cubed broad beans can be fed to livestock.
- Vicia faba varieties minor (horse beans) produces smaller seeds (250–350 g/1000 seeds) and is used mainly for livestock feeding.

20.4 Properties of Faba Beans

Dried faba bean harvesting, transportation, storage, and processing requires equipment designed for the properties of faba beans (Table 20.1). In designing separating, harvesting, sizing, and grinding machines, accouting for the size, shape, and mechanical behavior of faba beans is important. When designing for storage and transporting structures, the structure load and angle of repose is an important factor that is affected by bulk density and porosity (Table 20.1). It's also important to understand the coefficient of friction of the grain against various surfaces when designing conveying, transporting, and storing structures.

The total energy, crude protein, crude cellulose, and crude oil content (as percentage in dry matter) of faba bean are found as 18.87 MJ/kg, 29.63%, 6.39%, and 1.06%, respectively, and all elements determined in the research are listed in the text (Table 20.2).

20.5 Nutritional Value

Protein content in faba bean seeds ranges from 247–372 g/kg dry mass (20–41%) in genetic resources and from 270–320 g/kg dry mass in commercial varieties. Faba bean seeds contain 51– 68% of carbohydrates in total, the major proportion of which is starch (41–53%) (Punia et al., 2019). Starch is the major

TABLE 20.1

Physical Properties of Faba Beans

S. No.	Physical Properties	Values
1	Length	20.39 mm
2	Width	14.54 mm
3	Thickness	7.86 mm
4	Weight	1.31 g
5	Geometric mean diameter	13.25 mm
6	Sphericity	0.651 for 10.91% moisture content
7	Projected area	2.79 cm^2
8	Volume	1,210 mm^3
9	1,000 grain mass	1,349.34 g
10	Bulk density	608.17 kg/m^3
11	Kernel density	1,248 kg/m^3
12	Porosity	51.48%
13	Terminal velocity	4.94 m/s
14	Rupture strength	310.83 and 542.38 N

TABLE 20.2

Physical Properties as the Moisture Content Increased From 9.89% to 25.08% (Dry Basis)

S. No.	Physical Properties	Values
1	Average length	18.40–19.77 mm
2	Width	12.54–13.66 mm
3	Thickness	7.00–8.02 mm
4	Geometric mean diameter	11.68–13.01 mm
5	Unit mass of the faba bean	1.147–1.301 g
6	Sphericity	63.47–65.78%
7	Thousand grain mass	1140.15–1332.67 g
8	Bulk density	419.59–381.60 kg/m^3
9	True density	1151.33–1206.21 kg/m^3
10	Angle of repose	13.94–18.58°
11	Volume	0.998–1.099 cm^3
12	Porosity	63.09–67.21%
13	Surface area	4.29–5.31 cm^2
14	Static and dynamic coefficients of friction	0.517–0.822 and 0.394–0.681 (rubber)
		0.346–0.510 and 0.300–0.384 (plywood)
		0.340–0.480 and 0.271–0.366 (mild steel)
		0.338–0.450 and 0.286–0.309 (chipboard)
15	0.326 to 0.387 and 0.249 to 0.298	galvanized metal
16	Rupture force	314.17–185.10 N
		242.2–205.56 N
		551.43–548.75 N
		(for X-, Y- and Z-axes)
17	Specific deformation	6.81–36.76%
		5.93–44.41%
		14.90–49.89%
		(for X-, Y- and Z-axes)
18	Rupture energy	203.83–681.56 N mm
		135.63–651.03 N mm
		217.93–1090.6 N mm
		(for X-, Y- and Z-axes)

faba bean seed component; it has a mean content of 423 g/kg dry mass and varies in negative correlation with protein content. A detailed analysis on 12 genotypes revealed low fat and sugar content in the seeds (19 g/kg and 41 g/kg dry mass, respectively).

In comparison with cereal, faba bean seeds have high lysine content (19.8 g/kg dry mass) and low methionine, cysteine, and tryptophan content (2.6 g/kg, 3.7 g/kg, and 2.7 g/kg dry mass). The energy values of faba bean for ruminant (UFL and UFV) are high, comparable to cereal energy values and explained by the high starch content of the seeds. Crude fiber content in faba beans ranges from 5.0–8.5% and dietary fiber from 15–30%, which seems to depend on the seed variety, although hemi-cellulose seems to be the major component (60%). Faba beans are a good source of dietary minerals, such as phosphorus, potassium, calcium, Sulphur, and iron. Calcium content ranges from 120–260 mg/100 g dry mass. More than 40–60% of the phosphorus present is unavailable as phytates.

Studies indicate that the dry matter digestibility of faba beans is somewhat lower than soy bean meal and solubility of the protein is also lower in faba beans compared to soy bean meal. The fiber is higher and fat content is lower in faba beans versus soy bean meal. The faba bean is about 25% protein and is higher in energy than soy beans (Table 20.3). Most results suggest that substituting two parts faba beans for one part soy beans and one part cereal grain gives equal or better rates of gain.

TABLE 20.3

Comparative Nutrient Content of Faba Bean, Barley, and Soy Bean Meal

Nutrient Content	Faba Bean	Barley Grain	Soy Bean Meal
	% Dry Matter		
Crude protein	27.0	11.0	45.0
Digestible protein	22.6	8.8	41.8
Calcium	0.15	0.08	0.37
Phosphorus	0.50	0.35	0.67
Lysine	1.5	0.4	3.3
Methionine-Cystine	0.5	0.2	1.6

20.5.1 Nutrients and Antinutritional Factors in Faba Beans as Affected by Processing

Antinutritional compounds of faba beans are easily soluble in water, which means a soaking process can eliminate these compounds with the discarded soaking solution. Some of the metabolic reactions affect the soluble carbohydrate content. Trypsin inhibitors and volatile compounds are deactivated by thermal treatments, such as cooking, because of their heat-sensitive nature. Some other soluble compounds are discarded with cooking water. Dry-heating has been suggested as a simple process for reducing heat labile compounds while leaving water-soluble compounds unmodified, although heat and pressure conditions could also modify the final content. Germination has been suggested as an effective treatment to remove antinutritional factors in legumes, mobilizing secondary metabolic compounds that act as reserve nutrients (e.g. phytates and α-galactosides).

20.6 Processing

Faba beans are perhaps the most trouble-free pulse crop to harvest. If planted early in the season and dried at the appropriate time, they can be the last crop harvested. Largely unaffected by early snows, faba beans will remain standing through the most adverse fall weather while retaining quality. When harvesting for seed, choose an area of the paddock where there has been minimal disease, pest, and weed infestation. Germination rates are improved if seeds are harvested at 12–14% moisture and stored in mesh silos, aerated, or immediately graded and bagged.

20.6.1 Maturity

Depending on type and variety, seeding date, and seasonal moisture (drought to excess moisture), faba bean crops require a growing season of approximately 110–130 days. A color change of 90% or greater of the whole plant indicates it is physiologically mature and ready for harvest. As faba bean plants mature, the lower leaves darken and drop, and the bottom pods turn black and dry progressively up the stem. Faba beans are sensitive and responsive to moisture, and thus swathing or baling is not recommended. A faba bean crop can shatter as more pods are allowed to mature to the black stage or left standing until complete dry down, this is especially true if the crop is affected by drought or wind.

20.6.2 Seed Moisture Content

The crop is ready for harvesting at 18–20% seed moisture content if you have a clean sample and can aerate the seed. Seed should be aerated down to 16% moisture for safe storage. Dry for faba beans is <16% seed moisture content, tough is between 16.1–18% seed moisture content, and damp is >18% seed moisture content.

20.6.3 Field Monitoring

Preharvest field monitoring will help determine which harvest system to consider, if more than one is available, and will greatly assist in determining when to begin harvest operations. Monitoring the fields means checking plants in numerous locations for uniformity of stages of maturity. Most fields will not be 100% uniform in topography—there could be greener conditions in lower, wetter areas and further advanced plants on higher areas. A decision to begin harvesting will hinge on a majority of the field meeting certain criteria. Do not sacrifice the quantity and quality of your crop waiting for smaller greener areas to reach the proper stage. Harvesting too early will result in immature seeds. Harvesting too late when the pods are dry and brittle may result in shatter losses and will increase the risk of poorer quality seed due to adverse weather. The decision to start the harvest process will depend on three factors: Crop maturity (stage of uniformity), seed moisture content, and presence of weed growth. Other considerations may include weather patterns and marketing considerations.

20.6.4 Presence of Weed Control

Waiting for green weed growth to dry down will jeopardize quality and yields. Swathed green weeds are unlikely to dry sufficiently in a few days, so combining will be delayed. Green weed material in a straight-cut operation will cause extra wetness in the threshing areas of the combine, resulting in moisture on the seed coat and dirt adhering to this moisture (earth tag).

20.6.5 Preharvest Aids (Glyphosate)

Various chemical harvest management tools are available to aid in the preparation for combining. It's important to select the right product for the right crop and the intended outcome. Crop desiccation/dry down and the preharvest perennial weed products are not the same. Harvest aid products vary in speed of activity, efficacy, and preharvest intervals. Applying glyphosate, a systemic herbicide, for preharvest weed control is advisable, but not for desiccation. Glyphosate can be sprayed when the seed has less than 30% moisture content. At this stage, faba bean stems are green to brown, pods are yellow to brown, and 80–90% of the leaves have dropped. Timing your preharvest herbicide (glyphosate) can be a challenge. Applying glyphosate too early can reduce yield and seed size, and late-season application may result in levels of glyphosate in the seed that exceed maximum allowable levels. Glyphosate is registered for preharvest applications on faba beans and may be used to control perennial weeds such as quack grass, Canada thistle, sow thistle, common milkweed, toadflax, and dandelion.

20.6.6 Desiccation

The goal of desiccants or harvest aids is to make sure the crop is dry and goes through the combine efficiently. Diquat is a registered desiccant for faba beans. Diquat should be applied when most plants are ripe and dry. At this stage, pods are fully filled and the bottom pods will be tan or black in color. Always read and follow label directions prior to application. Timing of your desiccant (Diquat) application in faba beans can be a challenge. Spraying too early can decrease yield and seed size while spraying too late, especially after a frost, can greatly delay the dry down of the crop. Remember, Diquat has a much faster dry down period while glyphosate aids in perennial weed control. Spray only as many acres at one time as can be combined in two or three days after dry down. If the entire crop will take more than two or three days to combine, stagger the desiccant application so not all the crop is ready at the same time. Use proper rates, high water volume, and spray at the correct crop stage.

20.6.7 Swathing

Faba beans may shatter if left standing until full maturity or left standing until complete dry down—this is especially true if the crop is affected by drought. The crop should be swathed when about 25% of the plants in the field have the lowest one to three pods turning dark. By this time, the uppermost pods should be fully developed and the middle pods will be turning light green. At this stage, the moisture content of the most mature seeds may be over 40% and the seeds in the upper part of the plant may have moisture content over 60% (moisture content of the whole plant will be 70% or greater at this stage). A light narrow swath should be used as the crop may take up to 3 weeks to dry once cut.

20.6.8 Combining

Most producers will start combining at 18–20% moisture to reduce shattering and cracking, and typically use aeration to dry down to 16%. Combining in the early morning can reduce seed damage if the seed moisture content is lower than 18–20%. Combine settings for faba beans are similar to peas. Bean concaves are recommended and some producers have had success removing concave wires. The concaves should be open wide, although if roping is occurring with rotor combines, they may need to be tightened up. Cover plates for the first two concaves may also be considered with rotor combines.

20.7 Health Benefits

The European Union produced most of the high-tannin content faba beans (T+). Tannin decreases the energy content slightly, so their uses are limited for consumers. T+ faba beans are introduce in pig diets up to 350 g/kg on the basis of this anti nutritional factor. Faba bean seeds offer high protein to poultry nutrition. When faba beans are included at levels as high as 250 g/kg in broiler diets, they can replace soy bean meal in a large part. However, in different animal species, it has been observed that extreme diets containing approximately 50% faba beans significantly reduced absorption of zinc (Zn) and manganese (Mn), and a diet with faba beans as the only source of protein brought about an impairment in growth, muscle mass and liver weight in growing rats.

Removing the tannins from the seeds increases the energy value and the protein digestibility of faba bean for pigs and poultry. Removing VC increases the energy value of faba bean in broiler chickens, and increases the possible use of faba beans in the diet of laying hens. For faba beans, the more promising progress is to remove both tannins and VC, since the effect of the two antinutritional factors on the nutritional value for broiler chicken is additive. As far as human consumption is concerned, faba beans are considered beneficial. Besides protein and energy supply, the effects of a faba bean-enriched diet in humans has shown a significant decrease in plasma LDL-cholesterol levels. Other health benefits are recognized from their contribution in fibre, particular vitamins, and minerals. In contrast, G6PD-deficient individuals, particularly deficient small children and older subjects, may experience severe red blood cell destruction even after consumption of small amounts of fresh beans. G6PD deficiency is widespread in large numbers worldwide due to its protective effect against severe malaria. It is used for kidney stones, liver malfunction, and eye diseases. Thus, faba beans can play a part in human and animal nutrition.

20.8 Uses of Faba Bean

Although faba beans are consumed less in Western countries, they are one of the main sources of protein and energy for much of Africa, Asia, and Latin America. The longevity of their storage life, their ease of transportation, and their low cost are attractive points for farmers. Faba beans are a good alternative to expensive meat and fish protein. Faba beans are consumed as fresh faba beans, fresh faba bean kernels, and dried faba bean grain. Moreover, faba beans have been used in medicinal capacities (Etemadi et al., 2019). Faba bean seed size is an important trait in determining market and consumption form. Large-seeded varieties (broad beans) are widely used for food, either as a fresh green vegetable or de-hulled dry seeds. Varieties with small- to medium-size seeds are mostly used for animal feed. Faba beans can also be used in the bakery industry—a combination of faba bean and wheat flour improves the nutritional properties of bread. In Spain, small faba bean seeds (<12 mm) are currently highly accepted in the industry. Small seed genotypes are generally preferred by the frozen faba bean and canning industries;

the ability to use a microwave oven encourages the consumption of this legume, because seeds are much more easily cooked, and bags can be stored for up to 10 days at 5°C.

20.8.1 Livestock Feed

The faba bean does not possess any components toxic to animal or humans. Broken grains are mixed into animal diets and vegetative parts of the plant are used for the litter in animal houses. It is possible to feed the bean to all types of livestock or poultry, provided it is cracked or crushed. No further processing is required. Canadian research showed no significant difference in milk production when cows were fed grain rations containing either faba beans or soy bean meal as the protein supplement.

Pelleting has a very positive and significant effect on the AMEn value of faba beans fed to young chicks (Lacassagne et al., 1988). The increase in the AMEn value that can be attributed to pelleting is about 1.23 MJ/kg (d.m.) (12% of the AMEn value of unpelleted faba bean). Lucbert and Castaing (1986) showed that the use of faba bean in broiler diets leads to a slight increase in the feed conversion ratio, due to an increase in feed intake and a decrease in daily growth.

20.8.2 Forage/Silage

Faba bean plants make high-quality silage. In a 3-year experiment in Rosemont, MN, horse beans sown at 180 pounds/acre produced 4,370 pounds/acre of dry forage containing 10.5% protein. A mixture of 60 pounds of horse beans and 64 pounds of oats produced 5,613 pounds/acre of dry forage containing 10.1% protein. This and other data suggest that an oat/faba bean mixture for silage might be superior in production of protein per acre than oats alone.

20.8.3 Human Food

The seed coat of faba beans requires more chewing than that of the common baked bean varieties used in the United States, but the seed can be baked after it is softened in water. The large broad bean seeds are often preferred; the seed coats are often removed by hand before eating. Skinned beans are cooked, salted, and used for sandwich filling in North Africa. In Egypt and other Mid-Eastern countries, faba beans are eaten as a staple food by many strata of the society, and the increasing population of Middle-Eastern groups in the United States may be a potential market for faba beans.

Faba beans have been an essential staple food for millennia, widespread in the Mediterranean area including continental areas such as modern Iraq, Syria, Iran, and in northwest India, Pakistan, and southern China. The high content of digestible proteins and starch in their seeds mainly explain this extensive food use; more recent works have demonstrated the health benefits of faba bean seeds, because, like other pulses, they are a good source of fiber, and particular vitamins, and minerals (Ofuya and Akhidue, 2005; Champ, 2002). However, their VC constituents may cause health problems in particular situations.

20.9 Faba Bean Products

Legume-fortified pasta offers a broader spectrum for people wishing to improve the nutritional quality of their diet (enrichment in proteins and fiber, vitamins, minerals, and amino acid complementarity). However, substitution of durum wheat semolina with a high level (35% dry basis) of split pea or faba bean flour required an adaptation of the pasta making process at a pilot-scale.

20.10 Future of Faba Bean Products in India

Faba beasn are one of the best crops among the grain legumes with more than 7.0 grain yield potential and the addition of atmospheric nitrogen to the soil as added advantages. Being such an incredible crop, it's important to make it more widely used by and acceptable to consumers. Despite their good qualities, the major set backs for faba beans are their antinutritional elements, taste, and aroma. The availability of a zero tannin cultivar is a boon for expansion and inclusion in the daily diet, especially populations that depends upon vegetable protein.

Alternative protein sources are required to meet the nutritional requirements of the growing world population. These proteins will need to be produced in a sustainable manner, with little detrimental effect on the environment and be both economically and agriculturally advantageous. In addition, the food produced should be beneficial to combat the worldwide increase in chronic disorders (diabetes, cardiovascular disease, obesity, and cancer) and acceptable to both the food industry and consumers. Research on the nutritional and agronomic properties of legumes, along with advances in food processing and production, suggest that faba beans will become an important agricultural commodity. Several agronomic features of faba beans, including its symbiotic nitrogen fixation, may help address future agricultural challenges. Nutritionally, the high fiber, richness, and diversity of bioactive compounds point to faba bean as having a potential role in maintaining human health and disease prevention. It is clear that, where appropriate, faba beans could replace soy protein in both feed and food products, as well as partially replacing meat and meat-based products in the human diet. It is now important that research and technological innovation focuses on producing faba bean products of optimum nutritional value and consumer acceptability.

REFERENCES

Askar A, Faba beans (Vicia faba L.) and their role in the human diet. *Food and Nutrition Bulletin*, 8(3)1986, The United Nations University.

Caracuta, V., Barzilai, O., Khalaily, H., Milevski, I., Paz, Y., Vardi, J., Regev, L. and Boaretto, E., 2015. The onset of faba bean farming in the Southern Levant. *Scientific Reports*, 5, 14370.

Caracuta, V., Weinstein-Evron, M., Kaufman, D., Yeshurun, R., Silvent, J. and Boaretto, E., 2016. 14,000-year-old seeds indicate the Levantine origin of the lost progenitor of faba bean. *Scientific Reports*, 6, 37399.

Champ, M. M. J., 2002. Non-nutrient bioactive substances of pulses. *British Journal of Nutrition*, 88(S3), 307–319.

Duc, G., Aleksić, J. M., Marget, P., Mikic, A., Paull, J., Redden, R. J., Sass, O., Stoddard, F. L., Vandenberg, A., Vishnyakova, M. and Torres, A. M., 2015. Faba bean. In *Grain Legumes*. Springer, New York, NY, 141–178.

Etemadi, F., Hashemi, M., Barker, A. V., Zandvakili, O. R. and Liu, X., 2019. Agronomy, nutritional value, and medicinal application of faba bean (*Vicia faba* L.). *Horticultural Plant Journal*, 5(4), 170–182.

Etemadi, F., Hashemi, M., Zandvakili, O., Dolatabadian, A., Sadeghpour, A., 2018d. Nitrogen contribution from winter-killed faba bean cover crop to spring-sown sweet corn in conventional and no-till systems. *Agronomy Journal*, 110, 455–462.

FAO STAT (2009). Production stat: crops. FAO statistical databases. (FAO stat), food and agriculture organization of the United Nations (FAO), http://faostat.fao.org.

Galloway, J. N., Dentener, F. J., Caone, D.G., Boyer, E. W., Howarth, R. W., Seitzinger, S. P., Asner, G. P., Cleveland, C. C., Green, P. A., Holland, E. A., Karl, D. M., Michaels, A. F., Porter, J. H., Townsend, A. R., Voosmarty, C. J., 2004. Nitrogen cycles: Past, present, and future. *Biogeochemistry*, 70, 153–226.

Hu, Y., Schmidhalter, U., 2005. Drought and salinity: A comparison of their effects on mineral nutrition of plants. *Journal of Plant Nutrition and Soil Science*, 168, 541–549.

Lacassagne, L., Francesch, M., Carré, B. and Melcion, J. P., 1988. Utilization of tannin-containing and tannin-free faba beans (*Vicia faba*) by young chicks: Effects of pelleting feeds on energy, protein and starch digestibility. *Animal Feed Science and Technology*, 20(1), 59–68.

Lucbert, J. and Castaing, J., 1986. Utilization of sorghum with different concentrations of tannins in the feeding of broiler chickens. In *Proceedings of the 7th European Poultry Conference*, 472–476.

Multari, S., Stewart, D., Russell, W. R., 2015. Potential of fava bean as future protein supply to partially replace meat intake in the human diet. *Comprehensive Reviews in Food Science and Food Safety*, 14, 511–522.

Mekkei, M. E., 2014. Effect of intra-row spacing and seed size on yield and seed quality of faba bean (*Vicia faba* L.). *International Journal of Agriculture and Crop Sciences*, 7, 665–670.

Ofuya, Z. M. and Akhidue, V., 2005. The role of pulses in human nutrition: A review. *Journal of Applied Sciences and Environmental Management*, 9(3), 99–104.

Punia, S., Dhull, S. B., Sandhu, K. S., & Kaur, M. 2019. Faba bean (*Vicia faba*) starch: Structure, properties, and in vitro digestibility—A review. *Legume Science*, *1*(1), e18.

Singh, A. K., Verma, N., Singh, C. S., Kumar, S., Gupta, A. K., 2012. Faba Bean (*Vicia faba* L.): A potential leguminous crop of India, In: Singh, A. K., Bhatt, B. P. (Eds.), Ethnobotany of Faba Bean (Vicia Faba L.). ICAR, RC for ER, Patna, 431–450.

Weaver, T. (2015) Profitable pulses—it's all about the harvest window. Grains Research and Development Corporation website. Available from: https://grdc.com.au/resources-and-publications/groundcover/ground-cover-issue-118-sep-oct-2015/profitable-pulses-its-all-about-the-harvest-window

21

Pigeon Peas Possess Significant Protein and Starch Fractions with Potential as a Functional Ingredient in Food Applications

Aderonke Ibidunni Olagunju and Olufunmilayo Sade Omoba
Federal University of Technology, Akure, Ondo State, Nigeria

CONTENTS

21.1 Introduction

Pigeon peas (*Cajanus cajan*) are a grain legume of the tropic and subtropic regions belonging to the *Fabaceae* family. Pigeon pea ranks sixth in area and production compared to other grain legumes such as beans, peas, and chickpeas (FAOSTAT, 2015) and is cultivated in several regions, especially Asia, Africa, the Caribbean, and South America (Ecuador, Colombia, Venezuela, Peru, and Bolivia). The seed crop forms a good portion of the human diet in many countries owing to its high nutritive value and biological properties. Pigeon peas are abundant in protein (varying from 18–25% to as high as 32%), carbohydrates (50–60%), minerals, and B-complex vitamins (Olawuni et al. 2012; Aggarwal et al., 2015). It also possesses medicinal properties owing to the presence of a number of polyphenols and flavonoids (Singh, 2016).

The major component of pulse seeds in general are carbohydrates (starch), although greater attention has been given to their protein content because of its contribution to food quality and presence of health-promoting compounds. In a couple of research articles, the isolation and characterization of pulse starch has been considered and the legume/pulse starch has been of great interest to nutritionists because of the health-promoting attribute as a dietary control for diabetes (Punia et al., 2019, 2020a). This quality is attributable to the low digestibility and glycemic index values found in cereal and root/tuber starches. Extraction of pulse starch can also contribute to reduced overexploitation of food crops (cereals, roots, and tubers) for starch while increasing the utilization and industrialization of certain underutilized crops. The application of pigeon pea components as functional ingredients in food is associated with the fact that the food industry has been on an intensive search for ingredients with physiologically active components toward the development of foods that promote health and well-being. Interestingly, pigeon peas meet the requirement considering the richness of the seed in polyphenolic compounds, bioactive components and antioxidants, high essential and hydrophobic amino acids, as well as dietary starch components (Uchegbu & Ishiwu, 2016; Rinthong & Maneechais, 2018; Olagunju et al., 2018).

21.2 Pigeon Pea Protein Isolation Techniques

Protein isolation is immensely important and relevant in the development of new class of formulated food. Alkaline isoelectric precipitation is the widely used method for isolation of protein from legumes. Different techniques have been employed to produce pigeon pea protein concentrates/isolates, namely isoelectric precipitation (IEP) and alkaline extraction micellization (Mwasaru et al., 1999; Pazmiño et al., 2018). Modified micellization and isoelectric precipitation techniques were employed by Mwasaru et al. (1999) to isolate protein from pigeon peas. The study modified the conventional isoelectric precipitation by adjusting pH values of pigeon pea meal ranging from 8.5, 9.5, 10.5, 11.5, and 12.5. The protein content of the IEP isolate ranged from 78.1–83.4% with a pH of 8.5 having the highest protein content whereas, protein content of micelle isolate was 82.8%. However, several other isolation techniques have been employed to extract high quality protein fractions from pigeon pea. Adenekan et al. (2018) prepared pigeon pea protein isolate using four different methods, namely the methanol precipitation method, water extraction method, ammonium sulfate extraction method, and acetone precipitation method. Results revealed that high purity protein isolates were recovered as the isolate had a high protein content (91.83%) comparable to 90.65% reported by Olawuni et al. (2012). The high protein content of isolate showed that the solvents used (water, methanol, ammonium sulfate, and acetone) are good precipitants of protein from pigeon peas (Adenekan et al., 2018). They also reported that the isolation techniques employed were gentle and did not affect the amino acid profile of the isolate.

21.3 Pigeon Pea Protein Functionality and Application

There is an increasing global demand for high-quality plant proteins and alternative protein sources for application in the food industry, hence the need for research into both conventional and new sources of protein (Abdel-Rahman et al., 2011; Schlegel et al., 2019). Different studies evaluated the proximate composition of pigeon pea and reported fairly high protein content averaging 19–26% (Tiwari et al., 2008; Akande et al., 2010; Nwanekezi et al., 2017; Adenekan et al., 2018; Adepoju et al., 2019). The variation in observed values may be attributed to differences in the geographical location of pigeon pea seed used in each study. Wild species of pigeon peas have prospects to be a very promising source of high protein, nonetheless, some high-protein genotypes (as high as 32% protein) were earlier developed (Singh & Bains, 1988).

Pulse seeds have two types of protein i.e. metabolic or storage forms of protein. Often times, the protein exists as storage form and represents a larger percentage of the total seed protein. For instance, storage protein in lentils comprises up to 80% of total seed and the seed protein comprises of 7S and 11S globulin type protein (Joshi et al., 2017). Considering another example of pulse, pea seeds contain an average of 15–30% protein. Water-insoluble globulins and the water-soluble albumins form the major fractions of pea seed protein. Globulin fraction accounts for approximately 50–80% of total pea seed protein and are divided into the 7S (vicilin and convicilin) and 11S (legumin) fractions. However, vicilin is the predominant globulin in pea, with a varying composition of 26–52% of total seed protein while legumin accounts for 7–25% (Tzitzikas et al., 2006).

Similar to other pulse seeds, pigeon pea protein exists in the storage form as globulin and the fraction represents more than half (54–60%) of the total protein content (Saxena & Sawargaonkar, 2015; Kesari et al., 2017). The prominent protein in pigeon peas have been identified as 7S vicilin type globulin (Krishnan et al., 2017; Pazmiño et al., 2018). The 7S fraction consists mainly of low to intermediate molecular weight polypeptide with bands between 10 kDa and 100 kDa. According to Muangman et al. (2011), the observed molecular weight for crude protein isolate is in the range of 6–66 kDa, the protein was extracted by suspending pea powder in extraction solvents. Krishnan et al. (2017) found proteins with a molecular weight of 47 kDa and 64 kDa as the most prominent proteins in pigeon peas, which represent the two subunits of 7S vicilin. Likewise, Pazmiño et al. (2018) identified bands of 50–70 kDa in mature pigeon pea protein concentrate. In contrast to pigeon pea protein, lupin reportedly comprises two major protein types namely β-conglutin and α-conglutin representing 7S globulin, vicilin-like protein and 11S globulin, legumin-like protein, respectively (Duranti et al., 1981; Schlegel et al., 2019; Vogelsang-O'Dwyer et al., 2020). The molecular weight distribution of pigeon pea protein by different authors is similar to the molecular weight of polypeptides in other pulse seeds. Zehadi et al. (2015) reported an approximate band size of 25–100 kDa for lentil protein hydrolysate; the specific bands reported were 25–37 kDa, 35–45 kDa, 48 kDa, and 63 kDa identified as basic subunit, acidic subunit, vicilin, and convicilin, respectively. Lupin protein isolate showed a polypeptide weight distribution ranging from low (10–23 kDa) to intermediate (27–36 kDa) and high (41–84 kDa) molecular weight (Schlegel et al., 2019).

Solubility is an important property of proteins; it determines the crystallization of proteins. Essentially, protein solubility is related to pH, which is achieved when equilibrium exists between hydrophilic and hydrophobic interactions and solubility determines optimal applicability of protein in the food system. Minimum and maximum solubility of proteins exist; usually, minimum solubility occurs near an isoelectric point and maximum solubility above an isoelectric point. The solubility and isoelectric pH of pigeon peas are comparable to those of other related pulse crops. Most pulses possess similar solubility patterns with optimum solubility at both acidic and alkaline pH regions, while the isoelectric point of low solubility is usually at a low pH. The isoelectric point of rapeseed proteins was reportedly in the range from pH 4.5–5.5 (Ghodsvali et al., 2005). In contrast, a study by Ahmed et al. (2011) reported high solubility for different legume flours at extreme levels of pH and an isoelectric pH of 4.0. However, the minimum solubility of pigeon peas was reported in an earlier study by Oshodi and Ekperigin (1989) to be pH 5.0, whereas, according to Mwasaru et al. (2000), the minimum and maximum solubility of pigeon peas is 2.0 and 12.0, respectively, while the isoelectric point is pH 4.5. Similar values were also reported by Pazmiño et al. (2018). High protein solubility is often recorded at both acidic and alkaline pH regions; it is a guide to optimum protein isolation.

Pigeon pea proteins have a significant content of essential amino acids and are particularly rich in lysine, valine, threonine, and phenylalanine; similar to other pulse seeds, it is limiting in sulphur-containing amino acids, namely methionine and cysteine (Amarteifio et al., 2002). Equally, they contain a high amount of hydrophobic amino acids, reportedly shows a relationship with antioxidant activity, making the protein a good substrate for the production of Angiotensin-I converting enzyme (ACE) inhibitory peptides (Valdez-Ortiz et al., 2012; Pihlanto & Mäkinen, 2013). The balanced amino acid profile of pigeon peas qualifies them as good candidate for cereal supplementation to complement the limiting amino acids (especially lysine) in cereals.

Pigeon pea protein can be used to increase protein levels in food products as well serve as a useful ingredient to provide technological improvements in food systems (Pazmiño et al., 2018). Exploiting the protein component is a versatile approach to increase pigeon pea production and utilization. Some authors have evaluated the protein content of protein concentrate and isolate extracted from pigeon peas via different methods, they reported a high value of 76% for protein concentrate and 81–92% of recovered protein isolate (Olawuni et al., 2012; Adenekan et al., 2018; Ratnayani et al., 2019b; Olagunju et al., 2021). The protein content is higher than those reported for chickpea protein isolate (Paredes-López et al., 1991), but relative to 89–94% reported for lupin protein isolate (Jayasena et al., 2011).

Different techniques are employed in pulse processing to obtain a number of products and byproducts including flour or proteins (isolate, concentrate, hydrolysate, peptide). These processing techniques include, but are not limited to de-hulling, germination, fermentation, enzymatic hydrolysis, etc. Processing has shown the capability of increasing bioactivity and functionality of proteins.

- **Fermentation:** Pigeon pea seeds were fermented and used as pasta ingredient. The incorporation into semolina increased the total antioxidant capacity of the pasta product (Torres et al., 2006).
- **Germination:** According to Uchegbu and Ishiwu (2016), germination of pigeon peas increased the total phenolic content and 2,2-diphenyl-1-picryl-hydrazyl-hydrate (DPPH) radical scavenging ability by 30% and 39% respectively, exhibited inhibitory activity against alpha (α)-amylase

and α-glucosidase, as well showed hypoglycemic potential when administered to diabetic rats by reducing blood glucose levels.

*   **Hydrolysis:** Tempe peptide from hydrolyzed pigeon pea exhibited higher ACE activity than peptide from cooked pigeon peas (Putra et al., 2021).

A number of successful applications of pigeon pea for food formulation/development have been reported. Pigeon pea and oat mixture was fermented with *Lactobacillus reuteri* to produce a functional cream with improved nutrient profile and excellent sensory characteristics (Barboza et al., 2012). Pigeon pea isolate has been incorporated into wheat flour (at levels 15%, 20%, and 25%) and used in the development of protein-rich biscuits (protein content between 15.6–25.7%) without any noticeable change in desirable organoleptic properties (Hassan et al., 2013).

Parent protein from different pulse seeds have been subjected to natural processing and digestion processing in order to generate bioactive peptides with biological activities. These processing techniques (fermentation, germination, enzymatic hydrolysis, gastrointestinal digestion, etc.) generate oligopeptides with a specific amino acid sequence and molecular weight distribution. The peptides possess health-promoting properties useful in treating and/or reducing the onset of diseases. Research involving *in vitro* and *in vivo* studies have reported an array of nutraceutical properties exhibited by pigeon pea protein and its peptides, which are dependent on the protein substrate, the proteolytic enzymes used for hydrolysis, and the condition of hydrolysis process (Pihlanto & Mäkinen, 2013).

21.4 Bioactive Properties of Pigeon Pea Protein

Free radicals are a consequence of normal cell metabolism or external activities such as respiration, pollution, smoking, radiation etc; they attack important macromolecules causing cell damage and homeostatic disruption (Lobo et al., 2010). Excessive accumulation of free radicals in the body with inability of the cells to adequately destroy the excess free radicals formed results in oxidative stress (Pham-Huy et al., 2008). Free radicals have been implicated to adversely affect lipids, proteins, and DNA, thereby leading to oxidative stress and contributing to aging and the incidence of chronic diseases including diabetes mellitus, cardiovascular diseases, respiratory diseases, neurodegenerative diseases, cancer, etc (Marcadenti & Coelho, 2015; Phaniendra et al., 2015). However, antioxidants have capability to inhibit cellular damage via free radical scavenging activities. Antioxidants could be either endogenous, produced within the body (via enzyme system) or exogenous, received from outside the body (especially through diet). The endogenous nonprotein (glutathione) and protein (superoxide dismutase, catalase, glutathione peroxidase) antioxidant molecules act as the first line of defense against oxidative stress on the body. They confer protection by preventing the generation of oxidants, scavenging free radicals, or inhibiting oxidation reactions.

Endogenous antioxidants are easily damaged by excessive production of free radicals. Exogenous antioxidants refer to those from outside the human body, they exists as synthetic and natural antioxidants. Natural antioxidants include vitamins (C and E), minerals (zinc and selenium), and carotenoids; they are usually obtained from natural sources (especially fruits, vegetables, and legumes) and play vital role in preventing oxidative stress and DNA damage (Guerra-Araiza et al., 2013). The natural antioxidants, on the other hand, possess great potential as a preservative in place of synthetic antioxidants (Peschel et al., 2006). Pulses are important sources of antioxidants owing to their rich polyphenol content. Specifically, pigeon peas are a potential source of antioxidants since its high polyphenolic content has been documented (Al-Saeedi & Hossain, 2015; Chugh & Ritu, 2018, Cheboi et al., 2019). ACE is a vital enzyme in blood pressure regulation. In a renin angiotensin system, ACE releases Angiotensin II (a vasoconstrictor hormone), thereby promoting elevation of blood pressure. The inhibition of ACE activity using an inhibitory agent results in control of blood pressure; this approach has been well exploited in the control and management of hypertension. The production of dietary sources of ACE inhibitory agents will contribute immensely to the functional food market and health sector. Hydrophobic amino acid (HAA) has a direct correlation with ACE inhibitory activity (Pripp et al., 2004; Udenigwe & Aluko, 2011; Chalé et al., 2014; Teymoori et al., 2017), thus, the relative abundance of HAA in a protein sample suggests it could possess

ACE inhibitory potential and antihypertensive property. Interestingly, pigeon pea protein contains high levels of hydrophobic amino acid (Oshodi et al., 2009; Olagunju et al., 2018) and studies have confirmed the assertion.

Olagunju et al. (2018) evaluated the antioxidative capacity *(in vitro* and *in vivo)* and the antihypertensive properties of pigeon pea hydrolysate and reported that enzyme-hydrolyzed pigeon pea protein exhibited significant radical scavenging ability, ferric reducing property, as well the ability to inhibit linoleic acid oxidation. The same study showed the antihypertensive properties of pigeon pea hydrolysate owing to the superior angiotensin converting enzyme and renin inhibitory activities of pancreatin and pepsin-pancreatin hydrolysate in addition to an instantaneous systolic blood pressure lowering effect when administered to spontaneously hypertensive rats. This qualifies pigeon pea protein as a promising source of bioactive peptides with potential as a functional ingredient in food. To further confirm the bioactive potential of pigeon pea protein, Ratnayani et al. (2019a) reported that peptide fraction of germinated pigeon peas has the potency as an ACE inhibitory nutraceutical. In another study, Ratnayani et al. (2019b) subjected pigeon peas to gastrointestinal digestion followed by centrifugal ultrafiltration; they reported higher ACE inhibitory activity (86%) in low molecular weight ultrafiltered peptide (<3 kDa) compared to 3–10 and >10 kDa fraction (51–60%). The amino acid composition was dominated by hydrophobic amino acids with high content of proline and leucine, which are contributors to high ACE inhibitory activity of peptide fraction (Ryan et al., 2011).

In addition, thermoase-hydrolyzed pigeon pea protein and ultrafiltered peptides were reported to exhibit significant antidiabetic properties (Olagunju et al., 2021). This study observed a significant inhibition against key carbohydrate hydrolyzing enzymes (α-amylase and α-glucosidase). In view of the few studies on pigeon pea protein fraction and peptides, it is safe to submit that pigeon pea protein possesses significant bioactive properties with beneficial health implications.

21.5 Pigeon Pea Starch Functionality and Application

Pulses are important source of carbohydrates, composed of an average of 65–72% starch and 10–20% dietary fiber (Pehrsson et al., 2013). Starch is an insoluble polymeric carbohydrate produced by higher plants as well the major storage form of energy in various plants. Starch comprises two polymers namely amylose (straight-chain unit) and amylopectin (branched-chain unit). Pulse starches are characterized by a high amylose content (24–65%), the most significant component of carbohydrates in the human diet has been identified to be the starch fraction (Mahadevamma et al., 2004). Starch has found application as key ingredient in food and non-food industries by contributing functionality in terms of thickening, stabilization, texturization, gelation, encapsulation, and shelf-life extension (Ayucitra, 2012; Punia, 2019, 2020). Although, only a few studies exist on the starch and carbohydrate component of pulse crops compared to cereal and root crops, the subject has received tremendous research attention.

21.5.1 Isolation of Starch from Pigeon Pea Seeds

The high carbohydrate content of pigeon pea seed qualifies the pulse as an important potential source of edible starch (Olagunju et al., 2020a). The legume seed comprises a mixture of constituents, majorly starch, protein, and fine fiber; extraction of starch from this mixture may be a difficult procedure due to co-precipitation of the protein and fine fiber constituents (Ratnayake et al., 2001). In addition, the insoluble flocculent protein and fiber, decreases sedimentation and co-settles with the starch to form a brownish deposit. Starch is reportedly isolated from legumes/pulses by wet extraction or dry processing techniques. The wet milling process is a commonly employed technique for the isolation of starch from legumes/pulses, the process entails de-hulling of the seeds, steeping of de-hulled seeds in distilled water or sodium chloride solution (to facilitate solubilization of salt soluble proteins), wet milling, repeated centrifugation, and repeated treatment of sediment with 0.1–1.0 M sodium chloride solution (to further solubilize proteins and remove non-starch carbohydrates). The starch slurry is thereafter subjected to a hydrochloric acid treatment before the recovery of the sediment by centrifugation of sedimentation process. The starch sediment is dried, pulverized, and sieved to obtain pea starch of high purity. Dry milling is another method of starch isolation from legumes; the technique involves air classification and rinsing of air classified starch has been suggested to achieve removal of protein impurities (Reichert & Youngs, 1978). The purity of isolated starch is however, lower than that obtained via wet milling (Wani et al., 2016).

Over the years, different researchers have isolated starch from pigeon pea seeds, they however, reported variations in yield ranging from 28–49% (Singh et al., 1989; Hoover et al., 1993; Olagunju et al., 2020a; Acevedo et al., 2020). Narina et al. (2014) reported 7–12% amylose content for starch from different cultivars of pigeon pea. They equally observed complete resistance of the starch to digestion by pancreatic amylases and categorized the starch as resistant starch type III (RS3). In recent studies, Acevedo et al. (2020) and Olagunju et al. (2020a) reported a higher amylose content (24% and 26%, respectively) for native pigeon pea starch whereas, Sandhu and Lim (2008) and Hoover et al. (1993) reported 28% and 29%, respectively, for total amylose content of native pigeon pea starch, which was resistant to hydrolysis by porcine pancreatic alpha amylase. Different from the aforementioned reports, Singh et al. (1989) reported a significantly higher amylose content (47%) for whole seed pigeon pea starch. The significant variations in amylose contents may be attributable to varietal differences; overall, pigeon pea starch has a high amylose content, which confers positive benefits on the pulse starch.

Considering the fact that amylose digests slowly, and corroborating earlier reports of starch resistance to amylase enzyme, it suffice to say that there may be higher availability of resistant starch (RS) in the starch material thus, predicting lower digestibility of starch. Moreover, Sandhu and Lim (2008) as well as Narina et al. (2012), published that resistant starch of pigeon pea was comparatively higher than those of other pulse seeds (chickpeas, lentils, and field peas). Resistant and slowly digestible starches contribute to a low glycemic index in starch-based food products, a feature that is beneficial in the management of diabetes and hyperlipidemia (Rizkalla et al., 2002). Acevedo et al. (2020) reported low content (10%) of rapidly digesting starch (RDS) for pigeon pea starch and considered them useful for sufferers of type I diabetes. Thus, pigeon pea starch may be a useful functional ingredient in functional foods with potential health benefits.

Pulse starch granules generally have a distinct shape that appears round, oval, or ellipsoidal different from the irregular, spherical, lenticular, polyhedral, hexagonal granule morphology found in cereal starches (Ashogbon & Akintayo, 2013). Similar to other pulse seed starch, pigeon pea starch granule morphology was reported to be oval and or elliptical (Lawal, 2011; Narina et al., 2014; Acevedo et al., 2020). Pulse starch has comparable functional properties such as gelling, emulsion, and stabilizing capabilities hence, can be used in various food industrial applications. It has been used as a thickener in soups, thickener and texture improver in bread, fat replacer, stabilizer, mouthfeel improver, and stabilizer in yogurt (Messina, 1999; Mlyneková et al., 2014; Saleh et al., 2020). Despite the significant functionalities of pulse starch (in this case, pigeon pea starch), native starch is accompanied with several limitations in attaining specific industrial requirements. Hence, starch is subjected to different modification processes, specifically physical (thermal and nonthermal) and chemical (esterification, oxidation, succinylation, hydroxylpropylation, and cross linking) with the aim of improving the starch physicochemical, pasting, and techno-functional properties (Punia et al., 2020b).

Legume starches have occupied an important place in noodle preparation in several countries of the world, and pigeon pea starch is not left out as it has been used in preparation of noodles. The functionality of pigeon pea starch noodles compared favorably with noodles from mung bean starch and could even serve as a potential starch source as it possessed superiority in making transparent noodles (Singh et al., 1989). Studies have shown that legume/pulse starch have superior functional properties than cereal/tuber starch (Wani et al., 2016; Acevedo et al., 2020). Pigeon pea starch has higher peak viscosity and stability than other legume starches, thus, they may find application as thickeners, binders, and fillers in both the food and pharmaceutical industries (Ashogbon, 2014; Acevedo et al., 2020). Pulse starches (chickpeas and Turkish beans) have also been evaluated for their potential as a fat replacer or promising stabilizer in yogurt (Saleh et al., 2020). Similarly, pigeon pea starch has found application in this regard (Olagunju et al., 2020b). The study reported the ability of acetylated pigeon pea starch in promoting water holding capacity by about 85% and reducing whey syneresis (up to 25%) during a 28-day refrigerated storage period.

21.5.2 Modification of Pigeon Pea Starch and its Positive Impact on Starch Functionality

Modification of starch is carried out with the purpose of extending its applications in food products. Modification extends the properties of starch for food use and the process involves altering the structure and affecting the hydrogen bonding in a controllable manner (Murphy, 2000). The structure, reactivity, and functionality of the native starch can be modified by physical (pregelatinization, annealing, heat moisture treatment, microwave oven methods and other nonthermal treatments), chemical esterification, etherification, oxidation, cross-linking, and enzymatic methods or a combination of these techniques.

Physical modification methods are simple and inexpensive because chemical agents are not employed. However, chemical modification involves the exploitation of hydroxyl groups present in the starch structure to bring about the desired results for possible utilization of modified starch in various food and non-food applications. The modification process exposes reactive functional groups which then cause modification of the reactivity of starch when it comes in contact with water, oil, enzymes, or chemical reagents. Starch modification, therefore, brings about altered physicochemical properties, structural changes and results in the improvement of functional properties (such as water holding capacity, swelling capacity, gelling ability, viscosity, solubility), reduction in values of pasting parameters (such as peak, trough and final viscosities, breakdown value), decrease in values of thermal properties (gelatinization temperatures, enthalpy of gelatinization, retrogradation), and improved digestibility (starch and enzyme) of starch. The significant changes may be attributed to the weakening of the granular structure of the starch due to chemical modifications (Wojeicchowski et al., 2018). Lawal (2011) and Olagunju et al. (2020a; 2020b), carried out different chemical modification (specifically hydroxypropylation and acetylation) on pigeon pea starch and they reported significant changes/improvement in physicochemical, functional, pasting, and thermal properties of modified starch. Different modification techniques confer specific properties to native starches. Some commonly used modification processes are hereby discussed.

21.5.2.1 Hydrothermal Modification

Annealing and high moisture treatments involve the application of heat and water at low temperatures and a high water-to-starch ratio, it subsequently results in a physical reorganization of the starch granules (Simsek et al., 2012). This technique is widely employed to modify starches while the integrity of the granules is preserved. The methods differ from other physical modification process because it occurs below the gelatinization temperature of starch granules, thus, ultimately helping to preserve the granular structure (Xiong & Fei, 2017). Heat-moisture treated pigeon pea starch increased the pasting temperature and susceptibility of starch to the α-amylase enzyme (Hoover et al., 1993).

21.5.2.2 Acetylation

Acetylation is a process involving the use of predetermined volume of acetic acid or acetic anhydride in starch slurry i.e. esterification of the starch polymer with an acetyl group and part of the hydroxyl groups on anhydro-glucose units substituted with acetyl groups, consequently forming esters (starch acetates) (Babic et al., 2007). Acetylation affects the physicochemical properties of the starch positively; the extent depends on the acetyl group distribution and the degree of substitution (DS) of hydroxyl group with acetyl group (Chen et al., 2004). The number of acetyl groups incorporated into the starch molecule depend on the reactant concentration, pH of reaction mixture, reaction time, and presence of catalysts (Ačkar et al., 2015).

Acetylated starch with a low DS commonly finds application in the food industry; the standard maximum DS of acetylated starches for food application is 0.2 (Han et al., 2012). Acetylated starch functions to improve consistency, texture, and stability when incorporated to food systems (Bello-Perez et al., 2010). Starch acetates with a higher DS on the other hand, find application in extensive non-food applications, including the pharmaceutical industry (as tablet binders), biodegradable packaging materials, coatings, adhesives, etc. (Ačkar et al., 2015).

Acetylation of starch works to bring about significant changes in physicochemical properties such as swelling capacity, water absorption power and oil absorption capability, pasting behavior and thermal properties (Yadav & Patki, 2015).

21.5.2.3 Oxidation

Oxidation is also a chemical modification technique, oxidation of starch can be performed using different oxidizing agents such as hydrogen peroxide and ozone (Dias, et al., 2011; Klein et al., 2014; Oladebeye et al., 2018). Nonetheless, the widely used oxidizing agent to produce oxidized starch for industrial application is sodium hypochlorite (Vanier et al., 2012; Zhang et al., 2012). Sodium hypochlorite is known to enable oxidation at specific sites of the starch molecule by targeting the C2, C3, C6 glycosidic link only (Okekunle et al., 2020). Starch oxidation involves the application of sodium hypochlorite to a starch slurry under a controlled temperature and pH, the hydroxyl groups in the starch molecule are primarily converted/oxidized to carbonyl and then to carboxyl (Garrido et al., 2014). The process is associated with depolymerization of starch molecules mainly by cleaving amylose and amylopectin molecules at α-(1 $\rightarrow$ 4)-glycosidic linkages. The level of starch oxidation is defined by the carbonyl and carboxyl group contents and the degree of depolymerization in the starch granule (Kuakpetoon & Wang, 2006). Also, the intensity of oxidation reaction depends on the pH of the reaction mixture, time of reaction, and the oxidizing agent employed (Sangseethong et al., 2010). Oxidized starch possesses low viscosity, high stability, clarity, film-forming, and binding properties, and these associated properties have increased the industrial food application of oxidized starch (Sánchez-Rivera et al., 2005).

21.6 Potential Use of Pigeon Pea Starches in the Food Industry

Considering the available research reports, pigeon pea starch has shown potential and has been adjudged as a functional ingredient in industrial applications. However, compared to some other pulse starch, there is scant report regarding on its industrial application (food and nonfood industry). Studies have showed the physicochemical, functional, and thermal properties of both native and modified pigeon pea starches (Singh et al., 1989; Lawal, 2011; Hoover et al. ; 1993; Ashogbon & Oluwafemi, 2018; Acevedo et al., 2020; Olagunju, et al., 2020 a, b). The properties displayed exhibited their potential for use in food systems and a few studies have explored the food application potential.

For instance, pigeon pea starch was reportedly used in preparing noodles, and the noodle sensory qualities did not significantly differ from mung bean dhal starch noodles (Singh et al., 1989). Also, both native and modified pigeon pea starch were used as stabilizers in set yogurt; the starches had a positive influence on the water-holding capacity of the yogurt and prevented whey syneresis during refrigerated storage (Olagunju et al., 2020b). Pigeon pea starch has also been used in nonfood applications, specifically in the pharmaceutical industry. The study evaluated the compressional, mechanical, and disintegration properties of paracetamol tablets, comparing pigeon pea starch and plantain starch with conventional corn starch. Results showed that formulations containing pigeon pea starch exhibited the highest bond strength and lowest brittleness, thus, pigeon pea starch may be useful in producing strong tablets with minimal lamination tendency (Dare et al., 2006). Pigeon pea starch was also employed in nonfood applications in a study by Katta et al. (2013). The comparative cohesiveness efficiency of the *Cajanus cajan* seed starch and established binders (potato starch, methyl cellulose, and PVP) was studied. *Cajanus cajan* seed starch compared favorably with other established binders used as they all exhibited complete drug release after 40 minutes. The extracted *Cajanus cajan* seed starch was reported to be a good granulating agent and may be useful as an alternative and effective natural polymer binder in the manufacture of tablets. Pigeon pea starch has similar (in some cases better) properties with other types of starches, hence, it may serve as a substitute or alternative for other starch types, increasing utilization, exploitation, and industrial applications of pigeon pea seed product.

21.7 Conclusion

Pigeon pea is an inexpensive source of protein and carbohydrates as well as a potent source of bioactive peptide and dietary starch. The protein shows competitive nutritional and health-promoting potentials. Also, the starch fraction possesses superior functional and rheological properties, making it a better nontraditional starch source/option for use in industrial food and nonfood applications. Modified starches have numerous possibilities to produce new functional and value added properties as demanded by the food industries.

REFERENCES

Abdel-Rahman, S. M., Eltayeb, A. O. A., Azza, A. A., Feria, M. A. (2011). Chemical composition and functional properties of flour and protein isolates extracted from Bambara groundnut (*Vigna subterranean*). African Journal of Food Science, 5, 82–90.

Acevedo, B. A., Villanueva, M., Chaves, M. G., Avanza, M. V., Ronda, F. (2020). Starch enzymatic hydrolysis, structural, thermal and rheological properties of pigeon pea (*Cajanus cajan*) and dolichos bean (Dolichos lab-lab) legume starches. International Journal of Food Science & Technology, 55, 712–719. doi: 10.1111/ijfs.14334.

Ačkar, D., Babić, J., Jozinović, A., Miličević, B., Jokić, S., Miličević, R., Rajič, M., Šubarić, D. (2015). Starch modification by organic acids and their derivatives: A review. Molecules, 20, 19554–19570.

Adenekan, M. K., Fadimu, G. J., Odunmbaku, L. A., Oke, E. K. (2018). Effect of isolation techniques on the characteristics of pigeon pea (*Cajanus cajan*) protein isolates. Food Science & Nutrition, 6, 146–152.

Adepoju, O. T., Dudulewa, B. I., Bamigboye, A. Y. (2019). Effect of cooking methods on time and nutrient retention of pigeon pea (*Cajanus cajan*). African Journal of Food Agriculture, Nutrition & Development, 19, 14708–14725.

Aggarwal A, Nautiyal U, Negi D. (2015). Characterization and evaluation of antioxidant activity of *Cajanus cajan* and *Pisum sativum*. International Journal of Recent Advances in Science & Technology, 2, 21–26.

Ahmed, S. H., Ahmed, I. A., Eltayeb, M. M., Ahmed, S. O., Babiker, E.E. (2011). Functional properties of selected legumes flour as influenced by pH. Journal of Agricultural Technology, 7, 2091–2102.

Akande, K. E., Abubakar, M. M., Adegbola, T. A., Bogoro, S. E., Doma, U. D. (2010). Chemical Evaluation of the Nutritive Quality of Pigeon Pea [*Cajanus cajan* (L.) Millsp.]. International Journal of Poultry Science, 9, 63–65.

Al-Saeedi, A. H., Hossain, M. A. (2015). Total phenols, total flavonoids contents and free radical scavenging activity of seed crude extracts of pigeon pea traditionally used in Oman for the treatment of several chronic diseases. Asian Pacific Journal of Tropical Disease, 5, 316–321.

Amarteifio, J. O., Munthali, D. C., Karikari, S. K., Morke, T. K. (2002). The composition of Pigeon Pea (*Cajanus cajan*) grown in Bostwana. Plant Foods for Human Nutrition, 57, 173–177.

Ashogbon, A. O. (2014). Comparative study of the physicochemical properties of starch blends-pigeon pea and rice starches versus bambara groundnut and cassava starches. Journal of Current Chemical & Pharmaceutical Sciences, 4, 142–151.

Ashogbon, A. O., Akintayo, E. T. (2013). Morphological and functional properties of starches from cereal and legume: A comparative study. International Journal of Biotechnology and Food Science, 1, 72–83.

Ashogbon, A. O., Oluwafemi, A. D. (2018). Physicochemical properties of microwaved starch blends from cocoyam (*Xanthosoma sagittifolium*) and pigeon pea (*Cajanus cajan*) for industrial applications. Journal of Chemical Society of Nigeria, 43, 189–204.

Ayucitra, A. (2012). Preparation and characterisation of acetylated corn starches. International Journal of Chemical Engineering and Applications, 3, 156–159.

Babic, J., Subaric, D., Ackar, D., Kovacevic, D., Pilizota, V., Kopjar, M. (2007). Preparation and characterization of acetylated tapioca starches. Deutsche Lebensmittel-Rundschau, 103, 580–585.

Barboza, Y., Márquez, E., Parra, K., Piňero, M. P., Medina, L. M. (2012). Development of a potential functional food prepared with pigeon pea (Cajanus cajan), oats and Lactobacillus reuteri ATCC 55730. International Journal of Food Sciences and Nutrition, 63, 813–820.

Bello-Perez, L. A., Agama-Acevedo, E., Zamudio-Flores, P. B., Mendez-Montealvo, G., Rodriguez Ambriz, S. L. (2010). Effect of low and high acetylation degree in the morphological, physicochemical and structural characteristics of barley starch. Food Science & Technology, 43, 1434–1440.

Chalé F. G. H., Ruiz J.C.R., Fernández J. J. A., Ancona D. A. B., Campos M. R. S. (2014). ACE inhibitory, hypotensive and antioxidant peptide fractions from *Mucuna pruriens* proteins. Process Biochemistry, 49, 1691–1698.

Cheboi, J. J., Kinyua, M. G., Kimurto, P. K., Kiplagat, O. K., Nganga, F., Ghimire, S. R. (2019). Biochemical composition of pigeon pea genotypes in Kenya. Australian Journal of Crop Science, 13, 1848–1855.

Chen, Z., Schols, H. A., Voragen, A. G. J. (2004). Differently sized granules from acetylated potato and sweet potato starches differ in the acetyl substitution pattern of their amylase populations. Carbohydrate Polymers, 56, 219–226.

Chugh, C., Ritu. (2018). Variability in polyphenols, antioxidants and mineral composition in different genotypes of pigeon pea (*Cajanus cajan*) grown in India. Chemical Science Review & Letters, 7, 165–169.

Dare, K., Akin-Ajani, D. O., Odeku, O. A., Itiola, O. A., Odusote, O. M. (2006). Effects of pigeon pea and plantain starches on the compressional, mechanical, and disintegration properties of paracetamol tablets. Drug Development & Industrial Pharmacy, 32, 357–365.

Dias, A. R. G., Zavarezea, E. R., Helbig, E., de Mouraa, F. A., Vargas, C. G., Ciacco, C. F. (2011). Oxidation of fermented cassava starch using hydrogen peroxide. Carbohydrate Polymers, 86, 185–191.

Duranti, M., Faoro, F., Harris, N. (1981). The unusual extracellular localization of conglutin Y in germinating *lupinus albus* seeds rules out its role as a storage protein. Journal of Plant Physiology, 143, 711–716.

FAOSTAT (2015). Food and Agriculture Organization of the United Nations Statistics Division. Agricultural production: Pigeon pea. FAOSTAT website: Available from: http://www.fao.org. Assessed August 26, 2020.

Garrido, L. H., Schnitzler, E., Zortéa, M. E. B., de Souza Rocha, T., Demiate, I. M. (2014). Physicochemical properties of cassava starch oxidized by sodium hypochlorite. Journal of Food Science & Technology, 51, 2640–2647.

Ghodsvali, A., Khodaparast, M. H., Vosoughi, M., Diosady, L. L. (2005). Preparation of canola protein materials using membrane technology and evaluation of meals functional properties. Food Research International, 38, 223–231.

Guerra-Araiza, C., Álvarez-Mejía, A. L., Sánchez-Torres, S., Farfan-García, E., Mondragón-Lozano, R., Pinto-Almazán, R., Salgado-Ceballos, H. (2013). Effect of natural exogenous antioxidants on aging and on neurodegenerative diseases. Free Radical Research, 47, 451–462.

Han, F., Liu, M., Gong, H., Lu, S., Zhang, B. (2012). Synthesis, characterization and functional properties low substituted acetylated corn starch. International Journal of Biological Macromolecules, 50, 1026–1034.

Hassan, H. A., Mustafa, A. I., Ahmed, A. R. (2013). Effect of incorporation of decorticated pigeon pea (*Cajanus cajan*) protein isolate on functional, baking and sensory characteristics of Wheat (*Triticum aesitivum*) biscuit. Advance Journal of Food Science and Technology, 5, 976–981.

Hoover, R., Swamidas, G., Vasanthan T. (1993). Studies on the physicochemical properties of native, defatted, and heat-moisture treated pigeon pea (*Cajanus cajan* L) starch. Carbohydrate Research, 246, 185–203.

Jayasena, V., Chih, H. J., Nasar-Abbas, S. M. (2011). Efficient isolation of lupin protein. Food Australia, 63, 306–309.

Joshi, M., Timilsena, Y., Adhikari, B. (2017). Global production, processing and utilization of lentil: A review. Journal of Integrative Agriculture, 16, 2898–2913.

Katta, R., Indranil, G., Bharath, S., Deveswaran, R., Basavaraj, B. V., Varadharajan, M. (2013). Studies on cohesiveness of isolated starch from the seeds of *Cajanus cajan*. International Journal of Pharmaceutical & Phytopharmacological Research, 3, 63–67.

Kesari P, Sharma A, Katiki M, Kumar P, Gurjar B, Tomar S, Sharma, A. K., Kumar, P. (2017). Structural, functional and evolutionary aspects of seed globulins. Protein & Peptide Letters. 24, 267–277.

Klein, B., Vanier, N. L., Moomand, K., Pinto, V. Z., Colussi, R., Zavareze, E. R., Dias, A. R. G. (2014). Ozone oxidation of cassava starch in aqueous solution at different pH. Food Chemistry, 155, 167–173.

Krishnan, H.B., Natarajan, S. S., Oehrle, N.W., Garrett, W. M., Darwish, O. (2017). Proteomic analysis of Pigeon pea (Cajanus cajan) seeds reveals the accumulation of numerous stress-related proteins. Journal of Agriculture & Food Chemistry, 65, 4572–4581.

Kuakpetoon, D., Wang, Y. J. (2006). Structural characteristics and physicochemical properties of oxidized corn starches varying in amylose content. Carbohydrate Research, 341, 1896–1915.

Lawal, O. S. (2011). Hydroxypropylation of pigeon pea (*Cajanus cajan*) starch: Preparation, functional characterizations and enzymatic digestibility. LWT—Food Science & Technology, 44, 771–778.

Lobo, V., Patil, A., Phatak, A., Chandra, N. (2010). Free radicals, antioxidants and functional foods: Impact on human health. Pharmacognosy Reviews, 4, 118–126.

Mahadevamma, S., Shamala, T. R., Tharanathan, R. N. (2004). Resistant starch derived from processed legumes: *In vitro* and *in vivo* fermentation characteristics. International Journal of Food Science & Nutrition, 55, 399–405.

Marcadenti, A., Coelho, R.C.L. (2015). Dietary antioxidant and oxidative stress: Interaction between vitamins and genetics. Journal of Nutrition Health & Food, 3, 1–7.

Messina J. (1999). Legumes and soybeans grown foods: Overview of their nutritional profiles and health. American Journal of Clinical Nutrition, 70, S439–S450.

Muangman, T., Leelamanit, W., Klungsupaya, P. (2011). Crude proteins from pigeon pea (*Cajanus cajan* (L.) Millsp) possess potent SOD-like activity and genoprotective effect against H_2O_2 in TK6 cells. Journal of Medicinal Plants Research 5, 6977–6986.

Murphy, P. (2000). Starch. In: Handbook of Hydrocolloids. Phillips, G. O., Williams, P.A. (Eds.), Abington, England: Woodhead Publishing, 41–65.

Mwasaru, M. A., Muhammad, K., Bakar, J., Man, Y. B. (1999). Effects of isolation technique and conditions on the extractability, physicochemical and functional properties of pigeon pea (*Cajanus cajan*) and cowpea (*Vigna unguiculata*) protein isolates. I. Physicochemical properties. Food Chemistry, 67, 435–443.

Mwasaru, M.A., Muhammad, K., Bakar, J., Man, Y. B. (2000). Influence of altered solvent environment on the functionality of pigeon pea (*Cajanus cajan*) and cowpea (*Vigna unguiculata*) protein isolates. Food Chemistry, 7, 157–165.

Mlyneková, Z., Chrenková, M., Formelová, Z. (2014). Cereals and legumes in nutrition of people with celiac disease. International Journal of Celiac Disease 2, 105–109.

Narina, S. S., Bhardwaj, H. L., Hamama, A. A., Burke, J. J., Phatak, S. C., Xu, Y. (2014). Seed protein and starch qualities of drought tolerant pigeon pea and native tepary beans. Journal of Agricultural Science, 6, 247–259.

Narina, S. S., Xu, Y., Hamama, A. A., Phatak, S. C., Bhardwaj, H. L. (2012). Effect of cultivar and planting time on resistant starch accumulation in pigeon pea grown in Virginia. International Scholarly Research Network. doi:10.5402/2012/576471.

Nwanekezi E. C., Ubbaonu C. N., Arukwe D. C. (2017). Effect of combined processing methods on the proximate and mineral composition of pigeon pea (*Cajanus cajan*) flour. International Journal of Food Science & Biotechnology, 2, 73–79.

Okekunle, M. O., Adebowale, K. O., Olu-Owolabi, B. I., Lamprecht, A. (2020). Physicochemical, morphological and thermal properties of oxidized starches from Lima bean (*Phaseolus lunatus*). Scientific Afican, 8, e00432. doi: 10.1016/j.sciaf.2020.e00432.

Oladebeye, A. O., Oshodi, A. A., Amoo, I. A., Karim, A. A. (2018). Gaseous ozonation of pigeon pea, lima bean, and jack bean starches: Functional, thermal, and molecular properties. Starch, 70, 1700367. doi: 10.1002/star.201700367.

Olagunju, A. I., Omoba, O.S., Enujiugha, V. N., Alashi, A. M., Aluko, R. E. (2018). Antioxidant properties, ACE/renin inhibitory activities of pigeon pea hydrolysates and effects on systolic blood pressure of spontaneously hypertensive rats. Food Science & Nutrition, 6, 1879–1889.

Olagunju, A. I., Omoba, O. S., Enujiugha, V. N., Alashi, A. M., Aluko, R. E. (2020b). Technological properties of acetylated pigeon pea starch and its stabilized set-type yoghurt. Foods, 9, 957. doi: 10.3390/foods9070957.

Olagunju, A. I., Omoba, O. S., Enujiugha, V. N., Alashi, A. M., Aluko, R. E. (2021). Thermoase-hydrolysed pigeon pea protein and its membrane fractions possess in vitro bioactive properties (antioxidative, antihypertensive, and antidiabetic). Journal of Food Biochemistry, 45, e13429. doi: 10.1111/jfbc.13429.

Olagunju, A. I., Omoba, O. S., Enujiugha, V. N., Wiens, R. A., Gough, K. M., Aluko, R. E. (2020a). Influence of acetylation on physicochemical and morphological characteristics of pigeon pea starch. Food Hydrocolloids, 100. doi: 10.1016/j.foodhyd.2019.105424.

Olawuni, I. A., Ojukwu, M., Eboh, B. (2012). Comparative study on the physicochemical composition of pigeon pea (*Cajanus cajan*) flour and protein isolate. International Journal of Agriculture and Food Science, 2, 121–126.

Oshodi, A. A., Ekperigin, M. M. (1989). Functional properties of pigeon pea (*Cajanus cajan*). Food Chemistry, 34, 187–191.

Oshodi, A. A., Olaofe, O. Hall, G. (2009). Amino acid and mineral composition of pigeon pea (*Cajanus cajan*). International Journal of Food Science & Nutrition, 43, 187–191.

Paredes-López, O., Ordorica-Falomir, C., Olivares-Vázquez, M. R. (1991). Chickpea protein isolates: Physicochemical, functional and nutritional characterization. Journal of Food Science, 56, 726–729.

Pazmiño A., Vásquez G., Carrillo W. (2018). Pigeon pea protein concentrate (*Cajanus cajan*) seeds grown in Ecuador functional properties. Asian Journal of Pharmaceutical and Clinical Research, 11, 430–435.

Pehrsson, P. R., Roseland, J. M., Khan, M. (2013). Composition of foods raw, processed, prepared USDA national nutrient database for standard reference, Release 26 Documentation and User Guide. doi: 10.13140/RG.2.1.2026.2645

Peschel, W., Sánchez-Rabaneda, F., Diekmann, W., Plescher, A., Gartzía, I., Jimenez, D., Lamuela-Raventos, R. M., Buxaderas, S., Codina, C. (2006). An industrial approach in the search of natural antioxidants from vegetable and fruit wastes. Food Chemistry, 97, 137–150.

Pham-Huy, L. A., He, H., Pham-Huy, C. (2008). Free radicals, antioxidants in disease and health. International Journal of Biomedical Science, 4, 89–96.

Phaniendra, A., Jestadi, D. B., Periyasamy L. (2015). Free radicals: Properties, sources, targets, and their implication in various diseases. Indian Journal of Clinical Biochemistry 30, 11–26.

Pihlanto, A., Mäkinen, S. 2013. Antihypertensive properties of plant protein derived peptides. In: Bioactive Food Peptides in Health and Disease. Hernández-Ledesma, B., Hsieh, C.-C. (Eds). United Kingdom: InTech Open, 45–72.

Pripp, A. H., Isaksson, T., Stepaniak, L., Sorhaug, T. (2004). Quantitative structure-activity relationship modelling of ACE-inhibitory peptides derived from milk proteins. European Food Research Technology, 219, 579–583.

Punia, S., Dhull, S. B., Sandhu, K. S., Kaur, M., Purewal, S. S. (2020a). Kidney bean (*Phaseolus vulgaris*) starch: A review. Legume Science, 2, e52.doi.10.1002/leg3.52.

Punia, S., Sandhu, K. S., Dhull, S. B., Siroha, A. K., Purewal, S. S., Kaur, M., Kidwai, M. K. (2020b). Oat starch: Physico-chemical, morphological, rheological characteristics and its applications—A review. International Journal of Biological Macromolecules, 154, 493–498.

Punia, S., Dhull, S. B., Sandhu, K. S., Kaur, M. (2019). Faba bean (*Vicia faba*) starch: Structure, properties, and in vitro digestibility—A review. Legume Science, 1 (1), e18, doi.10.1002/leg3.18.

Punia, S. (2020). Barley starch modifications: Physical, chemical and enzymatic—A review. International Journal of Biological Macromolecules, 144, 578–585.

Punia, S. (2019). Barley starch: Structure, properties and in vitro digestibility-A review. International Journal of Biological Macromolecules, 155, 868–875.

Putra, I. D., Marsono, Y., Indrati, R. (2021). Effect of simulated gastrointestinal digestion of bioactive peptide from pigeon pea (Cajanus cajan) tempe on angiotensin-I converting enzyme inhibitory activity. Nutrition & Food Science, 51, 244–254.

Ratnayake, W. S., Hoover, R., Shahidi, F., Perera, C., Jane, J. (2001). Composition, molecular structure, and physico-chemical properties of starches from four field pea (*Pisum sativum* L.) cultivars. Food Chemistry, 74, 189–202.

Ratnayani, K., Suter, I. K., Antara, N. S., Putra, I. N. (2019a). Angiotensin converting enzyme (ACE) inhibitory activity of peptide fraction of germinated pigeon pea (*Cajanus cajan* [L.] Millsp.). Indonesian Journal of Chemistry, 19, 900–906.

Ratnayani, K., Suter, I. K., Antara, N. S., Putra, I. N. (2019b). Effect of in vitro gastrointestinal digestion on the angiotensin converting enzyme (ACE) inhibitory activity of pigeon pea protein isolate. International Food Research Journal, 26, 1397–1404.

Reichert, R. D., Youngs, C. G. (1978). Nature of the residual protein associated with starch fractions from air classified pea. Cereal Chemistry, 55, 469–480.

Rinthong, P., Maneechais, S. (2018). Total phenolic content and tyrosinase inhibitory potential of extracts from Cajanus cajan (L.) Millsp. Pharmacognosy Journal, 10S, 109–112.

Rizkalla, S. W., Bellisle, F., Slama, G. (2002). Health benefits of low glycaemic index foods, such as pulses, in diabetic patients and healthy individuals. British Journal of Nutrition, 88, S255–S262.

Ryan, J. T., Ross, R. P., Bolton, D., Fitzgerald, G. F., Stanton, C. (2011). Bioactive peptides from muscle sources: Meat and fish. Nutrition, 3, 765–791.

Saleh, A., Mohamed, A. A., Alamri, M. S., Hussain, S., Qasem, A. A., Ibraheem, M. A. (2020). Effect of different starches on the rheological, sensory and storage attributes of non-fat set yogurt. Foods, 9, 61. doi:10.3390/foods9010061

Sánchez-Rivera, M. M., García-Suárez, F. J. L., Velázquez del Valle. M., Gutierrez-Meraz, F., Bello-Pérez. L. A. (2005). Partial characterization of banana starches oxidized by different levels of sodium hypochlorite. Carbohydrate Polymers, 62, 50–56.

Sandhu, K. S., Lim, S-T. (2008). Digestibility of legume starches as influenced by their physical and structural properties. Carbohydrate Polymers, 71, 245–252.

Sangseethong, K., Termvejsayanon, N., Sriroth, K. (2010). Characterization of physicochemical properties of hypochlorite- and peroxide-oxidized cassava starches. Carbohydrate Polymers, 82, 446–453.

Saxena, K. B., Sawargaonkar, S. L. (2015). Genetic enhancement of seed proteins in Pigeon pea methodologies, accomplishments, and opportunities. International Journal of Science Research, 4, 3–7.

Schlegel, K., Sontheimer, K., Hickisch, A., Wani, A. A., Eisner, P., Schweiggert-Weisz, U. (2019). Enzymatic hydrolysis of lupin protein isolate-changes in the molecular weight distribution, technofunctional characteristics, and sensory attributes. Food Science & Nutrition, 7, 2747–2759.

Simsek, S., Ovando-Martínez, M., Whitney, K., Bello-Pérez, L.A. (2012). Effect of acetylation, oxidation and annealing on physicochemical properties of bean starch. Food Chemistry, 34, 1796–1803.

Singh, I. P. (2016). Nutritional benefits of pigeon pea. In: Biofortification of Food Crops. Singh, U., Praharaj, C. S., Singh, S. S., Singh, N. P. (Eds.). Springer New Delhi, 73–81. doi: 10.1007/978-81-322-2716-8-7.

Singh, T., Bains, G. H. (1988). Legumes: Type and uses. In: A handbook of Food of Plant Origin, 4th ed. New York: Blackie Academic & Professional, 214–217.

Singh, U., Voraputhaporn, W., Rao, P.V., Jambunathan, R. (1989). Physicochemical characteristics of pigeon pea and mung bean starches and their noodle quality. Journal of Food Science, 54, 1293–1297. doi: 10.1111/j.1365-2621.1989.tb05976.x

Teymoori, F., Asghari, G., Mirmiran, P., Azizi, F. (2017). Dietary amino acids and incidence of hypertension: A principle component analysis approach. Science Reports, 7, 16838.

Tiwari, B. K., Tiwari, U., Mohan R. J., Alagusundaram, K. (2008). Effect of various pre-treatments on functional, physiochemical, and cooking properties of pigeon pea (*Cajanus cajan* L). Food Science & Technology International, 14, 0487–0495.

Torres, A., Frias, J., Granito, M., Vidal-Valverde, C. (2006). Fermented pigeon pea (Cajanus cajan) ingredients in pasta products. Journal of Agricultural & Food Chemistry, 54, 6685–6691.

Tzitzikas, E. N., Vincken, J. P., De Groot, J., Gruppen, H., Visser, R. G. (2006). Genetic variation in pea seed globulin composition. Journal of Agricultural & Food Chemistry, 54, 425–33.

Uchegbu, N. N., Ishiwu, C. N. (2016). Germinated pigeon pea (*Cajanus cajan*): A novel diet for lowering oxidative stress and hyperglycemia. Food Science & Nutrition, 4, 772–777.

Udenigwe C. C., Aluko R.E. (2011). Chemometric analysis of the amino acid requirements of antioxidant food protein hydrolysates. International Journal of Molecular Sciences, 12, 3148–3161.

Valdez-Ortiz, A., Fuentes-Gutiérrez, C. I., Germán-Báez, L. J., GutiérrezDorado, R., Medina-Godoy, S. (2012). Protein hydrolysates obtained from Azufrado (sulphur yellow) beans (Phaseolus vulgaris): Nutritional, ACE-inhibitory and antioxidative characterization. Food Science & Technology, 46, 91–96.

Vanier, N. L., Zavareze, E. R., Pinto, V. Z., Klein, B., Botelho, F. T., Dias, A. R., Elias, M. C. (2012). Physicochemical, crystallinity, pasting and morphological properties of bean starch oxidized by different concentrations of sodium hypochlorite. Food Chemistry, 131, 1255–1262.

Vogelsang-O'Dwyer, M., Bez, J., Petersen, I. L., Joehnke, M. S., Detzel, A., Busch, M., Krueger, M., Ispiryan, L., O'Mahony, J. A., Arendt, E. K., Zannini, E. (2020). Techno-functional, nutritional and environmental performance of protein isolates from blue Lupin and white Lupin. Foods, 9, 230. doi.10.3390/foods9020230.

Wani, I. A., Sogi, D., Hamdani, A. M., Gani, A., Bhat, N. A., Shah, A. (2016). Isolation, composition and physicochemical properties of starch from legumes: A review. Starch-Starke, 68, 1–12.

Wojeicchowski, J. P., Siqueira, G. L., Lacerda, L. G., Schnitzler, E., Demiate, I. M. (2018). Physicochemical, structural and thermal properties of oxidized, acetylated and dual-modified common bean (*Phaseolus vulgaris* L.) starch. Food Science & Technology, Campinas, 38, 318–327.

Xiong, Z. H., Fei, P. (2017). Physical and chemical modification of starches: A review. Critical Reviews in Food Science & Nutrition, 57, 2691–2705.

Yadav, D. K., Patki, P. E. (2015) Effect of acetyl esterification on physicochemical properties of chick pea (Cicer arientinum L.) starch. Journal of Food Science & Technology, 52, 4176–4185.

Zehadi, M. J. A., Masamba, K., Li, Y., Chen, M., Chen, X., Sharif, H. R., Zhong, F. (2015). Identification and purification of antioxidant peptides from lentils (Lens Culinaris) hydrolysates. Journal of Plant Science, 5, 123–132.

Zhang, Y.-R., Wang, X.-L., Zhao,G.-M., Wang, Y-Z. (2012). Preparation and properties of oxidized starch with high degree of oxidation. Carbohydrate Polymers, 87, 2554–2662.

22

Kidney Beans: Nutritional Properties, Biofunctional Components, and Health Benefits

Arashdeep Singh, Antima Gupta and Savita Sharma
Punjab Agricultural University, Ludhiana, Punjab, India

CONTENTS

22.1 Introduction

Kidney beans, often classified as common beans, are scientifically known as *Phaseolus vulgaris,* belong to the *Leguminosae* class, family *Phaseoleae*, and subfamily *Papilionoideae.* Kidney beans are annual monocarpic plants that are typically cultivated in a monocrop system and harvested mechanically. Kidney beans are primarily cultivated in temperate, tropical, and subtropical regions in most of the developed and devolving countries around the world. For most of the world's population, kidney beans are a most important grain legume in the daily human diet. Kidney bean pods can be harvested at both the stages before or after physical maturity. Green pods harvested before maturity are used as vegetables, while dry beans after maturity are used as a source of daily dietary proteins in many developing countries. After harvesting at maturity, the leftover plant and leaves were used as animal fodder (Hayat et al., 2014; Ganesan & Xu 2017).

On the basis of chronological, archeological, botanical, and linguistic evidences available, kidney beans were reported to be originated, domesticated, and cultivated in central Peru and western Mexico of the Central American region around 8,000 years ago. From there, they were taken to southwestern Europe, the Mediterranean region, eastern Africa, and Asia. At present, kidney beans are cultivated and consumed throughout the world (Hayat et al., 2014; Ganesan & Xu 2017). As per the Food and Agricultural Organization of the United Nations, (FAO), India is the leading producer of dry beans in the world, producing 5.8 million tons, followed by Myanmar, Brazil, the United Statesand Mexico, which produce 4.9, 3.0, 1.3, and 1.2 million tons, respectively, during 2013–2017 (FAOSTAT, 2019). Kidney beans, also termed the "king of nutrition" and "poor man's meat," is popularly known as "rajma" in the Indian subcontinent and is consumed primarily for its savory texture and health benefits (Parmar et al., 2014). Kidney beans were named because of their shape and color similar to human kidneys. Kidney beans are large-sized, firm-textured grains and the color of kidney beans ranges from very light red to very dark, almost purple, with glossy skin. Kidney beans make up a major part of the diet in many countries and they were consumed traditionally in cooked form either alone or in combination with other legumes such as rice or in canned form (Yasmin et al., 2008, Wani et al., 2010; 2013; 2015).

22.2 Nutritional Value

Kidney beans are the most important source of daily dietary nutrients for the population of many developed and developing countries around the world as it is rich and inexpensive source of protein and carbohydrates (Wani et al., 2010; Punia et al., 2020). Nutritionally, kidney beans are rich sources of protein that accounts for 15–30% and 50–60% of complex carbohydrates on dry matter basis. Being low in fat content that ranges from 0.5–1.5%, they are rich source of minerals, folates and dietary fibers. Of the total fat present in kidney beans, total saturated, monosaturated, and polysaturated fatty acid content is 0.106, 0.056, and 0.403 g/100 on dry weight basis, respectively. Proteins, which are the second largest component of the grains, exhibit fairly good amino acid composition, which helps to make the kidney bean is a good candidate for the nutritious foods formulation (Mundi & Aluko, 2012), even so, it is low in methionine and tryptophan and sulfur-containing amino acids. Lysine, leucine, aspartic acid, glutamic acid, and arginine are predominantly present in kidney beans. Red kidney has 10.2 g glutamic acid, 9.5 g aspartic acid, 7.2 g leucine, 7 g lysine, 6.9 g arginine, 5.2 g glycine, 4.6 g phenylalanine, 4.4 g alanine, 4.1 g valine, .7 g isoleucine, 3.4 g threonine, 3.3 g proline, 3.1 g tyrosine, 3.1 g serine, 3 g histidine, 1.7 g methionines and 1.2 g cysteine per 100 of protein on dry matter basis (Shehzad et al.,2015). Phaseolin and vicilin are the main storage protein of the kidney bean, and out of these, the major portion is contributed by Phaseolin which is also known as G1 globulin. Phaseolin is an oligomeric protein that consists primarily of alpha- (α), beta- (β), and gamma (γ)-phaseolin with molecular weight ranges from 43–53 kDa (Yin et al., 2010). Vicilin often referred to as phaseolin, is a 7S globulin that comprises 3–5 subunits. Legumin, an 11–12S, globulin, usually exists with vicilin (Wani et al., 2015). Glycoprotein I and glycoprotein II are the other minor and major storage proteins present in kidney bean.

Of the 50–60% of total carbohydrates present in kidney beans, starch accounts for about 35–40% while dietary fiber accounts for 15–20% out of which 40% are soluble and 60% are insoluble dietary fiber (Shehzad et al., 2015). Kidney bean starch has higher amount of resistant starch content and lower amount of slow digestible and rapidly digestible starch content. Kidney bean starch molecules are round, ovoid, and irregular in shapes and higher pasting viscosity in comparison to cereal starches. The ash content in kidney bean varies from 2.5–5.9%. Iron (Fe) content in kidney bean ranges from 0.064–1.324 mg/kg, manganese (Mn) content from 0.1–27.1 mg/kg, copper (Cu) content from 0.15–49.5 mg/kg, sodium (Na) from 1.5–32.15 mg/kg, zinc (Zn) from 0.006–1.88 mg/kg, potassium (K) from 3.65–242.45 mg/kg, magnesium (Mg) from 1.9–179.1 mg/kg, and calcium (Ca) content ranges from 12.0–432.0 mg/kg (Parmar et al., 2014). Kidney beans are also a rich source of ascorbic acid (4.5 mg/100 g), Niacin (2.11 mg/100 g) along with folic acid 394.1 µg and Vitamin A. Kidney beans also contain substantial amounts of many antinutritional components that exert both beneficial and harmful biological properties such as lectins, phytic acid, tannins, trypsin inhibitors, alpha-amylase inhibitors, saponins, phytoheamagglutinins, oligosaccharides such as raffinose, stachyose, and verbascose, which diminish the bioavailability and digestibility of proteins, carbohydrates, and minerals (Shimelis & Rakshit, 2007).

22.3 Bioactive Compounds

Antioxidant activity in terms of 2,2-diphenyl-1-picryl-hydrazyl-hydrate (DPPH) inhibition activity among 48 accessions of kidney beans grown in India showed wide variations. Parmar et al., 2014 reported that antioxidant activity of cotyledons among 48 accessions of kidney bean varied from 0.69–21.36%, while bran from 48 accessions of kidney bean ranges between 2.14–94.24%. Cotyledons from dark-colored beans exhibit higher antioxidant activity in comparison to cotyledons form white or light-colored kidney beans. Similarly, the total phenolic content also varied with the color of the kidney bean and it ranges from 0.07–7.10 mg GAE/g among 48 accessions of kidney bean studied (Parmar et al., 2014).

Zhao et al. (2014) in their study reported that the black kidney bean and red kidney bean had total phenolic content of 32.9 mg GAE/g and 27.1 mg GAE/g extract, also showing that dark-colored beans showed higher bioactive compounds in comparison to light-colored kidney beans. In another study, different kidney bean varieties showed wide variations among total phenolic and total flavonoid content that ranges from 0.25–35.11 mg GAE/g and 0.19–7.05 mg RE/g, respectively (Kan et al., 2016). The seed coat from red kidney bean also had total flavonoid content of 26.35 mg CE/100 g (Gan et al., 2016), while red kidney beans in whole showed a total flavonoid content of 3.39 mg CE/g (Xu et al., 2007). The total phenol content among different cultivars of kidney beans vary from 2.70–4.59 mg GAE/g, while that of ortho-Diphenols varies from 2.50–6.69 mg GAE/g and for flavonoids it ranges from 0.80–4.33 mg CE/g (Carbas et al., 2020).

Antioxidant activity was 5.9 μmol Trolox/g in kidney beans, 5.5 μmol Trolox/g in red kidney beans, 212 μmol TE/g in red kidney bean seed coat, 288 μmol TE/g in violet red kidney bean coat, 71.1 μmol TE/g in big speckled kidney beans, 118 μmol TE/g in big speckled kidney bean coat (Wang et al., 2015; Gan et al., 2016). Kan et al., (2016) also showed wide variations in the antioxidant activity of different morphological fractions of 26 kidney bean varieties. The antioxidant activity in 26 varieties of seed coat varies between 4.00–491.33 μmol Trolox/g, while in whole kidney beans it varies between 1.07–7.48 μmol Trolox/g. Black and red kidney bean also showed antioxidant activity of 602 U/g and 516 U/g (Zhao et al., 2014), while white kidney beans showed antioxidant activity of 217.97 μmol TE/g (García-Lafuente et al., 2014). Aldaric derivatives of ferulic, hydroxycinnamic acids, p-hydroxybenzoic acid, p-coumaric, and sinapic acids are main phenolic acids identified in white and red kidney bean and they account for nearly about 50% of the total phenolic content (García-Lafuente et al., 2014; Dueñas et al., 2015).

Total anthocyanin in 16 Korean kidney bean cultivars was found to be in range of 0.00–2.78 mg/g and the major content among the anthocyanins compounds were of cyanidin 3,5-diglucoside, delphinidin 3-glucoside, cyanidin 3-glucoside, petunidin 3-glucoside, and pelargonidin 3-glucoside (Choung et al., 2003). Among the seed coats from 26 kidney bean cultivars, cyaniding content ranges from 0–1.44 mg/g, pelargonidin from 0–0.71, petunidin from 0–0.41, malvidin from 0–0.27, delphinidin from 0–4.45 and the total anthocyanindin from 0–5.84 mg/g (Kan et al., 2016). Pelargonidin 3-O-glucoside, petunidin 3-O-(6″-acetyl-glucoside), pelargonidin 3,5-O-diglucoside, delphinidin 3-O-glucosyl-glucoside, cyanidin-3-O-glucoside, and malvidin-3-O-glucoside are the predominant anthocyanins reported in kidney beans (Dueñas et al., 2015). Among the flavanones, naringenin and hesperitin were most abundantly present in white kidney beans as flavanon-glycoside and they were present in the concentration of 11.30 μg/g and 0.14 μg/g, respectively (García-Lafuente et al., 2014). The hard-to-cook phenomena in the kidney bean also exerts a profound effect on its total phenolic content. The total phenolic content in easy-to-cook (ETC) grains among different accessions of kidney beans vary from 1.00–2.29 mg GAE/g, while in hard-to-cook (HTC) grains, total phenolic content was higher and ranges between 1.18–2.87 mg GAE/g (Parmar et al., 2017). Catechin, chlorogenic acid, protocatechuic acid, and p-coumaric acids were found to be predominantly present in both HTC and ETC grains from different kidney bean accessions. Gallic acid was higher in ETC kidney bean grains, while HTC kidney bean grains showed higher content of chlorogenic acid (Parmar et al., 2017). Among the light-colored kidney bean, *p*-Coumaric acid derivatives, ferulic acid derivatives, sinapic acid, ferulic acid, quercetin 3-*O*-glucoside, quercetin 3-O-rutinoside, quercetin 3-O-(6″-O-malonyl) glucoside, kaempferol 3-O-glucoside, quercetin, kaempferol, quercetin 3-Oxylosylglucoside, and kaempferol 3-O-xylosylglucoside were the major phenolic acids and flavonoids present, while among dark red kidney beans, *p*-Coumaric acid derivatives, ferulic acid derivatives, sinapic acid, ferulic acid, quercetin diglycosides, kaempferol diglycosides, quercetin, and kaempferol were present. Quercetin diglycosides, kaempferol diglycosides, *p*-coumaric acid derivatives, ferulic acid derivatives, sinapic acid, ferulic acid, quercetin 3-O-xylosylglucoside, kaempferol 3-O-xylosylglucoside, quercetin, and kaempferol were present in pink kidney beans (Lin et al., 2008). Montcalm dark red kidney beans contained 3′, 4′, 5, 7-Tetrahydroxyflavonol 3-O-β-D-glucopyranosyl (2→1) O-β-D-xylopyranoside, quercetin 3-O-β-D-glucopyranoside, kaempferol 3-O-β-D-glucopyranoside (Beninger & Hosfield, 1999). Delphinidin 3-O-glucoside, pelargonidin 3-O-glucoside, cyanidin 3-O-(6″-malonyl) glucoside, cyanidin 3, 5-diglucoside were present in red kidney beans while black kidney bean contained Delphinidin 3-O-glucoside and petunidin 3-O-glucoside (Choung et al., 2003). In white kidney beans, Naringenin derivatives, hesperitin derivatives, feruroyl aldaric acid, coumaroyl aldaric acid, sinapic aldaric acid, sinapic derivatives, sinapic acid, and ferulic acid were the major phenolic acids and flavonoids present (Garcıa-Lafuente et al., 2014).

22.4 Antinutrient Factors and Processing Methodologies to Reduce them

The kidney bean and its various botanical fractions contain a wide range of antinutrient components such as phytic acid, tannins, saponins, trypsin inhibitors, phytoheamagglutinins and α-galactosides, which hinder and limit the nutrient utilization, digestibility, and bioavailability. Tannin content in the seed coat of 26 varieties of kidney beans ranged from 10.52–57.13 mg/PAE (Kan et al., 2016), while in the study of Gan et al. (2016), tannin content in red kidney bean seed coat was 46.7 mg CE/g, violet red and big speckled kidney bean coats ranged 57.9–79.5 mg CE/g, and small speckled kidney bean seed coat ranged from 28.1–44.7 mg CE/g. Xu et al., (2007 reported 2.87 mg CE/g of tannin content in red kidney bean. The HTC phenomena in the kidney bean also exerts profound effect on its total tannins and phytic acid content. The tannin content in ETC grains among different accessions of kidney beans vary 0.03–1.26 mg/g, while in HTC grains tannin content was lower and ranged from 0.03–1.18 mg/g. The phytic acid content in ETC grains among different accessions of kidney bean varies 1.99–2.21 mg/g, while in HTC grains, tannin content was lower and ranges between 1.76–2.13 mg/g (Parmar et al., 2017). Shang et al., (2016) in their study on diversity of four antinutritional factors in 56 common bean cultivars reported that the lectin content varies from 0.025–10.850 mg/g and the content of saponins and phytic acid ranges between 0.966–7.856 and 2.065–5.120 mg/g, respectively. They also reported that the activity of trypsin inhibitors ranges between 0.016–15.947 mg/g. Soaking, cooking, autoclaving, germination, fermentation, and canning are some of the processing technologies used to decrease the antinutritional factors in the kidney beans. During the processing of kidney beans, phytic acid, saponin and tannin content after household cooking was found to be 627.33 mg/100 g, 106.02 mg/100 g, and 6.78 mg/100 g, while after canning changed to 386.32 mg/100 g, 118.6 mg/100 g, and 5.44 mg/100 g, respectively, showing that canning helps to reduce the phytic acid and tannin content while the content of saponins increased (Margier et al., 2018).

Kidney beans have 0.2 mg/kg of cyanide content, which was reduced up to 25% by cooking, 20% by germination, 7.7% by soaking in water, and 8.7% by soaking in a citric acid solution (Yasmin et al., 2008). Raw kidney bean have a tannin content of 6.1 mg/g, which was significantly reduced by 81–92% cooking of grains soaked in water, citric acid, and sodium bicarbonate solutions, while 63%, 68%, and 69% reduction was observed while only soaking with citric acid, soaking with sodium bicarbonate solution, and through germination, respectively. Similarly, the phytic acid content, which was 6.1 mg/g in raw kidney bean seed was significantly reduced by soaking and germination from 5.7–43%. Shimelis & Rakshit (2007) reported that hydration, autoclaving, germination, and cooking singly or in combinations effectively help to remove various antinutritional factors. They reported that 25% reduction in water soaking, 27% in soaking with sodium bicarbonate, 33–76% during 24–96 hours of sprouting, 34% with cooking, 70% with soaking and cooking, 68% with soaking in sodium bicarbonate and cooking, 72% with autoclaving, 75% with soaking in water and sodium bicarbonate and autoclaving, 97–100% with sprouting for 24-96 hours along with autoclaving. Similarly, the phytic acid, which was 23.51 mg/g in unprocessed seeds, was reduced by 18% in water soaking, 14% in soaking with sodium bicarbonate,

73–92% during 24–96 hours of sprouting, 25% with cooking, 64% with soaking and cooking, 65% with soaking in sodium bicarbonate and cooking, 66% with autoclaving, 65% with soaking in water and sodium bicarbonate and autoclaving, 98–100% with sprouting for 24–96 hours along with autoclaving.

22.5 Health Benefits of Kidney Beans

Epidemiological and clinical studies have shown positive effects of kidney bean consumption in lowering the risk of coronary heart diseases and cardiovascular diseases, management of obesity, diabetes, antiradical activity, antifungal activities, hypolipidemic activities, antioxidant and anti-inflammatory activities, cardioprotective activity, antihepatotoxic activity, hypoglycemic effects, and chemoprotective effects among many others (Nolan et al., 2020). The health benefits of kidney beans are primarily due to presence of dietary fiber, more resistant starch, various phenolic acids, flavonoids, tannins, enzyme inhibitors such as protease and amylase inhibitors, and lectins, etc. (Ganesan & Xu, 2017).

Studies of Amarowicz et al. (2008) showed that phenolic acids p-coumaric, ferulic and sinapic acids, quercetin, kaempferol, procyanidins B-2 and B-3, and tannins present in kidney beans exert antiradical activity when used at dosage level of 62.5–500 ug/mL for 36–48 hours in *Brochothrix thermosphacta*, *Staphylococcus aureus*, *Listeria monocytogenes* Scott A, *Salmonella typhimurium*, E. *coli* O157: H7, *Pseudomonas fragi*, and *Lactobacillus plantarum*. Total phenolic and anthocyanins present in kidney beans showed antidiabetic and antiobesity activity, at dosage levels of 50 g/body weight when subjected to overweight subjects using a randomized, double-blind, placebo-controlled clinical trial during 60 days trial period (Rondanelli et al., 2011). The Phenolic acids and bioactive peptide fractions present in different botanical fractions of kidney bean showed antidiabetic activity during *in vitro* studies when used at dosage levels of <1 kDa, 1–3.5 kDa, 3.5–5 kDa, 5–10 kDa, and 10 kDa during 24 hours of experimental studies (Oseguera-Toledo et al., 2011). Kyznietsova et al., (2015) also found that phenolic, tannins, and anthocyanins present in kidney beans showed potential antidiabetic activity in Wistar albino rats when fed at dosage level of 200 mg/kg of body weight for 30 days. Similarly, the phenolics present in white kidney beans also showed antidiabetic activity in Wister albino rats when administered at dosage level of 50 mg/kg of body weight for 7 days due to the alpha-amylase inhibitor present in kidney beans (Tormo et al., 2004; Tormo et al., 2006).

Presence of phenolic acids such as chlorogenic acid, gallic acid, p-hydroxy benzoic acid, caffeic acid, protocatechuic acid, p-coumaric acid, rosmarinic acid, ferulic acid, sinapic acid, and ellagic acid and flavonoids such as epicatechin, catechin, gallo-catechin gallate, epigallocatechin gallate, quercetin, hesperidin, and rutin in kidney beans showed antidiabetic, hypolipidemic, and cardioprotective activity in male Wister rats when administered orally at dosage levels of 0.4 g/kg, 0.8 g/kg, and 1.2 g/kg of body weight up to 21 days (Pérez-Ramírez et al., 2018). Kidney bean lectins and polyphenols showed potential antifungal activity against *Fusarium oxysporum, Coprinus comatus*, and *Rhizoctonia solani* when used at concentration of 20–200 ul/mL for 24 hours (Ye et al., 2001). Daniell et al., (2012) reported that phenolic present in kidney beans also showed antihepatotoxic effect in Sprague-Dawley rats at dosage levels of 7.5%, 15%, 30%, or 60% w/w during 7 days. Anti-inflammatory activity of white and red bean was also reported by García-Lafuente et al., (2014) in macrophage cell line RAW264.7 due to phenolic-rich extract that contains ferulic, coumaric, sinapic acid, catechin, malvidin 6-O-glucoside, and quercetin present in them at the 20 uL dosage level. Antimutagenic activity of (+)-catechin present in kidney bean phenolic extract was also reported by Cardador-Martinez et al., (2002) against aflatoxin B1in *salmonella typhimurium* strains TA98 and TA100 at dosage levels of 2.5 ug, 5 ug, 10 ug, 12.5 ug, 15 ug, and 25 ug. Zhu et al., (2012) reported that the presence of quercetin, kaempferol, p-coumaric acid, ferulic acid, p-hydroxybenzoic acid, and vanillic acid in kidney bean helps to modulate cardiovascular risk factors in Sprague-Dawley rats and a diet-induced obesity in C57Bl/6 mice when administered in the diet at dosage levels of 7.5%, 15%, 30%, or 60% w/w for 7 days.

Antioxidant and anti-inflammatory activities of flavonoids present in fermented kidney beans were also documented by Gabriele et al. (2016) as it down-regulates LOX-1, CHOP, and ICAM-1 in HMEC-1 lines stimulated by ox-LDL. Phenolic acids, flavonoids, and anthocyanins present in dark and white kidney bean showed various antioxidant and anti-inflammatory activities in C57BL/6 mice and helped to reduce colonic mucosal damage and inflammation in response to dextran sodium sulfate. Phenolics and tannins present in kidney beans showed chemoprotective effects on breast cancer in a dose-dependent manner by inhibition of rat mammary carcinogenesis along with tissue lipid metabolism in female Sprague-Dawley rats at dosage levels of 0, 7.5%, 15%, 30%, or 60% w/w for 46 days (Thompson et al., 2008; Mensack et al., 2012). Similarly, Campos-Vega et al., 2012 reported that kidney bean tannins had chemoprotective effects on colon cancer as they modifies protein expression associated with apoptosis, cell cycle arrest, and proliferation in Human HT-29 cell lines. Fantini et al. (2009) reported that flavonoids present in kidney bean exert a hypoglycemic and antiobesity effect in male Wister rats when administrated at dosage level of 50 mg/kg, 200 mg/kg, or 500 mg/kg body weight for 21 days.

Song et al., (2016) reported that when 48 male C57BL/6J mice were fed with 50 mg/kg per day of white kidney bean extract along with a high fat diet (45%) for 98 days, the mice showed reduction in weight gain, visceral fat, total, HDL and LDL cholesterol, serum adiponectin, glucose and insulin while they showerd an increased the *Verrucomicrobia* and *Actenobacteria* (phylum level) and gut microbiota, which include *bifidobacterium, lactobacillus,* and *akkermansia* genus level in comparison to a high fat diet (HFD) alone. Shi et al., (2020) in their study on Sprague-Dawley rats fed with HFD-induced obesity reported that when the diet was intervened with white kidney bean extract at 1.0% and 1.5%, it decreased the body weight, food intake, HFD-induced intra-abdominal fat accumulation, HFD-induced increase in serum triglycerides and LDL cholesterol, and increased total short chain fatty acids (SCFAs), acetic acid, propionic, and isobutyric acid content. Micheli et al., (2019) also reported that when a HFD was intervened with 500 mg/kg of white kidney bean extract, C57BL/6 mice having HFD-induced metabolic syndrome showed reduction in weight, total and LDL cholesterol, plasma glucose, insulin tolerance, plasma insulin, triglycerides, oxidative stress markers and protein carbonylation in plasma, while it increased the cardiac antioxidant enzymes such as catalase reductase and glutathione reductase in comparison to HFD. Tormo et al., (2006) also reported reduction in weight gain and blood glucose levels when white kidney bean extract was induced into the diet at 100 mh/kg of body weight in nondiabetic and type 2 diabetic male Wistar rats. Reduction in body weight, glycemia, alkaline phosphatase, serum urea, cardiac oxidative stress markers such as superoxidase dismutase, catalase, malondialdehyde, and cardiac collagen deposition was also observed when diabetes-induced male Wistar rats were feed a diet containing 100–1500 mg/kg of white kidney bean extract in standard extruded chow and water ad libitum (Oliveira et al., 2014). Reduction in blood glucose above baseline levels, food intake, body weight, and glycemia with white kidney bean extract was also reported by Preuss et al. (2007) and Carai et al. (2011).

22.6 Functional Properties

Legume flours can be used to improve the nutritional value of foods such as baby foods, bakery, and extruded products. However, their use as an ingredient in these products depends on their functional properties, which affect sensory characteristics and the physical behavior of foods (Kaur and Sandhu, 2010). These properties of legume flours have been related to the proportion of as proteins, lipids, carbohydrates and other components such as pectin compounds (Kaur and Sandhu, 2010; Sharma et al., 2019).

22.6.1 Protein Solubility

Protein solubility is one of the most crucial functional properties as it influences other related functional properties such as emulsifying, foaming and gelation, and whipping properties (Singh et al. 2017a). The structure of protein, extent of denaturation, temperature, pH, ionic strength, method of preparation, post extraction treatments, and protein-protein and protein-solvent interactions are significantly influence by the solubility of protein (Mizubuti et al., 2000;). Kidney bean protein primarily shows a U-shaped curve and the highest solubility at low acidic and high alkaline pH values, while their solubility markedly decreases near the isoelectric point, pH 4–5. The protein solubility of flours from different varieties of kidney beans range between 6.8–7.8% at pH 4.0, which increases and ranges from 73.3–80.3% at pH 2.0 and 95.3–97.8% at 10.0 pH (Wani et al., 2013).

The protein solubility of protein isolates from different cultivars and accessions of kidney beans ranges from 65.8–78.8% at pH 2 and 65.8–78.5% at pH 7.0, while at pH 9.0 it increased significantly and ranges from 75.1–94.6%. The protein solubility of kidney bean protein isolates at pH 5.0 ranges between 3.4–7.4% (Shevkani et al., 2015; He et al., 2018). Studies by Wani et al. (2015) reported that protein solubility of isolates from different kidney bean cultivars is 73.22–87.41% at pH 2.0, 1.63–5.60% at pH 5.0, and 86.38–93.26% at pH 10.0 while hydrolyzation of protein isolates increased protein solubility, which ranges between 77.24–86.40% at pH 2.0, 2.63– 9.80% at pH 5.0 and 96.21–99.77% at pH 10.0. Protein solubility of kidney bean protein isolates at pH values of 7.0 is 68.2, which increased to 78.0% at pH 8.0 and then to 81.4% at pH 9.0 (Shevkani & Singh, 2015). While study of Tang and Ma (2009) and Tang et al. (2009) reported protein solubility of 91% for native kidney bean protein isolate at pH 7. He et al., (2018) reported that PEGylation of black kidney bean protein isolate with succinimidyl carbonate, succinimidyl succinate, and succinimidyl propionate conjugation increased its solubility in comparison to its native counterpart at pH range of 2–12. The albumin and globulin fractions of kidney bean proteins varied significantly for their solubility, in which globulin showed minimum values near pH 4–6, whereas albumin has more than 65% solubility and it ranges up to 85% at all the pH ranges (Mundi & Aluko, 2012). The phaseolin-rich protein from the kidney bean has a higher protein solubility of around 99.0% at pH 2.0 and 10.0 in comparison to the native protein solubility of native protein isolate. However, at the isoelectric pH of 4–6.0, the protein solubility of native kidney protein isolate was higher in comparison to the phaseolin-rich protein fraction of kidney protein (Yin et al., 2010).

22.6.2 Water Absorption Capacity

Water absorption capacity (WAC), a very important functional property of flour and protein, is the power to physically hold water against gravity. Polar amino acid residues and carbohydrates such as polysaccharides are hydrophilic and have an affinity for water molecules Singh and Sharma 2017) and differences in the WAC of flour and proteins from pulses could be due to the differences in the content of these hydrophilic parts (Ghadivel and Prakash, 2006). Flours with higher WAC have a great potential to be used in the preparation of bakery products as higher WAC improves the dough-handling characteristics and also maintains the freshness of bread (Kaur et al. 2020; Singh et al 2017b).

Kidney bean flours were reported to have a WAC in the range of 1.07–2.7 g/g (Kaur and Singh, 2007; Siddiq et al., 2010; Aguilera et al., 2011; Wani et al., 2013; Du et al., 2014). Wani et al. (2013) reported that the WAC of kidney bean flour from different Indian cultivars varied from 2.6–2.7 g/g, however for the different kidney beans cultivars grown in the Indian temperate climate, WAC varied significantly between the flours and ranged 1.38–1.53 g/g (Wani et al., 2020). Higher WAC provides information about the presence of higher proportions of hydrophilic residues in the carbohydrate and protein constituents of kidney bean flours (Wani et al., 2013).

The WAC values of kidney bean protein isolates in different studies range from1.6–6.16 g/g (Mundi & Aluko, 2012; Shevkani et al., 2015; Wani et al., 2015). The water absorption capacity of native protein isolated from different Indian kidney bean cultivars varied from 5.34–5.85 g/g (Wani et al., 2015). Hydrolyzation of protein isolates for 30 minutes increased their WAC from 5.48–6.16 g/g, while proteolysis for 60 minues further increased the WAC of the isolates from 5.64–6.30 g/g. Protein isolates from different kidney beans have WACs of 2.07 g/g and 2.33 g/g at 25°C and 70°C and it increased with an increase in intensity of pressure reaching to 2.56 g/g and 3.21 g/g, at 25°C and 70°C, respectively (Ahmed et al., 2018). PEGylation of black kidney bean protein isolate with succinimidyl carbonate, succinimidyl succinate, and succinimidyl propionate conjugation decreased the WAC of kidney bean protein isolates. Native kidney bean protein isolates have WAC of 6.05 g/g, which decreased with PEGylation of isolates and ranges between 4.70–5.50 g/g (He et al. 2018). The albumin fraction of kidney bean protein has a higher WAC of 3.4 mL/g, while the globulin fraction has a lower WAC of 2.56 mL/g (Mundi & Aluko 2012). The differences in WAC among native kidney bean protein isolates are primarily due to diversity among their protein conformations and the difference in number and nature of water-binding sites on protein molecule structure (Chou & Morr, 1979).

22.6.3 Oil Absorption Capacity

Oil absorption capacity (OAC) is also an important functionality and it has been attributed to the physical entrapment of oil. OAC helps in flavor retention and enhanced the mouthfeel of the product where fat absorption is desired (Surasani et al. 2019) and it also reflects the emulsifying potential of the flour or protein. The OAC of flour or protein is influenced by the content and size of protein, composition of amino acid, conformation of protein, its surface hydrophobicity, and liquidity of the oil. Variations in the presence of many nonpolar side chains of proteins result in binding the hydrocarbon side chains of fat, thus resulting in variation in the OAC of flour and proteins isolates (Adebowale and Lawal, 2004). Legume flours are potentially useful in structural interaction, especially flavor retention, palatability, and shelf life, particularly in bakery or meat products. Flours with high OAC are potentially useful in food products for flavor retention, improvement of palatability, and extension of shelf life, particularly in bakery and meat products where oil absorption is desirable (Kaur and Singh 2007).

The OAC of flour from kidney bean cultivars of temperate climates are found to be 1.12–1.13 g/g (Wani et al., 2020) and for different Indian kidney bean cultivars, it ranges between 2.2–2.3g/g (Wani et al., 2013). Du et al. (2014) reported the OAC of red kidney bean cultivars is 1.20 g/g. OAC of native protein isolated from different Indian kidney bean cultivars varied from 5.82–6.54 g/g (Wani et al., 2015). Hydrolyzation of protein isolates for 30 minutes increased their OAC, which ranges from 6.79–7.39 g/g, while proteolysis for 60 minutes further increased the OAC of the isolates and it ranges between 7.62–7.99 g/g. Native kidney bean starch has fat absorption capacity of 3.6 g/g while the PEGylation of black kidney bean protein isolate with succinimidyl carbonate, succinimidyl succinate, and succinimidyl propionate conjugation slightly increased fat absorption capacity of isolates and ranges between 3.7–4.0 g/g (He et al., 2018). According to the studies of Mundi & Aluko (2012), the oil absorption capacities for the albumin fraction of kidney bean flour is higher i.e. 2.37 mL/g, while that of the globulin fraction is lower i.e. 1.87 mL/g, due to more globular structure of globulin protein.

22.6.4 Emulsifying Properties

Pulse flours mainly comprise carbohydrates, proteins, lipids, and minerals (Ma et al., 2013; Siddiq et al., 2010). Emulsifying properties are very important properties of the food proteins that play an important role in development of emulsion-based traditional or novel food products. Other than proteins, carbohydrates (starch and fiber) also improve the emulsifying properties by acting as bulky barriers between oil molecules. The interaction of proteins and carbohydrates also enhance the emulsifying properties to different extents (Aluko et al., 2009). Emulsion characteristics of the flour and proteins were primarily influenced by the source of protein, its levels, size, shape, solubility, processing conditions, temperature, pH, ionic strength, hydrophilic/hydrophobic ratio, level of shear, mixing ratio, level of coalescence and spatial arrangement of droplets, viscosity, and time (Dickinson, 2003; McClements, 2005). Presence of components other than proteins (possibly carbohydrates) also influence the emulsification properties of protein-containing products like pulse flours. To evaluate the emulsifying properties of legume flours and their protein counterparts, emulsion capacity (EC), emulsion stability (ES), emulsifying activity index (EAI), and emulsifying stability index (ESI) are often used.

Wani et al. (2013) reported that the EAI of kidney bean flour from different Indian cultivars at pH 3 ranged from 13.2–25.2 m²/g, at pH 5.0 ranged from 6.0–8.9 m²/g, and at pH 7.0 it ranged from 14.6–18.9 m²/g. This showed that at isoelectric pH EAI of the kidney bean flour was lowest while pH 3.0 and 7.0 it was higher. The ESI of kidney bean flour from different Indian cultivars at pH 3.0 varies from 15.5–33.6 min, at pH 5.0 it ranged from 23.7–73.9 min, and at pH 7.0 it ranged from 19.4–26.4 min. Different kidney beans cultivars grown in Indian temperate climates have an EC that ranges between 77.0–79.0%, and their ES ranges between 60.0–74.0% (Wani et al., 2020). EC and ES of 45.6–60.5% and 48.2–62.3%, respectively have been reported for dry bean flours (Siddiq et al., 2010). Du et al. (2014) reported that red kidney bean flour has emulsion activity of 82.46% and has ES of 86.54%.

Protein isolates from different kidney bean cultivars exhibit emulsion activity that ranges between 15.8–26.6 m^2/g and ES that ranges between 21.28–78.9 min (Shevkani et al., 2015). They also reported that emulsion properties were pH-dependent and better emulsifying properties were observed at acidic and alkaline pH, while poor at pI (pH 4.0 and 5.0). The EAI of the untreated kidney bean protein isolates was 24.2 m^2/g and it increased with an increase in pressure level up to 400 mPA, and after that, it decreased. As the level of pressure increased from 200–400 mPA, the emulsion activity index increased from 39.0–40.0 m^2/g and at 600 mPA, it decreased to 33.9 m^2/g. Similarly, ESI of the untreated kidney bean protein isolates was 68.3 min and it increased with an increase in pressure level up to 400 mPA, and after that, it decreased. As the level of pressure increased from 200–400mPA, the emulsion stability index increased from 123–142.3 min and at 600 mPA, it decreased to 80.8 min (Ahmed et al., 2018).

Native kidney bean protein isolate has EAI of 54 m^2/g while that PEGylation of black kidney bean protein isolate with succinimidyl carbonate, succinimidyl succinate, and succinimidyl propionate conjugation slightly increased emulsion activity index of isolates and t ranges between 60–64 m^2/g (He et al., 2018). The ESI of native protein isolate is 78% and PEGylation of black kidney bean protein isolate increased the emulsion stability index and it varied from 80–84%. Kidney bean protein isolates are reported to have EAI and ESI of 23.7–41.7 m^2/g and 30.9–55.1 min, respectively (Tang and Ma, 2009). Differences in the emulsifying ability of protein may be related to their solubility and conformational stability.

The EAI of native protein isolated from different Indian kidney bean cultivars at pH 3.0 varied between 6.45–19.16 m^2/g, which decreased when pH was 5.0 and at this pH ranges from 4.15–8.24 m^2/g. With an increase in pH of the suspension to 7.0, the EAI of native protein isolates increased and ranges between 15.71–48.92 m^2/g (Wani et al. 2015). Furthermore, at pH 7.0, the EAI of protein isolates hydrolyzed for 30 minutes ranges between 50.71–68.15 m^2/g, which after 60 minutes of hydrolyzation varies between 42.79–64.09 m^2/g, which was lower in comparison 30 minutes of hydrolyzation. The ESI of native protein isolated from different Indian kidney bean cultivars at pH 3.0 varied between 12.13–19.60 min, which increased when pH was 5.0 and at this pH ranges from 14.33–25.0. With an increase in pH of the suspension to 7.0, the EAI of native protein isolates decreased and it ranges between 13.03–16.99 min (Wani et al., 2015). Hydrolyzation of protein isolates had significant influence on the ESI. At pH 3.0 after 30 minutes of proteolysis, the ESI of protein isolated from different Indian kidney bean cultivars ranges between 12.49–22.32 min and after 60 minutes of proteolysis the values varied from 12.49–17.66. At pH 5.0, the ESI of protein isolates after 30 minutes of proteolysis varies from 19.61–20.71 min, which after 60 minutes of proteolysis ranges between 12.28–18.65 min. At pH 7.0, the ESI of protein isolates after 30 minutes of proteolysis varies from 13.60–15.45 min, which after 60 minutes of proteolysis ranges between 12.40–14.98 min (Wani et al., 2015). This shows that hydrolyzation for 30 minutes enhanced the ESI while hydrolyzation for 60 minutes decreases it. Phaseolin-rich protein fraction form kidney bean protein showed EAI of 45 m^2/g at pH 3.0, which increased to 215 m^2/g at pH 7.0 and then to 250 m^2/g at pH 9.0. At isoelectric pH the EAI of both the kidney bean protein isolate and its phaseolin-rich protein fraction exhibits the lowest values (Yin et al., 2010). The albumin fraction of kidney bean protein isolate had the highest emulsion activity at all the pH in comparison to globulin. The emulsion stability for both the albumin and globulin fractions were highest at pH 5.0, while it was lowest at acidic and alkaline pH (Mundi & Aluko 2012), however, among both, the globulin fraction exhibits higher ES.

22.6.5 Foaming Properties

Foams are a colloidal system having tiny air bubbles dispersed in a continuous phase (Damodaran, 2005). Flours produce foam when whipped mainly because of the surface-active proteins. These reduce surface tension at air-liquid interface, avoiding coalescence of the bubbles and give stabilized foam. Foam capacity (FC) and foam stability (FS) generally depend on the interfacial film formed by the proteins that keep air bubbles in suspension and slows down the rate of coalescence. Foam is stabilized by an adsorbed layer of protein at the air-water interface. Foam is a two-phase system consisting of air cells separated by a thin continuous liquid layer called the lamellar phase (Zayas, 1997). The basic requirement of proteins to be a good foaming agent is their ability to absorb rapidly at the air-water interface during bubbling and to undergo rapid conformational change and rearrangement at the interface. These factors are essential for better foamability. The ability of proteins to form a cohesive viscoelastic film via intermolecular interactions is of prime importance for the FS (Adebowale et al., 2005). Foaming properties such as foaming capacity and stability dependent upon concentration of flour in suspension, level and type of protein, size of protein, presence of polysaccharide, carbohydrates and pH of the suspension.

Du et al. (2014) reported that red kidney bean flour has a FC of 70% and has FS of 75% after 120 minutes. Wani et al. (2013) reported that FC (WAC) of kidney bean flour from different Indian cultivars varied from 82.1–122.2% at pH 2.0 and increased with increase in pH. At pH 10.0, the FC ranges 103.8–142.0%. Similarly, the half-life of foam also varied significantly with increase in pH. At pH 2.0 the half-life of foam from different kidney bean flour ranges between 1.77–6.70 hours and at pH 6.0 it ranges between 8.60–12.0 hours. The FC of the flours from different kidney bean cultivars grown in Indian temperate climates were 225.0% and their FS ranges between 65–70% (Wani et al., 2020).

FC of protein isolates ranged between 136.7–244.9% for kidney beans (Tang and Ma, 2009). Tang and Ma (2009) reported FS of native kidney bean protein isolates as 87.8%, while it was 74.2–80.6% for heat-treated kidney bean protein isolates. Shevkani et al. (2015) reported that protein isolates from different red kidney bean cultivars range between 83–121% and has FS from 90–95%. They also reported that foaming properties were pH-dependent and better foaming properties were observed at acidic and alkaline pH, while poor at pI (pH 4.0 and 5.0).

Native kidney bean protein isolate has a FC of 7.5 ml, while that PEGylation of black kidney bean protein isolate with succinimidyl carbonate, succinimidyl succinate, and succinimidyl propionate conjugation slightly increased FC of isolates, ranging between 8.3–8.8 ml (He et al., 2018). The foaming stability index of native protein isolate is 28% and PEGylation of black kidney bean protein isolate increased the emulsion stability index and it varied from 32–36%. FC of the untreated kidney bean protein isolates at pH 7 was 76.7% and it decreased with increase in pressure level. As the level of pressure increased from 200–600 mPA, the FC decreased from 61.5–42.1% due to partial denaturation by unfolding protein molecule structure. The FS measured after 60 minutes decreased linearly with the pressure treatment. Similarly the FS of the untreated kidney bean protein isolates at pH 7 after 60 minutes was 54.5% and it decreased linearly with increase in pressure level. As the level of pressure increased from 200–600 mPA, the FS decreased from 50.5–40.5% (Ahmed et al. 2018). The FA of flours makes them useful in food systems requiring aeration for textural and leavening properties (Kaur and Singh 2007), while FS is important since the usefulness of whipping agents depends on their ability to maintain the whip as long as possible (Lin et al., 1974). Flours having high foaming ability are useful for various products such as whipped cream, cakes, muffins, mousses, marshmallows, ice-creams, etc.

The FC of native proteins isolated from different Indian kidney bean cultivars at pH 2.0 ranges between 107.34–113.46% and at pH 6.0 it decreased and ranges between 76.0–102.0%, which further increased to 108.0–126.0 when pH of the suspension was maintained at 10.0, showing the higher FC at acid and alkali pH then at pH round 4-6.0. A similar trend for foaming capacity of hydrolyzed protein isolates was also observed. Also, an increase in hydrolysis time from 30 minutes to 60 minutes at each pH significantly increased the foaming capacity of protein isolates. At pH 2.0 after 30 minutes of proteolysis, the FC of protein isolated from different kidney bean cultivars ranges between 100.0–132.0%, which after 60 minutes of proteolysis ranges between 118.0–132.0%. At pH 4.0, after 30 minutes of proteolysis, the foaming capacity of protein isolated from different kidney bean cultivars ranges between 54.0–113.0%, which after 60 minutes of proteolysis ranges between 109–132.0%. At pH 10.0, foaming capacity after 30 minutes of proteolysis for proteins isolated from different kidney bean cultivars ranges between 92.0–156.70%, which after 60 minutes of proteolysis ranges between 101.0–142.0%. FS calculated as half-life of foam of native protein isolates at pH 2–10.0 ranges between 3.65≥12.0 hours, while for hydrolyzed proteins, the half-life of foam value ranges between 2.47≥12.0 hours. The proteins isolates hydrolyzed for 30 minutes had higher foam half-life in comparison to protein isolates hydrolyzed for 60 minutes. Furthermore, the FS of all the protein isolates was maximum at around pH 4–6.0 white it was lower for acid and alkali pH. The albumin fraction of kidney bean protein isolate had the highest FC (100%) in comparison to globulin, which had FC of 76% at pH 9.0. At pH 3.0 the FC of albumin fraction is 68%, while it was 60% for globulin fraction, while at pH 5.0 both globulin and albumin fractions showed lowest FC. Similarly, the higher FS was found in globulin fraction compared to the albumin at all the tested pH (Mundi & Aluko 2012).

22.7 Utilization of Kidney Beans

22.7.1 Bread

The suitability of red kidney bean flour for bread formulation and quality bread production was studied by Ramzy & Putra (2018). Moderately coarse, moderately fine, and standard red kidney bean flour was used at the 15% and 20% level in bread development and compared with control white bread. Their results revealed that the utilization of red kidney bean flour of any particle size had no influence on oven spring, baking loss, loaf specific volume, and crumb and crust moisture, however, the level of red kidney bean flour did have an impact on these properties. Red kidney bean flour with lower particle size improves the crumb area cell and cell density. Manonmani et al. (2014) reported that red kidney bean flour at 5–25% can be substituted in the whole wheat flour for the development of functional bread. From their studies it was concluded that the optimized level of substitution of red kidney bean flour for bread formulation was 15% as it give higher sensory scores and consumer acceptability. Increasing the levels of red kidney bean flour increased the hardness of the bread along with increased gumminess, chewiness, and redness values of the brea while decreasing its lightness. The addition of red kidney bean flour enhanced the protein content and fiber content of the bread in comparison to wheat flour bread.

Kidney bean flour has a great potential to use in the formulation and development of rice flour based gluten free extruded snacks. Extruded snack prepared with rice flour (80%) and kidney bean flour (20%) were highly accepted by the taste panellists while the extruded snacks prepared with higher levels of kidney bean starch had lower overall acceptability and expansion ratio and higher bulk density. Red kidney bean flour can be successfully incorporated into wheat flour at 20% to prepare nutritionally rich functional bread that is pack house of nutrients, proteins, dietary fiber, minerals and exhibits better sensory and textural characteristics in comparison to control bread (Bhol & Bosco, 2014). White kidney bean flour was used at 10–25% levels in the development wheat based bread. White bean flour supplementation increased the content of protein and crude fiber content in the bread and the highest sensory scores were received for the bread with 15% kidney beans flour (Ukeyima et al., 2019).

22.7.2 Muffins, Cakes, and Cupcakes

Legume processing industry by-products, such as broken legumes, can be extruded to make extruded flour and successfully used in the formulation for the development of gluten-free cake mixes and ultimately gluten-free cakes (Gomes et al., 2015). The cakes can be prepared with up to a 75% replacement of the gluten-free mix with extruded kidney bean flour and the resultant cakes have lower specific volume, tenderness, and good sensorial properties, however they exhibit darker color in comparison to the control. Flour from extruded kidney beans can be recommended as a vital food ingredient to partially or wholly substitute for wheat flour from nutritional, technological, and sensory viewpoints, which can help to value addition of kidney bean by-products. Red kidney bean flour has a great potential to be utilized as gluten-free ingredient for the development of various functional foods. Chompoorat & Phimphimol (2019) reported that red kidney bean flour has been the most promising ingredient for making gluten-free cupcakes, as cupcakes made from red kidney bean flour had the same overall likeness in comparison to commercial gluten-free cupcakes and they also have higher amounts of protein and dietary fiber; consumption of one cupcake offers 8% of the daily recommended protein intake. Bassinello et al. (2020) reported that the extruded kidney bean flour (75%) along with a rice grits-based gluten free cake has improved protein content, enhanced *in vitro* starch, and protein digestibility, dietary fiber, iron, zinc, and riboflavin, without the proper essential amino acids content and the cake possess richness of good fiber both soluble and insoluble.

Kidney bean flour has the potential to be used as fat replacer in cake formulation and in this regard, Trisnawati & Sutedja (2014) reported that steamed kidney bean paste can be used successfully along with mung bean paste to replace the margarine level in the cake, which helps to reduce the amount of fat in the cake and ultimately help control obesity. Flour from kidney beans up to 40% can be used for the development of peach blossom cake using Chinese edible Peach blossom flower, with good taste, flavor appearance, and a fine structure and enhanced nutritional quality (Wang et al., 2015). Kidney bean starch in native and gelatinized form can be successfully used as a fat replacer in wheat flour-based sponge cake at a 30% level. The addition of both the native and gelatinized starches makes cakes with higher specific volume, firmness, springiness, and cohesiveness in comparison to whole flour muffins (Bajaj et al., 2019). The addition of native and gelatinized starches from kidney beans darkens the crust color of the cake, while the cakes have a spongy texture and better sensory characteristics. Thermal treatment of kidney bean flour by boiling and drying results in production of the precooked flour, which can be used for preparation of gluten-free cupcakes. Chompoorat et al. (2018) reported that boiling kidney bean flour 20 minutes followed by drying at 80°C for 3 hours resulted in the production of highly acceptable gluten-free cupcakes with better consumer-acceptance scores in comparison to cupcakes made from kidney bean flour alone. Red kidney bean flour is a nutrient-rich ingredient and can be utilized potentially in formulation and enrichment of bakery products. Chompoorat et al. (2020) reported that red kidney bean flour up to 75% along with 25% rice flour can be successfully used in the preparation of gluten-free cupcakes that exhibit softer crumb texture with higher overall sensory scores in comparison to the cupcakes made with higher levels of kidney bean flour.

Protein isolates from kidney bean at a 10% level of protein can be replaced with the starch batter to prepare starch-based gluten-free muffins. The addition of protein isolates from kidney bean improves the specific volume of the muffins along with increasing the muffins' firmness, springiness, and cohesiveness with respect to muffins without isolates (Shevkani & Singh 2014). The addition of protein isolates also darkens the color of the crumb and crust and the protein isolates can be successfully incorporated to prepare gluten-free muffins. Gelatinized retrograded starch and extruded kidney bean starches were used at 10% and 20% in the preparation of muffins (Sharma et al., 2016). The addition gelatinized retrograded starch and extruded kidney bean starches decrease the height and volume of muffins. The hardness of muffins prepared with extruded kidney bean starches is higher in comparison to gelatinized retrograded starch muffins and also with control muffins. The addition of gelatinized retrograded starch and extruded kidney bean starches decreased the gumminess and springiness of the muffins while it increased the resistant-starch content. Muffins prepared with gelatinized retrograded starch exhibited better sensory scores in comparison to extruded kidney bean starch muffins.

22.7.3 Cookies and Biscuits

White kidney bean flour can be used up to 10% along with rice bran in wheat-based composite flour cookies, which are rich in proteins, minerals, vitamins, and total essential amino acids and can be used to feed children and adolescents (Amira et al. 2018). Kidney bean flour can also be used up to 15% in the development of antioxidant-rich wheat-based cookies with thyme and parsley. Kidney bean supplementation increased the content of protein and crude fiber content (Amer et al., 2009). Cookies made with 20% kidney bean flour exhibited higher protein and crude fiber content in comparison to wheat flour cookies with higher sensory scores and lower glycemic index (Eme et al., 2018). Sparvoli et al. (2016), while assessing the contribution of bean flour on biscuit quality, observed that supplementation of wheat flour with kidney bean flour from 12–29% enhance the protein and fiber content, had a better total amino acid score, low starch content, and low glycemic index, and biscuits prepared with 14% bean flour were liked mostly by the consumers while increase in bean flour levels decrease the acceptability of the biscuits. Adding red kidney bean flour at a10–30% level in wheat flour for biscuits increased the protein and fiber content of the biscuits as well as enhanced their antioxidant activity and total phenol content. Increasing the level of kidney bean flour decreases the spread ratio of the biscuits as well as the sensory scores, and results revealed that the best quality biscuits with highest sensory score were prepared using 10% kidney bean flour (Roy et al., 2020). Nineteen percent of germinated kidney bean flour along with 31% of germinated chickpea flour can used to formulate the germinated wheat flour-based composite cookies, having 12.32% of protein and 5.64% crude fiber, soft texture, low hardness, and slightly dark color with higher sensory scores (Sibian & Riar 2020b). The germinated kidney bean chickpea flour cookies had better *in vitro* protein and starch digestibility and enhanced total essential amino acid content in comparison to wheat flour cookies, and these cookies remais shelf-stable for 3 months at room temperature when packed in aluminum-laminated packs. Studies of Sibian & Riar (2020a) reported that flour from germinated kidney bean (15%) along with chickpea (35%) and triticale can be utilized for the development of novel composite flour cookies with good nutritional properties, enhanced *in vitro* starch and protein digestibility and essential amino acid score. These cookies remain acceptable up to 90 days of storage. Gelatinized retrograded starch and extruded kidney bean starches were used at 10% and 20% in the preparation of cookies (Sharma et al., 2016). The addition of extruded kidney bean starches resulted in the cookies with a higher spread ratio and hardness in comparison to gelatinized retrograded starch incorporated cookies, however, when compared with control, the addition of gelatinized retro-graded starch and extruded kidney bean starches increase the spread ratio, resistant starch content and decrease the hardness of the cookie. Cookies with gelatinized retrograded starch and extruded kidney bean starches exhibited better sensory scores.

22.7.4 Pasta/Noodles/Vermicelli

Consumers are diverting toward healthier lifestyles and are shifting their food patterns toward low gly-cemic foods to meet their nutritional requirements (Kanchana et al., 2020) and red kidney bean flour being a low GI food can be successfully incorporated in wheat flour in 2:1 for the development of value-added, low glycemic index, high protein healthy noodles, which remain acceptable to consumers even after 3 months of storage. Different dry beans, such as white kidney beans and dark red kidney beans, were used for the development of gluten-free pasta to increase the dry bean consumption (Hooper et al., 2019). Results revealed that gluten-free pasta made with dark red and white kidney bean flour was nutrition-ally superior to wheat pasta in terms of higher protein, dietary fiber, resistant starch, and better protein digestibility. The kidney bean pasta had a higher corrected amino acid score as well as lower total starch content in comparison to wheat pasta. On the basis of consumer acceptability data, it was observed that 36% consumers willingly buy the light colored kidney bean pastas and their results suggested that fresh kidney bean pastas have much commercial potential as a healthy gluten-free pasta option in the United States. Romero & Zhang (2019) examined great northern, navy, red kidney, and garbanzo beans for the development of gluten-free pasta and revealed that flours from garbanzo and navy beans are potentially superior sources for the development of gluten-free pasta as they form a compact network that does not differ much in its textural attributes among other bean flours and their resultant pasta. Kidney bean paste at a 30% level, along with wheat flour and eggs, results in the development of noodles rich in protein and minerals that can be used to eliminate the prevalence of malnutrition in Ecuador (Meneses 2002).

Starches isolated from kidney beans have an important practical application in the formulation and development of vermicelli. Chang et al. (2018) reported that vermicelli prepared from starches isolated from kidney beans lowered swelling and viscosity while cooking in comparison to mung bean starch vermicelli. Mung bean starch vermicelli exhibits higher cooking loss and weak tensile and shear strength while kidney bean starch vermicelli showed lower cooking loss, higher tensile strength, breaking rate, and elasticity, thus kidney bean starch can replace high-cost mung bean starch for vermicelli production, which also increases the application of kidney bean starch in vermicelli processing. Zhu et al. (2017) pre-pared the thin noodles from kidney bean starch and compared them with mung bean starch thin noodles and revealed that noodles prepared from kidney bean starch had higher transparency and tenacity while they showed low gruel loss in boiling water. Cooking quality of kidney bean starch thin noodles was as good as mung bean, but they are slightly harder, have more elasticity, and exhibit higher shear deforma-tion rate. Although the sensory scores showed that since the noodles made from kidney bean starch were not as superior as mung bean, the starch from kidney beans had a much higher potential to be used in bean noodles. Lii & Chang (1981) prepared the starch noodles from red kidney bean starch and com-pared them with mung bean starch noodles and observed that red kidney bean starch noodles exhibited fairly good quality singly and in 50:50 with mung bean starch, even though not equivalent to alone mung bean starch noodles. The kidney bean starch noodles had higher cooked water loss and lower tensile strength in comparison to mung bean starch noodles, while still they can be used in noodle preparation. Gelatinized retrograded starch and extruded kidney bean starches were used at 10% and 20% in the preparation of noodles (Sharma et al., 2016). The addition gelatinized retrograded starch and extruded kidney bean starches decrease minimum cooking time, water uptake ratio, and gruel solid loss content of the noodles in comparison to noodles prepared from control flour. The hardness, cohesiveness, gummi-ness, and adhesiveness of the noodles decreased with gelatinized retrograded starch and extruded starch in comparison to control noodles. Addition of gelatinized retrograded starch and extruded kidney bean starches increased the resistant starch content in the noodles and noodles with higher levels of modified starches exhibits highest overall acceptability scores.

Flour form rajma (kidney beans) can be successfully incorporated into durum semolina at 20–30% levels to enhance its nutritional quality and minimize its starch digestibility (Kumar & Prabhasankar 2015). Rajma flour incorporation decreased the *in vitro* digestibility, lowered the content of rapidly digestible starch while increasing the content of slowly digestible starch. Pasta made with rajma flour had higher protein content and both soluble and insoluble dietary fiber content. Firmness and gruel solid loss content of pasta increased with increase in level of rajma flour. They concluded that incorporation of rajma flour up to 20–30% can be done in semolina for development of low GI pasta without affecting its sensory quality. The authors also reported that this protein-rich product can also be beneficial to tackle protein malnutrition and provide the adequate nutrient to consumers suffering from protein deficiency.

22.7.5 Other Products

Red kidney bean flour has a great potential to supplement the roasted maize flour for the development of Aadun, a popular local maize snack from Nigeria (Adeyanju et al., 2016). Irrespective of the maize flour variety used, the addition of kidney bean flour from 15–26% increased the protein content, minerals such as sodium, potassium, calcium, magnesium, iron, and zinc, and vitamins such as Vitamin B1, B2 and B3. Red kidney bean flour at a 15% level can be used in the preparation of acceptable snacks with improved nutritional value and functionality of traditional products. Light red kidney beans can used to make por-ridge that has better overall acceptability in comparison to traditional porridge i.e. *Sosuma* (Nyombaire et al., 2011). A low-cost extruder can be used for the products using native legumes.

Kidney bean flour can be incorporated up to 15% in wheat flour to prepare highly acceptable protein and dietary fiber-rich chapattis with sensory scores as good as chapatti from whole wheat flour (Wani et al., 2016). The chapattis prepared with kidney bean and wheat flour have perfect amino acid ratios and these two flours can be combined to enhance the cereal protein's essential amino acid deficiency. Addition of kidney bean flour helps to maintain the quality of chapatti for longer periods owing to the lower setback viscosity of composite flour than wheat flour. A red kidney bean-based tempeh-like product was prepared by Gomez & Kothary (1979), which was fermented using *Rhizopus oligosporus* by soaking, de-hulling, and boiling of beans for 20 minutes followed by inoculation and incubation at ambient temperatures for 48 hours. The tempeh had good acceptability and higher protein and fiber content respective to its native counterpart and can be stored for 14 days in refrigerated conditions after steaming.

Sharma et al., (2020) used kidney bean nanostarch to form starch-based composite films at 0.5–10% levels. The addition of kidney bean nanostarch particles increased the thickness, solubility, and burst strength of the films while they decreased their moisture content and water vapor transmission rate. The results suggested that kidney bean starch could be used for the development of packaging films as it improves the film forming properties of starch. Kidney bean protein isolates can be blended with chitosan for the development of self-supported films at acidic pH that can be used in antimicrobial packaging and also as a carrier of antimicrobial compounds for food packaging (Ma et al., 2013). Fan et al., (2014) also prepared kidney bean protein isolates blended with chitosan films using an ultrasonic treatment to pro-duce flexible films that can be loaded with niacin to form flexible films with antimicrobial activity. Kidney bean protein isolates can also be used for the preparation of films. The films prepared at pH 8.0 had higher transparency and lower lightness, while heating of the protein isolate improved tensile strength and water resistance of the films and decreased the water solubilization (Shevkani & Singh 2015).

22.8 Conclusion

The kidney bean is an important source of daily dietary nutrients for the most of population in both developed and developing countries due to its high protein content. The low fat and higher resistant starch content makes the kidney bean a promising health-food ingredient for the consumer who requires essential food ingredients beyond basic food nutrition. Rich in various phenolic, flavonoids, and bioactive compounds, kidney beans exert various health-promoting aspects such as antidiabetic, antiradical, antifungal, hypolipidemic, antioxidant, anti-inflammatory, and antihepatotoxic activities. In addition, kidney beans also showed profound functional properties such as water and oil absorption, emulsification, and foaming due to the higher solubility. The kidney beans also finds a place in conventional food products as its flour, protein, and starch components have been utilized in formulation of muffins, breads, cakes, cookies, and pasta products beyond use in traditions diets. Therefore, the kidney bean seed shows a real potential for the grain world market due to its nutritional and agricultural properties.

REFERENCES

Adebowale, K. O., & Lawal, O. S. (2004). Comparative study of the functional properties of bambarra groundnut (Voandzeia subterranean), jack bean (Canavalia ensiformis) and mucuna bean (Mucuna pruriens) flours. Food Research International, *37*(4), 355–365.

Adebowale, K. O., Olu-Owolabi, B. I., kehinde Olawumi, E., & Lawal, O. S. (2005). Functional properties of native, physically and chemically modified breadfruit (Artocarpus artilis) starch. Industrial Crops and Products, *21*(3), 343–351.

Adeyanju, B. E., Enujiugha, V. N., & Bolade, M. K. (2016). Effects of addition of kidney bean (Phaseolus vulgaris) and alligator pepper (Aframomum melegueta) on some properties of 'aadun' (a popular local maize snack). Journal of Sustainable Technology, *7*(1), 45–58.

Aguilera, Y., Estrella, I., Benitez, V., Esteban, R. M., & Martín-Cabrejas, M. A. (2011). Bioactive phenolic compounds and functional properties of dehydrated bean flours. Food Research International, *44*(3), 774–780.

Ahmed, J., Al-Ruwaih, N., Mulla, M., & Rahman, M. H. (2018). Effect of high pressure treatment on functional, rheological and structural properties of kidney bean protein isolate. LWT Food Science and Technology, *91*, 191–197.

Aluko, R. E., Mofolasayo, O. A., & Watts, B. M. (2009). Emulsifying and foaming properties of commercial yellow pea (Pisum sativum L.) seed flours. Journal of Agricultural and Food Chemistry, *57*(20), 9793–9800.

Amarowicz, R., Dykes, G. A., & Pegg, R. B. (2008). Antibacterial activity of tannin constituents from Phaseolus vulgaris, Fagoypyrum esculentum, Corylus avellana and Juglans nigra. Fitoterapia, *79*(3), 217–219.

Amer, T. A., LL Louz, S., & Doweidar, M. M. (2009). Biochemical and technological evaluation of biscuits supplemented with nutritional plant sources rich in antioxidants. Journal of Food and Dairy Sciences, *34*(4), 2953–2967.

Amira, S. S., Saly, A. A., Abbas, S., & Mona, M. M. (2018). Preparation and evaluation of biscuits supplemented with some natural additives for children and adolescents feeding. Suez Canal University Journal of Food Sciences, *5*(1), 69–90.

Bajaj, R., Singh, N., & Kaur, A. (2019). Effect of native and gelatinized starches from various sources on sponge cake making characteristics of wheat flour. Journal of Food Science and Technology, *56*(2), 1046–1055.

Bassinello, P. Z., Bento, J. A. C., Gomes, L. D. O. F., Caliari, M., & Oomah, B. D. (2020). Nutritional value of gluten-free rice and bean based cake mix. Ciência Rural, *50*(6).

Beninger, C. W., & Hosfield, G. L. (1999). Flavonol glycosides from Montcalm dark red kidney bean: Implications for the genetics of seed coat color in *Phaseolus vulgaris* L. Journal of Agricultural and Food Chemistry, *47*(10), 4079–4082.

Bhol, S., & Bosco, S. J. D. (2014). Influence of malted finger millet and red kidney bean flour on quality characteristics of developed bread. LWT—Food Science and Technology, *55*(1), 294–300.

Campos-Vega, R., García-Gasca, T., Guevara-Gonzalez, R., Ramos-Gomez, M., Oomah, B. D., & Loarca-Piña, G. (2012). Human gut flora-fermented nondigestible fraction from cooked bean (Phaseolus vulgaris L.) modifies protein expression associated with apoptosis, cell cycle arrest, and proliferation in human adenocarcinoma colon cancer cells. Journal of Agricultural and Food Chemistry, *60*(51), 12443–12450.

Carai, M. A., Fantini, N., Loi, B., Colombo, G., Gessa, G. L., Riva, A., Bombardelli, E. and Morazzoni, P. (2011). Multiple cycles of repeated treatments with a *Phaseolus vulgaris* dry extract reduce food intake and body weight in obese rats. British Journal of Nutrition, *106*(5), 762–768.

Carbas, B., Machado, N., Oppolzer, D., Ferreira, L., Queiroz, M., Brites, C., Rosa, E. A. and Barros, A. I. (2020). Nutrients, antinutrients, phenolic composition, and antioxidant activity of common bean cultivars and their potential for food applications. Antioxidants, *9*(2), 186.

Cardador-Martinez, A., Castano-Tostado, E., & Loarca-Pina, G. (2002). Antimutagenic activity of natural phenolic compounds present in the common bean (Phaseolus vulgaris) against aflatoxin B 1. Food Additives & Contaminants, *19*(1), 62–69.

Chang, R., Shuzhen, L. I. U., Lijuan, Z. H. U., & Caiqiong, Z. H. O. U. (2018). Morphological structure, physicochemical properties analysis and application in processing vermicelli of the kidney bean starch grown in Qianjiang, China. *Journal of Food Engineering and Technology*, *7*(2), 77–77.

Chou, D. H., & Morr, C. V. (1979). Protein-water interactions and functional properties. Journal of the American Oil Chemists' Society, *56*(1), A53–A62.

Chompoorat, P., & Phimphimol, J. (2019). Development of a highly nnutritional and functional gluten free cupcake with red kidney bean flour for older adults. Food and Applied Bioscience Journal, *7*(3), 16–26.

Chompoorat, P., Kantanet, N., Hernández Estrada, Z. J., & Rayas-Duarte, P. (2020). Physical and dynamic oscillatory shear properties of gluten-free red kidney bean batter and cupcakes affected by rice flour addition. Foods, *9*(5), 616.

Chompoorat, P., Rayas-Duarte, P., Hernández-Estrada, Z. J., Phetcharat, C., & Khamsee, Y. (2018). Effect of heat treatment on rheological properties of red kidney bean gluten free cake batter and its relationship with cupcake quality. Journal of Food Science and Technology, *55*(12), 4937–4944.

Choung, M. G., Choi, B. R., An, Y. N., Chu, Y. H., & Cho, Y. S. (2003). Anthocyanin profile of Korean cultivated kidney bean (Phaseolus vulgaris L.). Journal of Agricultural and Food Chemistry, *51*(24), 7040–7043.

Damodaran, S. (2005). Protein stabilization of emulsions and foams. Journal of Food Science, *70*(3), R54–R66.

Daniell, E. L., Ryan, E. P., Brick, M. A., & Thompson, H. J. (2012). Dietary dry bean effects on hepatic expression of stress and toxicity-related genes in rats. British Journal of Nutrition, *108*(S1), S37–S45.

Dickinson, E. (2003). Hydrocolloids at interfaces and the influence on the properties of dispersed systems. Food hydrocolloids, *17*(1), 25–39.

Du, S. K., Jiang, H., Yu, X., & Jane, J. L. (2014). Physicochemical and functional properties of whole legume flour. LWT—Food Science and Technology, *55*(1), 308–313.

Dueñas, M., Martínez-Villaluenga, C., Limón, R. I., Peñas, E., & Frias, J. (2015). Effect of germination and elicitation on phenolic composition and bioactivity of kidney beans. Food Research International, 70, 55–63.

Eme, P. E., Mbah, B. O., Oly-alawuba, N. M., & Nwambeke, E. (2018). Physicochemical, sensory evaluation and glycemic effect of biscuit produced from Acha (Digitaria exilis)/red kidney bean (Phaseolus vulgaris)/unripe plantain (Musa paradisiacal) blends on normoglycemic adults. Nigerian Journal of Nutritional Sciences, *39*(2), 17–26.

Fan, J. M., Ma, W., Liu, G. Q., Yin, S. W., Tang, C. H., & Yang, X. Q. (2014). Preparation and characterization of kidney bean protein isolate (KPI)–chitosan (CH) composite films prepared by ultrasonic pretreatment. Food Hydrocolloids, *36*(1), 60–69.

Fantini, N., Cabras, C., Lobina, C., Colombo, G., Gessa, G. L., Riva, A., Donzelli, F., Morazzoni, P., Bombardelli, E. and Carai, M. A., (2009). Reducing effect of a Phaseolus vulgaris dry extract on food intake, body weight, and glycemia in rats. Journal of Agricultural and Food Chemistry, *57*(19), 9316–9323.

FAOSTAT (2019), "Crops – Beans, dry – Production quantity, years 1988 to 2017 – Export and import quantities, years 2012 to 2016; Food supply quantity, bean, kg/capita, year 2013", *FAO Statistics Database*, Food and Agriculture Organisation of the United Nations (FAO), http://faostat.fao.org

Gabriele, M., Pucci, L., La Marca, M., Lucchesi, D., Della Croce, C. M., Longo, V., & Lubrano, V. (2016). A fermented bean flour extract downregulates LOX-1, CHOP and ICAM-1 in HMEC-1 stimulated by ox-LDL. Cellular & Molecular Biology Letters, *21*(1), 10.

Gan, R. Y., Deng, Z. Q., Yan, A. X., Shah, N. P., Lui, W. Y., Chan, C. L., & Corke, H. (2016). Pigmented edible bean coats as natural sources of polyphenols with antioxidant and antibacterial effects. LWT—Food Science and Technology, 73, 168–177

Ganesan, K., & Xu, B. (2017). Polyphenol-rich dry common beans (Phaseolus vulgaris L.) and their health benefits. International Journal of Molecular Sciences, *18*(11), 2331.

García-Lafuente, A., Moro, C., Manchon, N., Gonzalo-Ruiz, A., Villares, A., Guillamon, E., Mateo-Vivaracho, L. (2014). *In vitro* anti-inflammatory activity of phenolic rich extracts from white and red common beans. Food Chemistry, *161*(15), 216–223.

Gomes, L. D. O. F., Santiago, R. D. A. C., Carvalho, A. V., Carvalho, R. N., Oliveira, I. G. D., & Bassinello, P. Z. (2015). Application of extruded broken bean flour for formulation of gluten-free cake blends. Food Science and Technology, *35*(2), 307–313

Gomez, M. I., & Kothary, M. (1979). Studies on the production of red-kidney-bean tempeh. Journal of Plant Foods, *3*(3), 191–198.

Hayat, I., Ahmad, A., Masud, T., Ahmed, A., & Bashir, S. (2014). Nutritional and health perspectives of beans (Phaseolus vulgaris L.): An overview. Critical Reviews in Food Science and Nutrition, *54*(5), 580–592.

He, Q., Sun, X., He, S., Wang, T., Zhao, J., Yang, L., … & Sun, H. (2018). PEGylation of black kidney bean (Phaseolus vulgaris L.) protein isolate with potential functironal properties. Colloids and Surfaces B: Biointerfaces, *164*, 89–97.

Hooper, S. D., Glahn, R. P., & Cichy, K. A. (2019). Single varietal dry bean (Phaseolus vulgaris L.) pastas: Nutritional profile and consumer acceptability. Plant Foods for Human Nutrition, *74*(3), 342–349.

Kan, L., Nie, S., Hu, J., Liu, Z., & Xie, M. (2016). Antioxidant activities and anthocyanins composition of seed coats from twenty-six kidney bean cultivars. Journal of Functional Foods, 26, 622–631.

Kanchana, R., SanciaVaz, D. R., Antao, A., & Ravedar, T. S. S. (2020). Sensory evaluation of nutritionally potential high protein low glycemic index noodles. Journal of Scientific Research, *64*(2).

Kaur, J., Singh, B., Singh, A., & Sharma, S. (2020). Geometric, physical and functional properties of selected pulses and millets for the formulation of complementary food products. International Journal of Chemical Studies, *8*(4), 2854–2858.

Kaur, M., & Sandhu, K. S. (2010). Functional, thermal and pasting characteristics of flours from different lentil (Lens culinaris) cultivars. Journal of Food Science and Technology, *47*(3), 273–278.

Kaur, M., & Singh, N. (2007). A comparison between the properties of seed, starch, flour and protein separated from chemically hardened and normal kidney beans. Journal of the Science of Food and Agriculture, *87*(4), 729–737.

Kumar, S. B., & Prabhasankar, P. (2015). A study on starch profile of rajma bean (Phaseolus vulgaris) incorporated noodle dough and its functional characteristics. Food chemistry, *180*, 124–132.

Kyznietsova, M. Y., Halenova, T. I., Savchuk, O. M., Vereschaka, V. V., & Ostapchenko, L. I. (2015). Carbohydrate metabolism in type 1 diabetic rats under the conditions of the kidney bean pods aqueous extract application. Fiziolohichnyi Zhurnal (Kiev, Ukraine: 1994), *61*(6), 96–103.

Lii, C. Y., & Chang, S. M. (1981). Characterization of red bean (Phaseolus radiatus var. Aurea) starch and its noodle quality. Journal of Food Science, *46*(1), 78–81.

Lin, L. Z., Harnly, J. M., Pastor-Corrales, M. S., & Luthria, D. L. (2008). The polyphenolic profiles of common bean (Phaseolus vulgaris L.). Food Chemistry, *107*(1), 399–410.

Lin, M. J. Y., Humbert, E. S., & Sosulski, F. W. (1974). Certain functional properties of sunflower meal products. Journal of Food Science, *39*(2), 368–370.

Ma, W., Tang, C. H., Yang, X. Q., & Yin, S. W. (2013). Fabrication and characterization of kidney bean (Phaseolus vulgaris L.) protein isolate–chitosan composite films at acidic pH. Food Hydrocolloids, *31*(2), 237–247.

Manonmani, D., Bhol, S., & Bosco, S. J. D. (2014). Effect of red kidney bean (Phaseolus vulgaris L.) flour on bread quality. Open Access Library Journal, *1*(1), 1–6.

Margier, M., Georgé, S., Hafnaoui, N., Remond, D., Nowicki, M., Du Chaffaut, L., Amiot, M. J. and Reboul, E. (2018). Nutritional composition and bioactive content of legumes: Characterization of pulses frequently consumed in France and effect of the cooking method. Nutrients, *10*(11), p.1668.

McClements, D. J. (2005). Food Emulsions: Principles, Practices, and Techniques. CRC press.

Meneses, I. (2002). Elaboration of Noodles, Enriched with Kidney Bean in the Communities of Cuambo and the Rinconada of the Canton Ibarra Province of Imbabura. Thesis, Universidad Tecnica del Norte, Facultad de Ciencias de la Salud, Escuela de Nutricion y Salud Communitaria.

Mensack, M. M., McGinley, J. N., & Thompson, H. J. (2012). Metabolomic analysis of the effects of edible dry beans (Phaseolus vulgaris L.) on tissue lipid metabolism and carcinogenesis in rats. British Journal of Nutrition, *108*(S1), S155–S165.

Micheli, L., Lucarini, E., Trallori, E., Avagliano, C., De Caro, C., Russo, R., Calignano, A., Ghelardini, C., Pacini, A. and Di Cesare Mannelli, L. (2019). Phaseolus vulgaris L. Extract: Alpha-amylase inhibition against metabolic syndrome in mice. Nutrients, *11*(8), p.1778.

Mizubuti, I. Y., Júnior, O. B., de Oliveira Souza, L. W., & Ida, E. I. (2000). Response surface methodology for extraction optimization of pigeon pea protein. Food Chemistry, *70*(2), 259–265.

Mundi, S., & Aluko, R. E. (2012). Physicochemical and functional properties of kidney bean albumin and globulin protein fractions. Food Research International, *48*(1), 299–306.

Nolan R, Shannon OM, Robinson N, Joel A, Houghton D and Malcomson FC (2020) It's no has bean: A review of the effects of white kidney bean extract on body composition and metabolic health. Nutrients,12, 1398.

Nyombaire, G., Siddiq, M., & Dolan, K. D. (2011). Physico-chemical and sensory quality of extruded light red kidney bean (Phaseolus vulgaris L.) porridge. LWT—Food Science and Technology, *44*(7), 1597–1602.

Oliveira, R. J., de Oliveira, V. N., Deconte, S. R., Calábria, L. K., da Silva Moraes, A., & Espindola, F. S. (2014). Phaseolamin treatment prevents oxidative stress and collagen deposition in the hearts of strepto-zotocin-induced diabetic rats. Diabetes and Vascular Disease Research, *11*(2), 110–117.

Oseguera-Toledo, M. E., de Mejia, E. G., Dia, V. P., & Amaya-Llano, S. L. (2011). Common bean (Phaseolus vulgaris L.) hydrolysates inhibit inflammation in LPS-induced macrophages through suppression of NF-κB pathways. Food Chemistry, *127*(3), 1175–1185.

Parmar, N., Singh, N., Kaur, A., & Thakur, S. (2017). Comparison of color, anti-nutritional factors, miner-als, phenolic profile and protein digestibility between hard-to-cook and easy-to-cook grains from dif-ferent kidney bean (Phaseolus vulgaris) accessions. Journal of Food Science and Technology, *54*(4), 1023–1034.

Parmar, N., Virdi, A. S., Singh, N., Kaur, A., Bajaj, R., Rana, J. C., Agrawal, L. and Nautiyal, C. S., (2014). Evaluation of physicochemical, textural, mineral and protein characteristics of kidney bean grown at Himalayan region. Food Research International, *66*, 45–57.

Pérez-Ramírez, I. F., Becerril-Ocampo, L. J., Reynoso-Camacho, R., Herrera, M. D., Guzmán-Maldonado, S. H., & Cruz-Bravo, R. K. (2018). Cookies elaborated with oat and common bean flours improved serum markers in diabetic rats. Journal of the Science of Food and Agriculture, *98*(3), 998–1007.

Preuss, H. G., Echard, B., Bagchi, D., & Stohs, S. (2007). Inhibition by natural dietary substances of gastroin-testinal absorption of starch and sucrose in rats 2. Subchronic studies. International Journal of Medical Sciences, *4*(4), 209.

Punia, S., Dhull, S. B., Sandhu, K. S., Kaur, M., & Purewal, S. S. (2020). Kidney bean (Phaseolus vulgaris) starch: A review. Legume Science, *2*(3), e52.

Punia, S., Dhull, S. B., Sandhu, K. S., & Kaur, M. (2019). Faba bean (Vicia faba) starch: Structure, properties, and in vitro digestibility—A review. Legume Science, *1*(1), e18.

Ramzy, R. A., & Putra, A. B. N. (2018). Evaluation of white bread physical characteristics substituted by red kidney bean flour with different particle sizes and concentrations. The Journal of Microbiology, Biotechnology and Food Sciences, *9*(3), 610.

Romero, H. M., & Zhang, Y. (2019). Physicochemical properties and rheological behavior of flours and starches from four bean varieties for gluten-free pasta formulation. Journal of Agriculture and Food Research, *1*, 100001.

Rondanelli, M., Giacosa, A., Orsini, F., Opizzi, A., & Villani, S. (2011). Appetite control and glycaemia reduction in overweight subjects treated with a combination of two highly standardized extracts from Phaseolus vulgaris and Cynara scolymus. Phytotherapy Research, *25*(9), 1275–1282.

Roy, M., Haque, S. M. N., Das, R., Sarker, M., Al Faik, M. A., & Sarkar, S. (2020). Evaluation of physicochemical properties and antioxidant activity of wheat-red kidney bean biscuits. World Journal of Engineering and Technology, *8*(4), 689–699.

Shang, R., Wu, H., Guo, R., Liu, Q., Pan, L., Li, J., Hu, Z. and Chen, C., 2016. The diversity of four antinutritional factors in common bean. Horticultural Plant Journal, *2*(2), 97–104.

Sharma, I., Sinhmar, A., Thory, R., Sandhu, K. S., Kaur, M., Nain, V., Pathera, A.K. & Chavan, P. (2020). Synthesis and characterization of nano starch-based composite films from kidney bean (Phaseolus vulgaris). Journal of Food Science and Technology, 1–8.

Sharma, S., Singh, A., & Singh, B. (2019). Effect on germination time and temperature on techno-functional properties and protein solubility of pigeon pea (Cajanus cajan) flour. Quality Assurance and Safety of Crops & Foods, *11*(3), 305–312.

Sharma, S., Singh, N., & Katyal, M. (2016). Effect of gelatinized-retrograded and extruded starches on characteristics of cookies, muffins and noodles. Journal of Food Science and Technology, *53*(5), 2482–2491.

Shehzad, A., Chander, U. M., Sharif, M. K., Rakha, A., & Ansari, A. Shuja. MZ (2015). Nutritional, functional and health promoting attributes of red kidney beans: A review. Pakistan Journal of Food Sciences, *25*(4), 235–246.

Shevkani, K., & Singh, N. (2014). Influence of kidney bean, field pea and amaranth protein isolates on the characteristics of starch-based gluten-free muffins. International Journal of Food Science & Technology, *49*(10), 2237–2244.

Shevkani, K., & Singh, N. (2015). Relationship between protein characteristics and film-forming properties of kidney bean, field pea and amaranth protein isolates. International Journal of Food Science & Technology, *50*(4), 1033–1043.

Shevkani, K., Singh, N., Kaur, A., & Rana, J. C. (2015). Structural and functional characterization of kidney bean and field pea protein isolates: A comparative study. Food Hydrocolloids, *43*, 679–689.

Shi, Z., Zhu, Y., Teng, C., Yao, Y., Ren, G., & Richel, A. (2020). Anti-obesity effects of α-amylase inhibitor enriched-extract from white common beans (Phaseolus vulgaris L.) associated with the modulation of gut microbiota composition in high-fat diet-induced obese rats. Food & Function, *11*(2), 1624–1634.

Shimelis, E. A., & Rakshit, S. K. (2007). Effect of processing on antinutrients and in vitro protein digestibility of kidney bean (Phaseolus vulgaris L.) varieties grown in East Africa. Food Chemistry, *103*(1), 161–172.

Sibian, M. S., & Riar, C. S. (2020a). Optimization and evaluation of composite flour cookies prepared from germinated triticale, kidney bean and chickpea. Journal of Food Processing and Preservation, e14996.

Sibian, M. S., & Riar, C. S. (2020b) Formulation and characterization of cookies prepared from the composite flour of germinated kidney bean, chickpea, and wheat. Legume Science, *2*(3): e42. doi: 10.1002/leg3.42

Siddiq, M., Ravi, R., Harte, J. B., & Dolan, K. D. (2010). Physical and functional characteristics of selected dry bean (Phaseolus vulgaris L.) flours. LWT-Food Science and Technology, *43*(2), 232–237.

Singh, A., & Sharma, S. (2017). Bioactive components and functional properties of biologically activated cereal grains: a bibliographic review. Critical Reviews in Food Science and Nutrition, *57*(14), 3051–3071.

Singh, A., Sharma, S., & Singh, B. (2017a). Effect of germination time and temperature on the functionality and protein solubility of sorghum flour. Journal of Cereal Science, *76*, 131–139.

Singh, A., Sharma, S., & Singh, B. (2017b). Influence of grain activation conditions on functional characteristics of brown rice flour. Food Science and Technology International, *23*(6), 500–512.

Song, H., Han, W., Yan, F., Xu, D., Chu, Q., & Zheng, X. (2016). Dietary Phaseolus vulgaris extract alleviated diet-induced obesity, insulin resistance and hepatic steatosis and alters gut microbiota composition in mice. Journal of Functional Foods, *20*, 236–244.

Sparvoli, F., Laureati, M., Pilu, R., Pagliarini, E., Toschi, I., Giuberti, G., … & Bollini, R. (2016). Exploitation of common bean flours with low antinutrient content for making nutritionally enhanced biscuits. Frontiers in Plant Science, *7*, 928.

Surasani, V. K. R., Singh, A., Gupta, A., & Sharma, S. (2019). Functionality and cooking characteristics of pasta supplemented with protein isolate from pangas processing waste. LWT-Food Science and Technology, *111*, 443–448.

Tang, C. H., & Ma, C. Y. (2009). Heat-induced modifications in the functional and structural properties of vicilin-rich protein isolate from kidney (Phaseolus vulgaris L.) bean. Food Chemistry, *115*(3), 859–866.

Tang, C. H., Xiao, M. L., Chen, Z., Yang, X. Q., & Yin, S. W. (2009). Properties of cast films of vicilin-rich protein isolates from Phaseolus legumes: Influence of heat curing. LWT-Food Science and Technology, *42*(10), 1659–1666.

Thompson, M. D., Thompson, H. J., Brick, M. A., McGinley, J. N., Jiang, W., Zhu, Z., & Wolfe, P. (2008). Mechanisms associated with dose-dependent inhibition of rat mammary carcinogenesis by dry bean (Phaseolus vulgaris, L.). The Journal of Nutrition, *138*(11), 2091–2097.

Tormo, M. A., Gil-Exojo, I., de Tejada, A. R., & Campillo, J. E. (2006). White bean amylase inhibitor administered orally reduces glycaemia in type 2 diabetic rats. British Journal of Nutrition, *96*(3), 539–544.

Tormo, M. A., Gil-Exojo, I., de Tejada, A. R., & Campillo, J. E. (2004). Hypoglycaemic and anorexigenic activities of an α-amylase inhibitor from white kidney beans (Phaseolus vulgaris) in Wistar rats. British journal of nutrition, *92*(5), 785–790.

Trisnawati, C. Y., & Sutedja, A. M. (2014). Utilization of mung bean and red kidney bean as fat replacer in rice cake. In: Proceeding International Conference Food for a Quality Life. Department of Food Science and Technology. Jakarta: IPB, 1–8.

Ukeyima, M. T., Dendegh, T. A., & Isusu, S. E. (2019). Quality characteristics of bread produced from wheat and white kidney bean composite flour. European Journal of Nutrition & Food Safety, 263–272.

Wang, B. H., Lei, S., Gu, L. H., & Qu, L. L. (2015). Optimization of peach blossom cake formula based on fuzzy mathematics sensory evaluation and orthogonal experiment. Food Science and Technology, (10), 46.

Wani, I. A., Andrabi, S. N., Sogi, D. S., & Hassan, I. (2020). Comparative study of physicochemical and functional properties of flours from kidney bean (Phaseolus vulgaris L.) and green gram (Vigna radiata L.) cultivars grown in Indian temperate climate. Legume Science, *2*(1), e11.

Wani, I. A., Sogi, D. S., Sharma, P., & Gill, B. S. (2016). Physicochemical and pasting properties of unleavened wheat flat bread (Chapatti) as affected by addition of pulse flour. Cogent Food & Agriculture, *2*(1), 1124486.

Wani, I. A., Sogi, D. S., Shivhare, U. S., & Gill, B. S. (2015). Physico-chemical and functional properties of native and hydrolyzed kidney bean (Phaseolus vulgaris L.) protein isolates. Food Research International, *76*, 11–18.

Wani, I. A., Sogi, D. S., Wani, A. A., & Gill, B. S. (2013). Physico-chemical and functional properties of flours from Indian kidney bean (Phaseolus vulgaris L.) cultivars. LWT—Food Science and Technology, *53*(1), 278–284.

Wani, I. A., Sogi, D. S., Wani, A. A., Gill, B. S., & Shivhare, U. S. (2010). Physico-chemical properties of starches from Indian kidney bean (Phaseolus vulgaris) cultivars. International Journal of Food Science & Technology, *45*(10), 2176–2185.

Xu, B. J., Yuan, S. H., & Chang, S. K. C. (2007). Comparative analyses of phenolic composition, antioxidant capacity, and color of cool season legumes and other selected food legumes. Journal of Food Science, *72*(2), S167–S177.

Yasmin, A., Zeb, A., Khalil, A. W., Paracha, G. M. U. D., & Khattak, A. B. (2008). Effect of processing on anti-nutritional factors of red kidney bean (Phaseolus vulgaris) grains. Food and Bioprocess Technology, *1*(4), 415–419.

Ye, X. Y. Ng, T. B. Tsang, P. W. Wang, J. (2001). Isolation of a homodimeric lectin with antifungal and antiviral activities from red kidney bean (Phaseolus vulgaris) seeds. Journal of Protein Chemistry, 20, 367–375.

Yin, S. W., Tang, C. H., Wen, Q. B., & Yang, X. Q. (2010). Functional and conformational properties of phaseolin (Phaseolus vulgris L.) and kidney bean protein isolate: A comparative study. Journal of the Science of Food and Agriculture, *90*(4), 599–607.

Zayas, J. F. (1997). Foaming properties of proteins. In Functionality of Proteins in Food (pp. 260–309). Springer, Berlin, Heidelberg.

Zhao, Y., Du, S. K., Wang, H., & Cai, M. (2014). In vitro antioxidant activity of extracts from common legumes. Food Chemistry, 152 (1), 462–466.

Zhu, L., Liu, S., Xie, Y., & Zhou, C. (2017). Kidney bean starch thin noodle processing and its characteristics. Food and Fermentation Industries, *43*(6), 225–231.

Zhu, Z., Jiang, W., & Thompson, H. J. (2012). Edible dry bean consumption (Phaseolus vulgaris L.) modulates cardiovascular risk factors and diet-induced obesity in rats and mice. British Journal of Nutrition, *108*(S1), S66–S73.

23

Black Gram: Bioactive Components for Human Health and Their Functions

Barinderjit Singh, Reetu and Gurwinder Kaur
Department of Food Science and Technology, I. K. Gujral Punjab Technical University, Kapurthala, Punjab, India

CONTENTS

23.1 Introduction

Black gram (*Vigna mungo* [L.] Hepper) is an Indian-originated fast-growing warm-season legume belonging to the family *Fabaceae* and the genus *Vigna*. The best suitable season for black gram cultivation is from September to October and March to April. Black gram is grown in all parts of India. India produces 70% of the world's production of black gram (Deepalakshmi and Kumar 2004). It has a well-developed taproot, and the leaves are trifoliate with ovate leaflets (Samad et al. 2013). The fruit is a narrow, cylindrical, erect hairy pod having 4–10 ellipsoid black or mottled seeds (Jansen 2006; Choon et al. 2010). The seed coat color ranges from dark black to dull grey. Black gram is a good source of carbohydrates (51.5%), protein (20–25%), dietary fiber, and many vitamins and minerals (Girish, Pratape, and Prasada Rao 2012). However, the seed's nutritional composition generally varies with cultivar, seed origin, and agronomic practices (Suneja et al. 2011; Indira 2003). Black gram seeds are mainly a staple food and a very popular pulse due to it being an essential ingredient in many Indian recipes.

The de-hulled and split seeds are a typical dish in South Asia. The whole or split seed of black gram is mostly consumed after boiling. There is extensive use of its de-husked seed in south India after grounding to prepare dosa, idli, papadum, and vada, whereas dal makhani is a famous dish of north India (Jansen 2006; Reddy, Salunkhe, and Sathe 1982). In Japan, it is generally consumed after overnight soaking as fresh bean sprout salad. Black gram sprouts are known for their high digestibility and lack of flatulence induction (Fery 2002; Jansen 2006). This pulse is also referred to as the "king of pulses." Besides human use, different parts of black gram crops, such as seeds, pods, leaves, stems, seed, and pod husks, are also used as fodder, green manure, and a dry season cover crop in rice or wheat (Fuller 2004; Jansen 2006; Parashar 2006). Black gram is very frequently used in traditional medicine to cure diarrhea. Black gram is also known for its antidiabetic, antihypertensive, anticancer, antimicrobial, and antioxidant properties (Shaheen et al. 2012). The antioxidant activities of black gram are due to phenolic acids, flavonoids, carotenoids, and tocopherols. These compounds have a beneficial role in protecting seeds against oxidative damage and microbial infections (Troszyńska et al. 2002).

23.2 Bioactive Components from Black Gram

The black gram seed fractions contain many different bioactive components, such as phenolic acids, flavonoids, phytoestrogen, tocopherols, fatty acids, carotenoids, proteins, amino acids, polysaccharides, and enzymes (Table 23.1). However, a small amount of bioactive components are present in black gram like other foods. However, their chemical structure makes them beneficial toward reducing or preventing many diseases like diabetes, inflammation, allergy, hyperpigmentation, cancer, and cardiovascular disease in the human body.

23.2.1 Phenolic Acids

A significant variation in phenolic acids are present in a different fraction of black gram, as shown in Table 23.1. Phenolic acids are concentrated in the seed coat (Preet and Punia 2000). These phenolic acids are nonflavonoid polyphenolic compounds that can be further classified as hydroxybenzoic acid and hydroxycinnamic acid derivatives. Gallic, protocatechuic, vanillic, genistic, and syringic acids are the primary hydroxybenzoic acids present in black gram. However, ferulic acid is major hydroxycinnamic acid derivative found in black gram that has good antioxidant, anti-inflammatory, antidiabetics, anticancer, antiaging, antiapoptotic, hepatoprotective, radioprotective, neuroprotective, antiatherogenic, hypotensive, and pulmonary protective properties (Girish, Pratape, and Prasada Rao 2012; Girish, Kumar, and Prasada Rao 2016; Dhumal, Chaudhari, and Chavan 2019; Srinivasan, Sudheer, and Menon 2007). Besides this, the ferulic acid is more stable over longer periods of time than vitamin C (Srinivasan, Sudheer, and Menon 2007). The black gram seed coat is enriched with these many phenolic acids, whereas the plumule fraction lacks vanillic, syringic, and caffeic acids (Dhumal, Yele, and Ghodekar 2013). Most phenolic acids in legumes are linked via ester, ether, or acetal bonds to cellulose, protein, lignin, or smaller organic molecules, such as glucose, quinic acid, and maleic acid, to be released with alkali, acid, and enzymatic hydrolysis (Amarowicz and Pegg 2019).

23.2.2 Flavonoids

The flavonoids are one of the classifications of polyphenols. The flavonoids are further classified into flavonols, flavones, isoflavones, flavanones, anthocyanidins, and flavanols. Three different flavonols glycosides, namely robinin, kaempferol 3-rutinoside, and kaempferol 7-rhamnoside, are present in an appreciable amount in the leaves of black gram (Mato and Ishikura 1993; Dhumal, Chaudhari, and Chavan 2019). Kaempferol appears to be the most predominant flavonol in black gram (Onyilagha, Islam, and Ntamatungiro 2009). Mature leaves of a black gram plant can also form a flavonol named quercetin glycosides, except for seedling leaves. Hypocotyl and stem tissues of black gram form all three quercetin glycosides, thus, their flavonol composition is quite different from that of young leaves (Mato and Ishikura 1993). C-glycosyl flavones such as vitexin and isovitexin are found higher in black gram seed husks than other fractions that exhibited anticancer activities by protecting DNA and erythrocytes

TABLE 23.1

Bioactive Components in Black Gram

Components	Black Gram Fraction	Concentration	Reference
Phenolic Acid			
Gallic acid	Whole seed, seed coat, flour, aleurone layer, plumule	0.22–9.77 mg/100 g	Girish, Pratape, and Prasada Rao 2012
Protocatechuic acid	Whole seed, seed coat, germ, aleurone layer, plumule, flour	0.29–8.5 mg/100 g	
Vanillic acids	Whole seed, plumule, flour	0.05–19.5 mg/100 g	
Genistic acid	Whole seed, aleurone layer, plumule, flour	1.0–88.2 mg/100 g	
Syringic acid	Whole seed, aleurone layer, flour	0.06–1.56 mg/100 g	
Ferulic acid	Whole seed, germ, plumule, flour	15.2–684.0 mg/100 g	
Caffeic acid	Whole seed, flour	0.02–1.34 mg/100 g	
p-Coumaric acid	Whole seed	Traces	Sharma 1979
Flavonoids			
Vitexin	Husk	2.55–536.5 µg/g	Girish, Kumar, and Prasada Rao 2016
Isovitexin		5.06–518.61 µg/g	
Cyclokievitone	Whole seed	1.0 µg/g	Adesanya, O'Neill, and Roberts 1984
Kievitone		19.8 µg/g	
2'-Hydroxydaidzein		0.7 µg/g	
5-Deoxykievitone		0.9 µg/g	
Isoferreirin		2.0 µg/g	
Dalbergioidin		5.6 µg/g	
2'-Hydroxydihydrodaidzein		0.1 µg/g	
Aureol		2.1 µg/g	
Kievitone hydrate		1.2 µg/g	
Genistein		3.6 µg/g	
2'-Hydroxygenistein		3.4 µg/g	
4'-Methylkievitone		1.0 µg/g	
Cyclokievitone hydrate		3.0 µg/g	
5-Deoxykievitone hydrate		0.1 µg/g	
Daidzein		88.0 mg/100 g	Sharma 1979
Total Dietary Fiber			
Total dietary fibre	Whole seed	39.9%	Girish, Pratape, and Prasada Rao 2012
	Cotyledon	24.4%	
	Germ	44.8%	
	Seed coat	78.5%	
	Plumule fraction	71.9%	
Fucose	Whole seed	0.07 g/100 g	Bravo, Siddhuraju, and Saura-Calixto 1999
Arabinose		6.22 g/100 g	
Xylose		2.21 g/100 g	
Mannose		0.31 g/100 g	
Galactose		2.14 g/100 g	
Glucose		7.57 g/100 g	
Total neutral sugars		18.6 g/100 g	
Klason lignin		6.60 g/100 g	
Uronic acids		5.69 g/100 g	

(Continued)

TABLE 23.1 (Continued)

Components	Black Gram Fraction	Concentration	Reference
Phytoestrogen			
Secoisolariciresinol	Whole seed	105.0 μg/100 g	Mazur et al. 1998; Kurzer and
Matairesinol		70.8 μg/100 g	Xu 1997
Coumestans		Present in trace quantity	
Protein			
Total protein content	Soaked seed	25.4 g/100 g	
Albumin		3.06 g/100 g	Mahajan, Malhotra, and
Globulin		15.9 g/100 g	Singh 1988
Prolamine		0.25 g/100 g	
Glutelin		5.31 g/100 g	
Amino Acids			
Aspartic acid	Soaked seed	0.28–2.79 g/100 g	Mahajan, Malhotra, and
Threonine		0.04–0.42 g/100 g	Singh 1988
Glutamic acid		0.90–5.64 g/100 g	
Serine		0.02–0.38 g/100 g	
Proline		0.34–2.09 g/100 g	
Glycine		0.12–1.38 g/100 g	
Alanine		0.30–0.88 g/100 g	
Valine		0.05–0.48 g/100 g	
Cysteine		Traces–0.19 g/100 g	
Methionine		0.01–0.44 g/100 g	
Isoleucine		0.03–1.03 g/100 g	
Leucine		0.13–2.33 g/100 g	
Tyrosine		0.08–0.69 g/100 g	
Phenylalanine		0.01–0.38 g/100 g	
Lysine		0.17–2.04 g/100 g	
Histidine		0.17–1.61 g/100 g	
Arginine		0.02–0.62 g/100 g	
Enzyme			
Peroxidase	Whole seed, germ, Plumule, aleurone rich husk	16500–1155000 U/g	Ajila and Prasada Rao 2009
Polyphenol oxidase		24–160 U/g	
Protease		102–2124 U/g	
Xylanase		0.4–2.0 U/g	
α-amylase		4–16 U/g	
Tocopherols			
γ-tocopherol	Whole seed	6.58 mg/100 g	Gopala Krishna, Prabhakar, and
α-Tocopherol		0.03 mg/100 g	Aitzetmüller 1997
δ-Tocopherol		0.15 mg/100 g	
Carotenoids			
Total carotenoids	Whole seed	0.052 mg/100 g	Girish, Pratape, and Prasada
	Cotyledon	0.042 mg/100 g	Rao 2012
	Germ	0.034 mg/100 g	
	Seed coat	0.415 mg/100 g	
	Plumule fraction	0.128 mg/100 g	

Components	Black Gram Fraction	Concentration	Reference
Minerals			
Calcium	Whole seed	375.0–485.4 mg/100 g	
Magnesium		208.5–263.8 mg/100 g	Zia-Ul-Haq et al. 2014
Phosphorous		440.9–500.2 mg/100 g	
Potassium		1600.0–1646.1 mg/100 g	
Manganese		2.39–3.32 mg/100 g	
Copper		3.92–4.51 mg/100 g	
Iron		5.89–6.55 mg/100 g	
Sodium		227.0–284.1 mg/100 g	
Zinc		1.94–2.50 mg/100 g	
Phytic Acid			
Phytic acid	Whole seed	633–657 mg/100 g	Kataria, Chauhan, and Punia 1989
Phytate phosphorus	Whole seed	5.2–5.5 mg/100 g	Reddy, Balakrishnan, and Salunkhe 1978
Trypsin Inhibitor Activity			
Trypsin inhibitor activity	Whole grain	905 TIU/g	Chitra and Sadasivam 1986
	Cotyledon	896.85 TIU/g	
	Husk	8.14 TIU/g	

g = gram, µg = microgram, mg = milligram, U = unit, TIU = trypsin inhibitor unit.

from oxidative damage (Girish, Kumar, and Prasada Rao 2016). The isoflavones, namely cyclokievitone, pterocarpan glycinol, 2'-hydroxydaidzein, 5-deoxykievitone, isoferreirin, daidzein, 2'-hydroxydihydro-daidzein, coumestan aureol, kievitone, 4'-0-methylkievitone, cyclokievitone hydrate, and 5-deoxykievi-tone hydrate, are also present in black gram (Adesanya, O'Neill, and Roberts 1984; Dhumal, Chaudhari, and Chavan 2019). The black seedcoats and the purple-red hypocotyls of black gram contain three kinds of anthocyanin: delphinidin 3-glucoside, delphinidin 3-p-coumaroyl glucoside, and cyanidin 3-gluco-side. Black gram also contains leucocyanidin and leucodelphinidin, with the former as the predominant component (Ishikura and Iwata 1981; Dhumal, Chaudhari, and Chavan 2019). Black gram also contains a significant amount of tannin content, which can be reduced with soaking and processing methods. The long duration of thermal processing will break down the tannin into gallic acid, thus increasing the food's antioxidant activity (González, Torres, and Medina 2010). Proanthocyanidin, delphinidin, and cyanidin are present in the seeds. These compounds are present in the black gram seed coat at higher concentrations (Dhumal, Chaudhari, and Chavan 2019).

23.2.3 Dietary Fiber

Black gram has higher total dietary fiber than other legumes such as moth beans, horse gram, green gram, haricot beans, and chickpeas (Bravo, Siddhuraju, and Saura-Calixto 1999). The seed coat and plu-mule fractions of black gram are higher in fiber content than cotyledon fractions. The intake of a dietary fiber-enriched diet can increase bile acid synthesis and excretion and reduce cholesterol biosynthesis and absorption, leading to lower blood cholesterol (Naumann et al. 2019; Soliman 2019). Black gram con-tains a higher quantity of insoluble dietary fiber as compared to soluble dietary fiber. Insoluble dietary fiber plays a significant role in the human body to collect water and increase the stool bulk in the large intestine. This prevents constipation by increasing the fecal bulk, rendering feces softer, and shortening bowel transit time (Rose et al. 2015). The consumption of a fiber-enriched diet can reduce the absorption of salt, toxins, and other potential carcinogens (Soliman 2019). Black gram dietary fiber is fermented in the large intestine, which can change the microflora present in the gut and inhibit the development of colorectal cancer by decreasing the colonic pH with more short-chain production fatty acids such as acetate, propionate, and butyrate (Lanza et al. 2006). The major composition of black gram dietary fiber is fucose, arabinose, xylose, mannose, galactose, glucose, total neutral sugars, uronic acids, and lignin (Bravo, Siddhuraju, and Saura-Calixto 1999).

23.2.4 Phytoestrogens

Phytoestrogens are naturally occurring plant compounds that are structurally and/or functionally like mammalian estrogens and their active metabolites. Most phytoestrogens, however, are phenolic compounds such as isoflavones and coumestans. However, isoflavones are most abundant in soy beans, but black gram also contains these significant phytoestrogens (Kurzer and Xu 1997). Besides this, another phytoestrogen like lignin such as secoisolariciresinol (SECO) and matairesinol (MAT) are present in black gram seed at a sufficient enough amounts known for their antiestrogenic, anticarcinogenic, antiviral, antifungal, and antioxidant activities (Mazur et al. 1998).

23.2.5 Proteins and Amino Acids

The black gram seed contains an appreciable amount of different proteins such as albumins (12%), globulins (63%), glutelins (21%), and prolamins (1%) (Mahajan, Malhotra, and Singh 1988). Globulins are the major storage proteins in vacuoles of black gram cotyledon that can protect from cardiovascular disease (Harris and Chrispeels 1975). Globulins can be further fractionated into legumin and vicilin-type proteins. The legumin proteins form 80% of the total globulins, and the vicilin forms 20% of the total globulins. The black gram amino acid profile is influenced by various processing methods, such as boiling, soaking, blanching, and roasting. In general, boiling decreases the majority of essential and nonessential amino acids, while roasting raises the black gram amino acid profile. Boiling black gram decreases the concentration of leucine, tryptophan, arginine, lysine, methionine, and threonine, while roasting increase the amino acid composition of leucine, isoleucine, phenylalanine, valine, aspartic acid, glutamic, alanine, histidine, tyrosine, tryptophan, methionine, cysteine, serine, and arginine except for aliphatic amino acids i.e. proline and glycine (Nwosu, Anyaehie, and Ofoedu 2019). Tyrosine and cysteine content in black gram is substantially reduced by boiling, roasting, soaking, and blanching, while processing techniques increase histidine in black grams. Roasting treatment is recommended to improve the nutritional consistency and incorporate it into food formulation to preserve amino acid composition in black gram preparation (Nwosu, Anyaehie, and Ofoedu 2019).

23.2.6 Enzymes

The various enzymes, such as peroxidase, chitinase, cysterine, and endopeptidase, are present in the different factions of black gram (Dhumal, Chaudhari, and Chavan 2019; Ajila and Prasada Rao 2009; Kandukuri et al. 2012; Yamauchi, Akasofu, and Minamikawa 1992). The aleurone layer of black gram is an excellent resource to isolate and purify the peroxidase enzyme with good broader specificity characteristics and application in the analytical and biomedical industry (Ajila and Prasada Rao 2009). Besides this, the peroxidase enzyme can also reduce aromatic amines and phenolic compounds present in wastewater and characterize disease status in experimental pathology (Dhumal, Chaudhari, and Chavan 2019). A novel chitinase can be purified from seeds of black gram. The purified chitinase can be tested to determine the chitin contents of the stored cereals (Preety, Sharma, and Hooda 2018). The cotyledons of dark-grown black gram seedlings contain enzymes named acid phosphatase and α-amylase, whereas black gram seeds sprouted under normal conditions have the presence of the principal antioxidant enzyme catalase (Dhumal, Chaudhari, and Chavan 2019; Kandukuri et al. 2012; Tamura, Minamikawa, and Koshiba 1982; Koshiba and Minamikawa 1981).

23.2.7 Tocopherol and Fatty Acids

Black gram seeds contain lipophilic compounds such as tocopherol and tocotrienol. The oil fraction of Asian black gram contained high amounts of γ-tocopherols, Δ-tocopherols, and unsaturated fatty acids like linolenic (Gopala Krishna, Prabhakar, and Aitzetmüller 1997; Zia-Ul-Haq et al. 2014; Dhumal, Chaudhari, and Chavan 2019). Naturally arising tocopherols are used for oil and fat equilibrium against oxidative degradation, which proposes their usage in pharmaceutical, biomedical, and nutritional products (Zia-Ul-Haq et al. 2014). Black gram seeds are a rich source of α-linolenic acid that reduces the risk of cholesterol-related heart diseases by consuming oils containing more unsaturated fatty acids (Jandacek 2017; Gopala Krishna, Prabhakar, and Aitzetmüller 1997).

23.2.8 Carotenoids

Carotenoids have good potential to neutralize the free radicals present in the human body. These compounds are present in black grams at a higher amounts than mung bean, peas, beans, and soy beans (Girish, Pratape, and Prasada Rao 2012; Fordham, Wells, and Chen 1975; Kantha and Erdman 1987). The carotenoids content of legumes can decrease with germination and processing methods such as thermal processing, drying, and gamma-ray irradiation (Fordham, Wells, and Chen 1975; Kantha and Erdman 1987). Among different milling fractions of black gram, the seed coat and aleurone layer are the carotenoid's richest sources. Carotenoids have many health-promoting functions such as immune-enhancement, provitamin A activity, decreased risk of cataract formation, and inhibition of inflammation, cancer, and age-related ophthalmic diseases, improving brain-cognitive functions, and prevention of cardiovascular disease and macular degeneration (Dutta, Chaudhuri, and Chakraborty 2005; Bungau et al. 2019). However, the presence of high fiber in the diet can reduce the bioavailability of the carotenoids (Dutta, Chaudhuri, and Chakraborty 2005).

23.2.9 Minerals

Black gram seed contains many essential minerals such as calcium (Ca), magnesium (Mg), phosphorous (P), potassium (K), manganese (Mn), copper (Cu), iron (Fe), and zinc (Zn), which play a vital role in various metabolic processes of the human body (Zia-Ul-Haq et al. 2014). The low sodium-potassium ratio of black gram is beneficial to the hypertension patient, whereas high zinc can reduce liver fibrosis and triglyceride levels and increase the high-density lipoproteins (HDL) concentration in the human body (Perez and Chang 2014; Ranasinghe et al. 2015).

23.2.10 Other Bioactive Components

The other bioactive components found in black gram are saponins, trypsin inhibitors, and phytic acids (Kakati et al. 2010). These compounds have both potential benefits and an adverse effects on health that also depend on these compounds' concentration intake. Saponins and phytic acid have the potential to treat hypercholesterolemia by preventing cholesterol from being consumed or converted into bile acids that improve fecal excretion and lowering plasma cholesterol. The phytic acid is known for its good antioxidant activities and antidiabetic and anticancer properties. Different processing methods such as cooking, soaking, and autoclaving can reduce the saponins and phytic acid of black gram. Trypsin inhibitors, as they are heat-labile, may be irrelevant in cooked pulses. The trypsin inhibitor activity (TIA) increased as black gram seeds matured and decreased during the initial germination phase (Kakati et al. 2010). The reduction of TIA in cotyledons can boost the breakdown of cotyledon reserve proteins necessary for plant growth.

23.3 Health Benefits

23.3.1 Black Gram and Cancer

Cancer is the second leading cause of death. It is caused by physical carcinogens (UV and ionizing radiation), chemical carcinogens (tobacco smoke, aflatoxin, and arsenic), and biological carcinogens (infections caused by certain viruses, bacteria, or parasites) (Parsa 2012). The high intake of alcohol, an unhealthy diet, tobacco, stress, infection, and cytotoxic or carcinogenic compounds produce a higher amount of free radicals in the human body that can lead to oxidative damage to DNA strands (Pham-Huy, He, and Pham-Huy 2008; Phaniendra, Jestadi, and Periyasamy 2015; Sharifi-Rad et al. 2020). Therefore, many antioxidants that enrich black gram components, such as carotenoids, phenolic acids, and flavonoids, can be beneficial to scavenge these free radicals and reduce cancer risk by restoring the damage caused by free radicals (Rajagopal, Pushpan, and Antony 2017). Beta (β)-carotene content increases in black gram after the germination process that increases the free radical scavenging activity. The isolation of the glycosyl flavones like vitexin and isovitexin from black gram can be further used to reduce the HeLa cells and cervical carcinoma cells due to high antioxidant activity (Girish, Kumar, and Prasada Rao 2016). This shows that black gram glycosyl flavones exhibited anticancer activity.

Colorectal cancer is the third most widely diagnosed cancer in males and the second in females with 1.8 million new cases in 2018 (The International Agency for Research on Cancer 2018). Hyperinsulinemia increases the colorectal cancer risk by promoting colon cell proliferation and reducing apoptosis (Godsland 2010). The lower activity of the intestinal microflora is also positively associated with colon cancer (Han et al. 2018; Hibberd et al. 2017). However, dietary fiber is used by intestinal microflora for fermentation to increase their activity and production of short-chain fatty acids to inhibit colorectal cancer (Pericleous, Mandair, and Caplin 2013; Venegas et al. 2019). Dietary fibers present in black gram also lead to diluting fecal carcinogens associated with a reduced risk of colorectal cancer. Dietary fiber contributes to reducing constipation by solidifying the stool production, accelerating the passage rate through the intestine, and altering the colonic metabolism of bile acids and minerals, increasing the bulk of the colonic content and increasing the water content (Prado et al. 2019). Both these pathways lead to the prevention of bowel cancer (Rajagopal, Pushpan, and Antony 2017).

Tocopherols and tocotrienols are another very promising cancer prevention agent that prevent colon, prostate, mammary, and lung cancer, but cannot be synthesized in humans and animals. Therefore, they need to be obtained from dietary sources such as black gram seeds. The major dietary tocopherols in the black gram supplement diet are γ- and α-tocopherols. Among these, γ-tocopherol is more effective in the prevention of cancer. Tocotrienols, vitamin-E isomers have more significant anticancer activity *in vitro* than tocopherols (Ju et al. 2009). Minor constituents present in black grams, such as lectin, oligosaccharides, saponins, phytates, and protease inhibitors, also have anticancer potential.

23.3.2 Black Gram and Diabetes

Diabetes mellitus is a group of metabolic diseases caused by a deficiency or ineffective insulin production by the pancreas, resulting in increased or decreased glucose concentration in the blood (American Diabetes Association 2009). There is an intimate relationship between the human diet and the risk of diabetes (Sami et al. 2015; Beigrezaei et al. 2019; Adeva-Andany et al. 2019). The black gram seed coat is rich in dietary fiber, saponin, and a group of polyphenolic antioxidants responsible for free-radical scavenging effects, which are beneficial to maintain blood glucose levels (Paul et al. 2011). Therefore, the black gram extracts help prevent obesity and type 2 diabetes due to low glycemic index and high fiber content. Regular intake of black gram fibers help regulate the absorption of nutrients, which maintains blood glucose concentration (Cryer 2005). The hypoglycemic action of black gram fiber may have been due to increased glucose in liver glycogen synthesis and decreased gluconeogenesis (Boby and Leelamma 2003).

Further, the cooking of black grams can reduce the glycemic index and lower blood glucose levels due to carbohydrate modification (Singhal, Kaushik, and Mathur 2014). Another bioactive component, such as saponins present in black grams also have good antidiabetic activity. Saponin produces the antidiabetic activities by different mechanisms such as activation of glycogen synthesis, which modulates insulin signaling, suppresses gluconeogenesis, regenerates insulin action, and suppressesf disaccharide activity (Barky and Hussein 2017; Elekofehinti 2015). Phenolic acids, flavonoids, and saponin are also known for having an excellent free-radical scavenging activity that will reduce the late diabetic complications produced due to DNA damage (Girish, Pratape, and Prasada Rao 2012).

23.3.3 Black Gram and Cardiovascular Disease (CVD)

Cardiovascular disease (CVD) is one of the world's most severe chronic diseases correlated with human lifestyles, eating habits, and lack of physical activity. Sometimes other factors responsible for increasing CVD are increased blood pressure, hypercholesterolemia, glucose, lipids, obesity, lack of physical activity, extensive use of tobacco, and alcohol (Buttar, Li, and Ravi 2005). Black gram constituents like polysaccharides (fiber), saponins, and protein (globulin), help increase cholesterol absorption by the degradation of cholesterol to bile acids and their fecal excretion possess lipid-lowering activity. Dietary fiber reduces the risk of cardiovascular diseases through various mechanisms such as improving serum lipid concentration, lowering blood pressure, and reducing inflammation (Indira 2003).

Saponins in black gram exhibit various improvements to cardiovascular activities. Saponin can penetrate cell and plasma membranes to cause positive inotropic effects in isolated cardiac muscles. Saponins inhibit cholesterol absorption from the intestinal lumen and reduce plasma cholesterol concentration by lowering low-density lipoprotein (LDL) cholesterol concentration without changing HDL cholesterol concentration levels. However, this may prevent hypercholesterolemia (Soetan 2008). There is a good association between the consumption of polyphenol-enriched foods and a reduction in CVD. Black gram polyphenols can alter the hepatic cholesterol absorption, triglyceride biosynthesis, and lipoprotein secretion and inflammation (Jesch and Carr 2017). The antiatherogenic in black gram inhibit the peroxyl radical-induced DNA strand breakage, inhibit platelet aggregation, and protect LDL from oxidative damage (Chikane et al. 2011). Polyphenols also decrease atherosclerosis development by decreasing the activity of many enzymes, such as secretory phospholipase, myeloperoxidase, and sphingomyelinase, to improve endothelial function and blood pressure (Cheng et al. 2017). Flavones such as vitexin and isovitexin present in black gram is very beneficial in reducing the total cholesterol by removing the free fatty acid from circulation and increasing the lipoprotein lipase activity (Kaur, Somaiya, and Patel 2015).

23.3.4 Black Gram and Immune System

Immune system imbalance can cause or trigger many diseases in humans because the pathogeneses of many human diseases involve immune function (Nicholson 2016). Immune dysfunction in the mucosa can reduce gut flora and lead to diarrhea (Zhang et al. 2015). A well-balanced diet can enhance immunity by interfering with proinflammatory cytokine synthesis and modifying the immune cell regulation and gene expression (Wu et al. 2019; Ding, Jiang, and Fang 2018). Black gram polyphenols can produce a significantly positive effect on the level of interleukin-2I and decrease the release of interleukin-1 β and interleukin-6. They can also modulate immune responses by affecting mechanisms such as regulatory DNA methylation, histone modification, and micro-RNA-mediated post-transcriptional repression that alters the expression of gene encoding key immune factors. Polyphenols also inhibit antigen-specific IgE antibody formation by two mechanisms: one is by affecting the formation of the allergen-IgE complex, and the second is by affecting the binding of this complex to its receptors (Ding, Jiang, and Fang 2018).

Quercetin is a key flavonoid in black gram with an antiallergic effect due to high free-radical scavenging capacity, inhibition of histamine release, reduction of proinflammatory cytokines, and leukocyte production (Mlcek et al. 2016). Besides this, ferulic acid inhibits tumor necrosis factor production with lipopolysaccharides, which enhance the antitumor immune activity (Ding, Jiang, and Fang 2018). Black gram resveratrol is another beneficial compound for the immune system that can improve central cellular components of innate and adaptive immunity. Stilbenes exert an immunomodulatory effect by decreasing the expression of activating receptors on immune cells and increasing immunosuppressive cytokines production. Thus, polyphenols enhance the immunomodulatory process, intestinal mucosa immunity, and antitumor immune activity (Ding, Jiang, and Fang 2018).

The nonstarchy polysaccharides of black gram also help induce macrophages by cytokines such as interferon-gamma and bacterial endotoxins such as lipopolysaccharides (Wismar 2010). Activated macrophages kill the invading bacteria or infected cells that lead to immunomodulatory properties. Black gram dietary fiber has intimate contact with the immune system of the intestine. Immune-enhancing effects of dietary fibers assess the different mechanisms by the change in gut microflora, impacting the immune system (Schley and Field 2002; Venegas et al. 2019).

23.3.5 Black Gram and Hepatotoxicity

Liver diseases are among the most severe health problems associated with oxidative stress and aging (Kim, Kisseleva, and Brenner 2015). The liver's primary function is to take up, store, and provide nutrients to other organs and eliminate unwanted substances. During the elimination process, there is a chance of accumulation of toxic materials inside the hepatocytes that lead to liver infection. Excess consumption of certain drugs like antibodies, acetaminophen, chemotherapeutic agents, and exposure to some chemicals, such as peroxidises, oils, aflatoxins, carbon tetrachloride, and ethanol, increases the susceptibility of the liver to a variety of disorders like jaundice and hepatitis, which are two major hepatic disorders that lead to mortality (Liguori et al. 2018; Kim, Kisseleva, and Brenner 2015; Li et al. 2015).

The alkaloids of black gram exhibit good hepatoprotective activity prevention by blocking secretion of hepatic triglycerides into the plasma (Satyanarayana et al. 2012). The hepatoprotective activity of black gram may be due to its antioxidant activity, which may be due to flavonoids and phenolic acids (Sangale and Patil 2017). The vitexin and isovitexin present in black gram extracts the active sites of reactive oxygen species that have a greater affinity to cause cellular damage (Kumari, Rao, and Padmaja 2012). Another flavonoid, such as quercetin present in black gram, has good hepatoprotective potential by inhibiting liver fibrosis against several diseases by scavenging free radical generation and inhibiting acrylamide (ACR)-induced injury in hepatic tissues. The tannins can also reduce serum lipid levels, produce liver necrosis, reduce blood pressure, and modulate immune responses (Chung et al. 1998).

The liver is one of the body's critical organs that works to maintain homeostasis by producing coagulation factors and inhibitors, synthesizes thrombopoietin, and through vitamin-K dependent modulation of clotting factors (Palta, Saroa, and Palta 2014). Hepatoxicity negatively correlates with coagulation disorders due to the reduced synthesis of clotting and inhibitor factors (Soultati and Dourakis 2006). Essential minerals present in black grams such as Fe, Cu, Zn, K, Mg, Mn, and Ca not only act as antioxidants but also play a vital role in various liver metabolic processes. Zinc supplementation reduces liver fibrosis and triglycerides levels, resulting in reduced hepatocyte injury by decreasing LDL concentration and increasing HDL concentration. Iron may promote scar tissue formation in the liver (Guo, Chen, and Ko 2013; Li et al. 2015).

23.3.6 Black Gram and Inflammation

Inflammation is a localized protective reaction of cells or tissues of the body to any kind of noxious stimuli such as allergic or chemical irritation, injury, or infections (Jain, Pandey, and Shukla 2015; Chen et al. 2018; Rajagopal, Pushpan, and Antony 2017). Many conventional medicines and steroids are available to treat inflammation, which may have a side effect on many body parts due to prolonged administration (Bakasatae et al. 2018). However, the side-effect problem can be overcome using bioactive compounds of black gram such as various isoflavones such as genistein, protensin, formonoetin, daidzein, and glycitein, which all have good anti-inflammation effects (Venkidasamy et al. 2019; Yu et al. 2016). Supplementation of black gram increases the activity of the enzymes such as sulphuroxide dismutase, catalase, GPx, and lipid peroxidation, leading to its anti-inflammatory and antioxidant activity (Rajagopal, Pushpan, and Antony 2017). However, isoflavone content increases in black gram with germination and decreases during the cooking process (Deorukhkar and Ananthanarayan 2020; Sharma 1981; Eum et al. 2020).

23.4 Conclusion

Black gram is one of the essential staple crops that contain a good amount of carbohydrates, proteins, dietary fiber, and many vitamins and minerals. Besides these, many health-promoting and disease-preventing bioactive components are known for their antioxidant, hypoglycaemic, hypolipidemia, anti-cancer, and immunomodulatory hepatoprotective and anti-inflammatory properties. These compounds are phenolic acid, flavonoids, phytoestrogens, carotenoids, tocopherols, and various enzymes such as peroxidase, α-amylase, and polyphenol oxidase. However, potential health benefits depend on the intake concentration of these components. Therefore, black gram's bioactive components have potential applications in the analytical and biomedical industries.

REFERENCES

Adesanya, S. A., M. J. O'Neill, and M. F. Roberts. 1984. "Induced and Constitutive Isoflavonoids in *Phaseolus Mungo* L. Leguminosae." *Zeitschrift Fur Naturforschung - Section C Journal of Biosciences* 39 (9–10): 888–93.

Adeva-Andany, M. M., E. Rañal-Muíño, M. Vila-Altesor, C. Fernández-Fernández, R. Funcasta-Calderón, and E. Castro-Quintela. 2019. "Dietary Habits Contribute to Define the Risk of Type 2 Diabetes in Humans." *Clinical Nutrition ESPEN* 34: 8–17.

Ajila, C. M., and U. J.S. Prasada Rao. 2009. "Purification and Characterization of Black Gram (*Vigna Mungo*) Husk Peroxidase." *Journal of Molecular Catalysis B: Enzymatic* 60 (1–2): 36–44.

Amarowicz, R., and R. B. Pegg. 2019. "Leguminous Seeds as a Source of Phenolic Acids, Condensed Tannins, and Lignans." In: M. Á. Martín-Cabrejas (Ed.), *Legumes: Nutritional Quality, Processing and Potential Health Benefits*. UK: The Royal Society of Chemistry, 19–48.

American Diabetes Association. 2009. "Diagnosis and Classification of Diabetes Mellitus." *Diabetes Care* 32 (Suppl 1): S62–67.

Bakasatae, N., N. Kunworarath, C. Takahashi Yupanqui, S. P. Voravuthikunchai, and N. Joycharat. 2018. "Bioactive Components, Antioxidant, and Anti-Inflammatory Activities of the Wood of Albizia Myriophylla." *Revista Brasileira de Farmacognosia* 28 (4): 444–50.

Barky, A. El, and S. A. Hussein. 2017. "Saponins and Their Potential Role in Diabetes Mellitus." *Diabetes Management* 7: 148–58.

Beigrezaei, S., R. Ghiasvand, A. Feizi, and B. Iraj. 2019. "Relationship between Dietary Patterns and Incidence of Type 2 Diabetes." *International Journal of Preventive Medicine* 10 (July): 122.

Boby, R. G., and S. Leelamma. 2003. "Blackgram Fiber (Phaseolus Mungo): Mechanism of Hypoglycemic Action." *Plant Foods for Human Nutrition* 58 (1): 7–13.

Bravo, L., P. Siddhuraju, and F. Saura-Calixto. 1999. "Composition of Underexploited Indian Pulses. Comparison with Common Legumes." *Food Chemistry* 64 (2): 185–92.

Bungau, S., M. M. Abdel-Daim, D. M. Tit, E. Ghanem, S. Sato, M. Maruyama-Inoue, S. Yamane, and K. Kadonosono. 2019. "Health Benefits of Polyphenols and Carotenoids in Age-Related Eye Diseases." *Oxidative Medicine and Cellular Longevity* 2019: 1–22.

Buttar, H. S., T. Li, and N. Ravi. 2005. "Prevention of Cardiovascular Diseases: Role of Exercise, Dietary Interventions, Obesity and Smoking Cessation." *Experimental and Clinical Cardiology* 10 (4): 229–49.

Chen, L., H. Deng, H. Cui, J. Fang, Z. Zuo, J. Deng, Y. Li, X. Wang, and L. Zhao. 2018. "Inflammatory Responses and Inflammation-Associated Diseases in Organs." *Oncotarget* 9 (6): 7204–18.

Cheng, Y. C., J.M. Sheen, W. L. Hu, and Y. C.hiang Hung. 2017. "Polyphenols and Oxidative Stress in Atherosclerosis-Related Ischemic Heart Disease and Stroke." *Oxidative Medicine and Cellular Longevity* 2017: 1–16.

Chikane, M. R., D. V. Parwate, V. N. Ingle, S. Chhajjed, and A. G. Haldar. 2011. "In Vitro, Antioxidant Effect of Seed Coats Extracts of Vigna Mungo." *Journal of Pharmacy Research*, 4 (3): 656–57.

Chitra, R., and S. Sadasivam. 1986. "A Study of the Trypsin Inhibitor of Black Gram (*Vigna Mungo* (L.) Hepper)." *Food Chemistry* 21 (4): 315–20.

Choon, S. Y., S. H. Ahmad, P. Ding, U. R. Sinniah, and A. A. Hamid. 2010. "Morphological and Chemical Characteristics of Black Gram (*Vigna Mungo* L.) Sprouts Produced in a Modified Atmosphere Chamber at Four Seeding Densities." *Pertanika Journal of Tropical Agricultural Science* 33 (2): 179–91.

Chung, K. T., T. Y. Wong, C. I. Wei, Y. W. Huang, and Y. Lin. 1998. "Tannins and Human Health: A Review." *Critical Reviews in Food Science and Nutrition* 38 (6): 421–64.

Cryer, Philip E. 2005. "Mechanisms of Hypoglycemia-Associated Autonomic Failure and Its Component Syndromes in Diabetes." *Diabetes* 54 (12): 3592–3601.

Deepalakshmi, A. J., and C. R. A. Kumar. 2004. "Creation of Genetic Variability for Different Polygenic Traits in Blackgram {*Vigna Mungo* (L.) Hepper} through Induced Mutagenesis." *Legume Research* 27 (3): 188–92.

388 *Handbook of Cereals, Pulses, Roots*

Deorukhkar, A., and L. Ananthanarayan. 2020. "Effect of Thermal Processing Methods on Flavonoid and Isoflavone Content of Decorticated and Whole Pulses." *Journal of Food Science and Technology* 1–9.

Dhumal, J. S., S. R. Chaudhari, and M. J. Chavan. 2019. "A Review Bioactive Components of *Vigna Mungo*." *Journal of Drug Delivery and Therapeutics* 9 (4-s): 748–54.

Dhumal, J. S., S. U. Yele, and S. N. Ghodekar. 2013. "Evaluation of Immunomodulatory Activity of *Vigna Mungo* (L) Hepper." *Pharmacy and Phytotherapetics*, 9–14.

Ding, S., H. Jiang, and J. Fang. 2018. "Regulation of Immune Function by Polyphenols." *Journal of Immunology Research* 2018: 1–8.

Dutta, D., U. R. Chaudhuri, and R. Chakraborty. 2005. "Structure, Health Benefits, Antioxidant Property and Processing and Storage of Carotenoids." *African Journal of Biotechnology* 4 (13): 1510–20.

Elekofehinti, O. O. 2015. "Saponins: Anti-Diabetic Principles from Medicinal Plants—A Review." *Pathophysiology* 22 (2): 95–103.

Eum, H. L., Y. Park, T. G. Yi, J. W. Lee, K. S. Ha, I. Y. Choi, and N. I. Park. 2020. "Effect of Germination Environment on the Biochemical Compounds and Anti-Inflammatory Properties of Soybean Cultivars." *PLOS ONE* 15 (4): 1–14.

Fery, F. L. 2002. "New Opportunities in *Vigna*." In: J. Janick and A. Whipkey (Eds.), *Trends in New Crops and New Uses*. Alexandria, VA: ASHS Press, 424–28.

Fordham, J. R., C. E. Wells, and L. H. Chen. 1975. "Sprouting of Seeds and Nutrient Composition of Seeds and Sprouts." *Journal of Food Science* 40: 552–56.

Fuller, M F. 2004. *The Encyclopedia of Farm Animal Nutrition*. Wallingford, UK: CABI Publishing.

Girish, T. K., V. M. Pratape, and U. J. S. Prasada Rao. 2012. "Nutrient Distribution, Phenolic Acid Composition, Antioxidant and Alpha-Glucosidase Inhibitory Potentials of Black Gram (*Vigna Mungo* L.) and Its Milled by-Products." *Food Research International* 46 (1): 370–77.

Girish, T. K., K. A. Kumar, and U. J. S. Prasada Rao. 2016. "C-Glycosylated Flavonoids from Black Gram Husk: Protection against DNA and Erythrocytes from Oxidative Damage and Their Cytotoxic Effect on HeLa Cells." *Toxicology Reports* 3: 652–63.

Godsland, I. F. 2010. "Insulin Resistance and Hyperinsulinaemia in the Development and Progression of Cancer." *Clinical Science* 118 (5): 315–32.

González, M. J., J. L. Torres, and I. Medina. 2010. "Impact of Thermal Processing on the Activity of Gallotannins and Condensed Tannins from *Hamamelis Virginiana* Used as Functional Ingredients in Seafood." *Journal of Agricultural and Food Chemistry* 58 (7): 4274–83.

Gopala Krishna, A. G., J. V. Prabhakar, and K. Aitzetmüller. 1997. "Tocopherol and Fatty Acid Composition of Some Indian Pulses." *Journal of the American Oil Chemists' Society* 74 (12): 1603–6.

Guo, C. H., P. C. Chen, and W. S. Ko. 2013. "Status of Essential Trace Minerals and Oxidative Stress in Viral Hepatitis C Patients with Nonalcoholic Fatty Liver Disease." *International Journal of Medical Sciences* 10 (6): 730–37.

Han, S., J. Gao, Q. Zhou, S. Liu, C. Wen, and X. Yang. 2018. "Role of Intestinal Flora in Colorectal Cancer from the Metabolite Perspective: A Systematic Review." *Cancer Management and Research* 10: 199–206.

Harris, N., and Maarten J. Chrispeels. 1975. "Histochemical and Biochemical Observations on Storage Protein Metabolism and Protein Body Autolysis in Cotyledons of Germinating Mung Beans." *Plant Physiology* 56 (2): 292–99.

Hibberd, A. A., A. Lyra, A. C. Ouwehand, P. Rolny, H. Lindegren, L. Cedgård, and Y. Wettergren. 2017. "Intestinal Microbiota Is Altered in Patients with Colon Cancer and Modified by Probiotic Intervention." *BMJ Open Gastroenterology* 4 (1): 1–14.

Indira, M. 2003. "Black Gram (*Vigna Mungo*)—A Hypolipidemic Pulse." *Indian Journal of Natural Products and Resources* 2 (5): 240–42.

Ishikura, A., and M. Iwata. 1981. "Flavonoids of Some Vigna Plants in Leguminosae." *The Botanical Magazine* 94: 197–205.

Jain, P., R. Pandey, and S. S. Shukla. 2015. "Inflammation." In: P. Jain, R. Pandey, and S. S. Shukla *(Eds.), Inflammation: Natural Resources and Its Applications*. Springer India, 5–14.

Jandacek, R. J. 2017. "Linoleic Acid: A Nutritional Quandary." *Healthcare* 5 (2): 25.

Jansen, P. C. M. 2006. "*Vigna Mungo* (L.) Hepper. Record from Protabase." In Brink, M. & Belay, G. (Eds.), *Plant Resources of Tropical Africa 1. Cereals and Pulses*. PROTA Foundation, Wageningen, Netherlands. Leiden, Netherlands:Backhuys Publishers, 206–8.

Jesch, E. D., and T. P. Carr. 2017. "Food Ingredients That Inhibit Cholesterol Absorption." *Preventive Nutrition and Food Science* 22 (2): 67–80.

Ju, J., S. C. Picinich, Z. Yang, Y. Zhao, N. Suh, A. N. Kong, and C. S. Yang. 2009. "Cancer-Preventive Activities of Tocopherols and Tocotrienols." *Carcinogenesis* 31 (4): 533–42.

Kakati, P., S. C. Deka, D. Kotoki, and S. Saikia. 2010. "Effect of Traditional Methods of Processing on the Nutrient Contents and Some Antinutritional Factors in Newly Developed Cultivars of Green Gram [*Vigna Radiata* (L.) Wilezek] and Black Gram [*Vigna Mungo* (L.) Hepper] of Assam, India." *International Food Research Journal* 17 (2): 377–84.

Kandukuri, S. S., A. Noor, S. S. Ranjini, and M. A. Vijayalakshmi. 2012. "Purification and Characterization of Catalase from Sprouted Black Gram (*Vigna Mungo*) Seeds." *Journal of Chromatography B: Analytical Technologies in the Biomedical and Life Sciences* 889–890: 50–54.

Kantha, S. S., and J. W. Erdman. 1987. "Legume Carotenoids." *Critical Reviews in Food Science and Nutrition* 26 (2): 137–55.

Kataria, A., B. M. Chauhan, and D. Punia. 1989. "Antinutrients in Amphidiploids (Black Gram × Mung Bean): Varietal Differences and Effect of Domestic Processing and Cooking." *Plant Foods for Human Nutrition* 39 (3): 257–66.

Kaur, G., R. Somaiya, and S. Patel. 2015. "Preventive and Curative Potential of Vigna Mungo against Metabolic Syndrome in Acute and Chronic Rat Models." *Journal of Biological Sciences* 15 (2): 85–91.

Kim, I. H., T Kisseleva, and D. A. Brenner. 2015. "Aging and Liver Disease." *Current Opinion in Gastroenterology* 31 (3): 184–91.

Koshiba, T., and T. Minamikawa. 1981. "Purification by Affinity Chromatography of Alpha-Amylase—A Main Amylase in Cotyledons of Germinating *Vigna Mungo* Seeds." *Plant and Cell Physiology* 22 (6): 978–87.

Kumari, M. K., B. G. Rao, and V. Padmaja. 2012. "Role of Vitexin and Isovitexin in Hepatoproctective Effect of Alysicarpus Monilifer Linn. against CCl 4 Induced Hepatotoxicity." *Phytopharmacology* 3 (2): 273–85.

Kurzer, M. S., and X. Xu. 1997. "Dietary Phytoestrogens." *Annual Review of Nutrition* 17: 353–81.

Lanza, E., T. J Hartman, P. S Albert, R. Shields, M. Slattery, B. Caan, E. Paskett, et al. 2006. "High Dry Bean Intake and Reduced Risk of Advanced Colorectal Adenoma Recurrence among Participants in the Polyp Prevention Trial." *The Journal of Nutrition* 136 (7): 1896–1903.

Li, S., H. Y. Tan, N. Wang, Z. J. Zhang, L. Lao, C. W. Wong, and Y. Feng. 2015. "The Role of Oxidative Stress and Antioxidants in Liver Diseases." *International Journal of Molecular Sciences* 16 (11): 26087–124.

Liguori, I., G. Russo, F. Curcio, G. Bulli, L. Aran, D. Della-Morte, G. Gargiulo, et al. 2018. "Oxidative Stress, Aging, and Diseases." *Clinical Interventions in Aging* 13: 757–72.

Mahajan, R., S. P. Malhotra, and R. Singh. 1988. "Characterization of Seed Storage Proteins of Urdbean (*Vigna Mungo*)." *Plant Foods for Human Nutrition* 38 (2): 163–73.

Mato, M., and N. Ishikura. 1993. "Flavonol Changes in Seedlings of *Vigna Mungo* During Growth." *Journal of Plant Physiology* 142 (6): 647–50.

Mazur, W. M., J. A. Duke, K. Wähälä, S. Rasku, and H. Adlercreutz. 1998. "Isoflavonoids and Lignans in Legumes: Nutritional and Health Aspects in Humans." *The Journal of Nutritional Biochemistry* 9 (4): 193–200.

Mlcek, J., T. Jurikova, S. Skrovankova, and J. Sochor. 2016. "Quercetin and Its Anti-Allergic Immune Response." *Molecules* 21 (5): 1–15.

Naumann, S., U. Schweiggert-Weisz, J. Eglmeier, D. Haller, and P. Eisner. 2019. "In Vitro Interactions of Dietary Fibre Enriched Food Ingredients with Primary and Secondary Bile Acids." *Nutrients* 11 (6): 1–18.

Nicholson, L. B. 2016. "The Immune System." *Essays in Biochemistry* 60 (3): 275–301.

Nwosu, J. N., M.A. Anyaehie, and C. E. Ofoedu. 2019. "Effect of Different Processing Techniques on the Amino Acid Profile of Black Gram." *IOSR Journal of Environmental Science, Toxicology and Food Technology* 13 (11): 79–84.

Onyilagha, J. C., S. Islam, and S. Ntamatungiro. 2009. "Comparative Phytochemistry of Eleven Species of Vigna (Fabaceae)." *Biochemical Systematics and Ecology* 37 (1): 16–19.

Palta, S., R. Saroa, and A. Palta. 2014. "Overview of the Coagulation System." *Indian Journal of Anaesthesia* 58 (5): 515–23.

Parashar, S. M. P. 2006. "Post Harvest Profile of Black Gram." *Department of Agriculture and Cooperation*, Ministry of Agriculture, Directorate of Marketing and Inspection, Government of India.

Parsa, N. 2012. "Environmental Factors Inducing Human Cancers." *Iranian Journal of Public Health* 41 (11): 1–9.

Paul, T., N. H. M .R. Mozumder, M. A. Sayed, and M. Akhtaruzzaman. 2011. "Proximate Compositions, Mineral Contents and Determination of Protease Activity From Green Gram." *Bangladesh Research Publications Journal* 5 (3): 207–13.

Perez, V., and E. T. Chang. 2014. "Sodium-to-Potassium Ratio and Blood Pressure, Hypertension, and Related Factors." *Advances in Nutrition* 5 (6): 712–41.

Pericleous, M., D. Mandair, and M. E. Caplin. 2013. "Diet and Supplements and Their Impact on Colorectal Cancer." *Journal of Gastrointestinal Oncology* 4 (4): 409–23.

Pham-Huy, L. A., H. He, and C. Pham-Huy. 2008. "Free Radicals, Antioxidants in Disease and Health Lien." *International Journal of Biomedical Science* 4 (2): 89–96.

Phaniendra, Alugoju, D. B. Jestadi, and L. Periyasamy. 2015. "Free Radicals: Properties, Sources, Targets, and Their Implication in Various Diseases." *Indian Journal of Clinical Biochemistry* 30 (1): 11–26.

Prado, S. B. R do, V. C. Castro-Alves, G. F. Ferreira, and J. P. Fabi. 2019. "Ingestion of Non-Digestible Carbohydrates from Plant-Source Foods and Decreased Risk of Colorectal Cancer: A Review on the Biological Effects and the Mechanisms of Action." *Frontiers in Nutrition* 6: 1–17.

Preet, K., and D. Punia. 2000. "Proximate Composition, Phytic Acid, Polyphenols and Digestibility (in Vitro) of Four Brown Cowpea Varieties." *International Journal of Food Sciences and Nutrition* 51 (3): 189–93.

Preety, S. Sharma, and V. Hooda. 2018. "Purification and Analytical Application of Vigna Mungo Chitinase for Determination of Total Fungal Load of Stored Cereals." *Applied Biochemistry and Biotechnology* 186 (1): 12–26.

Rajagopal, V., C. K. Pushpan, and H. Antony. 2017. "Comparative Effect of Horse Gram and Black Gram on Inflammatory Mediators and Antioxidant Status." *Journal of Food and Drug Analysis* 25 (4): 845–53.

Ranasinghe, P., W. S. Wathurapatha, M. H. Ishara, R. Jayawardana, P. Galappatthy, P. Katulanda, and G.R. Constantine. 2015. "Effects of Zinc Supplementation on Serum Lipids : A Systematic Review and Meta-Analysis." *Nutrition & Metabolism* 12 (26): 1–16.

Reddy, N. R., D. K. Salunkhe, and S. K. Sathe. 1982. "Biochemistry of Black Gram (*Phaseolus Mungo* L.): A Review." *Critical Reviews in Food Science and Nutrition* 16 (1): 49–114.

Reddy, N .R., C. V. Balakrishnan, and D. K. Salunkhe. 1978. "Phytate Phosphorus and Mineral Changes during Germination and Cooking of Black Gram (*Phaseolus Mungo*) Seeds." *Journal of Food Science* 43 (2): 540–43.

Rose, C., A. Parker, B. Jefferson, and E. Cartmell. 2015. "The Characterization of Feces and Urine: A Review of the Literature to Inform Advanced Treatment Technology." *Critical Reviews in Environmental Science and Technology* 45 (17): 1827–79.

Samad, M. A., N. Sarker, J. K. Sarker, A. K. Azad, and A. C. Deb. 2013. "Assessment of Variability in Twenty-Four Lines of Blackgram (*Vigna Mungo* L.)." *International Journal of Biosciences (IJB)* 3 (8): 307–12.

Sami, W., T. Ansari, N. S. Butt, M. Rashid, and A. Hamid. 2015. "Effect of Diet Counseling on Type 2 Diabetes Mellitus." *International Journal of Scientific & Technology Research* 4 (8): 112–18.

Sangale, P., and R. Patil. 2017. "Hepatoprotective Activity of Alkaloid Fractions from Ethanol Extract of *Murraya Koenigii* Leaves in Experimental Animals." *Journal of Pharmaceutical Sciences and Pharmacology* 3 (1): 28–33.

Satyanarayana, T., B. G. Gangarao, C. K. V. L. S. N. Anjana Male, and G. Surendra. 2012. "Hepatoprotective Activity of Whole Plant Extract of *Vigna Mung* Linn Against Carbon Tetrachloride Induced Liver Damage Model." *International Journal of Pharmacy and Biological Sciences* 2 (3): 256–63.

Schley, P. D., and C. J. Field. 2002. "The Immune-Enhancing Effects of Dietary Fibres and Prebiotics." *British Journal of Nutrition* 87 (S2): S221–30.

Shaheen, S., N. Harun, F. Khan, R. A. Hussain, S. Ramzan, S. Rani, Z. Khalid, M. Ahmad, and Muhammad Zafar. 2012. "Comparative Nutritional Analysis between *Vigna Radiata* and *Vigna Mungo* of Pakistan." *African Journal of Biotechnology* 11 (25): 6694–6702.

Sharifi-Rad, M., N. V. Anil Kumar, P. Zucca, E.M. Varoni, L. Dini, E. Panzarini, J. Rajkovic, et al. 2020. "Lifestyle, Oxidative Stress, and Antioxidants: Back and Forth in the Pathophysiology of Chronic Diseases." *Frontiers in Physiology* 11: 1–21.

Sharma, R. D. 1979. "Isoflavones and Hypercholesterolemia in Rats." *Lipids* 14 (6): 535–40.

Sharma, R. D. 1981. "Isoflavone Content of Bengalgram (*Cicer Arietinum*) at Various Stages of Germination." *Journal of Plant Foods* 3 (4): 259–64.

Singhal, P., G. Kaushik, and P. Mathur. 2014. "Antidiabetic Potential of Commonly Consumed Legumes: A Review." *Critical Reviews in Food Science and Nutrition* 54 (5): 655–72.

Soetan, K. O. 2008. "Pharmacological and Other Beneficial Effects of Anti-Nutritional Factors in Plants —A Review." *African Journal of Biotechnology* 7 (25): 4713–21.

Soliman, G. A. 2019. "Dietary Fiber, Atherosclerosis, and Cardiovascular Disease." *Nutrients* 11 (5): 1155.

Soultati, A., and S. P. Dourakis. 2006. "Coagulation Disorders in Liver Diseases." *Seminars in Liver Disease* 9 (1): 29–42.

Srinivasan, M., A. R. Sudheer, and V. P. Menon. 2007. "Ferulic Acid: Therapeutic Potential through Its Antioxidant Property." *Journal of Clinical Biochemistry and Nutrition* 40 (2): 92–100.

Suneja, Y., S. Kaur, A. K. Gupta, and N. Kaur. 2011. "Levels of Nutritional Constituents and Antinutritional Factors in Black Gram (*Vigna Mungo* L. Hepper)." *Food Research International* 44 (2): 621–28.

Tamura, T., T. Minamikawa, and T. Koshiba. 1982. "Multiple Forms of Acid Phosphatase in Cotyledons of *Vigna Mungo* Seedlings." *Journal of Experimental Botany* 33 (137): 1332–39.

The International Agency for Research on Cancer. 2018. "Latest Global Cancer Data: Cancer Burden Rises to 18.1 Million New Cases and 9.6 Million Cancer Deaths in 2018." *World Health Organization*, France. Available from: https://www.who.int/cancer/PRGlobocanFinal.pdf. Accessed on December 16, 2020.

Troszyńska, A., I. Estrella, M.L. López-Amóres, and T. Hernández. 2002. "Antioxidant Activity of Pea (*Pisum Sativum* L.) Seed Coat Acetone Extract." *LWT—Food Science and Technology* 35: 158–64.

Venegas, D. P., M. K. De La Fuente, G. Landskron, M. J. González, R. Quera, G. Dijkstra, H. J. M. Harmsen, K. N. Faber, and M. A. Hermoso. 2019. "Short Chain Fatty Acids (SCFAs) Mediated Gut Epithelial and Immune Regulation and Its Relevance for Inflammatory Bowel Diseases." *Frontiers in Immunology* 10: 1–16.

Venkidasamy, B., D. Selvaraj, A. S. Nile, S. Ramalingam, G. Kai, and S. H. Nile. 2019. "Indian Pulses: A Review on Nutritional, Functional and Biochemical Properties with Future Perspectives." *Trends in Food Science and Technology* 88: 228–42.

Wismar, R. 2010. "Regulation of Immune Responses by Non-Starch Polysaccharides: Induction of Distinct Phenotypes in TLR-Triggered Dendritic Cells and Adjuvant Properties." Ph.D. thesis, Technical University of Denmark.

Wu, D., E. D. Lewis, M. Pae, and S. N. Meydani. 2019. "Nutritional Modulation of Immune Function: Analysis of Evidence, Mechanisms, and Clinical Relevance." *Frontiers in Immunology* 10: 1–19.

Yamauchi, D., H. Akasofu, and T. Minamikawa. 1992. "Cysteine Endopeptidase from *Vigna Mungo*: Gene Structure and Expression." *Plant and Cell Physiology* 33 (6): 789–97.

Yu, J., X. Bi, B. Yu, and D. Chen. 2016. "Isoflavones: Anti-Inflammatory Benefit and Possible Caveats." *Nutrients* 8 (6): 1–16.

Zhang, Y. J., S. Li, R. Y. Gan, T. Zhou, D. P. Xu, and H. B. Li. 2015. "Impacts of Gut Bacteria on Human Health and Diseases." *International Journal of Molecular Sciences* 16 (4): 7493–7519.

Zia-Ul-Haq, M., S. Ahmad, S.A. Bukhari, R. Amarowicz, S. Ercisli, and H.Z.E. Jaafar. 2014. "Compositional Studies and Biological Activities of Some Mash Bean (*Vigna Mungo* (L.) Hepper) Cultivars Commonly Consumed in Pakistan." *Biological Research* 47 (1): 1–14.

24

Rice Beans: An Underutilized Legume with Nutritional Potential and Health Promoting Compounds

Neha Sharma, Gurkirat Kaur and Arashdeep Singh
Punjab Agricultural University, Ludhiana, Punjab, India

CONTENTS

24.1 Introduction

Rice beana [*Vigna umbellata* (Thumb.)] are a nonconventional and underutilized bean and an important crop for the generation of livelihood for poor rural and tribal farmers of South and Southeast Asia. Rice beans are also known as climbing mountain beana, *mambi* bean, oriental bean, and *Beziamah* in the Assamese language (Saini and Chopra, 2012). Rice beans have a rich genetic diversity and high agricultural and nutritional potential in terms of being able to grow well in comparatively poor soils in hot and humid climates and their resistance to storage pests and serious diseases (Chandel et al., 1978; Singh et al., 1980). Rice beans are a native crop of South and Southeast Asia, mainly grown in India, Philippines, China, Myanmar, Malaysia, Korea, Indonesia, Fiji, Sri Lanka, Mauritius, Sierra Leone, Ghana, Zaire, Tanganyika, Jamaica, Haiti, and Mexico, and also to a limited extent in the West Indies, United States, Queensland (Australia), and East Africa. In India, it is cultivated mainly as a rainfed crop in the Northeastern Hills, West Bengal, Sikkim Hills, Western and Eastern Ghats Hills, Chhota-Nagpurregion and parts of Odisha in the Eastern peninsular tract, and the Kumaon Hills (Uttarakhand), and Chamba region (Himachal Pradesh) in the Western Himalaya.

About 16 different grade colors, mainly black, red, cream, violet, purple, maroon, brown, chocolate, or mottled grains with greenish, brownish, or ash grey backgrounds, are available in rice bean grains (Arora et al., 1980), which indicates the grain is a rich source of bioactive compounds. In Northeast India, it is grown mainly in the tribal regions of Assam, Meghalaya, Manipur, Mizoram, Arunachal Pradesh, and Nagaland. It is an important crop of shifting cultivation (1.7 million ha area) or kitchen gardens (Sharma and Kawatra, 1995) for reasons of attributes such as quick maturity, freedom from major insect and disease problems, and producing easily cooked, good-tasting seeds. Rice bean seeds have a smooth shiny surface, are slender to oblong in shape, 6–8 mm in length, 3–5 mm in width, 3–4 mm thick; rounded at both ends and with concave, straight, or protruding hilum (de Carvalho and Vieira,1996; Bepary et al., 2018). The proportion of cotyledon, testa, and embryo in rice beans range from 88–90%, 7–9%, and 0.3–0.5%, respectively. The world production of rice beans alone is about 1,407 metric tons from an area of 1,804 hectares with productivity of 2.25 q/ha. Very limited information is available on the marketing of rice beans in international trade. In the international trade of rice beans, Japan is a major importer whereas Thailand, Myanmar, China, and Madagascar are major exporting countries. In 1998–2000, about 1,100 tons were exported annually. In recent years, Thailand has exported rice beans to Japan where it used in lieu of azuki bean (*Vigna angularis* (Wild.) Ohwi & Ohashi (Tomooka et al., 2006).

24.2 Nutritional Potential of Rice Beans

The proximate composition is presented in Table 24.1 and detailed description of each nutritional component is described next.

24.2.1 Proteins

In comparison to animal-based protein rich foods, plant-based foods have absolute importance as 57% of total global protein supply is contributed by these against 18% by meat, 10% by dairy, 6% by fish, and shellfish and by 9% other animal products (Henchion et al., 2017). Amongst various plant-based foods, the cheapest source of protein are legumes providing 14–35% of total proteins, moreover the associated risks with consumption of animal-based products is also reduced with their intake. Rice beans are one of the cheapest high-quality protein sources with 14–26% protein, which is quite comparable to other commonly consumed and marketed legumes like chickpea (*Cicer arietinum*, 18.77%), kidney beans (*Phaseolus vulgaris*, 19.91%), pigeon peas (*Cajanuscajan*, 20.27%), dry peas (*Pisumsativum*, 20.43%), black gram (*Vigna mungo*, 21.97%), lentil (*Lens culinaris*, 22.49%), and green gram (*Vignaradiata*, 22.53%) (Pattanayak et al., 2019).

The major storage protein fraction of rice beans consist of albumins and globulins 13.11–15.56 g/100 g and 6.13–7.47 g/100 g rice bean flour, respectively. Prolamins and glutins are also present in appreciable amounts 1.60–1.97 g/100 g seed flour and 1.77–2.22 g/100 g seed flour, respectively. The reported *in vitro* protein digestibility of rice beans vary from 82–88.50% higher than mung bean (80%) and cow peas (74%), making it an excellent option to improve the overall quality of diet.

24.2.2 Fats

Rice beans are considered a low-fat food. The range of saturated and unsaturated fats varies considerably in the reported studies (Kaur and Kapoor, 1992; Saikia et al., 1999; Katoch, 2013; Bepary et al., 2017). Crude fat content varied from 0.46–2.3%, saturated fats varied from 27.4–55.0%, and unsaturated fats (mono and poly unsaturated) varied from 36–73.2%. With respect to fatty acids, linolenic acid is predominant in rice beans (39.89–44.36%) followed by linoleic acid (17.24–18.98%), oleic acid (15.62–17.91%), palmitic acid (14.23–16.88%), and stearic acid (4.36–5.87%).

TABLE 24.1

Proximate Composition of the Rice Bean Legume

Component	Bepary et al., (2017)	Du et al., (2014)	Pawar et al., (2012)	Kalidass and Mohan, (2012)	Baruah et al., (2018)
Moisture	12.57 ± 0.012	9.8 ± 0.0	10.3 ± 0.07	5.56 ± 0.02	3.58 ± 0.00
Carbohydrates	57.67 ± 1.84	-	61.6 ± 0.08	54.23 ± 1.6	77.54 ± 0.0037
Proteins	19.52 ± 0.04	25.68 ± 1.19	17.5 ± 0.05	26.12 ± 0.50	29.09 ± 0.50
Fats	1.34 ± 0.10	1.58 ± 0.04	0.51 ± 0.005	9.18 ± 0.05	1.72 ± 0.38
Crude Fiber	5.56 ± 0.12	-	5.59 ± 0.001	4.74 ± 0.0	-
Ash	3.34 ± 0.04	4.25 ± 0.16	4.5 ± 0.04	4.04 ± .004	-

24.2.3 Carbohydrates

In legumes, the carbohydrate content generally ranges from 50–60%, and starch is the most abundant carbohydrate. The starch from legumes contains about 60–70% amylopectin and 30–40% amylose content (Punia et al., 2019; 2020). Legume starch is generally referred as nonstructural carbohydrate, which is not efficiently utilized by monogastric animals, hence it being called resistant starch. Starch constitutes around 52–57% of total dry matter present in rice bean seeds with 35.20% and 64.80% of amylose and amylopectin, respectively (Kaur and Kapoor, 1990). Rice beans have average starch digestibility (32.86 mg maltose release/gm of meal), which is low in comparison to other commonly consumed pulses such as faba beans (42 mg maltose release/g of meal) (Kaur and Kapoor, 1990; Saharan et al., 2002). The low glycemic index makes rice beans suitable for consumption by diabetic patients and those with an elevated risk of developing diabetes. Rice beans also have total soluble sugars (5.0–5.60 g/100 g) and nonreducing sugars (4.70–5.30 g/100 g) and less starch (50–55 g/100 g). The oligosaccharide (flatulence-producing saccharides such as raffinose, stachyose, verbascose, and ajugose) content in rice beans is lower than other pulses (Bepary and Wadikar, 2019).

24.2.4 Vitamins

Rice bean grain is rich in water-soluble vitamins (mg/100 g) such as thiamine (0.261 ± 0.164), ascorbic acid (0.903 ± 0.404), niacin (1.141 ± 0.713), pyridoxine hydrochloride (1.242 ± 1.593), pyridoxal-5- phosphate (3.353 ± 0.650), pantothenic acid (3.450 ± 2.609), folic acid (0.139 ± 0.071), and riboflavin (0.083 ± 0.038) (Bepary et al., 2019). The niacin content in rice beans was also higher in comparison to *Cajanus cajan, Dolichos lablab, Dolichos biflorus, Phaseolus mungo, Vignacatjan*g, and other *Vigna species* (Rajyalakshmi and Geervani, 1994).

24.2.5 Minerals

The total ash content of different rice bean varieties varies from 3.00–7.43 g/100 g, comparatively higher than other beans like pinto beans, navy beans, black beans, lentils, cow peas, and chickpeas. Rice bean seeds contain appreciable quantities of sodium, potassium, calcium, magnesium, phosphorous, zinc, copper, and iron. Zinc (2.45–24.18 mg/100 g), potassium (610.44–1752.77 mg/100 g), calcium (111.51–168.00 mg/100 g), and iron (4.00–9.25 mg/100 g) are major mineral constituents in rice bean seeds (Saharan et al., 2002; Katoch, 2013).

24.3 Nutraceutical Potential of Rice Beans

Rice beans are a potential underutilized legume holding potential to be served as a nutraceutical food owing to the presence of different bioactive components. In spite of the scant literary evidences on the role of rice beans for disease prevention, its prospects of imparting health benefits are fairly extensive. This legume is a powerhouse of recuperative properties due to the richness of proteins, dietary fiber, macro-, and micronutrients, and the beneficial phytochemicals essential for maintaining human health. The overall potential of rice bean legumes in health promotion and the quantities of health-promoting factors present are tabulated in Table 24.2.

24.3.1 Polyphenols

The abundance in bioactive compounds like polyphenols advocates the use of rice beans as an excellent functional food. Phenolic acids and flavonoids including p-coumaric acid, ferulic acid, sinapic acid, catechin, epicatechin, vitexin, isovitexin, and quercetin are the major antioxidants present in rice bean seeds. The pigmented seeds of legumes contain higher total phenolic content than the light-colored seeds. Among the hydroxycinnamic acids, ferulic acid is present in higher amounts in rice bean seeds, ranging from 11.57–78.32 µg/g followed by p-coumaric acid and sinapic acid. Rice beans stand parallel to adzuki

TABLE 24.2

Bioactive Compounds Present in Rice Beans and their Associated Health Benefits

Bioactive Compounds	Content	Health Benefits	References
Dietary fiber	1.74–8.52 g/100 g	Reduce constipation, modulation of blood glucose level, cholesterol reduction, prebiotic effects, prevention of certain cancers, cardiovascular diseases (CVD), diverticulosis, obesity, and lower blood pressure	(Redgwell and Fischer, 2005)
α-amylase inhibitor and α-glucosidase inhibitor	7,529 (IU/g) to 10,766 (IU/g)	Management of diabetes mellitus by Inhibiting α-amylase and α-glucosidase enzymes that are responsible for catalyzing the hydrolysis of α-1, 4-glycosidic linkage in sugar polymers, which is important for converting complex sugar polymers into simpler units	Katoch and Jamwal(2013)
Saponins	0.30–0.70%	Anti-inflammation, antiviral, hepatoprotective activity. hypocholesterolemic, anticoagulant, anticarcinogenic, hepatoprotective, hypoglycemic and antioxidant activity	(Kitagawa et al., 1983)
Advanced glycation end products (AGEs) inhibitors	Glycation end products formation inhibition activity (34.11–75.75%)	Inhibit accumulation of AGEs and modification of proteins	Yao et al (2012)
Phenolic acids		Risk factors for menopause, coronary heart disease, and anticarcinogenic, possess free radical scavenging activity against superoxide radicals, hydroxyl radicals, and nitric oxide	Yao et al (2012)
Vitexin	401.84 µg/g		
Isovitexin	190.24 µg/g		
Total flavanoids	3.27 mg GAE		
p-coumaric acid	5.67–39.72 µg/g		
Ferulic acid	11.57–78.32 µg/g		
Sinapic acid	8.16–31.08 µg/g		
Catechin	24.76–182.64 µg/g		
Quercetin	10.77–35.46 µg/g		

beans in having similar amounts of p-coumaric acid (31.30 µg/g). These hydroxycinnamic acid-derived compounds are acknowledged for their free radical scavenging capacity, metal-chelating activity, and an ability to inhibit lipid peroxidation and prooxidative enzymes. Bhagyawant et al. (2019) investigated the antioxidant potential of different rice bean genotypes and reported that the 2,2-diphenyl-1-picryl-hydrazyl-hydrate (DPPH) free-radical scavenging activity was observed in a range of 9.36–18.57%. Yao et al. (2012) evaluated the antioxidant potential of rice bean extracts by measuring their DPPH radical scavenging activities (µM Trolox Equivalent[TE]/g) and the antioxidant values ranged between 39.87 µM TE/g and 46.40 µM TE/g, which is comparable to the other well-stablished legumes.

24.3.2 Dietary Fiber

Rice bean seeds are a rich source of dietary fiber. Dietary fiber refers to all polysaccharides that are resistant to digestion and absorption in the small intestine but complete and partial fermentation in large intestine (Prosky and DeVries, 1992). Major dietary fiber includes cellulose, hemicelluloses, pectin, arabinoxylans, beta-glucan, glucomannans, plant gums, mucilages, and hydrocolloids, which are principally found in the plant cell wall (Cummings and Stephen, 2007). In rice bean seeds, the fiber content has been found higher in seed coat fractions (12.60%) in comparison with other seed components (Katoch, 2011). In humans, fibers primarily act on gastrointestinal tract, affecting different physiological effects like alteration of the gastrointestinal transit time, satiety changes, influence on the levels of body cholesterol, after-meal serum glucose and insulin levels, flatulence, and alteration in nutrient bioavailability (Lajolo et al., 2001). The main bioactive functions that have been attributed to dietary fibers are reduce constipation, modulation of blood glucose level, cholesterol reduction, prebiotic effects, prevention of certain cancers (Redgwell and Fischer, 2005), cardiovascular diseases (CVD), diverticulosis, obesity, and lower blood pressure (Brand et al., 1990). Similarly, Sharma and Kawatra (1995) also reported that soluble fiber decreases serum cholesterol, reducing the risk of heart attack and colon cancer. Insoluble dietary fiber is required for normal, lower intestinal function in humans (Anderson et al., 1994).

24.3.3 Advanced Glycation End Product (AGE) Inhibitors

The nonenzymatic binding of sugars to proteins alters the structure and function of proteins and leads to formation and accumulation of advanced glycation end products (AGEs). The accumulation of AGEs in the body contributes to the onset and progression of diabetic complications, osteoporosis, and lifestyle-related diseases such as arteriosclerosis. *In vivo* glycation represents a risk factor for accelerated aging and is known as glycation stress, a topic of recent interest. Mitigation of glycation stress to inhibit accumulation of AGEs is probably playing important role in prevention of diseases. The inhibitory effects of most plant extracts on the formation of AGEs is mainly contributed by phenolic antioxidants present in them. As free radicals are involved in the formation of AGEs, it is reasonable to expect that phenolic antioxidants could inhibit the formation of AGEs and subsequent inhibition of modification of proteins, which is a major mechanism for mediating their antiglycation activities (Katoch, 2020).

24.3.4 Saponins

Saponins consist of triterpenoid and steroidal glycosides and present in the range of 0.30–0.70% in rice beans. These compounds are believed to have anti-inflammation, antiviral, and hepatoprotective activity. Metabolic risk factors like hypertension, cholesterol abnormalities, central obesity, and an increased risk for blood clotting further intensify the risk of cardiovascular diseases. As saponins have hypocholesterolemic, anticoagulant, anticarcinogenic, hepatoprotective, hypoglycemic, and antioxidant activity, these saponins from different sources including rice beans could provide a remedy for the treatment of cardiovascular diseases.

24.4 Antinutritional Factors in Rice Beans

24.4.1 Trypsin and Chymotrypsin Inhibitors

Trypsin inhibitors are known to impair the activities of protein digesting enzymes that lower the body's ability to consume protein and thereby reduces the growth. The trypsin and chymotrypsin inhibitors also have very high anticarcinogenic activities on human colon cancer (Clemente and Domoney, 2006). The trypsin inhibitor activities (TIA) of in rice beans were reported by different authors and are widely varied. The TIA value depends on extraction procedure, nature, and concentration of the substrate. The TIA of 17 genotypes of rice beans was found to range from 55.2–163.98 TIU/g (Malhotra et al., 1988, Saharan et al., 2002).

24.4.2 α-Amylase Inhibitors

The α-amylase inhibitors interfere with hydrolysis of starch and glycogen by inhibiting the human amylase. Among the pulses, field beans and chickpeas contain very low concentrations of α-amylase whereas winged beans, adzuki beans, soy beans, lima beans, lentils, and peas are devoid of amylase inhibitors (Grant et al., 1995). As reported by Katoch and Jamwal, (2013), α-amylase inhibitor activities of rice beans were observed to vary from 7.529–10.766 IU/g. High intake of amylase inhibitor starch blockers were reported to cause symptoms like diarrhea, nausea, and vomiting because of the reduced or no diges-tion and enlargement of the small intestine and of the pancreas. (Grant et al., 1998)

24.4.3 Allergens

Legumes are counted as an important source of food allergens containing highly abundant proteins and storage proteins (e.g., 7S and 11S globulins). The clinical manifestations of the allergy to legumes range from oral allergy syndrome, urticaria, angioedema, rhinitis, to asthma (Kasera et al., 2013). The allergenic proteins are present in lentils, common beans, mung beans, chickpeas, and peas, etc. (Riascos et al., 2010). There is little scientific evidence available on allergenicity of rice beans. Andersen and Hollington (2007) reported rice bean as a nonallergic food based on conventional consumer perception evidence.

24.4.4 Phytic Acid

The phytic acid content of rice varies from 1.88–8.77% (Kaur and Kapoor, 1992; Chau et al., 1997; Saharan et al., 2002). The phytic acid content of rice beans was found higher lower than lentils, soy beans and cow peas. Phytic acid contents of rice beans was reported higher than faba bean, lablab, green gram, and black gram but lower than *P. angularis* (Kaur and Kapoor, 1992; Chau et al., 1997; Saharan et al., 2002). It is often regarded as an antinutrient because of strong mineral, protein, and starch binding prop-erties thereby decreasing their bioavailability. Soaking and cooking of legumes results in a significant reduction in phytic acid. On the other hand, phytic acid can be helpful in the prevention kidney stone formation, heart disease (Jenab and Thompson, 1998), and heavy metal toxicity (Kumar et al., 2010).

24.4.5 Phenolic Compounds

As revealed by Katoch (2019), the polyphenolic content of 30 different varieties of rice bean crops were found to be varied from 0.57–0.80%. This wide range of polyphenol content may be due to variety, geographical origin, and method of extraction and analysis. The polyphenol content of rice beans was observed lower than cow peas and mung beans, but higher than faba beans (Malhotra et al., 1988, Kaur and Kapoor, 1992; Khabiruddin et al., 2002). In another study, it found that the polyphenol content of rice beans was highest among lentils, soy beans, black gram, and cow peas. The variation in varieties is because the dark and pigmented varieties contain more polyphenols compared to other varieties. Yao et al. (2012) reported the total phenolic contents from 123.09 ± 10.35–843.75 ± 30.15 µg/g, which was composed of three phenolics acids (p-coumaric acid, ferulic acid, and sinapic acid) and five flavonoids (catechin, epichatechin, vitexin, isovitexin, and quercetin).

24.4.6 Tannin

The tannin content of rice beans is lower than some of the traditional pulses, varying from 0.24–1.55% (Kalidass and Mohan, 2012, Katoch, 2013). Katoch (2013) reported the total, condensed, and hydrolys-able tannins as 1.37–1.55%, 0.76–.8%, and 0.56–0.79%, respectively, for 16 genotypes of rice bean. Phenolic compounds can reduce the activity of digestive enzymes such as α-amylase, trypsin, chymo-trypsin, and lipase and decrease the absorption of nutrients such as vitamin B12 (Duranti, 2006) and reduce the bioavailability minerals such as iron, zinc, and calcium (Gilani et al., 2005).

24.4.7 Saponins

Saponins (plant glycosides) have the hemolytic ability to break down red blood cells. The saponin con-tent of 22 genotypes of rice beans varies from 1.2–3.1% (Kaur and Kapoor, 1992, Saharan et al., 2002, Katoch, 2013). The saponin content of rice beans (2.28%) was higher than lentils, black gram, and cow peas, but lower than soy beans. Saponins are also believed to affect chymotrypsin and trypsin activity, which would affect the absorption of protein (Bepary et al., 2016).

24.4.8 Oxalates

Oxalic acid binds strongly to divalent cations like calcium and magnesium and form water soluble salts. The higher intakes of oxalate-rich foods lead to the lowered of absorption of calcium (Noonan and Savage, 1999). The deposition of insoluble oxalate salts, particularly calcium oxalate, results in crystal nephropathies. Patients suffering from kidney ailments should not exceed the intake of oxalate 50–60 mg/day (Massey et al., 2001). Oxalate binds with nutrients, rendering them inaccessible to body. The formation of calcium oxalate crystal leads to disturbed calcium-to-potassium (Ca:P) ratio in the body and results in increased mobilization of minerals from bones to recover from hypocalcemic conditions. The excessive mobilization of calcium and phosphorus from bone causes secondary hyperparathyroidism or osteodystrophy fibrosa. Bepary et al. (2017) evaluated oxalate content in 10 rice bean genotypes ranging from 23.17–34.12 mg/100 g.

24.4.9 Flatulance-Causing Factors

Oligosaccharides have a low molecular weight, are nonreducing, and contain water-soluble sugars that constitute approximately 53% of the total soluble sugars. The oligosaccharides e.g., raffinose, stachyose, and verbascose, are nondigestible carbohydrates that cause flatulence due to the fermentation of undigested carbohydrates by ruminal microbes. Raffinose family oligosaccharides (RFOs) are α-(1-6)-linked galactosides and raffinose is the representative of this group. The other structural homologues of raffinose are stachyose (tetrasaccharide) and verbascose (pentasaccharide). Raffinose is a trisaccharide containing galactose linked by α-(1–6) bond to glucose unit of sucrose. Stachyose is a tetrasaccharide containing a galactose linked by α-(1–6) to the terminal glucose of raffinose. Most of the legumes generally contain up to 50 mg/g of total oligosaccharides.

24.5 Postharvest Processing of Rice Bean Legumes

Postharvest processing involves a number of various practices such as judging maturity index, harvesting, threshing, drying, storing, processing and marketing of food grains. Every processing step determines the quality and safety of the final product and also decides the processing losses. These steps are discussed in detail next.

24.5.1 Judgment of Maturity Index

For a properly matured and developed rice bean pod, it takes 90–120 days to reach harvesting stage, but varies depending on varietal, geographical, and environmental factors (Acharya, 2008). Correct judgement of the maturity index is very crucial for both a qualitative and quantitative harvest. Late harvesting after attaining physiological maturity renders the beans difficult to cook. Formation of acceptable pod coat color (green to yellow to brown), thousand kernels weight, moisture content, protein and phytic acid contents, and crop duration are some of the factors that determine maturity of pods or grains. Rice bean pods are prone to shattering when left on the plant for prolonged periods, hence they require immediate harvesting. Shattering occurs owing to the formation of lignin deposition in fiber cap cells, the ventral suture tissue, and the inner sclerenchyma cells of pod valves, which helps the pods twist under low relative humidity conditions (Rau et al., 2019).

24.5.2 Harvesting

Harvesting of rice beans is defined as taking away the economic parts such as pods of the plant from the field when it reaches harvest maturity. Harvesting can be done either manually or by mechanical means. Harvesting of rice bean pods is carried out by pulling out mature plants with the roots or cutting near the base of the mature plant. This harvesting practice is less labor-intensive than harvesting by picking individual pods from plants.

24.5.3 Predrying

In order to achieve an effective threshing, the moisture content of rice bean pods are lowered by sun drying. Predrying is a unit operation of postharvest practices wherein harvested rice bean plants are dried to reduce moisture content for effective threshing. This practice is necessary when harvested plants have plenty of green foliage, nonuniform pod maturity, and high moisture (Lal and Verma, 2007). For the maximum shattering effect of pods, the harvested plants are dried for 1–2 weeks in the sun. During the predrying period, the harvested plants need protection from rain, dew, nocturnal humidity, and direct ground contact. Drying reduces the moisture content of pods, which facilitates twisting forces in the pod wall so that the pod shatters. Farmers dry the plants either in the backyard or on a roof top. The drying place should be free from food safety hazards. Careless drying results in contamination with physical hazards such as stone, cattle droppings, hair, and feathers.

24.5.4 Threshing and Cleaning

The practice of detachment and separation of rice bean kernels from the harvested plants is called threshing. Optimum moisture requirements are necessary in order to facilitate easy removal of chaffs during winnowing. After predrying, manual threshing is practiced by small and marginal farmers wherein dried pods are beaten with bamboo and wooden sticks or sometimes by trampling cattle. Mechanical threshing is done in the case of large harvests. As soon as the threshing operation is completed, the grains are cleaned to remove chaffs, very light seeds, broken grains, glumes, and other foreign materials by winnowing or dropping the grain from a bucket against the natural wind.

24.5.5 Drying

The moisture content of threshed rice beans is high and must be decreased to 9–12% for efficient grain storage. High moisture (18% or higher) results in microbial spoilage and pest attacks and has high enzymatic activity during storage, which shortens storage life and reduces grain quality and market price (Mohan et al., 2011). Drying is considered as an important postharvest practice wherein excess moisture of the rice bean grains is removed by applying heat by natural sun radiation or from artificial hot air. Rice beans require 2–3 days of sun drying to remove moisture for safe storage. During the peak sunny period of the day, the relative humidity (RH) of the air remains low (below 70%) and temperature remains high, which facilitates the quick evaporation of moisture from the grains. The rate of drying is not uniform in the sun drying process because of the frequent changes in solar radiation intensity during drying days. A fast rate of drying creates cracks on grains, increases integument permeability, reduces germination percentage, and changes the seed coat color (Scariot et al., 2017).

The efficiency of the solar drying practice of rice beans can be increased by the introduction of a small hand-held moisture meter to monitor moisture, a hygrometer to monitor air RH, and a tarpaulin to prevent contamination. Rice bean grains that are harvested in the monsoon season have poor grain quality and a short storage life because of the low solar radiation and high grain moisture (Joshi et al., 2008). This problem can be overcome by using artificial dryers with temperature control devices. In an artificial dryer, blowing hot air continuously transfers the heat and moisture away from grain surface. The efficiency of artificial dryer depends on dryer type as well as drying purpose. Drying at high temperature of high moisture rice bean grains should avoided to prevent severe damage to the grain.

24.5.6 Storage

Rice bean grains are stored at different levels of the food chain for different purposes. Most Southeast Asian farmers store rice bean grains either for home consumption or as seed for future cultivation. Other storage are handled by marketing agencies and processors. Rice beans can be stored for 3–4 years under cool and dry conditions. As other pulses, rice beans are more difficult to store than cereal grains because a slight variation in the storage conditions can cause drastic changes in the physicochemical properties, seed color, insect infestation, and microbial spoilage, with development of hard-to-cook seed. These changes ultimately influence the final quality of the processed grains. Six factors that determine the qualitative and quantitative changes of rice bean grains in storage are temperature, RH, oxygen, light, moisture content of grain, and rice bean variety. In combination, these factors have severe effects causing postharvest loss. Grain moisture content is one of the prime factors that affect grain quality. Cotyledon and testa cracking was observed in canned beans prepared from grains stored at low moisture. Grains storage at high initial moisture is vulnerable for off-flavor and off-color development, with reduced hydration capacity, increased cooking time, and the encouragement of microbial growth and pest infestation (Uebersax and Muhammad, 2013). A moisture content of 5–14% is considered as safe storage when the grains are stored at a temperature range between above freezing point and below 30°C. Within in this moisture range, the storage life is doubled with 1% decreases in moisture content (Lal and Verma, 2007). The surrounding relative humidity of the storage system can change the moisture level of the grains during storage as the grain has a tendency to reach equilibrium moisture content. Grains storage at high initial moisture (above 18%) and high RH (above 70%) promote the growth of molds, including aflatoxin-producing mold and pest infestation by bruchids. Marketing agencies and processors pack the rice beans in gunny sacks stored in a well-ventilated warehouses.

24.6 Tackling Antinutritional Factors in Rice Beans

24.6.1 Soaking

Soaking is one of the prime methods in consumption of pulses where soaking temperature, time period, composition of soaking solution (water, acidic or basic), and pulse type affect the level of different antinutrients. For highly intact seeds or grains, salt (0.25–1%) is often added to soften the cotyledon, which ultimately reduces the cooking time. Soaking also improves the structural and textural profile of leguminous seeds.

Soaking of rice bean grains in water for 12–16 hours can reduce the antinutritional factors such as tannin by 30.65%, polyphenolic compounds by 10–52%, trypsin inhibitor by 9–17%, amylase inhibitor by 11%, phytic acid by 6–33%, saponin by 3–9%, raffinose by 9.5–14%, and stachyose by 5–16.4% (Kaur and Kapoor, 1990; Chau and Cheung, 1997; Kaur and Kawatra, 2000; Saharan et al., 2002; Pawar et al., 2012). Most of the antinutritional factors are reduced by leaching out into the soaking water and enzymatic activities during soaking; the loss of saponin is not caused by leaching but rather by hydrolysis of the glycosidic linkage between the sapogenin and its sugar during soaking. The flavonoids (52%) and antioxidant activity (145%) are increased whereas oxalate (55%), hydrogen cyanide (HCN, 54.9%), and verbascose (49.5%) are decreased in soaked bean as compared to native beans (Oboh et al., 2000; Boateng et al., 2008; Okudu and Ojinnaka, 2017).

24.6.2 Effect of Germination

Germination or sprouting is one of the important household processing practices to improve taste and flavor. For germination of rice beans, the grains are soaked overnight (6–12 hours), and then are allowed to develop rootlets under humid conditions at room temperature for 24–48 hours (Kaur and Kawatra, 2000).

Germination of rice beans can reduce non-nutritional components and increase digestibility (starch and protein), bioavailability of minerals (calcium and iron), and antioxidant activity. Germination of rice beans for 40–48 hours can decrease polyphenol by 35%, phytate by 26.98–55%, tannin by 63%, saponin by 18.68%, trypsin inhibition activity by 42%, amylase inhibition by 57%, raffinose by 15%, stachyose by 75%, *in vitro* starch digestibility by 212%, *in vitro* protein digestibility by 12%, HCl extractability of calcium by 8.2%, and HCl extractability of iron by13%, as compare to native grain (Chau and Cheung, 1997; Saharan et al., 2001; Kaur and Kawatra, 2000; Kaur, 2015). Germination can diminish flavonoids by 5.34% anverbascose by 51.21% (Pal et al., 2015; Oboh et al., 2000). The citric acid (1%) pretreatment followed by germination for 24 hours followed by steam-cooking for 10 minutes can decrease phytate by 19% and increase total phenol by 105%, and 2,2-diphenyl-1-picryl-hydrazyl-hydrate (DPPH) assay by 111% as compared to 6-hour soaked grain (Sritongtae et al., 2017), with increased total polyphenol and B-vitamins.

24.6.3 Roasting

Roasting causes little loss of polyphenol (16–48%), tannins (16.54%), saponins (22.22%), phytic acid (26.04%), raffinose (24.71%), stachyose (24.27%), verbascose (24.29%), and oxalate (11.81%), and little increase of flavonoids (15%), whereas it causes more loss of trypsin inhibitor activity (TIA, 82.04%), α-amylase inhibitor activity (AIA, 78.9%), and hydrogen cyanide (HCN, 66.93%), hence improving *in-vitro* protein digestibility (IVPD)(16.96%) and *in*-vitro starch digestibility (IVSD) (16.47%) (Siddhuraju et al., 1996; Siddhuraju and Becker, 2007; Sharma and Punia, 2017).

24.6.4 Extrusion Processing

The extrusion process can lessen the functional components such as polyphenols (45.89%), flavonoids (39.37%), and antioxidant activities (5.71%) but improve *in vitro* protein digestibility, IVPD (21.87%), *in vitro* starch digestibility, IVSD (128%) (Alonso et al., 2001). The extrusion process greatly decreases the content of saponin (84.5%), TIA (86.12%), AIA (100%), oxalate (94.44%), HCN (78.18%), and raffinose (65.62%), with less impact on phytic acid (20.75%), stachyose (18.8%), and verbascose (20.83%) (Alonso et al., 2001; Anuonye et al., 2012).

24.6.5 Flaking

Hydrothermal flaking can cause loss of polyphenols (27.85–75.47%), phytate (40–88%), and tannin (70–94%), and improve IVSD (several times over), IVPD (some), in vitro iron bioaccessibility (IVIB) (122–923%), and in vitro calcium bioaccessibility IVCB (70%) (Itagi et al., 2012; Wu et al., 2018). The loss of total polyphenols results from the loss of the bran layer from grains and thermal dissociation of conjugated phenolic into moderate reduction of hydroxycinnamic acid derivatives (Randhir et al., 2008). Flaking can decrease free-radical scavenging activity, and flaking followed by toasting can increase free-radical scavenging activity. Protein digestibility is also increased, but as compared to starch less so, as starch is complexed with protein, thereby partly decreasing its susceptibility to hydrolysis.

24.7 Value Addition of Rice Bean Flour

Underutilized crops are a valuable source food as well as income for indigenous people. They also play a significant role in maintaining the productivity and stability of traditional agroecosystems. These little-known crops contribute to food security and play vital roles in the nutrition that enrich the diet of the rural population. Most of the underutilized crops are often available only in the local markets and are practically unknown in other parts of the world. A large number of these crops have the potential to resist the adverse growing conditions with excellent nutritive and therapeutic value and can satisfy the demands of health-conscious consumers. The value addition of underutilized crops is an effective approach for their utilization. Most of consumers these days are demanding new, delicious, nutritious, and attractive food products to meet their food and nutritional requirements. The value addition and utilization of underutilized crops that have excellent nutritive value and therapeutic properties is a possible way for meeting the current require-ment of the consumers. Furthermore, the generation of different food products by means of value addition would be useful in promoting and commercializing the underutilized crops in global market. Among the underutilized legumes, rice beans have high production potential, balanced nutritional profile with high protein content, essential amino acids, vitamins, and minerals in comparison to other well-established legumes. Low fat content and a relatively high proportion of healthy, unsaturated fatty acids make rice beans a wholesome food over other traditionally consumed pulses. Therefore, it is worthwhile for promoting rice beans as a potential pulse crop for the development of different value-added products and their popularization among the masses.

For the development of high-quality rice bean products, in-depth knowledge of the physicochemical properties of rice bean flour is required. Knowledge of the physical properties also helps in the designing of the efficient and economical machinery for product development. For example, physical properties like porosity, density, and angle of repose directly affects the production, packaging, storage, and distribution of the developed products.

Hydration parameters such as swelling capacity and swelling index are directly linked to starch gelatinization. High swelling capacity and the swelling index of items denotes delayed gastric emptying and concurrently increases stomach distension, which triggers the feeling of fullness in the stomach. The swelling capacity and swelling index of different rice bean genotypes ranged from 0.46–0.79 ml/seed and 5.50–9.93, respectively (Kaur et al., 2013). Hydration properties such as hydration capacity and hydration index are two absolute factors that determine the quality of baked food items. Kaur et al. (2013) observed that the hydration capacity and hydration index of different rice bean genotypes ranged from 0.02–0.19 g/seed and 0.25–2.33, respectively. Cooking is known to be a fundamental processing method for legume consumption, as it increases digestibility, inactivates antinutritional factors, increases nutrient biological value, and confers sensorial quality. The cooking quality of pulses is a function of the cooking time. Cooking, in other words, means the time taken between the beginning of the boil and when the seeds are ready to eat. This means that at least 90% of them are soft enough to masticate. Rice bean seeds generally take 30–44 min for cooking; however varietal differences in cooking time have also been reported (Kaur et al., 2013).

24.7.1 Fortified Rice Bean Multimix

Fortification of foods with underutilized legumes is one of the effective strides in tackling malnutrition. Different global organizations like the Food and Agricultural Organization of the United Nations (FAO) and World Health Organization (WHO) focus on the potential utilization of orphan legumes in the preparation of the value-added products with commonly consumed food ingredients making use of their nutrient strengths. In this effort, Baruah et al. (2018) developed food multimix (FMM) using raw and malted rice beans, millet (Konidhan), flax seed, rice (luit variety), and tomato powder. Two food multi-mixes were developed; the first one (FMM-I) was formulated based on energy density value between 1,512.00–1,890.00 kJ (360–450 kcal) per 100 g of sample and further by mixing all the ingredients at appropriate amount. Subsequently, FMM II was formulated by inoculating probiotic bacteria, viz., *Lactobacillus plantarum* and *Lactobacillus rhamnosus* in FMM-I both individually and in combination in different test samples. The study revealed that the FMM developed from malted rice beans had appreciable level of nutrients. The FMM also had good physical properties in terms of bulk density, viscosity, and water-holding and fat-holding capacities, and these properties make these food mixes suitable to be used for the preparation of different value-added products like cakes, cookies, and savory items.

24.7.2 Rice Bean Curry

Curry is a traditional recipe of Southeast Asia that is eaten with rice and roti. Rice bean seeds are usually taken as a soup or as a pulse (dhal) with rice. Dhal (soup, curry) is the common use of pulses and is prepared by different kinds of pulses. In the rice-based food system, dhal is preferred by most of the people in the morning meal with pickle and vegetables. People use both de-hulled and unhulled seeds, directly to make dhal. However, the ways of preparing dhal from de-hulled and unhulled seeds are different. The unhulled beans are normally soaked and cooked, but the de-hulled bean is not soaked before cooking. Mostly, it is mixed with other pulses such as black gram. The rice bean dhal is also known as Dhal Mori and Dhal Khatti and is prepared by mixing unsplit rice bean seeds with others pulses and by adding mango powder in the recipe, respectively. For preparation of curry, grains soaked overnight are cooked and seasoned with spices, salt, and condiments in gravy form. In some places, the unsoaked grains are pressure-cooked to prepare curry.

24.7.3 Nuggets

Soaked grains are utilized for the preparation of number of other food products. Nuggets (barian) are also prepared from the paste of soaked and ground rice bean seeds with black gram. Rice bean seeds are soaked overnight, drained, and ground with ginger, chilies, and garlic paste without the addition of water, followed by adding spices. Plated on a greased sheet, these nuggets are sun dried and stored in air tight containers for future use.

24.7.4 Sundal

Parvathi and Kumar (2006) developed this recipe from rice beans. It is prepared by boiling the whole grain that has been soaked for 12 hours with water and salt until a soft texture is reached. Cooked grains are seasoned with onion, chili, curry leaves, and coconut scraps; the resultant products are used as snacks. Sensory attributes are reported as appearance, 3.7, color, 4.0, taste, 3.7, flavor, and OAA, 3.7.

24.7.5 Rice Bean Coix Gruel

This product is also from China where it is used medicinally to remove swelling and dampness, cure constipation, calm the mind, energize the heart, and refresh the spleen. It is the best food for patients with obesity or edema and for postpartum women. It prepared by boiling the soaked bean and coix seed initially with a hot flame, after which it is cooked slowly until the rice beans look like blowing sand and the coix seed explodes. The resultant clear soup is seasoned with spices and beaten for smoothness (Wei et al., 2015).

24.7.6 Eromba

Eromba is in the traditional cuisine of Manipur. It is made from brinjal, cabbage, potato, bean, onions, and chili and served with rice beans. To prepare *eromba*, brinjal, cabbage outer leaves, soaked rice beans, and potatoes are boiled to softness and then these boiled vegetables are mashed into a paste. The water extract of roasted dry red chili is added into the paste and mixed uniformly after adding salt.

24.7.7 Rice Bean Kheer

For preparation of kheer, de-hulled rice beans soaked for 3–4 hours are cooked until they reach a soft texture. The required amounts of jaggey, coconut milk, cardamom powder, and salt are added after mashing the cooked dhal, cooked to a slurry-like texture, and served as dessert.

24.7.8 Khichadi

Khichadi is a national dish of India and is also a popular dish in Pakistan, Nepal, and Bangladesh. It is prepared mainly from rice and legumes. Selection of cereals and pulses depends on the availability of these grains in the respective areas. In rice bean growing areas, *khichadi* is made from rice beans and rice. Split de-hulled rice beans are cooked with rice, spices, salt, and condiments (Andersen et al., 2009).

24.7.9 Pakora

Pakora is a popular fried snack of the Indian subcontinent. Gram flour is generally used to make the batter. Kaur and Mehta (1993) used rice bean flour for *pakora* preparation and found more crude fiber and calcium than in Bengal gram *pakora*. Seed coats of rice bean grains soaked overnight are removed; the grain is allowed to dry for 3–4 days, and the dried de-hulled grain is then milled into flour. Rice bean flour and spinach leaves are mixed with water and then beaten for uniformity, followed by a second mixing after adding chopped onion, garlic, ginger, coriander leaves, green chilies, and salt. The shaped paste (0.5 inches thick, 2 inches in diameter) is deep fried. Kaur and Mehta (1993) reported the nutritional composition of *pakora* as moisture, 15.5%; crude protein, 13.1%; fat, 16.8%; ash, 2.3%; crude fiber, 2.1%; calcium, 138.5 mg/100 g; iron, 6.7 mg/100 g; zinc, 3.1 mg/100 g; copper, 0.6 mg/100 g; while amino acids (g/16g N) were methionine,1.0; cystine, 1.0; tryptophan, 0.9; and lysine, 4.0.

24.7.10 Sepuvadi

Sepuvadi is a traditional dish of Himachal Pradesh. It is commonly prepared from lentils and black gram with leafy vegetables and other spices, but in some rice-bean growing areas it is also prepared from de-hulled rice beans. Sharma (2014) prepared rice bean-based *sepuvadi* and found that it had more fat, crude fiber, and zinc than black gram *sepuvadi* but with a lower overall acceptability (6.7) compared to black gram (7.3). Wet grinding of soaked split dhal produces the paste. The paste is mixed with the desired quantity of garam masala and salt and shaped into large-sized balls that are then steam-cooked until the core of the balls is cooked properly. Rectangular pieces are made from the cooked balls and deep fried.

24.7.11 Papad

Papad is another traditional food product of the Indian subcontinent popular because of its crispness and taste. It is generally prepared from de-hulled black gram flour, but Sharma (2014) prepared rice bean *papad* and reported it had more crude fiber and zinc, but overall acceptability was lower (6.5) than for black gram *papad* (7.8). To prepare rice bean *papad*, rice bean whole flour is sieved and blended with desired quantities of Ajwan, cumin seed, sodium bicarbonate, chili powder, and salt. Water is added into blended flour to make a dough that can be held for 2 hours after covering with wet muslin cloth. Small dough balls are formed from soft dough made by hammering with a heavy pestle. Dough balls are shaped into a circular dish shape and dried in hot air at 60°C, then deep fried in oil.

24.7.12 Boondi

Boondi is a dessert or savory type of food available in the Indian subcontinent in various forms. It generally made from chickpea flour. Sharma (2014) prepared *boondi* from rice bean flour and found more crude fiber, ash, calcium, and zinc than in Bengal gram flour. For preparation, rice bean flour, baking powder, salt, garam masala, and chili powder are blended uniformly; the flour blend is transformed into a lump-free batter after adding the desired quantity of water. The ready batter passes through a slotted spoon into heated oil where it forms small pearl-like balls after frying. The frying is continued to a light brown color. *Boondi* is ready for use after draining the excess oil.

24.7.13 Bhujia

Bhujia is a popular crispy Indian snack usually made from moth bean, besan, and spices. Sharma (2014) developed rice bean-based *bhujia* and reported more crude fiber, ash, calcium, and zinc than with Bengal gram flour. For development of rice bean *bhujia*, the desired quantity of rice bean flour is kneaded with other ingredients such as boiled potatoes, salt, turmeric powder, garam masala, chili powder, and citric acid into a smooth dough that is then covered for 15–20 min. The rolled dough is allowed to pass through a vermicelli-making machine fitted with a fine net; the machine's piston pushes the raw *bhujia* into already heated oil. The raw *bhujia* is fried till it turns golden brown in color and is ready for use after removing excess oil.

24.7.14 Muruku

*Muruku*or *chakli* is a popular snack product of the Indian subcontinent, especially southern India and Sri Lanka. This crunchy savory product is also available in Malaysia, Singapore, and Fiji and is usually made from rice flour and urad dal. Parvathi and Kumar (2006)) developed rice bean-based *muruku* in the ratio 50:50 rice and de-hulled rice beans with a sensory acceptability of 3.7 of 4.

24.7.15 Ladoo

Ladoo is a popular sweet product of the Indian subcontinent usually served at festivals and religious ceremonies. This sphere-shaped sweet product is made from sugar, ghee/butter/oil, and flour. Sharma (2014) developed a rice bean-based *ladoo* that had more crude fiber, ash, calcium, and zinc than Bengal gram flour *ladoo*. For making of *ladoo*, rice bean flour is roasted in ghee unil golden brown and the aroma is released. The cooled roasted flour is mixed with powdered sugar and shaped into small ball-shaped *ladoo*. Parvathi and Kumar (2006) also developed a de-hulled rice bean-based *sweetened ball* with 3.6 sensory acceptability of 4.

24.7.16 Ojojo

Ojojo is a crispy-crust traditional fried product of Nigeria that is usually prepared from water yam with salt and spices. Okoye et al. (2018) developed rice bean-based *ojojo* by replacing water yam flour by 10–50%. To prepare *ojojo*, rice bean grain is soaked, de-hulled, dried, and milled into flour. A thick batter is made from the desired quantity of water yam, rice bean flour, and water. Onions (chopped or ground), seasoning cubes, fresh pepper, salt, and garlic (ground) are added to the batter and mixed uniformly. The batter formed into ball shapes and deep fried in ground nut oil until golden brown in color and served after draining off oil. *Ojojo* prepared by replacing 50% water yam flour with rice bean flour was reported to have maximum overall acceptability (7.2) and higher amounts of nutritional attributes such as protein, ash, crude fiber, calcium, magnesium, vitamin B12, vitamin C, and phosphorus than the *ojojo* prepared from 100% water yam flour. The nutritional attributes of this *ojojo* were reported as moisture, $18.56 \pm 0.02\%$; crude protein, $10.84 \pm 0.92\%$; fat, $14.78 \pm 0.04\%$; ash, $2.81 \pm 0.14\%$; crude fiber, $2.92 \pm 0.14\%$; carbohydrate, $43.25 \pm 0\%$; phosphorus, $83.04 + 0.03$ mg; calcium, 75.05 ± 0.0 mg; magnesium, 91.01 ± 0.02 mg; vitamin B12, 35.51 ± 0.01 mg; and vitamin C, 39.26 ± 0.37 mg per 100 g.

24.7.17 Cake

Cake is a baked product usually served as a sweetened dessert. It is commonly made from soft wheat flour, sugar, eggs, margarine/butter, milk, and baking powder. Sharma (2014) prepared a rice bean-based cake with greater quantities of protein, crude fiber, ash, calcium, iron, and zinc than a cake made from wheat flour with comparable sensory acceptability.

24.7.18 Cookies

Kii et al. (2013) developed cookies by blending rice bean flour from 0–30% with wheat flour and found a significantly increased texture and calcium content with an insignificant change in sensory quality.

24.7.19 Extruded Snacks

Lamichhane et al. (2013) developed extruded snacks from 75% maize flour, 10% sorghum flour, and 15% rice bean flour with particle size 1,090 μm. For extrusion, these flours are adjusted to moisture content of 17% and extruded at 135°C barrel temperature and 153 rpm screw speed. The extruded products have an expansion ratio of 4.0, bulk density of 80 kh/m³, water-soluble index of 21.48 ± 0.82, water absorption index of 9.77 ± 0.52, and measured contents of moisture, 0.667%; crude protein, 2.97%; fat, 1.87%; ash, 2.54%; crude fiber, 12.06%; carbohydrates, 79.86%; and beta-glucan, 6.52 ± 0.61. The sensory quality of extrudates were color, 7.5; taste, 7.0; texture, 7.5; and OAA, 7.5.

24.8 Conclusion and Future Prospects

The rice bean pulse crop represents a convincing replacement to various staple crops with a variety of favorable agronomic and nutritional qualities. Rice beans are identified with excellent nutritive value, mainly credited to the high-quality bioavailable proteins, excellent amino acid composition, and substantial level of different micronutrients. It can also exemplify an alternative or a supplement to uncommon food sources for economically demoted people. Moreover, the presence of low fats and essential amino acids, predominantly lysine and methionine, advocates their effective utilization in the nutritional security programs. The nutraceutical potential of rice beans is also promising owing to the presence of appreciable quantities of bioactive compounds like polyphenols and their role in preventing and treatment of disease like diabetes and obesity. Although a number of processing methods are available, their optimization still requires proper intervention. Modern processing methods can be accompanied by traditional knowledge and will certainly create a path for their commercial exploitation. Effective employment of of rice beans for the production of value-added products can also promote their commercial utilization. But most of these products are prepared using traditional methods of processing and require immediate consumption owing to the lack of marketing and proper packaging systems. Moreover, deficiency of proper nutritional profiling and storage studies also hampers their commercial viability. Hence, there is an immediate requirement for the production of ready-to-eat and ready-to-cook rice bean-based food products.

REFERENCES

Acharya, Bipin Kumar. "*Cultivation and use of ricebean. A case study of Dang District*, Nepal." Master's thesis, The University of Bergen (2008).

Alonso, R., L. A. Rubio, M. Muzquiz, and F. Marzo. "The effect of extrusion cooking on mineral bioavailability in pea and kidney bean seed meals." *Animal Feed Science and Technology* 94, no. 1-2 (2001): 1–13.

Andersen, P. and P. A. Hollington, "Food security through rice bean research in India and Nepal (FOSRIN). Report-3: Nutritional qualities of rice bean, Universitat Bergen, Norway, Bangor University, Wales, UK (2007), 1–10.

Anderson, J. W., A. E. Jones, and S. Riddell-Mason. "Ten different dietary fibers have significantly different effects on serum and liver lipids of cholesterol-fed rats." *The Journal of Nutrition* 124, no. 1 (1994): 78–83.

Anuonye, J. C., A. A. Jigam, and G. M. Ndaceko. "Effects of extrusion-cooking on the nutrient and antinutrient composition of pigeon pea and unripe plantain blends." *Journal of Applied Pharmaceutical Science* 2, no. 5 (2012): 158–162.

Arora, R. K., K. P. S. Chandel, B. S. Joshi, and K. C. Pant. "Rice bean: Tribal pulse of eastern India." *Economic Botany* (1980): 260–263.

Baruah, D. K., M. Das, and R. K. Sharma. "Nutritional and microbiological evaluation of ricebean (Vigna umbellata) based probiotic food multi mix using Lactobacillus plantarum and Lactobacillus rhamnosus." *Journal of Probiotics & Health* 6 (2018): 200.

Bepary, R. H., and D. D. Wadikar. "HPLC profiling of flatulence and non-flatulence saccharides in eleven ricebean (Vigna umbellata) varieties from North-East India." *Journal of Food Science and Technology* 56, no. 3 (2019): 1655–1662.

Bepary, R. H., D. D. Wadikar, and P. E. Patki. "Engineering properties of rice-bean varieties from North-East India." *Journal of Agricultural Engineering* 55, no. 3 (2018): 32–42.

Bepary, R. H., D. D. Wadikar, and P. E. Patki. "Analysis of eight water soluble vitamins in ricebean (Vigna umbellata) varieties from NE India by reverse phase-HPLC." *Journal of Food Measurement and Characterization* 13, no. 2 (2019): 1287–1298.

Bepary, R. H., D. D. Wadikar, S. Borah Neog, and P. E. Patki. "Studies on physico-chemical and cooking characteristics of rice bean varieties grown in NE region of India." *Journal of Food Science and Technology* 54, no. 4 (2017): 973–986.

Bhagyawant, S. S., D. T. Narvekar, N. Gupta, A. Bhadkaria, A. K. Gautam, and N. Srivastava. "Chickpea (Cicer arietinum L.) Lectin Exhibit Inhibition of ACE-I, α-amylase and α-glucosidase Activity." *Protein and peptide letters* 26, no. 7 (2019): 494–501.

Boateng, J., M. Verghese, L. T. Walker, and S. Ogutu. "Effect of processing on antioxidant contents in selected dry beans (*Phaseolus spp. L.*)." *LWT Food Science and Technology* 41, no. 9 (2008): 1541–1547.

Brand, J. C., B. J. Snow, G. P. Nabhan, and A. S. Truswell. "Plasma glucose and insulin responses to traditional Pima Indian meals." *The American Journal of Clinical Nutrition* 51, no. 3 (1990): 416–420.

Chau, C-F., and P. C-K. Cheung. "Effect of various processing methods on antinutrients and in vitro digestibility of protein and starch of two Chinese indigenous legume seeds." *Journal of Agricultural and Food Chemistry* 45, no. 12 (1997): 4773–4776.

Chau, C-F., P. C-K. Cheung, and Y-S. Wong. "Effects of cooking on content of amino acids and antinutrients in three Chinese indigenous legume seeds." *Journal of the Science of Food and Agriculture* 75, no. 4 (1997): 447–452.

Clemente, A. and C. Domoney. "Biological Significance of Polymorphism in Legume Protease Inhibitors from the Bowman-Birk Family." *Current Protein & Peptide Science* 7, no. 3 (2006), 201–216. doi:10.2174/138920306777452349

Cummings, J. H., and A. M. Stephen. "Carbohydrate terminology and classification." *European Journal of Clinical Nutrition* 61, no. 1 (2007): S5–S18.

de Carvalho, N. M., and R. D. Vieria. Rice bean [*Vignaumbellata* (Thunb.) Ohwi & Ohashi]. In: E. Nkowolo and J. Smartt (Eds.), *Legumes and Oilseeds in Nutrition*. London: Chapman & Hall, (1996): 222–228.

Dhillon, P. K., and Tanwar B. "Rice bean: A healthy and cost-effective alternative for crop and food diversity." *Food Security* 10, no. 3 (2018): 525–535.

Du, S-K, H. Jiang, X. Yu, and J-L. Jane. "Physicochemical and functional properties of whole legume flour." *LWT Food Science and Technology* 55, no. 1 (2014): 308–313.

Duranti, M. "Grain legume proteins and nutraceutical properties." *Fitoterapia* 77, no. 2 (2006): 67–82.

Gilani, G. S., K. A. Cockell, and E. Sepehr. "Effects of antinutritional factors on protein digestibility and amino acid availability in foods." *Journal of AOAC International*, 88, no. 3 (2005): 967–987

Grant G, J. E. Edwards and A. Pusztai "α-Amylase inhibitor levels in seeds generally available in Europe" *Journal of Science and Food Agriculture*.67 (1995) :235–238. doi: 10.1002/jsfa.2740670214.

Henchion, M., M. Hayes, A. M. Mullen, M. Fenelon, and B. Tiwari. "Future protein supply and demand: strategies and factors influencing a sustainable equilibrium." *Foods* 6, no. 7 (2017): 53.

Itagi, H. N., B. V. Rao, S. Rao, P. A. Jayadeep, and V. Singh. "Functional and antioxidant properties of ready-to-eat flakes from various cereals including sorghum and millets." *Quality Assurance and Safety of Crops & Foods* 4, no. 3 (2012): 126–133.

Jenab, M. and L. U. Thompson. "The influence of phytic acid in wheat bran on early biomarkers of colon carcinogenesis." *Carcinogenesis* 19, no. 6 (1998): 1087–1092.

Joshi, K. D., B. Bhandari, R. Gautam, J. Bajracharya, and P. A. Hollington. "Ricebean: A multipurpose underutilised legume." *Proceedings from the International Symposium on New Crops and Uses: Their role in a rapidly changing world*. University of Southampton, UK: Centre for Underutilised Crops, (2008): 234.

Kalidass, C. and V. R. Mohan. "Nutritional composition and antinutritional factors of little-known species of vigna." *Tropical and Subtropical Agroecosystems* 15, no. 3 (2012).

Kasera, R., A. B. Singh, S. Lavasa, K. Nagendra, and N. Arora. "Purification and immunobiochemical characterization of a 31 kDa cross-reactive allergen from Phaseolus vulgaris (kidney bean)." *PLOS ONE* 8, no. 5 (2013): e63063.

Katoch, R. "Nutritional and anti-nutritional constituents in different seed components of rice bean (Vigna umbellata)." *Indian Journal of Agricultural Biochemistry* 24, no. 1 (2011): 65–67.

Katoch, R. "Nutritional potential of rice bean (*Vigna umbellata*): An underutilized legume." *Journal of Food Science* 78, no. 1 (2013): C8–C16.

Katoch, R. *Ricebean: Exploiting the Nutritional Potential of an Underutilized Legume.* Springer Nature, 2020.

Katoch, R., and A. Jamwal. "Characterization of α-amylase inhibitor from rice bean with inhibitory activity against midgut α-amylases from Spodoptera litura." *Applied Biochemistry and Microbiology* 49, no. 4 (2013): 419–425.

Kitagawa I, H., K. Wang, M. Saito and M. Yoshokawa. (1983) "Saponin and sapogenol. XXXII. Chemical constituents of the seeds of *Vigna angularis* (Willd.) Ohwiet Ohashi. (2). Azukisaponins I, II, III and IV". *Chemical and Pharmaceutical Bulletin* 31, no. 2 (1983):674–682

Kaur, M. "Chemical composition of ricebean (Vigna umbellata): Effect of domestic processing." *Indian Journal of Applied Research* 5, no. 4 (2015): 311–313.

Kaur, D. and A. C. Kapoor. "Nutrient composition and antinutritional factors of rice bean (Vigna umbellata)." *Food Chemistry* 43, no. 2 (1992): 119–124.

Kaur, M., and B. L. Kawatra. "Effect of domestic processing on flatus producing factors in ricebean (Vigna umbellata)." *Food/Nahrung* 44, no. 6 (2000): 447–450.

Kaur, A., P. Kaur, N. Singh, A. S. Virdi, P. Singh and J. C. Rana. "Grains, starch and protein characteristics of rice bean (*Vigna umbellata*) grown in Indian Himalaya regions." *Food Research International.* 54, no. 1 (2013):102–110. doi: 10.1016/j.foodres.2013.05.019.

Khabiruddin, M. D., S. N. Gupta, and C. S. Tyagi. "Nutritional composition of some improved genotypes of ricebean (Vigna umbellata)." *Forage Research* 28, no. 2 (2002): 104–5.

Kii, S. V. M., A. Wijanarka and T. K. A. Puruhita. "Effect of ricebean flour (*Vigna umbellata*) and wheat flour mixing variations on physical properties, organoleptic properties and calcium levels of cookies production." *Medika Respati*, 8, no.1 (2013).

Korus, J., D. Gumul, and K. Czechowska. "Effect of extrusion on the phenolic composition and antioxidant activity of dry beans of Phaseolus vulgaris L." *Food Technology and Biotechnology* 45, no. 2 (2007): 139–146

Kumar, V., A. K. Sinha, H. P. S. Makkar, and K. Becker. "Dietary roles of phytate and phytase in human nutrition: A review." *Food Chemistry* 120, no. 4 (2010): 945–959.

Lajolo F. M., F. Saura-Calixto, E. Penna, W. Menezes. Fibra dietética en Iberoamérica: tecnología y salud: obtención, caracterización, efecto fisiológico y aplicación en alimentos. Livraria Varela, São Paulo (2001).

Lal, R. R. and Verma, P. *Post Harvest Management of Pulses.* Kanpur: Indian Institute of Pulses Research, (2007).

Lamichhane, B., P. Ojha, and B. Paudyal. "Formulation and quality evaluation of extruded product from composite blend of maize, sorghum and ricebean." *Journal of Food Science and Technology Nepal* 8 (2013): 40–45.

Malhotra, S., D. Malik, and K. Singh Dhindsa. "Proximate composition and antinutritional factors in rice bean (Vigna umbellata)." *Plant Foods for Human Nutrition* 38, no. 1 (1988): 75–81.

Massey, L. K., R. G. Palmer, and H. T. Horner. "Oxalate content of soybean seeds (Glycine max: Leguminosae), soyfoods, and other edible legumes." *Journal of Agricultural and food Chemistry* 49, no. 9 (2001): 4262–4266.

Mohan, R. J., Sangeetha, A., Narasimha, H. V., & Tiwari, B. K. Post-harvest technology of pulses. In: B. K. Tiwari, A. Gowen, and B. McKenna (Eds.), *Pulse Foods: Processing, Quality and Nutraceutical Applications.* London: Academic Press–Elsevier, (2011).

Noonan, S. C. and G. P. Savage. "Oxalate content of foods and its effect on humans." *Asia Pacific Journal of Clinical Nutrition* 8, no. 1 (1999): 64–74.

Oboh, H. A., M. Muzquiz, C. Burbano, C. Cuadrado, M. M. Pedrosa, G. Ayet, and A. U. Osagie. "Effect of soaking, cooking and germination on the oligosaccharide content of selected Nigerian legume seeds." *Plant Foods for Human Nutrition* 55, no. 2 (2000): 97–110.

Okoye, E. C., E. U. Njoku, and G. R. Ugwuanyi. "Proximate composition, micronutrient contents and acceptability of 'Ojojo' from the blends of water yam and Rice bean flours." *International Journal of Food Sciences and Nutrition* 3, no. 5 (2018): 5–10.

Okudu, H. O., and M. C. Ojinnaka. "Effect of soaking time on the nutrient and antinutrient composition of Bambara groundnut seeds (Vigna subterranean)." *African Journal of Food Science and Technology* 8, no. 2 (2017): 25–29.

Pal, R. S., A. Bhartiya, R. A. Kumar, L. Kant, J. P. Aditya, and J. K. Bisht. "Impact of dehulling and germination on nutrients, antinutrients, and antioxidant properties in horsegram." *Journal of Food Science and Technology* 53, no. 1 (2016): 337–347.

Parvathi, S., and V. J. F. Kumar. "Value added products from rice bean (Vigna umbellata)." *Journal of Food Science and Technology* 43, no. 2 (2006): 190–193.

Pattanayak, A., S. Roy, S. Sood, B. Iangrai, A. Banerjee, S. Gupta, and D. C. Joshi. "Rice bean: A lesser known pulse with well-recognized potential." *Planta* (2019): 1–18.

Pawar, V. D., M. K. Akkena, P. M. Kotecha, S. S. Thorat, and V. V. Bansode. "Effect of presoak treatment on cooking characteristics and nutritional functionality of rice bean." *Journal of Food Legumes* 25, no. 4 (2012): 321–325.

Prosky, L., and J. W. DeVries. *Controlling Dietary Fiber in Food Products.* Van Nostrand Reinhold, 1992.

Punia, S., Dhull, S. B., Sandhu, K. S., and Kaur, M.. Faba bean (Vicia faba) starch: Structure, properties, and in vitro digestibility—A review. *Legume Science*, 1, no. 1 (2019): e18.

Punia, S., Dhull, S. B., Sandhu, K. S., Kaur, M., and Purewal, S. S. Kidney bean (Phaseolus vulgaris) starch: A review. *Legume Science*, 2 no. 3 (2020): e52.

Rajyalakshmi, P., and P. Geervani. "Nutritive value of the foods cultivated and consumed by the tribals of South India." *Plant Foods for Human Nutrition* 46, no. 1 (1994): 53–61.

Randhir, R., Y-I. Kwon, and K. Shetty. "Effect of thermal processing on phenolics, antioxidant activity and health-relevant functionality of select grain sprouts and seedlings." *Innovative Food Science & Emerging Technologies* 9, no. 3 (2008): 355–364.

Rau, D., M. L. Murgia, M. Rodriguez, E. Bitocchi, E. Bellucci, D. Fois, D. Albani et al. "Genomic dissection of pod shattering in common bean: Mutations at non-orthologous loci at the basis of convergent phenotypic evolution under domestication of leguminous species." *The Plant Journal* 97, no. 4 (2019): 693–714.

Redgwell, Robert J., and Monica Fischer. "Dietary fiber as a versatile food component: An industrial perspective." *Molecular Nutrition & Food Research* 49, no. 6 (2005): 521–535.

Riascos, J. J., A. K. Weissinger, S. M. Weissinger, and A. W. Burks. "Hypoallergenic legume crops and food allergy: factors affecting feasibility and risk." *Journal of Agricultural and Food Chemistry* 58, no. 1 (2010): 20–27.

Saharan, K., N. Khetarpaul, and S. Bishnoi. "Processing of newly released ricebean and fababean cultivars: Changes in total and available calcium, iron and phosphorus." *International Journal of Food Sciences and Nutrition* 52, no. 5 (2001): 413–418.

Saharan, K., N. Khetarpaul, and S. Bishnoi. "HCl-extractability of minerals from ricebean and fababean: Influence of domestic processing methods." *Innovative Food Science & Emerging Technologies* 2, no. 4 (2001): 323–325.

Saharan, K., N. Khetarpaul, and S. Bishnoi. "Variability in physico-chemical properties and nutrient composition of newly released ricebean and fababean cultivars." *Journal of Food Composition and Analysis* 15, no. 2 (2002): 159–167.

Saikia, P., C. R. Sarkar, and I. Borua. "Chemical composition, antinutritional factors and effect of cooking on nutritional quality of rice bean [Vigna umbellata (Thunb; Ohwi and Ohashi)]." *Food Chemistry* 67, no. 4 (1999): 347–352.

Saini, R. and A. R. Chopra. "In-vitro plant regeneration via somatic embryogenesis in rice-bean Vigna umbellata (Thunb.) Ohwi and Ohashi: An underutilized and recalcitrant grain legume." *Journal of Environmental Research and Development* 6, no. 3 (2012): 452–457.

Scariot, M. A., G. Tiburski, F. W. Reichert Júnior, L. L. Radünz, and M. R. R. Meneguzzo. "Moisture content at harvest and drying temperature on bean seed quality." *Pesquisa Agropecuária Tropical* 47, no. 1 (2017): 93–101.

Sharma, S. *"Nutritional quality, functional properties and value addition of rice bean (Vignaumbellata)."* PhD thesis, Chaudhary Sarwan Kumar Himachal Pradesh Krishi, Vishvavidyalaya, Palampur, India, 2014.

Sharma, M., and A. Kawatra. "Effect of dietary fibre from cereal brans and legume seedcoats on serum lipids in rats." *Plant Foods for Human Nutrition* 47, no. 4 (1995): 287–292.

Sharma, R., and D. Punia. "Effect of traditional processing methods on protein digestibility and starch digestibility of field pea (Pisum sativum)." *International Journal of Food and Nutritional Sciences* 6, no. 3 (2017): 82.

Siddhuraju, P. and K. Becker. "The antioxidant and free radical scavenging activities of processed cowpea (Vigna unguiculata (L.) Walp.) seed extracts." *Food Chemistry* 101, no. 1 (2007): 10–19.

Siddhuraju, P., K. Vijayakumari, and K. Janardhanan. "Chemical composition and protein quality of the little-known legume, velvet bean (Mucuna pruriens (L.) DC.)." *Journal of Agricultural and Food Chemistry* 44, no. 9 (1996): 2636–2641.

Singh, S. P., B. K. Misra, K. P. S. Chandel, and K. C. Pant. "Major food constituents of rice-bean (Vigna umbellata)." *Journal of Food Science and Technology, India* 17, no. 5 (1980): 238–240.

Sritongtae, B., T. Sangsukiam, M. R. Morgan, and K. Duangmal. "Effect of acid pretreatment and the germination period on the composition and antioxidant activity of rice bean *(Vigna umbellata).*" *Food Chemistry*, 227 (2017): 280–288. doi:10.1016/j.foodchem.2017.01.103

Tomooka, N., A. Kaga, and D. A. Vaughan. "The Asian Vigna (Vigna subgenus Ceratotropis) biodiversity and evolution." *Plant Genome: Biodiversity and Evolution* 1, no. part C (2006): 87–126.

Uebersax, M. A, and M. Siddiq. "Postharvest storage quality, packaging and distribution of dry beans," In: *Dry Beans and Pulses: Production, Processing and Nutrition*. John Wiley & Sons, (2013).

Walker, ARP. "Dietary fiber in the prevention and treatment of disease," In: G. A. Spilled (Ed.), *CRC Handbook of Dietary Fiber in Human Nutrition,* 3rd ed. CRC Press, (2001).

Wei, Y., J. Yan, F. Long, and G. Lu. "Vignaumbellata (Thunb.) Ohwiet Ohashi or Vignaangularis (Willd.) OhwietOhashi赤小豆 (Chixiaodou, Rice Bean)," In: *Dietary Chinese Herbs*. Vienna: Springer (2015): 551–559.

Wu, G., J. Ashton, A. Simic, Z. Fang, and S. K. Johnson. "Mineral availability is modified by tannin and phytate content in sorghum flaked breakfast cereals." *Food Research International* 103 (2018): 509–514.

Yao, Y., X-Z. Cheng, L-X. Wang, S-H. Wang, and G. Ren. "Major phenolic compounds, antioxidant capacity and antidiabetic potential of rice bean (Vigna umbellata L.) in China." *International Journal of Molecular Sciences* 13, no. 3 (2012): 2707–2716.

25

Adzuki Beans (Vigna Angularis): Nutritional and Functional Properties

Prachi Jain, C. Lalmuanpuia, Antima Gupta and Arashdeep Singh
Bhaskaracharya College of Applied Sciences, Delhi University, India

CONTENTS

25.1 Introduction

Adzuki bean (*Vigna angularis var. angularis*) is a diploid leguminous crop with a long history of cultivation in eastern Asia, China, Japan, Taiwan, and South Korea. It is also known as small red beans. It has been predominantly consumed as traditional desserts, puddings, snacks, and sweet cuisines. It is high in protein and fiber, and contains a good amount of vitamins, minerals, essential amino acids, essential fatty acids, and phytochemicals such as polyphenols and phytates. Many different varieties of adzuki beans are found throughout the world, and their nutritional composition varies due to factors like harvest time, climate, soil quality, grain size, and region of cultivation.

Adzuki beans have been reported to have several pharmacological properties owing to its bioactive compounds. It has been used as an herbal medicine since the Tang dynasty in China to control body weight. In China, it is used to treat dropsy and diuretic functions, while in Japan it is utilized as an ingredient in traditional confections, such as, youkan, wagashi, manju, and amana (Gohara *et al.*, 2016). Adzuki beans were found to be effective against edema, diarrhea, vomiting, etc., as reported in *Ben Cao Gang Mu,* the first herbal book in Chinese history from the 16th century (Luo *et al.*, 2016).

Recently, antioxidant properties of the adzuki bean were discovered as its most remarkable property as it contains many polyphenols (Amarowicz *et al.*, 2008. Nowadays the beans are also gaining popularity due to its antidiabetic and weight-reducing effects. The beans are high in α-glucosidase inhibitory activity, which delays the digestion and absorption of carbohydrates and consequently results in suppression of the postprandial hyperglycemia (Zhenxung *et al.* 2016). Over the past decade, the recognition of *Vigna angularis* as a therapeutic food has significantly drawn global interest. As a result, literature on its nutritional composition, health benefits and processing quality has grown.

25.2 Nutritional Composition

The appealing nature of adzuki beans is due to its desirable nutrient profile. Different nutritional parameters were analyzed for 100 g of edible portion of dried seed and the results were as follows: water, 15%; energy, 324 kcal; protein, 21.1 g; fat, 1.0 g; carbohydrate, 59.5 g; fiber, 3.9 g; ash, 3.4 g; calcium, 82g; iron, 6.4 mg; thiamine, 0.45 mg; riboflavin, 0.15 mg; and niacin, 2.2 mg (Doughty and Walker *et al.*, 1982).

These beans are important sources of manganese, iron, and zinc, which are essential in many metabolic reactions as they function as cofactors for different enzymes. Adzuki beans also contribute to antioxidant activities due to a number of phenolic compounds, which are further discussed in this chapter (Gohara *et al.*, 2016).

25.2.1 Carbohydrates

In Adzuki beans *(Vigna angularis L.)*, stachyose (nonglycemic) is analyzed as the major oligosaccharide while sucrose and raffinose have also been identified. The amylose and resistant starch contribute to about 11.08–26.19% and 19.92–26.90% of total starch, while the highest resistant starch content is found in Jingnong 6 variety. Resistant starch is valued because it is not degraded by amylase/pullinase activity and is therefore nondigestible (Agarwal and Chauhan, 2019). Many clinical studies conducted in humans show that resistant starch can have potent health benefits, including improved insulin sensitivity, lower blood sugar levels, higher satiety levels, and various benefits for digestion (Nugent *et al.*, 2015; Punia *et al.,* 2019a; Punia *et al.*, 2020). The adzuki starch serves as desirable food agents with increased paste stabilities and reduced pasting temperature owed to its various properties like large granule size, low gelatinization temperatures, high amylopectin content and low crystalline stability (Agarwal and Chauhan, 2019).

Heat moisture treatment (HMT) can produce changes in the pasting and digestibility of red azuki starch. The HMT can produce changes in the granular structure at molecular and crystalline level. This rearrangement of the starch molecules results in increased thermal stability. HMT also results in formation of slow digesting carbohydrates by converting fractions of rapidly digestible starch and resistant starch into that of slowly digestible starch (Wang *et al.,* 2017).

25.2.2 Fat

The beans are often cited as the "weight loss bean" due to its lower calorie and fat content. The reported fat content of the beans is as low as 0.4–2.1% while in other legumes it can be as high as 40% as in groundnuts (Yousif *et al.*, 2007). The low fat content of beans is advantageous over many similar low-fat

TABLE 25.1

Major Phospholipids in the Oils Obtained From Adzuki Beans

Phospholipids	mg/100 g of Beans	Relative Weight%
Phosphatidylethanolamine	360.49 (+/−) 17.3	25.8
Phosphatidylcholine	632.8 (+/−) 30.7	45.3
Phosphatidylinositol	300.4 (+/−) 14.8	21.5
Others	103.4 (+/−) 4.2	7.4

Source: Yoshida *et al.* 2008.

legumes such as chickpeas because of the high polyunsaturated fat content, which exerts many health benefits in the human body (Yoshida *et al.*, 2010).

The analysis of lipid profile in the adzuki bean through thin layer chromatography (TLC) revealed phospholipids as the key lipid component found in adzuki beans, holding an approximate share of about 63.5 (wt.%). Three classes of phospholipids were noticeable namely phosphatidylcholine 45.3 (wt.%), phosphatidylethanolamine 25.8 (wt.%), and phosphatidylinositol 21.5 (wt.%) with the highest number of saturated fatty acids (Yoshida *et al.*, 2008). These phospholipids are found in cell membranes and are actively involved in biological and chemical reactions including energy generation in the cell (Table 25.1 Yoshida *et al.* 2008)

The fatty acid distributions of other lipid components were as follows: triacylglycerol (21.2 wt.%), steryl esters (7.5 wt.%), and hydrocarbons (5.1 wt.%). The free fatty acids diacylglycerols (1, 3-DAG and 1, 2-DAG) and monoacylglycerols were present in minor proportions 0.2–1.1(wt. %). Unsaturated fatty acids were mostly condensed in the sn-2 position while the sn-1 or sn-3 position in the oils of adzuki beans was occupied by saturated fatty acids (Yoshida *et al.*, 2008).

25.2.3 Proteins

The nutritional influences of legume proteins are already known. Plenty of research has been conducted to find cheaper sources of protein and replace animal protein with vegan protein. Techniques like dialysis and isoelectric precipitation were applied to study the protein profile of adzuki beans. In the previously conducted studies, the albumin fractions and globulin fractions accounted for 73.3% and 10.4% of the total extractable proteins, respectively. Albumin being the major adzuki protein contributes toward the unique water solubility of the beans, making them favorable for use in food applications. Vicilin-like or 7S globulin protein is referred to as major storage adzuki protein. (Tjahjadi *et al.*, 2006). This 7S protein which is the major globulin protein of the adzuki beans is classified as a glycoprotein, which consists of α- and β1-subunits. The α-subunit has a greater molecular weight of 55 kDa and β1-subunits have 35 kDa (Chen *et al.*, 1984). The minor 11S fraction, also known as legumin protein, makes up only 10% of the adzuki globulin protein (Meng and Ma., 2002). The 11S subunits have molecular weights of 40 and 20 kDa, respectively.

Typically the first, and at times, the second most limiting amino acids of the bean (Table 25.2 Doughty and Walker (1982) methionine and tryptophan, were found somewhat in higher concentration in albumin rather than globulin (Bhatty., 1982). Additionally, the lysine and threonine content of albumin protein were also higher than those of protein globulin. Amino acids like aspartic acid and glutamic acid were concentrated more in the latter protein part (Sakakibara *et al.*, 1979).

Adzuki beans are labeled as "gluten-free" because they lack gluten protein. This property makes them highly useful in the preparation of gluten-free snacks. As more people are diagnosed with celiac disease, the outlook for adzuki beans is also broadening (Agarwal and Chauhan, 2019).

25.2.4 Antioxidants

Adzuki beans contain 11 different types of phenolic compounds that are resistant to oxidative damage and prevent cell degeneration. These phenolic compounds are culpable for anti-inflammatory,

TABLE 25.2

Essential Amino Acid Content of Adzuki Bean

Amino Acid	Content(g/16 g N)
Leucine	7.8
Lysine	7.0
Phenylalanine	5.4
Valine	5.4
Isoleucine	4.5
Threonine	3.8
Tyrosine	3.4
Methionine	1.8
Cystine	1.1

Source: Doughty and Walker (1982).

antiallergic, and anticarcinogenic activities. Phenolic compounds consist of phenolic acid, tannins, and flavonoids among which phenolic acid is the primary phenolic compound of adzuki beans. Polyphenolic substances are abundant in adzuki beans. A 100 g of dry adzuki beans enclosed an average of 370 mg of polyphenol, with levels ranging from 56–737 mg depending on the bean variety and harvesting time. These beans are dark skin seeds that contain pigments of anthocyanidin consisting of pelargonidin, delphinidin, cyanidin, petunidin, and malvidin (Khang *et al.*, 2016). In the present study, four bound phenolic acids: caffeic acid, syringic acid, ferulic acid, and p-coumaric acid and two free phenolic acids, ferulic acid and caffeic acid, were found in the beans of which caffeic acid is found to be predominant in most of the bean variety (Shi *et al.*, 2016).

Catechin, a phenolic substance with hydroxyl group at C-3 also known as flavan-3-ols (Singh *et al.,* 2017), is frequently found in many plants and was also found to be the major phenolic acid of adzuki beans. Catechin is a key contributor to the antioxidant properties of the beans. It efficiently scavenges reactive oxygen species (ROS) and nitrogen species counting singlet oxygen, peroxides, etc. (Khang *et al.*, 2016). Procyanidingallate, procyanidin dimers, and procyanidin trimers were in the range 12.4 µg/g, 16.0–213 µg/g, and 41.8–42.4 µg/g, respectively, in adzuki crude extract. Procyanidin are oligomers of molecules of catechin and epicatechin (Singh *et al.,* 2017).

Tannins are present in the testa region of legumes and are important during oxidative stress or unfavorable environmental conditions faced by the seed (Shahidi and Ambigaipalan, 2015). Condensed tannins are proanthocyanidins, which release catechin and anthocyanin during heat treatments in alcoholic solutions. The condensed tannin content (CTC) in the adzuki seed coat was 13.8 mg catechin equivalent (CE)/g (Gan *et al.*, 2016).

Tocopherols are generally termed as Vitamin E and are natural antioxidants found in the food system. Among the different forms of tocopherols, δ-tocopherol (77.6%) was the major component while β-tocopherol (22.4%) was found to be minor. However, the most common forms α- and β-tocopherols were not detected in adzuki beans, which are otherwise present in other legumes (Grela and Gunter, 1995). The adzuki bean tocopherols distribution pattern is unique and different from other legumes (Yoshida *et al.*, 2002).

25.2.5 Mineral

The ability of legume and vegetable plants to build up high metal levels from the soil (Cobb *et al.*, 2000; Kumar *et al.*, 2007) depends on metal distribution, availability, and concentration in plant species. For the mineral composition analysis, the atomic absorption spectrophotometer was used. The beans were found to be a valuable source of elements mainly potassium, phosphorus, calcium, magnesium, etc. (Table 25.3 Agarwal and Chauhan, 2019)

TABLE 25.3

The Mineral Contents of the Adzuki Bean

Minerals	ppm (dry basis)
Potassium	12,915
Phosphorus	4,787
Magnesium	1,530
Calcium	705
Iron	60
Zinc	35
Manganese	14
Boron	13
Copper	11
Aluminium	11

Source: Agarwal and Chauhan 2019.

25.3 Functional Properties of Adzuki Beans

25.3.1 Anti-nutritional Factor

The adzuki bean contains antinutritional factors that can be protein or nonprotein in nature. The most characterized antinutritional factor of the beans is trypsin inhibitor, which is a type of protease inhibitor. It affects the protein digestibility by reducing enzyme activity, which degrades or digests protein in the body. Trypsin inhibitor units (TIU) were observed in the range of 2,881.12–3,510.07 TIU/g. Another study conducted by Wati *et al.* (2009) indicated a range of 8,490–12,354 TIU/g. However, these differences in ranges can be due to varying environmental conditions, different extraction methods, etc.

Phytic acid, which is also a prominent antinutritional factor occurring in legumes, was found to be in the range of 368.96–507.92 mg/100 g (Sharma *et al.*, 2018). Saponins (mainly azuki saponins II and VI) are also reported in adzuki beans (Kinjo, Imagire *et al.,* 1998) Saponins are nonvolatile compounds in which polar and water-soluble sugar molecules (pentose, hexose, or uronic acid) are coupled with nonpolar, fat-soluble saponegin (or aglycone). The strong surface active property of saponins is due to its amphiphilic nature (Heng *et al.*, 2006: Singh, Kaur *et al.*, 2017).

25.3.2 Antimicrobial

Extracts from black, red, and green adzuki beans were assessed for their antimicrobial activity (Hori *et al.*, 2006). The beans were found to be effective against many different species of bacteria. Extracts from black beans inhibited the growth of *Escherichia coli* (*E. coli*), *bacillus subtilis, Pseudomonas aeruginosa*, and *Enterococcus faecalis. E. coli and P. aeruginosa* were inhibited by red and green extracts, respectively. Authors of the study signified polyphenols were responsible for the antimicrobial activity (Liu and Xu, 2016)

25.3.3 Anti-inflammatory

Adzuki methanol extracts trimmed the number of mast cells present in the skin, ratio of eosinophil present in peripheral leukocytes, and relative mRNA expression of inflammatory cytokines in the spleen, all of which contributes to the anti-inflammatory effect of adzuki beans (Liu and Xu, 2016). The 1% adzuki bean extract with incubated adipocytes was shown to decrease the inflammatory responses in the cell without affecting its viability in a study conducted by Okada *et al.,* 2012.

25.3.4 Antiallergic Effect

Adzuki beans are helpful in alleviating allergic symptoms. The intracellular elevation of [Ca2+] and passive cutaneous anaphylaxis reaction was suppressed by an extract made of 40% ethanol elute (Itoh *et al.* 2012). It is also reported that extracts of *vigna angularis* decreased intracellular calcium levels, which possibly further exerted inhibitory effects on the release of histamine (Kim *et al.,* 2013).

25.3.5 Antidiabetic Effect

Alpha-glucosidase is an important enzyme involved in the glucose absorption occurring in the intestine. The inhibition of this enzyme can be a way to reduce glucose absorption. Yao *et al.* (2011) investigated 15 legumes along with adzuki beans in which the ethanol extract of adzuki beans exhibited the highest rate (64.33%) of inhibition. Sreerama *et al.*, 2012 showed that the ethanol extracts from black adzuki beans had higher inhibitory effects than red adzuki extracts.

The levels of Blood glucose, serum insulin levels, urinary glucose, creatinine ratio, liver triacylglycerol, and total cholesterol of spontaneously diabetic KK-Ay mice, type 2 diabetes was reported to be lowered by 40% ethanol fraction of hot water extract made by using diaion HP-20 column (Itoh *et al.*, 2009).

25.3.6 Renal Protection

A study reported that white and red adzuki bean seed coats, preferably red adzuki bean seed coat, provided protection against macrophages in cis-platin-induced rat kidney damage. It also reduced the damage of fibrotic areas in the kidney (Sato, Hori *et al.*, 2005).

Sato,Yamate *et al.* (2005) also reported that adzuki bean seed coats that contain dietary fiber and proanthocyanidins improve interstitial fibrosis and suppress the action of infiltrating macrophages in the damaged kidney.

25.3.7 Hepatoprotective Effect

The hepatoprotective effect of adzuki beans was lower than mung bean extracts but still considerable. Adzuki bean extracts help in reducing oxidative damage of the liver by serving as a prophylactic agent. The water extracts of adzuki beans suppressed serum glutamate-oxalate-transaminase and serum glutamate-pyruvate-transaminase activities in acetaminophen-induced hepatotoxicity rats (Wu *et al.*, 2001). Moreover, the extracts were also beneficial in the recovery from D-galactosamine-induced liver damage (Ohba *et al.*, 2005).

25.3.8 Antihypertensive Effect

The blood pressure-adjusting effect of adzuki beans associates itself with increasing the production of nitric oxide and decreasing the expressions of endothelial nitric oxide synthase in hypertensive rats (Mukai and Sato, 2009). The polyphenol content of the adzuki bean seed coat also reduces the oxidative stress and inflammation during the progression of hypertension. (Mukai and Sato, 2011).

25.3.9 Antihyperlipidemic Effect

Incorporation of adzuki beans in the diet significantly lowers the levels of serum triglycerides and total serum cholesterol concentration. This effect of adzuki bean is associated with resistant starch, which further lowers serum cholesterol by enhancing activity of the hepatic low-density lipid-receptor mRNA and cholesterol 7alpha-hydroxylase mRNA (Han *et al.*, 2005).

The adzuki bean polyphenols were also reported to reduce the level of low density lipoprotein (LDLP), very low density lipoprotein (VLDP), and intermediate density lipoprotein (IDLP). Adzuki bean juice contained total polyphenol 880 mg per 150 g while 1,960 mg polyphenol per 150g was present in the concentrated juice part (Maruyama *et al.*, 2008). So, adzuki bean juice might be beneficial to prevent hyperlipidemia in individuals.

25.3.10 Antiobesity Effect

The antiobesity effect of the adzuki bean diet is well known. Experiments performed by many scientists concluded that these beans help in reducing final body weight and decreasing weight gains with zero toxicity in the experimented rats. Moreover, it reduces abdominal fat and epididymal fat of the rats (Matsumoto *et al.*, 2002. This antiobesity effect is related to the inhibition of the action of pancreatic lipase. Pancreatic lipase is the chief enzyme involved in the hydrolysis of dietary fat. It catalyses around 50–60% of dietary fat present in the digestive system (Mukherjee, 2003). The phenolic extracts of both red and black adzuki beans exert inhibition properties on pancreatic lipase.

25.3.11 Anticancer Effects

Xu and Chang (2012) investigated adzuki beans and found that they inhibited the proliferation of nine human cancer cell lines, including, gastric adenocarcinoma cell AGS, squamous carcinoma cell CAL 27, hepatocellular carcinoma cell HepG2, colorectal adenocarcinoma cell SW 480, prostate carcinoma cell DU145, colorectal adenocarcinoma cell Caco-2, ovary adenocarcinoma cell SK-OV-3, leukemia cell HL-60, and breast adenocarcinoma cell MCF-7. It was reported that ethanol fraction (40%) of hot-water extract from adzuki bean adsorbed by diaion HP-20 resin exerts a inhibitory effects on adhesion, invasion, and metastasis of murine B16 melanoma cells *in vitro* (Itoh *et al.*, 2012).

25.3.12 Anti-Alzheimer's Effect

Alzheimer is a neurological disorder that results in motor impairment. It is believed to be caused by the accumulation and aggregation of amyloid proteins (Aβ) in the brain. It was reported that adzuki beans can successfully delay the progression of the disease (Miyazaki *et al.*, 2019). The polyphenols present in adzuki beans inhibits the aggregate formation of beta amyloid proteins in the brain (Porat *et al.*, 2004; Sato *et al.*, 2005; Mukai and Sato, 2009) and also suppresses the oxidative stress due to Aβ42 proteins (Luo *et al.*, 2016). However, the authors suggested that adzuki beans may delay the progression of Alzheimer's disease, but further studies on the mechanism associated with it is required (Miyazaki *et al.*, 2019).

25.4 Uses of Adzuki Bean in Food Products

25.4.1 Adzuki Bean Extracts as a Natural Antioxidant in Pork Sausages

The high amount of unsaturated fats in meat makes them highly susceptible to lipid oxidation. Moreover, mincing or grinding of meat exposes it to metal catalysts, which further increase chances of oxidative damage (Morrissey *et al.*, 1998). Commercially produced processed meats contain antioxidants to prevent lipid oxidation. In case of uncured meats, it is synthetic phenolic antioxidants, such as BHT and BHA, while in uncured meat it is sodium nitrite.

Adzuki bean extract contains many polyphenols, such as anthocyanin and catechins, which exert antioxidant effects (Ariga *et al.*, 1988; Ariga & Hamano, 1990). In the experiments, it was found that 0.2% adzuki extract, when used in both cooked and uncooked sausages, was able to prevent lipid oxidation.

Color values (CIE lab L*, a*, b* and hue angle) did not show any considerable changes in both the samples, however in uncured sausages, the lightness denoted L* and redness symbolized as a* increased respectively due to the dark color of the extracts. In cured sausages, the a* value increased with a decrease in L* and b* (yellowness) values.

The sensory analysis revealed negligible differences in taste, flavor, and overall acceptability of both the sausages added with 0.2% adzuki extract. Thus, the use of adzuki extract in modulation with respective preservatives provides successful protection from lipid oxidation and can be researched more to increase consumer acceptability and product quality (Jayawardana *et al.,* 2011).

25.4.2 Effects of Frying, Roasting, and Boiling Adzuki Beans and their Potential Use in Bakery Products like Biscuits

Adzuki beans are more popularly used in processed form rather than in raw form. The beans are mainly subjected to processing methods like roasting, frying, and boiling, which plays an important role in development of flavor in the final product (Molteberg *et al.,* 1996). The total amount of aroma compounds was found to be 0.0012% (w/w) in the beans (Lee, Mitchell, and Shibamoto, 2000) along with identification of 142 volatile compounds, such as hexanal which is responsible for the beany flavors in the beans. The most common method of changing unacceptable flavors of beans to acceptable forms is by heating methods (Chigwedere *et al.,* 2019). Cooked adzuki beans have a sugary flavor due to the presence of maltol (Tokitomo and Kobayashi, 1988).

In raw, fried, boiled, and roasted samples of adzuki beans, 100 volatile compounds were identified. Processing methods like roasting and frying increased the number of volatile compounds while boiled adzuki beans showed a decrease in volatile compounds. Alcohols and aliphatic aldehydes, like 2-nonenal, 1-octanol, and 1-nonanol, were found in higher concentrations in raw beans than in cooked adzuki beans. Aromatic aldehydes and alcohols are formed as a result of the chemical reaction between amino acids, sugars, and phenolic reactants in the cooked beans. Roasted beans have the highest amount of aromatic aldehydes and alcohols (Ma *et al.,* 2016). The chief volatile compound found in cooked adzuki beans is 1-octen-3-ol (Buttery, 1975).

Aliphatic ketones like 3,5-octadien-2-one are detected in lower concentrations in raw beans and their concentration further decreases during cooking. Two esters, dibutyl phthalate and methyl salicylate, were found in cooked beans while methyl benzoate, methyl linoleate, and methyl palmitate were present in raw beans. Both ketones and esters generally exhibit pleasant odors, however, they do not contribute to the aroma due to their high threshold odor detection values (Mottram, 1994).

In general, it can be concluded that overall acceptability of aroma increases in cooked beans than in raw beans. It changes from a grassy odor to nutty and roasted in fried and roasted beans. Among all the cooked forms, the roasted form depicted the best flavor properties and highest number of volatile compounds as reported by Bi and Wang., 2020.

Biscuits prepared from adzuki bean flour had a volatile concentration of 4,787.06 ± 204.51 μg kg⁻¹ while the biscuits that contained a mixture of adzuki bean flour and millet flour had a volatile concentration of 6,694.43 ± 246.60 μg kg⁻¹. This difference in volatile material concentration was attributed to a difference in fat content. In general, the fat content in millet is 5.6 times more than adzuki beans, which results in more lipid degradation and thus more volatile concentration.

In the experiment, a total of five different types of aromas were reported, namely caramel-like, roasted bean-like, cream-like, nutty, and popcorn-like. The biscuits supplemented with adzuki bean flour were typically "roasted bean like," whereas biscuits containing both millet and adzuki flour had more "popcorn and nutty like" aroma. The biscuits with millet-adzuki combinations have higher intensities of all five aromas and were most acceptable as reported by Bi and Wang., 2020.

25.4.3 Adzuki Bean Flour as a Meat Extender and Fat Replacer in Meat Products

The high protein and fiber content of adzuki beans makes them highly suitable for replacement of corn flour and fat by different proportions of adzuki bean flour (ABF) in meat products. Reduced-fat meatballs are one such example of successfully produced meat products using ABF. This replacement also resulted in reduction of calories and better physicochemical properties.

Increase in cooking yield was observed, which indicates higher water uptake and thus more weight of the meatballs. In the experiment, it was found that meatballs with 100% ABF had a higher cooking yield value than one with zero % of ABF (control sample) (Dzudie *et al.,* 2002; Serdarogˇlu *et al.,* 2005). This greater absorption of water was reported due to good hydration properties of the adzuki seed coat (Sefa-Dedah and Stanley., 1979).

Incorporation of ABF in meatballs also results in desirable textural changes. The high protein and starch content of ABF contributes to greater hardness and chewiness (Huda *et al.*, 2010). However, springiness and cohesiveness were not much affected by ABF content (Modi *et al.*, 2009). In the study conducted by authors Huda *et al.* (2010) and Serdarog˘lu *et al.* (2005) it was reported that with increasing incorporation of ABF and decreasing fat content there was a decrease in lightness and yellowness of the meatball samples. The authors suggested that higher fat content dilutes the myoglobin, which results in lighter color. Huda *et al.* (2010) also suggested that this color dilution is a result of increasing protein and fat content.

ABF has considerable water-binding properties when used as a meat extender. The moisture content of meatballs significantly increased until an increase in ABF amounted up to 75% (Dzudie *et al.,* 2002). However, the moisture content in meatballs decreased with the addition of ABF to 100%. This further concluded that the water-binding ability highly depends on fat content and up to a certain amount of flour. A 100% replacement of corn flour and fat significantly decreased the water-binding capacity (Serdarog˘lu *et al.*, 2005).

As earlier discussed, ABF is a good source for protein and thus its increased amounts (w/w) 25%, 50%, and higher will subsequently have greater protein content. The protein values were even higher than meatballs produced with wheat, sago, or cassava flour, ranging 13.1–13.3% (w/w) (Ikhlas *et al.*, 2011). Similar protein content was found in meatballs made with potato starch 14.44% (w/w) and tapioca starch 13.55% (w/w) Purnomo and Rahardiyan (2008). However, meatballs made with common bean flour 18.49–19.78% (w/w) had higher protein content than meatballs made with ABF (Dzudie *et al.*, 2002).

ABF did not impart any typical or odd flavor or odor to the product. On the basis of texture, there was no considerable effect up to 75% (w/w) addition. Total replacement by ABF produced the darkest color in the meatballs. The findings in the study indicated that ABF is a potential fat replacer and meat extender up to 50–75% (w/w) (Aslinah *et al.,* 2018).

25.4.4 Use of Adzuki Bean in Production of Slow Digesting Red Rice

Slow-digesting starch is important for controlling obesity and digestive problems. Starch digestibility is dependent on many factors like its granular structure, amylose and amylopectin ratio, interactions with other food components, etc. (Magallanes-Cruz *et al.*, 2017; Punia *et al.*, 2019b). Proanthocyanidins are oligomers of flavan-3-ols and can interact with polysaccharides (Renard *et al.*, 2017) as well as decrease starch digestibility by suppression of α-amylase activity (Kawakami *et al.,* 2010). Adzuki beans contain 1,019 mg/100 g of procyanidin dimers (B-type) (Ariga and Asao, 1981). Adzuki-meshi, which is prepared by cooking adzuki beans and nonglutinous rice together, reportedly slows starch digestion in rice (Takahama *et al.* 2019; Hirota and Takahama, 2017). The mechanism of slowing digestion is cause when the amylose molecules of rice starch combine with procyanidin and its oxidation products; this combination then covers the surface of cooked rice, thus slowing the digestion process (Morina *et al.,* 2020).

25.4.5 Adzuki Bean Paste as a Confectionery Ingredient

The use of bean paste as a confectionery ingredient originated from Asia. Bean paste-containing products made from cooked beans and sugar are important traditional confectionery products in Japan. Bean paste qualities include paste yield, color, stickiness, smoothness, aroma, and flavor. It is used as a filling of sweet bread or rice cakes, or small pieces of sweets, collectively called wagashi (Okada *et al.*, 2011). The color of bean paste is determined by the seed coat color of the beans. Red bean paste is made mainly from red adzuki beans, and white bean paste is made from lima beans, white common beans and white adzuki beans (Kato *et al.*, 2000).

25.5 Conclusion

Adzuki beans are a brilliant source of numerous essential compounds. Several studies reported that these grains possess antithrombogenic, hypocholesterolemic, antiallergic, antiobesity, and many more effects in the body. The ratio of polyunsaturated fatty acids to monounsaturated fatty acids and saturated fatty acids were also appropriate. Belonging to the legume family, these beans are also a fair source of protein. The presence of phenolic compounds in beans indicates its antioxidant activity, which is an important property in the immune systems of humans. The antioxidant content was comparable to other beans and further adds to the nutritional quality of the adzuki beans. The tocopherol content of the beans was higher than found in any other beans. Moreover, the beans were also a prominent source of vitamins and minerals. The traditional use of adzuki beans in the eastern dietary pattern is potentially promising due to its intrinsic characteristics. Moreover, there is a wide scope for clinical trials and in research areas to dig deeper into the nutritional and functional properties of adzuki beans.

REFERENCES

Agarwal, S., & Chauhan, E. S. (2019). Adzuki beans–Physical and nutritional characteristics of beans and its health benefits. *International Journal of Health Science Research*, 9(4): 304–310.

Amarowicz, R., Estrella, I., Hernández, T., & Troszyńska, A. (2008). Antioxidant activity of extract of azuki bean and its fractions. *Journal of Food Lipids,* 15: 119–136.

Ariga, T., & Asao, Y. (1981). Isolation, identification and organoleptic astringency of dimericproanthocyanidins occurring in adzuki bean. *Agricultural and Biological Chemistry*, 45:2709–2712.

Ariga, T., & Hamano, M. (1990). Radical scavenging action and its mode in procyanidins B-1 and B-3 from azuki beans to peroxyl radicals. *Agricultural & Biological Chemistry*, 54:2499–2504

Ariga, T., Koshiyama, I., & Fukushima, D. (1988). Antioxidative properties of procyanidins B-1 and B-3 from azuki beans in an aqueous system. *Agricultural & Biological Chemistry*, 52:2717–2722.

Aslinah, L. N. F., Mat Yusoff, M., & Ismail-Fitry, M. R. (2018). Simultaneous use of adzuki beans (Vigna angularis) flour as meat extender and fat replacer in reduced-fat beef meatballs (beboladaging). *Journal of Food Science and Technology*, 55(8):3241–3248.

Barac, M. B., Pesic, M. B., Stanojevic, S. P., Kostic, A. Z., & Bivolarevic, V. (2014). Comparative study of the functional properties of three legume seed isolates: Adzuki, pea and soy bean. *Journal of Food Science and Technology*, 52(5):2779–2787.

Bhatty, R. S. 1982. Albumin proteins of eight edible grain legume species: Electrophoretic patterns and amino acid composition. *Journal of Agricultural and Food Chemistry,* 30:620.

Bi, S., Wang, A., Lao, F., Shen, Q., Liao, X., Zhang, P., & Wu, J. (2020). Effects of frying, roasting and boiling on aroma profiles of adzuki beans (Vigna angularis) and potential of adzuki bean and millet flours to improve flavor and sensory characteristics of biscuits. *Food Chemistry*, 127878.

Buttery, R. G. (1975). Nona-2, 4, 6-trienal, an unusual component of blended dry beans. *Journal of Agricultural and Food Chemistry*, 23(5):1003–1004.

Chen, T. H. H., Gusta, L. V., Tjahjadi, C., & Breene, W. M. (1984). Electrophoretic characterization of adzuki bean (Vignaangularis) seed proteins. *Journal of Agriculture and Food Chemistry,* 32:396–399.

Chigwedere, C. M., Tadele, W. W., Yi, J., Wibowo, S., Kebede, B. T., Van Loey, A. M., …Hendrickx, M. E. (2019). Insight into the evolution of flavor compounds during cooking of common beans utilizing a headspace untargeted fingerprinting approach. *Food Chemistry*, 275:224–238.

Cobb G. P., Sands K., Waters M., Wixson B. G., & Dorward-King E. (2000). Accumulation of heavy metals by garden vegetables. *Journal of Environmental Quality,* 29:934–939.

Doughty, J., Aykroyd, W. R., & Walker, A. F. (1982). *Legumes in human nutrition*. Italy: Food and Agriculture of the United Nations.

Dzudie T., Scher J., & Hardy J. (2002) Common bean flour as an extender in beef sausages. *JournalFood Engineering,* 52:143–147.

Gan, R-Y., Deng, Z-Q., Yan, A-X., Shah, N. P.., Lui, W-Y., Chan, C-L., & Corke, H. (2016). Pigmented edible bean coats as natural sources of polyphenols with antioxidant and antibacterial effects, *Lebensmittel-Wissenschaft & Technologie*, 73:168–177.

Gohara A. K., Souza A. H., Gomes S. T., Souza, N., Visentainer J. V., & Matsushita M. (2016). Nutritional and bioactive compounds of adzuki bean cultivars using the Chemometric approach. *Food Science and Technology*, 40(1): 104.

Grela, E. R., & Gunter, K. D. (1995). Fatty acid composition and tocopherol content of some legume seeds. *Animal Feed Science Technology*, 52:325–331.

Han, K. H., Iijuka, M., Shimada, K., Sekikawa, M., Kuramochi, K., Ohba, K., Ruvini, L., Chiji, H., & Fukushima, M. (2005). Adzuki resistant starch lowered serum Cholesterol and hepatic 3-hydroxy-3-methylglutaryl-CoA mRNA levels and increased Hepatic LDL-receptor and cholesterol 7alpha-hydroxylase mRNA levels in rats fed a Cholesterol diet. *British Journal of Nutrition*, 94(6): 902–908.

Heng, L., Vincken, J.-P., van Koningsveld, G., Legger, A., Gruppen, H., van Boekel, T., Voragen, F. (2006). Bitterness of saponins and their content in dry peas. *Journal of the Science of Food and Agriculture*, 86(8):1225–1231.

Hirota S. & Takahama U. (2017). Inhibition of pancreatic induced digestion of cooked rice starch by adzuki (Viganaangularis) bean flavonoids and the possibility of a decrease in the Inhibitory effects in the stomach. *Journal of Agricultural and Food Chemistry*, 65(10):2172–2179.

Hongwei, W., Zhaoyuan, W., Li, X., Chen, L., & Zhang, B. (2017). Multi-scale structure, pasting and digestibility of heat moisture treated red adzuki bean starch. *International Journal of Biological Macromolecules*, 102.

Hori, Y., Sato, S., & Hatai, A. (2006). Antibacterial activity of plant extracts from adzuki Beans (Vigna angularis) in vitro. *Phytotherapy Research*, 20(2): 162–164.

Huda, N., Shen, Y. H., Huey, Y. L., Ahmad, R, & Mardiah, A. (2010) Evaluation of physico-chemical properties of Malaysian commercial beef meatballs. *American Journal of Food Technology*, 5(1):13–21

Ikhlas, B., Huda, N., & Noryati, I. (2011) Chemical composition and physicochemical properties of meatballs prepared from mechanically deboned quail meat using various types of flour. *International Journal of Poultry Science*, 10(1):30–37.

Itoh, T., Hori, Y., Atsumi, T., Toriizuka, K., Nakamura, M., Maeyama, T., Ando, M.,Tsukamasa, Y., Ida, Y., & Furuichi, Y. (2012). Hot water extract of adzuki (Vigna NBC Angularis) suppresses antigen-stimulated degranulation in rat basophilic leukemia RBL-2H3 cells and passive cutaneous anaphylaxis reaction in mice. *Phytotherapy Research*, 26(7): 1003–1011.

Itoh, T., Kobayashi, M., Horio, F., & Furuichi, Y. (2009). Hypoglycemic effect of hot water extract of adzuki (Vigna angularis) in spontaneously diabetic KK-A(y) mice. *Nutrition*, 25(2):134–141.

Jayawardana, B. C., Hirano, T., Han, K.-H., Ishii, H., Okada, T., Shibayama, S., … Shimada, K. (2011). Utilization of adzuki bean extract as a natural antioxidant in cured and uncured cooked pork sausages. *Meat Science*, 89(2):150–153.

Kato, J. (2000). Studies on characteristics for food processing of adzuki beans and common beans, and factors for their variation [English abstract]. *Hokkaido Central Agricultural Experiment Station Report*, Iwate University.

Kawakami K., Aketa S., Nakanami M., Iizuka S., & Hirayama M. (2010). Major water-soluble polyphenols, proanthocyanidins, in leaves of persimmon (Diospyros kaki) and their a-amylase inhibitory activity. *Bioscience, Biotechnology, and Biochemistry*, 74(7):1380–1385.

Khang, D. T., Dung, T. N., Elzaawely, A. A., & Xuan, T. D. (2016). *Phenolic Profiles and Antioxidant Activity of Germinated Legumes. Foods*, 5(2):27.

Kim, H., Kim, S., Kim, D., Oh, H., Rho, M., & Kim, S. (2013) Vignaangularis inhibits mast cell-mediated allergic inflammation. *International Journal of Molecular Medicine*, 32(3):736–742.

Kinjo, J., Imagire, M., Udayama, M., Arao, T., & Nohara, T. (1998). Structure-hepatoprotective relationships study of soyasaponins I-IV Having soyasapogenol B as aglycone. *Planta Medica*, 64(3): 233–236.

Kumar N. J. I., Soni H., & Kumar R. K. (2007). Characterization of heavy metals in vegetables using inductive coupled plasma Analyzer (ICPA). *Journal of Applied Sciences and Environmental Management*, 11:75–79.

Lee, K. G., Mitchell, A. E., & Shibamoto, T. (2000). Determination of antioxidant properties of aroma extracts from various beans. *Journal of Agricultural and Food Chemistry*, 48(10):4817–4820.

Liu, R., & Xu, B. (2016). Bioactive compositions and health promoting effects of adzuki bean. In: J. N. Gavial (Eds.), *Recent Progress in Medicinal Plants Volume 44, Phytotherapeutics III*. Studium Press, 23–43.

Luo, J., Cai, W., Wu, T., & Xu, B. (2016) Phytochemical distribution in hull and cotyledon of adzuki bean (*Vigna angularis* L.) and mung bean (*Vigna radiate* L.), and their contribution to antioxidant, anti-inflammatory and anti-diabetic activities. *Food Chemistry*, 201:350–360.

Ma, Z., Boye, J. I., Azarnia, S., & Simpson, B. K. (2016). Volatile flavor profile of Saskatchewan grown pulses as affected by different thermal processing treatments. *International Journal of Food Properties*, 19(10):2251–2271.

Magallanes-Cruz P. A., Flores-Silva P. C., & Bello-Perez L. A. (2017). Starch structure influences its digestibility. *Journal of Food Science*, 82:2016–2023.

Maruyama, C., Araki, R., Kawamura, M., Kondo, N., Kigawa, M., Kawai, Y., Takanami,Y., Miyashita, K., & Shimomitsu, T. (2008). Adzuki bean juice lowers serum Triglyceride concentrations in healthy young women. *Journal of Clinical Biochemistry and Nutrition*, 43(1):19–25.

Matsumoto, Y., & Ono, A. (2002). Effect of dietary adzuki bean (Phaseolus angularis) on serum lipid concentrations in adult rats. *Kawasaki Journal of Medical Welfare*, 8:2, 49–55.

Meng, G.-T., & Ma, C-Y. (2002). Characterization of globulin from Phaseolusangularis (red bean). *International Journal of Food Science and Technology*, 37:687–695.

Miyazaki H, Okamoto Y, Motoi A, Watanabe T, Katayama S, Kawahara SI, Makabe H, Fujii H, & Yonekura S. (2019). Adzuki bean (*Vigna angularis*) extract reduces amyloid-β aggregation and delays cognitive impairment in Drosophila models of Alzheimer's disease. *Nutrition Research and Practice*, 13(1):64–69.

Modi V. K., Yashoda K. P., & Naveen S. K. (2009). Effect of carrageenan and oat flour on quality characteristics of meat kofta. *International Journal of Food Properties*, 12:228–242.

Molteberg, E. L., Magnus, E. M., Bjørge, J. M., & Nilsson, A. (1996). Sensory and chemical studies of lipid oxidation in raw and heat-treated oat flours. *Cereal Chemistry*, 73(5):579–587.

Morina, F., Hirota, S., & Takahama, U. (2020). Contribution of amylose-procyanidin complexes to slower starch digestion of red-colored rice prepared by cooking with adzuki bean. *International Journal of Food Sciences and Nutrition*, 71(6):1–11.

Morrissey, P. A., Sheehy, P. J. A., Galvin, K., Kerry, J. P., & Buckley, D. J. (1998). Lipid stability in meat and meat products. *Meat Science*, 49:573–586.

Mottram, D. (1994). Meat flavour. In: J. R. Piggott, & A. Paterson (Eds.), *Understanding Natural Flavors*. Boston: Springer, 140–163.

Mukai, Y., & Sato, S. (2009). Polyphenol-containing adzuki bean (*Vigna angularis*) extract attenuates blood pressure elevation and modulates nitric oxide synthase and Caveolin-1 expressions in rats with hypertension. *Nutrition Metabolism and Cardiovascular Diseases*, 19(7):491–497.

Mukai, Y., & Sato, S. (2011). Polyphenol-containing adzuki bean (*Vigna angularis*) seed coats attenuate vascular oxidative stress and inflammation in spontaneously hypertensive rats. *The Journal of Nutritional Biochemistry*, 22(11):16–21.

Mukherjee, M. (2003). Human digestive and metabolic lipases-a brief review. *Journal of Molecular Catalysis b: Enzymatic*, 22:369–376.

Nugent A. P. (2015) Health properties of resistant starch, *Nutrition Bulletin*, 30(1):27–54.

Ohba, K., Nirei, M., Watanabe, S., Han, K.H., Hashimoto, N., Shimada, K., Sekikawa,M., Chiji, H., & Fukushima, M. (2005). Effect of an adzuki bean extract on hepaticAnti-oxidant enzyme mRNAs in D-galactosamine-treated rats. *Bioscience, Biotechnology, and Biochemistry*, 69(10):1988–1991.

Okada T. Wagashi, Igarashi O., Ichishima E., Oga K., Kobayashi A., Tajima M. et al. eds, *Encyclopedia of Food [in Japanese]*. Maruzen, Tokyo, pp. 817–818 (2011).

Okada K, Tomoko, & Ito (2012). Anti-obesity role of adzuki bean extract containing polyphenols: In vivo and in vitro effects. *Journal of the science of food and agriculture*, 92, 2644–2651.

Porat, Y., Mazor, Y., Efrat, S., & Gazit, E. (2004). Inhibition of islet amyloid polypeptide fibril formation: A potential role for heteroaromatic interactions. *Biochemistry*, 43:14454–14462.

Punia, S., Siroha, A. K., Sandhu, K. S., & Kaur, M. (2019a). Rheological behavior of wheat starch and barley resistant starch (type IV) blends and their starch noodles making potential. *International Journal of Biological Macromolecules*, 130:595–604.

Punia, S., Dhull, S. B., Sandhu, K. S., & Kaur, M. (2019b). Faba bean (*Vicia faba*) starch: Structure, properties, and in vitro digestibility—A review. *Legume Science*, 1(1):e18.

Punia, S., Dhull, S. B., Sandhu, K. S., Kaur, M., & Purewal, S. S. (2020). Kidney bean (Phaseolus vulgaris) starch: A review. *Legume Science*, 2(3): e52.

Purnomo, H., & Rahardiyan, D. (2008). Indonesian traditional meatball. *International Food Research Journal*, 15(2):101–108.

Renard, C. M., Watrelot, A. A., & Le Bourvellec, C. (2017). Interactions between polyphenols and polysaccharides, mechanisms and consequences in food processing and digestion. *Trends in Food Science Techology*, 60:43–51.

Sakakibara, M., Aoki, T., & Noguchi, H. (1979). Isolation and characterization of 7s protein-I of phaseolusangularis (*Adzuki bean*). *Agricultural and Biological Chemistry*, 43:1951.

Sato, S., Hori, Y., Yamate, J., Saito, T., Kurasaki, M., & Hatai, A. (2005). Protective effect of dietary azuki bean (*Vigna angularis*) seed coats against renal interstitial fibrosis of rats induced by cisplatin. *Nutrition*, 21(4):504–511.

Sato, S., Yamate, J., Hori, Y., Hatai, A., Nozawa, M., & Sagai, M. (2005). Protective effect of polyphenol-containing adzuki bean (*Vigna angularis*) seed coats on the renal cortex in streptozotocin-induced diabetic rats. *The Journal of Nutritional Biochemistry*, 16(9):547–553.

Sefa-Dedah, S., & Stanley, D. W. (1979). Textural implications of the microstructure of legumes. *Food Technology* 33:77–83.

Serdarogˇlu, M., Yildiz-Turp, G., & Abrodıˊmov, K. (2005). Quality of low fat meatballs containing legume flours as extenders. *Meat Science,* 70:99–105.

Shahidi, F., & Ambigaipalan, P. (2015). Phenolics and polyphenolics in foods, beverages and spices: Antioxidant activity and health effects. *Journal of Functional Foods*, 18 (Part B): 820–897.

Sharma, S., Verma, R., Singh, N., & Dhaliwal, Y. S. (2018). Assessment of anti nutritional factors and antioxidants in three genotypes of adzuki beans. *Journal of Pharmacognosy and Phytochemistry,* 8(1): 1376–1378.

Shi, Z., Yao, Y., Zhu, Y., & Ren, G. (2016). Nutritional composition and biological activities of 17 Chinese adzuki bean (Vigna angularis) varieties. *Food and Agricultural Immunology*, 28(1):78–89.

Singh, B., Singh, J. P., Singh, N., & Kaur, A. (2017). Saponins in pulses and their health promoting activities. *Food Chemistry,* 233:540–549.

Singh, B., Singh, J. P., & Singh, N. (2017). Phenolic composition and antioxidant potential of grain legume seeds. *Food Research International*, 101:1–16.

Sreerama, Y. N., Takahashi, Y., & Yamaki, K. (2012). Phenolic antioxidants in some Vigna species of legumes and their distinct inhibitory effects on glucosidase and Pancreatic lipase activities. *Journal of Food Science*, 77(9):C927–933.

Takahama, U., Hirota, S., & Yanase, E. (2019). Slow starch digestion in the rice cooked with adzuki bean: Contribution of procyanidins and the oxidation products. *Food Research International*, 119:187–195.

Tjahjadi, C., Lin, S., & Breene, W. M. (2006). Isolation and characterization of adzuki bean (Vigna angularis cv Takara) proteins. *Journal of Food Science*, 53: 1438–1443.

Tokitomo, Y., & Kobayashi, A. (1988). Odor of cooked Japanese adzuki beans. *Journal of the Agricultural Chemical Society of Japan*, 62(1):17–22.

Wati, R. K., Theppakorn, T., & Rawdkeun, S. (2009) Extraction of trypsin inhibitor from three legume seeds of the Royal Project Foundation. *Asian Journal of Food and Agro Industry,* 2:245–254.

Wu, S. J., Wang, J. S., Lin, C. C., & Chang, C. H. (2001). Evaluation of hepatoprotective activity of legumes. *Phytomedicine*, 8(3):213–219.

Xu, B. J., & Chang, S. K. C. (2012). Comparative study on antiproliferative properties and cellular antioxidant activities of commonly consumed food legumes against nine human cancer cell lines. *Food Chemistry*, 134:1287–1296.

Yao, Y., Cheng, X., Wang, L., Wang, S., & Ren, G. (2011). A determination of potential glucosidase inhibitors from adzuki beans (*Vigna angularis*). *International Journal of Molecular Science.* 12(10):6445–6451.

Yoshida, H., Hirakawa, Y., Abe, S., & Mizushina, Y. 2002. The content of tocopherols and oxidative quality of oils prepared from sunflower (*Helianthus annuus* L.) seeds roasted in a microwave oven. *European Journal of Food Science and Technology,* 104:116–122.

Yoshida, H., Tomiyama, Y., Yoshida, N., Shibata, K., & Mizushina, Y. (2010). Regiospecific profiles of fatty acids in triacylglycerols and phospholipids from adzuki beans (*Vigna angularis*). *Nutrients,* 2(1):49–59.

Yoshida, H., Yoshida, N., Tomiyama, Y., Saiki, M., & Mizushina, Y. (2008). Distribution profiles of tocopherols and fatty acids of phospholipids in adzuki beans (*Vigna angularis*). *Journal of Food Lipids.* 15:2, 209–221.

Yousif, A. M., Kato, J., & Deeth, H. C. (2007). Effect of storage on the biochemical structure and processing quality of adzuki bean (*Vigna angularis*), *Food Reviews International*, 23(1): 1–33.

26

Lupine: A Versatile Legume with Enhanced Nutritional Value

Prabhjot Singla, Sucheta Sharma and Arashdeep Singh
Punjab Agricultural University, Ludhiana, Punjab, India

CONTENTS

26.1 Introduction

Lupine (*Lupinus*) is a cool-season legume, believed to originate in Egypt approximately 2,000 years ago, then spreading worldwide due to its beneficial effects. It is a nonstarch leguminous seed with high protein and relatively low oil content and is cultivated in different parts of the world for its nutritional quality and adaptability to marginal soils and climates. It is used as green manure and forage and can fix atmospheric nitrogen for crop rotations with cereal and oil seed crops. Lupine is rich in dietary fiber, phytochemicals, antioxidants, phytosterols, vitamins, and minerals. It contains lower levels of antinutrients, namely phytates, flatus-inducing raffinose series oligosaccharides, protease inhibitors, lectins, and saponins as compared to other legumes. Lupine proteins have good emulsifying power, binding and foaming properties, and are used for development of animal-free protein foods or food products suitable for celiac patients. It can reduce transit time in human digestion and has beneficial effects on stool bulking. Lupine seeds are gaining importance as potential health promoters as they exhibit biological properties. Lupine fiber and proteins show hypoglycemic and hypocholesterolemic effects (Beyer et al., 2015). *Lupinus* seeds are detoxified and being used to combat nutrient deficiencies. Its flour is used to fortify cereal-based foods, pasta, bread, and emulsified meat products to increase nutritional value, aroma, and modify the texture of the end products. Consumption of lupine-enriched foods may be advantageous on food appetite, energy balance, blood lipid improvement, and glycemic index reduction (Prusinski, 2017). This chapter will discuss the history of the lupine crop, nutritional diversity among genotypes, and their health benefits and uses as food or feed ingredients.

26.2 History of the Lupine Crop

Lupine plants belong to genus *Lupinus* of family *Leguminosae* and subfamily *Papilonoideae*. The name was derived from the Latin for "wolf," and, according to the Encyclopaedia Britannica (2016), there are two reasons for the name i.e. (i) the plant is known for "wolfing" minerals from the ground and (ii) the use of different plant-part extracts for epilepsy treatment (Dendle, 2001). This might be due to lupine's capacity to concentrate various metals from the soil, including manganese. As epileptic fits can be triggered by manganese depletion, providing manganese tends to lower the seizure rate (Dendle, 2001). The *Lupinus* genus consists of approximately 300 annual and perennial herbs, soft-woody shrubs, and small tree species distributed from the semidesert and subtropical to subarctic climate conditions and from Alpine to sea-level ecosystems.

Lupine species are divided into two groups: (1) the "Old World" group of 12–13 lupine species that are annual, herbaceous, and large seeded and belong to European and Eastern and North African regions, and (2) the "New World" group of approximately 280 of the total species that are annual and herbaceous perennials including few shrubs and are distributed from South Argentina and Chile to Alaska. Among all the species, *L. albus* L., or white lupine (WL); *L. angustifolius* L., or narrow-leafed lupine (NLL), also known as Australian sweet lupine or blue lupine; and *L. luteus* L., or yellow lupine (YL, from Old World species) and *L. mutabilis* or Andean lupine/bitterlupine (of the New World species) are being produced on a commercial scale. Some of the other *Lupinus* species are cultivated to be used as forage, green manure, ornamentals, or for land stabilization.

Lupines can be characterized on the basis of differences in geographic distribution, chromosomal, and morphological polymorphisms with large variation in protein content of seed (Islam and Ma, 2016). The Old World species are taxonomically divided into two classes based on seed texture; (1) smooth-seeded species *Malacospermae* with chromosome number (2n) in the range of 40–52. *L. albus*, *L. luteus*, *L. angustifolius*, and *L. hispanicus* belong to the class of smooth-seeded species; and (2) *Scabrispermae* consisting of rough-seeded species with chromosome number (2n = 32–42) and include *L. pilosus*, *L. cosentinii*, and *L. digitalis species*. In contrast New World species with the predominant chromosome number (2n = 48; and basic chromosome number of x = 6) are not well taxonomically defined.

Lupine seeds show good germination within an average of 12 days after sowing (Singla et al., 2017). The status of some lupine genotypes, days to 50% flowering, days to first podding, days to maturity, and seed weight have been listed in Table 26.1. The flowers formed high above the leaves, producing a spectacular raceme on which blue, yellow, or white blossoms occur (Figure 26.1). The roots are relatively long, usually nodular, and may grow down to 3 meters. The pods are hairy, normally flat and 4–10 cm in length. Depending on the species and genotypes, pods contain seeds that may vary in color (white, gray, and brown), size (a few millimeters to 1 cm in diameter) and shape (oval, round, and flat) (Figure 26.2).

Lupine occupies about 3% of total pulse production (in tons) in the world (FAO, 2015). In the last 30 years, Australia is the major lupine producer at nearly 70% of the total world's production followed by Poland as the second largest producer of lupine at present. There is about 17.6% lupine production in European countries and 5.1% in the United States (FAO, 2015). Yellow lupine species are widely grown in Australia whereas white and blue lupines are cultivated in European countries and South America.

Lupine is grown mostly on deep, coarse-textured (<10% clay) and mildly acid soils. The modern cultivars of NLL, a deep-rooted and acid-tolerant (due to its association with *Bradyrhizobium)* species, yield well under these circumstances. YL is tolerant to waterlogging; brown leaf spot, Eradu patch diseases, and soil aluminum that can limit plant growth in acidic soils and are better in comparison to NLL or WL. WL is sown in autumn due to its tolerance to soil freezing, whereas YL and NLL are sown in spring. NLL has tolerance against anthracnose (*Colletotrichum gloeosporioides*). *L. mutabilis* with a thin seed

TABLE 26.1

Agronomic Characteristics of Different Lupine Genotypes Grown in Northwest India in 2016–17

Species	Genotype	Color	Days to 50% Flowering	Days to First Podding	Days to Maturity	Seed Weight (g)
Lupinus albus	Kiev Mutant	Yellowish white	83	98	147	0.28
	Andromeda	Yellowish white	87	102	148	0.30
Lupinus luteus	Pootalong	Yellowish white	89	100	145	0.12
	Wodjil	Yellowish white	89	99	145	0.11
Lupinus angustifolius	Chitick	Yellowish white with brown dots	87	104	149	0.11
	Jenabillup	Creamish white with single brown line	83	100	146	0.12
	Tanjil	White with brown dots	80	100	150	0.11
	Mandelup	White with brown dots	80	98	152	0.12
	Quilinock	White with brown dots	85	100	150	0.12
	Merrit	White with brown dots	86	100	148	0.11
	Merry	Creamish white	87	106	147	0.14

Sources: Singla et al., 2017, Parmdeep et al., 2017.

coat is tolerant to water logging and possesses high phosphorus-use efficiency. Its grain quality matches to soybeans as it contains about 42% protein and 18% oil content. But its use in cropping systems is limited as it is susceptible to frost, has late maturity, and gives low and unstable yields (Carton et al., 2020).

Lupines can act as a cover crop, improve soil fertility by fixing atmospheric nitrogen (N), provide N for other crops, and the crop residues can be used as feed or manure (Lambers et al., 2013). Australian dairy farmers prefer lupines as a supplementary feed source due to their lower cost and easy storage and handling as compared to oilseed proteins. Lupine crops in Australia require phosphorus fertilizer, and NLL show higher response to phosphorus as compared to yellow or white lupines. Lupines have a high potential to mobilize phosphorus and micronutrients and play a role in the removal of heavy metals and hydrocarbons from soils through the accumulation or promotion of microorganisms that are able to detoxify these contaminants (Lambers et al., 2013).

FIGURE 26.1 Plants of *Lupinus luteus, Lupinus angustifolius,* and *Lupinus albus* with blue, yellow, or white blossoms on raceme (see Singla et al., 2017).

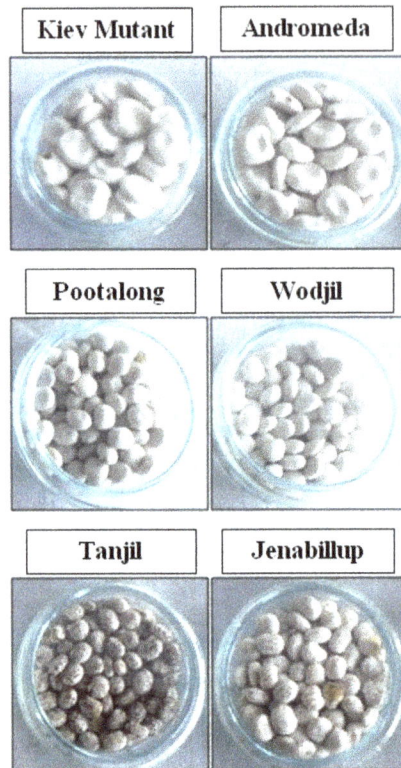

FIGURE 26.2 Seeds of different genotypes belonging to *Lupinus albus, Lupinus luteus,* and *Lupinus angustifolius* (see Singla et al., 2017).

Wild lupines or "bitter" lupines contain high alkaloid content whereas sweet lupines have low alkaloid content. All modern varieties belonging to *L. angustifolius* species are known as sweet lupines by food safety organizations, thus making this crop an important part of agriculture and food systems in the 20th century.

26.3 Nutrients in Lupines

Lupine grains contain high protein and relatively low but high-quality oil. They can compete with cereals and oil seed meals as energy and protein sources, respectively. Lupines are also a rich source of dietary fiber, vitamins, minerals, and phytochemicals including antioxidants and phytosterols (Khan et al., 2015) and have the highest protein and dietary fiber contents among the pulses, beans, lentils, and peas. Lupine seeds deserve greater interest as a result of their chemical composition and increased availability in many countries in recent years. Human consumption of lupines has increased as it is regarded as a beneficial food ingredient and being recommended as staple food by health organizations and dieticians (Kohajdova et al., 2011). The nutritional composition of lupine has been given in Table 26.2.

26.3.1 Protein

Protein plays an important role in humans as they are a source of energy, act as major building blocks in the body, and maintain different body functions. The genus *Lupinus* typically contains 36–52% protein and it varies from 28–48% between species and cultivars (Trugo et al., 2003; Capraro et al., 2008; Singla et al., 2017) because of the characteristics of the growing environment (Martinez-Villaluenga et al., 2006).

TABLE 26.2

Nutrient Composition of the Lupine

Nutrients (%)	L. albus	L. luteus	L. angustifolius	L. mutabilis	References
Crude Protein	30.1–38.2	24.3–49.2	21.0–28.9	46.6	Trugo et al., 1988; Singla et al., 2017
Crude fat	9.5–10.1	4.8–9.1	6.6–8.0	15.8	Parmdeep et al., 2017
Crude fiber	14	17.6	17.6	9.0	Muzquiz et al., 1982
Oligosaccharides	7–8	na	8–9	na	Trugo et al., 2003 Capraro et al., 2008
Nonstarch polysaccharides	18	na	47–51	na	
Ash	3.1	5.3	3.2	3.6	

na: not available.

Most of the cultivated species, especially YL, possess relatively higher protein than soybeans and protein content increases after de-hulling. The amino acid profile of lupine protein is also different from soybeans as it is characterized with significantly higher arginine, glutamic, and aspartic acid content and a low content of lysine, threonine, tryptophan, and sulphur-containing amino acids cysteine and methionine (Stanek et al., 2006; Suchy et al., 2006; Pisarikova et al., 2008). Methionine, cysteine, valine, and tryptophan are the main limiting amino acids in lupines (Villacres et al., 2020). Lysine, branched chain amino acids isoleucine and leucine and aromatic amino acids (phenylalanine and tyrosine) contents in lupines are almost similar to the Food and Agricultural Organization standards for amino acids of ideal reference protein appropriate for adults and therefore, may be a good complement of wheat flour (Doxastakis et al., 2002). Lupine has a biological value of 91% in comparison to egg proteins and is an important protein source for human beings. Yellow lupine and *L. mutabilis* contain higher protein content than WL followed by NLL. Yellow lupines also have higher S-amino acids than WL or NLL. The comparison of amino acid profile of different lupine species with soybean has been presented in Table 26.3.

The main lupine seed proteins are albumin and globulins present in the ratio of approximately 1:9. Albumins are the functional proteins especially the metabolic enzymes. Globulins constitute 80–90% of the storage proteins in lupines whereas glutelins and prolamines are present in minor amounts (Gulewicz et al., 2008). Globulins are the major lupine storage proteins that contain sufficient levels of nearly all of the essential amino acids, thus making these proteins an important protein source. Globulins consist

TABLE 26.3

Comparison of Amino Acid Profile of Different Lupine Species with Soy Beans

Amino Acid (mg/g)	L. albus[a]	L. luteus[a]	L. angustifolius[a]	L. mutabilis[b]	Glycine Max[c]
Valine	39	32	35	41	19.4
Leucine	75	79	71	71	34.7
Isoleucine	44	37	40	57	19.7
Threonine	36	32	34	38	16.3
Lysine	47	49	46	58	23.7
Histidine	22	29	27	26	10.5
Methionine	7	7	7	9	5.9
Cystine	16	24	18	11	8
Phenylalanine	35	39	337	38	22.5
Tyrosine	46	27	34	40	13.5
Tryptophan	8	10	8	8	5.7

Sources: [a]Gross (1988); [b]Gross et al (1989); [c]Carrera et al. (2011).

of alpha, beta, gamma and delta (α, β, γ and δ)-conglutin (Foley et al., 2011), known as 11S, 7S and 7S basic globulin, respectively, and are the major storage proteins; δ-conglutin, a 2S sulphur-rich albumin is present in minor amounts. The α-conglutin is hexameric protein made up of subunits linked by disulfide bonds. The β-conglutins consist of three monomers and is a trimeric protein. The γ-conglutin is a sulphur-rich dimeric protein consisting of heterogeneous disulfide linked subunits whereas δ-conglutin is a low-molecular-weight monomeric protein. These seed proteins are nutritionally important and are therapeutic in nature against blood glucose and cholesterol (Sedláková et al., 2016).

26.3.2 Fat

Lupine seed contains approximately 5–20% of crude oil (Uzun et al., 2007; Parmdeep et al., 2017) and its quantity is quite low in comparison to oilseeds. However, oil quality in terms of unsaturated fatty acids/saturated fatty acids and omega (ω)-3/ω-6 polyunsaturated fatty acids is very favorable. Lupine oil contains 10% saturated and 90% unsaturated fatty acids (Hamama and Bhardwaj, 2004). The proportion of oleic (18:1) is 32–50%, linoleic (18:2) 17–47% and linolenic (18:3) acid 3–11% of total unsaturated fatty acids. Lupine oil is an important source of energy, essential fatty acids and vitamins A, D, and E. The vitamin E content in lupine oil is less in comparison to sunflower and rapeseed oil but almost equal to soybeans (Lampart-Szczapa et al., 2003). Lupine species differ in total lipid content (Borek et al., 2009). Yellow lupine contains about 6% lipids whereas white and Andean lupine exhibited about 7–14% and 20% oil, respectively. Seeds of *L. mutabilis* contain the highest oil (20%) content among all the *Lupinus* species and values are comparable to soybean seed with 12–26% oil (Borek et al., 2009). White lupine has greater oil content than YL or NLL.

The fatty acid distribution in the oil fraction is also variable between species. Loredo-Davila et al. (2012) separated the lupines into bitter and sweet groups depending on the presence of mono unsaturated fatty acids (MUFA), total saturated fatty acids (TSFA), and poly unsaturated fatty acids (PUFA). The sweet lupine group contained the highest amount of MUFA as compared to TSFA and PUFA whereas in case of bitter group, the PUFA were of higher concentration than TSFA and MUSA. Oleic acid is the major fatty acid for *L. albus,* while linoleic predominates in cotyledonary tissues in *L. luteus* and *L. angustifolius*. *L. angustifolius* contained 19% saturated, 33% MUFA and 48% PUFA with linolenic acid contributing 6% of the total lipids (Kouris-Blazos and Belski 2016). The level of storage lipids was also dependent on the species as well as growing conditions and phosphatidylcholine was the major phospholipid present in seed storage organs (Borek et al., 2009).

26.3.3 Carbohydrates

Lupine is a nonstarch leguminous seed and its soluble nonstarch polysaccharides content ranges from 30–40% (Erbas et al., 2005; Parmdeep et al., 2017). Total carbohydrate content of *L. albus* seeds is nearly 48% as found in other legumes. It contains higher amounts of soluble sugars in comparison to other legumes, except soybeans (Martinez-Villaluenga et al., 2006). Mature dry lupine seeds contain 1.5% sucrose and 10.9% oligosaccharides (Gorecki et al., 1997) that are important for development of desiccation tolerance in orthodox seeds and also improve seed storability (Piotrowicz-Cieslak 2005). Sucrose and glucose are synthesized immediately after anthesis, and oligosaccharides are deposited at the onset of the drying phase of seed ripening in developing seeds of lupine species *L. albus* cv. Ultra and *L. angustifolius* cv. Unicrop (Saini and Lymbery1983). Lupine seed contains raffinose series oligosaccharides (RFO) including raffinose, stachyose, verbacose, and ajucose as main α-galactosides and their contents are highest among all types of legumes (Martinez-Villaluenga et al., 2006). These RFOs are important dietary fiber fractions in lupine seeds that have an important role in osmotic regulation in the gastrointestinal tract (GIT), but they also cause flatulence. Stachyose was the predominant sugar in the cotyledons and the seed coat of fully ripened seeds of *L. reaxus, L. exaltatus,* and *L. mexicanus* (Ruiz-Lohpez et al., 2000). In dry seeds, 80% of total carbohydrates are present in cotyledons with 19% as soluble sugars and rest as structural polysaccharides. The soluble sugars accounted for 7% of the total amount present in seed coats. Parmdeep et al. (2017) reported 9.02–21.69 mg/g sucrose and 29.70–55.49 mg/g starch in different lupine genotypes.

Dietary fibers constitute 34–40% of the total seed weight of sweet lupine (Clark and Johnson 2002; Hall et al., 2005; Smith et al., 2006). Polysaccharides made of arabinose, galactose, and uronic acid are present in the cotyledon part of lupine seed, but there is no starch. Lupine hulls mainly consist of cellulose (79%), hemicellulose (14%), lignin (4%), and pectic polymers (30%) (Ciesiolka et al., 2005; Parmdeep et al.,2017). The hull contains 85–95% dietary fiber with glucose, uronic acid, and xylose as the three main monosaccharides units of nonstarch polysaccharides (Dandanell and Aman 1993; Evans and Cheung, 1993). *L. albus* contains about 2.8–2.9 times more total dietary fibers as oats (Mohamed and Rayas-Duarte 1995). Dietary fibers are categorized into insoluble and soluble fibers. Lupine cotyledon contains about 21.5% and 22% of soluble and insoluble fibers, while hull contained 86.2% and 1% of insoluble and soluble fibers. Insoluble fiber acts as a fat replacer in the digestive system and helps in bowel health due to its high water-binding capacity and viscosity in comparison to other insoluble fibers. It has been reported that *Lupinus* reduce transit time in human digestion and have beneficial effects on stool bulking (Johnson et al., 2006). It reduces glucose in nondiabetics (Hall et al., 2005). Dietary fiber is not digested in humans and it is either fermented or added to bulk in the colon part of the GIT (Dhingra et al., 2012). Dietary fibers are hypocholesterolemic, anti-inflammatory and anticarcinogenic and improve glucose tolerance in diabetes (Jenkins et al., 2002; Yang et al., 2014).

26.3.4 Minerals and Vitamins

The mineral composition of lupine is similar to other legumes in relation to the major elements, except that all of the species are low in calcium. However, some differences have been reported in the amount of trace elements. Ash content is related to a high content of micronutrients, like phosphorous and potassium, other macronutrients, like iron, and lower levels of essential minerals, like calcium and magnesium, in lupine seeds (Ortega-David et al., 2010). The mineral composition of lupine seeds is given in Table 26.4.

L. albus species unusually contain high amounts of manganese (896 mg/kg). Lupine can be fortified in ferritin to get a good iron source in the diet and the stability of plant ferritin in some pasta and other bakery products has been confirmed experimentally (Zielinska-Dawidziak 2015). Thiamine, niacin, riboflavin, and fat-soluble tocopherols are the main vitamins in lupine seeds (Trugo et al., 2003). Chandra-Hioe and Arcot (2015) reported that the content of vitamin B12 in lupine meal can be increased by microbial biosynthesis.

TABLE 26.4

Mineral Composition of Lupine Seeds

Mineral (mg/g dw)	L. albus[a]	L. luteus[a]	L. angustifolius[a,b]	L. mutabilis[a,c]	References
Calcium	200	210	232	147	[a]Petterson,1998;
Copper	0.5	0.9	0.5	1	[b]Petterson and Crosbie, 1990;
Iron	2.6	9.3	6.1	5.9	[c]Villacres et al., 2000
Potassium	na	na	Na	1265	
Magnesium	na	na	Na	285	
Manganese	83.5	8.6	2.1	3.2	
Phosphorus	360	610	321	753	
Zinc	3	5.6	3.6	3.5	

na = not available.

26.4 Bioactive Compounds

The hull fraction of the lupine seed contains a major proportion of polyphenols (Khan et al., 2015; Luo et al., 2016). Unlike other species, lupine contains higher amounts of flavonoids than phenolic acids.

Earlier, it was thought that phenolics act as an antinutrient because of their negative impact on the absorption and digestibility of nutrients. Now, these compounds are known to have anticancerous, antioxidant, and antimutagenic properties, reduce oxidative stress, and inhibit chronic diseases including high blood pressure, type 2 diabetes, and inflammation and cardiovascular diseases (Arnoldi et al., 2015). Flavonoids are the secondary metabolites with the highest antioxidant activity due to their chemical structure. These are the phenolic compounds that defend the plants against ultraviolet radiation damage and pathogens (Manach et al., 2004). Lupine flour comprises 55% trans-ferulic, 17% p-hydroxybenzoic, 15% of syringic and 13% of trans-p-coumaric acids. Whereas, the lupine hull contains 12.6% of the whole seed phenolics constituting mainly of 60% of trans-ferulic and 40% p-hydroxybenzoic acids (Sosulski and Dabrowski, 1984).

Lupines contain color imparting pigments β-carotene, lutein, and zeaxanthin (Entisar and Hudson 1979). Wang et al. (2008) reported high levels of carotenoids in Australian sweet lupine seeds as compared to white and narrow leafed lupines. Carotenoids transmit color to the oil fraction, and the yellow color of lupine flour can be imparted into bakery products, thus reducing the need for butter and egg yolk as colorants (Kohajdova et al., 2011; Krawczyk et al., 2015). Australian sweet lupine contains 2 mg/kg, 15 mg/kg, and 15 mg/kg of β-carotenoid, lutein, and zeaxanthin, respectively. Carotenoids can prevent development of some chronic diseases in human beings including cancer, cardiovascular diseases, and other biological activities including antioxidant activities, influence on immune system, control of cell growth and differentiation and stimulation effects on gap junction communication (Wang et al., 2008). Lupine contains lupeol, a triterpene alcohol as a component of the lipidic fraction that helps in improving the renewal of epidermal tissue (Msika et al., 2006). Guemes-Vera et al. (2012) studied the carotenoids content in different lupine species. The content of carotenoids (both isochromic red and yellow fraction) was significantly higher for *L. barkeri* as compared to *L. montanus* and *L. albus*. The difference in carotenoid content among lupine samples was due to interspecies variation with characteristic pigments including lutein, zeaxanthin, and β-carotenoid (Cerletti et al., 1978).

26.5 Antinutritional Factors

Plants contain certain compounds that are biologically active without any nutritional importance. Some of these compounds may not be desirable for human or animal nutrition, and are known as antinutritional factors. Lupine also contains such compounds, including phytic acid, oligosaccharides, trypsin inhibitors, lectins, and saponins in lower quantities than other legumes (Fernandez-Orozco et al., 2008, Martinez-Villaluenga et al., 2006, Pastor-Cavadaet al., 2009).

Lupine seeds contain various alkaloids belonging to the quinolizidine group as a main antinutritional component (Sujak et al., 2006). These include lupineine, lupanine, hydroxylupanine, lupineidine, anagirine, monolupine, termophsine, puziline, sparteine, and angustifoline (Maknickiene and Razukas, 2007) and their presence limits its consumption (De Cortes-Sanchez et al., 2005) because of their strong bitter taste and toxicity in high doses that can result in trembling, convulsions, and, potentially, respiratory arrest. The content of lupine alkaloids differs due to soil type, growing season, and among different cultivars (Bhardwaj and Hamama, 2012). These are synthesized only by the green tissues but are also present in lupine seeds;large seeded lupines contain higher amounts than small seeds. Wild lupines are known to contain toxic alkaloids lupanine and sparteinene, several antinutritive factors, and must be properly processed before consumption (Muzquiz et al., 1993). The sweet lupine contains up to 0.09% of total alkaloids. Some sweet lupine mutants with low or no alkaloid content (0.01–0.05% versus 1–8% of landraces) have been reported from European countries (Gresta et al., 2010). They possess specific plant antiviral, antifungal, and antibacterial properties, thus improving resistance to insects and other pests, inhibiting the germination of grass seeds and other weeds that grow in their vicinity, deterring snails and some insects from eating bitter lupine greens, thus improving their selective survival at times of heavy pest infestation (Dubois et al., 2019; Romeo et al., 2018). Table 26.5 showed the main bioactive and antinutritional substances found in lupine seeds.

In lupine, the protease inhibitors and saponins are present in very low amounts as compared to pea and soybeans (Samtiya et al., 2020). Different *Lupinus* spp. exhibit almost negligible trypsin inhibitor activity

TABLE 26.5

The Main Antinutritional Substances Found in Lupine Seeds

Antinutrients (%)	L. albus	L. luteus	L. angustifolius	L. mutabilis
Alkaloids[a]	0.04	0.1	0.06	1.27
Lupanine[b]	70	60	70	64.4
Sparteine[b]	na	30	na	12.6
Angustifoline[b]	na	na	10	2.3
Phytic acid[c]	0.36–1.42	2.72	1.45	2.74
Saponin[c]	0.9–1.44	1.22	0.9	1.7
Tannins[d]	0.01–0.08	0.02	0.01	0.06
Trypsin inhibitor[d]	0.01	0.03	0.01	na
Stachyose[c]	0.8	1	0.3	1.3
Raffinose[c]	0.8	1	0.6	0.9

Sources: [a]Sujak et al., 2006; [b]Petterson, 1998; [c]Muzquiz et al.,1989; [d]Petterson and Mackintosh, 1994.

in comparison to "very intense" in soybean (43–84 trypsin inhibitor units/mg) and high in common bean (Guillamón et al., 2008). Trypsin inhibitors inhibit activity of trypsin and chymotrypsin, induce secretion of large quantities of pancreatic enzymes, and finally result in the reduction of dietary protein digestion and absorption (Lima et al., 2019). This can lead to lower retention of sulphur and nitrogen, resulting in impaired growth of animals (Shi et al., 2017). The protease inhibitor can alter the water-holding, gel-formation capacity and foaming and whipping ability of food products (Karami and Akbari-Adergani, 2019).

There are two types of trypsin inhibitors—namely Kunitz trypsin inhibitor (KTI) and Bowman-Bick inhibitor (BBI). KTI is about 20 kDa polypeptide containing two disulphide bridges and inhibits trypsin only. BBI, with a molecular mass of 6-8 kDa, contains a number of disulphide bonds and has the ability to inhibit trypsin and chymotrypsin simultaneously by binding at independent sites and is also called a double-headed inhibitors (Indarte et al., 2017). It has been suggested that KTI and BBI inhibitors are anticarcinogenic and suppress different stages of carcinogenesis (Kennedy 1993). Trypsin inhibitor is either absent or the activity is very low ranging from 0.1–0.2 mg/g in white lupines (Erbas et al., 2005). Shoeneberger et al. (1982) reported lower trypsin inhibitor activity in wild lupine seeds as compared to *L. mutabilis* and potato cultivars. These inhibitors are mainly located in cotyledons (Embaby 2010) and trypsin inhibitor activity is higher in sweet lupine than the bitter one. Different lupine species contain trypsin inhibitor activity in the range of 1.78–4.25 TI units/mg (Guillamón et al., 2008).

Saponins are derivatives of triterpenoids steroidal aglycones and possess the ability to lower plasma cholesterol, anticancer activity or act as an inhibitor of viral replication (Roopashree and Naik 2019). Saponins are absent in *L. albus* but small quantities ranging from 57–470 mg/kg may be present in lupine species *L. luteus*, *L. mutabilis*, and *L. angustifolius*. However, these values are still low compared with soybeans that contain saponins in the range of 2,000–5,000 mg/kg.

Phytic acid can chelate divalent metal ions such as zinc (Zn), calcium (Ca), and iron (Fe), and reduces their bioavailability in animals. It may also form nonspecific complexes with protein molecules through metal ion binding mechanisms or by charge neutralization depending upon the charge on the protein. It is also found to inhibit the action of number of digestive enzymes by affecting the solubility and function of protein (Dersjant-Li et al., 2015). In last few years, it is considered to have some beneficial role as an antioxidant by complexing iron, lowering free radical generation and peroxidation of membranes (Quirrenbach et al., 2009). Phytate can play a role in protection against colon cancer (Abdulwaliyu et al., 2019), lower blood glucose, exhibit antidiabetic properties (Khayata et al., 2013), and prevent renal stone development (Fakier et al., 2019) in humans. The role of phytate in the regulation of various cellular functions such as DNA repair, chromatin remodeling, endocytosis, nuclear mRNA export, and hormonal signaling in plant and seed development have also reported (Dieck et al., 2012). Ruiz-Lohpez et al. (2000) reported phytate content in the range of 11.1–1.856 g/kg in seeds of three *Lupinus* species, including

L. exaltatus, L. reexus, and *L. mexicanus* from Jalisco, Mexico. The phytate content of lupine as about 500 mg/g and the values are similar to that found in peas and soybeans, but less than in rapeseed and canola meal (Carter and Hauler 1999). Variation in phytate content in lupine by number of authors has been reported (Trugo et al., 1993; Mohamed and Rayas-Duarte 1995; Saastamoinen et al., 2013). The average phytate content in white lupine seed is around 0.8 g/100 g and can be further lowered by fermentation, soaking, cooking, and dehulling or extrusion of lupine seeds (Da Silva et al., 2005). Martinez-Villaluenga et al. (2006) recorded 0.025–0.044% of phytic acid in white lupin seeds, whereas Saastamoinen et al. (2013) observed a value of 0.063%. Its content may be reduced even further through fermentation. The phytate-to-Zn ratio was 4:4 in wild species *L. digitatus* and 5:3 in *L. pilosis* P23030, which indicated excellent bioavailability of Zn. The phytate-to-Zn molar ratio in lupins is generally lower and phytate-to-calciummolar ratio is similar or better than in other legume seeds (Trugo et al., 1993).

Lectins are found at lower levels in lupines than in many other legumes. A negligible amount of tannin has been reported in lupines (Lampert-Szczapa et al., 2003). Lupine seeds also show lipoxygenase (LOX) activity. Lipoxygenases are used in the food and baking industry in bread making for aroma production (Baysal and Demirdovan, 2007), for improved dough rheology, and also as a bleaching agent (Szymanowska et al., 2009). Lupine seeds mainly show LOX-1 activity and exhibit lower LOX activity than other legumes such as peas and soybeans (Szymanowska et al., 2009). LOX type-2 has been identified in white lupines (Olias and Valle 1988). Stephany et al. (2015) reported significant variation (50–1,004 units/mg protein) in LOX activity in seeds of different species and varieties of sweet lupine. Removal of hulls and flaking of seeds can increase LOX activity.

26.6 Health Benefits

Lupine-enriched foods increase satiety, reduce energy intake, and decrease blood pressure, blood glucose, and cholesterol levels (Figure 26.3) thus making lupine an important constituent of a healthy menu.

26.6.1 Antiobesity

It has been suggested that a high-fiber diet compared to low-fiber diet, or high-protein diet compared to high-sugar diet, has an effect on increasing satiety. Thus, lupine has the ability to decrease food and energy intake as it has high protein and fiber content. Lee et al. (2006) found that lupine kernel fiber increased satiety by 20%. The possible reason behind the increased satiety was the high water-binding capacity of lupine fiber in the upper GIT and its fermentation to short-chain fatty acids in the colon (Johnson et al., 2006). Turnbull et al. (2005) observed that under *in vitro* conditions, lupine fibers showed greater water binding and viscosity than soy, pea hull, cellulose, or wheat fiber. The delayed stomach emptying and increased gastric distension that triggers the signals of fullness to the brain, prolonged small intestine transit time, and absorption rate of nutrients are the result of high water-binding capacity of lupine fiber (Turnbull et al., 2005). The other probable cause of reduced appetite and energy intake was the suppression of the ghrelin hormone in the stomach for more than 3 hours from the lupine flour. Ghrelin is a peptide hormone that acts as a powerful appetite stimulant produced in the cells lining

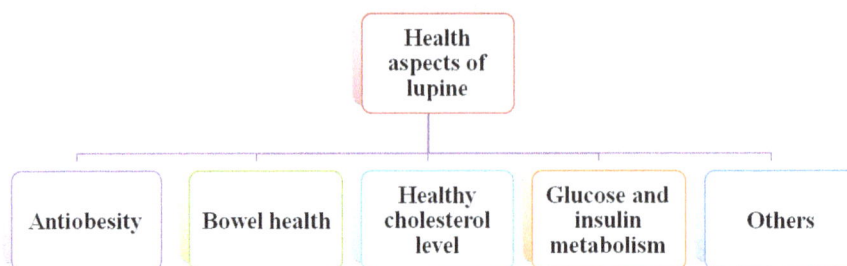

FIGURE 26.3 Health aspects of lupine.

the stomach. The intake of protein-rich lupines stimulate the cholecystokinin production in GIT that can induce delayed emptying of the stomach, help in ghrelin regulation (Koliaki et al., 2010), and send signals of fullness to the brain (Paddon-Jones et al., 2008). Ingestion of high protein and high dietary fiber lupine bread resulted in decreased post-meal ghrelin levels that could increase satiety and lowered short-term energy intake in comparison to wheat bread (Lee et al., 2006).

26.6.2 Bowel Health

Lupine kernel fiber, which is classified as a "prebiotic" by the authors, increases the levels of potentially beneficial *Bifidobacterium* spp. in the feces while reducing levels of the potentially pathogenic *Clostridia* group (Smith et al., 2006). Lupine RFOs were found to be an excellent carbon source for *Bifidobacteria* (Gulewicz et al., 2002). These bacteria did not metabolize RFOs to gaseous products, but prevented the excessive growth of detrimental microflora through acid production by reducing fecal pH. Diets enriched with lupine kernel fiber have been reported to improve bowel function, lower pH of fecal matter, and increase butyric acid levels in feces (Johnson et al., 2006).

26.6.3 Healthy Cholesterol Levels

It is important to maintain healthy cholesterol levels because cholesterol levels are a measure of heart health. Lammi et al. (2015) found that the activity of 3-hydroxy-3-methyl-glutaryl-CoA reductase was hindered by the peptides obtained from the hydrolysis of lupine proteins. Three-hydroxy-3-methyl-glutaryl-CoA reductase is the rate limiting enzyme of mevalonate pathway that produces cholesterol and other isoprenoids. Also, the higher the levels of low-density lipoprotein (LDL) or "bad cholesterol" in the blood, the greater risk for heart disease. It has been shown that genes involved in lipid synthesis were down-regulated by lupine proteins and resulted in reduced LDL cholesterol levels (Bettzieche et al., 2008). The high water-binding capacity of lupine fiber also resulted in the formation of short-chain fatty acids and are thought to reduce LDL levels. Moreover, it increased the viscosity in GIT that reduced the rate of diffusion of bile acids, thus inhibiting cholesterol reabsorption (Zacherl et al., 2011).

It was observed from immune-blotting experiments that the levels of sterol regulatory element binding protein 2 (SREBP2) were up-regulated by lupine peptides that further regulated the LDL receptor. This can increase the uptake ability of the Hep G2 (a human liver carcinoma cell line) cells for LDL, resulting in hypocholesterolemic effects (Lammi et al., 2015). In contrast to this, phytosterols in lupin kernels can lead to down-regulation of LDL cholesterol, and prevent the cholesterol absorption as they compete for cholesterol in bile salt micelles, resulting in decreased solubilization of cholesterol. Lupine oil consists of greater levels of phytosterols (2.4%) in comparison to other known phytosterol-rich oil such as soybean (0.4%) and rapeseed (1.2%) (Hamama and Bhardwaj, 2004).

26.6.4 Glucose and Insulin Metabolism

The glycemic index (GI) is a number that defines how rapidly a specific carbohydrate food raises blood glucose levels. It offers a way to tell slower acting "good carbs" from the faster-acting "bad carbs." In this regard, since lupine doesn't have an effective available carbohydrate, it cannot have a measurable GI. Lupines lower the glycemic load of any meal due to their very low available carbohydrate. Low glycemic-load foods act as a part of the diet that protects against type 2-diabetes (Buyken et al., 2010). In addition, the presence of γ-conglutin in lupines that stimulates insulin activity may be responsible for lowering glycemic response. Moreover, from the previous studies, it has been shown that the presence of higher protein content, higher dietary fiber content, and phytochemicals could decline the processes of digestion of starch and glucose absorption. Oligosaccharides, saponins, phytic acid, and tannins may also show glycemic-lowering properties. The presence of arginine, phenylalanine, and stearic acid in Australian sweet lupine might increase insulinaemia after consumption of lupine bread (Hall et al., 2005).

26.6.5 Other Benefits

The replacement of refined sugars by an increased protein content in the diet might have a favorable influence on blood pressure. Laurant et al. (1995) showed that arginine-rich lupine proteins improve vascular function and attenuate blood pressure in animal models. Nitrogen oxide, synthesized from arginine, is a strong endothelial relaxation factor and improves the vessel tonus, thus, decreasing blood pressure. Moreover, lupine proteins may also improve blood pressure due to their hypocholesterolemic properties.

Lutein is found in the macula of eyes that filters short or blue light wavelengths. Blue light can lead to the production of free radicals due to oxidative stress and damages the macula of the eyes (Zhao et al., 2018). People with a low level of lutein and zeaxanthin pigments in their blood can develop age-related macular degeneration. As lupines contain a good amount of lutein so, as a part of a diet can delay the start and development of the effects of macular disease (Fryirs et al., 2008).

Scarafoni et al. (2008) identified a biologically active protein from lupine seeds—the Bowman-Birk serine-proteinase inhibitor. The Bowman-Birk serine proteinase inhibitor can suppress both initiation and progression stages of carcinogenesis in addition to its effects on other pathological conditions that include degeneration of nerves, angiogenesis, rheumatoid arthritis, and heart ailments (Clemente and Arques 2014).

Australian sweet lupine sprouts are a rich source of isoflavones, the natural antioxidants that may lower the risk of cardiovascular disease when taken in the diet (Pabich and Materska 2019). The daidzein and genistein may help in conserving calcium and improve bone health thus reducing the risk for osteoporosis.

26.7 Functional Properties of Lupine Flour and its Products

Functional properties are the most important physical characteristics of flour, protein concentrate, and isolates for their utilization on different food formulations. Lupine flour and its different protein fractions exhibit different functional properties, which help to determine their potential in food products at different levels. The protein from lupines showed wide variations in the protein solubility profile (19–33%). However, the highest solubility of the proteins was found at pH 10 and 12 (Lqari et al., 2002). The solubility of lupine protein primarily depends on pH of the solution, ionic strength of the solution, and the temperature at which the proteins are solubilized (Villarino et al., 2016). Water absorption capacity (WAC) of lupine flour from different cultivars varied from 1.2–2.4 g/g, which was similar to the water absorption capacity of soybean flour (2.0–2.4). The WAC of protein isolate and concentrates showed many variations in comparison to lupine flour. The WAC of isolates and concentrates from different species of lupine ranges between 0.5–6.0 g/g. This variation in the WAC of protein isolate and concentrates was primarily due to the variation in extraction pH and temperature, drying conditions, and protein content in the product (Villarino et al., 2016). Oil absorption capacity, which determines the mouth feel and acceptability of the product, in lupine flour ranges between 1.5–1.7 g/g, which was much higher in comparison to oil absorption capacity of soybean flour. Interestingly, protein isolates and concentrates from lupines showed significantly higher oil absorption capacity in comparison to lupine flour and soy flour. Protein isolates and concentrates from lupine flour showed oil absorption capacity that varied between 2.9–3.9 g/g. Furthermore, authors have reported that oil absorption capacity of defatted protein concentrate is higher in comparison to full-fat protein concentrates. Processing conditions such as acid and alikali extraction, pH of solubilizing media, ionic strength, and presence of undesirable compounds such as sugars, phenols, fibers, and fat influence the oil absorption capacity of lupine flour and its different protein products (Sathe et al., 1982; Lqari et al., 2002).

Emulsifying properties such as emulsion capacity, emulsion activity, and emulsion stability are of prime importance in the utilization of protein flour and their products in different food products. Lupine flour exhibits an emulsifying capacity of 55.1 g/g, while its protein concentrate showed an emulsifying capacity of 89.9 g/g at 2% concentration. The emulsifying capacity of protein isolates from lupines ranges between 370–570 mL/g of isolate (Sathe et al., 1982). The emulsion capacity of the flour showed maximum values at a particular strength, and beyond this strength increasing its concentration does not enhance the emulsion capacity of lupine flour and its protein isolates. Emulsifying properties vary mostly with changes in the pH (Villarino et al., 2016). Emulsion-forming properties are another most important functional property of lupine flour (Pollard et al., 2002) and emulsification capacity varies as the lupine is de-hulled, milled, soaked, germinated, fermented, and toasted. α-, β-conglutin are the major fractions of the lupine protein that exhibit emulsifying properties (Sironi et al., 2005). Lupine protein fractions have been categorized as E (emulsifying) fraction (α-, β-conglutin) and F (foaming) fraction (γ-conglutin-rich) owing to their significant functional properties in food formulations and product preparation (Sironi et al., 2005; Wäsche et al., 2001). Consumer preference and sensorial properties of lupine are far better than peas, beans, and lentils; lupine flour mainly used in the preparation of various sweet and savory food products.

Foaming properties, such as foaming capacity and foaming stability of the lupine flour, range between 130–180% as the concentration of lupine flour increased from 2–10% (Sathe et al., 1982). This increase in the foaming capacity with increase in the flour concentration was due to an increase in the protein content. Defatting also improved the foaming capacity of the protein concentrates and isolates, and the addition of salt and sugar improved the foaming capacity of the protein products. The foaming stability of lupine flour after 120 minutes ranges between 92.4–79.2% at 2% and 3% of flour concentration. This also showed that maximum foaming stability is at particular flour concentration, while beyond this concentration it declines (Villarino et al., 2016). The least gelation concentration of lupine flour ranges between 6–14%, while for protein isolates it ranges between 8–10%. Furthermore, the variation in the least gelation concentration was primarily due to the change in pH and compositional changes (Sathe et al., 1982). Thermal processing of whole grain lupine results in the flour with active lipoxygenase enzyme, which will reduce its storage stability. This can be overcome in the milling of lupine by skipping thermal processing, producing whole lupine flour with intact lipoxygenase (Yoshie-Stark et al., 2004). This inactivated lipoxygenase enzyme in the flour further helps in providing a bleaching effect to the formulation in which it is primarily used as bread improver. Owing to the higher water absorption capacity of lupine flour, it helps in extending the shelf life of bread and other bakery products as it helps to lower the rate of staling. In comparison to flour from other nonconventional protein sources, the flour from lupines exhibits the highest water absorption capacity (Raikos et al., 2014).

26.8 Utilization of Lupine Flour in Food Products

In the current scenario, consumers are more health conscious and they look for foods that provide additional health benefits beyond basic nutrition such as functional foods with whole grains and nonconventional grains as an alternative to refined products. Also being vegetarian, gluten-free, and genetic modified organism-free are more points that are encouraging for the further utilization of lupine in a daily diet. In regards to the conventional proteins, lupines can be used as a high-quality substitute for conventional legume protein, which helps in its utilization as protein enrichment of various foods for their utilization by the populations of many developing countries.

Germinated lupine is a powerhouse of bioactive compounds and is rich in antioxidants that can be used in several food formulations. With advancements in the food formulations and utilization of nonconventional ingredients, lupines are gaining popularity for the preparations of various food products such as bread, cookies, cake, pasta, noodles, muffins, and drinks. Legumes have interesting nutritional properties and their inclusion in the diet is encouraged; their incorporation into bakery products could be a good method for increasing consumption. This could be utilized for the development of composite blends from locally produced lupines at a small-scale industry level as value-add products (Ahmed 2014). Owing to the rich source of nutrients, proteins, nonstarchy polysaccharides, and micronutrients, lupine flour was considered as an excellent raw material for supplementation in different food products (Pollard et al., 2002). Due to the enhanced functional properties of lupine flour, it was utilized as an egg replacement in different bakery products such as cakes, pancakes, bread, and pasta (Dervas et al., 1999). Owing to the yellow color of lupine flour and lupine protein concentrate, it provides a considerable appeal in pasta and noodles (Doxastakis et al., 2002; Guémes-Vera et al., 2008). Lupine flour can be incorporated into wheat flour to improve the nutritional value of the final product with little or no detriment in product sensory quality, and lupine fiber can also be used as a source of dietary fiber (Clark and Johnson, 2002).

Due to the presence of good-quality highly digestible proteins, health promoting dietary fiber, low digestible carbohydrates, high levels of bioavailable nutrients and minerals (calcium, magnesium, and iron), high essential amino acids, no cholesterol and low levels of antinutrients, lupine flour in extensively used in bakery products and breads, substituting wheat flour with 50% lupine flour is available in market. Incorporation of lupine flour in the bread formulation helps to increase its nutritional quality in terms of protein, dietary fiber, carotenoids, and peptides. Globulins and albumins comprising of cysteine and tyrosine residues are present in lupine and availability of cysteine and tyrosine residues in lupine proteins primarily help in the development of crosslinks between lupine and gluten proteins in the wheat-flour-substituted bread, thus helping to provide the required structure and strength to dough and bread (Shoup et al., 1966; Sujak et al., 2006). Lupine flour upto 20% and bran upto 10% can be used as a substitute in wheat flour cookies formulated with 0.4% xylanase enzyme. Both lupine flour and bran improved the minerals such as calcium, copper, magnesium, and phosphorus protein content (Bilgiçli and Levent 2014). Maghaydah et al. (2012) reported that lupine flour alone, or in different proportions with rice flour or corn starch, with different hydrocolloids can be employed in the preparation of acceptable cookies, which are most desirable and is considered suitable for all ages due to its low manufacturing cost, convenience, long shelf life and good eating quality.

Studies reported that lupine flour and fiber can be used at 10% levels in the development of bread and cakes with acceptable sensory scores. Although the addition of lupine flour and fiber results in a weaker structural and gluten network, their addition results in good viscoelastic behavior of the dough and in the protein-rich baked product. Lupine flour and fiber can be utilized as hypoglycemic agents in different bakery products such as bread and cake (Ahmed, 2012). Bread quality has been improved when 10% lupin flour is substituted for refined wheat flour, possibly due to lupin-wheat protein cross-linking assisting bread volume and the high water-binding capacity of lupin fiber delaying staling. Above 10% substitution appears to reduce bread quality due to the low elasticity of lupin proteins and the high WBC of its dietary fiber interrupting gluten network development (Doxastakis et al., 2002; Pollard et al., 2002). Sweet lupine flour and its different products such as sweet lupine protein isolate and concentrate can be used at 6% and 3% levels, respectively, in wheat flour formulation for the preparation of bread (Mubarak, 2001). Lupine flour and its protein products improve the sensory properties and consumer acceptability of bread along with increased protein content and total essential amino acids, especially lysine and *in-vitro* protein digestibility. The addition of lupine protein isolate to bread along with bread gum significantly improved the protein content and corrected the protein digestibility score, having a positive effect on the loss of available lysine. However, it did decrease the bread volume and form heterogeneous crumbs structure (López, 2014).

Lupines can also be used in the development of gluten-free products due to their higher water absorption capacity and functionality. Lupine flour can be used in combination with rice flour in a 20:80 ratio for the development of pasta that exhibits similar sensory characteristics in comparison to control along with of protein and fiber content and thus lupine flour can be used for partial substitution of rice flour by lupine flour and could be a reliable alternative for gluten-free products. Lupine flour can also be used in the preparation of wheat flour-based instant noodles; the addition of 20% lupin flour enhanced the nutritional value of instant noodles by increasing protein by 42% and dietary fiber by approximately 200% without affecting the sensory properties of instant noodles (Jayasena et al., 2010). Semolina can be substituted with 20% lupine flour for the development of nutrient-rich pasta without affecting its color, texture, and sensory properties. The addition of lupine flour results in a substantial increase in the protein and dietary fiber contents of pasta (Jayasena and Nasar-Abbas, 2012). Whole and defatted lupine meal can also be used as functional ingredients to enrich the noodles at 10% level of enrichment with improved flavor and overall acceptability scores (Mahmoud et al., 2012).

26.9 Conclusion

Lupines have a number of advantages above all other pulses and can become one of the most important and sustainable protein sources to feed the growing global population. In addition, lupines can fix atmospheric nitrogen, helping to reduce the demand for fossil-based nitrogenous fertilizers. The nutritional properties such as high protein, nondigestible starch, and unsaturated fats make lupines an importance source of nutrition. Thus, lupines seem to be a realistic sustainable alternative to soybean that can be used without the loss in the quantity and quality of the food products. Lupines also possess biological properties that include cholesterol- and glucose-lowering capacities, and antiatherogenic, hypotensive, and hypoglycemic activities. In addition, with physical properties such as emulsification and foaming, lupine ingredients find use in a variety of food products, including baked goods and meat alternatives. Therefore, the lupine seed shows a real potential for the grain world market due to its nutritional and agricultural properties.

REFERENCES

Abdulwaliyu I, Arekemase S O, Adudu J A, Batari M L, Egbule M N and Okoduwa S I R (2019). Investigation of the medicinal significance of phytic acid as an indispensable anti-nutrient in diseases. *Clinical Nutrition Experimental*, 28:42–61.

Ahmed A R A (2012). *Technological and Nutritional Studies on Sweet Lupine Seeds and its Applicability in Selected Bakery Products*. Ph. D thesis, Technical University of Berlin.

Ahmed A R A (2014). Influence of chemical properties of wheat-lupine flour blends on cake quality, *American Journal of Food Science and Technology*, 2:67–75.

Arnoldi A, Boschin G, Zanoni C, Lammi C (2015). The health benefits of sweet lupin seed flours and isolated proteins. *Journal of Functional Foods*, 18:550–563.

Baysal T and Demirdoven A (2007). Lipoxygenase in fruits and vegetables: A review. *Enzyme Microbial Technology*, 40(4):491–96.

Bettzieche A, Brandsch C, Weibe K, Hirche F, Eder K and Stang G I (2008). Lupin protein influences the expression of hepatic genes involved in fatty acid synthesis and triacylglycerol hydrolysis of adult rats. *British Journal of Nutrition*, 99(5):952–62.

Beyer H, Schmalenberg A K, Jansen G, Jurgens H U, Uptmoor R, Broer I, Huckauf J, Dietrich R, Michel V, Zenk A and Ordon F (2015). Evaluation of variability, heritability and environmental stability of seed quality and yield parameters of *L. angustifolius*. *Field Crop Research*, 174: 40–47.

Bhardwaj H L and Hamama A A (2012). Cultivar and growing location effects on white lupin immature green seeds. *Journal of Agricultural Sciences*, 4 (2): 135–38.

Bilgiçli N and Levent H (2014). Utilization of lupin (*Lupinus albus L.*) flour and bran with xylanase enzyme in cookie production, *Legume Research*, 37 (3): 264–271.

Borek S, Pukacka S, Michalski K and Ratajczak L (2009). Lipid and protein accumulation in developing seeds of three lupine species: *Lupinus luteus* L., *Lupinus albus* L., and *Lupinus mutabilis* Sweet. *Journal of Experimental Botany*, 60(12): 3453–3466.

Buyken, A E, Mitchell, P, Ceriello, A and Brand-Miller, J (2010). Optimal dietary approaches for prevention of type 2 diabetes: A life-course perspective. *Diabetologia*, 53, 406–418.

Capraro J, Magni Ch, Fontanesi M, Budelli A and Duranti M (2008). Application of two-dimensional electrophoresis to industrial process analysis of proteins in lupin-based pasta. *LWT Food Science and Technology*, 41(6): 1011–1017.

Carrera C S, Reynoso C M, Funes G J, Martínez M J, Dardanelli J and Resnik S L (2011). Amino acid composition of soybean seeds as affected by climatic variables. *Pesquisa Agropecuaria Brasileira*, 46 (12): 1579–1587.

Carter C G and Hauler R C (1999). Evaluation of phytase in Atlantic salmon, *Salmo salar* L., feeds containing fish meal and different plant meal. In: Fish Meal Replacement In Aquaculture Feeds Development For Atlantic Salmon, Fisheries Research and Development Corporation, 34–45.

Carton N, Naudin C, Piva G and Corre-Hellou G (2020). Intercropping winter lupin and triticale increases weed suppression and total yield. *Agriculture*, 10:316.

Cerletti P, Fumagalli A and Venturin D (1978). Protein composition of seeds *Lupinus albus*. *Journal of Food Sciences* 43: 1049–1411.

Chandra-Hioe M V and Arcot J (2015). Microbial biosynthesis to enhance vitamin B12 contents in Lupin flour. In: Capraro J, Duranti M, Magni Ch, Scarafoni A (Eds), Proceedings of the XIV International Lupin Conference. Milan, Italy June 21–26, 2015, pp. 84–86.

Ciesiolka D, Gulewicz P, Martinez-Villaluenga C, Pilarski R, Bednarczyk M and Gulewicz K (2005). Products and biopreparations from alkaloid-rich lupin in animal nutrition and ecological agriculture. *Folia Biologica*, 53(1): 59–66.

Clark R and Johnson S (2002). Sensory acceptability of food with added lupine (*Lupinus angustifolius*) kernel fiber using pre-set criteria. *Journal of Food Sciences,* 67: 356–62.

Clemente A and Arques Mdel C (2014). Bowman-Birk inhibitors from legumes as colorectal chemo preventive agents. *World Journal of Gastroenterology,* 20(30):10305–10315.

da Silva LG, Trugo LC, Terzi SDC, Couri S (2005). Low phytate lupin flour based biomass obtained by fermentation with a mutant of Aspergillus niger. *Process Biochemistry.* 40: 951–954.

Dandanell DY and Aman P (1993). Chemical composition of certain dehulled legume seeds and their hulls with special reference to carbohydrates. *Swedish Journal of Agricultural Research,* 23: 133–39.

De Cortes-Sanchez M, Altares P, Pedrosa M M, Burbano C, Cuadrado C, Goyoaga C, Muzquiz M, Jimenez-Martinez C and Davila-Ortiz G (2005). Alkaloid variation during germination in different lupin species. *Food Chemistry,* 90(3): 347–55.

Dendle, P (2001). Lupines, manganese, and devil-sickness: An Anglo-Saxon medical response to epilepsy. *Bulletin of the History of Medicine,* 75: 91e101.

Dersjant-Li Y, Awati A, Schulze H and Partridge G (2015). Phytase in non-ruminant animal nutrition: A critical reviewon phytase activities in the gastrointestinal tract and influencing factors. *Journal of the Science of Food and Agriculture,* 95: 878–896.

Dervas, G, Doxastakis, G, Hadjisavva-Zinoviadi, S and Triantafillakos, N (1999). Lupin flour addition to wheat flour doughs and effect on rheological properties. *Food Chemistry,* 66:67–73.

Dhingra D, Michael M, Rajput H and Patil R T (2012). Dietary fiber in foods: A review. *Journal of Food Science and Technology,* 49(3):255–266.

Dieck C B, Boss W F and Perera I Y (2012). A role for phosphoinositides in regulating plant nuclear functions. *Frontiers in Plant Science,* 3: 50.

Doxastakis, G, Zafiriadis, I, Irakli, M, Marlani, H, and Tananaki, C (2002). Lupin, soya and triticale addition to wheat flour doughs and their effect on rheological properties. *Food Chemistry,* 77:219–227. doi: 10.1016/S0308-8146(01)00362-4

Dubois O, Allanic C, Charvet C L, Guégnard F, Février H, Théry-Koné I et al. (2019). Lupin (*Lupinus spp.*) seeds exert anthelmintic activity associated with their alkaloid content. *Scientific Reports,* 9: 9070.

Embaby H E (2010). Effect of soaking, dehulling and cooking methods on certain antinutrients and in vitro protein digestibility of bitter and sweet lupin seeds. *Food Science and Biotechnology,* 19: 1055–62.

Encyclopædia Britannica website (2016). Lupine. Available at:https://www.britannica.com/plant/lupine. Accessed on September 17, 2020.

Entisar A E and Hudson B J F (1979). Identification and estimation of carotenoids in the seed of four lupin species. *Journal of the Science of Food and Agriculture,* 30: 1168–70.

Erbas, M, Certel, M and Uslu, M K (2005). Some chemical properties of white lupin seeds (Lupinus albus L.). *Food Chemistry,* 89: 341–345.

Evans A J and Cheung P C K (1993). The carbohydrate composition of cotyledons and hulls of cultivars of *Lupinus angustifolius* from Western Australia. *Journal of the Science of Food and Agriculture,* 61(2): 189–94.

Fakier, S, Rodgers, A, Jackson, G (2019). Potential thermodynamic and kinetic roles of phytate as an inhibitor of kidney stone formation: Theoretical modelling and crystallization experiments. *Urolithiasis,* 47, 493–502.

Food and Agriculture Organization of the United Nation (FAO) (2015). FAOSTAT: Crop production data. Available at: http://faostat.fao.org/site/567/default:aspx Accessed on September 25, 2020.

Fernandez-Orozco R, Frias J, Munoz R, Zielinski H, Piskula M K, Kozlowska H and Vidal-Valverde C (2008). Effect of fermentation conditions on the antioxidant compounds and antioxidant capacity of *Lupinus angustifoliuscv.*Zapaton. *European Food Research and Technology,* 227: 979–88.

Foley, R C, Gao, L-L, Spriggs, A, Soo, L Y C, Goggin, D E, Smith, P M C, Atkins, C A, and Singh, K B (2011). Identification and characterisation of seed storage protein transcripts from *Lupinus* angustifolius. *BMC Plant Biology,* 11.

Fryirs C, Eisenhaur B, Duckworth S (2008). Luteins in lupins—An eye for health. In: Palta J A, Berger J D (Eds.), *Proceedings of the 12th International Lupin Conference. CSIRO Plant Industry,* Wembley, Western Australia.

Gorecki R J, Piotrowicz-Cieslak A I, Lahuta L B and Obendorf R L (1997). Soluble carbohydrates in desiccation tolerance of yellow lupin seeds during maturation and germination. *Seed Science Research,* 7(2): 107–116.

Gresta, F, Abbate, V, Avola, G, Magazzù, G, Chiofalo, B (2010). Lupin seed for the crop-livestock food chain. *Italian Journal of Agronomy,* 4:333–340.

Gross R (1988). Lupins in human nutrition. *Proceedings of the 5th International Lupin Conference,* Poland, 51–63.

Gross R, Koch F, Malaga I, de Miranda A F, Schoeneberger H and Trugo L C (1989). Chemical composition and protein quality of some local Andean food sources. *Food Chemistry,* 34:25–34.

Guemes-Vera N, Martinez-Herrera J, Hernandez-Chavez J F, Yanez-Fernandez J and Totosaus A. (2012). Comparison of chemical composition and protein digestibility, carotenoids, tannins and alkaloids content of wild Lupinus varieties flour. *Pakistan Journal of Nutrition,* 11: 676–82.

Guémes-Vera, N, Pena-Bautista, R J, Jimenez-Martinez, C, Davila-Ortiz, G and Calderon-Dominguez, G (2008). Effective detoxification and decoloration of Lupinus mutabilis seed derivatives, and effect of these derivatives on bread quality and acceptance. *Journal of the Science of Food and Agriculture,* 88:1135–1143.

Guillamón E, Pedrosa M M, Burbano C, Cuadrado C, Sánchez M de C and Muzquiz M (2008). The trypsin inhibitors present in seed of different grain legume species and cultivar. *Food Chemistry,* 107: 68–74.

Gulewicz P, Martínez-Villaluenga C, Frias J, Ciesiołka D, Gulewicz K and Vidal-Valverde C (2008). Effect of germination on the protein fraction composition of different lupin seeds. *Food Chemistry,* 107: 830–844.

Gulewicz, P, Szymaniec, S, Bubak, B, Frias, J, Vidal-Valverde, C, Trojanowska, J and Gulewicz, D (2002). Biological activity of alpha-galactoside preparations from *Lupinus* angustifolius L. and Pisum sativum L. seeds. *Journal of Agricultural and Food Chemistry,* 50(2):384–389.

Hall R S, Thomas S J and Johnson S K (2005). Australian sweet lupin flour addition reduces the glycaemic index of a white bread breakfast without affecting palatability in healthy human volunteers. *Asia Pacific Journal of Clinical Nutrition,* 14(1): 91–97.

Hamama A A and Bhardwaj H L (2004). Phytosterols, tri-terpene alcohols, and phospholipids in seed oil from white lupin. *Journal of the American Oil Chemists' Society,* 81: 1039–1044.

Indarte, M, Lazza, C M, Assis, D.et al. (2017). A Bowman–Birk protease inhibitor purified, cloned, sequenced and characterized from the seeds of *Maclura pomifera* (Raf.) Schneid. *Planta,* 245:343–353. https://doi.org/10.1007/s00425-016-2611-6

Islam S. and Ma W (2016). Lupine.In: Caballero B, Finglas P M, and Toldrá F (Eds.), *Encyclopedia of Food and Health.* Oxford, UK: Academic Press, 579–585.

Jayasena V and Nasar-Abbas S M (2012). Development and quality evaluation of high-protein and high-dietary-fiber pasta using lupin flour. *Journal of Texture Studies,*43: 153–163.

Jayasena V, Leung P and Nasar-Abbas S M (2010). Effect of lupin flour substitution on the quality and sensory acceptability of instant noodles. *Journal of Food Quality,* 33: 709–727.

Jenkins D J A, Kendall C W C, Augustin L S A, Martini M C, Axelsen M, Faulkner D, Vidgen E, Parker T, Lau H, Connelly P W, Teitel J, Singer W, Vandenbroucke A C, Leiter LA and Josse RG (2002). Effect of wheat bran on glycemic control and risk factors for cardiovascular disease in type 2 diabetes. *Diabetes Care,* 25:1522–1528.

Johnson S K, Chua V, Hall R S and Baxter A L (2006). Lupin kernel fiber foods improve bowel function and beneficially modify some putative faecal risk factors for colon cancer in men. *British Journal of Nutrition,* 95: 372–78.

Karami Z and Akbari-Adergani B (2019). Bioactive food derived peptides: a review on correlation between structure of bioactive peptides and their functional properties. *Journal of Food Science and Technology,* 56(2):535–547.

Kennedy A R (1993). Cancer prevention by protease inhibitors. *Preventive Medicine,* 22: 796–811.

Khan M K, Karnpanit W, Nasar-Abbas S M, Huma Z E, and Jayasena V (2015). Phytochemical composition and bioactivities of lupin: A review. *International Journal of Food Science and Technology,* 50:2004–2012.

Khayata W, Bashor G, and Sadek M (2013). The effect of phytic acid on rate of starch digestibility in *Vicia faba* and white wheat bread *in vitro. International Journal of Pharmaceutical Sciences and Research,* 5(1): 161–64.doi: 10.13040/IJPSR.0975-8232.5(1).161–64

Kohajdova, Z, Karovicova, J and Schmidt, S (2011). Lupin composition and possible use in bakery-A review. *Czech Journal of Food Science,* 29 (3):203–211.

Koliaki C, Kokkinos A, Tentolouris N, and atsilambros N (2010). The effect of ingested macronutrients on postprandial ghrelin response: A critical review of existing literature data. *International Journal of Peptides,* 2010.

Kouris-Blazos A and Belski R (2016). Health benefits of legumes and pulses with a focus on Australian sweet lupins. *Asia Pacific Journal of Clinical Nutrition,* 25(1):1–17.

Krawczyk M, Przywitowski M and Mikulski D (2015). Effect of yellow lupin (*L. luteus*) on the egg yolk fatty acid profile, the physicochemical and sensory properties of eggs, and laying hen performance. *Poultry Science,* 94: 1360–1367.

Lambers H, Clements J C and Nelson M N (2013). How a phosphorus-acquisition strategy based on carboxylate exudation powers the success and agronomic potential of lupines (Lupinus, Fabaceae). *American Journal of Botany,*100: 263–288.

Lammi C, Zanoni C, Scigliuolo G M, D'Amato A and Arnoldi A (2015). Molecular investigation of the mechanism of action through which lupin peptides induce hypocholesterolemic effects on HepG2 cells. In: Capraro J, Duranti M, Magni Ch, Scarafoni A (Eds): *Proceedings of the XIV International Lupin Conference.* Milan, Italy. June 21–26, 91–94.

Lampart-Szczapa E, Siger A, Trojanowska K, Nogala-Kalucka M, Walecka M and Pacholek B (2003). Composition and antibacterial activities of lupin seeds extracts. *Nahrung Foods,* 47: 286–90.

Laurant P, Demolombe B and Berthelot A (1995). Dietary L-arginine attenuates blood pressure in mineralocorticoid-salt hypertensive rats. *Clinical and Experimental Hypertension,* 17(7):1009–1024.

Lee Y P, Mori T A, Sipsas S, Barden A, Puddey I B, Burke V,Hodgson J M (2006). Lupin-enriched bread increases satiety and reduces energy intake acutely. *American Journal of Clinical Nutrition,* 84:975–980.

Lima V C O D, Piuvezam G, Maciel B L L and Morais A H D A (2019). Trypsin inhibitors: Promising candidate satietogenic proteins as complementary treatment for obesity and metabolic disorders? *Journal of Enzyme Inhibition and Medicinal Chemistry,* 34:405–419.

López E P (2014). Influence of the addition of lupine protein isolate on the protein and technological characteristics of dough and fresh bread with added Brea Gum. *Food Science and Technology,* 34(1): 195–203.

Loredo-Davila S, Espinosa-Hernandez V, Goytia-Jimenez M A, Diaz- Ballote L, Soto-Hernandez R M and Marrone P G (2012). Fatty acid methyl ester profile from *Lupinus* in the identification of sweet and bitter species from this gender with oil of *Lupinus uncinatus* Schlecht seeds. *Journal of Nutrition and Food Sciences,* 2: 158–61.

Lqari H, Vioque J, Pedroche J and Millan F (2002). Lupinus angustifolius protein isolates: Chemical composition, functional properties and protein characterization. *Food Chemistry.* 76:349–356.

Luo J, Cai W, Wu T and Xu, B (2016). Phytochemical distribution in hull and cotyledon of adzuki bean (*Vigna angularis* L.) and mung bean (*Vigna radiate* L.), and their contribution to antioxidant, anti-inflammatory and anti-diabetic activities. *Food Chemistry,* 201:350–360.

Maghaydah S, Abdul-Hussain S, Ajo R, Tawalbeh Y and Elsahoryi N (2012). Effect of lupine flour on baking characteristics of gluten free cookies. *Advance Journal of Food Science and Technology,* 5(5): 600–605.

Mahmoud E A M, Nassef, S L and Basuny, A M M (2012). Production of high protein quality noodles using wheat flour fortified with different protein products from lupine. *Annals of Agricultural Science,* 57(2):105–112.

Maknickiene Z and RaZukas A (2007). Narrow-leaved forage lupine (*Lupinus angustifolius* L.) breeding aspects. *Zemes Ukio Mokslai,* 14: 27–31.

Manach C, Scalbert A, Morand C, Remesy C and Jimenez L (2004). Polyphenols: Food sources and bioavailability. *American Journal of Clinical Nutrition,* 79(5): 727–47.

Martinez-Villaluenga C, Sironi E, Vidal-Val- verde C and Duranti M. (2006). Effects of oligosac- charide removing procedure on the protein profiles of lupin seeds. *European Food Research and Technology,* 223: 691–696.

Mohamed A A and Rayas-Duarte P (1995). Composition of *Lupinus albus. Cereal Chemisry,* 72: 643–47.

Msika P, Piccirilli A and Paul F (2006). Peptideextract of lupin and pharmaceutical or cosmetic or nutritional composition comprising the same. US Patent 7029713.

Mubarak A E (2001). Chemical, nutritional and sensory properties of bread supplemented with lupin seed (*Lupinus albus*) products. *Nahrung,* 45(4):241–245.

Muzquiz M, Burbano C, Cuadrado C and De la Cuadra C (1993). Determination of heat-resistant anti-nutritional factors in legumes. *Prod Rotec Veg*, 8(1): 351–61.

Muzquiz M, Burbano C, Gorospe M J and Rodenas I (1989). A chemical study of Lupinus hispanicus seed-toxic and antinutritional components. *Journal of the Science of Food and Agriculture*, 47: 205–214.

Muzquiz M, Rbdenas I, Villaverde J and Casinello M (1982). Valoracioncuantitativa de 10s alcaloidesensemillas del genero Lupinus (L.). In: Lopez Bellido L, Fuentes M, and Milne D A (Eds.), *Actas II Conf. Int. Lupino*, May 1982, Torremolinos, 196–206.

Olias J M and Valle M (1988). Lipoxygenase from lupin seed: Purification and characterization. *Journal of the Science of Food and Agriculture*, 45: 165–74.

Ortega-David E, Rodriguez A, David A and Zamora Burbano A (2010). Characterization of lupine seeds (*Lupinus mutabilis*) planted in Andes of Colombia. *Acta Agronomica*, 59: 111–18.

Pabich M and Materska M (2019). Biological effect of soy isoflavones in the prevention of civilization diseases. *Nutrients*, 11(7):1660.

Paddon-Jones D, Westman E, Mattes R D, Wolfe R R, Astrup A and Westerterp-Plantenga M (2008). Protein, weight management, and satiety. *The American Journal of Clinical Nutrition* 87:1558S–1561S.

Parmdeep, Sharma S and Singh S (2017). Comparison of cell wall constituents, nutrients and anti-nutrients of lupin genotypes. *Legume Research*, 40 (3): 478–484.

Pastor-Cavada E, Juan R, Pastor J E, Alaiz M and Vioque J (2009). Analytical nutritional characteristics of seed proteins in six wild *Lupinus* species from Southern Spain. *Food Chemistry*, 117: 466–69.

Petterson D S (1998). Composition and food uses of lupins. In: Gladstones J S, Atkins C A and Hamblin J (Eds.) *Lupins as Crop Plants: Biology, Production and Utilization*. Wallingford, UK: CAB International, 353–84.

Petterson D S and Crosbie G B (1990). Potential for lupins as food for humans. *Food Australia*, 42: 266–268.

Petterson D S and Mackintosh J B (1994). The chemical composition of lupin seed grown in Australia. Perth, Western Australia: Proceedings of the first Australian Lupin Technical Symposium.

Piotrowicz-Cieslak A I (2005). Changes in soluble carbohydrates in yellow lupin seed under prolonged storage. *Seed Science Technology*, 33(1): 141–45.

Pisarikova B, Zraly Z, Bunka F and Trckova M (2008). Nutritional value of white lupine cultivar Butan in diets for fattening pigs. *VeterinárniMedicina*, 53: 124–134.

Pollard N J, Stoddard F L, Popineau Y, Wrigley C W and MacRitchie F (2002). Lupin flours as additives: Dough mixing, bread making, emulsifying, and foaming. *Cereal Chemistry*, 79: 662–669.

Prusinski J. (2017). White Lupin (*Lupinus albus* L.)—Nutritional and health values in human nutrition—A review. *Czech Journal of Food Science*, 35(2): 95–105.

Quirrenbach H R, Kanumfre F, Rosso N D and Carvalho Filho M A (2009). Behaviour of phytic acid in the presence of iron (II) and iron (III). *Food Science and Technology*, 29(1): 24–32.

Raikos V, Neacsu M, Russell W and Duthie G (2014). Comparative study of the functionalproperties of lupin, green pea, fava bean, hemp, and buckwheat flours as affected by pH. *Food Science and Nutrition*, 2(6): 802–810.

Romeo F V, Fabroni S, Ballistreri G, Muccilli S, Spina A and Rapisarda P (2018). Characterization and anti-microbial activity of alkaloid extracts from seeds of different genotypes of *Lupinus* spp. *Sustainability*, 10: 788–800.

Roopashree K M and Naik D (2019). Saponins: Properties, applications and as insecticides. *Trends in Biosciences*, 12(1): 1–14.

Ruiz-Lopez M A, Garcia-Lopez P M, Castaneda-Vazquez H, Zamora N J F, Garzon-De laMora P, Banuelos Pineda J, Burbano C, Pedrosa M M, Cuadrado C and Muzquiz M (2000). Chemical composition and antinutrient content of three *Lupinus* species from Jalisco, Mexico. *Journal of Food Composition and Analysis*, 13: 193–99.

Saastamoinen M, Eurola M and Hietaniem V (2013). The chemical quality of some legumes, peas, fava beans, blue and white lupins and soybeans cultivated in Finland. *Journal of Agricultural Science and Technology*, 3(2B): 92–100.

Saini H P and Lymbery J (1983). Soluble carbohydrates of developing lupin seeds. *Phytochemistry*, 22: 1367–70.

Samtiya M, Aluko R E and Dhewa T (2020). Plant food anti-nutritional factors and their reduction strategies: An overview. *Food Production, Processing, and Nutrition*, 2(1):1–14.

Sathe S K, Deshpande S S and Salunkhe D K (1982). Functional properties of lupin seed (*Lupinus* mutabilis) proteins and protein concentrates. *Journal of Food Science*, 47: 491–497.

Scarafoni A, Consonni A, Galbusera V, Negri A, Tedeschi G, Rasmussen P (2008). Identification and characterization of a Bowman-Birk inhibitor active towards trypsin but not chymotrypsin in *Lupinus* albus seeds. *Phytochemistry,* 69: 1820–1825.

Sedláková K, Straková E, Suchý P, Krejcarová J, Herzig I (2016). Lupin as a perspective protein plant for animal and human nutrition—A review. *Acta Veterinaria Brno*, 85: 165–175.

Shi L, Mu K, Arntfield S D and Nickerson M T (2017). Changes in levels of enzyme inhibitors during soaking and cooking for pulses available in Canada. *Journal of Food Science and Technology*, 54(4):1014–1022.

Shoeneberger H, Gross R, Cremerand H D and Elmadfa I (1982). Composition and protein quality of *Lupinus* mutabilis. *Journal of Nutrition,* 112: 70–76.

Shoup F K, Pomeranz Y and Deyoe CW (1966). Amino acid composition of wheat varieties andflours varying widely in bread-making potentialities. *Journal of Food Science,* 31:94–101.

Singla P, Sharma S and Singh S (2017). Amino acid composition, protein fractions and electrophoretic analysis of seed storage proteins in lupins. *Indian Journal of Agricultural Biochemistry*, 30(1):33.

Sironi E, Sessa F and Duranti M (2005). A simple procedure of lupin seed protein fractionation for selective food applications. *European Food Research and Technology*, 221: 145–150.

Smith S C, Choy R, Johnson S K, Hall R H, Wildeboer-Veloo A C M and Welling G W (2006). Lupin kernel fiber consumption modifies fecal microbiota in healthy men as determined by rRNA gene fluorescent in situ hybridization. *European Journal of Nutrition,* 45: 335–41.

Sosulski F W and Dabrowski J D (1984). Composition of free and hydrolysable phenolic acids in the flours and hulls of ten legume species. *Journal of Agricultural and Food Chemistry,*32(1): 131–33.

Stanek M, Bogusz J, Sobotka W and Bieniaszewski (2006). Nutritive value of new varieties of yellow lupin (*Lupinus luteus*) and narrow-leaved lupin (*Lupinus angustifolius*). *Annals of Animal Science*, 2/1: 206–210.

Stephany, M, Bader-Mittermaier, S, Schweiggert-Weisz, U and Carle, R (2015). Lipoxygenase activity in different species of sweet lupin (*Lupinus L.*) seeds and flakes. *Food Chemistry,* 174: 400–406.

Suchy P, Straková E, Večerek V, Šerman V and Mas N (2006). Testing of two varieties of lupin seeds as substitutes for soya extracted meal in vegetable diets designed for young broilers. *Acta Veterinaria Brno,* 75(4): 495–500.

Sujak A, Kotlarz A and Strobel W (2006). Compositional and nutritional evaluation of several lupin seeds. *Food Chemistry,* 98: 711–719.

Szymanowska U, Jakubczyk A, Baraniak B and Kur A (2009). Characterisation of lipoxygenase from pea seeds (*Pisum sativum* var. Telephone L.). *Food Chemistry,* 116: 906–10.

Trugo L C, Donangelo C M, Duarte Y A and Tavares C L. (1993). Phytic acid and selected mineral composition of seed from wild species and cultivated varieties of lupin. *Food Chemistry,* 47(4): 391–94.

Trugo LC, Almeida DCF and Gross R. (1988). Oligosaccharide contents in the seeds of cultivated lupins. *Journal of the Science of Food and Agriculture,* 45: 21–24.

Trugo L C, von Baer D and von Baer E (2003). Lupin. In: Benjamin, C. (Ed.), Encyclopedia of Food Sciences and Nutrition. Oxford: Academic Press, 3623–3629.

Turnbull C M, Baxter A L and Johnson S K (2005). Water-binding capacity and viscosity of Australian sweet lupin kernel fiber under in vitro conditions simulating the human upper gastrointestinal tract. *International Journal of Food Sciences and Nutrition,* 56:87–94.

Uzun B, Arslan C, Karhan M and Toker C (2007). Fat and fatty acids of white lupin (*Lupinus albus* L.) in comparison to sesame (*Sesamum indicum L.*). *Food Chemistry,* 102: 45–49.

Villacres E, Quelal M B, Jacome X, Cueva G and Rosell C M (2020). Original article effect of debittering and solid-state fermentation processes on the nutritional content of lupine (*Lupinus mutabilis* Sweet). *International Journal of Food Science and Technology*, 55:2589–2598.

Villacres E, Caicedo C and Peralta E (2000). Diagnostico del processamiento artesanal, comercializacion y consumo del chocho. In: Caicedo C and Peralta E. (Eds.), ZonificacionPotencial, Sistemas de Produccion y Procesamiento Artesanal del Chocho (Lupinus mutabilis Sweed) en Ecuador. INIAP-FUNDACIT, Quito, Ecuador, pp. 24–41.

Villarino C B J, Jayasena V, Coorey R, Chakrabarti-Bell S and Johnson S K (2016). Nutritional, health, and technological functionality of lupin flour addition to bread and other baked products: Benefits and challenges, *Critical Reviews in Food Science and Nutrition*, 56:5, 835–857.

Wang N, Hatcher D W and Gawalko E J (2008). Effect of variety and processing on nutrients and certain anti-nutrients in field peas (*Pisum sativum*). *Food Chemistry,* 111: 132–38.

Wasche A, Muller K and Knauf U (2001). New processing of lupin protein isolates and functional properties. *Nahrung Food,* 45(6): 393–95.

Yang J, Xiao A and Wang C (2014). Noval development and characterization of dietary fiber from yellow soybean hulls. *Food Chemistry,* 161: 367–75.

Yoshie-Stark Y, Bez J, Wada Y and Wäsche A (2004). Functional properties, lipoxygenase activity, and health aspects of *Lupinus albus* protein isolates. *Journal of Agricultural and Food Chemistry,* 52(25): 7681–9.

Zacherl, C, Eisner, P and Engel, K H (2011). In vitro model to correlate viscosity and bile acid-binding capacity of digested water-soluble and insoluble dietary fibers. *Food Chemistry,* 126:423–428.

Zhao Z C, Zhou Y, Tan G and Li J (2018). Research progress about the effect and prevention of bluelight on eyes. *International Journal of Ophthalmology,* 11(12):1999–2003.

Zielinska-Dawidziak M (2015). Lupin ferritin application for enrichment food in iron. In: Capraro J, Duranti M, Magni Ch, Scarafoni A (Eds), Proceedings of the XIV International Lupin Conference. Milan, Italy. June 21–26, 2015, 87–90.

27

Mung Beans: *Bioactive Compounds and Their Potential Health Benefits*

Barinderjit Singh, Gurwinder Kaur and Reetu
Department of Food Science and Technology, I. K. Gujral Punjab Technical University, Kapurthala, Punjab, India

CONTENTS

27.1 Introduction

The mung bean (*Vigna radiata*) is a legume that can be cultivated in any soil across tropical and subtropical regions of Thailand, Burma, China, Indonesia, West India, South America, Africa, and Malaysia for human consumption (Kahraman et al. 2014; Nair, Kavrekar, and Mishra 2013; Walde et al. 2005). *Vigna radiate* subsp. *radiata (cultivated), Vigna radiata* subsp. *sublobata (wild),* and *Vigna radiata* subsp. *glabra (wild)* are three major subspecies of *Vigna radiata* (Heuzé et al. 2015). While cultivated types are more erect, wild types tend to lie face down. The plant is slightly hairy with multibranched stems and a well-defined root system (Göhl 1982). The leaves are alternate, trifoliated, 5–18 cm long and 3–15 cm wide with elliptical to ovate leaflets. The mung bean flowers are papilionaceous, pale yellow, or green in color. The long hairy cylindrical-shaped pod contains 7–20 ellipsoid, or cube-shaped seeds. The color of seeds such as green, golden, yellow, brown, olive, or black are used to differentiate between various mung bean varieties (Lambrides and Godwin 2007; Mogotsi 2006). Generally, cultivated types are green or golden and, depending on the texture layer, can be shiny or dull (Lambrides and Godwin 2007). However, green, black, and brown colored mung beans are available in India (Mogotsi 2006).

The milling fraction of the whole mung bean is a seed coat (12.1%), cotyledon (85.6%), and embryo (2.3%) (Adsule, Kadam, and Salunkhe 1986). Mung beans are an excellent source of protein (20–40% d.b.), carbohydrates (61.8–64.9%), vitamins including thiamin, riboflavin, and niacin, and minerals including calcium, phosphorus, iron, sodium, and potassium (Wiryawan et al. 1995). The different milling fractions contain an even distribution of these nutrients. The cotyledons and seed coat have more starch and crude fiber, whereas proteins and lipids are the primary nutrients in the embryo (Adsule, Kadam, and Salunkhe 1986). The mung bean is good source of the digestible protein for vegetarian populations. Worldwide, it is cultivated for human food and feed for cattle (Sharma and Rao 2009), sheep (Garg et al. 2004), and goats (Khatik, Vaishnava, and Gupta 2007). Mung beans are consumed fresh as a vegetable or in dry form with rice and chapati (Mogotsi 2006). The mung bean can also be used for preparation of vermicelli, noodles, soups, snacks, and bread (Heuzé et al. 2015). Sprouted seeds are also eaten raw or cooked all over the world (Bayode and Okhonlaye, Ojokoh 2020). Mung beans provide complete nutrition during pregnancy and the early growing years of children. Mung beans help in reducing oxidative stress, weight loss, regulating blood pressure, lowering atherosclerosis risk, maintaining bowel activity, controlling blood sugar levels, and possess antimicrobial activity. Since ancient times, the Chinese have used mung beans due to their benefits for gastrointestinal problems, skin moisturizing, detoxification activities, and decreasing heat stroke (Kahraman et al. 2014; Min 2001). Chinese consume mung bean noodles, which are transparent, tasteless, and quickly cooked. They also consume sprouted seeds of mung bean as a fresh salad. Many other traditional mung bean products are consumed in India, such as boiled dry bean dhal, bean cake, papad, fried green gram dhal, soups, porridge, confections, curries, alcoholic beverages, and fresh green beans (Vuong et al. 2014). Whole mung beans are also used to prepare a rice-mung bean food that meets children's nutritional requirements when weaning from a liquid diet to a soft semisolid diet (Adsule, Kadam, and Salunkhe 1986).

27.2 Bioactive Compounds of the Mung Bean

The mung bean is known for its antioxidant (Kanatt, Arjun, and Sharma 2011), anti-inflammatory, anti-hypertensive, antidiabetic (Lee, Yoon, and Lee 2012), anticancer (Lin, Humbert, and Sosulski 1974), lipid regulation metabolism (Tang et al. 2014), and antimicrobial properties (Jaya Prakash Priya et al. 2012; Kanatt, Arjun, and Sharma 2011) due to presence of different bioactive components such as polyphenols, polysaccharides, protein, peptides, phytoestrogen, tocopherols, tocotrienols and other constituents like phytic acid, trypsin inhibitor enzyme, saponin (Figure 27.1).

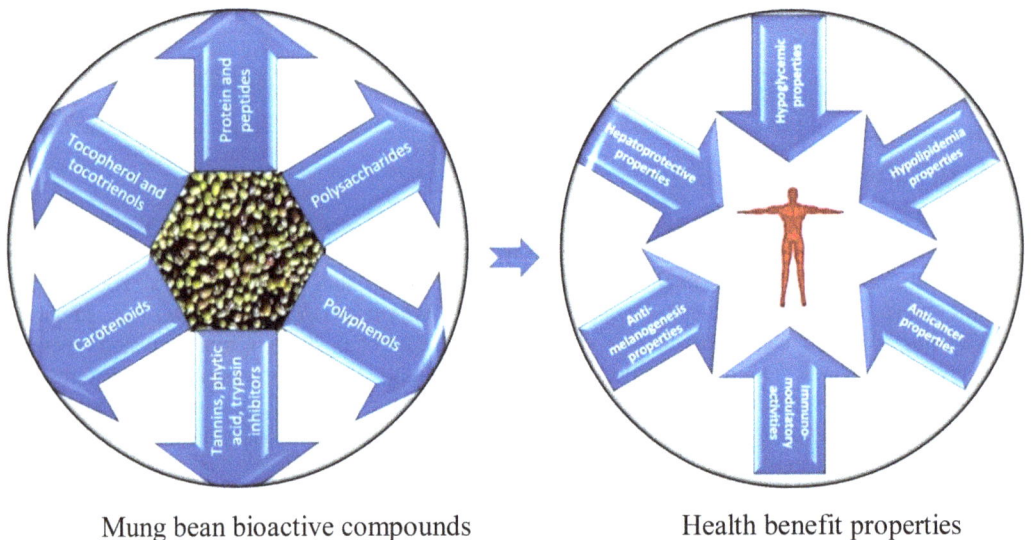

Mung bean bioactive compounds   Health benefit properties

FIGURE 27.1   Mung bean bioactive compounds and their health benefit properties.

27.2.1 Polyphenols

Mung bean fractions contain many different polyphenolic compounds like phenolic acids, flavonoids, anthocyanin, and tannins. However, polyphenols are more abundant in the seed coat of the mung bean. The polyphenolic compounds profile can be affected by factors such as genotype, growing environmental factors, seed coat color, extraction conditions, and analysis methods (Singh et al. 2017). Genotype and growing environmental conditions can significantly change the polyphenolic content and their antioxidant activities (Satya, Saha, and Bhattacharya 2011; Yao et al. 2013). Some extraction factors, including extraction method, solvent concentration, temperature, time, pH, selection of solvent, particle size, and solid-to-solvent ratio, may significantly alter the extraction efficiency (Singh, Sharma, and Sarkar 2012). The extraction of polyphenolic compounds can be done by liquid-solid extraction. It is a separation process involving removing a solute component from the solid by using a liquid solvent.

Various extraction methods have been reported for polyphenol extraction: maceration, soxhlet extraction, ultrahigh-pressure extraction, microwave extraction, maceration, accelerated solvent extraction, supercritical fluid extraction, and heat-reflux extraction. The selection of solvent and its concentration is the most critical factor for extracting polyphenols from mung beans (Singh et al. 2017). The polar solvents (methanol, ethanol, acetone, and water) are more suitable for extracting these polyphenols due to their polar in nature. However, the literature also highlighted that aqueous solvent is more effective in extracting polyphenols than the pure solvent (Singh et al. 2017). The solvent's polarity will change with the changing concentration of the solvent, which will change the profile of the polyphenols. The extraction temperature also affects the quality and quantity of the polyphenols in the extract. The high temperature may increase the solutes' solubility, whereas most heat-labile phytochemicals can also be lost due to higher temperatures. The polyphenol yield can also be increased with increasing extraction time; however, the longer duration can also degrade the polyphenols (Singh, Sharma, and Sarkar 2012; Singh et al. 2017). The efficiency of the extraction of polyphenolic compounds from mung beans also depends on the solid-to-solvent ratio. The nature of the solid-to-solvent ratio determines the amount and type of extracted phenolics and the antioxidant activity of the extracts. However, the solvent concentration and type, extraction time, and temperature are the significant variables that may affect the extraction of polyphenolic compounds and their antioxidant properties. However, these variables, such as solvent type and concentration, duration of extraction, and temperature of extraction, need to be optimized (Singh, Sharma, and Sarkar 2012; Singh et al. 2017).

Most polyphenolic compounds are present in bound form with different macromolecules such as sugar, starch, polysaccharides, protein, and fatty compounds (Shahidi and Yeo 2016; Santos-Zea, Villela-Castrejón, and Gutiérrez-Uribe 2019). These fatty acids interferences can be reduced by mixing the sample with a nonpolar solvent for 24 hours before the extraction of the polyphenols. However, the breakdown of the bond between polyphenols and the macromolecules cannot be broken down by simple polar solvent extraction. Therefore, the bound polyphenols can be released by using exogenous enzymes and alkali-acid hydrolysis (Brglez Mojzer et al. 2016). The different exogenous enzymes, such as alpha-amylase, alpha-glucosidase, pepsin, proteases, hemicellulose, and viscozymes, can break down polyphenol-macromolecule linkages, which will result in a higher yield. The higher extraction of polyphenolic compounds with enzymes is also due to the modification or synthesis of new complex compounds from different biomaterials under the right conditions of pH, temperature, and time for each enzyme (Meyer, Jepsen, and Sørensen 1998; Arcan and Yemenicioğlu 2010; Urbano et al. 2007). On the other hand, the polar solvent crude extract can also be treated with alkali and acid to break down the linkage before analysis of the sample in high liquid chromatography as these glycosides need to be converted into aglycones with hydrolysis before their evaluation. Base hydrolysis led to the breakdown of ester bond linkage between phenolic acid and the cell wall, whereas acid hydrolysis broke the glycosidic bond and caused the solubilization of sugars (Fazary and Ju 2007; Verma, Hucl, and Chibbar 2009; Liu 2007; Bhanja, Kumari, and Banerjee 2009). Base and acid hydrolysis also caused degradation of some polyphenolic compounds like CA, FA, and pCoA (Gao and Mazza 1994). These polyphenols have good corelation with antioxidant activity (Li, Lu, and Yue 2014; Sandhu et al., 2016; Punia and Sandhu 2015). The mung bean polyphenols have high free-radical scavenging activity, ferric reducing power, oxygen, and peroxyl radical absorbing capacity (Xu and Chang 2012; Chai et al. 2019).

27.2.1.1 Phenolic Acid

Phenolic acid contains one carboxylic acid functionality in phenols. These compounds are divided into derivatives of cinnamic acid and derivatives of benzoic acids. They differ in the position of the hydroxyl group on the aromatic ring of two different phenolic acid (Kumar and Goel 2019). The hydroxycinnamic acids are caffeic acid, ferulic acid, p-coumaric acid, chlorogenic acids, and sinapic acid, whereas the hydroxybenzoic acids are gallic acid, protocatechuic acid, vanillic acid, syringic acid, and gentisic acid (Hou et al. 2019). Among these phenolic acids, sinapic acid is the major hydroxycinnamic acid, while syringic acid is the predominant hydroxybenzoic acid present in mung beans (Singh et al. 2017; Hou et al. 2019). Among the phenolic acids, the hydroxycinnamic acids are the predominant phenolic acid present in mung beans. The phenolic acid profile of mung beans can vary with the extraction method, extraction temperature, extraction duration, type, and concentration of the solvent (Singh et al. 2017). Phenolics are important bioactive compounds that impart health benefits for the human body; mostly bound phenolics are absorbed into the blood stream after fermentation of food in the lower parts of gastrointestinal tract (Wang et al. 2020; Adebo and Medina-Meza 2020). Ferulic, chlorogenic, p-coumaric, and gentisic are primary phenolic acids present in bound form. However, caffeic and synergic acids are available in both free and bound forms.

27.2.1.2 Flavonoids

Flavonoids are classified based on the central pyran ring's oxidation state as flavonols, flavones, flavanones, and isoflavones. The chemical structure and degree of hydroxylation and polymerization can change the antioxidant properties of different types of flavonoids (Kumar and Pandey 2013). Flavanols contain a saturated three-carbon chain with a hydroxyl group in C3 and exist in both monomer and polymer forms. Mung beans contain flavonols (quercetin, quercetin-3-0-glucoside, myricetin, and kaempferol), flavanols (catechin), flavones (luteolin, vitexin, isovitexin, and isovitexin-6-O-α-L-glucoside), and isoflavonoids (dulcinoside) (Hou et al. 2019; Kang et al. 2015; Jeong et al. 2016; Lee et al. 2011; Liu et al. 2014). Luteolin is the predominant flavonoid present in mung bean fractions (Singh et al. 2017). Vitexin and isovitexin are produced in larger amounts during the germination of the mung bean (Jeong et al. 2016). In mung beans, flavonoids also play an important role in the nitrogen-fixing symbiosis as signal molecules in the Nod factor transduction pathways to the formation of nodule primordial (Tang et al. 2014). However, flavonoids exhibit various biological activities such as antioxidant, anticancer, anti-inflammatory, antifungal, antibacterial, and antiviral activities. A wide range of health-promoting effects are associated with flavonoids and are an indispensable component of various nutraceutical, pharmaceutical, medicinal and cosmetic applications (Tang et al. 2014). Effectively, flavonoids inhibit lipid peroxidation that is responsible for different disorders, such as atherosclerosis, diabetes, hepatotoxicity, inflammation, and aging. Anthocyanins are water-soluble color compounds that are attached to the sugar moieties in plant structure (Martín 2017). They are a complex form of flavonoids that are synthesized via the phenylpropanoid pathway. The black mung bean's pigmentation is due to the high amount of total anthocyanin, i.e., 437.7–810.6 μg/g. The peonidin-3-glucoside (5.43–8.42 μg/g), pelargonidin-3,6-malonylglucoside (8.53–12.5 μg/g), pelargonidin-3-glucoside (129.2–350.7 μg/g) and cyanidin-3-glucoside (256.3–476.5 μg/g) are the major anthocyanins responsible for pigmentation (Hou et al. 2019). These anthocyanins are known for many health benefits such as hypolipidemia, hypoglycemic and anticancer properties, and antiviral and antimicrobial activities.

27.2.1.3 Stilbene

Stilbenes are naturally occurring polyphenolic compounds with aromatic groups bonded to each end of a carbon-carbon double bond. Trans-stilbene is the simplest example of the stilbene that is easily found in mung bean fractions. The conventional method of extraction is more suitable to extract trans-stilbene, whereas the mung bean ultrasound treatment reduces the concentration of trans-stilbene (Singh et al. 2017). Mung beans also contain hydroxylated stilbene derivatives like resveratrol that exhibit anticarcinogenic, antiaging, vasorelaxant, neuroprotective, anti-inflammatory cardioprotective, and phytoestrogenic properties (Ganesan and Xu 2018). The mung bean hull contains a higher quantity of resveratrol, which can be increased by using ultrasound extraction in an acetone solvent. However, other polar solvents such as methanol, water, and ethanol can adversely affect the extraction of resveratrol (Singh et al. 2017).

27.2.2 Polysaccharides

Mung beans contain both soluble and insoluble dietary fibers; however, the insoluble dietary fiber is available in a higher amounts (Liyanage et al. 2018). Cellulose is the predominant insoluble dietary fiber present in mung beans (Adsule, Kadam, and Salunkhe 1986). A small quantity of lignin is also reported in the literature. The mung bean has higher total dietary fiber compared to chickpeas. The dietary fiber can improve gut health by producing short-chain fatty acids during colonic fermentation that will reduce systemic inflammation (Perry and Ying 2016). Dietary fiber can also help diabetic patients regulate insulin production, reducing the predicted glycemic index (Perry and Ying 2016). The dietary fibers can also be useful for weight management by decrease the macronutrients, promoting satiety, and modifying the secretion of gut hormones (Perry and Ying 2016). The mung bean is also known for the presence of arabinogalactan, which plays a vital role in macrophage activation. Macrophages impart a significant role in immunomodulatory properties by killing pathogens via phagocytosis (Dai et al. 2014). The alkali-extractable polysaccharides from mung bean can be used as a functional ingredient for preparing functional food due to good antioxidant properties and immunoregulatory effects on macrophages (Zhu et al. 2016). The milling industry of mung beans produced hull waste that can further be processed to extract xylitol. The xylitol-enriched diet can reduce serum glucose, cholesterol, and triglyceride levels (Mushtaq et al. 2014). Therefore, the production of xylitol from mung beans resolves the environmental issue and can produce high-valued functional ingredients.

27.2.3 Protein and Peptides

Protein isolate can be produced from mung beans due to their high protein content. Like soy bean protein, mung bean protein isolate is made up of 8S globulins (Nakatani et al. 2018). This protein isolate can be prepared by enzymatic hydrolysis. These protein isolates can be used to regulate blood pressure and cardiovascular functions by inhibiting angiotensin-1-converting enzyme activities (Li et al. 2006). This bioactive property of mung beans is due to the presence of peptides released upon enzymatical digestion (Hou et al. 2019; Li et al. 2006; Watanabe et al. 2017). Therefore, the mung bean isolates not only contain amino acids like asparagine, glutamine, serine, histidine, glycine, threonine, alanine, arginine, tyrosine, cystine, valine, methionine, phenylalanine, isoleucine, leucine, lysine, proline, and tryptophan for normal growth and maintenance in the human body, but also have bioactive peptides that are beneficial toward chronic diseases (Li et al. 2006).

27.2.4 Lipophilic Compounds

Mung bean seeds also contain lipophilic-based antioxidant components such as carotenoids, tocopherol, and tocotrienols that are reduced with the germination of the seeds (Jaleel et al. 2009; Zia-Ul-Haq, Ahmad, and Iqbal 2008). Epidemiological studies have highlighted that adequate carotenoid supplementation reduces the risk of many chronic diseases mediated by reactive oxygen species (Fiedor and Burda 2014). The mung bean contains a higher amount of γ-tocopherol and α-tocotrienols than other respective isomers (Zia-Ul-Haq, Ahmad, and Iqbal 2008). The γ-tocopherol has good potential to reduce the platelet aggregation, low-density lipoprotein (LDL) oxidation, and delaying intra-arterial thrombus formation. However, the tocotrienols are beneficial toward the inhibition of cholesterol biosynthesis due to higher antioxidant activities (Zia-Ul-Haq, Ahmad, and Iqbal 2008). The mung bean oil has different polyunsaturated fatty acids such as omega-6 and omega-3 acids; however, the ratio of omega-6 to omega-3 is 1.70 (Ramesh et al. 2011). It is recommended that if the ratio of omega-6 to omega-3 in the diet is less than 4, the diet is beneficial for human health (Simopoulos 2002). Therefore, mung bean oil can be used as a functional ingredient for preparing functional foods.

27.2.5 Other Compounds

The mung bean also contain other bioactive components such as tannins, phytic acid, trypsin inhibitors, and hemagglutinin. These compounds showed the beneficial affect toward promoting digestion, biological function, and eliminating toxins from the human body (Mubarak 2005; Tang et al. 2014). However, the high percentage of these compounds can produce negative impacts on the human body by reducing the bioavaiablity of minerals such as calcium, iron, phosphorus, and zinc (Dahiya et al. 2015). The mung bean contain less phytic acid compared to other legumes such as soy beans, kidney beans, pinto beans, black-eyed peas, chickpeas, lentils, and adzuki beans (Xu and Chang 2012). In addition to this, different processing methods such as fermenattaion, germination, de-hulling, boiling, roasting and thermal processing further reduce the content of trypsin inhibitors, hemagglutinin, tannins, and phytic acid due to degradation of these compounds (Barakoti and Bains 2007; Hemalatha, Platel, and Srinivasan 2007; Dahiya et al. 2015).

27.3 Potential Health Benefits of Bioactive Compounds

27.3.1 Hypoglycemic Properties

The mung bean and its extract have excellent hypoglycemic properties. Hypoglycemia, also known as low blood sugar, slows the absorption of glucose. Various studies showed that low glycemic index food intake helps prevent the risk of diabetes mellitus and obesity (Yeap et al. 2012; Hou et al. 2019). Diabetes mellitus is a heterogeneous, multi-causal metabolic syndrome characterized by chronic hyperglycemia with partial or total insulin secretion deficiency and decreased hormone sensitivity in peripheral tissues (American Diabetes Association 2009). There are many factors such as obesity, higher age, impaired glucose tolerance, less physical activity, high-calorie diet, high glycemic index foods, genetics, race, and other chronic diseases, which can trigger diabetes (Sami et al. 2015). Diabetics can be controlled by using dietary strategies. Mung bean extracts help inhibit the activity of α-amylase and α-glucosidase activity, which reduces intestinal absorption of carbohydrates and reduces body weight by promoting the effect of insulin (Liyanage et al. 2018). In addition to the regulation of LDL cholesterol, the fermented mung bean can provide health benefits toward lowering blood sugar levels and promoting weight management (Yeap et al. 2012).

27.3.2 Hypolipidemia Properties

Hyperlipidemia is a disease in which elevated amounts of lipids are found in the blood, most of which are associated with high-fat diets, a sedentary lifestyle, obesity, diabetes, and genetic causes like mutations in the gene microsomal triglyceride transfer protein (Tang et al. 2014). However, some secondary factors such as hyperthyroidism, cancer, hematologic, undernutrition, malabsorption, hepatitis C infection, and other inflammatory problems can trigger the hyperlipidemia in the human body (Elmehdawi 2008). Many studies highlighted the direct relationship between the mung bean and lowered the lipid levels in plasma. The mung bean-supplemented diet can reduce the apparent cholesterol absorption, total cholesterol triglycerides, and nonhigh-density lipoprotein cholesterol due to the increase in bile acids' production mRNA3-hydroxy-3-methyl glutaryl coenzyme A reductase, and sterol excretion (Udem, Ezeonuegbu, and Obidike 2011).

27.3.3 Anticancer Properties

Cancer is a category of disease linked to the development of abnormal, uncontrolled cells due to many extrinsic (chemicals, infectious organisms, radiation, and smoking) and intrinsic factors (hormones, immune conditions, inherited and metabolic mutations) (Chanda and Nagani 2013). Many conventional treatments for cancer, such as surgery, radiotherapy, hormonal therapy, and chemotherapy, have either relapse or many treatment-induced side effects (Palumbo et al. 2013). Consumption of the mung bean and its sprouts can reduce several cancers, including colon, breast, and prostate (Ganesan and Xu 2018), so it can be a potent anticancer agent granting new anticancer therapy (Hafidh et al. 2012). Dose-dependent antiproliferative effects of mung beans are exerted against different cell lines, such as the digestive system, ovary, and breast. Among all types of cancer in women, breast cancer still represents the highest percentage. A high intake of mung beans helps to protect against breast cancer during adolescence (Yeap et al. 2013).

Fermented mung beans, on the other hand, prevent the development of colon cancer due to the presence of different bioactive components such as polysaccharides, polyphenols, saponin, and protease inhibitors (Hou et al. 2019). These compounds have good antioxidant and anticarcinogenic properties. Antioxidant properties of mung bean polyphenols reduce the metastatic proliferation of the cell and inhibit the development of cancer. Mung bean sprout extract stimulates factor-alpha and interferon-gamma tumour necrosis, thus stimulating immunity mediated by cells (Yeap et al. 2013). The extract of mung bean sprouts also induces cell cycle arrest and apoptosis on colon cancer cells due to the presence of trypsin inhibitors in the extract (Rao et al. 2018).

27.3.4 Immunomodulatory Activity

The immune system is a complex network of cells, tissues, organs, and substances that work together to help the body combat infections and other diseases. Two large subsystems, innate and adaptive immunity, are divided into immune processes (Hou et al. 2019). Phagocytes, natural killer cells, and the complement cascade are the main constituents of innate immunity. Adaptive immunity has recognition mechanisms unique to the antigen, and it has an immunological memory. Mung bean nonstarch polysaccharides such as fiber can help to activate macrophages (Hou et al. 2019). Mung bean extract includes polyphenols, gallic acid, vitexin, and isovitexin that reduce the activity of murine macrophages by preventing proinflammatory gene expression, including interleukin (IL)-Iβ, IL-6, IL-I2β, tumor necrosis factor, and inducible NO synthase, and decrease lipopolysaccharide stimulant macrophages without cytotoxicity (Dai et al. 2014). Mung bean sprouts have high vitamin-C content that makes them a powerful stimulant for the body's white blood cells to fight off infections and diseases. Fermented mung bean induced splenocyte proliferation and enhanced serum interleukin-α and interferon-γ (Ali et al. 2016).

27.3.5 Antimelanogenesis Properties

The human body produces melanin pigment in melanosomes by the interaction of melanocytes with the endocrine, immune, inflammatory, and central nervous systems; this process is known as melanogenesis (Videira, Moura, and Magina 2013). Many factors such as ultraviolent radiation, drugs, hormones, and inflammation can alter the synthesis of melanin pigments like pheomelanin and eumelanin, which can cause a change in human skin color. Pheomelanin and eumelanin pigments are derived from the amino acid L-tyrosine, and their synthesis involves melanogenic enzymes such as tyrosinase, a copper-dependent enzyme, and a rate-limiting enzyme of melanogenesis (Kondo and Hearing 2011). However, the change in the free radical-antioxidant balance in the human body can alter the mRNA concentration of tyrosinase and melanin production, which can lead to severe skin problems like hyperpigmentation. This problem can be overcome by using polyphenols containing mung bean extract due to high antioxidant potential. These compounds have the capability to reduce the rate of melanin production by supressing the tyrosinase enzyme activities that will improve skin health and color (Kao et al. 2013; Chai et al. 2019).

27.3.6 Hepatoprotective Properties

The main functions of the liver are in the metabolism, secretion, and storage of many nutrients and purifying the body by clearing the toxins and harmful substances. The high exposure of chemotherapeutics, drugs, and environmental toxins can also increase the risk of liver intoxication (Venkat 2017). The hepatoprotective properties of the mung beans due the presence of vitexin and isovitexin. These compounds known for having good antioxidant properties that inhibit the accumulation of hepatic lipids

(Liu et al. 2014). Therefore, the mung bean extracts act as a potential hepatoprotective agent in dietary supplements (Wu et al. 2001). High-dose fermented and germinated mung bean extracts are the most effective extracts to revert the NO level elevation after induction with ethanol. Thus, by suppressing the liver's NO production, fermented and germinated mung beans show their potential properties as a hepatoprotective agent. The hepatoprotective effect of fermented and germinated mung beans at high dose may be due to flavonoids and phenolic acid bioactive compounds, which are positively detected in fermented a germinated product. The increased GABA content and amino acids in mung bean after germination and fermentation processes possess a better hepatoprotective effect than normal mung beans (Mohd Ali et al. 2013).

27.4 Conclusion

In summary, mung beans have demonstrated they are an excellent source of not only major nutrients but also a significant number of bioactive components such as polyphenols, polysaccharides, protein, peptides, phytoestrogen, tocopherols, tocotrienols, phytic acid, trypsin inhibitor enzyme, and saponin. The concentration of these compounds can vary with the extraction methods or conditions (solvent type and concentration, extraction duration, and extraction temperature) and different processing methods (soaking, germination, fermentation, boiling, roasting, and thermal processing). However, these bioactive components possess various health benefits properties including hypoglycemic, hypolipidemic, anticancer, immunomodulatory, antimelanogenesis, and hepatoprotective activities. Therefore, a mung bean-enriched diet can be a natural and cost-effective solution for curing and preventing different chronic diseases.

REFERENCES

Adebo, O. A., and I. G. Medina-Meza. 2020. "Impact of Fermentation on the Phenolic Compounds and Antioxidant Activity of Whole Cereal Grains: A Mini Review." *Molecules* 25 (4): 1–19.

Adsule, R. N., S. S. Kadam, and D. K. Salunkhe. 1986. "Chemistry and Technology of Green Gram (*Vigna Radiata* [L] Wilczek)." *CRC Critical Reviews in Food Science and Nutrition* 25 (1): 73–105.

Ali, N. M., S. K. Yeap, H. M. Yusof, B. K. Beh, W. Y. Ho, S. P. Koh, M. P. Abdullah, N. B. Alitheen, and K. Long. 2016. "Comparison of Free Amino Acids, Antioxidants, Soluble Phenolic Acids, Cytotoxicity and Immunomodulation of Fermented Mung Bean and Soybean." *Journal of the Science of Food and Agriculture* 96 (5): 1648–1658.

American Diabetes Association. 2009. "Diagnosis and Classification of Diabetes Mellitus." *Diabetes Care* 32 (SUPPL. 1): S62–S76.

Arcan, I., and A. Yemenicioğlu. 2010. "Effects of Controlled Pepsin Hydrolysis on Antioxidant Potential and Fractional Changes of Chickpea Proteins." *Food Research International* 43 (1): 140–147.

Barakoti, L., and K. Bains. 2007. "Effect of Household Processing on the In Vitro Bioavailability of Iron in Mungbean (*Vigna Radiata*)." *Food and Nutrition Bulletin* 28 (1): 18–22.

Bayode, A. A., and A. Okhonlaye, Ojokoh. 2020. "Microorganisms Associated with the Fermentation of Gari Fortified with Sprouted Mung Beans Flour." *South Asian Journal of Research in Microbiology* 5 (4): 1–8.

Bhanja, T., A. Kumari, and R. Banerjee. 2009. "Enrichment of Phenolics and Free Radical Scavenging Property of Wheat Koji Prepared with Two Filamentous Fungi." *Bioresource Technology* 100 (11): 2861–2866.

Brglez Mojzer, E., M. K. Hrnčič, M. Škerget, Ž. Knez, and U. Bren. 2016. "Polyphenols: Extraction Methods, Antioxidative Action, Bioavailability and Anticarcinogenic Effects." *Molecules* 21 (7): 1–38.

Chai, W. M., Q. M. Wei, W. L. Deng, Y. L. Zheng, X. Y. Chen, Q. Huang, C. Ou-Yang, and Y. Y. Peng. 2019. "Anti-Melanogenesis Properties of Condensed Tannins from: *Vigna Angularis* Seeds with Potent Antioxidant and DNA Damage Protection Activities." *Food and Function* 10 (1): 99–111.

Chanda, S., and K. Nagani. 2013. "*In Vitro* and *in Vivo* Methods for Anticancer Activity Evaluation and Some Indian Medicinal Plants Possessing Anticancer Properties: An Overview" *Journal of Pharmacognosy and Phytochemistry* 2 (2): 140–152.

Dahiya, P. K., A. R. Linnemann, M. A. J. S. Van Boekel, N. Khetarpaul, R. B. Grewal, and M. J. R. Nout. 2015. "Mung Bean: Technological and Nutritional Potential." *Critical Reviews in Food Science and Nutrition* 55 (5): 670–688.

Dai, Z., D. Su, Y. Zhang, Y. Sun, B. Hu, H. Ye, S. Jabbar, and X. Zeng. 2014. "Immunomodulatory Activity *in Vitro* and *in Vivo* of Verbascose from Mung Beans (*Phaseolus Aureus*)." *Journal of Agricultural and Food Chemistry* 62 (44): 10727–10735.

Elmehdawi, R. 2008. "Hypolipidemia: A Word of Caution." *Libyan Journal of Medicine.* 3 (2): 84–90.

Fazary, A. E., and Y. H. Ju. 2007. "Feruloyl Esterases as Biotechnological Tools: Current and Future Perspectives." *Acta Biochimica et Biophysica Sinica* 39 (11): 811–828.

Fiedor, J., and K. Burda. 2014. "Potential Role of Carotenoids as Antioxidants in Human Health and Disease." *Nutrients* 6 (2): 466–488.

Ganesan, K., and B. Xu. 2018. "A Critical Review on Phytochemical Profile and Health Promoting Effects of Mung Bean (*Vigna radiata*)." *Food Science and Human Wellness* 7 (1): 11–33.

Gao, L., and G. Mazza. 1994. "Rapid Method for Complete Chemical Characterization of Simple and Acylated Anthocyanins by High-Performance Liquid Chromatography and Capillary Gas-Liquid Chromatography." *Journal of Agricultural and Food Chemistry* 42 (1): 118–125.

Garg, D. D., R. S. Arya, T. Sharma, and R.K. Dhuria. 2004. "Effect of Replacement of Sewan Straw (*Lasirus Sindicus*) by Moong (*Phaseolus Aureus*) Chara on Rumen and Haemato-Biochemical Parameters in Sheep." *Veterinary Practitioner* 5 (1): 70–73.

Göhl, B. 1982. Livestock Feed in the Tropics. FAO, Production and Animal Health Division, Rome, Italy.

Hafidh, R. R., A. S. Abdulamir, F. A. Bakar, F. A. Jalilian, F. Abas, and Z. Sekawi. 2012. "Novel Molecular, Cytotoxical, and Immunological Study on Promising and Selective Anticancer Activity of Mung Bean Sprouts." *BMC Complementary and Alternative Medicine* 12: 1–24.

Hemalatha, S., K. Platel, and K. Srinivasan. 2007. "Influence of Germination and Fermentation on Bioaccessibility of Zinc and Iron from Food Grains." *European Journal of Clinical Nutrition* 61 (3): 342–348.

Heuzé, V., G. Tran, D. Bastianelli, and F. Lebas. 2015. "Mung Bean (*Vigna radiata*)." Feedipedia, a Program by INRA, CIRAD, AFZ, and FAO. Available at: https://www.feedipedia.org/node/235 (Accessed November 06, 2020).

Hou, D., L. Yousaf, Y. Xue, J. Hu, J. Wu, X. Hu, N. Feng, and Q. Shen. 2019. "Mung Bean (*Vigna radiata* L.): Bioactive Polyphenols, Polysaccharides, Peptides, and Health Benefits." *Nutrients* 11 (6): 1–28.

Jaleel, C. A., K. Jayakumar, Z. Chang-Xing, and M. M. Azooz. 2009. "Antioxidant Potentials Protect *Vigna radiata* (L.) Wilczek Plants from Soil Cobalt Stress and Improve Growth and Pigment Composition." *Plant Omics* 2 (3): 120–126.

Jaya Prakash Priya, A., S. L. G. Yamini, F. Banu, S. Gopalakrishnan, P. Dhanalakshmi, E. Sagadevan, A. Manimaran, and P. Arumugam. 2012. "Phytochemical Screening and Antibacterial Activity of *Vigna radiata* L. against Bacterial Pathogens Involved in Food Spoilage and Food Borne Diseases." *Journal of Academia and Industrial Research* 1 (6): 355–359.

Jeong, Y. M., J. H. Ha, G. Y. Noh, and S. N. Park. 2016. "Inhibitory Effects of Mung Bean (*Vigna radiata* L.) Seed and Sprout Extracts on Melanogenesis." *Food Science and Biotechnology* 25 (2): 567–573.

Kahraman, A., M. Adali, M. Onder, N. Koc, and C. Kaya. 2014. "Mung Bean [*Vigna radiata* (L.) Wilczek] as Human Food." *International Journal of Agriculture and Economic Development* 2 (2): 9–17.

Kanatt, S. R., K. Arjun, and A. Sharma. 2011. "Antioxidant and Antimicrobial Activity of Legume Hulls." *Food Research International* 44 (10): 3182–3187.

Kang, I., S. Choi, T. J. Ha, M. Choi, H. R. Wi, B. W. Lee, and M. Lee. 2015. "Effects of Mung Bean (*Vigna radiata* L.) Ethanol Extracts Decrease Proinflammatory Cytokine-Induced Lipogenesis in the KK-Ay Diabese Mouse Model." *Journal of Medicinal Food* 18 (8): 841–849.

Kao, Y. Y., T. F. Chuang, S. H. Chao, J. H. Yang, Y. C. Lin, and H. Y. Huang. 2013. "Evaluation of the Antioxidant and Melanogenesis Inhibitory Properties of Pracparatum Mungo (*Lu-Do Huang*)." *Journal of Traditional and Complementary Medicine* 3 (3): 163–170.

Khatik, K. L., C. S. Vaishnava, and L. Gupta. 2007. "Nutritional Evaluation of Green Gram (*Vigna radiata* L.) Straw in Sheep and Goats." *Indian Journal of Small Ruminants* 13 (2): 196–198.

Kondo, T., and V. J. Hearing. 2011. "Update on the Regulation of Mammalian Melanocyte function Function and Skin Pigmentation." *Expert Review of Dermatology* 6 (1): 97–108.

Kumar, N., and N. Goel. 2019. "Phenolic Acids: Natural Versatile Molecules with Promising Therapeutic Applications." *Biotechnology Reports* 24: 1–10.

Kumar, S., and A. K. Pandey. 2013. "Chemistry and Biological Activities of Flavonoids: An Overview." *The Scientific World Journal* 2013: 1–15.

Lambrides, C. J., and I. D. Godwin. 2007. "Mungbean." In: *Pulses, Sugar and Tuber Crops*, edited by Kole, C. 69–90. Springer Berlin Heidelberg, Berlin.

Lee, C. H., S. J. Yoon, and S. M. Lee. 2012. "Chlorogenic Acid Attenuates High Mobility Group Box 1 (HMGB1) and Enhances Host Defense Mechanisms in Murine Sepsis." *Molecular Medicine* 18 (11): 1437–1448.

Lee, S. J., J. H. Lee, H. H. Lee, S. Lee, S. H. Kim, T. Chun, and J. Y. Imm. 2011. "Effect of Mung Bean Ethanol Extract on Pro-Inflammtory Cytokines in LPS Stimulated Macrophages." *Food Science and Biotechnology* 20 (2): 519–524.

Li, F. M., Z. G. Lu, and M. Yue. 2014. "Analysis of Photosynthetic Characteristics and UV-B Absorbing Compounds in Mung Bean Using UV-B and Red LED Radiation." *Journal of Analytical Methods in Chemistry* 2014: 1–5.

Li, G. H., J. Z. Wan, G. W. Le, and Y. H. Shi. 2006. "Novel Angiotensin I-Converting Enzyme Inhibitory Peptides Isolated from Alcalase Hydrolysate of Mung Bean Protein." *Journal of Peptide Science* 12 (8): 509–514.

Lin, M. J. Y., E. S. Humbert, and F. W. Sosulski. 1974. "Certain Functional Properties Of Sunflower Meal Products." *Journal of Food Science* 39 (2): 368–370.

Liu, R. H. 2007. "Whole Grain Phytochemicals and Health." *Journal of Cereal Science* 46 (3): 207–219.

Liu, T., X. H. Yu, E. Z. Gao, X. N. Liu, L. J. Sun, H. L. Li, P. Wang, Y. L. Zhao, and Z. G. Yu. 2014. "Hepatoprotective Effect of Active Constituents Isolated from Mung Beans (*Phaseolus Radiatus* L.) in an Alcohol-Induced Liver Injury Mouse Model." *Journal of Food Biochemistry* 38 (5): 453–459.

Liyanage, R., C. Kiramage, R. Visvanathan, C. Jayathilake, P. Weththasinghe, R. Bangamuwage, B. C. Jayawardana, and J. Vidanarachchi. 2018. "Hypolipidemic and Hypoglycemic Potential of Raw, Boiled, and Sprouted Mung Beans (*Vigna radiata* L. Wilczek) in Rats." *Journal of Food Biochemistry* 42 (1): 1–6.

Martín, J. 2017. "Anthocyanin Pigments: Importance, Sample Preparation and Extraction." In: *Phenolic Compounds — Natural Sources, Importance and Applications*, edited by Navas, M. J. 117–152. IntechOpen, Rijeka.

Meyer, A. S., S. M. Jepsen, and N. S. Sørensen. 1998. "Enzymatic Release of Antioxidants for Human Low-Density Lipoprotein from Grape Pomace." *Journal of Agricultural and Food Chemistry* 46 (7): 2439–2446.

Min, L. 2001. "Research Advance in Chemical Composition and Pharmacological Action of Mung Bean." *Shanghai University of Traditional Chinese Medicine* 5: 18.

Mogotsi, K. K. 2006. "*Vigna radiata* (L.) R. wilczek." In: *Plant Resources of Tropical Africa 1: Cereals and Pulses*, edited by M. Brink and G. Belay, 208–213. PROTA Foundation, Wageningen/Backhuys Publishers/CTA, Wageningen, Netherlands.

Mohd Ali, N., H. Mohd Yusof, K. Long, S. K. Yeap, W. Y. Ho, B. K. Beh, S. P. Koh, M. P. Abdullah, and N. B. Alitheen. 2013. "Antioxidant and Hepatoprotective Effect of Aqueous Extract of Germinated and Fermented Mung Bean on Ethanol-Mediated Liver Damage." *BioMed Research International* 2013: 1–9.

Mubarak, A. E. 2005. "Nutritional Composition and Antinutritional Factors of Mung Bean Seeds (*Phaseolus aureus*) as Affected by Some Home Traditional Processes." *Food Chemistry* 89 (4): 489–495.

Mushtaq, Z., M. Imran, Salim-ur-Rehman, T. Zahoor, R. S Ahmad, and M. U. Arshad. 2014. "Biochemical Perspectives of Xylitol Extracted from Indigenous Agricultural By-Product Mung Bean (*Vigna radiata*) Hulls in a Rat Model." *Journal of the Science of Food and Agriculture* 94 (5): 969–974.

Nair, S. S, V. Kavrekar, and A. Mishra. 2013. "*In Vitro* Studies on Alpha Amylase and Alpha Glucosidase Inhibitory Activities of Selected Plant Extracts." *European Journal of Experimental Biology* 3 (1): 128–132.

Nakatani, A., X. Li, J. Miyamoto, M. Igarashi, H. Watanabe, A. Sutou, K. Watanabe, et al. 2018. "Dietary Mung Bean Protein Reduces High-Fat Diet-Induced Weight Gain by Modulating Host Bile Acid Metabolism in a Gut Microbiota-Dependent Manner." *Biochemical and Biophysical Research Communications* 501 (4): 955–961.

Palumbo, M. O., P. Kavan, W. H. Miller, L. Panasci, S. Assouline, N. Johnson, V. Cohen, et al. 2013. "Systemic Cancer Therapy: Achievements and Challenges That Lie Ahead." *Frontiers in Pharmacology* 4 (May): 1–9.

Perry, J. R. and W. Ying. 2016. "A Review of Physiological Effects of Soluble and Insoluble Dietary Fibers". *Journal of Nutrition & Food Sciences* (6)1–6

Punia, S., and K. S. Sandhu. 2015. "Functional And Antioxidant Properties Of Different Milling Fractions Of Indian Barley Cultivars." *Journal of Food Science & Technology* (7) 19–27

Ramesh, C. K., A. Rehman, B. T. Prabhakar, B. R. Vijay Avin, and S. J. Aditya Rao. 2011. "Antioxidant Potentials in Sprouts vs. Seeds of *Vigna radiata* and *Macrotyloma uniflorum*." *Journal of Applied Pharmaceutical Science* 1 (7): 99–103.

Rao, S., K. Chinkwo, A. Santhakumar, and C. Blanchard. 2018. "Inhibitory Effects of Pulse Bioactive Compounds on Cancer Development Pathways." *Diseases* 6 (3): 72.

Sami, W., T. Ansari, N. S. Butt, M. Rashid, and A. Hamid. 2015. "Effect of Diet Counseling on Type 2 Diabetes Mellitus." *International Journal of Scientific & Technology Research* 4 (8): 112–118.

Sandhu, K. S., S. Punia, and M. Kaur. 2016. "Effect of Duration of Solid State Fermentation by Aspergillus Awamorinakazawa on Antioxidant Properties of Wheat Cultivars." *LWT Food Science and Technology* 71: 323–328.

Sandhu, K. S., and S. Punia. 2017. "Enhancement of Bioactive Compounds in Barley Cultivars by Solid Substrate Fermentation." *Journal of Food Measurement and Characterization*, 11 (3): 1355–1361.

Santos-Zea, L., J. Villela-Castrejón, and J. A. Gutiérrez-Uribe. 2019. "Bound Phenolics in Foods." In: *Bioactive Molecules in Food*, edited by Mérillon, J. M., and Ramawat, K. G. 973–989. Springer International Publishing, Cham.

Satya, P., A. Saha, and P. M. Bhattacharya. 2011. "Tolerance to Chilling Stress in Germinating Mung Bean (*Vigna radiata* L. wilczek) Is Associated with Increased Phenolics And Peroxidase Activity." *Genetics and Plant Physiology* 1 (3–4): 139–149.

Shahidi, F., and J. D. Yeo. 2016. "Insoluble-Bound Phenolics in Food." *Molecules* 21 (9): 1–22.

Sharma, V., and L. J. M. Rao. 2009. "A Thought on the Biological Activities of Black Tea." *Critical Reviews in Food Science and Nutrition* 49 (5): 379–404.

Simopoulos, A. P. 2002. "The Importance of the Ratio of Omega-6/Omega-3 Essential Fatty Acids." *Biomedicine and Pharmacotherapy* 56 (8): 365–379.

Singh, B., H. K. Sharma, and B. C. Sarkar. 2012. "Optimization of Extraction of Antioxidants from Wheat Bran (*Triticum* spp.) Using Response Surface Methodology." *Journal of Food Science and Technology* 49 (3): 294–308.

Singh, B., N. Singh, S. Thakur, and A. Kaur. 2017. "Ultrasound Assisted Extraction of Polyphenols and Their Distribution in Whole Mung Bean, Hull and Cotyledon." *Journal of Food Science and Technology* 54 (4): 921–932.

Tang, D., Y. Dong, H. Ren, L. Li, and C. He. 2014. "A Review of Phytochemistry, Metabolite Changes, and Medicinal Uses of the Common Food Mung Bean and Its Sprouts (*Vigna radiata*)." *Chemistry Central Journal* 8 (1): 1–9.

Udem, S. C., U. C. Ezeonuegbu, and R. I. Obidike. 2011. "Experimental Studies on the Hypolipidemic and Haematological Properties of Aqueous Leaf Extract of Cleistopholis Patens Benth. & Diels. (Annonacae) in Hypercholesterolemic Rats." *Annals of Medical and Health Sciences Research* 1 (1): 115–11521.

Urbano, G., S. Frejnagel, J. M. Porres, P. Aranda, E. Gomez-Villalva, J. Frías, and M. López-Jurado. 2007. "Effect of Phytic Acid Degradation by Soaking and Exogenous Phytase on the Bioavailability of Magnesium and Zinc from *Pisum sativum*, L." *European Food Research and Technology* 226 (1–2): 105–111.

Venkat, S. 2017. "A Review on Hepatoprotective Activity." *International Journal of Current Research* 9 (6): 51876–51881.

Verma, B., P. Hucl, and R. N. Chibbar. 2009. "Phenolic Acid Composition and Antioxidant Capacity of Acid and Alkali Hydrolysed Wheat Bran Fractions." *Food Chemistry* 116 (4): 947–954.

Videira, I. F. d. S., D. F. L. Moura, and S. Magina. 2013. "Mechanisms Regulating Melanogenesis." *Anais Brasileiros de Dermatologia* 88 (1): 76–83.

Vuong, Q. V., C. D. Goldsmith, T. T. Dang, V. T. Nguyen, D. J. Bhuyan, E. Sadeqzadeh, C. J. Scarlett, and M. C. Bowyer. 2014. "Optimisation of Ultrasound-Assisted Extraction Conditions for Phenolic Content and Antioxidant Capacity from *Euphorbia Tirucalli* Using Response Surface Methodology." *Antioxidants* 3 (3): 604–617.

Walde, S. G., J. Tummala, S. M. Lakshminarayan, and M. Balaraman. 2005. "The Effect of Rice Flour on Pasting and Particle Size Distribution of Green Gram (*Phaseolus radiata*, L. wilczek) Dried Batter." *International Journal of Food Science and Technology* 40 (9): 935–942.

Wang, Z., S. Li, S. Ge, and S. Lin. 2020. "Review of Distribution, Extraction Methods, and Health Benefits of Bound Phenolics in Food Plants." *Journal of Agricultural and Food Chemistry* 68 (11): 3330–3343.

Watanabe, H., Y. Inaba, K. Kimura, S. I. Asahara, Y. Kido, M. Matsumoto, T. Motoyama, et al. 2017. "Dietary Mung Bean Protein Reduces Hepatic Steatosis, Fibrosis, and Inflammation in Male Mice with Diet-Induced, Nonalcoholic Fatty Liver Disease." *Journal of Nutrition* 147 (1): 52–60.

Wiryawan, K. G., J. G. Dingle, A. Kumar, J. B. Gaughan, and B. A. Young. 1995. "True Metabolisable Energy Content of Grain Legumes: Effects of Enzyme Supplementation." In: *Recent Advances in Animal Nutrition in Australia*, edited by Rowe, J. B. and Nolan, J. V. pp. 196. University of New England, Armidale.

Wu, S. J., J. S. Wang, C. C. Lin, and C. H. Chang. 2001. "Evaluation of Hepatoprotective Activity of Legumes." *Phytomedicine* 8 (3): 213–219.

Xu, B., and S. K. C. Chang. 2012. "Comparative Study on Antiproliferation Properties and Cellular Antioxidant Activities of Commonly Consumed Food Legumes against Nine Human Cancer Cell Lines." *Food Chemistry* 134 (3): 1287–1296.

Yao, Y., X. Yang, J. Tian, C. Liu, X. Cheng, and G. Ren. 2013. "Antioxidant and Antidiabetic Activities of Black Mung Bean (*Vigna radiata* L.)." *Journal of Agricultural and Food Chemistry* 61 (34): 8104–8109.

Yeap, S. K., N. Mohd Ali, H. Mohd Yusof, N. B. Alitheen, B. K. Beh, W. Y. Ho, S. P. Koh, and K. Long. 2012. "Antihyperglycemic Effects of Fermented and Nonfermented Mung Bean Extracts on Alloxan-Induced-Diabetic Mice." *Journal of Biomedicine and Biotechnology* 2012: 1–7.

Yeap, S. K., H. Mohd Yusof, N. E. Mohamad, B. K. Beh, W. Y. Ho, N. M. Ali, N. B. Alitheen, S. P. Koh, and K. Long. 2013. "*In Vivo* Immunomodulation and Lipid Peroxidation Activities Contributed to Chemoprevention Effects of Fermented Mung Bean against Breast Cancer." *Evidence-Based Complementary and Alternative Medicine* 2013: 1–7.

Zhu, Y., Y. Yao, Y. Gao, Y. Hu, Z. Shi, and G. Ren. 2016. "Suppressive Effects of Barley β-Glucans with Different Molecular Weight on 3T3-L1 Adipocyte Differentiation." *Journal of Food Science* 3(81): 786–793.

Zia-Ul-Haq, M., M. Ahmad, and S. Iqbal. 2008. "Characteristics of Oil from Seeds of 4 Mungbean [*Vigna radiata* (L.) Wilczek] Cultivars Grown in Pakistan." *Journal of the American Oil Chemist's Society* 85 (9): 851–856.

28

Nutraceutical and Health Benefits of Pulses

Nikita Wadhawan and Chavan Sagar Madhukar
College of Technology and Engineering, Udaipur, Rajasthan, India

Gaurav Wadhawan
Pacific Medical College & Hospital, Udaipur, India

CONTENTS

28.1 Introduction

The *portmanteau* "nutraceutical" came from the words "nutrition" and "pharmaceutical" and means a food, or part of a food, that provides medical or health benefits, including the prevention and/or treatment of a disease. Dr Stephen DeFelice coined the term "Nutraceutical" from "Nutrition" and "Pharmaceutical" in 1989 about the food items that not only supplement the diet but should also aid in the prevention and/ or treatment of a disease and/or a disorder. The term nutraceuticals is applied broadly to foods and food constituents that provide specific health or medical benefits, including the prevention and treatment of diseases (Ekta K. Kalra 2003). These may include food additives, vitamins, and mineral supplements, herbs, phytochemicals, and probiotics, etc. Nutraceuticals can also be defined as "food by virtue of physiologically active components, provide health benefits beyond basic nutrition." The American Dietetic Association (ADA) define these as foods "that include whole foods as fortified, enriched or enhanced foods that have potentially beneficial effect on health when consumed as part of a varied diet on a regular basis at effective levels." In 1994, the Japanese Ministry of Health and Welfare highlighted three conditions that the term must satisfy: First, they are foods, not capsules, tablets or powders and are derived from naturally occurring ingredients. Second, they can and should be consumed as part of the daily diet. Finally, they should have a particular function when ingested, serving to regulate a particular body process such as enhancement of the biological defense mechanisms, prevention of a specific disease and control of physical and mental conditions or slowing down the aging process (ADA, 2009).

Pulses are food components providing rich nutrients and hence are often recommended as a staple food by health organizations and dieticians (Rizkalla et al., 2002). Pulses are also recognized as a food of choice with significant potential health benefits. They contain complex carbohydrates (dietary fibers, resistant starch, and oligosaccharides), protein with a good amino acid profile (high lysine), important vitamins and minerals (B vitamins, folates and iron), as well as antioxidants and polyphenols. Pulse grains are an excellent source of protein, carbohydrates, dietary fiber, vitamins, minerals and photochemicala (phenolic acid, anthocyanins) (Tharanathan and Mahadevamma, 2003) and their consumption and production is increasing worldwide. Researchers have demonstrated that pulses can prevent or manage chronic health issues such as diabetes, cardiovascular disease, and obesity and contribute to overall health and wellness. Pulses contain a number of bioactive substances including enzyme inhibitors, lectins, phytates, oligosaccharides, and phenolic compounds that play metabolic roles in humans or animals that frequently consume these foods (Campos-Vega et al., 2010). These effects may be regarded as positive, negative or both (Champ, 2002). Some of these substances have been considered as antinutritional factors due to their effect on hindering nutrient absorption and affecting the diet quality. Frequent legume consumption (four or more times compared with less than once a week) has been associated with a 22% and 11% lower risk of coronary heart disease (CHD) and cardiovascular disease (CVD), respectively (Flight and Clifton, 2006). In an earlier study on 9,632 participants free of CVD at their baseline examination in the First National Health and Nutrition Examination Survey (NHANES 1) Epidemiological Follow-up Study (NHEFS), Bazzano et al. (2001) found that legume consumption was significantly and inversely associated with risk of CHD and CVD.

28.2 Ethical Issues in the Use of Nutraceuticals

Discussions of ethics are now a common feature of academic curricula and the underpinnings for research societies' and journals' "codes of conduct." Consequently, a primary objective is to provide ethical perspectives and background with a focus on data acquisition and translation, publication and communication, monitoring, and the marketing of research findings. Examples of historical, cultural, and political events that have influenced nutrition and health policies and practices in relation to nutraceutical use are discussed next.

28.2.1 Tailoring Products to Domestic Tastes and Preferences

These might include vegetarianism, Halal, or Hindu dietary practices, traditional remedies, flavor and formulation preferences reflecting social and cultural diversity, or reluctance to see functional benefits in staple foods. Yogurt, for example, is a dietary mainstay in India, and is often homemade. Educational programs on diet, nutrition, and disease risk-reduction will help to clear a path for premium-priced nutraceutical brands. Mainstream food companies, with their long experience of consumer markets and varying tastes and textures, will have the competitive edge here. Pharmaceutical companies entering the nutraceuticals market may want to access these capabilities through joint ventures, research partnerships, or acquisitions.

28.2.2 The Right Price Point

Premium pricing of nutraceuticals in India may be viable for the urban middle class, but not for the rural poor. It will require more attention to scientific evidence and professional endorsement that substantiates health claims. A comprehensive approach to the Indian market calls for a split-level strategy, with tailored branding, formulation, positioning, pricing, and distribution. There may be opportunities to penetrate poorer segments of the market on a platform of social responsibility, in tandem with nongovernment organizations or government fortification programs. As things stand, nutraceuticals fall under the Food Safety & Standards Act. They are treated broadly as food products, subject to manufacturing/ quality standards and restrictions on claims to prevent or mitigate disease.

28.2.3 Market Players Say the Lack of Carved-Out Nutraceutical Regulations Leads To Confusion and Expediency

Any food supplement in a "drug-like" oral formulation is categorized as a medicine, with corresponding approval requirements, even if it is only making structure-function claims (Santini, 2018). More recently, the Indian government announced that proposed sector-specific regulations—the Food Safety and Standards (Food or Health Supplements, Nutraceuticals, Foods for Special Dietary Uses, Foods for Special Medical Purpose, Functional Foods, and Novel Foods) Regulations, 2015—were at the final notification stage. These establish conditions for the manufacture, formulation, approval, and sale of nutraceuticals, as well as for health claims, RDAs, scientific evidence and labeling. There are lists of permissible additives, vitamins, minerals, amino acids, plant ingredients; etc. To the dismay of many manufacturers, the Food Safety and Standards Authority said in March 2016 that the draft regulations would apply to all nutraceuticals launched in India after 2011 until the rules were finalized (Yates, 2016).

28.3 Nutraceutical Properties of Pulses

Nutraceuticals as defined earlier are a food or part of a food that provides medical or health benefits, including the prevention or treatment of a disease (Nazri et al. 2014). Legumes are well known for the presence of different bioactive compounds such as saponins, tannins, flavonoids, isoflavones, lectins, phytic acid, etc., which is important for their nutraceutical properties (Campos-Vega et. al. 2010). Several nutraceuticals found in legumes are listed among the top 200 list of the American Nutraceutical Association (Murphy and Hendrich,). Highly pigmented and dark-colored legume seeds have higher levels of phenolic and flavonoid content, which helps in its antioxidant activity (2002; Pieta, 2000; Yeh and Yen, 2003) Legume seeds contain enzyme inhibitors like alpha-amylase, alpha-glucosidase, and alpha-aminobutyric acid (GABA) for which it can be used as a nutraceutical molecule (Lajolo, Mancini Filho, and Menezes, 1984). Green legume seeds are also a good source of nutraceuticals (Mathers, 2002). Legume seeds are normally consumed after processing thereby increasing the bioavailability of nutrients by inactivating trypsin, growth inhibitors, and hemagglutinins (Nachbar and Oppenheim, 1980). Different species of legumes are involved in the treatment of various diseases like coronary heart diseases, cardiovascular diseases, cancer, diabetes, etc. Legume seeds also contain resistant proteins and carbohydrates that play an effective role in human health (Barman et al., 2018). The importance of legumes in the human diet is expected to increase in the near future to meet the demand for protein and other nutrients in the increasing world population, and also to reduce the risk related to animal food source consumption. Molecules present in legume seeds that are considered toxic or unhealthy may also provide positive effects on human health in the prevention and treatment of certain diseases if consumed in a limited scale and in a proper way. Hence these can play an excellent role in the nutraceutical and antioxidant properties (Barman et al., 2018) of seed legumes (Table 28.1).

TABLE 28.1

Legumes Possessing Nutraceutical Properties

Name of the Legume	Scientific Name	Role as Nutraceutical
Black soy bean	*Glycine max* L.	Treatment of diabetes, hypertension, antiaging, cosmetology, blood circulation, etc. due to presence of active peptide compounds (Morris, 2003).
Pigeon pea	*Cajanus cajan* L. Millspaugh	It has both nutritional and medicinal properties. Scorched seeds can relieve headache and vertigo when added to coffee while fresh seeds help urinary incontinence in males. Immature seeds are used in the treatment of kidney ailments. Pigeon pea seed husks possess an effective antioxidant and antihyperglycemic activity that may be a potential organic resource for the development of nutraceutical for hyperglycemic individuals (Crandall and Duren, 2013)
Mung bean	*Vigna radiata*	It contains different bioactive compound that help lower the risk of various diseases (Morris, 2003).
Cow pea	*Vigna unguiculata*	It is an important leguminous food rich in protein, carbohydrates, minerals, and water soluble vitamins like thiamine, riboflavin, and niacin (Shinde et al., 2014).
Rice bean	*Vigna umbellate*	It also contains a number of bioactive compounds such as phytate, alpha-galactosides, and trypsin inhibitors that can act as antioxidant,anticancer, and antidiabetic agents (Ajit, 2013).
Black gram	*Vigna mungo*	Used as a diet during fever, cooling astringent, poultice for abscesses, affection for cough and liver, and also recommended for treating diabetes (Rajasekaran et al., 2008)
Lentil	*Lens culinaris*	Due to the high content of dietary fiber and low glycemic response of lentil seeds, it is highly recommended for patients suffering from cardiovascular diseases and diabetes. Several bioactive compounds present in lentil seeds such as phytic acids, lectins, saponins, etc. show anticarcinogenic, antimutagenic, antioxidative and antihyperglycemic activities (Rajasekaran A, 2008).
Chickepeas	*Cicer arietinum*	Chickpea seed oil contains different sterols, tocopherols, and tocotrienols. These phytosterols have been reported to exhibit antiulcerative, antibacterial, antifungal, antitumor and anti-inflammatory properties coupled with a lowering effect on cholesterol levels (Anderson and Major, 2002).
Lupines		Different potential health benefits of lupines have been investigated, particularly in the area of dyslipidemia, hyperglycemia, and hypertension prevention (Anderson and Major, 2002).

28.4 Nutrient Bioactive Compounds

Several bioactive compounds may exhibit a wide range of beneficial effects on human health that may contribute to thier nutraceutical properties. A vital challenge today is to change our current diets to be healthier and more sustainable. In this context, the promotion of pulse consumption is widely acknowledged and promoted. Phytates are abundant in all pulses, assayed from 413 mg/100 g in brown lentils to 714 mg/100 g in green lentils. Saponin content is present around 120 mg/100 g (mean value). Great variations were seen for tannins (from 1.7 mg/100 g in household-cooked white beans to 16.6 mg/100 g in household-cooked chickpeas) and total polyphenols (from 15 mg/100 g in household-cooked white beans to 284 mg/100 g in household-cooked green lentils). The method of preparation could significantly influence pulse bioactive compound composition. For instance, phytate contents were lower in canned kidney beans and chickpeas than in household-cooked grains (−38.5% and −24%, $p < 0.05$). Conversely, polyphenol content was higher in canned white beans and chickpeas compared to household-cooked white beans and chickpeas (+24.4% and +10.5%, respectively, $p < 0.05$). The saponin content moderately varied depending on the preparation method: it was significantly higher in canned kidney beans than in household-cooked kidney beans (+12%), whereas it was significantly lower in canned chickpeas compared to household-cooked chickpeas (−4.1%). In a well-balanced diet, pulse bioactive compounds may contribute to reduce the risk of chronic diseases (Rajasekaran A, 2008).

28.4.1 Polyphenols

The potential health benefits of common beans are attributed to the presence of secondary metabolites such as phenolic compounds that possess antioxidant properties (Cardador-Martinez et al., 2002; Punia et al., 2020a). The major polyphenolic compounds of pulses consist mainly of tannins, phenolic acids and flavonoids. Pulses with the highest polyphenolic content have dark and highly pigmented grains, such as red kidney beans (*Phaseolus vulgaris*) and black gram (*Vigna mungo*). Condensed tannins (proanthocyanidins) have been quantified in hulls of several varieties of field beans (*Vicia faba*) and are also present in pea seeds of colored-flower cultivars. Tannin-free and sweet seeds have been selected among broad beans, lentils, and lupines (Smulikowska et al., 2001). Pulses vary based on their total phenolic contents and antioxidant activities (Table 28.2). Lentils have the highest phenolic, flavonoid, and condensed tannin content (6.56 mg gallic acid equivalents and malvidin, while kaempferol and its 3-*O*-glycosides were present in pinto beans. Light red kidney beans had traces of quercetin 3-*O*-glucoside and its malonates, but pink and dark red kidney beans contained the diglycosides of quercetin and kaempferol. Small red beans contained kaempferol 3-*O*-glucoside and pelargonid in 3-*O*-glucoside, while flavonoids were undetected in alubia, cranberry, great northern, and navy beans. Total anthocyanin content in whole grain and seed coat ranged from 37.7–71.6 mg g^{-1} and 10.1–18.1 mg g^{-1}, respectively, among 15 black bean cultivars grown in Mexico. The anthocyanins in seed coats of beans were identified as delphinidin 3-glucoside 65.7%, petunidin 3-glucoside 24.3%, and maldivin 3-glucoside 8.7% (Salinas-Moreno et al., 2005). Chickpeas also contain a wide range of polyphenolic compounds, including flavonols, flavone glycosides, and oligomeric and polymeric proanthocyanidins (Sarma et al., 2002). Total phenolic content in chickpea ranges from 0.92 to 1.68 mg gallic acid equivalents g^{-1} (Xu and Chang, 2007; Zia-Ul-Haq et al., 2007). Lignans, diphenolic compounds with a 2,3-dibenzylbutane skeleton, have both estrogenic and antiestrogenic properties (Orcheson et al., 1998). The plant lignans, secoisolariciresinol (SEC) and matairesinol (MAT), are converted to the metabolites enterodiol (ED) and enterolactone (EL), known as the mammalian lignans, in the gastrointestinal tract. Most studies have only looked at the isoflavonoid content of legumes; only one study (Wiltold et al., 1998) has analyzed the SEC and MAT content. The concentrations of SEC lignans in legumes ranged from 0–240 mg 100 g^{-1} and 13–273 mg 100 g^{-1} in soy beans with trace or no MAT detected (Mazur et al., 1998).

TABLE 28.2

Phenolic Content of Pulses

Legume	Total Phenolic Content (mg Gallic Acid Equivalents g^{-1})	Total Flavonoid Content (mg Catechin Equivalents g^{-1})	Condensed Tannin Content (mg Catechin Equivalents g^{-1})
Green pea	1.53	0.08	0.26
Yellow pea	1.67	0.18	0.42
Chickpea	1.81	0.18	1.05
Lentil	6.56	1.30	5.97
Red kidney	4.98	2.02	3.85
Black bean	5.04	2.49	3.40

28.4.2 Isoflavones

Flavones and isoflavones have been isolated from various plants, though the isoflavones are largely reported from the *Fabaceae/Leguminosae* family. According to the US Department of Agriculture survey on iso-flavone content, lentils do not contain significant amounts of these isoflavones (USDA, 2002). Chickpeas contain daidzein, genistein, and formononetin (0.04 mg 100 g^{-1}, 0.06 mg 100 g^{-1}, and 0.14 mg 100 g^{-1}, respectively) and approximately 1.7 mg 100 g^{-1} biochanin A. Soy beans have significantly higher levels of daidzein and genistein (47 mg 100 g^{-1} and 74 mg 100 g^{-1}, respectively) but contain less formononetin and biochanin A compared to chickpeas, 0.03 mg 100 g^{-1} and 0.07 mg 100 g^{-1}, respectively. Total isoflavones in *L. muta-bilis* range from 9.8–87 mg 100 g^{-1}, 16.1–30.8 mg 100 g^{-1}, and 1.3–6.1 mg 100 g^{-1} fresh weight of sample (expressed as genistein) in seed coat, cotyledon and hypocotyl fractions, respectively (Lena et al., 2009).

Barcelo and Munoz (1989) identified isoflavones such as genistein, 29-hydroxigenistein, luteon and wighteone in sprouted hypocotyls of *L. albus* CV multolupa, suggesting that these compounds are related to cell wall lignification. This may explain luteone (a tetrahydroxyisoflavone), which was detected in immature seeds of *L. luteus* (Fukui et al., 1973). Dini et al. (1998) detected two genistein derivatives, mutabilin (glycosylated form) and mutabilein (aglycon form), in seeds of *L. mutabilis*. Furthermore, formononetin, genistein, and the phytoestrogen secoisolariciresinol were found in seeds of *L. mutabilis* (23 mg 100 g^{-1}, 2,420 mg 100 g^{-1}, and 3.1 mg 100 g^{-1}, respectively) (Wiltold et al., 1998). Soy is the most widely available source of isoflavones used therapeutically. Effects of the isoflavone supplement on hormonal states in young premenopausal women suggest that isoflavones influence not only estrogen receptor-related functions but the hypothalamo-hypophysis-gonadal axis (Duncan et al., 1999).

28.4.3 Phytosterols

Pulses contain small quantities of phytosterols, of which b-sitosterol, campesterol, and stigmasterol are the most common (Benveniste, 1986). These compounds are also abundant as sterol glucosides and esterified sterol glucosides, with b-sitosterol representing 83% of the glycolipids in defatted chickpea flour (Sanchez-Vioque et al., 1998) reported that the total phytosterol content in the pulses ranged from 134 mg 100 g^{-1} (kidney beans) to 242 mg 100 g^{-1} (peas) while total b-sitosterol content ranged from 160 mg 100 g^{-1} (chickpeas) to 85 mg 100 g^{-1} (butter beans). Chickpeas and peas contained high levels of campesterol (21.4 mg 100 g^{-1} and 25.0 mg 100 g^{-1}, respectively). Stigma-sterol content is higher in butter beans (86 mg 100 g^{-1}) as is the squalene content in peas (1.0 mg 100 g^{-1}). The role of different bioactive compounds in management of various diseases is as described next.

28.4.3.1 Cardiovascular Diseases

Legume consumption four times or more per week compared with less than once a week was associated with a 22% lower risk of CHD (relative risk, 0.78; 95% confidence interval, 0.68–0.90) and an 11% lower risk of CVD (Crandall and Duren, 2001). It has been reported that legume seed consumption lowers low-density lipoprotein (LDL) cholesterol by partially interrupting the enterohepatic circulation of the bile acids and increasing the cholesterol saturation by increasing the hepatic secretion of cholesterol. It has been examined that 30% neutral detergent fiber of black gram in the diet can reduce cholesterol levels compared to cellulose on binding with bile acids (Anderson and Major, 2002).

28.4.3.2 Diabetes

Legume seeds play a significant role in the treatment of diabetes. They have high content of fiber and oligosaccharides, which help maintain the glycemic levels in the blood (Rizkalla et al., 2002).

28.4.3.3 Cancer

Legume seeds contain several nutrients and bioactive compounds like fiber, oligosaccharides, phenolic compounds, and antioxidants that show anticarcinogenic activity. It has been found that adzuki bean

(*Vigna angularis*) has differentiation/maturation inducing activity for dendritic cells and apoptosis inducing activity for human leukemia U937 cell. Hence legume seeds help in the treatment of different types of cancer (Rajasekaran A, 2008). There is a reduced risk of prostate cancer, breast cancer, colon cancer, and pancreatic cancer associated with the consumption of legume seeds (Anderson and Major, 2002).

28.4.3.4 Hepatotoxicity

Legume seeds are a healthy food for liver functionality. GABA present in legume seeds is a potent hepatoprotective agent (Rajasekaran et al., 2008).

28.4.3.5 Osteoporosis

Legume seeds are a good source of calcium and protein, helping to build strong bones—hence legume seeds can help reduce the risk of osteoporosis. Isoflavones daidzein and genistein prevent the breakdown of bone (Shinde et al., 2014).

28.4.3.6 Postprandial Hyperglycemia

Legume seed husks possess potent antioxidant and antihyperglycemic activity. Methanolic extract of seed husks potentially mitigates development of postprandial hyperglycemic spikes and glycemic load close to the clinically used drug acarbose (Rizkalla et al., 2002).

28.4.3.7 Anticarcinogenic Effects

Various studies have reported results for intakes of pulses and cancer risk; among the potential protective components against cancer that can be present in pulses are protease inhibitors, saponins, phytosterols, isoflavones, and phytates (Van Loo et al., 1995; Trinidad et al., 2010).

28.4.3.8 Weight Control and Obesity

Dry legumes can help maintain a healthy body weight, thanks to their great satiety effect, which helps in limiting daily food intake. Studies performed with healthy subjects showed an increase in stool weight when they included soy bean or pea fiber in the diet (Smarta, 2012). A specific direct action of grain legume alpha-amylase protein inhibitors has been considered for its potential use in the prevention and therapy of obesity and diabetes (Shinde et al., 2014).

28.4.4 Protein and Amino Acids

Pulses present both environmental and nutritional benefits and are often recommended in sustainable diets (Rajasekaran A, 2008). Their environmental benefit is related to their ability to restore soil nitrogen without adding fertilizer. Diversifying crop rotations with pulses is thus a green solution to enhance system productivity. Moreover, health organizations such as the Food and Agriculture Organization of the United Nations (FAO) recommend pulses as staple foods to fulfill the basic protein and energy requirements of the human diet (Borkar et al., 2015). Indeed, pulse grains are i) a low-fat source of proteins and carbohydrates and ii) of interest as a gluten-free food category. They exhibit complementary amino acid profiles to those of cereals in well-balanced semivegetarian or plant-based diets (Milner, 2004). Mature seeds of pea and faba beans contain 18–20% protein and lupine and soy beans contain 35–45% protein. Most of the proteins found in legume seeds are storage proteins, which are of 7S and 11S globulins based on their sedimentation coefficient (Trinidad et al., 2010). Some proteins in legume seeds show antifungal and antiviral activity and hence act as anti-HIV and antidiabetic agent. These proteins contain several essential amino acids that are beneficial to human health. The proteins present in *Vigna species* show antifungal and antiviral activity. Ground bean lectins inhibit the hemagglutinating activity by polygalacturonic acid but not galacturonic acid and simple monosaccharides. It decreases the viability of hepatoma (HepG2), leukemia (L1210), and leukemia (M1) cell and also elects a mitogenic response from mouse splenocytes. Due to presence of all these properties, these proteins in pulsed act as an excellent drug for the treatment of AIDS patients with no adverse effects as compared to synthetic drugs (Barac et al., 2005; Duranti, 2006; Aluko, 2008).

28.4.5 Carbohydrates: Starch, Soluble Sugars, and Dietary Fiber

The presence of carbohydrates in grain legume seeds amounts up to 60%, which include oligosaccharides like alpha-galactosides and complex molecules such as starch and fibers (Navarro et al., 2018; Punia et al., 2019, 2020b). Starch is a storage polysaccharide made up of amylose and amylopectin in the ratio roughly of 1:3, but in wrinkled peas, the ratio reaches 3:1. Lupines and soy beans contain the lowest amounts of starch (about 1–2%) whereas in peas and faba beans, starch accounts for about 50% of the dry seed weight. From a nutritional viewpoint, this polysaccharide can be classified according to its hydrolysis degradation in animal model systems as rapid digestion starch (RDS), slow digestion starch (SDS), and resistant starch. The latter is not hydrolyzed by human amylases, but it can be fermented by the microorganisms present in the colon as if it is fiber.

28.5 Non-Nutrient Bioactive Compounds

Some bioactive compounds present in pulses can display ambivalent properties. It has been suggested that low doses of phytates could reduce the risk of colon cancer because of the phytate antioxidant effect and phytate prebiotic activity due to their ability to bind enzymes such as amylases, so that a portion of the undigested starch reaches the intestine (Pusztai et al. 1997; Ewen et al. 1998). Conversely, at higher doses, phytates act by chelating different cations, thereby decreasing the bioavailability of the minerals present in the bolus. Furthermore, it is acknowledged that saponins interfere with the absorption of dietary lipids, cholesterol, and bile acids, thus displaying interesting lipid-lowering properties, as well as vitamins A and E in chicks. Saponins have additionally been shown to bind to the cells of the small intestine, which can affect the *in vitro* absorption of nutrients across the intestinal wall (Fu and Renxiao, 2010). Finally, tannins, well known for their antioxidant capacity (Ajit, 2013) have been found to interfere with digestion by displaying antitrypsin and antiamylase activity (MacCarty, 2005; Obiro et al., 2008). Tannins are likely responsible for the trypsin inhibitor activity of faba beans and have also been reported to interfere with iron and zinc absorption (FDA, 1999).

28.5.1 Phytic Acids

Phytic acid (myoinositol hexakisphosphate or IP6) is a major phosphorus storage form in plants and its salts known as phytates. Phytic acid constitutes 1–3% of cereal grains, pulse seeds, and nuts, especially wholegrain cereals and legumes, which have a high content of phytate (Sandberg, 2003). Phytates are located in the protein bodies in the endosperm portion of pulses. Phytate occurs as a mineral complex, which is insoluble at the physiological pH of the intestine (Rochfort and Panozzo, 2007). InsP6 and InsP5 account for more than 95% of the total inositol in raw, dry pulses, measuring 1.9 mmol kg^{-1} and 1.36–2.52 mmol kg^{-1} in green split peas and black-eye peas, respectively, accounting for 16% of total inositol phosphates. The most abundant inositol phosphate in raw, dry pulses was InsP6, accounting for 83% of total inositol phosphates, 77% in chickpeas and 88% in black beans. Oomah (2001) reported that phytic acid (InsP6) represents 75% of the total phosphorus in several Canadian bean varieties. Varietal and agronomic factors, alone and in combination, often result in wide variation in phytate content of mature pulse seeds (Dintzis and Dintzis, 1992; Mason et al., 1993). Chen (2004) reported InsPn contents in beans with only InsP6 and InsP5 detected in all beans. There was a wide variation in the InsP6 or InsP5 content among different types of raw dry black beans or red kidney beans. InsP6 content (kg^{-1}, adjusted by moisture) in raw dry beans ranged from 5.87 mmol to 14.86 mmol in mung beans and black beans, respectively. InsP6 was the predominant inositol phosphate of the total InsPn determined in raw dry beans, ranging from 63.9% in red kidney beans to 97.5% in pinto beans.

28.5.2 Phenolic Compounds

Phenolic acids are derivatives of benzoic acid (gallic, syringic, and vanillic acid) or of cinnamic acid (caffeic and quinic acids). Ferulic, p-coumaric, caffeic, and vanillic acids have been quantified in different pulse species. The total phenolic acid content of common bean (*P. vulgaris* L.) has been found to be 30 mg 100 g^{-1}, with ferulic acid as the prevalent compound, followed by p-coumaric acid (Luthria and Pastor-Corrales, 2006; Carbonaro et al., 2001\. Phenolic acids are expected to have a positive effect on prevention of LDL cholesterol oxidation. They may also inhibit carcinogenesis in the breast and liver.

28.5.3 Flatulence Factors

Most legumes contain relatively high amounts of both dietary fiber and resistant starches. The soluble oligosaccharides found in legumes are not digestible by human intestinal enzymes alone. Instead, oligosaccharides, such as raffinose and stachyose, are broken down by bacterial fermentation in the intestines (Gulewicz et al., 2000). Although some rectal gas is due to the ingestion of air, the majority of flatulence is produced from bacterial fermentation (Carnovale et al., 2001). The byproducts of this degradation are hydrogen, carbon dioxide, methane, and sometimes sulfur, depending upon the bacteria. Normal intestinal processes move these gases out of the body in the form of flatus. Removal or alteration of the oligosaccharide content of legumes will reduce the amount of gas produced (Carbonaro et al., 2001). However, it is not clear if changing the oligosaccharide component will alter the health benefits of legumes.

28.5.4 Enzyme Inhibitors

An enzyme inhibitor is a molecule that binds to an enzyme and decreases its activity. By binding to enzymes' active sites, inhibitors reduce the compatibility of substrates and enzymes, leading to the inhibition of the formation of the enzyme substrate complex, which prevents the catalysis of reactions and decreasing (at times to zero) the amount of product produced by a reaction. It can be said that as the concentration of enzyme inhibitors increases, the rate of enzyme activity decreases, and thus, the amount of product produced is inversely proportional to the concentration of inhibitor molecules (Srinivasan, 2020). Legumes have natural components, such as lectins, amylase, and trypsin inhibitors that may adversely affect their nutritional properties.

Much information has already been obtained on their antinutritional significance and how to inactivate them through proper processing. Chronic ingestion of residual levels is unlikely to pose risks to human health. On the other hand, the ability of these molecules to inhibit some enzymes such as trypsin, chymotrypsin, disaccharides, and α-amylases, to selectively bind to glycol-conjugates and to enter the circulatory system, may be a useful tool in nutrition and pharmacology. Trypsin inhibitors have also been studied as cancer risk reducing factors (Srividya 2010). These components seem to act as plant defense substances. However, increased contents may represent an impairment of the nutritional quality of legumes because these glycoprotein and the sulfur-rich protease inhibitors have been shown to be poorly digested and to participate in chemical reactions during processing, reducing protein digestibility (Milner, 2004).

Elimination or reduction of antinutritional factors can be achieved by classical breeding, molecular biology techniques, and by several technological treatments, often used in combination. The most commonly employed post-harvesting treatments are: dry and moist thermal treatment, extrusion cooking, steaming, soaking, germination, fermentation, de-hulling, and enzymatic treatment (Champ, 2002). In particular, germination can be used to reduce the content of phytates, tannins, and protease inhibitors, while de-hulling is effective in decreasing the levels of tannins. A phytate-degrading enzyme has been isolated and characterized from faba bean seeds and used for phytate removal during processing (Greiner et al., 2001).

28.6 Conclusion

The key nutritional role of grain pulses is unquestionable, due to the massive presence of macro- and micronutrients. Pulses supply significant amounts of protein and calories for both rural and urban populations of developing and developed countries. These pulses contain up to 60% carbohydrates (mainly starch). Pulses are also a good source of major and minor (polyphenols, vitamins, minerals) compounds that may have important metabolic and/or physiological effects. More recent evidence provides potential information of their impact on health, so these secondary metabolites are currently marketed as functional foods and nutraceutical ingredients. In the frame of a reappraisal of the effects that grain legume components may have on human health and wellness, widely accepted claims on their beneficial activity in the prevention and treatment of various diseases have been made. Altogether, these claims strongly support the regular dietary intake of grain legumes as one of the ways to a healthy life. Nevertheless, many efforts and further studies are still needed to disclose the mechanism(s) underlying the legume proteins/peptides effects, to identify and characterize novel biological activities often "hidden" inside the polypeptide chains, and to establish clear dose-response relationships in order to calibrate the preparation and use of nutraceutically enhanced foods. Similar direction is needed for other bioactive compounds in pulses (phenolic acids, anthocyanins, vitamins, etc.). As parallel or further steps, the biotechnological approaches can be extremely useful as cognitive tools and in the design of novel, more effective biologically active molecules.

REFERENCES

ADA (American Dietetic Association). 2009. Position of the American Dietetic Association: Functional foods. *Journal of the American Dietetic Association.* 109 (4), 735–746.

Ajit S. 2013. Nutraceuticals—Critical supplement for building a healthy India. Ernst and Young, FICCI task force on nutraceuticals.

Aluko R. E., 2008. Determination of nutritional and bioactive properties of peptides in enzymatic pea, chickpea, and mung bean protein hydrolysates. *Journal of AOAC International.* 91, 947–956.

Anderson J. W. and Major A. W. 2002. Pulses and lipaemia, short- and long-term effect: Potential in the prevention of cardiovascular disease. *The British Journal of Nutrition.* 88, 5263–5271. doi: 10.1079/BJN2002716.

Barac M. B., Stanojevic S. P., Pesic M. B. 2005. Biologically active components of soybean and soy protein products—A review. *Acta Periodica Technologica.* 36, 155–168.

Barman A., Marak C. M., Barman R. M. and Sangma C. S. 2018. Nutraceutical properties of legume seeds and their impact on human health. *Legume Seed Nutraceutical Research.* http://dx.doi.org/10.5772/intechopen.78799.

Bazzano L. H. J., Ogden L. G., Loria C., Vupputuri S., Myers L., and Whelton P. K. 2001. Legume consumption and risk of coronary heart disease in US men and women: NHANES I Epidemiologic Follow-up Study. *Archives of Internal Medicine,* 161, 2573–2578.

Benveniste P. 1986. Sterol biosynthesis. *Annual Review Plant Physiolgy.* 37, 275–308. Copyright © 1986 by Annual Reviews Inc.

Borkar N., Saurabh S. S., Rathore K. S., Pandit A., Khandelwal K. R. 2015. An Insight on Nutraceuticals. *PharmaTutor.* 3 (8), 13–23. Available from: https://www.researchgate.net/publication/326301043_Nutraceuticals (Accessed on January, 11 2021)

Campos-Vega R., Loarca-Piña, G., and Oomah B. D. 2010. Minor components of pulses and their potential impact on human health. *Food Research International.* 43 (2), 461–482.

Carbonaro, M., Grant, G., Pusztai, A. 2001. Evaluation of polyphenol bioavailability in rat small intestine. *Europen Journal of Nutrition.* 40, 84–90.

Cardador-Martínez A., Loarca-Piña G., Oomah B. D. 2002. Antioxidant Activity in Common Beans *(Phaseolus vulgaris* L.). *Journal of Agricultural and Food Chemistry.* 50 (24), 6975–80.

Champ, M. M. 2002. Non-nutrient bioactive substances of pulses. *British Journal of Nutrition.* 88 (Suppl. 3), S307–S319.

Chen L., Hebrard G., Beyssac E., Denis S., Subirade M. 2010. *In vitro* study of the release properties of soy-zein protein microspheres with a dynamic artificial digestive system. *Journal of Agricultural and Food Chemistry.* 58, 9861–9867.

Crandall K. M. and Duren S. E. 2001. Nutraceuticals: what are they and do they work? In: Pagan, J. D. (Ed.), *Advances in Equine Nutrition II.* Versailles, KY: Kentucky Equine Research, 29–36.

Cuatrecasas P., Tell G. P. E. 1973. Insulin-like activity of concanavalin A and wheat germ agglutinin-direct interactions with insulin receptors. *Proceedings of the National Academy of Science.* 70, 485–489. doi: 10.1073/pnas.70.2.485.

Dintzis H. M. and Dintzis R. Z. 1992. Profound specific suppression by antigen of persistent IgM, *IgG, and IgE antibody production. Proceedings of the National Academy of Sciences of the United States of America.* 89 (3) 1113–1117.

D Dini , L Del Mastro, A Gozza, R Lionetto, O Garrone, G Forno, G Vidili, G Bertelli, M Venturini. 1898. The role of pneumatic compression in the treatment of postmastectomy lymphedema. A randomized phase III study. *Annual Oncology.* 9(2), 187–190. doi:10.1023/a:1008259505511

Duranti M., 2006. Grain legume proteins and nutraceutical properties. *Fitoterapia.* 77, 67–82.

Duncan A. M., Merz B. E., Xu X., Nagel T. C., Phipps W. R., Kurzer M. S. 1999. *Journal of Clinical Endocrinology and Metabolism.* 84 (1), 192–7.

Ekta K. Kalra. 2003. Nutraceutical—Definition and introduction. *AAPS PharmSciTeach*, 5 (3), 27–28.

Ewen S. W. B., Bardocz S., Grant G., Pryme I. F., Pusztai A. 1998. The effects of PHA and misletoe lectin binding to epithelium of rat and mouse gut. In: Bardocz, S., Pfuller, U., Pusztai, A. (Eds.), *COST 98— Effects of Antinutrients in the Nutritional Value of Legume Diets*, Vol. 5. Office Official Publications of the European Commission, Luxembourg, 221–225.

Flight I. and Clifton P. 2006. Cereal grains and legumes in the prevention of coronary heart disease and stroke: A review of the literature. *European Journal of Clinical Nutrition.* 60, 1145–1159.

Food and Drug Administration, 1999. *Federal Register.* 64 (206), 57699–57733.

Fu L. and Renxiao W. 2010. Hemolytic mechanism of dioscin by molecular dynamics simulations. *Journal of Molecular Modeling.* 16, 107–118.

Fukui H., Egawa H., Koshimizu K. and Mitsui T. (1973). A new isoflavone with antifungal activity from immature fruits of Lupinus luteus. *Agricultral Biology Chemistry.* 37, 417–421.

Greiner R., Muzquiz M., Burbano C., Cuadrado C., Pedrosa M. M., Goyaga C., 2001. Purification and characterization of a phytate-degrading enzyme from germinated faba bean (*Vicia faba* var. Almeda). *Journal of Agricultural and Food Chemistry.* 49, 2234–2240.

Gulewicz P., Ciesiolka D., Frias J. 2000. Simple method of isolation and purification of alpha-galactosides from legumes. *Journal of Agricultural and Food Chemistry.* 48, 3120–3123.

Kabir S. M. S. 2016, Ethical Approaches in Research. *In: Basic Guidelines for Research: An Introductory Approach for All Disciplines*, 1st ed. Bangladesh: Book Zone Publication.

Lena Ga´lvez Ranilla, Maria Ine´ s Genovese, Franco Maria Lajolo (2009) Isoflavones and antioxidant capacity of Peruvian and Brazilian lupin cultivars. *Journal of Food Composition and Analysis.* 22 397–404

Lajolo F. M., Mancini Filho J., and Menezes E. W. 1984. Effect of a bean (Phaseolus vulgaris) a-amylase inhibitor on starch utilization. *Nutrition Reports International*, 30, 45–54.

Luthria D. L. and Pastor-Corrales M. A., 2006. Phenolic acid content of fifteen dry edible beans (*Phaseolus vulgaris* L.) varieties. *Journal of Food Composition and Analysis.* 19, 205–211.

MacCarty M. F., 2005. Nutraceutical resources for diabetes prevention. *Medical Hypotheses.* 64, 151–158.

Mason W. L., McConell G., Hargreaves M. 1993. Carbohydrate ingestion during exercise: Liquid vs solid feedings. *Medicine and Science in Sports Exercise.* 25 (8), 966–9.

Mathers J. C. 2002. Pulses and carcinogenesis: Potential for prevention of colon, breast and other cancers. *British Journal of Nutrition*, 88, 272–279.

Mazur, W. M., Duke, J. A., Wahala, K., Rasku, S. and Adlercreutz, H. (1998). Isoflavones and lignans in legumes: Nutritional and health aspects in human. *Journal of Nutritional Biochemistry.* 9:193–200.

Milner J. A., 2004. Molecular targets for bioactive food components. *Journal of Nutrition.* 134, 2492S–2498S.

Morris B. 2003. Bio-functional legumes with nutraceutical, pharmaceutical and industrial uses. *Economic Botany.* 57, 254–261. doi: 10.1017/S1479262113000397

Murphy P. A. and Hendrich S. 2002. Phytoestrogens in foods. *Advances in Food and Nutrition Research.* 44, 195–246.

Nachbar M. S. and Oppenheim J. D. 1980. Lectins in the United States diet. A survey of lectins in commonly consumed foods and a review of the literature. *American Journal of Clinical Nutrition.* 33, 2338–2345.

Navarro D. M. D. L, Bruininx E. M. A. M., de Jong L, Stein H. H. 2018. Analysis for low-molecular-weight carbohydrates is needed to account for all energy-contributing nutrients in some feed ingredients, but physical characteristics do not predict in vitro digestibility of dry matter. *Journal of Animal Science.* 96 (2), 532–44. https://doi.org/10.1093/jas/sky010.

Nazri H., Baradaran A., Shirzad H, Kopaei M. R. 2014. New concepts in nutraceuticals as alternative for pharmaceuticals. *International Journal of Preventive Medicine.* 5 (12), 1487–1499.

Obiro W. C., Zhang T., Jiang B. 2008. The nutraceutical role of *Phaseolus vulgaris* alpha-amylase inhibitors. *British Journal of Nutrition.* 100, 1–12.

Orcheson, L. J., Rickard, S. E., Seidl, M. M. & Thompson, L. U. 1998 Flaxseedand its mammalian lignan precursor cause lengthening or cessation of estrous cycling in rats. *Cancer Letter.* 125: 69–76.

Oomah BD. 2001. Flaxseed as a functional food source. *Journal of Science Food Agric.* 81, 889–894

Pieta P. 2000. Flavonoids as antioxidants. *Journal of Natural Products.* 63, 1037–1042.

Pulses website. (n.d.) Future of Food—Pulses. Available from: https://pulses.org/future-of-food/pulses-your-health (Accessed on 2021).

Punia S., Dhull S. B., Sandhu K. S., and Kaur M. 2019. Faba bean (Vicia faba) starch: Structure, properties, and in vitro digestibility—A review. *Legume Science.* 1 (1), e18.

Punia S., Sandhu K. S. and Kaur M. 2020a. Quantification of phenolic acids and antioxidant potential of wheat rusks as influenced by partial replacement with barley flour. *Journal of Food Science and Technology.* 57 (10), 3782–3791.

Punia S., Dhull S. B., Sandhu K. S., Kaur M. and Purewal S. S. 2020b. Kidney bean (Phaseolus vulgaris) starch: A review. *Legume Science.* 2 (3), e52.

Pusztai A., Grant G., Bardocz S., Bainter K., Gelencser E., Ewen S. W. B. 1997. Both free and complexed trypsin inhibitors stimulate pancreatic secretion and change duodenal enzyme levels. *American Journal of Physiology.* 272, G340–G350.

Rajasekaran A, Sivagnanam G, Xavier R. 2008. Nutraceuticals as therapeutic agents: A review. *Research Journal of Pharmacy and Technology.* 1 (4), 328–340.

Rochfort S. and Panozzo J. 2007. Phytochemicals for health, the role of pulses. *Journal of Agricultural and Food Chemistry.* 55, 7981–7994.

Rizkalla S. W., Bellisle F., and Slama G. 2002. Health benefits of low glycaemic index foods, such as pulses, in diabetic patients and healthy individuals. *British Journal of Nutrition.* 88, Suppl. 3, S255–S262.

Sánchez-Vioque R., Clemente A., Vioque J., Bautista J. and Millán F. 1998. Neutral lipids of chickpea flour and protein isolates. *Journal of the American Oil Chemists' Society.* 75, 851.

Sandberg A-S. 2003. Bio-availability of minerals in legumes. *British Journal of Nutrition.* 88, Suppl 3(S3), S281–5.

Santini A., Cammarata S. M., Capone G., Ianaro A., Tenore G. C., Pani L., and Novellino E. 2018. Nutraceuticals: Opening the debate for a regulatory framework. *British Journal of Clinical Pharmacology.* 84 (4), 659–672.

Sarma B. K., Singh D. P., Mehta S., H. B. Singh. 2002. Plant growth-promoting rhizobacteria-elicited alterations in phenolic profile of Chickpea (*Cicer arietinum*) infected by *Sclerotium rolfsii. Journal of Phytopathology.* 150, 277–282

Smulikowska S., Pastuszewska B., Swiech E., Ochtabinska A., Mieczkowska A., Nguyen V. C. and Buraczewska L. 2001. Tannin content affects negatively nutritive value of pea for monogastrics. *Journal of Animal and Feed Sciences.* 10 (3), 511–523.

Shinde N, Bangar B, Deshmukh S, Kumbhar P. 2014. Nutraceuticals: A Review on current status. *Research Journal of Pharmacy and Technology.* 7 (1).

Smarta R. B. 2012. Regulatory perspective of nutraceuticals in India. Interlink Marketing Consultancy Pvt. Ltd. Report; 1–12. (www.interlinkconsultancy.com)

Srinivasan B. 2020. Words of advice: Teaching enzyme kinetics. *The FEBS Journal.* 288 (7). doi:10.1111/febs.15537

Srividya A. R., Venkatesh N., Vishnuvarthan V. J. 2010. *Nutraceutical as medicine. International of Advances in Pharmacutical Sciences.* 1(2), 133–145.

Tharanathan R. N and Mahadevamma S. 2003. Grain legumes—A boon to human nutrition. *Trends in Food Science & Technology.* 12 (14).

Trinidad T. P., Mallillin A. C., Loyola A. S., Sagum R. S., Encabo R. R., 2010. The potential health benefits of legumes as a good source of dietary fibre. *British Journal of Nutrition.* 103, 569–574.

Van Loo J., Cummings J., Delzenne N., 1995. Functional food properties of non-digestible oligosaccharides: A consensus report from the ENDO project (DGXII AIRII-CT94-1095). *British Journal of Nutrition.* 81, 121–132.

Witold M. Mazur, James A. Duke, Kristiina Wahala, Sirpa Rasku, and Herman Adlercreutz 1998. Isoflavonoids and lignans in legumes: Nutritional and health aspects in humans. *Nutritional Biochemistry.* 9:193–200, Elsevier Science Inc. Avenue of the Americas, New York, NY 10010.

Xu B. J. and Chang S. K. C. 2007. A comparative study on phenolic profiles and antioxidant activities of legumes as affected by extraction solvents. *Journal of Food Science.* 72 (2), S159–66.

Yeh C. T. and Yen, G. C. 2003. Effects of phenolic acids on human phenolsulfotransferases in relation to their antioxidant activity. *Journal of Agricultural and Food Chemistry*, 51, 1474–1479.

Yates M. 2016. Nutraceuticals in India: A Challenging Opportunity. Reuters Events website. Available from: https://www.reutersevents.com/pharma/column/nutraceuticals-india-challenging-opportunity (acessed on nov 12,2020).

Salinas-Moreno Y., Rojas-Herrera L., Sosa-Montes E. and Pérez-Herrera P. 2005. Anthocyanin composition in black bean (Phaseolus vulgaris L.) varieties grown in México. *Agrociencia.* 39 (4), 385–394.

USDA. (2002). Iowa State University Isoflavones Database. United States Department of Agriculture [Web page]. March 15, 2002. Available at: http://www.nal.usda.gov/fnic/foodcomp/Data/isoflav/isoflav.html. [Google Scholar]

Zia-Ul-Haqq M., Iqbal S., Ahmad S., Imran M., Niaz A. and Bhanger M. I. 2007. Nutritional and compositional study of Desi chickpea (Cicer arietinum L.) cultivars grown in Punjab, Pakistan. *Food Chemistry*, 105, 1357–1363.

Section III

Oilseed Crops

29

Sesame: An Emerging Functional Food

Sneh Punia
Chaudhary Devi Lal University, Sirsa, India
Clemson University, Clemson, SC, USA

Anil Kumar Siroha
Chaudhary Devi Lal University, Sirsa, Haryana, India

Manoj Kumar
*ICAR - Central Institute for Research on Cotton Technology,
Mumbai, India*

CONTENTS

29.1 Introduction

Sesame (*Sesamum indicum L.*), also known as beniseed, belongs to order *Tubiflorae* and family *Pedaliaceae*. Amongst 37 species of sesamum, *S. indicum* is the dominant species and called the queen of oil seeds. The total worldwide production of sesame in 2018 was 6,015,573 tons and occupies 14% of the total world export of the crop. It is cultivated mainly in India, Sudan, China, and Burma. India ranks third in production and accounts for 42% of world's total sesame area. In India, sesame occupies an area of nearly 1,730,000 ha with total production of 746,000 tons (FAO, 2018). Sesame is an erect branched annual plant and generally known as *sesame* in English, *tila* in Ayurveda, and *til* and *kunjad* in Unani.

Sesame is used as a condiment and is grown primarily for its oil and protein. Sesame is a nutritionally important source of oil (50–60%), protein (18–25%), carbohydrates and ash (El Khier et al., 2008) and also contains important minerals (calcium, phosphorus, and iron) and vitamins (niacin and

TABLE 29.1

Nutritional Composition of Sesame Seeds

Components	Availability	References
Oil	50–60%	El Khier et al. (2008)
Unsaturated		
Linolic acid	42.7%	Gharby et al. (2017)
Oleic acid	41.9%	Gharby et al. (2017)
Saturated		
Palmitic acid	8.47%	Crews et al. (2006)
Stearic acid	12.9%	Crews et al. (2006)
Protein	18–25%	El Khier et al. (2008)
α-globulin	80%	Johnson et al. (1979)
β-globulin	20%	Johnson et al. (1979)
Ash	10.46%	Sharma and Singh (2016)
Moisture	6.94%	Sharma and Singh (2016)
Fat	0.66%	Sharma and Singh (2016)
Crude fiber	2.31%	Sharma and Singh (2016)

thiamine) (Ojiako et al., 2010). The sesame oil comprises unsaturated fatty acids that contain glycerides of oleic acid and linoleic acid. Other components are saturated fatty acids (myristic acid, palmitic acid, stearic acid, and arachidonic acid). The unsaponifiable matter includes tocopherols and the lignans (sesamin, sesamolin, sesamol, and sesaminol), which make sesame oil stable toward oxidation, enhanced shelf-life, and improved flavor and taste of foods. Sesame oil is used for preparing margarine, for cooking purposes, and for preparing tehina and Halva (Sankar et al., 2005). Factors such as climate, cultivars, soil type and maturation of the plant affects the quantity and quality of the sesame oil. Sesame seeds contain 35–40% protein, which is a byproduct after oil extraction by alkaline or salt extraction and isoelectric precipitation methods (Onsaard et al., 2010). The sesame protein is heat stable and rich in methionine and tryptophan, therefore, may be incorporated into low methionine-protein meal to form a balance. Both methionine and tryptophan are reported to be essential amino acids and possess considerable properties to become valuable for food applications. Hemalatha and Prasad (2003) reported that some seed proteins are metabolically active while others are inactive. The protein isolates, which are an excellent source of protein and contain a low amount of antinutrients, are used in various food applications. As reported by Orruño and Morgan (2005), depending on the basis of their solubility, the majority of sesame seed proteins are storage proteins and comprise globulins (67.3%), albumins (8.9%), glutelins (6.9%), and prolamins (1.3%). The nutritional composition of sesame are presented in Table 29.1.

29.2 Sesame Oil

The role of dietary fats and oils in human nutrition is one of the most complex areas to examine nutrition science. Dietary fats and oils can help promoting cardiovascular disease, diabetes, obesity etc. Due to superior oil quality, high smoke point, and greater stability than other vegetable oils, sesame oil is reported to have several health benefits. Sesame oil is good source of oleic acid (monounsaturated) and linoleic acid (polyunsaturated) accounting more that 80% of the total fatty acids. Palmitoleic, stearic, linolenic and arachidic acids were reported to be minor fatty acid components present in sesame seed oil (Gharby et al., 2017) and exhibit saturated fatty acid (20%) mainly composed of palmitic (8.47%) and stearic acids (12.9%) (Crews et al., 2006). Although sesame oil contains 85% unsaturated fatty acids, it possesses a pleasant odor and good taste and is resistant to oxidative rancidity. Uzun et al. (2002)

Raw sesame seed
↓
Cleaning and drying
↓
Roasting
↓
Crushing (in rotary machines and expellers)
↓

Raw sesame oil   Sesame cake
↓
Solvent extraction
↓
Filtration
↓
Distillation
↓
Extracted oil

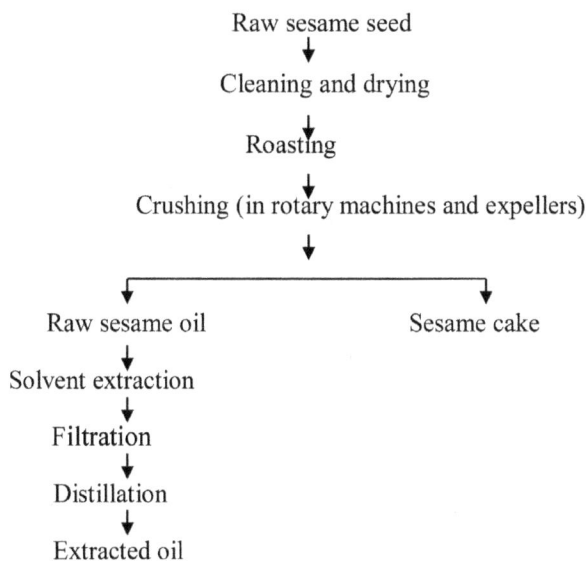

FIGURE 29.1 Sesame oil extraction.

reported that sesame oil contains oleic acid (47%), linoleic acid (39%), palmitic acid (9.0%), stearic acid (4.1%), and arachidic acid (0.7%).

For food and pharmaceuticals purposes, the industrialization of sesame oil is important. For an extraction process, the quality of the final product and environmental prospects are the desirable considerations (Figure 29.1). The extraction of sesame oil includes milling, then extraction using solvents, and recovery of the solvents by the distillation process. After cleaning, the seeds are pressed in expellers and rotary machines and the crude sesame oil is further subjected to a solvent extraction process with a suitable organic solvent (i.e., hexane, haptane isopropanol, etc.). Corso et al. (2010) reported an alternative procedure called the super critical extraction process, which was capable of minimizing organic solvents. This process generally uses carbon dioxide (temperature of 313–333K with 19–25 MPa pressure) and propane solvent (temperature of 303–333K with 8–2 MPa pressure). Hamada et al. (2009) reported that as the pressure and flow rate of CO_2 is increased, the yield of extracted oil is also increased, whereas time for extraction is reduced. Sesame seed yields approximately 50% oil with 32.3% lignan content (Prasad et al., 2012).

Sesame offers a combined source of omega-3, omega-6, and omega-9, however, omega-3 (α-linolenic, ALA) and omega-6 (linoleic acid, LA) are essential fatty acids, whereas omega-9 (oleic acid) is a nonessential fatty acid. Interestingly, humans are capable of synthesizing omega-9, but the mechanism of metabolism is not understood yet. Fundamental polyunsaturated fatty acids (PUFAs) are ALA (omega-3) and LA (omega-6), which further form the long chain PUFAs. The delta (Δ)-6 destruase enzyme is used for metabolism of ALA as well as LA, therefore, conversion of ALA acid to EPA competes with the conversion of LA to AA. Sesame oil is source of linoleic (LA, omega-6) and linolenic (ALA, omega-3) acids (Park et al., 2010). LA is the simplest n-6 fatty acid of biologically active fatty acids and may be converted to DPA which are more biologically active (Figure 29.2). During conversion of LA to DPA, arachidonic acid (AA) is also formed. AA through the steps outlined in the synthesis of eicosanoid. Yui et al. (2015) reported that AA is an abundant PUFA of the membrane phospholipids and responsible for synthesis of eicosanoids. Eicosanoids generally regulate brain functions as well as immune and inflammatory responses (Hallahan & Garland, 2005).

ALA (omega-3) is biologically active fatty acids and converted to eicosapentonic acid (EPA) and Dicosahexonic acid (DHA) (Figure 29.3). ALA, as triglycerides is taken from diet and digested in small intestine and then absorbed into the bloodstream and transported to body tissues (Kaur et al., 2014). Further they undergo for esterification, oxidation and to form longer chain products (Figure 29.3).

LA

↓ desaturase

gamma-linoleic acid

↓ elongase

Dihomo-gamma-linoleic acid (DGLA)

↓ desaturase

Arachidonic acid (AA)

↓ elongase

Docosatetraenoic (DTA)

↓ elongase

Tetracosatetraoenoic (TTA)

↓ desaturase

Tetracosapentaenoic (TPA)

↓ oxidation

Docosapentaenoic (DPA)

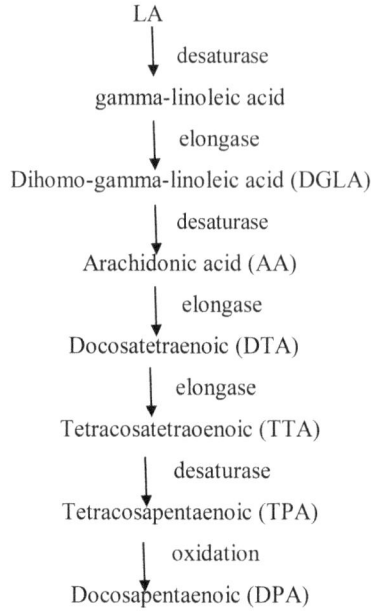

FIGURE 29.2   Metabolism of linoleic acid (omega-6).

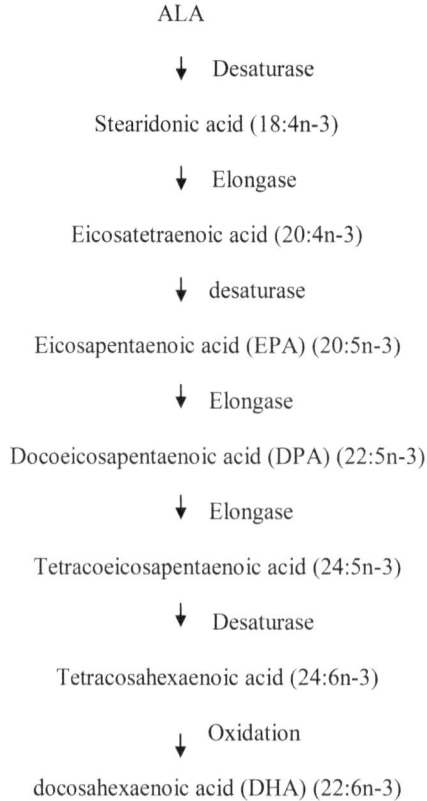

ALA

↓ Desaturase

Stearidonic acid (18:4n-3)

↓ Elongase

Eicosatetraenoic acid (20:4n-3)

↓ desaturase

Eicosapentaenoic acid (EPA) (20:5n-3)

↓ Elongase

Docoeicosapentaenoic acid (DPA) (22:5n-3)

↓ Elongase

Tetracoeicosapentaenoic acid (24:5n-3)

↓ Desaturase

Tetracosahexaenoic acid (24:6n-3)

↓ Oxidation

docosahexaenoic acid (DHA) (22:6n-3)

FIGURE 29.3   Metabolism of ALA fatty acids into EPA and DHA.

Mammals are not capable enough to synthesize longer chain ALA, but get shorter chain ALA to form eicosapentaenoic acid (EPA) and then docosahexaenoic acid (DHA). As reported by Domenichiello et al. (2015), competition between both n-6 and n-3 fatty acids for desaturase makes conversion of ALA to EPA and DHA inefficient in humans, but the DHA conversion is sufficient for brain functioning. First, ALA is converted into stearidonic acid by Δ6-desaturase, followed by eicosatetraenoic acid. Further desaturation by Δ5-desaturase yields EPA and docosapentaenoic acid (DPA). According to Sprecher (2000), through the elongation and desaturation of DPA into tetracosapentaenoic and tetracosahexenoic acids, finally DHA by peroxisomal oxidation is synthesized.

29.3 Functional Components

The stability of sesame oil to oxidation is reported to be high because of presence of phenolic compounds and their antioxidant potential (Pathak et al., 2014; Sharma et al., 2021). Phenolics like phenolic acid, sesamol, and chlorine containing naphthoquinone occur in sesame in small amounts (Lyon, 1972; Shimoda et al., 1997; Hasan et al., 2000). Sesamol, sesamolin, and gamma tocopherol present in sesame oil gives its high oxidative stability and many health benefits (Rangkadilok et al., 2010; Kochhar, 2002). The associated health benefits of sesame meal and sesame oil consumption i.e. anti-inflammatory, hypo-cholesterolemic, reduction of cardiovascular disease and oxidative stress-related diseases, and antimutagenic effects have been reported (Gouveia et al., 2016; Lazarou et al., 2007; Chen et al., 2005). A study conducted by Sani et al. (2013) reported flavonoids from *Sesamum indicum* seed oil are responsible for its antioxidant property and also inhibit the replication of human colon cancer cells.

29.3.1 Lignans

Lignans present in sesame make sesame oil more stable to autoxidation than other vegetable oils (Gertz et al., 2000) and contribute to its long shelf life (Chung, 2004; Suja et al., 2004). Lignans are formed by the coupling of two molecules of p-hydroxyphenyl propane by a bond between β-positions in the propane side chains. Currently, 16 types of lignans are isolated from sesame that are further categorized into glycosylated water-soluble lignans and oil-soluble lignans. Sesamin, sesamolin, sesamolinol, pinoresinol, and sesaminol are the main oil-soluble lignans present in sesame (Pathak et al., 2014). The major glycosylated lignans are sesaminol triglucoside, pinoresinol tri-glucoside, sesaminol monoglucoside, pinoresinol monoglucoside, two isomers of pinoresinol diglucoside, and sesaminol diglucoside (Hemalatha, 2004). Sesamin, episesamin, sesaminol, and sesamolin are major lignans present in sesame (Gokbulut, 2010).

By inhibiting absorption and cholesterol synthesis, these components help lower blood lipids, arachidonic acid levels, and cholesterol levels. Moazzami et al. (2007) reported the highest amount of sesamin (167–804 mg/100 g) followed by sesamolin (48–279 mg/100 g), sesaminol (32–298 mg/100 g), and sesamolinol (58 mg/100 g) in sesame seed. Liu et al. (2006) evaluated 1,520 μmol of sesamin in the total lignan content of 2,180 μmol/100 g. The amount of sesamin and sesamolin in Indian sesame were observed higher when compared with 65 cultivars of sesame from Texas, US., (1.63 mg/g for sesamin and 1.01 mg/g for sesamolin) (Moazzami & Kamal-Eldin, 2006) and 403 cultivars of sesame seeds of Korea (2.09 mg/g for sesamin and 1.65 mg/g for sesamolin) (Kim et al., 2006). Ide et al. (2015) reported lignan consumption resulted in physiological activity to promote health.

In addition to sesamin (a major lignan), sesamolin, sesaminol, and sesamolinol, have been reported in sesame (Namiki, 1995). Metabolism of sesame lignans in the intestine depends on the individual intestinal microflora (Figure 29.4). As explained by Tomimori et al. (2017), after oral intake of sesame lignan, sesamin is metabolised to enterodiol (ED) and enterolactone (EL) by the action of intestinal microflora and transferred to body tissues and fluids. During fermentation in the colon, enterolactone is the major end-product and is reported to be inversely correlated with the risk for developing various chronic diseases. Further, they are absorbed up to 90% followed by metabolization in the liver and then excretion from the liver and kidneys.

Sesame lignan
↓
Small intestine ⟶ Liver
↓ Microflora
Colon
↓
Feces

→ Tissues
→ Kidney
↓
Urine

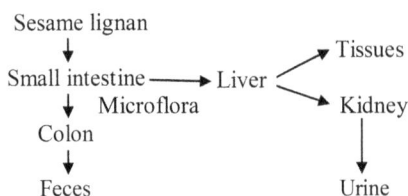

FIGURE 29.4 Metabolism of lignin.

29.3.2 Tocopherols

Apart from other compounds, sesame seed and oil also contain tocopherol, especially γ-tocopherol, (Williamson et al., 2008) and possesses health benefits. Generally, tocopherols are lipophilic plant phenolics and possess strong antioxidant activity and nutritional potential (Brigelius-Flohé et al., 2002). Besides being a source of natural antioxidants, tocopherols prevent oxidation of oil, act as scavengers of free radicals, and help prevent lung and oral cancer, Alzheimer's disease, and diseases related to the nervous system (Pasias et al., 2018). Haji et al. (2008) reported γ-tocopherol content of 563–1,095 mg/kg and 293–569 mg/kg in sesame oil and seed, respectively. Pathak et al. (2014) reported that γ-tocopherol is the major tocopherol of sesame with lesser amounts of α- and δ- tocopherols. Williamson et al. (2008) evaluated α-, δ-, and γ-tocopherol in 11 genotypes of sesame seeds with values ranging between 0.034–0.175 μg/g, 0.44–3.05 μg/g, and 56.9–99.3 μg/g, respectively. Υ-tocopherol is reported to be a free-radical remover among the isomers of vitamin E and possesses unique properties that are beneficial in sustaining human health and disease prevention. They are believed to prevent many diseases such as heart disease and cancer. In addition, they have strong anti-inflammatory activity, anticarcinogenic effect (Ju et al., 2010), and modulate the expression of proteins involved in cholesterol metabolism (Wallert et al., 2014). Both α- and δ-tocopherols help reduce platelet aggregation, low density lipid oxidation, delay in thrombus formation, and reduce photo carcinogenesis (Balan et al., 2009). A study conducted by Cooney et al. (2001) reported that dietary intake of sesame seeds in humans has been associated with increasing plasma gamma tocopherol.

The mechanism of absorption of tocols is shown in Figure 29.5. In the human diet, tocols are generally in the form of dl-α-tocopheryl acetate (Jakobsen et al., 1995) and metabolized into free alcohols, or with fat and thereafter absorption takes place (Gallo-Torres, 1970). Borel et al. (2001) reported that lipid-rich foods, bile acids, and pancreatic enzymes are necessary for the absorption process. They further concluded that tocols are firstly hydrolyzed with gastric lipase in the stomach and are thereafter digested by pancreatic and other digestive enzymes secreted in the intestine (Burton et al., 1988). Absorption of

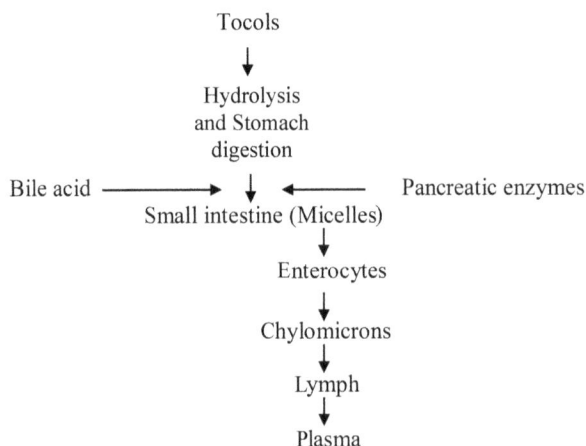

Tocols
↓
Hydrolysis
and Stomach
digestion
Bile acid ⟶ ↓ ⟵ Pancreatic enzymes
Small intestine (Micelles)
↓
Enterocytes
↓
Chylomicrons
↓
Lymph
↓
Plasma

FIGURE 29.5 Metabolism of tocols.

tocols mainly occurs at the middle and upper parts of the intestinal tract. This process is accelerated by triglycerides,whereas it is inhibited by PUFA (Traber et al., 1985). Further, micelles are transferred into enterocyte from where they reach into lymph within chylomicrons and then into plasma where they are processed by lipoprotein lipase (Bjorneboe et al., 1986). It is reported that α- and β-tocopherol are transferred directly to tissues through chylomicron and mediated by lipoprotein lipase (LPL) (Traber et al., 1985). As a result of LPL activity, tocols and lipids are circulated into adipose and muscular tissue and, to some extent, the brain also. It is believed that the biological activities of tocols from its antioxidant action, and α-tocopherol is reported to have the highest biological activity and antioxidant activity of other tocols.

29.3.3 Phenolic Acids

Sesame seeds contain several phenolic compounds (ferulic, vanillic, cinnamic, and p-coumaric acids, among others) (Mohdaly et al., 2013; Othman et al., 2015), which have been reported to display important biological properties. Hassan (2012) reported that sesame contains a significant amount of catechin, chlorogenic, ellagic, and benzoic acids.

During consumption of food, phenolic compounds are present either in glycosidic form or as aglycones associated with a number of sugars. Glucose is found to be a frequently used sugar, whereas fructose is rarely used (Di Carlo et al., 1999). During absorption of phenols, sugar molecules are firstly removed by the action of β-glucosidase enzyme and lactase phlorizin hydrolase (LPH) enzyme followed by absorption in the duodenum. For absorption, β-glucosidase and LPH enzymes release aglycone into the lumen by a diffusion mechanism. Phenolic compounds, which are associated with rhamnose, are hydrolyzed by ramnosidases (Manach et al., 2004). Hydroxycinnamic acids (HCAs) are common in the human diet, but their potential health benefits depend on their bioavailability. Generally, HCAs are esterified to organic acids, sugars, and fats and HCAs metabolism is done by the large intestine microbiome, although most of the absorption has been completed in the small intestine (almost one-third) (Clifford, 2000; Erk et al., 2014). After the absorption process, HCA undergoes some degree of phase II metabolism at the enterocyte level, enters the blood stream by the portal vein, then enters the liver where it goes through to more phase II metabolism, undergoes conjugation, and is transported again to bloodstream until it is secreted in urine (Stalmach et al., 2010). Landete (2011) reported that ellagitannins are resistant to the action of LPH and may not be absorbed in the small intestine. LPH reaches the large intestine and is cleaved and metabolized by microbiota into many hydroxyphenyl acetic acids. A fraction of conjugated products are transported back into the intestinal tract where they converted into deconjugated compounds by gut microbial enzymes before being reabsorbed again (Cardona et al., 2013). The unabsorbed metabolites are excreted through feces.

29.4 Antinutritional Factors

Antinutritional factors (ANFs) are present naturally in edible seeds, and when consumed, reduce the absorption of nutrients i.e. proteins, vitamins, and minerals in the digestive tract (Soetan & Oyewole, 2009). ANFs have an impact on the gastrointestinal tract, affect the metabolic system, cause many health-related problems in humans, and also limit the use of sesame seeds in food production. Phytate, phenolic compounds, lectin, enzyme inhibitors, saponins, and oxalates are the major antinutritional compounds present in sesame seeds. Sesame seed is normally considered to be free from antinutritional constituents and thus, used in a wide variety of foods. However, the seed coat contains oxalates (3–5%) and phytates (2.25–3.5%) that form insoluble complexes with minerals (calcium, magnesium, zinc, and iron) and ultimately interfere with their utilization (Adegunwa et al., 2012).

29.4.1 Phytates

The presence of high amounts of phytic and oxalic acids present in sesame hinders its utilization in food products (Konietzny & Greiner, 2003). Phytic acid is an important source of phosphorus and has a strong chelating ability to form protein complexes, thereby contributing to antinutritional effects (Urbano et al., 2000). Besides antinutritional properties, dietary phytates have some health benefits i.e. reducing cholesterol levels, helping to prevent cancer (Steer & Gibson, 2002), and reducing lipid peroxidation (Serna-Saldivar et al., 2015). In 1975, de Boland et al. (1975) compared sesame and soy beans for their phytate content and concluded that sesame seeds are reported to be a rich source of phytates (5.18%) as compared to soy bean meal and soy bean protein (1% and 1.5%), respectively. Toma et al. (1979) evaluated the phytin content with values of 4.7%, 5.2%, 4.7%, and 5.1%, respectively in whole, de-husked, roasted de-husked, and roasted sesame seeds. Makinde and Akinoso (2014) reported phytate in de-husked white cultivar and black cultivar with values of 30 mg/100 g and 25.07 mg/100 g. Phytates reduce mineral bioavailability and impairs protein digestibility by forming phytic-protein complexes and depresses absorption of nutrients because of damage of intestinal pyloric caeca region (Francis et al., 2001).

29.4.2 Oxalic Acid

Sesame seed contains 2.2% oxalic acid (Manikantan et al., 2015). Oxalic acid salts in calcium and magnesium are insoluble crystals and are not absorbed in the human body (Leeson et al., 2001). Instead, these salts are excreted into urine (by the kidneys) where they are combined with calcium to form an insoluble salt that leads to kidney stones (Massey et al., 2001). To overcome the drawbacks of oxalates of sesame, decortication of the seed helps in removing the oxalate content.

29.4.3 Tannins

Tannins, a phenolic derivative of flavones, form complexes with protein molecules and reduce the bioavailability of amino acids in the human body (Lampart-Szczapa et al., 2003). Tannins are the main ANF present in sesame (*Sesamum indicum*) meal (Mukhopadhyay & Bandyopadhyay, 2003). The tannin content of sesame seed is reduced with processing time. Jimoh et al. (2011) reported that moist heat treatment is the appropriate method to reduce the tannin content is sesame.

29.5 Sesame as an Ingredient

Sesame is gaining interest to become a functional foods for human nutrition and its utilization in food products because of its high PUFAs, protein, and bioactive compounds. The utilization of sesame in food products is shown in Table 29.2. Flaxseed products are stable for a long time at ambient temperature despite of generous amount of linolic acid.

Nonculinary applications of sesame include their utilization in soap, cosmetics, lubricants, and medicines as an ingredient. It is very convenient to use sesame as tahini (sesame paste) and halvah (sweet sesame paste) as it provides a nutty and crunchy taste. Decorticated sesame seeds supply functional components at a low cost. Sesame is also being incorporated in edible films to improve their transparency and tensile strength (Rodrigues et al., 2018; Sharma & Singh, 2016). Sesame, as an excellent source of protein, may be utilized in imitation dairy goods used for lactose intolerant infants and adults, vegetarians, and for those who want cholesterol-free dairy products. Additionally, sesame-based dairy products may overcome the problems of antinutrients and flatulence associated with soy-based dairy goods (Quasem et al., 2009). Similarly, bread with black sesame flour was substituted with wheat flour at 0%, 5%, 10%, 15%, and 20%, and these bread were reported to nutritionally superior in protein, fat, and crude fiber than wheat bread (Makinde & Akinoso, 2014).

29.6 Conclusion

During the recent decades, awareness toward the role of functional compounds in human health and disease prevention has been unremittingly increasing. Sesame seed is a reservoir of nutritional components with numerous beneficial effects along with health promotion in humans. Sesame seeds are a

TABLE 29.2

Sesame as a Functional Ingredient in Various Food Products

Sesame as an Ingredient	Processing Method	Products	Features	References
Sesame oil + tamarind kernel xyloglucans	Sonication, high shear	Film	Film has lower water vapor permeability and good tensile properties	Rodrigues et al. (2018)
Sesame meal	Microwave	Edible films	Higher transparency	Sharma and Singh (2016)
Sesame cake + cassava starch	Thermal expansion process	Baked foams	Starch-based foams incorporated with sesame cake can be an alternative for packing dry foods and foods with low moisture content, reducing the EPS use	Machado et al. (2017)
Tahini (sesame paste)	Milling	Sesame paste	Serves as a functional food for treating type 2 diabetes	Bahadoran et al. (2015)
Sesame + wheat flour	Baking	Bread	Nutritionally superior in protein, fat, and crude fiber than wheat bread	Makinde and Akinoso (2014)
Sesame cake + corn grits	Extrusion	Extrudates	An alternate way to improve nutritional and sensory characteristics	Carvalho et al. (2012)
Sesame + wheat flour	Baking	Biscuits	Increased protein content over the traditional product	Gandhi and Taimini (2009)
Sesame	Boiling, pressure cooking, and steaming	Sesame milk	Imitation of dairy goods used for lactose-intolerant infants and adults	Quasem et al (2009)
Sesame	Milling	Paste	Sesame paste is shelf stable with respect to chemical deteriorative reactions	Çiftçi et al. (2008)
Sesame	Roasting	Sesame paste	An excellent source of dietary fiber and could be used as an ingredient for functional foods	Elleuch et al. (2007)

rich source of protein, oil, biologically active, and health-promoting bioactive compounds such as sesamin, sesamolin, tocopherols, PUFAs, phytosterols, phytates, etc., which may be used as a supplemental functional food formulations that are effective in reducing the risk of cardiovascular disease. Sesame is a combined source of essential fatty acids (omega-3 [ALA] and omega-6 [LA]). The presence of phenylpropanoid compounds, namely lignans along with tocopherols and phytosterols, provide defense mechanisms against reactive oxygen species and increases the shelf life of oil by preventing oxidative rancidity. The superior functional components of sesame will help to strongly promote the use of sesame seeds in a daily diet world-wide.

REFERENCES

Adegunwa, M. O., Adebowale, A. A., & Solano, E. O. (2012). Effect of thermal processing on the biochemical composition, antinutritional factors and functional properties of beniseed (*Sesamum indicum*) flour. *American Journal of Biochemistry and Molecular Biology,* 2:175–182.

Bahadoran, Z., Mirmiran, P., Hosseinpour-Niazi, S., & Azizi, F. A. (2015). Sesame seeds-based breakfast could attenuate sub-clinical inflammation in type 2 diabetic patients: A randomized controlled trial. *International Journal of Nutrition and Food Sciences*, 4:1–5.

Balan, V., Rogers, C. A., Chundawat, S. P., da Costa Sousa, L., Slininger, P. J., Gupta, R., & Dale, B. E. (2009). Conversion of extracted oil cake fibers into bioethanol including DDGS, canola, sunflower, sesame, soy, and peanut for integrated biodiesel processing. *Journal of the American Oil Chemists' Society*, 86:157–165.

Bjorneboe, A., Bjornboe, E. A., Bodd, E., Hagen, B. F., Kveseth, N., & Drevon, C. A. (1986). Transport and distribution of et-tocopherol in lymph, serum and liver cells in rats. *Biochimica et Biophysica Acta*, 889:310–315.

Borel, P., Pasquier, B., Armand, M., Tyssandier, V., Grolier, P., Alexandre-Gouabau, M. C., Andre, M., Senft, M., Peyrot, J., Jaussan, V., & Lairon, D. (2001). Processing of vitamin A and E in the human gastrointestinal tract. *American Journal of Physiology-Gastrointestinal and Liver Physiology*, 280:G95–G103.

Brigelius-Flohé, R., Kelly, F. J., Salonen, J. T., Neuzil, J., Zingg, J. M., & Azzi, A. (2002). The European perspective on vitamin E: Current knowledge and future research. *The American Journal of Clinical Nutrition*, 76:703–716.

Burton, G. W., Ingold, K. U., Foster, D. O., Cheng, S. C., Webb, A., & Hughes, L. (1988). Comparison of free α-tocopherol and α-tocopheryl acetate as sources of vitamin E in rats and humans. *Lipids*; 23:834–840.

Cardona F., Andrés-Lacueva C., Tulipani S., Tinahones F. J., & Queipo-Ortuño M. I. (2013). Benefits of polyphenols on gut microbiota and implications in human health. *Journal of Nutritional Biochemistry*, 24: 1415–1422.

Carvalho, C. W. P., Takeiti, C. Y., Freitas, D. D. G. C., & Ascheri, J. L. R. (2012). Use of sesame oil cake (*Sesamum indicum* L.) on corn expanded extrudates. *Food Research International*, 45: 434–443.

Chen, P. R., Chien, K. L., Su, T. C., Chang, C. J., Liu, T. L., Cheng, H., & Tsai, C. (2005). Dietary sesame reduces serum cholesterol and enhances antioxidant capacity in hypercholesterolemia. *Nutrition Research*, 25:559–567.

Chung CH, editor. (2004). Molecular strategy for development of value-added sesame variety. In: Proceedings of International Conference on Sesame Science, East Asian Society of Dietary Life. Seoul; p. 15–39

Çiftçi, D., Kahyaoglu, T., Kapucu, S., & Kaya, S. (2008). Colloidal stability and rheological properties of sesame paste. *Journal of Food Engineering*, 87:428–435.

Clifford, M. N. (2000). Chlorogenic acids and other cinnamates–nature, occurrence, dietary burden, absorption and metabolism. *Journal of the Science of Food and Agriculture*, 80:1033–1043.

Cooney, R. V., Custer, L. J., Okinaka, L., & Franke, A. A. (2001). Effects of dietary sesame seeds on plasma tocopherol levels. *Nutrition and Cancer*, 39:66–71.

Corso, M. P., Fagundes-Klen, M. R., Silva, E. A., Cardozo Filho, L., Santos, J. N., Freitas, L. S., & Dariva, C. (2010). Extraction of sesame seed (*Sesamun indicum* L.) oil using compressed propane and supercritical carbon dioxide. *Journal of Supercritical Fluids*, 52:56–61.

Crews, C., Hough, P., Godward, J., Brereton, P., Lees, M., Guiet, S., & Winkelmann, W. (2006). Quantitation of the main constituents of some authentic grape-seed oils of different origin. *Journal of Agricultural and Food Chemistry*, 54:6261–6265.

de Boland, A. R., Garner, G. B., & O'Dell, B. L. (1975). Identification and properties of phytate in cereal grains and oilseed products. *Journal of Agricultural and Food Chemistry*, 23:1186–1189.

Di Carlo, G., Mascolo, N., Izzo, A. A., & Capasso, F. (1999). Flavonoids: Old and new aspects of a class of natural therapeutic drugs. *Life Sciences*, 65:337–353.

Domenichiello, A. F., Kitson, A. P., & Bazinet, R. P. (2015). Is docosahexaenoic acid synthesis from α-linolenic acid sufficient to supply the adult brain? *Progress in Lipid Research*, 59:54–66.

El Khier, M. K. S., Ishag, K. E. A., & Yagoub, A. E. A. (2008). Chemical composition and oil characteristics of sesame seed cultivars grown in Sudan. *Research Journal of Agriculture and Biological Sciences*, 4:761–766.

Elleuch, M., Besbes, S., Roiseux, O., Blecker, C., & Attia, H. (2007). Quality characteristics of sesame seeds and by-products. *Food Chemistry*, 103:641–650.

Erk, T., Hauser, J., Williamson, G., Renouf, M., Steiling, H., Dionisi, F., & Richling, E. (2014). Structure and dose absorption relationships of coffee polyphenols. *Biofactors*, 40:103–112.

Food and Agriculture Organization of the United Nations (FAO). (2018). Available from FAOSTAT Statistics database-agriculture, Rome, Italy (http://faostat.fao.org/beta/en/#data/QC 2020). Retrieved on August, 18, 2020.

Francis, G., Makkar, H. P., & Becker, K. (2001). Antinutritional factors present in plant-derived alternate fish feed ingredients and their effects in fish. *Aquaculture*, 199:197–227.

Gallo-Torres, H.E. (1970). Obligatory role of bile for the intestinal absorption of vitamin E. *Lipids*, 5:379–384.

Gandhi, A. P., & Taimini, V. (2009). Organoleptic and nutritional assessment of sesame (*Sesame indicum*, L.) biscuits. *Asian Journal of Food & Agro-industry*, 2:87–92.

Gertz, C., Klostermann, S., & Kochhar, S. P. (2000). Testing and comparing oxidative stability of vegetable oils and fats at frying temperature. *European Journal of Lipid Science and Technology*, 102:543–551.

Gharby, S., Harhar, H., Bouzoubaa, Z., Asdadi, A., El Yadini, A., & Charrouf, Z. (2017). Chemical characterization and oxidative stability of seeds and oil of sesame grown in Morocco. *Journal of the Saudi Society of Agricultural Sciences*, 16:105–111.

Gokbulut, C. (2010). Sesame oil: Potential interaction with P450 isozymes. *Journal of Pharmacology and Toxicology*, 5:469–472.

Gouveia, L. A., Cardoso, C. A., de Oliveira, G. M., Rosa, G., & Moreira, A. S. (2016). Effects of the intake of sesame seeds (*Sesamum indicum* L.) and derivatives on oxidative stress: A systematic review. *Journal of Medicinal Food*, 19:337–345.

Haji, M. M., Oveysi, M., Sadeghi, N., Janat, B., Baha, A. Z., & Mansouri, S. (2008). Gamma tocopherol content of Iranian sesame seeds. *Iranian Journal of Pharmaceutical Sciences*, 7:135–139.

Hallahan, B., & Garland, M. R. (2005). Essential fatty acids and mental health. *The British Journal of Psychiatry*, 186:275–277.

Hamada, N., Fujita, Y., Tanaka, A., Naoi, M., Nozawa, Y., Ono, Y., & Ito, M. (2009). Metabolites of sesamin, a major lignan in sesame seeds, induce neuronal differentiation in PC12 cells through activation of ERK1/2 signaling pathway. *Journal of Neural Transmission*, 116:841–852.

Hasan, A. F., Begum, S., Furumoto, T., & Fukui, H. (2000). A new chlorinated red naphthoquinone from roots of *Sesamum indicum*. *Bioscience, Biotechnology, and Biochemistry*, 64:873–874.

Hassan, M. A. (2012). Studies on Egyptian sesame seeds (*Sesamum indicum* L.) and its products 1-physicochemical analysis and phenolic acids of roasted Egyptian sesame seeds (*Sesamum indicum* L.). *International Journal of Dairy Technology*, 7:195–201.

Hemalatha, K. P. J., & Prasad, D. S. (2003). Changes in the metabolism of protein during germination of sesame (*Sesamum indicum* L.) seeds. *Plant Foods for Human Nutrition*, 58:1–10.

Hemalatha, S. (2004). Lignans and tocopherols in Indian sesame cultivars. *Journal of the American Oil Chemists' Society*, 81:467.

Ide, T., Azechi, A., Kitade, S., Kunimatsu, Y., Suzuki, N., Nakajima, C., & Ogata, N. (2015). Comparative effects of sesame seeds differing in lignan contents and composition on fatty acid oxidation in rat liver. *Journal of Oleo Science*, 64:211–222.

Jakobsen, K., Engberg, R. M., Andersen, J. O., Jensen, S. K., Henckel, P., Bertelsen, G., Skibsted, L. H., & Jensen, C. (1995). Supplementation of broiler diets all-rac-α- or a mixture of natural RRR-α-, ϒ-, δ-tocopheryl acetate. 1. Effect on Vitamin E status of broilers in vivo and at slaughter. *Poultry Science*, 74:1984–1994.

Jimoh, W. A., Fagbenro, O. A., & Adeparusi, E. O. (2011). Effect of processing on some minerals, antinutrients and nutritional composition of sesame (*Sesamum indicum*) seed meals. *Electronic Journal of Environmental, Agricultural & Food Chemistry*, 10:1858–1864.

Johnson, L. A., Suleiman, T. M., & Lusas, E. W. (1979). Sesame protein: A review and prospectus. *Journal of the American Oil Chemists' Society*, 56:463–468.

Ju, J., Picinich, S. C., Yang, Z., Zhao, Y., Suh, N., Kong, A. N., & Yang, C. S. (2010). Cancer-preventive activities of tocopherols and tocotrienols. *Carcinogenesis*, 31:533–542.

Kaur, N., Chugh, V., & Gupta, A. K. (2014). Essential fatty acids as functional components of foods-a review. *Journal of Food Science and Technology*, 51:2289–2303.

Kim, K. S., Lee, J. R., & Lee, J. S. (2006). Determination of sesamin and sesamolin in sesame (*Sesamum indicum* L.) seeds using UV spectrophotometer and HPLC. *The Korean Journal of Crop Science*, 51:95–100.

Kochhar, S. P. (2002). Sesame, rice-bran and flaxseed oils. In F.D. Gunstone (Ed.), Vegetable Oils in Food Technology, 297–326. Oxford: Blackwell Publishing, CRC Press.

Konietzny, U., & Greiner, R. (2003). Phytic acid: Nutritional Impact. In: *Encyclopedia of Food Science and Nutrition*, B. Caballero, L. Trugo, P. Finglas Eds., Elsevier, London, UK, 4555–4563.

Lampart-Szczapa, E., Siger, A., Trojanowska, K., Nogala-Kalucka, M., Malecka, M., & Pacholek, B. (2003). Chemical composition and antibacterial activities of lupin seeds extracts. *Nahrung/Food*, 47: 286–290.

Landete, J. M. (2011). Ellagitannins, ellagic acid and their derived metabolites: A review about source, metabolism, functions and health. *Food Research International*, 44:1150–1160.

Lazarou, D., Grougnet, R., & Papadopoulos, A. (2007). Antimutagenic properties of a polyphenol-enriched extract derived from sesame seed perisperm. *Mutation Research/Genetic Toxicology and Environmental Mutagenesis*, 634:163–171.

Leeson, S., Summers, J. D., & Caston, L. J. (2001). Response of layers to low nutrient density diets. *Journal of Applied Poultry Research*, 10:46–52.

Liu, Z., Saarinen, N. M., & Thompson, L. U. (2006). Sesamin is one of the major precursors of mammalian lignans in sesame seed (*Sesamum indicum*) as observed in vitro and in rats. *Journal of Nutrition*, 136:906–912.

Lyon, C. K. (1972). Sesame: Current knowledge of composition and use. *Journal of the American Oil Chemists' Society*, 49:245–249.

Machado, C. M., Benelli, P., & Tessaro, I. C. (2017). Sesame cake incorporation on cassava starch foams for packaging use. *Industrial Crops and Products,* 102:115–121.

Makinde, F. M., & Akinoso, R. (2014). Physical, nutritional and sensory qualities of bread samples made with wheat and black sesame (*Sesamum indicum Linn*) flours. *International Food Research Journal*, 21:1635–1640.

Manach, C., Scalbert, A., Morand, C., Rémésy, C., & Jiménez, L. (2004). Polyphenols: Food sources and bioavailability. *The American Journal of Clinical Nutrition*, 79:727–747.

Manikantan, M. R., Sharma, R., Yadav, D. N., & Gupta, R. K. (2015). Selection of process parameters for producing high quality defatted sesame flour at pilot scale. *Journal of Food Science and Technology*, 52:1778–1783.

Massey, L. K., Palmer, R. G., & Horner, H. T. (2001). Oxalate content of soybean seeds (Glycine max: Leguminosae), soyfoods, and other edible legumes. *Journal of Agricultural and Food Chemistry*, 49:4262–4266.

Moazzami, A. A., & Kamal-Eldin, A. (2006). Sesame seed is a rich source of dietary lignans. *Journal of the American Oil Chemists' Society*; 83:719.

Moazzami, A. A., Haese, S. L., & Kamal Eldin, A. (2007). Lignan contents in sesame seeds and products. *European Journal of Lipid Science and Technology*, 109:1022–1027.

Mohdaly, A. A., Ramadan-Hassanien, M. F., Mahmoud, A., Sarhan, M. A., & Smetanska, I. (2013). Phenolics extracted from potato, sugar beet, and sesame processing by products. *International Journal of Food Properties*, 16:1148–1168.

Mukhopadhyay, N., & Bandyopadhyay, S. (2003). Extrusion cooking technology employed to reduce the antinutritional factor tannin in sesame (*Sesamum indicum*) meal. *Journal of Food Engineering*, 56:201–202.

Namiki, M. (1995). The chemistry and physiological functions of sesame. *Food Reviews International*, 11:281–329.

Ojiako, O. A., Igwe, C. U., Agha, N. C., Ogbuji, C. A., & Onwuliri, V. A. (2010). Protein and amino acid compositions of Sphenostylis stenocarpa, Sesamum indicum, Monodora myristica and Afzelia africana seeds from Nigeria. *Pakistan Journal of Nutrition*, 9:357–361.

Onsaard, E., Pomsamud, P., & Audtum, P. (2010). Functional properties of sesame protein concentrate from sesame meal. *Asian Journal of Food and Agro-Industry*, 3:420–431.

Orruño, E., & Morgan, M.R.A. (2005). Purification and characterization of the 7s globuling storage protein from sesame (*Sesamum indicum* L.). *Food Chemistry*, 100:926–934.

Othman, S. B., Katsuno, N., Kanamaru, Y., & Yabe, T. (2015). Water-soluble extracts from defatted sesame seed flour show antioxidant activity in vitro. *Food Chemistry*, 175:306–314.

Park, Y. W., Chang, P. S., & Lee, J. (2010). Application of triacylglycerol and fatty acid analyses to discriminate blended sesame oil with soybean oil. *Food Chemistry*, 123:377–383.

Pasias, I. N., Kiriakou, I. K., Papakonstantinou, L., & Proestos, C. (2018). Determination of vitamin E in cereal products and biscuits by GC-FID. *Foods*, 7:3.

Pathak, N., Rai, A. K., Kumari, R., & Bhat, K. V. (2014). Value addition in sesame: A perspective on bioactive components for enhancing utility and profitability. *Pharmacognosy Reviews*, 8:147–155.

Prasad, M. N. N, Sanjay, K. R., Prasad, D. S., Vijay, N., Kothari, R., & Nanjunda Swamy, S. (2012). A review on nutritional and nutraceutical properties of sesame. *Journal of Nutrition & Food Sciences,* 2:1–6.

Quasem, J. M., Mazahreh, A. S., & Abu-Alruz, K. (2009). Development of vegetable based milk from decorticated sesame (*Sesamum Indicum*). *The American Journal of Applied Sciences*, 6:888–896.

Rangkadilok, N., Pholphana, N., Mahidol, C., Wongyai, W., Saengsooksree, K., Nookabkaew, S., & Satayavivad, J. (2010). Variation of sesamin, sesamolin and tocopherols in sesame (*Sesamum indicum* L.) seeds and oil products in Thailand. *Food Chemistry*, 122:724–730.

Rodrigues, D. C., Cunha, A. P., Silva, L. M., Rodrigues, T. H., Gallão, M. I., & Azeredo, H. M. (2018). Emulsion films from tamarind kernel xyloglucan and sesame seed oil by different emulsification techniques. *Food Hydrocolloids*, 77:270–276.

Sani, I., Sule, F. A., Warra, A. A., Bello, F., Fakai, I. M., & Abdulhamid, A. (2013). Phytochemicals and mineral elements composition of white *Sesamum indicum* L. seed oil. *International Journal of Traditional and Natural Medicines*, 2:118–130.

Sankar, D., Sambandam, G., Rao, M. R., & Pugalendi, K. V. (2005). Modulation of blood pressure, lipid profiles and redox status in hypertensive patients taking different edible oils. *Clinica Chimica Acta*, 355:97–104.

Serna-Saldivar, S. O., Gutiérrez-Uribe, J. A., & García-Lara, S. (2015). Phytochemical profiles and nutraceutical properties of corn and wheat tortillas. In: Rooney, L. W. Serna-Saldivar, S. O. (Eds.), *Tortillas,*. AACC International Press, USA, pp 65–96.

Sharma, L., & Singh, C. (2016). Sesame protein based edible films: Development and characterization. *Food Hydrocolloids*, 61:139–147.

Sharma, L., Saini, C. S., Punia, S., Nain, V., & Sandhu, K. S. (2021). Sesame (Sesamum indicum) Seed. In: *Oilseeds: Health Attributes and Food Applications,* 305–330. Springer, Singapore.

Shimoda, M., Nakada, Y., Nakashima, M., & Osajima, Y. (1997). Quantitative comparison of volatile flavor compounds in deep-roasted and light-roasted sesame seed oil. *Journal of Agricultural and Food Chemistry*, 45:3193–3196.

Soetan, K., & Oyewole, O. (2009). The need for adequate processing to reduce the anti-nutritional factors in plants used as human foods and animal feeds: A review. *African Journal of Food Science*, 3:223–232.

Sprecher, H. (2000). Metabolism of highly unsaturated n-3 and n-6 fatty acids. *Biochimica et Biophysica Acta*, 1486:219–231.

Stalmach, A., Steiling, H., Williamson, G., & Crozier A. (2010). Bioavailability of chlorogenic acids following acute ingestion of coffee by humans with an ileostomy. *Archives of Biochemistry and Biophysics*, 501:98–105.

Steer, T. E., & Gibson, G. R. (2002). The microbiology of phytic acid metabolism by gut bacteria and relevance for bowel cancer. *International Journal of Food Science & Technology*, 37:783–790.

Suja, K. P., Jayalekshmy, A., & Arumughan, C. (2004). Free radical scavenging behavior of antioxidant compounds of sesame (*Sesamum indicum* L.) in DPPH system. *Journal of Agricultural and Food Chemistry*, 52:912–915.

Toma, R. B., Tabekhia, M. M., & Williams, J. D. (1979). Phytate and oxalate contents in sesame seed (Sesamum indicum 1). *Nutrition Reports International*, 20(1): 25–31.

Tomimori, N., Rogi, T., & Shibata, H. (2017). Absorption, distribution, metabolism, and excretion of [^{14}C] sesamin in rats. *Molecular Nutrition & Food Research,* 61:1600844.

Traber, M. G., Olivecrona, T., & Kayden, H. J. (1985). Bovine milk lipoprotein lipase transfers tocopherol to human fibroblasts during triglyceride hydrolysis in vitro. *The Journal of Clinical Investigation*, 75:1729–1734.

Urbano, G., Lopez-Jurado, M., Aranda, P., Vidal-Valverde, C., Tenorio, E., & Porres, J. (2000). The role of phytic acid in legumes: Antinutrient or beneficial function? *Journal of Physiology and Biochemistry*, 56:283–294.

Uzun, B., Ülger, S., & Çağirgan, M. İ. (2002). Comparison of determinate and indeterminate types of sesame for oil content and fatty acid composition. *Turkish Journal of Agriculture and Forestry*, 26:269–274.

Wallert, M., Schmölz, L., Galli, F., Birringer, M., & Lorkowski, S. (2014). Regulatory metabolites of vitamin E and their putative relevance for atherogenesis. *Redox Biology*, 2:495–503.

Williamson, K. S., Morris, J. B., Pye, Q. N., Kamat, C. D., & Hensley, K. (2008). A survey of sesamin and composition of tocopherol variability from seeds of eleven diverse sesame (*Sesamum indicum* L.) genotypes using HPLC-PAD-ECD. *Phytochemical Analysis,* 19:311–322.

Yui, K., Imataka, G., Nakamura, H., Ohara, N., & Naito, Y. (2015). Eicosanoids derived from arachidonic acid and their family prostaglandins and cyclooxygenase in psychiatric disorders. *Current Neuropharmacology*, 13:776–785.

30

Flaxseed: An Underrated Superfood with Functional Properties

Alok Mishra and Amrita Poonia
Department of Dairy Science and Food Technology, Institute of Agricultural Sciences, Banaras Hindu University, Varanasi. (U.P) India

CONTENTS

30.1 Introduction

Flaxseed (*Linum usitatissimum*) is an age-old known oilseed derived from the blue-flowered flax plant that chiefly grows in cold climates (Table 30.1; Figure 30.1). It is believed to be the oldest agronomic crop cultivated by humans (Kaur et al., 2018a). Spanning more than 300 species, the flax plant has been in use for food and textile fiber for about 5,000 years (Singh et al., 2011). Flaxseed is produced mostly in the Northern hemisphere and in about 50 nations worldwide. Canada is the largest producer with a total production of 940,000 metric tons in 2015–16. In 2014–15, Canada exported 80% of its total flax production to China (50%), European Union (23%), and the United States (21%) (Flax Council of

TABLE 30.1

Scientific classification and vernacular naming of flaxseed

Classification	Vernacular Names
Kingdom: Plantae	English: Flax (flaxseed for food; linseed for industrial purposes)
Division: Magnoliophyta	Sanskrit: Nilapushpika, Atasi, Ksuma, Budrapatni, Tailottama, Parvathi, Masrina
Class: Magnoliopsida	Hindi: Alsi, Tisi
Subclass: Rosidae	Tamil: Alci, Ali vittu, Alishivirai
Order: Linales	Kannada: Alashi, Agase beeja, Athasi gida
Family: Linaceae	Malayalam: Akasi, Cheru-chanattinte-vitta
Genus: *Linum*	Telugu: Ali, Atasi, Avishi
Species: *usitatissimum* (~ very useful)	

Canada, 2020). India also ranks high in global flax production and its chief producing states are Madhya Pradesh, Chhattisgarh, Maharashtra, and Bihar (Singh et al., 2011; Ayelign and Alemu, 2016). Flaxseed used to be a staple food crop in ancient times and was a native crop to India (Shakir and Madhusudhan, 2007). The global market of flaxseed is growing at a compound annual growth rate of 12% and the market was expected to be worth $1.95 million USD by 2021 (Technavio, 2018).

Flaxseeds are oval, flat, and pointed at the tip with their color varying from yellow to dark brown, and measure approximately 2.5 mm × 5.0 mm × 1.5 mm (Singh et al., 2011). Flaxseed has a true hull or seed

FIGURE 30.1   Flax: (A) Flower, (B) fruit, (C) dried plant, (D) seeds. (Photos Courtesy: Flax Council of Canada, 2020.)

coat (testa), a fine endosperm, two embryos, and an embryonal axis (Morris, 2007). The envelope or testa of the seed contains about 15% of mucilage (Katare et al., 2012). The two basic varieties of flaxseeds are yellow and golden; both possess almost similar nutritional values and equal amounts of short-chain omega (ω)-3 fatty acids (Amin and Thakur, 2014). On an average, brown Canadian flaxseed comprise fat (41%), total dietary fiber (28%), protein (20%), moisture (7.7%), and ash (3.4%) (Yograj et al., 2017). The protein and oil content of flaxseed are inversely proportional, i.e., the former decreases with increase in oil content (Katare et al., 2012).

The texture of flaxseed is crisp and chewy possessing a pleasant nutty taste (Ganorkar and Jain, 2013). Germinated flaxseed (sprouts) is edible, having a slightly spicy flavor. Flaxseed is chemically stable until milled or ground because milling and grinding expose its oil content, which can go rancid at room temperature within a week. In such cases, it should be refrigerated or stored in air-tight containers (Bernacchia et al., 2014). Traditionally, flaxseed has been used as a remedy for constipation (Tarpila et al., 2005). Flaxseed can be consumed as a whole grain, which may be germinated or roasted. Flaxseed grits, powder, isolates, and roasted grains can be incorporated into a number of products, such as ready-to-eat breakfast cereals, snacks, drinks, crackers, biscuits, cakes, soups, and salad dressings (Bernacchia et al., 2014; Ayelign and Alemu, 2016).

Despite having a long history of use as a food, feed, and industrial fiber, the functional and health benefits of flaxseed are not known to many, hence, it remains an underrated oilseed. In recent years, the consumption of flaxseed beyond its oil has been increased, primarily due to the raised consciousness among consumers toward their health and a search for functional food alternatives (Yograj et al., 2017).

Flaxseed has gained popularity as a functional food as it is a good source of polyunsaturated fatty acid (PUFA), digestible proteins, and lignans (Dzuvor et al., 2018). Among vegetarian foods, flaxseed is one of the sources of α- linolenic acid (ALA, an omega [ω]-3 fatty acid) and mucilage (a soluble dietary fiber) (Ganorkar and Jain, 2013). The presence of phytoestrogens and phenolic compounds add to the functional value of flaxseed (Tarpila et al., 2005).

Flaxseed is considered a "superfood." The functional components of flaxseed are generally recognized as safe (GRAS) for human consumption. The health preventive and bioactive properties of these flaxseed components have been well studied. For example, the lipids, lignans, and fiber in flaxseed have been shown to have hypolipidemic, antiatherogenic, postprandial glycemic and insulinemic responses, and anticholesterolemic and anti-inflammatory properties (Repin et al., 2017; Martinchik et al., 2012; Fodje et al., 2009). Moreover, other flaxseed components, such as proteins and peptides, have been shown to induce certain desirable biologically active properties such as antioxidant, anti-inflammatory, antihypertensive, immune suppression/enhancement, glucose absorption control, etc. in living body systems (Dzuvor et al., 2018).

Several health benefits have been linked with the consumption of flaxseed as oil or as an enriched product. Studies suggest that flaxseed consumption may prevent many diseases including obesity disorders, cardiovascular diseases (CVDs), and cancer. However, the presence of toxic compounds may impose adverse health effects and some health risks are also associated with the effects of lignans in men and in pregnant women (Bernacchia et al., 2014)

In this chapter, the functional/bioactive ingredients derived from flaxseed are described, while further elaborating on the processing techniques used for extraction or isolation of these ingredients.

30.2 Nutritional Profile of Flaxseed

Flaxseed is widely acclaimed for its high ALA content in addition to mucilage, lignans, and high-quality proteins. It is a rich source of various phytochemicals, putting it in the functional food arena (Table 30.2). Oil constitutes a major part of the flaxseed, ranging from 35–45% on dry basis (Daun et al., 2003; Singh et al., 2011). The composition, however, varies according to the genetic makeup, growth conditions, processing, and method of analysis (Morris, 2007).

TABLE 30.2

Composition of Flaxseed

Nutritive or Bioactive Compound	Unit	Amount (per 100 g)	Components	Unit	Amount (per 100 g)
Total lipids (fats)	g	42.16	Fatty acids, total polyunsaturated	g	28.730
			Fatty acids, total monounsaturated	g	7.527
			Fatty acids, total saturated	g	3.663
Carbohydrate (by difference)	g	28.88	Neutral arabinoxylan fraction	mg	1.2
			Acidic rhamnogalacturonan fraction	mg	0.4
Dietary Fiber (total)	g	27.3	Insoluble fiber	g	12.8–17.1
			Soluble fiber	g	4.3–8.6
Protein	g	18.29	Glutamic acid	g	19.6
			Aspartic acid, arginine	g	9.2
			Glycine, leucine	g	5.8
			Serine, valine	g	4.5
Minerals	g	2.4	Potassium	mg	831
			Phosphorus	mg	622
			Magnesium	mg	431
			Calcium	mg	236
Vitamins	mg	n/a	γ-tocopherol	mg	522
			Folic acid	mg	112
			Biotin	mg	6
			Niacin	mg	3.2
Phenolic compounds			Lignans	mg	10–2600
			Secoisolariciresinol	mg	165
			Total flavonoids	mg	35–70
Essential fatty acids	g	n/a	α-linolenic acid	g	22.8
			Linoleic acid	g	5.9
Carotene	μg	30			
Moisture	g	6.5–7.5			
Energy	Kcal	534			

Sources: Flax council of Canada; 2020; Singh et al., 2011; USDA, 2016; Kaur et al., 2018a.

30.2.1 Lipids

Of all the valued functional compounds present in flaxseed, it is α-linoleic acid that makes it stand out (Gebauer et al., 2006). Oil from flaxseed is rich in unsaturated fatty acids (91%), constituting high PUFA content (73%) and moderate monounsaturated fatty acid content (18%). Only a small fraction (9%) is saturated fatty acid (Dubois et al., 2007). Out of all the lipids available in flaxseed, the major component is α-linolenic acid (~53%), followed by oleic acid (~19%), linoleic acid (~17%), palmitic acid (~5%), and stearic acid (~3%) (Bernacchia et al., 2014). It also provides an excellent ω-6:ω-3 fatty acid ratio nearly 0.3:1 (Pellizzon et al. 2007). For this reason, vegans must include flaxseed into their diet, which is devoid of marine foods (El-Beltagi et al., 2007). Flaxseed contains approximately 23 g ALA per 100 g (USDA, 2016). Although flaxseed is rich in ALA, its bioavailability depends upon the type and form of product ingested. ALA has greater bioavailability in oil than in milled seed followed by whole seed (Austria et al., 2008). Flaxseed oil also contains considerable amounts of vitamin E (tocopherol and tocotrienol), carotenoid (β-carotene and lutein), and phytosterols (Daun et al., 2003; Goyal et al., 2014).

Considerate amount of eicosapentaenoic acid (EPA) and docosahexaenoic acid (DHA) is also present in flaxseed. EPA and DHA are known to impart various health benefits on their consumption even in small amounts. Clinical trial reports suggest that they can help control the development of atherosclerosis, rheumatoid arthritis, inflammation, and asthma (Goyal et al., 2014; Kajla et al., 2015). ALA is known to be anticarcinogenic and aids in suppressing the development of malignant tumors and their metastases

(Dzuvor et al., 2018). Antiulcer activity of flaxseed oil and mucilage has been suggested in the study on rat models of ethanol-induced gastric ulcers (Dugani et al., 2008). Flaxseed lipids are also known to provide some degree of protection against liver disorders (Kajla et al., 2015). Flaxseed consumption also helps in significant lowering of bad cholesterol levels (LDL) without much effect on HDL levels, nullifying lipid peroxidation (Shakir and Madhusudhan, 2007).

30.2.2 Proteins

Flaxseed protein (FP) is a high-value vegetable protein source having an amino acid score comparable to soy beans without having any gluten (Hongzhi et al., 2004; Oomah, 2001). However, FP is not considered as a complete protein due to some of the limiting amino acids that need to be supplemented from other protein sources (Bernacchia et al., 2014). A protein content of about 21% and 34% is usually present in the flaxseed grain and paste, respectively. The protein content varies with the genetic and environmental factors (Chung et al., 2005). A low protein and higher oil content are generally obtained in colder climates (Bernacchia et al., 2014).

FP is constituted of approximately 80% globulins, which are a predominant salt-soluble-fraction with high molecular weight (11-12S, 18.6% nitrogen), and 20% albumin, which is a water-soluble basic component with low molecular weight (1.6-2S; 17.7% nitrogen) (Hall et al., 2006; Chung et al., 2005). Linin (around 65% of the total FP) is the major high molecular weight fraction of flaxseed protein, whereas, colinin (about 42% of the total FP) is the major low molecular weight component (Bekhit et al., 2018). Flaxseed (whole, meals, isolates) is a considerable source of sulphur amino acids (methionine and cysteine); glutamic acid/glutamine and arginine; branched-chain amino acids (valine, leucine, isoleucine, BCAA); and aromatic amino acid (phenylalanine and tyrosine). The limiting amino acids are lysine, tyrosine, and threonine (Bernacchia et al., 2014; Goyal et al., 2014; Gopalan et al., 2004).

The beneficial proteins and bioactive peptides in flaxseed contain various nutraceutical properties (Dzuvor et al., 2018). The amino acids cysteine and methionine exhibit antioxidant characteristics (Oomah, 2001). FPs have been reported to demonstrate antifungal properties against *Alternaria solani*, *Candida albicans*, and *Aspergillus flavus*; antineurodegenerative properties by inhibiting nitric oxide synthesis; antihypertensive properties by obstructing the transformation of angiotensin I to angiotensin II; and plasma glucose lowering abilities (Xu et al., 2008a, b; Marambe et al., 2008; Omoni and Aluko, 2006; Bhathena et al., 2003). Flaxseed is a considerable source of biopetides. Approximately 25 kinds of cyclolinopeptides have been reported from flaxseed (Dzuvor et al., 2018). These biopeptides possess various known physiological activities, such as antioxidant capacity, antibacterial activity, antidiabetic effect, and inhibition of angiotensin-converting enzyme (ACE) (Wu et al., 2019).

A relatively high biological value (77.4%) of FPs has been reported (Martinchik et al., 2012). The digestibility of FPs is also very high. However, the digestibility of a protein depends on its form of ingestion. The coefficient of digestibility value flaxseed protein extracts is around 89.6%. The extraction of oil from flaxseed and further removal of mucilage results in a concentration of proteins with improved *in-vitro* digestibility (Dzuvor et al., 2018; Marambe and Wanasundara, 2017).

30.2.3 Carbohydrates

Carbohydrate content is considerably low in flaxseed, contributing very little carbohydrates upon ingestion (Bernacchia et al., 2014; Katare et al., 2012). It is the hull of the flaxseed where most of the carbohydrate is concentrated (Bekhit et al., 2018). Out of total flaxseed carbohydrates (~29%), the bulk is indigestible and only 1–2% is in the form of soluble sugar/starch (Dzuvor et al., 2018). The soluble indigestible fiber is generally flaxseed mucilage and flaxseed gum. A study reported that flaxseed mucilage accounts for 60–64% of the carbohydrate fraction depending on the fractionation method used for separation of mucilage (Tirgar et al., 2017). The insoluble part of flaxseed carbohydrate consists of non-starch polysaccharides, mainly cellulose and lignan (Vaisey-Genser and Morris, 2003). Flaxseed mucilage is composed of the heterogeneous polysaccharides neutral arabinoxylan (75–85%) and acidic rhamnogalacturonan (15–25%). Arabinoxylan consists of arabinose, galactose, and xylose. Rhamnogalacturonan consists of L-rhamnose, L-fucose acid, D-galactose, and D-galacturonic acid (Bernacchia et al., 2014).

Of note is that the exact composition of monosaccharides, carbohydrate yield, and quality varies significantly among flaxseed cultivars in different parts of the world (Ho et al., 2007; Warrand et al., 2005).

The antioxidant properties of oligosaccharides from flaxseed cake induce antitumor properties (Gutiérrez et al., 2010). Anticancerous properties have also been reported by many authors based on the antiradical activity of the flaxseed saccharides, which check the oxidation of proteins, lipid, or DNA. Antimicrobial properties due to chito-oligosaccharides were reported in cake prepared using flaxseed. The activity was found against various pathogenic bacteria and fungi including *Fusarium graminearum*, *Aspergillus flavus*, *Candia albicans*, and *Penicillium chrysogenum* (Xu et al., 2008a, b).

30.2.4 Dietary Fiber

Flaxseed contains appreciable amounts of dietary fiber in the form of insoluble and soluble (mucilage) dietary fiber. Dietary fiber from flaxseed can be extracted using an aqueous solution at pH 12, followed by centrifugal separation to get the supernatant (Gutiérrez et al., 2010). The solution obtained is sticky mucilage that can be dried using a vacuum drier, spray drier, or lyophilization to obtain a shelf-stable powder (Bekhit et al., 2018). Various health benefits have been attributed to flaxseed mucilage. It serves as a prebiotic, i.e., good for the healthy growth of gut flora (HadiNezhad et al., 2013). Studies also suggest that it has antiulcer properties and can significantly reduce the number of gastric ulcers and their length (Dugani et al., 2008).

30.2.5 Phenolic Compounds

The three types of phenolic compounds present in flaxseed are phenolic acids, flavonoids, and lignans, which contribute a small but significant fraction of the total flaxseed composition (Dzuvor et al., 2018). Phenolic acids are simpler compounds whereas the lignans are complex. The phenolic acid content in flaxseed ranges from 790–1,030 mg/100 g depending on the cultivar and environment (Bekhit et al., 2018). The phenolic acids usually present in defatted flaxseed are hydroxybenzoic acid, ferulic acid, chlorogenic acid, gallic acid, and coumaric acid (Mazza, 2008; Beejmohun et al., 2007). Major flavonoids present in flaxseeds are flavone-C-glycosides and flavone-O-glycosides (Mazza, 2008). Lignans are concentrated in the seed coat of the flaxseeds. The major lignan present in flaxseed is secoisolariciresinol diglucoside (SDG), amounting to nearly 11.7–24.1 mg/g and 6.1–13.3 mg/g in defatted flaxseed flour and whole flaxseed, respectively. Flaxseed happens to be the richest source of SDG among all foods (Bekhit et al., 2018; Johnsson et al., 2002). Other lignans reported in minor levels are matairesinol and pinoresinol. Plant lignans provide the precursors needed for the development of mammalian lignans, which have positive health effects. Gut bacteria converts SDG to lignans: enterodiol and enterolactone. These provide health benefits due to their weak estrogenic or antiestrogenic as well as antioxidant effects (Adlercreutz, 2007). Some health benefits reported include protection against chronic diseases like cancer, diabetes, kidney disease, cardiac disease, and hepatic diseases (Prasad, 2009; Thompson and Cunnane, 2003; Stavro et al., 2003). Flax lignans have shown promising effects in reducing the growth of cancerous tumors, especially hormone-sensitive tumors of the breast, endometrium, and prostate (Goyal et al., 2014). The *in vitro* and *in vivo* anti-inflammatory and antioxidant activities of flaxseed lignans have also been investigated (Korkina et al., 2011).

30.2.6 Micronutrients

Flaxseed is a considerable source of vitamins and minerals as well. It contains abundant quantities of niacin and all forms of tocopherols (α, β, and γ). Vitamin E (usually γ-tocopherol) amounts to ~0.039% (Dzuvor et al., 2018). Vitamin E is antioxidant in nature and helps in scavenging free radicals from the body. Besides, vitamin E may protect against cancer by preventing the formation of carcinogenic nitrosamines from nitrites in the stomach (Winter, 2013). It may also help in lowering blood pressure by promoting secretion of sodium in the urine (Meagher et al., 2001). Among minerals, potassium is abundant (~5.6% to 9.2%) followed by phosphorous (~0.65%), magnesium (~0.40%), calcium (~0.25%), and small quantities of sodium (~0.027%) (Dzuvor et al., 2018; Morris, 2007). The high potassium indicates flaxseeds provide protection against stroke, help promote free radical scavenging, and inhibit platelet accumulation (Dzuvor et al., 2018). Strikingly, a mere 30 g portion of the seed constitutes 7– 30% of the recommended dietary allowance (RDA) for these minerals (Singh et al., 2011).

30.2.7 Antinutritive Compounds

Various antinutritive compounds have been reported in flaxseed. These are cyanogenic glycosides, cadmium, trypsin inhibitors, linatine, and phytic acid. However, consumption of flaxseed leading to food poisoning or critical toxicity has not been reported in literature (Daun et al., 2003).

Hall et al. (2006) reported small amounts of cyanogenic glycosides (264–354 mg/100 g) and linamarin (acetone–cyanohydrin-β–glucoside, 10–11.8 mg/100 g). According to Mazza (2008), whole flaxseed contains 2.5–5.5 mg/g cyanogenic glycosides. Its major components were identified as linustatin (2.07 mg/g) and neolinustatin (1.74 mg/g) (Park et al., 2005). Ingestion of these compounds may be lethal to humans at a dosage of ~100 mg and above (Bernacchia et al., 2014). Cyanogenic glucosides cause chronic effects on the nervous system, which are usually observed in populations ingesting cyanate rich foods (Tarpila et al., 2005). However, these compounds can be eliminated by subjecting flaxseed to thermal and mechanical processing like roasting, cooking, boiling, pelleting, and autoclaving (Guerrero-Beltrán et al., 2009; Park et al., 2005). Moreover, certain detoxifying enzymes like β-glycosidases may be used. They release hydrogen cyanide that may evaporate using steam (Yamashita et al., 2007; Feng et al., 2003).

The presence of cadmium in flaxseed makes it potentially toxic to humans. It can accumulate in the kidneys, causing renal dysfunction, pulmonary emphysema, glycosuria, and aminoaciduria. It can also compromise mineral reabsorption, which may lead to osteomalacia (Bernacchia et al., 2014).

Small amounts of trypsin inhibitor activity (TIA) has been reported in flaxseed, which has 42–51 units of TIA in laboratory-prepared flaxseed meal (Bernacchia et al., 2014). Trypsin inhibitors slow the digestion and absorption of proteins by the inhibition of proteases (Puvača et al., 2012).

Linatine (gammaglutamyl-1-amino-D-proline) is not a high-risk antinutrient for humans. It was found responsible for vitamin B_6 deficiency symptoms in flaxseed-fed chicks. However, the studies reported that human consumption of flaxseed up to 50 g/per day did not affect vitamin B_6 levels or metabolism (Bekhit et al., 2018). Thus, it is advisable to include enough vitamin B_6 to overcome the deficit due to linatine (Daun et al., 2003).

Phytic acid, which reduces the bioavailability of micronutrients, is also present in flaxseed (~2.3–3.3%) (Kajla et al., 2015; Mazza, 2008). The chelating nature of phytic acid is responsible for the deficiency of zinc, calcium, copper, magnesium, and iron (Bekhit et al., 2018). Recently, some authors have reported anticancerous, antioxidant, hypolipidemic, and hypocholesterolemic properties of phytic acid (Daun et al., 2003; Mazza, 2008).

30.3 Traditional Uses of Flaxseed in Disease Manifestation

Many authors have reported that flaxseed consumption in humans has been prevalent since historic times. Flaxseed oil has been used in food preparations and its health-based applications have been reported as well. Flaxseed oil brings mental and physical benefits by fighting fatigue and controlling the aging process. Utilization of flaxseed as a remedy to cough and abdominal pain has been in practice (Goyal et al., 2014). The Indian system of medicine, Ayurveda, identifies various properties of flaxseed such as *Vranahrit* (wound healing), *Balya* (improves tensile strength or elasticity of the skin), *Tvagdoshahrit* (removes skin blemishes), *Madhura* (balances the skin pH), *Grahi* (improves moisture holding capacity of skin), *Picchaila* (lubricous), and useful in *Vata* (skin) disorders including dryness, undernourishment, and lack of luster/glow (Goyal et al., 2014). Preparations of flaxseed were used for the treatment of gastrointestinal disorders as an enveloping and wound-healing agent (Ivanov et al., 2011). Diuretic uses of flaxseed oil in the treatment of kidney disorders were prevalent in the middle ages. Flaxseed can be a remedy in cough, pain, and inflammation (Moghaddasi, 2011). Flaxseed in combination with sweet clover was recommend as an antitumoral. A mixture of flaxseed with soda and figs can be used in the treatment of freckles and a mixture with garden cress and honey may treat nail disorders (Tolkachev and Zhuchenko, 2004).

30.4 Functional and Health Aspects of Flaxseed Consumption

The presence of the aforementioned nutrients indicates the various functional and health benefits of flax-seed consumption (Figure 30.2; Table 30.3). Most of the beneficial aspects of flaxseed are due to three major components: (1) high ALA content; (2) high composition of soluble and insoluble dietary fiber; and (3) the highest concentration of lignan among all plant-based foods (Singh et al., 2011). Flaxseed may prove to be beneficial in various critical diseases such as diabetes, cancer, CVDs, obesity, and renal and bone disorders (Katare et al., 2012).

30.4.1 Role in Reducing Tumors and Cancer

Flaxseed can protect from cancers of the breast, colon, prostate, thyroid, and ovaries by preventing the formation of tumors and additionally diminishing blood vessel cell development. The tumor-reducing effect of flaxseed can be attributed to the high content of SDG lignan (Mason and Thompson, 2014; Truan et al., 2012; Chen et al., 2011; Saggar et al., 2010). SDG can attenuate tumorigenesis by decreasing the cell proliferation and angiogenesis. It also increases the apoptosis via modulation of the estrogenic receptor (ER) and growth-factor signaling pathways (Saggar et al. 2010; Chen et al. 2009). Studies relate the breast tumor-reducing property to the high SDG lignan found in flaxseed (Chen et al., 2011). The anticarcinogenic potential of flax-lignans against hormone-induced cancer (e.g., prostate cancer) has been reported (Waldschläger et al., 2005). A more than 50% reduction in size and 37% reduction in the number of mammary tumors were reported in carcinogenic-treated rats due to flaxseed lignans (Katare et al., 2012). The antioxidant activity of flaxseed lignans is the reason suggested by researchers to help

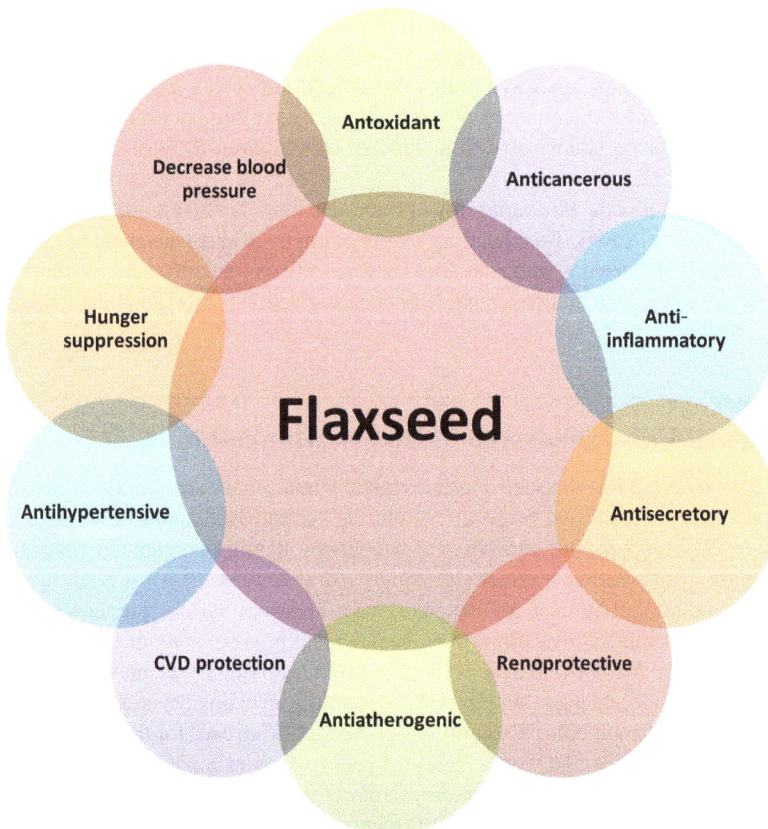

FIGURE 30.2   Various health benefits and functional aspects of flaxseed.

TABLE 30.3

Studies on Functional and Health Implications of Flaxseed and their Products

Flax Form	Supplementation Method	Functional or Health Implications	Reference
Ground flaxseed	Baked products	• 9–16% decrease in HDL levels in male subjects • Decrease in short-lived LDL levels after 5 weeks of consumption	Bloedon et al., 2008
Ground flaxseed	Flax powder, bread, and muffins	• 10% decrease in LDL levels • 7% decrease in total cholesterol level	Patade et al., 2008
Extracted oil	Oil capsules	• 8 µg/mL increase in blood plasms ALA	Kaul et al., 2008
Lignan isolate	Low-fat muffin	• Reduction in C-reactive protein, a strong predictor of CHD	Hallund et al., 2008
Lignan isolate	Baked products	• Dose-dependent reduction in total plasma cholesterol and blood glucose level	Zhang et al., 2008
Lignan isolate	Capsules	• Moderate improvement in glycemic control	Pan et al., 2007
Flaxseed meal	Clinical feed	• Antioxidant potential indicated by restoration of hepatic enzymes	Rajesha et al., 2006
Ground flaxseed	Muffins and bread	• Decrease in bone resorption rate	Arjmandi et al., 2005

maintain more early stages of cancer (Hall et al., 2006). Lignans affect the metabolism of estrogenic metabolism, which may reduce the risk of ovarian cancer (McCann et al., 2007). The chemopreventive effect activity of flaxseed may be contributed by the antioxidant activity mediated by flavonoids such as herbacetin 3, 7-Odimethyl ether. The presence of phytoestrogen and dietary fiber in flaxseed also imparts cancer protective effects (Katare et al., 2012; Hu et al., 2007).

30.4.2 Role in CVD Treatment

The presence of adequate amounts of ALA, lignans, phytoestrogens, and dietary fiber makes flaxseed an ideal candidate to fight CVDs. Dietary flaxseed modifies cardiovascular risk factors by improving the lipid profile in hyperlipidemic patients (Truan et al., 2012). The hypocholesterolemic activity of flaxseed was reported in many animal model studies on rats, mice, and rabbits (de França Cardozo et al., 2012; Kristensen et al., 2012; Park and Velasquez, 2012; Leyva et al., 2011; Mani et al., 2011). Consumption of flaxseed oil for 28 days resulted in reduced high-density lipoprotein (HDL) levels in human serum (Gillingham et al., 2011). A significant decrease in systolic and diastolic blood pressure levels were observed in middle-aged men having dyslipidemia after a 12-week dietary supplementation with flaxseed oil (8 g/day) (Paschos et al., 2007). Enterolactone, produced by intestinal microflora is associated with atherosclerosis protective effects (Fuchs et al., 2007). Dietary flaxseed may also offer protection against ischemic heart disease by improving vascular relaxation responses and by inhibiting the incidence of ventricular fibrillation (Adolphe et al., 2010). An animal model study has shown potent antiatherogenic effects in LDL-receptor-deficient-mice administered with 10% flaxseed-supplemented diet for 24 weeks (Dupasquier et al., 2007). Lignans, being selective estrogen receptor modulators (SERMs), affect various body metabolisms, including cardiovascular responses. Thus, flaxseed may help lower blood plasma LDL (Zhang et al., 2008; Regitz-Zagrosek et al., 2007). Flaxseed fiber may be beneficial in various cases such as hypocholesterolemia, constipation, gastrointestinal motility, glucose intolerance, and fermentation (Kristensen et al., 2013, 2012; Mani et al., 2011).

30.4.3 Role in Diabetes Treatment

The protective effects of flaxseed administration against diabetes were evident due to ω-3 fatty acids, lignans, and dietary fiber (Goyal et al., 2014). Flaxseed lignan and soluble fiber may lower the glycemic response of blood glucose, which may lead to prolonged satiety and help fight the problem of obesity (Pan et al., 2009, 2007). The presence of ω-3 fatty acid, lignans, and fibers in flaxseed may have a protective effect against the risk of diabetes and lifestyle diseases (Adlercreutz, 2007). Kapoor et al. (2011) observed that postprandial blood glucose was decreased by 7.9% and 19.1%, respectively, on supplementation of flaxseed powder (15 g/day and 20 g/day) for 2 months in diabetic females. Supplementation of the diet of type 2 diabetics with 10 g of flaxseed powder for a period of 1 month reduced fasting blood glucose by 19.7% and glycated hemoglobin by 15.6% (Mani et al., 2011). Utilization of flaxseed for glycemic control can also be associated with reduced risks of obesity and dyslipidemia because these are the risk factors behind development of diabetes resistance to insulin (Wu et al. 2010; Morisset et al., 2009). SDG (flaxseed lignan) has been found to reduce the high-fat-diet-induced accumulation of visceral and liver fat. It has positive effect on hypercholesterolemia, hyperleptinemia, and hyperinsulinemia (Fukumitsu et al., 2010).

30.4.4 Role in Nephrology

Consumption of ω-3 fatty acids has various health benefits because of their anti-inflammatory properties that protect from the destruction of kidneys in adults. Consumption of flaxseed and flaxseed products increases the concentrations of ALA, which may reduce the manifestation of chronic kidney diseases (Gopinath et al., 2011). Flaxseed meal has been found more effective than soy protein in reducing proteinuria and renal histologic abnormalities. Moreover, flax lignans and oils have a positive effect on reducing the progression of renal injury including polycystic kidney disease (PKD) (Ogborn et al., 2006; Velasquez et al., 2003).

30.4.5 Other Health Benefits

Laxative effects of flaxseed fibers in the young and in elderly adults have been related to the soluble fiber of flaxseed (Vaisey-Genser and Morris, 2003; Morris, 2007). Non-starchy soluble dietary fiber of flaxseed form viscous solutions. It prolongs the stomach emptying process, resulting in diminution of nutrient absorption in the small intestine. Thus, it can help prevent obesity-related diseases (Singh et al., 2011). Consumption of ALA has been linked to the regulation of leptin protein encoded by obese genes. The secretion of neuropeptide-Y (NPY), a potent appetite stimulator is repressed by diet-induced obesity. Leptins bind to the NPY neurons and decrease their activity. The signals sent by the leptins to the hypothalamus induce a feeling of satiety and reduce craving for high calorie diet. Supplementation of 10% flaxseed in a regular diet resulted in elevated leptin protein levels that may be attributed to higher plasma and adipose levels of flaxseed contributed ALA. Thus, the positive correlation suggested between leptin expression and ALA content may lower the risk of atherosclerosis (Goyal et al., 2014; McCullough et al., 2011).

30.5 Techniques for Extraction

Functional ingredients can be extracted from flaxseed by applying modern techniques to incorporate them in various health foods and nutraceuticals. The demarcation of nutraceuticals from functional foods must be understood before going further. Functional food is very much identical to conventional food as it is part of a regular diet. Such a food demonstrates physiological benefits and leaps ahead of basic nutritional functions to reduce the chances of chronic disease. In contrast, nutraceuticals are isolated or purified products sold as medicine or as potential therapeutic agents without being associated with regular foods. It's worth noting that nutraceuticals can form a part of functional foods, but the functional foods must provide essential nutrients necessary for normal growth, development, and maintenance (Goyal et al., 2014).

30.5.1 Extraction of Lipids

Flaxseed, principally, is a kind of oilseed that produces triglyceride oil rich in ω-3 fatty acids. Several techniques are used for oil extraction from flaxseed (Table 30.4). These can be divided into conventional

TABLE 30.4

Comparison of Different Techniques Used in the Extraction of Flaxseed Compounds

Method of Extraction	Comments	Advantages	Disadvantages
Extraction of Lipids			
Cold pressing	• Gentle mechanical oil extraction • Application of pressure and shear force using screw press or oil expeller	• Low equipment cost • Low energy requirement • High quality consumable oil • No use of chemicals	• Low yield • Low content of vitamins, phospholipids, phytosterols, antioxidants • High residual oil in cake
Solvent extraction	• Depends on factors such as nature of solvent, reaction time, temperature, and solid/solvent ratio.	• Cheap and high yield as compared to mechanical process	• Long extraction time • Large amounts of organic solvent waste generated
Microwave- and ultrasound-assisted extraction	• These techniques aid in solvent penetration and increase the solvent-seed contact surface	• Improved oil extraction efficiency • Shorter extraction times • Reduced solvent consumption • Decrease in loss of bioactive	• Use of organic solvents is hazardous • High cost input
Sub/Supercritical fluids extraction	• A gas at supercritical state is used as solvent (carbon dioxide, propane, toluene, and ethane)	• High quality of extract • Eco-friendly approach • Industrial feasibility and scalability	• High setup costs
Extraction of Proteins			
Isoelectric precipitation	• Most used technique in protein recovery • Involves manipulating the pH of the protein solution to reach the solute pH that achieves precipitating of the proteins.	• High yield	• Irreversible denaturation of protein by mineral acid • Potential loss of functional properties • Reduction of nutritional quality • Concentration of antinutritional components such as phytic acid
Partial enzyme hydrolysis-assisted extraction	• Flaxseed meal is treated with proteases such as papain and further processed	• Improve functional properties • Ease of control • Helps in extraction of conjugated proteins such as glycoproteins	• Increases overall processing cost • Extensive hydrolysis is undesirable
Micellization	• Precipitation with salts such as ammonium sulphate • Also called salting out	• Native state of protein is preserved • Nonprotein components are removed • Very high protein yield (93%)	• Ammonium sulphate precipitation can be onerous
Extraction of Carbohydrates			
Solid-liquid extraction	• Involves the mass transfer of solutes from a solid matrix into a solvent • Water used as primary solvent	• Simplest technique • Reasonable yields with little capital costs	• Time consuming • Laborious purification process • Slow mass transfer rate

(Continued)

TABLE 30.4 (Continued)

Method of Extraction	Comments	Advantages	Disadvantages
Pressurized fluid extraction	• Based on the principles of Soxhlet • Organic solvent used	• Rapid mass transfer of compounds • Enhanced performance of solvents due to high pressure and high temperature	• High capital cost • Removal of residues puts under labor
Ionic liquids (ILs) and natural deep eutectic solvents (NADES)	• Choline chloride IL and choline chloride NADES are commonly used in carbohydrate extractions	• Cost-effective • Relatively nontoxic • Easy to prepare • NADES are not only safer than IL but also cheaper	• More research is required to better identify and understand the toxicity

Source: Dzuvor et al., 2018.

(mechanical cold processing and solvent extraction) and novel (supercritical fluid extraction, ultrasonic assisted extraction, and microwave assisted extraction) (Dzuvor et al., 2018).

30.5.2 Extraction of Proteins

Proteins are extracted from seeds before their implementation in food to enhance their digestibility, reduce levels of antinutritive compounds, improve techno-functional properties, and enhance sensory characteristics like taste and color (Rommi et al., 2015). There are three classes of flaxseed proteins: globulin, glutelin, and albumin (Malomo and Aluko, 2015). Various seed protein fractionation techniques (traditional as well as novel) are available. Several techniques have been proposed for protein-isolate preparation from flaxseed (Dzuvor et al., 2018). Before the onset of extraction process, some prerequisites are performed to facilitate the recovery of protein from the raw flaxseed. These are extraction of mucilage, removal of flax gum, milling, and defatting of flaxseed (Dzuvor et al., 2018; Gutiérrez et al., 2010).

Conventional methods of flaxseed protein isolation include isoelectric/alkaline precipitation, ammonium sulphate precipitation (micellization), and acid pretreatment with ultrafiltration (Hadnađev et al., 2017). Novel techniques of protein extraction include ultrasound assisted extraction, pulse-electric field technique, enzyme-assisted extraction, pressurized low-polarity water extraction, high-voltage electrical discharge, and pressurized low-polarity water extraction (Gutiérrez et al., 2010). Novel methods offer improved yield, functionality, and sustainability in protein production (Hadnađev et al., 2017).

Protein isolation using these techniques from the defatted flax seed paste or cake increases the purity of protein and raises the amino acid content (Teh et al., 2014). Furthermore, the enzymatic hydrolysis improves the bioactivities of the flax protein isolates, which can be used in the production of bioactive peptides (Teh et al., 2016).

30.5.3 Extraction of Carbohydrates

The flaxseed surface contains mucilage as the part of its waxy coat. These hygroscopic carbohydrates are a mixture of neutral and acidic polysaccharide fractions. Their extraction and purification are a tedious and time-consuming process that is of commercial value only when it is coupled with an oil extraction process (Elboutachfaiti et al., 2017; Dzuvor et al., 2018).

30.6 Flaxseed Utilization in Functional Food Development

In urban societies, the increased rate of obesity and other lifestyle diseases are a major concern. Many people are looking for healthy alternatives that are low in calories but have a high punch of nutrients. Flaxseed products are increasingly used as functional food or nutraceuticals in many societies

FIGURE 30.3 Utilization of flaxseed as (A) whole seeds, (B) extracted oil, (C) garnishing, (D) incorporated in food product. (Photos courtesy: Flax Council of Canada, 2020.)

throughout the world (Figure 30.3) (Ogborn et al., 2006; Lemay et al., 2002; Watkins et al., 2001). The consumption of 9 g per day of flaxseed can provide sufficient ω-3 fatty acid (Goyal. et al., 2014). Flaxseed can be used in food preparations as a whole or in the form of roasted, ground, milled, pow-dered, or as extracted oil. Various products have been formulated using flaxseed or its oil such as baked foods, milk and dairy products, breakfast cereals, extruded snacks, soups, and waffles (Kaur et al., 2019, 2018b-c, 2017, 2013; Aliani et al., 2012; Ivanov et al., 2011; Bilek and Turhan, 2009; Manthey and Sinha, 2008; Sinha and Manthey, 2008; Pohjanheimo et al., 2006 a, b). Studies suggest increased levels of protein, dietary fiber, and antioxidant activity because of the flaxseed addition in various forms (Kaur et al., 2018a; Cameron and Hosseinian, 2013; Khouryieh and Aramouni, 2012; Hao and Beta, 2012a, b).

A slight modification in flaxseed may provide more enhanced benefits. As discussed earlier, roasting causes detoxification of flaxseeds. However, inferior digestibility of roasted flaxseed proteins has been reported. Such a high-heat treatment has negative effects on physical properties like water absorption and dough stickiness (Kaur et al., 2019; Pourabedin et al., 2017; Marpalle et al., 2015). Germination of flaxseeds increases the content and bioavailability of certain nutrient components such as amino acids, vitamins, and enzymes. It also reduces the antinutrients such as phytic acid, enzyme inhibitors, and insoluble fiber (Bekhit et al., 2018; Marpalle et al., 2015).

Baked food products have wide consumer base that make them ideal candidates for food fortification trials and public interventions (Gat and Ananthanarayan, 2015; Kadam and Prabhasankar, 2010). The regular ingredients used deprive these products of the protective health components such as ALA and dietary fiber (Ganorkar and Jain, 2014; Fardet, 2010). The challenges associated with the addition of flaxseed in bakery products are oxidative instability at high temperatures and during storage (Kaur et al., 2018b, c). Various baked products incorporating flaxseed have been prepared such as cookies, biscuits, buns, breads, pizza, bagels, and patties (Kaur and Kaur, 2018b, c, 2017, 2013). Blood levels of ALA were found to be raised upon the consumption of dietary flaxseeds being incorporated into baked goods like breads or muffins in ground or milled form (Kaur and Kaur, 2018b). Other than baked goods, flaxseed applications in dairy-based, extruded, snack, and fermented products has been reported. A brief account of some recent applications of functional food development using flaxseed and their products is elicited in Table 30.5.

Flaxseed byproduct, which is the residual paste obtained after the extraction of oil, has a potential to be used in human food. Flaxseed paste contains 34% protein, which can be further processed to prepare protein concentrate to make it suitable for addition in conventional food products. Thus, a product with high protein content having certain desirable functional properties can be obtained (Goyal et al., 2014; Wang et al. 2010a, b). Other than protein, this residual paste also contains more lignan and fiber than the whole flaxseed. Thus, it can be a beneficial nutritious source of nonoil flaxseed with a rich composition of protein, lignan, and fiber (Bekhit et al., 2018).

The incorporation of milled flaxseed in food products affects their overall quality due to its gums and proteins. Flaxseed has been reported to enhance viscosity; stabilize emulsions; increase absorption thereby improving loaf volume, oven spring, and keeping quality of bread; and affect shear rate. Thus, it has promise as a thickening and improving agent in baked foods (Goyal. et al., 2014; Wang et al., 2011; Koca and Anil, 2007). The effects on sensory qualities and shelf life of flaxseed-fortified foods is less studied. While not all the products have shown good results, a few received lower sensory acceptability and neutral health impacts, the majority of the studies reported positive health benefits of flax-enriched food products (Goyal. et al., 2014; Conforti and Davis, 2006; Pohjanheimo et al., 2006a). Flaxseed addition provides a nutty cereal-like flavor with a slight bitter tang or rancid flavor due to lipid oxidation of unstable ALA (Bekhit et al., 2018). An enhanced dark color due to flaxseed flour addition is generally observed (Conforti and Davis, 2006). Flaxseed mucilage has been reported to increase water-retention in many products such as meat, bread dough, etc. (Chen et al., 2009). The addition of flaxseed may inhibit the structure formation in wheat-based products by interfering with wheat gluten, thus, causing a textural problem in such products (Shearer and Davies, 2005). The inhibition effect of flaxseed on some spoilage-causing bacteria has also been reported which may help extend the shelf life of fortified food products (Bekhit et al., 2018; Xu et al., 2008a, b).

30.7 Challenges in Flaxseed Consumption

The functional properties and health benefits of flaxseed have been thoroughly investigated, verified, and established over a long period of time by both traditional wisdom and scientific studies. However, the presence of antinutrients (as discussed previously) limits the bioavailability and bioaccessibility of the valued nutrients present in flaxseed; some processing and treatment can overcome these challenges. Of all the antinutrients present in flaxseed, cyanogenic glycosides are of the most concern and must be removed (Dzuvor et al., 2018). Various simple techniques and measures can be adapted for this purpose, which are listed in Table 30.6. These approaches are dependent on the type and intended use of the final

TABLE 30.5

Application of Flaxseed and its Products in Functional Food Development

Food Product	Flaxseed Component & Its Dosage (w/w)	Observed Results	Reference
Dahi	• Microencapsulated flaxseed oil powder • 1–3%	• Improved fatty acid profile of dahi • No major effect on sensorial properties	Goyal et al., 2016
Wheat bread	• Flaxseed flour • Full fat (10%) • Defatted (15%) • Full fat + defatted (10% + 15%)	• Acceptable product obtained • Improvement in fat, fiber, and protein levels • Decrease in serum LDL, VLDL, total cholesterol • Increase in serum HDL	Mervat et al., 2015

Functional bread	• Raw and roasted ground flaxseed flour
	• 5%, 10%, and 15%
		• Increase in ω-3 levels
		• With increase in flaxseed levels:
		  • more water absorption leading to dough stickiness
		  • increase in crumb softness
			Marpalle et al., 2015

Whole wheat bread	• Roasted brown flaxseed flour • 10%	• 1.51 g ALA and 0.15 g ω-6 (linoleic acid) content per 100 g bread • Lower protein digestibility compared to control	Marpalle et al., 2015
Omega-3 rice paper	• Whole flaxseed flour • 10%	• High antioxidant value compared to traditional rice paper	Cameron and Hosseinian, 2013
Whole wheat bread	• Germinated or nongerminated flaxseed flour • 0%, 5%, 10%, 15%, and 20%	• Combination yielded unique and healthy fatty acid profile • Germination resulted in 57% increase in linoleic acid and 20% decrease in linolenic acid • Decrease in linolenic acid content adds to the stability of the product	Kaur et al., 2013
Cookies	• Whole flaxseed flour • 0–18%	• Decrease in cookie dough stickiness with flaxseed flour • 6% and 12% supplementation was found acceptable	Khouryieh and Aramouni, 2012
Unleavened flat bread	• Full-fat and partially defatted flaxseed flour • 4–20%	• Increase in fiber and essential fatty acid content • 12% full-fat and 16% defatted flaxseed flour enriched bread were found acceptable	Hussain et al., 2012
Gluten free cookies	• Flaxseed meal • 21%	• Prepared using irradiation (0.5 kGy, 1.0 kGy, and 1.5 kGy) • No significant difference compared to control was observed	Rodrigues et al., 2012
Cheese	• Flaxseed oil • 1%	• High retention of flax oil (5.2 mg/g) was observed • No effect on shelf stability • Improved fatty acid profile	Bermúdez-Aguirre and Barbosa-Cánovas, 2012

TABLE 30.6

A Comparison of Different Techniques for Detoxification of Cyanogenic Glycosides in Flaxseed

Technique	Key Element	Mechanism	Comments
Enzymatic methods	Commercial enzymes such as linamarase, glycosidases, xylanase, and cellulase	Degradation of cyanogenic glycosides to HCN followed by volatilization of HCN	Relatively cheaper approach. Enzyme used should be safe, reproducible, cost-effective, and suitable for large scaling.
Heat treatment	Oven, steam, microwave, or sun heating	Heating above 26°C (boiling point of HCN) volatilizes the cyanogen. The higher the temperature, the higher the volatilization rate of HCN.	Cost effective method, but high temperatures can result in nutritional loss
Solvent extraction	Methane and hexane can remove cyanogenic glycosides	Two-phase solvent extraction system comprising alkanol-ammonia-water/hexane can be used for detoxification	Incomplete detoxification along with reduced fat, protein, fiber, and lignan content is observed

Sources: Bekhit et al., 2018; Dzuvor et al., 2018.

product, i.e., whole seed or flour as a cheaper protein source, ingredient in proprietary foods, functional foods, or nutraceuticals (Bekhit et al., 2018).

30.8 Summary

Flaxseed is a nutrient-rich oilseed with many functional properties responsible for proven health benefits. Low carbohydrates, absence of gluten, adequate protein with better amino acid scores, and richness in ω-3 fatty acids make flaxseed a highly nutritious food. Moreover, the presence of lignans, phytoestrogens, mucilage, vitamins, and minerals add to the functional properties of flaxseed. In this regard, flaxseed meets countless requirements of health-conscious consumers looking for a healthy food alternative. Also, flaxseed offers a cheap source of protein in the form of residual cakes from the oil extraction. However, the presence of antinutrients pose a limitation in flaxseed consumption, which can be easily overcome by relevant detoxification approaches depending on the intended use of the final product. Thus, flaxseed has enough potential to be put under the category of superfood and can be used in many ways, in many products, and for many purposes.

REFERENCES

Adlercreutz, H. 2007. Lignans and human health. *Critical Reviews in Clinical Laboratory Sciences 44*: 483–525.

Adolphe, J. L., Whiting, S. J., Juurlink, B. H., Thorpe, L. U., & Alcorn, J. 2010. Health effects with consumption of the flax lignan secoisolariciresinol diglucoside. *British Journal of Nutrition 103*: 929–938.

Aliani, M., Ryland, D., & Pierce, G. N. 2012. Effect of flax addition on the flavor profile and acceptability of bagels. *Journal of Food Science 77*: S62–S70.

Al-Okbi, S. Y. 2005. Highlights on functional foods, with special reference to flaxseed. *Journal of Natural Fibers 2*: 63–68.

Amin, T., & Thakur, M. 2014. Linum usitatissimum L. (Flaxseed)—A multifarious functional food. *Online International Interdisciplinary Research Journal 4*: 220–238.

Arjmandi, B. H., Lucas, E. A., Khalil, D. A., Devareddy, L., Smith, B. J., McDonald, J., ... & Mason, C. 2005. One year soy protein supplementation has positive effects on bone formation markers but not bone density in postmenopausal women. *Nutrition Journal 4*: 8.

Austria, J. A., Richard, M. N., Chahine, M. N., Edel, A. L., Malcolmson, L. J., Dupasquier, C. M., & Pierce, G. N. 2008. Bioavailability of alpha-linolenic acid in subjects after ingestion of three different forms of flaxseed. *Journal of the American College of Nutrition 27*: 214–221.

Ayelign, A., & Alemu, T. 2016. The Functional nutrients of flaxseed and their effect on human health: A review. *European Journal of Nutrition & Food Safety 6*: 83–92.

Beejmohun, V., Fliniaux, O., Grand, É., Lamblin, F., Bensaddek, L., Christen, P., ... & Mesnard, F. 2007. Microwave-assisted extraction of the main phenolic compounds in flaxseed. *Phytochemical Analysis 18*: 275–282.

Bekhit, A. E. D. A., Shavandi, A., Jodjaja, T., Birch, J., Teh, S., Ahmed, I. A. M., ... & Bekhit, A. A. 2018. Flaxseed: Composition, detoxification, utilization, and opportunities. *Biocatalysis and Agricultural Biotechnology 13*: 129–152.

Bermúdez-Aguirre, D., & Barbosa-Cánovas, G. V. 2012. Fortification of queso fresco, cheddar and mozzarella cheese using selected sources of omega-3 and some nonthermal approaches. *Food Chemistry 133*: 787–797.

Bernacchia, R., Preti, R., & Vinci, G. 2014. Chemical composition and health benefits of flaxseed. *Austin Journal of Nutrition and Food Sciences 2*: 1045.

Bhathena, S. J., Ali, A. A., Haudenschild, C., Latham, P., Ranich, T., Mohamed, A. I., ... & Velasquez, M. T. 2003. Dietary flaxseed meal is more protective than soy protein concentrate against hypertriglyceridemia and steatosis of the liver in an animal model of obesity. *Journal of the American College of Nutrition 22*: 157–164.

Bilek, A. E., & Turhan, S. 2009. Enhancement of the nutritional status of beef patties by adding flaxseed flour. *Meat Science 82*: 472–477.

Bloedon, L. T., Balikai, S., Chittams, J., Cunnane, S. C., Berlin, J. A., Rader, D. J., & Szapary, P. O. 2008. Flaxseed and cardiovascular risk factors: Results from a double blind, randomized, controlled clinical trial. *Journal of the American College of Nutrition 27*: 65–74.

Cameron, S. J., & Hosseinian, F. 2013. Potential of flaxseed in the development of omega-3 rice paper with antioxidant activity. *LWT Food Science and Technology 53*: 170–175.

Chen, J., Saggar, J. K., Corey, P., & Thompson, L. U. 2009. Flaxseed and pure secoisolariciresinol diglucoside, but not flaxseed hull, reduce human breast tumour growth (MCF-7) in athymic mice. *The Journal of Nutrition 139*: 2061–2066.

Chen, J., Saggar, J. K., Corey, P., & Thompson, L. U. 2011. Flaxseed cotyledon fraction reduces tumour growth and sensitises tamoxifen treatment of human breast cancer xenograft (MCF-7) in athymic mice. *British Journal of Nutrition 105*: 339–347.

Chung, M. W. Y., Lei, B., & Li-Chan, E. C. Y. 2005. Isolation and structural characterization of the major protein fraction from NorMan flaxseed (Linum usitatissimum L.). *Food Chemistry 90*: 271–279.

Conforti, F. D., & Davis, S. F. 2006. The effect of soya flour and flaxseed as a partial replacement for bread flour in yeast bread. *International Journal of Food Science and Technology 41*: 95–101.

Daun, J. K., Barthet, V. J., Chornick, T. L., & Duguid, S. 2003. Structure, composition, and variety development of flaxseed. In: L. U. Thompson & S. C. Cunnane (Eds.), *Flaxseed in Human Nutrition, 6–45*. AOCS Publishing.

de França Cardozo, L. F. M., Boaventura, G. T., Brant, L. H. C., Pereira, V. A., Velarde, L. G. C., & Chagas, M. A. 2012. Prolonged consumption of flaxseed flour increases the 17β-estradiol hormone without causing adverse effects on the histomorphology of Wistar rats' penis. *Food and Chemical Toxicology 50*: 4092–4096.

Dubois, V., Breton, S., Linder, M., Fanni, J., & Parmentier, M. 2007. Fatty acid profiles of 80 vegetable oils with regard to their nutritional potential. *European Journal of Lipid Science and Technology 109*: 710–732.

Dugani, A., Auzzi, A., Naas, F., & Megwez, S. 2008. Effects of the oil and mucilage from flaxseed (linum usitatissimum) on gastric lesions induced by ethanol in rats. *Libyan Journal of Medicine 3*: 1–5.

Dupasquier, C. M., Dibrov, E., Kneesh, A. L., Cheung, P. K., Lee, K. G., Alexander, H. K., ... & Pierce, G. N. 2007. Dietary flaxseed inhibits atherosclerosis in the LDL receptor-deficient mouse in part through antiproliferative and anti-inflammatory actions. *American Journal of Physiology–Heart and Circulatory Physiology 293*: H2394–H2402.

Dzuvor, C. K. O., Taylor, J. T., Acquah, C., Pan, S., & Agyei, D. 2018. Bioprocessing of functional ingredients from flaxseed. *Molecules 23*: 2444.

El-Beltagi, H. S., Salama, Z. A., & El-Hariri, D. M. 2007. Evaluation of fatty acids profile and the content of some secondary metabolites in seeds of different flax cultivars (Linum usitatissimum L.). *General and Applied Plant Physiology 33*: 187–202.

Elboutachfaiti, R., Delattre, C., Quéro, A., Roulard, R., Duchêne, J., Mesnard, F., & Petit, E. 2017. Fractionation and structural characterization of six purified rhamnogalacturonans type I from flaxseed mucilage. *Food Hydrocolloids 62*: 273–279.

Fardet, A. 2010. New hypotheses for the health-protective mechanisms of whole-grain cereals: What is beyond fibre? *Nutrition Research Reviews 23*: 65–134.

Feng, D., Shen, Y., & Chavez, E. R. 2003. Effectiveness of different processing methods in reducing hydrogen cyanide content of flaxseed. *Journal of the Science of Food and Agriculture 83*: 836–841.

Flax Council of Canada. 2020. Retrieved from: https://flaxcouncil.ca/ (accessed on 12 September, 2020).

Fodje, A. M., Chang, P. R., & Leterme, P. 2009. In vitro bile acid binding and short-chain fatty acid profile of flax fibre and ethanol co-products. *Journal of Medicinal Food 12*: 1065–1073.

Fuchs, D., Piller, R., Linseisen, J., Daniel, H., & Wenzel, U. 2007. The human peripheral blood mononuclear cell proteome responds to a dietary flaxseed-intervention and proteins identified suggest a protective effect in atherosclerosis. *Proteomics 7*: 3278–3288.

Fukumitsu, S., Aida, K., Shimizu, H., & Toyoda, K. 2010. Flaxseed lignan lowers blood cholesterol and decreases liver disease risk factors in moderately hypercholesterolemic men. *Nutrition Research 30*: 441–446.

Ganorkar, P. M., & Jain, R. K. 2013. Flaxseed-a nutritional punch. *International Food Research Journal 20*: 519–525.

Ganorkar, P. M., & Jain, R. K. 2014. Effect of flaxseed incorporation on physical, sensorial, textural and chemical attributes of cookies. *International Food Research Journal 21*.

Gat, Y., & Ananthanarayan, L. 2015. Physicochemical, phytochemical and nutritional impact of fortified cereal-based extrudate snacks. *Nutrafoods 14*: 141–149.

Gebauer, S. K., Psota, T. L., Harris, W. S., & Kris-Etherton, P. M. 2006. n-3 fatty acid dietary recommendations and food sources to achieve essentiality and cardiovascular benefits. *The American Journal of Clinical Nutrition 83*: 1526S–1535S.

Gillingham, L. G., Gustafson, J. A., Han, S. Y., Jassal, D. S., & Jones, P. J. 2011. High-oleic rapeseed (canola) and flaxseed oils modulate serum lipids and inflammatory biomarkers in hypercholesterolaemic subjects. *British Journal of Nutrition 105*: 417–427.

Gopalan, C., Sastri, R., & Balasubramanian, S. C. 2004. *Nutritive Value of Indian Foods*. National Institute of Nutrition. Hyderabad: ICMR.

Gopinath, B., Harris, D. C., Flood, V. M., Burlutsky, G., & Mitchell, P. 2011. Consumption of long-chain n-3 PUFA, α-linolenic acid and fish is associated with the prevalence of chronic kidney disease. *British Journal of Nutrition 105*: 1361–1368.

Goyal, A., Sharma, V., Sihag, M. K., Singh, A. K., Arora, S., & Sabikhi, L. 2016. Fortification of dahi (Indian yoghurt) with omega-3 fatty acids using microencapsulated flaxseed oil microcapsules. *Journal of Food Science and Technology 53*: 2422–2433.

Goyal, A., Sharma, V., Upadhyay, N., Gill, S., & Sihag, M. 2014. Flax and flaxseed oil: An ancient medicine & modern functional food. *Journal of Food Science and Technology 51*: 1633–1653.

Guerrero-Beltrán, J. A., Estrada-Girón, Y., Swanson, B. G., & Barbosa-Cánovas, G. V. 2009. Pressure and temperature combination for inactivation of soymilk trypsin inhibitors. *Food Chemistry 116*: 676–679.

Gutiérrez, C., Rubilar, M., Jara, C., Verdugo, M., Sineiro, J., & Shene, C. 2010. Flaxseed and flaxseed cake as a source of compounds for food industry. *Journal of Soil Science and Plant Nutrition 10*: 454–463.

HadiNezhad, M., Duc, C., Han, N. F., & Hosseinian, F. 2013. Flaxseed soluble dietary fibre enhances lactic acid bacterial survival and growth in kefir and possesses high antioxidant capacity. *Journal of Food Research 2*: 152–152.

Hadnađev, M. S., Hadnađev-Dapčević, T., Pojić, M. M., Šarić, B. M., Mišan, A. Č, Jovanov, P. T., & Sakač, M. B. 2017. Progress in vegetable proteins isolation techniques: A review. *Food and Feed Research 44*: 11–21.

Hall, C., Tulbek, M. C., & Xu, Y. 2006. Flaxseed. *Advances in Food and Nutrition Research 51*: 1–97.

Hallund, J., Tetens, I., Bügel, S., Tholstrup, T., & Bruun, J. M. 2008. The effect of a lignan complex isolated from flaxseed on inflammation markers in healthy postmenopausal women. *Nutrition, Metabolism and Cardiovascular Diseases 18*: 497–502.

Hao, M., & Beta, T. 2012a. Development of Chinese steamed bread enriched in bioactive compounds from barley hull and flaxseed hull extracts. *Food Chemistry 133*: 1320–1325.

Hao, M., & Beta, T. 2012b. Qualitative and quantitative analysis of the major phenolic compounds as antioxidants in barley and flaxseed hulls using HPLC/MS/MS. *Journal of the Science of Food and Agriculture* 92: 2062–2068.

Ho, C. H., Cacace, J. E., & Mazza, G. 2007. Extraction of lignans, proteins and carbohydrates from flaxseed meal with pressurized low polarity water. *LWT Food Science and Technology* 40: 1637–1647.

Hongzhi, Y., Zhihuai, H. T., & Hequn, T. 2004. Determination and removal methods of cyanogenic glucoside in flaxseed. In: *Proceedings from the 2004 ASAE Annual Meeting.* American Society of Agricultural and Biological Engineers, St. Joseph, Michigan: 04066.

Hu, C., Yuan, Y. V., & Kitts, D. D. 2007. Antioxidant activities of the flaxseed lignan secoisolariciresinol diglucoside, its aglycone secoisolariciresinol and the mammalian lignans enterodiol and enterolactone in vitro. *Food and Chemical Toxicology* 45: 2219–2227.

Hussain, S., Anjum, F. M., Butt, M. S., Alamri, M. S., & Khan, M. R. 2012. Biochemical and nutritional evaluation of unleavened flat breads fortified with healthy flaxseed. *International Journal of Agriculture and Biology* 14: 190–196.

Ivanov, S., Rashevskaya, T., & Makhonina, M. 2011. Flaxseed additive application in dairy products production. *Procedia Food Science* 1: 275–280.

Johnsson, P., Peerlkamp, N., Kamal-Eldin, A., Andersson, R. E., Andersson, R., Lundgren, L. N., & Åman, P. 2002. Polymeric fractions containing phenol glucosides in flaxseed. *Food Chemistry* 76: 207–212.

Kadam, S. U., & Prabhasankar, P. 2010. Marine foods as functional ingredients in bakery and pasta products. *Food Research International* 43: 1975–1980.

Kajla, P., Sharma, A., & Sood, D. R. 2015. Flaxseed-a potential functional food source. *Journal of Food Science and Technology* 52: 1857–1871.

Kapoor, S., Sachdeva, R., & Kochhar, A. 2011. Efficacy of flaxseed supplementation on nutrient intake and other lifestyle pattern in menopausal diabetic females. *Studies on Ethno-Medicine* 5: 153–160.

Katare, C., Saxena, S., Agrawal, S., Prasad, G. B. K. S., & Bisen, P. S. 2012. Flax seed: A potential medicinal food. *Journal of Nutrition & Food Sciences* 2: 120.

Kaul, N., Kreml, R., Austria, J. A., Richard, M. N., Edel, A. L., Dibrov, E., ... & Pierce, G. N. 2008. A comparison of fish oil, flaxseed oil and hempseed oil supplementation on selected parameters of cardiovascular health in healthy volunteers. *Journal of the American College of Nutrition* 27: 51–58.

Kaur, A., Sandhu, V., & Sandhu, K. S. 2013. Effects of flaxseed addition on sensory and baking quality of whole wheat bread. *International Journal of Food Nutrition and Safety* 4: 43–54.

Kaur, M., Singh, V., & Kaur, R. 2017. Effect of partial replacement of wheat flour with varying levels of flaxseed flour on physicochemical, antioxidant and sensory characteristics of cookies. *Bioactive Carbohydrates and Dietary Fibre* 9: 14–20.

Kaur, P., Sharma, P., Kumar, V., Panghal, A., Kaur, J., & Gat, Y. 2019. Effect of addition of flaxseed flour on phytochemical, physicochemical, nutritional, and textural properties of cookies. *Journal of the Saudi Society of Agricultural Sciences* 18: 372–377.

Kaur, P., Waghmare, R., Kumar, V., Rasane, P., Kaur, S., & Gat, Y. 2018a. Recent advances in utilization of flaxseed as potential source for value addition. *Oilseeds and fats, Crops and Lipids* 25: A304.

Kaur, R., & Kaur, M. 2018b. Microstructural, physicochemical, antioxidant, textural and quality characteristics of wheat muffins as influenced by partial replacement with ground flaxseed. *LWT Food Science and Technology* 91: 278–285.

Kaur, R., Kaur, M., & Purewal, S. S. 2018c. Effect of incorporation of flaxseed to wheat rusks: Antioxidant, nutritional, sensory characteristics, and in vitro DNA damage protection activity. *Journal of Food Processing and Preservation* 42: e13585.

Khouryieh, H., & Aramouni, F. 2012. Physical and sensory characteristics of cookies prepared with flaxseed flour. *Journal of the Science of Food and Agriculture* 92: 2366–2372.

Koca, A. F., & Anil, M. 2007. Effect of flaxseed and wheat flour blends on dough rheology and bread quality. *Journal of the Science of Food and Agriculture* 87: 1172–1175.

Korkina, L., Kostyuk, V., De Luca, C., & Pastore, S. 2011. Plant phenylpropanoids as emerging anti-inflammatory agents. *Mini Reviews in Medicinal Chemistry* 11: 823–835.

Kristensen, M., Jensen, M. G., Aarestrup, J., Petersen, K. E., Søndergaard, L., Mikkelsen, M. S., & Astrup, A. 2012. Flaxseed dietary fibers lower cholesterol and increase fecal fat excretion, but magnitude of effect depend on food type. *Nutrition & Metabolism* 9: 8.

Kristensen, M., Savorani, F., Christensen, S., Engelsen, S. B., Bügel, S., Toubro, S., … & Astrup, A. 2013. Flaxseed dietary fibers suppress postprandial lipemia and appetite sensation in young men. *Nutrition, Metabolism and Cardiovascular Diseases 23*: 136–143.

Lemay, A., Dodin, S., Kadri, N., Jacques, H., & Forest, J. C. 2002. Flaxseed dietary supplement versus hormone replacement therapy in hypercholesterolemic menopausal women. *Obstetrics & Gynecology 100*: 495–504.

Leyva, D. R., Zahradka, P., Ramjiawan, B., Guzman, R., Aliani, M., & Pierce, G. N. 2011. The effect of dietary flaxseed on improving symptoms of cardiovascular disease in patients with peripheral artery disease: Rationale and design of the FLAX-PAD randomized controlled trial. *Contemporary Clinical Trials 32*: 724–730.

Malomo, S. A., & Aluko, R. E. 2015. A comparative study of the structural and functional properties of isolated hemp seed (*Cannabis sativa* L.) albumin and globulin fractions. *Food Hydrocolloids 43*: 743–752.

Mani, U. V., Mani, I., Biswas, M., & Kumar, S. N. 2011. An open-label study on the effect of flax seed powder (*Linum usitatissimum*) supplementation in the management of diabetes mellitus. *Journal of Dietary Supplements 8*: 257–265.

Manthey, F. A., Sinha, S., Wolf-Hall, C. E., & Hall III, C. A. 2008. Effect of flaxseed flour and packaging on shelf life of refrigerated pasta. *Journal of Food Processing and Preservation 32*: 75–87.

Marambe, H. K., & Wanasundara, J. P. D. 2017. Protein from flaxseed (Linum usitatissimum L.). In: S. R. Nadathur, J. P.D. Wanasundara, & L. Scanlin (Eds.), *Sustainable Protein Sources,* 133–144. Academic Press.

Marambe, P. W. M. L. H. K., Shand, P. J., & Wanasundara, J. P. D. 2008. An in-vitro investigation of selected biological activities of hydrolysed flaxseed (Linum usitatissimum L.) proteins. *Journal of the American Oil Chemists' Society 85*: 1155–1164.

Marpalle, P., Sonawane, S. K., LeBlanc, J. G., & Arya, S. S. 2015. Nutritional characterization and oxidative stability of α-linolenic acid in bread containing roasted ground flaxseed. *LWT Food Science and Technology 61*: 510–515.

Martinchik, A. N., Baturin, A. K., Zubtsov, V. V., & Molofeev, V. 2012. Nutritional value and functional properties of flaxseed. *Voprosy Pitaniia 81*: 4–10.

Mason, J. K., & Thompson, L. U. 2014. Flaxseed and its lignan and oil components: Can they play a role in reducing the risk of and improving the treatment of breast cancer? *Applied Physiology, Nutrition, and Metabolism 39*: 663–678.

Mazza, G. 2008. Production, processing and uses of Canadian flax. In: *First CGNA International Workshop,* 3–6. Temuco, Chile.

McCann, S. E., Wactawski-Wende, J., Kufel, K., Olson, J., Ovando, B., Kadlubar, S. N., … & Freudenheim, J. L. 2007. Changes in 2-hydroxyestrone and 16α-hydroxyestrone metabolism with flaxseed consumption: modification by COMT and CYP1B1 genotype. *Cancer Epidemiology and Prevention Biomarkers 16*: 256–262.

McCullough, R. S., Edel, A. L., Bassett, C. M., LaVallée, R. K., Dibrov, E., Blackwood, D. P., … & Pierce, G. N. 2011. The alpha linolenic acid content of flaxseed is associated with an induction of adipose leptin expression. *Lipids 46*: 1043–1052.

Meagher, E. A., Barry, O. P., Lawson, J. A., Rokach, J., & FitzGerald, G. A. 2001. Effects of vitamin E on lipid peroxidation in healthy persons. *JAMA 285*: 1178–1182.

Mervat, E.–D., Mahmoud, K. F., Bareh, G. F., & Albadawy, W. 2015. Effect of fortification by full fat and defatted flaxseed flour sensory properties of wheat bread and lipid profile laste. *International Journal of Current Microbiology and Applied Sciences 4*: 581–598.

Moghaddasi, M. S. 2011. Linseed and usages in Human life. *Advances in Environmental Biology,* 1380–1393.

Morisset, A. S., Lemieux, S., Veilleux, A., Bergeron, J., Weisnagel, S. J., & Tchernof, A. 2009. Impact of a lignan-rich diet on adiposity and insulin sensitivity in post-menopausal women. *British Journal of Nutrition 102*: 195–200.

Morris, D. H. 2007. *Flax: A Health and Nutrition Primer.* Winnipeg: Flax Council of Canada.

Ogborn, M. R., Nitschmann, E., Bankovic-Calic, N., Weiler, H. A., & Aukema, H. M. 2006. Effects of flaxseed derivatives in experimental polycystic kidney disease vary with animal gender. *Lipids 41*: 1141–1149.

Omoni, A. O., & Aluko, R. E. 2006. Effect of cationic flaxseed protein hydrolysate fractions on the in vitro structure and activity of calmodulin-dependent endothelial nitric oxide synthase. *Molecular Nutrition & Food Research 50*: 958–966.

Oomah, B. D. 2001. Flaxseed as a functional food source. *Journal of the Science of Food and Agriculture 81*: 889–894.

Pan, A., Sun, J., Chen, Y., Ye, X., Li, H., Yu, Z., ... & Demark-Wahnefried, W. 2007. Effects of a flaxseed-derived lignan supplement in type 2 diabetic patients: A randomized, double-blind, cross-over trial. *PLOS ONE 2*: e1148.

Pan, A., Yu, D., Demark-Wahnefried, W., Franco, O. H., & Lin, X. 2009. Meta-analysis of the effects of flaxseed interventions on blood lipids. *The American Journal of Clinical Nutrition 90*: 288–297.

Park, E. R., Hong, J. H., Lee, D. H., Han, S. B., Lee, K. B., Park, J. S., ... & Kim, M. C. 2005. Analysis and decrease of cyanogenic glucosides in flaxseed. *Journal of the Korean Society of Food Science and Nutrition 34*: 875–879.

Park, J. B., & Velasquez, M. T. 2012. Potential effects of lignan-enriched flaxseed powder on bodyweight, visceral fat, lipid profile, and blood pressure in rats. *Fitoterapia 83*: 941–946.

Paschos, G. K., Zampelas, A., Panagiotakos, D. B., Katsiougiannis, S., Griffin, B. A., Votteas, V., & Skopouli, F. N. 2007. Effects of flaxseed oil supplementation on plasma adiponectin levels in dyslipidemic men. *European Journal of Nutrition 46*: 315–320.

Patade, A., Devareddy, L., Lucas, E. A., Korlagunta, K., Daggy, B. P., & Arjmandi, B. H. 2008. Flaxseed reduces total and LDL cholesterol concentrations in Native American postmenopausal women. *Journal of Women's Health 17*: 355–366.

Pellizzon, M. A., Billheimer, J. T., Bloedon, L. T., Szapary, P. O., & Rader, D. J. 2007. Flaxseed reduces plasma cholesterol levels in hypercholesterolemic mouse models. *Journal of the American College of Nutrition 26*: 66–75.

Pohjanheimo, T. A., Hakala, M. A., Tahvonen, R. L., Salminen, S. J., & Kallio, H. P. 2006a. Flaxseed in breadmaking: Effects on sensory quality, aging, and composition of bakery products. *Journal of Food Science 71*: S343–S348.

Pohjanheimo, T., Hakala, M., & Kallio, H. 2006b. Effect of baking process and storage on volatile composition of flaxseed breads. In: W. Bredie, & M. Petersen (eds), *Developments in Food Science (Vol. 43)* 339–342. Elsevier.

Pourabedin, M., Aarabi, A., & Rahbaran, S. 2017. Effect of flaxseed flour on rheological properties, staling and total phenol of Iranian toast. *Journal of Cereal Science 76*: 173–178.

Prasad, K. 2009. Flaxseed and cardiovascular health. *Journal of Cardiovascular Pharmacology 54*: 369–377.

Puvača, N., Stanaćev, V., Milić, D., Kokić, B., & Čabarkapa, I. 2012. Limitation of flaxseed usage in animal nutrition. In: *XV International Feed Technology Symposium. COST—"Feed for Health" joint Workshop Proceedings,* 58–63. Novi Sad, Serbia: Institute of Food Technology.

Rajesha, J., Murthy, K. N. C., Kumar, M. K., Madhusudhan, B., & Ravishankar, G. A. 2006. Antioxidant potentials of flaxseed by in vivo model. *Journal of Agricultural and Food Chemistry 54*: 3794–3799.

Regitz-Zagrosek, V., Wintermantel, T. M., & Schubert, C. 2007. Estrogens and SERMs in coronary heart disease. *Current Opinion in Pharmacology 7*: 130–139.

Repin, N., Kay, B. A., Cui, S. W., Wright, A. J., Duncan, A. M., & Goff, H. D. 2017. Investigation of mechanisms involved in postprandial glycemia and insulinemia attenuation with dietary fibre consumption. *Food & Function 8*: 2142–2154.

Rodrigues, F. T., Fanaro, G. B., Duarte, R. C., Koike, A. C., & Villavicencio, A. L. C. 2012. A sensory evaluation of irradiated cookies made from flaxseed meal. *Radiation Physics and Chemistry 81*: 1157–1159.

Rommi, K., Ercili-Cura, D., Hakala, T. K., Nordlund, E., Poutanen, K., & Lantto, R. 2015. Impact of total solid content and extraction pH on enzyme-aided recovery of protein from defatted rapeseed (Brassica rapa L.) press cake and physicochemical properties of the protein fractions. *Journal of Agricultural and Food Chemistry 63*: 2997–3003.

Saggar, J. K., Chen, J., Corey, P., & Thompson, L. U. 2010. The effect of secoisolariciresinol diglucoside and flaxseed oil, alone and in combination, on MCF-7 tumor growth and signaling pathways. *Nutrition and Cancer 62*: 533–542.

Shakir, K. F., & Madhusudhan, B. 2007. Effects of flaxseed (Linum usitatissimum) chutney on gamma-glutamyl transpeptidase and micronuclei profile in azoxymethane treated rats. *Indian Journal of Clinical Biochemistry 22*: 129.

Shearer, A. E., & Davies, C. G. 2005. Physicochemical properties of freshly baked and stored whole-wheat muffins with and without flaxseed meal. *Journal of Food Quality 28*: 137–153.

Singh, K. K., Mridula, D., Rehal, J., & Barnwal, P. 2011. Flaxseed: A potential source of food, feed and fiber. *Critical Reviews in Food Science and Nutrition 51*: 210–222.

Sinha, S., & Manthey, F. A. 2008. Semolina and hydration level during extrusion affect quality of fresh pasta containing flaxseed flour. *Journal of Food Processing and Preservation 32*: 546–559.

Stavro, P. M., Marchie, A. L., Kendall, C. W., Vuksan, V., & Jenkins, D. J. 2003. Flaxseed, fiber, and coronary heart disease: Clinical studies. In: L. U. Thompson, & S. C. Cunnane (Eds.), *Flaxseed in Human Nutrition, 288-300*. Champaign, Illinois: AOCS Press.

Tarpila, A., Wennberg, T., & Tarpila, S. 2005. Flaxseed as a functional food. *Current Topics in Nutraceutical Research 3*: 167–188.

Technavio. (2020). *Flax Seeds Market by Product and Geography - Forecast and Analysis 2020-2024*. Retrieved from https://www.technavio.com/report/flax-seeds-market-industry-analysis.

Teh, S. S., Bekhit, A. E. D. A., Carne, A., & Birch, J. 2016. Antioxidant and ACE-inhibitory activities of hemp (Cannabis sativa L.) protein hydrolysates produced by the proteases AFP, HT, Pro-G, actinidin and zingibain. *Food Chemistry 203*: 199–206.

Teh, S. S., Bekhit, A. E. D., Carne, A., & Birch, J. 2014. Effect of the defatting process, acid and alkali extraction on the physicochemical and functional properties of hemp, flax and canola seed cake protein isolates. *Journal of Food Measurement and Characterization 8*: 92–104.

Thompson, L. U., & Cunnane, S. C. 2003. *Flaxseed in Human Nutrition*. Illinois: AOCS Publishing.

Tirgar, M., Silcock, P., Carne, A., & Birch, E. J. 2017. Effect of extraction method on functional properties of flaxseed protein concentrates. *Food Chemistry 215*: 417–424.

Tolkachev, O. N., & Zhuchenko, A. A. 2004. Biologically active substances of flax: Medicinal and nutritional properties (a review). *Pharmaceutical Chemistry Journal 34*: 360–367.

Truan, J. S., Chen, J. M., & Thompson, L. U. 2012. Comparative effects of sesame seed lignan and flaxseed lignan in reducing the growth of human breast tumors (MCF-7) at high levels of circulating estrogen in athymic mice. *Nutrition and Cancer 64*: 65–71.

USDA. (2016). Retrieved June 10, 2020, from United States Department of Agriculture: https://ndb.nal.usda.gov/ndb/foods/show/3716?manu=&fgcd=

Vaisey-Genser, M., & Morris, D. H. 2003. Introduction: History of the cultivation and uses of flaxseed. In: A. D. Muir, & N. D. Westcott (Eds.), Flax, 13–33. London: CRC Press.

Velasquez, M. T., Bhathena, S. A. M. J., Ranich, T., Schwartz, A. M., Kardon, D. E., Ali, A. L. I. A., … & Hansen, C. T. 2003. Dietary flaxseed meal reduces proteinuria and ameliorates nephropathy in an animal model of type II diabetes mellitus. *Kidney International 64*: 2100–2107.

Waldschläger, J., Bergemann, C., Ruth, W., Effmert, U., Jeschke, U., Richter, D. U., … & Briese, V. 2005. Flaxseed extracts with phytoestrogenic effects on a hormone receptor-positive tumour cell line. *Anticancer Research 25*: 1817–1822.

Wang, B. O., Wang, L. J., Li, D., Adhikari, B., & Shi, J. 2011. Effect of gum Arabic on stability of oil-in-water emulsion stabilized by flaxseed and soybean protein. *Carbohydrate Polymers 86*: 343–351.

Wang, B., Li, D., Wang, L. J., & Özkan, N. 2010a. Effect of concentrated flaxseed protein on the stability and rheological properties of soybean oil-in-water emulsions. *Journal of Food Engineering 96*: 555–561.

Wang, B., Li, D., Wang, L. J., Adhikari, B., & Shi, J. 2010b. Ability of flaxseed and soybean protein concentrates to stabilize oil-in-water emulsions. *Journal of Food Engineering 100*: 417–426.

Warrand, J., Michaud, P., Picton, L., Muller, G., Courtois, B., Ralainirina, R., & Courtois, J. 2005. Structural investigations of the neutral polysaccharide of Linum usitatissimum L. seeds mucilage. *International Journal of Biological Macromolecules 35*: 121–125.

Watkins, B. A., Devitt, A. A., & Feng, S. 2001. Designed Eggs Containing Conjugated Linoleic Acids and Omega-3 Polyunsaturated Fatty Acids. In: A. P. Simopoulos (Edd.), *Nutrition and Fitness: Metabolic Studies in Health and Disease, 162–182*. Switzerland: Karger Publishers.

Winter, R. 2013. *Vitamin E: Your Protection Against Exercise Fatigue, Weakened Immunity, Heart Disease, Cancer, Aging, Diabetic Damage, Environmental Toxins*. Three Rivers Press.

Wu, H., Pan, A., Yu, Z., Qi, Q., Lu, L., Zhang, G., … & Tang, L. 2010. Lifestyle counselling and supplementation with flaxseed or walnuts influence the management of metabolic syndrome. *The Journal of Nutrition 140*: 1937–1942.

Wu, S., Wang, X., Qi, W., & Guo, Q. 2019. Bioactive protein/peptides of flaxseed: A review. *Trends in Food Science & Technology 92*: 184–193.

Xu, Y., Hall Iii, C., & Wolf-Hall, C. 2008. Antifungal activity stability of flaxseed protein extract using response surface methodology. *Journal of Food Science 73*: M9–M14.

Xu, Y., Hall III, C., Wolf-Hall, C., & Manthey, F. 2008. Fungistatic activity of flaxseed in potato dextrose agar and a fresh noodle system. *International Journal of Food Microbiology 121*: 262–267.

Yamashita, T., Sano, T., Hashimoto, T., & Kanazawa, K. 2007. Development of a method to remove cyanogen glycosides from flaxseed meal. *International Journal of Food Science & Technology 42*: 70–75.

Yograj, S., Gupta, G., Gupta, V., Bhat, A. N., & Arora, A. 2017. Flaxseed—A shield against diseases? *Anatomy Physiology and Biochemistry International Journal 1*: 001–005.

Zhang, W., Wang, X., Liu, Y., Tian, H., Flickinger, B., Empie, M. W., & Sun, S. Z. 2008. Dietary flaxseed lignan extract lowers plasma cholesterol and glucose concentrations in hypercholesterolaemic subjects. *British Journal of Nutrition 99*: 1301–1309.

31

Soy Bean Processing and Utilization

Nikita Wadhawan, Sagar M. Chavan, N.K. Jain and Seema Tanwar
College of Technology and Engineering, Udaipur, Rajasthan, India

CONTENTS

31.1 Introduction

Soy beans (*Glycine max.*) often called the "golden miracle bean," are a leguminous crop and rich in proteins. Soy beans originated in Southeast Asia, first domesticated by Chinese farmers around 1100 B.C. By the first century A.D., soy beans were grown in Japan and many other countries. In 1904, the famous

American scientist George Washington Carver discovered that soy beans are a valuable source of protein and oil. The United States, Brazil, Argentina, China, and India are the top producers of soy beans in the world. The total production of soy beans worldwide is about 276.4 million tons from a total acreage of about 111.27 million ha. In India, soy beans are produced abundantly in Madhya Pradesh, Maharashtra, Rajasthan, Andhra Pradesh, Karnataka, Chhattisgarh, and Gujarat. In India, too, unprecedented growth has been witnessed in soy beans. The mean area has increased from 0.03 m ha. in 1970 to 9.30 m ha. in 2010. The mean national productivity has also increased from 0.43 t/ha in 1970 to 1.36 t/ha in 2010. The total production of soy beans in India, during the year 2013, was reported to be 11.95 million tons from an area of about 12.2 m ha. (Source: Anonymous, 2014). It is also reported that India is the secondary center for domestication of crops in the world after China. At present, the country stands fifth in area and production of soybean in world after the US, Brazil, Argentina, and China.

31.1.1 Indian Soy Bean Crop Calendar

- **Mainly grown in kharif season**: Monsoon
- **Sowing**: June–July
- **Harvest**: October–December

31.2 Composition and Physical Properties of Soy Beans

Unlike other legumes, soy beans have a unique chemical composition. They have a variety of nutritional and health-promoting components (Kökten et al, 2013.) like isoflavones, tocopherol, lecithin, etc., in addition to containing 20% oil and 40% protein. Whole soy beans or bean cotyledons are utilized for commercial soy products. Soy bean composition, highlighted in Table 31.1, includes varying amounts of protein (38–42%) and fat (18%) of which 85% is unsaturated and high in linoleic and linolenic acids (a precursor to omega-3 fatty acids), 23% oleic acid and 16% palmitic acid, which is saturated. Most fatty acids in soy beans and their derivatives are unsaturated and therefore susceptible to oxidation (Penalvo *et al.*, 2004). Other components include varying concentrations of high levels of minerals, including iron, calcium, and zinc and vitamins including α-tocopherol, niacin, pyridoxine, and folacin. Soy beans are also a good source of antinutrient factors such as saponins, phospholipids, protease inhibitors, phytates, and trypsin inhibitors. Soy bean oil is also a good source of omega-3 and omega-6 fatty acids, similar to those found in fish oils, and is cholesterol free. Soy beans are an excellent source of both soluble and insoluble dietary fiber. Soluble fiber may help lower serum cholesterol and control blood sugar. Insoluble fiber increases stool bulk, may prevent colon cancer, and can help relieve symptoms of some digestive

TABLE 31.1

Nutrient Composition of Soy Beans (per 100 g)

Constituent	Quantity	Constituent	Quantity
Energy	1,866 kJ (446 kcal)	Tryptophan	0.391 g
Protein	40.49 g	Methionine	0.547 g
Carbohydrates	20.16 g	Threonine	1.766 g
Dietary fiber	7.3 g	Leucine	1.309 g
Fat	19.94 g	Lysine	2.706 g
Arginine	1.153 g	Histidine	1.097 g
Alanine	1.715 g	Aspartic acid	0.112 g
Proline	1.379 g	Serine	0.357 g
Vitamin B6	0.377 mg (29%)	Vitamin C	6.0 mg (10%)
Oil	18%		

Source: Baisya, 2008.

disorders. Soy beans have more than two times the amount of most of the minerals, especially calcium, iron, phosphorus, and zinc than any other legume and a very low sodium content. Soy beans have all the important vitamins and are a very good source for B complex vitamins and vitamin E. Soy foods are being recognized as having potential roles in the prevention and treatment of chronic diseases, cancer, heart disease, and kidney disease (Messina, 1995). The beans have almost 40% protein, making soy beans higher in protein than any other legumes and many animal products. The protein in just 250 g of soy beans is equivalent to the amount of protein in 3 liters of milk, 1 kg of mutton, or 24 eggs. The quality of soy protein is virtually equivalent in quality to that of milk and egg protein. Thus, one can say that soy beans provide health benefits along with high-quality proteins at much cheaper costs.

The physical properties of soy beans are dependent on its moisture content. The moisture content of the beans range from 8.7–25.0% db. The various physical parameters of the grain demonstrate that the length of grain increases from 6.32–6.75 mm, width from 5.23–5.55 mm, thickness from 3.99–4.45 mm, geometric mean diameter from 5.09–5.51 mm, sphericity from 0.806–0.816, the surface area from 0.813–0.952 cm^2, the volume of grain from 0.091–0.113 cm^3 and thousand grain mass from 0.110–0.127 kg. While the kernel density decreases from 1216 kg/m^3 to 1124 kg/m^3, bulk density from 735 kg/m^3 to 708 kg/m^3, and porosity from 0.40 to 0.37.

31.3 Health Benefits of Soy Beans

Soy beans are the world's most important and in-demand seed legume. They contribute about 25% of the global edible oil and about two-thirds of the world's protein concentrate. It is equivalently considered as a replacement for meat, cheese, bread, and oil. Other proven health benefits of soy beans have been shown to help in reducing the risk of different reproductive cancers, combating osteoporosis, alleviating menopause symptoms, and managing blood sugar levels in type 2 diabetic patients. With ongoing scientific evidence on the health benefits of soy beans in reducing the burden of degenerative diseases, these beans are becoming an ingredient in functional foods. Because of sophisticated soy processing technologies, the end product containing this protein ingredient is more palatable and more acceptable. The use of 30–35 g of carefully processed soy beans in a daily diet can help people to have better health. Owing to its high nutritional significance, both for humans and animals, soy beans are also called as "cow of the field" or "gold from the soil."

31.3.1 Export and Import

Trade in soy beans and soy bean products is quite high and soy bean prices in India are greatly affected by the international trade situation and prices. As mentioned before, more than 90% of soy beans produced in India are used for oil extraction and for manufacturing oilcake and oil meal, which in turn is exported in great quantities. As India has a large deficit in edible oil, this deficiency is met through imports, which mainly consist of palm oil and soya oil. Exports constituted more than 50% of soy bean production in the country during the early-to-mid-1990s. Since 1997–1998, however, the proportion of production for export sharply declined. However, in recent years, India has exported only 30% of its soy bean production.

31.4 Soy Bean Processing

Raw soy beans should not be consumed due to its antinutritional factors, i.e. digestion problems due to trypsin inhibitors. However, sprouted soy beans contains some antinutritional parts that, too, can cause health hazards. Hence, it is necessary to consume well-processed soy beans to take advantage of their beneficial aspects like a selection of good quality soy bean grain (healthy, cleaned, washed, and dried grain). Antinutritional factors of soy beans are inactivated by soaking, steaming (3 liters water per kg of soy beans), and blanching (hot water blanching or steam blanching).

31.5 Soy Bean Products

Soy beans contain 18% oil and 38% protein. A small amount of soy beans are used in many nonfood products. Soy beans can be processed into many soy products, such as soy dhal and soy milk, as well as soy flour, soy beverage (soy milk), tofu (soy paneer), soy yogurt, bakery products, and dairy products by using different processing and machineries.

Soy bean oil is used in cooking and frying foods. Margarine is a product made from soy bean oil. Salad dressings and mayonnaise are made with soy bean oil, as are some baked breads, crackers, cakes, cookies, and pies. Some foods are packed in soy bean oil (tuna, sardines, etc.).

Products made from soy beans can easily meet the protein requirements of a vegetarian diet. Soy foods have a number of health benefits such as, cancer prevention, cholesterol reduction, combating osteoporosis, and menopause regulation. Human studies suggest that as little as one serving of soy foods each day may be protective against many types of cancers.

31.5.1 Soy Dal Analogue and Use

Soy-based dal analogue is prepared from 49.5% soy + 49.5% wheat + turmeric and an emulsifier mix. The dal analogue is protein rich (30%), nutritious, and scores higher than tur dal (22% protein). Not only can it meet the growing consumption requirement, it can save consumers some money since it is less expensive than tur dal. Soy dal analogue is made on a single-screw or twin-screw extruder. When using a single-screw machine, a combination is used: the first machine acts as gelatinizer or high-temperature short time (HTST) process, while the second machine forms a dense product that needs drying to remove excess moisture. These two functions of cooking/forming can be achieved in one twin-screw extruder equipped with special profiled screw segments that provide necessary mechanical shear and heat for cooking the ingredients and forming a dense product after the material exits the die. Soy dhal is made from soy beans that have been firstly cleaned and graded and then dehulled.

31.5.2 Soy Flour and Use

Soy flour is made from roasted soy beans that have been ground into a fine powder by using grain mill. In use, soy bean flour takes one part and the other flour takes nine parts.

The steps followed in preparation of soy flour is, cleaning and grading, de-husking and splitting, blanching of split grains (boiling water for 30 minutes at around 100°C), drying in a tray dryer for 4–5 hours at around 80°C, and milling in a grain mill.

31.5.2.1 Types of Soy Flour

31.5.2.1.1 Full-Fat Soy Flour (FFSF)

Soy flour contains 40% protein and 20% oil. Enzyme-active full fat soy flour has a wide variety of applications, including its use up to 3% in bread making, which helps to improve whiteness in blend. It is a good emulsifier and stabilizer. It helps in homogenizing milk for cakes and improves mixing tolerance when mixed with ready flour mixes.

31.5.2.1.2 Defatted Soy Flour (DFSF) (Toasted)

Defatted soy flour contains more than 50% protein and less than 1% fat. It is produced by milling desolventized soy beans (without husks) obtained from a solvent extraction plant. Defatted soy flour has a variety of application in making various products apart from its use in soy milk and milk products, including fortification in cereals for enhanced protein content of processed foods, enhancement of protein content of baked goods, improved crumb body. Its reduced moisture content in cookies increases their shelf life. Defatted soy flour can enhances the protein content in baby food and acts as the best base material for fermentation in making antibiotics.

31.5.2.1.3 Defatted Soy Flour (Untoasted)

The advantages and applications of defatted soy flour include its maintaining the balance of essential amino acids required for overall development in human bodies, its use as a good agent for surface sealing and protein fortification, and its addition of moistness and longer shelf life in baked goods.

31.5.3 Soy Milk and Use

Grinding, soaking and straining soy beans creates a mild-tasting liquid known as soy milk. Soy milk is usually a suitable replacement for dairy milk. Vanilla and chocolate soy milk are often sold alongside unflavored soy milk, which are all typically packaged in aseptic containers. A 1-cup serving of soy milk has 104 calories, 6 grams of protein and 3.5 grams of fat, on average. Fortified soy milk is a good source of calcium, iron, vitamin B-12, and vitamin D. Soy beverage (soy milk) is a water soluble extract from whole soy beans. It is an off-white emulsion or suspension containing water soluble proteins, carbohydrates, and lipids. It resembles dairy milk in appearance. However, it is a healthy drink as it is cholesterol and lactose free, so represents an alternative to dairy milk.

The omega-3 and omega-6 fatty acids as well as the powerful phytoantioxidants in soy can effectively protect blood vessels from lesions and hemorrhage as it improves the fluidity and flexibility of blood vessels. It is naturally lower in sugar content than regular milk and, hence, promotes weight loss. It is a rich source of phytoestrogen that can inhibit the production of testosterone in men and cut significantly the risk of prostate cancer. It helps in prevention of osteoporosis, which is an age- and hormone-related disease. It prevents postmenopausal syndromes in women, thus reducing the risk of health problems both physical and psychological. It also contains phytochemicals, which have proven health benefits.

About 6–8 kg of soymilk can be obtained from 1 kg of raw soy bean splits. There are different machineries used in soy milk processing like power-operated paddles and cleaner-graders, power operated dehullers, containers for soaking, grinder cookers, boilers, and filtering units. Soy *gulab jamun,* soy *shrikhand,* and soy curd products are also prepared from soy milk.

31.5.3.1 Types of Soy Milk

Commercial soy beverages can be classified according to their composition such as high-solids soy milk (bean-to-water ratio of 1:5), dairy-like soy extract (bean-to-water ratio of 1:7), and lower-solids soy beverage (bean-to-water ratio of 1:20). Depending upon the processing parameters and water-to- soy bean ratio, soy milk would have a typical solids content around 8–10%. Within this, protein is 3.6%, fat 2.0%, carbohydrates 19.9%, and ash 0.5%.

31.5.4 Soy Tofu and Use

Soy bean curd—or tofu—is created by curdling soy with a coagulant. Tofu, which has minimal flavor, can absorb seasonings and flavorings easily. Firm tofu is dense and useful in stir fries or soups. Soft tofu is creamier and works in place of yogurt in smoothies. A one-half-cup serving of firm tofu has 88 calories, more than 10 grams of protein, and 5 grams of fat. Creamy desserts using tofu are common in grocery stores, as are plain blocks of tofu with varying firmness. Most Asian markets carry fresh tofu, which has a smoother texture and flavor.

Soy paneer, also called tofu, is one of the most popular nonfermented nutritional products prepared from soy milk. The high nutritional value and low cost make it a suitable solution for malnutrition problems for poor populations. Hence, tofu would be an ideal substitute in terms of price as well as nutritious values.

Tofu is a high-protein food often used as a meat or cheese substitute. It is sold as ready-to-eat cakes that resemble paneer or soft white cheese. The preparation of tofu involves extraction of soymilk and then coagulation of this extract to form curd. The curd is then pressed to form tofu cakes (Figure 31.1).

Typical wet composition of tofu is 85% moisture, 15% protein, and 4.2% lipids. The remaining constituents are carbohydrates and minerals. The typical dry composition is made up of 50% protein, 27% fat,

Soybeans

▲

Soaking in cold water (1:3 soybean:water ratio) for 4–6 hours

▲

Draining of water and removal of cotyledons

▲

Grinding with hot water (1:8 soybean:water ratio)

Soy paste

▲

Filtration and pressing (muslin cloth)

▲ → Okara

Soymilk

▲

Heating (up to boiling)

▲

Cooling (to 25–30°C)

Adding coagulant (0.2% citric acid)

▲

Decanting

▲ →Whey

▲ Solid

▲

Pressing (15–20 minutes)

Dipping in chilled water

▲

Soy paneer (tofu)

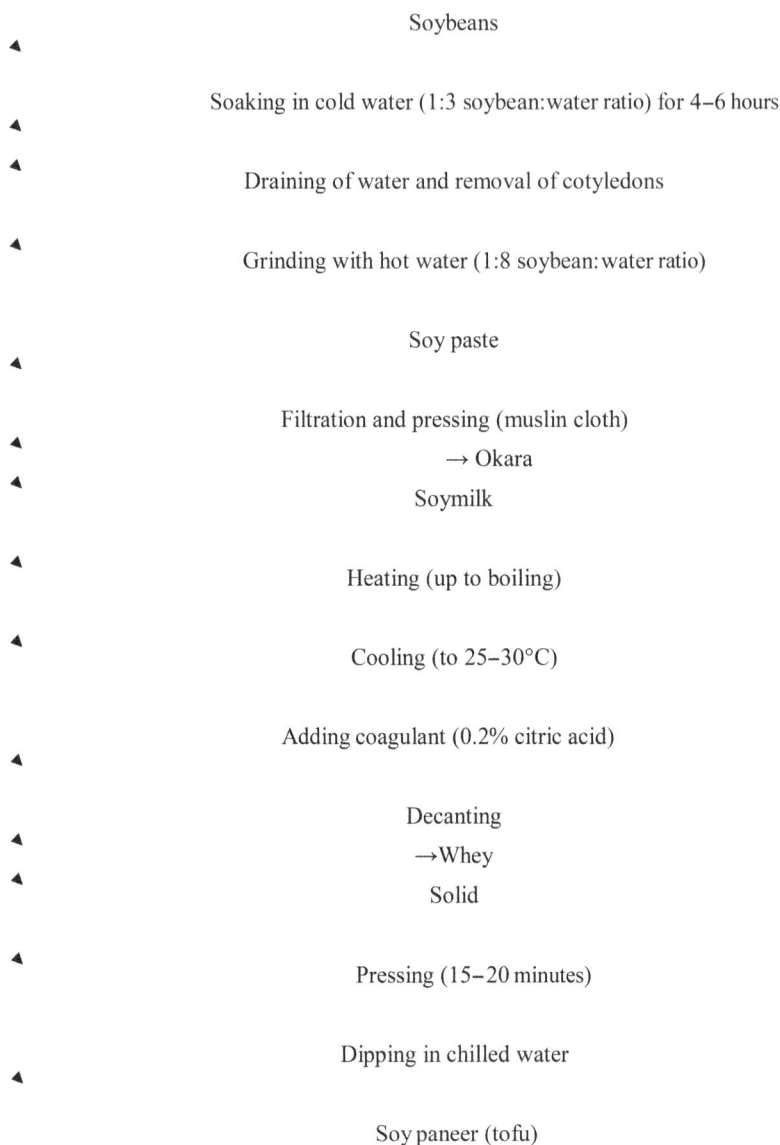

FIGURE 31.1 Process flow chart for the preparation of Tofu.

and 23% carbohydrates and minerals. From one kilogram of soy bean, approximately 2 kg of tofu can be obtained.

31.5.4.1 Types of Tofu

Tofu can be categorized as silken or pressed tofu. Silken tofu production involves soy extract being finely filtered and heated before cooling to a temperature of 65–70°C. Calcium sulphate/magnesium chloride of low concentration is added to the extract. A fine, smooth, and firm curd forms after 30–60 minutes. This curd is left unbroken.

    In pressed tofu, a coagulant is vigorously stirred into the hot soy extract. The curd is broken and pressed. Pressed tofu contains about 22% protein and 61.6% moisture. There are different machineries used in tofu processing like power-operated paddles and cleaner-graders, power-operated de-hullers, containers for soaking, grinder cookers, boilers, filtering units, paneer pressing devices, and cooling units.

31.5.5 Soy Yogurt and Use

While milk-based yogurt has long been consumed in many countries, soy yogurt, also known as sogurt, is a relatively new product made from soy milk. It is produced through the fermentation of soy milk by different cultures of bacteria to form a soft, fragile, custard like texture, generally containing 12–14% total solids and possessing a clean tart flavor. It is an easy substitute for sour cream or cream cheese.

31.5.5.1 Types of Soy Yogurt

There are several types of sogurts. Sogurt can be produced in the form of a highly viscous texture, a softer gel, or in frozen form as a dessert or drink. Generally, they can be classified as a set-type soy yogurt, stirred-type soy yogurt, and drink-type soy yogurt.

31.5.6 Soy Sauce

Soy sauce is one of the most common soy products available. This dark brown liquid with a salty taste is made from fermented soy beans that have undergone a fermenting process. Shoyu, tamari, and teriyaki are common varieties of soy sauce and are typically available in different levels of darkness. A 1-teaspoon serving of tamari has 4 calories and 335 milligrams of sodium. Vegetable, meat, and tofu dishes often call for soy sauce, but it is even used in some cookie recipes. Soy sauce products can be used in many different ways in the kitchen including marinating, adding saltiness, darkening foods, changing appearance, and enhancing and balancing flavors.

31.5.7 Lecithin and Use

Lecithin has been utilized by the food industry in a variety of forms to serve many different functions. Lecithin is nature's best emulsifier, helping to blend materials that do not mix spontaneously or easily when combined. Soy lecithin is often used in many bakery applications such as bread, buns, and tortillas. It can extend the shelf life and softness of bread products. In industrial applications, it is used as an antioxidant in the automotive industry in cleaning and penetrating products, as a chelating agent for multivalent metals. Lecithin modifies and stabilizes the crystal structure of fats, mono- and diglycerides, and is used for this crystallization control in cosmetics and pharmaceuticals. Lecithin can be used in dust control by applying a spray of lecithin diluted 10–30% in oil, which helps small particles from becoming airborne. It also is used as an emollient in shampoo products, as it is easily dispersible and improves shine and body.

31.5.8 Soy Oil and Use

According to The United Soybean Board, most margarines, shortenings, and salad dressings contain soy bean oil (Güzeler and Yildirim, 2016). In addition, most of the "vegetable oil" you see in the grocery store is pure soy bean oil. The American Heart Association lists soy beans oil as a safe fat for maintaining health and longevity. Additionally it is important in terms of its medicinal properties such as antioxidant and cholesterol reducing agent (Khamar and Jasrai, 2015). A 1-teaspoon serving of soy bean oil has 40 calories, 4.5 grams of fat, and less than 1 gram of saturated fat. Soy bean oil is mostly flavorless, making it a nonintrusive ingredient in most dishes.

The majority of soy beans are processed for oil. The oil is primarily used for making mayonnaise, salad dressings, shortenings, imitation chocolates, creams, cheese, frozen desserts, etc. The oil may be refined for cooking, other edible or industrial uses. The high-protein fiber that remains after processing is used as animal feed for the poultry and pork industries. Biodiesel fuel for diesel engines can be produced through transesterification of soy beans. This process removes the glycerin that acts as a fuel. Moreover, this energy is renewable and environmental friendly, as soy oil produces an ecofriendly solvent. Other product and byproducts are soy crayons, soy ink, soy-based lubricants, hydraulic fluid, foams for furniture, and foot wear.

In the past it was ground nut oil and mustard oil that were the main edible oils being consumed. Now, soy bean oil is the most widely consumed oil in the world. The oil is, however, sensitive to oxidation and rapid development of undesirable flavors, which can be improved by hydrogenation, winterization, and transesterification. Soy oil is a perfect alternative to olive and canola oil. It is used to stir fry meals, dress salads make dip, and in baking. With time, the price of ground nut and mustard oils rose very high, which restricted a large part of the population from their use. In addition to the cost factor, health awareness has increased. The amount of fat found in ground nut and mustard oils was much higher than soy bean oil. These were the major two reasons for the shift of demand to soy bean oil.

31.5.9 Soy Bean Meal and Hull

Soy bean meal is livestock feed and a major protein product in the world. Around 90–95% of the total output of soy bean meal is being used for livestock feed in the world. It is safe ingredient for cattle and poultry and contains minerals, vitamins, and all essential amino acids. Soy hulls are a byproduct of soy bean meal and oil processing used extensively as animal feed.

31.6 Overview of Modern Processing and Industrial Uses for Soy Beans

Modern soy bean processing consists of a series of unit processes that in total determine the costs of processing and quality of final products. This section reviews a sequential discussion of each unit process and outlines the processes employed and their interrelationships and interdependencies. Modern emphasis in soy bean processing has tended to concentrate on energy conservation and savings, increased production, and environmental considerations. These are real-world concerns and merit emphasis, but not at the expense of the basic technical goals in processing. First and foremost, the overall technical goal should be production of products of acceptable quality. Secondly, such production should be at the lowest possible cost that allows meeting the first goal. The overall technical goal should be production of products of acceptable quality. Such production should be at the lowest possible cost that allows meeting the first goal.

31.6.1 Extraction Processes

The extraction processes for soy beans, which involve continuous screw presses, solvent extraction, or, in some cases, a combination of the two.

31.6.2 Mechanical Extraction

In the United States, less than 1% of soy beans are processed by continuous screw presses (expellers); the process of a flow diagram combined with solvent extraction is, soy bean storage, size reduction or flaking, cooking, continuous screw press or hydraulic, crude oil press cake, ± cake breaker, and solvent extraction. The soy bean meal (press cake) from pressing operations contains 4–5% residual oil, and the crude soy bean oil is similar to that from solvent extraction, with generally a lower phosphatide content.

31.6.3 Solvent Extraction

The solvent extraction process breaks down into the three steps: preparation, extraction, and desolventizing. The major differences in soy bean solvent extraction processes occur in the preparation steps. A commonality in all the preparation processes is the choice of whether to de-hull. One variation in the conventional system is the introduction of an expander after flaking or other method of size reduction (grinding). Another variation is the introduction of "hot de-hulling," which replaces the tempering step in conventional preparation and has some other advantages. A final variation is the so-called Alcon process, which is a method of cooking the flakes prior to extraction.

31.6.4 Extraction

The preparation process affects both the efficiency of oil extraction and the quality of the oil. The efficiency of extraction is based on the residual extract in the soy bean meal in relation to production rates. The scientist (Kock and Amer, 1983) has shown that there is a correlation between the temperature of extraction and undesirable enzymatic action. From this observation, he patented the so-called Alcon process.

31.6.5 Desolventizing

The desolventizing of soy bean meal and soy bean oil is the final step in solvent extraction. The advent of the expander process required some changes in the desolventizing process, because of less residual solvent in the collets from the extractor and more phosphatides in the oil. Use of the Alcon process also required changes in operation of the desolventizing equipment, especially for the much higher gum content in crude oil. The desolventizing of the miscella can have an adverse effect on the quality of the extracted oil. If the temperature in desolventizing is too high (in excess of 115°C), the content of nonhydratable phosphatides may increase.

31.6.5.1 Soy Bean Meal Processing and Utilization

Extracted soy bean flakes normally exit a continuous extractor with a residual hexane content of 29–35% hexane on an "as is" basis. The removal and recovery for reuse of the residual hexane, and the cooking, drying, cooling, and sizing of the extracted meal is required for its principal end use in livestock feeds. Desolventizing processes for other uses of the extracted flakes, such as in edible products or industrial uses. The desolventizing section of the extraction process is one of the more dangerous spots in the process because the solids leave the enclosed solvent-handling area and enter a process area not designed for solvent vapors. The principal market for toasted soy bean oil meal is incorporation into livestock feeds as the principal source of protein. The dominance of soy bean oil meal is due to its high nutritional value and favorable cost of production.

31.6.5.2 Soy Protein Processing and Utilization

The term "soy protein" means processed, edible dry soy bean products. Many types are produced for use in human and pet foods, milk replacers and starter feeds for young animals. Most soy proteins are sold in bulk as ingredients for commercial meat processing, baking, and for remanufacturing into grocery store, fast food, institutional, and restaurant convenience foods. The crude soy flour sold in the United States primarily as health flour was initially made from whole soy beans processed from a hydraulic press or expeller cakes, and later from solvent-defatted soy bean meals. In spite of the fact that the beans are excellent nutritional staples, they are not consumed raw due to its hard texture and beany/undesirable flavor. However, the strong beany flavor that also limits the market growth of soy flours may undergo a "debittering" processes to remove the objectionable compounds. In processing all soy proteins, it is essential to start with thoroughly cleaned, sound, mature, yellow soy beans sorted to uniform size.

31.6.5.3 Degumming and Lecithin Processing and Utilization

Degumming is the process for removal of phosphatides from crude soy bean and other vegetable oils. The phosphatides are also called gums and lecithin. The latter term is also the common name for phosphatidyl choline, but common usage refers to the array of phosphatides present in all crude vegetable oils. Although all crude vegetable oils contain gums, soy bean oil is currently the major source of commercial lecithin because it contains the largest amount of gums and is also the world's leading vegetable oil. In conventional solvent extraction, only half of the phosphatides present in soy beans are extracted. Use of preparation processes such as the Alcon process or expanders will change the array of phosphatides in the crude oil by increasing the phosphatidyl choline content by about 30–40%, increasing the total extracted phosphatides. Some plants use steam condensates for degumming rather than deionized water; however, the absence of iron in the condensate should be ensured. Some plants also use excess stripping steam in the last stages of miscella desolventizing as a means for hydration of phosphatides, but this is more difficult to control than simple water addition. This chapter illustrates the flowsheet for degumming soy bean oil and crude lecithin production. Lecithin has unique release properties and as such has been used in pan frying formulations and in pan greases for baking. In addition, it is also used industrially as a release agent for ready removal of both wooden and metal concrete casting forms.

31.7 Cost Estimates in Soy Bean Processing

Estimation of total costs provides guidance for developing the capital cost and operating expense estimates necessary for use in project feasibility studies, evaluation of alternatives, funding requests, and construction budgeting. The intent of this effort is toward handling major project activities, which fall outside the domain of plant sustenance, improvements, debottlenecking, and nominal incremental capacity increases. Although these latter projects generally lack the complexity, capital exposure/risk, and demand on resources to follow formalized project development procedures, they do go through the same project stages and thought processes as a major project. Among the many critical factors that determine the ultimate profitability of an oilseed-, edible oil-, or soy food-processing plant are its design, location, and capacity. These are also the factors that most affect the capital costs, the subsequent associated operating expenses, the conversion efficiency, and capability (product yield/product mix) of that facility. While this is true of all chemical processing and manufacturing plants, it is of particular importance in the processing of agricultural raw materials and the utilization of their products. These businesses fall into the realm of "commodity businesses," which are typified as market-driven, high-volume, low-margin operations that produce generic (nonunique) products from universally available raw materials. Commodity businesses require low life-cycle costs—that is, the one-time capital costs plus the "lifetime" value of the direct operating expenses and indirect commercial/administrative expenses—for success. In order to achieve low life-cycle costs for a proposed new facility (whether a green field site, a new product addition at an existing site, or a major expansion of an existing facility), it is fundamental for evaluating the business and technology that both the capital cost and the operating expense be estimated appropriately from the initial project conception to project startup (Table 31.2).

Table 31.3 indicates projected estimating factors for a soy bean plant and oil refinery. All of these estimating factors can be applied to either the major or total equipment costs. It should be pointed out that the equipment costs include both major and minor equipment (i.e., extractors and pumps, diverters, and piping specialties) but not supplies (pipe, fittings, valves, conduit), which are associated with the installation factor. Each engineering firm organizes "factored data" a little differently, so it is sometimes very difficult to compare and reconcile differences reported in the literature. Further, factors can change significantly based on supply of materials and labor from area to area. A general rule is that major equipment cost × 2.5 to 3.5 = Total Project Cost.

31.8 Future of Soy Products in India

In India, health consciousness is increasing among the general public and this increased consciousness is a boon for the soy bean producers and processors. Through intended efforts, soy bean use is gaining acceptance and with time this acceptance is gaining in India. Soy beans are mainly used in the form of textured vegetable protein, commonly known as soy badi or soy nuggets. Even chapattis made from soy-fortified wheat flour is used. Biscuits made with wheat flour mixed with soy bean flour are also available in the market. Slowly but steadily, these products are penetrating the diets of Indians. In reference

TABLE 31.2

Checklist for Project Cost Estimates

Direct Project Costs

1.0 Site Work
- Clearing and structure demolition
- Grading, fill
- Rail road track-age: on-site, off-site
- Roadways: on-site
- Underground piping: fire protection loop, utility run-ins, sewer
- Security fencing
- Truck staging and parking
- Storm-water drainage reservoir
- Foundations: piling, caissons

2.0 Buildings
- Process buildings
- Support buildings: shops, part storage, offices, laboratory, guard house, welfare

3.0 Storage, Reclamation, and Receiving/Shipping Facilities
- Liquid tank farms
- Solids bins, silos, steel tanks, outside flat storage
- Unloading facilities: marine, truck dumpers, rail pits

4.0 Utilities
- Steam generation facility, including water treatment and reserve fuel handling as required
- Power generation/primary supply
- Cooling water
- Fire-fighting water
- Refrigeration
- High-temperature systems

5.0 Instrumentation and Controls
- Central control system
- Field-mounted controls

6.0 Environmental Controls
- Air pollution control
- Wastewater treatment

7.0 Construction
- Civil/structural
- Piping and insulation
- Mechanical: equipment setting
- Electrical: supply, distribution, MCCs

Indirect Project Costs

1.0 Engineering and design
- Process, mechanical, civil/structural, electrical, instrumentation
- Programming
- Startup services

2.0 Procurement
3.0 Site Surveys, Topography and Soil Studies, Environmental Permits
4.0 Construction Management
- Receiving, warehousing, and control

5.0 Miscellaneous Equipment
- Mobile equipment: track mobiles, diesels, trucks, etc.
- Shop/lab equipment and tools
- Office/welfare furnishings

6.0 Taxes
7.0 Freight
8.0 Land
9.0 Contingencies

TABLE 31.3

Estimating Factors Based on Major Equipment Costs

Capital Cost Category	Typical, %	Range, %
Major equipment = A	30%	27.5–34.5
Minor equipment = 5% A	1.5%	1.0–2.0
Instrumentation hardware = 16.7% A	5.0%	3.0–7.0
Freight = 3.3% A	1.0%	0.5–1.5
Total equipment = 1.25% A	37.5%	32.0–45.0
Buildings and storage structures	12.5%	10.0–25.0
Utilities	3.5%	2–7.5
Civil/site work	4.5%	3.5–7.5
Mechanical	6.5%	4.0–8.0
Piping and protective coverings	6.0%	3.0–7.0
Electrical and instrumentation	7.5%	5.0–9.0
Land	2.0%	0–5.0
Maintenance (parts, tools, mobile equipment)	1.5%	1.0–3.0
Engineering, procurement	10%	8.0–12.0
Expenses (permits, training, etc.)	1%	0.5–1.5
Contingency and escalation	7.5%	5.0–10.0
Total project cost (2.9 to 3.6) x A	100%	

to soy milk and tofu, the fact is that they are not easily available everywhere in the country and thus are confined only to local areas of production. India being a country consisting mainly of a vegetarian population has a very great potential for soy products, primarily due to its higher nutritional value with an affordable price. Soy bean oil is extensively used in India. Because of its lower price and high nutritional value, it has gained an important place in the Indian diet and household and as a strong sector of India. The productivity of soy beans have more than doubled from 1970 to 2011–12, from 426 kg/ha to 1,264 kg/ha. Still, there are several reasons that India could not stand out from the rest of the world. The reviews suggest that the low adoption of improved production technologies by farmers with dependence on only oil cakes for export may cause not so favorable results or may cause stagnation in productivity of the crop. Hence, some needful strategies have to be identified and enforced to make India the leader in legume production and processing.

REFERENCES

Güzeler N and Yildirim C. (2016). The Utilization and Processing of Soybean and Soybean Products. Journal of Agricultural Faculty of Uludag University 30 (Special Issue): 546–553, 546.

Khamar R, Jasrai Y. T. 2015. Soybean Oil, Soy Germ Oil and DOD of Soybean Oil – Good source of Nutraceuticals. Journal of Medicinal and Aromatic Plants Research 1(1):1–5.

Agarwal D. K., Billore S. D. and Sharma A. N. 2013. Soybean-Introduction, Improvements and Utilization in India- problems & prospects. Agricultural Research 2(4), 293–300.

Anonymous. 2014. Food and Agriculture Organization of the United Nations Statistics division. (http://faostat.fao.org/site/567/DesktopDefault.aspx?PageID=567#ancor)

Baisya, R. K. 2008. Changing Face of Processed Food Industry in India. An e Books India, New Delhi.

Besin Değerlerinin Belirlenmesi. Tr. Doğa ve Fen Derg. 2 (2): 7–10.

Cabrera-Orozco, A., Jiménez-Martínez, C., & Dávila-Ortiz, G. (2013). Soybean: Non-nutritional factors and their biological functionality. Soybean-bio-active compounds, 387–410.

Chavan, S. M. 2015. Process development for dehydration of tofu. An unpublished M. Tech. Thesis, submitted to Maharana Pratap University of Agriculture and Technology, Udaipur, Rajasthan.

Dijkstra, J. 2016. Soybean Oil. In Encyclopedia of Food and Health.

Erickson, D. R. 1995. Overview of Modern Soybean Processing and Links between Processes. Practical Handbook of Soybean Processing and Utilization, 56.

Guzeler, N. and Yildirium, C. 2016. The utilization and processing of soybean and soybean products. Journal of Agricultural Faculty of Uludag University 30 (special issue): 546–553.

Jain, N. K., Chavan, S. M. and Wadhawan, N. 2020. E-Compendium on Process Technology for Legumes and Oil seed (FT 222), MPUAT, Udaipur

Kock, M. J. and Amer, J. 1983. Oil Chem. Soc. 60: 210.

Kökten, K, Boydak, E, Kaplan, M, Seydoşoğlu, S, Kavurmacı, Z (2013). Bazı Soya Fasulyesi (*Glycine max* L.) Çeşitlerinden Yapılan Silajların

Messina, M. 1995. Modern applications for an ancient bean: soybeans and the prevention and treatment of chronic disease. The Journal of Nutrition, 125 (3 Suppl): 567–69S.

Penalvo, J. L., Matallana, M. C. and Torija, M. E. 2004. Chemical composition and nutritional value of traditional soymilk. The Journal of Nutrition. 134 (5): 1254S.

Richard, J. Fiala. Chapter 6- Cost Estimates for Soybean Processing and Soybean Oil Refining. Practical Handbook of Soybean Processing and Utilization, 519-535

Section IV

Roots and Tubers

32

Roots and Tubers: Functionality, Health Benefits, and Applications

Adeleke Omodunbi Ashogbon
Adekunle Ajasin University, Akungba-Akoko, Ondo State, Nigeria

CONTENTS

32.1 Introduction

Roots and tubers are predominantly cultivated in tropical and subtropical regions of the world, especially in Africa, South America, and parts of Asia. A polysaccharide, starch is the most important biopolymer in roots and tubers, closely followed by protein, and lipids, in that order. In most of the poor countries in the world where roots and tubers are planted, they are utilized mostly for consumption, and there is little attempt to extract the significant starch in them for food and nonfood applications in the industry.

Some of the roots and tubers that are cultivated for edible purposes are cassava (*Manihot esculenta*), potato (*Solanum tuberosum*), sweet potato (*Ipomoea batatas*), true yams (*Dioscorea* species [*D. alata, D. cayenensis, D. spicata, D. bulbifera, D. esculenta, D. abbyssinia*]), arrow root (East Indian arrow-root [*Tacca leontopetaloides*]), Queensland arrow root (*Canna edulis*), Indian arrowroot (*Hutchenia caulina*), West Indian arrowroot (*Maranta arundinacea*), kudzu (*Pueraria hirsuta*), buffalo gourd (*Cucurbita foetidissima*), lotus (*Nelumbo nucifera*), ginger (*Zingiber officinale*), and the edible aroids of the *Araceae* family consisting of five genera (*Colocassia, Xanthosoma, Amorphallus, Alocassia,* and *Cytosperma*).

Generally, starchy roots and tubers that are nutritionally and economically significant are aroids, cassava, potatoes, sweet potatoes, and yams. A crucial agronomic benefit of root and tuber crops as staple foods is their auspicious adaptation to various soil and environmental conditions and a variety of farming systems with minimum agricultural inputs in term of fertilization and agrochemicals. Nevertheless, root and tuber crops possess a bulky nature with high moisture content ranging from 60–90%, making them to be linked with high transportation cost, short shelf life, and restricted market brink in developing countries where they are produced (Chandrasekara, 2017). The yearly worldwide production of roots and tubers is approximately 845 million tons (FAOSTAT, 2014). The principal producer of roots and tubers is Asia, closely followed by Africa, Europe, and the US. Cassava, potatoes, and sweet potatoes constitute about 90% of the global production (FAOSTAT, 2014). Generally, roots and tubers supply limited amount of proteins ranging from 1–2% on a dry weight basis, nevertheless, potatoes and yams comprise higher quantities of proteins among the tubers (Chandrasekara, 2017). Cassava, sweet potatoes, potatoes, and yams supply some vitamin C, and the yellow varieties of cassava, yam, and sweet potatoes comprise β-carotene, but generally roots and tubers are deficient in most other vitamins and minerals but contain high amounts of dietary fiber (FAO, 1990).

Roots and tubers of food crops are mainly for consumption in the tropic and subtropic regions of the world. The starches extracted from them are solely utilized in the food and nonfood industries. There native starches are sometimes biologically, chemically, and physically modified to better their functionalities for optimal utilization in the industries (Punia, 2020; Punia et al., 2020). The imperfect nature of single modification resulted into consideration of dual modification and other multiple modifications such as triple and quadruple modification. Generally, native starches are rather inert and hydrophilic, not resistance to thermal processing conditions, easily retrograde and insolubility in aqueous medium (Punia et al., 2019; Dhull et al., 2020). That is the pivotal reason for modification.

Starches from roots and tubers are easily extracted and purified compared to starches from other botanical sources because of their low content of proteins and lipids. Characteristically, root and tuber starches possessed large granules and high swelling power unlike legume starches with restricted swelling and cereal starches that overwhelming possessed small granules. The high swelling power of root and tuber starches, especially that of potato starch granules, has been ascribed to its high possession of phosphate monoester. The latter is located on C6 of the starch chains and the repulsion between them resulted in weakening and easy percolation of water into the starch granules. In the study of roots and tubers, the starches are very important as raw materials for utilization in the industries due to their sustainability. On the other hand, the phytochemicals are vital for the benefits of health. The bioactive phytochemicals are especially potentially pivotal in the management of some health ailments such diabetics, obesity, cancer, and other cardiovascular affliction.

It is difficult to imagine the food security position in some tropical and subtropical regions of the world without roots and tubers. The advantage inherent in them is their ability to grow under very harsh environmental conditions with varying climatic conditions and little or no input in terms of fertilization and agrochemicals. Some of the roots and tubers are resistant to pest due to the phytochemicals produced by them and some of the phytochemicals are capable of attracting pollinating insects. The aim of this chapter is to lay the foundation for more detail discussion about roots and tubers. The attention here is a brief deliberation on the starches and phytochemicals from roots and tubers. The functionality, applications, health benefits of the roots and tubers and products derived from them are also discussed.

32.2 Important Roots and Tubers

32.2.1 Aroids

The edible aroid root crops belonging to the family *Araceae* included five genera (*Colocassia, Xanthosoma, Amorphallus, Alocassia,* and *Cytosperma*). *Alocassia, Xanthosoma* and *Colocassia* are of the tribe *Colocasiae* and in the subfamily *colocasiodeae*. The classification of the edible aroid genera into species has caused puzzlement. *Colocassia* is part of the subtribe *colocasinae*. *Colocassia* species can be divided as follows: (i) *C. esculenta* (L) Schott var *escluenta* (produces a large corm and is also named taro, dasheen, coco, tannia); and (ii) *C. esculenta* (L) Schott var antiquorum (produces a small central corm surrounded by numerous side cormels and is also known as eddoe). Both i and ii are collectively referred to as "old" cocoyams. *Xanthosoma* is part of the subtribe caladinae. *Xanthosoma sagittifolium* (L) Schott is normally considered as the sole cultivated species. Other closely related species include X. *brasillense*, X. *atravirens*, X. *violaceum*, X. *robustrum*, X. *auriculatum*, X. *roseum*, and

FIGURE 32.1 Cocoyam corms.

X. varacu. The *Xanthosoma* species are collectively known as the "new" cocoyams. The above expositions on classification and subclassification of edible aroid root crops are due to Hoover, 2001.

Aroid plants or the Arum family has characteristic corms, which are rich in polysaccharides including the starch and the mucilage (Mijinyawa et al., 2018). A vast production of *Colocasia esculenta* (L.) Schott or taro has been documented worldwide since the 1980s. Nigeria (3, 45 million metric tons [MMT]), China (1.8 MMT), Cameroon (1.6 MMT), Ghana (1.3 MMT), and Papua New Guinea (0.3 MMT) are major producers of taro and Nigeria is leading the other countries (Mijinyawa et al., 2018). It is broadly utilized as food substance (Nagata et al., 2015; Andrade et al., 2015; Kaushal et al., 2015). Taro is an erect herbaceous perennial root crop that is broadly cultivated in tropical and subtropical regions of the world. The origin of taro is uncertain, however it is documented that it was originated and first domesticated in Southeast Asia (Hunt et al., 2013). The crop has been largely produced in Africa even though the time of its spread to the region is largely unknown and is now cultivated in Nigeria, Cameroon, Ghana, and Burkina Faso where it has gained high significance (Chair et al., 2016). Cocoyam corms are shown in Figure 32.1.

Despite the immense nutritional potential of cocoyam corm, its exploitation is not economically feasible unless it is first processed to diminish its high calcium oxalate levels (Coronell-Tovar et al., 2019). The compound negatively affects the corm palatability, conferring acridity and a bitter-astringent taste (Owusu-Darko et al., 2014), and it also possess antinutritional and toxic characteristics, which are hazardous to humans when ingested in high concentration. The methodologies involve in the minimization or elimination of high concentration of calcium oxalate are cooking, soaking, oven drying, sun drying, and fermentation (Igbabul et al., 2014). Among all of these methods, water soaking for 72 hours at 28°C under constant shaking was found to be the most efficient method as it reduces the oxalate content of cocoyam corm by 77% (Coronell-Tovar, 2015).

32.2.2 Cassava

Cassava (*Manihot esculenta* Crantz: Euphorbiaceae) is a woody shrub grown in tropical and subtropical regions of the world (Hsieh et al., 2019). It is occasionally called tapioca, yucca, or manioc and is presently the sixth most pivotal global food crop providing nourishment for more than 800 million people in Africa, Asia and Latin America (Prochnik et al., 2012). Cassava has been categorized as a perfect 21st century "climate change" crop because it grows resiliently under adverse conditions, is drought tolerant, adapts to marginal soil, and possesses a rich store of starch in its tuberous root (average 84.5%, dry basis). It has been calculated that the domestication of cassava materialized between 5,000 and 8,000 years B.C. in the Amazon region (Allem, 2002). The Caribbean and northern South America people are historically accepted to have been the earliest cultivators of cassava (Henry and Hershey, 2020). It was estimated (FAOSTAT, 2018) that there was an increase in global cassava root production (million tons) from 2007 to 2016; and the production of cassava in 2016 from 104 countries around the world was estimated to be 281.9 million tons (FAOSTAT, 2018). In 2016, Africa's cassava production has extended to 57% of the global output, pursued by 32% for Asia, and 10% for South America (FAOSTAT, 2016). The four countries contributing to the majority of cassava world production of 20.6%, 11.2%, 7.6%, and 7.5% are Nigeria, Thailand, Brazil, and Indonesia, respectively. Cassava is presently considered as one of the world's fastest expanding staple crops (FAO, 2016).

It is to be noted that cassava produced in Africa is primarily consumed to provide food energy. In disparity, most cassava in Asia and South America is processed into starch for industrial uses (Hsieh et al., 2019). The most valuable part of cassava is the storage root and its consists of nutrients such as vitamin C, carotenoid, calcium, potassium, iron, magnesium, copper, zinc, and manganese (Iyer et al., 2010). The constituents of the cassava starch from its root includes ash, protein, lipid, phosphorus, and fiber content, ranging between 0.03–0.29%, 0.06–0.75%, 0.01–1.2%, 0.0029–0.0095%, and 0.11–1.9%, respectively (Zhu, 2015a). Some other bioactive compounds such as cyanogenic glucosides (linamarin and lotaustralin), noncyanogenic glucosides, hydroxycoumarins such as scopoletin, terpenoids, and flavonoids have been documented in cassava roots (Blagbrough et al., 2010).

Typical mature cassava roots contain on average composition of 30–40% dry matter, 30–35% carbohydrate, 1–2% fat, 1–2% fiber, and 1–2% protein, with trace quantities of vitamins and minerals (Wang and Gao, 2020). The toxicity of some species of cassava is unacceptable due to the presence of cyanogenic glycosides, hence the need for effective detoxification especially the ones utilized for human consumption or as animal feed. Cassava roots are shown in Figure 32.2.

Cassava is a tuberous shrub known for its ability to thrive under harsh climatic (El Sharkawy, 2014) and unfavorable abiotic environmental conditions. Its broad cultivation in most countries of sub-Saharan Africa, Asia, South America and the Caribbean makes it a significant staple in these regions (Abass et al., 2018; Aristizabal et al., 2017). Cassava is pivotal to the food industry due to its high carbohydrate content, low cost, and unique functional properties of its flour and starch. When cassava root was

FIGURE 32.2 Cassava roots.

subjected to three cooking methods, it was found that the steam cooking method retained more phenolic compounds and antioxidant activity, followed by microwaving, and boiling (De Lima et al., 2016).

32.2.3 Potato

Potato (*Solanum tuberosun* L.) is a very widespread crop in Argentina and worldwide (Calliope et al., 2020). It is the most valuable crop globally after wheat and rice, with 368,168,914 million tons of fresh weight tubers produced in 1.76 billion hectares during 2018 (FAOSTAT, 2018). Potato is rich in carbohydrates, with weighty amounts of protein and a good balance of amino acids (Samaniego et al., 2020). These authors asserted the richness of potato tuber in dietary fiber content when consumed with the peel and indicated it is prevalent in antioxidant components. Antioxidants are substances that generally delay degradation by oxidation.

The main antioxidants in potatoes are phenolic compounds, vitamin C, flavonoids, and carotenoids (Camire et al., 2009). The four types of polyphenolics in potatoes are phenolic acids, flavonoids, flavan-3-ols, and anthocyanin components (Samaniego et al., 2020); and the most plentiful is the chlorogenic acid, accounting for approximately 90% of total polyphenols content (TPC). The chlorogenic acids are found in higher concentration than cryto-chlorogenic, neochlorogenic, and caffeic acids in various potato cultivars (Lachman et al., 2013). In a nutshell, the concentration of the phytochemicals present in the potato tuber can be impacted by a number of factors such as genotype, environment, postharvest management, and processing conditions (Samaniego et al., 2020). Other inclusive factors could be the planting altitude, fertilization, and anthropogenic interference. Potato tubers are displayed in Figure 32.3.

32.2.4 Sweet Potato

Sweet potato (*Ipomoea batatas* L.) is a tuberous perennial plant in the family *Convolvulaceae* (morning glory). It is the seventh most pivotal food crop in the world (Shekhar et al., 2015) but it is underutilized in food and nonfood applications (Guo et al., 2019). Sweet potato is a higher yield staple crop in the world because of its ease of cultivation; the price is lower than other crops (Guo et al., 2020). The yearly output is approximately 117 million tons in China, accounting for around 90% of total global production (Bao et al., 2017).

FIGURE 32.3 Potato tubers.

FIGURE 32.4 Sweet potato roots.

In the sweet potato, the total phenolics in yellow and purple varieties amount to 2.2 and 5.2 times more than in the white-fleshed variety, respectively (Satheesh and Solomon, 2019). Sweet potato is the seventh most significant food crop globally in annual production (Song et al., 2014), which is rich in starch, vitamins, minerals, and protein. Furthermore, sweet potatoes also contain some functional ingredients, such as anthocyanin, polyphenols, and dietary fiber. Sweet potato roots are shown in Figure 32.4.

32.2.5 Yam

Yam (*Dioscorea* spp.) is one of the most salient root and tuber, which belong to the monocotyledonous family *Dioscoreaceae* and is a staple food for people of West Africa, Southeast Asia, and the Caribbean regions. The most usually cultivated species are D. *alata* (water yam), *D. cayenensis* (yellow yam), *D. esculenta* (lesser yam), *D. opposita* (Chinese yam), *D. rotundata* (white yam), and *D. trifida* (cush-cush yam), whereas those less cultivated but of local importance are *D. abyssinica, D. bulbifera, D. dumetorum, D. persimilis, D. pseudojaponica,* and D. *septemloba* (Zhu, 2015b). This author asserted that yam tubers from diverse species differ in shapes, with flesh colors from white to yellow and purple. Yam tubers are displayed in Figure 32.5. The cultural, economic and nutritional significance of the yam tuber in the tropical and subtropical regions of the world must be noted.

FIGURE 32.5 Yam tubers.

Phytochemical analysis of *D. bulbifera* shows that it contains saponins, tannins, flavonoids, sterols, polyphenols, and glycosides (Ghosh, 2015). The phytochemical compositions vary according to the geographical location and the extraction solvents utilized. *Dioscorea bulbifera* has a wide spectrum of biological activities due to its chemical diversity, though it is on the verge of extinction (Ikiriza et al., 2019). In yam species, the total phenolic content (TPC) of D. *cayenensis* is often documented as low, whereas that D. *alata* seems to be richest in this compound (Price et al., 2018). Within a species, the changeability in TPC was reported to be almost six times higher in the purple variety than in the noncolored variety (Njoh Ellong et al., 2015).

Yam is a generic name for the plants of twining climbers in the genus *Dioscorea* of the monocot family *Dioscoreaceae*, and is recognized as one of the medicine-food homologous crops because of its nutritional composition and pharmacological value (Zhang et al., 2014). Yam tubers or rhizomes are rich in starch and provide a staple food for millions of people in Asia, Africa, and South America, and play pivotal role in food security (Shao et al., 2020). The global production of yams reached 72.58 MMT in 2018, and the top producers are Nigeria, Ghana, and Cote D'Ivoire with production quantities over 47.5, 7.9, and 7.3 MMT, respectively (FAOSTAT, 2020).

32.3 Chemical Composition of Roots and Tubers

32.3.1 Phytochemicals

There are various definitions, classifications, and subclassifications of phytochemicals in the literature. What is generally common to all the definitions is that phytochemicals are chemicals from plants. Phytochemical is a collective term for plant chemicals with varied structure and function (Huang et al., 2016). In plants, they serve various functions for protection and reproduction, such as color and odor for protection and insect attraction for pollination, phytoalexins for pathogen defense, hormonal functions for growth and signaling, antifeedants and toxins for insect protection, and allelochemicals for defense against herbivory (Saxena et al., 2013). Some phytochemicals are colored, while others are odoriferous or vice versa. Unsaturation or the presence of multiple bonds in conjugated form in phytochemicals enhances the manifestation of color or odor. These similarities in the chemistries of color and odor are worth-noting. Furthermore, the presence of chromophores as functional groups on aromatic rings brings about color and the addition of auxochromes resulted in increment of the intensity of the color. Some phytochemicals produce activity in biological systems, including humans; they are termed "bioactive phytochemicals" (Huang et al., 2016). Biological activity may manifest when phytochemicals are consumed by humans.

In variable quantity, the following phytochemicals; phenolics, flavonoid, vitamin C, carotenoid, provitamin A carotenoid, and lutein are found in edible parts of potato, cassava, sweet potato, yam, taro, and cocoyam (Adepoju et al., 2018; Chandrasekara, 2017; Cornago et al., 2011; Dako et al., 2016; De Oliveira et al., 2019; Dilworth et al., 2012; Hamouz et al., 2016; Labot et al., 2018; Price et al., 2018; Satheesh and Solomon, 2019).

Generally, phytochemicals have been classified into six major categories based on their chemical structures and characteristics. These categories include carbohydrates, lipids, phenolics, terpenoids, alkaloids, and other nitrogen-containing compounds (Campos-Vega and Oomah, 2013). There are subclassifications of the main classification. For example, the carbohydrate can be divided into monosaccharide, disaccharide, oligosaccharide, and polysaccharide. Glucose, fructose, and galactose are the most important examples of monosaccharide. The disaccharide has sucrose, maltose, lactose and cellobiose as their main examples. The main pivotal components of the polysaccharides are cellulose and starch. The lipids could be monounsaturated, polyunsaturated (presence of multiple bonds), saturated fats, and fatty acids. The terpenoids, like the phenolic acids, alkaloids and other nitrogen-containing metabolites possesse various examples with diverse structures and functions in relation to nutrition and health benefits.

In other classifications of phytochemicals, emphases are placed on the polyphenols and flavonoids. The polyphenols are further divided into polyphenolic acids and non-flavonoids. The phenolic acids or polyphenolic acids are aromatic and have at least one benzene ring to which at least one hydroxyl group is directly attached. The principal examples of the phenolic acids are hydroxybenzoic acid and hydroxy-cinnamic acid. The manifestation of color in roots and tubers is mainly due to their carotenoid content. The color produced by the carotenoids is due to the extensive conjugation in their structures.

Tropical tubers such as sweet potato, taro, cocoyam, and some species of yam (D. *alata*) are richer in TPC than potato and cassava (Ronaldo, 2020). The latter has the lowest TPC. Yellow and orange fleshed varieties contain more total phenolics and also more provitamin A carotenoids than noncolored fleshed ones. Total phenolics and total flavonoids are generally considered to be the main contributors to antioxidant activity among tubers; D. *alata*, taro, and sweet potato have the highest antioxidant activity, whereas cocoyam has the lowest (Rinaldo, 2020). The contribution of ascorbic acid to antioxidant activity is much lower than phenolics in starchy food.

32.3.2 Polysaccharides

The polysaccharide can be broadly classified into starchy polysaccharide and nonstarchy polysaccharide. The nonstarchy polysaccharide constituents in plants consist mainly of cellulose and inulin. Starch is given a special and pivotal position in this write-up because it is the most important carbohydrate in the human diet. There is need for extensive discussion of starch being the most important polysaccharide in roots and tubers.

Roots and tubers are mainly cultivated due to their starchily-rich tubers for consumption in Africa, especially in the West Africa region. In contrast, these roots and tubers are solely grown in Asia and South America for exportation and industrial utilization. Starch is the main constituent of roots and tubers. This is closely followed by proteins, thirdly by lipids as far as biopolymers are concerned. Starch is a condensation polymer and second most abundant organic compound in the biosphere, closely following cellulose. The importance of starch as an industrial raw material lies in its abundant availability, non-toxicity, renewability, biodegradability, versatility, and cheapness. The native starches are significant in the food industry and other nonfood industries such as cosmetics, textiles, pharmaceuticals, and pulp and paper.

Generally, the extraction, isolation, and purification of native root and tuber starch are easier than that from other botanical sources such as cereal, legume, and green fruit. This has been attributed to the low protein and lipid content of roots and tubers compared to other plant origins.

Starches occupy a unique position among the polysaccharides. The ubiquitous hydroxyl groups on starch facilitate chemical modifications. The sources of various starches are multifarious: cereal, root and tuber, legume and green fruit (Ashogbon and Akintayo, 2014). The modification of starches is realizable by physical, chemical, enzymatic, biotechnological or a combination of these processes. Physical modification is favored because it is simple, cheap and safe; it requires no chemicals or biological agents when compared to other methods of starch modification (Ashogbon, 2018a). Chemical modification of starch is most commonly utilized because it can be easily regulated and major actions are well comprehended. On the other hand, enzymolysis is currently the most desirable and green method since it possesses higher substrate selectivity, product specificity, mild reaction conditions, higher yields, and low production of coproducts and byproducts.

The individuality and uniqueness of each starch is undeniable and no two starches are identical. The similarity between starches from the same botanical sources is not contestable but the discrepancies between them are also real. For instance, despite the similarity between both corn and sorghum starches (cereal starches), there are many obvious differences. The differences between the same botanical sources of starches grown in the same soil or different soil and under various climatic conditions with the influences of different inputs and anthropogenic factors are known. These contradictions in the study of starch chemistry have been well-documented (Ashogbon, 2018b).

Generally, modifications are carried out to influence the functionalities of various starches and the properties improved involve, but not limited to one or more of the following: emulsion stabilization; film formation; flavor release; shelf stability of product; stability of acids to heat resistant, shear, tackiness; temperature required to cook; viscosity (hot paste and cold paste); hydration rate; moisture retention and control in product; mouth feel of product; oil migration control in product; paste texture/consistency;

sheen of product; adhesion; and clarity of solutions/pastes (BeMiller, 2016). The composition of starch granules are mainly amylose (AM) and amylopectin (AP), with infinitesimal quantities of lipids, protein, and ionic salts. The AM is linear with randomly limited branching. In contrast, AP is a bigger molecule that is highly branched, but the branching in AP is nonrandom. The physicochemical and functional properties of starches are due to these polymeric components of the starch granules and the influences of lipids and proteins are also significant.

There is a need to modify starches for their acceptance in some applications of the food and non-food industries (Ashogbon et al., 2020). The industrial utilization of native starches is restricted due to their flawed nature, such as cold water insolubility, and tendency to easily retrograde and undergo syneresis, and therefore, form rickety paste and gels (Ashogbon and Akintayo, 2014). The functionality of starch can be modified through physical, chemical, and biological means, or a combination of these (Ashogbon, 2020a, Chen, Kaur, and Singh, 2018; Haq et al., 2019; Masina et al., 2017; Vanier et al., 2017). Classification of starches based on physical modification depends on whether the molecular integrity of the starches are destroyed or preserved after the modification. They are mainly hydrothermal treatments (annealing [ANN] and heat-moisture treatment [HMT]), pregelatinization, and nonthermal processes. During the hydrothermal processes of ANN and HMT, the starch granules are preserved because the treatments take place before the gelatinization process, but above the glass transition temperature. This enables a lot of flexibility and mobility of starch chains especially in the amorphous domain constituted by amylose and unstable region of the amylopectin and thus allows for alterations in the totality of the starch system. In contrast, during pregelatinization, the starches are subjected to high temperatures for a short period of time using the methods of spray drying, extraction cooking, and drum drying. Consequently, the starch granules are totally destroyed, with the resultant reduction in polymerization and average molecular weight. The nonthermal processes (strictly not absolutely nonthermal) include microwaving, high hydrostatic pressure, pulsed electric field, and ultrasonic treatment (Ashogbon, 2020b). They possess various effects on the physicochemical, functional, and morphological properties of the starch granules. The biological modification can be subdivided into enzymatic and genetic modification. In disparity, the chemical modification is usually achieved by derivatization such as esterification, etherification, oxidation, acid hydrolysis, and cross-linking. The whole process of classification and subclassification of starch is depicted in Figure 32.6.

Physically modified starches are simple and not exorbitant because they can be produced without chemicals or biological agents. In contrast, chemical modification is possible due to ubiquitous hydroxyl groups at C2, C3, and C6 of the repeat units that constitute the starch chains (Ashogbon and Akintayo, 2014).

Native starches (NSs), single, and dual modification of various starches have their significance in terms of altered functionalities of starch properties in order to satisfy the growing interest of consumers in both the food and nonfood industries. It is rational and logical to state the problems inherent in the NSs, single, and dual modifications of different starches in order to justify the necessity for triple modification of various starches. In the first place, NSs usage in food applications are limited due to intrinsic factors such as insolubility in cold water, lower gelatinization temperature, retrogradation and loss of viscosity after cooking (Goel et al., 2020). Some specific examples of the problems linked with NSs are: unmodified sweet potato products possess unacceptable hardness and poor transparency that in turn reduce consumer acceptability (Guo et al., 2019); NS granules have low surface areas and pore volumes, these could be enhanced by enzymic treatment such as amyloglucosidase (AG), α-amylase (AA), cyclodextrin glycosyltransferase, and branching enzyme (BE) to produce porous starches (Benavent-Gil and Rosell, 2017); the high hydrophilicity, solubility, and low tensile strength of NS films restrict their usage for food packaging, therefore, there is a need for their modification to develop films of better physical, optical, morphological, and barrier properties (Fonseca et al., 2015); other problems are the narrow window between glass transition temperature and the disordering temperature of NSs, its hygroscopicity, brittleness, sensitivity to enzymes, and insolubility in most organic solvents. Furthermore, the native starch of *hedychium coronarium* (popularly called ginger lily) has a high tendency for retrogradation, which is unfit for certain industrial utilization and phosphorylation of the NS has been suggested as a means of mitigating this shortcoming (De Oliveira et al., 2019). Poor water dispensability, high paste viscosity, and low fluidity limit the wholesome utility of native kudzu starch (Li et al., 2019). There are

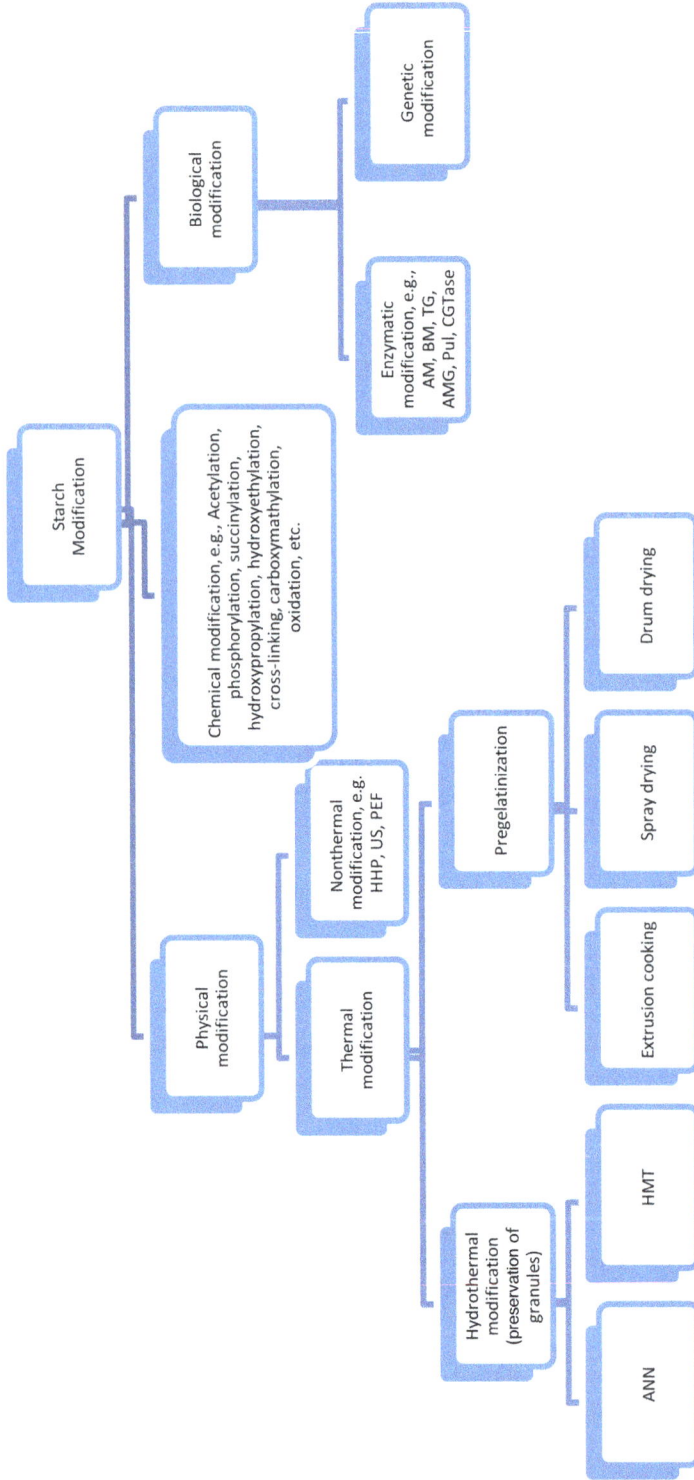

FIGURE 32.6 Classification of starch modification.

AM, α-amylase; **AMG**, amyloglucosidase; **ANN**, annealing; **BM**, β-amylase; **HHP**, high hydrostatic pressure; **HMT**, heat-moisture treatment; **PEF**, pulsed electric field; **PUL**, pullulanase; **TG**, transglucosidase; **US**, ultrasonication problems associated with the utilization of cassava flour and starch in the bakery industry. Cassava flour produces poorly-structured bakery products as a result of its innate gluten-free nature and its composition of minute sulfur-containing amino acids. On the other hand, cassava starch is unable to withstand the thermal conditions during baking thus eliciting inferior baked products (Dudu et al., 2019). In other cases, sweet potato starch has limitations including weak thermal stability and poor solubility, which inhibit its utilization in the production process (Das et al., 2010). Potato starch has restrictions such as low solubility in cold water, low shear and thermal resistance, high tendency to retrograde, narrow peak viscosity range, and poor process tolerance (Li et al., 2015). Additionally, in the production of biodegradable starch films, the problems inherent in potato starch are their hydrophilicity due to the many hydroxyl groups on its starch structure and poor mechanical properties. The application of potato starch is also limited due to its high viscosity at low concentration, low hygroscopicity, dispersity, acid and vulnerability (Guo et al., 2019). Another restriction to the industrial utilization of potato starch is its easy thermal decomposition, low shear strength (Heo et al., 2017), and poor acid resistance. All the limitations stated above for NSs that inhibit their utilization in the food and nonfood industries necessitate the need for one form of modification or the combinations of various modifications (dual or multiple).

The second stage is the problem associated with some single modification of various starches. The succinylation process is a chemical esterification process of modifying starch with desirable qualities such as high solubility in water, better thickening power, high viscosity, increased paste clarity, retarded retrogradation, and freeze-thaw stability (Moin et al., 2016). However, succinylated starches are unstable during shearing at high temperature (Ackar et al., 2015), therefore the need exists for new modifications to mitigate these shortcomings. Another example is that acid modification is widely nonspecific, randomly attacking the α-1, 4 and α-1, 6 linkages and leading to the production of undesirable products (coproducts and byproducts) apart from the net product.

The problem with dual modified starches were that some porous starches (PSs) produced by dual-enzyme modification had inherent shortcomings such as low resistance to shear, heat, and light sensitivity to thermal decomposition, which limit their applications (Xie et al., 2019a). In order to further enhance the adsorption capacity of PSs produced by dual-enzyme modification, the enzymatic hydrolysis modification is often combined with other treatments such as moderate gelatinization, freeze-thawing, and HMT to improve the efficiency of preparation (Wang et al., 2016; Xie et al., 2019b). Sometimes, these PSs are treated with cross-linking (Xie et al., 2019a) or repeated HMT (Xie et al., 2019b) to improve their adsorptive capacity.

32.4 Health Benefits of Roots and Tubers

In relation to the potato tuber, the positive impact of its antioxidant components on human health has been documented (Vinson, 2019; Xing et al., 2019). These researchers revealed the possibilities of improving the quality of high antioxidant diets. Therefore, there is a continuous search for high antioxidant natural compounds such as proteins, hydrolyzates, peptides, polyphenols (PPs), carotenoids, and flavonoid components (Vilcacundo et al., 2018a, b: Carrillo et al., 2016). A relationship has been developed between PP consumption and the prevention of cardiovascular diseases in humans from epidemiological studies. Polyphenols in potato tubers were found to avert oxidative stress, cell damages, DNA, protein damage, and lipid peroxidation processes (Abbas et al., 2016; Moga et al., 2016; Zhao et al., 2020), as well as Alzheimer's and Parkinson's (neurodegerative diseases) (Abbas et al., 2016; Zhao et al., 2020).

Sweet potato tubers are pivotal because of their unique nutritional and functional properties due to the possession of bioactive carbohydrates, proteins, lipids, carotenoids, anthocyanin, conjugated phenolic acids, and minerals represent versatile nutrients (Wang et al., 2016). These authors asserted that this unique composition of sweet potato contributes to their different health benefits, such as antioxidative, hepatoprotective, anti-inflammatory, antitumor, antidiabetic, antimicrobial, antiobesity, and antiaging effects. Polyphenolic acids (chlorogenic, isochlorogenic, caffeic, cinnamic, and hydroxycinnamic acids) serve as functional ingredients of health-promoting foods toward preventing cancer, diabetes, and inflammation (Lu et al., 2020). It has been indicated that prolonged consumption of sweet potato-based foods could contribute positively to human health (Wang et al., 2016).

The usage of cocoyam or taro increases efforts to diversify food and support food security (Venema et al., 2020). Furthermore, to potentially lower blood glucose concentration, taro contains resistant starch, which reaches the colon and can contribute to the adjustment of the composition and/or activity of gut microbiota (Kovatcheva-Datchary., 2009), and through this increases human health, e.g., by increasing the production of butyrate (Rose et al., 2010). It must be noted that resistant starches are starches that escape digestion in the small intestine and reach the large intestine or colon. In the latter, the so-called resistant starch is degraded to short chain fatty acids (SCFAs) and the consequential benefits in humans include the help in alleviating cancer, diabetes, obesity, and other cardiovascular ailments.

Native taro starch showed a lower glycemic load than native wheat starch, and modified taro starch could be used as a substitute for refined foods by diabetics and people suffering from other glucose metabolic diseases (Surono et al., 2020). Furthermore, taro contains anthocyanins such as cyanidin-3-glucoside, pelargonidin-3-glucoside and cyanidin-3-chemnode, which are reported to have antioxidative and anti-inflammatory properties (Kaushal et al., 2015).

In a nutshell, the major classes of phytochemicals found in roots and tubers such as alkaloids, phenolics, terpenoids, and tannins have the potential to prevent diseases and act as antimicrobial, anti-inflammatory, antioxidant, anticancerous, and detoxifying, immunity-potentiating, and neuropharmacological agents (Koche et al., 2016).

32.5 Applications of Roots and Tubers

The products from cassava roots not only serve for human consumption, but are also utilized as animal feed and raw materials for paper, textiles, alcoholic drinks, and bioethanol (Falade and Akingbala, 2010). Cassava starch has been used as food packaging materials (Zhao et al., 2018) and as filler in pharmaceutical tablets. Its derivatives such as dextrins, glucose, and high fructose syrups have become the main products of the cassava industry.

Food application of cassava flour and starch has been broadly exploited as bakery additives in bread and other bakery products (Charoenkul et al., 2011), and as thickeners and structure enhancers in other food systems (Steenkamp and McCrindle, 2014). Unlike some cereals such as wheat, cassava flour is gluten-free and thus confers no allergic effects such as celiac disease when consumed (Dudu et al., 2019). Furthermore, cassava flour and starch could be viable replacements for wheat flour and starch in countries where cassava is the principal staple food.

Potato starch is broadly utilized in the fields of food, medicine, textile and paper making due to its low gelatinization transition temperature, high transparency, and viscosity of its paste (Cao and Gao, 2019). Furthermore, it is applied in other multifarious industries such as coating, chemical, and biodegradable materials (Zhao et al., 2018). Potato flour can be used as a substitute for wheat flour in confectionery and pastry, and as a thickener and stabilizer in ice creams, jellies, soups, and sauces. Additionally, in the pharmaceutical industry, potato starch is employed as a raw material for the production of dextrose, an excipient for tablets and pills and other products of the industry. Potato starch is also used in the configuration of inflatable or pregelatinized starch flakes for the fabrication of pulp, coated paper, and kraft paper. These expositions on the utilization of potato starches are due to Chavez et al. (2020).

Some commercial sweet potato-related products can be divided into a few wide categories such as noodles (pasta) (Lee et al., 2006; Wu et al., 2006), pickled vegetables (El Sheikha & Montet, 2014), beverages (Panda et al., 2009), dairy products (Mohapatra et al., 2007), major condiments (i.e., red vinegar) (Terahara et al., 2003), food additives (Bindumole et al., 2000), and sugar syrups (Johnson et al., 2009). Liquefaction and saccharification of sweet potato have been utilized for the production of glucose and high-fructose syrup for the food and beverage industries (Dominque et al., 2013; Johnson et al., 2010). Apart from being used in the manufacture of modified starches and bioethanol, sweet potato starch has been widely applied to produce noodles, chips, bakery foods, snack foods, confectionery products and starch syrup (Guo, 2018).

After an extensive study of yam starches, the following conclusions were reached on its applicability. There was large variability in the physicochemical and functional properties of yam starches, hence their heterogeneous utilizations in the food and nonfood industries. The small granule size of *D. dumetorum* compared to other yam starches displayed that it could find application as a fat replacer in foods, because starches with small granules demonstrate fat mimetic characteristics that produce agreeable textures in frozen desserts, cookies, and other low-fat and fat-free food formulations. They can also be employed as stabilizers in baking powder, aerosols, face (or dusting) powder in cosmetics and textile industries. Yam starches can be utilized as thickeners, binders, and gelling agents in food especially those with high swelling power such as *D. rotundata* and *D. bulbifera*. The high retrogradation and syneresis tendency of some yam starches will be a demerit in the frozen product industry but could be applied in products where retrogradation is merited, such as noodles, soups, and sauces. Heterogeneous gelation properties of the yam starches could also be utilized in various food applications; those with high gelling ability could be applied in food like jams and jellies and in foods where Hooke's law is obeyed and desired while those with low gelling ability will be very beneficial as ingredient in processing interdependent diet. The above expositions on the potential industrial utilizations of yam starches are due to Otegbayo et al. (2014).

Water yam (*Dioscorea alata*) starch has been established to be employable to thicken yogurts, producing a transparent, creamy texture, with a sweet taste, flavor, and consistency to create an overall acceptable product (Tortoe et al., 2019). Research attributed this to the low moisture content, water activity, and low acidity of the water yam starch that favored quality and appropriate shell life. Furthermore, research indicated the restricted water absorption capacity, swelling power, and lightness in color of the water yam starch paste as being supportive in the development of a new products, especially as congeals for yogurt production.

The functional, physical and thermal properties of the spherical aggregates of taro starch granules showed that this material could offer a good potential for the microencapsulation of bioactive compounds (Hoyos-Leyva et al., 2018). One of the finest attributes of taro is the small size of its starch granules. As a result of this, it has found attention and attraction in the production of food such as gluten-free products, noodles, extruded snacks, and breads, as well as nonfood applications including bioethanol and pharmaceutical usage (Singla et al., 2020). Taro starch microcapsules are better than that of rice starch in usage as wall material or matrix for protection and growth of *Lactobacillus paracasei* subsp. *paracasei*. This advantage is based on the small size of taro starch granules that proposed protection and maintenance of probiotic strains viability during storage at both refrigeration and optimum temperature (Alfaro-Galarza et al., 2020). The gluten-free nature of taro starch unlike some cereal starches such as wheat starches makes it suitably fit and desirable for bread production along with mixing of hydrocolloids and enzymes (Calle et al., 2020).

The low amylose content of taro starch in combination with high amylose loaded cassava starch is a suitable selection to cast edible film (Hamim et al., 2018). Alpha-amylase modified taro starch is preferable as a better stabilizer in ice-cream; its showed appropriate foam quality and better absolute viscosity and optimum overrun performance than guar gum stabilized ice-cream (More et al., 2017). According to the later authors, the low cost, optimum texture, and low melting properties of taro starch compared to other traditional food additives gives it better storage stability. In a nutshell, taro corms starch is considered an inexpensive option for the food industries due to its various potentialities in food as stabilizer, emulsifier, fat substitute, and as filler agent (Singla et al., 2020). Furthermore, with the incorporation of resistant starch in taro, its utilization in baby foods, package material novelties, and geriatric foods is favored.

32.6 Conclusion

Roots and tubers are mostly cultivated in the tropical and subtropical regions of the world for their edible starchy polysaccharide. Starch is the most pivotal carbohydrate in the human diet. The uniqueness of roots and tubers is their ability to grow under various harsh environmental conditions. The extracted native starch from roots and tubers are utilized in the food and nonfood industries. The nonfood industries, such as pharmaceuticals, cosmetic, pulp and paper, textile, chemical, and petroleum have benefitted in the usage of different starches. Native starches with poor functionality need their physicochemical and functional properties modified to meet the need of some modern industries. The native starches are physically, chemically, and biologically modified or their combination undergoes dual, triple, or quadruple modifications for better functionality.

Apart from their role as an energy provider, starchy roots and tubers supply a number of helpful nutritional and health benefits such as antioxidative, hypoglycemic, hypocholesterolemic, antimicrobial, and immunomodulatory activities. Processing affects the bioactivities of component compounds. Roots and tubers may serve as functional foods and nutraceutical ingredients to help in the prevention of noncommunicable chronic diseases such as obesity, diabetes, cancer, aging, and heart-related problems, as well as maintain wellness. Plants yielding starchy roots and tubers are pivotal to nutrition and health. They perform a significant role in the diets of people in developing countries in their utilization for animal feed and for manufacturing of starch, alcohol, fermented foods, and beverages.

REFERENCES

Abbas, M., F. Saeed, F. M. Anjum, M. Afzaal, T. Tufail, M. S. Bashir, A. Ishtiaq, S. Hussain, and H. A. R. Suleria. 2016. Natural polyphenols: An overview. *International Journal of Food Properties* 20(8): 1689–1699.

Abass, A., W. Awoyale, B. Alenkha, N. Malu, B. Asiru, V. Manyong, and N. Sanginga. 2018. Can food technology innovation change the status of a food security crop? A review of cassava transformation into 'bread' in Africa. *Food Review International* 34(1): 87–102.

Ackar, D., J. Babic, A. Jozinovic, B. Milicevic, S. Jokic, M. Milicevic, et al. 2015. Starch modification by organic acids and their derivatives: A review. *Molecules* 20: 19554–19570.

Adepoju, O. T., O. Boyejo, and P. O. Adeniji. 2018. Effects of processing methods on nutrient and antinutrient composition of yellow yam (*Dioscorea cayenensis*) products. *Food Chemistry* 238: 160–165.

Alfaro-Galarza, O., E. O. Lopez-Villegas, N. Rivero-Perez, D. Tapia-Maruri, A. R. Jimenez-Aparicio, H. M. Palma-Rodriguez, and A. Vargas-Torres. 2020. Protective effects of the use of taro and rice starches as wall material on the viability of encapsulated *Lactobacillus paracasei subsp. paracaasei. LWT Food Science and Technology* 117. doi:10.1016/j.lwt.2020.108686

Andrade, L. A., C. A. Nunes, and J. Pereira. 2015. Relationship between the chemical components of taro rhizome mucilage and its emulsifying property. *Food Chemistry* 178: 331–338.

Allem, A. C. 2002. The origins of taxonomy of cassava. In: Cassava: Biology, Production and Utilization, CABI Publishing, 1–16.

Aristizabal, J., J. A. Garcia, and B. Ospina. 2017. Refined cassava flour in bread making: A review. *Ingenieriae Investigacion* 37(1): 25–33.

Ashogbon, A. O. 2020a. Dual modification of various starches: Synthesis, properties and applications. *Food Chemistry* 342: 128325. doi:10.1016/j.foodchem.2020128325

Ashogbon, A. O. 2020b. Limited quadruple modification of various starches in the literature; Why? *Starch/Starke* 73(3–4): 2000126. doi:10.1002/star.202000126

Ashogbon, A. O. 2018a. Current research addressing physical modification of starch from various botanical sources. *Global Nutrition and Dietetics* 1(1): 1–7.

Ashogbon, A. O. 2018b. Contradictions in the study of some compositional and physicochemical properties of starches from various botanical sources. *Starch/Starke* 70: 1–7.

Ashogbon, A. O. and E. T. Akintayo. 2014. Recent trend in the physical and chemical modification of starches from different botanical sources: A review. *Starch/Starke* 66: 41–57.

Ashogbon, A. O., E. T. Akintayo, A. O. Oladebeye, A. D. Oluwafemi, A. F. Akinsola, and O. E. Imanah. 2020. Developments in the isolation, composition, and physicochemical properties of legume starches. *Critical Review in Food Science and Nutrition*. doi: 10.1080/10408398.2020.1791048

Bao, P. E., L. B. Bao, P. Ngoc, N. N. Thanh, and P. Van. 2017. Impact of heat-moisture and annealing treatments on physicochemical properties and digestibility of starches from different colored sweet potato varieties. *International Journal of Biological Macromolecules* 105: 1071–1078.

BeMiller, J. N. 2016. Starch: Modification. In: Encyclopedia of Food Grains, 2nd ed., Academic Press, 223–286. doi:10.1016/B978-0-12-394437-5.00147-9

Benavent-Gil, Y., and C. M. Rosell. 2017. Morphological and physicochemical characterization of porous starches obtained from different botanical sources and amylolytic enzymes. *International Journal of Biological Macromolecules* 101: 587–595.

Bindumole, V. R., and C. Bulagopalan. 2001. Saccharification of sweet potato flour for ethanol production. *Journal of Root Crops* 27(1): 89–93.

Blagbrough, I. S., S. A. I. Bayoumi, M. G. Romaw, and J. R. Beeching. 2010. Cassava: An appraisal of its phytochemical and its biotechnological prospects—A review. *Photochemistry* 71(17-18): 1940–1951.

Calle, J., Y. Benavent-Gil, and C. M. Rosell. 2020. Development of gluten-free bread from Collocasia esculenta flour blended with hydrocolloids and enzymes. *Food Hydrocolloids* 198. doi:10.1016/j.foodhyd.2020.105243

Calliope, S., J. Wagner, and N. Samman. 2020. Physicochemical and function characterization of potato starch (*Solanum tuberosum* spp. Andigenum) from the Quebrada De Humahuaca, Argentina. *Starch/Starke* 72: 1900069. doi:10.1002/star.201900069

Camire, M. E., S. Kubow, and D-J. Donnelly. 2009. Potatoes and human Health. *Critical Reviews in Food Science and Nutrition* 49: 823–840.

Campos-Vega, R., and B. D. Oomah. 2013. Chemistry and Classification of Phytochemicals, Handbook of Plant Food Phytochemicals, John Wiley & Sons Ltd.

Cao, M., and Q. Gao. 2019. Effects of high-voltage electric field treatment on physicochemical properties of potato starch. *Journal of Food Measurement and Characterization* 13: 3069–3076. doi:10.1007/s11694-019-00229-x

Carrillo, W., J. A. Gomez-Ruiz, B. Miralles, M. Rames, D. Barrio, and I. Recio. 2016. Identification of anti-oxidant peptides of hen egg-white lysozymes and evaluation of inhibition of lipid peroxidation and cytotoxicity in the Zebrafish model. *European Food Research and Technology* 242: 1777–1785.

Chair, H., R. E. Traore, M. F. Duval, R. Rivallan, A. Mukherjee, et al. 2016. Genetic diversification and dispersal of taro (*Colocasia esculenta* (L.) Schott). *PLOS ONE* 11: 1–9.

Chandrasekara, A. 2017. Roots and tubers as functional foods. In: Bioactive Molecules in Food, Springer, 1–29.

Charoenkul, N., D. Uttapap, W. Pathipanawat, and Y. Takeda. 2011. Physicochemical characteristics of starches and flours from cassava varieties having different cooked root textures. *LWT Food Science and Technology* 44(8): 1774–1781.

Chen, Y.-F., L. Kaur, and J. Singh. 2018. Chemical modification of starch. In: Starch in Food, Part 1, Analyzing and Modifying Starch, Woodhead Publishing, 283–321. doi:10.1016/B978-0-08-100868-3.00007-x

Cornago, D. F., R. G. O., Rumbaoa, and I. M. Geronimo. 2011. Philippine yam (Dioscorea spp.) tubers phenolic content and antioxidant capacity. *Philippine Journal of Science* 140(2): 145–152.

Coronell-Tovar, D. C. 2015. Elaboracion de harina a partir del cormo de la yautia (Xanthosoma spp.) de alimentos (Master's dissertation). Programa de Ciencia y Tecnologia de Alimentos, Universidad de Puerto Rica, Mayaguez, Puerto Rico.

Coronell-Tovar, D. C., R. N. Chavez-Jauregui, A. Bosques-Vega, and M. L. Lopez-Moreno. 2019. Characterization of cocoyam (*Xanthosoma* spp.) corm flour from the Nazareno cultivar. *Food Science and Technology, Campinas* 39(2): 349–357.

Dako, E. N., Retta, and G. Desse. 2016. Comparison of three sweet potato (Ipomoea Batatas (L.) Lam) varieties on nutritional and anti-nutritional factors. *Global Journal of Science Frontier Research* 16(4): 1–11.

Das. A. B., G. Singh, and C. S. Riar. 2010. Effect of acetylation and dual modification on physicochemical, rheological and morphological characteristics of sweet potato (*Ipomoea Batatas*) starch. Carbohydrate Polymers 80: 725–732.

De Lima, A. C. S., J. D. da Rocha Viana, L. B. de Sousa Sabino, L. M. R. da Silva, N. K. Y. da Silva, and P. H. M. da Sousa. 2016. Processing of three different cooking methods of cassava: Effects of in vitro bio-accessibility of phenolic compounds and antioxidant activity. *LWT Food Science and Technology* 76 (Part B): 254–258. Doi: 10.1016/j.lwt.2016.07.023.

De Oliveira, A. F., J. Machado Soares, E. C. da Silva, P. S. Loubet Filho, C. Jordao Candido, L. A. do Amaral, et al. 2019. Evaluation of chemical, physical and nutritional composition and sensory acceptability of different sweet potato cultivars. *Ciencias Agrarias*, 40(3): 1113–1127.

Dilworth, L., K. Brown, R. Wright, M. Oliver, S. Hall, and H. Asemota. 2012. Antioxidants, minerals and bioactive compounds in tropical staples. *African Journal of Food Science and Technology* 3(4): 90–98.

Dhull, S. B., Punia, S., Kumar, M., Singh, S., & Singh, P. (2020). Effect of different modifications (physical and chemical) on morphological, pasting, and rheological properties of black rice (Oryza sativa L. Indica) starch: A comparative study. *Starch–Stärke*, 2000098.

Dominque, B., P. N., Gichuhi, V. Rangari, and A. C. Bovell-Benjamin. 2013. Sugar profile, mineral content, and rheological and thermal properties of an isomerized sweet potato starch syrup. *International Journal of Food Science* 2013: 1–8.

Dudu, O. E., A. B. Oyede, S. A. Oyeyinka, and Y. Ma. 2019. Impact of steam-heat-moisture treatment on structural and functional properties of cassava flour and starch. *International Journal of Biological Macromolecules* 126: 1056–1064.

El Sharkawy, M. A. 2014. Global warming: Causes and impacts on agro-ecosystems productivity and food security with emphasis on cassava comparative advantage in the tropics/subtropics. *Photosynthetica* 52(2): 161–178.

El Sheikha, A. F., and D. Montet. 2014. African fermented foods: historical roots and real benefits. In: Microorganisms and Fermentation of Traditional Foods, Food Biology Series. Ray, R. C. and Montet, D. (Eds.). Science Publishers Inc., CRC Press, Boca Raton, Floride, USA, 248–282.

Falade, K. O., and J. O. Akingbala. 2010. Utilization of cassava for food. *Food Reviews International* 27(1): 51–83.

Food and Agricultural Organization of the United Nations (FAO). 1990. Roots, Tubers, Plantains and Bananas in Human Nutrition. Food and Nutrition Series, NO. 24. Rome: FAO.

FAO. 2016. Food outlook October, 2016. Rome: FAO. Available from: http://www.fao.org/3/i6198e/i6198e.pdf, Published by Trade and Markets Division of FAO, Rome – Italy.

FAOSTAT. 2014. Agricultural data. Rome, Italy. Food and Agricultural Organization Corporate Statistical Database online. Rome: FAO. Available from http://www.fao.org/faostat/en /#home. Assessed on the 26th of May, 2021.

FAOSTAT. 2016. Food and Agriculture Organization Corporate Statistical Database online. Rome: FAO. http://faostat.fao.org/beta/en/#data/QC

FAO. 2018. FAOSTAT. Food and Agriculture Organization of the United Nations http://faostat.fao.org/beta/en/#data/QC. Assessed on the 26th of May, 2021.

FAOSTAT. 2020. Food and Agriculture Organization Corporate Statistical Database Online. Rome: FAO. Available online at http://faostat.fao.org/beta/en/#data/QC. Assessed on the 26th of May, 2021

Fonseca, I. M., J. R. Goncalves, S. L. M. Halal, V. Z. Pinto, A. R. G. Dias, A. C. Jacques, and E. A. Zavareze. 2015. Oxidation of potato starch with different sodium hypochlorite concentrations and its effects on biodegradable films. *LWT Food Science and Technology* 60: 714–720.

Ghosh, S. 2015. Photochemistry and therapeutic potential of medicinal plant: *Discorea bulbifera*. *Medicinal Chemistry* 5(4): 160–172.

Goel, C., A. D. Semwal, A. Khan, S. Kumar, and G. K. Sharma. 2020. Physical modification of starch changes in glycemic index, starch fractions, physicochemical and functional properties of heat-moisture treated buckwheat starch. *Journal of Food Science and Technology* 57: 2941–2948. doi:10.1007/s13197-020-04326-4

Guo, L. 2018. Sweet potato starch modified by branching enzymes, β-amylase and transglucosidase. *Food Hydrocolloids* 83: 182–189.

Guo, L., H. Tao, B. Cui, and S. Janaswamy. 2019. The effects of sequential enzyme modifications on structural and physicochemical properties of sweet potato starch granules. *Food Chemistry* 277: 504–514.

Hamim, F. A., S. A. Ghani, F. Zainuddin, and H. Ismail. 2018. Taro powder (Colocasia esculenta) filler reinforced recycled high density polyethylene/ethylene vinyl acetate composites: Effect of different filler loading and high density polyethylene grafted glycolic acid as compatibilizer, in: S. M. Sapuan, H. Ismail, E. S. Zainudin (Eds.), *Natural Fiber Reinforced Vinyl Ester and Vinyl Polymer Composites*, Woodhead Publishing, 225–247.

Hamouz, K., K. Pazderu, J. Lachman, J. Cepl, and Z. Kotikova. 2016. Effect of cultivar, flesh color, locality and year on carotenoid content in potato tubers. *Plant Soil and Environment* 62(2): 86–91.

Haq, F., H. Yu, L. Teng, M. Haroon, R. U. Khan, S. Mehmood, Bilal-Ul-Amin, A. Khan, and A. Nazir. 2019. Advances in chemical modifications of starches and their applications. *Carbohydrate Research* 476: 12–35.

Henry, G., and C. Hershey. 2020. Cassava in South America and the Caribbean. CABI Publishing, Wallingford, UK.

Heo, H., Y.-K. Lee, and Y.-H. Lee. 2017. Effect of cross-linking on physicochemical and in vitro digestibility properties of potato starch. *Emirate Journal of Food and Agriculture* 29(6): 463–469.

Hoyos-Leyva, J., L. A. Bello-Perez, E. Agama-Acevedo, and J. Alvarez-Ramirez. 2018. Potential of taro starch spherical aggregates as wall material for spray drying microencapsulation: Functional, physical and thermal properties. *International Journal of Biological Macromolecules* 120: 237–244.

Hoover, R. 2001. Composition, molecular structure, and physicochemical properties of tuber and root starches: A review. *Carbohydrate Polymers* 45: 253–267.

Hsieh, C-F., W. Liu, J. K. Whaley, and Y-C. Shi. 2019. Structure, properties, and potential applications of waxy tapioca starches—A review. *Trends in Food Science & Technology* 83: 225–234.

Huang, Y., D. Xiao, B. M. Burton-Freeman, and I. Edirisinghe. 2016. Chemical changes of bioactive phyto-chemicals during thermal processing. In: *Reference Module in Food Science*. Elsevier, 1–9.

Hunt, V. H., H. M. Moot, and P. J. Mathews. 2013. Genetic data confirms field evidence for natural breeding in a wild taro (*Colocasia esculenta*) population in northern Queensland, Australia. *Genetic Resource and Crop Evolution* 60: 1695–1707.

Igbabul, B. D., J. Amove, and I. Twadue. 2014. Effect of fermentation on the proximate composition, anti-nutritional factors and functional properties of cocoyam (Colocasia esculenta) flour. *African Journal of Food Science* 5(3): 67–74.

Ikiriza, H. P. E. Ogwang, E. L. Peter, O. Hedman, C. U. Tolo et al. 2019. *Dioscorea bulbifera*, a highly threat-ened African medicinal plant, A review. *Cogent Biology* 5: 1–6.

Iyer, S., D. S. Mattinson, and J. K. Fellman. 2010. Study of the early events leading to cassava root postharvest deterioration. *Tropical Plant Biology* 3(3): 151–165.

Johnson, R., G. Padmaja, and S. N. Moorthy. 2009. Comparative production of glucose and high fructose syrup from cassava and sweet potato roots by direct conversion techniques. *Innovative Food Science & Emerging Technologies* 10: 616–620.

Johnson, R., S. N. Moorthy, and G. Padmaja. 2010. Production of high fructose syrup from cassava and sweet potota flours and their blends with cereal flours. *Food Science and Technology International* 16(3): 251–258.

Kaushal, P., V. Kumar, and H. K. Sharma. 2015. Utilization of taro (Colocasia esculenta): A review. *Journal of Food Science and Technology* 52:27–40.

Koche, D., R. Shirsat, and M. Kawale. 2016. An overview of major classes of phytochemicals: Their types and role in disease prevention. *Hislopia Journal* 9(1/2): 1–11.

Kovatcheva-Datchary, P., M. Egert, A. Maathuis, M. Stojanovic, A. A. de Graaf, H. Smidt, W. M. de Vos, and K. Venema. 2009. Linking phylogenetic identities of bacteria to starch fermentation in an in vitro model of the large intestine by RNA-based stable isotope probing. *Environmental Microbiology* 11(4): 914–926.

Lachman, J., K. Hamouz, J. Musilova, K. Hejtmankovo, Z. Kotikoya, et al. 2013. Effect of peeling and three cooking methods on the content of selected phytochemicals in potato tubers with various color of flesh. *Food Chemistry* 138: 1189–1197.

Labot, V., R. Malapa, and T. Molisale. 2018. Development of HP-TLC method for rapid quantification of sugars, catechins, phenolic acids and saponins to assess yam (Dioscorea spp.) tuber flour quality. *Plant Genetic Resources*, 1–11.

Lee, W. S., I. C. Chen, C. H. Chang, and S. S. Yang. 2012. Bioethanol production from sweet potato by coim-mobilization of saccharolytic molds *Saccharomyces cerevisiae*. *Renewable Energy* 39: 216–222.

Li, H., B. Cui, S. Janaswamy, and L. Guo. 2019. Structural and functional modification of kudzu starch modi-fied by branching enzyme. *International Journal of Food Properties* 22(1): 952–966.

Li, W. H., X. L. Tian, L. P. Liu, P. Wang, G. L. Wu, et al. 2015. High pressure induced gelatinization of red adzuki bean starch and its effects on starch physicochemical and structural properties. *Food Hydrocolloids* 45: 132–139.

Lu, P., X. Li, S. Janaswamy, C. Chi, L. Chen, Y. Wu, and Y. Liang. 2020. Insights on the structure and digest-ibility of sweet potato starch: Effect of postharvest storage of sweet potato roots. *International Journal of Biological Macromolecules* 145: 694–700.

Masina, N., Y. E. Choonara, P. Kumar, L. C. D. Toit, M. Govender, S. Indermun, and V. Pillay. 2017. A review of the chemical modification techniques of starch. *Carbohydrate Polymers* 157: 1226–1236.

Mijinyawa, A. A., G. Durga, and A. Mishra. 2018. Isolation, characterization and microwave assisted surface modification of *Colocasia esculenta* (L.) Schott mucilage by grafting polylactide. *International Journal of Biological Macromolecules* 119:1090–1097.

Moga, M. A., O. G. Dimienescu, C. A. Arvatescu, A. Mironescu, I. Dracea, and I. Ples. 2016. The role of natural polyphenols in the prevention and treatment of cervical cancer—An overview. *Molecules* 21: 1055.

Mohapatra, S., S. H. Panda, A. K. Sahoo, P. S. Sivakumar, and R. C. Ray. 2007. β-carotene rich sweet potato curd: Production, nutritional and proximate composition. *International Journal of Food Science and Technology* 42: 1305—1314.

Moin, A., T. M. Ali, and A. Hasnain. 2016. Effect of succinylation on functional and morphological properties of starches from broken kernels of Pakistani basmati and irri rice cultivars. *Food Chemistry* 191: 52–58.

More, P. R., R. V. Solunke, M. I. Talib, and V. R. Parate. 2017. Stabilization of ice-cream by incorporating α-amylase modified taro starch. *International Journal of Creative Research and Thoughts*. 410–416. DOI: 10.1727/IJCRT.17172

Nagata, C. L. P., L. A. Andrade, and J. Pereira. 2015. Optimization of taro mucilage and fat levels in sliced breads. *Journal of Food Science and Technology* 52: 5890–5897.

Njoh Ellong, E., C. Billard, D. Petro, S. Adenet, and K. Rochefort. 2015. Physicochemical, nutritional and sensorial qualities of Boutou yam (*Dioscorea alata*) varieties. *Journal of Experimental Biology and Agricultural Sciences* 3(2): 138–150.

Otegbayo, B. D., Oguniyan, and O. Akinwumi. 2014. Physicochemical and functional characterization of yam starch for potential industrial applications. *Starch/Starke* 66: 235–250.

Owusu-Darko, P. G. A., Paterson, and E. L. Omenyo. 2014. Cocoyam (corms and cormels) an underexploited food and feed resource. *Journal of Agricultural Chemistry and Environment* 03(01): 22–29.

Prochnik, S., P. R. Marri, B. Desamy, P. D. Rabinowicz, C. Kodira, M. Mohiuddin, et al. 2012. The cassava genome: Current progress, future directions. *Tropical Plant Biology* 5(1): 88–94.

Punia, S. (2020). Barley starch modifications: Physical, chemical and enzymatic—A review. *International Journal of Biological Macromolecules* 144: 578–585.

Punia, S., K. S. Sandhu, S. B. Dhull, A. K. Siroha, S. S. Purewal, M. Kaur, and M. K. Kidwai. (2020). Oat starch: Physico-chemical, morphological, rheological characteristics and its applications—A review. *International Journal of Biological Macromolecules* 154: 493–498.

Punia, S., K. S. Sandhu, S. B. Dhull, and M. Kaur. (2019). Dynamic, shear and pasting behaviour of native and octenyl succinic anhydride (OSA) modified wheat starch and their utilization in preparation of edible films. *International Journal of Biological Macromolecules* 133: 110–116.

Ronaldo, D. 2020. Carbohydrate and bioactive compounds composition of starchy tropical fruits and tubers, in relation to pre and postharvest conditions: A review. *Journal of Food Science* 82(2): 249–259.

Rose, D. J., K. Venema, A. Keshavarzian, and B. R. Hamaker. 2010. Starch-entrapped microspheres show a beneficial fermentation profile and decrease in potentially harmful bacteria during in vitro fermentation in faecal microbiota obtained from patients with inflammatory bowel disease. *The British Journal of Nutrition* 103(10): 1514–1524.

Price, E. J., R. Bhattacharjee, A. Lopez-Montes, and P. O. Fraser. 2018. Carotenoid profiling of yams: Clarity, comparisons and diversity. *Food Chemistry* 259: 130–138.

Samaniego, I., S. Espin, X. Cuesto, V. Arias, A. Rubio, et al. 2020. Analysis of environmental conditions effect in the phytochemical composition of potato (*Solanum tuberosum*) cultivars. *Plants (MDPI)* 9: 815. doi:10.3390/plants90708

Satheesh, N. and F. Solomon. 2019. Review on nutritional composition of orange-fleshed sweet potato and its role in the management of Vitamin A deficiency. *Food Science and Nutrition* 7: 1920–1945.

Saxena, M., J. Saxena, R. Nema, D. Singh, and A. Gupta. 2013. Photochemistry of medicinal plants. *Journal of Pharmacognosy and Photochemistry* 1: 168–182.

Shao, Y., L. Mao, W. Guan, X. Wei, F. Xu, Y. Li, and Q. Jiang. 2020. Physicochemical and structural properties of low-amylose Chinese yam (*Dioscorea opposite* Thunb.) starches. *International Journal of Biological Macromolecules* 164: 427–433.

Shekhar, S., D. Mishra, A. K. Buragohain, S. Chakraborty, and N. Chakraborty. 2015. Comparative analysis of phytochemicals and nutrient availability in two contrasting cultivars of sweet potato (*Ipomoea batatas* L.). *Food Chemistry* 173: 957–965.

Singla, D., A. Singh, S. B. Dhull, P. Kumar, T. Malik, and P. Kumar. 2020. Taro starch. Isolation, morphology, modification and novel applications concern. *International Journal of Biological Macromolecules* 163: 1283–1290.

Song, H. Y., S. Y. Lee, S. J. Choi, K. M. Kim, S. K. Jin, and G. J. Han. 2014. Digestibility and physicochemical properties of granular sweet potato starch as affected by annealing. *Food Science & Biotechnology* 23: 23–31.

Steenkamp, V., and C. McCrindle. 2014. Production, consumption and nutritional value of cassava (*Manihot esculenta*, Crantz) in Mozambique: An overview. *Journal of Agriculture and Biotechnology and Sustainable Development* 6(3): 29–38.

Surono, I. S., J. Verhoeven, and K. Venema. 2020. Low glycemic load after digestion of native starch from the indigenous tuber Belitung taro (Xanthosoma sagittifolium) in a dynamic in vitro model of the upper GI tract (TIM-1). *Food & Nutrition Research* 64: 4623. Doi: 10.29219/fnr.v64.4623

Tortoe, C., P. T. Akonor, and J. Ofori. 2019. Starches of two water yam (Dioscorea alata) varieties used as congeals in yogurt production. *Food Science & Nutrition* 7: 1053–1062.

Vanier, N. L., S. L. M. El Halal, A. R. G. Dias, and E. D. R. Zavareze. 2017. Molecular structure, functionality and applications of oxidized starches: A review. *Food Chemistry* 221: 1546–1559.

Venema, K., J. Verhoeven, I. S. Surono, P. Waspodo, and A. Simatupang. 2020. Differential glucose bioaccessibility from native and modified taro-starches in the absence or presence of beet juice. *CYTA-Journal of Food* 18(1): 670–674.

Vilcacundo, R. B., D. A. Barrio, L. Pinuel, P. Boeri, A. Tombari, A. Pinto, J. Welbaum, H. Hernandez-Ledesma, and W. Carrillo. 2018a. Inhibition of lipid peroxidation of kiwicha (Amaranthus caudatus) hydrolyzed protein using zebrafish larvae and embryos. *Plants* 7: 69.

Vilcacundo, R., B. Miralles, W. Carrillo, and B. Hernandez-Ledesma. 2018b. In vitro chemo-preventive properties of peptides released from quinoa (*Chenopodum quinoa* Wild) protein under simulated gastrointestinal digestion. *Food Research International* 105: 403–411.

Vinson, J. A. 2019. Intracellular polyphenols: How little we know. *Journal of Agriculture and Food Chemistry* 67: 3865–3870.

Wang, S., and P. Gao. 2020. Botanical sources of starch: In: S. Wang (Ed.), Starch Structure, Functionality and Applications in Foods, Springer Nature, Singapore Pte Ltd, 9–27. doi:10.1007/978-981-15-0622-2_2

Wang, S., S. Nie, and F. Zhu. 2016. Chemical constituents and health effects of sweet potato. *Food Research International* 89: 90–116.

Xie, Y., B. Zhang, M.-N. Li, and H.-Q. Chen. 2019a. Effects of cross-linking with sodium trimetaphosphate on structural and adsorptive properties of porous wheat starches. *Food Chemistry* 289: 187–194.

Xie, Y., M. N. Li, H. Q. Chen, and B. Zhang. 2019b. Effects of the combination of repeated heat-moisture treatment and compound enzymes hydrolysis on the structural and physicochemical properties of porous wheat starch. *Food Chemistry* 274: 351–359.

Xing, L., H. Zhang, R. Qi, R. Tsao, and Y. Mine. 2019. Recent advances of the health benefits and molecular mechanisms associated with green tea polyphenols. *Journal of Agriculture and Food Chemistry* 67: 1029–1043.

Zhang, B. et al. 2018. Comparison of structural and functional properties of starches from the rhizome and bulbil of Chinese yam (Dioscorea opposite Thunb.). *Molecules* 23(2).

Zhang, W., H. Chen, J. Wang, Y. Wang, L. Xing, and H. Zhang. 2014. Physicochemical properties of three starches derived from potato, chestnut, and yam as affected by freeze-thaw treatment. *Starch/Starke* 66: 353–360.

Zhao, K., B. Li, M. Xu, L. Jing, M. Gou, Z. Yu, et al. 2018. Microwave pretreated esterification improved the substitution degree, structural and physicochemical properties of potato starch esters. *LWT Food Science and Technology* 90: 116–123.

Zhao, D., J. E. Simon, and Q. A. Wu. 2020. A critical review on grape polyphenols for neuroprotection: Strategies to enhance bioefficacy. *Critical Review in Food Science and Nutrition* 60: 597–626.

Zhu, F. 2015a. Composition, structure, physicochemical properties, and modification of cassava starch. *Carbohydrate Polymers*, 122, 456–480.

Zhu, F. 2015b. Isolation, composition, structure, properties, modifications, and uses of yam starch. *Comprehensive Reviews in Food Science and Food Safety* 14(4): 357–386. doi:10.1111/1541-4337.12134

33

Potatoes: Processing, Properties, and Application

Oseni Kadiri
Edo State University Uzairue, Auchi, Edo State, Nigeria

Babatunde Olawoye
First Technical University, Ibadan, Oyo State, Nigeria

Timilehin David Oluwajuyitan
Federal University of Technology, Akure, Nigeria

CONTENTS

33.1 Introduction

Potatoes belong to the *Solanaceae* family and originated from the Andean mountain region of South America. About 5,000 varieties of potato exist worldwide. Potatoes range in color, shape, size, flavor and starch content. Some common potato varieties includes the White Rose, Russet Burbank, Katahdin, Red Pontiac, and Red LeSoda (Zaheer & Akhtar, 2016). Other varieties that feature a beautiful deep violet flesh and purple-grey skin also exist (Zaheer & Akhtar, 2016). However, the distribution and popularity of potato cultivars differ with geographical region. The International Potato Center (CIP) holds an ISO-accredited collection of germplasm of potatoes (CIP, 2008). In the developed world, potatoes being a staple food account for 130 kcal of energy per person per day when compared with 41 kcal for a person living in developing countries.

In 2010, the total world potato production was estimated at 324 million metric tons. About one-third of the global potato production is accountable to India and China (Zaheer & Akhtar, 2016) and is expected to soar higher in the coming years. Though there has been a slight decrease in global potato production between the year 2000 and 2010 from 327 million to 324 million metric tons; its production is continually increasing in developing countries, which now account for about 55% of total potato production. Bangladesh, a developing country, had its total potato production increased from about 3 million to 8 million metric tons between 2000–2010, while potato production decreased considerably in developed countries like Russia, Poland and the United States during the same period (Geohive, 2011).

Potato is a food product rich in carbohydrates and widely consumed worldwide. It is a versatile crop served in a variety of ways. Starch makes up 60–80% dry matter of the potato tuber while about 80% and 20% consist of moisture and dry matter of the freshly harvested potato tuber, respectively.

After wheat and rice, potato stands as the third most essential crop in the world (CIP, 2016). It is consumed by more than a billion people around the globe with total production exceeding 300 million metric tons (CIP, 2016). An estimated 125 countries are involved in potato cultivation throughout the world.

Potato can be propagated via vegetative reproduction, making it agriculturally unique since a potato or piece of potato can be used to produce a new plant. This plant produces between 5–20 tubers and 100–400 botanical seeds which are genetic clones of the original plant (CIP, 2016). About 180 wild potato species and 4,000 varieties of native potatoes exist (CIP, 2016). The potato makes a significant contribution of potassium, dietary fiber, and vitamin C to the human diet (McGill *et al.* 2013).

Nutritional deficiencies are not a common occurrence in countries whose population majorly depends on potatoes as its staple food (McCay *et al.,* 1987). Aside from its rich starch content, potatoes also consist of an appreciable number of bioactive molecules and secondary metabolites which are of importance in several processes. These compounds are of importance in the diet because of their beneficial health effect on humans (Katan & De Ross, 2004). Likewise, potatoes consist of higher phytonutrient content which can be improved upon through improved breeding processes and biotechnology approaches (Nzaramba *et al.,* 2007).

The year 2008 was declared as the International Year of the Potato by a United Nations General Assembly Resolution (UNGAR) to increase awareness of the potential role potatoes can make in defeating hunger (UNGAR, 2011). Consumption of fried potato snacks is a common practice among all age groups across the world. Recent literature reviewed the impact of nutrient and bioactive compounds present in the potato on human health (Hijmans & Spooner, 2001).

In a recent study, it was reported that the consumption of French fries, oven-baked fries, and white potatoes could be an essential source of magnesium, calcium, vitamin E, and fiber to children and adolescents. In arriving at their conclusion, these authors used the data provided from 2003–2006 by the National Health and Nutrition Examination Survey (NHANES) (Freedman & Keast, 2011).

33.2 Global Situation of Potatoes

Sweet potatoes, a potato variety, are an essential crop in economies of developing countries but of less importance in the developed world (FAO, 1990). About 80–85% of global sweet potato production comes from China (Khan, 2016). Other countries in Asia, Africa, and Latin America account for the remaining global production of sweet potatoes (Khan, 2016). They are mostly referred to as a perishable and bulky commodity with limited export potential due to cross-border transactions and high transport costs. There is still a low volume of international trade in potatoes and potato products as output trade hovers around 6%. High transportation and storage costs are some major obstacles to a wider international marketplace.

Sweet potatoes are among the world's most under-exploited food crops (Grant, 2003). With more than 133 million metric tons in annual production, sweet potatoes currently rank as the fifth most important food crop on a fresh-weight basis in developing countries after rice, wheat, maize, and cassava (Grant, 2003). Sweet potatoes are normally categorized as a "famine relief" or "food security" crop. In the last decade, the uses of sweet potato have diversified substantially in developing countries (Grant, 2003). The sweet potato is a suitable substitute for wheat in bread-making as shown by the FAO statistics, which reported the suitability of sweet potato to climates not suitable for wheat production (Katan & De Roos, 2004). This has the advantage of reducing dependency on wheat or wheaten flour importation.

The global situation of potato as a food commodity is that it is grown widely throughout the world. Nevertheless, only about 1% of potato production enters world trade, with France, the United Kingdom, Netherlands, and Canada being the major countries importing this crop (Katan & De Roos, 2004). The United States accounts for 35% of world's exportation of sweet potato, while China, Israel, France, Indonesia, and Netherlands/France accounts for 12%, 9%, 7%, 6%, and 5%, respectively. Most potato products are consumed domestically while a small percentage goes into animal feed and industrial purposes. Aside from its role as a staple food crop to some parts of the world such as the Philippines, Tonga Island, Solomon Island, and Papua New Guinea, the sweet potato plays several diverse roles in the human diet (Sosinski *et al.*, 2001). The potato and its products are mainly consumed as a luxury food, while it plays the role of a nutraceutical in Japan and other parts of the world.

33.3 Dietary and Nutritional Composition of Potatoes

Table 33.1 presents the dietary and nutritional composition of potatoes. From literature, it could be deduced that potatoes are rich sources of dietary fiber ranging between 1.80–8.10 g/100 g based on verities/cultivars (Burgos *et al.*, 2020; Jaggan *et al.*, 2020). The dietary fiber contained in potatoes is higher than white rice (0.3 g/100 g) and green plantain (1.2 g/100 g) as reported by Burgos *et al.* (2020) and Oluwajuyitan *et al.* (2020), respectively. It is also worthy of knowing that potatoes contribute to more than 26.2% of the daily fiber consumption in men and women living in the United States (NHANES, 2010). Dietary fiber is known as the nondigestible carbohydrate that promotes health benefits such as reducing the risk of colon cancer, high blood cholesterol levels, coronary heart disease, and high blood glucose level (type 2 diabetes) (Reynolds *et al.*, 2019; Sampaio *et al.*, 2020). Hence, the consumption of potatoes is encouraged in patients with cardiovascular diseases.

Ash content represents the incombustible portion after a sample is burned in the furnace at a high temperature. It is used as an index in measuring the mineral value of food samples. The ash content of potato ranged between 1.00–5.00 per 100 g (Kaplan *et al.*, 2018; Tejeda *et al.*, 2020). It is also within the same range compared to that of other staple foods such as rice (0.73 g/100 g), wheat (2.35 g/100 g), and sorghum (2.29 g/100 g), respectively (Verma & Srivastav 2017; Czaja *et al.*, 2020; Tasie & Gebreyes, 2020).

Protein is reported to be the second most abundant macronutrient present in potatoes and it ranged between 6.00–15.00 g/100 g (Bonierbale *et al.*, 2010; Kaplan *et al.*, 2018). The crude protein content of potatoes is higher when compared to other roots and tuber stables (2.0–4.00 g/100 g) (Beals, 2019), but lower compared with those of beans. In addition to the building and repair of worn-out tissue, proteins help in lowering blood pressure and maintaining weight loss (Skytte *et al.*, 2019; Daliri *et al.*, 2019). The protein quality of a food product is usually evaluated based on its biological value, which is referred to as the amount (%) of protein bioavailability in a given food sample for the body to utilize, usually in respect to amino acids. Using egg as a reference (<100%), the biological value of potatoes (90%) is relatively higher compared with other plant protein sources such as soy beans (84%) (McGill *et al.*, 2013; Beals, 2019). It is known that some plants are missing or lacking in one or more essential amino acid, however,

TABLE 33.1

Dietary and Nutritional Composition of Potatoes

Parameters	Composition	References
Dietary fiber (g/100 g)	1.80–8.10	Burgos *et al.* (2020); Jaggan *et al.* (2020)
Total ash (g/100 g)	1.00–5.00	Kaplan *et al.* (2018); Tejeda *et al.* (2020)
Crude protein (g/100 g)	6.00–15.00	Bonierbale *et al.* (2010); Kaplan *et al.* (2018)
Crude fat (g/100 g)	<1.00	Bellumori *et al.* (2020); Tejeda *et al.* (2020)
Carbohydrate (g/100 g)	9.00–35.00	Bellumori *et al.* (2020); Tejeda *et al.* (2020)
Energy (Kcal/100 g)	86.00–141.00	de Haan *et al.* (2019); Tejeda *et al.* (2020)
Potassium (mg/100 g)	150–1386	Nassar *et al.* (2012); Burgos *et al.* (2020)
Zinc (mg/100 g)	0.40–12.67	de Haan *et al.* (2019); Jaggan *et al.* (2020)
Iron (mg/100 g)	0.29–26.25	de Haan *et al.* (2019); Jaggan *et al.* (2020)
Vitamin C	8.00–110.80	Burgos *et al.* (2020); Kourouma *et al.* (2020)

potatoes exhibit a peculiar attribute by containing almost all essential amino acids (Gorissen *et al.* 2018; Beals, 2019).

Potatoes contained negligible amount of crude fat as it is less than 1.00 g/100 g, however, this is reported to contribute to its attractive flavor and appealing taste (Bonierbale *et al.*, 2010; Bellumori *et al.*, 2020). Potato is known as a carbohydrate-rich (9.00–35.00 g/100 g) food and is widely consumed as a cheap source of calories containing (86.00–141.00 kcal/100 g) energy (de Haan *et al.*, 2019; Bellumori *et al.*, 2020; Tejeda *et al.*, 2020). Observed calories values could be attributed to the protein and carbohydrate content obtained and this varies based on crop genotype and species (Bellumori et al., 2020). Carbohydrate quality of food samples is measured using the glycemic index—an index that shows the rate at which carbohydrates are broken down in the bloodstream (Venn & Green, 2007). Potatoes are classified as a low glycemic index food (<50%), which is due to a higher proportion of amylopectin (70–80%) than amylose (20–30%), thus potatoes are slowly digested, increase satiety, lower blood glucose levels, and reduce cholesterol levels, making them a suitable food for diabetic and hyperlipidemia patients (Miao *et al.*, 2015; Burgos *et al.*, 2020).

Micronutrients are found in appreciable quantities in potatoes and this includes potassium, zinc, iron, and vitamin C (de Haan *et al.*, 2019; Burgos *et al.*, 2020; Jaggan *et al.*, 2020). Research has shown that potatoes contain 12% of the recommended daily allowance (RDA) for potassium and 33% of the RDA of vitamin C (Navarre *et al.*, 2009). Potassium plays an important role in nervous system functioning as well as control of high blood pressure and decreased risk of stroke (Bethke & Jansky, 2008. Meanwhile, zinc and iron help in wound healing and boosting of the immune system while Vitamin C acts as an antioxidant that helps in scavenging of free radicals from the body, which can cause tissue damage and serves as a cofactor in the synthesis of collagen needed to support cardiovascular function, wound healing, and maintain cartilage, bones, and teeth (Naidu, 2003; Burgos *et al.*, 2020).

33.4 Beneficial Bioactive Compounds in Potatoes

To guide against invasion from pests, insects, microbes, animals, and humans, plants generally synthesize bioactive compounds that act as their natural defense. Aside from this function, bioactive compounds also serve to mitigate some degenerating diseases in humans, such as cancer, high blood pressure, atherosclerosis, diabetes, inflammation, Alzheimer's disease, etc. (Ezekiel *et al.*, 2013). Potato,specifically, the sweet potato,contains some edible parts such as the stem, leaves, and root, which contains various biologically active compounds. The presence of these biological active compound (phenols, flavonoids, peptides, starch, dietary fibre etc.) in potatoes had spurred the interest of researchers in the area of bioactive compounds as well as human nutritionists. The next sections will discuss some updates on and review the biologically active components present in potato leaves, stems and roots as well as their benefit to human health.

33.4.1 Phenols

Phenolic compounds are bioactive compounds that have a benzene ring in their structure and consist mainly of groups of compounds such as phenolic acid, flavonoids, and stilbenes. Different varieties of potatoes, which include orange, red and yellow flesh potatoes, have been reported to consist of various concentration of phenolic compounds. According to several researchers (Burlingame *et al.*, 2009; de Albuquerque *et al.*, 2019; Ezekiel *et al.*, 2013), potato was ranked after oranges and apples as the third most important source of phenols. The phenol composition in sweet potatoes according to Sun *et al.* (2014) are made up of a different concentration of polymeric and monomeric phenols, coumarins, tannins, flavanones, carotenoid, and anthocyanins. Although these compounds are found in potatoes in a small amount unlike macromolecules, theyplayed an essential role in the determination of the organoleptic properties of potatoes as well as providing different arrays of health-promoting characteristics such as antimicrobial and antioxidant properties (Wang *et al.*, 2016; Kadiri *et al.*, 2017). Discussed in the next section are the health-inducing polyphenols found in sweet potatoes.

33.4.2 Phenolic Acids

Phenolic acids are aromatic acid compounds composed of a phenolic ring and a carboxylic acid. Phenolic acid exists naturally in two forms: hydroxybenzoic and hydroxycinnamic acids derived from benzoic and cinnamic acid, a nonphenolic moiety (Kadiri & Olawoye, 2015). A higher concentration of phenolic acids had been reported in potato peels compared to the flesh by Friedman *et al.* (2018). These researchers said this was due to the pigmentation of the potato peel in different colors such as purple, orange, yellow, etc. The composition and concentration of phenolic acids in sweet potatoes vary according to varieties and geographical location of cultivation or production. Liquid chromatography-mass spectroscopy analysis of a purple flesh sweet potato root revealed it has in composition, trans-4,5-dicaffeolyqunic acid, caffeoylquinic acid, and caffeoylquinic acid derivatives such as chlorogenic acid (Ezekiel *et al.*, 2013). Lee *et al.* (2013) in their research reported a chlorogenic acid of 30–900 ug/g and 100–4,000 ug/g in fresh potato pulp and skin, respectively. However, a higher concentration of sinapic acid and coumaric acid, a hydroxycinnamic acid, as well as p-anisic acids and benzoic acid, hydroxybenzoic acids had been reported in other parts of potato such as peel and leaves. The phenolic acids found in white-fleshed cultivars of potatoes were reported to be three-to-four times less than that of orange or purple flesh potatoes. This was attributed to the pigmentation found in orange or purple potatoes (Lee *et al.*, 2013).

33.4.3 Flavonoids

Flavonoids are natural secondary metabolite products of plants and horticultural commodities with low molecular weights whose structures is polyphenolic in nature. Generally, they are composed of a 15-carbon structure that is made up of two phenyl rings as well as a heterocyclic ring. Flavonoids exhibit many health-promoting characteristics and play an important role in the management of degenerating diseases such as diabetes, cancers, atherosclerosis as well as Alzheimer's disease (Ezekiel *et al.*, 2013). It is made up of several subgroups that include isoflavone, chalcones, flavonols, and flavones. The flavonoid content of the sweet potato varies significantly among varieties. The flavonoid content of potatoes in increasing order of abundance according to de Albuquerque *et al.* (2019) were naringenin, kaempferol, erodictyol, epicatechin, and catechin, respectively. Although, the flavonol content of potatoes is not as high as the flavonoid, however, rutin (flavonol) is found in purple and orange flesh potatoes. Anthocyanins, another subgroup of flavonoids had been reported in a significant amount (35 mg/100 g) in the flesh of (orange, yellow or purple) potato (de Albuquerque *et al.*, 2019). Also, quercetin, a dominant flavonoid, was identified by Burlingame *et al.* (2009) in pigmented potatoes (orange and purple). The anthocyanins had been reported to impart a different color to sweet potatoes, however, the predominant color was purple. A report by Reyes *et al.* (2005) revealed that pigmented sweet potatoes could serve as a potential colorant and as antioxidants in the food industry. In an *in vitro* analysis, Chu *et al.* (2000) revealed that extracted flavones and flavonoids from orange and purple flesh sweet potato exhibits high radical scavenging activities against oxidative radicals.

33.4.4 Carotenoids

Carotenoids are hydrophobic compounds synthesized from isoprenoids in the plastids. Recent evidence revealed that the onset of many health degenerating diseases can be reduced through the consumption of foods or food products rich in carotenoids. Sweet potatoes, especially those with pigmented flesh, are excellent sources of carotenoids. Grace *et al.* (2014) revealed a higher carotenoid content in the peel of orange-fleshed potatoes (281.9 ug/g), being a less common variety of sweet potato, than in other parts of the potato as well as from other varieties of potato. Yellow-fleshed sweet potatoes was also reported to contain 2.0 ug/g of betacarotene, the predominant carotenoid found in this variety.

33.4.5 Dietary Fiber

Dietary fiber referrs to the groups or collection of the constituents of plant that are resistant to enzymatic digestion in the gastrointestinal tract but are fermented into a short-chain fatty acid in the large intestine. These group of plant constituents include pectin, lignin, cellulose, and hemicellulose (Olawoye *et al.,* 2020b). Although, they are not digested in the gastrointestinal tract, their fermented products convey health benefits such as prevention of colon cancer as well as diabetes mellitus. The dietary fiber present in sweet potatoes varies and it is dependent on the cultivar as well as the growing conditions. Sweet potatoes had been reported to contain 20 mg/g of dietary fiber, a value lower than whole-grain cornmeal (73 mg/g) (Ogutu & Mu, 2017).

33.5 Processing Techniques of Potatoes

Potatoes are processed into different consumables using different processing techniques, which vary from country, region, and culture. In developed nations, potatoes are usually processed and consumed in the form of French fries. Processing techniques include peeling, cutting into smaller shapes, blanching (using hot water to prevent a browning reaction), draining, par-frying, freezing, and frying. This process is usually accompanied by the formation of acrylamide and also involved the development of desirable and appealing color, aroma and flavor (Ngobese & Workneh, 2018; Jaggan *et al.*, 2020). Potatoes can be cooked with their peel for about 15–25 minutes. The tubers are then peeled and sliced (optional) into smaller sizes on consumption with sauces. Potato is also processed into flour and consumed as dough meal. Processing involves the washing of tubers to remove foreign materials, peeling, cutting into pieces, drying using either electric oven/sun drying and milling into powder. Prepared dough is usually consumed with vegetable soups. Potato protein fractions can also be extracted and used in the production of potato juice concentrate characterized with high nutritional value and antioxidant properties (Kowalczewski *et al.,* 2019). Literature has shown that processing techniques help reduce the antinutritional composition of potatoes (Navarre *et al.,* 2009) and improve nutrient composition (Ngobese & Workneh, 2018. Potatoes can also be processed into starch, which can be used as animal feed and in the production of glucose syrups, noodles and organic acids (Akoetey *et al.,* 2017).

33.5.1 Effects of Processing on Potatoes

Most foods, especially starchy foods, are usually not consumed in their natural state as they are not easily digestible, hence the need to process the food before consumption. Processing, however, influences some changes on the nutritional and organoleptic properties of potatoes. Some processing methods such as cooking, frying, and other thermal processing increase palatability and digestibility and also improves the keeping quality of the potato. While some processing methods are beneficial, some are detrimental to the nutritional, functional, and organoleptic properties of sweet potatoes as well as products obtained therein. The next sections discuss the effects of processing on the nutritional, functional, and organoleptic properties of sweet potatoes.

33.5.2 Effects of Processing on Potato Starch

In sweet potatoes, the starch content usually defines the carbohydrate composition and its composition in sweet potato root ranges between 18–22%. Sweet potato starch is formed by small granules and is made up of 70–80% amylopectin (a highly branched form of starch) and approximately 20–30% amylose, a linear nonbranched form of starch (Akanbi *et al.*, 2019). The properties of the starch are markedly influenced by many factors such as morphology, botanical origin, and processing methods among others. Several researchers had reported different processing methods of potato starch and their effects on the functionality of the starch (Akanbi *et al.*, 2019; Furrer *et al.*, 2018; Liu *et al.*, 2019). Thermal processing on potato starch has been reported to lead to the pasting of the starch, which in turns increases significantly the digestibility of the starch. The increase in starch digestibility resulted in a rapid increase in the blood glucose response, which invariably led to high glycemic index and loads an undue effect in the management of diabetes mellitus. According to (Yang *et al.*, 2016), the annealing of potato starch, a thermal modification of starch in which the starch is subjected to high moisture content within the range of 40–60% and heated at a low temperature (60–80°C) resulted in structural changes in the starch crystallinity and hence promotes the susceptibility of the starch to enzymatic hydrolysis. However, when the starch was subjected to baking and other dry thermal treatment, there was the stabilization of the resistant starch in the product coupled with the complex of the starch with protein through the action of heat and water, resulting in a decrease in the susceptibility of the starch to enzymatic digestion (Mishra *et al.*, 2008). The heating of potato starch at an elevated temperature followed by subsequent cooling caused the retrogradation of the starch amylose. Unlike thermal processing, starch modification using chemicals was found to significantly improve the water and oil holding capacity, swelling power, emulsion stability and other functional properties of potato starch (Ogutu& Mu, 2017; Punia, 2020; Punia *et al.*, 2020; Dhull *et al.*, 2020). The digestibility of the starch, on the order hand, decreases due to the substitution of the hydroxyl group of the starch with other functional groups, making it unsusceptible to enzyme hydrolysis. According to Chi *et al.* (2019), high-pressure processing of potato starch between 400–600 Mpa caused more than a 50% increase in resistant starch. This increase was attributed to temperature rise as a result of adiabatic effect during pressurization. This invariably might lead to the recrystallization of starch coupled with starch nuclei formation and, in turn, resulted in increased resistant starch. Potato starch, when subjected to frying at high temperatures, resulted in excessive formation of brown color (milliard reaction) and acrylamide, an undesirable product associated with cancer formation.

33.5.3 Effects of Processing on Potato Protein and Lipids

Although present in a minute amount, sweet potatoes contain 1.73–9.14 g/100g of protein. The protein present in sweet potatoes is water-soluble and hence is usually discarded along with wastewater during potato processing to produce starch. Different forms of proteins are found in sweet potatoes and they include salt-soluble protein (Ipomoein) and globular protein (Sporamin), which is the most abundant soluble protein in sweet potato roots (Xu *et al.*, 2009). The concentration of the proteins in sweet potatoes vary and are therefore affected by cultivar, geographical location, stage of maturity, and processing methods. Upon processing, the structural and functional properties of protein changes and therefore affects the utilization of the protein. Thermal processing, such as cooking, frying, and baking caused protein denaturation and made it susceptible to digestion by proteolytic enzymes. In addition to protein denaturation during thermal processing, most potato proteins are lost during prolonged cooking due to the water solubility of the proteins (Sun *et al.*, 2020). Major changes occur to amnio acids during thermal processing, especially during baking and other dry heat treatments. Amino acids, especially lysine, are lost when they form a milliard reaction with the carboxylic acid end of sugar in the presence of water. During frying, baking, and roasting of potatoes, asparagine, which is the most abundant amino acid in potatoes, react with sugar to form acrylamide, a compound known for its ability to induce cancerous cells. Also, there is a reduction in asparagine during yeast fermentation of potato flour as a result of the conversion of asparagine to aspartic acid by asparaginase (Tian *et al.*, 2016). Natural lipids present in potatoes are typically low across various cultivar. The majority of lipids in the form of fatty acids present in potatoes include linolenic, linoleic, and palmitic acids. Although fewer changes occur to the natural lipids during the processing of potatoes, the addition of fat to potatoes increases palatability. The added fat form complexes with starch, improving the functional properties as well as reducing the digestibility of the starch.

33.5.4 Effects of Processing on Vitamins and Minerals

Among root and tuber crops, sweet potatoes have been reported to contain a substantial amount of vitamins and minerals. Vitamins and minerals are regarded as micronutrients because they are required in a small amount for the functionality of the body system. However, vitamin and mineral composition in sweet potatoes varies with respect to varieties as well as their processing conditions. There had been a report on the effects of processing on micronutrients. The report of Tian *et al.* (2016) revealed that 20–30% of vitamin C was lost in unpeeled sweet potato, however, peeling the potato before boiling resulted in greater loss in vitamin C. Vitamin C being the most heat-labile among the vitamins, leached easily into cooking or boiling water. Lachman *et al.* (2013) compared the extent of vitamin C loss in fried and boiled potatoes and observed that frying resulted in a 50–56% loss compared to 20–28% loss in boiled potatos. The higher temperature at which the frying occurs brings about a significant loss in the vitamin C content of the sweet potato. In contrast, vitamin A, a fat-soluble and heat-stable vitamin, is insignificantly affected by cooking. However, there have been some reports on the loss of vitamin A in boiled orange-fleshed sweet potatoes. The loss was attributed to the destruction of betacarotene, a precursor of vitamin A. One critical reaction that resulted in the loss of vitamin A activity is the isomerization of betacarotene to form neobetacarotene during canning of sweet potatos (Tian *et al.*, 2016). This isomerization of betacarotene led to the development of an off-flavor during the storage of sweet potato at ambient temperature. Thiamine, a vitamin that falls under the vitamin B group is a thermolabile vitamin and its reduction due to thermal processing had been reported. The major change that occurs to minerals during processing is leaching during thermal processing. Processing involving milling and canning of sweet potatoes was found to increase the iron content of the potato significantly (Tian *et al.*, 2016). The increase is obviously due to leaching of the metal can into the canned product as well as the wearing of metals into the milled potato.

33.5.5 Effects of Processing on Phytochemicals

Earlier in this chapter, sweet potatoes were reported to be an excellent source of phytochemicals such as polyphenols, carotenoid, flavonoids, and anthocyanin. Although these phytochemicals are relatively low in sweet potatoes compared to fruits and vegetables, they, however, play an important role as an antioxidant (Burgos *et al.*, 2013). Before consumption or utilization, sweet potatoes are usually processed, causing some changes to the physical, chemical, and functional properties of the phytochemicals. Thermal processing had been reported to alter the phytochemicals in potatoes. Narwojsz *et al.* (2020) reported a decrease in the polyphenol content of potatoes due to the thermal degradation of the phytochemical. However, Burgos *et al.* (2013) observed a 10.67% increase in antioxidant activities of the potato during roasting. The increase was attributed to the polymerization of the polyphenols with macromolecules, especially protein to form a complex that exhibits antioxidant properties. Carotenoid and anthocyanin decrease during cooking as a result of leaching of these phytochemicals into the cooking water.

33.6 Byproducts from Potato

Potatoes are one of the most commonly consumed root and tuber crop throughout the world. In some countries, potatoes are utilized as an ingredient in some freshly consumed traditional cuisines, however, in recent times, the fresh consumption of potato had decline owing to the steady growth of varieties of potato products such as potato starch, fries, dried, frozen, and canned potato. To produce these products requires a large volume of potatoes, resulting in the generation of high amounts of potato byproducts, usually regarded as potato waste. The following sections discuss the byproducts generated during the harvesting and processing of potatoes.

33.6.1 Potato Peels

Potato peel is regarded as the outer coat or epidermal layer removed during the processing of potatoes. The waste generated as a result of potato peeling varies according to the method of peeling (lye, abrasion, or steam peeling) as well as the types of potato products and ranges between 14–40% of fresh potato weight. Among the products of potato processing, potato flour, starch, and canned potatoes generated the largest volume of potato peel. This large volume of peel waste generated during processing is of environmental concern due to its disposal, as most of the peel waste is disposed in landfills (Sampaio *et al.*, 2020). The chemical composition of the peel is dependent on the peeling procedure, as peeling obtained through the lye procedure contained starch ranging between 500–560 mg/g, crude protein ranging from 40–56 mg/g, crude fiber of 76 mg/g, ash of 69 mg/g, and lipids of 1 mg/g (Ncobela *et al.*, 2017). Peeling using a steaming procedure, however, contained less chemical components compared to the lye procedure. The chemical composition of the potato peel makes it suitable for utilization as animal feed. Apart from its chemical composition, potato peel also contains phytochemical whose antioxidant and antimicrobial activities are of health benefits to humans. The phytochemicals present in potato peel differ from cultivar to cultivar as more phenolic compounds are associated with orange- or purple-fleshed potato than yellow- or white-fleshed potato (Sampaio *et al.*, 2020). The study oft he phenolic compounds present in potato peels revealed that chlorogenic acid and its derivatives are the most abundant molecules followed by caffeic acid. The antioxidant and antimicrobial activities of the phytochemicals present in the peel make it a suitable raw material in the formulation of functional food as well as in the pharmaceutical industry. In their research, Kadiri *et al.* (2019) extracted polyphenol from orange-fleshed potato peel to produce a functional noodle. The inclusion of the polyphenol from the peel increases the antioxidant properties of their formulated noodles.

33.6.2 Potato Pulp

Potato pulp is a byproduct obtained from the production of potato starch and is made up of cellulosic material of the potato tuber. It contains potato skin and residual cells with a small amount of starch mixed with potato liquid. It is rich in undigestible food materials such as cellulose, hemicellulose, and pectin as well as a small amount of starch (Serena & Knudsen, 2007). Due to its chemical composition, potato pulp is usually utilized as feed for livestock. Other forms of utilization include processing into single-cell protein through the fermentation of the pulp.

33.6.3 Potato Hash

Potato hash is the byproduct obtained from chip and snack production. It is composed of a large amount of potato peel, starch, and rejected fresh potato tuber (Ncobela *et al.*, 2017). It has high moisture content and low water-soluble carbohydrate. Its composition makes it fit only for the production of animal feed.

33.7 Potential Use of Byproducts from Potatoes

Potatoes byproducts are typically regarded as waste products; however, they are rich sources of energy, carbohydrates, fiber, and vitamins, but low in protein (Ncobela *et al.*, 2017). Generation of these byproducts begins from the farm, during harvesting, transportation, and processing, which includes potato leaves, vines, damaged tubers, chunks of tuber, trimmings, peels, and nutrient-rich wastewater (Akoetey *et al.*, 2017; Ncobela *et al.*, 2017).

Potatoes leaves, vines, and peels have been reported to contain high amounts of bioactive polyphenols and antioxidant compounds, and as such, these byproducts can find better uses in the production of value-added products than animal feed. Better uses include supplementation of the human diet/encapsulation of these bioactive compounds, thereby promoting growth and better well-being in animals and humans (Sun *et al.*, 2014; Akoetey *et al.*, 2017; Ncobela *et al.*, 2017).

Wastewater from potato processing (e.g. starch) has been reported to contain a high amounts of bio-chemical oxygen demand, nitrogen, and phosphorus. This may be a suitable medium for the production of biofertilizer via cultivation of *Paenibacillus polymyxa* (Xu *et al.*, 2014; Akoetey *et al.*, 2017). Likewise, the sediment leftover after starch extraction is rich in dietary fiber and can be used in food production to help reduce constipation and the risk of cardiovascular diseases such as diabetics and hypertension (Takamine *et al.*, 2000; Akoetey *et al.*, 2017; Oluwajuyitan *et al.*, 2020).

33.8 Potato Products

Recently, potato consumption as a food has drastically shifted from fresh potatoes to processed food products. Potatoes are one of the most consumed roots and tubers, and are processed into varieties of added-valued products including potato flour, potato starch, potato chips, potato flakes, pan-fried potatoes, dehydrated potatoes, to mention just a few.

33.8.1 Potato Flour

Potato flour is one of the most common products obtained from potato processing. The processing of potato to either mashed or chip forms, which can later be processed into flour could serve as an alternative or substitute for expensive wheat flour especially for countries with comparative disadvantage in the production and cultivation of wheat grain. The flour obtained can serve as a dough conditioner in bread, as well as in the confectionery industry where it can be substituted in small quantities and used in the production of gluten-free products (Ahn *et al.*, 2013). The utilization of potato flour in the formulation of gluten-free diets will help in the management of celiac disease, an immune-mediated disease that arises as a result of the inability of the consumer of wheat products to ingest gluten (Olawoye *et al.*, 2020a). Also, due to the sweetness of the potato, the flour can serve as a natural sweetener, flavoring, and colorant in the production of food products. The production of flour from potato tubers involves several processing steps or unit operations. The first processing step involves the raw material preparation, such as cleaning, washing, brushing, and peeling followed by slicing. Next, the slice potato is dried, milled, and packaged. The type of milling operation as well as the particle size of the flour determine the type of flour obtained as well as the products it can be used for. Of all these unit operations, the brushing and washing steps are the most important as they help determine the quality of the potato flour. During the production of potato flour and starch, peeling can be excluded as the skin of the potato root is very thin and has little influence on the color of the flour.

33.8.2 Potato Starch

Starch is one of the most-studied among all macromolecules owing to its usefulness and functionality in many food products. The functionality and physical properties and their applications in various food products vary with respect to the biological origin of the starch (Olawoye *et al.*, 2020b). Starch, however, exists in two different polymers which are amylose, a linear chain polymer joined together by an alpha 1-4 glycosidic bond, and amylopectin, a branched-chain polymer that, along with the linear chain, is branched at a alpha 1-6 glycosidic linkage (Olawoye & Gbadamosi, 2020). The ratio of both polymers (i.e. amylose:amylopectin) determines the functionality as well as the application of the starch. Starch is used in numerous applications in the food industry such as confectionaries, soups, thickeners, and meats, as well as in nonfoods such as paint, textile, plastics, fuel, and adhesives. Potatoes are an excellent source of starch being one of the most highly cultivated root and tuber crops. Its starch is made up of 20–25% amylose while the amylopectin composition is 75–80% (Gbadamosi *et al.*, 2020). The production of starch from potatoes can either be of two methods: dry and wet. In the dry method, starch is obtained by dissolving potato flour in water, the slurry obtained is sieved using various mesh sizes to obtain potato starch. The limitation of this method is that the starch yield is low compared to the wet method, which is the general method of starch extraction (Kadiri *et al.*, 2019). Starch production using a wet method of extraction involves the washing of the fresh potato tubers. This washing step is very critical to the quality of the starch. Different impurities such as soil and small stones are removed during the washing process. Following washing, the potato is milled into mash from which the starch is extracted. The starch can be extracted using distilled water, sodium chloride, or sodium metabisulfite. Potato starch extraction using sodium metabisulfite is well detailed in the report from Akanbi *et al.* (2019). After extraction, the starch slurry is always subjected to drying at 30–40°C, milled, and packaged. Potato starch has found applications as functional noodles (Gbadamosi *et al.,* 2020), in biscuit production, bread, biofuel, biofilm, etc.

33.8.3 Potato Chips or French Fries

French fries or potato chips are added-value products obtained from the processing of fresh potatoes. In the snack food industry, potato chips or French fries are considered as one of the most important products. They are produced through the frying of a slices of potatoes in vegetable oil. During frying, the vegetable oil used replaces the water present in the potato slices and can be done with the potato slices first dipped into glucose solution (Furrer *et al.,* 2018). This invariably affects the color of the final products due to the browning reaction that occurs from the excess glucose solution. Maintenance of the color of the chips is one of the major problems confronting the potato chip industry and hence, the standard of the product. This is difficult because the color of the chips is dependent on the chemical composition of the potato tuber, which in turns depends on uncontrollable environmental conditions. The oil uptake constitutes about 40% of the product and this invariably affects the properties as well as the shelf-life of the chips. Depending on geographical location and consumer preference, the frying conditions (volume of oil, type of frying equipment, and recipes) differ. Recently, a low-fat potato is preferred as chips with excess frying oil are associated with some degenerating diseases such as atherosclerosis and hypertension (Ngobese & Workneh, 2018). The production of potato chips involves washing, brushing, peeling, and slicing of potato tubers, which is then fried, seasoned and packaged. Seasoning of the chips is dependent on consumer's preference, but usually contain salt, pepper, and flavoring, etc. Several frying methods and equipment have been used during frying including batch and continuous frying methods. In batch frying, the potato slices are put inside a fryer, and, after a predetermined frying time, are discharged from the fryer. In continuous frying, the frying operation is continuous and there is limited uptake of oil by the chips.

33.8.4 Dehydrated Potato

Drying is an ancient method of food processing and preservation, and when applied to potato tissues, it causes physical and structural modifications. The physical modification of the tissue caused by drying is the shrinkage and deformation of the tissue. Water loss and segregation during drying leads to damage, rigidity, and cell wall disruption which in turns reduces the weight and volume of the dehydrated potatoes (Troncoso *et al.,* 2009). Products of dehydrated potatoes include potato flakes and granules. The production of dehydrated potato involves cleaning, peeling, slicing, precooking, and cooling. The cooled potato is then cooked and mashed followed by drying using either a drum dryer for flake or air-dryer for potato granule production. Depending on consumer preference, the color, texture and shelf life of dehydrated potatoes can be improved with additives. According to Kakade *et al.* (2011), potato flakes contains 5–10% moisture content, which makes their cell was intact.

33.9 Miscellaneous Uses of Potato

The processing of potatoes into its flour not only improves their functionality but also extend their shelf life. Potato flour has improved function as a thickener and flavor improver. The uses of potato flour as a combined flavoring-thickener have been reported by Marwaha & Sandhu (1996). Food processes like steaming, baking, deep-frying, boiling, and baking incorporate potato flour with other flours. For instance, incorporating potato flour with wheat during the baking process improves bread freshness.

Potato flour exhibits unique functional properties that determine its suitability for use in some specific product formulations. Its starch-rich content improves its functionality and suitability for food use applications. Processing methods can influence the properties of potato flour, some of which include modification types, the severity of heat treatment, and the presence of other components like protein, fiber, etc. Modification of the structural characteristics of starch results in specific potato flour functionality. Factors such as drying techniques, pretreatments, parboiling, peeling, blanching, variety, drying temperatures, and processing steps are among some documented factors that can influence the quality and acceptability of food products developed from potato flour (Jangchud *et al.,* 2003; Van Hal, 2000; Yadav *et al.,* 2006; Olatunde *et al.,* 2016).

The functional properties reflect the complex linkage between the composition, molecular conformation of the structure, and the physicochemical characteristics of food components (Ojo *et al.,* 2017; Kaur & Singh, 2006). Such properties decide the production and use of food ingredients for different foods, and also regulate the processing and storage of these items (Olatunde *et al.,* 2016; Adebowale *et al.,* 2012).

Potato starch is a low-cost biopolymer readily available for the production of spherulites (Ziegler *et al.,* 2003), which can find use as drug encapsulants. Starch spherulites are produced using steam jet cooking or differential scanning calorimetry (DSC). The procedure using steam jet cooking has been described by Kaur & Singh (2009). Due to its ease of use and high sensitivity, differential scanning calorimetry is most suitable for spherulite production and starch study.

The rate of spherulite formation is dependent on the cooling rate, fatty acid solubility in water, and amount used relative to starch (Fanta *et al.,* 2008). Spherulite formation and application was extensively discussed by Kaur & Singh (20016).

Several studies have reported the use of waste derived from the potato-processing industry in the production of biodegradable packaging trays, which are environmentally friendly and comparable in terms of performance to conventional plastics (Murphy *et al.,* 2004). British potato products generate around 17,000 tons yearly that has the potential of being converted to biodegradable plastics (Kaur & Singh, 2016). It has been established that starch extracted from waste potato materials is better than starch extracted from the tuber (Murphy *et al.,* 2004). Studies have reported the conversion of potato waste to high-value products, such as polylactic acid and xanthan, which constitute a challenge in waste management (Robertson, 2006). Polylactic acid is a material required for the production of several biodegradable materials such as films, sheet, board, paint, and fiber because of the low-energy requirement during production when compared to petroleum-based production. Though production of these biodegradable materials is cost-effective when produced from petroleum origin, there is the need to develop technology that uses a cost-effective system in the utilization of biodegradable materials from potato and its waste.

33.10 Postharvest Storage

Postharvest storage/handling of potatoes is an important aspect to be considered since potatoes contain a high percentage of moisture (<50%) content depending on its cultivars and, as such, they are highly perishable, prone to microbial attack and spoilage. Poor handling of potatoes such as exposure to excessive sunlight, high temperature, and bruising of tubers could increase the level of toxic substances (glycoalkaloids) within the tuber with the onset of greening and sprouting of potatoes tubers (Musita *et al.,* 2019). Consumption of these damaged tubers are harmful to human health however; potatoes tubers can be stored in good conditions until consumption without the aforementioned damages if the following are put in proper consideration:

1. **Maturity stage of potatoes (early/late):** Potatoes are harvested either early or late for sales. Maturity is defined when potatoes tubers have fully reached both physiological and morphological conditions. When harvested early, crops are usually still increasing in size and yet to reach full (physiological and morphological) maturity and as such, they potatoes possess soft skin (epidermis) that is prone to abrasion/mechanical injury and increased respiration rate. These attributes increase the susceptibility of early potatoes to spoilage and increased postharvest losses while the reversed is observed in late- harvest potatoes (Bishop & Clayton, 2009; Pinhero *et al.,* 2009).

2. **Preharvest conditions of potatoes:** Several preharvesting conditions affect the soil output and this includes moisture stress (water holding capacity of the soil), nutrient status of the soil, and the incidence of pests and diseases. The higher the water holding capacity and nutritional status, the better the potatoes harvested. However, the higher the incidence of pests and diseases, the higher the postharvest losses (Pinhero *et al.,* 2009).

3. **Harvest and handling conditions:** Conditions under which potato tubers are harvested also affect the quality of the potatoes. The susceptibility of tubers to external damage during harvesting varies based on the cultivar, stage of maturity of the crop, soil and weather conditions, harvester skills, and design of harvesting and handling equipment (Gottschalk & Ezhekiel, 2006; Pinhero *et al.,* 2009).

4. **Health of the crop:** Potatoes tubers are usually affected by pests on the farm before and after harvesting, and this can increase during transportation to market if the potatoes have sustained mechanical injuries, which gives room for entry of pathogenic organisms. Even if a small quantity of tubers are affected, disease can spread quickly to others tubers within a short time, which can increase the rate of postharvest loses. For example, the inoculum of *Fusarium* may gain access into tubers via wounds caused by mechanical injuries. It is recommended that tubers be examined for pests and diseases before storage.

5. **Biochemical changes:** Potato tubers are affected by several biochemical changes such as respiration, water loss, sprouting, the incidence of pests, disease, and temperature. Tuber respiration involves the production of respiratory heat and this directly increases the storage temperatures and temperatures of the ventilation systems. If this heat is not removed, it could cause blackheart, rot, and increase the rate of potato deterioration. Potato tubers that lost large amounts of water can succumb to mechanical damage, such as bruising and blue/black discoloration, greater peeling losses, and reduction of culinary quality leading to economic loss. Damaged and diseased tubers sprout earlier than healthy tubers. Stored potatoes may be affected by microorganisms such as fungus or bacteria. These microbial infections may have occurred in the field before harvesting or due to infection following harvest (Gottschalk & Ezhekiel, 2006; Pinhero *et al.,* 2009).

33.11 Summary and Future Perspectives

Today, potatoes are grown globally. Though there has been a steady decline in potato production in developed countries, there has been an increase in production output in developing countries over the years (Geohive, 2011). For instance, potato production decreased from 190 million to 155 million tons in developed countries and increased from 89–159 million tons in developing countries in the years 1990–2006. Consumption of potatoes was observed to also increase dramatically in low-income and food-deficient areas. This should be expected due to the increasing prices of other staple foods such as wheat, rice, and corn.

Potatoes are mostly used as snacks, or a term called "convenience" foods such as French fries and chips, in developed countries. The carcinogen acrylamide is produced when fresh potatoes are converted into snacks. This is due to the temperature at which snacks are produced that initiate the Maillard reaction linked to this carcinogen compound formation. There has been considerable research globally to develop processing methods that can help eliminate or significantly reduce acrylamide in edible food portions, though no single process has been identified to completely remove this compound. A combination of processes can however reduce acrylamide to nondetectable levels. Acrylamide levels can be reduced in food fortified with natural antioxidants in the form of additives/ingredients. Potatoes processed into flour, baked, or boiled have not been reported to have acrylamide content to date. However, moderation in potato snack consumption can go a long way in reducing and mitigating these deleterious effects.

REFERENCES

Adebowale, A. A., Adegoke, M. T., Sanni, S. A., Adegunwa, M. O., & Fetuga, G. O. (2012). Functional properties and biscuit making potentials of sorghum-wheat flour composite. *American Journal of food technology*, 7(6), 372–379.

Ahn, J. H., Baek, H. R., Kim, K. M., Han, G. J., Choi, J. B., Kim, Y., & Moon, T. W. (2013). Slowly digestible sweetpotato flour: Preparation by heat-moisture treatment and characterization of physicochemical properties. *Food Science and Biotechnology*, 22(2), 383–391.

Akanbi, C. T., Kadiri, O., & Gbadamosi, S. O. (2019). Kinetics of starch digestion in native and modified sweetpotato starches from an orange fleshed cultivar. *International Journal of Biological Macromolecules*, 134, 946–953.

Akoetey, W., Britain, M. M., & Morawicki, R. O. (2017). Potential use of byproducts from cultivation and processing of sweet potatoes. *Ciência Rural*, 47(5).

Beals, K. A. (2019). Potatoes, nutrition and health. *American Journal of Potato Research*, 96(2), 102–110.

Bellumori, M., Chasquibol Silva, N. A., Vilca, L., Andrenelli, L., Cecchi, L., Innocenti, M., ... & Mulinacci, N. (2020). A Study on the biodiversity of pigmented Andean potatoes: Nutritional profile and phenolic composition. *Molecules*, 25(14), 3169.

Bethke, P. C., & Jansky, S. H. (2008). The effects of boiling and leaching on the content of potassium and other minerals in potatoes. *Journal of Food Science*, 73(5), H80–H85.

Bishop, C. F. H., Clayton, R. C., 2009. Physiology. In: Pringle, R.T., Bishop, C.F.H., Clayton, R.C. (Eds.), *Potato Postharvest*, UK: CABI, 1–29.

Bonierbale, M., Zapata, G. B., zum Felde, T., & Sosa, P. (2010). Composition nutritionnelle des pommes de terre. *Cahiers de Nutrition et de Diététique*, 45(6), S28–S36.

Burgos, G., Amoros, W., Muñoa, L., Sosa, P., Cayhualla, E., Sanchez, C., Díaz, C., & Bonierbale, M. (2013). Total phenolic, total anthocyanin and phenolic acid concentrations and antioxidant activity of purple-fleshed potatoes as affected by boiling. *Journal of Food Composition and Analysis*, 30(1), 6–12.

Burgos, G., Zum Felde, T., Andre, C., & Kubow, S. (2020). The potato and its contribution to the human diet and health. In: *The Potato Crop*, Springer, Cham, 37–74.

Burlingame, B., Mouillé, B., & Charrondière, R. (2009). Nutrients, bioactive non-nutrients and anti-nutrients in potatoes. *Journal of Food composition and Analysis*, 22(6), 494–502.

Chi, C., Li, X., Lu, P., Miao, S., Zhang, Y., & Chen, L. (2019). Dry heating and annealing treatment synergistically modulate starch structure and digestibility. *International Journal of Biological Macromolecules*, 137, 554–561.

Chu, Y.-H., Chang, C.-L., & Hsu, H.-F. (2000). Flavonoid content of several vegetables and their antioxidant activity. *Journal of the Science of Food and Agriculture*, 80(5), 561–566.

CIP—International Potato Center. (2008). ISO accreditation a world-first for CIP genebank.

CIP—International Potato Center. (2016). BNFB Facts on biofortification. Available at: https://cipotato.org/bnfb/facts/ Accessed on December, 21,2020.

Czaja, T., Sobota, A., & Szostak, R. (2020). Quantification of ash and moisture in wheat flour by Raman Spectroscopy. *Foods*, 9(3), 280.

Daliri, E. B. M., Ofosu, F. K., Chelliah, R., Park, M. H., Kim, J. H., & Oh, D. H. (2019). Development of a soy protein hydrolysate with an antihypertensive effect. *International Journal of Molecular Sciences*, 20(6), 1496.

de Albuquerque, T. M. R., Sampaio, K. B., & de Souza, E. L. (2019). Sweet potato roots: Unrevealing an old food as a source of health promoting bioactive compounds—A review. *Trends in Food Science & Technology*, 85, 277–286.

de Haan, S., Burgos, G., Liria, R., Rodriguez, F., Creed-Kanashiro, H. M., & Bonierbale, M. (2019). The nutritional contribution of potato varietal diversity in Andean food systems: A case study. *American Journal of Potato Research*, 96(2), 151–163.

Dhull, S. B., Punia, S., Kumar, M., Singh, S., & Singh, P. (2020). Effect of different modifications (physical and chemical) on morphological, pasting, and rheological properties of black rice (Oryza sativa L. Indica) starch: A comparative study. *Starch-Stärke*, 2000098.

Ezekiel, R., Singh, N., Sharma, S., & Kaur, A. (2013). Beneficial phytochemicals in potato—A review. *Food Research International*, 50(2), 487–496.

Fanta, G. F., Felker, F. C., Shogren, R. L., & Salch, J. H. (2008). Preparation of spherulites from jet cooked mixtures of high amylose starch and fatty acids. Effect of preparative conditions on spherulite morphology and yield. *Carbohydrate Polymers*, *71*(2), 253–262.

FAO (1990). *Roots, Tubers, Plantains and Bananas in Human Nutrition*. Rome: Food and Agriculture Organization.

Freedman, M. R., & Keast, D. R. (2011). White potatoes, including French fries, contribute shortfall nutrients to children's and adolescents' diets. *Nutrition Research*, *31*(4), 270–277.

Friedman, M., Huang, V., Quiambao, Q., Noritake, S., Liu, J., Kwon, O., Chintalapati, S., Young, J., Levin, C. E., Tam, C., Cheng, L. W., & Land, K. M. (2018). Potato peels and their bioactive glycoalkaloids and phenolic compounds inhibit the growth of pathogenic trichomonads. *Journal of Agricultural and Food Chemistry*, *66*(30), 7942–7947.

Furrer, A. N., Chegeni, M., & Ferruzzi, M. G. (2018). Impact of potato processing on nutrients, phytochemicals, and human health. *Critical Reviews in Food Science and Nutrition*, *58*(1), 146–168.

Gbadamosi, S. O., Kadiri, O., & Akanbi, C. T. (2020). Quality characteristics of noodles produced from soybean protein concentrate and sweet potato starch: a principal component and polynomial cubic regression model approach. *Journal of Culinary Science & Technology*, *19*(3), 1–21.

Geohive website (2011).

Gorissen, S. H., Crombag, J. J., Senden, J. M., Waterval, W. H., Bierau, J., Verdijk, L. B., & van Loon, L. J. (2018). Protein content and amino acid composition of commercially available plant-based protein isolates. *Amino acids*, *50*(12), 1685–1695.

Gottschalk, K., & Ezhekiel, R. (2006). Storage. In: *Handbook of Potato Production, Improvement, and Postharvest Management* (489–522).New York London, Oxford: Food Products Press.

Grace, M. H., Yousef, G. G., Gustafson, S. J., Truong, V.-D., Yencho, G. C., & Lila, M. A. (2014). Phytochemical changes in phenolics, anthocyanins, ascorbic acid, and carotenoids associated with sweetpotato storage and impacts on bioactive properties. *Food Chemistry*, *145*, 717–724.

Grant, V. (2003). Select markets for taro, sweet potato and yam. A report for the Rural Industries Research and Development Corporation (RIRDC). Publication No 0 3/052 RIRDC project No UCQ-13A.

Hijmans, R. J., & Spooner, D. M. (2001). Geographic distribution of wild potato species. *American Journal of Botany*, *88*(11), 2101–2112.

Jaggan, M., Mu, T., & Sun, H. (2020). The effect of potato (Solanum tuberosum L.) cultivars on the sensory, nutritional, functional, and safety properties of French fries. *Journal of Food Processing and Preservation*, e14912.

Jangchud, K., Phimolsiripol, Y., & Haruthaithanasan, V. (2003). Physicochemical properties of sweet potato flour and starch as affected by blanching and processing. *Starch-Stärke*, *55*(6), 258–264.

Kadiri, O., & Olawoye, B. (2015). Underutilized indigenous vegetable (UIV) in Nigeria: A rich source of nutrient and antioxidants—A review. *Annals. Food Science & Technology*, *16*(2), 236–247.

Kadiri, O., Akanbi, C. T., Olawoye, B. T., & Gbadamosi, S. O. (2017). Characterization and antioxidant evaluation of phenolic compounds extracted from the protein concentrate and protein isolate produced from pawpaw (Carica papaya Linn.) seeds. *International journal of food properties*, *20*(11), 2423–2436.

Kadiri, O., Gbadamosi, S. O., & Akanbi, C. T. (2019). Extraction kinetics, modelling and optimization of phenolic antioxidants from sweet potato peel vis-a-vis RSM, ANN-GA and application in functional noodles. *Journal of Food Measurement and Characterization*, *13*(4), 3267–3284.

Kakade, R. H., Das, H., & Ali, S. (2011). Performance evaluation of a double drum dryer for potato flake production. *Journal of Food Science and Technology*, *48*(4), 432–439.

Kaplan, M., Ulger, S., Kokten, K., Uzun, S., Oral, E. V., Ozaktan, H., ... & Kale, H. (2018). Nutritional composition of potato (Solanum tuberosum L.) Haulms. *Progress in Nutrition*, *20*(Suppl. 1), 90–95.

Katan, M. B., & De Roos, N. M. (2004). Promises and problems of functional foods. *Critical Reviews in Food Science and Nutrition*, *44*, 369–377.

Kaur, L., & Singh, J. (2016). Novel applications of potatoes. In: *Advances in Potato Chemistry and Technology* (627–649), Academic Press.

Kaur, M., & Singh, N. (2006). Relationships between selected properties of seeds, flours, and starches from different chickpea cultivars. *International Journal of Food Properties*, *9*(4), 597–608.

Khan, S. H. (2016). Sweet potato (Ipomoea batatas (L.) Lam as feed ingredient in poultry diets. *World's Poultry Science Journal*, *73*(1), 77–88. doi: 10.1017/S0043933916000805

Kourouma, V., Mu, T. H., Zhang, M., & Sun, H. N. (2020). Comparative study on chemical composition, polyphenols, flavonoids, carotenoids and antioxidant activities of various cultivars of sweet potato. *International Journal of Food Science & Technology*, 55(1), 369–378.

Kowalczewski, P. Ł., Olejnik, A., Białas, W., Rybicka, I., Zielińska-Dawidziak, M., Siger, A. … & Lewandowicz, G. (2019). The nutritional value and biological activity of concentrated protein fraction of potato juice. *Nutrients*, 11(7), 1523.

Lachman, J., Hamouz, K., Musilová, J., Hejtmánková, K., Kotíková, Z., Pazderů, K., … & Cimr, J. (2013). Effect of peeling and three cooking methods on the content of selected phytochemicals in potato tubers with various colour of flesh. *Food Chemistry*, 138(2-3), 1189–1197.

Lee, M. J., Park, J. S., Choi, D. S., & Jung, M. Y. (2013). Characterization and quantitation of anthocyanins in purple-fleshed sweet potatoes cultivated in Korea by HPLC-DAD and HPLC-ESI-QTOF-MS/MS. *Journal of Agricultural and Food Chemistry*, 61(12), 3148–3158.

Liu, Y., Yang, L., Ma, C., & Zhang, Y. (2019). Thermal behavior of sweet potato starch by non-isothermal thermogravimetric analysis. *Materials (Basel)*, 12(5).

Marwaha, R. S., & Sandhu, S. K. (1996). Annual Scientific Report. Central Potato Research Station, Indian Council of Agricultural Research, Jalandhar, India, 6.

McCay, C. M., McCay, J. B., & Smith, O. (1987). The nutritive value of potatoes. In: W. F. Talburt, & O. Smith (Eds.), *Potato Processing* (pp. 287–332). New York: Van Nostrand Reinhold.

McGill, C. R., Kurilich, A. C., & Davignon, J. (2013). The role of potatoes and potato components in cardio-metabolic health: A review. *Annals of Medicine*, 45(7), 467–473.

Miao, M., Jiang, B., Cui, S. W., Zhang, T., & Jin, Z. (2015). Slowly digestible starch—A review. *Critical Reviews in Food Science and Nutrition*, 55(12), 1642–1657.

Mishra, S., Monro, J., & Hedderley, D. (2008). Effect of processing on slowly digestible starch and resistant starch in potato. *Starch-Stärke*, 60(9), 500–507.

Murphy, R. J., Bonin, M., & Hillier, W. R. (2004). Life cycle assessment of potato starch based packaging trays. *Report prepared (draft) for STI Project Sustainable GB Potato Packaging, Imperial College London, London, UK*.

Musita, C. N., Okoth, M. W., & Abong, G. O. (2019). Postharvest handling practices and perception of potato safety among potato traders in Nairobi, Kenya. *International Journal of Food Science*, 1–8.

Naidu, K. A. (2003). Vitamin C in human health and disease is still a mystery? An overview. *Nutrition Journal*, 2(1), 7.

Narwojsz, A., Borowska, E. J., Polak-Śliwińska, M., & Danowska-Oziewicz, M. (2020). Effect of different methods of thermal treatment on starch and bioactive compounds of potato. *Plant Foods for Human Nutrition*, 75(2), 298–304.

Nassar, A. M., Sabally, K., Kubow, S., Leclerc, Y. N., & Donnelly, D. J. (2012). Some Canadian-grown potato cultivars contribute to a substantial content of essential dietary minerals. *Journal of Agricultural and Food Chemistry*, 60(18), 4688–4696.

Navarre, D. A., Goyer, A., & Shakya, R. (2009). Nutritional value of potatoes: Vitamin, phytonutrient, and mineral content. In: *Advances in Potato Chemistry and Technology* (pp. 395–424). Academic Press.

Ncobela, C. N., Kanengoni, A. T., Hlatini, V. A., Thomas, R. S., & Chimonyo, M. (2017). A review of the utility of potato by-products as a feed resource for smallholder pig production. *Animal Feed Science and Technology*, 227, 107–117.

Ngobese, N. Z., & Workneh, T. S. (2018). Potato (Solanum tuberosum L.) nutritional changes associated with French fry processing: Comparison of low-temperature long-time and high-temperature short-time blanching and frying treatments. *LWT Food Science and Technology*, 97, 448–455.

NHANES. (2010). National Health and Nutrition Examination Survey. *Centers for Disease Control and Prevention*.

Nzaramba, M. N., Bamberg, J. B., & Miller, J. C., Jr. (2007). Effect of propagule type and growing environment on antioxidant activity and total phenolic content in potato germplasm. *American Journal of Potato Research*, 84, 323–330.

Ogutu, F. O., & Mu, T.-H. (2017). Ultrasonic degradation of sweet potato pectin and its antioxidant activity. *Ultrasonics Sonochemistry*, 38, 726–734.

Ojo, M. O., Ariahu, C. C., & Chinma, E. C. (2017). Proximate, functional and pasting properties of cassava starch and mushroom (Pleurotus Pulmonarius) flour blends. *American Journal of Food Science and Technology*, 5(1), 11–18.

Olatunde, G. O., Henshaw, F. O., Idowu, M. A., & Tomlins, K. (2016). Quality attributes of sweet potato flour as influenced by variety, pretreatment and drying method. *Food Science & Nutrition, 4*(4), 623–635.

Olawoye, B., & Gbadamosi, S. O. (2020). Digestion kinetics of native and modified cardaba banana starch: A biphasic approach. *International Journal of Biological Macromolecules, 154*, 31–38.

Olawoye, B., Gbadamosi, S. O., Otemuyiwa, I. O., & Akanbi, C. T. (2020a). Gluten-free cookies with low glycemic index and glycemic load: Optimization of the process variables via response surface methodology and artificial neural network. *Heliyon, 6*(10), e05117. https://doi.org/10.1016/j.heliyon.2020.e05117

Olawoye, B., Gbadamosi, S. O., Otemuyiwa, I. O., & Akanbi, C. T. (2020b). Improving the resistant starch in succinate anhydride-modified cardaba banana starch: A chemometrics approach. *Journal of food processing and preservation, 44*(9), e14686. https://doi.org/10.1111/jfpp.14686

Oluwajuyitan, T. D., Ijarotimi, O. S., & Fagbemi, T. N. (2020). Nutritional, biochemical and organoleptic properties of high protein-fibre functional foods developed from plantain, defatted soybean, rice-bran and oat-bran flour. *Nutrition & Food Science, 50*(6), 1–21. https://doi.org/10.1108/NFS-06-2020-0225

Pinhero, R. G., Coffin, R., & Yada, R. Y. (2009). Post-harvest storage of potatoes. In: *Advances in potato chemistry and technology* (pp. 339–370). Academic Press.

Punia, S. (2020). Barley starch modifications: Physical, chemical and enzymatic—A review. *International Journal of Biological Macromolecules, 144*, 578–585.

Punia, S., Sandhu, K. S., Dhull, S. B., Siroha, A. K., Purewal, S. S., Kaur, M., & Kidwai, M. K. (2020). Oat starch: Physico-chemical, morphological, rheological characteristics and its applications-A review. *International Journal of Biological Macromolecules, 154*, 493–498.

Reyes, L. F., Miller, J. C., & Cisneros-Zevallos, L. (2005). Antioxidant capacity, anthocyanins and total phenolics in purple-and red-fleshed potato (*Solanum tuberosum* L.) genotypes. *American Journal of Potato Research, 82*(4), 271.

Reynolds, A., Mann, J., Cummings, J., Winter, N., Mete, E., & Te Morenga, L. (2019). Carbohydrate quality and human health: A series of systematic reviews and meta-analyses. *The Lancet, 393*(10170), 434–445.

Robertson, T. (2006). Study seeks new uses for potato waste. Minnesota Public Radio. Available from: http://news.minnesota.publicradio.org/features/2005/01/27_robertsont_potatowaste/. Accessed on December, 21,2020.

Sampaio, S. L., Petropoulos, S. A., Alexopoulos, A., Heleno, S. A., Santos-Buelga, C., Barros, L., & Ferreira, I. C. F. R. (2020). Potato peels as sources of functional compounds for the food industry: A review. *Trends in Food Science & Technology, 103*, 118–129.

Serena, A., & Knudsen, K. E. B. (2007). Chemical and physicochemical characterisation of co-products from the vegetable food and agro industries. *Animal Feed Science and Technology, 139*(1), 109–124.

Skytte, M. J., Samkani, A., Petersen, A. D., Thomsen, M. N., Astrup, A., Chabanova, E., ... & Larsen, T. M. (2019). A carbohydrate-reduced high-protein diet improves HbA 1c and liver fat content in weight stable participants with type 2 diabetes: A randomised controlled trial. *Diabetologia, 62*(11), 2066–2078.

Sosinski, B., He, L., Cervantes-Flores, J., Pokrzywa, R. M., Bruckner, A., & Yencho, G. C. (2001). Sweetpotato genomics at North Carolina State University. *I International Conference on Sweetpotato Food and Health for the Future, 583, 51-60.*

Sun, H., Mu, T., Xi, L., Zhang, M., & Chen, J. (2014). Sweet potato (Ipomoea batatas L.) leaves as nutritional and functional foods. *Food Chemistry, 156*, 380–389.

Sun, N., Wang, Y., Gupta, S. K., & Rosen, C. J. (2020). Potato tuber chemical properties in storage as affected by cultivar and nitrogen rate: Implications for acrylamide formation. *Foods, 9*(3).

Takamine, K., Abe, J. I., Iwaya, A., Maseda, S., & Hizukuri, S. (2000). A new manufacturing process for dietary fibre from sweet potato residue and its physical characteristics. *Journal of Applied Glycoscience, 47*(1), 67–72.

Tasie, M. M., & Gebreyes, B. G. (2020). Characterization of nutritional, antinutritional, and mineral contents of thirty-five sorghum varieties grown in Ethiopia. *International Journal of Food Science, 2020.*

Tejeda, L., Mollinedo, P., Aliaga-Rossel, E., & Peñarrieta, J. M. (2020). Antioxidants and nutritional composition of 52 cultivars of native Andean potatoes. *Potato Research*, 1–10.

Tian, J., Chen, J., Ye, X., & Chen, S. (2016). Health benefits of the potato affected by domestic cooking: A review. *Food Chemistry, 202*, 165–175.

Troncoso, E., Pedreschi, F., & Zúñiga, R. N. (2009). Comparative study of physical and sensory properties of pre-treated potato slices during vacuum and atmospheric frying. *LWT Food Science and Technology, 42*(1), 187–195.

UNGAR—United Nations General Assembly Resolution. (2005). Session 60 International Year of the Potato 2008, Resolution 191. Available at: https://digitallibrary.un.org/record/563735?ln=en.

Van Hal, M. (2000). Quality of sweet potato flour during processing and storage. *Food Reviews International*, *16*(1), 1–37.

Venn, B. J., & Green, T. J. (2007). Glycemic index and glycemic load: Measurement issues and their effect on diet–disease relationships. *European Journal of Clinical Nutrition*, *61*(1), S122–S131.

Verma, D. K., & Srivastav, P. P. (2017). Proximate composition, mineral content and fatty acids analyses of aromatic and non-aromatic Indian rice. *Rice Science*, *24*(1), 21–31.

Wang, S., Nie, S., & Zhu, F. (2016). Chemical constituents and health effects of sweet potato. *Food Research International*, *89*, 90–116.

Xu, S., Bai, Z., Jin, B., Xiao, R., & Zhuang, G. (2014). Bioconversion of wastewater from sweet potato starch production to Paenibacillus polymyxa biofertilizer for tea plants. *Scientific Reports, 4*(1), 1–7.

Xu, X., Li, W., Lu, Z., Beta, T., & Hydamaka, A. W. (2009). Phenolic content, composition, antioxidant activity, and their changes during domestic cooking of potatoes. *Journal of Agricultural and Food chemistry*, *57*(21), 10231–10238.

Yadav, A. R., Guha, M., Tharanathan, R. N., & Ramteke, R. S. (2006). Influence of drying conditions on functional properties of potato flour. *European Food Research and Technology*, *223*(4), 553–560.

Yang, Y., Achaerandio, I., & Pujolà, M. (2016). Classification of potato cultivars to establish their processing aptitude. *Journal of the Science of Food and Agriculture*, *96*(2), 413–421.

Zaheer, K., & Akhtar, M. H. (2016). Potato production, usage, and nutrition—a review. *Critical Reviews in Food Science and Nutrition*, *56*(5), 711–721.

Ziegler, G. R., Nordmark, T. S., & Woodling, S. E. (2003). Spherulitic crystallization of starch: influence of botanical origin and extent of thermal treatment. *Food Hydrocolloids*, *17*(4), 487–494.

34

Sweet Potato Starch: Properties and Its Bioactive Components

Prixit Guleria
Maharshi Dayanand University, Rohtak (Haryana), India

Roshanlal Yadav
Bhaskaracharya College of Applied Sciences, University of Delhi, Dwarka (New Delhi), India

Baljeet Singh Yadav
Maharshi Dayanand University, Rohtak (Haryana), India

CONTENTS

34.1   Introduction

Sweet potato (*Ipomoea batatas* L.) is one of the most important food crops that belongs to the family *Convolvulaceae* and is cultivated all over the world. It is a tuberous rooted perennial, usually grown as an annual crop, which is now grown extensively in a wide range of environments. Sweet potato roots are tubers and vary in different shapes, sizes, and colors, depending on the area of cultivar and the environment in which they are cultivated (Takeiti et al., 2011). Sweet potato varieties are classified as soft-fleshed or firm-fleshed. The soft-fleshed varieties have orange flesh and are sweet in taste, often called yams. The firm-fleshed varieties have yellow, purple, or even white flesh (Farley and Drost, 2010). Sweet potato is the sixth most important food crop in the world after rice, wheat, potatoes, maize, and cassava. The variety of sweet potato tuber and its growing location might affect starch content, nutritional composition, and phytochemicals. Sweet potato roots contain a high amount of carbohydrates ranging between 25–30%. Also, sweet potato roots are rich in vitamins, minerals, and dietary fibers (Takeiti et al., 2011).

Sweet potato also contains some antinutritional compounds, including oxalic acid, trypsininhibitor, and furano-terpenoid compounds (Woolfe 1992; Shireen et al., 2001). Sweet potatoes are also good sources of ascorbic acid (vitamin C) and contain moderate amounts of thiamine (B1), riboflavin (B2), and niacin as well aspyridoxine and its derivatives (B6), pantothenic acid (B5), and folic acid. They also contain satisfactory quantities of vitamin E. Many endogenous amylolytic enzymes are found in sweet potato tubers like α-amylase, β-amylase, and starch phosphorylase (Hagenimana et al., 1992). These enzymes are important for the breakdown of starch into simpler sugars during storage and processing (Walter et al., 1976). Starch is the main biochemical component present in sweet potato tubers. The use of starch in the food industry primarily depends upon the starch content and its properties (Moorthy et al., 2012). In Asia, about 45% of the sweet potato supply is used for animal feed, and nearly 50% is used for human consumption (Molina, 2005). Sweet potato roots are widely used for consuming directly with simple cooking and roasting. They can be boiled, steamed, baked, fried, chipped, candied, canned, frozen, and made into flour or starch for industrial alcohols (Tian et al., 1991). Sweet potato flour and starches are also used in the production of various types of food products like soup, sauces, and noodles (Ahmed et al., 2010; Zhang and Oates, 1999; Lin et al., 2005). In some countries, sweet potato tubers are also used for the production of food products like noodles, candy, desserts, and flour by farm households to extend the availability and increase the value of the crop.

34.2 Sweet Potato Starch

Starch, the most important polysaccharide, is found in cereal grain seeds (e.g. sorghum, rice, corn, and wheat), tubers (e.g. potato), roots (e.g. sweet potato, cassava, and arrowroot), legume seeds (e.g. lentils, peas, beans), and fruits (e.g. green bananas). The amount of starch in sweet potato roots depends upon the area of cultivation and variety. Sweet potato tubers contain about 80–90% carbohydrates, mostly starch, which is a good raw material for the food industry (Benjamin, 2007). Sweet potato starch contains 19.0–22.9% amylose content, which varies with variety and area of a cultivar (Ngoc et al., 2017). Starch contains two major high-molecular weight components, amylose, and amylopectin. Amylose is a relatively long, linear polymer α-glucan containing about99% (1→ 4)-α- and 1%(1→6)-α-linkages while amylopectin is a much larger molecule and a heavily branched structure built from about 95% (1→4)-α- and 5% (1→6)-α-linkages. Both structures' polymers play an important role in the functionality of native and modified starches (Punia, 2020; Punia et al., 2020). The amylose content of sweet potato starch varies between 8.5–38% (Tian et al., 1991; Takeda et al., 1986). The physicochemical properties of sweet potato starch, such as gelatinization, retrogradation, water absorption, and pasting properties, are affected by the amylose and amylopectin ratio of sweet potato starch (Collado et al., 1999; Black et al., 2000). These properties can be developed by controlling the rate of heating during cooking, which activates endogenous amylolytic enzymes of the sweet potato root to convert a portion of the starch into dextrin (Hoover, 1967). Starch granules are found in different sizes, shapes, and dimensions that depend on their botanical source and growing and harvest conditions. Most of the tuber and root starches are simple granules, whereas the cassava and taro starches are blended into compound and simple granules (Hoover, 2001). Sweet potato starch granules size ranges between 2–42 μm and are round, oval, and polygonal in shape (Tian et al., 1991; Hoover, 2001). Both A and C types of X-ray patterns have been found by various researchers in sweet potato starch (Gallant et al., 1982; Zoebel, 1988; Jane et al., 1999).

34.3 Starch Isolation

Starch is generally extracted from fresh tubers and the time of harvesting is also very important. A delay in the processing of sweet potatoes results in the conversion of starch in the tubers into sugar and fiber and thus affects the yield and quality of starch. The starch isolation method affects the physicochemical properties of the starch (Moorthy et al., 2012; Lii and Chang, 1978). Wet milling methods are generally used by various researchers for the extraction of sweet potato starch. Extraction involves the washing of sweet potatoes and de-skinning with the help of a peeler for complete removal of skin. The sweet potatoes are then cut into small pieces and homogenized to make a slurry with distilled water. The slurry is then passed through double-layered cheesecloth and allowed to settle for a minimum of 3–4 hours at room temperature. The precipitated starch is washed with distilled water two to three times for complete removal of impurities and dried at 50°C for 12 hours in a hot-air oven (Wickramasinghe et al., 2009; Zheng et al., 2016).The starch paste should be clear and free from off-colors for better acceptability in the food industry. The color of starch is an important criterion for its application in the food and textile industries. Kallabinski and Balagopalan (1994 used an enzymatic technique to isolate sweet potato starch by using pectinase and cellulose enzymes. This technique gives a 20% higher yield than the conventional wet method. Radley (1976a) added lime, diluted acetic acid, and lactic acid during starch extraction and observed an increase in starch yield. An alkali steeping method with 0.2% NaOH for sweet potato starch isolation was used by Lee and Lee (2016) and Song et al. (2014). Sajeev et al. (2005) used a multipurpose mobile starch extraction plant developed by Central Tuber Crops Research Institute for starch exaction from sweet potatoes. Starch extract by this method was 21% and the manual wet method was about 17.7%.

34.4 Physicochemical Properties

The amylose content in sweet potato starch by different researchers has been reported between 15.9%–25.3%. (Table 34.1). The amount of amylose in sweet potato starch is one of the most important factors influencing the cooking and textural qualities of sweet potato starch-based products (Collado et al., 1999). Swelling power and starch solubility provide evidence of noncovalent bonding between starch molecules and therefore allow comparison of relative bond strength at specific temperatures (Moorthy, 2002). The swelling power of starch is related to the gelatinization of starch, reflecting the breaking of hydrogen bonds in the crystalline regions, uptake of water by hydrogen bonds, and water absorption by nonstarch polysaccharides and protein (Thitipraphunkul et al., 2003). The swelling power and solubility of starch are temperature-dependent and increase with an increase in temperature due to the weakening of internal associative forces maintaining the granular structure (Peroni et al., 2006). Swelling and solubility of sweet potato starch varied with botanical source and also was influenced by the chemical composition of starch, amylose-amylopectin ratio, granular morphology, and structural characteristics of amylose and amylopectin, lipid and phosphate contents in starch (Hoover, 2001; Ratnayake et al., 2002). The swelling power and solubility of sweet potato starches not varies with varieties but also varies with different temperatures. The solubility of four varieties of sweet potato starch was found to vary between 15–98% at a temperature range between 70–90°C (Nuwamanya et al,2011). Moorthy (2002) reported swelling volume and solubility of sweet potato starch in the range of 42–71 mg/ml and 25–48%, respectively. Various studies have shown that the solubility of sweet potato starch was less than potato and cassava but higher than maize starch (Rasper, 1969; Delpeuch and Favier, 1980). Kusumayanti et al. (2015) reported swelling power and solubility of sweet potatoes from various varieties as not significantly different and ranged from 3.40–3.67 (g/g) and 8.61–9.57 (%) respectively.

34.5 Pasting Properties

Pasting properties of starch can be measured with a rapid visco-analyzer (RVA) (Hoover et al., 1996; Sandhu et al., 2007; Shaikh et al., 2015;Yadav et al., 2013) or by using a dynamic rheometer (Punia et al., 2019a; Siroha and Sandhu, 2018). The pasting parameters including the pasting temperature (PT), peak viscosity (PV), hot paste viscosity (HPV), breakdown (BD), and setback (SB). Pasting properties are shown in Table 34.1. The pasting properties of starches are governed by granule-granule interactions, size and shape of granules, and their distribution and amylose content. Also, the viscosity behavior of starches changes with its concentration, shear rate, and temperature (Morikawa and Nishinari, 2002). Swelling power and amylose content are the major factors affecting the pasting properties of flour and starch samples. The PV of starches represents the ability of starch granules to swell freely before the disintegration of starch granules at a particular concentration. Starches that swell at a higher temperature

TABLE 34.1

Physicochemical and Functional Properties Sweet Potato Starches

Properties Range References		
Amylose content (%)	22–25	Shiotahi et al., 1991
	15.9	Aprianita et al., 2013
	17.2–19	Takeda et al., 1986
	19.4–22.8	Chiang and Chen, 1988
	16.2–25.3	Katayama et al., 2011
	16.2–23.4	Nabubuya et al., 2012
Pasting temperature (°C)	82–95	Moorthy et al., 2012
	57.1–80.3	Katayama et al., 2011
	79.75–82.73	Sajeev et al., 2012
	67–75	Suganuma and Kitahara, 1997
	78.9 (purple color)	Lee and Lee, 2016
	66.6	Aprianita et al., 2013
Solubility (%)	60–79	Seog et al., 1987
	19.5–39.5	Moorthy and Balagopalan, 1999
	15–98	(Nuwamanya et al, 2011).
	7.18–13.65	Moorthy et al., 2010
Swelling volume (ml/g)	17.75–19.50	Moorthy and Balagopalan, 1999
	27.5–33.3	Chiang and Chen, 1988
	63–95	Seog et al., 1987
	32.5–50	Moorthy et al., 2010
T_{onset} (°C)	61.3	Collado et al., 1999
	64.9 (Yulmi)	Song et al., 2014
	67.1 (commercial sweet potato starch)	Song et al., 2014
	62.9	Aprianita et al., 2013
	67–75	Chiang and Chen, 1988
	67.3	Valetudie et al., 1995
	65.6–68.21	Kitada et al., 1988
T_{max} (°C)	70.2–77	Collado et al., 1999
	73.5	Aprianita et al., 2013
	72.8–74.3	Kitada et al., 1988
	73–79	Chiang and Chen, 1988
	72.7	Valetudie et al., 1995
T_{end} (°C)	80.5	Aprianita et al., 2013
	79.6	Valetudie et al., 1995
	86.20 (purple color)	Lee and Lee, 2016
	81.6	Lee and Yoo, 2009
ΔH (J/g)	12.6	Lee and Yoo, 2009
	9.3 (Yulmi)	Song et al., 2014
	12.0 (commercial sweet potato starch)	Song et al., 2014
	11.3	Aprianita et al., 2013
	13.6	Valetudie et al., 1995

are less resistant to breakdown during cooling and exhibit a significant decrease in viscosity after attaining PV. An increase in viscosity during the cooling phase not only indicates the inverse relationship between viscosity and temperature but also shows the tendency of various starch components to retrograde as the temperature decreased (Adebowale and Lawal, 2003). A lot of literature is available on the pasting properties of sweet potato starch and their genotype variation. PT observed by different researchers ranged between 66.6–95°C and is presented in Table 34.1. Jangchud et al. (2003) reported starch-pasting temperature 80.5°C for orange-fleshed tubers and 74.8°C for purple-fleshed tubers. PT provides the information of estimated minimum cooking time for a particular food material and the energy costs that may be involved during cooking (Shimelis et al., 2006; Ikegwu and Okechukwu, 2010). High PT of flour and starches has been associated with higher amylose content and high resistance toward swelling (Ikegwu and Okechukwu, 2010). Several studies show found that sweet potato starch does not show at 4–6% concentration (Tian et al., 1991) but Lii and Chang(1978) observed a moderate peak and high SB on cooling with a 7% starch concentration. Varietal differences in pasting properties of white, yellow, and purple sweet potato starches were observed by Ngoc et al. (2017). They observed that the white sweet potato starch had significantly higher PT but lower peak and final viscosity (FV) and breakdown than yellow and purple sweet potato starches. The paste of the white sweet potato starch exhibited a greater gel consistency and hot paste stability and the lower FV and SB of this starch paste indicated that this starch paste had high resistance against retrogradation. Yellow sweet potato starch paste had the lowest pasting temperature and highest peak and final viscosities and breakdown as compared to white and purple sweet potato starch. These results showed that the paste of yellow sweet potato starch had a weak resistance against retrogradation caused by high amylose content in starches. Hung and Morita (2005) reported the same results in their findings. Lee and lee (2016) studied the pasting profile of white, orange, and purple-fleshed sweet potato starches. They observed that the pasting temperature of the white-fleshed sweet potato starch had the highest and orange-fleshed sweet potato starch showed the lowest PT (76.9°C), but a sharper rise in viscosity and yielding the highest PV. The highest BD was observed in the orange-fleshed sweet potato starch. When the temperature was decreased to 50°C, a sharp rise in pasting viscosity was observed in all starch samples, indicating a gelling tendency of the starch paste during cooling. A large number of intermolecular hydrogen bonds will be formed during the cooling cycle, resulting in an increase in viscosity and formation of gels at low temperatures.

34.6 Thermal Properties

Starch gelatinization is an important functional property of starch and can be determined by a differential scanning calorimeter (DSC). Thermal properties of starch vary with starch source, the composition of starch granules, size and shape of starch granules, the ratio of amylose and amylopectin, and their chain length distribution (Singh and Kaur, 2004; Perez et al., 2005). Starch-to-water ratio, excess amount of water, and the heating rate are very important in the starch gelatinization and determination of the gelatinization temperature range of starches (Calzetta-Resio et al., 2000). Gelatinization temperature is a measure of the cooking quality of starch and an important parameter in food processing industries for the making of various starch-based products. Starches with low gelatinization temperature ranges have good cooking quality (Waters et al., 2006). The gelatinization transition temperatures T_O (onset), T_P (peak), T_C (end set), and the enthalpy of gelatinization (ΔH gel) are obtained from thermo grams and influenced by the molecular architecture of the crystalline region (Wani et al., 2016). Gelatinization temperature range is calculated by the difference between T_C and T_O, and peak height index is measured as the ratio of the enthalpy to the ratio between T_P and T_O. Collado et al. (1999) obtained considerable variations in all the DSC parameters of all cultivars of sweet potato. They found that the T_O was 64.6°C and mean range was 61.3–70°C, T_P was 73.9°C and mean range was 70.2–77°C, T_C was 84.6°C and mean range 80.7–88.5°C, and the mean gelatinization range was 20.1°C with a range of 16.1–23°C. Garcia and Walter (1998) found a range between 58–64°C for two varieties of sweet potato cultivated at different locations. Noda et al. (1998) reported variations in DSC parameters for 51 varieties of sweet potato and observed range between 55.7–73.1°C, 61.3–77.6°C, and 12.7–16.8J/g for T_O, T_P and ΔH. Zhu et al. (2011) found variations in enthalpy values (7.6–13.2J/g) of 11 varieties of sweet potato. They reported that increase in short outer chains of amylopectin reduced the packing efficiency of double helices within the crystalline region, resulting in a low gelatinization temperature and enthalpy. The gelatinization enthalpy depends upon several factors like the degree of crystallinity, amylose content, and intermolecular bonding. The ΔH enthalpy value of sweet potato starch ranges between 10–18.6J/g (Tian et al., 1991;

Garcia and Walter, 1998; Collado et al., 1999) and 11.8–13.4J/g (Noda et al., 1992). Effect of variety and environmental conditions also affect the ΔH enthalpy value (Noda et al., 1996; Garcia and Walter, 1998; Noda et al., 1992).Lee and Lee (2016) studied the DSC profile of white sweet potato starch and showed a T_O temperature of 58.02°C and a T_P temperature of 76.94°C. White sweet potato starch exhibited a higher peak gelatinization temperature and enthalpy than the orange- or purple-fleshed sweet potato starches. This result concludes that the white sweet potato starch had a more thermally stable granular structure and a more highly ordered structure (Huang et al., 2010).

34.7 Starch Modification

Less use of native starches in industries is due to its functional and physicochemical properties (Jyothi et al., 2005). Starches are generally modified for use in foods to stabilize and provide good thickening and gelling properties. Modification of starch is generally achieved by acetylation, hydroxypropylation, oxidation, acid hydrolysis, cross-linking, and dual modification (Singh et al., 2007). These techniques are, however, limited due to issues concerning consumers safety and the environment. Chemical modification involves the introduction of functional groups into the starch molecules, resulting in marked changes in starch physicochemical properties (Majzoubi et al., 2009; Hung and Morita, 2005). Such modification of native granular starches profoundly changes the proximate compositions, gelatinization, retrogradation, and pasting characteristics. Chemical modification is intended to facilitate intra- and intermolecular bonds at random locations in the starch granule for their stabilization. The chemical and functional properties achieved by modified starches depend upon the starch source, reaction conditions (reactant concentration, pH, reaction time, and the presence of a catalyst), type of substituent, degree of substitution (DS), and the distribution of the substituents in the starch molecule (Hirsch and Kokini, 2002; Wang and Wang, 2002). Various researchers modified sweet potato starch by physical, chemical, and enzymatic techniques. Lee and Yoo (2009) modified sweet potato starch by acetic anhydride. The same treatment was also given by Shon and Yoo (2006) and Wurzburg (1964) for sweet potato starch. Babu, et al. (2016) modified sweet potato starch by pullulanase enzyme and studied its debranching effect on starch digestibility. Song et al., (2014) modified sweet potato starch by physical modification (annealing) and studied its effect on physicochemical properties and digestibility.

34.8 Antioxidant Activity

Antioxidant activity is the "restriction of oxidation of lipids, proteins, DNA or several other molecules that take place by inhibiting the propagation stage in oxidative chain reactions." (Sivonova et al., 2007). There are two main categories of antioxidants: primary, which directly scavenges free radicals, and secondary, which act indirectly by stopping the generation of free radicals through Fenton's reaction (Huang et al., 2005). The antioxidant activity of sweet potato (*Ipomoea batatas L.*) is due to the presence of various phytochemicals such as phenolics, anthocyanin, carotenoids, saponin, alkaloids, and tannins. In sweet potatoes, the antioxidant activities have been mostly accredited to their betacaroteneand anthocyanin contents (Figure 34.1). These constituents may differ with varieties and also impart in the development of skin and flesh color. Betacarotenoids take part in the development of orange flesh color while anthocyanins are responsible for purple flesh color in sweet potatoes (Swamy and Omwenga, 2014; Lila, 2004). The anthocyanins, such as peonidin and cyanidin, are present in purple sweet potatoes. In a study conducted by Park et al. (2016) peonidin and cyanidin were only observed in purple-fleshed sweet potatoes but not in the white and orange varieties. Luteolin, a flavonoid, was detected in orange and purple varieties, however, it was not found in the white. Besides these phyto-constituents, the sweet potato is rich in some phenolic acids such as cinnamic, caffeic, chlorogenic, hydroxycinnamic, and isochlorogenic acids. In previous studies, it was concluded that these phenolic acids have great potential for antioxidant activity of foods (Robbins, 2003).

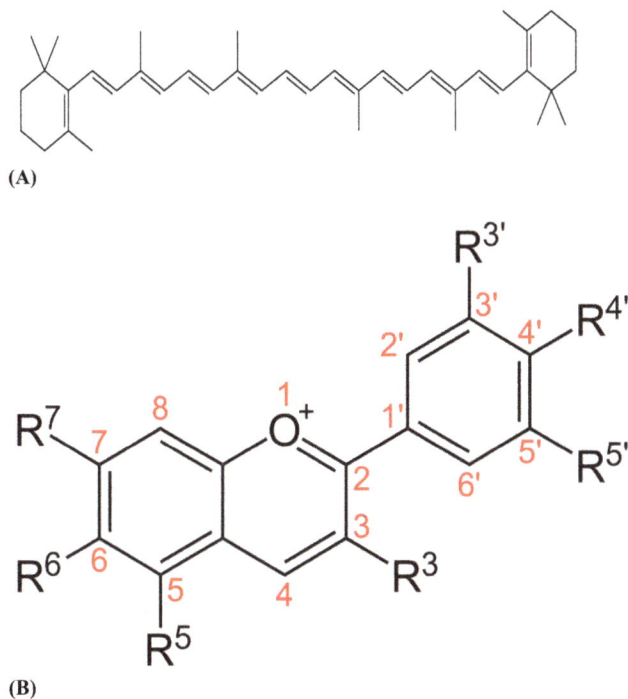

FIGURE 34.1 Basic structure of beta carotenoids (A) and anthocyanin (B).

34.8.1 β-carotene and Anthocyanin

β-carotene is an organic, red-orange colored pigment abundant in sweet potato tubers, especially orange-fleshed tuber. It is an element of the carotenes (terpenoids) having a long chain of carbon molecules (40 carbons). It was reported up to 276.98 μg/g in orange-fleshed sweet potato tubers, which is a significantly high content and helps in the fulfillment of vitamin-A deficiency (Tumuhimbise et al., 2009). It has been reported that processing techniques such as maceration and thermal treatments recover β-carotene from root tubers, which might be due to the breakdown of the microstructure of plant tissue, which helps in the release of nutrients (Tumuhimbise et al., 2009). Anthocyanins are members of the flavonoid group, which are water-soluble pigments mainly available in purple-fleshed sweet potato, and take part in the development of the antioxidant activity. High anthocyanin content in purple- and red-fleshed sweet potato tubers has been reported in numerous studies (Oki et al., 2002). The highest anthocyanin (0.53 mg/g fw) was reported by Teow et al. (2007 in purple-fleshed sweet potato. In another study, Cevallos-Casals and Cisneros-Zevallos (2003) investigated 1.82 mg/g (fw) anthocyanin content in a red-fleshed sweet potato variety. Both the components are abundantly present in sweet potatoes and very much responsible for antioxidant activity.

34.8.2 Free Radical Scavenging Activity

Several methods are available to determine the antioxidant activities of natural compounds in foods. Two assays 2,2-diphenyl-1-picrylhydrazyl (DPPH) and 2, 2'-azinobis(3-ethylbenzothiazoline-6-sulfonic acid(ABTS) are generally used in the determination of antioxidant activities. Antioxidants can prevent the oxidation reaction by quenching free radicals, in other words, antioxidants act as scavengers of free radicals reactive oxygen species within the cell (Devasagayam et al., 2004). To determine the scavenging effect of plant extract, a powder composed of stable free-radical molecules i.e. DPPH and ABTS are generally used. Phenolic compounds are the primary contributors to the DPPH radical scavenging ability of sweet potatoes (Rumbaoa et al., 2009). A significant positive correlation between phytochemicals and free radical scavenging activity indicates that a high amount of phytochemicals i.e phenolics and flavonoids in sweet potato is responsible for elevated free-radical scavenging activity (Grudzinska et al., 2016). The DPPH activity of a different variety of sweet potato was measured in the range of 1.9–17.9 μmol TE/g dw (Cartier et al., 2017). In a study conducted by Kim et al. (2019) the DPPH and ABTS radical-scavenging activities of sweet potato cultivars were measured in the range from 1.91–36.29 mg TE/g ER and from 17.77–62.20 mgTE/g of extract, respectively. Oxygen radical absorbance capacity (ORAC) is one more assay used to calculate the antioxidant potential. This is based on the inhibition of the peroxyl-radical induced oxidation initiated by thermal decomposition of azo-compounds, such as 2, 20-azobis (2-amidino propane) dihydrochloride (AAPH). The advantage of the ORAC method is its ability to assay both hydrophilic and lipophilic antioxidants, which results in better measurements of total antioxidant activity (Prior et al., 2003). The lipophilic ORAC values of different samples of sweet potato were in the range of 0.85–2.87 μmol TE/g fw, (Wu et al., 2004).

34.8.3 Ferric Reducing Antioxidant Power (FRAP)

The FRAP method depends on the reduction of ferric 2,4,6-Tri(2-pyridyl)-s-triazine (TPTZ) (Fe [III]-TPTZ) to ferrous TPTZ (Fe [II[-TPTZ) by a reductant (antioxidant) at a lower pH value. The intense blue color is formed and noted at wavelength 593 nm when Fe^{3+}TPTZ complex is reduced to the ferrous form. Generally, the reducing properties are highly linked with the presence of compounds that exert their action by breaking the free radical chain by donating a hydrogen atom. The FRAP values of various samples of sweet potato ranged from 8.55–28.57 mmol FE/g (Kim et al., 2019; Padda and Picha 2007). In another study conducted by Tang et al. (2015), the FRAP value was observed in the range of 17.5–274 mmol FE/g.

34.8.4 Polyphenolic Profile

Sweet potato tubers are an abundant source of polyphenolic compounds and these polyphenols possess potent antioxidant activities. Chlorogenic acid is the major phenolic, which was first identified by Rudkin and Nelson (1947) in the root tissue of sweet potatoes. A strong correlation of chlorogenic acid and antioxidant activity of several sweet potato cultivars was measured in some early studies (Oki et al., 2002; Huang et al., 2004). The concentration and types of phenolics may vary and mainly depends upon the variety of crops and sometimes upon environmental factors. Various phenolics namely, caffeic acid, chlorogenic acid, and isomers of dicaffeoylquinic acid were also recognized as the chief phenolic acids in numerous cultivars of sweet potato (Guan et al. 2006; Walter and Schade 11981). The level of caffeic acid varies with variety and it is mainly present in the periderm tissue of root tuber (Harrison et al., 2003). Interestingly, caffeic acid has great retention capacity inside the root tuber after steaming (Tudela et al., 2002). Flavonoids are one of the important categories of polyphenolic compounds, which also are responsible for antioxidant activity. Numerous flavonoids such as peonidins, cyanidins (Islam et al., 2002; Terahara et al., 1999), rhamnocitrin, kaempferol, tiliroside, astragalin, and rhamnetin (Luo and Kong, 2005) as well as quercetin and rutin (Guan et al., 2006) are also found in different varieties of sweet potato. The esters of caffeic acid and quinic acid are the caffeoylquinic acids, which are one of the significant phenolic acids present in root tubers. The different isomer of caffeoylquinic acid i.e 1,3-di-O-caffeoylquinic acid, 3,5-di-O-caffeoylquinic acid, and 4-O-caffeoylquinic acid acquire potent antioxidant activities and these have been measured in sweet potatoes in significant quantity (Blum-Silva et al., 2015). The total phenol content was determined to be in the range of 2.73–12.46 g/100 g dw in various samples of sweet potato (Islam et al., 2002; Sun et al., 2014). In another study of sweet potato, the phenol content was estimated in the range 13.70–72.41 mg GAE/g extract residue (Song et al., 2005; Kim et al., 2019).

34.9 Health Benefits

The sweet potato possesses protective abilities against colon cancer and vascular diseases due to the presence of vital components such as fiber, cellulose, and hemicellulose (Woolfe, 1992). Betacarotene is a precursor of vitamin A, which is plentiful in orange-fleshed sweet potato, hence its greatly use to maintain adequate level of vitamin A in the body (Gurmu et al., 2014). Carotenoids possess antioxidant potential and take part in the reduction or inhibition of mutagenesis in cells (Mohanraj and Sivasankar, 2014). An abundant quantity of vitamin B6 in root tubers has been reported in previous studies, which may help in reducing the levels of blood amino acid i.e. homocysteine, which may impart in cardiovascular diseases (Ayeleso et al., 2016). Although the human body has a defense mechanism system against the generation of free radicals, extreme production of free radicals can cause oxidative damage in the cells (Silalahi, 2001). The free radicals damage the activity of DNA, lipids, and protein, which may cause neurodegenerative disorders, aging, cancer, and atherosclerosis (Ames et al., 1993; Aruoma, 1998). Polyphenolic compounds available in sweet potatoes may help in averting chronic illnesses such as cardiovascular diseases, neurodegenerative diseases, cancers, and diabetes (Surh, 2003; Scalbert et al. 2005). Some specific phytochemicals such as chlorogenic acid and quercetin have been reported in sweet potatoes that help in the inhibition of cancer and also defend the heart (Mohanraj and Sivasankar, 2014).

34.10 Conclusion

Sweet potato roots are directly consumed by people all over the world by simple cooking, and its starch is mostly used for the production of various types of food products. Sweet potato tubers contain about 80–90% carbohydrates, mostly starch, which is a good raw material for the food industry. The physicochemical properties of sweet potato starch such as gelatinization, retrogradation, water absorption, and pasting properties are affected by the amylose and amylopectin ratio of sweet potato starch. Sweet potato roots are rich in vitamins, minerals, and dietary fibers. Sweet potato tubers also possess various antioxidant properties. The antioxidant activity of sweet potato tubers is due to the presence of various phytochemicals such as phenolics, anthocyanin, carotenoids, saponin, alkaloids, and tannins. The sweet potato tubers possess protective abilities against various types of cancers and vascular diseases and their polyphenolic compound contents may help in averting chronic illnesses such as cardiovascular diseases, neurodegenerative diseases, and diabetes.

REFERENCES

Adebowale, K. O. and Lawal, O. S. 2003. Microstructure, physicochemical properties, and retrogradation behavior of mucuna bean (*Mucunapruriens*) starch on heat moisture treatments. *Food Hydrocolloids* 17: 265–272.

Ahmed, M., Sorifa, A. M. and Eun, J. B. 2010. Effect of pretreatments and drying temperatures on sweet potato flour. *International Journal of Food Science and Technology* 45: 726–732.

Ames, B. N., Shigenaga, M. K. and Hagen, T. M. 1993. Oxidants, antioxidants, and the degenerative diseases of aging. *Proceedings of the National Academy of Sciences* 90: 7915–7922.

Aprianita, A., Vasiljevic, T., Bannikova, A. and Kasapis, S. 2013. Physicochemical properties of flours and starches derived from traditional Indonesian tubers and roots. *Journal of Food Science and Technology.* 51: 3669–3679

Aruoma, O. I. 1998. Free radicals, oxidative stress, and antioxidants in human health and disease. *Journal of the American Oil Chemist's Society* 75: 199–212.

Ayeleso, T. B., Ramachela, K. and Mukwevho, E. 2016. A review of therapeutic potentials of sweet potato: Pharmacological activities and influence of the cultivar. *Tropical Journal of Pharmaceutical Research* 15: 2751–2761.

Babu, A. S., Parimalavalli, R., Jagannadham, K. and Rao, J. S. 2016. Fat mimicking properties of citric acid-treated sweet potato starch. *International Journal of Food Properties* 19: 139–153.

Benjamin, A. C. B. 2007. Sweet potato: A review of its past, present, and future role in human nutrition. *Advances in Food and Nutrition Research* 52: 1–59.

Black, C. K., Panozzo, J. F., Wright C. L. and Lim, P. C. 2000. Survey of white salted noodle quality characteristics in heat landraces. *Cereal Chemistry* 4: 468–472

Blum-Silva, C. H., Chaves, V. C., Schenkel, E. P., Coelho, G. C. and Reginatto, F. H. 2015. The influence of leaf age on methylxanthines, total phenolic content, and free radical scavenging capacity of Ilex para-guariensisaqueous extracts. *Revista Brasileira de Farmacognosia* 25: 1–6.

Calzetta-Resio, A. N., Tolaba, M. P. and Suarez, C. 2000. Some physical and thermal properties of amaranth starch. *International Journal of Food Science and Technology* 6: 371–378.

Cartier, A., Woods, J., Sismour, E., Allen, J., Ford, E., Githinji, L. and Xu, Y. 2017. Physiochemical, nutritional, and antioxidant properties of fourteen Virginia-grown sweet potato varieties. *Journal of Food Measurement and Characterization* 11: 1333–1341.

Cevallos-Casals, B. A. and Cisneros-Zevallos, L. 2003. Stoichiometric and kinetic studies of phenolic anti-oxidants from Andean purple corn and red-fleshed sweetpotato. *Journal of Agricultural and Food Chemistry* 51: 3313–3319.

Chiang, W. C. and Chen, K. L. 1988. Comparison of physicochemical properties of starch and amylolytic enzyme activity of various sweet potato varieties. *Shih Pin Ko Hsueh (Taipei)* 15: 1–11

Collado, L. S., Mabesa, R. C. and Corke, H. 1999. Genetic variation in the physical properties of sweet potato starch. *Journal of Agriculture and Food Chemistry* 47: 4195–4201.

Delpeuch, F. and Favier, J. C. 1980. Characteristics of starches from tropical food plants; α-amylase hydrolysis, swelling, and solubility patterns. *Annals of Technology in Agriculture* 29: 53–67.

Devasagayam, T. P. A., Tilak, J. C., Boloor, K. K., Sane, K. S., Ghaskadbi, S. S. and Lele, R. D. 2004. Free radicals and antioxidants in human health: Current status and future prospects. *Journal of the Associations of Physicians India* 52: 794–804.

Farley, J. and Drost, D. 2010. Sweet potatoes in the garden. Home gardening. Utah State University.

Gallant, D. J., Bewa, H., Buy, Q. H., Bouchet, B., Szylit, O. and Sealy, L. 1982. On ultrastructural and nutritional aspects of some tropical tuber crops. *Starch/Stärke* 34: 255–262.

Garcia, A. M. and Walter, W. M. 1998. Physico-chemical characterization of starch from Peruvian sweet potato selections. *Starch/Stärke* 50: 331–337.

Grudzinska, M., Czerko, Z., Zarzynska, K. and Boowska-Komenda, M. 2016. Bioactive compounds in potato tubers: Effects of farming system, cooking method, and flesh color. *PLOS ONE*, 11: e-0153980.

Guan, Y., Wu, T., Lin, M. and Ye, J. 2006. Determination of pharmacologically active ingredients in sweet potato (*Ipomoea batatas* L.) by capillary electrophoresis with electrochemical detection. *Journal of Agricultural and Food Chemistry* 54: 24–28.

Gurmu, F., Hussein, S. and Laing, M. 2014. The potential of orange-fleshed sweet potato to prevent vitamin A deficiency in Africa. *International Journal of Vitamin and Nutrition Research* 84: 65–78.

Hagenimana, V. L. P., VezinaL. P. and Simard, R. E. 1992. Distribution of amylases within sweetpotato (*Ipomoea batatas* L.) root tissues. *Journal of Agriculture and Food Chemistry* 40: 1777–1783.

Harrison, H. F., Peterson, J. K., Snook, M. E., Bohac, J. R. and Jackson, D. M. 2003. Quantity and potential biological activity of caffeic acid in sweet potato (*Ipomoea batatas* L.) storage root periderm. *Journal of Agricultural and Food Chemistry* 51: 2943–2948.

Hirsch, J. B. and Kokini, J. L. 2002. Understanding the mechanism of cross-linking agents (POCl3, STMP, and EPI) through swelling behavior and pasting properties of crosslinked waxy maize starches. *Cereal Chemistry* 79: 102–107.

Hoover, M. W. 1967. An enzyme activation process for producing sweet potato flakes. *Food Technology* 21: 322–325.

Hoover, R. 2001. Composition, molecular structure and physicochemical properties of tuber and root starches: A review. *Carbohydrate Polymers* 45: 253–267.

Hoover, R., Swamidas, G., Kok, L. S. and Vasanthan, T. 1996. Composition and physicochemical properties of starch from pearl millet grains. *Food Chemistry* 56: 355–367.

Huang, C. C., Lai, P., Chen, I. H., Liu, Y. F. and Wang, C. C. R. 2010. Effects of mucilage on the thermal and pasting properties of yam, taro, and sweet potato starches. *LWT Food Science and Technology* 43: 849–855.

Huang, D. J., Chun-Der, L. I. N., Hsien-Jung, C. H. E. N. and Yaw-Huei, L. I. N. 2004. Antioxidant and anti-proliferative activities of sweet potato (*Ipomoea batatas* [L] Lam Tainong 57.) constituents. *Botanical Bulletin- Academia Sinica Taipei* 45: 179–186.

Huang, D., Ou, B. and Prior, R. L. 2005. The chemistry behind antioxidant capacity assays. *Journal of Agricultural and Food Chemistry* 53: 1841–1856.

Hung, P. V. and Morita, N. 2005. Physicochemical properties and enzymatic digestibility of starch from edible canna (*Canna edulis*) grown in Vietnam. *Carbohydrate Polymers* 61: 314–321.

Ikegwu, O. J. and Okechukwu, P. E. 2010. Physicochemical and pasting characteristics of flours and starch from achibrachystegiaeurtcoma seed. *Journal of Food Technology* 8: 58–66.

Islam, M. S., Yoshimoto, M., Yahara, S., Okuno, S., Ishiguro, K. and Yamakawa, O. 2002. Identification and characterization of foliar polyphenolic composition in sweetpotato (*Ipomoea batatas L.*) genotypes. *Journal of Agricultural and Food Chemistry* 50: 3718–3722.

Jane, J., Chen, Y. Y., Lee, L. F., McPherson, A. E., Wong, K. S., Radosavljevic, M. and Kasemsuwan, T. 1999. Effects of amylopectin branch chain length and amylose content on the gelatinization and pasting properties of starch. *Cereal Chemistry* 76:629–637.

Jangchud, K., Phimolsiripol, Y. and Haruthaithanasan, V. 2003. Physico-chemical properties of sweet potato flour and starch as affected by blanching and processing. *Starch/Stärke* 55: 258–264.

Jyothi, A. N., Rajasekharan, K. N., Moorthy, S. N. and Sreekumar, J. 2005. Synthesis and characterization of low DS succinate derivatives of cassava (*Manihoresculentacrants*) starch. *Starch/Stärke* 57: 556–563.

Kallabinski, J. and Balagopalan, C. 1994. Enzymatic starch extraction from tropical root and tuber crops. In: *Symposium on Tropical Root Crops in developing economy Acta Horticulturae* 1: 83–88.

Katayama, K., Kitahara, K., Sakai, T., Kai, Y. and Yoshinag, M. 2011. Resistant starch and digestible starch contents in sweet potato cultivars and line. *Journal of Applied Glycoscience* 58: 53–59.

Kim, M. Y., Lee, B. W., Lee, H. U., Lee, Y. Y., Kim, M. H., Lee, J. Y., Lee, B. K., Woo, K. S. and Kim, H. J. 2019. Phenolic compounds and antioxidant activity in sweet potato after heat treatment. *Journal of the Science of Food and Agriculture* 99: 6833–6840.

Kitada, Y., Sasaki, M., Yamazoe, Y. and Nakazawa, H. 1988 Measurements of the thermal behaviour and amylose content of kuzu and sweet potato starches. *Nippon Kogyo Gokkaishi* 35: 135–140.

Kusumayanti, H., Handayani, N. A. and Santosa, H. 2015. Swelling power and water solubility of cassava and sweet potatoes flour. *Procedia Environmental Sciences* 23: 164–167.

Lee, B. H. and Lee, Y. T. 2016. Physicochemical and structural properties of different colored sweet potato starches. *Starch/Stärke* 61: 407–413.

Lee, H. L. and Yoo, B. 2009. Dynamic rheological and thermal properties of acetylated sweet potato starch. *Starch/Stärke* 61: 407–413.

Lii, C. Y. and Chang, S. M. 1978. Studies on the starches in Taiwan, sweet potato, cassava, yam and arrowroot starches. In: *Proceedings of the National Scientific Council. ROC2 Part A: Physics Science Engineering* 2: 146–423.

Lila, M. A. 2004. Anthocyanins and human health: An in vitro investigative approach. *Journal of Biomedicine and Biotechnology* 5: 306–313.

Lin, Y. P., Tsen, J. H. and King, V. A. E. 2005. Effects of far-infrared radiation on the freeze-drying of sweet potato. *Journal of Food Engineering* 68: 249–255.

Luo, J. G. and Kong, L. Y. 2005. Study on flavonoids from leaf of Ipomoea batatas. Zhongguo Zhong yao za zhi= Zhongguo zhongyao zazhi= *China Journal of Chinese Materia Medica* 30: 516–518.

Majzoubi, M., Radi, M., Farahnaki., A., Jamalian, J. and Tongdang, T. 2009. Physico-chemical properties of phosphoryl chloride cross-linked wheat starch. *Iranian Polymer Journal* 18: 491–499.

Mohanraj, R. and Sivasankar, S. 2014. Sweet Potato (*Ipomoea batatas L.*)—A valuable medicinal food: A review. *Journal of Medicinal Food* 17: 733–741.

Molina, La. 2005. *Sweet Potato, Treasure for the Poor*. Peru: InternationalPotato Center.

Moorthy, S. N. 2002. Physicochemical and functional properties of tropical tuber starches: A review. *Starch/Stärke* 54: 559–592.

Moorthy, S. N. and Balagopalan, C. 1999. Physiochemical properties of enzymatically separated starch from sweet potato. *Tropical Science* 39: 23–27.

Moorthy, S. N., Naskar, S. K., Shanavas, S., Radhika, G. S. and Mukharjee. 2010. Physico chemical characterization of selected sweet potato cultivars and their Starches. *International Journal of Food Properties* 13:1280–1298.

Moorthy, S. N., Sajeev, M. S. and Shanavas, S. 2012. Sweet potato starch: Physio-chemical, functional, and thermal and rheological characteristics. *Fruit Vegetable and Cereal Science and Biotechnology* 6: 124–133.

Morikawa, K. and Nishinari, K. 2002. Effects of granular size and size distribution on oftuber and root starches: A review. *Carbohydrate Polymers* 3: 253–267.

Nabubuya, A., Namutebi, A., Byaruhanga, Y., Narvhus, J. and Wicklund, T. 2012. Potential use of selected sweet potato (Ipomeabatatas L.) varieties as defined by chemical and flour pasting characteristics. *Food and Nutrition Sciences* 3: 889–896.

Ngoc, L. B. B., Trung, P. T. B., Hoa, P. N. and Hung, P. V. 2017. Physicochemical properties and resistant starch contents of sweet potato starches from different varieties grown in Vietnam. *International Journal of Food Science and Nutrition* 2: 53–57.

Noda, T., Takahata, Y., Nagarata, T. 1992. Development changes in properties of sweet potato starch. *Starch/ Stärke* 44: 405–409.

Noda, T., Takahata, Y., Sato, T., Ikoma, H. and Mochida, H. 1996. Physiochemical properties of starch from purple and orange fleshed sweet potato roots at two major level of fertilizer. *Starch/ Stärke* 48: 395–399. Noda, T., Takahata, Y., Suda, I., Morishita. T. and Ishiguro, K. 1998. Relationship between chain length distribution of amylopectin and gelatinization properties within the same botanical origin for sweet potato and buckwheat. *Carbohydrate Polymers* 37: 153–158.

Nuwamanya, E., Baguma, Y., Wembabazi, E. and Rubaihayo, P. 2011. A comparative study of the physico-chemical properties of starch from root, tuber and cereal crops. *African Journal of Biotechnology* 10: 12018–12030.

Oki, T., Masuda, M., Furuta, S., Nishiba, Y., Terahara, N. and Suda, I. 2002. Involvement of anthocyanins and other phenolic compounds in radical scavenging activity of purple-fleshed sweet potato cultivars. *Journal of Food Science* 67: 1752–1756.

Padda, M. S. and Picha, D. H. 2007. Antioxidant activity and phenolic composition in 'Beauregard'sweet potato are affected by root size and leaf age. *Journal of the American Society for Horticultural Science* 132: 447–451.

Park, S. Y., Lee, S. Y., Yang, J. W., Lee, J. S., Oh, S. D., Oh, S., Lee, S. M., Lim, M. H., Park, S. K., Jang, J. S. and Cho, H. S. 2016. Comparative analysis of phytochemicals and polar metabolites from colored sweet potato (*Ipomoea batatas* L.) tubers. *Food Science and Biotechnology* 25: 283–291.

Perez, E., Schultzb, F.S. and Pacheco-de-Delahaye, E. 2005. Characterization of some properties of starches from Xanthosomasagittifolium (*Tannia*) and Colocassiaesculenta (*Taro*). *Carbohydrate Polymers* 60: 139–145.

Peroni, F. H. G., Rocha, T. S. and Franco, C. M. L. 2006. Some structural and physicochemical characteristics of tuber and root starches. *International Journal of Food Science and Technology* 12: 505–513.

Prior.,R. L., Hoang., H. A., Gu, L., Wu., X., Bacchiocca, M., Howard., L., Hampsch-Woodill., M., Huang, D., Ou, B. and Jacob, R. 2003. Assays for hydrophilic and lipophilic antioxidant capacity (oxygen radical absorbance capacity (ORACFL) of plasma and other biological and food samples. *Journal of Agricultural and Food Chemistry* 51: 3273–3279.

Punia, S., Siroha, A. K., Sandhu, K. S. and Kaur, M. 2019a. Rheological and pasting behavior of OSA modi-fied mungbean starches and its utilization in cake formulation as fat replacer. *International Journal of Biological Macromolecules* 128:230–236.

Punia, S. (2020). Barley starch modifications: Physical, chemical and enzymatic—A review. *International Journal of Biological Macromolecules*, *144*, 578–585.

Punia, S., Sandhu, K. S., Dhull, S. B., Siroha, A. K., Purewal, S. S., Kaur, M., & Kidwai, M. K. (2020). Oat starch: Physico-chemical, morphological, rheological characteristics and its applications—A review. *International Journal of Biological Macromolecules*, *154*, 493–498.

Radley, J. A. 1976a. Starch Production Technology, Applied Science Publishers Ltd., London, 189–229.

Rasper, V. 1969. Investigations on starches from major starch crops grown in Ghana. II. Swelling and solubility patterns and amyloelastic susceptibility. *Journal of the Science of Food and Agriculture* 20: 642–646.

Ratnayake, W. S., Hoover, R. and Warkentin, T. 2002. Pea starch: Composition, structure and properties — Review. *Starch/Stärke* 54: 217–234.

Robbins, R. J. 2003. Phenolic acids in foods: An overview of analytical methodology. *Journal of Agricultural and Food Chemistry* 51: 2866–2887.

Rudkin, G. O. and Nelson, J. M. 1947. Chlorogenic acid and respiration of sweet potatoes. *Journal of the American Chemical Society* 69: 1470–1475.

Rumbaoa, R. G. O., Cornago, D. F. and Geronimo, I. M. 2009. Phenolic content and antioxidant capacity of Philippine potato (*Solanumtuberosum*) tubers. *Journal of Food Composition and Analysis* 22: 546–550.

Sajeev, M. S. and Balagopalan, C. 2005. Performance evaluation of a multi-purpose mobile starch extraction plant for small scale processing of tuber crops. *Journal of Root Crops* 31: 105–110.

Sajeev, M. S., Sreekumar, J., Vimala, B., Moorthy, S. N. and Jyothi, A. N. 2012. Textural and gelatinization characteristics of white, cream and orange fleshed sweet potato tubers (*Ipomoea batatas*L.). *International Journal of Food Properties* 15: 912–931.

Sandhu, K. S., Singh, N. and Lim, S. T. 2007. A comparison of native and acid thinned normal and waxy corn starches: Physicochemical, thermal, morphological and pasting properties. *LWT Food Science and Technology* 40: 1527–1536.

Scalbert, A., Manach, C., Morand, C., Rémésy, C. and Jimenez, L., 2005. Dietary polyphenols and the prevention of diseases. *Critical Reviews in Food Science and Nutrition* 45: 287–306.

Seog, H. M., Park, Y. K, Nam, Y. J., Shin, D. H. and Kim, J. P. 1987. Physicochemical properties of several sweet potato starches. *Hanguk Nanghwa Hakhoechi* 30: 179–185.

Shaikh, M., Ali, T. M. and Hasnain, A. 2015. Post succinylation effects on morphological functional and textural characteristics of acid-thinned pearl millet starches. *Journal of Cereal Science* 63: 57–63.

Shimelis, E., Meaz M. and Rakshit, M. 2006. Physicochemical properties, pasting behaviour and functional characteristics of flour and starches from improved bean (*Phase-olus-vulgaris l.)* varieties in East Africa. *International Journal of Agricultural Engineering* 8: 1–18.

Shiotahi, I., Nishimura, A., Yamanaka, S., Taki, M. and Yamada, T. 1991. Starch properties of the sweet potato diploid Ipomoea trifida (H.B.K.) Doni and tetraploid hybrids. *Starch/Stârke* 43: 133–137.

Shireen, K. F, Pace, R. D, Egnin, M., Prakash, C. S. 2001. Effects of dietary proteins and trypsin inhibitors on growth and lipid metabolism in hamsters. *Malaysian Journal of Nutrition* 7: 1–14.

Shon, K. J. and Yoo, B. 2006. Effect of acetylation on rheological properties of rice starch. *Starch/Stärke* 58: 177–185.

Silalahi, J. 2001. Free radicals and antioxidant vitamins in degenerative diseases. *Journal of Indonesian Medical Association* 51: 16–21.

Singh, J., Kaur, L. and McCarthy, O. J. 2007. Factors influencing the physico-chemical, morphological, thermal and rheological properties of some chemically modified starches for food applications—A review. *Food Hydrocolloids* 21: 1–22.

Singh, N. and Kaur, L. 2004. Morphological, thermal, rheological and retrogradation properties of potato starch fractions varying in granule size. *Journal of the Science of Food and Agriculture* 84: 1241–1252.

Siroha, A. K. and Sandhu, K. S. 2018. Physicochemical, rheological, morphological, and in vitro digestibility properties of cross-linked starch from pearl millet cultivars. *International Journal of Food Properties* 21: 1371–1385.

Sivonova, M., Tatarkova, Z., Durackova, Z., Dobrota, D., Lehotsky, J., Matakova, T. and Kaplan, P. 2007. Relationship between antioxidant potential and oxidative damage to lipids, proteins and DNA in aged rats. *Physiological Research* 56: 757–764.

Song, H. Y., Lee, S. Y., Choi, S. J., Kim, K. M., Kim, J. S., Han, G. J. and Moon, T. W. 2014. Digestibility and physicochemical properties of granular sweet potato starch as affected by annealing. *Food Science Biotechnology* 23: 23–31.

Song, J., Chung, M. N., Kim, J. T., Chi, H. Y. and Son, J. R. 2005. Quality characteristicsand antioxidative activities in various cultivars of sweet potato. *Korean Journal of Crop Science* 50: 141–146.

Suganuma, T. and Kitahara, K. 1997. Sweet potato starches its properties and utilization in Japan. *International Workshop on Sweet Potato production systems towards 21st century*, Kyushu National Agricultural Experimental Station, Miyazaki, 285–294.

Sun, H., Mu, T., Xi, L., Zhang, M. and Chen, J. 2014. Sweet potato (*Ipomoea batatas L.*) leaves as nutritional and functional foods. *Food Chemistry* 156: 380–389.

Surh, Y. J. 2003. Cancer chemoprevention with dietary phytochemicals. *Nature Reviews Cancer* 3: 768–780.

Swamy, A. T. and Omwenga, J. 2014. Analysis of phytochemical composition of white and purple sweet potato (*Ipomoea batatas (L.) Lam*) root. *Indian Journal of Advance Plant Research* 1: 19–22.

Takeda, Y., Tokunaga, N., Takeda, C. and Hizukuri, S.1986. Physicochemical properties of sweet potato starches. *Starch/Stärke* 38: 345–350.

Takeiti, C. Y., Oliveira, R. A. and Park, K. J. 2011. Sweet potato: Production, morphological and physicochemical characteristics, and technological process. *Fruit, Vegetable and Cereal Science and Biotechnology.* Global Science Books5: 1–18.

Tang, Y., Cai, W. and Xu, B. 2015. Profiles of phenolics, carotenoids and antioxidative capacities of thermal processed white, yellow, orange and purple sweet potatoes grown in Guilin, China. *Food Science and Human Wellness* 4: 123–132.

Teow, C. C., Truong, V. D., McFeeters, R. F., Thompson, R. L., Pecota, K. V. and Yencho, G. C. 2007. Antioxidant activities, phenolic and β-carotene contents of sweet potato genotypes with varying flesh colours. *Food Chemistry* 103: 829–838.

Terahara, N., Shimizu, T., Kato, Y., Nakamura, M., Maitani, T., Yamaguchi, M. A. and Goda, Y. 1999. Six diacylatedanthocyanins from the storage roots of purple sweet potato, Ipomoea batatas. *Bioscience Biotechnology and Biochemistry* 63: 1420–1424.

Thitipraphunkul, K., Uttapap, D., Piyachomkwan, K. and Takeda, Y. 2003. A comparative study of edible canna (*Cannae edulis*) starch from different cultivars. Part I. Chemical composition and physicochemical properties. *Carbohydrate Polymers* 53: 317–324.

Tian, S. J., Rickard, J. E. and Blanshard, J. M. V. 1991. Physicochemical properties of sweet potato starch. *Journal of the Science of Food and Agriculture* 57:459–491.

Tudela, J. A., Cantos, E., Espin, J. C., Tomas-Barberan, F.A. and Gil, M. I. 2002. Induction of antioxidant flavonol biosynthesis in fresh-cut potatoes- Effect of domestic cooking. *Journal of Agricultural and Food Chemistry* 50: 5925–5931.

Tumuhimbise, G. A., Namutebi, A. and Muyonga, J. H. 2009. Microstructure and in vitro beta carotene bioaccessibility of heat processed orange fleshed sweet potato. *Plant Foods for Human Nutrition* 64: 312–318.

Valetudie,J. C., Colonna, J., Bouchet, B. and Gallant, D. J. 1995. Gelatinization of sweet potato, tannia and yam starches. *Starch/Stärke* 47, 298–306.

Walter, J. W. M. and Schadel, W. E. 1981. Distribution of phenols in "Jewel"sweet potato roots. *Journal of Agricultural and Food Chemistry* 29: 904–906.

Walter, JR.W. M., Purcell, A. E. and Hoover M. W. 1976. Changes in amyloid carbohydrates during preparation of sweet potato flakes. *Journal of Food Science* 41: 1374.

Wang, Y. J. and Wang, L. 2002. Characterization of acetylated waxy maize starches preparedunder catalysis by different alkali and alkaline-earth hydroxides. *Starch/Stärke* 54: 25–30.

Wani, I. A., Sogi, D. S., Hamdani, A. M., Gani, A., Bhat, N. A. and Shah, A. 2016. Isolation, composition, and physicochemical properties of starch from legumes: A review. *Starch/Stärke* 68: 834–845.

Waters, D. L., Henry, R. J., Reinke, R. F. and Fitzgerald, M. A. 2006. Gelatinization temperature of rice explained by polymorphisms in starch synthase. *Plant Biotechnology Journal* 4: 115–122.

Wickramasinghe, H. A. M., Takigawa, S., Matsura-Endo, G., Yamauchi, H. and Noda, T. 2009. Comparative analysis of starch properties of different root and tuber crops Sri Lanka. *Food Chemistry* 112: 98–103.

Woolfe, J. A. 1992. Sweet Potato, an Untapped Food Resource. Cambridge, U.K.: Cambridge University Press.

Wu, X., Beecher, G. R., Holden, J. M., Haytowitz, D. B., Gebhardt, S. E. and Prior, R. L. 2004. Lipophilic and hydrophilic antioxidant capacities of common foods in the United States. *Journal of Agricultural and Food Chemistry* 52: 4026–4037.

Wurzburg, O. B. 1964. Acetylation. In: Methods in Carbohydrate Chemistry (Ed. R. L. Whistler) New York: Academic Press, 64: 286–288.

Yadav, B. S., Guleria, P. and Yadav, R. B. 2013. Hydrothermal modification of Indian water chestnut starch: Influence of heat-moisture treatment and annealing on the physicochemical, gelatinization and pasting characteristics. *LWT Food Science and Technology* 53: 211–217.

Zhang, T. and Oates, C. G. 1999. Relationship between a-amylase degradation and physico-chemical properties of sweet potato starches. *Food Chemistry* 65: 157–163.Zheng, Z., Li, Q. B., Lin, L., Tundis, R., Loizzo, M. R., Zheng, B. and Xiao, J. 2016. Characterization and prebiotic effect of the resistant from purple sweet potato. *Molecules* 21: 932–42.

Zhu, F., Yang, X., Cai, Y., Bertoft, E. and Corke, H. 2011. Physicochemical properties of sweet potato starch. *Starch/Stärke* 63: 249–259.

Zoebel, H. F. 1988. Molecules to granules - A comprehensive starch review. *Starch/Stärke* 40: 44–50.

35

Structure and Properties of Lotus Seed Flour and Starch

Sukriti Singh
Maharishi Markenedeshvar (deemed to be) University, Mullana, India
Guru Nanak Dev University, Amritsar, India

CONTENTS

35.1 Introduction

Nelumbo nucifera, a perennial aquatic plant belonging to the *Nelumbonaceae* (water lilly) family is a rhizome popular for its religious, ornamental, medicinal, and nutritional values. The *Nelumbonaceae* family comprises only one genus *Nelumbo* with two species: *Nelumbo lutea Pear.* (American lotus) widely growing from North America to Northern South America and *Nelumbo nucifera Gaertn.* (*Asian lotus*) cultivated from Asia to Northern Australia (Ming et al., 2013; Zhu et al., 2017). It has several common names (e.g, Chinese water lily, water lily, Indian lotus, and sacred lotus) and various botanical names (Nelumbium *nelumbo* [L.] *Druce, Nelumbium speciosum Willd,* and *Nymphaea nelumbo L.*). Countries such as India, China, and Egypt have been known to honor the lotus for its beauty, purity, and perfection (Karki et al., 2013) and has been considered a spiritual symbol by Hindus, Buddhists, and Egyptians (Tungmunnithum et al., 2018). Every part of the lotus—seeds, leaves, stamens, flowers, and rhizomes—have been consumed as functional foods and medicine for more than 2,000 years (Guo, 2009; Velusami et al., 2013; You et al., 2014). Lotus herbal tea prepared from dry lotus leaves in Taiwan, China, Korea, and India are consumed to decrease body fat index and treat diarrhea, insomnia, and nervous disorders. Lotus seed, which is eaten either fresh or dry matured is rich in protein, minerals, amino acids, and unsaturated fatty acids (Bhat & Sridhar, 2008). In folk remedies common in India and China, lotus seeds are used as a cooling agent, diuretic, and as an antidote for the treatment of tissue inflammation (Liu et al., 2004). The rhizome from lotus contains lisensinine, an alkaloid effective in treating fever, sunstroke, stomach problems, and arrhythmia (Ling et al., 2005; Lee et al., 2005). Lotus seeds can be consumed raw, roasted, or in ground form after peeling. In traditional medicine, use of lotus seeds for the treatment of hypertension, cancer, palpitations, insomnia has been recorded (Chen et al., 2007). Lotus is also known for its unique features such as seed longevity, floral thermoregulation, and leaf ultrahydrophobicity (Lin et al., 2019). Extracts of various pharmacologically active constituents, such as flavonoids, tannins glycosides, alkaloids, steroids, and polyphenols, have been isolated from the lotus plant and have been studied for their anti-inflammatory, antiproliferative, antidiabetic, antiobesity, antiangiogenic, and anticancer properties (Chen et al., 2012; Man et al., 2012; Fang et al., 2014; Yamini et al., 2019).

Lotus seeds have been under investigation as a nonconventional source for its protein content as high as 24% in mature desiccated seed (Zheng et al., 2013), starch (31.2%) with no characteristic taste or odor (Fatima et al., 2018), and minerals and vitamins (calcium, phosphorus potassium, iron, thiamine, riboflavin, niacin, and ascorbic acid Sheikh, [2014]). Various studies have been conducted regarding the utilization of lotus seed flour, rhizome powder, lotus starch in the food industry owing to its unique properties and phytochemicals (Hu & Skibsted, 2002; Linshang et al., 2015; Thanushree et al., 2017; Antarkar et al., 2019; Saeed et al., 2020). This chapter summarizes the various aspects of lotus flour and starch for its better utilization in the food industry and provides insight into the unique characteristics that make it an economically benefiting plant.

35.2 Lotus Flour Isolation

Flour from lotus seeds or roots has been studied as a substitute to conventional flours in its application in the food industry, especially in value-added products. Flour from lotus seed has high amylose content and gelatinization temperature and lower swelling point in comparison to flour from the rhizome (Jirukkakul & Sengkhamparn, 2019). Flour obtained after gelatinizing lotus seeds were found to have high antioxidant content as compared to raw seed flour. Lotus seeds have been reported to be rich in proteins (10–14%), making it suitable as an essential food ingredient to combat malnutrition in developing countries (Sridhar & Bhat, 2007). Flour from lotus does not contain any gluten so can be used to develop gluten-free products for celiac disease patients (Zhang et al., 2015). Singthong and Meesit (2017) extracted flour from lotus seeds by boiling seeds in 2% (W/V) sodium hydroxide solution (75 ± 5°C for 2 minutes) and seeds were milled and ground using a 100 mesh sieve to obtain whole flour. The dry grinding method involved the mechanical removal of hard seed coats to obtain seed plumule, which was further ground in an electronic grinder (1000 rpm) and the obtained flour was sieved using a 100 mesh sieve (Nawaz et al., 2020).

35.3 Chemical Composition

The lotus flours are mainly composed of carbohydrates but also have high protein content that emphasizes their value as a vital source of nutrients. Lotus seeds are ovoid, black with a greyish tinge in color, and possessed a tough testa. The cotyledon portion is white with the presence of a green-colored embryo. Flour obtained from lotus seed is yellowish in color (L* = 88.01–89.35, a* = −0.22–(−1.00) and b* = 10.80–14.38) as reported by Singthong and Meesit (2017). The moisture, fat, and fiber content of lotus flour ranges from 4.46–9.68%, 1.99–2.41%, and 1.70–4.87%, respectively (Bhat et al., 2009; Singthong and Meesit, 2017; Antarkhar et al., 2019). The low moisture content of the flour is desirable for better storage stability and baking quality due to slowed respiration and activity of microorganisms (Butt et al., 2004). Fat content is responsible for amylose-lipid complex formation during flour processing and is responsible for resistance starch production. The fiber content of lotus flour is higher than conventional sources of flour (wheat, rice, and maize). High fiber content is beneficial to human health as they minimize the risks of hypertension and illnesses by decreasing cholesterol levels (Rachkeeree et al., 2018). Ash content of lotus flour was found to be 3.30–4.03% (Bhat et al., 2009). Flour characterized by a better ash levels is typically less purified and is widely used as an index of flour purity. The amylose content of lotus flour ranges from 21.27–24.09%. Amylose content of flour affects the baking quality of flour as the ratio of amylose to amylopectin (i.e. amylose content) is an important characteristic in food processing, affecting the pasting property of flour (Araki et al., 2016). Amylose content affects wheat flour thermal and pasting properties as studies showed that wheat flour with low amylose content showed unique flour- starch pasting properties (Graybosch et al., 2000). High gas formation during fermentation and a decreased gas retention coefficient is associated with low flour amylose. Porous crumb structure, large volume, and high moisture content were observed in the bread prepared using the low amylose flour even after 7 days of storage at 4°C (Lee et al., 2001). Minerals detected in the study of lotus flour were calcium (318 mg/100 g), magnesium (43.90 mg/100 g), potassium (23.77 mg/100 g) and iron (16 mg/100 g) (Bhat and Sridhar, 2008; Bhat et al., 2009). High levels of potassium in lotus flour will benefit patients suffering from hypertension and patients with nutritional deficiency caused due to excessive excretion of potassium through body fluids (Siddhuraju and Becker, 2001). The presence of the powerful antioxidant selenium (1.10 mg/100 g) was also detected in flour (Bhat et al., 2009; Singthong and Meesit, 2017). These minerals play a significant role in improving the nutritional value of lotus flour, as iron and selenium are found to strengthen the immune system, whereas zinc and magnesium are known to prevent immunological dysfunctions, growth retardation, muscle degeneration, bleeding disorders, and congenital malformations (Chaturvedi et al., 2004). Lotus seeds are found to be free from heavy metals such as nickel, copper, lead, chromium, and cobalt. Since wheat flours used in the baking industry are usually deficient in some of the essential minerals (e.g., calcium and iron) (Khalil and El-Adawy, 1994), the fortification of wheat flours with lotus seed flour will definitely improve the dietary requirements.

35.4 Functional Properties

The functional properties include foaming, solubility, hydration (water-binding), and emulsification (Chandra and Samsher, 2013). The application of flours from different sources in food systems depends greatly on the information about their physiochemical and functional properties. These functional properties study the hydrophilic/hydrophobic nature of flour and hydrogen bond formed between starch and protein molecules during processing (Lawal, 2004). A functional property represents complex interactions among the structural and molecular conformations of various food components that are important for the maintenance of texture, flavor, taste, and consistency of the food products (Kaur and Singh, 2005; Siddiq et al., 2009). In the food industry, food products undergo various stages of processing and are stored at variable temperatures that may vary the functional properties and the quality of food materials by affecting the intrinsic properties of proteins (Shad et al., 2013).

The swelling power and solubility index of flours from different lotus seed cultivars ranged from 10.54–1.20 and 28.20–36.29%, respectively (Singthong and Meesit, 2017). Swelling power is a temperature-dependent property and is followed by the solubilization of starch granule constituents (Doporto et al., 2011). Water solubility determines the using ability of flour as a stabilizer in dairy, bakery, and emulsified/ground meat food industries (Singh and Singh, 2004). The water and oil absorption capacity of lotus flour ranges from 0.93–1.59 (ml/g) and 1.50–1.55 (ml/g), respectively. The water absorption capacity is the indicator of flour's ability to associate with the limited amount of water available. High water-absorption capacity means a high amount of fiber and starch in the flour. High water-absorption capacity and a high solubility index of lotus seed flours will help it to produce better aqueous flour dispersion in food applications (Torre-Gutiérrez et al., 2008). The water and oil absorption capacity values are the indicators of protein present. Flour's ability to absorb and retain water or oil that improves flavor retention, mouth feel, and reduces fat and moisture losses in baked and extended meat products (Kaur et al., 2007; Bhat & Sridhar, 2008). Bulk density of the lotus flour was recorded as 0.59–0.88% (Hussain et al., 2016; Bhat and Sridhar, 2008). High bulk density of flour is a desirable quality for various food preparations whereas, flours with low bulk density are found to be suitable in formulating high nutrient density foods such as complementary foods (Akpata and Akubor, 1999). The foaming capacity of the flour is protein dependent as protein dispersion is responsible for lowering surface tension at the water-air interface. Foaming properties help in maintaining the structure of creamy foods like ice cream and soups (Shad et al., 2011). The foaming capacity of lotus flour was found to be ranging from 22.66–49.99% (Singthong and Meesit, 2017). The emulsifying activity of lotus flour is found to be 52.00–61.17% and emulsifying stability was found in the range of 79.98–83.81% (Singthong and Meesit, 2017; Hussain et al., 2016; Bhat and Sridhar, 2008). Emulsifying capacity may be a measure of the quantity of emulsified oil (hydrophobic polar liquid) per grams of emulsifying oil (hydrophilic polar liquid) (Alvarez et al., 2009). High emulsion activity and stability make lotus flour desirable for preparing comminuted meats. Another important parameter that affects the stability of the gel system and is responsible for the storage stability of food products at low temperatures is syneresis. Lotus flour has found to have high syneresis and low gel stability, making it less desirable for the food system, which requires freezing or refrigeration.

Least gelation concentration is considered as the gelling ability of flour, which provides a structural matrix for holding water and other water-soluble materials like sugars and flavors. Ratios between lipids, carbohydrates, and proteins vary at least gelation concentration, which helps the flour to work as a binder or to provide consistency to semisolid products. Lotus flour least gelation concentration was found to be 18.0 ± 2.0% (Bhat and Sridhar, 2008).The increasing concentration of proteins in the flour facilitates the gelation properties, which may be due to the enhanced interaction among the binding forces (Lawal, 2004). Gelation capacity is affected by the water requirement between protein and starch in the flour.

35.5 Thermal Properties

A differential scanning calorimeter (DSC) is a thermal analysis apparatus that measures changes in the physical properties of flour with a change in temperature (Haynie, 2008). These measurements are able to provide quantitative and qualitative information regarding physical and chemical changes that involve endothermic (energy-consuming) and exothermic (energy-producing) processes, or changes in heat capacity as most of the food systems are subjected to heating or cooling during processing. DSC showed a single symmetrical endotherm during the gelatinization of lotus flour. Onset, peak, and conclusion temperatures of gelatinization (T_o, T_p, and T_c) for the different cultivars of lotus seed flours were studied by Singthong and Meesit (2017) and found them ranging from 71.62–75.37°C, 76.28–79.04°C, and 80.48–82.37°C, respectively. The flour containing high amylose content gelatinizes at a high temperature (Huijbrechts et al., 2008). The same phenomenon was found for gelatinization temperature in lotus and rice flour (72.86–80.15°C), which was higher than tapioca starch (62.99–79.47°C), corn starch (68.41–77.14°C), or waxy rice starch (60.49–71.67°C). The high temperature ensures complete gelatinization and pasting, which is desirable in products that need delayed pasting, like canned foods (Torre-Gutiérrez et al., 2008). The gelatinization enthalpy of lotus seed flour was found to be lower than waxy rice flour (5.98 J/g), wheat flour (5.45 J/g), or rice flour (5.34 J/g).

35.6 Pasting Properties

Pasting properties are the indicator of flour's ability to form a paste; starch granules present in the flour gelatinize at various temperatures for different flour (Rincon & Padilla 2004). Pasting temperatures of the different lotus seed flours ranged from 80.31–83.47°C (Singthong and Meesit, 2017), flour from rhizome shows high pasting temperature in comparison to lotus seed flour (89.75°C) (Hussain et al., 2016). Peak viscosity for flour from rhizome and seeds ranges from (1,409–2,607 cP) and final viscosities ranged from 2,232.50–2,456 cP (Singthong and Meesit, 2017; Hussain et al., 2016). The breakdown values of flours from seeds were found to be lower (151 cP) than flour from the rhizome (461 cP). Low breakdown stability shows paste stability. The setback of lotus seed flours ranged from 907.33–1,396.08 cP (Singthong and Meesit, 2017), and the setback found for rhizome flour (842 cP) was low, which indicated a lower tendency to retrograde. The high pasting temperature of lotus flour is the indicator of flour stability during various heating processes and low viscosity values (hold, final, bulk, and setback viscosities) of lotus seed flours indicate it will remain stable during the cooling process

35.7 Lotus Starch Isolation

Lotus rhizome and seeds attract attention for a number of years owing to their high starch content (500g/ kg dry basis) and amylose content (40% w/w), as a nonconventional source of starch (Guo, 2009). Starch from lotus seeds and rhizome found its application like other starches as a thickening, gelling, and stabilizing agents in the food industry (Guo et al., 2015). In China, lotus starch is commercially available and is used in traditional confectionery, fast foods, and food additives (Zhong et al., 2007; Man et al., 2012). Lotus rhizomes can be divided into two classes: First class is characterized as lotus rhizome with low starch and crude fiber content, high sugar and water content (responsible for its crispiness); the second class is characterized by low water content and high starch (Wattebled et al., 2002). High starch will give a soft texture to the rhizome and increases the viscosity of food products. A significant difference between the structural and functional properties of lotus starch with other starches has been studied (Man et al., 2012; Yu et al., 2013; Zhang et al., 2014). The use of native starches is limited due to their high resistance and viscosity, low light transmittance, and susceptibility to thermal decomposition, which make them less desirable in the food industry. Modification of native starches by using the enzymatic treatment, chemical, or physical treatment changes the properties of starch, making it suitable for industrial use (Singh et al., 2007). Various modifications such as gamma irradiation (Punia et al., 2020), microwave treatment (Nawaz et al., 2020), oxidation and cross-linking (Sukhija et al., 2016), and ultra-high pressure (Guo et al., 2015) on lotus seed and rhizome starches have been done to see the change in characteristics of lotus starch and make it a preferable candidate in starch-based formulations.

Isolation methods for rhizome and seed starch from lotus are different due to the difference in their composition and physical structure (Man et al., 2012; Guo et al., 2015). For isolating starch from lotus rhizomes, rhizomes are to be washed, brushed, peeled, and sliced in small pieces immediately. These small pieces are then homogenized with cold water in a household blender. The slurry obtained is then filtered through gauze and the fibrous residue is again homogenized, while the filtrate is collected in the glass beaker and further filtered using a 100, 200, and 300 mesh sieve and centrifuged. The process of centrifugation separation is repeated several times to remove the yellow gel layer on top. White starch granule pellets are carefully scraped and, finally, the starch is collected and air-dried (Yu et al., 2013). Lotus seeds are removed from the kernel and steeped in water containing 0.16% sodium hydrogen sulphite for 12 hours at 50°C. The seeped seeds are then ground using a laboratory blender. The ground slurry is then sieved through a muslin cloth and the filtrate slurry is allowed to stand for 1 hour, the supernatant is to be removed using suction, and settled starch is centrifuged three to four times to collect the white layered starch. The dried starch is then vacuum dried at 50°C for 6–7 hours (Punia et al., 2020).

35.8 Chemical Composition

Amylose content plays a significant part in the internal quality of starch and is responsible for determining the properties of starch, which render starch suitable for industrial applications. Amylose content of lotus starch has been found varying depending upon its growing condition, method of quantification, and genetics. Amylose content of lotus rhizome and seed starch ranges from 18.5–23.9% and 27.7–38.9%, respectively (Ito et al., 1996; Man et al., 2012). Other components like protein and ash content also show variation in starches from rhizome and seeds (Table 35.1).

35.9 Granular Structure

35.9.1 Polymorph and Crystallinity

X-ray diffraction interprets the distribution of crystal sizes and lattice disorder found in crystalline samples like starches, so it is an effective method in studying the structure of starches (Yu et al., 2013). X-ray diffraction shows three types of polymorphic forms: A type (cereals), B type (tubers and roots), and

TABLE 35.1

Chemical Composition of Lotus Starch From Rhizome and Seeds

Source	Moisture (%)	Amylose (%)	Protein (%)	Ash (%)	Reference
Seed	-	20.2 ± 0.18	-	-	Punia et al., 2020
Seed	-	38.89 ± 0.68	-	-	Man et al., 2012
Rhizome	-	23.86 ± 0.28	-	-	Man et al., 2012
Rhizome	-	20.60 ± 0.40	-	-	Sukhija et al., 2016
Rhizome	9 ± 0.05	21.16 ± 0.29	0.16 ± 0.01	0.4 ± 0.02	Syed and Singh,2013
Rhizome	10.08 ± 0.27	20.06 ± 0.40	0.15 ± 0.02	0.2 ± 0.1	Sukhija et al., 2015
Rhizome	12.5 ± 0.1	30.0 ± 0.05	0.14 ± 0.05	0.1 ± 0.1	Gani et al., 2013

C type (pulses), among which A-type and B-type polymorphic forms are common among native starches (Yu et al., 2013). Studies on lotus rhizome starch regarding the polymorphic form of starch have been found to be inconclusive, as studies show the presence of all the three forms: A-type polymorph (Yu et al., 2013; Sukhija et al., 2016), B-type (Geng et al., 2007; Gani et al., 2013) and C-type (Man et al., 2012; Yu et al., 2013; Nawaz et al., 2020). The discrepancy may be attributed to environmental factors, genetic factors, or misidentification of correct species. Various techniques have been used to locate the position of granules showing different polymorph form and found that the A-type polymorph is mostly located in the periphery hilum region of rhizome and B-type is located at the opposite direction (Cai et al., 2014). On calculating the degree of crystallinity, variations have been found as it was calculated as 35–37% (Yu et al., 2013), 28% (Sukhija et al., 2016), or 23–29% (Man et al., 2012). These discrepancies can be attributed to quantification methods and genetics. Lotus seed starches showed the A-type polymorph (Man et al., 2012; Cai and Wei, 2013; Punia et al., 2020). The degree of crystallinity has been found to be 29% for lotus seed starches (Man et al., 2012)

35.9.2 Morphology

A scanning electron microscope helps in characterizing starch on the basis of granule size and shape (Figure 35.1). Lotus rhizome starches are found to be oval to elongate in shape and varying in size. Lotus starch showed a bimodal size distribution (Geng et al., 2007). The size of starch granules from the rhizome is found to be 10–50μm (long axis) and 10–35μm (short axis) (Sukhija et al., 2016). Lotus

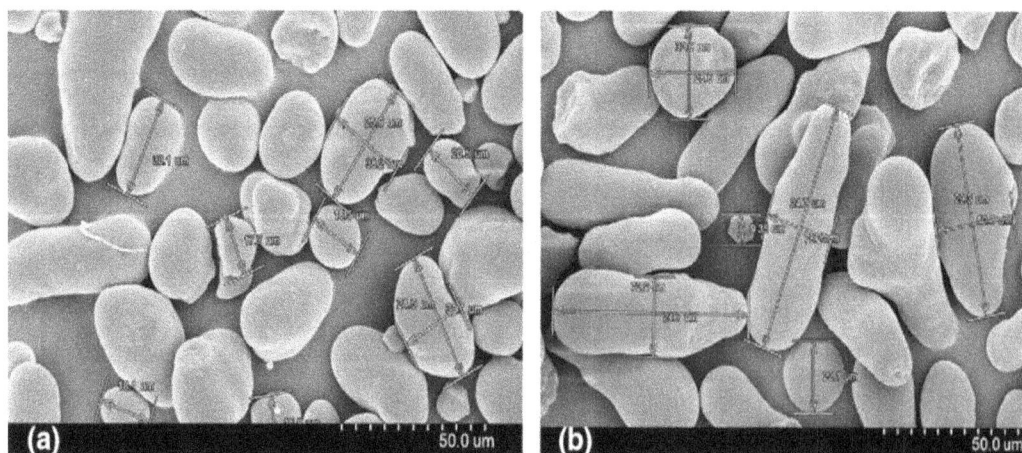

FIGURE 35.1 Scanning electron microscope from rhizome starch from two cultivars (a) Meirenhong and (b) Wawalian. (Reprinted with permission from Yu et al., 2013).

TABLE 35.2

Pasting Properties of Lotus Rhizome and Seed Starches

Plant Tissue	Instrument	Starch Conc. (%)	Peak Viscosity	Breakdown Viscosity	Setback Viscosity	Reference
Rhizome	RVA	6	1,136 cP	733 cP	295 cP	Yu et al., 2013
Rhizome	RVA	6	8,060 cP	5,320 cP	997 cP	Syed and Singh, 2013
Rhizome	RVA	10	2,718 cP	428 cP	1,103 cP	Gani et al., 2013
Rhizome	RVA	10.7	6,214 cP	4,123 cP	976 cP	Sukhija et al., 2016
Rhizome	BV	6	950 BU	352 BU	438 BU	Akuzawa and Kawabata, 2003
Seed	Rheometer	8	398 mPa.s	88 mPa.s	401 mPa.s	Punia et al., 2020
Seed	RVA	10	1,337 mPa.s	93 mPa.s	670 mPa.s	Guo et al., 2015
Seed	BA	6	610 BU	-	-	Ito et al., 1996

RVA = Rapid visco analyzer, BA = Brabender amylograph, BV = Brabender viscograph, BU = Brabender unit.

seed starches are found to be mostly oval in shape. The size of starch granules from seeds is found to be 9.7µm (long axis) and 6.7µm (short axis) (Man et al., 2012). The ratio of long/short axis in starch granules affects the shape, 1.1 (round shape), 1.1–1.4 (oval shape), and a ratio above 1.4 (elongated shape) (Yu et al., 2013). Various treatments on starches showed significant changes in shape and size. Starch exposed to irradiation (10–20 kGy) showed the appearance of dents and cavities and rupturing of few granules at a high dose (Punia et al., 2020). Lotus seed resistant starch shows the complete destruction of the oval shape and a block structure appeared (Zhang et al., 2014).

35.10 Rheological Properties

Pasting properties are the measure of the change in starch properties when subjected to heating and cooling (50–95°C) to a programmed cycle under a constant shearing force (Zhu, 2015). Pasting properties of lotus rhizome and seeds from various researchers showed significant differences; these differences can be attributed to the different botanical sources, environmental factors, amylose content, granule size, molecular structure, amylose leaching, and amylopectin molecular structure (Table 35.2). Several different instruments such as Rapid Visco Analyzer (RVA), micro Visco amylograph, and Brabender Visco amylograph have been used for studying pasting properties. Among these instruments, the RVA has been found to be most effective as it requires a smaller sample size and output is fast (Cozzolino et al., 2016). These instruments have been shown to have variable rotational speed, spindle geometry, and temperature time programs, the extent of shear, and viscosity units (Gani et al., 2013). Interpreting the data of pasting properties that have been studied using different instruments can be difficult due to a lack of correlation between them. A slight change in the concentration of starch slurry also gives variable data, which is difficult to compare (Gani et al., 2013; Yu et al., 2013). Pasting properties of lotus rhizome and seed starch have been compared with other starches and found to be higher than cassava, maize, sweet potatoes, and arrowroot starch (Akuzawa & Kawabata, 2003; Geng et al., 2007; Sukhija et al., 2016).

35.11 Starch Modifications

Inherent weaknesses in native starch such as high viscosity, retrogradation tendency, low shear strength andease of thermal decomposition, (crystallization and aging of gels) limit its use in other applications. Physical, chemical, or enzymatic modifications of starch alter the properties of starch, which make them more suitable for food industries. Lotus rhizome and seed starches have been subjected to various modifications (Table 35.3)

TABLE 35.3

Modifications of Starches Isolated From Lotus Rhizome and Seeds

Starch Source	Modification Type	Experimental Condition	Major Finding	Reference
Chemical	Modification			
Rhizome	Oxidation	Sodium hypochlorite (2.5% active chlorine w/w)	Oxidation caused reduction in amylose content and improved its solubility, significant increase in transition temperatures, but decrease in enthalpy	Sukhija et al., 2016
Rhizome	Cross-linking	Starch (100 g) in water (150 g), sodium trimetaphosphate (3 g) was mixed at pH 10. Added to it 0.1 N NaOH (45°C for 2 hours)	Cross-linking starch showed lower peak, trough, and breakdown and setback viscosity.	Sukhija et al., 2016
Rhizome	Acid hydrolysis	Starch (2 g) was hydrolyzed with 100 mL HCl (2.2 M) at 40°C for 96 hours	Starch Polymorph changed from C- to A-type. NMR analysis showed that acid hydrolysis increased the proportion of double helix and decreased that of amorphous region	Cai et al., 2014
Rhizome	Hydroxypropylation	Starch (50 g) + water (110 ml) + Na_2So_4 (10 g) was mixed with propylene oxide (5 ml) at pH 11.3 and kept at 35°C for 2 hours. (Terminated by adjusting pH 5.3)	Hydroxypropylation increased swelling factor and decreases pasting viscosity, retrogradation, gel firmness and shearing stability	Gunaratne and Corke, 2007
Rhizome	Acetylation	Starch (117.5 g) + water (235 ml) (pH = 8). Acetic anhydride (2.03g) was added dropwise for 30 minutes (Terminated by adjusting pH at 6.5)	Acetyltion increased swelling, solubility, light transmittance, and decreased pasting viscosity	Sun et al., 2016)
Physical	**Modification**			
Rhizome	Microwave irradiation	Starch was microwaved at different treatment times (1, 2, 3, 4, and 5 minutes at low intensity of 200 W). Treatment was discontinued for each minute after 30 seconds followed by through mixing	Microwaved starch showed increased starch crystallinity, water holding capacity and decreased amylose content, swelling capacity	Nawaz et al., 2020
Seed	Microwave irradiation	Starch slurry (30% dry basis) was microwave treated (2.4 w/g, 4.0 w/g, 6.4 w/g, and 8.0 w/g) until fully gelatinized	Microwave treatment showed decrease in amylose leaching swelling power, molecular properties and increase in resistant starch, slowly digestible starch, and crystalline region.	Zeng et al., 2016

(Continued)

TABLE 35.3 (Continued)

Starch Source	Modification Type	Experimental Condition	Major Finding	Reference
Rhizome	γ radiation	Starch was subjected to γ radiations at different doses (5 kGy, 10 kGy, 15 kGy, and 20 kGy)	Irradiation increased solubility, light transmittance and decrease in swelling power, pasting properties, and gelatinization temperature	Punia et al., 2020
Rhizome	γ radiation	Starch was subjected to γ radiations at different doses (5 kGy, 10 kGy, and 20 kGy)	Irradiation increased amylose leaching, gel transmittance, carboxyl content and water holding capacity and decrease in swelling power, pasting viscosities and degree of crystallinity	Gani et al., 2013
Seed	High pressure	Starch (15% w/w) was vacuum packed and subjected at pressure of 100–600 MPa for 30 minutes (at 1 MPa). Starch was also subjected at 600 MPa for 1 minute	Starch showed increase in granule size, change in polymorph type and decrease in swelling power, solubility at high temperatures, starch retrogradation, and gelatanization temperature.	Guo et al., 2015

35.11.1 Chemical Modifications

Lotus rhizome starch has been subjected to oxidation, cross-linking, and dual modification (cross-linked oxidized and oxidized cross-linked) (Sukhija et al., 2016). Modifications showed a decrease in viscosity, amylose content, and an increase in swelling power, solubility, and light transmittance as compared to native starches. Starch granules showed rough surfaces in the case of oxidized and dual modified starches and smooth surfaces in case of native and cross-linked starches. The difference was also found in dual modifications as oxidized-cross linked starch showed low viscosity values and lower thermal resistance, whereas cross-linked oxidized starch showed better pasting and thermal properties making it more suitable for food preparations such as thickeners, stabilizers, emulsifying agents, and biodegradable coatings.

Acid hydrolysis of lotus rhizome and seed starch at various conditions was studied (Srichuwong et al., 2005; Kaur et al., 2011; Man et al., 2012; Cai et al., 2014). Acid treatment showed a decrease in the swelling power, solubility, water-binding capacity, and pasting viscosity of lotus rhizome starch, while an increase in the paste syneresis (Kaur et al., 2011). Acid hydrolysis of lotus rhizome starch showed that small granules of starches are more susceptible to hydrochloric acid (HCl) due to their higher surface-to-volume ratio and lower contents of ordered structure in the external region observed in Fourier transform infrared spectroscopy (FTIR) (Lin et al., 2015). Lotus rhizome starch when subjected to hydroxypropylation increased the swelling power and α-amylase susceptibility, while reducing its acid tolerance, pasting viscosity, retrogradation, shearing stability, gelatinization temperatures, and gel firmness (Gunaratne and Corke, 2007). Acetylated lotus rhizome starch showed an increase in swelling power, solubility, light transmittance, freeze-thaw stability, and a decrease in pasting viscosities (Sun et al., 2016).

35.11.2 Physical Modifications

Lotus rhizome and seed starch has been subjected to microwave treatment (Zeng et al., 2016; Nawaz et al., 2020). Microwave treated rhizome starch showed a decrease in oil holding capacity, swelling capacity, amylose content, and an increase in water holding capacity and crystallinity. Microwave treatment causes disruption of morphology and changes the C-type crystalline structure (C_A to C_B- type) in lotus starch. Seed starch showed a reduction in amylose leaching, swelling power, and molecular properties with increase in microwave power. Changes in the crystalline region and starch-water interaction due to changed water distribution of starch granules was also observed (Zeng et al., 2016).

Lotus rhizome and seed starches were subjected to gamma (γ)-radiations (Gani et al., 2013; Punia et al., 2020). With an increase in dose (20 kGy), a significant increase was seen in amylose leaching, carbonyl content, water absorption capacity, and gel transmittance and decreases in pasting properties, apparent amylose content, and syneresis. Gamma-irradiation induces granular damages and degrades starch chains through free radical reactions (Zhu, 2016). Granule polymorph remains the same, however, upon irradiation, some dents and hollows were observed (Gani et al., 2013). Lotus seed starch was subjected to high-pressure up to 600 MPa (Guo et al., 2013; Guo et al., 2015). High pressure increased peak viscosity, but decreased breakdown viscosity and set back viscosity. A change in polymorph from C to B type was also observed under high pressure but the granular structure showed no change. Lack of amylose leaching under high pressure may be the reason behind reduced retrogradation.

35.12 Food Applications

There have been only a few reports on the applications of lotus flour and starches (Table 35.4). This can be attributed to the widely explored commercially successful sources such as maize, wheat, rice, potato, and cassava. Although, there have been a lack of comparative studies of lotus flour and starches from those of the available sources, it should be explored for its effective utilization in food industries.

35.13 Conclusion

Nelumbo nucifera is an underutilized plant that has gained popularity due to its nutritional, medicinal, and historical importance. Lotus is a good source of starch, minerals, dietary fiber, and dietary proteins.

TABLE 35.4

Utilization of Lotus Flour and Lotus Starch

Plant Source	Uses	Major Finding	References
Seed Flour	Bread (10–40% substitution)	Lotus flour substituted breads had firm texture, lower loaf volume, increased protein and fiber content. Hedonic scale showed a 10% substitution of lotus flour gives same result as control.	Singthong & Meesit, 2017
Seed flour	Noodles (5–15% substitution)	5% pregelatinized lotus flour can be substituted in the noodle recipe rendering the noodles with a low-fat, high- fiber, ash, and phenolic content.	Jirukkakul & Sengkhamparn, 2019
Rhizome starch	Pudding ingredients	Puddings with acetylated lotus rhizome starch were found to have higher sensory scores, improved color, mouth melting, and smoothness compared to native starch pudding.	Sun et al., 2016
Seed starch	Edible films	Edible starch films formed from lotus seed starch (nisin [0.02%] and glycerol [1.5%] added) showed reduced rates of browning, softening and microbial growth for freshly cut pineapple.	Lin et al., 2011

Flour and starch obtained from lotus have been found to exhibit appreciable functional properties in comparison to conventional grains or legumes, indicating its potential utility in various food industries to improve the nutritional value of the product and to develop value-added products. Its negligible fat content contributes to being a very interesting constituent for fat-free diets. The development of flour blends with lotus could improve the utilization of indigenous food crops in those countries where wheat is not the main crop. Appropriate processing methods of lotus seeds are essential to exploit the nutritional and medicinal potential.

REFERENCES

Akpata, M. I. and Akubor, P. I. 1999. Chemical composition and selected functional properties of sweet orange (*Citrus sinensis*) seed flour. *Plant Foods for Human Nutrition* 54: 353–362.

Akuzawa, S. and Kawabata, A. 2003. Relationship among starches from different origins classified according to their physicochemical properties. *Journal of Applied Glycoscience* 50:121–126.

Alvarez, G., Poteau, S., Argillier, J. F., Langevin, D. and Salager, J. L. 2009. Heavy oil water interfacial properties and emulsion stability: Influence of dilution. *Energy Fuels* 23: 294–99.

Antarkar, S., Gabel, S., Tiwari, S., Mahajan, S. and Azmi, R. M. 2019. Evaluation of nutritional and functional properties of partially substituted whole wheat flour with taro root and lotus seed flour (Composite flour). *Pharma Innovation* 8:125–128.

Araki, E., Ashida, K., Aoki, N., Takahashi, M. and Hamada, S. 2016. Characteristics of rice flour suitable for the production of rice flour bread containing gluten and methods of reducing the cost of producing rice flour. *Japan Agricultural Research Quarterly* 50: 23–31.

Bhat, R. and Sridhar, K. R. 2008. Nutritional quality evaluation of electron beam-irradiated lotus *(Nelumbo nucifera)* seeds. *Food Chemistry* 107: 174–184.

Bhat, R., Sridhar, K. R., Alias, A. K., Young, C. C. and Arun, B. A. 2009. Influence of gamma radiation on the nutritional and functional qualities of lotus seed flour. *Journal of Agricultural and Food Chemistry* 57: 9524–9531.

Butt, M. S., Nasir, M., Akhtar, S. and Sharif, M. K. 2004. Effect of moisture and packaging on the shelf life of wheat flour. *Internet Journal of Food Safety* 4: 1–6.

Cai, C. and Wei, C. 2013. In situ observation of crystallinity disruption patterns during starch gelatinization. *Carbohydrate Polymers* 92: 469–478.

Cai, J., Cai, C., Man, J., Yang, Y., Zhang, F. and Wei, C. 2014. Crystalline and structural properties of acid-modified lotus rhizome C-type starch. *Carbohydrate Polymers* 102: 799–807.

Chandra, S. and Samsher. 2013. Assessment of functional properties of different flours. *African Journal of Agricultural Research* 8: 4849–4852.

Chaturvedi, U., Shrivastava, R. and Upreti, R. 2004). Viral infections and trace elements: A complex interaction. *Current Science*, 87: 1536–1554.

Chen, S., Fang, L., Xi, H., Guan, L., Fang, J., Liu, Y. and Li, S. H. 2012. Simultaneous qualitative assessment and quantitative analysis of flavonoids in various tissues of lotus (*Nelumbo nucifera*) using high-performance liquid chromatography coupled with triple quad mass spectrometry. *Analytica Chimica Acta* 724:127–135.

Chen, Y., Fan, G., Wu, H., Wu, Y. and Mitchell A. 2007. Separation, identification and rapid determination of liensine, isoliensinine and neferine from embryo of the seed of *Nelumbo nucifera* Gaertn. by liquid chromatography coupled to diode array detector and tandem mass spectrometry. *Journal of Pharmaceutical and Biomedical Analysis* 43: 99–104.

Cozzolino, D., Degner, S. and Eglinton, J. 2016. The use of the rapid visco analyser (RVA) to sequentially study starch properties in commercial malting barley (*Hordeum vulgare* L.). *Journal of Food Measurement and Characterization* 10: 474–479.

Doporto, M. C., Mugridge, A., García, M.A. and Viña, S. Z. 2011. Pachyrhizus ahipa (Wedd.) Parodi roots and flour: Biochemical and functional characteristics. *Food Chemistry* 126: 670–1678.

Fang, F. F., Du, S. G., Dai, X. M., Guo, X. L., Chen, H. W. and Luo, L. P. 2014. Rapid analysis alkaloids in lotus seeds by extractive electrospray ionization mass spectrometry. *Chemical Journal of Chinese University* 4: 730–735.

Fatima, T., Iftikhar, F. and Hussain, S. Z. 2018. Ethno-medicinal and pharmacological activities of lotus rhizome. *Journal of Pharmaceutical Innovations* 7: 238–241.

Gani, A., Gazanfar, T., Jan, R., Wani, S. M. and Masoodi, F. A. 2013. Effect of gamma irradiation on the physicochemical and morphological properties of starch extracted from lotus stem harvested from Dal lake of Jammu and Kashmir, India. *Journal of the Saudi Society of Agricultural Sciences* 12:109–115.

Geng, Z., Chen, Z. and We, Y. 2007. Physicochemical properties of lotus (*Nelumbo nucifera Gaertn.*) and kudzu (*Pueraria hirsute Matsum.*) starches. *International Journal of Food Science and Technology* 42: 1449–1455.

Graybosch, R. A., Guo, G. and Shelton, D. 2000. Aberrant falling numbers of waxy wheats independent of α-Amylase activity 1. *Cereal Chemistry* 77:1–3

Gunaratne, A. and Corke, H. 2007. Functional properties of hydroxypropylated, cross-linked, and hydroxy-propylated cross-linked tuber and root starches. *Cereal Chemistry* 84: 30–37.

Guo, H. B. 2009. Cultivation of lotus (*Nelumbo nucifera Gaertn. ssp. nucifera*) and its utilization in China. *Genetic Resources and Crop Evolution* 56: 323–330.

Guo, Z. B., Liu, W. T., Zeng, S. X. and Zheng, B. D. 2013. Effect of ultra high pressure processing on the particle characteristics of lotus-seed starch. *Chinese Journal of Structural Chemistry* 32: 525–532.

Guo, Z., Zeng, S., Lu, X., Zhou, M., Zheng, M. and Zheng, B. 2015. Structural and physicochemical properties of lotus seed starch treated with ultra-high pressure. *Food Chemistry* 186: 223–230.

Haynie, D. T. 2008. Biological Thermodynamics. Cambridge, UK: Cambridge University Press.

Hu, M. and Skibsted, L.H. 2002. Antioxidative capacity of rhizome extract and rhizome knot extract of edible lotus (*Nelumbo nuficera*). *Food Chemistry* 76: 327–333.

Huijbrechts, A. M. L., Desse, M., Budtova, T., Franssen, M. C. R., Visser, G. M., Boeriu, C. G. and Sudhölter, E.J.R. 2008. Physicochemical properties of etherified maize starches. *Carbohydrate Polymers* 74:170–184.

Hussain, S., Shah, F., Hameed, O., Naik, H. R. and Reshi, M. 2016. Functional behavior of lotus rhizome harvested from high altitude Dal Lake of Kashmir. *Indian Journal of Ecology* 43:835–837.

Ito, T., Murase, M., Yamada, T. and Namiki, K. 1996. Properties of lotus-seed starch. *Journal of Applied Glycoscience* 43:7–13

Jirukkakul, N. and Sengkhamparn, N. 2019. Physicochemical properties and potential of lotus seed flour as wheat flour substitute in noodles. *Songklanakarin Journal of Science and Technology* 40:1354–1360

Karki, R., Jeon, E. R. and Kim, D. W. 2013. *Nelumbo nucifera* leaf extract inhibits neointimal hyperplasia through modulation of smooth muscle cell proliferation and migration. *Nutrition* 29: 268–275.

Kaur, M. and Singh, N. 2005. Studied on functional, thermal and pasting properties of flours from different chickpea (*Cicer arietinum* L.) cultivars. *Food Chemistry* 91: 403–411.

Kaur, M., Oberoi, D. P. S., Sogi, D. S. and Gill, B. S. 2011. Physicochemical, morphological and pasting properties of acid treated starches from different botanical sources. *Journal of Food Science and Technology* 48: 460–465.

Kaur, M., Sandhu, K. S. and Singh, N. 2007. Comparative study of the functional, thermal and pasting properties of flours from different field pea and pigeon pea cultivars. *Food Chemistry* 104:259–267.

Khalil, A. and El-Adawy, T. A. 1994. Isolation, identification and toxicity of saponin from different legumes. *Food Chemistry* 50:197–201.

Lawal, O. S. 2004. Composition, physicochemical properties and retrogradation characteristics of native, oxidised, acetylated and acid-thinned new cocoyam (*Xanthosoma sagittifolium*) starch. *Food Chemistry* 85: 205–218.

Lee, H. K., Choi, Y. M., Noh, D. O. and Suh, H.J. 2005. Antioxidant effect of Korean traditional lotus liquor (Yunyupju). *International Journal of Food Science and Technology* 40: 709–715.

Lee, M. R., Swanson, B. G. and Baik, B. K. 2001. Influence of amylose content on properties of wheat starch and breadmaking qualities of starch and gluten blends. *Cereal Chemistry* 78:701–706.

Lin, L., Huang, J., Zhao, L., Wang, J., Wang, Z. and Wei, C. 2015. Effect of granule size on the properties of lotus rhizome C-type starch. *Carbohydrate Polymers* 134: 448–457.

Lin, Y. Y., Zheng, B. D., Zeng, S. X., Zhang, F. and Wu, S.Z. 2011. Effects of edible lotus seed starch coating on quality of fresh-cut pineapple. *Journal of Fujian Agriculture and Forestry University* 40: 205–210.

Lin, Z., Zhang, C., Cao, D., Damaris, R. N. and Yang, P. 2019. The latest studies on lotus (*Nelumbo nucifera*)-an emerging horticultural model plant. *International Journal of Molecular Science* 20: 3680.

Ling, Z. Q., Xie, B. J. and Yang, E. L. 2005. Isolation, characterization, and determination of antioxidative activity of oligomeric procyanidins from the seedpod of *Nelumbo nucifera* Gaertn. *Journal of Agricultural and Food Chemistry* 53: 2441–2445.

Lingshang, L., Jun H., Lingxiao, Z., Juan, W., Zhifeng, W. and Cunxu, W. 2015. Effect of granule size on the properties of lotus rhizome C-type starch. *Carbohydrate. Polymer* 134: 448–457.

Liu, C. P., Tsai, W. J., Lin, Y. L., Liao, J. F., Chen, C. F. and Kuo, Y. C. 2004. The extracts from *Nelumbo nucifera* suppress cell cycle progression, cytokine genes expression, and cell proliferation in human peripheral blood mononuclear cells. *Life Sciences* 75:699–716.

Man, J., Cai, J., Cai, C., Xu, B., Huai, H. and Wei, C. 2012. Comparison of physicochemical properties of starches from seed and rhizome of lotus. *Carbohydrate Polymers* 88:676–683.

Ming, R., VanBuren, R. and Lin, Y. 2013. Genome of the long-living sacred lotus (*Nelumbo nucifera Gaertn.*). *Genome Biology* 10:14(5):R41

Nawaz, H., Akbar, A., Andaleeb, H., Shah, M., Amjad, A., Mehmood, A. and Mannan, R. 2020. Microwave-induced modification in physical and functional characteristics and antioxidant potential of *Nelumbo nucifera* rhizome starch. *Journal of Polymers and the Environment* 28: 2965–2976

Punia, S., Dhull, S. B., Kunner, P. and Rohilla, S. 2020. Effect of γ-radiation on physico-chemical, morphological and thermal characteristics of lotus seed (*Nelumbo nucifera*) starch. *International Journal of Biological Macromolecules* 157: 584–590.

Rachkeeree, A., Kantadoung, K., Suksathan, R., Puangpradab, R., Page, P. A. and Sommano, S. R. 2018. Nutritional compositions and phytochemical properties of the edible flowers from selected zingiberaceae found in Thailand. *Frontiens in Nutrition* 5:3.

Rincon, A. M. and Padilla, F. C. 2004. Physicochemical properties of breadfruit (*Artocarpus altilis*) starch from Margarita island, Venezuela. *Archivos latinoamericanos de nutricion* 8: 95–97

Saeed, S. M. G., Tayyaba, S., Ali, S.A., Tayyab, S., Sayeed, S. A., Ali, R., Mobin, L. and Naz, S. 2020. Evaluation of the potential of Lotus root *(Nelumbo nucifera)* flour as a fat mimetic in biscuits with improved functional and nutritional properties, *CyTA Journal of Food* 18: 624–634.

Shad, M., Nawaz, H., Hussain, M. and Yousuf, B. 2011. Proximate composition and functional properties of rhizomes of lotus (*Nelumbo nucifera*) from Punjab, Pakistan. *Pakistan Journal of Botany* 43.895–904.

Shad, M. A., Nawaz, H., Siddique, F., Zahra, J. and Mushtaq, A. 2013. Nutritional and functional characterization of seed kernel of lotus (*Nelumbo nucifera*): Application of response surface methodology. *Food Science and Technology Research* 19:163–172

Sheikh, S. A. 2014. Ethno-medicinal uses and pharmacological activities of lotus (*Nelumbo nucifera*). *Journal of Medicinal Plants Studies* 2: 42–46.

Siddhuraju, P. and Becker, K. 2001. Effect of various domestic processing methods on antinutrients and in vitro protein and starch digestibility of two indigenous varieties of Indian tribal pulse, *Mucuna pruriens* Var. utilis. *Journal of Agricultural and Food Chemistry 49*: 3058–3067

Siddiq, M., Nasir, M., Ravi, R., Dolan, K. D. and Butt, M. S. 2009. Effect of defatted maize germ flour addition on the physical and sensory quality of wheat bread. *LWT Food Science and Technology* 42:464–470.

Singh, J., Kaur, L. and McCarthy, O. J. 2007. Factors influencing the physico–chemical, morphological, thermal and rheological properties of some chemically modified starches for food applications—A review. *Food Hydrocolloids* 21:1–22.

Singh, J. and Singh, N. 2004. Effect of process variables and sodium alginate on extrusion behavior of nixtamalized corn grit. *International Journal of Food Properties* 7: 329–340.

Singthong, J. and Meesit, U. 2017. Characteristic and functional properties of Thai lotus seed (*Nelumbo nucifera*) flours. *International Food Research Journal* 24: 1414–1421.

Srichuwong, S., Isono, N., Mishima, T. and Hisamatsu, M. 2005. Structure of lintnerized starch is related to X-ray diffraction pattern and susceptibility to acid and enzyme hydrolysis of starch granules. *International Journal of Biological Macromolecules* 37: 115–121.

Sridhar, K. and Bhat, R. 2007. Lotus—A potential nutraceutical source. *Journal of Agricultural Technology* 3:143–155.

Sukhija, S., Singh, S. and Riar, C. S. 2016. Isolation of starches from different tubers and study of their physicochemical, thermal, rheological and morphological characteristics. *Starch/ Stärke* 68:160–168.

Sun, S., Zhang, G. and Ma, C. 2016. Preparation, physicochemical characterization and application of acetylated lotus rhizome starches. *Carbohydrate Polymers* 135: 10–17.

Syed, A. and Singh, S. 2013. Physicochemical, thermal, rheological and morphological characteristics of starch from three Indian lotus root (*Nelumbo Nucifera Gaertn*) cultivars. *Journal of Food Processing and Technology* S1:003.

Thanushree, M. P., Sudha, M. L. and Crassina, K. 2017. Lotus (*Nelumbo nucifera*) rhizome powder as a novel ingredient in bread sticks: rheological characteristics and nutrient composition. *Journal of Food Measurement and Characterization* 11: 1795–1803.

Torre-Gutiérrez, L., Chel-Guerrero, L. A. and Betancur-Ancona, D. 2008. Functional properties of square banana (*Musa balbisiana*) starch. *Food Chemistry* 106:1138–1144.

Tungmunnithum, D., Thongboonyou, A., Pholboon, A. and Yangsabai, A. 2018. Flavonoids and other phenolic compounds from medicinal plants for pharmaceutical and medical aspects: An overview. *Medicines* 5: 93.

Velusami, C. C., Agarwal, A. and Mookambeswaran, V. 2013. Effect of *Nelumbo Nucifera* petal extracts on lipase, adipogenesis, adipolysis, and central receptors of obesity. *Evidence Based Complement and Alternative Medicine 2013*: 145925.. doi:10.1155/2013/145925

Wattebled, F., Buléon, A., Bouchet, B., Ral, J.P., Liénard, L., Delvallé, D., Binderup, K., Dauvillée, D., Ball, S. and D'Hulst, C. 2002. Granule-bound starch synthase I. *European Journal of Biochemistry* 269: 3810–3820.

Yamini, R., Kannan, M., Thamaraisevi, S. P., Uma, D. and Santhi, R. 2019. Phytochemical screening and nutritional analysis of *Nelumbo nucifera* (Pink lotus) rhizomes to validate its edible value. *Journal of Pharmacognosy and Phytochemistry* 8: 3612–3616.

You, J. S., Lee, Y. J., Kim, K. S., Kim, S. H. and Chang, K. J. 2014. Ethanol extract of lotus (*Nelumbo nucifera*) root exhibits an anti-adipogenic effect in human pre-adipocytes and antiobesity and anti-oxidant effects in rats fed a high-fat diet. *Nutrition Research* 3:258–267.

Yu, H., Cheng, L., Yin, J., Yan, S., Liu, K., Zhang, F., Xu, B. and Li, L. 2013. Structure and physicochemical properties of starches in lotus (Nelumbo nucifera Gaertn.) rhizome. *Food Science and Nutrition* 1: 273–283.

Zeng, S., Chen, B., Zeng, H., Guo, Z., Lu, X., Zhang, Y., and Zheng, B. 2016. Effect of microwave irradiation on the physicochemical and digestive properties of lotus seed starch. *Journal of Agricultural and Food Chemistry* 64: 2442–2449.

Zhang, Y., Lu, X., Zeng, S., Huang, X., Guo, Z., Zheng, Y., Tian, Y. and Zheng, B. 2015. Nutritional composition, physiological functions and processing of lotus (*Nelumbo nucifera Gaertn.*) seeds: A review. *Phytochemistry Reviews* 14: 321–334.

Zhang, Y., Zeng, H., Wang, Y., Zeng, S. and Zheng, B. 2014. Structural characteristics and crystalline properties of lotus seed resistant starch and its prebiotic effects. *Food Chemistry* 155: 311–318.

Zheng, Y., Jagadeeswaran, G., Gowdu, K., Wang, N., Li, S., Ming, R. and Sunkar, R. 2013. Genome-wide analysis of microRNAs in sacred lotus, Nelumbo nucifera (Gaertn). *Tropical Plant Biology 6*: 117–130.

Zhong, G., Chen, Z. D. and We, Y. M. 2007. Physicochemical properties of lotus (*Nelumbo nucifera* Gaertn.) and kudzu (*Pueraria hirsute* Matsum.) starches. *International Journal of Food Science and Technology* 42: 1449–1455.

Zhu, F. 2015. Interactions between starch and phenolic compound. *Trends in Food Science and Technology* 43:129–143

Zhu, F. 2016. Impact of γ-irradiation on structure, physicochemical properties, and applications of starch. *Food Hydrocolloids* 52: 201–212.

Zhu, M., Liu, T., Zhang, C. and Guo, M. 2017. Flavonoids of lotus (*Nelumbo nucifera*) seed embryos and their antioxidant potential. *Journal of Food Science* 82: 1834–1841.

36

Antinutritional Factors and Their Minimization Strategies in Root and Tuber Crops

Uma Prajapati and Vikono Ksh
ICAR—Indian Agricultural Research Institute, New Delhi, India

Manoj Kumar
ICAR—Central Institute for Research on Cotton Technology, Mumbai, India

Alka Joshi
ICAR—Indian Agricultural Research Institute, New Delhi, India

CONTENTS

36.1 Introduction

Root and tuber crops are excellent sources of several nutrients and bioactive compounds. Except potatoes, their intake is limited for table purpose due to their earthy flavor (e.g. geosmine in beet root) and sensory and processing limitations (e.g. thick peel, high fiber, starchy nature). The plant constitutes numerous phenolic bioactives, which can play a central role in various bioactivities (antioxidant, antiinflammatory, antidiabetic, etc.) in the human body (Nishad et al., 2020; Kumar et al., 2019a,b; Kumar et al., 2020a,b; Kumar et al., 2018). Plants species synthesize a range of low- and high-molecular weight (Mw) metabolites, which protect plant species from herbivores, insects, pathogens and adverse growing condition (Harborne, 1989). Generally, these compounds have ill effects on human health and are labelled antinutritional factors (ANFs) when included in the human diet. Being very commonly consumed, it has become critical to discuss the ANF profiles of roots and tubers. Scientific studies take decades to observe and confirm the antinutritional effect of certain food compounds on human health. Thus, it requires *in vivo* studies on rats, livestock, and other domestic animals for evaluation of toxicological effects of ANFs present in roots and tubers. These evaluations can be carried out by continuously feeding the roots and tubers over a long period of time as a part of unvarying diet. However, intake of ANFs with food stuff varies largely according to the consumption pattern, composition, and processing of foods. Undesirable effects of consumption of certain ANFs include flatulence due to α-galactoside, anemia due to phytate-induced iron deficiency, and signs of nausea, vomiting, and diarrhea due to saponins. These compounds also possess anticancerous, antimicrobial, hypoglycaemic, and hypocholesterolemic properties in human cell lines. Therefore, ANFs are becoming of great interest in the field of biochemistry, medicine, pharmacology, and nutrition. The reduction of ANFs from food is an arduous operation as it requires the knowledge of chemical structure and its physical properties especially its stability under heating (Khokhar and Apenten, 2003). One can't design a processing protocol for keeping ANFs as the nutritional quality of roots and tubers is critical. Effects of already established processing protocols (soaking or cooking) of ANFs is currently the primary way to manage ANFs except in the use of biotechnological and breeding protocols. In this chapter, diverse aspects of ANFs are discussed within the context of root and tuber crops (Table 36.1). These ANFs will be outlined by commonly used roots in foods i.e. carrot, radish, parsnip, beet root, celeriac, and tubers including potato, cassava, yam, colocasia, and elephant foot yam. This chapter will also focus on the impact of some common food processing phenomenon on root and tuber ANFs. Physical, chemical, or biological methods employed to combat ANFs are soaking, cooking, fermentation, and pulsed electric field (Khokhar and Chauhan, 1986) will also be highlighted. The application of a single technique might not be effective, hence a combination of techniques are commonly employed. One study suggested soaking and cooking as an effective way

TABLE 36.1

Antinutrients Found in Different Root and Tuber Crops Consumed Worldwide Along With Their Mechanism and Health Complications

Antinutrients	Root Crops	Mechanism	Symptoms
Oxalate	Beet root, parsley, elephant foot yam	Influence calcium, iron, magnesium, and copper absorption	Kidney disease and stones, rheumatoid arthritis, gout
Glucosinolate	Radish, turnip	Thiocyanate inhibits assimilation of iodine by the thyroid gland	Goiter
Furanocoumarins	Parsnip	Furanocoumarins react with nucleobases in DNA under the influence of UV-A radiation giving rise to cross-links in DNA and leading to characteristic acute bullous lesions.	Causes redness, burning, and blisters on skin
Nitrate	Celeriac, potato	Formation of nitrosamines in gastrointestinal tract	Cancer in stomach, intestine, methemoglobinemia
Saponin	Potato, yam, asparagus	Surfactant activity, decomposing cellular membranes	Nausea, vomiting, diarrhea, dizziness, elevated blood pressure, cardiac arrhythmia
Tannins	Sweet potato	Binds proteins, inhibiting digestive enzymes and reducing iron bioavailability	Liver injury, esophageal cancers
Glycoalkaloid	Potato	Inhibition cholinesterases	Neurotoxicity
Cyanogenic glycosides	Elephant foot yam, cassava	Toxicity of cyanides with high affinity to cytochrome oxidase	Liver injury, esophageal cancers
Phytic acid	Sweet potato, elephant foot yam	Influence iron, magnesium, zinc, and copper absorption	Symptoms of iron and zinc deficiencies
Protease inhibitor	Potato, sweet potato, taro	Bind the inhibitor to enzyme, decreasing the nutritional value of proteins	Growth retardation, pancreatic hypertrophy

of reducing saponin content. Industrial processing like canning, toasting, fractionation, and isolation of protein concentrates have also been shown to be effective in reducing ANFs. However, adverse effects of processing should also be taken under consideration while focusing ANFs as this may introduce some undesirable compounds in foods such as volatile aldehydes, ketones, or peroxides as a result of lipid oxidation or it can change nature and concentration of some desirable compounds such as proteins and essential minerals. In the end, this chapter will also highlight the adverse and beneficial effects of ANFs along with the futuristic approach to deal with them.

36.2 Root and Tuber Crops

36.2.1 Root Crops

36.2.1.1 Beet Root

Beet root (*Beta vulgaris* L.) is grown for its taproot. It can be consumed both in raw or cooked form. It is also popular because of its highly bioactive compound which gives its root red violet chroma. Its root extracts are used for improving the food coloring and also flavor of tomato paste, sauces, desserts, jams, jellies, ice cream, candy, and breakfast cereals (Grubben and Denton, 2004). Betalin-rich chips and papaya candy has also been developed as an innovative fusion product with enhanced functionality (Joshi et al., 2019a, b). Beet roots consist of approximately 300–600 mg kg^{-1} of betalins and hence, can be used for the industrial production of natural colorant (Kanner et al., 2001). Betalin glycosides, known for their water solubility, remarkable pH stability, and generally recognized as safe (GRAS) status, can successfully replace synthetic pigments in the human diet (Joshi et al., 2019 a, b). Their bio-functionality is mainly governed by two pigments: purple betacyanin and yellow betaxanthins (Delgado-Vargas et al., 2000; Stintzing and Carle, 2004). Betacyanins compounds exhibit antioxidant and radical-scavenging activities (Escribano et al., 1998). Betaxanthins have been used as a food supplement to fortify processed food products with essential amino acids popularly known as "essential dietary colorant" (Leathers et al., 1992). Being a highly nutritious vegetable, it also contains some ANFs like oxalates and nitrates. Higher consumption of oxalates leads to the formation of kidney stones by interfering in the absorption of micronutrients and leading to abdominal pains, muscle weakness, nausea, and diarrhea. Even smaller concentrations of oxalate can lead to burning in the eyes, ears, mouth, and throat in people sensitive to this compound (Natesh et al., 2017).

36.2.1.2 Carrot

The carrots (*Daucus carota* L.) is a popular root vegetable from the *Apiaceae* family. It is found in variety of colors like orange, red, purple, white, black, and yellow (Iorizzo et al., 2020). Its bioactive richness is due to carotenoids and dietary fibers, with appreciable levels of several other functional components (Sharma et al., 2012). Purple and black carrots are rich sources of anthocyanins and can act as natural colorant. Apart from various useful compounds, it also constitutes cytotoxic compounds namely polyacetylenes (falcarinol and falcarindiol). Falcarinol shows antifungal activity while falcarindiol is the main compound responsible for bitterness in carrots (Garrod et al., 1978; Czepa and Hofmann, 2003).

36.2.1.3 Radish

The radish (*Raphanus sativus* L.) is an important crucifer vegetable whose roots and leaves are utilized for eating purposes. Radish roots contain 4-methylthio-3-butenyl glucosinolate (4-MBG) as a characteristic glucosinolate with the common name of glucoraphasatin (Carlson et al., 1985). Glucosinolates are hydrolyzed in to raphasatin in the presence of the myrosinase enzyme that gives the characteristic pungent flavor to the radish. Isothiocyanates derived from glucosinolates are also found in radishes and play an important role in cancer prevention (Ishida et al., 2011). Thiocyanate inhibits assimilation of iodine by the thyroid gland, which leads to its multiplication and swelling, known as a goiter (Eymar et al., 2016).

36.2.1.4 Parsnip

The parsnip (*Pastinaca sativa* L.), is a root vegetable belonging to family *Apiaceae*. The parsnip is a biennial plant, but is usually grown as an annual plant. It is popular for its cream colored long tuberous roots that can be consumed raw or cooked. Parsnips are loaded with plenty of vitamins and minerals especially potassium. Parsnips also contain soluble and insoluble dietary fibers and comprises cellulose, hemicellulose, and lignin. Presence of high fiber content in parsnips may help in preventing constipation and reduce blood cholesterol levels (Siddiqui, 1989). They also contain antioxidants such as falcarinol, falcarindiol, panaxydiol, and methyl-falcarindiol, which potentially have anticancer, anti-inflammatory, and antifungal properties (Christensen, 2011). The health benefits are counteracted by presence of the ANF furanocoumarins, which causes redness, burning, and blisters on skin when exposed to sunlight, a condition is known as phytophotodermatitis. Compounds responsible for the toxicity of parsnip extract are heat resistant and persist for several months under storage (Cain et al., 2010).

36.2.1.5 Turnip

The turnip (*Brassica rapa* subsp. Rapa) is commonly grown in temperate climates for its white and fleshy taproot. The word turnip is derived from the latin word *napus,* which means turned/rounded on a lathe and neep (Smillie, 2010). Turnip roots are low in calories although they contain valuable components like carbohydrates, proteins, vitamins, minerals, dietary fibers, and antioxidants (Scalzo et al., 2008). Phenolic compounds such as isorhamnetin 3, 7-O-diglucoside, kaempferol 3-O-sophoroside-7-O-glucoside, isorhamnetin 3-O-glucoside, and kaempferol 3-O-(feruloyl/caffeoyl)-sophoroside-7-O-glucoside are also found in turnips. Phenolic compounds can act as a natural pesticide and help prevent the risk of cardiovascular diseases, cancer, and inflammatory ailments when consumed as food. Tannins present in turnips affect iron (Fe) bioavailability and protein absorption by forming insoluble complexes with the protein as well as with minerals (Gemede and Ratta, 2014; Delimont et al., 2017). A small quantity of oxalic acid is also found in turnips that can cause stones in urinary tract (Lin and Harnly, 2010).

36.2.1.6 Parsley

Parsley (*Petroselinum crispum* L.) is a low-calorie highly nutritious vegetable belonging to family *Apiaceae*. Parsley is commonly used as snack or a vegetable in many soups, stews, and casseroles. It is a rich source of flavonoid, antioxidants (especially luteolin, apigenin), folate, and vitamins K, C, and A (Meyer et al., 2006). Parsley contains essential oils, including apiol and myristicin, which have antibacterial effects and fight potentially harmful bacteria, such as *Staphylococcus aureus* (Linde et al., 2016). Apigenin present in parsley regulates immune function by reducing inflammation and preventing cellular damage (Cardenas et al., 2016). Parsley also contains potentially harmful compounds such as oxalates. As mentioned earlier, intake of large amounts of soluble oxalates can increase the risk of kidney stones (Savage et al., 2000).

36.2.1.7 Celeriac

Celeriac is a popular root vegetable in France and Italy that can be eaten both raw and cooked. It is a closely related species of celery and grown as a winter vegetable. Its knobby baseball sized roots have a crisp texture and intense celery flavor. Celeriac is also grown for its stalks and leaves, which can be used for flavoring of soups and stews. It is a dense source of fibers, pyridoxine, ascorbic acid, vitamin K, antioxidants, and minerals (phosphorus, potassium and manganese), and hence referred to as a nutritional powerhouse (MacLeod and Ames, 1989). Terpenoid compounds, psoralens, alkaloids, choline-esterase inhibitor are the few ANFs present in celery and may cause skin disorders.

36.2.1.8 Sweet Potato

The sweet potato (*Ipomoea batatas* L.) is an herbaceous perennial vine, belonging to family *Convolvulaceae*. It is originated in Central America and is widely grown in tropical and subtropical countries. It is grown for its large, starchy, sweet-tasting tuberous roots (Woolfe, 1992). Its edible tuberous root consists of color ranges between yellow, orange, red, brown, purple, and beige. Sweet potato is considered as an "insurance crop" because of its year-round production and minimal crop loss due to adverse climatic conditions. Sweet potatoes play an important role in food security because of its availability for comparatively longer duration in a year. Sweet potatoes are a rich source of vitamin A, carotenoids, dietary fiber, minerals, and several bioactive compounds that contribute to its role in nutritional safety and prevention of various disorders (Chandrasekara and Kumar, 2016). Phytates, oxalates, and tannins are some of the ANFs detected in sweet potato that reduce the bioavailability of minerals and nutrients by making complexes with divalent ions (Oboh et al., 1989).

36.2.2 Tuber Crops

36.2.2.1 Potato

The potato (*Solanum tuberosum* L.) is a perennial plant belonging to the family *Solanaceae* and is grown for its tuber all over the world. Potatoes are the fourth most important crop after maize, wheat, and rice in the world. It is a rich source of carbohydrates, bioavailable proteins, potassium, and ascorbic acid (Hale et al., 2008). Consumption of pigmented potato also has several health beneficial effects. It has elevated antioxidant status, reduced inflammation and DNA damage through reduction of inflammatory cytokine and C-reactive protein concentrations (Kaspar et al., 2011). Potatoes also contains ANFs such as glycoalkaloids, from which most prevalent are solanine and chaconine compounds. These compounds are generally present to protect the plant from predators (Friedman et al., 1997). The consumption of excess of glycoalkaloids can cause headaches, diarrhea, cramps, and in severe cases, coma and death. In general, the glycoalkaloid concentration should be less than 200 ppm, however Morris and Lee (1984) reported that its safety limit should be less than 60 ppm. Nevertheless, cultivated varieties are least prone of causing glycoalkaloid poisoning. Exposure of potato tuber to light can cause greening which can also increase this ANF (Pavlista, 2001).

36.2.2.2 Cassava

Cassava (*Manihot esculenta*) is a tropical perennial shrub grown worldwide for its tubers. It is predominantly consumed in boiled form and is also used to extract starch, which can be used for food, animal feed, and industrial purposes. It is a rich source of carbohydrate content, hence it plays an important role as a staple crop for more than 500 million people in the world (Blagbrough et al., 2010). In addition, it is also a rich source of several bioactive compounds such as cyanogenic glucosides (linamarin and lotaustralin), noncyanogenic glucosides, and hydroxycoumarins (scopoletin, terpenoids, and flavonoids) (Chandrasekara and Kumar, 2016). Cassava is also a rich source of fiber, which assists in intestinal peristalsis and bolus progression. Overconsumption may lead to negative effects in humans by decreasing the nutrient absorption in the body (Montagnac et al., 2009). Cassava must be properly washed and cooked before consumption because it contains ANFs like cyanogenic glycoside, which can be converted to cyanide and cause acute cyanide intoxication, goiters, and even ataxia, partial paralysis, or in severe cases, death (Cereda and Mattos, 1996).

36.2.2.3 Yams

The yam (*Dioscorea* sp.) is a common name of all plant species belonging to the genus *Dioscorea* and family *Dioscoreaceae*. It is widely cultivated for its edible tubers. These are perennial herbaceous vines largely grown in temperate and tropical regions of Africa, South America, Asia, and Oceania (Lu et al., 2012). Major cultivable species of yams are *Dioscorea rotundata* (white yam), *D. cayennensis* (yellow yam), *D. alata* (greater yam), *D. esculenta* (lesser yam), and *D. polystachya* (Chinese yam) (Diop and Calverley, 1998). Yams are grown widely as a staple food, can be consumed in raw or cooked form, and its flour is used for various food preparations. It is a rich source of various bioactive compounds such as dioscin, choline, diosgenin, mucin, allantoin, polyphenols, and vitamins such as carotenoids and tocopherols (Bhandari et al., 2003). Several studies have shown that yams have a potential role in human health as they possess hypoglycemic, antimicrobial, and antioxidative properties. It also enhances digestive enzymes in the small intestine by stimulating the proliferation of gastric epithelial cells (Chandrasekara and Kumar, 2016). Yams also contain some toxins such as dihydrodioscorine that impart a bitter taste to the yam. These bitter compounds are water-soluble alkaloids that can lead to severe and distressing symptoms and sometimes even death. Its removal is an important aspect prior to consumption and can be done through fermentation, roasting, or cooking (Wanasundera and Ravindran, 1992).

36.2.2.4 Elephant Foot Yam

Elephant foot yam is a popular tuber crop belonging to the family *Araceae* and is native to South Asia. It is widely distributed in India, Bangladesh, Malaysia, Philippines, and Indonesia. In India, it is generally cultivated as an intercrop with ginger under coconut or banana (Santosa et al., 2016). The elephant foot yam tuber is largely eaten as a vegetable after boiling. In many countries, its tubers are used as a traditional food source (Ravi et al., 2009). Elephant foot yam tubers contain phosphorous, calcium, vitamin A, crude protein, fat, calcium and crude fiber. They are also used for the treatment of hemorrhoids, abdominal pain, and constipation. Pharmacologically, it has been proved to exhibit gastrokinetic, anti-hemorrhoidal, anticolitic, analgesic, central nervous system (CNS) depressant, anti-inflammatory, cytotoxic, antibacterial, and antifungal activities (Dey et al., 2017). Its tubers also contain ANFs such as oxalates and phytates. If elephant foot yams are not properly cooked, they can give an itchy sensation in the mouth and throat. The level of oxalic acid in the elephant foot yam is 1.3% (Peetabas et al., 2015).

36.3 Classification of ANFs

ANFs in plants are broadly classified on the basis of chemical nature, solubility, and thermostability. Although this classification does not encompass all the known groups of ANFs, it does present the list of those which are frequently found in human foods and animal feed.

36.3.1 On the Basis of Chemical Nature

- Proteins (such as α-amylase inhibitors and protease inhibitors) that are sensitive to normal processing temperatures
- Polyphenolic compounds (mainly condensed tannins), nonprotein amino acids, and galacto-mannan gums
- Antimetals (phytates and oxalates)
- Antivitamins (antivitamins A, D, E, and B12)
- Enzyme inhibitors (trypsin and chymotrypsin inhibitors, plasmin inhibitors, elastase inhibitors)
- Triterpenoids (saponins)
- Alkaloids (solanine and chaconine)

36.3.2 On the Basis of Solubility

- **Water soluble**: Polyphenols and tannins, oxalates (potassium and sodium salts), saponin, cyanogenic glycosides, α-galactosides, nitrates, glucosinolates
- **Water insoluble**: Phytates, oxalates (calcium, magnesium, iron), furanocoumarins, alkaloids

36.3.3 On the Basis of Thermostability

- **Heat stable**: Total free phenolics, tannins, L-Dopa, oxalates, cyanogens and nitrates
- **Heat labile**: Protease inhibitors
- **Controversial**: Saponin, phytates, phenolics

36.4 Characteristics of ANFs and Their Analysis

36.4.1 Characteristics of Phytic Acid

Phytic acid (myoinositol hexakisphosphate, IP6) is a ubiquitous plant compound first identified in 1855. It possesses a unique structure responsible for its characteristic properties. It has 12 replaceable protons that allow it to make complexes with multivalent cations and positively charged proteins. IP6 consists of two forms: phytate (calcium [Ca] salt of phytic acid) and phytin (Ca magnesium [Mg] salt of phytic acid). These are principal storage forms of phosphorus in many plant tissues and can exist in both forms according to physiological pH and metal ions present in the crop. Complex hydrolysis of IP6 results in to the formation of inositol and inorganic phosphates (Oatway et al., 2001). IP6 is present in cereal grains, nuts, oilseeds, legumes, pollen, spores, and tubers. In tuber crops 21–25% of total phosphorus (P) may occur as phytates (Ravindran et al., 1994). IP6 reduces the bioavailability of certain nutrients because of its characteristic property of making complexes directly or indirectly with minerals, proteins, and starch. Phosphate groups are negatively charged at a normal pH, which allows its interaction with positively charged minerals or proteins. Metal ions can bind with phosphate groups forming complexes that may or may not be water soluble while proteins bind directly with phytic acid through electrostatic charges. Zinc (Zn) forms the most stable insoluble complex with IP6 and hence is the most affected mineral (Rickard and Thompson, 1997). Therefore, consortia research projects (CRP-respective crops) were targeted around biofortification of Zn. This is the reason why plant-based fiber-rich diets can create risk of Zn vulnerability among vegetarian populations. IP6 also binds with several other minerals like Ca, sodium (Na), Fe, Mg, and manganese (Mn), and makes them unavailable for consumption in humans. The formation of complexes (phytic acid-mineral nutrient) alters solubility, functionality, digestion, and absorption of nutrients in gastrointestinal tract, hence it is considered an ANF.

36.4.1.1 Analysis of Phytic Acid

In tubers, phytic acid exists in the form of phytates. Tubers and root crop samples are mostly oven-dried at 60°C and pulverized or lyophilized prior to analysis. Charles et al. (2005) analyzed the phytic acid content of five genotypes of cassava. The sample was extracted using hydrochloric acid (HCl) for 2 hours at room temperature and centrifuged. A 3 mL aliquot of the filtered supernatant was drawn and diluted to 18 mL with distilled water and passed through a chloride anion exchange resin. Phytate was eluted with 0.7 M NaCl and determined calorimetrically. The phytic acid level was found in the range from 95–135 mg g^{-1}. The phytate levels in Jerusalem artichokes were determined spectrophotometrically at 500 nm (Judprasong et al., 2018). A rapid cost-effective methodology for phytate estimation was also standardized using ELISA-based technique in potato samples (Joshi et al., 2015). The method is again based on the colorimetric method in which the phytic acid (phytate) is precipitated with an acidic iron-III solution of known iron content. The decrease in iron content (determined calorimetrically with 2, 2-bipyridine in the supernatant) is a measure of the phytic acid content. Estimated phytate content in exotic potato cultivars were found to be 21.0–47.0 mg 100 g^{-1} (fwb) in German Butterball and Russet Norkotah potato varities, respectively (Phillippy et al., 2004). The detail of the spectrophotometer test is described here and can be performed with a very basic lab structure. For this, extract a 0.5–1.0 g sample with 10 mL of 0.65 N HCl and filter the extract through a Whatman No. 42 filter paper. Next, mix 0.5 mL of the aliquot sample extract in 1 mL ferric solution (0.02 g ammonium iron sulphate in 10 mL of 2N HCl and make up to 100 mL with distilled water) in test tubes and boil in a water bath for 30 minutes followed by immediate cooling in an ice-water bath for 15 minutes. Then, allow the reaction mixture to adjust to room temperature and add 2 mL of 2,2-bipyridine solution (1 g 2,2-bipyridine and 1 mL thioglycolic acid in 100 mL distilled water), and after 1 minute, take absorbance at 519 nm. The amount of phytate phosphorus in the samples can be calculated by comparing the absorbance with the standard graph (extra pure sodium phytate) (Haug and Lantzsch, 1983).

36.4.2 Characteristics of Oxalate

Oxalate is the conjugate base of oxalic acid found naturally in plants and humans and acts as ligand for metals. Oxalic acid also forms water-soluble salts with Na^+, K^+, and NH^{4+} ions and also binds with minerals (Ca^{2+}, Fe^{2+}, and Mg^{2+}) rendering them unavailable to animals. The pH level affects the existence of oxalates in cell sap; for example, at pH 2, it occurs as acid oxalate primarily as K oxalate while at pH 6 it exists as an Na oxalate and the insoluble Ca and Mg oxalates. At acidic pH, Ca oxalates dissolve freely while at neutral or alkaline pH it is insoluble (Noonan and Savage, 1999). Excess consumption of oxalates can lead to kidney stones. Conjugation of oxalic acid with divalent metallic cations such as Ca and Fe further leads to the formation of crystals of oxalates. A single crystal can work as nucleus around which more crystals can accommodate. These oxalate crystals are sometimes excreted through urine or can form large kidney stones that can obstruct the kidney tubules. Since 80% of kidney stones are formed from Ca oxalates, patients with gout, kidney disorders, rheumatoid arthritis, or certain forms of chronic vulvar pain (vulvodynia) are advised to avoid foods which contain high oxalic acid (Coe et al., 2005).

36.4.2.1 Analysis of Oxalate

A method involving digestion, precipitation and permanganate titration is extensively used for oxalate estimation (Adane et al., 2013; Ogbonna et al., 2017; Ramos et al., 2020). However, to eliminate the need of digestion, an ashing method was used (acid soluble ash). Gemede and Fekadu (2014) estimated oxalate in anchote samples by digesting the sample in deionized water and HCl at 100°C for 1 hour, followed by precipitation with excessive calcium ions, and then by titration against 0.05 M standard $KMnO_4$ solution to a faint pink color that needs to persist for at least 30 seconds. Buta (2020) also followed a similar procedure of digestion, oxalate precipitation, titrating with a standardized $KMnO_4$ solution, and the use of ultraviolet-visible (UV-V) spectroscopy (a method to estimate oxalate), and reported 140.45 ± 17.51 mg $100g^{-1}$ oxalate content in taro roots. Capillary electrophoresis and ion chromatography were also tried by Holmes and Kennedy (2000) in various food samples and it was found that like traditional method, both the methods were able to trace even a very low amount of oxalate i.e., 1.8 mg $100g^{-1}$. The ion chromatography was found to be more effective for oxalate estimation in comparison with capillary electrophoresis.

36.4.3 Characteristics of Furanocoumarins

Furanocoumarins are natural phototoxic compounds present in plants belonging to the *Apiaceae* family, which includes celery, parsnips, parsley, and carrots. Currently, more than 50 plant furanocoumarins are known (Søborg et al., 1996). On the basis of chemical structure, there are two subgroups of furanocoumarins, linear and angular. Linear furanocoumarins include psoralen, bergapten, xanthotoxin, trioxsalen, isopimpinellin, and bergamottin whereas angular furanocoumarins are angelicin, pimpinellin, sphondin, and isobergapten. Furanocoumarin levels increase as a plant undergoes stress conditions such as an insect or pathogen attack, mechanical damage, or unfavourable environmental condition (Schulzová et al., 2007). Furanocoumarins yield reactive intermediates under UV light irradiation, which leads to the formation of DNA adducts (Llano et al., 2003). In addition, this higher level of furanocoumarins also leads to mutagenic and carcinogenic effects. The lowest observed adverse effect level of furanocoumarins was estimated in the range $0.14–0.38$ mg kg^{-1} body weight. The average daily dietary intake of furanocoumarins was estimated to be 0.5 mg kg^{-1} body weight from vegetables (Søborg et al., 1996). Dietary intake of furanocoumarins is a potential health risk for consumers.

36.4.3.1 Analysis of Furanocoumarins

For furanocoumarins, high-performance thin liquid chromatography (HPLC) coupled with medium and high-resolution mass spectrometry has been extensively used for qualitative profiling while those coupled with ultraviolet, diode array detection (DAD), and fluorescence detection have been reported for quantitative analysis. These are also often combined with other techniques such as antioxidant activity evaluation, biological assay, pharmacokinetic studies, nuclear magnetic resonance spectroscopy (NMR), and Fourier transform infrared spectroscopy. Cook et al. (2017), quali-quantitatively analyzed 14 fura-nocoumarins in parsnip samples using a two-dimensional liquid chromatography (LC × LC) coupled to DAD followed by the 2D chromatograms with manual integration for concentration determination. Ten different furanocoumarin compounds were detected at very low concentrations ($<5~\mu g~g^{-1}$). Supercritical fluid chromatography (SFC) can also be explored as a novel green alternative for furanocoumarin esti-mation (Pfeifer et al., 2016).

36.4.4 Characteristics of Cyanogens

Cyanogens are glycosides of 2-hydroxyl nitriles and widely distributed among plants belonging to the *Rosaceae*, *Leguminosae*, *Graminae*, and *Araceae* families. Stress, particularly wounding, leads to the breakdown of cellular walls, causing the release of cyanogenic glycosides that come into contact with active β-glucosidase, which hydrolyze cyanogenic glucosides to yield 2-hydroxynitrile or cyanogens (Zagrobelny et al., 2004). Similarly, when the plant is consumed, glucosides present in it can be broken down in to sugar and aglycone in the presence of enzyme β-glucosidase. The aglycone in the presence of hydroxynitrile lyase produces cyanide and an aldehyde or ketone. Cyanide is an extremely toxic com-pound that often leads to death by inhibiting cytochrome oxidase that acts at the final step in the electron transport chain, and thus blocks ATP synthesis. Other symptoms may include faster and deeper respira-tion, disfunction of the central nervous system, faster irregular and weaker pulse, salivation and froth-ing at the mouth, muscular spasms, dilation of the pupils, bright red mucous membranes, and cardiac arrest (Bjarnholt and Møller, 2008). Apart from exhibiting toxic effects, cyanogens also serve as mobile nitrogen storage compounds in seeds, which is important at the time of germination. Cassava is one of the important sources of cyanogenic glucoside, hence, it is suggested that it should be detoxified prior to its consumption.

36.4.4.1 Analysis of Cyanogens

Silver nitrate titrimetric method (AOAC18 standard method) was employed by Gemede and Fekadu (2014) to determine cyanide content in tuber crops. Ogbonna et al. (2017) used an alkaline picrate method fol-lowed by spectrophotometric analysis for determination of cyanogenic glycoside in root tubers of *Tacca leontopetaloides* L. (yam) at 490 nm. The cyanide content in root tubers was found to be 0.18 ± 0.03 mg 100 g^{-1}. Cyanides were extracted from cassava samples using steam distillation (Mulualem et al., 2020). The distillate was added to silver nitrate and acidified with nitric acid. Using iron alum as an indicator, the distillate mixture was titrated with potassium thiocyanate. Another method involves hydrolysis of cyanide present in the samples by water soaking followed by steam distillation. The distilled fraction needs to absorb in steam at a neutral pH that forms cyanogen chloride with chloramine T. Cyanogen chloride can react with isonicotinic acid-pyrazolone to form a blue compound (λ max 638 nm). Thus, quantification can be done using Beer Lambert's law and the derivation of a standard equation. A stan-dard calibration equation was derived by Chen et al. (2018) for cyanogen's content ($y = 0.0814 \times -0.031$, $R = 0.9954$) in bitter almond oil, which is comparatively a difficult matrix for cyanogen estimation since hydrophobic in nature.

36.4.5 Characteristics of Saponins

Saponin is a widely occurring secondary metabolite in plant species that exhibits a broad range of benefi-cial and harmful effects on human health. The term saponin is derived from the Latin word *sapo*, which means "soap." Chemically, these are glycosides of high Mw that produce lather or foam when shaken with water. Saponins consist of a polycyclic aglycones attached to one or more sugar side chains. Aglycone part is also called as sapogenin, which can be either steroid (C27) or triterpene (C30). The foaming abil-ity of saponin is due to combination of sapogenin (hydrophobic) with sugar part (hydrophilic) (Kaushik et al., 2018). Toxic saponins are called sapotoxins, while saponins (steroid or triterpene glycoside com-pounds) possess some medicinal properties and antimicrobial potential. Saponins cause a bitter taste if present in higher concentrations and can affect nutrient absorption by inhibiting metabolic or digestive enzymes with mineral nutrients like zinc. The adverse effects of saponins involve strong hypocholesterolemic effects, hypoglycemia, impaired protein digestion, uptake vitamins and minerals in the gut, as well as lead to the development of a leaky gut (Akwaowo et al., 2000; Barky et al., 2017). Potato, yam, and asparagus are major sources of saponin among various root and tuber crops. These can be eliminated by soaking, rinsing, cooking, canning, and fermentation (Güçlü-Üstündağ and Mazza, 2007).

36.4.5.1 Analysis of Saponins

Saponin content in yams analyzed by following the spectrophotometric method of Brunner (1984), which involved extraction of a sample with isobutyl alcohol, followed by filtration (Mulualem et al., 2018). To this filtrate, 40% (saturated solution) of magnesium carbonate was added to obtain a clean colorless solution. After appropriate dilution, the reaction mixture was allowed to stand for 30 minutes for color development so that the analyte concentration (optical density [OD]) can lie in linear range of the concentration-OD graph at 350 nm. However, Ogbonna et al. (2017) analyzed root tuber samples of *Tacca leontopetaloides* L. (yam) by using a modified spectrophotometric method of Brunner. Lebot et al. (2019) developed an HPLC protocol for quantification of saponins in wild yam species. Total saponins varied between 37.36–129.97 mg g^{-1}. The peels of tubers were found to have higher saponin content than the flesh. Senanayake et al. (2012) evaluated saponin content in yam and tuber samples by extracting tissue with ethanol and diethyl ether. The solution was evaporated and the saponin content was measured on dry basis.

36.4.6 Characteristics of Polyphenols and Tannins

Polyphenols and tannins are naturally occurring metabolites widely distributed in the plant kingdom. These compounds possess a wide array of beneficial and adverse effects on human health. Beneficial effects may include radical scavengers or chelators of metals thus preventing low-density lipoprotein oxidation and DNA strand scission. Polyphenols and tannins are also found to control certain types of cancer, cardiovascular disease, and the process of aging (Salunke, 2006). There are many factors that affect the level of polyphenols in different plant species such as, germination, light, variety, degree of ripeness processing and storage; genetic factors can also influence levels (Ahn et al., 1989). Tannins are another major group of antioxidant polyphenol with multifunctional properties to human health. Chemically, tannins are oligomers of flavan-3-ols and flavan-3, 4-diols and possess antinutritional properties by impairing the digestion of various nutrients, hindering their absorption in human body (Ertop and Bektas, 2018). Tannins can bind to protein and their complexes may cause inactivation of digestive enzymes thus reducing protein digestibility (Salunkhe et al., 1990).

36.4.6.1 Analysis of Polyphenols and Tannins

Polyphenols have been widely reported to be estimated using HPLC (Champagne et al., 2011). Eight yam species (*Dioscorea* spp.) were analyzed for polyphenols using HPLC with toluene, ethyl acetate, and formic acid (4:6:1, v/v) as the mobile phase. The highest total phenolic acid concentrations (9.96 mg g^{-1} and 9.55 mg g^{-1}, respectively) were reported in *Dioscorea bulbifera* bulbils and *D. nummularia* tubers (Lebot et al., 2019). Sousa et al. (2015) determined polyphenols in yacon tuber flour (*Smallanthus sonchifolius*) using a Folin-Ciocalteu reagent and checking the absorbance at 750 nm. The total phenolic content can be measured by using a modified colorimetric Folin-Ciocalteu method given by Singleton and Rossi (1965) wherein the Folin-Ciocalteu reagent, a mixture of phosphotungstic and phosphomolybdic acids, is reduced to oxides of tungstene and molybdene and provides a bluish tinge to the reaction mixture. This can be quantified using a spectrophotometer at lambda (λ) max 765 nm. A rapid ELISA-based protocol was reported by Raigond et al. (2016) in which 31 samples can be estimated at once using 96-welled ELISA plate readers along with control. In this sample extract, 80% ethanol was reacted with Folin-Ciocalteu reagent (1:1) and diluted three times approximately before examination. Next was the addition of 5% sodium carbonate and incubation of the plate for 1 hour at room temperature to generate a colored mixture that can be measured calorimetrically and polyphenols can be calculated after comparing with standard curve.

Tannins in anchote and taro have been reported to be analyzed with the Burns method, which involves extraction of a sample with 1% HCl in methanol followed by the addition of vanillin-HCl reagent for color development. The developed color can then be spectrophotometrically analyzed at 500 nm using D-catechin as the standard (Adane et al., 2013; Gemede and Fekadu, 2014). Wild Polynesian arrowroot, yams, and cocoyam tubers have been analyzed for tannin content with the Folin-Dennis colorimetric method comprising of Folin-Dennis reagent and 20% sodium carbonate solution followed by spectrometric measurement at 260 nm (Adegunwa et al., 2011; Ogbonna et al., 2017; Mulualem et al., 2018).

36.4.7 Characteristic of α-amylase Inhibitors

Alpha (α)-amylase inhibitor is a proteinaceous ANF that inhibits an α-amylase by forming a tight stoichiometric 1:1 complex with α-amylase and further leads to steric blockage of the active site of the enzyme (König et al., 2003). Root and tuber crops like sweet potato, taro, and yam have been reported to contain α-amylase inhibitor. The presence of α-amylase inhibitor in the diet can lead to celiac disease and pancreatic hypertrophy (Rekha and Padmaja, 2002). Some studies have found its potential on treating obesity and diabetes based on its starch-blocking mechanism and its physicochemical properties (Obiro et al., 2008).

36.4.7.1 Analysis of α-amylase Inhibitors

Bhandari and Kawabata (2006) evaluated α-amylase inhibitor activity (AIA) according to the method of Alonso et al. (1998). A 1 g sample was extracted with 10 mL of deionized water for 12 hours at 4°C and the supernatants were tested for AIA by maintaining 1:1 of sample extract and α-amylase enzyme solution (0.003% in 0.2 m sodium phosphate buffer, pH 7.0, and containing 0.006 m NaCl) for 15 minutes at 37°C. Next, 0.5 mL of a 1% starch solution was added to this mixture. The enzymatic reaction can be stopped at the end of 3 minutes by the addition of 2 mL of a dinitrosalicylic acid reagent followed by severe boiling for at least 10 minutes. The colored reactant in the supernatant can be quantified at 540 nm. One unit inhibited of α-amylase activity was defined as one α-amylase inhibitory unit (IU), and expressed as IU per gram of dry matter (Bhandari and Kawabata, 2006). It can be quantified in comparison with control, as an ability to liberate one micromole of reducing groups (calculated as maltose) per minute at 37°C and pH 7.0 under the specified conditions, from soluble starch.

36.4.8 Characteristics of Protease Inhibitors

Protease inhibitors are a commonly occurring class of ANFs in plants. Biologically, these are molecules that inhibit the function of proteases, which hydrolyze proteins. Protease inhibitors are classified into two groups: one is a single-chain polypeptide (Kunitz type) that inhibit the enzyme activity of only trypsin, and the other is also a single-chain polypeptide, but of a different Mw (also called as Bowman-Birk inhibitors) and that inhibit the enzyme activity of both trypsin and chymotrypsin. Hence, protease inhibitors affect trypsin and chymotrypsin by irreversible binding in the human digestive tract. These are also resistant to digestive enzymes and the acidic pH of the stomach (Kaushik et al., 2010). Protease inhibitors are widely distributed in various plant species including seeds, roots, and tubers. The specific nature of these compounds makes processing easier by heat, which causes protein denaturation, though their residual activity might be seen in some processed products. The antinutritional activity of protease inhibitors is associated with growth inhibition and pancreatic hypertrophy (Giri and Kachole, 2004). There are also some potential benefits of protease inhibitors such as the lowered incidence of pancreatic cancer and prevention of the development of chemically induced cancer of the liver, lung, colon, oral area, and esophagus (Finotti et al., 2006).

36.4.8.1 Analysis of Protease Inhibitors

Wild yam tubers and sweet potato were analyzed for trypsin inhibitor activity (Bhandari and Kawabata, 2006; Senanayake et al., 2012) using a method proposed by Smith et al. (1980), which involves extraction of the protease inhibitors from the sample at pH 9.5 and mixing unfiltered suspensions with bovine trypsin. The activity of the unreacted trypsin is then measured by providing benzoylarginine p-nitroanilide (BAPNA) to the reaction base under standard conditions (37°C); the p-nitroaniline released is measured spectrophotometrically at 410 nm. This provides a linear measure of the residual trypsin activity, so that the amount of pure trypsin inhibited per unit weight of sample can be calculated. Trypsin inhibitor activity in samples can also be determined according to the method described by Raghuramulu et al. (2003). The activity of the trypsin enzyme was assayed using casein as a substrate. Kiran and Padmaja (2003) calculated trypsin inhibitor activity in sweet potato and taro tubers on dry weight basis.

36.4.9 Characteristics of Nitrates

Nitrate is a naturally occurring essential plant nutrient and is used as primary source of nitrogen for plants. It is available in soil in higher amounts and can be utilized as fertilizer. Nitrate accumulation in vegetables can be influenced by several factors such as fertilizer doses, time, and source of fertigation, species and variety of crops, water supply, light intensity, and temperature. Also, the slow growth of plants can lead to its accumulation as enough nitrates were not converted to proteins (Mirecki et al., 2015). Nitrates in food commodities can be converted into nitrite and nitrous acid. It also forms nitrosamines in combination with primary and secondary amines, which pose risks to human health (Thorup-Kristensen, 2006; Prasad and Chetty, 2008). The antinutritional effects of nitrate can be seen when its intake is high; in the gastrointestinal tract, nitrates lead to the formation of nitrosamines, which ultimately cause cancer in the stomach, intestine, bladder, mouth, as well as methemoglobinemia in children (Mozolewski and Smoczyński, 2004; Chetty and Prasad, 2009). According to the European Food Safety Authority (EFSA), the maximum daily nitrate intake should not exceed 3.65 mg kg^{-1} of body weight on average (Brkić et al., 2017).

36.4.9.1 Analysis of Nitrates

Nitrate content was determined by the cadmium column method. The nitrite is determined calorimetrically at 540 nm by diazotization of sulphanilic acid and subsequent coupling with N-(1-naphthyl)-ethylenediamine. The concentration of nitrate plus nitrite is determined similarly but after reduction of the nitrate to nitrite on a cadmium column (Sen and Donaldson, 1978). A similar method was used in Jerusalem artichokes (Judprasong et al., 2018). Kristl et al. (2016) extracted nitrate from lyophilized taro corm samples by following the procedure developed by Kobayashi et al. (2011) with slight modification. The nitrate concentration was analyzed by ion chromatography. The nitrate distribution was not uniform, with the upper portion of the corm reporting the highest concentration. In the edible parts, the nitrate content was relatively low in the range of 29 mg kg^{-1} and 313 mg kg^{-1} fresh weight (FW). Ion chromatography was also used for nitrate computation by Chung et al. (2011) in tubers. Ebrahimi et al. (2020) showed that the average nitrate content in potato was 123.26 mg kg^{-1} which was lower than the safe limit of nitrate.

36.4.10 Characteristics of Alkaloids

Alkaloids are nitrogen-containing secondary metabolites able to form salts with acid. These are distributed widely in nature and impart a bitter taste to food. Alkaloids can be isolated from roots, seeds, leaves, or bark of some plant species. Pyrrolizidine is a widely found alkaloids in nature. Other glycoalkaloids also found in common foods are solanine and tomatine. Solanine is present in potato tubers and glycoalkaloid acts as a natural pesticide. Glycoalkaloids, solanine, and chaconine present in potato tubers are hemolytically active and toxic to both fungi and humans (Saito et al., 1990). The antinutritional response of alkaloids is due to its action on the nervous system, disrupting or inappropriately augmenting electrochemical transmission. Consumption of high-dose of alkaloids can result in gastrointestinal disorder, neurological disorders, rapid heartbeat, paralysis, and in worst case, even death (Fernando et al., 2012). Some plant alkaloids, such as solasodine, are reported to cause infertility (Olayemi, 2010).

36.4.10.1 Analysis of Alkaloids

Senanayake et al. (2012) treated yam and tuber flour samples with 20% acetic acid in ethanol. The contents were concentrated and crude alkaloids were precipitated using concentrated ammonium. Dried precipitation gives an estimate of alkaloids content in the sample of dry weight basis. A similar gravimetric technique was adopted by Ogbonna et al. (2017) to evaluate alkaloid content in *Tacca leontopetaloides* L.

36.4.11 Characteristics of Glucosinolates

Glucosinolates are sulphur containing secondary metabolites mainly found in crop species belonging to the family *Brassicaceae*. Glucosinolate has three moieties: a β-thioglucose moiety, a sulfonated oxime moiety, and a variable aglycone side chain derived from an alpha-amino acid. There are more than 120 glucosinolates, which can be classified into three groups aliphatic, aromatic, and indolic glucosinolates (Fahey et al., 2001). The composition of glucosionlates varies among different plant genera and different organs. Glucosinolate and its derivatives have a wide range of biological functions, including antimicrobial and anticarcinogenic properties and is also used to control soil-borne pests and diseases in some crops through fumigation. Glucosinolates are hydrolyzed into an isothiocyanate, which give a unique and characteristic flavor to *Brassicaceae* family members. This breakdown is catalyzed by enzyme myrosinase, which is stored in the vacuoles of certain phloem cells and is released when a vegetable is chopped or chewed (Deng et al., 2015). Toxicity of glucosinolates are generated after hydrolysis by the endogenous enzyme myrosinase. Some of the hydrolytic compounds are goitrogenic and/or hepatogenic, although this hydrolysis also leads to the generation of the pungent volatile compound in radishes and horseradish, which are relished by Asian consumers. Hydrolytic toxic compounds can also lead to the enlargement of thyroid glands like in goiter. Since the thyroid is an endocrine gland, the hyposynthesis and utilization of thyroid hormones forces the thyroid gland cells to multiply and to generate more thyroid hormone to compensate this lack, causing an enlargement of thyroid gland (Embaby et al., 2010).

36.4.12 Analysis of Glucosinolates

Maldini et al. (2017) analyzed total glucosinolate content in radish samples by using a qualitative (Ultra high-performance liquid chromatograph -Electrospray ionization-Mass spectrometry) UHPLC-ESI–MS/MS technique in an ultra high-performance liquid chromatograph (UHPLC) system coupled with an ABSciex Q-Trap instrument. In a C18 column maintained at 47°C, a 5 ml sample was eluted at 300 mL min^{-1} by employing acidified water (0.1% formic acid) and acidified acetonitrile (0.1% formic acid) as mobile phases in gradient mode. Information dependent acquisition (IDA) was used to switch to mass spectrometry (MS). For MS, a 1 mg mL^{-1} standard solution of glucoraphanin in methanol: water 50:50 (v/v) was infused into the sample (10 mL min^{-1}). This was then followed by HPLC-ESI–MS/MS in a multiple reaction monitoring (MRM) mode for quantitative analyses. For the root samples, glucoraphasatin (56 mg 100 g^{-1} FW) was found to be the predominant glucosinolate followed by glucoraphanin (16 mg 100 g^{-1} FW) and methoxyglucobrassicin (7 mg 100 g^{-1} FW), respectively. In another study, Choquechambi et al. (2019) analyzed mashua samples for glucosinolate content using the HPLC technique according to the American Oil and Chemist's Society (AOCS) Official Method (AOCS, 1997).

36.5 Technological Intervention for Minimizing Antinutritional Factors

36.5.1 Physical Processing

36.5.1.1 Cooking

Cooking in water reduced oxalate content by more than 85%. The oxalate content reduced from 699.27 to 78.51 mg 100g^{-1} and 499.67 to 60.66 mg 100g^{-1} in *Xanthosoma sagittifolium* and *Pachyrhizus tuberosus* respectively, by cooking in water (Ramos et al., 2020). Adane et al. (2013) reported that boiled taro had reduced levels of tannin, phytate, and oxalate. The depletion of tannins and oxalate may be attributed to their solubility in boiling water while that of phytate may be due to the formation of complexes such as phytate-protein, phytate-protein-mineral, or inositol hexaphosphate hydrolyzed and penta- and tetra-phosphates. In another study, boiling resulted in a 65.7–82.1% reduction of oxalates of cocoyam tuber flours while in wild yam, boiling led to about 20% of phytate depletion (Iwuoha and Kalu, 1995; Bhandari and Kawabata, 2004). Cooking tubers resulted in retention of 29–59% amylase inhibitor activity (AIA) in sweet potato and 11–16% in taro (Rekha and Padmaja, 2002).

36.5.1.2 Oven-Drying

An oven drying of taro chips at 90°C and 100°C for 24 hours almost totally inactivated the α-amylase inhibitor (AI) while sweet potato chips retained about 0.8–10% activity under the same conditions (Rekha and Padmaja, 2002). The boiled-oven dried technique was found to be the most effective among all processing techniques, such as sun-dried, roasted-oven-dried, and fresh-oven-dried in removal of tannin content in yams (Adegunwa et al., 2011). Sweet potatoes oven-dried for 24 hours at 80°C and 90°C detected negligible trypsin inhibitor while no trypsin inhibitor was reported for those dried for 24 hours at 100°C (Kiran and Padmaja, 2003).

36.5.1.3 Microwave Baking

The AI of the sweet potato was completely inactivated in two of four cultivars while 19% and 29% was retained in the other two after 180 seconds of microwave baking. In contrast, after 180 seconds of microwave baking, only trace activity of the AI was present in all four cultivars of taro (Rekha and Padmaja, 2002). Kiran and Padmaja (2003) reported that microwave baking for 120 sec was effective in inactivation of the trypsin inhibitor of sweet potatoes. Further, microwave baking for 180 seconds ensured complete trypsin inhibitor inactivation in sweet potatoes and taro tuber pieces.

36.5.1.4 Flour Preparation

Kiran and Padmaja (2003) prepared flour from taro and sweet potato after pretreatment with oven drying, cooking, and microwaving. The residual trypsin inhibitor activity was completely eliminated in case of flour prepared from taro while those prepared from sweet potato retained only 5–12% trypsin inhibitor activity. The heat generated during size reduction and refinement of flour may have aided in the reduction of the trypsin inhibitor activity.

36.5.1.5 Pulsed Electric Field (PEF)

PEF can effectively reduce the oxalate content in oca tubers (*Oxalis tuberosa*) with no changes in overall tuber/tissue structure. Tubers softened above 0.5 kV cm^{-1} but unevenly with inner tuber cores softening more than the middle regions. Higher electric field strength of 0.8 kV cm^{-1} was required to cause a reduction of oxalates in the inner-tuber layers. At 1.2 kV cm^{-1}, oxalate content was found to be 42% and 50% of untreated tubers in the outer and inner regions, respectively. The reduction can be attributed to the disruption of the tonoplast membrane, which may have led to the leakage of the oxalate compounds into the PEF media inside the treatment chamber (Liu *et al.*, 2018).

36.5.2 Bioprocessing

36.5.2.1 Fermentation

Natural fermentation of taro at room temperature for 72 hours resulted in a 43.52%, 84.75%, and 35.79% reduction in tannins, phytate, and oxalate content, respectively, due to the action of enzymes during fermentation (Adane et al., 2013). Microbial fermentation of local cassava yam products was found to significantly deplete ($p < 0.05$) the level of cyanide, tannins, phytate, oxalate, and saponins by 86%, 73%, 72%, 61%, and 92%, respectively (Etsuyankpa et al., 2015). Batista et al. (2019) described a fermentation technique for 24 hours of an aqueous mixture of 40% (w/v) yam with phytase producing lactic acid bacteria. *L. lactis* CCMA 0415 fermentation showed 82% (9.79 mg 100 g^{-1}) reduction in phytate concentration while the other strains maintained the same concentration as that found in the substrate (54.92 mg 100 g^{-1}). Additionally, oxalate concentration in the fermented yam was not detectable.

36.5.3 Biotechnology

Biotechnological and conventional breeding interventions can reduce antinutrients systematically from the genetic background of the crop, which is out of the scope of this chapter. However, in bioprocessing, natural or artificial inoculums are used as a tool for processing.

36.6 ANFs and Human Health

36.6.1 Adverse Effects

- Furanocoumarins like psoralen, bergapten, and xanthotoxin have been documented to possess toxic effects. Excessive amounts may lead to dermatitis, blisters, and hyperpigmentation (Dugrand et al., 2013).

- Phytic acid inhibits the absorption of metals by forming complexes with the metal ions thereby reducing their bioavailability. They also form soluble complexes with proteins and reduce the bioavailability (Liener et al., 1980). They interact with enzymes like trypsin, pepsin, α-amylase, and β-galactosidase in the small intestine and stomach causing a decrease in their activity (Kies et al., 2006).

- Tannins affect the digestibility of protein and have been reported to lower the bioavailability of amino acids by forming tannin-protein complexes between the hydroxyl group of tannins and the carbonyl group of proteins (Raes et al., 2014). This results into a loss of body weight, loss of appetite, breathing issues, and cardiac complications. Yams and coco yam tubers contain tannins, which are phenolic compound that can reduce the bio-availability of iron by forming a complex with iron in the gastrointestinal lumen.

- Cyanide is known to bind to the ferric ion's cytochrome oxidase within the mitochondria and halts cellular respiration. It hampers oxidative phosphorylation, which is a process of utilization of oxygen during ATP production (Graham and Traylor, 2018).

- Oxalate content impairs calcium absorption, which may lead to the formation of kidney stones. Oxalate also forms complexes with magnesium, iron, and zinc, and decreases their bioavailability (Golden, 2009).

- At a high-dose intake, polyphenols may exert carcinogenic or genotoxic effects as they interfere with thyroid hormone biosynthesis and hamper iron absorption, leading to iron depletion too (Mennen et al., 2005).

- The chronic toxic effects of nitrate arise from those nitrates that are formed when nitrite is reduced by bacterial enzymes. N-nitroso compounds are formed when nitrite binds to other substances before or after ingestion. Both nitrite and N-nitroso are known to be toxic. Nitrite reacts with hemoglobin to form methemoglobin and nitrate. Due to this, the delivery of oxygen to tissue gets disrupted (Santamaria, 2006).

- Acute alkaloid poisoning may cause abdominal pain, ascites, nausea, vomiting, diarrhea, edema and very rarely jaundice and fever. Long term exposure usually leads to cirrhosis of the liver. It may also affect the liver, lungs (pulmonary hypertension), or cardiovascular system (cardiac right ventricular hypertrophy) (Koleva et al., 2012).

- The consumption of high amounts of amylase inhibitors and protease inhibitors may lead to growth retardation due the unavailability of essential carbohydrates and proteins. The presence of undigested starch and proteins in the colon may result in various indigestion problems and diarrhea.

- At high concentrations, saponin may alter the integrity of intestinal epithelial cells which, may cause many health complications. The absorption of vitamina A and E and lipids are also impaired. Saponins have also been reported to obstruct the activities of digestive enzymes such as amylase, glucosidase, trypsin, chymotrypsin, and lipase, which may lead to indigestion-related health disorders (Samtiya et al., 2020).

- The human intestinal mucosa lacks the hydrolytic enzyme α-galactosidase, making α-galactosides indigestible as these sugars cannot pass through the intestinal wall. Excess intake leads to its accumulation in the hindgut where it undergoes anaerobic fermentation and causes flatulence. Its presence also throws off the osmotic pressure balance of the small intestine, which impedes the digestion as well as the absorption of other nutrients (Martinez-Villaluenga et al., 2008).

36.6.2 Beneficial Effects

Although the detrimental effects of antinutrients are prevalent, it is interesting to note that these compounds also have beneficial effects when consumed at lower concentrations. Some of the benefits are listed and discussed next.

36.6.2.1 Antioxidant Activities

Oxidative stress has been reported to play a major role in the development of several chronic diseases including cancer, cardiovascular diseases, arthritis, diabetes, autoimmune and neurodegenerative disorders, and aging. This makes tackling oxidative stress a necessity. Free radicals such as dehydroascorbate (DHA) and monodehydroascorbate (MDA) have been reported to be nullified by sporamin, which is a type of trypsin inhibitor found in sweet potatoes (Hou and Lin, 1997). Sporamin accounts for 60–80% of total proteins in sweet potato (Shewry, 2003). Pereira et al. (2016) found that the antioxidant activity of yacon flour extract was highly correlated (0.88) with its tannin content. The yacon peel extract reported the highest antioxidant activity. This may be used as an alternative to tackle the detrimental consequences of oxidative stress. Dietary supplementation of phytates at low concentration (0.1%) has also been reported to exhibit antioxidant properties (Reddy and Sathe, 2001).

36.6.2.2 Antiulcerative Activities

A study of antiulcerative activity of sweet potato roots conducted on a rat model revealed that the sweet potato extract did not exhibit any toxic or deleterious effects by oral route up to 2,000 mg kg^{-1}. In treated rats, the superoxide dismutase, catalase, glutathione peroxidase, and glutathione reductase activities were significantly elevated indicating the ability of restoring enzyme activities compared to the control (Panda and Sonkamble, 2012).

36.6.2.3 Anticancer Activities

In a study conducted by Wang et al. (2011), it was revealed that H_2O_2-$CuSO_4$ induced damage of calf thymus DNA and protected human lymphoblastoid cells from $CuSO_4$. Induced DNA damage can be checked with the aqueous extract of yam (*Dioscorea alata*). This activity is attributed to the phenolic compounds, saponins, and mucilage polysaccharides present in yams. Antitumour activity has been described for flavonoids and phenolic acids extracted from yam. These extracts are known in China to promote health and longevity (Zhang et al., 2014). Madiwale et al. (2011) reported that purple-fleshed potatoes had a higher potential for suppressing proliferation and elevated apoptosis of HT-29 human colon cancer cell lines in comparison with white-fleshed potatoes. The anticancer influence was more pronounced for fresh potatoes than in stored tubers. The storage duration was found to have a strong positive correlation with antioxidant activity and percentage of viable cancer cells while a negative correlation existed with apoptosis induction. It was concluded that storage enhanced the antioxidant activity and phenolic content, but at the cost of its antiproliferative and proapoptotic activities.

36.6.2.4 Antimicrobial Activity

The phenolic compounds present in yam varieties have been reported to be potential antimicrobial agents. Methanolic extracts of *Dioscorea* yams (*Dioscore adumetorum* and *Dioscorea hirtiflora*) exhibited an antagonistic effect against *Staphylococcus aureus, E. coli, Bacillus subtilis, Proteus mirabilis, Salmonella typhi, Candida albicans, Aspergillus niger,* and *Penicillium chrysogenum.* The agar diffusion and pour plate methods were employed to assess this antimicrobial activity (Sonibare and Abegunde, 2012).

36.6.2.5 Hypoglycemic Activities

Alpha-amylase inhibitors are known to combat gastrointestinal and metabolic activities in diabetic patients (Boivin et al., 1988). Plasma glucose levels were observed to be lowered in diabetic patients upon ingestion of tablets containing extracts of sweet potato peels (Ludvik et al., 2002). In another study by Kusano and Abe (2000), oral administration of white-skinned sweet potato extracts were reported to reduce hyperinsulinemia in Zucker fatty rats by 23%, 26%, 60%, and 50%, after 3 weeks, 4 weeks, 6 weeks, and 8 weeks, respectively. Maithili et al. (2011), found that antidiabetic activity was detected in alloxan-induced diabetic rats when administered with ethanolic extract of tubers of *Dioscorea alata.* The creatinine levels decreased as a consequence of improved renal function by reduced plasma glucose levels and subsequent glycosylation of renal basement membranes.

36.6.2.6 Hypocholesterolemic Activity

Son et al. (2007) demonstrated that diosgenin, a steroidal saponin of yam (*Dioscorea*), exhibited anti-oxidative and hypolipidemic effects *in vivo.* The study revealed that there was reduced total cholesterol level, pancreatic lipase inhibitory activity, protective effect of liver under high-cholesterol diet, and protection against the oxidative damaging effects of polyunsaturated fatty acids in diabetic rats supplemented with either 0.1% or 0.5% diosgenin for 6 weeks. Incorporation of Taiwan yam (*Dioscorea alata*) in the diet (50% w/w) has been reported to abate plasma and hepatic cholesterol levels and also increased fecal steroid excretions in mice models (Chen et al., 2003).

36.7 Molar Ratio Concept

The molar ratio of antinutrient to mineral is an important criterion for predicting potential mineral bioavailability. Lower molar ratios are indicative of higher mineral bioavailability. It can be calculated by dividing the mole of antinutrient to the mole of minerals (Norhaizan and Norfaizadatul, 2009). In cereal crops, the molar ratios of phytate to calcium (Phy:Ca) ranged from 0.016–0.018 (Woldegiorgis et al., 2015). Phytate to iron (Phy:Fe) ratios of the formulated diets varied from 0.047–0.051. The phytate:iron molar ratios >0.15 is indicative of poor iron bioavailability (Siegenberg et al., 1991). The molar ratios of phytate to zinc of the formulated diets varied from 2.30–2.52. Bhandari and Kawabata (2004) demonstrated that bioavailability of dietary zinc might be reduced by phytic acid. Oxalate to calcium (Ox:Ca) ratio of the yam-based diet varied from 0.076–0.080. Oxalic acid and its salts can have deleterious effects on human nutrition and health, particularly by decreasing calcium absorption and aiding the formation of kidney stones (Bhandari and Kawabata, 2004). To date, critical levels/limits of antinutrient to nutrient ratios have not been established for any horticultural commodity as has been done for phytate to mineral (for iron, zinc and calcium) ratio for rice (Lee et al., 2015) and phytate to zinc ratio for wheat and pearl millet (Jou et al., 2012).

36.8 Safe Limits

It is important to be aware of the antinutrient compounds present in tuber and root crops to avoid the adverse effects associated with it and limit its intake accordingly. Safe limits of some of these antinutrient compounds as reported in literature are given in Table 36.2.

TABLE 36.2

Safe Limits for Commonly Found Antinutrients in Tuber and Root Crops

Antinutrients	Safe Limit	References
Furocoumarins	0.25 mg/kg bw (for phototoxic effect)	Dugrand et al., 2013
Phytic acid	10 mg/100 g	Mehrjardi et al., 2014
Oxalate	<10 mg serving per day	Horner et al., 2005
	Foods with >10 mg/125 mL (high-oxalate foods)	Gelinas and Seguin (2007)
Saponins	<5 mg/kg bw	Ndie and Okaka, 2018
α-galactosides	<3 g/day	Martinez-Villaluenga et al., 2008
Trypsin inhibitors	5 mg/g of protein in a finished product	Codex Alimentarius Commission, 1989
Amylase inhibitors	6 g/kg/day for a 70 kg man	Obiro et al. 2008
Nitrates	3.7 mg/kg bw per day	Karwowska and Kononiuk, 2020

* bw = body weight.

36.9 Commercial Applications

Plants naturally produce many secondary metabolites or antinutrients for several reasons including self-defense against other organisms, metal transportation agents, and as agents of symbiosis (Demain and Fang, 2000). These antinutrients are being explored and exploited enormously for use for the betterment of human health. Listed below are some of the commercial applications of antinutrients.

- Amylase inhibitors hamper carbohydrate digestion or absorption. They reduce calorie intake and promote weight loss and combat obesity. They are also known as starch blockers (Mahmood, 2016).
- Asparagus, commonly known as *Safed Musali*, has been identified for its antimicrobial potential due to saponins and tried by various workers as a biopreservative agent in dairy products (Saini, 2008).
- Polyphenols are used in the production of nutraceuticals and as preservatives. They are also incorporated as active ingredients in the development of functional foods (El Gharras, 2009; Martillanes et al., 2017).
- At low doses, α-galactosides have been found to be effective as prebiotics and beneficially stimulate the growth and activity of living bifidobacteria and lactobacilli in the human colon. This improves the digestive health. The recommended dose for obtaining this benefit has been reported to be 3 g day^{-1} (Martinez-Villaluenga et al., 2008).
- Alpha-galactosides can selectively stimulate the growth of bifidobacteria in the colonic microbiota and they are considered as model type prebiotics. Hence, α-galactosides could be recognized as functional food ingredients for which health claims may become authorized (Tomomatsu, 1994).
- Nitrates, at a concentration <200 ppm, are used in meat processing as curing agents. They also aid in the preservation process (Honikel, 2008).
- Saponins are used in the food industry for the manufacture of beverages and food ingredients. It has also found its use in the cosmetic industries for the production of shampoos, liquid detergents, toothpastes and extinguishers as an emulsifier and long-lasting foaming agent (Yukuyoshi et al., 2012).
- Protease inhibitors, such as trypsin inhibitors, are reported to be used as insecticide. They are either fed to the insects through bait or by the production of transgenic crops through gene stacking/multigene engineering (Singh et al., 2020).
- Tannins are used in the leather industry for dyeing purposes, as a clarifying agent for wine and beer in the food industry, as an antiscale formation agent in boilers, for reducing the viscosity of drilling mud in oil wells, and in the nutraceutical and pharmaceutical industries (Singh and Kumar, 2019).

36.10 Future Prospect

Most of the antinutrients' characterization, validation, and analysis protocols are focused around cereals, pulses, and oilseeds. Therefore, data pertaining to horticultural crops specifically for tuber and root crops are scant. Additionally, most of them are underutilized and the molar ratio concept is also limited in targeted arable crops. The time has come when the dependency for food and nutritional security is also going to be inclined toward horticultural produce. Biochemical attributes, physiology, storage requirements, postharvest management, and sensory profiles of horticultural produce are altogether different than cereals, pulses, and oilseed commodities through plant origin. Therefore, intensive research work is required to see the effect of various primary and secondary processing techniques (including minimal processing) and storage on antinutritional factors of horticultural crops including root and tuber crops.

36.11 Conclusion

Observing all horticultural crops at once, most of the tuber and root crops are generally thermally processed before consumption, therefore in comparison with other crops, their consumption is relatively safe. Sensory and processing limitations associated with root and tuber crops also limit their intake, thus saving the consumers from the risk associated with excess intake of antinutritional factors. Such self-limiting factors prove the suitability and superiority of tuber and root crop produce over other horticultural crops. Their unique ability to provide bulk to the diet, huge production, and easy availability and affordability also make them a suitable carrier for nutrient fortificants' delivery to the poor masses (vulnerable). Therefore, exhaustive research work is required not only from an ANF quantification point of view and minimization strategies, but to develop shelf stable, market-competitive, and versatile products. The research and development of products can help crops bifunctionality reach everyone—from the vulnerable as well as elite consumer class of society.

REFERENCES

Adane, T., A. Shimelis, R. Negussie, B. Tilahun, and G. D. Haki 2013. Effect of processing method on the Proximate composition, mineral content and antinutritional factors of Taro (*Colocasia esculenta*, L.) growth in Ethiopia. *African Journal of Food, Agriculture, Nutrition and Development* 13(2).

Adegunwa, M. O., E. O. Alamu, and L. A. Omitogun 2011. Effect of processing on the nutritional contents of yam and cocoyam tubers. *Journal of Applied Biosciences* 46: 3086–3092.

Ahn, J. H., B. M. Robertson, R. Elliot, R. C. Gutterridge, and C. W., Ford 1989. Quality assessment of tropical browse legumes, tannin content and protein degradation. *Animal Feed Science and Technology* 27: 147–156.

Akwaowo, E. U., B. A. Ndon, and E. U. Etuk 2000. Minerals and antinutrients in fluted pumpkin (*Telfairia occidentalis* Hook f.). *Food Chemistry* 70(2): 235–40.

Alonso, R., E. Orue, and F. J. F. C. Marzo 1998. Effects of extrusion and conventional processing methods on protein and antinutritional factor contents in pea seeds. *Food Chemistry* 63(4): 505–512.

AOCS. 1997. Determination of glucosinolate content in rapeseed and canola by HPLC. Official Method Ak 1-92. Official Methods and Recommended Practices of the AOCS. American Oil Chemist's Society: Champaign, IL, USA.

Barky, A. R., S. A. Hussein, and A. E. Alm-Eldeen 2017. Saponins and their potential role in diabetes mellitus. *Diabetes Management* 7(1): 148.

Batista, N. N., C. L. Ramos, L. D. F. Vilela, D. R. Dias, and R. F. Schwan 2019. Fermentation of yam (*Dioscorea* spp. L.) by indigenous phytase-producing lactic acid bacteria strains. *Brazilian Journal of Microbiology* 50(2): 507–514.

Bhandari, M. R., and J. Kawabata 2004. Assessment of antinutritional factors and bioavailability of calcium and zinc in wild yam (*Dioscorea* spp.) tubers of Nepal. *Food Chemistry* 85(2): 281–287.

Bhandari, M. R., and J. Kawabata 2006. Cooking effects on oxalate, phytate, trypsin and α-amylase inhibitors of wild yam tubers of Nepal. *Journal of Food Composition and Analysis* 19(6–7): 524–530.

Bhandari, M. R., T. Kasai, and J. Kawabata 2003. Nutritional evaluation of wild yam (*Dioscorea* spp.) tubers of Nepal. *Food Chemistry* 82(4): 619–623.

Bjarnholt, N., and B. L. Møller 2008. Hydroxynitrile glucosides. *Phytochemistry* 69(10): 1947–1961.

Blagbrough, I. S., S. A. Bayoumi, M. G. Rowan, and J. R. Beeching 2010. Cassava: An appraisal of its phytochemistry and its biotechnological prospects. *Phytochemistry* 71(17–18): 1940–1951.

Boivin, M., B. Flourie, R. A. Rizza, V. L. W. Go, and E. P. DiMagno 1988. Gastrointestinal and metabolic effects of amylase inhibition in diabetics. *Gastroenterology* 94(2): 387–394.

Brkić, D., J. Bošnir, M. Bevardi, A. G. Bošković, S. Miloš, D. Lasić, A. Krivohlavek, A. Racz, A. Mojsović-Ćuić, and N. U. Trstenjak 2017. Nitrate in leafy green vegetables and estimated intake. *African Journal of Traditional, Complementary and Alternative Medicinea* 14:31–41.

Brunner, J. H. 1984. Direct spectrophotometric determination of saponin. *Analytical Chemistry* 34(396): 1314–1326.

Buta, B. M. 2020. Evaluation of oxalate content in boyna and taro roots grown in Areka (Ethiopia). *World Scientific Research* 7(1): 12–16.

Cain, N., S. J. Darbyshire, A. Francis, R. E. Nurse, and M. J. Simard 2010. The biology of Canadian weeds. 144. *Pastinaca sativa* L. *Canadian Journal of Plant Science* 90(2): 217–240.

Cardenas, H., D. Arango, C. Nicholas, S. Duarte, G. J. Nuovo, W. He, O. H. Voss, M. E. Gonzalez-Mejia, D. C. Guttridge, E. Grotewold, and A. I. Doseff 2016. Dietary apigenin exerts immune-regulatory activity in vivo by reducing NF-κB Activity, halting leukocyte infiltration and restoring normal metabolic function. *International Journal of Molecular Sciences* 17(3): 323.

Carlson, D. G., M. E. Daxenbichler, C. H. VanEtten, C. B. Hill, and P. H. Williams 1985. Glucosinolates in radish cultivars. *Journal of the American Society for Horticultural Science* 110(5): 634–638.

Cereda, M. P., and M. C. Y. Mattos 1996. Linamarin: The toxic compound of cassava. *Journal of Venomous Animals and Toxins* 2(1): 06–12.

Champagne, A., G. Hilbert, L. Legendre, and V. Lebot 2011. Diversity of anthocyanins and other phenolic compounds among tropical root crops from Vanuatu, South Pacific. *Journal of Food Composition and Analysis* 24(3): 315–325.

Chandrasekara, A., and T. J. Kumar 2016. Roots and tuber crops as functional foods: A review on phytochemical constituents and their potential health benefits. *International Journal of Food Science* 2016:3631647. http://dx.doi.org/10.1155/2016/3631647

Charles, A. L., K. Sriroth, and T. C. Huang 2005. Proximate composition, mineral contents, hydrogen cyanide and phytic acid of 5 cassava genotypes. *Food Chemistry* 92(4): 615–620.

Chen, H. L., C. H. Wang, C. T. Chang, and T. C. Wang 2003. Effects of Taiwanese yam (*Dioscorea japonica* Thunb var. *pseudojaponica* Yamamoto) on upper gut function and lipid metabolism in Balb/c mice. *Nutrition* 19(7-8): 646–651.

Chen, J., L. Liu, M. Li, X. Yu, and R. Zhang 2018. An improved method for determination of cyanide content in bitter almond oil. *Journal of Oleo Science* 67(3): 289–294.

Chetty A. A., and S. Prasad 2009. Flow injection analysis of nitrate-N determination in root vegetables: Study of the effects of cooking. *Food Chemistry* 116:561–566.

Choquechambi, L. A., I. R. Callisaya, A. Ramos, H. Bosque, A. Mújica, S. E. Jacobsen, M. Sørensen, and E. O. Leidi 2019. Assessing the nutritional value of root and tuber crops from Bolivia and Peru. *Foods* 8(11): 526.

Christensen, L. P. 2011. Aliphatic C(17)-polyacetylenes of the falcarinol type as potential health promoting compounds in food plants of the Apiaceae family. *Recent Patents on Food, Nutrition and Agriculture* 3 (1): 64–77.

Chung, S. W., J. C. Tran, K. S. Tong, M. Y. Chen, Y. Xiao, Y. Y. Ho, and C. H. Chan 2011. Nitrate and nitrite levels in commonly consumed vegetables in Hong Kong. *Food Additives and Contaminants* 4(1): 34–41.

Codex Alimentarius Commission. (1989). Report of the Fifth Session of the Codex Committee on vegetable proteins. Ottawa, Canada February, 6-10.

Coe, F. L., A. Evan, and E. Worcester 2005. Kidney stone disease. *The Journal of Clinical Investigation* 115(10): 2598–2608.

Cook, D. W., M. L. Burnham, D. C. Harmes, D. R. Stoll, and S. C. Rutan 2017. Comparison of multivariate curve resolution strategies in quantitative LCxLC: Application to the quantification of furanocoumarins in apiaceous vegetables. *Analytica Chimica Acta* 961: 49–58.

Czepa, A., and T. Hofmann 2003. Structural and sensory characterization of compounds contributing to the bitter off-taste of carrots (*Daucus carota* L.) and carrot puree. *Journal of Agricultural and Food Chemistry* 51(13): 3865–3873.

Delgado-Vargas, F., A. R. Jiménez, and O. Paredes-López 2000. Natural pigments: Carotenoids, anthocyanins, and betalains—characteristics, biosynthesis, processing, and stability. *Critical Reviews in Food Science and Nutrition* 40(3): 173–289.

Delimont, N. M., M. D. Haub, and B. L. Lindshield 2017. The impact of tannin consumption on iron bioavailability and status: A narrative review. *Current Developments in Nutrition* 1(2): 1–12.

Demain, A. L., and A. Fang 2000. The natural functions of secondary metabolites. In: Fiechter A. (ed.), *History of Modern Biotechnology, vol. 69*. Springer, Berlin, Heidelberg, 1–39.

Deng, Q., K. G. Zinoviadou, C. M. Galanakis, V. Orlien, N. Grimi, E. Vorobiev, N. Lebovka, and F. J. Barba 2015. The effects of conventional and non-conventional processing on glucosinolates and its derived forms, isothiocyanates: extraction, degradation, and applications. *Food Engineering Reviews* 7(3): 357–381.

Dey, Y. N., M. M. Wanjari, D. Kumar, V. Lomash, S. N. Gaidhani, and A. K. Jadhav 2017. Oral toxicity of elephant foot yam (*Amorphophallus paeoniifolius*) tuber in mice. *Journal of Pharmacy & Pharmacognosy Research* 5(1): 55–68.

Diop, A., and D. J. B. Calverley 1998. Storage and Processing of Roots and Tubers in the Tropics. Food and Agriculture Organization of the United Nations. Available at: http://www.fao.org/docrep/X5415E/X5415E00.htm

Dugrand, A., A. Olry, T. Duval, A. Hehn, Y. Froelicher, and F. Bourgaud 2013. Coumarin and furanocoumarin quantitation in citrus peel via ultraperformance liquid chromatography coupled with mass spectrometry (UPLC-MS). *Journal of Agricultural and Food Chemistry* 61(45): 10677–10684.

Ebrahimi, R., A. Ahmadian, A. Ferdousi, S. Zandi, B. Shahmoradi, R. Ghanbari, S. Mahammadi, R. Rezaee, M. Safari, H. Daraei, and A. Maleki 2020. Effect of washing and cooking on nitrate content of potatoes (cv. Diamant) and implications for mitigating human health risk in iran. *Potato Research* 63:449–462.

El Gharras, H. (2009). Polyphenols: Food sources, properties and applications—A review. *International Journal of Food Science & Technology* 44(12): 2512–2518.

Embaby, H. E., R. A. Habiba, A. A. Shatta, M. M. Elhamamy, N. Morita, and S. S. Ibrahim 2010. Glucosinolates and other anti-nutritive compounds in canola meals from varieties cultivated in Egypt and Japan. African Journal of Food, Agriculture, Nutrition and Development 10(8).

Ertop, M. H., and M. Bektaş 2018. Enhancement of bioavailable micronutrients and reduction of antinutrients in foods with some processes. *Food and Health* 4(3): 159–165.

Escribano, J., M. A. Pedreño, F. García-Carmona, and R. Muñoz 1998. Characterization of the antiradical activity of betalains from *Beta vulgaris* L. roots. *Phytochemical Analysis* 9: 124–127.

Etsuyankpa, M. B., C. E. Gimba, E. B. Agbaji, I. Omoniyi, M. M. Ndamitso, and J. T. Mathew 2015. Assessment of the effects of microbial fermentation on selected anti-nutrients in the products of four local cassava varieties from Niger state, Nigeria. *American Journal of Food Science and Technology* 3(3): 89–96.

Eymar, E., C. Garcia-Delgado, and R. M. Esteban 2016. Food poisoning: Classification. In: B. Caballero, P. M. Finglas, and F. Toldrá (Eds.), *Encyclopedia of Food and Health, Academic Press*, 56–66.

Fahey, J. W., A. T. Zalcmann, and P. Talalay 2001. The chemical diversity and distribution of glucosinolates and isothiocyanates among plants. *Phytochemistry* 56(1): 5–51.

Fernando, R., M. D. P. Pinto, and A. Pathmeswaran 2012. Goitrogenic food and prevalence of goitre in Sri Lanka. *International Journal of Internal Medicine* 1(2): 17–20.

Finotti E., A. Bertone, and V. Vivanti 2006. Balance between nutrients and anti-nutrients in nine Italian potato cultivars. *Food Chemistry* 99: 698–701.

Friedman, M., G. M. McDonald, and M. Filadelfi-Keszi 1997. Potato glycoalkaloids: Chemistry, analysis, safety, and plant physiology. *Critical Reviews in Plant Sciences* 16(1): 55–132.

Garrod, B., B. G. Lewis, and D. T. Coxon 1978. Cis-heptadeca-1,9-diene-4,6-diyne-3,8-diol, an antifungal polyacetylene from carrot root tissue. *Physiological Plant Pathology* 13(2): 241–246.

Gelinas, B., and P. Seguin 2007. Oxalate in grain amaranth. *Journal of Agricultural and Food Chemistry* 55(12): 4789–4794.

Gemede, H. F., and H. Fekadu 2014. Nutritional composition, antinutritional factors and effect of boiling on nutritional composition of Anchote (*Coccinia abyssinica*) tubers. *Journal of Scientific and Innovative Research* 3(2): 177–188.

Gemede, H. F., and N. Ratta 2014. Antinutritional factors in plant foods: Potential health benefits and adverse effects. *International Journal of Nutrition and Food Sciences* 3(4): 284–289.

Giri A.P., and M. S. Kachole 2004. Amylase inhibitors of pigeon pea (*Cajanus cajan*) seeds. *Phytochemistry* 47: 197–202.

Golden, M. 2009. Proposed nutrient requirements of moderately malnourished populations of children. *Food and Nutrition Bulletin* 30(3): S267–S342.

Graham, J., and J. Traylor 2018. Cyanide toxicity. StatPearls. StatPearls Publishing, Treasure Island (FL).

Grubben, G. J. H., and O. A. Denton 2004. *Plant Resources of Tropical Africa, vol 2. Vegetables.* PROTA Foundation, Wageningen, Netherlands.

Güçlü-Üstündağ, Ö., and G. Mazza 2007. Saponins: Properties, applications and processing. *Critical Reviews in Food Science and Nutrition* 47(3): 231–258.

Hale, A. L., L. Reddivari, M. N. Nzaramba, J. B. Bamberg, and J. C. Miller 2008. Interspecific variability for antioxidant activity and phenolic content among Solanum species. *American Journal of Potato Research* 85(5): 332.

Harborne, J. B. 1989. *Biosynthesis and function of anti-nutritional factors in plants. Aspects of Applied Biology (UK)* 19: 21–28.

Haug, W., and H. J. Lantzsch 1983. Sensitive method for the rapid determination of phytate in cereals and cereal products. *Journal of the Science of Food and Agriculture* 34(12): 1423–1426.

Holmes, R. P., and M. Kennedy 2000. Estimation of the oxalate content of foods and daily oxalate intake. *Kidney International* 57(4): 1662–1667.

Honikel, K. O. 2008. The use and control of nitrate and nitrite for the processing of meat products. *Meat Science* 78(1-2): 68–76.

Horner, H. T., T. Cervantes-Martinez, R. Healy, M. B. Reddy, B. L. Deardorff, T. B. Bailey, I. Al-Wahsh, L. K. Massey, and R. G. Palmer 2005. Oxalate and phytate concentrations in seeds of soybean cultivars (*Glycine max* (L.) Merr.). *Journal of Agricultural and Food Chemistry* 53(20): 7870–7877.

Hou, W. C., and Y. H. Lin 1997. Dehydroascorbate reductase and monodehydroascorbate reductase activities of trypsin inhibitors, the major sweet potato (*Ipomoea batatas* [L.] Lam) root storage protein. *Plant Science* 128(2): 151–158.

Iorizzo, M., J. Curaba, M. Pottorff, M. G. Ferruzzi, P. Simon, and P. F. Cavagnaro 2020. Carrot anthocyanins genetics and genomics: Status and perspectives to improve its application for the food colorant industry. *Genes* 11(8): 906.

Ishida, M., T. Kakizaki, T. Ohara, and Y. Morimitsu 2011. Development of a simple and rapid extraction method of glucosinolates from radish roots. *Breeding Science* 61(2): 208–211.

Iwuoha, C. I., and F. A. Kalu 1995. Calcium oxalate and physico-chemical properties of cocoyam (*Colocasia esculenta* and *Xanthosoma sagittifolium*) tuber flours as affected by processing. *Food Chemistry* 54(1): 61–66.

Joshi, A., J. Prakash, V. R. Sagar, and B. S. Tomar 2019a. Betanin rich papaya candy (tutti-fruity): An innovative fusion product with enhanced functionality. *International Journal of Tropical Agriculture* 37(4): 459–465.

Joshi, A., P. Raigond, B. Kaundal, R. Kumar, and B. Singh 2015. ELISA based phytate estimation in Indian potato cultivars. *ICAR-CPRI Newsletter.*

Joshi, A., S. Sethi, V. R. Sagar, B. S. Tomar, P. Raigond and B. Singh 2019b. Betanins rich purple potato chips: An innovative fusion product with enhanced functionality. *Journal of Agricultural Engineering and Food Technology* 6(1): 1–3.

Jou, M. Y., X. Du, C. Hotz, and B. Lönnerdal 2012. Biofortification of rice with zinc: Assessment of the relative bioavailability of zinc in a Caco-2 cell model and suckling rat pups. *Journal of Agricultural and Food Chemistry* 60(14): 3650–3657.

Judprasong, K., N. Archeepsudcharit, K. Chantapiriyapoon, P. Tanaviyutpakdee, and P. Temviriyanukul 2018. Nutrients and natural toxic substances in commonly consumed Jerusalem artichoke (*Helianthus tuberosus* L.) tuber. *Food Chemistry* 238: 173–179.

Kanner J., S. Harel, and R. Granit 2001. Betalains—A new class of dietary cationized antioxidants. *Journal of Agricultural and Food Chemistry* 49:5178–5185.

Karwowska, M., and A. Kononiuk 2020. Nitrates/nitrites in food—Risk for nitrosative stress and benefits. *Antioxidants* 9(3): 241.

Kaspar, K. L., J. S. Park, C. R. Brown, B. D. Mathison, D. A. Navarre, and B. P. Chew 2011. Pigmented potato consumption alters oxidative stress and inflammatory damage in men. *The Journal of Nutrition* 141(1): 108–111.

Kaushik, G., P. Singhal, and S. Chaturvedi 2018. Food processing for increasing consumption: The case of legumes. In: A. M. Grumezescu and A. M. Holban (Eds.), *Food Processing for Increased Quality and Consumption,* Academic Press, 1–28.

Kaushik, G., S. Satya, and S. N. Naik 2010. Effect of domestic processing techniques on the nutritional quality of the soybean. *Mediterranean Journal of Nutrition and Metabolism* 3(1), 39–46.

Khokhar, S., and B. M. Chauhan 1986. Anti-nutritional factors in moth beans (*Vigna aconitifolia*): Varietal difference and effects of methods of domestic processing and cooking. *Journal of Food Science* 51(3): 591–594.

Khokhar, S., and R. K. O. Apenten 2003. Antinutritional factors in food legumes and effects of processing. In: V. R. Squires (Ed.), *The Role of Food, Agriculture, Forestry and Fisheries in Human Nutrition, vol* 4. EOLSS Publications, 82–116.

Kies, A. K., L. H. De Jonge, P. A. Kemme, and A. W. Jongbloed 2006. Interaction between protein, phytate, and microbial phytase. *In vitro studies. Journal of Agricultural and Food Chemistry* 54(5): 1753–1758.

Kiran, K. S., and G. Padmaja 2003. Inactivation of trypsin inhibitors in sweet potato and taro tubers during processing. *Plant Foods for Human Nutrition* 58(2): 153–163.

Kobayashi, N. I., K. Tanoi, A. Hirose, T. Saito, A. Noda, N. Iwata, A. Nakano, S. Nakamura, and T. M. Nakanishi 2011. Analysis of the mineral composition of taro for determination of geographic origin. *Journal of Agricultural and Food Chemistry* 59(9): 4412–4417.

Koleva, I. I., T. A. V. Beek, A. E. Soffers, B. Dusemund, and I. M. Rietjens 2012. Alkaloids in the human food chain-natural occurrence and possible adverse effects. *Molecular Nutrition and Food Research* 56(1): 30–52.

König, V., L. Vértesy, and T. R. Schneider 2003. Structure of the α-amylase inhibitor tendamistat at 0.93 Å. *Acta Crystallographica Section D: Biological Crystallography* 59(10): 1737–1743.

Kristl, J., A. Ivancic, A. Mergedus, V. Sem, M. Kolar, and V. Lebot 2016. Variation of nitrate content among randomly selected taro (*Colocasia esculenta* (L.) Schott) genotypes and the distribution of nitrate within a corm. *Journal of Food Composition and Analysis* 47: 76–81.

Kumar, M., Dahuja, A., Sachdev, A., Kaur, C., Varghese, E., Saha, S., and Sairam, K. V. S. S. (2019a). Valorisation of black carrot pomace: Microwave assisted extraction of bioactive phytoceuticals and antioxidant activity using Box-Behnken design. *Journal of Food Science and Technology*, 56(2): 995–1007.

Kumar, M., Dahuja, A., Sachdev, A., Kaur, C., Varghese, E., Saha, S., and Sairam, K. V. S. S. (2019b). Evaluation of enzyme and microwave-assisted conditions on extraction of anthocyanins and total phenolics from black soybean (*Glycine max* L.) seed coat. *International Journal of Biological Macromolecules*, 135: 1070–1081.

Kumar, M., Dahuja, A., Sachdev, A., Kaur, C., Varghese, E., Saha, S., and Sairam, K. V. S. S. (2020a). Black Carrot (*Daucus carota* ssp.) and black soybean (*Glycine max* (L.) Merr.) anthocyanin extract: A remedy to enhance stability and functionality of fruit juices by copigmentation. *Waste and Biomass Valorization*, 11(1): 99–108.

Kumar, M., Dahuja, A., Sachdev, A., Kaur, C., Varghese, E., Saha, S., and Sairam, K. V. S. S. (2018). Valorization of Black carrot marc: Antioxidant properties and enzyme assisted extraction of flavonoids. *Research Journal of Biotechnology*, 13(11): 12–21.

Kumar, M., Tomar, M., Punia, S., Amarowicz, R., and Kaur, C. (2020b). Evaluation of cellulolytic enzyme-assisted microwave extraction of *Punica granatum* peel phenolics and antioxidant activity. *Plant Foods for Human Nutrition*, 75(4): 614–620.

Kusano, S., and H. Abe 2000. Antidiabetic activity of white skinned sweet potato (*Ipomoea batatas* L.) in obese Zucker fatty rats. *Biological and Pharmaceutical Bulletin* 23(1): 23–26.

Leathers, R. R., C. Davin, and J. P. Zryd 1992. Betalain producing cell cultures of Beta vulgaris L. var. bikores monogerm (red beet). *In Vitro–Plant* 28(2): 39–45.

Lebot, V., B. Faloye, E. Okon, and B. Gueye 2019. Simultaneous quantification of allantoin and steroidal saponins in yam (Dioscorea spp.) powders. *Journal of Applied Research on Medicinal and Aromatic Plants* 13: 100200.

Lee, H. H., S. P. Loh, C. F. J. Bong, S. R. Sarbini, and P. H. Yiu 2015. Impact of phytic acid on nutrient bioaccessibility and antioxidant properties of dehusked rice. *Journal of Food Science and Technology* 52(12): 7806–7816.

Liener, I. E., R. J. Summerfield, and A. H. Bunting 1980. Advances in Legume Science. Royal Botanic Gardens, Kew.

Lin, L.Z., and J. M. Harnly 2010. Phenolic component profiles of mustard greens, yu choy, and 15 other Brassica vegetables. *Journal of Agricultural and Food Chemistry* 58(11): 6850–6857.

Linde, G. A., Z. C. Gazim, B. K. Cardoso, L. F. Jorge, V. Tešević, J. Glamočlija, M. Soković, and N. B. Colauto 2016. Antifungal and antibacterial activities of Petroselinum crispum essential oil. *Genetics and Molecular Research* 15(3).

Liu, T., D. J. Burritt, G. T. Eyres, and I. Oey 2018. Pulsed electric field processing reduces the oxalate content of oca (*Oxalis tuberosa*) tubers while retaining starch grains and the general structural integrity of tubers. *Food chemiStry* 245: 890–898.

Llano, J., J. Raber, and L. A. Eriksson 2003. Theoretical study of phototoxic reactions of psoralens. *Journal of Photochemistry and Photobiology A: Chemistry* 154(2–3): 235–243.

Lu, Y. L., C. Y. Chia, Y. W. Liu, and W. C. Hou 2012. Biological activities and applications of dioscorins, the major tuber storage proteins of yam. *Journal of Traditional and Complementary Medicine* 2(1): 41–46.

Ludvik, B. H., K. Mahdjoobian, W. Waldhaeusl, A. Hofer, R. Prager, A. Kautzky-Willer, and G. Pacini 2002. The effect of *Ipomoea batatas* (Caiapo) on glucose metabolism and serum cholesterol in patients with type 2 diabetes: A randomized study. *Diabetes Care* 25(1): 239–240.

MacLeod, G., and J. M. Ames 1989. Volatile components of celery and celeriac. *Phytochemistry* 28(7): 1817–1824.

Madiwale, G. P., L. Reddivari, D. G. Holm, and J. Vanamala 2011. Storage elevates phenolic content and anti-oxidant activity but suppresses antiproliferative and pro-apoptotic properties of colored-flesh potatoes against human colon cancer cell lines. *Journal of Agricultural and Food Chemistry* 59(15): 8155–8166.

Mahmood, N. 2016. A review of α-amylase inhibitors on weight loss and glycemic control in pathological state such as obesity and diabetes. *Comparative Clinical Pathology* 25(6): 1253–1264.

Maithili, V., S. P. Dhanabal, S. Mahendran, and R.Vadivelan 2011. Antidiabetic activity of ethanolic extract of tubers of *Dioscorea alata* in alloxan induced diabetic rats. *Indian Journal of Pharmacology* 43(4): 455.

Maldini, M., M. Foddai, F. Natella, G. L. Petretto, J. P. Rourke, M. Chessa, and G. Pintore 2017. Identification and quantification of glucosinolates in different tissues of *Raphanus raphanistrum* by liquid chromatography tandem-mass spectrometry. *Journal of Food Composition and Analysis* 61: 20–27.

Martillanes, S., J. Rocha-Pimienta, M. Cabrera-Bañegil, D. Martín-Vertedor, and J. Delgado-Adámez 2017. Application of phenolic compounds for food preservation: Food additive and active packaging. In: *Phenolic Compounds–Biological Activity. London, UK: IntechOpen*, 39–58.

Martinez-Villaluenga, C., J. Frias, and C. Vidal-Valverde 2008. Alpha-galactosides: Antinutritional factors or functional ingredients? *Critical Reviews in Food Science and Nutrition* 48(4): 301–316.

Mathew, S. and Abraham, E. 2006. In vitro antioxidant activity and scavenging effects of Cinnamomum verum leaf extract assayed by different methodologies. Food and Chemical Toxicology 44: 198–206.

Mehrjardi, M. M., A. Dehghani, G. J. Khaniki, F. S. Hosseini, B. Hajimohammadi, and N. Nazary 2014. Determination of phytic acid content in different types of bread and dough consumed in Yazd, Iran. *Journal of Food Quality and Hazards Control* 1(1): 29–31.

Mennen, L. I., R. Walker, C. Bennetau-Pelissero, and A. Scalbert 2005. Risks and safety of polyphenol consumption. *The American Journal of Clinical Nutrition* 81(1): 326S–329S.

Meyer, H., A. Bolarinwa, G. Wolfram, and J. Linseisen 2006. Bioavailability of apigenin from apiin-rich parsley in humans. *Annals of Nutrition and Metabolism* 50(3): 167–172.

Mirecki, N., Z. S. Ilic, L. Šunic, and A. Rukie 2015. Nitrate content in carrot, celeriac and parsnip at harvest time and during prolonged cold storage. *Fresenius Environmental Bulletin* 24: 3266–3273.

Montagnac, J. A., C. R. Davis, and S. A. Tanumihardjo 2009. Nutritional value of cassava for use as a staple food and recent advances for improvement. *Comprehensive Reviews in Food Science and Food Safety* 8(3): 181–194.

Morris S. C., and T. H. Lee 1984. The toxicity and teratogenicity of Solanaceae glycoalkaloids, particularly those of potato (*Solanum tuberosum*). *Food Technology in Australia* 36: 118–124.

Mozolewski, W., and S. Smoczyński 2004. Effect of culinary processes on the content of nitrates and nitrites in potato. *Pakistan Journal of Nutrition* 3(6): 375–361.

Mulualem, T., F. Mekbib, S. Hussein, and E. Gebre 2018. Analysis of biochemical composition of yams (*Dioscorea* spp.) landraces from Southwest Ethiopia. *Agrotechnology* 7(1): 1–8.

Mulualem, T., N. Semman, G. Etana, and S. Alo 2020. Evaluation of cassava (*Manihot esculenta* Crantz) genotypes for total cyanide content, storage tuber and starch yield in South Western Ethiopia. *International Journal of Biomedical Materials Research* 8(2): 14–19.

Natesh, H. N., L. Abbey, and S. K. Asiedu 2017. An overview of nutritional and antinutritional factors in green leafy vegetables. *Horticulture International Journal* 1(2): 00011.

Ndie, E. C., and J. C. Okaka 2018. Risk assessment of antinutrient consumption of plant foods of south eastern Nigeria. *Journal of Food Science and Nutrition* 1(2): 9–12.

Nishad, J., Dutta, A., Saha, S., Rudra, S. G., Varghese, E., Sharma, R. R., Tomar, M., Kumar, M. & Kaur, C. 2020. Ultrasound-assisted development of stable grapefruit peel polyphenolic nano-emulsion: Optimization and application in improving oxidative stability of mustard oil. *Food Chemistry*, 334, 127561.

Noonan, S. C., and G. P. Savage 1999. Oxalate content of foods and its effect on humans. *Asia Pacific Journal of Clinical Nutrition*. 8 (1): 64–74.

Norhaizan, M. E., and A. A. Norfaizadatul 2009. Determination of phytate, iron, zinc, calcium contents and their molar ratios in commonly consumed raw and prepared food in Malaysia. *Malaysian Journal of Nutrition* 15(2): 213–222.

Oatway, L., T. Vasanthan, and J. H. Helm 2001. Phytic acid. *Food Reviews International* 17(4): 419–431.

Obiro, W. C., T. Zhang, and B. Jiang 2008. The nutraceutical role of the *Phaseolus vulgaris* α-amylase inhibitor. *British Journal of Nutrition* 100(1): 1–12.

Oboh, S., A. Ologhobo, and O. Tewe 1989. Some aspects of the biochemistry and nutritional value of the sweet potato (*Ipomea batatas*). *Food Chemistry* 31(1): 9–18.

Ogbonna, A. I., S. O. Adepoju, C. I. C. Ogbonna, T. Yakubu, J. U. Itelima, and V. Y. Dajin 2017. Root tuber of *Tacca leontopetaloides* L. (kunze) for food and nutritional security. *Microbiology: Current Research* 1(1):7–13.

Olayemi, F. O. 2010. Review on some causes of male infertility. *African Journal of Biotechnology* 9(20).

Panda, V., and M. Sonkamble 2012. Phytochemical constituents and pharmacological activities of *Ipomoea batatas* l. (Lam)—A review. *International Journal of Research in Phytochemistry and Pharmacology* 2(1): 25–34.

Pavlista, A. D. 2001. G1437 Green potatoes: The problems and the solution. Historical Materials from University of Nebraska-Lincoln Extension. Paper 88. Available from: http://digitalcommons.unl.edu/extensionhist/88

Peetabas, N., R. P. Panda, N. Padhy, and G. Pal 2015. Nutritional composition of two edible aroids. *International Journal of Bioassays* 4: 4085–4087.

Pereira, J. A. R., M. C. Teixeira, A. A. Saczk, M. D. F. P. Barcelos, M. F. D. Oliveira, and W. C. D. Abreu 2016. Total antioxidant activity of yacon tubers cultivated in Brazil. *Ciência e Agrotecnologia* 40(5): 596–605.

Pfeifer, I., A. Murauer, and M. Ganzera 2016. Determination of coumarins in the roots of *Angelica dahurica* by supercritical fluid chromatography. *Journal of Pharmaceutical and Biomedical Analysis* 129: 246–251.

Phillippy, B. Q., M. Lin, and B. Rasco 2004. Analysis of phytate in raw and cooked potatoes. *Journal of Food Composition and Analysis* 17(2): 217–226.

Prasad, S., and A. A. Chetty 2008. Nitrate-N determination in leafy vegetables: Study of the effects of cooking and freezing. *Food Chemistry* 106(2): 772–780.

Raes, K., D. Knockaert, K. Struijs, and J. V. Camp 2014. Role of processing on bioaccessibility of minerals: Influence of localization of minerals and anti-nutritional factors in the plant. *Trends in Food Science & Technology* 37(1): 32–41.

Raghuramulu, N., K. Madhavan Nair, and S. Kalyanasundaram 2003. *A Manual of Laboratory techniques*, 2nd ed. National Institute of Nutrition, Indian Council of Medical Research, Hydrabad (India).

Raigond, P., Bandana, A. Joshi, S. Dutt, M. Tomar, A. Mehta, B. Singh, and S. K. Chakrabarti 2016. Potato Biochemistry: A Laboratory Manual. Technical Bulletin No. 102, ICAR—Central Potato Research Institute, Shimla, India.

Ramos, A. D. S., R. D. M. Verçosa, S. M. L. Teixeira and B. E. Teixeira-Costa 2020. Calcium oxalate content from two Amazonian amilaceous roots and the functional properties of their isolated starches. *Food Science and Technology* 40(3).

Ravi, V., C. S. Ravindran, and G. Suja 2009. Growth and productivity of elephant foot yam (*Amorphophallus paeoniifolius* [Dennst.] Nicolson): An overview. *Journal of Root Crops* 35(2): 131–142.

Ravindran, V., G. Ravindran, and S. Sivalogan 1994. Total and phytate phosphorus contents of various foods and feedstuffs of plant origin. *Food Chemistry* 50(2): 133–136.

Reddy, N. R., and Sathe, S. K. (Eds.). 2001. *Food Phytates*. CRC Press.

Rekha, M. R., and G. Padmaja 2002. Alpha-amylase inhibitor changes during processing of sweet potato and taro tubers. *Plant Foods for Human Nutrition* 57(3–4): 285–294.

Rickard, S. E., and L. U. Thompson 1997. Interactions and biological effects of phytic acid. In; Shahidi, F. (Ed.), Antinutrients and Phytochemicals in Food. American Chemical Society: Washington, DC, 294–312.

Saini, P. 2008. Preparation of shatavari (*Asparagus racemosus*) powder and its utilization as a preservative and therapeutic agent in burfi. Ph.D. thesis, GB Pant University of Agriculture and Technology, Pantnagar-263145, Uttarakhand.

Saito, K., M. Horie, Y. Hoshino, N. Nose, and H. Nakazawa 1990. High-performance liquid chromatographic determination of glycoalkaloids in potato products. *Journal of Chromatography A* 508: 141–147.

Salunke, B. K., 2006. Anti-nutritional constituents of different grain legumes grown in North Maharashtra. *Journal of Food Science* 43(5): 519–521.

Salunkhe, D. K., J. K. Chavan, and S. S. Kadam 1990. *Dietary Tannins: Consequences and Remedies.* Boca Raton, FL: CRC Press, 150–73.

Samtiya, M., R. E. Aluko, and T. Dhewa 2020. Plant food anti-nutritional factors and their reduction strategies: An overview. *Food Production, Processing and Nutrition* 2(1): 1–14.

Santamaria, P. 2006. Nitrate in vegetables: Toxicity, content, intake and EC regulation. *Journal of the Science of Food and Agriculture* 86(1): 10–17.

Santosa, E., A. D. Susila, A. P. Lontoh, A. Noguchi, K. Takahata, and N. Sugiyama 2016. NPK fertilizers for elephant foot yam (*Amorphophallus paeoniifolius* [Dennst.] Nicolson) intercropped with coffee trees. *Indonesian Journal of Agronomy* 43(3): 257–263.

Savage, G. P., L. Vanhanen, S. M. Mason, and A. B. Ross 2000. Effect of cooking on the soluble and insoluble oxalate content of some New Zealand foods. *Journal of Food Composition and Analysis* 13(3): 201–206.

Scalzo, R. L., A. Genna, F. Branca, M. Chedin, and H. Chassaigne 2008. Anthocyanin composition of cauliflower (*Brassica oleracea* L. var. botrytis) and cabbage (*B. oleracea* L. var. capitata) and its stability in relation to thermal treatments. *Food Chemistry* 107(1): 136–144.

Schulzová, V., J. Hajšlová, P. Botek, and R. Peroutka 2007. Furanocoumarins in vegetables: Influence of farming system and other factors on levels of toxicants. *Journal of the Science of Food and Agriculture* 87(15): 2763–2767.

Sen, N. P., and B. Donaldson 1978. Improved colorimetric method for determining nitrate and nitrite in foods. *Journal of the Association of Official Analytical Chemists* 61(6): 1389–1394.

Senanayake, S. A., K. K. D. S. Ranaweera, A. Bamunuarachchi, and A. Gunaratne 2012. Proximate analysis and phytochemical and mineral constituents in four cultivars of yams and tuber crops in Sri Lanka. *Tropical Agricultural Research Extension* 15(1): 32–36.

Sharma, K. D., S. Karki, N. S. Thakur, and S. Attri 2012. Chemical composition, functional properties and processing of carrot—A review. *Journal of Food Science and Technology* 49(1): 22–32.

Shewry, P. R. 2003. Tuber storage proteins. *Annals of Botany* 91(7): 755–769.

Siddiqui, I. R. 1989. Studies on vegetables: Fiber content and chemical composition of ethanol-insoluble and -soluble residues. *Journal of Agricultural and Food Chemistry* 37 (3): 647–650.

Siegenberg, D., R. D. Baynes, T. H. Bothwell, B. J. Macfarlane, R. D. Lamparelli, N. G. Car, P. MacPhail, U. Schmidt, A. Tal, and F. Mayet 1991. Ascorbic acid prevents the dose-dependent inhibitory effects of polyphenols and phytates on nonheme-iron absorption. *The American Journal of Clinical Nutrition* 53(2): 537–541.

Singh, A. P., and S. Kumar. 2019. Applications of Tannins in Industry. *In: Tannins-Structural Properties, Biological Properties and Current Knowledge.* IntechOpen.

Singh, S., A. Singh, S. Kumar, P. Mittal, and I. K. Singh 2020. Protease inhibitors: Recent advancement in its usage as a potential biocontrol agent for insect pest management. *Insect Science* 27(2): 186–201.

Singleton, V., and J. A. Rossi 1965. Colorimetry of total phenols with phosphor molybdic phoshphotungstic acid reagents, *American Journal of Enology and Viticulture* 16: 144–158.

Smillie, S. 2010. Are 'neeps' swedes or turnips. *The Guardian.* Available from: https://www.theguardian.com/lifeandstyle/wordofmouth/2010/jan/25/neeps-swede-or-turnip

Smith, C., W. V. Megen, L. Twaalfhoven, and C. Hitchcock 1980. The determination of trypsin inhibitor levels in foodstuffs. *Journal of the Science of Food and Agriculture* 31(4): 341–350.

Søborg, I., C. Andersson, and J. Gry 1996. TemaNord: Furanocoumarins in Plant Food. Nordic council of Ministers, 17–29.

Son, I. S., J. H. Kim, H. Y. Sohn, K. H. Son, J. S. Kim, and C. S. Kwon 2007. Antioxidative and hypolipidemic effects of diosgenin, a steroidal saponin of yam (*Dioscorea* spp.), on high-cholesterol fed rats. *Bioscience, Biotechnology, and Biochemistry* 71(12): 3063–3071.

Sonibare, M. A., and R. B. Abegunde 2012. In vitro antimicrobial and antioxidant analysis of *Dioscorea dumetorum* (Kunth) Pax and *Dioscorea hirtiflora* (Linn.) and their bioactive metabolites from Nigeria. *Journal of Applied Biosciences* 51: 3583–3590.

Sousa, S., J. Pinto, C. Rodrigues, M. Giao, C. Pereira, F. Tavaria, F. X. Malcata, A. Gomes, M. B. Pacheco, and M. Pintado 2015. Antioxidant properties of sterilized yacon (*Smallanthus sonchifolius*) tuber flour. *Food Chemistry* 188: 504–509.

Stintzing, F. C., and R. Carle 2004. Functional properties of anthocyanins and betalains in plants, food, and in human nutrition. *Trends in Food Science and Technology* 15(1): 19–38.

Thorup-Kristensen, K. 2006. Root growth and nitrogen uptake of carrot, early cabbage, onion and lettuce following a range of green manures. *Soil Use and Management* 22(1): 29–38.

Tomomatsu, H. 1994. Health effects of oligosaccharides. *Food Technology* 48:61–65.

Wanasundera, J. P. D., and G. Ravindran 1992. Effects of cooking on the nutrient and antinutrient contents of yam tubers (*Dioscorea alata* and *Dioscorea esculenta*). *Food Chemistry* 45(4): 247–250.

Wang, T. S., C. K. Lii, Y. C. Huang, J. Y. Chang, and F. Y. Yang 2011. Anticlastogenic effect of aqueous extract from water yam (*Dioscorea alata* L.). *Journal of Medicinal Plants Research* 5(26): 6192–6202.

Woldegiorgis, A. Z., D. Abate, G. D. Haki, and G. R. Ziegler 2015. Major, minor and toxic minerals and antinutrients composition in edible mushrooms collected from Ethiopia. *Journal of Food Processing & Technology* 6(3):1

Woolfe, J. A. 1992. *Sweet Potato: An Untapped Food Resource*. Cambridge, UK: Cambridge University Press and the International Potato Center (CIP).

Yukuyoshi, T., M. Masazumi, and Y. Masaji 2012. Application of Saponin containing plants in foods and cosmetics. In: *H. Sakagami (Ed.), Alternative Medicine. Intech Open*, 85–101.

Zagrobelny, M., S. Bak, A. V. Rasmussen, B. Jørgensen, C. M. Naumann, and B. L. Møller 2004. Cyanogenic glucosides and plant-insect interactions. *Phytochemistry* 65(3): 293–306.

Zhang, Z., W. Gao, R. Wang, and L. Huang 2014. Changes in main nutrients and medicinal composition of Chinese yam (*Dioscorea opposita*) tubers during storage. *Journal of Food Science and Technology* 51(10): 2535–2543.

Index

For Product Safety Concerns and Information please contact our EU
representative GPSR@taylorandfrancis.com
Taylor & Francis Verlag GmbH, Kaufingerstraße 24, 80331 München, Germany

www.ingramcontent.com/pod-product-compliance
Lightning Source LLC
Chambersburg PA
CBHW080348220326
41598CB00030B/4635